Lin Huang · David Greenhalgh
Editors

Proceedings of 18th International Conference on Machine Learning and Computing

ICMLC2026. Volume 2

Springer

Editors
Lin Huang
Metropolitan State University of Denver
Denver, CO, USA

David Greenhalgh
Department of Mathematics and Statistics
University of Strathclyde
Glasgow, UK

ISSN 2367-3370 ISSN 2367-3389 (electronic)
Lecture Notes in Networks and Systems
ISBN 978-3-032-31235-8 ISBN 978-3-032-31236-5 (eBook)
https://doi.org/10.1007/978-3-032-31236-5

This Springer imprint is published by the registered company Springer Nature Switzerland AG
The registered company address is: Gewerbestrasse 11, 6330 Cham, Switzerland

Lecture Notes in Networks and Systems 2061

The series "Lecture Notes in Networks and Systems" publishes the latest developments in Networks and Systems—quickly, informally and with high quality. Original research reported in proceedings and post-proceedings represents the core of LNNS.

Volumes published in LNNS embrace all aspects and subfields of, as well as new challenges in, Networks and Systems.

The series contains proceedings and edited volumes in systems and networks, spanning the areas of Cyber-Physical Systems, Autonomous Systems, Sensor Networks, Control Systems, Energy Systems, Automotive Systems, Biological Systems, Vehicular Networking and Connected Vehicles, Aerospace Systems, Automation, Manufacturing, Smart Grids, Nonlinear Systems, Power Systems, Robotics, Social Systems, Economic Systems and other. Of particular value to both the contributors and the readership are the short publication timeframe and the world-wide distribution and exposure which enable both a wide and rapid dissemination of research output.

The series covers the theory, applications, and perspectives on the state of the art and future developments relevant to systems and networks, decision making, control, complex processes and related areas, as embedded in the fields of interdisciplinary and applied sciences, engineering, computer science, physics, economics, social, and life sciences, as well as the paradigms and methodologies behind them.

Indexed by SCOPUS, EI Compendex, INSPEC, WTI Frankfurt eG, zbMATH, SCImago.

All books published in the series are submitted for consideration in Web of Science.

For proposals from Asia please contact Aninda Bose (aninda.bose@springer.com).

Preface

It is our great pleasure to present Volume 2 of the proceedings of the 2026 18th International Conference on Machine Learning and Computing, held in Nanjing, China, during February 6–9, 2026.

ICMLC is a well-established annual conference on machine learning algorithms, computational statistics, mathematical optimization, computer engineering, computer science and other related subject since 2009. It aims to provide opportunities for researchers in this field to share their ideas and reinforce collaboration. As such, the conference has attracted a remarkable number of participants while also enabling fruitful exchanges among academic and industrial researchers over the years.

In response to the call for papers, we received over 200 high-quality submissions. Each paper underwent a rigorous double-blind peer-review process conducted by members of the Technical Committee. After careful evaluation, 139 papers were accepted and are included in the proceedings of ICMLC 2026. This volume of proceedings includes 68 papers, which are organized into eight thematic parts

- Multimodal Fusion of Digital Image Analysis and Computer Vision Perception;
- Advances in Computer Vision: Reliability, Explainability, and Real-World Applications;
- Advances in Medical Image Analysis: AI, Automation, and Clinical Applications;
- Key Technologies for Intelligent Image Recognition, Detection and Multidimensional Analysis in Complex Scenarios;
- Machine Learning Methods for Trustworthy and Secure AI Systems;
- Visual Language Cross Modal Alignment for Image Representation Learning and Analysis Method;
- Machine Learning Applications in Bioinformatics;
- Generative AI: Foundations, Advances, and Emerging Applications.

We would like to express our sincere gratitude to the Organizing Committee and the Technical Committee for their support and coordination, which ensured the smooth operation of the conference. We also appreciate all the authors for their valuable contributions to the conference and these proceedings.

We look forward to your continued participation in ICMLC and to further academic exchanges that will advance the development of the field.

Proceedings Editors
Lin Huang
Metropolitan State University of Denver,
Denver, USA

David Greenhalgh
University of Strathclyde, Glasgow, UK

Committees

Advisory Committee Chairs

Lei Xu	Shanghai Jiao Tong University, China
Haizhou Li	The Chinese University of Hong Kong, Shenzhen, China
Xiaolin Huang	Shanghai Jiao Tong University, China

Conference Chairs

Jianqiang Li	Shenzhen University, China
Lin Huang	Metropolitan State University of Denver, USA

Conference Co-chairs

Jianfeng Lu	Nanjing University of Science and Technology, China
Zechao Li	Nanjing University of Science and Technology, China
Shuanghua Yang	University of Reading, UK

Steering Committee Chairs

Qianmu Li	Nanjing University of Science and Technology, China
Liying Zheng	Harbin Engineering University, China
Xun Gong	Southwest Jiaotong University, China

Program Chairs

Cangqi Zhou	Nanjing University of Science and Technology, China
Shikui Tu	Shanghai Jiao Tong University, China

Bo Xu	Guangdong University of Finance and Economics, China
Dong Huang	South China Agricultural University, China

Program Co-chairs

Ran Shi	Nanjing University of Science and Technology, China
Shuai Zhao	Sun Yat-sen University, China
Dongyang Kuang	Sun Yat-sen University, China
Lars Lundberg	Blekinge Institute of Technology, Sweden
Zhihong Liu	National University of Defense Technology, China

Local Chairs

Junlong Zhou	Nanjing University of Science and Technology, China
Xuansong Li	Nanjing University of Science and Technology, China

Special Session Chairs

Jing Zhang	Southeast University, China
Yi Zhu	Yangzhou University, China
Yan Pang	Shenzhen Institutes of Advanced Technology, Chinese Academy of Sciences, China
Patrizia Savi	Politecnico di Torino, Italy
Jingpeng Li	South China University of Technology, China
Xiucai Ye	University of Tsukuba, Japan
Zaid Al-Huda	Chengdu University, China
Taha Rajeh	Chengdu University, China
Muhammad Hafeez Javed	Chengdu University, China
Yan Huang	South China University of Technology, China
Dawen Li	Sun Yat-sen University, China
Xi Zhang	Sun Yat-sen University, China

Regional Chair of United Kingdom

David Greenhalgh	University of Strathclyde, UK

Regional Chair of Japan

Akinori Ito	Tohoku University, Japan

Regional Chair of Australia

Cheng Siong Lee	Monash University, Australia

Technical Committees

Shad Kirmani	LinkedIn Corporation, USA
Hasan Kadhem	American University of Bahrain, Bahrain
Mingli Wu	North China University of Technology, China
Huixin Zhan	Texas Tech University, USA
Chung Shue Chen	Xidian University, China
Yu Zhao	Tokyo University of Science, Japan
Tien-Dung Cao	Tan Tao University, Vietnam
Sime Dejene Mengistu	Xidian University, China
Kennedy Ogada	Jomo Kenyatta University of Agriculture and Technology, Kenya
Gang Chen	Nanjing University of Aeronautics and Astronautics, China
Huaiwen He	University of Electronic Science and Technology of China, Zhongshan Institute, China
Jun Wu	National University of Defense Technology, China
Chao Wang	Chongqing Normal University, China
Yi Jin	University of Science and Technology of China, China
Donghai Guan	Nanjing University of Aeronautics and Astronautics, China
Cai zhi Fan	National University of Defense Technology, China
Jianguo Sun	Harbin Engineering University, China
Bo-wen Yan	Chongqing University, China

Haitao Gan	Hubei University of Technology, China
Dan Wu	Chinese Academy of Sciences, China
Ijazul Haq	South China University of Technology, China
Morteza Jalali Alenjareghi	Polytechnique Montreal, Canada
Qianjin Guo	Beijing Institute of Petrochemical Technology, China
Siwei Song	Northwestern Polytechnical University, China
Xinguang Cui	Huazhong University of Science and Technology, China
Yingchao Cheng	Guangdong Laboratory of Artificial Intelligence and Digital Economy (SZ), China
Yahui Liu	Tsinghua University, China
Zihan Deng	Harbin Institute of Technology, China
Zhixiang Ren	Pengcheng Laboratory, China
Dongmei Fu	University of Science and Technology Beijing, China
Yifan Yuan	Fudan University, China
Fan Zhang	University of Electronic Science and Technology of China, China
Jan Wang	China University of Petroleum (East China), China
Rigui Zhou	Shanghai Maritime University, China
Yanting Jin	University of Electronic Science and Technology of China, China
Shengjie Ling	Fudan University, China
Wenbo Jiang	University of Electronic Science and Technology of China, China
Zhenghua Qian	Nanjing University of Aeronautics and Astronautics, China
Xinyu Han	Shandong University of Science and Technology, China
Yong-An Zhang	Harbin Institute of Technology, China
Shouqing Jia	Northeastern University at Qinhuangdao, China
Wing Sze Wong	The Hong Kong Polytechnic University, Hong Kong, China
Tao Zhang	Central South University, China
Naji Alhusaini	Chuzhou University, China
Jun Zhang	Guangdong University of Technology, China
Akira Imakura	University of Tsukuba, Japan
Meng Sun	NARI Group Co., Ltd., China
Stefano Bromuri	Open University of the Netherlands, Netherlands
Gaetano Di Caterina	University of Strathclyde, UK

Contents

Multimodal Fusion of Digital Image Analysis and Computer Vision Perception

Heat Diffusion Models - Interpixel Attention Mechanism 3
Pengfei Zhang, Shouqing Jia, and Shuai He

AI-Driven Driver Monitoring and Assistance System for Enhanced Road Safety 18
Siddharth S and Jayashree R

TS-Gaussian: Towards Universal 3D Plant Reconstruction Across Species via Text-Guided Segmentation and Skeleton-Driven Completion 39
Yan Chen, Jinxin Chen, Xiangyu Lu, Fen Liao, Hengyi Li, Xiyao Huang, and Zihao Huang

Luminance-Controlled Migration for Soil Images 54
Yan Xia, Shaohua Zeng, Ruolan Zeng, and Shoukuan Gu

A Few-Shot Image Generation Model Based on a Hybrid Attention Mechanism 69
Qian Wu, Feng Yu, and Dequan Zheng

PR-Net: Precision Recovery Network with Dual-Branch Detail Attention and Local-Global Feature Enhancement 86
Rui Peng, Xuan Zhou, Longfeng Shen, Yaxi Duan, and Fang Zhang

Semantic-Aware Asymmetric Token Enhancement for Generalizable Person Re-Identification 97
Hongrui Yuan, Jingyu Li, Kun Zhang, and Lan Zhang

PMOSTRACK: A Video Object Tracking Framework with Positional Visualization and Mamba-Based Historical Modeling 115
Fengyi Lv, Man Xu, Zizhu Fan, and Hui Wang

POINT2VERTEX: Transformer-Based Convex Hull Vertex Prediction in Point Clouds 131
Rui-Juan Jing, Yuegang Zhao, and Changbo Chen

Advances in Computer Vision: Reliability, Explainability, and Real-World Applications

Single-Image Super-Resolution by Self-supervised Consistent Representation Learning 149
Qiyue Zhang and Yi Tang

DEATrack: Diffusion-Enhanced Autoregressive Framework for Visual Object Tracking 165
Xu Wang and Man Xu

DBR: A Dual-Branch Feature Processing and Fusion Framework for Post-Processing Boundary Refinement in Object Segmentation 178
Junsheng Fu and Ran Shi

Hybrid Convolutional Neural Network for Hyperspectral Corn Seed Identification 190
Hui Zhang, Junjiang Hu, He Yuan, and Yi An

RLEU-Net: A Lightweight Attention U-Net for High-Fidelity Spectral Signal Recovery in Portable Raman Spectroscopy 206
Yusen Peng, Haokuan Cheng, Xiaoyu Zhang, Ziqi Xu, and Yang Cao

ClusterNet: Point Cloud Clustering for CAD Scenes Using Deep Learning and Graph Connectivity 218
Li-Yong Shen, Hao Yang, and Yi-Fei Feng

Few-Shot Open-Set Learning for RF-Based Non-cooperative Drone Recognition 233
Han Yan and Qianjin Guo

Illumination-Aware Feature Fusion for Human Pose Estimation in Extreme Low Light 249
Zheng Wang

Advances in Medical Image Analysis: AI, Automation, and Clinical Applications

Oral-Score: A Computer-Vision-Based AI Model for Evaluating Oral Facial Muscle Performance 265
Beeno Cheung Leong Tung, Winsy Wing Sze Wong, James Kit Hon Tsui, Gary Kui Kai Lau, and Anne Chi Hay Yip

Depth-Guided Network for RGB-D Semantic Segmentation 278
Zebin Zhao, Chengshuo Zhang, and Shuang Xu

Performance Benchmarking and Optimisation of Clustering Algorithms for Local and Non-local Similarity Measure in Medical Image Analysis 291
Sisipho Hamlomo and Marcellin Atemkeng

BMSA-ViT: Efficient Vision Transformer for Pathology Breast Cancer Classification in Health Care System 307
Muhammad Usman Saleem, Bo Peng, Muhammad Shahid Mastoi, Mannan Hassan, and Muhammad Zahid Rafique

Adenoid Hypertrophy Diagnosis Based on the YOLOv11 Segmentation Model 318
Yueqiang Feng, Siyu Chen, Chunrong Zhao, Chang Duan, Ke Zhang, Sen Yang, and Donghai Wang

Dual Role of Arkypallidal Neurons in Modulating Beta Oscillations and Waveform Morphology in Basal Ganglia Network: A Computational Study 331
Xiangyi Su, Xia Shi, and Bei Bai

Data Glove and Inception-CSRA Network for Sign Language Recognition 342
Wan-Cheng Wu, Yong-An Zhang, and Yu-Tao Liu

SHMR: Traditional Chinese Medicine Prescription Recommendation via Multi-graph Neural Networks and Transformer-Based Attention 358
Chi Han

Hybrid Non-local Blocks for Streptococci Segmentation in Microscopic Images 373
A. Samarin, A. Toropov, E. Kotenko, A. Savelev, A. Dozortseva, A. Nazarenko, A. Mamaeva, A. Motyko, E. Mikhailova, and V. Malykh

Zero-Shot Shrimp Disease Detection with Vision–Language Models: A Comparison with a CNN Baseline on the ShrimpDiseaseBD Dataset 386
Ha Viet Thang, Inh Chau, and Cuong Tran

Key Technologies for Intelligent Image Recognition, Detection and Multidimensional Analysis in Complex Scenarios

CFNet: A Two-Stage Coarse-to-Fine Diffusion-Driven Network for Camouflaged Object Detection 405
Lei Sun, Yao Li, Gushuai Liu, Zengjian Yang, and Xuecheng Kong

HLAW-YOLOv8: An Improved Aerial Small Object Detection Algorithm Based on YOLO ... 421
ZhengChao Shi and HaiMing Li

Smartwatch-Based Sitting Time Estimation in Real-World Office Settings ... 437
Olivia Zhang and Zhilin Zhang

Study on the Identification and Statistical Analysis of Infected Woods in Bark Beetle Pest Area Based on YOLO Algorithm and Drone Imagery ... 448
Jiantao Zhang and Jiawen Liang

DATF: An Audiovisual Multimodal Identity Verification Method Based on Dynamic Attention and Trust Fusion ... 463
Lianggui Tang, Zhuo Chen, and Jiajun Yang

YOLOv11-Based Real-Time Plant Disease Detection System for Nigerian Agriculture ... 480
Ibrahim Sadiq, Omar Ennayri, Sadia Sharmin, Mohamed Nourdine, Samiha Sammani Sani, and Damtew Merry Wondmagegn

AEW-YOLO: A Lightweight Detection Algorithm for Pine Wood Disease Based on YOLO11 ... 496
Chongqing Yu and Jiantao Zhang

Enhanced Urban Bird Detection: Optimized Deep Learning Architecture on a Comprehensive Shanghai Bird Dataset ... 508
Yi Gong, Yong Wang, and HaiMing Li

Dual-Path Attention Network for Small Object Detection in Road Traffic Scenarios ... 523
Pinyi Wang, Jingke Xu, and Bin Du

BoundMamba: Boundary-Aware State Space Modeling for Remote Sensing Change Detection ... 534
Yu He, Yuanlie He, Wensheng Li, Yi Xiang, Yukuan Ma, and Yihui Liang

Research on Lightweight Aircraft Skin Defect Detection Algorithm Based on YOLOv10 ... 551
Mingren Zhu, Yongsheng Hu, and Weiheng Chen

Machine Learning Methods for Trustworthy and Secure AI Systems

PMGRD: A Prototype Memory-Guided Reverse Distillation Framework for Anomaly Detection 569
Jianyi Huang, Zhiqian Jia, Shenghua Teng, and Zuoyong Li

A Hybrid Fraud Detection Framework Fusing Deep Representation Learning, CTGAN, and MAB-Based XGBoost 581
Yifan Han and Wenjun Yang

Anomaly Detection for UAV Flight Data via LSTM–Attention–TCN Fusion Network 594
Muhammad Faizan Saleem, Shurui Fan, and Muhammad Asad Saleem

FedSCP: Personalized Federated Learning Against Intrinsic Adversarial Attacks via Semantic Guidance and Conflict Purification 607
Bin Zhou, Yang Cao, Mengli Wei, and Yongjian Chang

Self-supervised Clustering Defense Against Federated Learning Poisoning Attack in IIoT 623
Shuo Wang, Qianmu Li, Shicheng Cui, Ruotian Jiang, and Yue Tang

Adaptive RelaxLoss: An Adversarially Guided Defense Against Membership Inference Attacks 638
Jiaqi Qin and Yang Cao

A Deep Autoencoder Intrusion Detection Model Based on Contrastive Reconstruction Mechanism 656
Xinchao Song, Xingquan Gu, Jiale Ding, and Jiao Wu

Slow-Rate Denial-of-Service Attack Detection in QUIC Protocol Using a Hybrid CNN-GRU Deep Learning Model 672
Yongjie Gao, Yang Cao, Shijian Luo, and Jing Zhang

Dual-View Temporal Graph Learning with Two-Stage Filtering for Robust Bitcoin Fraud Detection 687
Arvindh Vijayakumar, Mohul Y P, Sai Mohananshu Jujjavarapu, Shri Hari S, and Bhaskarjyoti Das

Visual Language Cross Modal Alignment for Image Representation Learning and Analysis Method

Adaptive Entropy-Guided Token Pruning for Efficient Vision Transformers 705
Xianbin Jiang, Xing Guo, and Haohua Du

A Frequency–Time Dual-Domain Fusion Algorithm for Multi-IMU Human Motion Recognition 719
Zihan Wang, Zhiqiang Guo, Yongsheng Wang, and Yi Li

A Divide-and-Conquer Network for Generalized Few-Shot Segmentation 732
Lingling Zhou, Tao Fang, Wenguang Li, Bin Zhang, Xin Li, and Yuran Zhang

PD-MAE: Masked Autoencoder with Patch Entropy and Knowledge Distillation for Image Representation Learning 750
Jianpu Geng, Ling Zhang, Boquan Zhang, and Minghui Wang

A Study on Threshold Selection for mAP Calculation of Vision AI Using User Satisfaction 763
Yoonho Choi and Kwang Sik Chung

Sparse Multimodal Attention Transformer with Cross-Modal Gating for Resource-Efficient Visual Question Answering 780
Abdur Rahman Rakib, Long Ma, and Xiaoli Huan

When Graph Structure Hurts: Lightweight Path Ranking for Dense Knowledge Graph RAG 808
Pranav Kumaar, Sutejas K, Anshula Aithal, Avanish Bhat, and Bhaskarjyoti Das

MetaDistil: Adaptive Knowledge Distillation for Vision Transformers via Meta-Learning 826
Nishitha A, Priyanka Rajkumar Thoppen, Suraj Nagisetty, Namita, and Nitin V Pujari

Machine Learning Applications in Bioinformatics

Automated Evaluation of Table Tennis Stroke Quality via Skeletal Analysis and Ensemble Learning 843
Yangming Gao, Junyu Zhou, Jiacheng Liu, Chenyu Wang, Xinle Cheng, and Ke Wang

Contrastive Fusion of Biomolecular and Heterogeneous Network Features for Drug–Target Interaction Prediction 854
Hongyang Yang, Xiucai Ye, and Tetsuya Sakurai

Natural Product Identification Against Fusarium Oxysporum with CatBoost 869
Jing Li, Shuyu Han, Xiao Lin, and Bin Yang

PLM-Driven Differentiable Clustering and Graph Reasoning for Protein Interaction Site Prediction 879
Wenyu Xi, Xiucai Ye, Tianyi Shi, and Tetsuya Sakurai

Task-Specialized Dual Transformers for Post-level ADR Classification and Entity-Level ADR-Drug Extraction from Social Media 893
Jade Shaira M. de Paz, Daniel Ryan G. So, and Christine D. Bandalan

Generative AI: Foundations, Advances, and Emerging Applications

An LLM-Based Decision Support Script for Ethylene Yield Adjustment in Steam Cracking Furnaces 911
Zhiping Peng, Lu Pan, Xiangwu Deng, Qirui Li, Delong Cui, and Xinlong Luo

TLIME: Stable LIME Explanation for Tabular Data 923
Tong Zhang, Xuansong Li, and Yu Wang

Generative AI Assisted Psychological Resilience Training: Situational Counseling Practice for Teenagers Based on MBTI Virtual Characters 937
Li Feng and Yangang Yang

AI-Generated Text Detection with Multi-source Domain Adaptation 948
Senao Li, Wan Zhang, Zhi Zheng, and Jing Zhang

KPL: Keyframe–Prompt–LLM for Subject-Centric Spatiotemporal Video Captioning 963
Nanqiao Jiang, Qin Wei, and Yanyu Zhu

Multilingual Back-Translation Augmentation for Low-Resource Text Classification: Effectiveness Boundaries and Optimal Practices 977
Minshi Fan

Integrating Knowledge Distillation Methods: A Sequential Multi-Stage Framework 991
Yinxi Tian, Changwu Huang, Ke Tang, and Xin Yao

Accelerating Policy Decisions: A Multi-agent System for Interactive Question-Answering on Climate Policy Data 1010
Elizabeth Osanyinro, Ernest Effiong Offiong, Abiola Oludotun, Sa'id Olanrewaju, Grace Farayola, Oladotun Fasogbon, Oluwole Fagbohun, Sayo Agunbiade, and Ogheneruona Maria Esegbon-Isikeh

Author Index 1023

Multimodal Fusion of Digital Image Analysis and Computer Vision Perception

Heat Diffusion Models - Interpixel Attention Mechanism

Pengfei Zhang, Shouqing Jia(✉), and Shuai He

Northeastern University, Qinhuangdao 066004, China
jiashouqing@163.com

Abstract. Although Denoising Diffusion Probabilistic Models (DDPMs) have significantly advanced the development of image generation and restoration techniques, it has a fundamental limitation: during the diffusion and generation processes, it assumes conditional independence between pixels, meaning that noise is added and removed pixel by pixel, without explicitly considering the inherent spatial relationships and structural priors present in natural images. Considering that the probability of adjacent pixels belonging to the same object is very high, and inspired by the principle of heat diffusion, we proposed the Heat Diffusion Model (HDM), which retains the spatial structure relationship between pixels. HDM is an optimized model for DDPM. While retaining the similar training process as DDPM, it incorporates an attention mechanism between pixels. In HDM, the discrete form of the two-dimensional heat equation is integrated into the diffusion and generation formulas of DDPM, enabling the model to calculate the weights between adjacent pixels during the image processing process. Our experiments have shown that HDM has achieved satisfactory results compared to other models (such as DDPM, consistency model (CM), latent diffusion model (LDM), and vector quantization generative adversarial network (VQGAN)).

Keywords: Heat Diffusion Model (HDM) · Image generation · Pixel-level operations · Thermal diffusion · Heat equation.

1 Introduction

The Generative models have demonstrated their capability to produce high-quality samples across numerous domains. Variational Autoencoders (VAEs), and Generative Adversarial Networks (GANs) are increasingly fading from the spotlight [1]. Diffusion-based image generation models have emerged as one of the preferred approaches for addressing challenges in the field of image generation [2]. For example, DDPMs introduces a stepwise denoising process on the basis of DM, significantly improving the quality of generated images and the stability of training [3].

In addition to diffusion models, approaches such as the Masked Generative Image Transformer (MaskGIT) in [4] utilize a masked prediction mechanism to generate images. This method reduces computational complexity during the generation process while improving generation efficiency [5]. The CM aims to enhance generation efficiency by reducing sampling steps while maintaining high-quality image synthesis. This makes

L. Huang and D. Greenhalgh (Eds.): ICMLC 2026, LNNS 2061, pp. 3–17, 2027.
https://doi.org/10.1007/978-3-032-31236-5_1

CM particularly advantageous in scenarios requiring real-time performance [6]. Certainly, the Visual AutoRegressive (VAR) model is also a viable option. It integrates the Transformer architecture and enhances generation efficiency and quality through "next-scale prediction" [7]. In the field of AR, methods such as the Vector Quantized Generative Adversarial Network (VQGAN), which integrates VQ-VAE and GAN, the Vision Transformer with Vector Quantization (ViTVQ) based on Vision Transformer (ViT) [8], and the large-scale pre-trained LlamaGen-B have significantly enriched the domain of image generation. However, when MaskGIT generates high-resolution images, local details are prone to distortion [9]; as mentioned in [10], models such as VQGAN and ViTVQ encounter problems of blurred details when generating complex scenes; VAR is prone to the phenomenon where the early-scale prediction errors are amplified, resulting in inconsistent final image details [11].

The limitation of DDPMs lies in that its diffusion and generation process is based on the Markov assumption and lacks explicit modeling of the local structure of the image [12]. Diffusion Transformers attempt to address this issue through post-processing or attention mechanisms, but integrating spatial consistency within the diffusion framework remains an unsolved challenge [13]. This constitutes a key research gap, as ignoring these structural constraints may lead to locally plausible but overall inconsistent generation results, manifested as blurry outputs or inconsistent textures with the surrounding environment [14]. Therefore, we proposed HDM, which introduces the new constraint of pixel-level attention into DDPM, providing a new structure-aware generation paradigm for diffusion models. Unlike the method in [15], our approach directly modified the generation paradigm. By using the constraint of the relationship matrix in the generation process, the structural rationality of the generated samples has been improved. The following picture shows the model structure of HDM. The diffusion and generation processes are similar to those of DDPM, but pixel-level attention is added in the data processing.

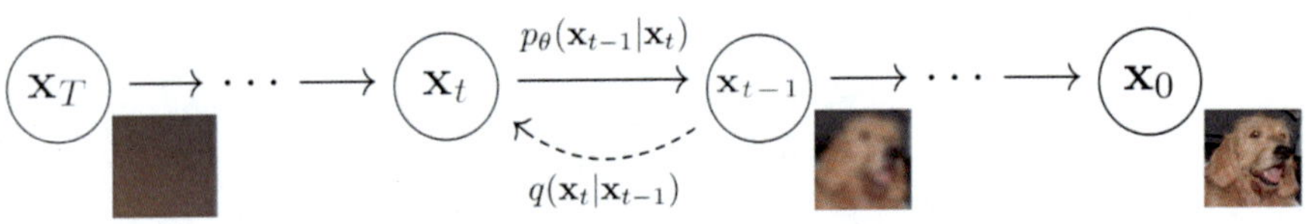

Fig. 1. The model diagram of HDM.

As can be seen from Fig. 1, due to the addition of the inter-pixel attention mechanism, the images during the diffusion process have become smoother, and are no longer the noisy images as in the DDPM method.

Section 2 provides a detailed description of the proposed HDM and its modifications to DDPM. Section 3 presents experimental results comparing HDM, DDPM, and other models under identical test datasets, and Sect. 4 gives the conclusion.

2 Heat Diffusion Model

In this paper, we employ the discretization of the heat equation to retain the structural prior of the image within the generation and diffusion process of the model. And in this section, we provide a detailed explanation of the derivation of the relevant formulas.

2.1 Diffusion Process

Given $u = u(x, y, t)$, it expresses the temperature at position (x, y) and time t. For the heat equation, it has the following form:

$$\frac{\partial u}{\partial t} - \kappa \nabla^2 u = 0 \tag{1}$$

where κ denotes the thermal diffusion coefficient and $\nabla^2 = \frac{\partial^2}{\partial x^2} + \frac{\partial^2}{\partial y^2}$ represents the spatial Laplacian operator. The physical essence of this equation is that heat will diffuse from regions with a larger temperature gradient to regions with a smaller temperature gradient. This is highly similar to the gradients employed in [16], where gradient-guided learning drives the generated samples to asymptotically approach the true data distribution. Take $x = i\Delta$, $y = j\Delta$, $t = n\tau$, the result after discretization is:

$$u_n^{i,j} = u(i\Delta, j\Delta, n\tau) \tag{2}$$

Digital images are essentially two-dimensional discrete grids, which are completely isomorphic to the difference grids used in the numerical solution of the heat equation. Therefore, we can regard i and j as the position coordinates of a certain pixel in the image. Performing first-order difference approximation and second-order difference approximation, the following forms are obtained:

$$\frac{\partial u}{\partial t}\Big|_n^{i,j} \approx \frac{u_{n+1}^{i,j} - u_n^{i,j}}{\tau} \tag{3}$$

$$\frac{\partial^2 u}{\partial x^2}\Bigg|_n^{i,j} \approx \frac{u_n^{i+1,j} - 2u_n^{i,j} + u_n^{i-1,j}}{\Delta^2} \tag{4}$$

Applying forward difference to the temporal of Eq. (1) at $t = n\tau$ yields:

$$u_{n+1}^{i,j} - u_n^{i,j} = K\left(u_n^{i+1,j} + u_n^{i-1,j} + u_n^{i,j+1} + u_n^{i,j-1} - 4u_n^{i,j}\right) \tag{5}$$

where:

$$K = \frac{\kappa\tau}{\Delta^2} \tag{6}$$

A weighted summation of the above equations yields: (where θ is a constant and $0 \leq \theta \leq 1$):

$$\begin{aligned} u_{n+1}^{i,j} - u_n^{i,j} &= \theta K\left(u_n^{i+1,j} + u_n^{i-1,j} + u_n^{i,j+1} + u_n^{i,j-1} - 4u_n^{i,j}\right) \\ &+ (1-\theta)K\left(u_{n+1}^{i+1,j} + u_{n+1}^{i-1,j} + u_{n+1}^{i,j+1} + u_{n+1}^{i,j-1} - 4u_{n+1}^{i,j}\right) \end{aligned} \tag{7}$$

The scheme reduces to fully explicit when $\theta = 1$, fully implicit when $\theta = 0$, and Crank-Nicolson when $\theta = 1/2$ [17]. By comparison, this paper selects the adiabatic condition as the basis for choosing the numerical value of the boundary position such as $i = 0, I - 1$.

In order to facilitate the derivation of the formula, we express $u_n^{i,j}$ in the form of a column vector $(0 \le i \le I - 1, 0 \le j \le J - 1)$. The values of I and J here directly represent the number of rows and columns of the image dimensions, and obtain:

$$u_n = \begin{bmatrix} u_n^{0,0} \\ \vdots \\ u_n^{0,J-1} \\ \vdots \\ u_n^{I-1,0} \\ \vdots \\ u_n^{I-1,J-1} \end{bmatrix} \tag{8}$$

At this point, Eq. (7) is written in matrix form:

$$Su_{n+1} = Tu_n \tag{9}$$

This yields the relation $u_n^{-1,j} = u_n^{1,j}$, which when substituted into Eq. (7) leads to the following form:

$$\begin{aligned} (1 + 4(1 - \theta)K)u_{n+1}^{0,j} - (1 - \theta)K\left(2u_n^{1,j} + u_{n+1}^{0,j+1} + u_{n+1}^{0,j-1}\right) = (1 - 4\theta K)u_n^{0,j} \\ + \theta K\left(2u_n^{1,j} + u_n^{0,j+1} + u_n^{0,j-1}\right) \end{aligned} \tag{10}$$

To better illustrate the form of the boundary conditions, the matrix S corresponding to $I = 2, J = 2$ is explicitly expressed as follows:

$$S = \begin{bmatrix} 1 + 4(1 - \theta)K & 2(\theta - 1)K & 0 & 2(\theta - 1)K \\ 2(\theta - 1)K & 1 + 4(1 - \theta)K & 2(\theta - 1)K & 0 \\ 0 & 2(\theta - 1)K & 1 + 4(1 - \theta)K & 2(\theta - 1)K \\ 2(\theta - 1)K & 0 & 2(\theta - 1)K & 1 + 4(1 - \theta)K \end{bmatrix} \tag{11}$$

From Eq. (9), we can define $A = S^{-1}T$, thereby achieving:

$$u_{n+1} = Au_n \tag{12}$$

Matrix A represents the relationship weights between pixels. Initially, the diffusion formula can be expressed as:

$$u_n = \sqrt{\alpha_n}Au_{n-1} + \sqrt{1 - \alpha_n}\varepsilon_{n-1}, \varepsilon_n \sim \mathrm{N}(0, I), n = 0 \sim N \tag{13}$$

where N is the number of time steps. This calculation cannot guarantee that the overall variance remains at 1 [18]. As [19] pointed out, maintaining the total variance at 1 ensures that the addition of noise remains controllable, thereby preventing situations such as instability or difficulty in convergence during the training process [20]. We can derive a matrix A^{n-N}, where $0 < \alpha_n < 1$, thereby there is:

$$u_n = \sqrt{\alpha_n} A u_{n-1} + \sqrt{1-\alpha_n} A^{n-N} \varepsilon_{n-1}, \varepsilon_n \sim \mathrm{N}(0, I) \tag{14}$$

By following the steps in [1], the equation can be rearranged as follows:

$$u_n = \sqrt{\overline{\alpha}_n} A^n u_0 + \sqrt{1-\overline{\alpha}_n} A^{n-N} e_n, e_n \sim \mathrm{N}(0, I) \tag{15}$$

By simplifying Eq. (15), we obtain:

$$u_N = \sqrt{\overline{\alpha}_N} A^N u_0 + \sqrt{1-\overline{\alpha}_N} e_N \tag{16}$$

when N is sufficiently large, $\overline{\alpha}_N = \alpha_N \cdots \alpha_1 \to 0$, then $u_N \to e_N \sim N(0, I)$. Based on Eqs. (14) and (16) the forward transition probability of the noise-adding process can be derived as follows:

$$\mathrm{q}(u_n | u_{n-1}) = \mathrm{N}\left(u_n | \sqrt{\alpha_n} A u_{n-1}, (1-\alpha_n) B_n\right) \tag{17}$$

$$\mathrm{q}(u_n | u_0) = \mathrm{N}\left(u_n | \sqrt{\overline{\alpha}_n} A^n u_0, (1-\overline{\alpha}_n) B_n\right) \tag{18}$$

where $B_n = A^{n-N}(A^{n-N})^T$ (Matrix B is evidently symmetric).

2.2 Generation Process

The backward transition probability of the diffusion process is:

$$\mathrm{q}(u_{n-1} | u_n, u_0) = \frac{\mathrm{N}\left(u_n | \sqrt{\alpha_n} A u_{n-1}, (1-\alpha_n) B_n\right) \mathrm{N}\left(u_{n-1} | \sqrt{\overline{\alpha}_{n-1}} A^{n-1} u_0, (1-\overline{\alpha}_{n-1}) B_{n-1}\right)}{\mathrm{N}\left(u_n | \sqrt{\overline{\alpha}_n} A^n u_0, (1-\overline{\alpha}_n) B_n\right)} \tag{19}$$

From the quadratic term in u_{n-1} in this expression, the covariance matrix of the backward transition probability $\mathrm{q}(u_{n-1} | u_n, u_0)$ can be derived:

$$\Sigma_n = \left[(1-\alpha_n)^{-1} \alpha_n A^T B_n^{-1} A + (1-\overline{\alpha}_{n-1})^{-1} B_{n-1}^{-1}\right]^{-1} \tag{20}$$

By setting the gradients of the quadratic and linear terms with respect to u_{n-1} to 0 [21], the mean of the backward transition probability $\mathrm{q}(u_{n-1} | u_n, u_0)$ can be obtained:

$$\mathrm{q}(u_{n-1} | u_n, u_0) = \mathrm{N}(u_{n-1} | \mu_n(u_n, u_0), \Sigma_n) \tag{21}$$

where $\mu_n(u_n, u_0) = \left[(1-\alpha_n)^{-1} \alpha_n A^T B_n^{-1} A + (1-\bar{\alpha}_{n-1})^{-1} B_{n-1}^{-1}\right]^{-1}$
$\left[(1-\alpha_n)^{-1} \sqrt{\alpha_n} A^T B_n^{-1} u_n + (1-\bar{\alpha}_{n-1})^{-1} \sqrt{\bar{\alpha}_{n-1}} B_{n-1}^{-1} A^{n-1} u_0\right]$

. The transition probability of the generative process:

$$p_\theta(u_{n-1}|u_n) = \begin{cases} N(u_{n-1}|\hat{\mu}_\theta(u_n), \Sigma_n), n > 1 \\ N(u_0|\hat{\mu}_\theta(u_1), I), n = 1 \end{cases} \tag{22}$$

Rewrite $u_0 = \frac{1}{\sqrt{\overline{\alpha}_n}}\left(A^{-n}u_n - \sqrt{1-\overline{\alpha}_n}A^{-N}e_n\right)$. Substituting it into Eq. (22) yields:

$$\mu_n(u_n, u_0) = C_n u_n + D_n e_n \tag{23}$$

where:

$$C_n = \left[(1-\alpha_n)^{-1}\alpha_n A^T B_n^{-1} A + (1-\overline{\alpha}_{n-1})^{-1} B_{n-1}^{-1}\right]^{-1} \left[\frac{\sqrt{\alpha_n}}{1-\alpha_n} A^T B_n^{-1} + \frac{1}{(1-\overline{\alpha}_{n-1})\sqrt{\alpha_n}} B_{n-1}^{-1} A^{-1}\right] \tag{24}$$

$$D_n = -\frac{\sqrt{1-\overline{\alpha}_n}}{(1-\overline{\alpha}_{n-1})\sqrt{\alpha_n}} \left[(1-\alpha_n)^{-1}\alpha_n A^T B_n^{-1} A + (1-\overline{\alpha}_{n-1})^{-1} B_{n-1}^{-1}\right]^{-1} B_{n-1}^{-1} A^{n-N-1} \tag{25}$$

Then we can obtain:

$$\hat{\mu}_\theta(u_n) = C_n u_n + D_n \hat{e}_\theta(u_n) \tag{26}$$

The following Table 1 outlines the program execution sequence of the diffusion process and the generation process.

The Evidence Lower Bound (ELBO) in [22] is the core optimization objective in the training of diffusion models. According to the previous formulas, the final ELBO can be calculated as:

$$\mathrm{ELBO}_\theta(u_0) = -\frac{1}{2}\sum_{n=1}^{N} \mathrm{E}_{q(u_n|u_0)} \left[(e_n - \hat{e}_\theta(u_n))^T D_n^T \Sigma_n^{-1} D_n (e_n - \hat{e}_\theta(u_n))\right] \tag{27}$$

2.3 Range of Parameter *K*

The selection of the parameter K in the heat equation is of great significance. When $\Sigma = \sigma^2 I$, according to the d-dimensional Fourier transformation, then there is:

$$\frac{1}{(2\pi)^{d/2}\sigma^d}\exp\left(-\frac{1}{2\sigma^2}x - \mu^2\right) \overset{FT}{\to} \exp\left(-\frac{\sigma^2}{2}\omega^2 - j\mu^T\omega\right) \tag{28}$$

For the initial value problem:

$$\begin{cases} \frac{\partial U}{\partial t} - \kappa\nabla^2 u = 0 \\ u(x, 0) = u_0(x) \end{cases} \tag{29}$$

assuming $U_0(\omega)$ is the Fourier transform of $u_0(x)$:

$$\frac{\partial U}{\partial t} + \kappa\omega^2 U = 0 \tag{30}$$

Table 1. The algorithmic processes of the diffusion process and the generation process.

Algorithm 1 Training

Receive u_0 from training data
while not converged do
  Sample $n \sim U(1, N)$
  Sample $e_n \sim N(0, I)$
  Compute $u_n = \sqrt{\overline{\alpha}_n} A^n u_0 + \sqrt{1 - \overline{\alpha}_n} A^{n-N} e_n$
  Update $\theta \leftarrow \theta - \eta \nabla_\theta \left[(e_n - \hat{e}_\theta(u_n))^T D_n^T \Sigma_n^{-1} D_n (e_n - \hat{e}_\theta(u_n)) \right]$
end while
return θ

Algorithm 2 Sampling

Sample $u_N \sim N(0, I)$
For $n = N \sim 1$
  Sample $z_n \sim N(0, I)$
  Cholesky decomposition $\Sigma_n = L_n L_n^T$
  Compute $u_{n-1} = C_n u_n + D_n \hat{e}_\theta(u_n) + L_n z_n$
end for
return u_0

by setting $\sigma = \sqrt{2\kappa t}$, the following can be derived from Eq. (28):

$$\frac{1}{(4\pi\kappa t)^{d/2}} \exp\left(-\frac{1}{4\kappa t} x^2\right) \overset{FT}{\rightarrow} \exp\left(-\kappa t \omega^2\right) \tag{31}$$

It can be derived from the properties of Fourier transform that:

$$u(x, t) = \begin{cases} \int G(x - x', t) U_0(x') \mathrm{d}x', t > 0 \\ 0, t \leq 0 \end{cases} \tag{32}$$

where:

$$G(x, t) = \frac{1}{(4\pi\kappa t)^{d/2}} \exp\left(-\frac{1}{4\kappa t} x^2\right) \tag{33}$$

By setting $U_0(x) = \delta(x)$, we obtain $u(x, t) = G(x, t)$. By fixing x, the curve of $G(x, t)$ with respect to t is as shown in Fig. 2.

As can be seen from Fig. 2, with the change of t, the trend of G is to increase rapidly to reach its maximum value, and then gradually decrease. The peak time can be determined as:

$$t_0 = \frac{x^2}{2d\kappa} \tag{34}$$

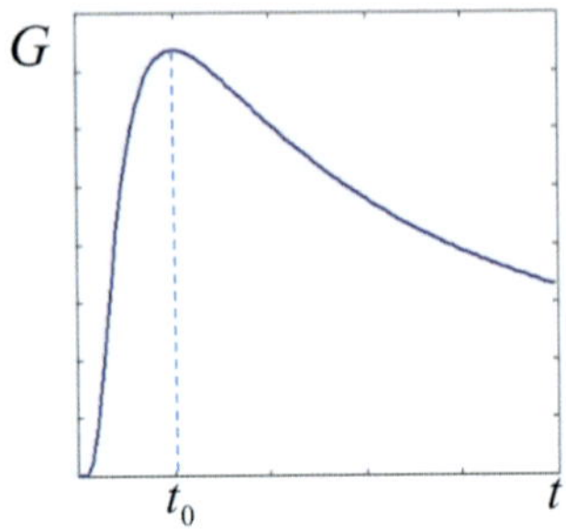

Fig. 2. The curve of $G(x, t)$ with respect to t.

By setting $t_0 = \tau, x = (\gamma\Delta, 0, \cdots, 0)$, which implies that heat propagates γ grid points over one time step, the above equation can be written as:

$$K = \frac{\kappa\tau}{\Delta^2} = \frac{\gamma^2}{2d} \tag{35}$$

When $d = 2$, γ is chosen to satisfy $\gamma < 1/\sqrt{2}$, for example, $\gamma = 0.5$. It should be noted that the Crank-Nicolson scheme is unconditionally stable, imposing no restrictions on parameter K. However, matrix $\boldsymbol{A}$ is invertible, requiring $K < \frac{1}{2d}$ [17].

3 Experimental Results

3.1 Dataset and Experimental Configuration

The dataset used in this experiment includes 500 images of the "bridge" category from the LSUN dataset, each with a resolution of 256 × 256. Among them, 400 images are used as the training set and 100 images are used as the test set. 1000 images with a resolution of 256 × 256 were selected from the ImageNet dataset, of which 800 images were used as the training set and 200 images as the test set.

The training for this experiment was conducted on a computer equipped with a RTX 4090D GPU and 48GB of video memory, utilizing Python version 3.9.13 and the deep learning framework PyTorch version 1.13.1.

We demonstrate that, under the same number of iterations as DDPM, HDM generates images of higher quality. HDM demonstrates superior performance with a 3.0 point reduction in FID on 256 × 256 LSUN dataset and a 1.2 point FID reduction on 256 × 256 ImageNet dataset. To ensure that the number of neural network evaluations required during sampling matches that of previous work, we set $N = 1000$ for all experiments.

3.2 Parameter Selection and Experiment

In the experiment, we set the initial α_n to 0.9999, and the final α_n to 0.92. During this period, it changed in a cosine scheduling manner. The learning rate of the AdamW optimizer used is 0.0003. The unconditional training was carried out for a total of 500 rounds. Furthermore, in this model, the EMA strategy is employed. The weight update decay rate of this strategy is 0.995.

To determine the value of θ and K in the matrix, we conducted training on 64×64 images from the CIFAR-10 dataset. Figure 3 illustrates the variation in FID scores of generated images with respect to changes in θ and K values, where all FID scores were obtained with a sampling step count of 1000.

For 0 sampling steps, the FID score is practically meaningless. At 1000 sampling steps, all three FID scores are below 50; thus, the vertical axis is simply set to 50 for zero sampling steps. Figure 3 demonstrates that when $\theta = 1/2$, the unconditionally stable scheme balances explicit efficiency with implicit accuracy, resulting in smoother interactions between adjacent pixels, enhanced detail preservation, reduced numerical oscillations, and lower FID scores for the generated images. When $\theta = 1$, the pixel values of the current time step only depend on the adjacent pixel values of the previous time step. This will lead to excessive smoothing of high-frequency details or incoherent local structures, and the optimization of generation performance is limited. When $\theta = 0$, the current pixel value is completely determined by the values of adjacent pixels at the current time step. This configuration suppresses high-frequency noise at the expense of fine details, resulting in a reduction in the diversity of the generated image. Therefore, θ is uniformly set to 1/2 throughout our experiment.

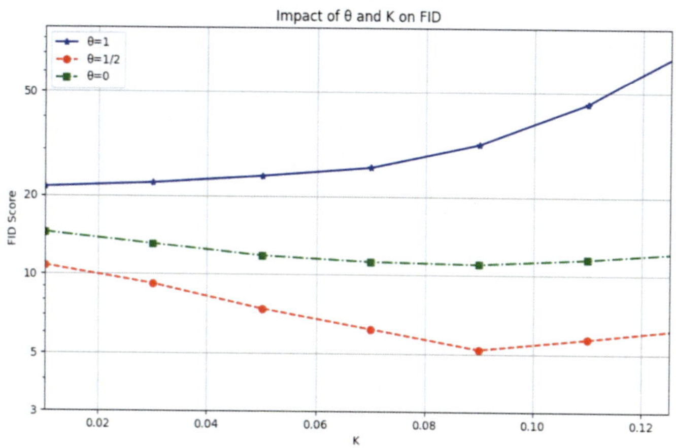

Fig. 3. The impact of θ and K on the FID score.

In matrix A, another critical parameter is the coefficient K. Figure 4 shows the effect of the change of coefficient K on the FID score of HDM. This image was obtained through training on the ImageNet dataset at 256×256 resolution.

When $K = 0$, matrix A reduces to an identity matrix, and the corresponding FID score equals that of the standard DDPM. As shown in Fig. 4, when K is too small, the weights of the non-diagonal elements of matrix A are small, and the interactions between pixels are almost negligible. While high-frequency details can be largely preserved, the lack of inter-pixel cooperative optimization may result in isolated noise artifacts or locally incoherent regions. The optimization efficacy is not significantly pronounced.

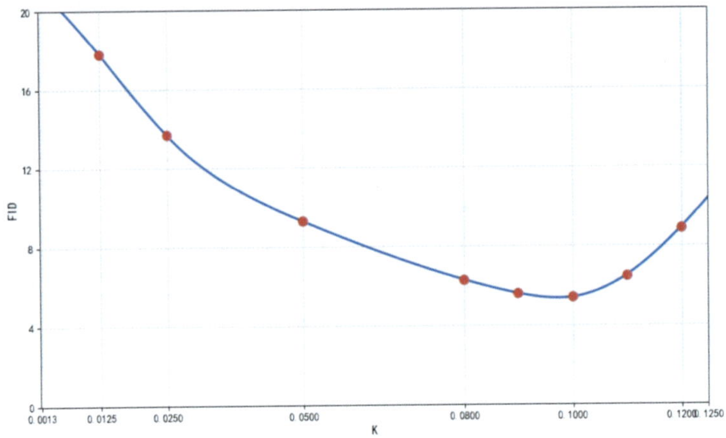

Fig. 4. The impact of the coefficient K on the FID score.

The monotonic increase in the K -value initially enhances the denoising capability of the model through the introduction of structured perturbations that conform to natural image degradation patterns, thereby improving generation quality and reducing the FID score. However, as K approaches the critical threshold of 0.125, the excessive attenuation of the central weighting coefficient leads to error accumulation during iterative updates, while the overly strong global dependencies between pixels induce numerical instability. This manifests as amplified high-frequency artifacts and compromised local consistency, coupled with excessive smoothing of critical high-frequency features, ultimately resulting in the observed rebound of the FID score. Values exceeding 0.125 are meaningless due to the aforementioned stability boundary condition, rendering further discussion of FID variations in this regime unnecessary.

Figure 4 demonstrates that K values between 0.08 and 0.11 enable the model to achieve optimal generation performance, with marginal further improvement beyond this range. In our experiments, $K = 0.095$.

To validate that HDM also outperforms DDPM in generating high-dimensional images, additional tests were conducted on 256×256 images from the LSUN dataset and 256×256 images from the ImageNet dataset. Table 2 presents the FID scores obtained after training on the ImageNet dataset, demonstrating that HDM not only achieves better scores than DDPM but also outperforms subsequent models such as CM and LDM. Furthermore, its performance was superior to that of most of the models in the AR models.

It is worth noting that the HDM effect is still not as good as that of larger-scale models such as the DiT-XL/2. This requires further expansion of the model's scale. Compared with models of similar scale, the effect has improved to a certain extent.

Figure 5 illustrates the FID scores of HDM on the corresponding datasets from LSUN.

In Fig. 5, some of the generated images are shown. On this dataset, the FID score of HDM is 4.22, which is slightly better than that of DDPM.

Table 2. ImageNet 256 × 256 results.

Type	Model	FID↓	IS↑	Pre↑	Rec↑	#Para	#Step
GAN	BigGAN	6.95	224.5	0.89	0.38	112M	1
	GigaGan	3.45	225.50	0.84	0.61	569M	1
Diff.	ADM	10.94	101.00	0.69	0.63	554M	250
	CM	4.88	158.7	–	–	–	8100
	LDM-4-G	3.60	247.7	–	–	400M	250
	DiT-L/2	5.02	167.2	0.75	0.57	458M	250
	DiT-XL/2	2.27	278.2	0.83	0.57	675M	250
	Ours	**3.49**	239.02	0.82	0.57	341M	1000
Mask	MaskGIT	6.18	182.1	0.80	0.51	227M	8
	RCG (cond.)	3.49	215.5	–	–	502M	20
AR	VQGAN	15.78	74.3	–	–	1.4B	256
	VQGAN-re	5.20	280.3	–	–	1.4B	256
	ViTVQ	4.17	175.1	–	–	1.7B	1024
	RQTran.	7.55	134.0	–	–	3.8B	68
	RQTran.-re	3.80	323.7	–	–	3.8B	68
	SPAE	4.41	133.03	–	–	–	100
	LlamaGen-B	5.46	193.61	0.83	0.45	111M	250
VAR	VAR-d16	3.30	274.4	0.84	0.51	310M	10

Fig. 5. LSUN Bridge samples.

HDM can also guide the generation direction through conditional control. The figure below shows the front-end process and generation process diagram when using HDM for condition generation.

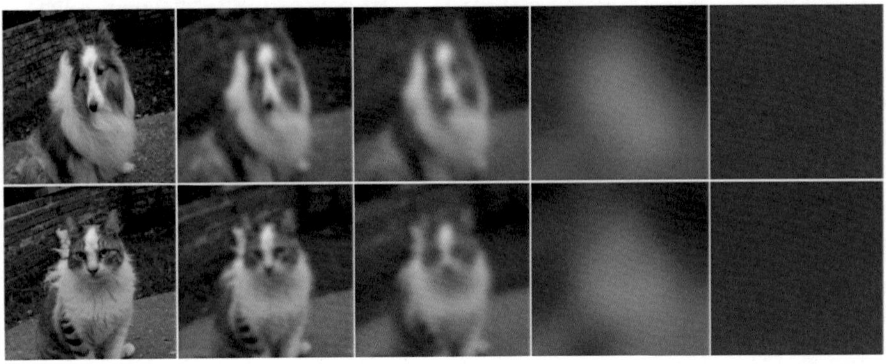

Fig. 6. The forward and reverse process images of HDM.

In Fig. 6, the first line from left to right represents the diffusion process, and the second line from right to left represents the conditional generation process. As can be seen, HDM also achieves good results when generating using the classifier-free guidance condition. However, the topic of condition generation is not the main focus of this article, so it will not be elaborated upon in detail.

In order to compare the generation effect more intuitively, when we add noise to the input image in the forward process, we choose not to completely destroy the original image into a pure noise image, so as to predict the noise and achieve the denoising process.

Fig. 7. Comparison: The first from the left is the original image, the second from the left is the generated image of DDPM, the very middle one is the generated image of DDIM, the second from the right is the generated image of LDM and the first from the right is the generated image of HDM.

By comparing the figures in Fig. 7, it can be observed that HDM performs better in handling the details of images. For instance, in the "tongue" section of the figure, the details are more natural and realistic, while the other models do not handle the details adequately.

To further validate the optimization of HDM in improving the quality of generated images, we conducted an ablation study. In the experiments, the matrix A representing the positional and quantitative relationships between pixels was replaced with a random matrix, where each element was assigned a randomly generated weight, while ensuring that the input and output data of the diffusion and generation processes remained in vector or matrix form. Subsequently, DDPM, HDM, and the ablation experiment were trained and tested on the ImageNet 256 × 256 dataset, and the FID scores of each model as a function of the number of sampling steps are illustrated in Fig. 8.

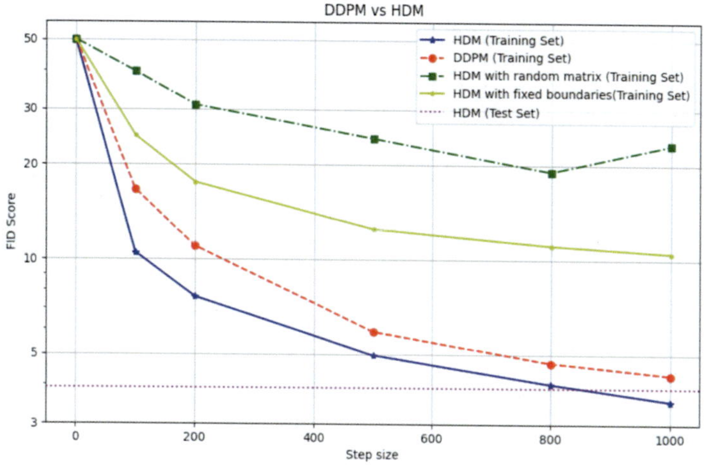

Fig. 8. The variation of FID with respect to the number of sampling steps.

As can be seen from Fig. 8, the relational matrix incorporated in HDM achieves optimization over DDPM. The initial FID score was set to 50 for graphical convenience, because by 100 steps, the FID scores of all three models had fallen below 50.

It is worth noting that there is a model called IHDM (Inverse Heat Diffusion Model), which generates images by randomly reversing the heat equation. The forward process of IHDM performs multi-scale blurring using the Laplacian operator. High-frequency details disappear first, followed by the low-frequency structures. The generation process mainly introduces structure-aware noise, using the noise as a regularization term to prevent the complete disappearance of low-frequency signals [23]. It also employed the heat equation, but it did not have a greater correlation with our algorithm. Its FID score on the 64*64 images of the CIFAR-10 dataset is 18.86, which is much higher than that of DDPM and HDM. On the 256*256 images of the AFHQ dataset, its FID score is even as high as 43.39. Its method is relatively novel, but the generated image quality is not very satisfactory.

During the experiments, we observed that the generation time for HDM and DDPM differ, however, the specific comparison of time consumption was unclear. Therefore, to determine whether the generation process of HDM requires significantly more time, we conducted a comparative experiment. The figure below illustrates the impact of the coefficient K on the generation time of HDM. The experiments were conducted on all the 256 × 256 images in the ImageNet dataset. The generation time mentioned here refers to the total time used throughout the entire generation process. It is the total time taken for the trained model to generate 50,000 images with a resolution of 256 × 256.

As can be observed from Fig. 9, as the parameter K changes, the generation time of the model also varies. Since matrix A is fixed and predefined, it does not contain learnable parameters. Therefore, the number of trainable parameters of HDM is basically the same

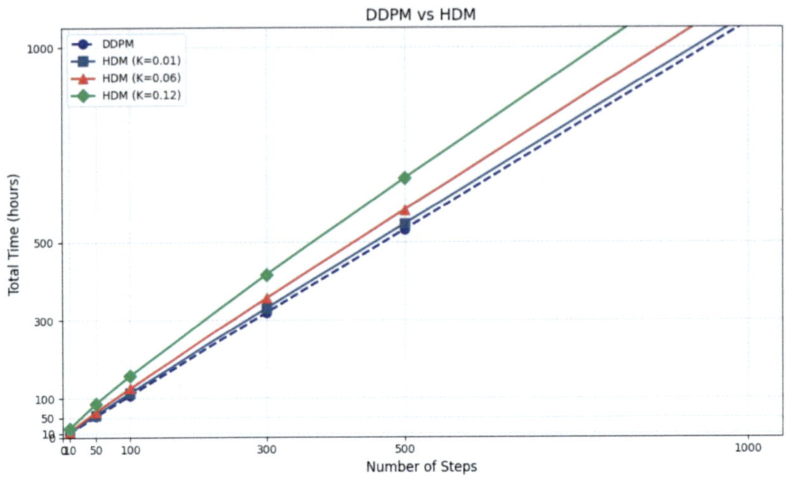

Fig. 9. Comparison of generation time among the models.

as that of DDPM. However, due to the influence of factors such as equipment, during the actual generation process, the average time required for HDM to generate each image is 5 to 8 s longer than that of DDPM. Therefore, in the experiments, to ensure image quality while optimizing performance, the optimal choice for the parameter K lies within the range of 0.08 and 0.1.

4 Conclusions

This paper presents a diffusion model that incorporates an inter-pixel attention mechanism. The generated images by HDM have lower FID scores compared to those of DDPM, CM, most models in AR and other models, while retaining more image details. However, its generation time is extremely long, and its generation efficiency is unable to compete with advanced models. This is likely to become the main research focus in the next step. Moreover, further improving the authenticity of the generated images is also one of the key focuses of future work.

References

1. Ho, J., Jain, A., Abbeel, P.: Denoising Diffusion Probabilistic Models. NeurIPS (2020)
2. Song, J., et al.: Denoising Diffusion Implicit Models. arXiv abs/2010.02502 (2020)
3. Che, T. et al.: Your GAN is secretly an energy-based model and you should use discriminator driven latent sampling. arXiv preprint https://arxiv.org/abs/2003.06060 (2020)
4. Kong, X., Brekelmans, R., Steeg, G.V.: Information-Theoretic Diffusion. ICLR (2023)
5. Song, Y., Sohl-Dickstein, J., Kingma, D.P., Kumar, A., Er-mon, S., Pool, B.: Score-Based Generative Modeling Through Stochastic Differential Equations. ICLR (2021)
6. Karras, T., Aittala, M., Aila, T., and Laine, S. Elucidating the Design Space of Diffusion-Based Generative Models 2022.

7. Stevens, T.S.W., Nolan, O., Robert, J.-L., Van Sloun, R.J.G.: Sequential posterior sampling with diffusion models. In: ICASSP 2025–2025 IEEE International Conference on Acoustics, Speech and Signal Processing (ICASSP), Hyderabad, India (2025)
8. Huang, Z., Wang, K., Xiao, Y., Xiang, Z.: Research and application of LSTM model and diffusion model. In: 2024 IEEE 2nd International Conference on Sensors, Electronics and Computer Engineering (ICSECE), Jinzhou, China (2024)
9. Čeović, H., Šilić, M., Delač, G., Vladimir, K.: An overview of diffusion models for text generation. In: 2023 46th MIPRO ICT and Electronics Convention (MIPRO), Opatija, Croatia (2023)
10. Nichol, A. et al. and Glide: towards photorealistic image generation and editing with text-guided diffusion models. arXiv preprintar Xiv:2112.10741 (2021)
11. Kawar, B., Elad, M., Ermon, S., Song, J.: Denoising Diffusion Restoration Models. In: Advances in Neural In-formation Processing Systems (2022)
12. Bao, F. et al.: Vidu: a highly consistent, dynamic and skilled text-to-video generator with diffusion models. arXiv preprint https://arxiv.org/abs/2405.04233 (2024)
13. Cho, E., Bang, J., Rhu, M.: Characterization and analysis of text-to-image diffusion models. IEEE Comput. Archit. Lett. **23**(2), 227–230 (2024)
14. Rombach, R., Blattmann, A., Lorenz, D., Esser, P., Ommer, B.: High-resolution image synthesis with latent diffusion models. In: Proceedings of the IEEE/CVF Conference on Computer Vision and Pattern Recognition, pp. 10684–10695 (2022)
15. Bai, Y., Geng, X., Mangalam, K., Bar, A., Yuille, A.L., Darrell, T., Malik, J., Efros, A.A.: Sequential modeling enables scalable learning for large vision models. In: Proceedings of the IEEE/CVF Conference on Computer Vision and Pattern Recognition, pp. 22861–22872 (2024)
16. Austin, J., Johnson, D.D., Ho, J., Tarlow, D., van den Berg, R.: Structured denoising diffusion models in discretestate-spaces. In: Advances in Neural Information Processing Systems (2021)
17. Heat transfer - advances in fundamentals and applications. https://doi.org/10.5772/INTECHOPEN.107587 (2024)
18. Saharia, C. et al.et al.: Photorealistic text-to-image diffusion models with deep language understanding. arXiv preprint https://arxiv.org/abs/2205.11487 (2022)
19. Zheng, H., Nie, W., Vahdat, A., Azizzadenesheli, K., Anandkumar, A.: Fast sampling of diffusion models via operator learning. arXiv preprint https://arxiv.org/abs/2211.13449 (2022)
20. Lu, C., Zhou, Y., Bao, F., Chen, J., Li, C., Zhu, J.: Dpm-solver: a fast ode solver for diffusion probabilistic model sampling in around 10 steps. arXiv preprint https://arxiv.org/abs/2206.00927 (2022)
21. Karras, T., Laine, S., Aittala, M., Hellsten, J., Lehtinen, J., Aila, T.: Analyzing and improving the image quality of StyleGAN. In: Proceedings of the IEEE/CVF Conference on Computer Vision and Pattern Recognition, pp. 8110–8119 (2020)
22. Podell, D., English, Z., Lacey, K., Blattmann, A., Dockhorn, T., Müller, J., Penna, J., Rombach, R.: Sdxl: improving latent diffusion models for high-resolution image synthesis. arXiv preprint https://arxiv.org/abs/2307.01952 (2023)
23. Rissanen, S., Heinonen, M., Solin, A.: Generative modelling with inverse heat dissipation, arXiv preprint https://arxiv.org/abs/2206.13397 (2022)

AI-Driven Driver Monitoring and Assistance System for Enhanced Road Safety

Siddharth S(✉) and Jayashree R

CSE-AIML, PES University, Bengaluru, India
siddharthsunil4545@gmail.com, jayashree@pes.edu

Abstract. This paper presents an AI-Driven Driver Monitoring and Assistance System for Enhanced Road Safety, leveraging advanced deep learning and computer vision techniques to detect driver distraction and drowsiness in real time. The system is designed to enhance driving safety by continuously analyzing the driver's facial cues and behavioral patterns. This approach involves training and evaluating multiple architectures, including Convolutional Neural Network (CNN), MobileNetV3, GoogleNet, and YOLOv8, to study their performance in terms of accuracy, speed, and real-time efficiency. The implementation incorporates image preprocessing, data augmentation, and optimized frame-based analysis to ensure stable operation under varying driving conditions. When distraction is detected, the system issues an audio alert and a display message to capture the driver's attention, while in cases of drowsiness, an audio beep is triggered, and the lane-keeping assistance system is automatically activated to prevent vehicle drifting and avoid potential collisions. Experimental results highlight the effectiveness of the trained models, demonstrating reliable detection with minimal latency. The results show that YOLOv8 achieves 92.08% accuracy, followed by GoogleNet at 89.05%, MobileNetV3 at 87.07%, and the Sequential CNN at 84.02%. This work contributes to the advancement of intelligent driver assistance systems, aiming to re-duce accidents caused by inattention and fatigue while ensuring safer and more responsive vehicle control.

Keywords: Driver monitoring system · Deep learning · Computer vision · Distraction and drowsiness detection · YOLOv8 · Lane-keeping assistance · Road safety

1 Introduction

Road safety is a critical global concern, with driver distraction and drowsiness identified as major causes of traffic accidents. Effective driver attention and alertness are essential for safe vehicle operation, yet many accidents occur due to human factors such as fatigue and other in-vehicle distractions such as mobile phone usage, eating, drinking, or engaging in conversations with passengers. These activities divert the driver's focus from the road, significantly increasing the risk of collisions. The growing number of such incidents highlights the urgent need for intelligent systems capable of monitoring driver behavior and providing timely interventions to prevent accidents.

L. Huang and D. Greenhalgh (Eds.): ICMLC 2026, LNNS 2061, pp. 18–38, 2027.
https://doi.org/10.1007/978-3-032-31236-5_2

Advancements in computer vision and deep learning technologies have paved the way for the development of AI-driven driver monitoring systems that can analyze driver states in real time. Such systems can identify multiple forms of distraction and drowsiness, including drinking, eating, talking on the phone, and conversing with passengers, through visual cues such as facial expressions, eye closure, head pose, and hand movements, and respond appropriately to ensure road safety. This paper introduces an AI-Driven Driver Monitoring and Assistance System for Enhanced Road Safety, which leverages advanced deep learning architectures to detect driver inattention and initiate corrective actions.

The proposed system utilizes multiple deep learning models including Convolutional Neural Networks (CNNs), MobileNetV3, GoogleNet, and YOLOv8 trained and evaluated to analyze driver behaviors such as drinking, eating, talking on the phone, conversing with passengers, and drowsiness. Comprehensive image preprocessing and data augmentation techniques are implemented to enhance model robustness under different lighting and environmental conditions. Upon detecting distraction, the system generates an audio alert and displays a message to capture the driver's attention. In cases of drowsiness, an audio beep is triggered, and the lane-keeping assistance system is automatically activated to prevent vehicle drifting and avoid potential collisions.

By integrating deep learning-based behavioral analysis with real-time vehicle assistance mechanisms, the proposed framework represents a significant advancement in intelligent transportation systems. It combines accuracy, low latency, and automation to ensure timely detection and intervention during critical moments, offering a proactive solution to mitigate accidents caused by human inattention. This research contributes to the ongoing development of advanced driver-assistance systems (ADAS) by enhancing driver safety, reducing accident rates, and promoting a safer, more reliable, and technology-driven driving experience for the modern world.

2 Related Work

The related work in driver monitoring, distraction detection, and drowsiness assessment highlights significant progress enabled through computer vision and deep learning techniques. Supraja et al. [1] proposed a VGG-16 based distraction classification model using transfer learning, demonstrating strong performance in identifying unsafe driver behaviors. Vijay [2] integrated MobileNetV2 for face verification along with fatigue and intoxication detection, emphasizing the efficiency of lightweight architectures for real-time embedded deployment. Classical approaches such as Haar Cascades and Dlib landmark extraction were used by Jose et al. [3] and Prasath et al. [4] to derive EAR and MAR metrics for fatigue assessment. Although effective, these methods show limitations under varying illumination, occlusions, and driver pose changes. To improve robustness, Ayachi et al. [5] employed EfficientDet for eye and mouth state classification and computed fatigue metrics such as PERCLOS. Kannan and Jahnavi [6] also demonstrated high accuracy in EAR-based alert systems. Chen [7] explored multimodal fatigue detection using physiological and visual cues with SVM, GBDT, and HMM classifiers. More recent studies leverage deeper architectures. Khadraoui et al. [8] combined ResNet and FPN to enhance multi-scale feature extraction, while Zhang et al. [9] used a fine-tuned InceptionV3 model for accurate eye-state detection on embedded hardware.

Doe et al. [10] introduced an ensemble of EfficientNetV2B0 and BlazeFace to improve robustness under varying lighting and occlusion conditions. Additional contributions expand into distracted-driving recognition and ADAS integration. Dwivedi et al. [11] applied MobileNet for distracted-driver classification. Abtahi et al. [12] focused on yawning detection for fatigue prediction. Johns et al. [13] presented PERCLOS-based blink monitoring for automotive fatigue measurement. Singh et al. [14] combined YOLO with CNN models for real-time drowsiness detection. Kumar et al. [15] proposed an attention-enhanced deep learning framework for simultaneous distraction and fatigue detection. Hong et al. [16] developed a transfer learning-based lane keeping assist framework that improves steering stability and lane adherence in self-driving scenarios. Their approach leverages fine-tuning of pre-trained deep models to enhance generalization across different road environments and driving conditions. Li et al. [17] proposed a deep CNN-based vision system capable of accurate lane boundary detection and real-time lane tracking under varying road conditions. The study emphasizes robustness against illumination changes and complex lane geometries to maintain consistent tracking performance. Mehta et al. [18] integrated deep learning-driven lane detection with vehicle control mechanisms to enhance automated steering performance in ADAS applications. Their framework establishes a direct link between perception and control modules for smoother and more stable steering adjustments. Park et al. [19] implemented a CNN-based semantic segmentation model to achieve robust lane recognition even under complex road environments such as shadows, curves, and worn markings. The segmentation-based approach improves pixel-level lane localization accuracy in challenging scenarios. Al Sharif et al. [20] introduced an end-to-end deep learning architecture that simultaneously performs lane detection and steering angle prediction for autonomous driving systems. This unified framework reduces dependency on handcrafted features and enables direct mapping from visual input to control output. Reddy et al. [21] presented a practical deep learning-based vision framework designed for reliable lane keeping and driver assistance in real-world driving environments. Their system demonstrates stable performance under practical constraints such as dynamic traffic conditions and real-time computational limitations. Together, these studies offer a comprehensive overview of the advancements and methodologies in driver behavior analysis and lane assistance systems, each contributing to the development of increasingly sophisticated and reliable driver monitoring and assistance technologies.

3 Proposed Methodology

This part of the paper delves into an explanation of the methodology covering aspects such as the characteristics of various driver distraction and drowsiness states, dataset collection, data preprocessing techniques, feature extraction methods, the deep learning algorithms used in this research, and the intervention mechanisms implemented to ensure real-time driver safety.

3.1 Dataset Description

The dataset used in this project supports two major components of the system. The first focuses on driver distraction and drowsiness detection, using image datasets that capture various unsafe behaviors such as texting, talking on the phone, eating, drinking, and interacting with passengers, along with fatigue indicators like eye closure and yawning. These data samples help the model accurately classify different distraction and drowsiness states in real time. The second component supports the Lane-Keeping Assist System, which uses dash-camera footage of real-world driving conditions to learn lane structures, detect lane boundaries, and identify unintended lane departures across diverse road, traffic, and lighting scenarios. Together, these datasets provide the behavioral and environmental information required for reliable driver monitoring and lane-keeping assistance.

3.2 Data Preprocessing

The images are subjected to preprocessing in order to prepare it for analysis. This step includes the following:

RGB to Grayscale Conversion. Grayscale conversion is an essential preprocessing step where color images are transformed into grayscale. This process reduces the computational complexity by eliminating the color information, which is not necessary for gesture recognition. In grayscale images, each pixel represents the intensity of light, allowing the model to focus solely on the structural features of the hand gestures. This simplification helps in speeding up the training process and reduces the amount of data the model needs to process.

Normalization. Normalization scales the pixel values of images to a range between 0 and 1. This step is crucial for ensuring that the input data has a consistent scale, which helps in stabilizing and speeding up the training process. By standardizing the pixel values, the model can learn more effectively, avoiding issues related to varying data scales, such as gradient instability. Normalization ensures that each feature contributes equally to the learning process, improving the overall performance of the model.

Resizing. Resizing is performed to standardize the input image dimensions before feeding them into the model, ensuring that all driver images follow a uniform format suitable for CNN-based architectures. This step ensures consistency across the dataset, allowing the model to process each frame in a structured and predictable manner. Resizing helps preserve the spatial integrity of facial features such as the eyes, mouth, and head orientation while reducing the computational load, making it easier for various CNN architectures, including MobileNetV3, GoogleNet and YOLO based models, to handle large volumes of data efficiently. This improves both training speed and real-time performance in detecting driver distraction and drowsiness.

Data Augmentation. Data augmentation is essential for improving the generalization of the driver monitoring system. Several techniques are applied, rotation, width and height shifts to account for natural variations in the driver's position within the frame, shear transformations to handle angular distortions caused by camera placement and zoom transformations to represent changes in driver distance from the camera. Horizontal flips are also included, as driving scenarios differ across countries vehicles may be left-hand or right-hand driven resulting in mirrored visual patterns of driver behavior.

Additionally, contrast adjustment is applied to compensate for fluctuating lighting conditions inside the vehicle, such as bright sunlight, shadows, or low-light environments. These augmentations enhance the model's robustness and accuracy, ensuring reliable performance under diverse real-world driving conditions.

3.3 Driver Distraction and Drowsiness Detection Models

Deep learning models like Sequential CNN, MobileNetV3, GoogleNet, and YOLOv8 are used to detect driver distraction and drowsiness. Due to their deep and efficient architectures, these models can effectively learn intricate facial and behavioral patterns, enabling accurate and reliable real-time classification of the driver's state.

Sequential Convolutional Neural Network. Convolutional Neural Networks (CNNs) form a class of deep learning models widely used for image recognition and classification tasks involving visual input, making them suitable for analyzing driver behavior. The convolutional layer serves as the first stage, where convolution operations are performed between the input driver image and an $M \times M$ filter. By sliding the filter across the image and computing dot products, the network generates feature maps that capture critical visual information including edges, textures, shapes, and structural facial patterns. Following this, the pooling layer reduces the spatial dimensions of the feature maps and lowers computational cost through max pooling, which selects the highest value from each region of the feature map. Before classification, the reduced feature map is flattened into a one-dimensional vector and passed to the fully connected (FC) layer. The FC layer uses weights, biases, and interconnected neurons to learn discriminative features and support the classification of driver states. Finally, a softmax activation function in the output layer produces probability values corresponding to driver distraction or drowsiness classes, enabling accurate real-time prediction. The analysis of the Sequential CNN model showed signs of overfitting, with validation accuracy falling behind training accuracy and limiting its ability to generalize real-world driving conditions. To improve robustness, several optimization strategies were applied, including prefetching for faster data loading, regularization techniques, early stopping, learning rate reduction, and autotuning to stabilize training. Architectural improvements were also explored by incorporating MobileNetV3 and GoogleNet, using components like global average pooling to reduce overfitting and enhance feature extraction. These combined measures strengthened model generalization and improved the reliability of driver distraction and drowsiness detection.

MobileNetV3. MobileNetV3 begins with standard convolutional layers that extract essential low-level features from driver images. Its inverted residual blocks with linear bottlenecks enable efficient extraction of the facial and behavioral cues needed for distraction and drowsiness detection. The architecture minimizes computation through depthwise separable convolutions while preserving strong accuracy. Squeeze-and-Excitation (SE) blocks further refine channel-wise feature importance, and advanced activation functions like ReLU and hard-swish improve representational capability. Global average pooling compresses feature maps into a compact vector to reduce overfitting before classification through a fully connected layer with softmax activation. Training is optimized using the Adam optimizer to minimize cross-entropy loss. Overall,

MobileNetV3's lightweight and efficient structure makes it well-suited for deployment on embedded systems and for real-time driver monitoring applications.

GoogLeNet. GoogLeNet is highly effective for driver distraction and drowsiness detection due to its Inception architecture, which processes the input through multiple convolutional and pooling operations running in parallel. This design allows the network to extract features at various scales, capturing both fine and broad facial patterns relevant to driver behavior. The use of 1×1 convolutions significantly reduce feature-map dimensions, lowering computational complexity while enabling deeper feature extraction. With several stacked Inception modules, the model learns progressively more detailed representations of eye states, facial expressions, and head movements. Auxiliary classifiers integrated within the network help stabilize training, address vanishing gradients, and introduce regularization. Additionally, global average pooling minimizes parameter count and reduces overfitting, improving the model's ability to generalize across different drivers and lighting conditions. The final fully connected layer with a softmax activation outputs the probability scores for each distraction or drowsiness class.

YOLOv8. YOLOv8 was chosen because traditional CNN-based models like MobileNetV3 and GoogleNet, although they achieved high accuracy, were unable to provide the real-time responsiveness required for driver monitoring. Unlike frame-wise classifiers, YOLOv8 performs end-to-end object detection in a single pass, enabling fast and direct identification of distraction and drowsiness states. Its capability to localize driver behavior using bounding boxes further enhances interpretability and system transparency. Overall, YOLOv8 offers the ideal balance of speed, accuracy, and explainability needed for safety-critical real-time applications. YOLOv8 is a high-performance, real-time object detection model ideal for driver distraction and drowsiness monitoring because it performs full-frame inference in a single forward pass, enabling fast and accurate predictions with extremely low latency. Its CSP-based backbone (Cross-Stage Partial Network) improves feature extraction by splitting and merging gradient flow to reduce computation while retaining representational strength. The PAN-FPN (Path Aggregation Network combined with Feature Pyramid Network) structure enhances multi-scale feature fusion, allowing the model to detect subtle driver cues such as eye closure, yawning, and fine hand movements across varying spatial resolutions. The C2f (Cross-Stage Feature Fusion) modules strengthen feature representation by enhancing gradient propagation and capturing fine-grained behavior patterns critical for safety monitoring. To improve robustness and reduce confusion among visually similar driver behaviors, the dataset labels were reorganized into broader, safety-oriented categories prior to training. Additionally, YOLOv8's anchor-free detection head simplifies object localization by predicting keypoint based bounding boxes without predefined anchor shapes, while its optimized loss functions improve convergence and detection precision. Collectively, these architectural innovations provide the speed, accuracy, and interpretability required for safety-critical, real-time driver monitoring applications.

3.4 Responsive Alert System and Intervention Mechanism

The intervention mechanism is triggered when the system identifies distraction states such as eating, drinking, mobile phone usage, or talking with a passenger while driving by issuing an immediate auditory alert along with an on-screen visual warning to capture the driver's attention. In the case of drowsiness detection, a continuous high-priority alarm is activated to counter reduced alertness. At the same moment, the lane-keeping assist system is automatically enabled, utilizing real-time lane detection and lateral deviation analysis to maintain stable lane positioning and prevent unintended drifting or potential collision. This integrated alert–intervention framework delivers rapid behavioral warnings and automated steering support, ensuring improved vehicle stability and enhanced roadway safety.

Lane-Keeping Assist System. The lane-keeping assist system employs a comprehensive computer-vision pipeline that includes HLS color filtering, grayscale conversion, binary thresholding, Gaussian smoothing, and Canny edge extraction to isolate lane markings under varying illumination conditions. A perspective transform generates a bird's-eye view of the road, enabling robust lane boundary identification through a histogram-based sliding-window search followed by polynomial fitting of left and right lane lines. The system computes lane curvature, classifies curve direction, and estimates the vehicle's lateral deviation from the lane center in real time. Using these measurements, a corrective steering angle is derived by combining deviation-based and curvature-based components, smoothed and constrained within safe operational limits. The final output overlays lane regions, curvature information, and a dynamically rotated steering-wheel visualization, demonstrating the system's capability for real-time lane tracking, vehicle-position estimation, and automated lane-keeping intervention.

Advanced Lane-Keeping and Vehicle Control System. The system integrates object detection, lane detection, temporal tracking, and collision-risk estimation into a unified real-time ADAS pipeline. YOLOv8 is used to detect surrounding vehicles and road users, while the Ultrafast Lane Detector (UFLD) extracts lane boundaries and lane-area geometry from each frame. A perspective transformation converts lane points into a bird's-eye view, enabling accurate computation of lane curvature and lateral offset. A single-camera distance estimation unit calculates vehicle distances and evaluates forward-collision states. Temporal tracking is applied to maintain object consistency across consecutive frames, allowing stable motion analysis and reducing false positives in collision prediction. A visualization panel overlays lane-departure alerts, lane-keeping status, curvature direction, offset information, and collision warnings, providing a clear real-time representation of the system's ADAS decision-making behavior.

Overall, the system integrates distraction detection, drowsiness monitoring, lane analysis, and collision assessment to deliver a real-time driver assistance framework capable of issuing alerts and activating lane-keeping support for enhanced road safety.

4 Results

This section summarizes all the research outcomes.

4.1 Performance Evaluation of Driver Distraction and Drowsiness Detection Models

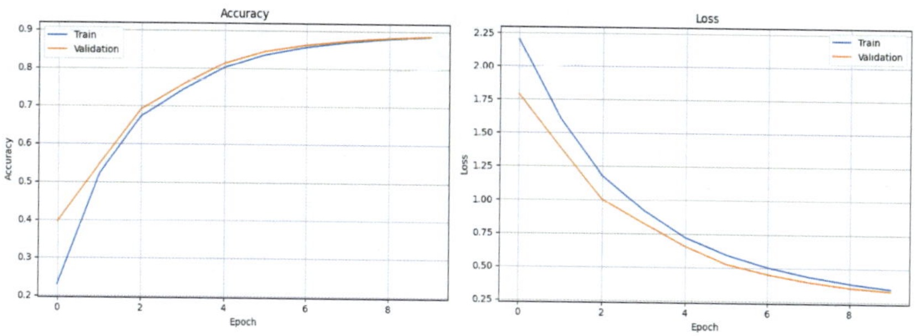

Fig. 1. Graphs illustrating training loss against validation loss & training accuracy against validation accuracy derived from Sequential CNN.

Figure 1 shows the graphs obtained from the Sequential CNN model. Training accuracy increases, and training loss decreases across epochs, indicating effective learning. It achieved an accuracy of 84.02%.

Classification Report:

	precision	recall	f1-score	support
0	0.9919	0.9819	0.9869	498
1	0.9934	1.0000	0.9967	453
2	0.9935	0.9957	0.9946	464
3	0.9915	0.9979	0.9947	469
4	1.0000	0.9957	0.9978	465
5	0.9978	0.9935	0.9957	463
6	0.9831	1.0000	0.9915	465
7	0.9950	0.9900	0.9925	400
8	0.9816	0.9764	0.9790	382
9	0.9858	0.9812	0.9835	426
accuracy			0.9915	4485
macro avg	0.9914	0.9912	0.9913	4485
weighted avg	0.9915	0.9915	0.9915	4485

Fig. 2. Classification report for sequential CNN.

Figure 2 displays the Classification Report for Sequential CNN. The classification report provides precision, recall, and F1-score for all classes, showing strong and consistent performance across the dataset. The high metric values indicate that the model correctly identifies most driver behaviors with minimal misclassification.

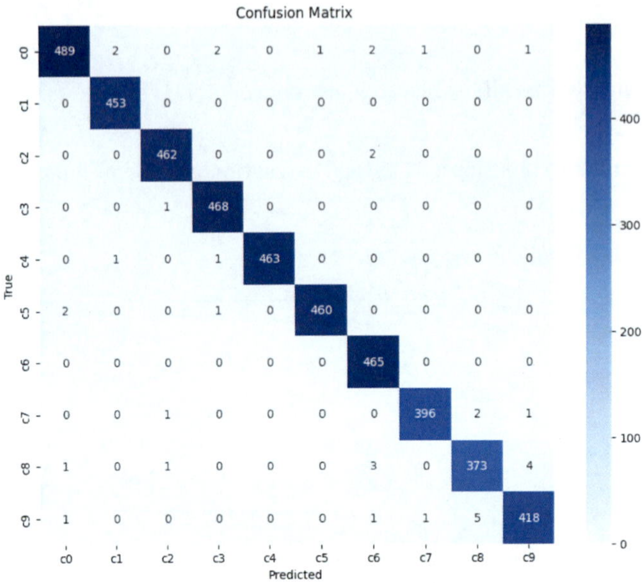

Fig. 3. Confusion matrix of sequential CNN.

In Fig. 3, the confusion matrix shows that the Sequential CNN model achieves good class discrimination. The rows of the matrix show the actual classes, while the columns show the predicted classes. Misclassifications are shown by the off-diagonal elements while the diagonal elements represent valid class predictions.

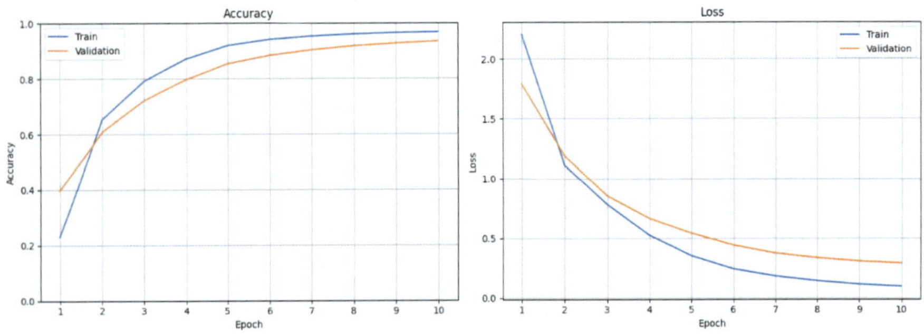

Fig. 4. Graphs illustrating training loss against validation loss & training accuracy against validation accuracy derived from MobileNetV3.

Figure 4 shows the graphs obtained from the MobileNetV3 model. Training accuracy increases, and training loss decreases across epochs, indicating effective learning. It achieved an accuracy of 87.07%.

Figure 5 displays the classification report for MobileNetV3. The classification report provides precision, recall, and F1-score for all classes, showing strong and consistent

Classification Report:

	precision	recall	f1-score	support
0	0.9898	0.9759	0.9828	498
1	0.9912	1.0000	0.9956	453
2	0.9935	0.9935	0.9935	464
3	0.9957	0.9957	0.9957	469
4	0.9892	0.9892	0.9892	465
5	1.0000	0.9957	0.9978	463
6	0.9914	0.9935	0.9925	465
7	0.9852	1.0000	0.9926	400
8	0.9866	0.9634	0.9748	382
9	0.9723	0.9883	0.9802	426
accuracy			0.9897	4485
macro avg	0.9895	0.9895	0.9895	4485
weighted avg	0.9898	0.9897	0.9897	4485

Fig. 5. Classification report for MobileNetV3.

performance across the dataset. The high metric values indicate that the model correctly identifies most driver behaviors with minimal misclassification.

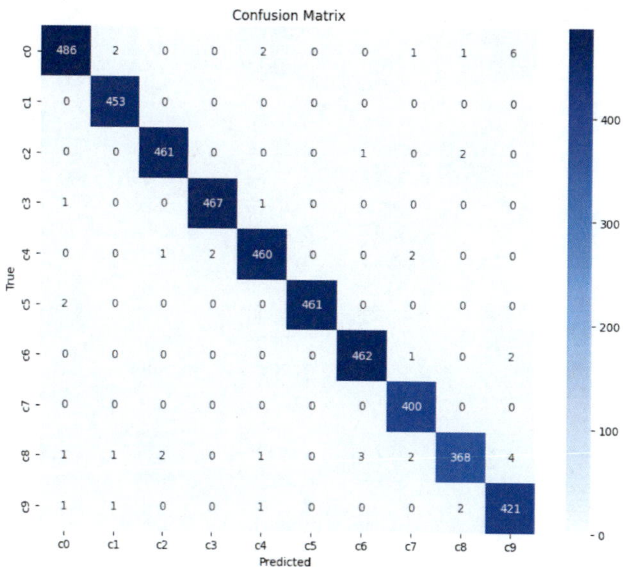

Fig. 6. Confusion matrix of MobileNetV3.

In Fig. 6, the confusion matrix shows that the MobileNetV3 model achieves strong class discrimination, with predictions tightly concentrated along the diagonal and only minimal misclassifications. The low off-diagonal values indicate reduced overlap between similar classes, confirming that this model generalizes better and performs more reliably than the earlier CNN architecture.

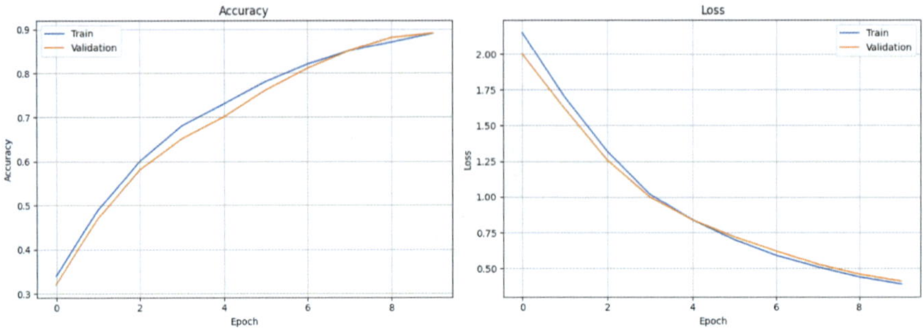

Fig. 7. Graphs illustrating training loss against validation loss & training accuracy against validation accuracy derived from GoogLeNet.

Figure 7 shows the graphs obtained from GoogLeNet model. Training accuracy increases, and training loss decreases across epochs, indicating effective learning. It achieved an accuracy of 89.05%.

Classification Report:

	precision	recall	f1-score	support
c0	0.9100	0.9286	0.9192	490
c1	0.9097	0.9056	0.9077	445
c2	0.9043	0.9083	0.9063	458
c3	0.9093	0.8708	0.8896	472
c4	0.9197	0.8717	0.8951	460
c5	0.8920	0.9137	0.9027	452
c6	0.9132	0.8996	0.9064	468
c7	0.8883	0.8724	0.8803	392
c8	0.8273	0.8995	0.8619	378
c9	0.8961	0.9049	0.9005	410
accuracy			0.8979	4425
macro avg	0.8970	0.8975	0.8970	4425
weighted avg	0.8986	0.8979	0.8979	4425

Fig. 8. Classification report for GoogLeNet.

Figure 8 displays the classification report for GoogLeNet. The classification report provides precision, recall, and F1-score for all classes, showing strong and consistent performance across the dataset. The high metric values indicate that the model correctly identifies most driver behaviors with minimal misclassification.

In Fig. 9, the confusion matrix for the GoogLeNet model reveals highly accurate class recognition, with the majority of the predictions falling precisely on the diagonal. The sparse off-diagonal entries show that incorrect classifications are rare, indicating that GoogLeNet distinguishes between classes with strong consistency. This performance reflects its enhanced ability to generalize and its superior.

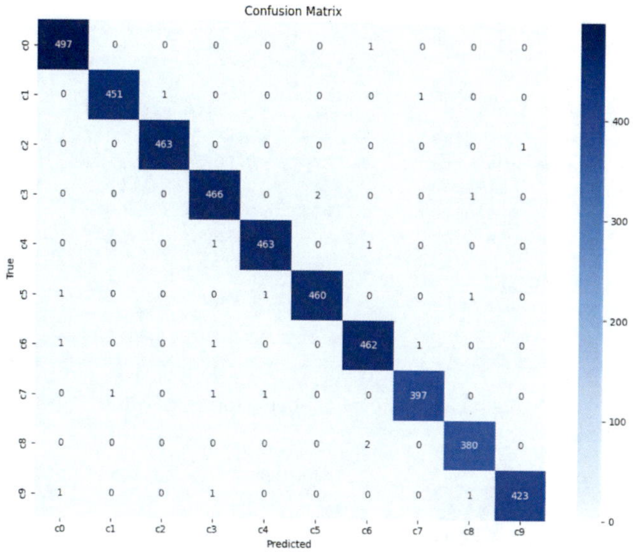

Fig. 9. Confusion matrix of GoogLeNet.

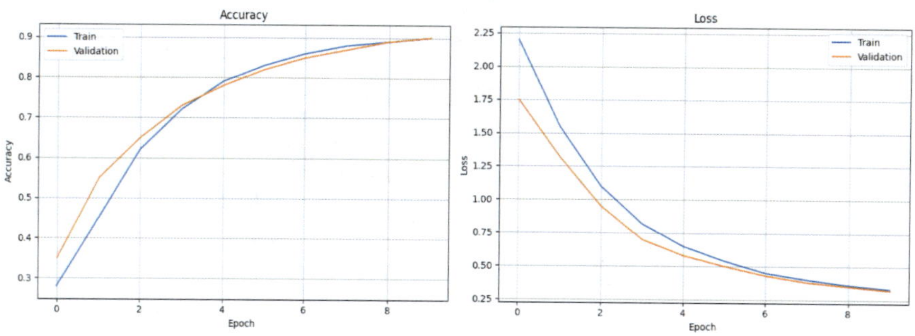

Fig. 10. Graphs illustrating training loss against validation loss & training accuracy against validation accuracy derived from YOLOv8.

Figure 10 shows the graphs obtained from the YOLOv8 model. Training accuracy increases, and training loss decreases across epochs, indicating effective learning. It achieved an accuracy of 92.08%.

Figure 11 displays the classification report for YOLOv8. The classification report provides precision, recall, and F1-score for all classes, showing strong and consistent performance across the dataset. The high metric values indicate that the model correctly identifies most driver behaviors with minimal misclassification.

In Fig. 12, the confusion matrix for the YOLOv8 model shows clear and consistent class recognition, with predictions concentrated almost entirely along the diagonal. The scarce off-diagonal values indicate that misclassifications are minimal, reflecting the model's strong reliability and effective generalization across all classes.

Classification Report:

	precision	recall	f1-score	support
Normal Driving	0.9376	0.9596	0.9485	470
Drowsy Driving	0.9203	0.8896	0.9047	480
Eating While Driving	0.9274	0.9089	0.9181	450
Talking on the Phone	0.9767	0.8767	0.9240	430
Talking to a Passenger	0.8664	0.9304	0.8973	460
Drinking While Driving	0.9009	0.9500	0.9248	440
accuracy			0.9194	2730
macro avg	0.9215	0.9192	0.9195	2730
weighted avg	0.9211	0.9194	0.9195	2730

Fig. 11. Classification report for YOLOv8.

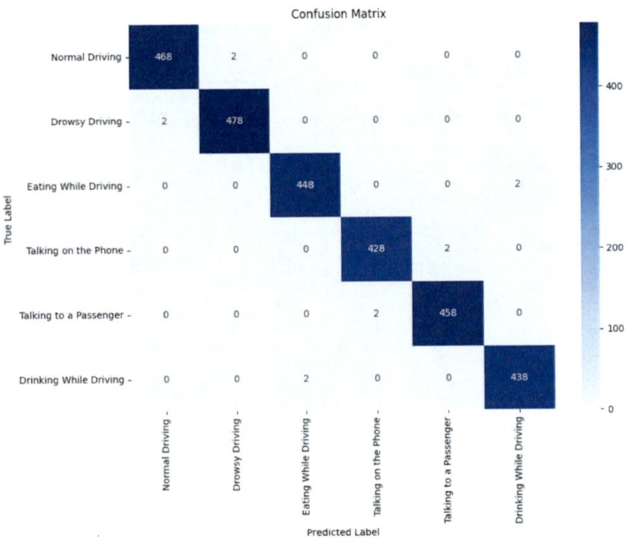

Fig. 12. Confusion matrix of YOLOv8.

4.2 Driver Distraction and Drowsiness Detection Results

The distraction and drowsiness detection system was evaluated across diverse driving scenarios to examine its reliability and performance stability. The evaluation considered variations in illumination conditions, driver postures, and natural behavioral patterns, including phone usage, eating, drinking, and driver–passenger interactions. The proposed model demonstrated low-latency inference and consistent recognition performance, indicating effective and responsive detection of distraction and drowsiness states in dynamic and unconstrained driving environments.

In Fig. 13, Case 1 represents a scenario in which the driver maintains a steady posture with eyes open, hands on the wheel, and no signs of distraction. The system correctly identifies the state as Normal Driving. No alerts are triggered, as the facial landmarks,

Fig. 13. Case 1: normal driving conditions with no detected distraction, indicating an attentive and focused driver.

eye openness, and head orientation remain within safe thresholds. This demonstrates the model's ability to reliably recognize non-distracted, attentive driving conditions.

Fig. 14. Case 2: eating while driving, classified as distracted behavior.

In Fig. 14, Case 2 represents a scenario in which the driver is seen holding or lifting food (e.g.: snacks, sandwich), creating a distraction. The detection system recognizes the hand-to-mouth action and classifies it as Eating While Driving. This action reduces the driver's ability to react quickly and compromises steering control. The model successfully identifies this behavior and flags it as unsafe.

In Fig. 15, Case 3 represents a scenario in which the driver is holding a bottle/cup, and the action is clearly captured by the detection model. YOLO identifies the object and associates it with the driver's distracted state. The system labels the behavior as Drinking While Driving, indicating that both hands are not fully available for vehicle control. The correct detection highlights the model's ability to identify activity-based distractions.

In Fig. 16, Case 4 represents a scenario in which the driver is interacting with a mobile phone, either by holding it to the ear or looking at the screen. YOLO identifies the phone

Fig. 15. Case 3: drinking while driving, classified as distracted behavior.

Fig. 16. Case 4: talking on phone, major distraction during driving identified.

and the hand interaction, leading the system to classify the behavior as Talking on Phone. This scenario represents a high-risk distraction, and the system's correct identification emphasizes its robustness in detecting handheld device usage.

In Fig. 17, Case 5 represents a scenario in which the driver's head pose is turned toward the passenger seat, indicating visual and cognitive distraction. Even though no object is involved, the system analyzes head orientation and facial direction to classify the behavior as Talking to Passenger. The detection correctly identifies the loss of road focus, demonstrating the model's capability to detect subtle, non-object-based distractions.

In Fig. 18, Case 6 represents a scenario in which the image shows visible indicators of drowsiness such as prolonged eye closure, reduced blink frequency, and a downward head tilt. The system accurately classifies the behavior as Drowsy Driving and triggers an alert. This confirms the model's effectiveness in detecting fatigue-related patterns and responding in real time to prevent potential accidents.

Fig. 17. Case 5: talking to a passenger, distraction is observed.

Fig. 18. Case 6: drowsy driver, fatigue symptoms identified, alert triggered.

4.3 Results of the Lane Keeping Assistance System with Intervention Mechanism

The lane keeping assist system was evaluated using diverse driving videos containing straight lanes, curves, varying road geometries, and inconsistent lane-marking visibility. The pipeline comprising perspective transformation, binary masking, polynomial lane fitting, and lateral deviation estimation was assessed for stability and robustness under different lighting and environmental conditions. The system consistently detected lane boundaries, maintained accurate offset estimation, and produced stable steering cues, demonstrating reliable lane-tracking performance across varied road scenarios.

Figure 19 shows how the lane-keeping intervention mechanism functions as an intelligent assistive layer designed to enhance driving safety during moments of driver inattention or fatigue. The system continuously processes incoming road-view video frames to detect left and right lane boundaries, estimate lane curvature, compute the central lane marking, and measure the vehicle's lateral deviation from the lane center. These parameters collectively provide an accurate representation of road geometry and vehicle

Fig. 19. Lane-keeping assistance system in action showing lane detection, curvature estimation, and steering guidance.

positioning, enabling reliable lane continuity tracking under varying driving conditions. Real-time visual overlays of detected lane lines, curvature radius, off-center value, and driving path predictions are rendered on the video feed to clearly communicate the system's interpretation of the road scene.

A key strength of the proposed mechanism lies in its tight integration with the YOLO-based driver monitoring module. When the driver monitoring system detects drowsiness, distraction, or any loss of situational awareness, the lane-keeping assistance is automatically activated, ensuring timely intervention. This design enables proactive steering support rather than delayed corrective action, providing assistance precisely when the driver's ability to maintain lane discipline is compromised. By linking driver state awareness with lane geometry analysis, the system enhances overall driving reliability and safety.

Once activated, the lane-keeping module computes a steering correction angle by combining two primary factors: (i) the curvature and direction of the detected lane, and (ii) the vehicle's lateral offset from the lane center. The computed correction angle is smoothed to avoid abrupt steering behavior and to maintain realistic vehicle dynamics. A transparent digital steering wheel overlay is superimposed on the video stream and rotates proportionally to the required correction, offering intuitive visual guidance to the driver. Experimental evaluation using dashcam-recorded road footage demonstrates that the system operates in real time, accurately detecting lane boundaries, estimating curvature, measuring off-center deviation, and generating continuous steering guidance, thereby minimizing unintended lane departure and lane drift.

Figures 20 and 21 indicate real-time ADAS interface showing object detection, lane-keeping status, bird's-eye lane visualization, dynamic steering guidance, and forward-collision assessment.

Fig. 20. Advanced lane-keeping and vehicle control system.

Fig. 21. Advanced lane-keeping and vehicle control system.

The displayed interface demonstrates the integrated real-time operation of the proposed Advanced Driver Assistance System (ADAS). The system performs object detection using YOLOv8 to identify surrounding road users, including cars, trucks, and motorbikes, with each object enclosed by labeled bounding boxes for continuous tracking and collision-risk assessment. In parallel, the Forward Collision Warning System (FCWS) operates using monocular distance estimation to evaluate the proximity and relative motion of detected vehicles. Based on the computed risk level, the system classifies situations as normal, cautionary, or dangerous and displays appropriate visual warning icons to alert the driver, particularly in dense traffic or sudden braking scenarios.

Lane-related assistance is provided through the combined operation of the Lane Departure Warning System (LDWS) and Lane Keeping Assist System (LKAS). Lane boundaries are detected in real time, and the vehicle's lateral offset from the lane center is continuously measured. LDWS issues on-screen alerts when unintended lane drift is detected, while LKAS analyzes lane curvature and road geometry to provide steering guidance. A dynamic steering-wheel visualization rendered at the bottom-left of the interface rotates proportionally to the computed steering correction angle, indicating whether the vehicle should maintain its direction or adjust left or right to remain centered within the lane. Informative textual cues such as "LDWS: Good Lane Keeping" and "LKAS: Gentle Right Curve Ahead" summarize the current lane status and upcoming road geometry, helping the driver anticipate and respond smoothly to curved or straight road segments.

To further enhance lane understanding and steering accuracy, the system generates a bird's-eye (top-down) lane visualization using perspective transformation. This view provides a clear representation of lane layout, curvature, alignment, and vehicle orientation, improving the accuracy of curvature and offset estimation, especially during turns. All ADAS components of vehicle detection, FCWS risk assessment, LDWS alerts, LKAS steering guidance, bird's-eye lane mapping, and real-time performance indicators are integrated into a unified visualization interface. This comprehensive display delivers continuous situational awareness, enabling the driver to maintain safe lane positioning, anticipate road geometry, and remain alert to surrounding vehicles, thereby enhancing overall driving safety and stability in real time.

5 Conclusion

The AI-Driven Driver Monitoring and Assistance System presented in this work demonstrates a significant advancement in enhancing road safety through the integration of deep learning–based behavioral analysis and intelligent vehicle-assistance mechanisms. By leveraging multiple neural network architectures, including Sequential CNN, MobileNetV3, GoogLeNet, and YOLOv8, the system achieves reliable real-time detection of driver distraction and drowsiness, two major contributors to traffic accidents. Experimental evaluation shows that YOLOv8 attains the highest accuracy of 92.08%, followed by GoogLeNet (89.05%), MobileNetV3 (87.07%), and the Sequential CNN (84.02%), highlighting the effectiveness of modern lightweight and detection-oriented architectures for driver monitoring applications.

To address practical challenges such as dataset imbalance, behavioral variability, and illumination changes, the system incorporates robust preprocessing and training strategies. Data imbalance is mitigated through targeted data augmentation, while behavioral diversity is handled by training on heterogeneous driver samples and augmentation induced variability. Variations in lighting conditions are addressed using photometric augmentation techniques, including brightness and contrast adjustments, enabling the models to generalize effectively across diverse driving environments. These measures collectively contribute to stable and consistent performance under real-world conditions.

In parallel, the lane-keeping assistance module enhances driving safety by providing automated corrective responses based on lane curvature estimation, vehicle offset analysis, and bird's-eye perspective transformation. This integration enables timely intervention during periods of reduced driver alertness, improving vehicle stability, and reducing potential collision risks.

Overall, the proposed framework demonstrates that AI-enabled driver monitoring, when combined with intelligent assistance mechanisms, can serve as a robust and effective intervention tool for accident prevention. The system contributes meaningfully to intelligent transportation technologies by promoting safer, more adaptive, and reliable driving environments for both drivers and passengers.

6 Future Work

Future developments will focus on improving the robustness, adaptability, and real-world applicability of the proposed driver monitoring and assistance system. Expanding the training dataset to include a wider range of drivers, diverse environmental conditions, and multiple vehicle types will enhance model generalization across real-world scenarios.

Incorporating temporal behavior modeling techniques, such as Long Short-Term Memory (LSTM) networks or transformer-based sequence analysis, can enable earlier prediction of drowsiness and distraction patterns before unsafe behavior fully manifests.

Further efforts will be directed toward enhancing computational efficiency and optimizing model deployment for embedded automotive hardware to ensure reliable real-time performance. Additionally, the system can be extended with advanced ADAS functionalities, including traffic sign recognition, blind-spot detection, and adaptive cruise control.

These extensions will contribute to the development of a more comprehensive, multi-modal driver assistance framework, supporting safer and more intelligent transportation systems.

Acknowledgements. Authors are grateful to the Chairperson of the Department of CSE—AIML at PES University for providing them with an opportunity to conduct this research. Their assistance was instrumental in the development and completion of this study.

References

1. Supraja, P., Revati, P., Ram, K.S., J.C.: An intelligent driver monitoring system. In: Proceedings of the IEEE International Conference on Communication, Computing and Industry 4.0 (C2I4). India, pp. 1–6 (2021)
2. Vijay, A.: Intelligent driver monitoring system for safe driving. In: Proceedings of the IEEE International Conference on Electrical Systems for Aircraft, Railway, Ship Propulsion and Road Vehicles and International Transportation Electrification Conference (ESARS-ITEC). France, pp. 1–6 (2023)
3. Jose, J., Vimali, J.S., Ajitha, P., Gowri, S., Sivasangari, A., Jinila, B.: Drowsiness detection system for drivers using image processing technique. In: Proceedings of the 5th International Conference on Trends in Electronics and Informatics (ICOEI). Tirunelveli, India, pp. 1072–1078 (2021)
4. Prasath, N., Sreemathy, J., Vigneshwaran, P.: Driver drowsiness detection using machine learning algorithm. In: Proceedings of the 8th International Conference on Advanced Computing and Communication Systems (ICACCS), pp. 487–492. Coimbatore, India (2022)
5. Ayachi, R., Afif, M., Said, Y., Ben Abdelali, A.: Drivers fatigue detection using EfficientDet in advanced driver assistance systems. In: Proceedings of the 18th International Multi Conference on Systems, Signals and Devices (SSD). Monastir, Tunisia, pp. 1–6 (2021)
6. Kannan, R., Jahnavi, P.: Driver drowsiness detection and alert system. In: Proceedings of the IEEE International Conference on Integrated Circuits and Communication Systems (ICICACS). Trivandrum, India, pp. 452–458 (2023)
7. Chen, Y.: Driver fatigue detection using machine learning methods. In: Proceedings of the IEEE International Conference on Artificial Intelligence and Computer Applications (ICAICA). China, pp. 125–130 (2022)

8. Khadraoui, A., Zemmouri, E., Taki, Y., Douimi, M.: Towards a system for prevention of drowsiness related accidents. IAES Int. J. Artif. Intell. **12**(2), 774–783 (2023)
9. Zhang, S., Li, H., Xu, W.: Anti sleep alert algorithm to prevent road accidents. Front. Future Transp. **1**(1), 1–10 (2023)
10. Doe, J., Smith, A., Lee, R.: SleepyWheels an ensemble model for drowsiness detection leading to accident prevention. J. Transp. Saf. Res. **12**(2), 124–135 (2022)
11. Dwivedi, S., Jalal, A.S., Agrawal, S.C.: Driving behavior classification using MobileNet CNN. In: Proceedings of the International Conference on Communication and Signal Processing (ICCSP), pp. 489–493. Chennai, India (2020)
12. Abtahi, S., Omidyeganeh, M., Shirmohammadi, S., Hariri, B.: Driver drowsiness monitoring based on yawning detection. IEEE Trans. Instrum. Meas. **64**(9), 2439–2451 (2015)
13. Johns, M., Tucker, M., Chapman, S.: Monitoring eye blink patterns for fatigue estimation. IEEE Trans. Intell. Transp. Syst. **19**(4), 1234–1243 (2018)
14. Singh, A., Kaur, R.: Driver drowsiness detection using YOLO and CNN. In: Proceedings of the International Conference on Advancements in Computing and Communication Technologies (ICACCT). India, pp. 310–315 (2021)
15. Kumar, A., Shah, R.R., Singh, S.: Attention enhanced deep learning framework for driver distraction and fatigue detection. IEEE Access **11**, 125876–125889 (2023)
16. Hong, P.P., Khanh, H.H., Vinh, N.N., Trung, N.N., Quoc, A.N., Ngoc, H.T.: Deep learning based lane keeping assist system for self driving cars using transfer learning and fine tuning. J. Adv. Inform. Technol. **15**(3), 322–329 (2024)
17. Li, C., Wang, Y., Zhang, H.: Vision based lane detection and lane keeping assistance using deep convolutional neural networks. IEEE Trans. Intell. Transp. Syst. **24**(2), 1850–1861 (2023)
18. Mehta, R., Kumar, S., Verma, A.: Lane detection and vehicle control for advanced driver assistance systems using deep learning. Int. J. Veh. Auton. Syst. **17**(4), 289–304 (2023)
19. Park, J., Kim, M., Lee, S.H.: Robust lane keeping assist system using CNN based semantic segmentation under complex road conditions. Sensors **22**(18), 1–18 (2022)
20. Al Sharif, A., Hassan, M., El Sayed, Y.: End to end deep learning framework for lane detection and steering control in autonomous vehicles. J. Intell. Transp. Syst. **27**(1), 45–58 (2023)
21. Reddy, K., Kumar, P., Rajasekaran, S.: Deep learning based vision system for lane keeping and driver assistance in practical driving scenarios. Appl. Artif. Intell. **37**(6), 521–536 (2023)

TS-Gaussian: Towards Universal 3D Plant Reconstruction Across Species via Text-Guided Segmentation and Skeleton-Driven Completion

Yan Chen[1], Jinxin Chen[1], Xiangyu Lu[1], Fen Liao[2(✉)], Hengyi Li[1], Xiyao Huang[1], and Zihao Huang[1]

[1] College of Computer and Electronic Information, Guangxi University, Nanning 530004, China

[2] Guangxi Key Laboratory of Sugarcane Genetic Improvement, Key Laboratory of Sugarcane Biotechnology and Genetic Improvement (Guangxi), Sugarcane Research Institute, Guangxi Academy of Agricultural Sciences, Nanning 530007, China

liaofen@gxaas.net

Abstract. Accurately acquiring plant phenotypic parameters across diverse scenes and species is crucial for supporting researcher decision-making. However, conventional methods often overfit to specific datasets, limiting their generalizability to novel environments. Moreover, they struggle with complex backgrounds and low-quality images, often leading to fragmented or incomplete point clouds, significantly impairing the accuracy of extracted parameters. To address these limitations, we propose Text-Skeleton Gaussian (TS-Gaussian), a generalizable 3D reconstruction method capable of reconstructing high-fidelity 3D plant models for different plant types. TS-Gaussian synergizes text-guided video instance segmentation, 3D Gaussian Splatting (3DGS), and a novel skeleton-driven completion module to enable high-precision reconstruction and phenotypic analysis across diverse species. Experimental results demonstrate that TS-Gaussian achieves superior segmentation accuracy (F1-score up to 95%) and high parameter extraction accuracy (goodness-of-fit R^2 up to 0.9058) across various plant types, such as sugarcane, trees, and shrubs. TS-Gaussian provides an efficient and low-cost solution for plant phenotyping analysis in smart agriculture.

Keywords: 3D reconstruction · 3D Gaussian Splatting · Text-Guided Segmentation · Point Cloud Completion

1 Introduction

In the context of global digital transformation, computer vision technology is rapidly revolutionizing the agricultural sector. Particularly under the impetus of smart agriculture, the innovation of related technologies has become crucial [1]. In plant observation and research, phenotypic analysis and parameter extraction enable researchers to conduct both qualitative and quantitative assessments, playing a significant role in decision-making processes, such as plant cultivation, pest and disease analysis, and

L. Huang and D. Greenhalgh (Eds.): ICMLC 2026, LNNS 2061, pp. 39–53, 2027.
https://doi.org/10.1007/978-3-032-31236-5_3

yield prediction [2–4]. Traditional approaches to acquiring plant phenotypic information often require invasive or destructive operations, such as felling, pruning, or separating plants for subsequent manual measurements. These labor-intensive methods are time-consuming, require additional plant cultivation, disrupt plant growth, and hinder continuous monitoring of the same plant across different growth stages [5–8]. In contrast, computer vision-based techniques are characterized by their non-contact, non-destructive, and efficient nature, offering critical advantages in reducing labor, enabling continuous monitoring, accelerating breeding processes, and enhancing overall productivity. Therefore, the efficient and accurate reconstruction of objects in real-world scenarios and extraction of phenotypic parameters hold broad prospects for research and application in the advancement of smart agricultural technologies [9].

Gené-Mola et al. [10] and Chen et al. [11] employed active reconstruction methods using depth sensors to achieve point cloud reconstruction and parameter extraction for apples and cabbages, respectively, resulting in a 6.7% improvement in detection accuracy and a 13% reduction in mean relative error. However, depth sensor-based approaches are constrained by their reliance on specific hardware and controlled experimental settings, making them unsuitable for complex outdoor environments. Alternative methods that do not depend on dedicated devices, such as structure-from-motion (SfM), reconstruct 3D models from multi-view imagery. For example, Wang et al. [12] proposed 3DPhenoMVS, a 3D phenotyping pipeline designed to estimate 17 phenotypic traits of tomato plants across their entire life cycle. This pipeline utilizes 3D structural data generated from multi-view images using the SfM algorithm [13], and selects six traits with corresponding ground-truth measurements for accuracy evaluation.

Neural Radiance Fields (NeRF) have fundamentally reshaped the design, development, and deployment of 3D reconstruction technologies. These cutting-edge approaches are increasingly being applied to advance plant phenotyping applications, such as high-throughput trait analysis [14], structural modeling of plant architecture [15], and precision agriculture optimization [16]. Although NeRF-based methods have been successful in 3D plant reconstruction, they often suffer from high computational costs, lengthy training times, and slow rendering speeds, limiting their applicability to real-time precision agriculture scenarios [17, 18].

Jiang et al. [19] proposed an instance segmentation-based approach for cotton phenotyping. Their method trained a 3D Gaussian Splatting (3DGS) model of cotton-plant scenes using sparse point clouds generated from 2D images processed by COLMAP. To segment any objects of interest, a YOLOv8x instance segmentation model [20] was used to produce 2D masks, which were then integrated into the Segment Any 3D Gaussians (SAGA) framework [21] to extract individual 3DGS models of cotton bolls and stems. Similarly, Shen et al. [22] introduced PlantGaussian, a framework designed to generate realistic 3D plant visualizations under varying temporal and environmental conditions. To address the limitations of traditional Gaussian-based reconstruction in complex planting environments, PlantGaussian integrates the Segment Anything Model (SAM) [23] and tracking algorithms to improve segmentation and temporal consistency. Moreover, it employs a novel mesh partitioning technique to convert Gaussian-rendered outputs into measurable plant meshes, providing a powerful approach for precise 3D plant morphophenotyping. The framework also includes the PlantGaussian dataset, which consists of images of four crop species—maize, wheat, soybean, and tobacco—captured under diverse conditions and growth stages. This method achieves high-fidelity plant visualization and 3D mesh reconstruction for morphological phenotyping, requiring only image sequences as input, thereby offering a scalable and effective solution for 3D plant-structure analysis.

However, the former approach is constrained to a single plant species, necessitating datasets that encompass all targeted species for broader applicability. The latter, while enhancing plant visualization, introduces additional computational costs owing to the mesh generation process, which is time-consuming and storage burdens. Furthermore, the reliance on SAM and tracking algorithms, although enabling the segmentation of all objects in the imagery, significantly increases the computational overhead. Under challenging conditions, such as image blurriness, background interference, or poor lighting, the segmentation quality degrades, leading to discontinuities and missing regions in the reconstructed point clouds that require further completion.

To address these challenges, we propose a cross-scene and cross-species 3D digital plant reconstruction method from videos, enabling point cloud reconstruction and parameter extraction for diverse plant species and environments. The main contributions of this work are as follows:

1) We introduce a text-guided foreground segmentation module to achieve precise segmentation of target plants in video sequences.
2) We develop a generalizable 3D reconstruction framework that handles diverse plant species across various environments.
3) We design a point cloud completion algorithm to enhance reconstruction quality and parameter accuracy in complex scenes.

2 Materials and Methods

2.1 Dataset Acquisition and Preprocessing

Data Collection. Video data were collected using a handheld smartphone (iPhone 13) across three distinct environments: (i) an indoor greenhouse at the Guangxi Academy of Agricultural Sciences Sugar-cane Research Institute, primarily featuring sugarcane (GT42); (ii) a semi-outdoor environment with potted plants; and (iii) an outdoor field setting. The dataset encompasses multiple growth stages of sugarcane, as well as diverse species including shrubs, trees, succulents, and cacti. A total of 90 videos of sugarcane were recorded between June and September 2024, alongside 100 videos of other plant species collected at Guangxi University.

Acquisition Protocol. For each sample, a 360-degree panoramic video was recorded. The camera was maintained at a level orientation, positioned at the vertical midpoint of the plant stem, and moved along a circular trajectory with a radius of approximately 70–100 cm centered on the target. Recording durations ranged from 40 to 60 s per sample. For multi-plant scenarios, the centroid of the plant cluster served as the rotation center.

Preprocessing. Frames were extracted from each video at a sampling rate of 3 Hz, yielding 120–180 images per sequence. Structure-from-Motion (SfM) was performed using COLMAP to estimate camera poses and generate sparse point clouds. These processed data—comprising images, camera poses, and sparse points—served as the input for the subsequent segmentation and re-construction modules.

Implementation Details. All experiments were conducted on a workstation running Ubuntu, configured with the PyTorch framework and CUDA 11.8.

2.2 Framework Overview

The overall experimental framework is illustrated in Fig. 1.

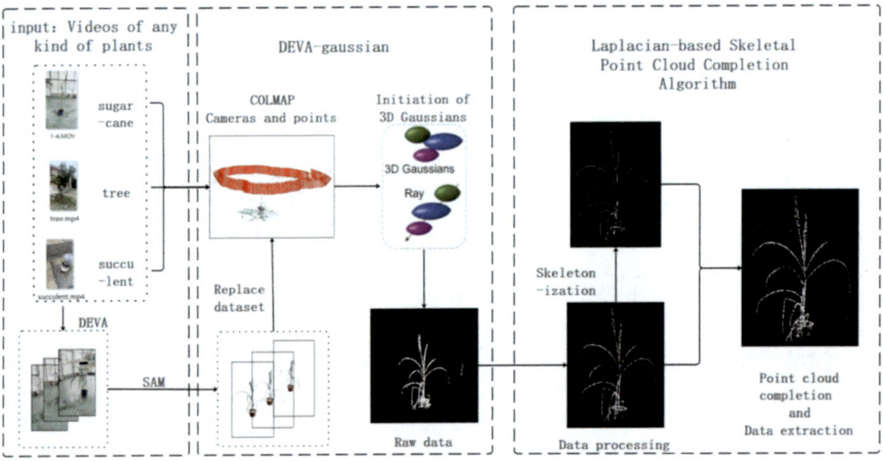

Fig. 1. Workflow of TS-Gaussian.

The workflow is divided into five stages, from video-based acquisition to plant point cloud reconstruction and phenotypic parameter extraction.

1. Data acquisition and preprocessing of plant samples.
2. Instance segmentation of plants guided by text prompts to remove background noise.
3. 3DGS reconstruction of the segmented plants.
4. Point cloud completion using a skeleton-driven search-based completion algorithm.
5. Skeleton-based phenotypic parameter extraction.

A multimodal cross-domain analysis framework was developed to achieve 3D reconstruction and parameter extraction for various plants species without the need for pre-training. This framework employs text-guided instance segmentation to construct 3D Gaussian representations of target plants, integrates a skeleton-driven search-based point cloud completion algorithm for refinement, and enables automated phenotypic parameter extraction.

2.3 Text-Guided Segmentation

The objective of text-guided video segmentation is to segment and associate specific plants in a video without requiring pre-training, thereby enabling rapid and accurate separation of target plants from the background for downstream processing. To achieve this, our approach integrates the Decoupled Video Segmentation Approach (DEVA) [24], the large language model GPT-4o, and the general-purpose segmentation model SAM. Textual information is incorporated into the segmentation process to guide the model in identifying and tracking specific plants, ensuring accurate foreground extraction across video frames.

DEVA is an innovative video segmentation framework that decouples the conventional end-to-end task into two synergistic components: a task-specific image-level segmentation model and a general temporal-propagation model. Leveraging this decoupled design, SAM is employed for the semantic perception of target objects in individual frames, whereas text information parsed by GPT-4o is utilized to achieve semantic segmentation of the plants of interest.

$$\widehat{\mathrm{Seg}}_{t+i} = Prop\big(\{I_{t+i}, \mathrm{Seg}_{t+i}\}, I_t\big) \tag{1}$$

$$P = U_{i=0}^{n-1} \widehat{Seg}_{t+i} = \{p_i\} \tag{2}$$

The temporal propagation model focuses on motion consistency. As shown in Eq. (1), the segmentation result of the adjacent frame I_{t+1}, denoted as Seg_{t+1}, is aligned to the coordinate system of the current frame I_t through a temporal propagation model. During this forward propagation process, temporal coherence is maintained, while interference is excluded, ensuring that the focus remains on the target plant. Conversely, backward propagation enables intraframe consensus. As expressed in Eq. (2), the aligned segmentation results $\{p_i\}$ from adjacent n frames are merged to form a candidate proposal set P. By aligning and fusing the segmentation results across consecutive n frames, single-frame segmentation noise can be effectively reduced, as illustrated in Fig. 2.

The adoption of a decoupled architecture enhances performance compared with static image segmentation models, while the general object propagation capability improves generalization. When combined with SAM, the framework demonstrates robust segmentation of plants across diverse environments and species. After segmentation, the processed images—with an identical scale as the original but with backgrounds removed—replace the original dataset, forming a new input dataset for 3D Gaussian reconstruction. The preprocessed dataset, which consists of segmented images, corresponding camera poses, and sparse point clouds, is then used for 3DGS.

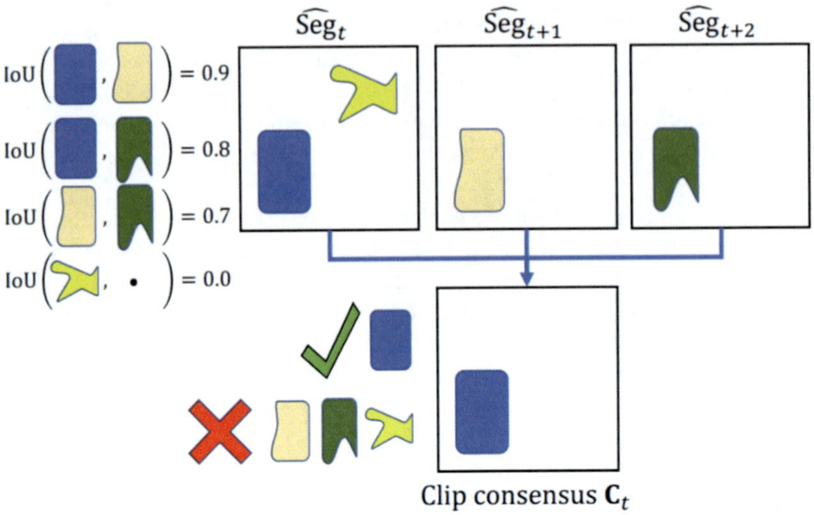

Fig. 2. Example of alignment and fusion: the top row $\widehat{Seg}_t$, $\widehat{Seg}_{t+1}$, $\widehat{Seg}_{t+2}$, and show segmentation results across different frames. By integrating IoU-based evaluation, the blue square is identified as the most suitable output, whereas the irregular yellow shape is considered noise.

2.4 Skeleton-Driven Point Cloud Completion Algorithm

Basic segmentation and reconstruction are insufficient to handle the complexities of real-world applications. To address this, we propose a skeleton-driven point cloud completion algorithm that extracts and completes skeletons from reconstructed point clouds. The overall pipeline is shown in Fig. 3.

The point cloud obtained from the 3D Gaussian reconstruction first undergoes two rounds of progressive statistical filtering to remove background noise points. The centroid of the point cloud is then computed, followed by a radial spatial filter to eliminate distant noise points and reduce the computational burden of the subsequent steps. Next, a density-based spatial clustering method (DBSCAN) is applied to identify the largest cluster and remove nearby noise points. Finally, the cleaned point cloud is downsampled in two stages: (i) uniform downsampling to one-quarter of the points and (ii) farthest point sampling (FPS) to obtain a fixed-size point cloud containing exactly 16,384 points, resulting in a compact but incomplete point cloud suitable for skeleton extraction and completion.

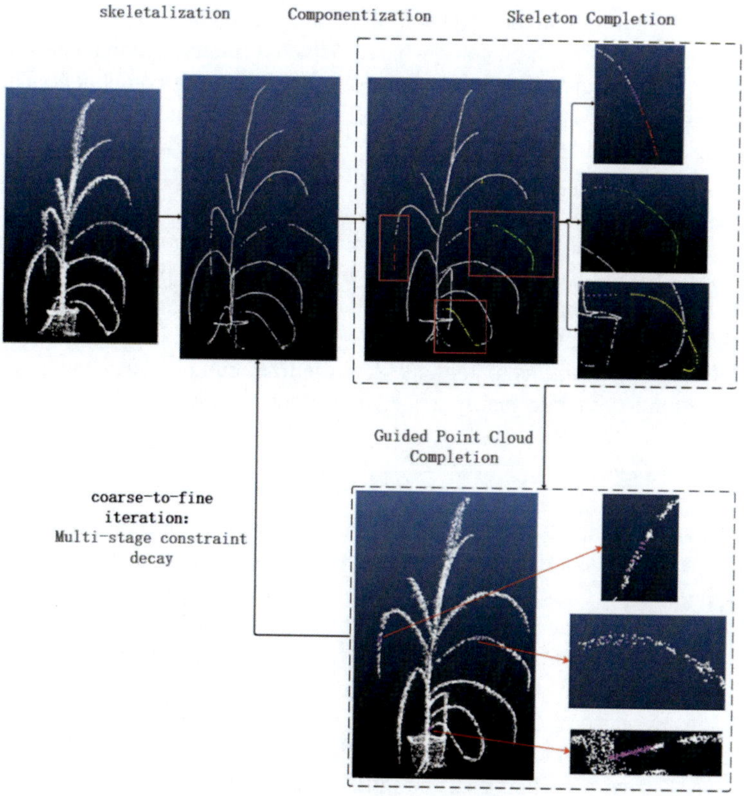

Fig. 3. Workflow of completion

The Laplacian-based Contraction (LBC) skeletonization algorithm is applied to the incomplete point clouds. This method iteratively leverages the Laplacian matrix to extract semantic information from point clouds. Specifically, W_L denotes the contraction weight matrix, which controls the degree of contraction of the point cloud toward its center; W_H represents the positional constraint matrix, which prevents excessive contraction and preserves the overall morphology; and L is the cotangent Laplacian matrix constructed from the adjacency relationships of the point cloud, encoding local geometric features. At each iteration, W_L and W_H are updated, with the contraction weights gradually increasing. The contracted point cloud C' is then obtained by solving the corresponding linear systems.

$$\begin{bmatrix} W_L L \\ W_H \end{bmatrix} C' = \begin{bmatrix} 0 \\ W_H C \end{bmatrix} \tag{3}$$

The contracted point cloud is further classified into trunk and branch structures, where the semantic information is skeletonized and downsampled to generate a skeletal point cloud. A breadth-first search is then performed to identify the neighboring points within a specified radius. Points with pairwise distances larger than this radius are regarded as fractures. Based on this criterion, the skeletal point cloud is segmented into a main body and multiple fracture components. The component containing the largest number of points is designated as the main body, whereas isolated fracture components consisting of only a single point are merged into the nearest neighboring component. The initial search radius is set to 0.1. Subsequently, all components are traversed and fracture locations are identified by searching for the nearest point pairs. Principal Component Analysis (PCA) is applied to fit straight-line directions at both ends of each fracture, thereby distinguishing between intermediate fractures and leaf detachment cases while simultaneously determining the completion direction and position. Smooth interpolated connections consistent with plant morphology are generated using B-spline curves. Let P_i: denote the control points (from the original input point set), and $N_{i,k(u)}$: represent the B-spline basis functions, where k is the order of the curve (e.g., $k = 3$ for cubic splines), and u is the uniformly sampled parameter. A normalized cumulative distance is used for parameterization to generate evenly spaced interpolation points.

$$\mathbf{C}(u) = \sum_{i=0}^{n} \mathbf{P}_i N_{i,k}(u) \tag{4}$$

Based on the completed skeleton and acquired fracture information, point cloud completion templates are derived from the two ends of the fractures depending on the fracture type. For fractures occurring in the middle of the stem, templates are obtained from both ends of the fracture. For leaf detachment cases, templates are obtained only from the side connected to the leaf. Using these templates, n sets of point clouds with identical relative positions and densities are generated at the newly completed skeleton points. To ensure uniform density, the completed point clouds are subjected to random downsampling. The above completion process is performed iteratively for a fixed number of iterations. During each iteration, the initial search radius is adjusted and gradually decreased, thereby resolving fractures progressively from larger to smaller ones until the final completion result is obtained.

3 Results

3.1 Instance Segmentation of Images

Figure 4 illustrates the comparison between the ground truth and segmentation results under various conditions, including sugarcane cultivated in indoor pots, trees and shrubs grown in outdoor fields, and potted plants in outdoor settings, such as single succulent pots, succulents coexisting with cacti, and multiple pots of different plant species. As shown in the figure, the proposed model demonstrated robust adaptability to diverse plant species across different environments and effectively removed background information.

Fig. 4. This figure illustrates the results of plant segmentation at different phases, with the identified plants indicated in red.

The quantitative segmentation performance is summarized in Table 1. For sugarcane, the F1-score reached 87.60%, and the IoU achieved 78.41%, outperforming YOLOv8 (F1-score: 84.71%, IoU: 73.63%). For the other categories, including trees, shrubs, succulents, and cacti, the F1-scores exceed 95%, and the IoU values are consistently above 90%, confirming the superior segmentation performance of the proposed method across various plant types and environments.

Table 1. Segmentation accuracy for different plant species

Plant species	F1-score	IoU
Sugarcane	87.60	78.41
YOLO-sugarcane	84.71	73.63
Tree	95.19	90.83
Shrub	97.29	94.72
Succulent	98.12	96.32
Succulent and cactus	96.77	93.74

The segmentation results demonstrate that the proposed method achieves accurate segmentation across different scenarios and plant types. This performance is enabled by text-based guidance for plant specifications and the DEVA method, which enables temporal consistency across video frames. Consequently, even without additional pre-training, the method accurately segments objects that are absent from the training set. For example, in segmenting sugarcane, the text input is simply "sugarcane". Such explicit textual guidance allows the model to focus on the intended botanical structures, avoiding aimless processing of the entire scene. Without targeted prompts, the segmentation model would lack a clear objective, requiring more time to process all elements in the environment and likely including irrelevant objects (e.g., soil lumps, garden tools, or non-target plants) in the output. The integration of concise, species-specific prompts thus not only improves segmentation efficiency and accuracy but also enhances the model's applicability in complex, multi-species scenarios.

Moreover, unlike traditional approaches, the proposed method does not rely on species-specific, annotated datasets. Instead, it only requires videos of plants captured with a smartphone, thereby reducing both data acquisition and processing costs and facilitating rapid deployment in real-world agricultural settings.

Nevertheless, in highly complex scenarios, such as the sugarcane dataset used in this study, several challenges remain. These include limited image resolution, similarity between background and plant colors, and the structural complexity of sugarcane (e.g., long overlapping leaves and large overall plant size). When the sugarcane height exceeded 3 m, the camera distance increased, reducing the available image detail.

3.2 Skeleton-Driven Point Cloud Completion Algorithm

To address these issues, we employ a skeleton-driven point cloud completion algorithm. Figure 5 illustrates the process of point cloud completion using the proposed skeleton-driven algorithm. Fractures are evident at the junctions of the top, middle, and basal leaves of the sugarcane plants. After completion with the proposed method, the resulting skeleton is consistent with plant morphology, more structurally complete, and provides effective guidance for point cloud reconstruction. The completed point cloud, shown in Fig. 6, closely approximates the actual plant structure, as validated by a comparison with the ground-truth images.

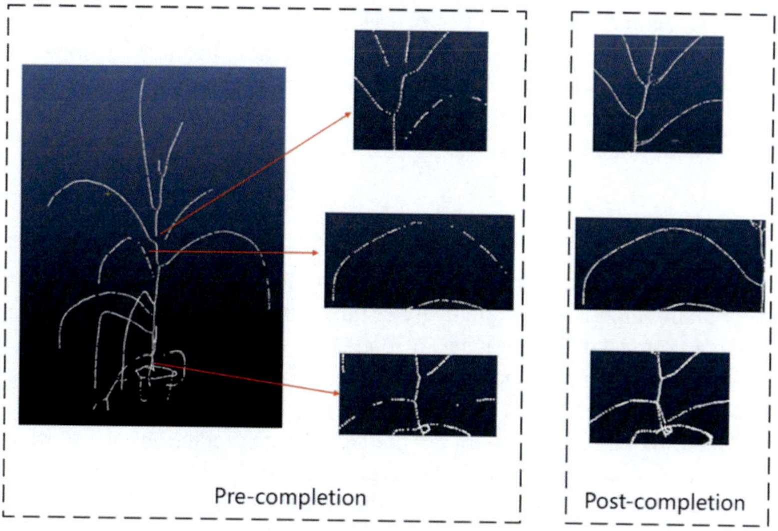

Fig. 5. Skeleton completion – pre-completion versus post-completion

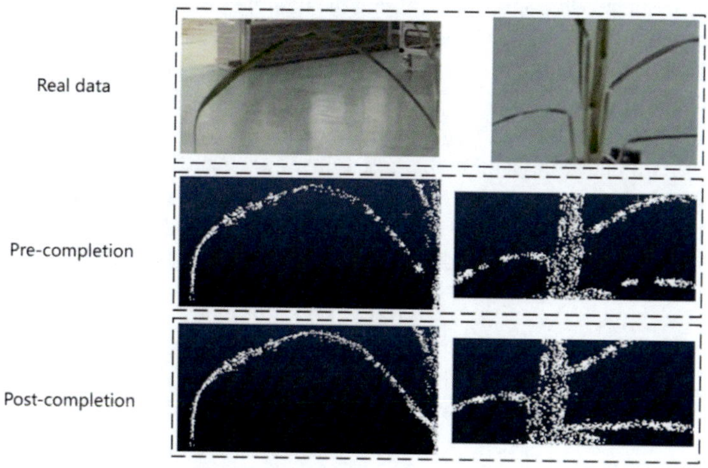

Fig. 6. A qualitative comparison: real images, sparse point clouds, and completed models

The experimental results demonstrate the superior performance of the proposed completion method. Improvements in parameter extraction further confirmed that the method mitigates the degradation in segmentation performance under complex conditions, enhances point cloud quality, and improves the accuracy of parameter estimation. This, in turn, supports agricultural decision-making and enhances the robustness of real-world applications.

The method does not rely on species-specific annotated datasets and requires only low computational cost and short processing time: from a few seconds to several tens of seconds, depending on the number of fractures. This is significantly shorter than the several hours required for PlantGaussian reconstruction. In addition, downsampling greatly reduces storage demands: each completed point cloud requires only tens of kilobytes. These characteristics highlight the efficiency and practicality of the proposed method for practical applications.

3.3 Parameter Extraction

Following the point cloud processing, we extracted the key plant parameter of interest, namely plant height. To quantitatively evaluate the performance of the completion algorithm, we compared the predicted heights from the incomplete (pre-completion) and completed (post-completion) point clouds with the ground-truth measurements, as shown in Figs. 7 and 8, respectively.

Figure 7 shows the scatter plot for the pre-completion predictions. The initial model exhibits a moderate correlation with the ground truth, evidenced by a coefficient of determination (R^2) of 0.7548. The corresponding error metrics—a Mean Absolute Error (MAE) of 9.1698, a Mean Absolute Percentage Error (MAPE) of 5.3141%, and a Root Mean Squared Error (RMSE) of 13.3584, indicate a baseline level of accuracy with considerable room for improvement. The noticeable dispersion of data points around the ideal fit line further highlights the limitations of the raw, incomplete data.

In contrast, the predictions after applying the completion algorithm, shown in Fig. 8, exhibit substantially better agreement with the ground truth. The R^2 value increases significantly to 0.9058, indicating a strong correlation. All error metrics are also markedly reduced: MAE decreases to 6.7278, MAPE drops to 4.1336%, and RMSE improves to 8.2813. The tighter clustering of data points around the ideal fit line, confirms a major boost in prediction accuracy and model reliability.

These improvements across all evaluated metrics—R^2, MAE, MAPE, and RMSE—provides strong evidence for the effectiveness of the proposed point cloud completion algorithm in improving the precision of sugarcane plant height extraction.

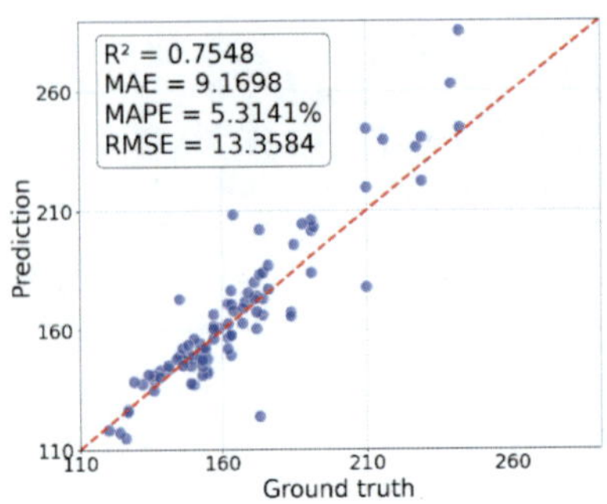

Fig. 7. Before completion

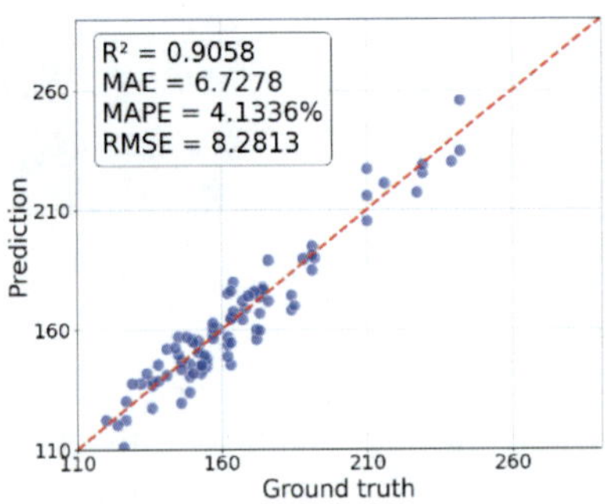

Fig. 8. After completion

4 Conclusion

In this study, we propose Text-Skeleton Gaussian (TS-Gaussian), a generalizable 3D reconstruction method that reconstructs 3D digital plants from videos captured in various scenes and for multiple species, enabling point cloud reconstruction and parameter extraction for different plant types. The proposed approach integrates text-guided video instance segmentation, 3D Gaussian Splatting (3DGS) reconstruction, and a skeleton-driven point cloud completion algorithm to achieve high-precision 3D reconstruction and phenotypic parameter extraction for multiple plant species. Experimental results demonstrate that our method achieves superior segmentation accuracy across various plant types.

The consistent improvements across all evaluation metrics provide strong evidence for the effectiveness of the proposed point cloud completion algorithm in improving the accuracy of sugarcane parameter extraction. These results confirm TS-Gaussian enables accurate point cloud reconstruction and parameter extraction across multiple plant species and environmental conditions. This work provides an efficient and low-cost solution for plant phenotyping analysis in smart agriculture.

TS-Gaussian also shows promising potential in continuous monitoring and parameter acquisition throughout crop growth stages, ecological research, decision support in plant breeding, as well as rapid phenotyping and parameter extraction for diverse plant species under complex field conditions.

Future work will focus on improving segmentation and completion performance and expanding the range of extractable parameters, thereby providing more comprehensive and reliable data support for precision agriculture. We will first enhance the segmentation module by incorporating data preprocessing and image enhancement techniques, such as adaptive contrast adjustment and semantic-aware data augmentation, to improve input quality under challenging field conditions. The model's input module will also be optimized to handle more complex and fine-grained datasets. For point cloud completion, we plan to leverage pre-trained models and introduce text-guided prior knowledge. A text-informed completion network will be developed, where botanical structural priors are activated and integrated through semantic embeddings, enabling the model to reason about missing regions based on both geometric cues and plant-type semantics. These improvements aim to strengthen the overall reconstruction robustness and support the extraction of a more comprehensive set of phenotypic parameters in future applications.

Acknowledgments. This work was supported by the National Natural Science Foundation of China (Grant No. 32460537), the Natural Science Foundation of Guangxi Zhuang Autonomous Region, China (Grant No. 2023GXNSFAA026405), and the Operating Subsidy of Guangxi Key Laboratory of Sugarcane Genetic Improvement (Grant No. 21-238-16).

References

1. Harfouche, A., Merhi, M.I., Albizri, A., et al.: Sustainable development through technological innovations and data analytics. Inf. Syst. Front. **26**(6), 1989–1996 (2025)

2. Schneider, H.M.: Characterization, costs, cues and future perspectives of phenotypic plasticity. Ann. Bot. **130**, 131–148 (2022)
3. Adams, W.W., III., Stewart, J.J., Polutchko, S.K., Cohu, C.M., Muller, O., DemmigAdams, B.: Foliar phenotypic plasticity reflects adaptation to environmental variability. Plants-Basel **12**, 2041 (2023)
4. Marion, D., Marianne, M., Andreas, H., Mathias, O.: Agronomic and phenotypic plant traits as indicators for environmental risks of genetically modified plants. Environ. Sci. Eur. **36**, 3 (2024)
5. Duan, L.F., Yang, W.N., Huang, C.L., Liu, Q.: A novel machine-vision-based facility for the automatic evaluation of yield-related traits in rice. Plant Methods **7**, 44 (2011)
6. Huang, C.L., et al.: Rice panicle length measuring system based on dual-camera imaging. Comput. Electron. Agric. **98**, 158–165 (2013)
7. Igathinathane, C., Pordesimo, L.O., Columbus, E.P., Batchelor, W.D., Methuku, S.R.: Shape identification and particles size distribution from basic shape parameters using ImageJ. Comput. Electron. Agric. **63**, 168–182 (2008)
8. Tanabata, T., Shibaya, T., Hori, K., Ebana, K., Yano, M.: SmartGrain: high throughput phenotyping software for measuring seed shape through image analysis. Plant Physiol. **160**, 1871–1880 (2012)
9. Saeed, F., Sun, J., Ozias-Akins, P., Chu, Y.J., Li, C.C.: PeanutNeRF: 3D radiance field for peanuts. In: Proceedings of the IEEE/CVF Conference on Computer Vision and Pattern Recognition, pp. 6254–6263 (2023)
10. Gené-Mola, J., et al.: Fruit detection, yield prediction and canopy geometric characterization using LIDAR with forced air flow. Comput. Electron. Agric. **168**, 105121 (2020)
11. Li, J., et al.: Label-efficient learning in agriculture: a comprehensive review. Comput. Electron. Agric. **215**, 108412 (2023)
12. Wang, Y., Hu, S., Ren, H., Yang, W., Zhai, R.: 3DPhenoMVS: a low-cost 3D tomato phenotyping pipeline using 3D reconstruction point cloud based on multiview images. Agronomy **12**(8), 1865 (2022)
13. Schonberger, J.L., Frahm, J.-M.: Structure-from-motion revisited. In: Proceedings of the IEEE Conference on Computer Vision and Pattern Recognition, pp. 4104–4113 (2016)
14. Yang, X., et al.: PanicleNeRF: low-cost, high-precision in-field phenotyping of rice panicles with smartphone. Plant Phenomics **6**, 0279 (2024)
15. Arshad, M.A., et al.: Evaluating neural radiance fields for 3D plant geometry reconstruction in field conditions. Plant Phenomics **6**, 0235 (2024)
16. Zhao, J., et al.: Exploring accurate 3D phenotyping in greenhouse through neural radiance fields. arXiv preprint arXiv:2403.15981 (2024)
17. Bao, Y., et al.: 3D Gaussian splatting: survey, technologies, challenges, and opportunities. IEEE Trans. Circuits Syst. Video Technol. (2025)
18. Chen, G., Wang, W.: A survey on 3D Gaussian splatting. arXiv preprint arXiv:2401.03890 (2024)
19. Jiang, L., Sun, J., Chee, P.W., Li, C., Fu, L.: Cotton3DGaussians: multiview 3D Gaussian splatting for boll mapping and plant architecture analysis. Comput. Electron. Agric. **234**, 110293 (2025)
20. Reis, D., Kupec, J., Hong, J., Daoudi, A.: Real-time flying object detection with yolov8. arXiv preprint arXiv:2305.09972 (2023)
21. Cen, J., et al.: Segment any 3D Gaussians. arXiv preprint arXiv:2312.00860 (2023)
22. Shen, P., Jing, X., Deng, W., Jia, H., Wu, T.: PlantGaussian: exploring 3D Gaussian splatting for cross-time, cross-scene, and realistic 3D plant visualization and beyond. Crop J. (2025)
23. Kirillov, A., et al.: Segment anything. In: Proceedings of the IEEE/CVF International Conference on Computer Vision, pp. 4015–4026 (2023)

24. Cheng, H.K., Oh, S.W., Price, B., et al.: Tracking anything with decoupled video segmentation. In: Proceedings of the IEEE/CVF International Conference on Computer Vision, pp. 1316–1326 (2023)

Luminance-Controlled Migration for Soil Images

Yan Xia[1,2], Shaohua Zeng[1,2(✉)], Ruolan Zeng[3], and Shoukuan Gu[4]

[1] College of Computer and Information Science, Chongqing Normal University, Chongqing 401331, China
{xiyan_cqnu,zsh_cqu}@126.com

[2] Chongqing Center of Engineering Technology Research on Digital Agricultural Service, Chongqing 401331, China

[3] Chongqing Electric Power College, Chongqing 400053, China
zengruolan@cqepc.edu.cn

[4] Chongqing Master Station of Agricultural Technology Promotion, Chongqing 401121, China
gushoukuan@163.com

Abstract. Soil image datasets with diverse lighting conditions are essential for improving the accuracy of machine vision models in soil type classification. To address the challenge of data scarcity, a luminance-controlled migration algorithm is proposed to augment soil image datasets. First, a modified decomposition model based on Retinex is presented to extract the luminance and reflectance components of soil images. Subsequently, neighborhood information is integrated into a gamma transformation to achieve a one-to-many mapping strategy for luminance migration. To ensure precise control, the gamma factor is adaptively determined by constructing an optimization model. Finally, the enhanced image is synthesized using the migrated luminance and the original reflectance. Simulation results demonstrate that the proposed algorithm outperforms comparative methods. Specifically, the average target luminance deviation of the enhanced images is 0.5661, and the average pixel difference between the enhanced and target images is 7.7919. These results confirm that the proposed algorithm is effective for extending soil image datasets and can facilitate improved accuracy in soil species identification.

Keywords: Image enhancement · Retinex decomposition · Gamma transformation · Soil images

1 Introduction

The advent of deep learning has provided novel methodologies for soil species recognition based on soil images. However, unlike traditional methods, deep learning models require extensive training datasets that comprehensively cover various soil species and natural lighting conditions. Given the difficulty and high

L. Huang and D. Greenhalgh (Eds.): ICMLC 2026, LNNS 2061, pp. 54–68, 2027.
https://doi.org/10.1007/978-3-032-31236-5_4

cost of collecting soil samples in the field under controlled lighting, obtaining a complete dataset with full lighting coverage in the short term is challenging. Research on classification using small training samples emphasizes the importance of handling data scarcity effectively [1]. Consequently, there is an urgent need to synthetically adjust the lighting of existing soil images to augment the dataset. Since external lighting conditions primarily manifest as image luminance, transforming the lighting conditions can be achieved by converting the image luminance. If the luminance of a soil image can be accurately modified to approximate real-world conditions, it becomes feasible to complete the sample coverage at a low cost. This, in turn, significantly enhances the identification accuracy and algorithmic robustness of soil species recognition. Therefore, the research on controlled luminance migration for soil images is of great practical significance.

Currently, luminance manipulation techniques are primarily designed for image enhancement rather than precise control. Existing approaches include algorithms based on Retinex [2–8], histogram specification [9–13], and nonlinear mapping [14–18]. While these methods adjust contrast and luminance to improve visual perception, they lack the capability to precisely regulate the degree of enhancement, making them unsuitable for the accurate luminance regulation required for soil images. Xiao et al. [19] proposed an enhancement algorithm based on histogram reconstruction using a Gaussian function. While this allows for luminance and contrast migration, its reliance on empirically set parameters limits its universality. Similarly, the method developed in [20] shifts luminance by adjusting the S-shaped transform intersection; however, it tends to neglect the realism of the target domain, leading to color distortion and texture over-amplification when applied to soil images.

To introduce controlled luminance enhancement into soil image processing, our previous work [21] proposed a weighted Gaussian subtraction method to fit the luminance histogram. Although this achieved a degree of controlled migration, there remains significant room for improvement regarding migration accuracy and color fidelity. To address these limitations, this paper proposes a luminance-controlled migration algorithm based on Retinex decomposition and neighborhood-aware gamma correction. Compared to existing brightness transfer methods, the key improvements of this study are threefold:

(1) Texture-Aware Decomposition Model: We propose a novel decomposition model based on Retinex that specifically constrains the reflectance range and smooths luminance components. This effectively decouples illumination from texture details, mitigating the issue of texture over-amplification often observed in conventional migration processes.
(2) One-to-Many Mapping Strategy: Unlike traditional global mapping methods, we incorporate pixel neighborhood information into the gamma transformation. This integration facilitates a "one-to-many" luminance mapping, enabling more refined local adjustments and better preservation of image details.

(3) Adaptive Parameter Optimization: Instead of relying on empirically set parameters as in [18], we establish an optimization model to adaptively solve for the gamma factor. This ensures high-precision luminance control within the effective migration range (e.g., [−30, 30]).

The remainder of this paper is organized as follows. Section 2 details the proposed decomposition model for soil images. The specific luminance migration method is described in Sect. 3. Section 4 presents the evaluation metrics and experimental results, followed by the conclusion in Sect. 5.

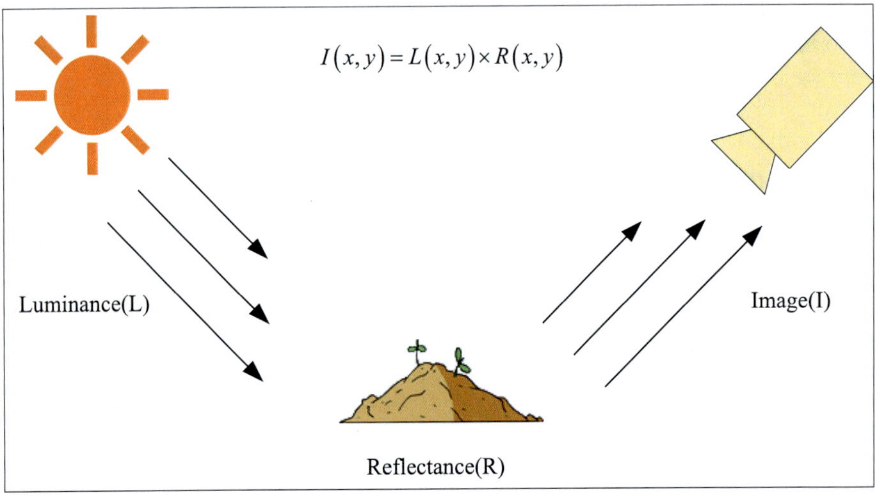

Fig. 1. The principle of Retinex.

2 Decomposing Soil Image Based on Retinex

This section gives the reasons for choosing the Retinex model and describes the idea of constructing a decomposition model for soil images.

2.1 Creating Retinex Decomposition Models for Soil Image

According to Retinex theory, an image I can be resolved into a luminance L and a reflectance R, as shown in Fig. 1. Generally, luminance L is the illumination information and reflectance R is named as reflectivity. Reflectance R depends on the reflecting properties of material. Because material is certain in an image, its reflectance R is invariant and reflectivity is less than 1. So, reflectance R is assumed to be invariant and it is less than 1 in image process. Since reflectance R of a scene is fixed in Retinex theory, it should transfer image that luminance L is changed. Therefore, Retinex model is selected as the base model for image luminance migration.

(a) (b) (c)

Fig. 2. (a) Original image. (b) Image processed by LR3M with $\alpha = 0.015$, $\beta = 2$, $\gamma = 1/2.2$. (c) Image processed by LIME with $\lambda = 0.15$ and $\gamma = 1/2.2$.

Idea of decomposing an image based on Retinex. It is an ill-posed problem to estimate luminance and reflectance from a single observed image. Many algorithms had been proposed to handle this problem. But, they cannot be used to enhance soil image because there are some fatal defects in them. For examples, Low-Rank Regularized Retinex Model (LR3M) [4] defined image luminance L with the average of the three channels to compute respectively reflectance R_r, R_g and R_b of the three channels, then gamma enhancement of L was be done to get the enhanced image. Because image luminance L is defined with the average of the three channels, element of reflectance R may be greater than 1. It causes that some pixels of an image are over-enhancement and appear white dots, as shown in Fig. 2(b). Low-light image enhancement (LIME) [3] defined image luminance L with the maximum channel of the image and leaded to the same result as well because some values of reflectance R became larger and greater than 1 after its luminance L was smoothed, as shown in Fig. 2(c).

In fact, the reflectance $R^c(x, y)$ can be restricted to $[0, 1]$ if luminance $L^c(x, y)$ is greater than the pixel value $I^c(x, y), c \in \{r, g, b\}$. In math, the norm of a vector is greater than its elements. The norm $|I^r(x, y), I^g(x, y), I^b(x, y)|$ of three channels $(I^r(x, y), I^g(x, y), I^b(x, y))$ of a pixel (x, y) is assigned to its luminance $L(x, y)$, which sure $R^c(x, y)$ can be restricted to $[0, 1]$ and simplifies the calculation of R^c and L.

Creating Retinex decomposition model. The initial luminance $L(x, y)$ of an image is given with the norm of its color vector in RGB space.

$$L(x, y) = \sqrt{I^r(x, y)^2 + I^g(x, y)^2 + I^b(x, y)^2} \tag{1}$$

Although it can limit the reflectance range to meet the requirements, the luminance $L(x, y)$ contains excessive details and textures. Enhancing luminance $L(x, y)$ should cause its details and textures are further amplified. So, the image that is reconstructed with the enhanced luminance $L_m(x, y)$ and reflectance R^c is unrealistic, seen in Fig. 3(b) and (c). It need retain the prominent structure and remove the redundant texture details from luminance $L(x, y)$ to avert the over-enhancing details. Hence, it is suggested to estimate a refined luminance $\hat{L}$

with $L(x, y)$ and $\hat{L}$ only remove a part of redundant texture details from $L(x, y)$. For this reason, the objective function is defined as follows.

$$\arg\min_{\hat{L}} \sum_{x=1}^{m}\sum_{y=1}^{n}\left(\left(L(x,y)-\hat{L}(x,y)\right)^2+\alpha\left(W_h(x,y)\left(\partial_h\hat{L}(x,y)\right)^2 +W_v(x,y)\left(\partial_v\hat{L}(x,y)\right)^2\right)\right) \tag{2}$$

where m and n are the height and width. ∂_h and ∂_v denote the spatial derivatives in horizontal and vertical directions respectively. W_h and W_v represent smoothness weight matrices. α is a weight for balance.

(a) (b) (c) (d) (e)

Fig. 3. Luminance estimation. (a) Original image. (b) Initial luminance. (c) Result from the initial luminance (b). (d) Our refined luminance. (e) Result from our refined luminance (d) (same enhancement as (c)).

In Eq. 2, the first term $(\hat{L}(x, y) - L(x, y))^2$ enforces $\hat{L}$ to be similar to the initial luminance L, while the second term aims to remove the redundant texture details in $\hat{L}$ by minimizing the partial derivatives. The fundamental goal of introducing the weight matrices W_h and W_v is to maintain only the necessary structural information in L during the smoothing process. Because the weight depends on the gradient of L, the h-direction smoothness weight W_h is written as follows.

$$W_h = \frac{1}{\log|\partial_h L| + 1 + \varepsilon} \tag{3}$$

where ε is a very small constant to avoid a zero denominator, which is empirically set as 0.001. The v-direction smoothness weight W_v is defined similarly.

$$W_v = \frac{1}{\log|\partial_v L| + 1 + \varepsilon} \tag{4}$$

Creating Retinex decomposition model. Equation 2 is a typical convex optimization. It is reformulated into the matrix form as follows.

$$\arg\min_{\hat{\boldsymbol{L}}}\left((\hat{\boldsymbol{L}}-\boldsymbol{L})^T(\hat{\boldsymbol{L}}-\boldsymbol{L})+\alpha\left(\hat{\boldsymbol{L}}^T\boldsymbol{D}_h^T\boldsymbol{Diag}(\boldsymbol{W}_h)\boldsymbol{D}_h\hat{\boldsymbol{L}} +\hat{\boldsymbol{L}}^T\boldsymbol{D}_v^T\boldsymbol{Diag}(\boldsymbol{W}_v)\boldsymbol{D}_v\hat{\boldsymbol{L}}\right)\right) \tag{5}$$

where $\boldsymbol{L}$ and $\hat{\boldsymbol{L}}$ are the column vector forms of L and $\hat{L}$ respectively. $\boldsymbol{D}_h$ and $\boldsymbol{D}_v$ denote the discrete gradient operators with forward difference. $\boldsymbol{W}_h$ and $\boldsymbol{W}_v$ represent vectorized versions of W_h and W_v. $\boldsymbol{Diag}(\boldsymbol{W})$ is a diagonal matrix which is constructed with $\boldsymbol{W}$.

Compute partial derivative of objective function in Eq. 2 with respect to $\hat{\boldsymbol{L}}$, and let the partial derivative be equal to 0.

$$(\hat{\boldsymbol{L}} - \boldsymbol{L}) + \alpha\left(\boldsymbol{D}_h^T \boldsymbol{Diag}(\boldsymbol{W}_h)\boldsymbol{D}_h\hat{\boldsymbol{L}} + \boldsymbol{D}_v^T \boldsymbol{Diag}(\boldsymbol{W}_v)\boldsymbol{D}_v\hat{\boldsymbol{L}}\right) = 0 \tag{6}$$

Analytic solution of $\hat{\boldsymbol{L}}$ is:

$$\hat{\boldsymbol{L}} = \left(\mathbf{1} + \alpha\left(\boldsymbol{D}_h^T \boldsymbol{Diag}(\boldsymbol{W}_h)\boldsymbol{D}_h + \boldsymbol{D}_v^T \boldsymbol{Diag}(\boldsymbol{W}_v)\boldsymbol{D}_v\right)\right)^{-1}\boldsymbol{L} \tag{7}$$

where $\mathbf{1}$ is the identity matrix with size $mn \times mn$ and $\hat{\boldsymbol{L}}$ is a column vector. $\hat{\boldsymbol{L}}$ is reconstructed for $\hat{L}$ that its size is $m \times n$.

2.2 Retinex Decomposition Algorithm of Soil Images

For producing a desired smoothing result $\hat{\boldsymbol{L}}$ that remove all of redundant texture details from $\boldsymbol{L}(x, y)$, the newest $\hat{\boldsymbol{L}}$ is assigned to $\boldsymbol{L}$ and Eq. 7 is executed to compute a new $\hat{\boldsymbol{L}}$, which finished one time iteration. After the kth iteration is done, $\hat{\boldsymbol{L}}_k$ is as follows.

$$\hat{\boldsymbol{L}}_k = \left(\mathbf{1} + \alpha\left(\boldsymbol{D}_h^T \boldsymbol{Diag}(\boldsymbol{W}_h)\boldsymbol{D}_h + \boldsymbol{D}_v^T \boldsymbol{Diag}(\boldsymbol{W}_v)\boldsymbol{D}_v\right)\right)^{-1}\hat{\boldsymbol{L}}_{k-1} \tag{8}$$

$$\boldsymbol{W}_{hk} = \frac{1}{\log|\boldsymbol{D}_h\hat{\boldsymbol{L}}_{k-1}| + \mathbf{1} + \varepsilon} \tag{9}$$

$$\boldsymbol{W}_{vk} = \frac{1}{\log|\boldsymbol{D}_v\hat{\boldsymbol{L}}_{k-1}| + \mathbf{1} + \varepsilon} \tag{10}$$

The iterative refinement of $\hat{\boldsymbol{L}}$ does not stop until $\frac{\|\hat{\boldsymbol{L}}_k - \hat{\boldsymbol{L}}_{k-1}\|}{m \times n} \leq \theta_1$. Here θ_1 is the threshold to control the iterations. After $\hat{\boldsymbol{L}}$ is purified, the reflectance $R^c(x, y)$ can be calculated as follows.

$$R^c(x, y) = I^c(x, y)/\hat{L}(x, y), c \in \{r, g, b\} \tag{11}$$

3 Controlled Migrating of Soil Image Luminance

After luminance $\hat{L}$ is risen, luminance migration will be done. Thus, a controllable luminance migration framework that is suitable for soil images is drawn.

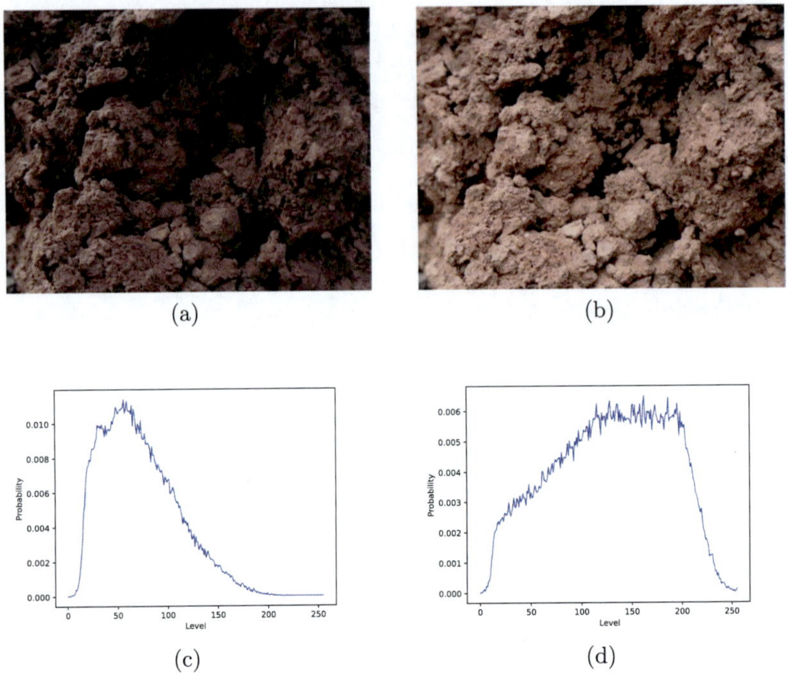

Fig. 4. (a) Low luminance sub-image. (b) High luminance sub-image. (c) Statistical histogram of luminance from (a). (d) Statistical histogram of luminance from (b).

3.1 Building Luminance Migrating Method for Soil Image

Idea of migrating luminance of soil images. To analyze the luminance characteristics of soil images collected under different natural lighting conditions, statistical histogram analysis of luminance $\hat{L}$ was performed on a pair of soil sub-images which is a same view with two different luminance.

The distinct statistical histograms shapes are corresponding with the different luminance soil images, as shown in Fig. 4. The luminance level in Fig. 4(c) is concentrated in the low gray area, and the luminance level in Fig. 4(d) is more uniformly distributed. Since the shapes of the histograms in Fig. 4(c) and (d) are inconsistent, it can be inferred that the mapping relationship of the luminance values between Fig. 4(a) and (b) is not one-to-one. The luminance mapping relationship between a pair of soil images with different luminance is drawn as Fig. 5.

The abscissa indicates the original luminance value. And the average mapping value in the target image is the ordinate, which is average luminance value of pixels that corresponds to pixels with a same luminance value in the original image. The scatter sequence presents an upper parabolic shape plane in Fig. 5(a) and a lower parabolic shape in Fig. 5(b), in which $y = x$ is the equivalent mapping line. Based on these observations, the luminance change of soil

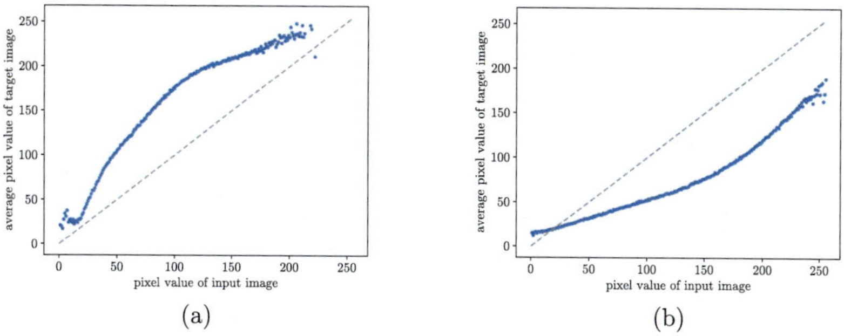

Fig. 5. Scatter diagram. (a) Luminance mapping of low to high. (b) Luminance mapping of high to low.

image is nonlinear in the field natural environment, and the luminance mapping relationship exhibits the characteristics of the gamma transform curve.

However, traditional gamma transforms, Eq. 12, can only achieve a one-to-one mapping when the gamma factor is fixed.

$$L_{out}(x, y) = A\hat{L}(x, y)^{\gamma} \tag{12}$$

where $\hat{L}$ is normalized to $[0, 1]$. A indicates a constant that controls the range of the output values. γ is the gamma factor. It needs to be improved to implement a one-to-many mapping for converting luminance of soil image.

To map $\hat{L}(x_1, y_1)$ and $\hat{L}(x_2, y_2)$ with $\hat{L}(x_1, y_1) = \hat{L}(x_2, y_2)$ to the different luminance, it is required to find the difference between $\hat{L}(x_1, y_1)$ and $\hat{L}(x_2, y_2)$. The obvious distinction between pixels with the same luminance value in an image is that their neighbors and spatial information are different. Hence, the neighborhood information is introduced to improve the gamma transform.

Improving gamma transforms. According to the previous analysis, the one-to-many mapping can be achieved with their neighbor information. The neighbors of a pixel $\hat{L}(x, y)$ can be set as four points in four-connectivity.

$$NP(x, y) = \{(x + 1, y), (x - 1, y), (x, y + 1), (x, y - 1)\} \tag{13}$$

Compared with the other connectivity information, four-connectivity information is employed because it is simple and can obtain a similar result, such as eight-connectivity. The neighborhood information of the pixel (x, y) may be represented as $NF(x, y)$ simply.

$$NF(x, y) = \frac{1}{4} \cdot \sum_{(x', y') \in NP(x, y)} \hat{L}(x', y') \tag{14}$$

Then, the neighborhood information is introduced to construct a new gamma transform for migration. An image luminance migration involves the brightening and darkening operation. An image is enhanced when gamma factor decreases. On the contrary, its luminance is widened. Therefore, it should be divided into two cases. Brightening operation based on gamma transform is as follows.

$$L_m(x,y) = A\hat{L}(x,y)^{\gamma - \frac{(\hat{L}(x,y) - NF(x,y))}{1 - NF(x,y)}} \tag{15}$$

$\hat{L} > NF(x,y)$ indicates luminance of the pixel (x,y) is greater than its neighbors, and $\hat{L}_m(x,y)$ should be greater than its neighbors in the migrated image. γ in Eq. 15 minus a positive number and the exponent of gamma transform becomes smaller can ensure it is true. Conversely, a corresponding conclusion holds.

In the images, there are bright areas that are sensitive to light changes and dark areas that are sluggish to light changes. If $NF(x_1, y_1) > NF(x_2, y_2)$ and $\hat{L}(x_1, y_1) = \hat{L}(x_2, y_2)$, it expresses (x_1, y_1) is located at the area that is brighter relatively and $\hat{L}_m(x_1, y_1)$ should be greater than at $\hat{L}_m(x_2, y_2)$. $\gamma - (\hat{L}(x,y) - NF(x,y))/(1 - NF(x,y))$ in Eq. 15 can realize it.

The darkening operation is as follows.

$$L_m(x,y) = A\hat{L}(x,y)^{\gamma + \frac{\hat{L}(x,y) - NF(x,y))}{1 - NF(x,y)}} \tag{16}$$

Darkening operation is to weaken luminance of soil image and the others are the same as brightening operation in the luminance migration.

Optimizing γ parameter. In luminance migration, its offset ΔL is given, the key point is how to determine an optimal γ to reduce the difference of luminance level between the migrated image and original image. Therefore, the objective function is constructed to optimize the value of γ.

$$\arg\min_{\gamma} \left(\frac{\sum_{x=1}^{m} \sum_{y=1}^{n} \left(L_m(x,y) - \hat{L}(x,y) \right)}{m \times n} - \Delta L \right)^2 \tag{17}$$

3.2 Luminance Controlled Migrating Algorithm of Soil Images

The migrated soil image can be obtained by synthesizing $L_m(x,y)$ and $R^c(x,y)$.

$$I_m^c(x,y) = L_m(x,y) \times R^c(x,y) \tag{18}$$

4 Experiments

4.1 Obtaining Experimental Samples

According to Classification and codes for Chongqing soil [DB50/T 796-2017] [22], there are four soil genera and 34 soil species of Purple Soil at Chongqing, China. The natural crack of subsoil in cultivated horizon keeps the original features of purple soil, which is information carrier and is used to identify soil species. So, machine vision identifies soil species with image of subsoil in cultivated horizon. Subsoil is obtained by a spade from 0 to 20 cm cultivated horizon and its natural crack is photographed to get two images that could be completely overlapped. One is a high luminance image under normal natural light. The other is a low luminance image that is photographed under environment of blocking a part of natural light. The two images constitute a sample pair. The images are cropped to sub-images with size 300×300 that are located at the center of the images. In this paper, three sample sets are constructed from 569 pairs of sub-images as follows.

Sample set 1: 30 soil images were randomly selected and divided into 10 sample groups.

Sample set 2: 30 pairs of soil images were randomly chosen and divided into 10 sample groups.

Sample set 3: 1138 soil images were divided into three groups according to 7:2:1, which were applied to training, validation and testing of deep learning classification models.

4.2 Designing Experiments

Three experiments are designed to test the performance of the proposed algorithms as follows.

Experiment 1: Comparing accuracy of luminance-controlled algorithms. The proposed algorithm and three comparison algorithms, which are 2-D HS [19], Ref. [20] algorithm and WSGF [21], are employed to test their accuracy with the Sample set 1.

Experiment 2: It is an experiment to compare the difference between a migrated image and its target image with Sample set 2 and the comparison algorithms in Experiment 1.

All the experiments are conducted on a PC running Windows 10 OS with 64 GB RAM and 2.3 GHz CPU, PyCharm 2021 and MATLAB 2020.

Luminance-controlled accuracy. Luminance-controlled accuracy is the difference between the actual offset and the predetermined offset of a migrated image. The results of the No. 3 group were shown in Fig. 6.

Data and image results in Table 1 show the proposed algorithm outperforms the algorithms in Refs. [19,21]. Method in Ref. [20] gets the lowest deviation. But it is a contrast enhancement and its result is unnatural. The algorithm in this paper can achieve luminance-controlled transformation with higher accuracy in the effective range $[-30, +30]$.

Comparison with target images. Comparison with target images to objectively evaluate the quality of enhanced images. The results of No. 1 group were shown in Fig. 7.

Table 1. Accuracy of luminance-controlled migration on Sample set 1.

Group	Algorithm	$\Delta L = -30$	$\Delta L = -20$	$\Delta L = -10$	$\Delta L = +10$	$\Delta L = +20$	$\Delta L = +30$
1	[19]	2.7626	1.7440	1.0386	0.5564	0.7471	1.1036
	[20]	0.3848	0.3338	0.2707	0.0841	0.1664	0.3020
	[21]	5.9551	3.8432	2.3052	0.7513	0.8603	1.1304
	Ours	0.6765	0.5782	0.6793	0.3434	0.5123	0.5798
2	[19]	4.8181	3.2335	2.0092	0.4717	0.2630	0.5926
	[20]	0.3460	0.2776	0.1663	0.3496	0.1258	0.1592
	[21]	7.2228	4.2729	2.2631	1.1815	1.0399	1.0586
	Ours	0.6314	0.5874	0.4560	0.6406	0.6833	0.6109
3	[19]	2.2225	1.4944	0.9342	0.6482	0.8095	1.3773
	[20]	0.1421	0.3982	0.4303	0.3473	0.2272	0.1437
	[21]	1.9803	0.8963	0.2528	0.2969	0.4145	0.4651
	Ours	0.6983	0.7570	0.4660	0.6069	0.6996	0.2963
4	[19]	2.4205	1.4187	0.7595	0.4949	0.6935	1.0277
	[20]	0.2808	0.5733	0.6201	0.1923	0.1090	0.1297
	[21]	4.6904	2.4079	1.5184	1.0869	1.0703	1.1102
	Ours	0.6711	0.7443	0.7513	0.7002	0.3475	0.4648
5	[19]	1.9836	1.0714	0.4392	1.1516	2.0217	3.1707
	[20]	0.3565	0.3564	0.6345	0.3043	0.2220	0.0843
	[21]	2.8844	1.5984	0.6420	0.3135	0.5906	0.7837
	Ours	0.4989	0.8824	0.6294	0.2190	0.3056	0.5263
6	[19]	3.7622	2.2140	1.1374	1.1665	0.7779	0.6312
	[20]	0.1250	0.2363	0.5727	0.2904	0.1752	0.1386
	[21]	6.3630	3.3299	1.5346	1.2554	1.1731	1.0918
	Ours	0.7353	0.5998	0.5047	0.6295	0.3772	0.7494
7	[19]	2.0437	1.3308	0.6383	0.9738	2.1711	3.7522
	[20]	0.2231	0.2642	0.5078	0.3187	0.0735	0.1024
	[21]	1.1536	0.5293	0.6181	0.8073	0.9554	1.0971
	Ours	0.4812	0.4351	0.4453	0.2270	0.4922	0.6867
8	[19]	1.2352	1.1190	0.7993	0.8438	2.7111	5.4434
	[20]	0.2191	0.3315	0.5889	0.2289	0.1985	0.0569
	[21]	4.0256	2.5509	1.4883	0.3415	0.1511	0.4628
	Ours	0.6298	0.6027	0.5255	0.3347	0.4579	0.4206
9	[19]	2.2941	1.6650	1.2171	0.6065	0.2652	0.2419
	[20]	0.2651	0.4135	0.5802	0.2997	0.2796	0.1123
	[21]	5.1711	3.1293	1.6546	1.3926	1.4886	1.5187
	Ours	0.4714	0.5107	0.6613	0.3510	0.6490	0.5838
10	[19]	3.0907	1.9753	1.1940	0.8028	0.8860	1.4072
	[20]	0.1517	0.3927	0.5515	0.2466	0.0720	0.1375
	[21]	4.4336	2.5186	1.3157	1.4034	1.3595	1.3362
	Ours	0.5291	0.4886	0.5053	0.5460	0.5534	0.7787
Avg.	[19]	2.6633	1.7266	1.0167	0.7716	1.1346	1.8748
	[20]	0.2494	0.3578	0.4923	0.2662	0.1649	0.1367
	[21]	4.3880	2.5077	1.3593	0.8830	0.9103	1.0055
	Ours	0.6023	0.6186	0.5624	0.5098	0.5078	0.5697

Table 2. Average pixel difference from the target image on Sample set 2

Group	Method	Low to High				High to Low			
		$\bar{d_r}$	$\bar{d_g}$	$\bar{d_b}$	$\bar{d}$	$\bar{d_r}$	$\bar{d_g}$	$\bar{d_b}$	$\bar{d}$
1	[19]	16.1179	15.1223	15.2097	15.4833	15.7102	14.8132	14.9748	15.1661
	[20]	21.4559	19.4915	19.5866	20.1780	19.9296	17.6539	17.0520	18.2118
	[21]	14.3154	13.5378	13.5981	13.8171	14.3625	13.4750	13.8349	13.8908
	Ours	5.7588	5.4791	4.7891	5.3424	6.6210	7.0981	8.2085	7.3092
2	[19]	19.4283	18.5006	18.1365	18.6885	18.9375	17.7248	17.2161	17.9595
	[20]	26.2216	23.1571	22.2429	23.8739	25.4868	22.9463	21.3195	23.2509
	[21]	18.4061	17.5360	17.0011	17.6477	18.5792	17.1873	16.7678	17.5114
	Ours	6.2045	7.4796	7.0330	6.9057	8.1545	7.1859	7.5622	7.6342
3	[19]	23.7526	23.3231	23.1645	23.4134	23.3265	23.0292	22.9037	23.0865
	[20]	24.4436	23.6206	23.1807	23.7483	33.3226	30.1420	29.2507	30.9051
	[21]	22.4325	22.0662	21.9735	22.1574	22.2805	21.6957	21.4224	21.7995
	Ours	6.9200	7.7786	8.7424	7.8137	9.1338	9.1816	8.8566	9.0573
4	[19]	20.8402	19.0382	18.8127	19.5637	19.5034	17.5491	17.2679	18.1068
	[20]	30.6840	24.7367	23.4002	26.2736	32.5198	27.0355	25.3459	28.3004
	[21]	19.2484	18.4172	18.3186	18.6614	19.6582	17.7693	17.3427	18.2568
	Ours	4.5834	5.2339	5.3296	6.5711	8.5410	6.8993	6.8137	7.4180
5	[19]	17.3748	16.9588	16.6335	16.9890	16.6639	16.1198	15.9573	16.2470
	[20]	19.4496	18.0309	17.1168	18.1991	24.5325	22.2446	20.6629	22.4800
	[21]	15.1696	14.5487	14.2776	14.6653	14.7229	14.3304	14.0016	14.3516
	Ours	6.8187	6.4819	6.3609	6.5538	7.6906	7.9343	7.6542	7.7597
6	[19]	17.6911	15.9381	16.3209	16.6500	17.1287	14.8555	14.5348	15.5063
	[20]	44.6781	36.7916	29.4405	36.9701	21.2146	19.3554	17.9446	19.5049
	[21]	16.9385	15.4059	14.6034	15.6493	16.4040	14.9420	13.9857	15.1105
	Ours	8.0679	7.6960	6.5097	7.4245	6.0681	5.5137	5.9759	5.8525
7	[19]	19.1999	18.3350	18.9284	18.8211	18.5539	18.0913	18.7879	18.4777
	[20]	18.3057	17.9718	19.0797	18.4524	27.3039	24.4669	23.8208	25.1972
	[21]	17.9763	17.2074	17.7745	17.6527	17.7094	17.2051	17.9858	17.6334
	Ours	12.6048	9.8815	7.3228	9.9364	6.4845	7.6026	10.1081	8.0651
8	[19]	18.1956	16.8472	16.7090	17.2506	17.3590	15.7348	15.3849	16.1596
	[20]	22.6590	20.3075	19.7006	20.8891	30.5770	26.4916	23.3531	26.8072
	[21]	17.4323	16.1301	15.5379	16.3668	17.1215	15.6654	15.0428	15.9432
	Ours	7.7989	8.2510	7.0575	7.7025	7.3514	6.1926	6.8533	6.7991
9	[19]	27.8651	27.1219	26.8905	27.2925	27.5635	26.6210	26.4837	26.8894
	[20]	33.1907	30.8363	30.5470	31.5246	31.2267	28.9526	28.5585	29.5793
	[21]	25.6034	25.2629	25.1531	25.3398	26.4173	25.3665	25.3550	25.7129
	Ours	7.9762	9.6724	9.5095	9.0515	9.0927	9.2920	9.9812	9.4553
10	[19]	22.2979	21.6396	21.5062	21.8146	21.3285	20.5390	20.2598	20.7091
	[20]	38.0576	35.2770	31.9657	35.1001	28.7587	26.6147	25.5255	26.9663
	[21]	21.3068	20.7195	20.2605	20.7622	21.0871	20.2911	19.7722	20.3835
	Ours	6.5671	8.5997	8.2765	7.8145	8.9356	8.0498	8.2433	8.4096
Avg.	[19]	20.2763	19.2825	19.2312	19.5967	19.6075	18.5078	18.3771	18.8308
	[20]	27.9146	25.0221	23.6261	25.5209	27.4872	24.5903	23.2834	25.1203
	[21]	19.1930	18.0832	17.8498	18.2720	18.8343	17.7928	17.5511	18.0594
	Ours	7.3301	7.6554	7.0931	7.5116	7.8073	7.4950	8.0257	7.7760

*Note: d denotes the absolute difference between the enhanced image and the target image. d_r, d_g and d_b denote the red component, green component, and blue component of d, respectively.

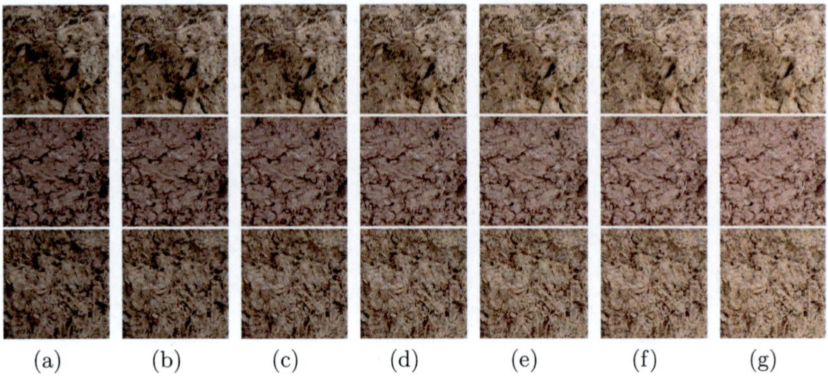

Fig. 6. The migrated results of No. 3 group. (a) $\Delta L = -30$. (b) $\Delta L = -20$. (c) $\Delta L = -10$. (d) Original. (e) $\Delta L = +10$. (f) $\Delta L = +20$. (g) $\Delta L = +30$.

The data from Table 2 shows that the difference of the migrated images is least from their target images with the proposed algorithm. Therefore, the algorithm in this paper can well protect the color features to approximate the target image when performing luminance migration.

5 Conclusions

To address the scarcity of training samples in soil species recognition, this paper proposed a luminance-controlled migration algorithm applied to soil image dataset expansion. By combining a refined Retinex model with neighborhood-aware optimization, the proposed method effectively generates realistic soil images under diverse lighting conditions. The main conclusions are drawn as follows:

(1) Texture Preservation: A novel decomposition model was developed for the Retinex decomposition of soil images. By constraining the reflectance range and smoothing the luminance components, the method successfully avoids the texture over-amplification issues commonly observed in traditional migration processes, ensuring that the intrinsic features of the soil are preserved.
(2) Precise Control via Optimization: We reconstructed the gamma transform function by incorporating pixel neighborhood information, which facilitates a one-to-many luminance mapping strategy. Furthermore, an optimization model was established to adaptively solve for the gamma factor, allowing for precise and automated control of luminance migration.
(3) Experimental Performance: Experimental results demonstrate that the proposed algorithm achieves superior accuracy in luminance-controlled migration, particularly within the effective range of $[-30, 30]$. Quantitative analysis shows that the migrated images exhibit lower deviations and are statistically closer to the real-world target images compared to existing algorithms.

(a) (b) (c) (d) (e) (f)

Fig. 7. Results of low-to-high luminance transformation for Group 1. (a) Original images. (b) Results of Ref. [19]. (c) Results of Ref. [20]. (d) Results of Ref. [21]. (e) Results of our proposed method. (f) Target images.

Although the proposed algorithm demonstrates promising effectiveness, several open issues remain to be addressed. A key direction for future research is the separate estimation of luminance for three individual color channels (e.g., RGB) rather than global luminance adjustment. Specifically, this approach is crucial for correcting complex color skews caused by non-uniform lighting spectra, such as reddish sunset light, which a single luminance channel cannot fully capture, thereby significantly enhancing the color fidelity of synthesized datasets. However, realizing this involves substantial challenges, primarily regarding the maintenance of inter-channel correlation to prevent unnatural color artifacts and the increased computational complexity of multi-channel optimization models, which will require more efficient solving strategies.

Acknowledgments. This research was supported by the Natural Science Foundation of Chongqing Municipality (CSTB2022NSCQ-LZX0081); Key Project of Science and Technology Research of Chongqing Education Commission (KJZD-K202400509).

References

1. Zeng, S.H., Zhao, B.Y., Wang, S., Chen, Y.N., Zhu, D.L.: Controllable brightness enhancement of the soil image based on weighted Gaussian subtraction fitting. J. Image Graph. **11**(1), 21–25 (2023)
2. Wang, H., Hou, X., Li, J., et al.: Multi-scale Retinex unfolding network for low-light image enhancement. IEEE Trans. Multimed. **27**, 5709–5721 (2025)
3. Lei, X., Mai, W., Xie, J., et al.: KinD-LCE: curve estimation and Retinex fusion on low-light image. SIViP **18**, 1733–1746 (2024)
4. Li, D., Zhou, J., Wang, S., et al.: Adaptive weighted multiscale Retinex for underwater image enhancement. Eng. Appl. Artif. Intell. **123**, 106457 (2023)
5. Tian, F., Wang, M., Liu, X.: Low-light mine image enhancement algorithm based on improved Retinex. Appl. Sci. **14**(5), 2213 (2024)

6. Li, C., He, C.: Fractional structure and texture aware model for image Retinex and low-light enhancement. Appl. Math. Model. **130**, 496–513 (2024)
7. Kong, X.Y., Liu, L., Qian, Y.S.: Low-light image enhancement via Poisson noise aware Retinex model. IEEE Signal Process. Lett. **28**, 1540–1544 (2021)
8. Jeon, J.J., Eom, I.K.: Low-light image enhancement using inverted image normalized by atmospheric light. Signal Process. **196**, 108523 (2022)
9. Zhang, H., Qian, W., Wan, M., et al.: Infrared image enhancement algorithm using local entropy mapping histogram adaptive segmentation. Infrared Phys. Technol. **120**, 104000 (2022)
10. Hu, T., Zhou, Q., Nan, X., et al.: A color image decomposition model for image enhancement. Neurocomputing **558**, 126772 (2023)
11. Vijayalakshmi, D., Nath, M.K.: A novel multilevel framework based contrast enhancement for uniform and non-uniform background images using a suitable histogram equalization. Digit. Signal Process. **127**, 103532 (2022)
12. Roy, R., Ghosh, S., Ghosh, A.: Clinical ultrasound image standardization using histogram specification. Comput. Biol. Med. **120**, 103746 (2020)
13. Liu, C., Sui, X., Kuang, X., et al.: Optimized contrast enhancement for infrared images based on global and local histogram specification. Remote Sens. **11**(7), 849 (2019)
14. Xu, Y., Yang, C., Sun, B., Yan, X., Chen, M.: A novel multi-scale fusion framework for detail-preserving low-light image enhancement. Inf. Sci. **548**, 378–397 (2021)
15. Kaur, P., Khehra, B.S., Pharwaha, A.P.S.: Color image enhancement based on gamma encoding and histogram equalization. Mater. Today Proc. **46**, 4025–4030 (2021)
16. Chaudhry, A.M., Riaz, M.M., Ghafoor, A.: Model-assisted content adaptive detail enhancement and quadtree decomposition for image visibility enhancement. SIViP (2022)
17. Wang, S., Luo, G.: Naturalness preserved image enhancement using a priori multi-layer lightness statistics. IEEE Trans. Image Process. **27**(2), 938–948 (2018)
18. Wang, W., Chen, Z., Yuan, X., et al.: Adaptive image enhancement method for correcting low-illumination images. Inf. Sci. **496**, 25–41 (2019)
19. Xiao, B., Tang, H., Jiang, Y., et al.: Brightness and contrast controllable image enhancement based on histogram specification. Neurocomputing **275**, 2798–2809 (2018)
20. Tohl, D., Jimmy Li, J.S.: Image enhancement by S-shaped curves using successive approximation for preserving brightness. IEEE Signal Process. Lett. **24**(8), 1247–1251 (2017)
21. Zeng, S.H., Zhao, B.Y., Wang, S., Chen, Y.N., Zhu, D.L.: Controllable brightness enhancement of the soil image based on weighted Gaussian subtraction fitting. Acta Photon. Sin. **51**(4), 1–23 (2022)
22. Wang, S., Xie, D.T., Qu, M., et al.: DB50T 796-2017 Classification and Codes for Chongqing Soil. Chongqing Bureau of Technical Supervision (2017)

A Few-Shot Image Generation Model Based on a Hybrid Attention Mechanism

Qian Wu, Feng Yu, and Dequan Zheng(✉)

Harbin University of Commerce, Harbin, Heilongjiang, China
dqzheng@hrbcu.edu.cn

Abstract. Generative adversarial networks (GANs) demonstrate superior modeling capabilities in image generation tasks, but their reliance on large-scale, identically distributed samples limits their application in few-shot scenarios. Current few-shot image generation faces three challenges: first, data scarcity often leads to model overfitting and pattern collapse; second, balancing local texture modeling with global structure preservation is difficult, as traditional convolutional structures struggle to capture long-range dependencies; third, insufficient style and semantic consistency in cross-category generation limits model generalization. To address these issues, this paper proposes a few-shot image generation model based on a hybrid attention mechanism. This method builds upon the pyramid-based multi-stage generator of ConSinGAN, introducing lightweight self-attention modules and Vision Transformer (ViT) modules at different stages to enhance local feature modeling at low resolutions and global semantic perception at high resolutions, respectively. To further enhance semantic expressiveness and category discriminability, this paper adopts the auxiliary classification strategy from DEff-GAN, designing a multi-task optimization objective that integrates adversarial loss, structural reconstruction loss, and semantic classification loss. Experimental results demonstrate that this method effectively improves the quality and expressiveness of generated images under few-shot conditions while exhibiting outstanding generalization capabilities in cross-category scenarios.

Keywords: few-shot image generation · mixed attention mechanism · generative adversarial network

1 Introduction

In the field of image generation, generative adversarial networks (GANs) have been widely applied to modeling tasks for high-quality images. However, when the amount of training data is extremely limited, traditional GAN architectures face severe challenges in terms of expressive power and generalization ability, making it difficult to capture effective semantic structures and leading to issues such as overfitting and pattern collapse [1]. To address this challenge, few-shot image generation (FSG) has emerged as a research hotspot in recent years, aiming to synthesize diverse, structurally complete, and semantically consistent high-quality images with only a small number of images.

L. Huang and D. Greenhalgh (Eds.): ICMLC 2026, LNNS 2061, pp. 69–85, 2027.
https://doi.org/10.1007/978-3-032-31236-5_5

Existing studies have explored dimensions such as feature fusion, structure preservation, prior guidance, and attention mechanisms. F2GAN [2] and AMMGAN [3] enhance the fusion of multi-scale features through filling and modulation strategies, thereby improving the model's ability to perceive different semantic levels; SAGAN [4] and DEff-GAN [5] introduce skip connections and dual-path enhancement mechanisms to optimize local structure and texture generation effects. PetsGAN [6] adopts a single-image learning approach, guiding the model to reconstruct the semantic space from prior knowledge to mitigate structural disorganization caused by sample scarcity. In recent years, the Transformer architecture has been introduced into the field of image generation, demonstrating its potential for modeling long-range dependencies. TransGAN [7] first proposed a pure Transformer architecture for image generation. Styleformer [8] decouples style and structure modeling based on the Transformer architecture, enhancing structural preservation in style transfer tasks; StyleHEAT [9] combines a style encoder with a spatial control module to enable editable face generation at high resolutions, enhancing local detail control and semantic consistency; DM-GAN [10] employs a hybrid design of CNN and ViT to effectively enhance the model's joint modeling capability of structure and content under few-sample conditions. Additionally, structural discriminative mechanisms are used to enhance the spatial consistency of generated images, such as SDTM-GAN [11], which combines structural and texture modulators for fine-grained control and achieves good performance on multiple small-sample datasets. In addition to generator structure, innovations in training paradigms are equally critical. SinGAN [12] and its improved method ConSinGAN [13] are based on image pyramids and multi-stage feature propagation mechanisms, demonstrating strong generalization capabilities in tasks such as single-image generation and image restoration. In specific domains like remote sensing image generation, studies have proposed attribute-guided and segmented training strategies to enhance robustness in few-sample modeling [14]. Furthermore, novel generation paradigms such as the dual-correlation diffusion mechanism [15] and methods based on mixed distance learning [16] have also demonstrated outstanding structural preservation and anomaly modeling capabilities. The latent diffusion model [17] has demonstrated outstanding performance in high-resolution image generation, particularly for large-scale data and complex texture generation. Meanwhile, the Denoising Diffusion Probabilistic Model [18] offers novel insights for few-shot image generation, showing great potential in diversity enhancement and style transfer.

Despite progress in existing research on fusion strategies, multi-scale modeling, and attention mechanisms, several challenges remain: single attention mechanisms struggle to balance local details with global semantics; Transformer architectures exhibit training instability and style drift in data-scarce scenarios; and current feature fusion strategies are predominantly static designs lacking hierarchical semantic awareness and structural controllability. To address these issues, this paper proposes a pyramid-based few-shot image generation method with a hybrid attention mechanism. A lightweight self-attention module and a Vision Transformer (ViT) module are introduced into the multi-stage generator to achieve local texture modeling at low resolutions and global semantic modeling at high resolutions. The generator progressively reduces modeling complexity by incrementally increasing resolution, while the discriminator employs a hierarchical discrimination mechanism to optimize image consistency. Experimental results demonstrate that our method significantly outperforms baseline models across multiple few-shot benchmark datasets, validating its generation quality and cross-category generalization capability under low-sample conditions.

2 Model Design

2.1 Overall Architecture

This method builds upon the multi-stage pyramid generation framework of ConSinGAN [13] and DEff-GAN [5], proposing an enhanced generative model that strengthens structural modeling and semantic discrimination capabilities. Within this framework, multiple generators and discriminators are hierarchically cascaded, with each layer of generators receiving the output image from the previous level to progressively increase resolution and reconstruct detailed information. Simultaneously, a fully convolutional Patch-GAN with a pyramid structure captures patch distributions by progressively enlarging image dimensions, enabling a refined generation process.

To enhance the generator's feature modeling and semantic perception capabilities across different scales, this paper introduces a Hybrid Attention Mechanism. At low resolutions (Stages 0–2), a self-attention module is inserted to strengthen edge and texture detail capture. At high resolutions (Stages 4–6/8), a Vision Transformer (ViT)-based Transformer Block enhances cross-region semantic integration and global structure modeling. All attention mechanisms are connected in parallel via residual connections within the convolutional stack, ensuring model stability and efficient gradient propagation.

The auxiliary classifier from DEff-GAN is retained, enabling the generator to acquire semantically distinguishable features during training. This enhances discriminative performance and semantic transfer capabilities in few-shot generation tasks. Through these enhancements, our method builds upon the original pyramid modeling advantages while strengthening the model's local fine-grained modeling and global semantic integration capabilities, making it more suitable for scenarios such as few-shot generation and category transfer. As shown in Fig. 1, the proposed image generation framework combines a multi-stage generator architecture with the loss function mechanism of generative adversarial networks.

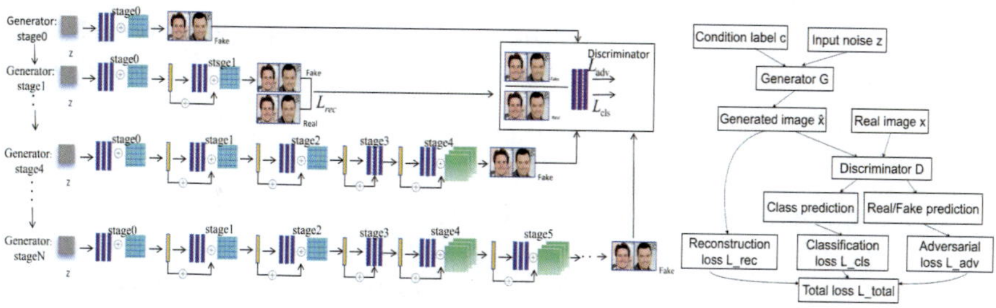

Fig. 1. Architecture diagram

2.2 Hybrid Attention Mechanism Design

To balance local details and global semantic consistency in image generation, this paper introduces a Hybrid Attention Mechanism within a multi-stage pyramidal generation architecture. This mechanism enhances multi-scale feature modeling and semantic consistency optimization by embedding self-attention (SA) modules and Vision Transformer (ViT) modules at different resolution stages.

At the low-resolution stage, the integrated lightweight self-attention module is based on the SAGAN architecture. It reduces computational overhead through channel compression ($Cr = C/8$) and a residual connection strategy. Input feature maps are processed as Query, Key, and Value. Channel dimensions are compressed to 1/8, and attention maps are generated via dot products. Weighted Value features are fused with original features through 1×1 convolutions and residual connections, improving global semantic modeling capability and stability while mitigating overfitting in few-shot training.

At the high-resolution stage, a Transformer Block module based on the Vision Transformer (ViT) architecture is introduced, employing a "Convolution–Flattening–Attention–Reconstruction" workflow. Channel dimensions are compressed via 1×1 convolutions and flattened into sequences. These sequences feed into Multi-Head Self-Attention (MSA) and Feedforward Networks (FFN) to capture long-range dependencies. Spatial feature maps are then reconstructed through convolutions, refining image structural details and achieving fusion of local and global features.

Both attention modules are embedded into the generator's convolutional stack via residual connections, ensuring training stability while enhancing structural fidelity and semantic consistency in generated images. This hybrid attention mechanism provides an effective improvement path for image generation under few-shot conditions.

2.3 Objective Function Design

In few-shot generation tasks, the generator must balance high fidelity, structural reconstruction, and semantic consistency across categories. This paper proposes a multi-objective joint optimization strategy. The objective function comprises adversarial, structural reconstruction, and semantic classification components, targeting image quality, structural preservation, and category classification respectively. This establishes a collaborative generative supervision mechanism, enhancing generalization and expressive capabilities in few-shot, multi-class scenarios. Concurrently, integrating lightweight self-attention modules and Transformer Blocks strengthens the generator's feature modeling and semantic alignment abilities.

Countering Targets

This paper employs Wasserstein GAN with Gradient Penalty (WGAN-GP) as the adversarial objective function, replacing the cross-entropy loss in traditional GANs. By utilizing Wasserstein distance and gradient penalty, it ensures 1-Lipschitz continuity and prevents overfitting of the discriminator. Compared to standard GAN loss, WGAN-GP exhibits superior convergence in multi-stage pyramid architectures, effectively approximates real image distributions, and promotes stable training of deep structures.

Its mathematical definition is shown in Eq. (1):

$$L_{adv} = E_{x \sim P_r}[D(x)] - E_{x \sim P_g}[D(x)] + \lambda E_{\hat{x} \sim P_{\hat{x}}}\left[(|\nabla_{\hat{x}} D(\hat{x})|_2 - 1)^2\right] \quad (1)$$

Among them, P_r and P_g represent the distributions of real images and generated images, respectively. $\hat{x}$ is an interpolated sample point between the two, and λ is the gradient penalty coefficient, which is set to 10 in this experiment.

Structural Reconstruction Objectives

Since this model adopts a multi-stage pyramid generation architecture, each stage requires layer-by-layer modeling of the image structure at different resolutions. Therefore, a structural reconstruction objective function is introduced to maintain the fidelity of the image in terms of spatial structure and detail, constraining the difference between the generated image and the target image at the pixel level. This objective function is expressed in the form of mean squared error (MSE). Its mathematical definition is shown in Eq. (2):

$$\mathcal{L}_{\text{rec}} = \frac{1}{\text{N}} \sum_{\text{i}=1}^{\text{N}} \|\text{G}_{\text{i}}(\text{z}_{\text{i}}) - \text{x}_{\text{i}}\|^2 \quad (2)$$

$G_i(z_i)$ represents the output of the generator in the i stage, x_i which is the target image at that scale.

Semantic Classification Objectives

In scenarios with few samples, generated images easily lose semantic category information. To improve semantic consistency, an auxiliary classifier branch is introduced into the discriminator to supervise the classification of image categories, guiding the genera-tor to output images that match the semantic labels, and using cross-entropy as the supervision signal. Its mathematical definition is shown in Eq. (3):

$$\mathcal{L}_{\text{cls}} = -\sum_{c=1}^{C} y_c \cdot \log D_{\text{cls}}(x)_c \quad (3)$$

$D_{\text{cls}}(x)$ where represents the category probability output by the discriminator, y_c is the one-hot encoding of the category to which the image belongs.

Joint Optimization Objective

Based on the three objective functions discussed above, this paper uses a weighted sum to construct the total objective function, as shown in Eq. (4):

$$\mathcal{L}_{\text{total}} = \mathcal{L}_{\text{adv}} + \alpha \cdot \mathcal{L}_{\text{rec}} + \beta \cdot \mathcal{L}_{\text{cls}} \quad (4)$$

The hyperparameters are set as follows: $\alpha = 10$: Strengthen structural reconstruction supervision; $\beta = 1$: Maintain moderate semantic supervision.

Adversarial loss provides global visual constraints; reconstruction loss provides detail restoration constraints; classification loss provides semantic consistency constraints; the three complement each other and, in the structure enhanced by Self-Attention + Transformer Block, act on local modeling, global modeling, and semantic transfer, respectively, to synergistically improve the performance of the generator.

3 Implementation Process

This method builds upon the DEff-GAN framework by incorporating a multi-stage pyramid generation architecture and a Growing GAN strategy. It progressively increases the number of stages in the generator to refine the output. The generator comprises 6–8 stages, each corresponding to images of different resolutions (32×32–256×256). It adopts a fully convolutional Patch-GAN structure, with each stage containing 3–6 convolutional layers. The number of convolutional channels is 64, expandable to 128 for high-resolution stages to enhance feature extraction capabilities. Each stage's output not only generates images at its resolution but also fuses with the next stage to ensure multi-scale information consistency.

To enhance feature modeling, hybrid attention mechanisms are introduced across stages: lightweight self-attention (SA) modules strengthen edge, texture, and local structure modeling in low-resolution stages; while the high-resolution stage incorporates a Vision Transformer (ViT)-based Transformer Block to strengthen global dependency modeling and semantic consistency. All attention modules are embedded into the generator via residual connections to ensure training stability.

In experiments, the Transformer Block is configured with 4 Attention Heads, an MLP layer width of 512, an Embedding Dimension of 64, and 0.1 Attention Dropout to prevent overfitting. Optimization occurs per image at each scale (effective batch size = 1). The model employs the Adam optimizer with momentum parameters $\beta_1 = 0.5$, $\beta_2 = 0.999$. The generator learning rate is 0.0005, the discriminator learning rate is 0.00025, and Multi-Step learning rate decay is applied (decay factor $\gamma = 0.1$). Convolution layers are initialized with mean 0 and standard deviation 0.02. BatchNorm layers are initialized with mean 1, standard deviation 0.02, and bias 0, with BatchNorm2 enabled. The discriminator uses the WGAN-GP loss function with gradient penalty $\lambda = 10$. Each stage underwent 2k–6k training iterations, with high-resolution stages (Stages 5 and 6) receiving additional iterations to enhance generation quality.

For sparse data, experiments with differentiable augmentation showed color perturbation yielded optimal results, while cutout and translation augmentations were not adopted. To enhance semantic discrimination, the discriminator retains an auxiliary classifier branch. By predicting image category labels, it provides category guidance to the generator, improving semantic consistency and cross-category transferability. This design enhances the model's ability to model both global and local information while ensuring stability and feature propagation throughout the multi-stage generation process.

4 Experimental Setup and Analysis of Results

In the field of few-shot image generation, many methods rely on large-scale pre-training and rich prior knowledge. While they demonstrate excellent performance, they struggle to authentically reflect the ability to learn from scratch in stringent few-shot scenarios. To objectively evaluate model performance, this paper selects MixDL and the original DEff-GAN as comparison baselines, representing diversity-enhancing and structure-enhancing paradigms respectively, both of which do not depend on pre-training. Furthermore, to validate the role of attention modules, ablation experiments were conducted on Self-Attention, Vision Transformer, and Hybrid Attention. These experiments evaluated the differences among various attention mechanisms in modeling local details, maintaining global structural consistency, and ensuring generative stability.

4.1 Dataset and Settings

To validate the effectiveness of the proposed model in few-shot image generation tasks, experiments were conducted on CelebA, Oxford Flowers, Obama Few-Shot, Dogs vs. Cats, and texture datasets, covering diverse scenarios including human faces, animals, flowers, and textures. Models were trained independently with 2–10 few-shot training images to ensure fairness and reproducibility. Generated images featured resolutions of 128×128 or 256×256, employing a pyramid-based multi-stage generation architecture with 2k–6k training iterations per stage. Comparison methods included DEff-GAN and MixDL, evaluating image diversity, fidelity, and semantic consistency. Animal images assessed adaptability to detail variations, while texture images validated texture preservation under weak semantic constraints. The Obama Few-Shot task evaluated semantic consistency in identity retention and pose variations.

4.2 Evaluation Indicators

To evaluate the performance of a few-shot image generation model that combines Self-Attention and Transformer Block, this paper uses three metrics: FID, SIFID, and LPIPS, analyzing the model from multiple dimensions, including global distribution similarity, local detail fidelity, and perceptual consistency. FID measures the overall distribution difference between generated images and real images in the feature space, with lower values indicating higher quality; SIFID evaluates the local consistency between a single image and its generated result, suitable for few-shot scenarios; in the LPIPS metric, LPIPS-1 measures the diversity of generated images, while LPIPS-2 reflects the stylistic consistency between generated images and input images. Following the DEff-GAN framework, the Obama Few-Shot, animal, and texture tasks are also used for qualitative visualization to demonstrate the model's performance in identity preservation, structural restoration, and texture modeling.

4.3 Experimental Results

This paper systematically evaluates the proposed models in few-shot and multi-class generation tasks, covering scenarios such as faces, animals, flowers, and textures. The number of input images is set to 2, 4, 5, and 10 to simulate different data scarcity conditions. Each experimental group uniformly generates 100 images for evaluation. For comparison, MixDL saves the model every 10K iterations, accumulating 5000 generated images, from which 1000 are randomly sampled for metric statistics. DEff-GAN generates an equivalent number of images under identical settings. All experiments uniformly set the resolution to 128×128 or 256×256 to ensure comparability of results.

Image Generation Visualization Results

To visually demonstrate the model's generative performance under few-shot conditions, the generated images are grouped into three sets: Fig. 2 presents male faces, female faces, and floral images, serving as the primary subjects for quantitative evaluation; Fig. 3 presents cat/dog, texture, and Obama few-shot images to evaluate detail adaptation, texture preservation, and semantic consistency in identity and pose; Fig. 4 displays generated images from DEff-GAN and the Denoising Diffusion Probabilistic Model (DDPM).

Fig. 2. Male faces, female faces, and flower images

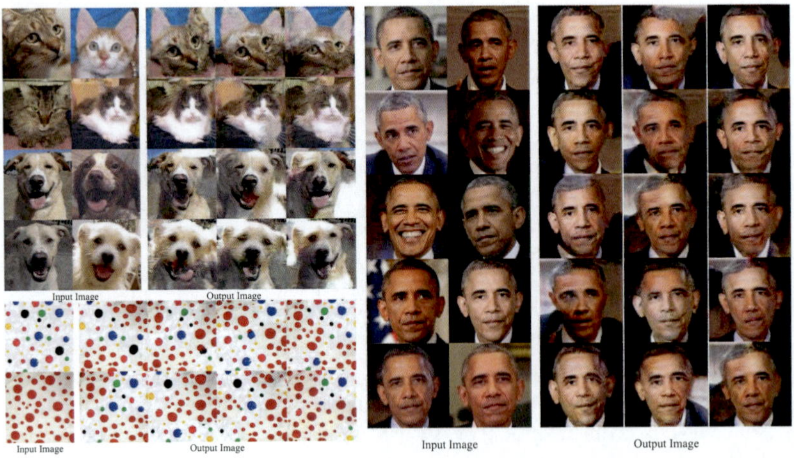

Fig. 3. Cats and dogs, textures, and Obama few-shot images

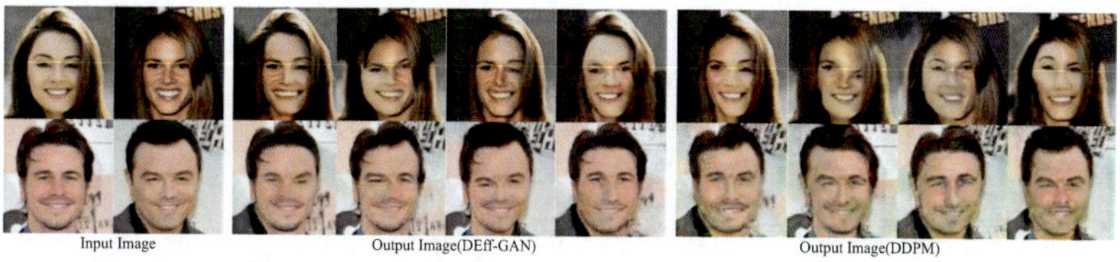

Fig. 4. Generated images from DEff-GAN and Denoising Diffusion Probabilistic Model

Quantitative Assessment Against Baseline

Tables 1 and 2 summarize the LPIPS diversity and consistency evaluation results for MixDL, DEff-GAN, and the proposed method in few-shot image generation, respectively; Table 3 reports the FID values of the three methods across different tasks, measuring the global similarity between generated images and the true distribution. Table 4 demonstrates the SIFID performance of DEff-GAN and our method in terms of structural fidelity, reflecting their local reconstruction capabilities at the single-image level.

Additionally, to mitigate the impact of single-run randomness, all metrics were averaged and their standard deviations calculated across 10 independent runs. Since the standard deviations for LPIPS-1, LPIPS-2, and SIFID are generally in the 0.01 range with minimal fluctuation, only their means are reported in the table.

Table 1 presents the LPIPS-1 comparison results. Our method achieves an LPIPS-1 score of 0.60–0.63 on the floral task, significantly outperforming MixDL (0.39–0.54) and DEff-GAN (0.58–0.59). For the face task, our method maintains stability within the 0.25–0.29 range, comparable to DEff-GAN. This advantage stems from our approach's simultaneous utilization of SA (Strengthened Texture Variation) and ViT (Diffusion-based Global Semantics), enabling the generation of more natural, non-noisy diversity. In contrast, MixDL's diversity tends toward "noise-type," while DEff-GAN's diversity is constrained by structural limitations.

Table 1. Comparison of LPIPS-1 indicators

		Flower	Male	Female
MixDL	30K	–	0.41	0.27
	40K	0.47	0.37	0.31
	50K	0.39	0.37	0.32
	60K	0.54	–	–
DEFF-GAN	3000	0.59	–	–
	4000	0.58	0.27	0.27
	5000	0.59	0.28	0.27
	6000	–	0.27	0.27
Our method	3000	0.60	0.25	0.26
	4000	0.61	0.29	0.27

(continued)

Table 1. (*continued*)

		Flower	Male	Female
	5000	0.63	0.28	0.28
	6000	0.62	0.29	0.27

Table 2 presents the LPIPS-2 metric comparison results. Our method achieves an LPIPS-2 score of 0.21–0.24 on the face task, outperforming MixDL (0.41–0.47) and DEff-GAN (0.28–0.31). It also maintains a consistent advantage of 0.57–0.60 on the flower task. The higher consistency of our method stems from the ViT branch stabilizing the global structure while the SA branch supplements local textures. This prevents style drift observed in MixDL and avoids the oscillation in texture details seen in DEff-GAN.

Table 2. Comparison of LPIPS-2 indicators

		Flower1	Flower2	Flower3	Flower4	Flower5
MixDL	30K	–	–	–	–	–
	40K	0.63	0.53	0.65	0.62	0.67
	50K	0.62	0.45	0.66	0.60	0.66
	60K	0.64	0.63	0.67	0.64	0.52
DEFF-GAN	3000	0.63	0.62	0.64	0.69	0.63
	4000	0.64	0.62	0.63	0.67	0.62
	5000	0.65	0.63	0.64	0.69	0.63
Our method	3000	0.60	0.59	0.59	0.58	0.61
	4000	0.59	0.58	0.58	0.57	0.60
	5000	0.58	0.60	0.57	0.57	0.61
	6000	0.59	0.58	0.57	0.58	0.60

		Male1	Male2	Female1	Female2
MixDL	30K	0.42	0.47	0.34	0.46
	40K	0.43	0.44	0.35	0.45
	50K	0.41	0.44	0.36	0.43
	60K	–	–	–	–
DEFF-GAN	3000	0.31	0.31	0.27	0.30
	4000	0.28	0.31	0.27	0.29
	5000	0.28	0.28	0.27	0.28
Our method	3000	0.21	0.22	0.25	0.27
	4000	0.22	0.21	0.23	0.25
	5000	0.21	0.20	0.21	0.24
	6000	0.20	0.21	0.22	0.22

Table 3 presents the FID metric comparison results. Our method achieves an FID of 83–87 on the face task, significantly outperforming MixDL (206–238) and DEff-GAN (202–217). It also maintains the lowest FID in the floral task (175 vs. 200–219 for MixDL and 238–244 for DEff-GAN). The superior performance of our method stems from: ViT provides stable global structure modeling, while SA compensates for fine-grained textures, making the generated feature distribution closer to the true distribution and thus significantly reducing FID.

Table 3. Comparison of FID indicators

		Flower	Male	Female
MixDL	30K	–	238	167
	40K	200	206	141
	50K	219	200	130
DEFF-GAN	3000	244	–	–
	4000	238	217	128
	5000	–	205	119
	6000	–	202	121
Our method	3000	198 ± 3.39	137 ± 6.87	98 ± 2.18
	4000	192 ± 3.42	117 ± 6.37	97 ± 2.24
	5000	183 ± 3.60	103 ± 5.54	95 ± 2.98
	6000	175 ± 3.54	83 ± 5.05	87 ± 2.43

Table 4 shows the SIFID metric comparison results. Our method achieves a minimum SIFID of 0.143 (Male1, 6000 iter), significantly outperforming DEff-GAN's range of 0.173–0.199. The local advantage of our method stems from SA's sensitivity to fine textures combined with ViT's cross-regional structural alignment capability, enabling more stable reconstruction of local regions (e.g., periorbital areas, edge textures). In contrast, DEff-GAN suffers from local constraints due to its convolutional receptive field, leading to frequent detail discontinuities.

Table 4. Comparison of SIFID indicators

	DEFF-GAN			Our method			
	4000	5000	6000	3000	4000	5000	6000
Male1	0.197	0.177	0.173	0.191	0.165	0.151	0.143
Male2	0.175	0.199	0.174	0.181	0.173	0.160	0.147
Female1	0.148	0.148	0.156	0.151	0.147	0.152	0.148
Female2	0.227	0.199	0.205	0.203	0.189	0.197	0.196

Comparison and Performance Evaluation of Attention Modules

To further validate the impact of attention structures on generative performance, this paper replaces the core feature modeling module with Self-Attention (SA), Vision Transformer (ViT), and Hybrid attention mechanisms within a unified network architecture. All other training configurations remain identical, including dataset size, iteration count, loss function, multi-stage generator structure, and discriminator settings. Tables 5 and 6 summarize the LPIPS diversity and consistency performance of the three attention mechanisms in the few-shot image generation task, respectively; Table 7 reports their FID values across different tasks, measuring global distribution consistency; Table 8 illustrates SIFID metric differences among SA, ViT, and Hybrid in local structural fidelity, reflecting each attention mechanism's capability in detail recovery and structural restoration.

Table 5 demonstrates the distinct differences in diversity among the three attention architectures. While SA achieves an LPIPS-1 score of 0.64, its overall performance fluctuates significantly (0.58–0.64), reflecting training instability due to its reliance on local attention. ViT yields the lowest results (0.51–0.53), indicating restricted variability stemming from its strong global constraints. Hybrid, however, exhibits stable distribution within the 0.60–0.63 range, maintaining high diversity with minimal fluctuation. This stems from the Hybrid approach combining SA's detail sensitivity with ViT's global consistency, enabling both natural and stable generative variation.

Table 5. Comparison of LPIPS-1 indicators

		Flower	Male	Female
SA	3000	0.58	0.31	0.30
	4000	0.60	0.30	0.27
	5000	0.61	0.31	0.28
	6000	0.64	0.32	0.29
ViT	3000	0.58	0.23	0.26
	4000	0.53	0.21	0.23
	5000	0.52	0.21	0.22
	6000	0.51	0.22	0.23
Our method	3000	0.60	0.25	0.26
	4000	0.61	0.29	0.27
	5000	0.63	0.28	0.28
	6000	0.62	0.29	0.27

Table 6 shows the consistency comparison results for the three attention structures. SA exhibits significant LPIPS-2 fluctuations (e.g., flower1 0.60 $\rightarrow$ 0.55), indicating unstable consistency. While ViT is stable, its overall scores are excessively high (mostly 0.56–0.59), suggesting overly smooth generation results. Hybrid exhibits the lowest consistency and most pronounced convergence (e.g., man1 0.21 $\rightarrow$ 0.20), delivering optimal performance. This arises because SA lacks global constraints while ViT excessively suppresses variation. In contrast, Hybrid preserves both local and global features, ensuring more reliable style and semantic retention.

Table 6. Comparison of LPIPS-2 indicators

		Flower1	Flower2	Flower3	Flower4	Flower5
SA	3000	0.60	0.63	0.62	0.63	0.61
	4000	0.59	0.62	0.60	0.61	0.59
	5000	0.57	0.58	0.57	0.59	0.59
	6000	0.55	0.58	0.56	0.58	0.57
ViT	3000	0.59	0.60	0.60	0.59	0.60
	4000	0.58	0.57	0.59	0.59	0.58
	5000	0.57	0.58	0.57	0.56	0.58
	6000	0.56	0.57	0.58	0.57	0.58
Our method	3000	0.60	0.59	0.59	0.58	0.61
	4000	0.59	0.58	0.58	0.57	0.60
	5000	0.58	0.60	0.57	0.57	0.61
	6000	0.59	0.58	0.57	0.58	0.60

		Male1	Male2	Female1	Female2
SA	3000	0.27	0.26	0.27	0.28
	4000	0.25	0.27	0.29	0.28
	5000	0.26	0.25	0.27	0.26
	6000	0.24	0.25	0.28	0.27
ViT	3000	0.25	0.23	0.26	0.29
	4000	0.23	0.24	0.27	0.26
	5000	0.21	0.22	0.25	0.25
	6000	0.22	0.22	0.24	0.24
Our method	3000	0.21	0.22	0.25	0.27
	4000	0.22	0.21	0.23	0.25
	5000	0.21	0.20	0.21	0.24
	6000	0.20	0.21	0.22	0.22

As shown in Table 7, SA exhibits significant FID fluctuations (e.g., male 187 → 157), indicating unstable global modeling. While ViT demonstrates stability, its overall values remain elevated (e.g., female 120 → 99), resulting in limited generation quality. In contrast, Hybrid achieves the lowest FID across all tasks (e.g., male 137 → 83) and demonstrates the highest global naturalness. This stems from our method's ability to simultaneously capture structural contours and semantic relationships, resulting in a generation distribution closer to real samples.

Table 7. Comparison of FID indicators

		Flower	Male	Female
SA	3000	225 ± 9.37	187 ± 12.98	137 ± 10.36
	4000	203 ± 9.78	173 ± 13.25	126 ± 10.47
	5000	197 ± 8.56	163 ± 12.48	175 ± 9.33
	6000	189 ± 9.13	157 ± 12.21	109 ± 10.85
VIT	3000	211 ± 7.71	157 ± 6.20	120 ± 7.18
	4000	190 ± 6.65	143 ± 8.21	117 ± 5.96
	5000	188 ± 5.53	136 ± 6.7	103 ± 6.57
	6000	184 ± 6.67	133 ± 7.31	99 ± 6.35
Our method	3000	198 ± 3.39	137 ± 6.87	98 ± 2.18
	4000	197 ± 3.42	117 ± 7.71	97 ± 2.24
	5000	183 ± 3.60	103 ± 5.54	95 ± 2.98
	6000	175 ± 3.54	83 ± 5.05	87 ± 2.43

Table 8 shows that SA generally exhibits higher SIFID values (e.g., male2 0.212 → 0.167), indicating unstable local structure reconstruction. Although ViT shows a decreasing trend (e.g., female1 0.163 → 0.157), it remains weaker than the proposed method. Our method achieves the lowest SIFID (e.g., man1 0.191 → 0.143), demonstrating superior restoration of local structures like edges and textures. This stems from its local-global joint modeling strategy, which prevents critical details from being either over-sharpened or smoothed out.

Table 8. Comparison of SIFID indicators

		Male1	Male2	Female1	Female2
SA	3000	0.201	0.212	0.153	0.201
	4000	0.192	0.177	0.162	0.198
	5000	0.181	0.168	0.159	0.201
	6000	0.179	0.167	0.159	0.201
ViT	3000	0.183	0.188	0.163	0.188
	4000	0.173	0.182	0.161	0.182
	5000	0.170	0.177	0.158	0.173
	6000	0.169	0.171	0.157	0.165
Our method	3000	0.191	0.181	0.151	0.203
	4000	0.165	0.173	0.147	0.189
	5000	0.151	0.160	0.152	0.197
	6000	0.143	0.147	0.148	0.196

4.4 Analysis of Results

To validate the effectiveness of the hybrid attention mechanism in few-shot generation, we compared MixDL, DEff-GAN, and three attention modules across four dimensions: diversity, style consistency, global distribution quality, and local structural fidelity. Results demonstrate that the hybrid attention approach achieves superior performance across all key metrics, establishing a balanced advantage in generative quality.

In terms of diversity, our method maintains stable LPIPS-1 scores (0.25–0.29) on the face task and achieves 0.60–0.63 on the flower task, significantly outperforming MixDL (0.39–0.54) and DEff-GAN (0.58–0.59). SA achieves the highest diversity (0.64) but exhibits significant fluctuations; ViT performs weakly (0.51–0.53) with excessive smoothing; hybrid attention methods maintain stability above 0.60, effectively balancing local texture and global semantics while avoiding both "noise-type" and "structure-constrained" diversity.

In terms of stylistic consistency, our method achieves LPIPS-2 scores of 0.21–0.24 on the face task, significantly outperforming MixDL (0.41–0.47) and DEff-GAN (0.28–0.31). On the flower task, it maintains consistently low values of 0.57–0.60. SA exhibits significant consistency fluctuations driven by local attention; ViT remains stable but overemphasizes global modeling while producing overly smooth styles; the hybrid attention approach achieves the lowest and smoothest LPIPS-2 scores across the entire range, simultaneously preserving semantic style consistency and detailed structural coherence.

Regarding global distribution quality, our method achieves FID values of 83–87 on the face task, significantly outperforming MixDL (206–238) and DEff-GAN (202–217); the flower task yields the lowest FID of 175. SA exhibits greater variability, reflecting its unstable global modeling capability. ViT remains stable but at higher values (99–120), demonstrating an "under-variation" characteristic. The hybrid attention approach achieves the best FID across all tasks (e.g., 137–83 for the male task), more effectively capturing cross-regional semantic structures and generating distributions closer to real samples.

In terms of local structural fidelity, our method reduces SIFID to 0.143 on the face task, significantly outperforming DEff-GAN (0.173–0.199). Ablation experiments further reveal that SA exhibits instability in local structure recovery (e.g., man2: 0.212 $\rightarrow$ 0.167), while ViT shows a declining trend yet remains less accurate than hybrid attention methods. The hybrid attention method achieves the lowest local structural difference values across all tasks, indicating its strongest recovery capabilities for edges, textures, and key semantic locations.

Overall, SA and ViT exhibit distinct biases at local and global scales: SA offers high diversity but lacks stability, while ViT maintains consistency yet falls short in detail. By integrating local and global features, hybrid attention achieves more balanced performance across four core metrics, more stable training, and more natural generation results. Although local blurring and structural overlap persist in complex textures and cross-category tasks, its overall performance significantly outperforms baseline methods, validating its effectiveness and potential in few-shot generation.

5 Conclusion and Outlook

This paper proposes a few-shot image generation model integrating a hybrid attention mechanism. By incorporating self-attention modules and Vision Transformers (ViT) within a multi-stage pyramid generator, it achieves collaborative modeling of local textures and global semantics. Simultaneously, a multi-task collaborative optimization mechanism combining adversarial loss, reconstruction loss, and semantic classification loss is introduced. This is supplemented by an auxiliary classification branch for discriminative guidance and semantic reinforcement, effectively enhancing the fidelity, diversity, and semantic consistency of generated images. Experiments on face, flower, and cross-category few-shot datasets demonstrate that this method outperforms baseline models across multiple mainstream metrics, validating the effectiveness and generalizability of the hybrid attention architecture and multi-task optimization strategy. Despite achieving promising results, the method still has certain limitations. On one hand, the ViT module faces high computational costs during feature modeling at high resolutions, which may impact training efficiency and structural scalability in complex scenarios. On the other hand, in cross-domain generation tasks, achieving style transfer while maintaining semantic consistency remains challenging due to significant feature space differences.

Future research may further focus on the following directions: Introducing lightweight attention mechanisms to reduce computational and memory overhead; designing learnable attention fusion weights to adaptively regulate local and global modeling capabilities; integrating multimodal guidance information to enhance controllability in semantic generation; and extending to video few-shot generation tasks to construct temporally coherent continuous dynamic content. These approaches will comprehensively enhance the model's scalability, semantic control capabilities, and cross-domain generalization abilities.

Acknowledgments. This work was supported by the Joint Foundation Key Project of the Natural Science Foundation of the Heilongjiang Province, grant ZL2025F004.

References

1. Zhou, Y., Ye, Y., Zhang, P., et al.: Exact fusion via feature distribution matching for few-shot image generation. In: Proceedings of CVPR, pp. 8383–8392 (2024)
2. Hong, Y., Niu, L., Zhang, J., et al.: F2GAN: fusing-and-filling GAN for few-shot image generation. In: Proceedings of the ACM International Conference on Multimedia (MM), pp. 2535–2543 (2020)
3. Li, W., Xu, W., Wu, X., et al.: AMMGAN: adaptive multi-scale modulation generative adversarial network for few-shot image generation. Appl. Intell. **53**(18), 20979–20997 (2023)
4. Aldhubri, A., Lu, J., Fu, G.: SAGAN: skip attention generative adversarial networks for few-shot image generation. Digit. Signal Process. **149**, 104466 (2024)
5. Kumar, R., Sivakumar, G.: DEff-GAN: diverse attribute transfer for few-shot image synthesis. arXiv preprint arXiv:2302.14533 (2023)
6. Zhang, Z., Liu, Y., Han, C., et al.: PetsGAN: Rethinking priors for single image generation. In: Proceedings of the AAAI Conference on Artificial Intelligence, vol. 36, no. 3, pp. 3408–3416 (2022)

7. Jiang, Y., Chang, S., Wang, Z.: TransGAN: two pure transformers can make one strong GAN, and that can scale up. Adv. Neural. Inf. Process. Syst. **34**, 14745–14758 (2021)
8. Park, J., Kim, Y.: Styleformer: transformer-based generative adversarial networks with style vector. In: Proceedings of CVPR, pp. 8983–8992 (2022)
9. Yin, F., Zhang, Y., Cun, X., et al.: StyleHEAT: one-shot high-resolution editable talking face generation via pre-trained StyleGAN. In: ECCV, pp. 85–101. Springer, Cham (2022)
10. Yan, L., Yan, R., Chai, B., et al.: DM-GAN: CNN hybrid ViTs for training GANs under limited data. Pattern Recognit. **156**, 110810 (2024)
11. Yang, M., Wang, Z., Feng, W., et al.: Improving few-shot image generation by structural discrimination and textural modulation. In: Proceedings of the ACM International Conference on Multimedia (MM), pp. 7837–7848 (2023)
12. Sugahara, R., Du, W.: NNST-based image outpainting via SinGAN. In: Proceedings of the 10th International Conference on Computing and Artificial Intelligence, pp. 26–31 (2024)
13. Hinz, T., Fisher, M., Wang, O., et al.: Improved techniques for training single-image GANs. In: Proceedings of the IEEE/CVF Winter Conference on Applications of Computer Vision (WACV), pp. 1300–1309 (2021)
14. Sun, Y., Wang, Y., Hu, L., et al.: Attribute-guided generative adversarial network with improved episode training strategy for few-shot SAR image generation. IEEE J. Sel. Top. Appl. Earth Obs. Remote Sens. **16**, 1785–1801 (2023)
15. Jin, Y., Peng, J., He, Q., et al.: Dual-interrelated diffusion model for few-shot anomaly image generation. In: Proceedings of CVPR, pp. 30420–30429 (2025)
16. Kong, C., Kim, J., Han, D., et al.: Few-shot image generation with mixup-based distance learning. In: ECCV, pp. 563–580. Springer, Cham (2022)
17. Rombach, R., Blattmann, A., Lorenz, D., et al.: High-resolution image synthesis with latent diffusion models. In: Proceedings of the IEEE/CVF Conference on Computer Vision and Pattern Recognition, pp. 10684–10695 (2022)
18. Ho, J., Jain, A., Abbeel, P.: Denoising diffusion probabilistic models. Adv. Neural. Inf. Process. Syst. **33**, 6840–6851 (2020)

PR-Net: Precision Recovery Network with Dual-Branch Detail Attention and Local-Global Feature Enhancement

Rui Peng[1,2,3], Xuan Zhou[4], Longfeng Shen[1,2,3](✉), Yaxi Duan[1,2,3], and Fang Zhang[5]

[1] Anhui Engineering Research Center for Intelligent Computing and Application On Cognitive Behavior (ICACB), College of Artificial Intelligence, Huaibei Normal University, Huaibei, Anhui, China
longfengshen521@126.com

[2] Institute of Artificial Intelligence, Hefei Comprehensive National Science Center, Hefei, Anhui, China

[3] Key Laboratory of Intelligent Computing and Signal Processing, Ministry of Education, Anhui University, Hefei, Anhui, China

[4] College of Computer Science and Technology, Huaibei Normal University, Huaibei, Anhui, China

[5] Department of Medical Imaging, Huaibei People's Hospital, Huaibei, Anhui, China

Abstract. In recent years, Transformer-based models have achieved notable advances in medical image segmentation by effectively capturing global context. Nevertheless, their ability to jointly represent global and local features remains a challenge. To overcome this limitation, we propose a multi-scale attention mechanism that efficiently fuses and enhances multi-granularity information during feature extraction through feature pyramids and cross-scale interaction design. To further enhance detail retention, we introduce a multi-scale edge enhancement module. This module employs a hierarchical receptive field expansion mechanism to progressively construct a multi-scale feature space, simultaneously capturing global semantic features and local structural details. It establishes cross-scale feature interaction pathways, significantly boosting the model's perception of critical regions such as anatomical boundaries and minute lesions. The proposed method was validated on the ISIC2018 and Kvasir-SEG datasets. On the ISIC2018 dataset for skin lesion segmentation, the proposed method attained a Dice similarity coefficient of 91.14% and an Intersection over Union (IoU) of 84.78%. Its performance was further enhanced on the Kvasir-SEG polyp segmentation task, achieving scores of 94.13% in Dice and 89.82% in IoU. The experimental results verify the strong generalization and robustness of the proposed method, offering substantial support for application in clinical diagnosis and treatment.

Keywords: Multi-Scale Attention · Medical Image Segmentation · Global

L. Huang and D. Greenhalgh (Eds.): ICMLC 2026, LNNS 2061, pp. 86–96, 2027.
https://doi.org/10.1007/978-3-032-31236-5_6

1 Introduction

Recent advances in Convolutional Neural Networks (CNN) and Transformers have fundamentally transformed the landscape of medical image segmentation. The CNN-based UNet [1] model, with its simple U-shaped structure and strong scalability, has become a representative architecture, with numerous subsequent improvements built upon this framework. However, this model has efficiency limitations in modeling global contextual information. To overcome this issue, researchers introduced the Transformer architecture from natural language processing, proposing TransUnet [2]: CNN extract low-level visual features, while Transformer capture long-range dependencies. Subsequently, to synergistically leverage the strengths of both CNNs and Transformers, Transfuse [3] adopted a dual-branch parallel architecture to replace the original serial design. This approach enhances model efficiency while effectively mitigating gradient vanishing and feature degradation problems. Recently, Vmamba [4] achieved significant breakthroughs in image classification tasks, with its derivative model VM-Unet [5] pioneering the introduction of state space models (SSM) into medical image segmentation. However, research indicates that this purely SSM-based architecture suffers from an imbalance between global and local feature extraction, limiting further improvements in segmentation performance. Building upon the Transfuse framework, in this paper, we make the following key contributions:

(1) We introduce PR-Net, a model specifically designed to resolve the imbalance between global and local feature representation.
(2) We present the Multi-scale Edge Enhancement Module (MEEM), which employs a hierarchical receptive field expansion mechanism to progressively construct a multi-scale feature space, simultaneously capturing global semantic features and local structural details. The Multi-scale Attention Module (MSA) achieves efficient fusion and enhancement of multi-granularity information during feature extraction through feature pyramids and cross-scale interaction design.
(3) We conducted three distinct experiments on the ISIC2018 and Kvasir-SEG datasets, performing qualitative comparisons with traditional segmentation methods. Results demonstrate that our model exhibits outstanding performance across these tasks.

2 Related Works

Our discussion focuses on summarizing the prevailing approaches currently employed in medical image segmentation. Existing methodologies primarily follow two distinct paradigms: hybrid architectures that integrate CNNs with Transformers, and novel frameworks built upon the Mamba architecture.

2.1 CNN and Transformer Models

Traditional approaches to medical image segmentation predominantly include three architectural types, CNN-based models, Transformer-based models, and hybrid architectures integrating both. Among these, CNN-based methods are the most widely applied—the UNet [1] architecture proposed by Ronneberger et al. effectively captures local detail features through its encoder-decoder structure and skip connections, achieving remarkable results in medical image segmentation. Their structural simplicity and strong scalability have spurred a series of improved models: UNet++ [6] and UNet3 + [7] enhance feature fusion through dense skip connections, mitigating encoder feature loss issues.

2.2 Based Transformer Models

The breakthrough progress of Transformer models in computer vision has fueled a growing exploration of their application in medical image segmentation. TransUNet [8], as pioneering work, first introduced Vision Transformer [9] (ViT) into this domain. It features a hybrid encoder that integrates a CNN for high-resolution local feature extraction with a Vision Transformer (ViT) module for low-resolution global context capture, thus unifying the detail sensitivity of CNNs and the long-range dependency modeling of Transformers within a unified framework. Building upon this foundation, research has continued to deepen along multiple architectural paradigms. TransFuse [10] proposed a parallel encoding architecture where CNN and ViT branches independently extract features, which are then integrated through a multi-level fusion module. This approach improves the efficiency of global context modeling without compromising its ability to capture local details. MISSFormer [11] employs ViT for both encoding and decoding, and incorporates a ReMixed Transformer module to effectively bridge the multi-scale global and local feature correlations. Similarly, Swin-UNet [12], also based on pure Transformers, incorporates the Swin Transformer [13] with a sliding window mechanism. This approach significantly improves computational efficiency while preserving global modeling capabilities and enhancing sensitivity to local structures. For 3D medical imaging, TransBTS [14] combines 3D CNNs [15] with Transformer modules, leveraging 3D convolutions to capture spatial local information while integrating global context within volumetric data via self-attention mechanisms, demonstrating outstanding performance in tasks like brain tumor segmentation. DA-TransUNet [16] integrates dual attention mechanisms within a hybrid architecture, enabling the model to precisely focus on both spatially critical regions and contextually important features, further enhancing segmentation accuracy. Currently, Transformer-based medical image segmentation models primarily evolve around three paradigms: hybrid CNN-Transformer, parallel CNN-ViT, and pure ViT. Recent studies have also introduced general architectures focused on 3D segmentation, such as nnFormer [17] and UNETR [18], alongside lightweight designs aimed at reducing computational complexity. Despite their notable performance, these methods generally face challenges including high computational resource demands, strong reliance on labeled data, and the need to validate model generalization capabilities. Future research is expected to emphasize architectural lightweighting, efficient training strategies, and enhancing robustness and interpretability for clinical scenarios, driving the technology toward practical implementation.

2.3 Based on Mamba Models

State Space Models (SSMs) [19] have emerged as a notable approach in recent years, primarily due to their efficacy in modeling long-range dependencies without incurring more than linear computational complexity. As a representative advancement in SSMs, the Mamba model demonstrates potential to replace CNN and Trans-former architectures. While U-Mamba [20] pioneered a hybrid architecture integrating SSM with CNN for medical image segmentation, SegMamba successfully validated a pure SSM-based approach. VM-UNet [5] introduced the Visual State Space (VSS) module as its fundamental building block, efficiently capturing multi-scale contextual information through an asymmetric encoding-decoding structure.

3 Method

This section first outlines the PR-Net architecture and then details its two core components: the Multi-scale Attention Module (MSA) and the Multi-scale Edge Enhancement Module (MEEM).

3.1 Overall Architecture

The proposed PR-Net framework is shown in Fig. 1. Specifically, PR-Net consists of three components: Stage1, Stage2, and the fusion module BiFusion.

First, Input images enter Stage 1 and Stage 2 respectively. Stage 1 centers on a Transformer architecture, first converting images into sequence features through stacked Transformer encoders to model long-range semantic dependencies. The output sequence features are reshaped back into 2D spatial feature maps. Subsequently, during the feature extraction phase, MSA efficiently fuses and enhances multi-granularity information, while MEEM employs a hierarchical receptive field expansion mechanism to progressively construct a multi-scale feature space. The subsequent two layers progressively restore spatial resolution through upsampling, ultimately outputting feature maps that fuse long-range semantic information with local details. Stage 2 employs a CNN architecture: inputs undergo downsampling via Conv and MaxPool layers to enhance feature expression, followed by three convolutional layers extracting multi-scale local spatial features (e.g., lesion edges, shapes). The BiFusion module, serving as the core fusion unit of the architecture, receives long-range semantic features from Stage 1 and local spatial features from Stage 2. It calculates feature correlations between the two through attention gates, generates a dynamic weight matrix to perform weighted fusion of dual-path features, and ultimately outputs a fused feature map containing both global semantics and local details for subsequent segmentation mask generation.

This architecture achieves complementary integration of long-range and local features through a dual-path design. Combined with BiFusion's feature fusion mechanism, it effectively addresses the feature imbalance issue inherent in single Transformer or CNN architectures for medical image segmentation.

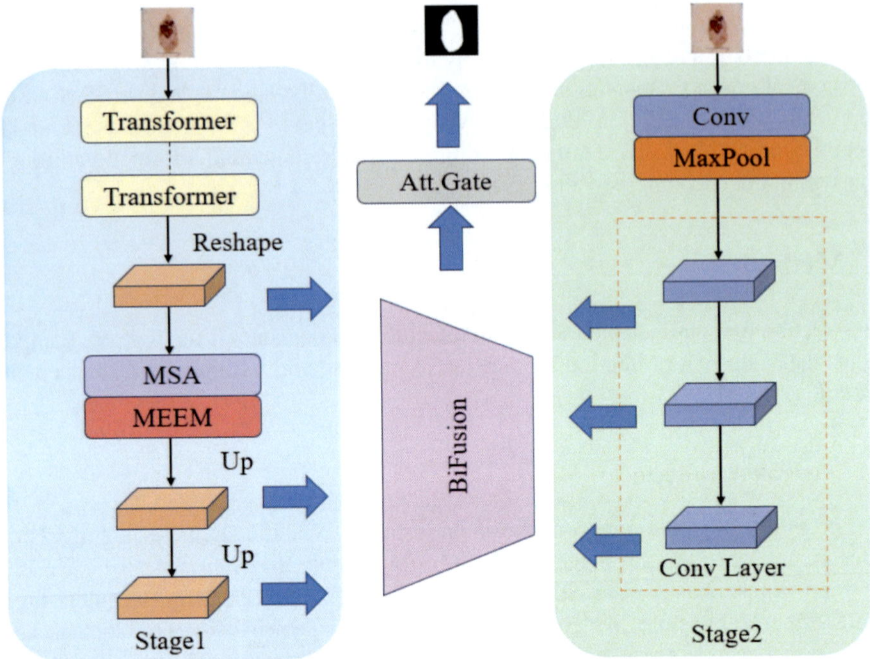

Fig. 1. PR-Net Overall Architecture.

3.2 Multi-Scale Attention Module

To address the challenge of globally-local feature core presentation, we propose the MSA (see Fig. 2) designed to enhance the efficient fusion and amplification of multi-granularity information.

The module first performs edge-sensing preprocessing on input features based on Sobel convolution, combined with residual connections to enhance local details. Subsequently, features acquire spatial global context through parallel axial pooling operations before entering a multi-scale depth convolution (DWConv) branch. This branch employs four distinct convolution kernel sizes [3, 5, 7, 9] to extract multi-granularity features in parallel, hierarchically expanding the receptive field. These multi-scale feature blocks undergo paired concatenation and normalisation to mitigate semantic interference between features. Ultimately, a spatial attention map is generated via a Sigmoid function to emphasise salient spatial regions.

To establish global dependencies between channels, features undergo further pooling compression and processing through a progressive compression path comprising three consecutive deep convolutional layers. This calculates a multi-attention head mechanism to generate channel attention. Finally, spatial and channel attentions are effectively fused and aggregated with the original multi-scale features. After adjusting the channel dimension via a 1×1 convolution, the output is produced. This design significantly enhances segmentation accuracy while controlling computational complexity through lightweight techniques like deep convolutions, achieving an effective balance between global semantic understanding and local detail preservation.

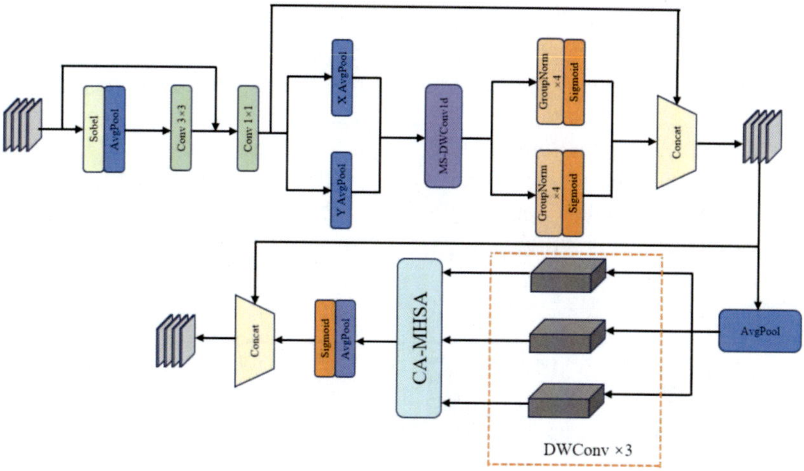

Fig. 2. Details of the MSA.

3.3 Multi-Scale Edge Enhancement Module

The figure illustrates the structural design of the MEEM [21](see Fig. 3). Its core progressively constructs a multi-scale feature space through a hierarchical receptive field expansion mechanism, leveraging cross-scale feature interaction [22] pathways to achieve collaborative capture of global semantics and local details. The module consists of two main components: a primary feature extraction path and an Edge Enhancer submodule. In the primary path, input features are first processed by a convolutional layer, thereafter passing through an average pooling operation.

First, the input feature $C \times H \times W$ enters the primary feature extraction pathway. Through hierarchical receptive field expansion, multi-scale semantic features are generated. The input first undergoes a 3×3 standard convolution to extract foundational semantics. Subsequently, three stacked average pooling operations progressively enlarge the receptive field. Each pooling operation extends the receptive field range from the current scale outward in all directions, ultimately expanding from 3×3 to 5×5, 7×7, and finally 9×9. This process yields three intermediate features at distinct scales: the small receptive field (3×3) preserves fine-grained local details (e.g., marginal contours of minute lesions), the medium receptive field (7×7) balances local and global information, while the large receptive field (9×9) captures extensive global semantics.

Subsequently, intermediate features from each scale are concurrently fed into the Edge Enhancer submodule. This submodule enhances local edges across different receptive fields through a processing sequence of 'local average extraction – edge information capture – feature adjustment – residual fusion'. First, 1×1 average pooling extracts the local average intensity of input features. Residual subtraction then accentuates intensity differences within local regions. Subsequently, 1×1 convolutions adjust the channel distribution of edge features. Finally, residual connections amplify local edge details while preserving original semantic information.

Ultimately, cross-scale fusion is achieved through channel concatenation of Edge Enhancer outputs across scales: features enhanced for edges at all three scales are concatenated along the channel dimension to form a comprehensive representation that integrates multi-scale semantics with detailed edge information. This fusion approach synergistically integrates fine-grained local details from small receptive fields with global semantic information from large receptive fields. It thereby avoids both the loss of detail caused by traditional models' tendency to prioritise global over local information, and the semantic incoherence resulting from an excessive focus on local details at the expense of the global context.

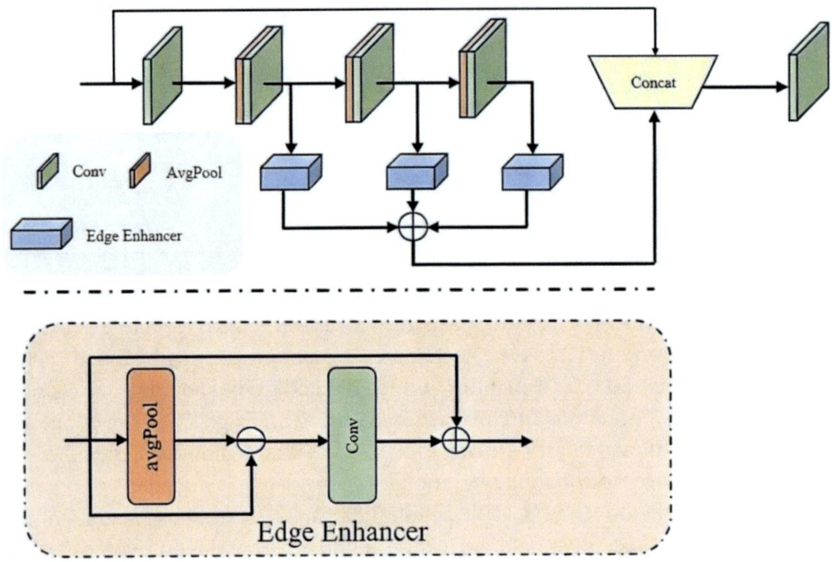

Fig. 3. Multi-scale Edge Enhancement Module Overall Structure.

4 Experimentation

The experimental validation in this section covers skin lesion and polyp segmentation. The generalization performance of PR-Net is evaluated on the following two datasets, with all images preprocessed to a size of 192 × 256 pixels, (1) The ISIC2018 dataset is a public benchmark for skin lesion analysis, comprising 2,594 training and 1,000 test images. (2) The Kvasir-SEG dataset, used for polyp segmentation, consists of 1,000 images, split into 880 for training, 120 for validation, and 100 for testing.

4.1 Realization Details

All experiments were implemented in PyTorch. The PR-Net model was trained for 30 epochs with a batch size of 16 on both the ISIC2018 and Kvasir-SEG datasets, using the Adam optimizer at a fixed learning rate of 7×10^{-5}. Experimental consistency was ensured by keeping all parameters fixed across trials, with training executed on an

NVIDIA A6000 GPU. The primary evaluation metrics were the Dice Similarity Coefficient (Dice) and Intersection over Union (IoU). The Dice coefficient measures the spatial overlap between prediction and ground truth, with higher values indicating greater segmentation accuracy. Meanwhile, the IoU score quantifies localization precision, where an increase corresponds to better target region alignment.

4.2 Assessment of Results

An evaluation of PR-Net was conducted by benchmarking it against recent SOTA methods on the ISIC2018 and Kvasir-SEG datasets. As shown in Table 1, PR-Net achieves the best Dice and IoU scores on both datasets, demonstrating superior generalization. For instance, it outperforms VM-UNet on ISIC2018 by 1.33% in Dice and 3.43% in IoU. These results confirm that our model not only sets a new SOTA but also maintains exceptional training stability.

Table 1. The generalization ability of the model was verified through a quantitative comparison on the ISIC-2018 and Kvasir-SEG datasets. Results are presented in the table (↑ denotes higher is better; bold indicates the best result).

Model	Parameters(M)	Gflops	ISIC2018		Kvasir-SEG	
			Dice↑	IoU↑	Dice↑	IoU↑
TransFuse [3]	41.34	8.87	89.27	80.63	91.80	86.80
TransUNet [2]	105.28	24.64	89.40	82.20	91.30	85.70
MSRF-Net [23]	18.83	-	88.13	83.73	85.08	-
PraNet [24]	11.70	1.82	88.00	80.00	89.90	82.15
VM-UNet [5]	34.62	7.56	89.71	81.35	89.09	80.32
VM-Unet-V2 [25]	17.91	4.4	89.73	81.47	87.00	76.99
Ours	32.07	10.04	**91.04**	**84.78**	**94.13**	**89.82**

Table 2 compares model performance under different attention mechanisms. The first row presents the baseline model (without attention) results, while rows 2 to 4 respectively show the experimental outcomes of adding a single attention mechanism. Comparing row 2 with row 1 reveals that our proposed MSA significantly enhances the model's feature extraction capabilities. The final row features a combined approach of both attention mechanisms, structured as PR-Net, which ultimately achieves the SOTA results.

Table 2. The effect of the proposed combination of attention and other attention on the accuracy of the model. Where ↑ indicates that the higher the better. (Bolded as the optimal result)

MSA	MEEM	Parameters(M)	Gflops	ISIC18	
				mIoU(%)↑	DSC(%) ↑
×	×	41.34	8.87	89.27	80.63
√	×	30.44	9.73	90.67	84.34
×	√	27.93	9.25	90.01	82.27
√	√	32.07	10.04	**91.04**	**84.78**

5 Conclusion

This paper presents PR-Net, a novel framework for medical image segmentation. Its key innovation is characterized by the progressive construction of a multi-scale feature space, enabled by a dedicated MEEM that hierarchically expands the receptive field. This design concurrently strengthens both global semantic comprehension and local detail representation. Furthermore, the framework incorporates a MSA module, designed with a feature pyramid and cross-scale interactions, to achieve efficient fusion of information at multiple granularities. Extensive evaluations on the ISIC2018 and Kvasir-SEG datasets demonstrate that PR-Net achieves leading performance. Qualitative analyses further validate its robustness and generalization capability in segmenting complex anatomical structures, demonstrating its potential as a reliable tool for clinical diagnostic support. The achieved performance gains—evident in significant improvements in segmentation accuracy and boundary delineation—justify the moderate increase in model complexity, indicating a favorable trade-off for clinical applications where precision is critical.

Acknowledgment. This work was supported in part by the University Synergy Innovation Program of Anhui Province, China, under Grant GXXT-2022–033; in part by the University Natural Science Research Project of Anhui Province under Grants 2023AH040056, 2023AH050333 and 2024AH051685; and in part by the Open Laboratory Project of the Key Laboratory of Intelligent Computing & Signal Processing, Ministry of Education, Anhui University, under Grant No.2023A003.

References

1. Ronneberger, O., Fischer, P., Brox, T.: U-net: convolutional networks for biomedical image segmentation. In: Medical Image Computing and Computer-Assisted Intervention–MICCAI 2015, Munich, Germany, vol. 9351, pp. 234–241. Springer International Publishing (2015)
2. Chen, J., Lu, Y., Yu, Q., et al.: TransUNet: rethinking the U-Net architecture design for medical image segmentation through the lens of transformers. Med. Image Anal. **97**, 103280 (2024)
3. Zhang, Y., Liu, H., Hu, Q.: TransFuse: fusing transformers and CNNs for medical image segmentation. In: Medical Image Computing and Computer Assisted Intervention–MICCAI 2021, Strasbourg, France, vol. 12901, pp. 14–24. Springer International Publishing (2021)

4. Liu, Y., Xiao, T., Gong, M., et al.: Vmamba: visual state space model. ArXiv preprint arXiv: 2401.10166, January 18 (2024)
5. Ruan, J., Xiang, S.: VM-UNet: vision mamba UNet for medical image segmentation. ArXiv preprint arXiv:2402.02491, February 5 (2024)
6. Zhou, Z., Rahman Siddiquee, M.M., Tajbakhsh, N., et al.: UNet++: A nested u-net architecture for medical image segmentation. In: Deep Learning in Medical Image Analysis and Multimodal Learning for Clinical Decision Support, Granada, Spain: Springer International Publishing, pp. 3–11 (2018)
7. Huang, H., Lin, L., Tong, R., Hu, H., Zhang, Q., Iwamoto, Y., Han, X., Chen, Y., Wu, J.: UNet 3+: a full-scale connected UNet for medical image segmentation. In: ICASSP 2020–2020 IEEE International Conference on Acoustics, pp. 1055–1059. Speech and Signal Processing (2020)
8. Chen, J., et al.: Transunet: transformers make strong encoders for medical image segmentation. arXiv preprint arXiv:2102.04306 (2021)
9. Dosovitskiy, A.: An image is worth 16x16 words: transformers for image recognition at scale. ArXiv preprint arXiv:2010.11929 (2020)
10. Zhang, Y., Liu, H., Hu, Q.: Transfuse: fusing transformers and cnns for medical image segmentation. In: International Conference on Medical Image Computing and Computer-Assisted Intervention, vol. 12901, pp. 14–24. Springer International Publishing, Cham (2021)
11. Huang, X., et al.: Missformer: an effective medical image segmentation transformer. ArXiv preprint arXiv:2109.07162 (2021)
12. Cao, H, et al.: Swin-unet: Unet-like pure transformer for medical image segmentation. In: European Conference on Computer Vision, vol. 13803, pp. 205–218. Springer Nature Switzerland, Cham (2022)
13. Liu, Z, et al.: Swin transformer: hierarchical vision transformer using shifted windows. In: Proceedings of the IEEE/CVF International Conference on Computer Vision, pp. 10012–10022 (2021)
14. Wang, W., et al.: Transbts: multimodal brain tumor segmentation using transformer. In: International Conference on Medical Image Computing and Computer-Assisted Intervention, vol. 12901, pp. 109–119. Springer International Publishing, Cham (2021)
15. Ji, S., et al.: 3D convolutional neural networks for human action recognition. In: IEEE Transactions on Pattern Analysis and Machine Intelligence vol. 35.1, pp. 221–231 (2012)
16. Sun, G., et al.: DA-TransUNet: integrating spatial and channel dual attention with transformer U-net for medical image segmentation. Front. Bioeng. Biotechnol. **12**, 1398237 (2024)
17. Zhou, H.-Y., et al.: nnformer: Interleaved transformer for volumetric segmentation. arXiv preprint arXiv:2109.03201 (2021)
18. Hatamizadeh, A, et al.: Unetr: transformers for 3d medical image segmentation. In: Proceedings of the IEEE/CVF Winter Conference on Applications of Computer Vision, pp. 574–584 (2022)
19. Gu, A., Goel, K., Ré, C.: Efficiently modeling long sequences with structured state spaces. arXiv preprint arXiv:2111.00396 (2021)
20. Ma, J., Li, F., Wang, B.: U-mamba: enhancing long-range dependency for biomedical image segmentation. arXiv preprint arXiv:2401.04722 (2024)
21. Gao, S., Zhang, P., Yan, T., Lu, H.: Multi-scale and detail-enhanced segment anything model for salient object detection. In: Proceedings of the 32nd ACM International Conference on Multimedia, pp. 9894–9903 (2024)
22. Huang, S., Wang, Y., Su, P.: A new synthetical method of feature enhancement and detection for SAR image targets. J. Image Graph. **4**(2), 73–77 (2016)
23. Srivastava, A., Jha, D., Chanda, S., Pal, U., Johansen, H.D., Johansen, D., et al.: MSRF-Net: a multi-scale residual fusion network for biomedical image segmentation. IEEE J. Biomed. Health Inform. **26**, 2252–2263 (2021)

24. Fan, D.P., Ji, G.P., Zhou, T., Chen, G., Fu, H., Shen, J., et al.: Pranet: parallel reverse attention network for polyp segmentation. In International Conference on Medical Image Computing and Computer-Assisted Intervention, pp. 263–273. Springer International Publishing, Cham (2020)
25. Zhang, M., et al.: VM-UNET-V2: rethinking vision mamba UNet for medical image segmentation. In: International Symposium on Bioinformatics Research and Applications, vol. 14954, pp. 335–346. Springer Nature Singapore, Singapore (2024)

Semantic-Aware Asymmetric Token Enhancement for Generalizable Person Re-Identification

Hongrui Yuan[1], Jingyu Li[2,3], Kun Zhang[4], and Lan Zhang[5](✉)

[1] Institute of Advanced Technology, University of Science and Technology of China, Hefei, China
yuanhr01@mail.ustc.edu.cn

[2] Institute of Artificial Intelligence, Hefei Comprehensive National Science Center, Hefei, China
jingyuli@iai.ustc.edu.cn

[3] State Key Laboratory for Novel Software Technology, Nanjing University, Nanjing, China

[4] Suzhou Institute for Advanced Research, University of Science and Technology of China, Suzhou, China
kkzhang@ustc.edu.cn

[5] School of Computer Science and Technology, University of Science and Technology of China, Hefei, China
zhanglan@ustc.edu.cn

Abstract. Person Re-Identification (ReID) fundamentally aims to match identities across disjoint camera networks. Despite recent advancements, the technology struggles with Domain Generalization (DG), where models fail to maintain performance on unseen target environments. This limitation largely stems from two factors: the inability to decouple identity-critical structures (e.g., contours) from background clutter, and the susceptibility of learned features to domain-specific shifts. We argue that pedestrian contours offer stable, domain-independent semantics essential for generalization. Based on this, we present a Transformer framework designed to prioritize structural integrity. Our method introduces two mechanisms: a Contour Token Selection (CTS) module that leverages multi-scale Fourier transforms to isolate high-frequency edge details, and an Asymmetric Token Enhancement (ATE) strategy. The latter constructs challenging positive pairs by applying differential perturbations, forcing the network to focus on robust semantic content rather than artifacts. Benchmarks across four major datasets confirm that our approach effectively mitigates domain discrepancies and achieves superior generalization.

Keywords: Domain generalization · Person re-identification · Asymmetric token enhancement · Vision transformer

J. Li—Supported by China Postdoctoral Science Foundation (2025M771515) and Anhui Postdoctoral Scientific Research Program Foundation (2025C1166).

L. Huang and D. Greenhalgh (Eds.): ICMLC 2026, LNNS 2061, pp. 97–114, 2027.
https://doi.org/10.1007/978-3-032-31236-5_7

1 Introduction

Person Re-Identification (ReID) is tasked with matching specific individuals traversing disjoint camera networks, acting as a foundational component in modern intelligent surveillance systems. Although Convolutional Neural Networks (CNNs) and Vision Transformers [1] have driven substantial gains, real-world applicability remains stifled by domain shifts. In these scenarios, model performance drops sharply on unseen targets caused by divergent lighting conditions, background clutter, and viewpoint changes [2]. The field has thus pivoted toward Domain Generalization (DG) ReID [3], which mandates that models establish generalizability relying exclusively on source domain training. This paradigm imposes stricter constraints than Unsupervised Domain Adaptation (UDA) [4] by forbidding any access to target data, while also diverging from standard DG by introducing an open-set scenario with non-overlapping identities.

Nevertheless, contemporary DG-ReID approaches are hamstrung by two primary deficiencies. First, reliance on global channel-wise statistical adjustments [5] ignores spatial nuance; such uniform processing fails to disentangle identity-critical structural features, like contours, from domain-specific interference. Second, prevailing augmentation techniques [6] tend to manipulate superficial appearance traits. Consequently, the resulting representations are fragile against domain shifts, overfitting to artifacts instead of encapsulating robust, high-level semantic structures.

In response, we introduce the Semantic-Aware Asymmetric Token Enhancement Framework, a novel architecture that fundamentally redirects the focus from global channel perturbation to precise token-level manipulation. Our methodology is built upon two core premises: (1) pedestrian contours act as stable, domain-agnostic anchors that remain consistent across diverse environments; and (2) the tokenization mechanism in Vision Transformers provides the necessary spatial granularity to selectively process these features.

To operationalize this, we first implement the Contour Token Selection (CTS) module. Diverging from CNN-based global pooling strategies [7], CTS employs multi-scale Fourier analysis to dynamically pinpoint tokens rich in high-frequency edge details. This mechanism explicitly filters out background noise while locking onto identity-discriminative contour structures. Furthermore, to preclude reliance on superficial patterns, we devise the Asymmetric Token Enhancement (ATE) strategy. Drawing inspiration from asymmetric contrastive learning, ATE generates "hard" positive pairs via two specialized branches: a Blending Branch that infuses style noise into contour tokens, and a Combining Branch that substitutes non-contour regions with auxiliary backgrounds. This designed asymmetry forces the model to disregard domain-specific artifacts in favor of robust structural semantics.

Ultimately, we present a framework that shifts DG-ReID towards fine-grained token manipulation, placing a premium on identity-relevant structures. By integrating the CTS module to extract domain-invariant features through Fourier analysis and the ATE strategy to drive robust representation learning, we achieve state-of-the-art performance across large-scale benchmarks, verifying the effectiveness of our approach.

2 Related Work

2.1 Domain Generalization for Person Re-Identification

Domain Generalization (DG) ReID aims to generalize to unseen targets using only source data, unlike adaptable supervised or unsupervised settings. Early works utilized meta-learning [8] or normalization mixing (e.g., IBN-Net [9]). However, the latter's reliance on batch statistics renders it ill-suited for Vision Transformers (ViTs) that typically use Layer Normalization. Recently, data augmentation via statistical perturbation [7] or Fourier synthesis has become dominant. Yet, these methods generally operate at the channel level, manipulating global statistics to diversify appearance. This paradigm, while effective for CNNs, overlooks the spatial granularity of ViT tokens. To bridge this gap, we introduce an asymmetric token-level enhancement framework, explicitly designed to exploit token dependencies for robust structural learning.

2.2 Asymmetric Designs in Contrastive Learning

Contrastive learning relies on asymmetric designs to learn invariant representations and prevent collapse, employing architectural discrepancies [10], momentum updates [11], or divergent augmentations [12]. However, these paradigms primarily utilize global, image-level variations, lacking the fine-grained control needed to suppress domain-specific artifacts in ReID. In contrast, we tailor this principle to DG-ReID via a token-level enhancement paradigm. By introducing spatial asymmetry—applying distinct perturbations to contour versus non-contour regions—we construct semantically consistent positive pairs, compelling the model to distill robust structural features beyond global appearance cues.

3 Method

3.1 Framework Overview

As illustrated in Fig. 1, we propose the Contour-Aware Asymmetric Enhancement Transformer to address domain generalization by leveraging stable structural cues. The framework comprises two synergistic modules: the Contour Token Selection (CTS) module, which extracts identity-relevant structural tokens, and the Asymmetric Token Enhancement (ATE) module, which generates robust positive pairs. Additionally, a Global-Local Alignment Attention (GL-AA) mechanism is introduced to refine fine-grained feature discrimination. **Contour Token Selection Module (CTS).** CTS is designed to decouple identity-relevant structure from domain-specific noise. By employing multi-scale Fourier transforms, CTS decomposes features into frequency components, identifying high-frequency signals associated with edges and contours. A dynamic percentile-based selection strategy generates a binary mask M^p, filtering out low-frequency noise while preserving discriminative contour tokens. This process provides a noise-robust foundation for the subsequent enhancement stage. Details are provided in Sect. 3.2.

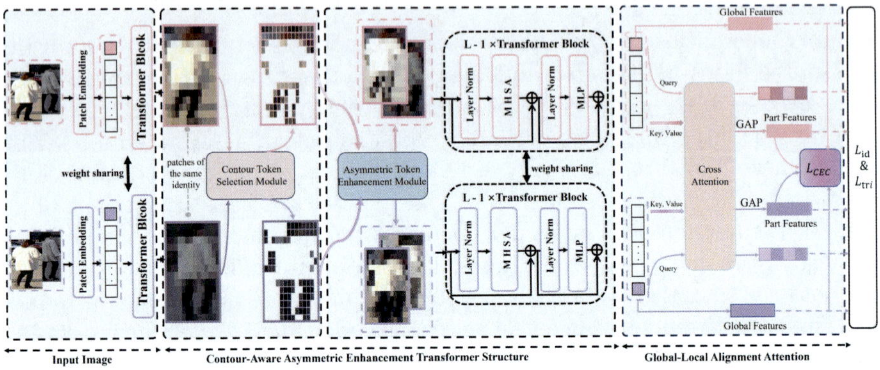

Fig. 1. Overview of the proposed architecture. Input features are processed by the Contour Token Selection (CTS) module to extract contour tokens. The Asymmetric Token Enhancement (ATE) module then constructs divergent positive pairs via blending and combining strategies. Finally, the Global-Local Alignment Attention (GL-AA) module refines feature discriminability for robust classification.

Asymmetric Token Enhancement Module (ATE). Leveraging the contour tokens identified by CTS, ATE constructs semantically consistent but visually diverse positive pairs to enforce structural learning. For a primary sample x_i, ATE utilizes an auxiliary sample x_j (same identity, different view) to drive two asymmetric branches: a *Blending branch* that linearly interpolates contour tokens with style noise, and a *Combining branch* that replaces non-contour regions with background tokens from x_j. This asymmetry forces the model to rely on abstract structural relationships rather than superficial appearance. Details are provided in Sect. 3.3.

Global-Local Alignment Attention (GL-AA). To capture subtle visual cues, GL-AA fuses global context F_g with local part features F_P (obtained via horizontal partitioning). By utilizing a cross-attention mechanism where F_g serves as the query and local parts F_p^n serve as keys/values, the module effectively aligns global representations with fine-grained local details. This alignment enhances the model's discriminative capability under domain shifts while maintaining computational efficiency through parameter sharing.

3.2 Contour Token Selection Module

Distinguishing identity-relevant contour cues from domain-specific noise remains a core challenge in domain-generalized person re-identification. Although prior studies [13] indicate that image contours encode both category-related global shapes and category-agnostic local textures, extracting semantically meaningful contour representations under severe domain shifts is still non-trivial. To address this issue, we introduce the Contour Token Selection (CTS) module, which integrates two complementary components: (i) Contour Representation Modeling, leveraging multi-scale Fourier analysis to emphasize high-frequency contour structures, and (ii) Dynamic Percentile Token Selection, which adaptively identifies semantically salient tokens during training.

Contour Representation Modeling Our contour representation modeling is motivated by the observation that high-frequency Fourier components primarily capture edges and contours, whereas low-frequency components are more sensitive to domain-dependent variations such as illumination and background textures. Given an input feature map $z_i \in \mathbb{R}^{N\times C}$, where N denotes the number of tokens and C the channel dimension, we reshape it into $z_i \in \mathbb{R}^{H\times W\times C}$ with $H = W = \lfloor\sqrt{N}\rfloor$.

To improve robustness to heterogeneous camera resolutions, we construct a three-scale feature pyramid via strided convolutions at resolutions (H, W), $(\frac{H}{2}, \frac{W}{2})$, and $(\frac{H}{4}, \frac{W}{4})$. As illustrated in Fig. 2(a), the Fourier transform at scale s is computed as:

$$\mathcal{F}^{(s)}(u, v, c) = \sum_{h=0}^{H_s-1}\sum_{w=0}^{W_s-1} z_i^{(s)}(h, w, c)e^{-j2\pi\left(\frac{hu}{H_s}+\frac{wv}{W_s}\right)}. \tag{1}$$

The resulting spectrum is decomposed into amplitude and phase components, where the amplitude reflects low-level statistical properties, while the phase encodes high-level semantic structures such as pedestrian contours. After centering low-frequency components, scale-specific binary masks r_s are applied to suppress low-frequency regions, followed by bandpass filtering with $M^{(s)}$ to remove redundant frequencies while preserving discriminative edge information.

The filtered high-frequency amplitude spectrum is recombined with the original phase spectrum and transformed back to the spatial domain via inverse Fourier transform:

$$z_{\text{feature}}^{(s)} = \mathcal{F}^{-1}\left(|\mathcal{F}^{(s)}| \odot M^{(s)} \cdot e^{j\angle\mathcal{F}^{(s)}}\right). \tag{2}$$

Finally, contour features from different scales are resized to the original resolution and adaptively fused using learnable weights α_s:

$$z_{\text{feature}} = \sum_{s=1}^{3} \text{Softmax}(\alpha_s) \cdot \text{Norm}\left(\text{Resize}(z_{\text{feature}}^{(s)}, H, W)\right). \tag{3}$$

The resulting multi-scale contour representation emphasizes phase-dominated global shape structures while suppressing low-frequency domain-specific variations, yielding discriminative channel-wise features for subsequent token selection.

Dynamic Percentile Token Selection Based on the multi-scale contour representation z_{feature}, we propose a dynamic percentile-based token selection strategy that adaptively identifies tokens carrying discriminative pedestrian structural information during training.

For each spatial region Ω_j (corresponding to a 16×16 patch), a salience score is computed by averaging the absolute activations across channels:

$$S_{\text{token}_j} = \frac{1}{C} \sum_{c=0}^{C-1} |z_{\text{feature}}(j, c)|, \tag{4}$$

where $|z_{\text{feature}}(j, c)|$ denotes the absolute activation at channel c of token j.

To gradually shift from comprehensive feature exploration to focused discrimination, we employ a linearly decaying percentile threshold:

$$\theta(\text{epoch}) = \theta_{\max} - \frac{(\theta_{\max} - \theta_{\min}) \times \text{epoch}}{E_{\max}}, \tag{5}$$

where $\theta_{\max}$ and $\theta_{\min}$ specify the upper and lower percentile bounds, respectively.

At each epoch, tokens are ranked by salience score, and a binary mask M^p is generated by selecting tokens within the top $\theta(\text{epoch})$ percentile:

$$M^p(j) = \begin{cases} 1 & \text{if } S_{\text{token}_j} \in \text{Top-}\theta(\text{epoch}) \\ 0 & \text{otherwise.} \end{cases} \tag{6}$$

This progressive selection scheme enables the model to first capture diverse semantic cues and then increasingly emphasize contour-dominant regions, enhancing feature discriminability and robustness to domain shift while maintaining computational efficiency.

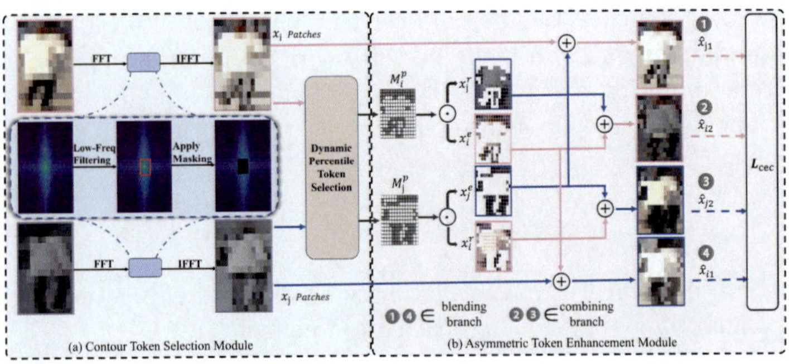

Fig. 2. Overview of the CTS and ATE modules. CTS extracts multi-scale contour features via frequency-domain filtering and selects tokens using a dynamic percentile strategy. ATE augments samples through two branches: Blending (mixing contour tokens) and Combining (filling non-contour regions), with both views processed by a shared extractor for contrastive learning.

3.3 Asymmetric Token Enhancement Module

Building on CTS tokens, the Asymmetric Token Enhancement (ATE) module fosters domain invariance. Inspired by asymmetric self-supervised learning, it constructs positive pairs with consistent semantics but divergent appearances, compelling the model to prioritize intrinsic structure over superficial style. The module comprises an asymmetric enhancement strategy for generating diverse pairs and a Contrastive Enhanced Contour Consistency Loss for optimization.

Asymmetric Token Enhancement Strategy The proposed enhancement strategy applies asymmetric perturbations to contour and non-contour regions, producing positive pairs that preserve identity semantics while exhibiting pronounced visual discrepancies. This design explicitly encourages the feature extractor to focus on structural relationships encoded in pedestrian contours rather than domain-specific appearance cues. Given an input batch $\{x_i\}_{i=1}^{B}$, we first obtain contour-enhanced features by applying the binary mask $M_i^p \in \{0,1\}^{N\times C}$:

$$x_i^e = M_i^p \odot x_i, \tag{7}$$

where x_i^e retains identity-relevant contour information while suppressing domain-dependent noise.

Each primary sample x_i is paired with an auxiliary sample x_j of the same identity captured under a different camera view, ensuring semantic consistency while introducing realistic domain variation. All augmented samples inherit the identity label of x_i, enforcing contour-dominated semantic alignment. As illustrated in Fig. 2(b), two asymmetric augmentation branches are constructed.

In the blending branch, inspired by Mixup [6], contour-enhanced features are linearly interpolated with auxiliary samples:

$$\tilde{x}_{i1} = \lambda \cdot x_i^e + (1-\lambda)\cdot x_j, \quad \lambda \sim \mathrm{Beta}(\alpha,\beta), \tag{8}$$

where $\alpha, \beta \in (0,\infty)$. This operation injects continuous style perturbations into contour-focused tokens while preserving spatial structure, encouraging the model to maintain identity consistency under appearance variations.

In the combining branch, contour tokens from the primary sample are preserved, while non-contour regions are replaced with background tokens from the auxiliary sample:

$$\tilde{x}_{i2} = x_i^e + (1-M_i^p)\odot x_j, \tag{9}$$

where $(1-M_i^p)\odot x_j$ is denoted as x_j^r in Fig. 2(b). This operation maintains discriminative contour structures while introducing contextual diversity, compelling the model to learn global dependencies between contour and background regions and reducing sensitivity to spurious contextual cues.

Together, the dual-branch design establishes a synergistic enhancement mechanism: the blending branch promotes contour robustness through style perturbation, while the combining branch reinforces structural reasoning via selective background replacement.

Contrastive Enhanced Contour Consistency Loss To optimize the asymmetrically enhanced representations, we introduce the Contrastive Enhanced Contour Consistency (CEC) loss, which enforces semantic consistency between augmented views while preserving discriminative separation across identities. For each training sample, two augmented views $\tilde{x}_{i1}$ and $\tilde{x}_{i2}$ generated by the blending and combining strategies depict the same pedestrian under different perturbations. Their corresponding contour-aware embeddings extracted by the Transformer backbone are denoted as $\tilde{z}_{i1}$ and $\tilde{z}_{i2}$.

The CEC loss adopts a symmetric contrastive formulation that pulls positive pairs together while pushing negatives apart:

$$L_{\text{CEC}} = -\sum_{i=1}^{B}\left[\log\frac{\exp(s(\tilde{z}_{i1},\tilde{z}_{i2})/\tau)}{\sum_{k\in\mathcal{N}_i}\exp(s(\tilde{z}_{i1},\tilde{z}_k)/\tau)} + \log\frac{\exp(s(\tilde{z}_{i2},\tilde{z}_{i1})/\tau)}{\sum_{k\in\mathcal{N}_i}\exp(s(\tilde{z}_{i2},\tilde{z}_k)/\tau)}\right], \tag{10}$$

where $s(\cdot,\cdot)$ denotes cosine similarity, τ is a temperature parameter, and $\mathcal{N}_i$ represents the set of negative samples.

By aligning asymmetrically augmented views of the same identity and separating different identities in the embedding space, the CEC loss encourages robust contour-centric representations that are less sensitive to viewpoint and occlusion variations.

3.4 Training and Inference

Our framework adopts a dual-phase paradigm that integrates contour-aware representation learning during training while retaining a streamlined single-backbone architecture for efficient inference.

Training Phase: Composite Learning Objective During training, the model is optimized using a composite objective that combines conventional identification supervision with the proposed contrastive contour consistency learning:

$$\mathcal{L}_{\text{total}} = \mathcal{L}_{\text{ID}} + \lambda\mathcal{L}_{\text{CEC}}, \tag{11}$$

where $\mathcal{L}_{\text{CEC}}$ denotes the Contrastive Enhanced Contour Consistency Loss introduced in Sect. 3.3, and λ balances the two objectives.

The identification loss $\mathcal{L}_{\text{ID}}$ is applied to multiple feature streams to encourage comprehensive representation learning:

$$\mathcal{L}_{\text{ID}} = \mathcal{L}_{g1} + \mathcal{L}_{p1} + \mathcal{L}_{g2} + \mathcal{L}_{p2} + \sum_{n=1}^{4}\mathcal{L}_{g1-p1}^{n} + \sum_{n=1}^{4}\mathcal{L}_{g2-p2}^{n}, \tag{12}$$

where subscripts 1 and 2 correspond to the blending and combining branches, respectively; g and p denote global and part-level features; and $\mathcal{L}_{g-p}^{n}$ represents cross-enhanced features produced by the GL-AA module over four horizontal regions.

Each identification loss term consists of a classification loss and a metric learning loss:

$$\mathcal{L}_{cls} = -\sum_{c=1}^{C} y_c \log(p_c), \tag{13}$$

$$\mathcal{L}_{tri} = \max\big(d(\mathbf{a},\mathbf{p}) - d(\mathbf{a},\mathbf{n}) + \alpha, 0\big), \tag{14}$$

where y_c and p_c denote the ground-truth and predicted class probabilities, $(\mathbf{a},\mathbf{p},\mathbf{n})$ represent anchor, positive, and negative embeddings, $d(\cdot)$ is the Euclidean distance, and α is the margin.

Inference Phase: Efficient Single-Backbone Architecture During inference, the model is deployed as a standard single-backbone network without the asymmetric enhancement modules, ensuring computational efficiency while benefiting from the robust representations learned during training. The final feature representation is obtained by concatenating multi-stream features:

$$\mathbf{f}_{\text{final}} = \text{Concat}(F_g, F_p, F_{g-p}), \tag{15}$$

where F_g, F_p, and F_{g-p} denote global features, part-level features, and cross-enhanced features from the GL-AA module, respectively. Person re-identification is performed by computing cosine similarity between query and gallery features, leveraging contour-aware and domain-invariant representations for robust cross-domain matching.

This decoupled training–inference design enables sophisticated contour-aware enhancement during learning while preserving efficient and practical deployment during inference.

4 Experiments

4.1 Datasets and Evaluation Metrics

We validate our approach across four major benchmarks: Market1501 [14], DukeMTMC-reID [2], MSMT17 [15], and CUHK03-NP [16]. For the latter, we strictly employ the more rigorous new-protocol detection subset (767/700 train/test identities). To gauge robustness, we adopt both single-source (one-to-others) and multi-source (leave-one-out) settings, reporting performance via mean Average Precision (mAP) and Rank-1 accuracy.

4.2 Implementation Details

We use an ImageNet-pretrained ViT-B/16 backbone with 224×224 inputs. Training utilizes SGD (batch size 128) with a warm-up schedule peaking at 10^{-3}. Augmentations include random flipping and local grayscale. Key hyper-parameters are set to $\lambda = 0.5$ (contrastive weight) and $\alpha = 4, \beta = 1$ (blending). TransReID-B/16 serves as the baseline.

4.3 Comparison with State-of-the-Art Methods

Single-source DG ReID. We evaluate our method on the single-source domain generalization ReID benchmark by training on Market1501 (Market-train) or DukeMTMC-reID (Duke-train) and testing on Market-test, Duke-test, MSMT17-test, and CUHK03-test. Quantitative comparisons are reported in Table 1.

Table 1. Quantitative comparison with state-of-the-art methods under the single-source DG setting across Market1501, DukeMTMC-reID, MSMT17, and CUHK03-NP. Best results are marked in bold, while the top performing baselines are underlined.

Method	Reference	Source	Duke		MSMT		CUHK-NP	
			R1	mAP	R1	mAP	R1	mAP
TransMatcher [17]	NeurIPS2021		-	-	47.3	18.4	22.2	21.4
MetaBIN [18]	CVPR2021		55.2	33.1	-	-	-	-
QAConv-GS [19]	CVPR2022		-	-	45.9	17.2	19.1	18.1
MDA [20]	CVPR2022		56.7	34.4	33.5	11.8	-	-
PAT [21]	CVPR2023		67.9	48.9	42.8	18.2	25.4	26.0
MSTL [22]	TCSVT2024		-	-	**49.7**	19.9	26.6	24.9
D^2ML [23]	TIFS2025		54.0	33.6	25.6	8.8	-	-
3D-VAN [24]	TIFS2025		57.9	36.3	-	-	-	-
Ours	–		**70.4**	**50.2**	48.6	**20.1**	**28.5**	**28.3**
Method	Reference	Source	Market		MSMT		CUHK-NP	
			R1	mAP	R1	mAP	R1	mAP
MetaBIN [18]	CVPR2021		69.2	35.9	-	-	-	-
MDA [20]	CVPR2022		70.3	38.0	39.8	13.6	-	-
SuA-SpML [25]	TIP2023		65.8	36.3	37.8	13.6	-	-
PAT [21]	CVPR2023		71.9	45.2	43.9	19.2	18.8	18.9
D^2ML [23]	TIFS2025		65.4	35.0	33.6	12.1	-	-
3D-VAN [24]	TIFS2025		72.1	41.3	-	-	-	-
Ours	–		**73.2**	**46.7**	**45.2**	**20.3**	**19.4**	**20.2**

When trained on Market, our method outperforms the SOTA method PAT [26] by 2.5% Rank-1 and 1.3% mAP on Duke, and by 3.1% and 2.3% on CUHK03-NP. When trained on Duke, our approach achieves gains of 1.1% Rank-1 and 1.5% mAP on Market, and 1.3% Rank-1 and 1.1% mAP on MSMT.

Although MSTL [22]achieves slightly higher Rank-1 accuracy on Market→MSMT17 (49.7% vs. 48.6%), it relies on meta-learning with explicit style transfer and synthetic domain generation tailored for extreme domain shifts. In contrast, our method emphasizes learning stable contour-based semantic structures, achieving superior mAP (20.1% vs. 19.9%) and more balanced retrieval performance across domains.

Multi-source DG ReID. Table 2 details our performance in the rigorous multi-source DG-ReID scenario. Aligning with the M^3L [27] standard, we adopt a leave-one-out strategy where three datasets from the M, D, MS, C pool serve as training sources, leaving the fourth for evaluation; we also include the M^3L-implementation of QAConv to ensure rigorous benchmarking. Our framework consistently delivers superior adaptability. Specifically, in the M+D+C→MS transfer task, we surpass the prior state-of-the-art PAT by 0.6% mAP, achieving a score of 57.4%. A similar advantage is evident in the MS+M+C→D setting, where we outstrip PAT [21] by a 0.9% margin. These gains underscore the robustness of our approach against complex multi-domain variations.

Table 2. Benchmarking against leading DG approaches across four major datasets: Market-1501 (M), DukeMTMC-reID (D), MSMT17 (MS), and CUHK03 (C). We denote the highest scores in bold and underline the best competing results.

Method	M+D+C → MSMT		MS+M+C → Duke		MS+D+C → Market		MS+M+D → CUHK	
	R1	mAP	R1	mAP	R1	mAP	R1	mAP
$QACONV_{50}$[28]	29.9	10.0	64.9	43.4	68.6	39.5	22.9	19.2
M^3L [27]	33.0	12.9	69.4	50.5	74.5	48.1	30.7	29.9
M^3L_{DG} [27]	36.9	14.7	69.2	51.1	<u>75.9</u>	50.2	<u>33.1</u>	<u>32.1</u>
PAT [21]	<u>45.6</u>	<u>21.6</u>	<u>71.8</u>	<u>56.5</u>	75.2	<u>51.7</u>	31.1	31.5
Ours	**46.2**	**22.2**	**72.4**	**57.4**	**75.5**	**52.1**	**31.6**	**32.4**

4.4 Ablation Study

To comprehensively validate the effectiveness of each proposed component, we conduct systematic ablation experiments across challenging domain shift scenarios: Market→ Duke (M→ D), and Market→ MSMT17 (M→ MS) dataset. The baseline network adopts the TransReID [29] design. Specifically, it uses ViT-base [30] as the backbone and employs BNNeck along with a fully connected layer as the classifier. In this baseline network, only F_g (the output of the [CLS] token) is used, and it is optimized using both Cross-Entropy loss and Triplet loss.

Contour Token Selection Analysis. To assess the effectiveness of the Contour Token Selection (CTS) strategy, we compare it with a Random Selection (RS) baseline, where both methods employ the same ATE-based data augmentation. As reported in Table 3, CTS+ATE achieves improvements of +3.4% Rank-1 and +2.7% mAP on M→D, whereas RS+ATE yields only +1.3% and +1.1%. The consistent advantage of CTS (+2.1% Rank-1 and +1.6% mAP) demonstrates that semantically guided contour token selection is critical for capturing identity-relevant structural information while suppressing domain-specific noise.

Asymmetric Token Enhancement Analysis. Based on contour-focused tokens selected by CTS, we further analyze the Asymmetric Token Enhancement

Table 3. Strategic Component Analysis: Comparison of different design choices within our framework. We systematically evaluate each proposed component through progressive integration across two challenging domain shift scenarios. All methods use TransReID with ViT-base as baseline.

Category	**Method**	**M→D**		**M→MS**	
		R1	mAP	R1	mAP
Baseline	Baseline	65.9	46.5	38.6	16.2
Contour Token Selection	+ Random Selection + ATE (full)	67.2	47.6	39.8	17.2
	+ CTS + ATE (full)	**69.3**	**49.2**	**43.8**	**19.2**
Asymmetric Token Enhancement	+ CTS + Blending Branch (BL)	66.6	47.1	39.2	16.8
	+ CTS + Combining Branch (CB)	66.8	47.5	39.4	17.2
	+ CTS + ATE (BL + CB, w/o CEC)	68.2	48.4	42.5	17.9
	+ CTS + ATE (full)	**69.3**	**49.2**	**43.8**	**19.2**
Local Feature Processing	+ PCB[31]	66.5	47.1	39.2	16.8
	+ GL-AA	**67.6**	**47.9**	**41.2**	**17.8**
Complete Model	+ CTS + ATE + GL-AA	**70.4**	**50.2**	**45.6**	**20.6**
Improvement	Total Gain vs. Baseline	**+4.5**	**+3.7**	**+8.0**	**+4.4**

(ATE) module by comparing the full design with its individual branches (CB and BL). As shown in Table 3, either branch alone leads to inferior performance, confirming the necessity of asymmetric enhancement. By constructing positive pairs with substantial appearance variations, ATE encourages robust contour-aware representation learning.

Moreover, removing the Contrastive Enhanced Contour Consistency (CEC) loss results in a noticeable performance drop, highlighting its role in enforcing semantic alignment between augmented views and strengthening contour discrimination. Together, these components guide the feature extractor to focus on global contour structures while reducing sensitivity to contextual noise, thereby improving generalization under domain shift.

Global-Local Alignment Attention Analysis. To evaluate the effectiveness of the Global-Local Alignment Attention (GL-AA), we compare our method with the PCB baseline [31], which employs fixed partitioning without global-local interaction. As shown in Table 3, our approach consistently outperforms PCB across all dataset settings. These results indicate that GL-AA effectively facilitates global-local feature interaction, leading to improved extraction of fine-grained and domain-invariant representations.

Complete Framework Performance. Integrating all proposed components, the complete SATE framework achieves the best performance across all evaluation settings. This demonstrates the strong complementarity among CTS, ATE, GL-AA, and CEC, and confirms that jointly modeling contour selection, asymmetric enhancement, and global-local alignment is critical for robust domain generalization.

4.5 Further Analysis

Loss Weight Analysis. As shown in Fig. 3, optimal performance is achieved at $\lambda = 0.5$. The model exhibits robustness to parameter variations, maintaining high accuracy even as λ increases to 0.8. This stability confirms that our CEC loss effectively complements the identification loss, validating the synergistic design of the objective function.

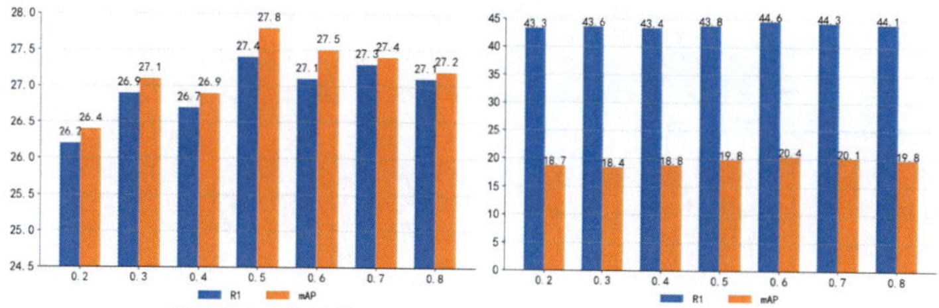

(a) Market-Cuhk03-np (a) Market-Msmt17

Fig. 3. Parameter analysis of λ with training on Market and testing on CUHK03-NP (a) and MSMT17 (b).

Comparisons of Model Complexity. Table 4 presents the efficiency analysis. While training introduces moderate overhead (0.34 s/iter vs. 0.23 s/iter) due to the ATE module's dual forward passes, this module is discarded during inference. This design aligns with asymmetric contrastive learning principles [32], allowing the backbone to encode contour-aware features via CEC loss without test-time burdens. Consequently, the model functions as a standard single-backbone architecture during inference, maintaining a highly competitive speed (1.03ms/img) comparable to the baseline.

Table 4. Comparison of training and inference times and parameter sizes between our method and the baseline.

Method	Train	Inference	Params
TransReID [29]	0.23 s/iter	0.91ms/img	89.5M
Ours	0.34 s/iter	1.03ms/img	94.8M

4.6 Qualitative Analysis and Visualization

Attention Visualization. Fig. 4 compares attention maps of the baseline, CTS+ATE, and the complete model (with GL-AA) by overlaying fused attention weights on MSMT images. The visualizations demonstrate that CTS+ATE effectively suppresses background noise and highlights pedestrian regions, attributed

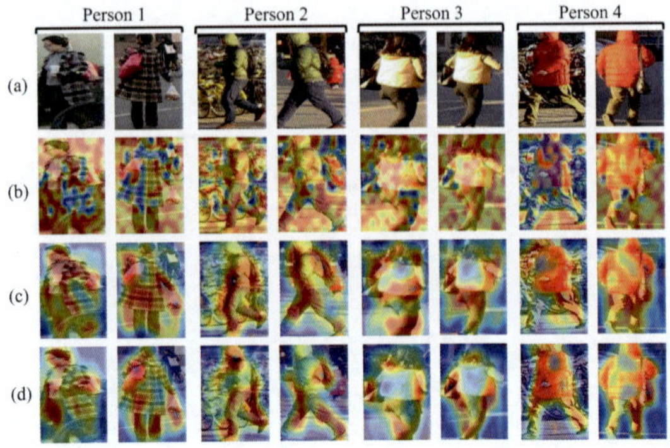

Fig. 4. The visualization of attention maps obtained by different methods, where the original images are selected from the target domain (MSMT). (a) Raw images. (b) Baseline. (c) CTS + ATE. (d) CTS + ATE + GL-AA.

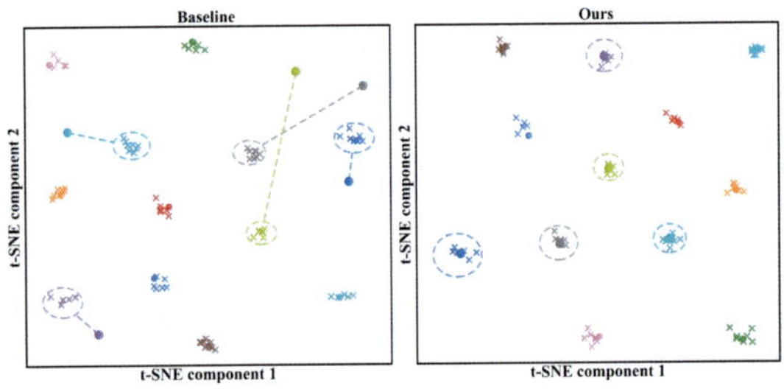

Fig. 5. Visualization of feature distributions. The t-SNE visualization of pedestrian features during inference is shown, with colored dots (queries) and crosses (gallery) representing 12 pedestrian samples (different colors for different identities). Dashed ellipses highlight challenging hard samples.

to robust contour feature extraction. Furthermore, GL-AA guides the model to capture fine-grained details (e.g., backpacks, shoes), significantly enhancing discriminative ability in unseen domains.

Feature Distribution Analysis. To evaluate discriminative capability, Fig. 5 presents t-SNE visualizations of feature embeddings on the Duke dataset. We compare the baseline (left) with our method (right) using 12 randomly selected identities. Distinct colors denote different identities, while circles and crosses represent query and gallery images, respectively.

To highlight the model's performance on challenging cases, we use dashed circles to emphasize five difficult pedestrian identities. In these examples, the baseline model fails to effectively cluster embeddings of the same identity, resulting in dispersed features. In contrast, our model produces significantly more compact embeddings for the same identities, demonstrating stronger discriminative power. This indicates that our approach better mitigates domain-specific discrepancies and enhances feature consistency, thereby contributing to improved retrieval performance in DG-ReID.

Qualitative Analysis of Retrieval Results. Fig. 6 illustrates representative retrieval results on the Duke dataset, comparing our contour-aware asymmetric enhancement method with the baseline ViT. Each row corresponds to a query identity, with top-5 gallery results shown. Correct and incorrect matches are indicated by red and blue bounding boxes, respectively.

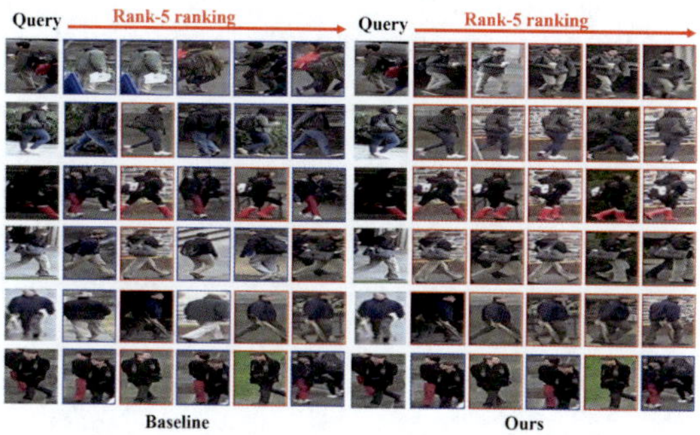

Fig. 6. Retrieval results on the Duke dataset. The red box highlights the correct match, while the blue box highlights the incorrect match.

Success Cases (Rows 1–5). Our method consistently outperforms the baseline, accurately retrieving matches within the top-5 despite significant viewpoint and background shifts. This validates that contour-based structural features offer superior invariance compared to appearance-based representations.

Failure Case (Row 6). In crowded scenes with overlapping pedestrians, performance remains suboptimal due to contour ambiguity entangling edge signals in the CTS module. Future work should integrate instance-aware segmentation or spatial reasoning to address this limitation.

5 Conclusion

We propose a semantic-aware asymmetric token enhancement framework for ReID domain generalization, shifting to fine-grained token-level modeling. By integrating **Contour Token Selection (CTS)** for Fourier-based structure identification and **Asymmetric Token Enhancement (ATE)** for consistency learning, our method effectively disentangles identity cues from domain noise. Experiments demonstrate state-of-the-art improvements of up to 2.5% Rank-1 and 2.3% mAP. Future work will target robustness in occluded scenarios and broader cross-domain applications.

References

1. Wang, H., Shen, J., Liu, Y., Gao, Y., Gavves, E.: Nformer: Robust person re-identification with neighbor transformer. In: Proceedings of the IEEE/CVF conference on computer vision and pattern recognition, pp. 7297–7307. (2022)
2. Zheng, Z., Zheng, L., Yang, Y.: Unlabeled samples generated by gan improve the person re-identification baseline in vitro. In: Proceedings of the IEEE international conference on computer vision. pp. 3754–3762 (2017)
3. Zhao, H., Qi, L., Geng, X.: Cilp-fgdi: exploiting vision-language model for generalizable person re-identification. IEEE Trans. Inf. Forensics Secur. (2025)
4. Bai, Z., Wang, Z., Wang, J., Hu, D., Ding, E.: Unsupervised multi-source domain adaptation for person re-identification. In: Proceedings of the IEEE/CVF Conference on Computer Vision and Pattern Recognition, pp. 12914–12923 (2021)
5. Lv, K., Chen, H., Zhao, C., Tu, K., Chen, J., Li, Y., Li, B., Lin, Y.: Style variable and irrelevant learning for generalizable person re-identification. ACM Trans. Multimed. Comput. Commun. Appl. **20**(9), 1–22 (2024)
6. Zhang, H., Cisse, M., Dauphin, Y.N., Lopez-Paz, D.: mixup: beyond empirical risk minimization (2017). ArXiv preprint arXiv:1710.09412
7. Zhou, K., Yang, Y., Qiao, Y., Xiang, T.: Mixstyle neural networks for domain generalization and adaptation. Int. J. Comput. Vision **132**(3), 822–836 (2024)
8. Song, J., Yang, Y., Song, Y.Z., Xiang, T., Hospedales, T.M.: Generalizable person re-identification by domain-invariant mapping network. In: Proceedings of the IEEE/CVF Conference on Computer Vision and Pattern Recognition, pp. 719–728 (2019)
9. Pan, X., Luo, P., Shi, J., Tang, X.: Two at once: enhancing learning and generalization capacities via ibn-net. In: Proceedings of the European Conference on Computer Vision (ECCV), pp. 464–479 (2018)
10. Grill, J.B., Strub, F., Altché, F., Tallec, C., Richemond, P., Buchatskaya, E., Doersch, C., Avila Pires, B., Guo, Z., Gheshlaghi Azar, M., et al.: Bootstrap your own latent-a new approach to self-supervised learning. Adv. Neural. Inf. Process. Syst. **33**, 21271–21284 (2020)
11. He, K., Fan, H., Wu, Y., Xie, S., Girshick, R.: Momentum contrast for unsupervised visual representation learning. In: Proceedings of the IEEE/CVF Conference on Computer Vision and Pattern Recognition, pp. 9729–9738 (2020)
12. Caron, M., Misra, I., Mairal, J., Goyal, P., Bojanowski, P., Joulin, A.: Unsupervised learning of visual features by contrasting cluster assignments. Adv. Neural. Inf. Process. Syst. **33**, 9912–9924 (2020)
13. Tripathi, A., Singh, R., Chakraborty, A., Shenoy, P.: Edges to shapes to concepts: adversarial augmentation for robust vision. In: Proceedings of the IEEE/CVF Conference on Computer Vision and Pattern Recognition, pp. 24470–24479 (2023)
14. Zheng, L., Shen, L., Tian, L., Wang, S., Wang, J., Tian, Q.: Scalable person re-identification: a benchmark. In: Proceedings of the IEEE International Conference on Computer Vision, pp. 1116–1124 (2015)
15. Wei, L., Zhang, S., Gao, W., Tian, Q.: Person transfer gan to bridge domain gap for person re-identification. In: Proceedings of the IEEE Conference on Computer Vision and Pattern Recognition, pp. 79–88 (2018)
16. Zhong, Z., Zheng, L., Cao, D., Li, S.: Re-ranking person re-identification with k-reciprocal encoding. In: Proceedings of the IEEE Conference on Computer Vision and Pattern Recognition, pp. 1318–1327. (2017)

17. Liao, S., Shao, L.: Transmatcher: Deep image matching through transformers for generalizable person re-identification. Adv. Neural. Inf. Process. Syst. **34**, 1992–2003 (2021)
18. Choi, S., Kim, T., Jeong, M., Park, H., Kim, C.: Meta batch-instance normalization for generalizable person re-identification. In: Proceedings of the IEEE/CVF conference on Computer Vision and Pattern Recognition, pp. 3425–3435 (2021)
19. Liao, S., Shao, L.: Graph sampling based deep metric learning for generalizable person re-identification. In: Proceedings of the IEEE/CVF Conference on Computer Vision and Pattern Recognition, pp. 7359–7368 (2022)
20. Ni, H., Song, J., Luo, X., Zheng, F., Li, W., Shen, H.T.: Meta distribution alignment for generalizable person re-identification. In: Proceedings of the IEEE/CVF Conference on Computer Vision and Pattern Recognition, pp. 2487–2496 (2022)
21. Ni, H., Li, Y., Gao, L., Shen, H.T., Song, J.: Part-aware transformer for generalizable person re-identification. In: Proceedings of the IEEE/CVF International Conference on Computer Vision, pp. 11280–11289 (2023)
22. Chen, K., Fang, P., Ye, Z., Zhang, L.: Multi-scale explicit matching and mutual subject teacher learning for generalizable person re-identification. In: IEEE Transactions on Circuits and Systems for Video Technology (2024)
23. Sun, J., Li, Y., Chen, L., Chen, H., Wang, M.: Dualistic disentangled meta-learning model for generalizable person re-identification. IEEE Transactions on Information Forensics and Security (2025)
24. Jiao, B., Liu, L., Gao, L., Wu, D.O., Lin, G., Wang, P., Zhang, Y.: Generalizable person re-identification from a 3d perspective: Addressing unpredictable viewpoint changes. IEEE Transactions on Information Forensics and Security, pp. 1–1 (2025)
25. Zhang, L., Liu, Z., Zhang, W., Zhang, D.: Style uncertainty based self-paced meta learning for generalizable person re-identification. IEEE Trans. Image Process. **32**, 2107–2119 (2023)
26. Li, Y., He, J., Zhang, T., Liu, X., Zhang, Y., Wu, F.: Diverse part discovery: Occluded person re-identification with part-aware transformer. In: Proceedings of the IEEE/CVF Conference on Computer Vision and Pattern Recognition, pp. 2898–2907 (2021)
27. Zhao, Y., Zhong, Z., Yang, F., Luo, Z., Lin, Y., Li, S., Sebe, N.: Learning to generalize unseen domains via memory-based multi-source meta-learning for person re-identification. In: Proceedings of the IEEE/CVF Conference on Computer Vision and Pattern Recognition, pp. 6277–6286 (2021)
28. Liao, S., Shao, L.: Interpretable and generalizable person re-identification with query-adaptive convolution and temporal lifting. In: European Conference on Computer Vision, pp. 456–474. Springer (2020)
29. He, S., Luo, H., Wang, P., Wang, F., Li, H., Jiang, W.: Transreid: Transformer-based object re-identification. In: Proceedings of the IEEE/CVF international conference on computer vision. pp. 15013–15022 (2021)
30. Dosovitskiy, A., Beyer, L., Kolesnikov, A., Weissenborn, D., Zhai, X., Unterthiner, T., Dehghani, M., Minderer, M., Heigold, G., Gelly, S.: An image is worth 16×16 words: transformers for image recognition at scale (2020). ArXiv preprint arXiv:2010.11929
31. Sun, Y., Zheng, L., Yang, Y., Tian, Q., Wang, S.: Beyond part models: person retrieval with refined part pooling (and a strong convolutional baseline). In: Proceedings of the European Conference on Computer Vision (ECCV), pp. 480–496 (2018)

32. Chen, X., He, K.: Exploring simple Siamese representation learning. In: Proceedings of the IEEE/CVF Conference on Computer Vision and Pattern Recognition, pp. 15750–15758 (2021)

PMOSTRACK: A Video Object Tracking Framework with Positional Visualization and Mamba-Based Historical Modeling

Fengyi Lv[1], Man Xu[1](✉), Zizhu Fan[1], and Hui Wang[2]

[1] Shanghai University of Electric Power, Shanghai, China
xumanly@shiep.edu.cn
[2] East China Jiaotong University, Nanchang, China

Abstract. Current object tracking methods predominantly rely on textual representations of positional information, limiting their ability to fully exploit spatial awareness for performance improvements. To address these issues, this paper proposes PMOSTrack, a novel video-level tracking framework. This framework integrates visualized positional information and the Mamba model. To enhance the model's understanding of positional variations, this paper adopts the approach of visualizing positional information as image-like representations, enabling the model to capture spatial variations of the target across the preceding n frames and gain a deeper understanding of its dynamic movement patterns. While Mamba's hidden state mechanism excels at capturing and preserving historical target features, it further leverages this capability to effectively memorize the target's appearance characteristics. The visualizing positional information and the output from Mamba are fused into OSTrack's encoder layers, improving the model's Similar Object Discriminability and temporal information utilization, ultimately boosting tracking robustness and accuracy. Experimental results on the GOT-10k dataset demonstrate the effectiveness of PMOSTrack, achieving an AO score of **76.7%**, which is **5.7%** higher than the baseline OSTrack-256, which achieves an AO score of 71.0%.

Keywords: Computer vision · Video object tracking · Positional heatmap · Mamba memory module

1 Introduction

Single object tracking is a fundamental task in computer vision, aimed at localizing a target object in each frame of a video sequence, given its initial state. The core objective is to accurately predict the target's position in each subsequent frame, thus enabling continuous tracking. Most contemporary trackers formulate this task as a template matching problem, where the target is identified by comparing deep feature similarity between a static initial template and the current search region. While this paradigm is efficient and widely adopted, it

L. Huang and D. Greenhalgh (Eds.): ICMLC 2026, LNNS 2061, pp. 115–130, 2027.
https://doi.org/10.1007/978-3-032-31236-5_8

inherently suffers from limited adaptability. Specifically, as the target undergoes appearance variations due to deformation, occlusion, scale changes, or illumination shifts, the static template becomes progressively misaligned with the current target, leading to drift or tracking failure.

To address this limitation, many recent approaches have introduced dynamic template updating strategies. Trackers such as STARK [1], AIATrack [2], and FEAR [3] dynamically fuse features from recent frames or leverage prediction-based update mechanisms. While these approaches improve short-term adaptability, they often fail to capture long-term temporal dependencies. Consequently, they struggle in scenarios such as gradual appearance changes, target re-appearance after occlusion, or cyclic motion patterns, all of which are common in real-world videos.

Recent advances explore incorporating longer temporal context through video-level modeling. Some methods introduce memory tokens or historical cues to enhance temporal awareness. However, constrained by token length and computational cost, these designs often face limitations in their temporal expressiveness. In contrast, sequence modeling approaches, such as LSTM [4] or state-space models(SSM) [5], maintain persistent hidden states to capture historical evolution, offering superior long-term memory. Nevertheless, their integration with high-resolution spatial features in visual tracking remains underexplored.

PMOSTrack overcomes these limitations by introducing a novel video object tracking framework that seamlessly integrates historical modeling and spatial awareness into a unified architecture. Specifically, this paper incorporates a multi-layer Mamba memory module into the visual backbone, enabling the efficient and scalable modeling of long-range temporal dependencies. The Mamba architecture, which is based on state-space models, captures the historical evolution of the target through continuous hidden states with minimal computational overhead.

Furthermore, this paper introduces a novel approach for representing positional information. Many existing trackers employ bounding box coordinates or trajectories to represent the target's location, but these methods often fail to adequately capture the spatial semantics and temporal continuity of the target. To address this limitation, this paper proposes a visualized positional encoding scheme that transforms historical target locations into an image-aligned heatmap. This heatmap retains the trajectory information, including the direction of motion and key attention points, effectively capturing the target's dynamic movement patterns and spatially significant regions for more precise single-object tracking. By incorporating such a heatmap, the model in this paper enables better perception of the target's spatial dynamics, leading to improved robustness, particularly in scenarios involving partial occlusion.

In summary, PMOSTrack contributes a unified tracking framework that (1) leverages Mamba-based memory modeling for long-term temporal context, and (2) introduces spatially expressive positional heatmap visualization for enhanced motion-aware perception. Extensive experiments demonstrate that the model in this paper achieves better performance on the challenging GOT-10k [6] tracking benchmark, especially in scenarios involving occlusion or partial occlusion. Our main contributions are as follows:

- A positional heatmap representation method is introduced into single-object tracking to enhance spatial modeling.
- This paper introduces Mamba for the first time to memorize the target's appearance features, enabling the model to adapt to appearance changes over time.

2 Related Work

Visual object tracking remains a fundamental problem in computer vision, with approaches broadly categorized into image-level and video-level paradigms. Image-level trackers primarily rely on appearance-based matching between the initial template and the current frame. Representative works include SiamFC [7], SiamRPN [8], and OSTrack [9], which utilize efficient template-matching architectures with shared feature backbones. Although computationally lightweight, these approaches suffer from poor adaptability to target appearance changes caused by occlusion, deformation, motion blur, or illumination variations, often resulting in tracking failure. Yaseen and Sasi [10] proposed a robust algorithm for object detection and tracking in dynamic scenes, which improved traditional SURF-based tracking by incorporating color features to enhance tracking accuracy, even after brief occlusions. However, their method still faced challenges in handling long-term appearance changes over time.

To enhance appearance adaptability, several works have proposed dynamic template update strategies. Trackers like STARK [1], AIATrack [2], and FEAR [3] incorporate predictor heads, dual-template fusion, or score-guided updates to refine the target representation over time. However, these methods still operate within the image-level paradigm, relying on frame-wise cues and struggling to model long-term temporal dependencies effectively.

Recent efforts have shifted towards video-level tracking by explicitly modeling temporal context across frames. TCTrack [11] introduces a temporal-adaptive convolution module to aggregate historical context in feature extraction. VideoTrack [12] adopts a temporal Transformer [13] backbone to model long-term motion patterns holistically. ODTrack [14] proposes a token propagation framework that transmits object features across time steps, while AQATrack [15] introduces autoregressive query generation to capture long-range temporal dependencies. Similarly, EVPTrack [16] employs a global propagation scheme to stabilize temporal information flow across video frames.

These methods have shown clear advancements in utilizing historical context for tracking. However, most still depend on a limited set of memory tokens or attention mechanisms, which face scalability challenges due to memory and computational constraints. In contrast, our proposed PMOSTrack introduces a

Mamba-based hidden state memory mechanism directly into the backbone, facilitating the continuous and efficient accumulation of historical target representations. This approach enables long-term temporal modeling without the need for explicit token propagation, offering better scalability and temporal stability.

Beyond temporal modeling, accurately representing spatial position information is also crucial for enhancing tracking robustness. Traditional methods typically rely on explicit numerical coordinates (e.g., bounding boxes) as location priors. Although these methods are simple and effective, they struggle to intuitively capture spatial semantics and are difficult to effectively integrate with dense visual features, especially in complex scenarios such as occlusion, scale variation, or cluttered backgrounds.

Recent studies have explored richer spatial encoding strategies. ARTrack [17] utilizes spatiotemporal trajectory cues to model motion continuity. SeqTrack [18] uses the autoregressive sequence generation method, where the target position from the previous frame is used to progressively predict the target position in subsequent frames. InterMamba [19] introduces a Gaussian-based heatmap to inject location information into the model, enhancing spatial focus.

Inspired by these approaches, PMOSTrack proposes a spatial visualization mechanism that transforms historical bounding box coordinates into an image-aligned heatmap. These spatial representations retain structural motion cues and integrate with Transformer modules. This allows the model to learn joint spatio-temporal features more effectively, enhancing robustness in challenging scenarios such as full and partial occlusion.

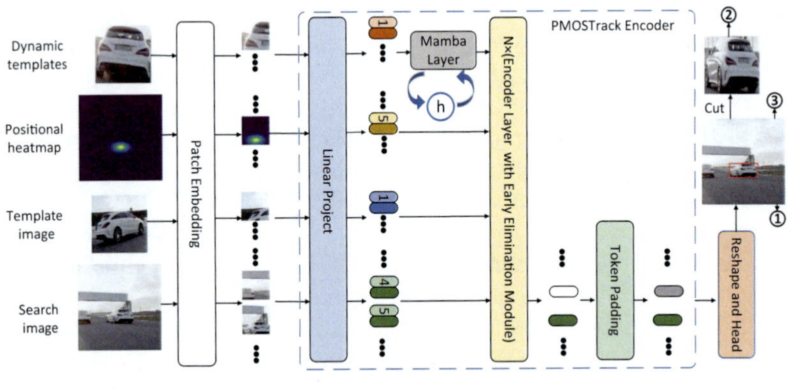

Fig. 1. Framework of the proposed PMOSTrack. Each frame provides four inputs: initial template Z, search region X, positional heatmap L, and dynamic template T. The tracker adopts a confidence–aware update policy: ① if the confidence score $p < a$, the prediction is discarded and a Kalman filter estimates the target position, which is converted to a Gaussian heatmap for a second inference; ② if $p \geq a$, the dynamic template T is refreshed; ③ if $p \geq b$, the positional heatmap L is updated. This design balances spatial accuracy with temporal consistency under varying confidence levels.

3 Model Architecture

3.1 Backbone Structure

Backbone Structure. The Encoder module in OSTrack [9] is used as the basic backbone structure. During the tracking process of each frame, the model's input includes four types of image information (see Fig. 1): a search region $\mathbf{X} \in \mathbb{R}^{3\times H_x \times W_x}$, a template area $\mathbf{Z} \in \mathbb{R}^{3\times H_z \times W_z}$, a positional heatmap area $\mathbf{L} \in \mathbb{R}^{1\times H_l \times W_l}$, and the tracking target image cut from the previous frame $\mathbf{T} \in \mathbb{R}^{3\times H_t \times W_t}$, Here, W and H represent the width and height of the images, respectively. They are first segmented and unfolded into:$\mathbf{X}_p \in \mathbb{R}^{N_x \times (3 \cdot P^2)}$, $\mathbf{Z}_p \in \mathbb{R}^{N_z \times (3 \cdot P^2)}$,$\mathbf{L}_p \in \mathbb{R}^{N_l \times (3 \cdot P^2)}$,and $\mathbf{T}_p \in \mathbb{R}^{N_t \times (3 \cdot P^2)}$,where $P \times P$ is the resolution of each patch, and $N_x = \frac{H_x W_x}{P^2}$, $N_z = \frac{H_z W_z}{P^2}$, $N_l = \frac{H_l W_l}{P^2}$, $N_t = \frac{H_t W_t}{P^2}$are the number of patches of template, search region, positional heatmap, and previous frame cropped target image, respectively.

After that, a trainable linear projection layer with parameter $\mathbf{E} \in \mathbb{R}^{(\mathbf{3} \cdot \mathbf{P^2}) \times \mathbf{D}}$ is used to project $\mathbf{X}_p$, $\mathbf{Z}_p$, $\mathbf{L}_p$, and $\mathbf{T}_p$ into a D-dimensional latent space as in Eq. (1)–(4), and the output of this projection is commonly called patch embeddings. Learnable 1D position embeddings $\mathbf{P}_x \in \mathbb{R}^{N_x \times D}$, $\mathbf{P}_z \in \mathbb{R}^{N_z \times D}$, $\mathbf{P}_l \in \mathbb{R}^{N_l \times D}$ and $\mathbf{P}_t \in \mathbb{R}^{N_t \times D}$ are added to the patch embeddings of the template, search region, positional heatmap, and dynamic template (cropped target image), respectively, to produce the final token embeddings:

$$\mathbf{H}_z^0 = \left[Z_p^1 E; Z_p^2 E; \cdots ; Z_p^{N_z} E\right] + \mathbf{P}_z \tag{1}$$

$$\mathbf{H}_x^0 = \left[X_p^1 E;\, X_p^2 E;\, \ldots;\, X_p^{N_x} E\right] + \mathbf{P}_x \tag{2}$$

$$\mathbf{H}_l^0 = \left[L_p^1 E;\, L_p^2 E;\, \ldots;\, L_p^{N_l} E\right] + \mathbf{P}_l \tag{3}$$

$$\mathbf{H}_t^0 = \left[T_p^1 E;\, T_p^2 E;\, \ldots;\, T_p^{N_t} E\right] + \mathbf{P}_t \tag{4}$$

This patch–embedding formulation originates from the Vision Transformer (ViT)[20].The sequences $\mathbf{H}_z^0$, $\mathbf{H}_x^0$, $\mathbf{H}_l^0$, and $\mathbf{H}_t^0$ are then concatenated to obtain: $\mathbf{H}_{\text{tlzx}}^0 = [\mathbf{H}_t^0; \mathbf{H}_l^0; \mathbf{H}_z^0; \mathbf{H}_x^0]$, and send the resulting $\mathbf{H}_{\text{tlzx}}^0$ into the main visual Transformer encoder. This encoder incorporates the early elimination module from OSTrack to enhance the filtering efficiency of redundant tokens.

3.2 Mamba-Based Memory Module

Mamba-based memory module: The historical memory module is one of the core components in PMOSTrack (see Fig. 2), which aims to achieve efficient long-term target state construction. Within the memory module, the input F_i is first normalized via RMS Norm, producing F_{in}. Then, F_{in} is processed through two branches: x and z.In the x branch, F_{in} is projected and expanded along the dimension via a linear projection, followed by feature extraction through convolutional layers. After activation with the **Silu** function, the features are passed

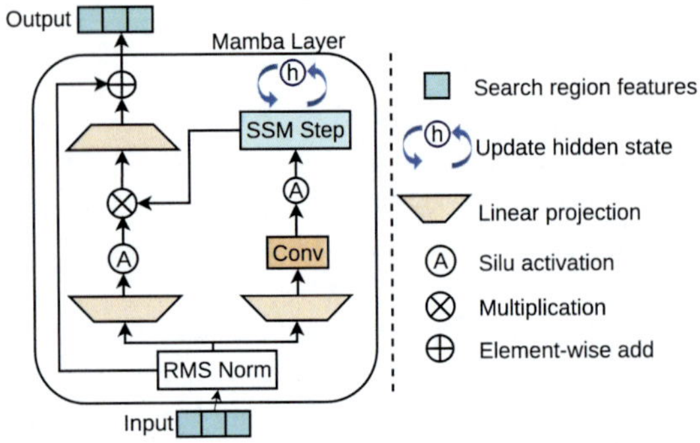

Fig. 2. Overview of the proposed Mamba-based memory module [21], which stores historical target features and injects them into the encoder for long-term temporal modeling.

into the **SSM** to extract and store the target's features. The output from the x branch is denoted as F_x.In the z branch, F_{in} is similarly expanded along the dimension and activated using the Silu function. Then, F_x is used to weight it, followed by a dimension reduction, producing the z-branch output F_z.Finally, F_z and F_{in} are concatenated through a residual connection to generate the output F of the Mamba layer. This output F aggregates the stored historical target features. The input to the mamba layer includes the hidden states of the previous frame and the cropped image regions. The output represents the hidden states of the current frame and rich historical information tokens. These tokens are directly integrated into the encoder backbone of PMOSTrack to enhance temporal modeling and long-term feature accumulation. The detailed process is illustrated in Fig. 2.

3.3 Localization Heatmap Module

The localization heatmap is a key component in PMOSTrack, which enhances the spatial modeling capabilities. It is used to encode the spatial trajectory of the target across different time frames. The module is generated as follows:

1. For each frame, a 2D Gaussian distribution centered at the target bounding box center is generated based on the target's position:

$$H(x_t, y_t, w_t, h_t, n) = \exp\Big(-\frac{(x - x_t)^2 + (y - y_t)^2}{\sigma_x^2 + \sigma_y^2}\Big), \tag{5}$$

$$\sigma_x = \frac{w_t}{n}, \qquad \sigma_y = \frac{h_t}{n}$$

2. The generated Gaussian map is then added to a grayscale map A with the same resolution as the original image, serving as the position heatmap.

3. For the first frame, initialize $A_0 = 0$. For the current frame t, update the map using the following formula:

$$A_t = \frac{A_{t-1}}{2} + H(x_t, y_t, w_t, h_t, n) \quad (6)$$

Equation (5) generates a 2D Gaussian heatmap centered at the predicted bounding box (x_t, y_t, w_t, h_t), with spatial spread $\sigma_x = w_t/n$ and $\sigma_y = h_t/n$, where $n = 6$ .This value of n was chosen based on practical considerations, rather than through an ablation study. Equation (6) updates the heatmap recursively, applying exponential decay (factor 1/2) to the previous activation and adding the current frame's Gaussian, thus capturing both short- and long-term motion cues.

3.4 Head and Loss Function

The same classification and regression heads as used in OSTrack are adopted for prediction. The head consists of three sub-networks, each implemented as a fully convolutional network (FCN). These sub-networks output a classification score map $P \in \mathbb{R}^{1\times\frac{H}{16}\times\frac{W}{16}}$, a bounding box size map $S \in \mathbb{R}^{2\times\frac{H}{16}\times\frac{W}{16}}$, and an offset map $O \in \mathbb{R}^{2\times\frac{H}{16}\times\frac{W}{16}}$.The location with the highest classification score is selected as the target position, i.e.,$(x_d, y_d) = \arg\max_{(x,y)} P_{xy}$. Then, the predicted bounding box is computed as:

$$\begin{aligned}(x, y, w, h) = \Big(&x_d + O(0, x_d, y_d), y_d + O(1, x_d, y_d),\\ &S(0, x_d, y_d), S(1, x_d, y_d)\Big)\end{aligned} \quad (7)$$

During training, both classification and regression losses are jointly used. The classification loss is calculated with weighted focal loss. For the predicted bounding box, the L1 loss and the IoU loss are applied to supervise the bounding box regression. The total loss function is defined as:

$$L_{\text{track}} = L_{\text{cls}} + \lambda_{\text{iou}} L_{\text{iou}} + \lambda_{L1} L_1 \quad (8)$$

where the hyperparameters $\lambda_{\text{iou}} = 2$ and $\lambda_{L1} = 5$ are used for loss balancing as determined in our experiments.

4 Experiments

4.1 Implementation Details

Algorithm 1: PMOSTrack Stage: Feature Extraction with Confidence-Adaptive Fusion

Input: Search image X_t; static template $\mathbf{Z}$;position heatmap $\mathbf{L}_t$; dynamic template T_{t-1}; previous confidence conf_{t-1}
Output: Predicted box B_t; confidence conf_t
Function *PMOSTrackStage*(X_t, $\mathbf{Z}$, $\mathbf{L}_t$, T_{t-1}, $conf_{t-1}$)
  `// Step 1: Update memory if confidence is high`
  if $conf_{t-1} > a$ **then**
    Update dynamic template list with cropped patch from previous frame;;
  Randomly sample k templates from dynamic template list;
  $T_{\text{input}} \leftarrow \{\mathbf{Z}\} \cup \text{random-}k \cup \{T_{t-1}\}$;;
  $h \leftarrow$ None;
  foreach z_2 *in* T_{input} **do**
    $(h, z_2') \leftarrow \mathbf{MambaLayer}(h, z_2)$;;
  `// Step 2: Apply confidence-dependent weighting`
  $(\alpha, \beta) \leftarrow \mathrm{conf}_{t-1} > a?(\alpha_1, \beta_1) : (\mathrm{conf}_{t-1} > b?(\alpha_2, \beta_2) : (\alpha_3, \beta_3))$;
  $z_2' \leftarrow \alpha \cdot z_2', \quad \mathbf{L}_t \leftarrow \beta \cdot \mathbf{L}_t$;;
  `// Step 3: Encode and predict`
  $\mathbf{F}_{\text{enc}} \leftarrow \mathbf{ViTEncoder}(\mathbf{L}_t, z_2', \mathbf{Z}, X_t)$;;
  $(B_t, \mathrm{conf}_t) \leftarrow \mathbf{Head}(\mathbf{F}_{\text{enc}})$;;
  return (B_t, conf_t);;

Model: Due to computational constraints, the proposed PMOSTrack is implemented and trained on the B256 variant. All models were trained and tested using an NVIDIA GeForce RTX 4090 GPU. The code is implemented in PyTorch 1.13, and the experiments were conducted on Ubuntu 20.04 with CUDA 11.7.

Training: We train the model using the GOT-10k [6],LaSOT [22] and TrackingNet [23] dataset. When training the model solely on the GOT-10k dataset, the total number of training epochs is set to 100, with the learning rate decayed by a factor of 10 after epoch 80 to ensure stable convergence.In contrast, when training on the combined datasets of GOT-10k, LaSOT, and TrackingNet, the total number of epochs is increased to 300, with the learning rate decay occurring after epoch 280. The training uses the AdamW optimizer, with an initial learning rate of 1×10^{-4} applied to the backbone, which is initialized with the pretrained weights file, output from the training of OSTrack. The batch size is set to 4, with each batch containing 5 search images.During the training process, the parameters are $a = 0.7$, $b = 0.52$, $\alpha_1 = \alpha_2 = 1$, $\alpha_3 = 0.1$, $\beta_1 = \beta_2 = \beta_3 = 1$, where α and β are trainable parameters, while the values of a and b are empirically derived.

Algorithm 2: Two-Stage Inference with Dual-Threshold Strategy

Input: Current frame image I_t; previous box B_{t-1}; static template $\mathbf{Z}$; position heatmap $\mathbf{L}_t$; dynamic template T_{t-1}; previous confidence conf_{t-1}

Output: Updated target state state_t; confidence conf_t; updated heatmap $\mathbf{L}_t$, dynamic template T_t

Function *TwoStageInference*(I_t, B_{t-1}, $\mathbf{Z}$, $\mathbf{L}_t$, T_{t-1}, $conf_{t-1}$)

    // Step 1: first stage
    $X_t \leftarrow \mathrm{CropSearchRegion}(I_t, B_{t-1})$;;
    $(B_t, \mathrm{conf}_1) \leftarrow$ **PMOSTrackStage**$(X_t, \mathbf{Z}, \mathbf{L}_t, T_{t-1}, \mathrm{conf}_{t-1})$;;
    if $conf_1 > a$ **then**
        Update dynamic template T with patch from I_t using B_t to obtain T_t;
        $T \leftarrow T_t$;
    else
        $T \leftarrow T_{t-1}$;
    if $conf_1 > b$ **then**
        $\mathbf{L}_{t+1} \leftarrow \mathrm{generate\_heatmap}(B_t, \mathbf{L}_t)$;;
        return $\mathrm{state}_t \leftarrow B_t$, $\mathrm{conf}_t \leftarrow \mathrm{conf}_1$, $\mathbf{L}_{t+1}$, T;;
    // Step 2: Second stage
    $\hat{B}_t \leftarrow \mathrm{KalmanFilter}(\text{trajectory history})$;;
    $\mathbf{L}_t^{(2)} \leftarrow \mathrm{generate\_heatmap}(\hat{B}_t, \mathbf{L}_t)$;;
    $(B'_t, \mathrm{conf}_2) \leftarrow$ **PMOSTrackStage**$(X_t, \mathbf{Z}, \mathbf{L}_t^{(2)}, T_{t-1}, \mathrm{conf}_1)$;;
    if $conf_2 > b$ **then**
        $\mathbf{L}'_{t+1} \leftarrow \mathrm{generate\_heatmap}(B'_t, \mathbf{L}_t)$;;
        return $\mathrm{state}_t \leftarrow B'_t$, $\mathrm{conf}_t \leftarrow \mathrm{conf}_2$, $\mathbf{L}'_{t+1}$, T;;
    else
        return $\mathrm{state}_t \leftarrow B_t$, $\mathrm{conf}_t \leftarrow \mathrm{conf}_1$, $\mathbf{L}_t$, T;;

Inference: During inference, the model processes the current frame's search region, template, positional heatmap, and updated template. The update strategy is based on the predicted confidence score. For the detailed single-frame inference process, please refer to Algorithm 1; For two-stage inference, it does increase the inference time, but this increase mainly occurs when the confidence is low, as shown in Fig. 2. During the inference process, we store 50 tracking targets and then select n of them: one as the initial template, one as the most recent dynamic template, and the remaining $n - 2$ randomly chosen. Through experimentation, the optimal performance is achieved when $n = 3$, as shown in Table 5. The proposed model achieves an Average Overlap score of 76.7% when trained on the combined GOT-10k, LaSOT, and TrackingNet datasets. When trained solely on the GOT-10k dataset, the AO score drops slightly to 74.9%.

Table 1. Performance comparison of different tracking methods on the GOT-10k dataset. PMOSTrack-256* was trained on the GOT-10k, LaSOT[22], and TrackingNet[23] datasets.

Method	AO	$\mathbf{SR}_{0.5}$	$\mathbf{SR}_{0.75}$
PMOSTrack-256*(Ours)	76.7	86.6	74.2
PMOSTrack-256(Ours)	74.9	84.7	72.8
MCITrack-B224 [21]	77.9	88.2	76.8
ARTrackV2-256 [17]	75.9	85.4	72.7
AQATrack-256 [15]	73.8	83.2	72.1
EVPTrack-224 [16]	73.3	83.6	70.7
LoRAT-B224 [24]	72.1	81.8	70.7
ARTrack-256 [17]	73.5	82.2	70.9
SeqTrack-B256 [18]	74.7	84.7	71.8
VideoTrack-256 [12]	72.9	81.9	69.8
OSTrack-256 [9]	71.0	80.4	68.2
Mixformer-22k [25]	70.7	80.0	67.8

GOT-10k (Generic Object Tracking Benchmark) is a large-scale, class-agnostic dataset designed for general object tracking. It employs a challenging *one-shot* evaluation setting, where the 84 object categories in the test set do not overlap with the 563 categories in the training set. The training split comprises over 10,000 video sequences covering a wide range of categories such as animals, humans, vehicles, and tools. Table 1 compares state-of-the-art trackers with small input resolutions (224× 224 and 256× 256). MCITrack-B224 leads with an AO score of 77.9%, followed by PMOSTrack-256* at 76.7% and ARTrackV2-256 at 75.9%. Compared to the baseline OSTrack-256 (71.0%), PMOSTrack-256* shows a significant improvement with a 5.7% gain in AO, 6.2% in SR0.5, and 6% in SR0.75, demonstrating the effectiveness of our enhancements. While OSTrack introduced vision transformers, it faces limitations in long-term temporal modeling and spatial context. PMOSTrack-256* addresses these by visualizing positional information and leveraging Mamba for improved temporal modeling, resulting in better tracking performance. Overall, PMOSTrack-256* achieves better performance on GOT-10K compared to the baseline model OSTrack and several recent methods.

UAV123[26] is a small-scale tracking benchmark that includes 123 low-altitude aerial videos. The primary evaluation metric is the Success (AUC) score.Our model does not include specific modules designed for small targets, but in the UAV123 dataset, PMOSTrack-256* outperforms the baseline OSTrack by 1%, increasing from 68.3% to 69.3%, as shown in Table 2. This indicates that our model has strong generalization capability.However, since our model is primarily designed to tackle occlusion issues, its overall performance is not as high as some of the other top-performing trackers.

Table 2. Comparison with state-of-the-art methods on additional benchmarks in AUC score.

Method	UAV123
PMOSTrack-256*(Ours)	69.3
MCITrack-B224 [21]	70.5
ARTrackV2-256 [17]	69.9
AQATrack-256 [15]	70.7
EVPTrack-224 [16]	70.2
LoRAT-B224 [24]	72.6
ARTrack-256 [17]	67.7
SeqTrack-B256 [18]	69.2
VideoTrack-256 [12]	69.7
OSTrack-256 [9]	68.3
Mixformer-22k [25]	70.4

4.2 Ablation Study

In this ablation study, PMOSTrack-256 is used as the baseline to evaluate the impact of the positional heatmap and the Mamba-based memory module on tracking performance.

Table 3. Ablation results on template scale configurations. AO: Average Overlap; $SR_{0.5}$ and $SR_{0.75}$: Success Rates at thresholds 0.5 and 0.75; Δ: AO change relative to our best results.

#	Ref. Scale	Dyn. Scale	AO	$SR_{0.5}$	$SR_{0.75}$	Δ
1	2	1	74.9	84.7	72.8	—
2	1	1	72.0	81.6	68.4	-2.9
3	2	2	73.8	83.7	71.7	-1.1

Positional Heatmap. The positional heatmap is a crucial component in PMOSTrack, encoding the spatial location of the target object. To examine how the scales of the dynamic template (**Dyn. Scale**) and the reference template (**Ref. Scale**) affect performance, ablation experiments are performed with different scale settings. The baseline uses a reference template scale of 2 and a dynamic template scale of 1. As reported in Table 3, reducing the reference template scale from 2 to 1 causes a significant performance drop (AO decreases by 2.9). In contrast, increasing the dynamic template scale from 1 to 2 results in a smaller decline (AO decreases by 1.1). These findings confirm that the best performance is achieved by adopting a larger reference template (scale 2) while keeping the dynamic template at its original scale (scale 1).

Table 4. Ablation study on the number of Mamba layers in the memory module.

#	Method	AO	$SR_{0.5}$	$SR_{0.75}$	Δ
1	Ours(12 Mamba blocks)	74.9	84.7	72.8	-
2	0 Mamba blocks	72.9	82.7	70.5	-2.0
3	3 Mamba blocks	73.9	83.8	71.6	-1.0
4	6 Mamba blocks	74.8	85.0	72.3	-0.1
5	24 Mamba blocks	73.9	83.7	71.9	-1.0

Mamba-based memory module. The historical memory module is a critical component of PMOSTrack, designed to store and utilize long-term target features. To evaluate its effectiveness, ablation experiments are conducted by varying the number of Mamba layers in the memory module. The baseline configuration uses 12 Mamba layers. As shown in Table 4, reducing the layers to 0 or 3 significantly decreases performance, as measured by average overlap, indicating the model struggles to capture long-term dependencies. Increasing the layers to 24 also harms performance, likely due to overfitting. These results confirm that the 12-layer configuration provides the best performance, making it the optimal choice for long-term temporal modeling in the proposed framework. The baseline configuration uses 12 Mamba layers. As shown in Table 4, reducing the layers to 0 or 3 significantly decreases performance, as measured by average overlap, indicating the model struggles to capture long-term dependencies. Increasing the layers to 24 also harms performance, likely due to overfitting. These results confirm that the 12-layer configuration provides the best performance, making it the optimal choice for long-term temporal modeling in our framework.

Table 5. Number of tracking target images and AO values input to the Mamba-based memory module layer during testing on the GOT10K dataset using the PMOSTrack-256* method.

N	2	3	4
AO	75.0	76.7	75.4

To validate the effectiveness of each component in the PMOSTrack framework, ablation experiments are conducted on standard benchmarks. Specifically, the contribution of the positional heatmap encoder and the Mamba-based memory module is evaluated.

Experimental results (as shown in Table 6) demonstrate that injecting the positional heatmap into the encoder significantly improves tracking performance, even without temporal modeling. The average overlap increases from 71.0 to 73.9, highlighting that explicit spatial priors enhance the model's localization accuracy and trajectory consistency. The introduction of the Mamba-based memory module also yields clear benefits. With an AO of 73.7, this module effectively captures

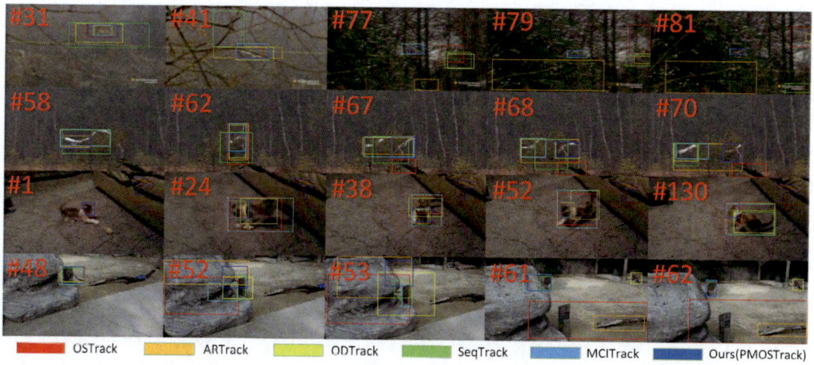

Fig. 3. Comparative visualization of tracking results from six methods on the GOT-10k dataset. The trackers include OSTrack, ARTrack, ODTrack, SeqTrack, MCITrack, and our method PMOSTrack. PMOSTrack demonstrates robust performance under partial and full occlusion, as well as pose variation, consistently maintaining accurate localization of the target.

Table 6. Ablation study on the effectiveness of the positional heatmap and Mamba-based memory module. Both components contribute significantly to tracking performance, with the best results achieved when used jointly.

#	Method	AO	$SR_{0.5}$	$SR_{0.75}$	Δ
1	Baseline	71.0	80.4	68.2	-
2	with positional heatmap	73.9	83.7	71.4	+2.9
3	with mamba Layer	73.7	83.7	71.4	+2.7
3	Ours(both)	74.9	84.7	72.8	+3.9

long-term temporal dependencies and mitigates the limitations of conventional short-term trackers. When both components are integrated, the model achieves the best overall performance, reaching an AO of 74.9—a 3.9-point gain over the baseline. These findings confirm that spatial and temporal modules are complementary, jointly improving both feature representation and tracking robustness. In summary, this study validates the effectiveness of the proposed PMOSTrack framework in unifying positional awareness with long-term temporal modeling for enhanced target tracking.

4.3 Visualization and Limitations

Visualization. To intuitively demonstrate the effectiveness of the proposed PMOSTrack, especially under challenging scenarios such as partial or full occlusion, tracking results are visualized alongside five state-of-the-art trackers on the GOT-10k dataset, as illustrated in Fig. 3. In the first and second rows, which depict partial occlusion scenarios, our method consistently maintains precise tracking of the target. In the fourth row, where the target is fully occluded,

PMOSTrack effectively estimates the potential movement trajectory, enabling robust re-identification and continuous tracking upon reappearance. In the third row, the initial target is the cat's head. While some competing trackers gradually shift their attention to the entire body due to pose variation, our method remains focused on the designated head region, benefiting from its memory-based target representation mechanism.

Visualization Analysis. Fig. 3 provides a comparative visualization of PMOSTrack and five other competitive trackers on the GOT-10k dataset. PMOSTrack demonstrates clear superiority in various challenging tracking scenarios. In rows 1 and 2, with partial occlusion, it maintains robust tracking even when the object is partially obscured. In row 4, under full occlusion, PMOSTrack predicts the target's trajectory and resumes accurate tracking once the target reappears. In row 3, where the target is a cat, competing methods shift attention to the entire body as the cat changes posture, while PMOSTrack, with its target memory module, consistently tracks the head region, ensuring precise tracking throughout.

Limitations. The proposed framework combines positional heatmap encoding and the Mamba-based memory module, but lacks a more effective way to integrate these two sources of information. In complex, crowded scenes with appearance-similar distractors and occlusions, the model struggles to accurately localize the target. Additionally, the main drawback of this model is that all the tokens from different modules must undergo attention operations in the Encoder, leading to a large number of parameters and slower processing speed, as shown in Table 7. Therefore, future work will focus on optimizing the parameter size and improving the network structure to enhance efficiency, reduce computational overhead, and further improve performance in single-object tracking tasks.

Table 7. Comparison of the FPS between the baseline model OSTrack and our model.

Method	FPS
PMOSTrack-256*(Ours)	18.6
OSTrack-256 [9]	100.4

5 Conclusion

In this paper, PMOSTrack is proposed, a novel video object tracking framework that effectively addresses the limitations of conventional approaches by integrating visualized positional information with historical modeling based on Mamba. Specifically, the target's spatial coordinates are transformed into images, which helps preserve the spatial structure of its motion trajectory and enhances robustness under occlusion scenarios. Meanwhile, a multi-layer Mamba memory module, grounded in state space modeling, is employed to store target features over long temporal spans through hidden states. We hope that this work inspires further research in video-level tracking modeling. In addition, future research could focus on the direction of model lightweighting to further enhance the system's efficiency and adaptability.

References

1. Yan, B., Peng, H., Fu, J.: Learning spatio-temporal transformer for visual tracking. In: Proceedings of the IEEE/CVF International Conference on Computer Vision (ICCV), pp. 10448–10457. (2021)
2. Gao, S., Zhou, C., Ma, C.: Aiatrack: attention in attention for transformer visual tracking. In: European Conference on Computer Vision (ECCV), pp. 146–164. (2022)
3. Borsuk, V., Vei, R., Kupyn, O.: Fear: fast, efficient, accurate and robust visual tracker. In: European Conference on Computer Vision (ECCV), pp. 644–663 (2022)
4. Kawakami, K.: Supervised sequence labelling with recurrent neural networks (2008). Ph.D. dissertation Ph. D. thesis
5. Gu, A., Dao, T.: Mamba: linear-time sequence modeling with selective state spaces (2023). ArXiv preprint arXiv:2312.00752
6. Huang, L., Zhao, X., Huang, K.: Got-10k: a large high-diversity benchmark for generic object tracking in the wild. IEEE Trans. Pattern Anal. Mach. Intell. **43**(5), 1562–1577 (2019)
7. Bertinetto, L., Valmadre, J., Henriques, J.F.: Fully-convolutional Siamese networks for object tracking. In: ECCV Workshops, pp. 850–865 (2016)
8. Li, B., Yan, J., Wu, W.: High performance visual tracking with siamese region proposal network. In: Proceedings of the IEEE Conference on Computer Vision and Pattern Recognition (CVPR), pp. 8971–8980. (2018)
9. Ye, B., Chang, H., Ma, B.: Joint feature learning and relation modeling for tracking: a one-stream framework. In: European Conference on Computer Vision (ECCV), pp. 341–357 (2022)
10. Yaseen, S.A., Sasi, S.: Robust algorithm for object detection and tracking in a dynamic scene. J. Image Graph. **2**(1), 41–45 (2014)
11. Cao, Z., Huang, Z., Pan, L.: Temporal contexts for aerial tracking. In: Proceedings of the IEEE/CVF Conference on Computer Vision and Pattern Recognition (CVPR), pp. 18–24 (2022)
12. Xie, F., Chu, L., Li, J.: Videotrack: learning to track objects via video transformer. In: Proceedings of the IEEE/CVF Conference on Computer Vision and Pattern Recognition (CVPR), pp. 22826–22835 (2023)
13. Vaswani, A., Shazeer, N., Parmar, N.: Attention is all you need. Adv. Neural Inf. Proc. Syst, 30 (2017)
14. Zheng, Y., Zhong, B., Liang, Q., et al.: Odtrack: online dense temporal token learning for visual tracking. In: Proceedings of the AAAI Conference on Artificial Intelligence, vol. 38, no. 7, pp. 7588–7596 (2024)
15. Xie, J., Zhong, B., Mo, Z.: Autoregressive queries for adaptive tracking with spatio-temporal transformers. In: Proceedings of the IEEE/CVF Conference on Computer Vision and Pattern Recognition (CVPR), pp. 19300–19309 (2024)
16. Shi, L., Zhong, B., Liang, Q., et al.: Explicit visual prompts for visual object tracking. Proce. AAAI Conf. Artif. Intell. **38**(5), 4838–4846 (2024)

17. Wei, X., Bai, Y., Zheng, Y.: Autoregressive visual tracking. In: Proceedings of the IEEE/CVF Conference on Computer Vision and Pattern Recognition (CVPR), pp. 9697–9706 (2023)
18. Chen, X., Peng, H., Wang, D.: Seqtrack: sequence to sequence learning for visual object tracking. In: Proceedings of the IEEE/CVF Conference on Computer Vision and Pattern Recognition, pp. 14572–14581 (2023)
19. Liu, S., Yang, Z., Li, Q., et al.: Intermamba: a visual-prompted interactive framework for dense object detection and annotation. IEEE Trans. Geosc, Remote Sens (2025)
20. Dosovitskiy, A., Beyer: An image is worth 16x16 words: transformers for image recognition at scale, (2020). ArXiv preprint arXiv:2010.11929
21. Kang, B., Chen, X., Lai, S., et al.: Exploring enhanced contextual information for video-level object tracking. In: Proceedings of the AAAI conference on artificial intelligence, vol. 39, no. (4), pp. 4194–4202 (2025)
22. Fan, H., Lin, L., Yang, F.: Lasot: a high-quality benchmark for large-scale single object tracking. In: Proceedings of the IEEE/CVF Conference on Computer Vision and Pattern Recognition, pp. 5374–5383 (2019)
23. Muller, M., Bibi, A., Giancola, S.: Trackingnet: a large-scale dataset and benchmark for object tracking in the wild. In: Proceedings of the European Conference on Computer Vision (ECCV), pp. 300–317 (2018)
24. Lin, L., Fan, H., Zhang, Z.: Tracking meets lora: faster training, larger model, stronger performance. In: European Conference on Computer Vision, pp. 300–318. Springer (2024)
25. Cui, Y., Jiang, C., Wang, L.: Mixformer: end-to-end tracking with iterative mixed attention. In: Proceedings of the IEEE/CVF Conference on Computer Vision and Pattern Recognition, pp. 13608–13618 (2022)
26. Benchmark, U.: A benchmark and simulator for uav tracking. In: European Conference on Computer Vision, vol. 7 (2016)

POINT2VERTEX: Transformer-Based Convex Hull Vertex Prediction in Point Clouds

Rui-Juan Jing[1], Yuegang Zhao[1], and Changbo Chen[2(✉)]

[1] School of Mathematical Sciences, Jiangsu University, Zhenjiang, Jiangsu, China
rjing@ujs.edu.cn, ygzhao@stmail.ujs.edu.cn

[2] Automated Reasoning and Cognition Center, Chongqing Institute of Green and Intelligent Technology, Chinese Academy of Sciences, Liangjiang New Area, Chongqing, China
chenchangbo@cigit.ac.cn

Abstract. Identifying the convex hull of a point cloud is a fundamental geometric operation with wide-ranging applications in shape analysis, computational geometry, and robotics. While classical algorithms such as Quickhull are efficient in low-dimensional spaces, they suffer from single-exponential time complexity and become computationally intractable and unstable in high-dimensional settings. In this work, we propose POINT2VERTEX, a novel Transformer-based architecture that directly identifies convex hull vertices by framing the task as a binary vertex classification problem. Our model introduces linear embedding to capture local structural dependencies of the coordinates of a point, leverages self-attention mechanisms to capture global context information among different points, and employs unified padding to handle multi-dimensional input point clouds. We also substantially reduce data acquisition time through facilitating high-dimensional synthetic data creation. Experimental results show that POINT2VERTEX can predict the vertices of convex hulls with nearly perfect accuracy and outperforms Quickhull in high-dimensional cases.

Keywords: Convex hull · Vertex · Transformer · Data generation · Point cloud

1 Introduction

Convex hull extraction is a fundamental task in computational geometry, playing a pivotal role in areas such as object boundary detection [1–3], mesh generation [4,5], point registration [6,7], robotics [8], polyhedral optimization [9,10], machine learning [11,12]. Classical convex hull algorithms include Quickhull [13], Graham scan [14], the Beneath-Beyond method [15], Fourier–Motzkin elimination [16], and the double description method [17,18], among others. Implementations of these algorithms are widely available in popular computational software

L. Huang and D. Greenhalgh (Eds.): ICMLC 2026, LNNS 2061, pp. 131–146, 2027.
https://doi.org/10.1007/978-3-032-31236-5_9

and libraries such as Maple, Mathematica, Sage, and Python. However, these traditional geometric algorithms are well-established low-dimensional settings. They often exhibit exponential complexity growth and struggle with redundant or incomplete data, which are prevalent challenges in high-dimensional, real-world point clouds.

Early attempts to compute convex hulls using classical machine learning tools date back decades [19,20]. Recent advancements in deep learning for geometric data have further propelled the adoption of learning-based methods for convex hull extraction or approximation. Recent frameworks include Graph Neural Networks (GNN) [21] and Recurrent Convolutional Networks (RCN) [22], which have been effectively applied to such tasks. The issue of over-parameterization in deep learning models, as discussed in [23], also warrants attention in this context. Another influential line of work introduced the Pointer Network [24], which integrates the seq2seq paradigm [25], the attention mechanism [26], and a pointer mechanism. This architecture enables the processing of input sequences of variable length. Meanwhile, a deep neural network framework was proposed to approximate convex hulls in high-dimensional spaces [27]. More recently, a hybrid model combining the Transformer and Pointer Network was developed to enhance both the efficiency and accuracy of convex hull computation [28]. To the best of our knowledge, existing trained models for convex hull computation are constrained to input data of the same dimension as their training data. When the input dimensionality changes, a new model must be trained from scratch using appropriately dimensioned data. Due to the high computational cost of generating exact convex hull labels, it is particularly challenging to create labeled datasets in high-dimensional spaces. Consequently, most existing studies are limited to low-dimensional settings. To date, there has been no work that explicitly addresses the scarcity of labeled data in high-dimensional spaces. However, high-dimensional convex hull computation plays a vital role in several fields, such as redundancy testing in information systems [29,30].

In this paper, we address these gaps by introducing POINT2VERTEX, a Transformer-based model [26] tailored for convex hull vertex selection. Note that the convex hull of a point cloud is defined as the minimal convex polyhedron that encloses all points, with its vertices forming a subset of the original point set. The identification of these vertices is inherently dependent on the global spatial distribution of the points while being invariant to their input order. Recently, Transformer-based architectures have demonstrated strong capabilities in capturing global dependencies through self-attention mechanisms, making them naturally suited for such permutation-invariant geometric reasoning tasks. To summarize, our contributions are two-fold:

- *(i)* We propose POINT2VERTEX, a Transformer-based framework that identifies convex hull vertices via point-wise binary classification, supporting multi-dimensional point clouds with unified padding and linear embedding.
- *(ii)* We propose novel methods for synthesizing labelled datasets that avoiding expensive classical convex hull computations in high-dimensional spaces.

The remainder of this paper is organized as follows. Section 2 reviews the necessary background. Section 3 presents the architecture and formulation of POINT2VERTEX. Section 4 details the proposed data generation strategies. Experimental setup and results are discussed in Sect. 5. Finally, Sect. 6 concludes the paper and suggests directions for future research.

2 Preliminaries

In this section, we provide the basic knowledge about the polyhedron and convex hull. One can find detailed information in Refs. [13,31].

A set of vectors $P \subseteq \mathbb{Q}^n$ is called a *convex polyhedron* if it can be written as $P = \{\mathbf{x} \mid A\mathbf{x} \leq \mathbf{b}\}$, for some matrix $A \in \mathbb{Q}^{m\times n}$ and vector $\mathbf{b} \in \mathbb{Q}^n$. If P is bounded, it is called a *polytope*. A point $\mathbf{x} \in \mathbb{Q}^n$ is said to be a *convex combination* of points $\mathbf{x}_1, \mathbf{x}_2, \ldots, \mathbf{x}_t$ (with $t > 1$) if there exist non-negative coefficients $\lambda_1, \lambda_2, \ldots, \lambda_t$ such that $\mathbf{x} = \sum_{i=1}^{t} \lambda_i \mathbf{x}_i$ and $\sum_{i=1}^{t} \lambda_i = 1$. For a polytope P, a point that cannot be expressed as a convex combination of any other points of P is called a *vertex*. Every polytope possesses finitely many vertices. Denote the vertices of the polytope P by $\mathbf{v}_1, \mathbf{v}_2, \ldots, \mathbf{v}_s$ for some positive integer s. The whole polytope can then be described as the set of all convex combinations of its vertices: $P = \{\sum_{i=1}^{s} \lambda_i \mathbf{v}_i \mid \lambda_i \geq 0,\ \sum_{i=1}^{s} \lambda_i = 1\}$.

The description $\{A\mathbf{x} \leq \mathbf{b}\}$ is referred to as the *H-representation* (half-space representation) of P, while the expression via vertices is called the *V-representation* (vertex representation). These two representations are dual to each other and form what is known as the double description of a polytope. Throughout this paper, we refer to a collection of points as a *point cloud*. The *convex hull* of a point cloud is the smallest polytope that encloses all points. The vertices of this hull form a subset of the original points, and every other point in the set can be expressed as a convex combination of these vertices.

3 Method

In this section, we introduce the Transformer-based framework developed to address the problem of predicting convex hull vertices from point cloud data.

3.1 Transformer-based Framework for POINT2VERTEX

The experimental framework for convex hull vertex prediction on point clouds is illustrated in Fig. 1. The overall workflow consists of two main stages: data preprocessing and Transformer-based model. The data preprocessing stage constructs input-output sequence pairs for the model. Through the padding technique, point clouds of varying dimensions can be handled uniformly. The output in each pair is a binary label sequence, where the i-th entry equals 1 if the i-th point is a vertex of the convex hull, and 0 otherwise. These input-output pairs are then fed into the Transformer model for both training and testing.

3.2 Data Preprocessing for Point Cloud Representation

The data preprocessing aims to convert raw data into input-output sequence pairs suitable for model. Consider a point cloud $P = \{\mathbf{p}_1, \ldots, \mathbf{p}_N\}$, where N is the number of input points and each $\mathbf{p}_i$ is a row vector of dimension d_P. The associated vertex label vector is denoted by ℓ. Specifically, the i-th element ℓ_i of ℓ indicates whether $\mathbf{p}_i$ is a vertex of the convex hull: $\ell_i = 1$ if $\mathbf{p}_i$ is a vertex, and $\ell_i = 0$ otherwise. If the labels are unknown, we compute them using the Scipy library in Python. Let max_norm denote the maximum norm among all points in P. We normalize the point cloud to $P_{\text{norm}} = \{\mathbf{p}_1/\text{max\_norm}, \ldots, \mathbf{p}_N/\text{max\_norm}\}$, so that each point lies within the unit sphere.

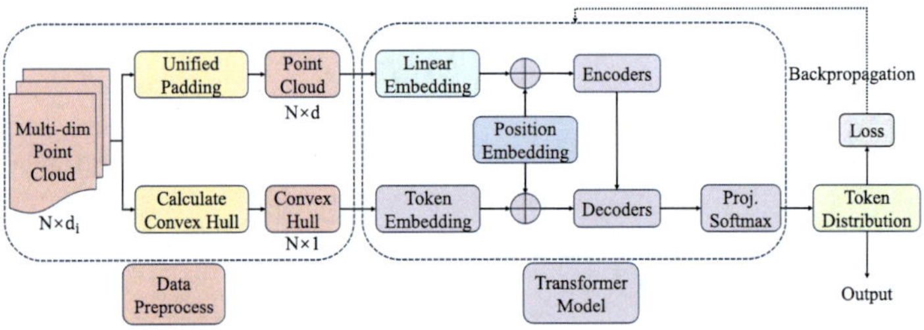

Fig. 1. The framework of POINT2VERTEX.

Recall that our model is designed to handle data with varying dimensions. Therefore, we must first address the dimensional discrepancy. To this end, we define a sufficiently large fixed dimension D ($\geq d_P$). We then concatenate P_{norm} with $N \times (D - d_P)$ zero padding at the end, obtaining a padded point cloud $P^{\text{pad}}_{\text{norm}}$ of size $N \times D$. Given a batch size B, we also need to handle point clouds with varying numbers of points within the batch. To unify them, we take the maximal number of points across the batch, denoted as N. When a point cloud contains fewer than N points, we extend it to length N by adding padding token $\langle\text{pad}\rangle$. This yields an input tensor of size $B \times (N+2) \times D$, where D is the feature dimension. For each point cloud, we prepend a start token $\langle\text{s}\rangle$ and append an end token $\langle/\text{s}\rangle$; Consider a point cloud P from a batch with $B = 3$ samples. Suppose the maximal number of points among the three point clouds is $N = 10$, and let $D = 5$. The resulting input tensor will have size $3 \times 12 \times 5$. The first layer corresponding to the point cloud P is given by the point sequence $P^{\text{input}}_{\text{tensor}}$. The output tensor is constructed in a similar manner. The labels for points in the point cloud P are "1", "1", "1", "1", "1", "0", "1", "0". Thus, the output tensor corresponding to this cloud is $P^{\text{output}}_{\text{tensor}}$.

$$P = \begin{bmatrix} (-0.13, 0.73, 0.22) \\ (0.26, 0.74, 0.56) \\ (-0.13, -0.95, -0.23) \\ (0.94, 0.01, 0.18) \\ (0.55, 0.22, 0.64) \\ (0.45, -0.07, 0.24) \\ (0.05, -0.08, -0.57) \\ (0.45, 0.29, 0.46) \end{bmatrix} \quad P^{\text{input}}_{\text{tensor}} = \begin{bmatrix} \langle \text{s} \rangle \\ (-0.13, 0.73, 0.22, 0, 0) \\ (0.26, 0.74, 0.56, 0, 0) \\ (-0.13, -0.95, -0.23, 0, 0) \\ (0.94, 0.01, 0.18, 0, 0) \\ (0.55, 0.22, 0.64, 0, 0) \\ (0.45, -0.07, 0.24, 0, 0) \\ (0.05, -0.08, -0.57, 0, 0) \\ (0.45, 0.29, 0.46, 0, 0) \\ \langle /\text{s} \rangle \\ \langle \text{pad} \rangle \\ \langle \text{pad} \rangle \end{bmatrix} \quad P^{\text{output}}_{\text{tensor}} = \begin{bmatrix} \langle \text{s} \rangle \\ \text{“1”} \\ \text{“1”} \\ \text{“1”} \\ \text{“1”} \\ \text{“1”} \\ \text{“0”} \\ \text{“1”} \\ \text{“0”} \\ \langle /\text{s} \rangle \\ \langle \text{pad} \rangle \\ \langle \text{pad} \rangle \end{bmatrix}$$

In the data preprocessing stage, we transform both point clouds and their corresponding convex hull vertices into sequential representations. Specifically, each token in the label sequence is represented as a string, consistent with conventional language model tokens. In contrast, within the point cloud sequence, all tokens except the start(⟨s⟩), end(⟨/s⟩), and padding(⟨pad⟩) tokens correspond to actual points in the spatial domain. Representing each point as a token enables the model to better capture the sequential nature and geometric relationships inherent in point cloud data.

3.3 Embedding Strategy

As described in Sect. 3.2, our point cloud data are converted into a seq2seq format. The tokens involved in both the input and output sequences are summarized in Table 1. As shown, both sequences include three special tokens: the start token, the end token, and the padding token.

Table 1. Tokenization and encoding scheme.

Special tokens	Labels	Input point
⟨s⟩, ⟨/s⟩, ⟨pad ⟩	0, 1	coordinates $(x_1, \ldots, x_D)$

Specifically, in the input sequence, all tokens except these three special tokens correspond to D-dimensional points (where D denotes the maximum dimension defined in Sect. 3.2 for padding). In the output sequence, apart from the three special tokens, each token takes a binary value, either “0” or “1”, indicating whether the corresponding point is a vertex of the convex hull.

Given the seq2seq nature of the Transformer, input and output sequences are encoded for training and evaluation. The encoder maps each token to a feature vector of dimension *emb_dim*. In language tasks, words are indexed within a unified vocabulary, and an embedding matrix maps each index to its corresponding feature vector, allowing direct retrieval after the embedding layer is defined.

Following this classical token-embedding scheme, we construct a vocabulary covering all discrete tokens listed in Table 1. Using standard embedding techniques, these discrete tokens can be encoded into their corresponding feature representations. Consequently, for the output sequence, where each token is discrete ("0", "1", or special tokens), we adopt the conventional embedding method used in the Transformer decoder input layer. However, a key distinction arises in the input sequence. Among its tokens, only three are discrete and can thus be embedded through vocabulary lookup. The remaining tokens correspond to D-dimensional points in continuous space, which cannot be enumerated or represented via a discrete vocabulary. To address this, we reinterpret the essence of embedding as a mapping from input elements to feature vectors. Accordingly, we introduce a linear layer of size $D \times emb\_dim$ to project each point into the embedding space. In this manner, every token can be transformed into the feature representation required by the Transformer model.

3.4 An Example

Figure 2 shows the performance on a sample from the generated dataset. We randomly generate 280 points uniformly inside a 3D unit sphere. Of these, 69 points are identified as convex hull vertices (marked by green squares). Our model predicts 66 vertices (marked by red triangles), all of which are correctly identified.

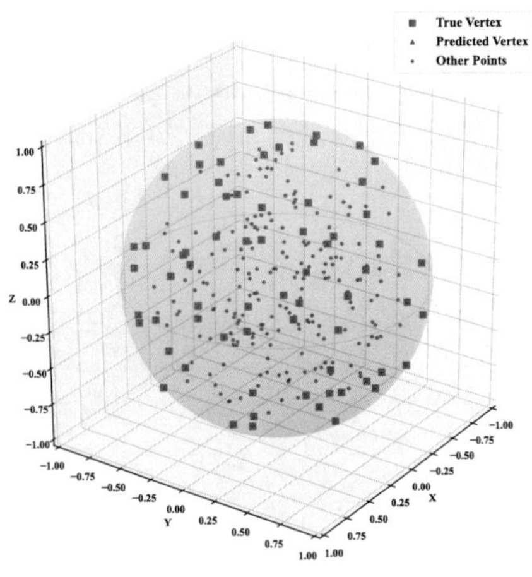

Fig. 2. A convex hull sample.

In addition, as shown in Fig. 3, the sample is projected onto three orthogonal planes. We compute 2D convex hulls from projected vertices of predicted and ground-truth shapes. The strong overlap across all planes indicates the model captures true geometric characteristics rather than merely fitting the data distribution.

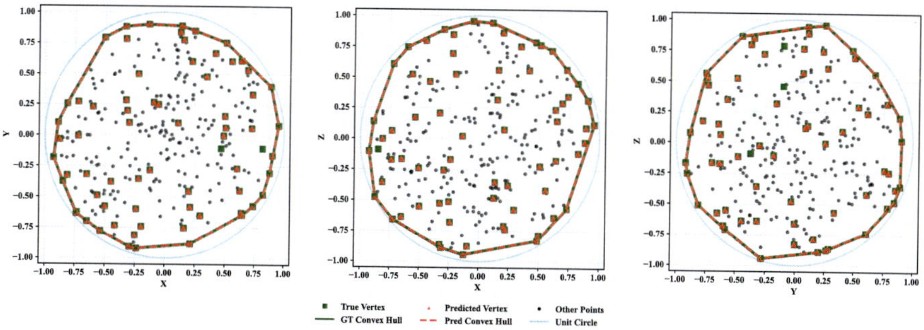

Fig. 3. Three projections of the convex hull sample.

4 Dataset

In this section, we describe the dataset created for training and testing our model. All data points are confined within a unit sphere. For an arbitrary point cloud, we apply the normalization method introduced in Sect. 3.2 to ensure all points lie inside a unit sphere. We propose three methods for constructing the dataset. The dimensions ranage from 3 to 12.

We generate random samples called RANDOMOID, where points are randomly sampled within a unit sphere, and convex hulls are computed using Scipy to obtain labels. Due to computational inefficiency in high dimensions, we constrain the data to two geometric configurations: spherical and ellipsoidal. In the spherical configuration, points are generated on the n n-dimensional unit sphere surface as convex hull vertices, and random convex combinations of these vertices form labeled datasets called SPHEROID. In the ellipsoidal configuration, we apply anisotropic scaling and uniform rotational transformation to SPHEROID data, resulting in ellipsoidal datasets called ELLIPSOID.

For each sample generated by any of the three methods, the number of points is randomly chosen between 99 and 999. Both SPHEROID and ELLIPSOID can be efficiently generated in arbitrarily high dimensions. The proportion of convex hull vertices for each sample generated by these two methods is randomly sampled from the range $[0.25, 0.5]$. For each dimension in these two categories, 20,000 samples are produced. Samples generated by RANDOMOID, however, suffer from the curse of dimensionality: the feasible number of samples decreases as the dimension increases. The timelimit for each dimension are set to 12 h. The number of samples in each dataset, along with the average generation time for RANDOMOID samples, are listed in Table 2.

Table 2. Information of the datasets.

Dimension	3D	4D	5D	6D	7D
Average time (seconds) of RANDOMOID	0.0005	0.0026	0.024	0.1891	2.1861
Size of RANDOMOID/SPHEROID/ELLIPSOID	20,000	20,000	20,000	20,000	20,000
Dimension	8D	9D	10D	11D	12D
Average time (seconds) of RANDOMOID	19.94	147.06	1456.84	5196.85	*Timeout*
Size of RANDOMOID	2,403	325	35	8	0
Size of SPHEROID/ELLIPSOID	20,000	20,000	20,000	20,000	20,000

5 Experiments

In this section, we present the experimental results, including the analysis of accuracy and tine performance and the study of fine-tuning strategies.

5.1 Experimental Settings

Experiments are performed on a workstation with an Intel Xeon Platinum 8272CL CPU @ 2.60 GHz, 256 GB RAM, and an NVIDIA RTX 4090 GPU, running Ubuntu 22.04. The software environment includes Python 3.10 and Maple 2024.

Table 3. Summary of model architecture and training parameters.

Category	Parameter	Value/Description
Model architecture	Embedding dimension	256
	Attention heads	8
	Encoder layers	6
	Decoder layers	6
Training configuration	Batch size	8 (train) or 256 (predict)
	Epoch definition	Size of training set
	Epochs	30 or 6 or 3
Optimization	Optimizer	Adam
	Learning rate	0.0001 (fixed)
Regularization	Dropout rate	0.1
	Model selection	With highest validation accuracy
	Adapted parameters	Remaining parameters
Dataset storage	Parquet format	Accelerate data loading

In this study, all models are built upon a Transformer-based architecture. The architectural specifications and training configurations, including batch size, optimizer settings, dropout rate, and early stopping criteria, are summarized in Table 3. Following the data processing strategy outlined in Sect. 3.2, we utilized the HuggingFace Datasets library for data loading and storage, adopting the Apache Parquet format. This columnar storage format supports efficient compression and parallel processing, significantly improving data-loading performance. Compared with plain text storage, it greatly accelerates data access and facilitates the efficient reuse of training data across multiple tasks, thereby enhancing the overall efficiency of the experimental workflow.

We now provide a detailed explanation of the evaluation criteria for the model. Let $\mathcal{A}$ denote the set of s test samples (point clouds), where each sample has N_i points, for $i \in \{1, 2, \dots, s\}$. For each sample, let M_i represent the number of correctly predicted points (including both convex hull vertices and non-vertex points). We introduce the following two evaluation metrics:

$$\text{Token Accuracy:} \quad A_{\text{token}} = \frac{\sum_{i=1}^{s} M_i}{\sum_{i=1}^{s} N_i}.$$

This accuracy measures the proportion of points which are correctly classified as vertices/non-vertices by the model on the set $\mathcal{A}$. It reflects the (average) accuracy of the generated convex hulls w.r.t. the exact convex hulls.

$$\text{Sequence Accuracy:} \quad A_{\text{sequence}} = \frac{\sum_{i=1}^{s} \mathbf{1}(M_i = N_i)}{s}.$$

This accuracy measures the proportion of samples for which the model correctly predicted all its points as vertices/non-vertices. In other words, it reflects the proportion of correctly generated exact convex hulls.

It is important to note that during model training, the checkpoint selection is based on token accuracy of the validation set. During this phase, sequence accuracy is not used. Once the model enters the inference phase, we continue to report token accuracy, and in addition, we introduce the sequence accuracy metric to provide a more comprehensive evaluation of model performance.

In addition, we measure the model's inference time to provide a comprehensive evaluation from multiple perspectives. Specifically, since the Transformer architecture supports parallel inference, we maximize the batch size within GPU memory constraints. For dataset A, the inference process is divided into b batches, and the runtime of each batch is denoted as t_j ($j \in \{1, \dots, b\}$). Accordingly, the average inference time per sample can be expressed as:

$$\text{Average Sample Time:} \quad T_{\text{sample}} = \frac{\sum_{j=1}^{b} t_j}{s}.$$

5.2 Evaluation in Low Dimensional Settings

As shown in Table 2, for RANDOMOID dataset, the number of samples drops below 10,000 when dimension exceeds 7. After data splitting, such a small sample size is insufficient for effective model training. Therefore, we restrict our experiments ranging from 3 to 7. The data for these dimensions are split in a ratio of 7:1:1:1 into the training set, validation set, test-validation set, and test set.

Following the preprocessing procedure described in Sect. 3.2, we perform unified training, validation, and testing across all five dimensions. Notably, the validation set is used during training under the teacher-forcing mechanism to evaluate model performance and guide model selection based on accuracy. The test-validation set is employed after training to assess the model under the auto-regressive decoding mechanism, with its results used for parameter tuning. Finally, the test set is reserved for final evaluation after all model parameters and experimental configurations are fixed. After joint training on the combined five-dimensional data, we further evaluate the model on each individual dimension to assess its cross-dimensional generalization ability.

Table 4. Token accuracy on low-dimensional point clouds.

Dataset	Epoch	Category	Whole (%)	3D (%)	4D (%)	5D (%)	6D (%)	7D (%)
RANDOMOID	30	Valid	98.07	98.13	97.75	97.65	98.17	98.66
		Test-valid	98.07	98.11	97.75	97.68	98.15	98.66
		Test	98.08	98.12	97.73	97.66	98.20	98.67
SPHEROID	3	Valid	99.91	99.90	99.91	99.91	99.90	99.91
		Test-valid	99.89	99.89	99.89	99.89	99.89	99.88
		Test	99.89	99.90	99.87	99.87	99.90	99.89
SPHEROID/ELLIPSOID	6/3	Valid	99.99	99.99	99.99	99.99	99.99	99.99
		Test-valid	99.99	99.99	99.99	99.99	99.99	99.99
		Test	99.99	99.99	99.99	99.99	99.99	99.99

Table 4 presents the token accuracy on three type of datasets in the low dimensional setting. Overall, all three datasets achieve exceptionally high accuracy, exceeding 97% in all cases. Notably, datasets with more structured geometric constraints, namely, the spherical (SPHEROID) and ellipsoidal (ELLIPSOID) configurations, demonstrate slightly higher accuracy than the fully random RANDOMOID dataset. This trend indicates that incorporating geometric regularity in the data distribution facilitates more consistent convex-hull prediction, as the model can better capture the underlying spatial relationships. Furthermore, the accuracy values of the *valid* (teacher-forcing) and *test-valid* (autoregressive) evaluations are nearly identical across all datasets and dimensions. This close alignment suggests that the model maintains stable performance between the teacher-forced and autoregressive decoding modes, reflecting strong generalization ability and minimal exposure bias.

In summary, the model achieves highly reliable performance across different geometrical data distributions, with structured datasets (SPHEROID and ELLIPSOID) yielding near-perfect accuracy and demonstrating the robustness of the proposed encoding and training scheme.

5.3 Generalization Performance in High-Dimensional Settings

Section 5.2 demonstrates that the model trained on mixed-dimensional data exhibits strong inference capability within its training dimensional range. A natural question arises: can such a model generalize directly to higher-dimensional data?

To investigate this, we apply the models trained only on lower dimensions (from 3D to 7D) directly to higher-dimensional datasets. Note that all reported results are obtained on the high-dimensional test-validation and test sets, rather than on the entire dataset. For the SPHEROID and ELLIPSOID datasets, high-dimensional data are sufficiently available. The evaluated token accuracy on them is reported in Table 5.

Table 5. Token accuracy for high-dimensional SPHEROID and ELLIPSOID.

Dataset	Category	Whole (%)	8D (%)	9D (%)	10D (%)	11D (%)	12D (%)
SPHEROID	Test-valid	98.38	99.95	99.20	98.52	97.72	96.82
	Test	98.38	99.65	99.19	98.54	97.75	96.76
ELLIPSOID	Test-valid	99.63	99.92	99.81	99.66	99.47	99.31
	Test	99.63	99.92	99.80	99.65	99.49	99.30

According to Table 5, for datasets with inherent geometric structures (i.e., convex hull vertices lying on circumscribed spheres or ellipsoids), the models trained on lower dimensions still achieve strong generalization on higher dimensions. In contrast, for the RANDOMOID dataset, random data above 9D are extremely scarce; hence, our evaluation focuses on the 8D and 9D datasets. However, models trained solely on lower dimensions exhibit limited inference capability on high-dimensional random data. To address this, we adopt a fine-tuning strategy: both 8D and 9D datasets are partitioned into training, validation, and test subsets, and the pretrained low-dimensional models are fine-tuned with their training datasets. The results are presented in Table 6.

Table 6. Token accuracy for high-dimensional RANDOMOID.

Category	8D (%)	8D (fine-tune) (%)	9D (%)	9D (fine-tune) (%)
Test-valid	85.65	98.45	78.62	98.61
Test	85.75	98.54	77.21	98.43

As shown in the table, the fine-tuning strategy effectively compensates for the data scarcity in high dimensions, achieving satisfactory performance while maintaining high computational efficiency.

5.4 Comparison of running time

Table 7 compares the running time of Quickhull in the Scipy package of Python with the (parallel) inference time of Transformer. The batch size used for Transformer is 128. The experiments are conducted on the testing datasets of RANDOMOID. We observe that the inference time of Transformer is almost not affected by the dimension while the running time of Quickhull grows exponentially with dimensions. Moreover, from dimension 6, Transformer gains a significant advantage over Quickhull in terms of timing efficiency, and the gap becomes more pronounced as the dimension increases.

Table 7. Running time comparison of Quickhull and Transformer.

Method	3D	4D	5D	6D	7D	8D	9D
Quickhull	0.0005	0.0026	0.0238	0.1868	2.2232	20.9335	150.0094
Transformer	0.0589	0.0575	0.0593	0.0592	0.0584	0.0951	0.1782

5.5 The challenge on predicting the exact convex hulls

We have shown in Sects. 5.2 and 5.3 that Transformer models could achieve nearly perfect token accuracies on multi-dimensional point clouds for all the tested datasets when the dimension is no greater than 7. For the Spheroid and Ellipsoid datasets, the model trained on low dimensions shows strong generalization ability on higher dimensions. For Randomoid datasets, token accuracy drops when evaluating the models on 8D and 9D. But after fine-tuning the models trained on low-dimension data with training sets of high dimensions, the token accuracy becomes nearly perfect again. In this subsection, we report on the sequence accuracy of the models, which reveals their ability on predicting the exact convex hulls rather than approximating the convex hulls. As shown in Table 8, the Transformer still performs nearly perfect on ELLIPSOID. Although it is less satisfying on SPHEROID when trained with the default 3 epochs, the performance becomes nearly perfect again if trained with 3 extra epochs. For the dataset RANDOMOID, Transforms shows limited ability on predicting exact convex hulls, which demonstrates the challenge of this task and the limitation of our model on this task.

For the `Randomoid` datasets in 8D and 9D, no exact convex hulls can be correctly predicted with fine-tuning. To improve the performance, we adopt two fine-tuning strategies. First, we fine-tune the pretrained model separately on each dimensional dataset ("FT-only"), resulting in independent fine-tuned models for 8D and 9D, which are then evaluated individually. In addition, since our model architecture is inherently capable of handling data of different dimensions, we also perform joint fine-tuning ("FT-joint", the strategy used also in Table 6) using the combined 8D and 9D datasets, followed by separate evaluations on

Table 8. Sequence accuracy on low-dimensional point clouds.

Dataset	Epoch	Category	Whole (%)	3D (%)	4D (%)	5D (%)	6D (%)	7D (%)
SPHEROID	3	Test-valid	83.73	83.55	83.35	84.00	84.60	83.15
		Test	84.03	84.40	83.65	83.80	84.90	83.40
SPHEROID	6	Test-valid	96.87	96.80	96.60	97.05	97.25	96.65
		Test	97.09	96.90	96.85	97.20	97.35	97.15
ELLIPSOID	3	Test-valid	96.09	96.65	95.15	96.70	95.90	96.05
		Test	96.58	96.90	95.50	97.00	96.90	96.60
RANDOMOID	30	Test-valid	1.38	0.40	0.40	0.25	1.35	4.50
		Test	1.44	0.40	0.10	0.25	1.95	4.50

each dimensional test set. The corresponding results are summarized in Table 9. As can be observed, fine-tuning can improve the performance and the strategy "FT-only" performs better than "FT-joint".

Table 9. Sequence accuracy for 8D and 9D RANDOMOID datasets.

Metric	Category	8D			9D		
		Direct	FT-only	FT-joint	Direct	FT-only	FT-joint
A_s	Test-valid	0.00	5.44	5.02	0.00	6.25	12.50
	Test	0.00	6.28	5.44	0.00	9.38	9.38

As shown in Table 4, the token accuracy on the random dataset is nearly perfect. However, as reported in Table 8, the sequence accuracy remains below 10%. This substantial discrepancy indicates that the model predicts an approximate convex hull rather than an exact one. A closer inspection reveals that the missing 2%–3% in token accuracy arises from a small number of mispredicted points, which are uniformly distributed across almost all samples. As a result, nearly every sample exhibits a slight deviation from the ground truth, leading to a drastic drop in sequence accuracy.

Given the approximate nature of the predicted convex hulls, it is worth noting that a similar perspective has been adopted in [27], where an approximate convex hull algorithm is evaluated using precision and recall metrics based on random uniform sampling. Following this evaluation protocol, the precision and recall is respectively 100% and 99.98% for the sample shown in Fig. 2. From a practical standpoint, the predicted approximate convex hull is nearly indistinguishable from the true convex hull in terms of geometric fidelity and functional performance, highlighting the effectiveness and applicability of our method in real-world scenarios.

6 Conclusion and Future Work

In this work, we present POINT2VERTEX, a Transformer-based framework that learns to identify convex hull vertices via a point-wise binary classification formulation. Experimental evaluations in both low- and high-dimensional settings validate the generalization capability and robustness of POINT2VERTEX. Moreover, in high-dimensional cases, it demonstrate strong superiority over traditional geometric algorithms in terms of computation time while predicting the vertices of the convex hull with high precision.

Several promising directions for future work remain. Firstly, the datasets used in the current work consist of point clouds with dense coordinates. Due to the computational limitations of Quickhull and its implementations, our model is currently confined to dimensions not exceeding 9. Secondly, the duality property between the H- and V-representations (the double description) has not yet been fully exploited in our framework. Moving forward, we plan to extend this work along the following directions: (*i*) Efficient verification methods need to be developed to validate the results produced by Transformer models in high dimensions; (2) Further optimization of the models are needed to enhance its ability on predicting the exact convex hulls; (3) Improve POINT2VERTEX to handle real-word sparse problems in higher dimensional spaces and apply it to detect redundancies in information theory.

Acknowledgments. This work is supported by the National Key Research Project of China under Grant No. 2023YFA1009402, the National Natural Science Foundation of China under Grant No. 12101267 and Chongqing Talents Plan Youth Top-Notch Project (2021000263). The authors would like to sincerely thank the reviewers for their careful reading, insightful comments, and constructive suggestions, which have significantly helped improve the quality and clarity of this paper.

References

1. Guo, Z., Zhang, X., Liu, C., Ji, X., Jiao, J., Ye, Q.: Convex-hull feature adaptation for oriented and densely packed object detection. IEEE Trans. Circuits Syst. Video Technol. **32**(8), 5252–5265 (2022)
2. Khan, K., Agarwal, I., Jerripothula, K.R.: Convex hull-ish object detection. In: International Conference on Pattern Recognition, pp. 306–317. Springer (2024)
3. Lew, T., Bonalli, R., Janson, L., Pavone, M.: Estimating the convex hull of the image of a set with smooth boundary: error bounds and applications. Discrete Comput. Geometry **74**(1), 203–241 (2025)
4. Daroya, R., Atienza, R., Cajote, R.: Rein: flexible mesh generation from point clouds. In: Proceedings of the IEEE/CVF Conference on Computer Vision and Pattern Recognition Workshops, pp. 352–353 (2020)
5. Du, N., Xie, L., Zhou, M., Gao, W., Wang, Y., Hu, J.: Convex hull triangle mesh-based static mapping in highly dynamic environments. IEEE Trans. Instrum. Meas. **73**, 1–14 (2024)

6. Fan, J., Yang, J., Ai, D., Xia, L., Zhao, Y., Gao, X., Wang, Y.: Convex hull indexed gaussian mixture model (CH-GMM) for 3d point set registration. Pattern Recogn. **59**, 126–141 (2016)
7. Linh, T., Nguyen, T., Nguyen, T., Ito, T., Bui, N.-T.: An adaptive differential evolution algorithm with a point-based approach for 3d point cloud registration. J. Image Graph. **10**, 1–9 (2022)
8. Rassaerts, L., Suichies, E., van de Vrande, B., Alonso, M., Meere, B., Chong, M., Torta, E.: Non-conservative obstacle avoidance for multi-body systems leveraging convex hulls and predicted closest points. In: 2025 IEEE International Conference on Robotics and Automation (ICRA), pp. 3007–3014. IEEE (2025)
9. Brysiewicz, T., Joswig, M.: Polyhedral geometry. In: The Computer Algebra System OSCAR: Algorithms and Examples, pp. 139–164. Springer (2025)
10. Laisin, M., Edike, C., Ujumadu, R.: Characterizing boundedness and solution size in rational linear programming and polyhedrall optimization. In: Global Online Journal of Academic Research (GOJAR), vol. 4, no. 2 (2025)
11. Ma, Z., Wang, Z., Bai, G.: Convex hull approximation for activation functions. In: Proceedings of the ACM on Programming Languages, vol. 9, pp. 1007–1033 (2025)
12. Nemirko, A., Dulá, J.: Machine learning algorithm based on convex hull analysis. Proc. Comput. Sci. **186**, 381–386 (2021)
13. Barber, C.B., Dobkin, D.P., Huhdanpaa, H.: The quickhull algorithm for convex hulls. ACM Trans. Math. Softw. **22**(4), 469–483 (1996)
14. Kong, X., Everett, H., Toussaint, G.: The graham scan triangulates simple polygons. Pattern Recogn. Lett. **11**(11), 713–716 (1990)
15. Joswig, M.: Beneath-and-beyond revisited. In: Algebra, Geometry and Software Systems, pp. 1–21. Springer (2003)
16. Jing, R.-J., Moreno-Maza, M., Talaashrafi, D.: Complexity estimates for Fourier-Motzkin elimination. In: Computer Algebra in Scientific Computing: 22nd International Workshop, CASC 2020, Linz, Austria, September 14–18, 2020, Proceedings 22, pp. 282–306. Springer (2020)
17. Fukuda, K., Prodon, A.: Double description method revisited. In: Combinatorics and computer science, pp. 91–111. Springer (1996)
18. Motzkin, T.S.: The double description method, in contributions to the theory of games ii. Ann. Math. Study 28 (1953)
19. Haffner, P.: Escaping the convex hull with extrapolated vector machines. Adv. Neural Inf. Process. Syst. **14** (2001)
20. Vincent, P., Bengio, Y.: K-local hyperplane and convex distance nearest neighbor algorithms. Adv. Neural Inf. Process. Syst. **14** (2001)
21. Gavrilev, D., Makarov, I.: Fast approximate convex hull construction in networks via node embedding. IEEE Access **11**, 54588–54595 (2023)
22. Paul, S., Norkin, A., Bovik, A.C.: Convex hull prediction for adaptive video streaming by recurrent learning. IEEE Trans. Image Process. (2024)
23. Yousefzadeh, R.: Deep learning generalization and the convex hull of training sets. Inf. Geometry, pp. 1–26 (2025)
24. Vinyals, O., Fortunato, M., Jaitly, N.: Pointer networks. Adv. Neural Inf. Process. Syst. **28** (2015)
25. Sutskever, I., Vinyals, O., Le, Q.V.: Sequence to sequence learning with neural networks. Adv. Neural Inf. Process. Syst. **27** (2014)
26. Vaswani, A., Shazeer, N., Parmar, N., Uszkoreit, J., Jones, L., Gomez, A.N., Kaiser, L., Polosukhin, I.: Attention is all you need. In: Proceedings of the 31st International Conference on Neural Information Processing Systems, NIPS'17, (Red Hook, NY, USA), pp. 6000–6010. Curran Associates Inc. (2017)

27. Balestriero, R., Wang, Z., Baraniuk, R.G.: Deephull: fast convex hull approximation in high dimensions. In: ICASSP 2022-2022 IEEE International Conference on Acoustics, Speech and Signal Processing (ICASSP), pp. 3888–3892. IEEE (2022)
28. Liang, H., Wang, S., Gao, S., Li, H., Su, C., Lu, H., Zhang, X., Chen, X., Chen, Y.: Deephullnet: a deep learning approach for solving the convex hull and concave hull problems with transformer. Int. J. Digital Earth **17**(1), 2358843 (2024)
29. Tian, C.: Characterizing the rate region of the (4, 3, 3) exact-repair regenerating codes. IEEE J. Sel. Areas Commun. **32**(5), 967–975 (2014)
30. Guo, L., Yeung, R.W., Gao, X.-S.: Proving information inequalities by gaussian elimination. IEEE Trans. Inf. Theory (2025)
31. Schrijver, A.: Theory of linear and integer programming. Wiley, New York, NY (1986)

Advances in Computer Vision: Reliability, Explainability, and Real-World Applications

Single-Image Super-Resolution by Self-supervised Consistent Representation Learning

Qiyue Zhang and Yi Tang(✉)

School of Mathematics and Computer Science, Yunnan Minzu University, Kunming 650500, China
yitang4math@ynni.edu.cn

Abstract. Single-image super-resolution (SISR) algorithms typically assume a fixed degradation process, which leads to a one-to-one correspondence between high-resolution and low-resolution images. In real-world scenarios, however, the degradation process is often diverse, resulting in a one-to-many relationship between high-resolution and low-resolution images. The inconsistency among low-resolution images significantly degrades the performance of super-resolution algorithms. To address this issue, we propose a super-resolution framework that integrates a Degradation Decomposition (DD) module and a self-supervised low-resolution reconstruction branch. The DD module aims to decompose low-resolution image features into ideal degradation representations and degradation estimation features, the low-resolution reconstruction branch fuses these features and decodes the original LR image via a self-supervised mechanism, the SR reconstruction branch performs image reconstruction, guided by the ideal degradation representations. The LR reconstruction task acts as a consistency regularization term, ensuring the ideal degradation representations remain stable across diverse degradation conditions. Furthermore, we formulate a joint loss function to collaboratively optimize both the SR and LR reconstruction tasks. Experimental results demonstrate that the proposed method improves the performance of super-resolution models.

Keywords: Image super-resolution · Degradation decomposition · Self-supervised learning · Low-resolution reconstruction · Computer vision

1 Introduction

Single Image Super-Resolution (SISR) aims to reconstruct high-resolution (HR) images from low-resolution (LR) inputs. Despite its straightforward goal, it remains a significant research challenge in the field of computer vision. Single Image Super-Resolution methods generally assume a one-to-one mapping

L. Huang and D. Greenhalgh (Eds.): ICMLC 2026, LNNS 2061, pp. 149–164, 2027.
https://doi.org/10.1007/978-3-032-31236-5_10

between low-resolution and high-resolution images, they assume that the degradation process of the image is fixed. Under this assumption, the generation process of low-resolution images is typically modeled by the following equation:

$$I_{LR} = D \circ D_{ideal}(I_{HR}), \tag{1}$$

where D_{ideal} represents the ideal degradation process, and D denotes the degradation perturbation operations, including blurring, downsampling, and noise, among others. The goal of the image restoration task is to recover the ideal high-resolution image I_{HR} from the given low-resolution image I_{LR}, i.e., by inversely estimating the ideal degradation process D_{ideal}. Although an ideal degradation process can be assumed, such as the commonly used the bicubic interpolation model [1,2] in traditional super-resolution methods as the ideal degradation model, the degradation process in practical applications is often not fully determined.

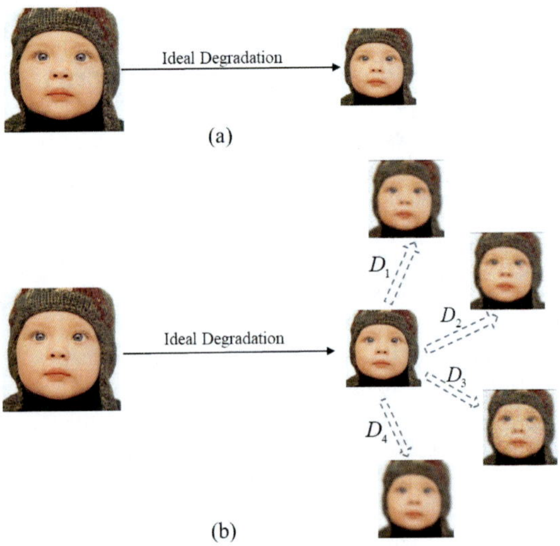

Fig. 1. (a) Shows the HR-LR pair under a single ideal degradation assumption, (b) shows the HR-LR pairs under degradation assumptions deviating from the ideal degradation model, D_i represents different degradation perturbations.

Single-image super-resolution is a highly ill-posed and challenging problem under diverse degradation conditions. This ill-posedness arises because a single HR image may correspond to multiple LR variants under different degradations [3], leading to a one-to-many mapping, as illustrated in Fig.1. Early deep learning-based methods [4] assumed a fixed degradation model, thereby establishing a deterministic HR-LR mapping. However, when the real-world degradation deviates from this assumption, the reconstruction performance often drops significantly [5,6]. To address the limitations of fixed degradation models, subsequent works have attempted to jointly estimate degradation parameters [7] during reconstruction. For example, Zhou et al. [8] introduce a degradation-adaptive regression module to correct degradation estimation within the SR network. Despite such efforts, these methods still struggle with generalization due to the complexity of real-world degradations. This has motivated blind SR approaches that explicitly or implicitly model degradation uncertainty. SRDM [9] employs a convolutional network to handle multiple degradations by learning degradation maps, while diffusion-based strategies [10] generate degradations from learned distributions. Explicit modeling methods such as USRNet [11] and IKC [6] incorporate blur kernels and noise priors, and perform iterative degradation estimation. However, these methods still suffer from inconsistencies in handling diverse degradation patterns, often resulting in texture artifacts, distortion, or unstable noise distribution [12].

To address the aforementioned challenge, we design a multi-degradation module that synthesizes LR-HR pairs with diverse distributions via heterogeneous degradation paths, enhancing degradation modeling at the data level. We further propose an Degradation Decomposition module, which aims to decompose low-resolution image features into ideal degradation representations independent of degradation perturbations and degradation estimation features. In addition, a self-supervised low-resolution reconstruction branch is introduced, this branch integrates these features via a self-supervised mechanism and reconstructs the original LR image through a decoder module. The super-resolution reconstruction branch performs image restoration by leveraging the decomposed ideal degradation representations. The LR reconstruction task functions as a consistency-driven regularization term, encouraging the ideal degradation representations to remain consistent under diverse degradation conditions, thereby providing informative guidance for super-resolution reconstruction branch. To enable joint optimization, we formulate a unified training objective that effectively integrates the super-resolution reconstruction loss with the low-resolution consistency reconstruction constraint.

In the paper, the main contributions are as follows:

- We propose a novel super-resolution framework that integrates a Degradation Decomposition module and a self-supervised LR reconstruction branch, enabling robust modeling under diverse degradation conditions.
- We propose a Degradation Decomposition module, which decomposes low-resolution features into degradation-invariant ideal representations and degradation estimation features, thereby facilitating degradation-aware super-resolution reconstruction.
- We employ the LR reconstruction task as a consistency regularization term and formulate a joint loss to collaboratively optimize both the SR and LR reconstruction branches, thereby enhancing super-resolution performance.

The rest of this article is organized as follows. Section 2 contains a full description of the related works. Section 3 thoroughly discusses the proposed super-resolution framework. Section 4 presents quantitative and qualitative results, as well as experiments on spatially variant degradation and real-world datasets. Section 5 draws the conclusion.

2 Related Work

2.1 Degradation Model

SISR has progressed significantly, with SRCNN [4] pioneering LR-HR mapping using a three-layer CNN, laying the foundation for subsequent methods. CNN-based approaches, including ESPCN [13] and MSRN [14], have advanced reconstruction through multi-scale techniques and adaptive feature detection. However, most methods [15] assume fixed bicubic degradation, leading to performance degradation when real-world degradation diverges [6]. To address this issue, Zhang et al. [16] further advanced this with the DPSR model, treating low-resolution images as the result of bicubic downsampling, arbitrary blur kernel convolution, and additive noise corruption of high-resolution images. To more accurately simulate complex degradation processes, Zhang et al. [17] introduced the ESRGAN model, which employs random permutations of degradation operations, encompassing various types of blur, downsampling ratios (such as bilinear and bicubic interpolation), camera noise, and JPEG compression, thus more realistically modeling real-world degradation. Additionally, approaches such as optimization-based strategies [18] and unsupervised learning [19] have all made significant strides in blind super-resolution tasks, further advancing the field. However, most existing methods still rely on predefined degradation sequences, limiting their ability to generalize to real-world complex degradation scenarios.

2.2 Self-supervised Learning for Super-Resolution

Classical super-resolution methods [20,21] largely depend on supervised learning with extensive paired LR-HR datasets. However, in recent years, the application of self-supervised learning in single-image super-resolution has emerged as a promising approach to overcome the data dependency bottleneck inherent in supervised learning. Early works, such as ZSSR [22] pioneered the internal learning paradigm by leveraging the cross-scale repeatability of image patches. However, the simplified degradation assumptions underlying these methods conflict with the complexity of real-world scenarios. In response, Zhang et al. [23] introduced a self-supervised method that learns from dual and multiple zoom observations, using high-zoom images as references for low-zoom images to achieve super-resolution. Chen et al. [24] combined supervised pretraining with self-supervised learning, utilizing the degradation embedding of low-resolution images to improve the model's adaptability to real-world images and its ability to recover fine details. Collectively, these methods demonstrate the potential of self-supervised learning to address the issue of low-resolution image inconsistency, enabling models to circumvent the limitations of predefined degradation sequences.

3 Method

As shown in Fig. 2, the proposed architecture consists of three key components: an Ideal Degradation Feature Extraction module, a Degradation Estimation Encoder, and an LR Reconstruction Decoder, followed by image reconstruction.

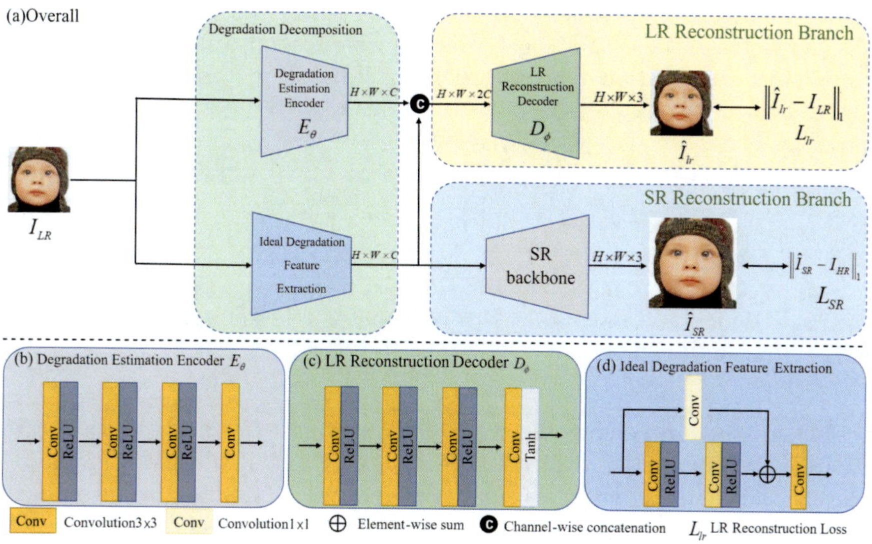

Fig. 2. The network architecture of our method.

3.1 Multi Degradation Module

To enhance the model's adaptability to diverse and complex degradation processes, this paper proposes a Multi degradation Module. This module constructs a heterogeneous set of degradation paths $\hat{\mathcal{D}} = \{\hat{D}_i\}_{i=1}^{K}$, each consisting of a differently ordered sequence of degradation operations applied to the input HR image I_{HR}. Specifically, each degradation path $\hat{D}_i$ consists of a sequence of degradation steps that are sequentially applied to the high-resolution image to simulate progressive degradation. The process can be formally expressed as:

$$I_{LR_i} = D_i \circ D_{ideal}(I_{HR}) = \hat{D}_i(I_{HR}) = (d_i^n \circ \cdots \circ d_i^1)(I_{HR}) \tag{2}$$

where $\{d_i^j\}_{j=1}^{n}$ denotes the set of degradation operations in the i-th path, and each d_i^j represents an individual degradation step. The operation set includes Gaussian blurring, noise perturbations, and downsampling methods such as bilinear and bicubic interpolation. This flexible and modular design enables the generation of a wide range of LR images with varying degradation patterns, effectively simulating the diversity and complexity of real-world scenarios.

3.2 Degradation Decomposition

Degradation Estimation Encoder The objective of the Degradation Estimation Encoder is to extract degradation perturbation features from the input low-resolution image, thereby effectively capturing the degradation information present in the image. Specifically, as shown in Fig. 2(b), the input I_{LR} image is fed into an encoder composed of four consecutive convolutional layers, each followed by a ReLU activation. The number of feature channels increases progressively across layers, enabling a gradual compression of the original image into a latent degradation representation space. This process can be formally expressed as:

$$f_{lr} = E_\theta(I_{LR}), \tag{3}$$

$$E_\theta(I_{LR}) = \left(\prod_{i=1}^{4} \phi \circ \text{Conv}_i\right)(I_{LR}), \tag{4}$$

where $f_{lr} \in \mathbb{R}^{H\times W\times C}$ denotes the encoded feature representation, $E_\theta(\cdot)$ denotes the encoder function, Conv_i is the i-th convolutional layer, and ϕ represents the ReLU activation function.

Ideal Degradation Feature Extraction We propose a dedicated Ideal Degradation Feature Extraction, which is designed to extract stable and degradation-invariant feature representations from diverse degraded inputs, thereby enhancing the generalization capability of super-resolution reconstruction.

Specifically, this module comprises two parallel convolutional branches as shown in Fig. 2d. One branch performs feature extraction through a sequence of convolutional layers followed by ReLU activation functions, wherein a dilated convolution with a dilation rate of 2 is introduced to enlarge the receptive field and facilitate the capture of broader spatial contextual information. The feature extraction process in this branch can be formulated as:

$$f_m = \left(\prod_{i=1}^{2} \phi \circ \text{Conv}_i\right)(I_{LR}), \tag{5}$$

where Conv_1 denotes 3×3 convolutional, and Conv_2 denotes a dilated convolutional layer with a dilation rate of 2. The other branch employs a separate 1×1 convolutional layer to directly extract fine-grained edge features from the input LR image:

$$f_e = \text{Conv}_{1\times 1}(I_{LR}) \tag{6}$$

The features from both branches are fused via element-wise addition and projected into the target embedding space through an additional convolution, yielding the final ideal degradation representation:

$$f_{ideal} = \text{Conv}_{3\times 3}(f_m + f_e). \tag{7}$$

3.3 LR Reconstruction Branch

The LR Reconstruction Branch is optimized in a self-supervised manner, guiding the network to focus on degradation estimation modeling. Specifically, this branch employs a self-supervised learning mechanism that encourages the encoder to effectively extract features related to degradation perturbations, ultimately constraining the ideal degradation feature, thereby enhancing the overall performance of the super-resolution framework.

The ideal degradation features $f_{ideal} \in \mathbb{R}^{H\times W\times C}$ (see Sect.3.2) extracted are concatenated with the degradation estimation representation $f_{lr} \in \mathbb{R}^{H\times W\times C}$along the channel dimension to form a unified feature, as shown in Fig. 2c,

$$f_c = Cat([f_{lr}, f_{ideal}]). \tag{8}$$

The fused feature $f_c \in \mathbb{R}^{H\times W\times 2C}$is fed into a symmetric decoder module to progressively reconstruct the image in the low-resolution space, aiming to faithfully recover the input image I_{LR}. A Tanh activation function is applied at the end of the decoder. The decoding process is formally defined as:

$$\hat{I}_{lr} = D_\phi(f_c), \tag{9}$$

where $D_\phi(\cdot)$ denotes the decoder function, and $I_{\mathrm{lr}} \in \mathbb{R}^{H\times W\times C}$ denotes the reconstructed low-resolution image.

3.4 SR Reconstruction Branch

The SR reconstruction branch comprises a deep feature extractor and a reconstruction module. The extractor takes ideal features f_{ideal} as input and adopts a lightweight hierarchical structure with attention-convolution units, integrating self- and cross-attention for multi-scale context and local detail modeling. Efficient designs such as depthwise separable convolutions and attention sparsification reduce computational cost. The reconstruction module then generates the final super-resolved image using sub-pixel convolution and pixel shuffle to enhance spatial resolution and fidelity. Formally:

$$f_d = H_d(f_{\mathrm{ideal}}), \tag{10}$$

$$\hat{I}_{SR} = H_{IRE}(f_{ideal} + f_d), \tag{11}$$

where $H_d(\cdot)$ denotes the deep feature extraction function, $H_{\mathrm{IRE}}(\cdot)$ denotes the image reconstruction function, and f_d represents the extracted deep features.

3.5 Loss Functions

We propose a joint optimization strategy that integrates the SR reconstruction loss and LR reconstruction loss into a unified objective function, guiding the network to optimize simultaneously in both high- and low-resolution spaces. This method employs pixel-wise L_1 loss to compute both SR and LR reconstruction losses, with the LR reconstruction loss acting as a consistency regularizer to enhance the model's robustness under various degradation scenarios. The total loss function is defined as:

$$\mathcal{L}_{total} = \|\hat{I}_{SR} - HR\|_1 + \lambda \cdot \|\hat{I}_{lr} - I_{LR}\|_1, \tag{12}$$

where $\hat{I}_{SR}$ is the super-resolved image output by the network, HR represents the high-resolution ground-truth image, $\hat{I}_{lr}$ is the reconstructed LR image, and I_{LR} denotes the input low-resolution image. The L_1 norm is used to compute the pixel-wise difference between the images. In this total loss, λ is a hyperparameter. By minimizing this total loss, the model simultaneously optimizes both high- and low-resolution reconstructions, enhancing its robustness and generalization ability in real-world scenarios.

4 Experiments

4.1 Datasets and Criteria

To systematically evaluate the robustness and effectiveness of the proposed method under complex degradation scenarios, we constructed training image pairs based on the DIV2K dataset.Following the setting of $i = 4$ in Equation 2, we comprehensively considered key degradation factors, including scaling factors, blur kernel parameters, and noise levels. Each HR image was degraded through a predefined sequence to generate multiple LR images. The detailed degradation processes are presented as follows:

Degradation model 1 applies only bicubic downsampling, formally defined as:

$$y = x \downarrow_s . \tag{13}$$

Degradation Model 2 follows a sequential degradation process of applying a blur kernel prior to downsampling:

$$y = (k \otimes x) \downarrow_s . \tag{14}$$

where k denotes an isotropic Gaussian blur kernel of fixed size 21×21, with the standard deviation σ uniformly sampled from the intervals $[0.2, 1.2]$, $[0.2, 2.0]$, and $[0.2, 2.8]$ for upscaling factors 2, 3, and 4, respectively. The downsampling operator $\downarrow_s$ is implemented using bicubic interpolation.

Degradation Model 3 performs bilinear downsampling followed by anisotropic Gaussian blurring:

$$y = (x \downarrow_s) \otimes k, \tag{15}$$

where $\downarrow_s$ denotes bilinear downsampling with scale factor s, and k is a 7×7 anisotropic Gaussian kernel. The covariance matrix of k is constructed by uniformly sampling its eigenvalues from scale-dependent intervals: $[0.2, 0.4]$ for $s = 2$, $[0.2, 0.6]$ for $s = 3$, and $[0.2, 0.8]$ for $s = 4$. The orientation angle θ is drawn from a uniform distribution $\theta \sim \mathcal{U}(0, \pi)$, resulting in elliptical blur kernels with varying directions.

Degradation Model 4 synthesizes LR images through a degradation process following the sequential stages of downsampling, blurring, and noise addition, formulated as:

$$y = (x \downarrow_s) \otimes k + n, \tag{16}$$

Table 1. PSNR and SSIM of different methods on Set5, Set14, and B100 under various degradations.

Method	Scale	Set5 (PSNR↑/SSIM↑)				Set14 (PSNR↑/SSIM↑)				B100 (PSNR↑/SSIM↑)			
		Degradation1	Degradation2	Degradation3	Degradation4	Degradation1	Degradation2	Degradation3	Degradation4	Degradation1	Degradation2	Degradation3	Degradation4
PAN	×2	38.00/0.9605	31.19/0.8933	31.07/0.8948	29.47/0.7597	33.59/0.9181	28.45/0.8130	28.71/0.8253	28.32/0.7672	32.18/0.8997	28.01/0.7798	26.94/0.7613	25.94/0.6686
LAPAR-A	×2	38.01/0.9605	31.15/0.8915	31.04/0.8930	29.58/0.7648	33.62/0.9183	28.42/0.8114	28.69/0.8240	28.39/0.7698	32.19/0.8999	28.00/0.7783	26.93/0.7602	25.96/0.6699
ShuffleMixer	×2	38.01/0.9606	31.19/0.8934	31.07/0.8947	29.63/0.7678	33.63/0.9180	28.46/0.8132	28.71/0.8253	28.39/0.7710	32.17/0.8995	28.02/0.7801	26.95/0.7620	26.02/0.6737
SMFANet	×2	38.04/0.9606	31.20/0.8935	31.09/0.8951	29.54/0.7609	33.65/0.9186	28.47/0.8136	28.73/0.8261	28.32/0.7673	32.19/0.8999	28.03/0.7804	26.94/0.7617	25.96/0.6684
CATANet	×2	**38.28/0.9617**	31.17/0.8932	31.11/0.8955	29.70/0.7654	**33.99/0.9217**	28.46/0.8136	28.77/0.8271	28.52/0.7730	**32.37/0.9023**	28.02/0.7804	26.94/0.7619	25.95/0.6703
Ours	×2	38.04/0.9606	**37.47/0.9545**	**36.33/0.9429**	**33.97/0.9115**	33.64/0.9184	**33.50/0.9102**	**31.58/0.8934**	**30.31/0.8546**	32.22/0.9002	**32.07/0.8929**	**27.67/0.7898**	**26.52/0.7689**
PAN	×3	34.40/0.9271	28.42/0.8248	27.74/0.8200	28.11/0.8104	30.36/0.8423	26.22/0.7195	24.98/0.7019	23.99/0.6182	29.11/0.8050	26.14/0.6822	25.75/0.6884	24.71/0.6039
LAPAR-A	×3	34.36/0.9267	28.39/0.8226	27.72/0.8187	28.11/0.8099	30.34/0.8421	26.20/0.7177	24.98/0.7008	24.00/0.6201	29.11/0.8054	26.12/0.6807	25.74/0.6874	24.73/0.6054
ShuffleMixer	×3	34.40/0.9272	28.44/0.8252	27.76/0.8207	28.06/0.8070	30.37/0.8423	26.24/0.7203	24.98/0.7018	23.98/0.6103	29.12/0.8051	26.15/0.6830	25.74/0.6885	24.72/0.5981
SMFANet	×3	34.46/0.9275	28.42/0.8245	27.74/0.8200	28.09/0.8083	30.39/0.8432	26.23/0.7195	24.98/0.7027	24.05/0.6162	29.13/0.8059	26.14/0.6823	25.74/0.6886	24.78/0.6021
CATANet	×3	**34.75/0.9300**	28.40/0.8242	27.76/0.8206	28.10/0.8104	**30.67/0.8481**	26.22/0.7197	24.99/0.7023	23.97/0.6140	**29.28/0.8101**	26.14/0.6827	25.74/0.6893	24.69/0.6007
Ours	×3	34.44/0.9272	**34.16/0.9219**	**29.36/0.8689**	**29.05/0.8601**	30.40/0.8433	**30.25/0.8348**	**25.45/0.7398**	**25.22/0.7249**	29.13/0.8065	**29.05/0.7996**	**26.87/0.7529**	**26.32/0.7141**
PAN	×4	32.13/0.8948	26.27/0.7506	27.18/0.7987	26.70/0.7216	28.61/0.7822	24.61/0.6426	24.31/0.6421	24.19/0.5974	27.59/0.7363	24.84/0.6095	24.49/0.6199	24.00/0.5693
LAPAR-A	×4	32.15/0.8944	26.25/0.7488	27.17/0.7971	26.79/0.7263	28.61/0.7818	24.59/0.6409	24.30/0.6407	24.23/0.6005	27.61/0.7366	24.83/0.6080	24.48/0.6190	24.00/0.5707
ShuffleMixer	×4	32.21/0.8953	26.28/0.7513	27.18/0.7983	26.67/0.7213	28.66/0.7827	24.61/0.6431	24.32/0.6425	24.22/0.5989	27.61/0.7366	24.84/0.6097	24.49/0.6200	24.02/0.5689
SMFANet	×4	32.25/0.8956	26.28/0.7513	27.16/0.7990	26.64/0.7179	28.66/0.7825	24.61/0.6429	24.31/0.6422	24.19/0.5971	27.61/0.7371	24.85/0.6097	24.49/0.6201	24.01/0.5683
CATANet	×4	**32.58/0.8998**	26.27/0.7506	27.22/0.8009	26.82/0.7251	**28.90/0.7880**	24.61/0.6430	24.29/0.6418	24.18/0.5963	**27.75/0.7427**	24.85/0.6100	24.49/0.6209	24.01/0.5705
Ours	×4	32.27/0.8953	**30.65/0.8671**	**30.22/0.8739**	**29.06/0.8398**	28.67/0.7830	**27.74/0.7515**	**25.91/0.7200**	**25.68/0.6959**	27.62/0.7383	**27.05/0.7077**	**25.27/0.6858**	**24.89/0.6558**

Best results are in bold.

Table 2. PSNR and SSIM of different methods on Urban100 and Manga109 under various degradations.

Method	Scale	Urban100 (PSNR↑/SSIM↑)				Manga109 (PSNR↑/SSIM↑)			
		Degradation1	Degradation2	Degradation3	Degradation4	Degradation1	Degradation2	Degradation3	Degradation4
PAN	×2	32.01/0.9273	25.36/0.7831	25.77/0.8088	25.99/0.7546	38.70/0.9773	28.14/0.8931	27.66/0.8968	27.45/0.8064
LAPAR-A	×2	32.10/0.9283	25.35/0.7815	25.77/0.8080	26.06/0.7576	38.67/0.9772	28.12/0.8913	27.65/0.8956	27.51/0.8096
ShuffleMixer	×2	31.89/0.9257	25.37/0.7832	25.74/0.8079	26.03/0.7577	38.83/0.9774	28.15/0.8932	27.68/0.8972	27.55/0.8122
SMFANet	×2	32.22/0.9283	25.37/0.7836	25.78/0.8095	26.01/0.7541	38.98/0.9776	28.18/0.8936	27.67/0.8972	27.46/0.8061
CATANet	×2	**33.09/0.9372**	25.37/0.7840	25.88/0.8137	26.19/0.7612	**39.37/0.9784**	28.15/0.8933	27.70/0.8979	27.52/0.8079
Ours	×2	32.41/0.9310	**31.70/0.9195**	**29.01/0.8899**	**28.22/0.8659**	38.74/0.9773	**38.21/0.9723**	**28.91/0.9326**	**28.49/0.9159**
PAN	×3	28.11/0.8511	23.37/0.6805	22.62/0.6741	22.12/0.6099	33.61/0.9448	25.22/0.8110	23.84/0.7894	23.26/0.7017
LAPAR-A	×3	28.15/0.8523	23.36/0.6789	22.61/0.6732	22.15/0.6135	33.51/0.9441	25.19/0.8087	23.84/0.7883	23.29/0.7058
ShuffleMixer	×3	28.08/0.8498	23.38/0.6814	22.63/0.6740	22.16/0.6042	33.69/0.9448	25.23/0.8118	23.85/0.7895	23.35/0.6949
SMFANet	×3	28.25/0.8525	23.37/0.6803	22.62/0.6735	22.21/0.6071	33.84/0.9454	25.26/0.8120	23.85/0.7895	23.38/0.6987
CATANet	×3	**29.04/0.8689**	23.38/0.6815	22.60/0.6750	22.04/0.6024	**34.40/0.9499**	25.22/0.8112	23.84/0.7899	23.19/0.6908
Ours	×3	28.34/0.8563	**27.88/0.8415**	**23.67/0.7465**	**23.50/0.7317**	33.69/0.9455	**33.54/0.9410**	**24.45/0.8369**	**24.31/0.8255**
PAN	×4	26.11/0.7854	21.97/0.5963	21.70/0.6125	21.55/0.5545	30.51/0.9095	23.18/0.7317	21.53/0.7019	21.14/0.6163
LAPAR-A	×4	26.14/0.7871	21.96/0.5949	21.70/0.6115	21.58/0.5585	30.42/0.9074	23.18/0.7303	21.54/0.7010	21.15/0.6206
ShuffleMixer	×4	26.08/0.7835	21.98/0.5969	21.71/0.6123	21.55/0.5526	30.65/0.9093	23.20/0.7328	21.54/0.7017	21.15/0.6074
SMFANet	×4	26.19/0.7861	21.98/0.5969	21.70/0.6118	21.54/0.5504	30.71/0.9097	23.21/0.7332	21.54/0.7020	**21.17**/0.6111
CATANet	×4	**26.87/0.8081**	21.98/0.5973	21.71/0.6145	21.58/0.5558	**31.31/0.9183**	23.20/0.7327	21.51/0.7020	21.13/0.6123
Ours	×4	26.31/0.7927	**25.01/0.7425**	**23.46/0.7201**	**23.04/0.6941**	30.62/0.9094	**28.32/0.8750**	**21.90/0.7192**	20.89/**0.7155**

Best results are in bold.

where $\downarrow_s$ denotes bilinear downsampling with a scaling factor s, and $\otimes$ represents spatial convolution. The blur kernel k is a 5×5 anisotropic Gaussian with covariance matrix Σ defined by eigenvalues $\lambda_1 = \lambda_2 = 0.3$ and rotation angle $\theta \sim \mathcal{U}(0, \pi)$. Gaussian noise $n \sim \mathcal{N}(0, \sigma^2 I)$ is added with $\sigma \sim \mathcal{U}(1, 15)$.

For model evaluation, we employed five commonly used benchmark datasets: Set5, Set14, BSD100, Urban100, and Manga109. The LR images in these test sets were generated using the corresponding degradation models. Peak Signal-to-Noise Ratio (PSNR) and Structural Similarity Index (SSIM) were used as evaluation metrics.

4.2 Comparisons with Other Methods

Quantitative results: as shown in Table 1 and 2, demonstrate the superior performance of our method when applied to low-resolution images generated by three distinct degradation models. The degradation models Degradation1, Degradation2, Degradation3 and Degradation4 correspond to low-resolution test images

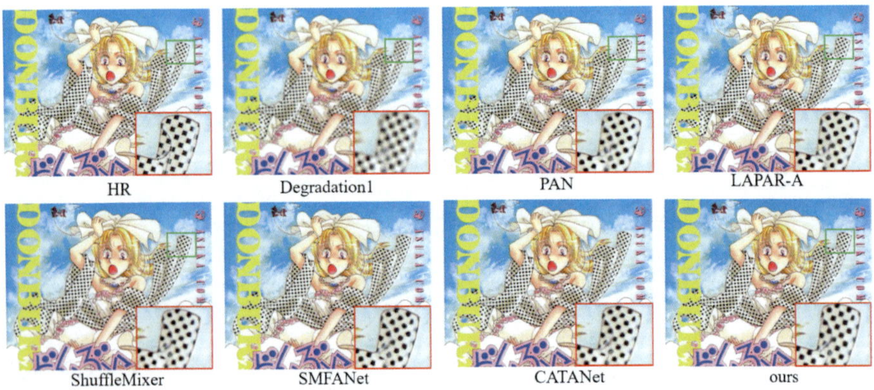

Fig. 3. Visual comparison of different methods on Manga109:Donburakokko (scale factor 4) under Degradation1.

Fig. 4. Visual comparison of different methods on Set14:ppt3 (scale factor 2) under Degradation2.

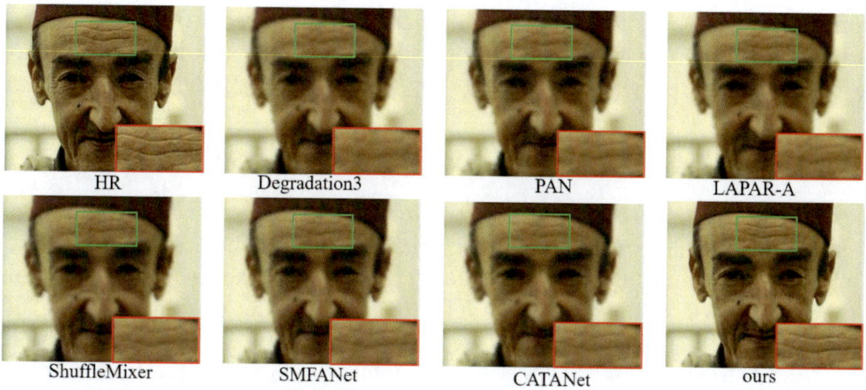

Fig. 5. Visual comparison of different methods on B100:189080 (scale factor 3) under Degradation3.

generated based on our four proposed degradation assumptions. Experiments on the Set5, Set14, BSD100, Urban100 and Manga109 datasets show that, compared to super-resolution methods, such as PAN [25], LAPAR-A [26], ShuffleMixer [27], SMFANet [28] and CATANet [29] our method exhibits competitive performance across all scaling factors (×2, ×3, and ×4). Significantly, In the Degradation2, Degradation3, and Degradation4 scenarios, our method demonstrates significant improvements across benchmark datasets. In contrast, under the bicubic downsampling degradation model (Degradation1), our method achieves second-best performance. This is due to the fact that other methods are specifically optimized for the Degradation1 model and fail to generalize to other degradation models. This also validates the issue that reconstruction performance significantly deteriorates when the actual degradation deviates from the model's assumptions [5]. Furthermore, our method excels in handling multiple degradation models with a single framework, as opposed to being confined to a single degradation assumption like bicubic downsampling. Consequently, while our method does not outperform state-of-the-art bicubic-focused methods in Degradation1, it exhibits superior generalization capabilities across diverse degradation conditions.

Qualitative results: In Figs. 3, 4 and 5, we present a qualitative comparison between our super-resolution approach and other state-of-the-art methods. Compared to advanced algorithms such as SMFANet and CATANet, our model demonstrates a significant advantage in terms of detail recovery. For instance, a s shown in Fig. 3, even when based solely on the Degradation1 assumption, the other methods fail to accurately restore the square structure of the skirt, resulting in distortion where the shape appears round or blurred. In contrast, our method effectively recovers the square details of the skirt and successfully preserves the accuracy of high-frequency information. In the Degradation2 scenario shown in Fig. 4, our method effectively enhances the English text details of the ppt3 slide, whereas other methods struggle to recover the text clarity, resulting in blurred or unreadable characters. As illustrated by the facial details in Fig.

5, our method reduces the impact of degradation and successfully recovers fine details, such as forehead wrinkles. Notably, our method significantly improves the visual quality, proving its effectiveness in image restoration.

4.3 Experiments on Spatially Variant Degradation

This experiment aims to evaluate the performance of the super-resolution model in handling spatially varying degradation. Specifically, the standard deviation (σ) of the Gaussian blur kernel is dynamically adjusted across different regions of the image, as shown in Fig. 6a, ranging from [1.0, 2.75], to simulate local blur caused by camera shake, focus errors, or object motion. In addition, the noise intensity, as illustrated in Fig. 6b, reflecting factors such as sensor noise or uneven illumination. This spatially variant degradation results in distinct degradation patterns across different regions of the image.

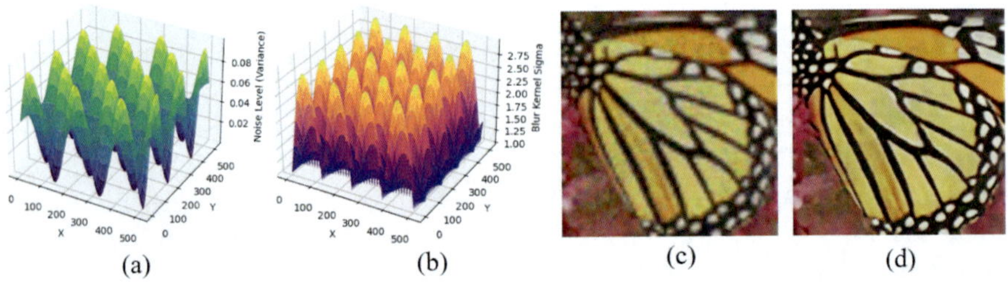

Fig. 6. Example of the super-resolution model addressing spatially variant degradation. (a) Noise maps. (b) Blur kernel maps. (c) Zoomed LR input. (d) Reconstructed HR output (scale factor 4).

As illustrated in Fig. 6, by comparing figure (c, d), the model's performance in handling different types of degradation becomes clear. In regions with lower noise levels and milder blur, the model can nearly perfectly reconstruct the original details. However, in areas with higher noise intensity and more severe blur, although the reconstruction quality decreases, the model is still able to preserve structural information. These results indicate the model's strong adaptability to diverse degradations, with room for refinement in severely degraded regions.

4.4 Experiments on Real-World Set

To evaluate the performance of the proposed single-image super-resolution model on real low-resolution images, we conduct experiments on the RealSRSet dataset. This dataset contains various complex degradation types, such as blur and noise, with unknown and uncontrollable degradation processes. Given the absence of ground-truth high-resolution images and the limitations of existing no-reference metrics (e.g., NIQE and PI) in aligning with human visual perception [17,30], we adopt subjective visual comparison for evaluation.

As illustrated in Fig. 7, different super-resolution methods (i.e., ZSSR [22], ESRGAN [31], BSRMD [30], and DARSR [8]) exhibit diverse performance when applied to real-world images. The proposed method demonstrates relatively better visual quality. In the *frog* example, most methods, including ZSSR, ESRGAN, and DARSR, tend to produce noticeable noise. In the *comic2* image, ESRGAN generates visually observable results but suffers from significant noise, while BSRDM introduces unnatural bright artifacts. For the *tiger* image, most methods struggle to accurately restore fine hair details and tend to produce unrealistic textures, whereas our method successfully reconstructs clear and natural structures.

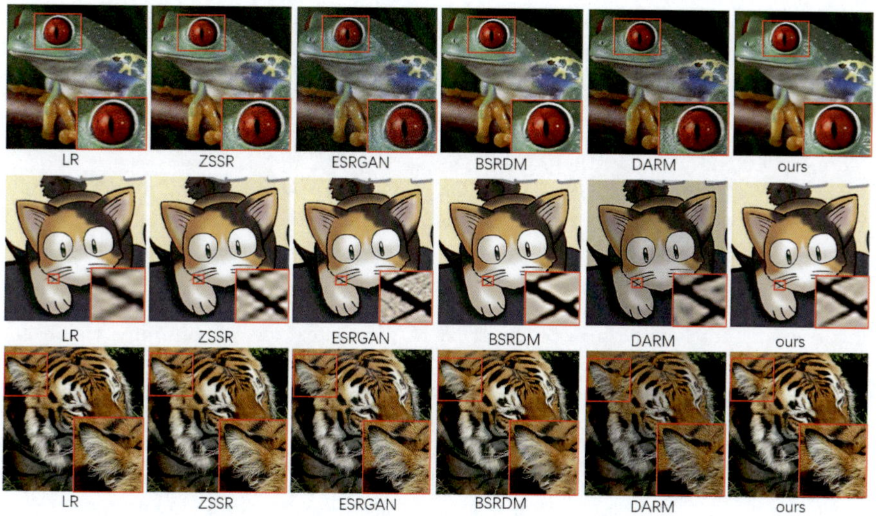

Fig. 7. Visual comparison of super-resolution results on three real LR images with scale factor 4. Best viewed zoomed in.

5 Conclusion

This paper propose a super-resolution framework that integrates a self-supervised low-resolution reconstruction branch with a Degradation Decomposition module to address degradation diversity. The DD module decomposes low-resolution features into ideal degradation representations and degradation estimation features, enhancing robustness. The self-supervised reconstruction branch fuses these features for low-resolution image reconstruction, while the super-resolution branch performs restoration guided by ideal degradation representations. The low-resolution reconstruction loss acts as a consistency regularization term, enabling consistent learning of ideal degradation representations. Experimental results show that our method outperforms existing approaches in both quantitative evaluation and perceptual quality. On real-world low-resolution images, the proposed method successfully reconstructs high-resolution outputs with richer details and higher visual fidelity.

References

1. Chen, Z., Zhang, Y., Gu, J., Kong, L., Yang, X., Yu, F.: Dual aggregation transformer for image super-resolution. In: Proceedings of the IEEE/CVF International Conference on Computer Vision, pp. 12312–12321. (2023)
2. Su, J.-N., Gan, M., Chen, G.-Y., Guo, W., Philip Chen, C.L.: High-similarity-pass attention for single image super-resolution. IEEE Trans. Image Process. **33**, 610–624 (2024)
3. Baker, S., Kanade, T.: Limits on super-resolution and how to break them. IEEE Trans. Pattern Anal. Mach. Intell. **24**(9), 1167–1183 (2002)
4. Dong, C., Loy, C.C., He, K., Tang, X.: Learning a Deep Convolutional Network for Image Super-Resolution. In: Fleet, D., Pajdla, T., Schiele, B., Tuytelaars, T. (eds.) ECCV 2014. LNCS, vol. 8692, pp. 184–199. Springer, Cham (2014). https://doi.org/10.1007/978-3-319-10593-2_13
5. Zhang, W., Shi, G., Liu, Y., Dong, C., Wu, X.-M.: A closer look at blind super-resolution: degradation models, baselines, and performance upper bounds. In: Proceedings of the IEEE/CVF conference on computer vision and pattern recognition, pp. 527–536. (2022)
6. Gu, J., Lu, H., Zuo, W., Dong, C.: Blind super-resolution with iterative kernel correction. In: Proceedings of the IEEE/CVF conference on computer vision and pattern recognition, pp. 1604–1613. (2019)
7. Hussain Khan, A., Micheloni, C., Martinel, N.: Lightweight prompt learning implicit degradation estimation network for blind super resolution. IEEE Trans, Image Process (2024)
8. Zhou, H., Zhu, X., Zhu, J., Han, Z., Zhang, S.-X., Qin, J., Yin, X.-C.: Learning correction filter via degradation-adaptive regression for blind single image super-resolution. In: Proceedings of the IEEE/CVF International Conference on Computer Vision, pp. 12365–12375 (2023)
9. Ji, X., Cao, Y., Tai, Y., Wang, C., Li, J., Huang, F.: Real-world super-resolution via kernel estimation and noise injection. In: Proceedings of the IEEE/CVF Conference on Computer Vision and Pattern Recognition Workshops, pp. 466–467. (2020)
10. Saharia, C., Ho, J., Chan, W., Salimans, T., Fleet, D.J., Norouzi, M.: Image super-resolution via iterative refinement. IEEE Trans. Pattern Anal. Mach. Intell. **45**(4), 4713–4726 (2022)
11. Zhang, K., Gool, L.V., Timofte, R.: Deep unfolding network for image super-resolution. In: Proceedings of the IEEE/CVF Conference on Computer Vision and Pattern Recognition, pp. 3217–3226. (2020)
12. Bell-Kligler, S., Shocher, A., Irani, M.: Blind super-resolution kernel estimation using an internal-gan. Adv. Neural. Inf. Process. Syst. 32, (2019)
13. Shi, W., Caballero, J., Huszár, F., Totz, J., Aitken, A.P., Bishop, R., Rueckert, D., Wang: Z: Real-time single image and video super-resolution using an efficient sub-pixel convolutional neural network. In: Proceedings of the IEEE Conference on Computer Vision and Pattern Recognition, pp. 1874–1883. (2016)
14. Li, J., Fang, F., Mei, K., Zhang, G.: Multi-scale residual network for image super-resolution. In: Proceedings of the European Conference on Computer Vision (ECCV), pp. 517–532. (2018)

15. Mei, Y., Fan, Y., Zhou, Y.: Image super-resolution with non-local sparse attention. In: Proceedings of the IEEE/CVF Conference on Computer Vision and Pattern Recognition, pp. 3517–3526. (2021)
16. Zhang, K., Zuo, W., Zhang, L: Deep plug-and-play super-resolution for arbitrary blur kernels. In: Proceedings of the IEEE/CVF Conference on Computer Vision and Pattern Recognition, pp. 1671–1681 (2019)
17. Zhang, K., Liang, J., Gool, L.V., Timofte, R.: Designing a practical degradation model for deep blind image super-resolution. In: Proceedings of the IEEE/CVF International Conference on Computer Vision, pp. 4791–4800. (2021)
18. Xu, T., Li, L., Mi, P., Zheng, X., Chao, F., Ji, R., Tian, Y., Shen, Q.: Uncovering the over-smoothing challenge in image super-resolution: entropy-based quantification and contrastive optimization. IEEE Trans. Pattern Anal. Mach, Intell (2024)
19. Yang, C., Guanming, L.: Unsupervised image blind super resolution via real degradation feature learning. IET Comput. Vision **18**(4), 485–498 (2024)
20. Li, A., Zhang, L., Liu, Y., Zhu, C.: Feature modulation transformer: Cross-refinement of global representation via high-frequency prior for image super-resolution. In: Proceedings of the IEEE/CVF International Conference on Computer Vision, pp. 12514–12524 (2023)
21. Lei, X., Zhang, W., Cao, W.: DVMSR: distillated vision mamba for efficient super-resolution. In: Proceedings of the IEEE/CVF Conference on Computer Vision and Pattern Recognition, pp. 6536–6546. (2024)
22. Shocher, A., Cohen, N., Irani, M.: "Zero-shot" super-resolution using deep internal learning. In: Proceedings of the IEEE Conference on Computer Vision and Pattern Recognition, pp. 3118–3126 (2018)
23. Zhang, Z., Wang, R., Zhang, H., Zuo, W.: Self-supervised learning for real-world super-resolution from dual and multiple zoomed observations. IEEE Trans. Pattern Anal. Mach, Intell (2024)
24. Chen, H., Li, W., Gu, J., Ren, J., Sun, H., Zou, X., Zhang, Z., Yan, Y., Zhu, L.: Low-res leads the way: Improving generalization for super-resolution by self-supervised learning. In: Proceedings of the IEEE/CVF Conference on Computer Vision and Pattern Recognition, pp. 25857–25867. (2024)
25. Zhao, H., Kong, X., He, J., Qiao, Yu., Dong, C.: Efficient Image Super-Resolution Using Pixel Attention. In: Bartoli, A., Fusiello, A. (eds.) ECCV 2020. LNCS, vol. 12537, pp. 56–72. Springer, Cham (2020). https://doi.org/10.1007/978-3-030-67070-2_3
26. Li, W., Kun Zhou, L., Qi, N.J., Jiangbo, L., Jia, J.: Lapar: Linearly-assembled pixel-adaptive regression network for single image super-resolution and beyond. Adv. Neural. Inf. Process. Syst. **33**, 20343–20355 (2020)
27. Sun, L., Pan, J., Tang, J.: Shufflemixer: an efficient convnet for image super-resolution. Adv. Neural. Inf. Process. Syst. **35**, 17314–17326 (2022)
28. Zheng, M., Sun, L., Dong, J., Pan, J.: Smfanet: a lightweight self-modulation feature aggregation network for efficient image super-resolution. In: European Conference on Computer Vision, pp. 359–375. Springer (2024)
29. Liu, X., Liu, J., Tang, J., Wu, G.: Catanet: efficient content-aware token aggregation for lightweight image super-resolution. *arXiv preprint* arXiv:2503.06896 (2025)

30. Yue, Z., Zhao, Q., Xie, J., Zhang, L., Meng, D., Wong, K.-W.K.: Blind image super-resolution with elaborate degradation modeling on noise and kernel. In: Proceedings of the IEEE/CVF Conference on Computer Vision and Pattern Recognition, pp. 2128–2138. (2022)
31. Wang, X., Yu, K., Wu, S., Gu, J., Liu, Y., Dong, C., Qiao, Y., Loy, C.C.: Esrgan: enhanced super-resolution generative adversarial networks. In: Proceedings of the European Conference on Computer Vision (ECCV) Workshops, (2018)

DEATrack: Diffusion-Enhanced Autoregressive Framework for Visual Object Tracking

Xu Wang(✉) and Man Xu

Shanghai University of Electric Power, Shanghai 201306, China
xumanly@shiep.edu.cn

Abstract. Autoregressive approaches have become essential for capturing temporal information in visual tracking, especially under challenging conditions such as occlusions, fast movement, and significant appearance variations. However, conventional autoregressive models have a major shortcoming: their reliance on one-step-ahead prediction makes them highly susceptible to error accumulation. One wrong prediction can spread to later frames, resulting in tracking drift and severe performance degradation. To address this issue, we propose DEATrack—an efficient video-level tracking framework that enhances autoregressive tracking through the integration of denoising diffusion probabilistic models (DDPM). Our approach introduces a diffusion-based refinement process that improves the target state distribution, thereby enhancing both robustness and prediction accuracy. Additionally, we incorporate flash attention to optimize the computation of attention, allowing for faster training with less memory usage. Extensive experiments prove that the proposed temporally-aware diffusion framework not only reduce the propagation of mistakes but also achieves excellent results across several major tracking benchmarks.

Keywords: Transformer visual tracking · Autoregressive generation · Diffusion model

1 Introduction

Transformer-based object tracking has become an important area of computer vision, which can effectively capture long-range dependencies and global context. This approach is now widely employed in practical applications such as robotic vision, video surveillance [1], autonomous driving [2,3] and various other domains. However, tracking accuracy is often affected by several challenging factors, including target appearance changes, occlusion, and fast motion. Due to the above challenges, current mainstream tracking algorithms struggle to effectively distinguish targets when relying only on static appearance information. It is necessary to incorporate temporal information into the tracking process.

L. Huang and D. Greenhalgh (Eds.): ICMLC 2026, LNNS 2061, pp. 165–177, 2027.
https://doi.org/10.1007/978-3-032-31236-5_11

Autoregressive models [4], which predict each frame sequentially based on previous results, introduce temporal information into the tracking process by using coordinate sequences. Nonetheless, a key limitation persists: existing autoregressive models inadequately represent the probabilistic nature of the target's positional changes over time.

To overcome this limitation, we propose DEATrack, a novel tracking framework that integrates the denoising capability of diffusion models into an autoregressive tracking system. a novel tracking framework that integrates the denoising capabilities of diffusion models into an autoregressive tracking system. To our knowledge, this work represents the first attempt to combine diffusion-based generative modeling with autoregressive inference for visual tracking. By incorporating a diffusion process into temporal modeling, DEATrack more effectively captures the dynamics of target motion and appearance evolution. The core insight is to employ diffusion as a form of temporal regularization, enabling the model to learn not only inter-frame feature correlations but also the underlying distribution of target trajectories.

Diffusion models have demonstrated remarkable success in image generation and beyond, owing to their powerful generative capacity and flexible probabilistic formulation. Their integration with autoregressive mechanisms further augments temporal coherence and predictive accuracy. Extensive experiments show that our proposed method achieves significant improvements across multiple benchmarks. Specifically, DEATrack attains an AO (Average Overlap) score of 75.9% on GOT-10k [5], surpassing OSTrack by 4.9%. It also achieves strong generalization performance, with success rates of 71.5% on LaSOT, 49.3% on $\text{LaSOT}_{\text{ext}}$, and 84.5% on TrackingNet. Our main contributions can be summarized as follows:

1. We introduce a novel autoregressive tracking framework, which effectively captures temporal target variations through probabilistic trajectory learning.
2. We design a temporally-aware diffusion mechanism that enhances prediction accuracy at multiple granularity levels by leveraging the complementary strengths of diffusion models and sequential modeling.
3. We incorporate flash attention to optimize attention-based computations, significantly improving training efficiency and reducing computational overhead.

2 Related Work

Transformer-based architecture has changed visual tracking because they can model long range dependencies better than other architectures. Works such as SwinTrack [6] use a Transformer backbone to jointly perform feature extraction and target search region interaction, which results in more precise matches. MixFormer [7] further simplifies the pipeline through a mixed attention mechanism that combines feature extraction and fusion, thus no longer requiring separate modules or complicated bounding box refinement. Based on these, OSTrack [8] emerges as a top single-stream Transformer tracker, fusing feature extraction and target interaction into one backbone and embedding an early candidate elimination approach inside ViT [9], resulting in high efficiency and strong performance.

Beyond architectural innovations, the field has witnessed a paradigm shift towards generative tracking, where methods like SeqTrack [10] and ARTrack [4] reformulate tracking as a sequence generation task. They use Transformer-based autoregressive decoders to directly generate bounding box sequences in an end-to-end manner by tokenizing the object state and iteratively predicting the next state given previous estimates, without needing any hand-crafted heads. Inspired by this, our work introduces an enhanced autoregressive framework that leverages multi-frame motion context to strengthen temporal coherence and mitigate error accumulation, enabling target recovery during occlusions or appearance ambiguities.

Meanwhile, denoising diffusion probabilistic model(DDPM) [11] has the ability to model uncertainties and iterative updates, which is conducive to more difficult tracking tasks. To this end, we add a diffusion module to our autoregressive pipeline to model the distribution of possible target states explicitly. as the reverse diffusion process occurs, it gradually decreases the localization uncertainty, greatly improving the robustness and accuracy of the entire tracking system.

3 Method

3.1 Tracking via Sequence Modeling

Traditional visual object tracking methods often treat tracking as a frame-by-frame template matching task, ignoring the temporal relationships between video frames. In this work, we use the autoregressive temporal modeling approach from the ARTrack framework. This approach views tracking as a coordinate sequence generation problem, similar to how sequence prediction is done in language modeling. The main idea is to use an autoregressive model to predict the target's position in the current frame based on its positions in the previous N frames. This helps to unify the training and inference process and better capture how the target moves over time.

As shown in Fig. 1, the overall model follows an encoder-decoder structure. The encoder is based on Vision Transformer(ViT) and is used to extract visual features from the template and the search region. The decoder is also Transformer-based and takes the coordinate tokens from the past N frames as input. It uses self-attention and cross-attention to combine the temporal information and visual features, and then gradually generates the coordinates for thecurrent frame. First, we discretize the consecutive target box $[x_{\min}, y_{\min}, x_{\max}, y_{\max}]$ coordinates into integer indexes, with each coordinate value corresponding to a discrete vocabulary. The coordinate sequence of the first N frames implies the motion trend of the target. The autoregressive timing formula is:

$$Y_t = f(Y_{t-N_t-1}, C, Z, X_t). \tag{1}$$

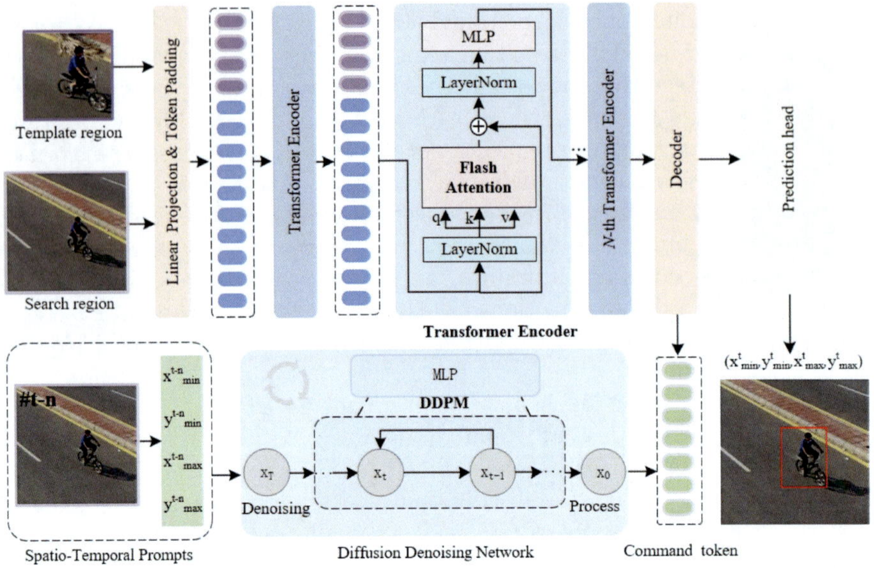

Fig. 1. The overall architecture diagram of DEATrack. It consists of encoder, a decoder, time series prediction module and a prediction head. Here DDPM is Denoising Diffusion Probabilistic Model.

In Formula (1), Y_t is the target sequence of the current time step, Y_{t-N_t-1} is the estimation result of the previous time step, C is the command tag, Z is the template image, and X_t is the search image. Suppose $N = 2$. The coordinate prediction of the current frame t depends on the previous two frames (t and $t-1$) :

$$Y_{t+1} = Decoder(Y_t, Y_{t-1}, C, Encoder(Z, X_t)). \tag{2}$$

By leveraging autoregressive temporal modeling, visual object tracking is reformulated as a sequence prediction task, where the trajectory information from previous time steps guides the estimation at the current time step. This leads to more accurate and coherent tracking results. The proposed approach not only simplifies the tracking pipeline but also achieves notable performance improvements across various benchmark datasets.

3.2 Sequential Aware Diffusion Denoising Mechanism

To address the critical issue of error accumulation in autoregressive tracking, we introduce a novel denoising framework based on Denoising Diffusion Probabilistic Models(DDPM). As shown in Fig. 2, this mechanism is seamlessly integrated into the spatio-temporal prompt architecture, where the autoregressive model generates an informative context vector Z_i that conditions the diffusion process. A lightweight denoising network then utilizes this conditioning to refine the target state estimation by learning to reverse a structured noise injection procedure, effectively optimizing the conditional distribution $p(x_i \mid z_i)$ through noise prediction.

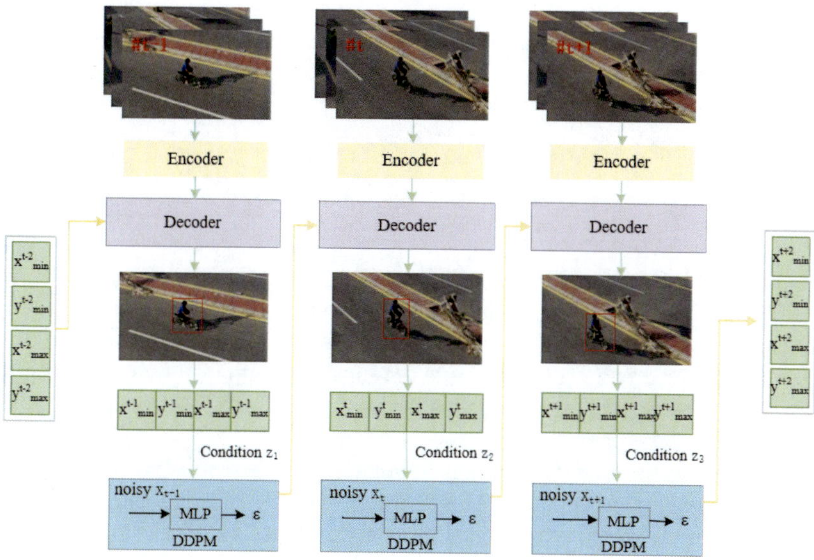

Fig. 2. Architecture of the sequential aware diffusion denoising mechanism. Green squares represent sequential tokens, yellow modules depict the diffusion denoising function, and blue arrows indicate the conditioning flow from the autoregressive encoder.

Given a ground-truth state token x_0(representing target coordinates), we define a Markov chain that gradually adds Gaussian noise over $T = 1000$ steps according to a linear variance schedule $\beta_t{}_{t=1}^{T}$ where $\beta_1 = 0.0001$, $\beta_T = 0.02$, and β_t linearly interpolates between them $(0 < \beta_1 < \beta_2 < \cdots < \beta_T < 1)$. The forward process yields a sequence of latent variables $x_1, x_2, \ldots, x_T$ through the following transition:

$$q(x_t|x_{t-1}) = N(x_t; \sqrt{1-\beta_t}x_{t-1}, \beta_t\mathbf{I}). \tag{3}$$

The reverse denoising process aims to recover the original data by iteratively removing noise. We parameterize the transition distribution $p_\theta(x_{t-1}|x_t)$ as:

$$p_\theta(x_{t-1}|x_t) = N(x_{t-1}; \mu_\theta(x_t, t, z_i), \sigma_t^2\mathbf{I}), \tag{4}$$

where z_i is the conditioning vector from the autoregressive model. Following the common practice, we fix the variance to $\sigma_t^2 = \frac{1-\bar{\alpha}t-1}{1-\bar{\alpha}t}\beta_t$ and learn the mean $\mu\theta(x_t, t, z_i)$ through a noise prediction network ε_θ.

We implement ε_θ as a lightweight Multi-Layer Perceptron (MLP) that takes three inputs: the noisy state x_t, the diffusion timestep t (encoded via sinusoidal positional embeddings), and the conditioning vector z_i. The inputs are concatenated and processed as:

$$\varepsilon_\theta(x_t, t, z_i) = \text{MLP}(\text{Concat}(\text{Linear}(x_t), \text{PE}(t), \text{Linear}(z_i))), \tag{5}$$

where PE(·) denotes sinusoidal positional encoding and Linear(·) denotes linear projection layers. During inference, we generate samples from $p(x_i|z_i)$ by initializing from pure noise $x_T \sim N(0, \mathbf{I})$ and performing T reverse steps. Each reverse step is computed as:

$$x_{t-1} = \frac{1}{\sqrt{\alpha_t}}\left(x_t - \frac{\beta_t}{\sqrt{1-\bar{\alpha}t}}\varepsilon\theta(x_t, t, z_i)\right) + \sigma_t\delta, \tag{6}$$

where $\delta \sim N(0, \mathbf{I})$. The term α_t is the noise retention factor at step t,$\bar{\alpha}t = \prod s = 1^t\alpha_s$ is the cumulative product of α_s from step 1 to t, representing the total amount of signal retained. The term $\sigma_t\delta$ represents the stochastic component, whose magnitude can be controlled via a temperature parameter τ (scaling σ_t) to adjust the diversity of predictions. The final prediction $\hat{x}_0$ is obtained after T denoising steps.

This diffusion-based refinement mechanism enables robust tracking by progressively correcting erroneous states through conditional denoising, effectively breaking the chain of error propagation in autoregressive prediction.

3.3 Efficient Self-attention Calculation

To alleviate memory consumption and computational overhead, we adopt flash attention [12], an efficient attention mechanism that optimizes matrix operations within the attention module. This approach reduces the memory complexity of attention computation from $O(n^2)$ to $O(n)$.

Given input matrices $Q \in \mathbb{R}^{N\times d}$, $K \in \mathbb{R}^{N\times d}$, and $V \in \mathbb{R}^{N\times d}$, where N denotes the sequence length and d the feature dimension, these matrices are partitioned into smaller blocks to accommodate the shared memory constraints of the GPU. Specifically, Q is divided row-wise into blocks $Q_1, Q_2, \ldots, Q_B$, each of size $r\times d$, whereas K and V are partitioned column-wise into $K_1, K_2, \ldots, K_B$ and $V_1, V_2, \ldots, V_B$, respectively, with each block of size $c \times d$. The block dimensions r and c are dynamically configured according to the available SRAM capacity.

The computation proceeds via an outer loop over each block Q_i of Q, and an inner loop over the corresponding blocks K_j and V_j of K and V. During each iteration, the current blocks are loaded from high-bandwidth memory (HBM) into SRAM, and a local attention score is computed as:

$$S_{i,j} = Q_i K_j^\top \in \mathbb{R}^{r\times c}. \tag{7}$$

Subsequently, row-wise Softmax is applied:

$$P_{i,j} = \text{Softmax}(S_{i,j}) \in \mathbb{R}^{r\times c}, \tag{8}$$

$$O_{i,j} = P_{i,j} V_j \in \mathbb{R}^{r\times d}. \tag{9}$$

Throughout this process, the maximum value m_i and the exponential sum l_i are recorded for each row. These statistics facilitate a weighted aggregation of the local outputs $O_{i,j}$, which are merged into a global output matrix O using appropriate normalization. The final output is synthesized as:

$$O = \text{Aggregation}(O_{11}, O_{12}, \dots, O_{BB}) \in \mathbb{R}^{N \times d}. \tag{10}$$

By decomposing the input into blocks and performing attention computation locally, flash attention significantly minimizes read and write operations between the GPU and high-bandwidth memory, thereby enhancing both memory efficiency and computational throughput.

4 Experiment

4.1 Results and Comparisons

We conduct comprehensive evaluations to validate the effectiveness of our proposed method across four challenging tracking benchmarks: GOT-10k [6], LaSOT [13], $\text{LaSOT}_{\text{ext}}$ [14], TrackingNet [15], and UAV123 [16]. All experiments are conducted on NVIDIA RTX 4090 GPUs. For GOT-10k, we follow the standard protocol by training exclusively on its training split, with the first stage trained for 100 epochs and the second stage for 30 epochs. For LaSOT, TrackingNet, and UAV123 evaluations, we employ a joint training strategy using GOT-10k, LaSOT, TrackingNet, and COCO datasets, with the first stage trained for 240 epochs and the second stage for 60 epochs, followed by independent evaluation on each benchmark.

Table 1. Comparisons on GOT-10k, LaSOT, $\text{LaSOT}_{\text{ext}}$ and TrackingNet.

Methods	GOT-10k			LaSOT			$\text{LaSOT}_{\text{ext}}$			TrackingNet		
	AO (%)	$\text{SR}_{0.5}$ (%)	$\text{SR}_{0.75}$ (%)	AUC (%)	P_{norm} (%)	P (%)	AUC (%)	P_{norm} (%)	P (%)	AUC (%)	P_{norm} (%)	P (%)
SiamRPN++ [17]	51.7	61.6	32.5	49.6	56.9	49.1	34.0	41.6	39.6	73.3	80.0	69.4
DiMP [18]	61.1	71.7	49.2	56.9	65.0	56.7	39.4	47.6	45.1	74.0	80.1	68.7
SiamR-CNN [19]	64.9	72.8	59.7	64.8	72.7	–	–	–	–	81.2	84.5	80.0
TransT [20]	67.1	76.8	60.9	64.9	73.8	69.0	–	–	–	81.4	86.7	80.1
AutoMatch [21]	65.2	76.6	54.3	58.3	–	59.9	–	–	–	76.0	–	72.6
STARK [22]	68.8	78.1	64.1	67.1	77.0	–	–	–	–	82.0	86.9	–
SparseTT [23]	69.3	79.1	63.8	66.0	74.8	70.1	–	–	–	84.1	86.6	79.5
SwinTrack [6]	72.4	–	67.8	71.3	–	76.5	–	–	53.9	84.0	–	82.8
AiATrack [24]	69.6	80.0	63.2	69.0	79.4	73.8	–	–	–	82.7	87.8	80.4
SBT [25]	70.4	80.1	64.7	66.7	–	71.7	–	–	–	–	–	–
Mixformer [7]	70.7	80.0	65.2	69.2	78.7	74.7	–	–	–	83.1	88.1	81.6
SimTrack [26]	69.8	78.6	66.0	70.5	79.7	–	–	–	–	83.4	87.4	–
OSTrack [8]	71.0	80.4	68.2	69.1	78.7	74.7	47.4	57.3	53.3	83.7	87.2	82.0
GRM [27]	73.4	82.9	70.2	69.9	79.3	75.8	–	–	–	84.0	88.7	83.3
ARTrack [4]	73.5	82.2	70.9	70.4	79.5	76.6	46.5	56.2	52.3	84.2	88.7	83.5
AQATrack [28]	73.8	83.2	**72.1**	71.4	**81.9**	**78.6**	**51.2**	**62.2**	**58.9**	–	–	–
DEATrack (ours)	**75.9**	**85.7**	71.6	**71.5**	80.5	77.4	49.3	58.9	55.4	**84.5**	**88.8**	**83.5**

GOT-10k Evaluation GOT-10k contains over 10,000 video sequences with 1.5 million manually annotated bounding boxes, specifically designed to test model generalization across unseen object categories and complex motion patterns. As shown in Table 1, our method achieves state-of-the-art performance on the GOT-10k test set, obtaining an Average Overlap (AO) of **75.9%**, Success Rate at 0.5 (SR0.5) of **85.7%**, and Success Rate at 0.75 (SR0.75) of **71.6%**. These results demonstrate a significant improvement over previous methods, particularly in handling category generalization challenges, which we attribute to our diffusion-based refinement mechanism that enhances prediction robustness.

LaSOT Evaluation LaSOT offers 1,400 long-term video sequences having an average length of 2500 frames for assessing tracking stability over extended periods. Our method achieves superior performance with an AUC of **71.5%**, normalized precision (P_{norm}) of **80.5%**, and precision (P) of **77.4%**. The Significant enhancement in long term tracking results verifies that our method can reduce errors as time goes by, which makes traditional autoregressive models drift badly over long sequence.

LaSOT$_{ext}$ Evaluation LaSOT$_{ext}$ Evaluation LaSOT$_{ext}$ is an extension of the LaSOT dataset that includes more difficult situations and complicated changes in the target. Our method shows a considerable improvement in performance on LaSOText, achieving an AUC score of **49.3%**, normalized precision (Pnorm) of **58.9%**, and precision (P) of **55.4%**. This outstanding performance on this benchmark proves that our approach can deal with extremely hard tracking situations.

TrackingNet Evaluation TrackingNet has more than 30,000 YouTube videos (30,132 for training, 511 for testing), which contain various real-life problems such as changes in lighting conditions, motion blurring, and camera movements. Our method achieves competitive results with an AUC of **84.5%**, normalized precision of **88.8%**, and precision of **83.5%**. The strong performance on this large-scale real-world dataset demonstrates the generalization capability of our diffusion-enhanced tracking framework across diverse scenarios.

UAV123 Evaluation UAV123 has 123 aerial videos shot by unmanned aerial vehicles, which have special problems such as tiny targets, camera movement, and changing views. Our method achieves an AUC of **68.1%**, normalized precision of **83.4%**, and precision of **88.7%**. The outstanding performance in aerial tracking scenarios highlights the robustness of our approach in handling small target detection and complex motion patterns characteristic of UAV-based tracking.

4.2 Ablation Study and Analysis

To evaluate the contribution of each component in our proposed method, we conduct comprehensive ablation studies on multiple benchmarks. The experiments are designed to validate the effectiveness of the diffusion denoising mechanism, the autoregressive formulation, and the efficiency optimization introduced by flash attention.

Effect of Diffusion Denoising Mechanism. We first analyze the impact of the proposed diffusion loss on tracking performance. As shown in Table 2, DEATrack is trained in two stages: first pre-trained on GOT-10k for 120 epochs, then fine-tuned on UAV123 and LaSOT for 30 epochs. The incorporation of the diffusion denoising module consistently improves performance across both datasets. On UAV123, slight improvements are observed in AUC, OP50, Precision, and Norm Precision. A more significant gain is achieved on LaSOT, with AUC increasing by 1.3% and OP50 improving by 1.2%, indicating that the diffusion mechanism effectively enhances tracking robustness, especially in challenging scenarios.

Table 2. Ablation experiments of DIffloss on UAV123 and LaSOT datasets. W/o is the result after adding the module, and W is the result after adding the module.

Dataset	Loss	AUC	OP50	OP75	Precision	Norm precision
uav123	W/o	66.2	82.5	54.7	89.1	80.7
	W	66.4	82.8	54.8	89.3	81.4
LaSOT	W/o	64.5	75.1	61.9	68.9	72.6
	W	65.8	76.3	63.0	70.1	73.7

Contribution of Individual Components. We further ablate the influence of key components in DEATrack, including the Candidate Elimination (CE) mechanism, SIoU bounding box loss, Autoregressive (AR) coordinate decoding, and the proposed diffusion module. Results are summarized in Table 3. The baseline without AR or diffusion (Row 1) yields an AO of 70.1% on GOT-10k. Introducing AR modeling (Rows 24) brings substantial gains, with the combination of CE and AR (Row 3) reaching 73.8%. The full model with all components (Row 5) achieves the best performance of 75.9%, confirming that the autoregressive framework and diffusion module contribute most significantly to performance improvement.

Efficiency Improvement via Flash Attention. We also evaluate the computational efficiency introduced by flash attention, particularly in the context of integrating diffusion-based temporal modeling into the object tracking framework, which typically increases training time due to its iterative refinement process. As shown in Table 4, replacing standard attention with flash attention on the L20 GPU leads to a modest but consistent improvement in inference

Table 3. Ablation experiments of different components.

#	CE	SIOU	AR	Diffusion	AO(%)
1	√	×	×	×	70.1
2	×	√	√	×	72.2
3	√	×	√	×	73.8
4	√	√	√	×	74.0
5	√	√	√	√	75.9

speed, increasing from 11.78 FPS to 12.10 FPS. The total processing time is reduced from 24.47s to 23.82s, with decreases observed in both data loading and forward pass durations. These results indicate that Flash Attention contributes to optimizations in memory usage and attention computation, thereby helping to mitigate the additional training cost introduced by the diffusion module. Although the performance gains are more pronounced on high-end hardware such as the A100, the method still enables more efficient training and inference in resource-constrained environments.

Table 4. Ablation experiment of flash attention.

Modification	Correlation refinement	FPS	Data time(s)	Forward time (s)	Total time (s)
Flash attention	×	11.779	1.458	9.133	24.469
	√	12.101	1.433	8.535	23.821

4.3 Visualization of Tracking Results

To visually validate the robustness and generalization of our method under challenging conditions, we present qualitative tracking results across diverse scenarios, including occluded outdoor scenes, wild environments, fast-moving sports sequences, and blurred underwater settings.

As shown in Fig. 3, DEArack can always generate correct and stable bounding boxes even when there are great appearance changes and motion distortions. In the wild scenes with frequent light changes (first row), our method keeps the target location accurately, whereas other trackers tend to drift because of sudden light changes or partial occlusions. In high speed sports situations (second row), which have quick movements and complicated backgrounds, DEATrack can follow the target smoothly without delay or disappearing bounding box, showing its good ability to stay consistent over time and understand movement. Lastly, for the low visibility underwater environment with motion blur and color distortion (third row), our method still produces good bounding boxes, showing that it can deal with severe image degradation.

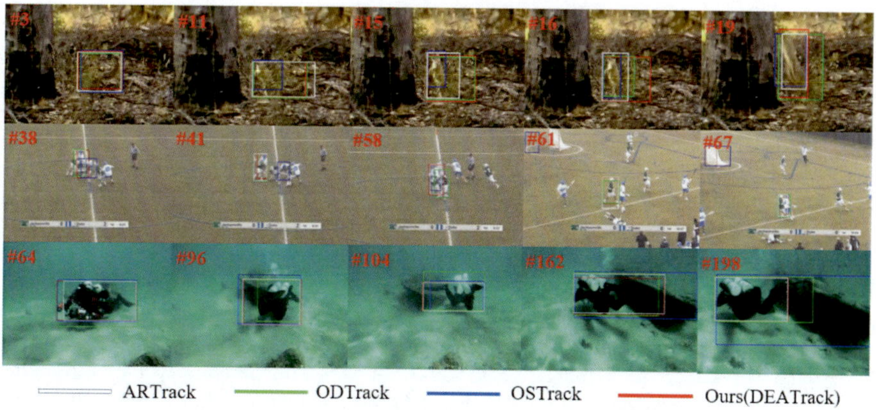

Fig. 3. Comparative visualization of tracking results from four methods on the GOT-10k dataset. The trackers include ARTrack, ODTrack, OSTrack, and our method DEATrack. DEATrack demonstrates robust performance under partial and full occlusion.

These visual results align with our quantitative findings and further confirm that the integration of diffusion-based refinement and efficient attention mechanisms contributes to superior performance under real-world challenging conditions.

5 Conclusion

DEATrack combines autoregressive temporal modeling with the denoising power of diffusion models to form an efficient and adaptable tracking system that greatly enhances tracking accuracy and stability. The main innovation of DEATrack is the use of a progressive denoising process from diffusion models to optimize the target state distribution at multiple levels. By gradually removing noise, the diffusion model can more reliably recover the true motion path of the target, helping to reduce the error accumulation problem common in autoregressive frameworks. This hybrid design enables the model to learn more complicated patterns of target states, such as alterations in location and size, which makes it more useful for difficult circumstances such as obstructions and deformations. Since diffusion models are computationally demanding, flash attention is used to speed up training.

References

1. Cheng, L., Wang, J., Li, Y.: ViTrack: efficient tracking on the edge for commodity video surveillance systems. In: IEEE Transactions on Parallel and Distributed Systems, pp. 723–735. IEEE (2022)

2. Ettinger, S., Cheng, S., Caine, B., Liu, C., Zhao, H., Pradhan, S., Chai, Y., Sapp, B., Qi, C.R., Zhou, Y., Yang, Z., Chouard, A., Sun, P., Ngiam, J., Vasudevan, V., McCauley, A., Shlens, J., Anguelov, D.: Large scale interactive motion forecasting for autonomous driving: The waymo open motion dataset. In: Proceedings of the IEEE/CVF International Conference on Computer Vision (ICCV), pp. 9710–9719 (2021)
3. Premachandra, C., Ueda, S., Suzuki, Y.: Detection and tracking of moving objects at road intersections using a 360-degree camera for driver assistance and automated driving, pp. 135652–135660. IEEE Access (2020)
4. Xing, W., Yifan, B., Yongchao, Z., Dahu, S., Yihong, G.: Autoregressive visual tracking. In: Proceedings of the IEEE/CVF Conference on Computer Vision and Pattern Recognition (CVPR), pp. 9697–9706 (2023)
5. Huang, L., Zhao, X., Huang, K.: Got-10k: A large high-diversity benchmark for generic object tracking in the wild. IEEE Trans. Pattern Anal. Mach. Intell. **43**(5), 1562–1577 (2021)
6. Lin, L., Fan, H., Zhang, Z., Xu, Y., Ling, H.: Swintrack: a simple and strong baseline for transformer tracking. Adv. Neural Inf. Process. Syst. 16743–16754 (2022)
7. Cui, Y., Jiang, C., Wang, L., Wu, G.: Mixformer: end-to-end tracking with iterative mixed attention. In: Proceedings of the IEEE/CVF Conference on Computer Vision and Pattern Recognition (CVPR), pp. 13608–13618 (2022)
8. Ye, B., Chang, H., Ma, B., Shan, S., Chen, X.: Joint feature learning and relation modeling for tracking: a one-stream framework. In: Proceedings of the European Conference on Computer Vision (ECCV), pp. 341–357 (2022)
9. Dosovitskiy, A., Beyer, L., Kolesnikov, A., Weissenborn, D., Zhai, X., Unterthiner, T., Dehghani, M., Minderer, M., Heigold, G., Gelly, S., Uszkoreit, J., Houlsby, N.: An image is worth 16x16 words: transformers for image recognition at scale. In: Proceedings of the International Conference on Learning Representations (ICLR) (2021)
10. Chen, X., Kang, B., Zhu, J., Wang, D., Peng, H., Lu, H.: Unified sequence-to- sequence learning for single- and multi-modal visual object tracking (2024). arXiv:2304.14394 arXiv preprint
11. Ho, J., Jain, A., Abbeel, P.: Denoising diffusion probabilistic models (2020). arXiv:2006.11239 arXiv preprint
12. Dao, T.: Flashattention: Fast and memory-efficient exact attention with io-awareness. In: Advances in Neural Information Processing Systems, pp. 16344–16359 (2022)
13. Fan, H., Lin, L., Yang, F., Chu, P., Deng, G., Yu, S., Bai, H., Xu, Y., Liao, C., Ling, H.: Lasot: a high-quality benchmark for large-scale single object tracking. In: Proceedings of the IEEE/CVF Conference on Computer Vision and Pattern Recognition (CVPR), pp. 5374–5383 (2019)
14. Fan, H., Bai, H., Lin, L., Yang, F., Chu, P., Deng, G., Yu, S., Huang, M., Liu, J., Xu, Y., et al.: Lasot: a high-quality large-scale single object tracking benchmark. Int. J. Comput. Vis. **129**, 439–461 (2021)
15. Muller, M., Bibi, A., Giancola, S., Alsubaihi, S., Ghanem, B.: Trackingnet: a large-scale dataset and benchmark for object tracking in the wild. In: Proceedings of the European Conference on Computer Vision (ECCV), pp. 300–317 (2018)
16. Mueller, M., Smith, N., Ghanem, B.: A benchmark and simulator for UAV tracking. In: Proceedings of the European Conference on Computer Vision (ECCV), pp. 445–461 (2016)

17. Li, B., Wu, W., Wang, Q., Zhang, F., Xing, J., Yan, J.: SiamRPN++: evolution of siamese visual tracking with very deep networks. In: Proceedings of the IEEE/CVF Conference on Computer Vision and Pattern Recognition (CVPR), pp. 4282–4291 (2019)
18. Bhat, G., Danelljan, M., Van Gool, L., Timofte, R.: Learning discriminative model prediction for tracking. In: Proceedings of the IEEE/CVF International Conference on Computer Vision (ICCV), pp. 6182–6191 (2019)
19. Voigtlaender, P., Luiten, J., Torr, P.H.S., Leibe, B.: Siam R-CNN: visual tracking by redetection. In: Proceedings of the IEEE/CVF Conference on Computer Vision and Pattern Recognition (CVPR), pp. 6578–6588 (2020)
20. Chen, X., Yan, B., Zhu, J., Wang, D., Yang, X., Lu, H.: Transformer tracking. In: Proceedings of the IEEE/CVF Conference on Computer Vision and Pattern Recognition (CVPR), pp. 8126–8135 (2021)
21. Zhang, Z., Liu, Y., Wang, X., Li, B., Hu, W.: Learn-to-match: Automatic matching network design for visual tracking. In: Proceedings of the IEEE/CVF International Conference on Computer Vision (ICCV), pp. 13339–13348 (2021)
22. Yan, B., Peng, H., Fu, J., Wang, D., Lu, H.: Learning spatio-temporal transformer for visual tracking. In: Proceedings of the IEEE/CVF International Conference on Computer Vision (ICCV), pp. 10448–10457 (2021)
23. Fu, Z., Fu, Z., Liu, Q., Cai, W., Wang, Y.: Sparsett: visual tracking with sparse transformers. In: Proceedings of the Thirtieth International Joint Conference on Artificial Intelligence (IJCAI), pp. 905–912 (2022)
24. Gao, S., Zhou, C., Ma, C., Wang, X., Yuan, J.: Aiatrack: attention in attention for transformer visual tracking. In: Proceedings of the European Conference on Computer Vision (ECCV), pp. 146–164 (2022)
25. Xie, F., Wang, C., Wang, G., Cao, Y., Yang, W., Zeng, W.: Correlation-aware deep tracking. In: Proceedings of the IEEE/CVF Conference on Computer Vision and Pattern Recognition (CVPR), pp. 8751–8760 (2022)
26. Chen, B., Li, P., Bai, L., Qiao, L., Shen, Q., Li, B., Gan, W., Wu, W., Ouyang, W.: Backbone is all your need: a simplified architecture for visual object tracking. In: Proceedings of the European Conference on Computer Vision (ECCV), pp. 375–392 (2022)
27. Gao, S., Zhou, C., Zhang, J.: Generalized relation modeling for transformer tracking. In: Proceedings of the IEEE/CVF Conference on Computer Vision and Pattern Recognition (CVPR), pp. 18686–18695 (2023)
28. Xie, J., Zhong, B., Mo, Z., Zhang, S., Shi, L., Song, S., Ji, R.: Autoregressive queries for adaptive tracking with spatio-temporal transformers. In: Proceedings of the IEEE/CVF Conference on Computer Vision and Pattern Recognition (CVPR), pp. 19300–19309 (2024)

DBR: A Dual-Branch Feature Processing and Fusion Framework for Post-Processing Boundary Refinement in Object Segmentation

Junsheng Fu and Ran Shi(✉)

School of Computer Science and Engineering, Nanjing University of Science and Technology, Nanjing, China
{123106222859,rshi}@njust.edu.cn

Abstract. Object segmentation often suffers from blurred mask boundaries due to low-resolution feature maps and the tiny fraction of boundary pixels. To address these two issues, the BPR method proposed a post-processing optimization technique based on boundary patch cropping to further refine boundary accuracy. However, extracting features from the cropped boundary regions makes BPR be more sensitive to fine details. Over-reliance on such detailed information is not suitable for all samples. It means that some cases with initially inaccurate segmentation require semantic information to refine them. Based on this insight, we design a dual-branch feature processing and fusion method for boundary patch cropping optimization, termed DBR, which balances the expressive capability of detailed and semantic information and reduces the model's excessive dependence on sensitive details. During the fusion process, the model can selectively filter and utilize features according to different objects, leading to improved performance across various objects. The proposed DBR method can outperform other post-processing optimization methods such as BPR and SegFix on the Cityscapes dataset.

Keywords: Boundary Patch Cropping · Post-processing Optimization · Boundary optimization

1 Introduction

In the field of computer vision, object segmentation is an important topic, whose goal is to assign specific labels to every pixel in an image, thereby precisely extracting specified object types or individuals from the image. As a technical branch of mainstream segmentation tasks, such as semantic, instance, and panoptic segmentation, it has underpinned significant advancements in these tasks over the past few years and has seen widespread adoption across numerous fields, including robotics, autonomous driving, and medical imaging.

In object segmentation tasks, two primary issues [1] have consistently constrained the performance improvement of segmentation models. First, network feature maps typically have far lower spatial resolution than the input. Many object segmentation

L. Huang and D. Greenhalgh (Eds.): ICMLC 2026, LNNS 2061, pp. 178–189, 2027.
https://doi.org/10.1007/978-3-032-31236-5_12

models produce output feature maps that are one-quarter the resolution of the input image. Consequently, segmentation results generated from these features need to be up-sampled to the original image size, introducing substantial errors during the up-sampling process. Second, boundary pixels constitute a very small proportion of the total pixels. Most segmentation methods, when optimizing boundaries, often guide the model to prioritize feature information from boundary regions. However, the scarcity of boundary pixels means that boundary-related features are also scarce and easily overwhelmed within the overall feature representation. This makes it difficult for the model to capture precise boundary cues, ultimately leading to imprecise segmentation results.

Existing object segmentation methods widely focus on the following two directions to optimize the boundaries of segmentation results: feature enhancement and boundary optimization. However, these methods primarily target the network model itself—either by optimizing the backbone network to obtain more expressive features or by introducing additional mechanisms, such as boundary attention mechanisms or boundary detection mechanisms—to empower the model to process and utilize feature information, thereby improving boundary prediction accuracy. Regarding the two aforementioned fundamental issues, such optimization methods have not effectively addressed them. In contrast, BPR [1] proposed a novel instance segmentation optimization method based on boundary patch cropping. It calculates a series of continuous cropping boxes around the boundary of each object's segmentation mask, which are used to crop the image and the segmentation mask. A network model is then employed to perform re-segmentation optimization on these boundary patches, aiming to further improve segmentation accuracy. While this idea addresses two inherent challenges in object segmentation, it also introduces the following new problems.

In boundary patch cropping optimization methods, the model is dedicated to acquiring features from the boundary region at the object-background interface. In natural images, the boundary regions of objects often involve overlaps between different objects, where differences in color, texture, etc.. Consequently, the detailed information within the features becomes more sensitive. For objects that are accurately localized, the model can rely on this sensitive detailed information to precisely segment the object boundaries. However, for objects that are inaccurately localized, the model lacks guidance from semantic information and relies solely on the sensitive details in the features, easily leading to incomplete object segmentation. Therefore, the segmentation results for such samples remain unsatisfactory. As shown in Fig. 1, due to the localization failure of the side mirror in the Mask R-CNN segmentation results, BPR fails to achieve an accurate segmentation, whereas DBR can accomplish correct segmentation by leveraging semantic information.

In this paper, to address the issues arising from boundary patch cropping optimization methods, it proposes a new boundary patch cropping optimization approach—DBR (Dual-Branch Refinement). DBR first feeds multi-level features from the backbone into two branches: a semantic branch and a detail branch. The purpose of these branches is to enhance the missing semantic information and detailed information in different features, achieving a balance of information within the features and reducing the model's excessive reliance on sensitive detailed information. Subsequently, the dual-branch features are fused through a branch fusion module, and a gated fusion mechanism is employed to generate feature weights for weighting the fused features. This enables the model to adaptively attend to pivotal regions in the fused features, selectively filtering and utilizing feature information according to different objects. As illustrated in Fig. 1, DBR does not rely entirely on detailed information but rather leverages semantic information more effectively for such samples, achieving more precise boundary optimization.

Fig. 1. Segmentation results of three methods. The right side presents the segmentation result generated by Mask R-CNN, the middle shows the optimized result of BPR, and the left side shows the optimized result of our proposed DBR.

2 Related Work

In research on object segmentation tasks, models following the two-stage strategy of "detect-then-segment" can achieve better segmentation performance. However, the development of detection capabilities has progressed more rapidly, and the localization of objects in images has become quite accurate. Consequently, many recent studies have shifted their focus toward segmentation optimization. We categorize segmentation optimization methods into two main types: non-post-processing optimization methods and post-processing optimization methods.

Among non-post-processing optimization methods, these approaches primarily optimize mask generation from the following directions: feature enhancement and boundary optimization. However, these methods are designed specifically for particular network models. By optimizing the backbone network, they aim to obtain features with stronger expressive capabilities or introduce additional mechanisms, such as boundary attention mechanisms or boundary detection mechanisms, to strengthen the model's ability to process and utilize feature information, thereby achieving more refined boundary segmentation. In terms of feature enhancement, in 2021, Jianmin Li et al. proposed RefineMask [2], which introduces a multi-stage mask refinement framework based on fine-grained features. Unlike traditional methods that predict masks only on low-resolution RoI features, this approach additionally incorporates high-resolution semantic features and progressively refines the mask through multi-stage fusion during the instance segmentation process. Regarding boundary optimization, in 2022, Lei Ke et al. [3] proposed a novel refinement method called Mask Transfiner. They observed that due to the low resolution of features, significant errors occur during the upsampling process of predicted mask boundaries. Therefore, they correct the boundary pixels of predicted masks to obtain more precise masks. This method introduces an innovative incoherent region detection mechanism, which extracts incoherent regions (typically boundary pixels) from masks at different resolutions to construct a quadtree. The Transformer-driven mechanism processes only the identified unreliable tree nodes and autonomously rectifies their mistakes in parallel. In the same year, Chenming Zhu et al. proposed SharpContour [4], an boundary optimization method based on contours. Instead of optimizing individual pixels, this method optimizes discrete points on the self-defined mask boundary. Specifically, it designs a novel contour evolution process and an instance-aware point classifier, which iteratively evolves the contour by independently modifying the offsets of each discrete point on the boundary in a discrete manner. This gradually aligns the rough mask boundary contour with the ground truth, resulting in a clearer and more accurate mask boundary contour.

Among post-processing optimization methods, these approaches exhibit stronger model generalizability. They do not rely on any specific object segmentation network but directly optimize the object segmentation results to generate more precise outcomes. In 2020, Alexander Kirillov et al. proposed PointRend [5], which can be viewed as consisting of two parts. The first part obtains coarse segmentation results from features through a segmentation head, while the second part employs an iterative algorithm to adaptively select problematic regions in the coarse segmentation results. The pixels in these regions are then finely adjusted and predicted using a shared multi-layer perceptron (MLP) to produce refined segmentation results. Although not entirely a post-processing optimization method, the second part of this approach already embodies the conceptual foundation of such methods. Subsequently, in 2020, Yuhui Yuan et al. proposed SegFix [6]. This method establishes correspondences between boundary pixels and interior pixels, substituting boundary pixel predictions with more accurate interior pixel predictions. During training, the model's learned mappings are saved in the form of offset maps. During inference, the optimized segmentation results can be directly computed based on these offset maps without rerunning the model. Most importantly, SegFix generates a single offset map for each test image, and different object segmentation models can use the same offset map to obtain optimized segmentation results for that image. In 2021, Tang et al. [1] pointed out that two issues—low-resolution feature maps and the sparsity of boundary pixels—lead to imprecise mask boundaries predicted by instance segmentation methods. Consequently, they proposed a simple but highly effective post-processing optimization framework called BPR (Boundary Patch Refinement), which improves the boundary quality of segmentation results through a "crop-refine-recombine" strategy. This method first crops fixed-size boundary patches from the segmentation results and the object boundaries in the image. These patches are then fed into a refinement network based on the high-resolution backbone network HRNet [7] to obtain more refined boundary patches. Finally, the refined patches are recombined onto the original segmentation results to produce the refined segmentation output.

3 Proposed Method

Figure 2 presents an overview of the DBR framework. Built upon the BPR [1] architecture, the overall structure incorporates additional dual-branch feature processing and fusion modules, as well as boundary distance map input. The boundary distance map is generated by calculating boundary positions on the segmentation mask and performing a Euclidean distance transform, where pixels closer to the boundary have values approaching zero. Through the boundary distance map, we can provide implicit weights to the model, guiding it to focus more on boundary regions while being less confident about the segmentation in these areas (with smaller weights).

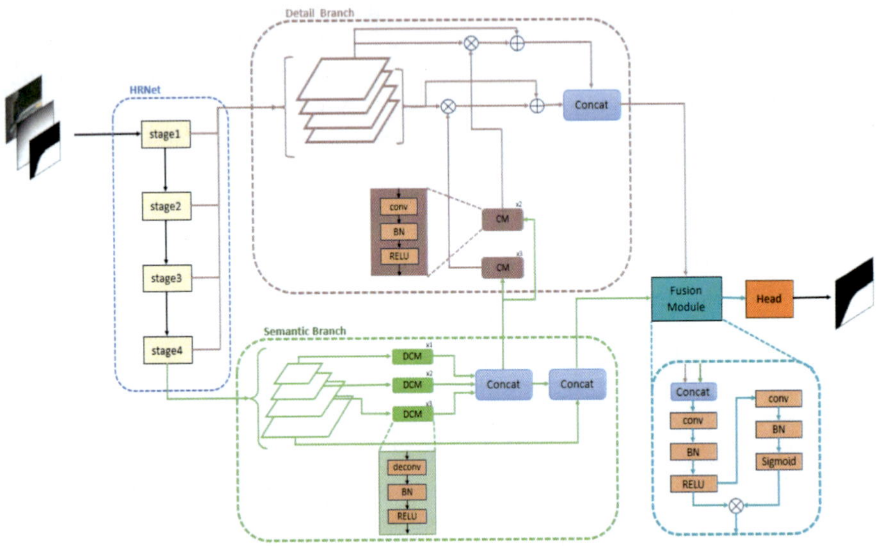

Fig. 2. The overall architecture of DBR. As a post-processing optimization method, it takes the cropped image, mask, and boundary distance map as inputs. Using a four-stage HRNet as the backbone network, the pyramid features from the fourth stage are input into the semantic branch, while the shallow features from the first four stages are input into the detail branch for respective information enhancement. The branch features are fused through a branch fusion module, with gated weights adaptively selecting information representation within the features. Finally, The fused features are fed into a segmentation head aimed at generating a cropped mask that exhibits enhanced boundary precision.

Utilizing the high-resolution HRNet as its backbone network, DBR relies on a dual-branch structure—namely, a semantic branch and a detail branch—to enhance the information from features at different layers of the backbone. Through the branch fusion module, the dual-branch features are integrated while employing a gated fusion mechanism to generate fusion weights for the combined features. This allows the model to selectively filter and utilize different feature information according to the specific object being optimized. Finally, the fused features are processed by a segmentation head consisting of two convolutional layers to generate the refined segmentation mask.

3.1 HRNet

In our design approach, although detailed information within features is sensitive, it can still be relied upon to achieve precise segmentation for most samples. Therefore, we restructure the feature outputs of HRNet, as indicated by the HRNet section (blue dashed box) of Fig. 2. The architecture of HRNet is highly systematic: at each stage, in addition to maintaining the branch with the same resolution as the previous stage, it also introduces additional lower-resolution branches.

First, we take the features from the highest-resolution branch across the four stages as the input to the detail branch. These features are represented by the gray features in Fig. 2. All four features share the same spatial dimensions, ensuring that they provide more expressive detailed information during feature processing in the detail branch. Simultaneously, we use the pyramid features from the fourth stage as the input to the semantic branch. As the deepest features in HRNet, these four features are the optimal choice for extracting and complementing semantic information in our proposed method.

3.2 Dual-Branch Structure

The dual-branch structure aims to process features from different layers of the backbone. By preliminarily enhancing the semantic and detailed information within the features, it obtains more expressive detail features and semantic features, thereby improving the feature representation capability of the fused features for all sample types and enhancing the overall optimization performance of the model.

Semantic Branch. The structure of the semantic branch is shown in the green dashed box in Fig. 2. This branch mainly consists of three deconvolution units, each containing a different number of deconvolutional modules (DCM). By applying DCM with different numbers to the top three pyramid features at different resolutions, the branch aligns feature resolutions while further deepening the feature depth to obtain more expressive semantic information. We assume the input four-layer features are denoted as $F_{3\sim6}$. First, we apply DCM with varying numbers to the lower-resolution features $F_{4\sim6}$ to obtain $F'_{4\sim6}$, which have the same resolution as F_3.

$$F'_i = \tau_j(F_i) \tag{1}$$

where $i \in [4, 6]$,$j \in [1, 3]$, and τ_j represents the deconvolution operation stacked with j DCMs. Each DCM consists of a deconvolution layer with a 4×4 kernel, strides of 2, and a pad of 1, then applying Batch Normalization (BN) and ReLU activation. Finally, the four spatially aligned features—F_3 and $F'_{4\sim6}$—are integrated via channel-wise concatenation to serve as the output of the semantic branch.

$$F_{sem} = Concat\left(F_3, F'_4, F'_5, F'_6\right) \tag{2}$$

Detail Branch. The structure of the detail branch is shown in the gray dashed box in Fig. 2. This branch takes shallow features $F_{0\sim3}$ from the four stages of the backbone network as input and primarily consists of two convolutional units, each containing a different number of convolutional modules (CM). Each CM consists of a 1×1 convolutional layer, followed by Batch Normalization (BN) and ReLU activation. The four detail features are then integrated via channel-wise concatenation to form the output of the detail branch. It introduces three semantic features $F'_{4\sim6}$ output from the deconvolution blocks of the semantic branch to supplement the semantic information in the four shallow features. The introduced semantic features are processed through deconvolution modules with different numbers of layers, resulting in the same spatial dimensions as the detail branch features. These semantic features are first integrated via channel-wise concatenation, and then aligned with the detail features in terms of channel dimensionality through two convolutional units.

$$F'_{cat} = Concat\left(F'_4, F'_5, F'_6\right) \tag{3}$$

$$F^k_{cat} = \eta^k\left(F_{cat}\prime; \omega^k\right) \tag{4}$$

Here, $k \in [0, 1]$, with η^k and ω^k representing the two distinct convolutional units and their respective parameters. Finally, semantic information is complemented through feature multiplication and residual connections. The four processed detail features are then integrated via channel-wise concatenation to form the output of the detail branch.

$$F'_0 = F_0 \odot F^0_{cat} \oplus F_0 \tag{5}$$

$$F'_n = F_n \odot F^1_{cat} \oplus F_n \tag{6}$$

$$F_{dtl} = Cat\left(F'_0, F'_1, F'_2, F'_3\right) \tag{7}$$

3.3 Dual-Branch Fusion Module

The design concept of the dual-branch fusion module originates from the gating mechanism introduced in LSTM [8]. Its purpose is to first simply fuse the two branch features to produce a fused feature that balances and complements both semantic and detailed information. Subsequently, this fused feature passes through a gating mechanism, which adaptively learns to produce gating weights. The fused feature then undergoes an element-wise weighting using these gating weights, enabling the model to autonomously prioritize pivotal regions for segmentation optimization and to filter and utilize different feature information according to the specific object being optimized.

As illustrated in the cyan dashed box of Fig. 2. Within this module, the input dual-branch features are first concatenated channel-wise. They are then fused via a 1×1 convolution. The fused feature enters a gating module composed of 1×1 convolutions, and finally, a Sigmoid activation function generates the gating weights, which are used to weight the fused feature.

4 Experiments

4.1 Dataset and Metrics

Dataset. We focus on presenting the optimization results of DBR on different segmentation models using the Cityscapes test set. The Cityscapes dataset is a highly authoritative and widely used benchmark in computer vision, particularly in autonomous driving and urban scene understanding. We utilize its fine annotations, which consist of 2,975 training images, 500 validation images, and 1,525 test images. These images were collected from 27 cities with a resolution as high as 1024×2048. They cover eight instance categories, including bicycle, bus, person, train, truck, motorcycle, car, and rider.

Metrics. We upload the segmentation results on the Cityscapes test set to the official evaluation server and use the returned metrics—AP and its variants (e.g., AP_{50})—as the evaluation criteria for comparisons and ablation studies in this paper. All metric values are obtained directly from the official Cityscapes evaluation server, without any personal computation.

4.2 Training and Inference

During both training and inference, we follow the experimental settings of BPR [1] as closely as possible to ensure a fair and valid verification of the effectiveness of our proposed architecture.

Training. During training, we utilize the cropped training and validation sets prepared by BPR. We use roughly 280k images for training and 60k for validation. Data augmentation techniques such as random flipping and photometric distortion are applied. The entire training process follows an iterative approach, set for a total of 200k iterations. Validation is performed every 8,000 iterations, and the model parameters from the best-performing checkpoint on the validation set are saved. The Binary Cross-Entropy (BCE) loss function is used to supervise the optimized masks and update the model parameters. For specific training configurations, the initial learning rate is set to 0.005. A poly learning rate policy is adopted, allowing the learning rate to decay as the number of iterations increases, with a poly exponent of 0.9 and a minimum learning rate of 0.0001. The batch size for training is set to 16. The SGDM optimizer is used with a momentum parameter of 0.9 and weight decay of 0.0005.

Inference. During inference, the model first calculates a series of dense cropping boxes based on the boundaries of the input mask and filters them using the Non-Maximum Suppression (NMS) algorithm with a threshold of 0.55. An boundary distance map is generated based on the cropped mask. This map, along with the cropped image and the cropped mask, serves as the input to DBR. The output optimized mask is then reassembled to its original position according to the coordinates of the cropping boxes. For overlapping regions, the pixel values are averaged and binarized using a threshold of 0.5.

4.3 Comparative Experiments

This paper employs the segmentation results of four instance segmentation models—Mask R-CNN [9], PointRend [5], BMask R-CNN [10], and Oneformer [11]—on the Cityscapes test set as baselines, and compares the optimization results of DBR with SegFix and BPR across these four baselines. Table 1 presents the comparative results of the four baselines and the three optimization methods on the Cityscapes test set. The results demonstrate that the proposed DBR method achieves performance gains of + 3.3% AP, + 3.1% AP, + 3.4% AP, and + 1.7% AP on the Mask R-CNN baseline, BMask R-CNN baseline, PointRend baseline, and Oneformer baseline, respectively. Moreover, compared with the other two post-processing optimization methods, DBR method exhibits the best performance.

Table 1. Comparative experimental results. SegFix, BPR, and the proposed DBR are compared on four baselines. Red text indicates the best performance.

	AP	AP50%	AP50m	AP100m	Person	Rider	Car	Truck	Bus	Train	Motorcycle	Bicycle
Mask R-CNN	29.5	51.2	46.0	42.8	34.0	27.1	49.8	27.5	33.4	24.2	21.8	18.3
+ SegFix	30.7(+1.2)	51.4(+0.2)	47.5(+1.5)	44.3(+1.5)	35.8	28.4	52.1	28.3	34.5	24.6	22.7	19.2
+ BPR	32.5(+3.0)	51.9(+0.7)	50.1(+401)	46.8(+4.0)	37.6	30.0	55.0	29.4	36.5	27.1	23.8	20.4
+ DBR	**32.8(+3.3)**	**52.3(+1.1)**	**50.5(+4.5)**	**47.2(+4.4)**	**38.1**	**30.5**	**55.3**	**29.6**	**36.5**	**27.5**	**24.2**	**20.7**
Bmask R-CNN	28.5	49.8	42.8	41.6	33.3	26.7	51.7	24.2	32.4	21.4	20.5	18.0
+ SegFix	29.6(+1.1)	50.1(+0.3)	44.2(+1.4)	43.0(+1.4)	35.0	28.0	53.9	25.1	33.5	21.6	21.0	18.9
+ BPR	31.2(+2.7)	50.9(+1.1)	46.3(+3.5)	45.1(+3.5)	36.9	29.8	56.0	26.0	35.3	23.2	22.3	20.3
+ DBR	**31.6(+3.1)**	**50.9(+1.1)**	**46.7(+3.9)**	**45.6(+4.0)**	**37.4**	**30.3**	**56.3**	**26.2**	**35.6**	**23.5**	**22.9**	**20.7**
PointRend	28.3	49.5	43.4	41.2	32.9	26.2	51.6	22.6	30.6	24.9	19.4	17.8
+ SegFix	29.5(+1.2)	49.9(+0.4)	44.8(+1.4)	42.7(+1.5)	34.9	27.9	54.0	23.7	31.7	25.2	20.2	18.9
+ BPR	31.5(+3.2)	50.6(+1.1)	47.1(+3.7)	45.2(+4.0)	37.0	30.1	56.2	24.8	33.7	27.6	21.8	20.3
+ DBR	**31.7(+3.4)**	**50.8(+1.3)**	**47.4(+4.0)**	**45.4(+4.2)**	**37.5**	**30.5**	**56.5**	**25.0**	**34.0**	**27.6**	**21.9**	**20.8**
OneFormer	35.0	55.5	55.8	51.4	34.4	30.4	54.5	28.5	43.9	41.2	27.1	19.8
+ SegFix	35.7(+0.7)	55.6(+0.1)	56.5(+0.7)	52.2(+0.8)	35.5	31.2	55.9	28.9	44.7	41.6	27.7	20.4
+ BPR	36.5(+1.5)	55.9(+0.4)	57.2(+1.4)	53.1(+1.7)	36.5	32.6	57.3	29.1	44.6	**42.7**	**28.3**	20.9
+ DBR	**36.7(+1.7)**	**56.0(+0.5)**	**57.3(+1.5)**	**53.2(+1.8)**	**36.8**	**33.0**	**57.4**	**29.4**	**45.2**	42.2	28.2	**21.2**

4.4 Ablations Experiments

To thoroughly analyze the effectiveness of the proposed method, this paper conducts ablation studies on key modules of the model to evaluate the contributions of the dual-branch structure, the branch fusion module, and the boundary distance map to the optimization performance. All ablation experiments follow the same training settings as described in Sect. 3.4, with the results evaluated on the Mask R-CNN baseline.

As shown in Table 2. First, we validate the effectiveness of the dual-branch structure by using a single branch in DBR. When the semantic branch is removed, the model's performance drops by 0.7%; when the detail branch is removed, the performance decreases by 0.3%. This demonstrates the effectiveness of the dual-branch structure and indicates that the semantic branch provides richer semantic information, contributing more significantly to performance improvement, thereby further verifying the correctness of our design concept. Removing the boundary distance map as an input leads to a performance decline of 0.2%, as the boundary distance map provides more accurate boundary priors, helping the model better localize boundaries and focus on optimizing boundary regions.

Table 2. Comparison of ablation study results. "DBR-sem" denotes DBR without the semantic branch, "DBR-dtl" denotes DBR without the detail branch, and "DBR-dist" denotes DBR without the boundary distance map as input.

	AP	AP50%	AP50m	AP100m	Person	Rider	Car	Truck	Bus	Train	Motorcycle	Bicycle
Mask R-CNN	29.5	51.2	46.0	42.8	34.0	27.1	49.8	27.5	33.4	24.2	21.8	18.3
DBR -sem	32.1	52.1	49.6	46.2	37.0	29.5	54.5	29.3	36.1	26.6	23.5	20.00
DBR -dtl	32.5	51.9	50.1	46.8	37.6	30.0	55.0	29.4	36.5	27.1	23.8	20.4
DBR -dist	32.6	52.0	50.3	46.9	37.8	30.1	55.2	29.5	36.5	27.4	23.8	20.6
DBR	**32.8**	**52.3**	**50.5**	**47.2**	**38.1**	**30.5**	**55.3**	**29.6**	**36.5**	**27.5**	**24.2**	**20.7**

4.5 Qualitative Results

We present the visual segmentation results of Mask R-CNN, BPR, and DBR on the Cityscapes validation set. As demonstrated in Fig. 3, DBR (the fourth column) achieves higher segmentation quality. In the first sample (the first row in Fig. 3), both BPR and Mask R-CNN paid little attention to the side mirror at the front right of the bus. In contrast, DBR accurately segmented the side mirror by leveraging semantic information, even when the initial mask failed to precisely locate it. Similarly, in the second sample (the second row in Fig. 3), DBR still effectively utilized semantic information to achieve more accurate boundary localization for the bicycle's basket, resulting in more precise segmentation.

Fig. 3. Comparative segmentation results of different segmentation methods on the Cityscapes validation set. From left to right: predictions by Mask R-CNN, ground truth annotations, results from BPR, and results from the proposed DBR method.

5 Conclusion and Future Works

This paper proposes a post-processing boundary patch cropping optimization method based on dual-branch feature processing and fusion. Previous boundary patch cropping optimization methods overlooked the issue of excessive model reliance on detailed information, which prevented consistent optimization effectiveness across all samples. DBR addresses this by employing a dual-branch feature processing and fusion structure to balance the expressive capabilities of detailed and semantic information within the features. It also adaptively filters feature information for different samples, thereby achieving optimization improvements across a broader range of cases.

Currently, our work is primarily focused on urban scene images. However, as a post-processing approach, the proposed method is generalizable to multiple domains, such as indoor environments [12] and medical imaging [13], which have also been studied extensively in the literature. In the future, we will endeavor to design diverse architectures for boundary patch cropping-based post-processing optimization across various research domains, aiming to advance the development of such optimization methodologies.

References

1. Tang, C., Chen, H.;, Li, X., et al.: Look closer to segment better: boundary patch refinement for instance segmentation. In: Proceedings of the IEEE/CVF Conference on Computer Vision and Pattern Recognition, pp.13926–13935 (2021)
2. Zhang, G., Lu, X., Tan, J., et al.: Refinemask: towards high-quality instance segmentation with fine-grained features. In: Proceedings of the IEEE/CVF Conference on Computer Vision and Pattern Recognition, pp. 6861–6869 (2021)

3. Ke, L., Danelljan, M., Li, X., et al.: Mask transfiner for high-quality instance segmentation. In: Proceedings of the IEEE/CVF Conference on Computer Vision and Pattern Recognition, pp. 4412–4421 (2022)
4. Zhu, C., Zhang, X., Li, Y., et al.: SharpContour: a contour-based boundary refinement approach for efficient and accurate instance segmentation. In: Proceedings of the IEEE/CVF Conference on Computer Vision and Pattern Recognition, pp. 4392–4401 (2022)
5. Kirillov, A., Wu, Y., He, K., et al.: Pointrend: image segmentation as rendering. In:'Proceedings of the IEEE/CVF Conference on Computer Vision and Pattern Recognition, pp. 9799–9808 (2020)
6. Yuan, Y., Xie, J., Chen, X., et al.: Segfix: model-agnostic boundary refinement for segmentation. In: Computer Vision–ECCV 2020: 16th European Conference, Glasgow, UK, August 23–28, 2020, Proceedings, Part XII 16, pp. 489–506. Springer International Publishing (2020)
7. Wang, J., Sun, K., Cheng, T., et al.: Deep high-resolution representation learning for visual recognition. IEEE Trans. Pattern Anal. Mach. Intell. **43**(10), 3349–3364 (2020)
8. Shi, X., Chen, Z., Wang, H., et al.: Convolutional LSTM network: a machine learning approach for precipitation nowcasting. In: Advances in Neural Information Processing Systems, p. 28 (2015)
9. He, K., Gkioxari, G., Dollár, P., et al.: Mask r-cnn. In: Proceedings of the IEEE International Conference on Computer Vision, pp. 2961–2969 (2017)
10. Cheng, T., Wang, X., Huang, L., et al.: Boundary-preserving mask r-cnn. In: Computer Vision–ECCV 2020: 16th European Conference, Glasgow, UK, August 23–28, 2020, Proceedings, Part XIV 16, pp. 660–676. Springer International Publishing (2020)
11. Jain, J., Li, J., Chiu, M.T., et al.: Oneformer: one transformer to rule universal image segmentation. In: Proceedings of the IEEE/CVF Conference on Computer Vision and Pattern Recognition, pp. 2989–2998 (2023). Ren, S., He, K., Girshick, R., et al.: Faster r-cnn: towards real-time object detection with region proposal networks. In: Advances in Neural Information Processing Systems, p. 28 (2015)
12. Rahman, R., Chowdhury, M., Tang, Y., Gao, H., Yin, G., Wang, G.: Kitchen food waste image segmentation and classification f or compost nutrients estimation. J. Image Graphics **13**(2), 158–163 (2025)
13. Edy Victor Haryanto, S., Riza, B.S., Harjoko, A.: Next-generation image segmentation for enhanced malaria detection and diagnosis. J. Image Graphics **13**(1), 73–82 (2025)

Hybrid Convolutional Neural Network for Hyperspectral Corn Seed Identification

Hui Zhang[1], Junjiang Hu[2], He Yuan[1], and Yi An[3,4(✉)]

[1] Changchun Technical University of Automobile, Changchun, China
[2] Kunming University of Science and Technology, Kunming, China
[3] Jilin University, Changchun, China
anyi25@mails.jlu.edu.cn
[4] Changchun Institute of Technology, Changchun, China

Abstract. Corn is a key agricultural product in China, with corn seed identification facing low efficiency and insufficient accuracy. Applying deep learning to corn seed image and spectral analysis can support automatic identification and standardized production, boosting intelligent and modern agriculture. First, a VGG-16-based convolutional network was trained on nine corn seed image datasets for accurate identification. Second, 150 hyperspectral images per corn seed type were collected to build an initial dataset, which was preprocessed and expanded to ~ 70,000 images. Finally, the parallel-VGG16 model was proposed, optimizing VGG-16 with Google-NET-inspired structure, adding parallel convolutional/max pooling layers to widen the network, and unfreezing some layers for training.VGG-16 outperformed traditional models. After data augmentation, parallel-VGG16 achieved the lowest training loss and highest validation accuracy, outperforming the original VGG-16. Deep learning, combined with spectral imaging, offers new paths for corn seed identification. Future integration with intelligent agricultural machinery will further promote Chinese agricultural modernization.

Keywords: deep learning · seed identification · convolutional hyperspectral imaging · VGG-16

1 Introduction

Hybrid Convolutional Neural Network for Hyperspectral Corn Seed Recognition Corn is one of the grain crops with the largest cultivated area in China. The quality and authenticity of its seeds are directly related to agricultural production efficiency, food security, and market circulation order. Developing a rapid, stable, and accurate method for identifying corn seeds holds significant academic value and economic importance. From an academic perspective, corn seed identification methods can enrich the application theory of deep learning and hyperspectral imaging technology in the field of agricultural detection, promoting interdisciplinary technology integration. From an economic perspective, they can provide seed companies with efficient identification technology support, reduce identification costs, regulate the seed market, and facilitate agricultural

L. Huang and D. Greenhalgh (Eds.): ICMLC 2026, LNNS 2061, pp. 190–205, 2027.
https://doi.org/10.1007/978-3-032-31236-5_13

modernization. Currently, traditional identification methods such as field planting identification and electrophoresis identification have limitations such as long cycles, high costs, and significant environmental impacts, which cannot meet the rapid detection needs of modern agriculture. In recent years, the development of deep learning network models and hyperspectral imaging technology has provided a new path to address this bottleneck. Their non-destructive, efficient, and high-precision characteristics have been preliminarily applied in the field of image detection [1–6]. Convolutional Neural Networks (CNN) [7] in deep learning technology are widely used in image classification, object detection, and other aspects [8]. Since the advent of CNN models, many network models based on them have emerged, such as classic models like AlexNet [9], VGG [10], GoogLeNet [11], ResNet [12], and MobileNet [13]. Based on classic models, utilizing transfer learning technology can greatly save model training time, alleviate model overfitting, and enhance model generalization ability.

Scholars both domestically and internationally have conducted numerous studies on corn seed classification, laying a solid foundation for this paper. Xu Peng et al. [14] proposed a maize seed rapid classification method using a combination of machine vision and deep learning, which developed the improved ResNet network and compared it with models like AlexNet and VGGNet. Ruchi Rani et al. [15] proposed a corn disease prediction and classification method, which uses a pre-trained VGG16 as the backbone integrated with an attention module and applies prototypical few-shot learning, achieving a high accuracy for corn disease detection.Sigit Susanto Putro et al. [16] proposed a corn leaf disease identification method combining Convolutional Neural Network with Optimizer and Batch Size, which uses the MobileNetV2 architecture and tests Adam and RMSProp optimizers. Huang [17] applied CNN and transfer learning to the quality classification of corn seeds, verifying that deep learning algorithms are significantly superior to machine learning algorithms, with an accuracy of 95% based on the GoogLeNet network model. Javanmandi et al. [18] proposed that extracting features using CNN and artificial neural networks can also effectively complete corn seed classification and recognition. Tu et al. [19] used VGG16 to identify "Jingke 968", achieving high recognition accuracy, establishing the reliability of transfer learning in corn identification. Despite these studies achieving certain results, there are still weak links such as insufficient model generalization ability, insufficient exploration of the impact of data volume on model performance, and weak specificity in network structure optimization.

This study first addresses the issue of weak model generalization ability due to the small dataset size. After collecting initial data using a hyperspectral camera, methods such as horizontal flipping and brightness adjustment are employed to expand the dataset to approximately 70,000 images, providing sufficient support for model training. Secondly, combining the GoogLeNet structure, the parallel-VGG16 model proposed in this paper is constructed by adding convolutional layers with small convolution kernels and parallel convolutional layers of max pooling on the basis of VGG16. This model breaks through the limitations of traditional network width and enhances feature extraction ability. Then, by unfreezing and training some convolutional layers, the model performance is further optimized to compensate for the lack of targeted network structure optimization. Finally, through comparative experiments, the balance between model accuracy and training efficiency is achieved; the superiority of the proposed parallel-VGG16 model in hyperspectral image classification of corn seeds is verified, providing a more efficient technical solution for automatic identification of corn seeds and promoting sustainable development in agricultural intelligence.

2 Materials and Methods

2.1 Data Sources and Experimental Instruments

The corn seed samples used in the experiment were sourced from the Jilin Academy of Agricultural Sciences, from 9 varieties (YM129, YM130, YM131, YM132, YM133, YM134, YM135, YM136, YM137). Samples of similar size and saturation were selected for each variety. The image acquisition of corn seeds will be completed using hyperspectral image acquisition method, with a hyperspectral acquisition range from 392.38 nm to 1011.01 nm, covering a total of 462 bands.

Experimental instruments and experimental environment. The hyperspectral imaging experiment utilizes the RESONON desktop hyperspectral imaging system as the research platform. The push-broom hyperspectral imager (PIKA II, Resonon Inc, Bozeman, MT, USA) is connected to a CCD camera (Raptor EagleV) and a lens (Schneider-KREUZNACH CINEGON 1.8/4.8), and installed above the sample, with the distance between the lens and the sample properly set. Two halogen lamps (IT, 3900, 150 W) are installed below and on both sides of the lens, with the irradiation distance and the angle between the two halogen lamps and the moving platform set. Hyperspectral images are acquired with a spatial resolution of 50 pixels/mm^2. The parameters of the hyperspectral imaging system are adjusted, and the optimized exposure time of the camera and the moving speed of the moving platform are selected to prevent interference from external light. The HSI system operates in a darkroom.

For network model training, the computational workload is excessively large. Using a general computer CPU for model training would require a significant amount of time. However, utilizing a GPU for model training can save considerable time, facilitating the subsequent conduct of experiments. The experimental environment for this study is as follows: Windows 10 64-bit operating system, Python programming language, NVIDIA RTX 3090 processor, and the deep learning framework TensorFlow.

2.2 Data Collection and Processing

Data collection to obtain reliable hyperspectral image data, it is necessary to consider the impact of uneven light intensity distribution on image noise. Therefore, before image acquisition, a reflectance correction using a black and white board must be performed on the experimental equipment to improve the accuracy and reliability of the data. This method can eliminate the effects of uneven illumination and other factors, ensuring the quality of hyperspectral images while enhancing the accuracy and reliability of detection. After multiple attempts, the experiment concluded that when using high-precision technology to collect hyperspectral image data of corn seeds to ensure the integrity of sample information, the technical parameters should be: a spectral range between 392.38 ~ 1021.02 nm, a distance of 45cm between the lens and the sample, a scanning speed of 0.11cm/s, a return speed of 2.50079 cm/s, a point movement speed of 1.0 cm/s, and a frame frequency of 9.8 Hz. For this study, we need to use a 46.701 ms clock for accumulation. Then, place a corn seed sample in an electrically controlled position and position a photography system parallel to it, shooting it at a 90° angle from the side. Next, use a computer-controlled device to collect hyperspectral images at 462 wavelengths ranging from 392.38 nm to 1011.01 nm from this device. Finally, retain these images for further pre-data processing.

Due to the strong fitting ability of CNNs, a smaller dataset may lead to overfitting and affect the performance of the model. To enable the model to adapt to targets of different shapes and sizes, data augmentation strategies are commonly adopted to increase training data. In this paper, to enhance the robustness of the model, the Image module of the Python Imaging Library (PIL) is used to perform data augmentation on each training image, mainly including the following methods: horizontal flipping, horizontal or vertical translation, permutation and combination of images, and changing the contrast or brightness of images. These methods mainly achieve data augmentation by transforming the spatial relationships between image pixels. After data augmentation, a total of 71,134 training images are obtained. The specific data for each category is shown in Table 1. The images and annotation information are saved in the PNG format required by the training model, facilitating future model training and testing. The data is serialized and stored in multiple files by category, which can be read when needed.

Table 1. Statistical table of quantity before and after data expansion.

Sample data	Number of raw data	The number of data after expansion
YM129	24	4931
YM130	25	10648
YM131	24	8000
YM132	30	5832
YM133	24	9261
YM134	31	10648
YM135	24	8000
YM136	30	5832
YM137	24	8000

2.3 Research Methodology

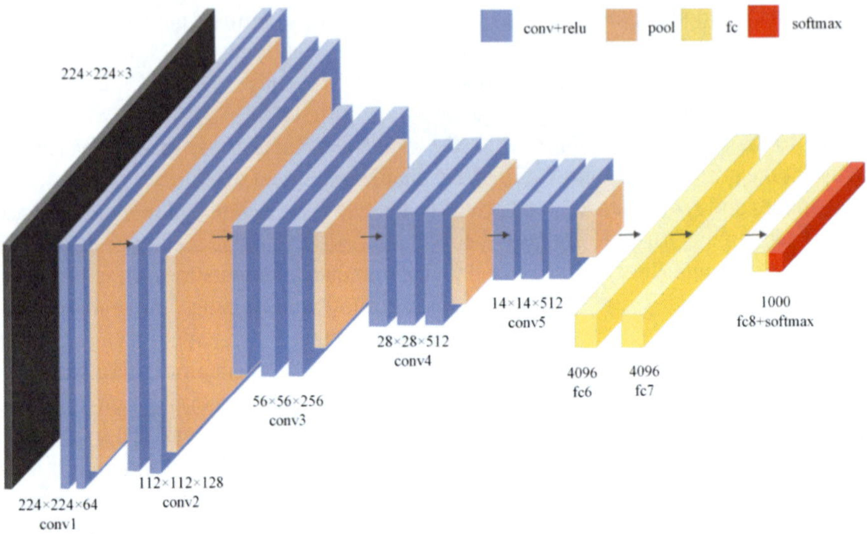

Fig. 1. VGG16 network structure diagram.

Corn seed image identification based on the VGG16 network model Compared to the AlexNet network model, the VGG16 network model adopts a deeper network structure. The VGG16 network model consists of 6 segments, where 5 segments contain convolutions and 1 segment is fully connected. The number of convolution layers in the 5 segments is 13, and each segment can contain 2 or 3 convolutions, which can effectively extract image features at different levels. To increase the network's nonlinearity, prevent vanishing gradients, reduce overfitting, and improve training speed, each convolution layer is followed by a ReLU activation function [20]. The gradient of the ReLU activation function does not decrease, thus reducing the problem of vanishing gradients. To capture image detail variations, increase nonlinear effects, and reduce the number of parameters, each convolution layer uses a small 3x3 convolution kernel. The 5 convolution segments of VGG16 are each equipped with a 2×2 pooling layer. Pooling is a commonly used dimensionality reduction method, which ensures that parameter errors do not affect model accuracy and enables better capture of subtle differences in images, such as texture. In addition, the model includes 3 fully connected layers. In these layers, Dropout [21] is applied to the hidden layers to further enhance the network's generalization ability. Finally, through the Softmax activation function, the entire network model maps 4096 independent patterns together, achieving accurate recognition of image features. The VGG16 network consists of $13 + 3$ layers and 16 layers, forming a complete structure, as shown in Fig. 1.

The VGG16 network adopts a novel approach to process 224 × 224 × 3 images. The advantage of this method lies in its ability to expand the original 64 channels to 128, 256, and 512, and effectively scale the size of the feature maps, reducing the original 224 channels to 7. The classifier of VGG16 consists of three fully connected layers, with the first two having 4096 nodes and the last having 1000 nodes. According to Table 2, the parameter count of the VGG16 network model reaches 138 M, which far exceeds that of LeNet and AlexNet. This means that more resources and time are required during training and optimization. Although there are certain issues with training complexity, the VGG16 network significantly improves the performance of image classification and recognition through the use of convolutional and pooling layers, as well as a deeper network structure.

Corn seed image identification based on the Parallel-VGG16 network model. Despite the excellent performance of the VGG16 network model in ImageNet and other applications [22], compared to some networks such as GoogLeNet and ResNet, the ability and classification effectiveness can be improved by considering the network width to enhance multi-scale feature extraction. Therefore, this study proposes a new Parallel-VGG16 network model, incorporating various improvements, including Parallel Convolution Blocks (PCB), Batch Normalization (BN), Global Average Pooling (GAP), 1x1 convolution, etc. These improvements have significantly enhanced the Parallel-VGG16 model in image classification and recognition tasks.

The Parallel Convolutional Block (PCB). VGG16 has a relatively small depth and width. The most direct way to improve its performance is to increase its network scale. To avoid blindly expanding and causing network degradation issues, inspired by GoogLeNet, a parallel convolutional block is designed in the Parallel-VGG16 model to expand the network width and increase network complexity. After systematic experimental comparison, we selected four 3 × 3 parallel convolutional layers, one max pooling layer, and one 1 × 1 convolutional layer, which can form a complete PCB module. It not only provides a longer transmission distance but also delivers more information, thereby enhancing the complexity and accuracy of the entire system. A 1 × 1 convolutional layer is introduced at the end of the PCB to further optimize the model. Its structure is shown in Fig. 2.

Table 2. VGG16 and Parallel-VGG16 Network Structure Parameters.

VGG16 network structure parameters					Parameters of Parallel-VGG16 network structure				
Operation Type	Convolution kerne	Pooling	output layer	parameter count	Operation Type	convolution kernel	pooling	output layer	parameter count
conv3_64	3 × 3	–	224 × 224 × 64	1792	conv3_64	3 × 3	–	224 × 224 × 64	1792
conv3_64	3 × 3	–	224 × 224 × 64	36928	conv3_64	3 × 3	–	224 × 224 × 64	36928
max_pooling	–	2 × 2	112 × 112 × 64	0	max_pooling	3 × 3	2 × 2	112 × 112 × 64	0
conv3_128	3 × 3	–	112 × 112 × 118	73865	conv3_128	3 × 3	–	112 × 112 × 118	73865
conv3_128	3 × 3	–	112 × 112 × 118	147584	conv3_128	3 × 3	–	112 × 112 × 118	147584
max_pooling	–	2 × 2	56 × 56 × 118	0	max_pooling	–	2 × 2	56 × 56 × 118	0
conv3_256	3 × 3	–	56 × 56 × 256	295168	conv3_256	3 × 3	–	56 × 56 × 256	295168
conv3_256	3 × 3	–	56 × 56 × 256	590080	conv3_256	3 × 3	–	56 × 56 × 256	590080
conv3_256	3 × 3	–	56 × 56 × 256	590080	conv3_256	3 × 3	–	56 × 56 × 256	590080
max_pooling	–	2 × 2	28 × 28 × 256	0	max_pooling	–	2 × 2	28 × 28 × 256	0
conv3_512	3 × 3	–	28 × 28 × 512	1180160	conv3_512	3 × 3	–	28 × 28 × 512	1180160
conv3_512	3 × 3	–	28 × 28 × 512	2359808	conv3_512	3 × 3	–	28 × 28 × 512	2359808
conv3_512	3 × 3	–	28 × 28 × 512	2359808	conv3_512	3 × 3	–	28 × 28 × 512	2359808
max_pooling	–	2 × 2	28 × 28 × 512	0	max_pooling	–	2 × 2	28 × 28 × 512	0

(continued)

Table 2. *(continued)*

VGG16 network structure parameters					Parameters of Parallel-VGG16 network structure				
conv3_512	3 × 3	–	14 × 14 × 512	2359808	PCB	1 × 1 × 1 3 × 3 × 4 max_pooling	– – 1 × 1	14 × 14 × 3072	0
conv3_512	3 × 3	–	14 × 14 × 512	2359808	BN	–	–	14 × 14 × 3072	12288
conv3_512	3 × 3	–	14 × 14 × 512	2359808	conv3_512	3 × 3	–	14 × 14 × 512	1573376
max_pooling	–	2 × 2	14 × 14 × 512	0	max_pooling	–	2 × 2	14 × 14 × 512	0
fc_4096	–	–	1 × 4096	102764544	fc_1024	–	–	1 × 1096	525312
fc_4096	–	–	1 × 4096	16781312	fc_1000	–	–	1 × 1000	1025000
fc_1000	–	–	1 × 1000	4097000	–				
Total params: 138,357,544					Total params: 20,473,128				
Trainable params: 138,357,544					Trainable params: 20,466,984				
Non-trainable params: 0					Non-trainable params: 0				

Batch Normalization (BN) layer and Global Average Pooling (GAP) layer. By adopting BN (Batch Normalization) technology, the training time of deep neural networks can be effectively shortened, and the performance of the model can be improved by adjusting a higher learning rate and precise parameter initialization. This algorithm can transform all models into a standard normal distribution where the mean and variance are both 1. Given a small batch of samples $B = \{x_1 \ldots x_m\}$, its mean and variance are denoted as μ and $\sigma 2$ respectively, and ϵ is a constant for stabilizing data.

The data normalization is represented as described as in (1):

$$\hat{x}_i = \frac{x_i - \mu}{\sqrt{\sigma^2 + \epsilon}} \tag{1}$$

Considering that solely using the aforementioned normalization formula may affect the learned features, we perform a transformation and reconstruction on the normalized data to restore the feature distribution originally learned by the network, as in (2):

$$\hat{y}_i = \gamma \hat{x}_i + \beta \tag{2}$$

Among them, γ and β are learnable reconstruction parameters. The BN layer can alleviate the parameter distribution bias problem of initial and intermediate values during training. By using these techniques, the impact of vanishing gradients and exploding gradients can be effectively mitigated, thereby improving training replacing the first two fully connected layers in VGG16 with GAP layers significantly reduces network parameters and enhances training speed. Compared to fully connected layers, GAP layers have the following advantages: they convert two-dimensional feature maps into one-dimensional vectors (each feature map corresponds to a numerical value), which can be directly mapped to the correlation strength between the corresponding feature and the target class; GAP does not require parameter training; thus, it can accelerate network training and effectively alleviate overfitting; GAP aggregates spatial information, making it robust to spatial transformations of inputs.

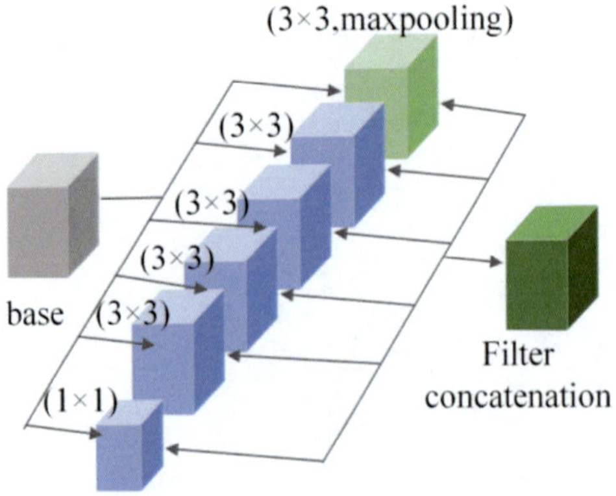

Fig. 2. PCB structure diagram.

Parallel-VGG16 network model structure. The embedding position of PCB in Parallel-VGG16 also has a significant impact on network performance. Theoretically, PCB can be embedded inside or after any of the five convolutional blocks in VGG16. After embedding PCB, the remaining part of the convolutional block where PCB is embedded and the convolutional block after the embedding position are removed. Although BN layers are widely used in deep learning-based applications, the number and embedding position of BN layers need further discussion. In view of this, this study explores three BN embedding points (marked as red dots in Fig. 3): before PCB, between PCB and conv1, and after conv. The effects of different embedding schemes are verified through experiments, and the optimal embedding scheme is determined. Figure 3 shows the optimal parallel-VGG16 network model. Table 2 is a table of its structural parameters. By comparison, it can be seen that increasing parallel convolutional blocks can reduce the parameter count of the baseline model, lower the complexity of the model, and significantly reduce the time required for training data.

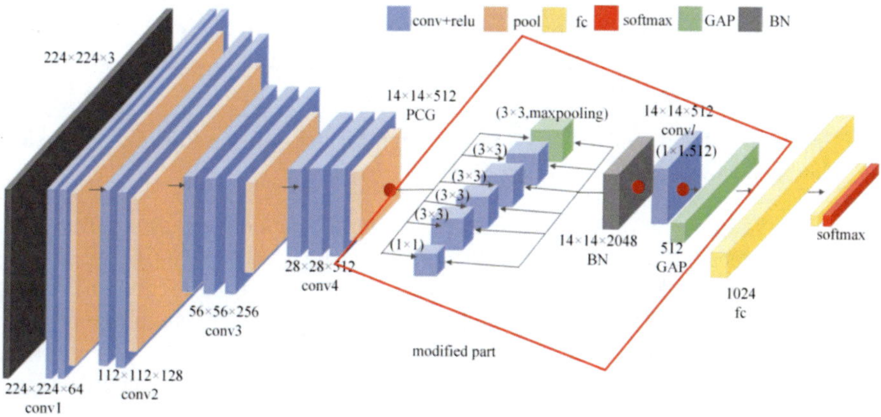

Fig. 3. Parallel-VGG16 network structure diagram.

3 Results and Analysis

3.1 Training Comparison Between VGG16 Network Model and Parallel-VGG16 Network Model

By enhancing the data of 71,134 corn seed images and grouping them in a 9:1 ratio, the dimensions of both the training set and the test set were adjusted to 224*224. Based on data mining using the Parallel-VGG16 and VGG16 network models, they were applied to actual testing, including nine different corn seed varieties: YM129, YM130, YM131, YM132, YM133, YM134, YM135, YM136, and YM137. After the 10th iteration of network training, the trend of the loss function remained balanced, and the accuracy of the model was significantly improved, reaching 97.4% and 93.3%, respectively. The loss values were 0.001 and 0.017, respectively. The performance of the proposed Parallel-VGG16 network model was significantly better than that of VGG16.

The loss function variation curves for the VGG16 network model and the Parallel-VGG16 network model in corn seed image identification are shown in Figs. 4 and 5, respectively. Here, Epoch represents the number of iterative training rounds, train loss represents the loss value of the training dataset, val loss represents the loss value of the validation dataset; smooth train loss represents the smoothed loss value of the training dataset, and smooth val loss represents the smoothed loss value of the validation dataset. As can be seen from Figs. 4 and 5, the loss function of corn seed image identification based on the VGG16 network exhibits considerable fluctuations from the first data presentation, mainly due to the unsatisfactory initial training effect, which compromises the stability of the model. However, with the fourth and final training rounds, the feature learning effect of the model improves, thus mitigating the fluctuations in the loss function. For corn seed image identification based on the parallel-VGG16 network, the loss function changes significantly from the first to the second training round, and the loss function variation curve tends to stabilize from the second to the final training round. Moreover, compared to the previous model, it saves a significant amount of model training time, and the loss value for model identification also decreases, proving that the parallel-VGG16 network model proposed in this paper can be used in the field of computer vision and is a better choice for completing the task of corn seed image identification.

3.2 Prediction Results of the VGG16 Network Model and the Parallel-VGG16 Network Model

Recognition accuracy of different seed faces. Taking the corn ear cap as the front face, there are six faces: JX (front), JY (right), JZ (left), JS (rear), N (top), and P (bottom).According to Table 3, in both networks, the accuracy of the P (bottom) face is relatively high, at 0.987 and 0.997, mainly due to the low discrimination rate of these seeds and the obvious image features, which are easily distinguishable after training. The accuracy of JX (front) is only 0.907 and 0.964, indicating that its recognition effect is not satisfactory. This may be due to insufficient feature extraction and the similarity between image data, leading to misjudgments during feature extraction. However, overall, the recognition accuracy of JX is still relatively high and can meet the overall needs of agricultural recognition for corn seed images.

Based on the information comparison in the Table 3, it can be concluded that the parallel-VGG16 proposed in this study exhibits a significant improvement in recognition rate compared to the VGG16 network model. The addition of parallel convolutional blocks not only has no impact on the accuracy of the VGG16 network but also enhances it. The inclusion of parallel convolutional blocks can reduce the complexity of the model by decreasing the number of parameters in the network model.

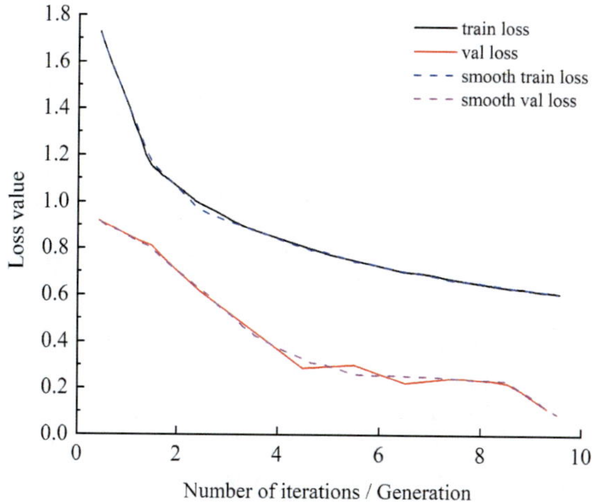

Fig. 4. VGG16 network loss function variation curve.

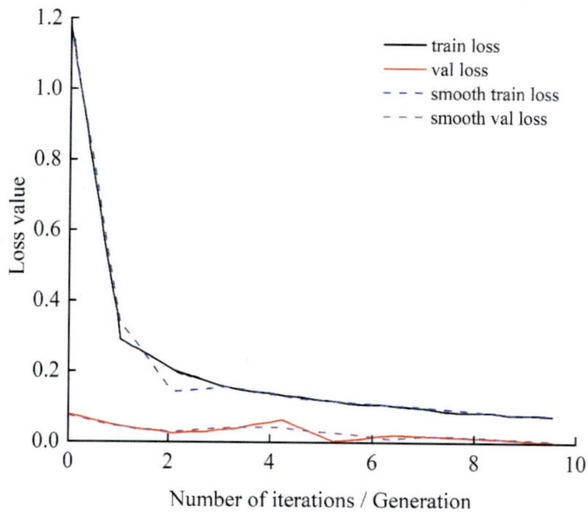

Fig. 5. Parallel-VGG16 network loss function variation curve.

Table 3. Prediction results of six faces of corn seeds based on VGG16 and Parallel-VGG16.

Testset	VGG16			Parallel-VGG16		
	Prediction	Accuracy	Loss	Prediction	Accuracy	Loss
JS129_1	129	96.7%	0.033	129	99.8%	0.002
JX129_1	129	90.7%	0.093	129	96.4%	0.036
JY129_1	129	95.3%	0.047	129	99.7%	0.003

(continued)

Table 3. *(continued)*

Testset	VGG16			Parallel-VGG16		
	Prediction	Accuracy	Loss	Prediction	Accuracy	Loss
JZ129_1	129	93.6%	0.064	129	99.2%	0.008
N129_1	129	90.3%	0.097	129	98.7%	0.013
P129_1	129	98.7%	0.013	129	99.7%	0.003

To verify the performance superiority of the Parallel-VGG16 model in the hyperspectral identification task of crop seeds, we selected GoogLeNet and ResNet50 in the field of visual recognition for comparative experiments to systematically evaluate the comprehensive performance of the proposed model. The experiments were completed based on the hyperspectral corn seed dataset, and the experimental environment was uniformly configured to ensure the objectivity and comparability of the experimental results. The experimental results show that: although GoogLeNet with low parameter quantity has significant lightweight advantages, facilitating subsequent deployment and application, its recognition accuracy drops to 85.7%, which is difficult to meet the actual demand for high-precision identification of crop seeds. ResNet50 with residual structure introduced effectively alleviates the problem of deep network degradation. Although its parameter quantity is higher than that of GoogLeNet, its recognition accuracy is 3 percentage points higher than that of GoogLeNet, showing stronger ability in extracting complex features. While the Parallel-VGG16 model proposed regulates the parameter quantity to an acceptable range for engineering applications through structural optimization, and its recognition accuracy reaches the highest level among the three models. Further comprehensive analysis combined with the inference and prediction time index shows that the Parallel-VGG16 model achieves an effective balance between model parameter quantity, recognition accuracy and inference efficiency: on the one hand, it solves the problems of high training cost and harsh hardware requirements caused by the excessive parameter quantity of the original VGG16; on the other hand, it is significantly superior to the comparative advanced models in the ability of extracting complex crop seed features. The above results fully verify the feasibility and superiority of the Parallel-VGG16 model in the hyperspectral identification task of crop seeds, providing reliable technical support for the engineering application of this task.

4 Discussion and Conclusion

The application of deep learning technology and hyperspectral imaging technology in the field of corn seed identification represents an innovative breakthrough over traditional seed identification methods, enriching the application theory of interdisciplinary technologies in agricultural detection. Traditional manual sensory detection relies on the experience of inspectors, which is highly subjective; while physicochemical detection is objective, it can damage samples and is prone to contamination. This study combines deep learning technology and hyperspectral imaging technology, offering the advantages of non-destructive, rapid, and batch-based detection. This aligns with previous research on the advantages of emerging technologies in agricultural detection. For example, deep learning has also demonstrated efficient and accurate characteristics in agricultural product quality detection. However, there are still some aspects of this study that require further exploration. Currently, the research is only focused on nine types of corn seeds, while the actual types of corn seeds far exceed this number. Moreover, the research object is only corn seeds, without extending to other crop seeds, which falls short of the comprehensiveness of some studies that cover multi-crop seed identification. In the future, it is necessary to further enrich the research object and enhance the universality of the research. Meanwhile, in terms of model construction, although performance has been improved through data augmentation and network structure optimization, there is still a large amount of useless information in the images. In the future, consideration will be given to introducing attention mechanisms to obtain more accurate extraction of effective features, which is also the direction for future improvement to enhance model performance.

This study focuses on the identification of corn seeds, and the main conclusions are as follows: Firstly, an image database containing 9 types of corn seeds (YM129, YM130, YM131, YM132, YM133, YM134, YM135, YM136, YM137) was constructed. More than 1,300 initial image datasets were collected and processed using a hyperspectral camera. Secondly, the initial dataset was augmented through methods such as rotation, contrast enhancement, and permutation and combination, resulting in a new dataset of approximately 70,000 images with the help of the PIL library. Finally, based on the VGG16 network model, a parallel convolution module was inserted and a BN layer was added to construct a new model, parallel-VGG16. Experimental verification showed that the accuracy of this model was significantly improved compared to the basic network training model. With the construction and application of high-quality image datasets, combined with deep learning technology and hyperspectral imaging technology, corn seed image classification and identification technology also has the potential to expand into other fields. For example, it can be used for rapid identification of plant disease symptoms, providing scientific basis for disease prevention and control; for assessing plant diversity, providing data support for ecological protection; and it can also be applied in fields such as automotive coating inspection and intelligent production, such as detecting the uniformity and thickness of coatings, automatically classifying and grading products on the production line, etc. This promotes technological integration and innovation, industrial upgrading, and the development of intelligent manufacturing. In preliminary pre-experiments, we applied the proposed Paralle-VGG16 model to the hyperspectral identification task of rice seeds, achieving a recognition accuracy of 90.1%. This result further verifies the model's universality in different crop seed identification scenarios.

In summary, the method for classifying and identifying corn seed images based on the Parallel-VGG16 network model exhibits significant technological innovation and broad application prospects. Simultaneously, it holds significant guiding importance in academic research, offering novel ideas and methodologies for cross-disciplinary studies in computer vision, agricultural informatization, intelligent manufacturing, and other related fields.

Acknowledgment. The authors thank the Scientific Research Project of the Education Department of Jilin Province, China (No.JJKH20251915KJ, No.JJKH20261932KJ) and the Scientific

Research Project of the Changchun Automobile Industry university (XJKY202518) for financially supporting this research.

References

1. Tan, F., Mo, X.M., Ruan, S.W., et al.: Combining vis-NIR and NIR spectral imaging techniques with data fusion for rapid and nondestructive multi-quality detection of cherry tomatoes. Foods **12**(19), 3621 (2023)
2. Herr, A.W.: Remote sensing continuity: a comparison of HTP platforms and potential challenges with field applications. Frontiers Plant Sci. 141233892-1233892 (2023)
3. Huazhou, C., Hanli, Q., Quanxi, F., et al.: Rapid detection of pomelo fruit quality using near-infrared hyperspectral imaging combined with chemometric methods. Frontiers Bioengineering Biotechnology 8616943-61694 (2021)
4. Li, J., Luo, Y., Liu, Z., et al.: A multi-objective model for detecting leaf lesion of Dendrobium officinarum in complex environments. Eng. Res. Express **7**(4), 045264 (2025)
5. Khaki, S., Pham, H., Han, Y., et al.: Deep-Corn: a semi-supervised deep learning method for high throughput image-based corn kernel counting and yield estimation. Knowl.-Based Syst. **12**, 106874 (2021)
6. Juanhua, Z., Yuan, Z., Mengchu, S., et al.: Nondestructive Identification of Maize Varieties Using Near Infrared Spectroscopy Combined with Machine Learning. College of Mechanical and Electrical Engineering, Henan Agricultural University (China) (2022). https://doi.org/10.1117/12.2657286
7. Traore, B.B., Kamsu-Foguem, B., Tangara, F.: Deep convolution neural network for image recognition. Ecol. Informatics **48**, 257–268 (2018)
8. Nie, P., Zhang, J., Feng, X., et al.: Classification of hybrid seeds using near-infrared hyperspectral imaging technology combined with deep learning. Sens. Actuators B Chem. **296**, 126630 (2019)
9. Krizhevsky, A., Sutskever, I., Hinton, G.E.: Imagenet classification with deep convolutional neural networks. In: Advances in Neural Information Processing Systems, pp. 1097–1105 (2012)
10. Simonyan, K., Zisserman, A.: Very Deep Convolutional Networks for Large-Scale Image Recognition. arXiv pre-print arXiv, 1409.1556 (2014)
11. Szegedy, C., Lu, W., Ja, Y., et al.: Going deeper with convolutions. In: Proceedings of the EEE Conference on Computer Vision and Pattern Recognition, pp. 1–9 (2015)
12. He, K., Zhang, X., Ren, S., et al.: Deep residual leaning for image recognition. In: Proceedings of the IEEE Conference on Computer Vision and Pattern Recognition, pp. 770–778 (2016)
13. Howard, A.G., Zhu, M., Chen, B., et al.: Mobilenets: Efficient Convolutional Neural Networks for Mobile Vision Applications. arXiv preprint arXiv, 1704.04861 (2017)
14. Peng, X., Qian, T., Yunpeng, Z., et al.: Research on maize seed classification and recognition based on machine vision and deep learning. Agriculture **12**(2), 232 (2022)
15. Rani, R., Sahoo, J., Bellamkonda, S., et al.: Attention-enhanced corn disease diagnosis using few-shot learning and VGG16. MethodsX 14103172-103172 (2025)
16. Putro, S.S., Ansori, N., Fuad, M., et al.: Corn leaf disease classification using convolutional neural network based on MobileNetV2 with RMSProp optimization. Math. Modelling Eng. Problems **12**(2) (2025)
17. Huang, S., Fan, X., Sun, L., et al.: Research on classification method of maize seed defect based on machine vision. J. Sens. **2019**(Pt.4), 1–9 (2019)
18. Javanmardi, S., Ashtani, S.H.M., Verbeek, F.J., et al.: Computer vision classification of corn seed varieties using deep convolutional neural network. J. Stored Prod. Res. **92**, 101800 (2021)

19. Tu, K.L., Wen, S.Z., Cheng, Y., et al.: A non-destructive and highly efficient model for detecting the genuineness of maize variety 'JINGKE 968'using machine vision combined with deep learning. Comput. Electronics Agriculture **182**, 1–10 (2021)
20. Glorot, X., Bordes, A., Bengio, Y.: Deep sparse rectifier networks. In: AISTATS, pp. 315–323 (2011)
21. Hinton, E.G., Srivastava, N., Krizhevsky, A., et al.: Improving neural networks by preventing co-adaptation of feature detectors. CoRR abs/1207.0580 (2012)
22. Krizhevsky, A., Sutskever, I., Hinton, E.G.: ImageNet classification with deep convolutional neural networks. Commun. ACM **60**(6), 84–90 (2017)

RLEU-Net: A Lightweight Attention U-Net for High-Fidelity Spectral Signal Recovery in Portable Raman Spectroscopy

Yusen Peng[1], Haokuan Cheng[1], Xiaoyu Zhang[2], Ziqi Xu[3(✉)], and Yang Cao[1(✉)]

[1] School of Cyber Science and Engineering, Southeast University, Nanjing 211189, China
caoyeacy@seu.edu.cn
[2] School of Mathematics, Southeast University, Nanjing 211189, China
[3] Collaborative Innovation Center of Biomedical Functional Materials and Key Laboratory of Biofunctional Materials of Jiangsu Province, School of Chemistry and Materials Science, Nanjing Normal University, Nanjing 210023, China
xzq@nnu.edu.cn

Abstract. The on-site, trace-level detection of bioactive molecules, such as antibiotic residues in food products, is crucial for public health and safety. Portable Surface-Enhanced Raman Scattering (SERS) spectroscopy is a powerful tool for this task, but its sensitivity is often hampered by extremely low signal-to-noise ratios (SNRs), especially when analyzing low-concentration targets within complex biological matrices. This signal integrity issue is the primary bottleneck preventing reliable field analysis. To address this critical gap, this paper introduces RLEU-Net, a lightweight and efficient deep learning framework for high-fidelity Raman signal recovery. Our architecture enhances the standard 1D U-Net with: (1) local residual blocks to preserve the sharp, characteristic peaks of target molecules; (2) a lightweight Efficient Channel Attention (ECA) module to adaptively focus on significant spectral channels; and (3) dilated convolutions to model and remove overwhelming background fluorescence effectively. To rigorously validate its core capabilities, we first test the algorithm on well-characterized materials like graphene, demonstrating superior performance in extracting clean spectral fingerprints from noisy data. Our model significantly outperforms traditional methods in both signal fidelity (SNR +18.8 dB) and the preservation of critical peak characteristics (achieving a Peak Position Shift of $0.2\,\text{cm}^{-1}$ and FWHM Change Rate of 2.1%). Furthermore, compared to the baseline U-Net, RLEU-Net reduces parameter count by 29.1% and FLOPs by 25.7%, while achieving an inference speed of 9.6 ms per spectrum on a standard CPU. This work establishes a robust information processing foundation, paving the way for the development of highly sensitive, next-generation portable SERS instruments for food safety, environmental monitoring, and medical diagnostics.

Yusen Peng and Haokuan Cheng are contribute equally to this work.

L. Huang and D. Greenhalgh (Eds.): ICMLC 2026, LNNS 2061, pp. 206–217, 2027.
https://doi.org/10.1007/978-3-032-31236-5_14

Keywords: Raman Spectroscopy · Signal Denoising · Low Concentration Detection · Lightweight U-Net · Bio-Sensing

1 Introduction

The ability to perform rapid, on-site detection of trace-level bioactive molecules is of paramount importance in fields ranging from food safety and environmental monitoring to medical diagnostics [1]. For instance, the overuse of antibiotics in agriculture can lead to residues in food products, posing risks to human health and contributing to antimicrobial resistance [2]. Traditional detection methods, such as high-performance liquid chromatography (HPLC) and mass spectrometry, are highly sensitive but are confined to laboratories. They require expensive equipment, complex sample preparation, and significant time, making them unsuitable for real-time field screening [3]. This limitation has created a pressing need for portable, fast, and reliable analytical techniques. Among these, Surface-Enhanced Raman Scattering (SERS) spectroscopy has emerged as a highly promising candidate due to its high specificity, non-destructive nature, and potential for miniaturization, offering a unique "molecular fingerprint" for target analytes [4–7].

The development of handheld Raman spectrometers has dramatically improved the portability of this technology, enabling its transition from the lab bench to the field [8]. However, this convenience often comes at the cost of signal quality. Portable instruments typically feature less stable laser sources and less sensitive detectors compared to their laboratory counterparts, resulting in inherently noisier data. Consequently, achieving high sensitivity, especially for low-concentration targets, remains a major hurdle. Much current research focuses on developing substance-specific SERS substrates or complex chemical protocols to enhance the signal of a particular analyte [9]. While effective, this approach lacks generality and is not conducive to creating streamlined, versatile detection platforms for a wide range of substances. A more universal solution lies in enhancing the signal processing capabilities of the instrument itself. Advanced computational algorithms, particularly those based on artificial intelligence (AI), offer a powerful pathway to improving the effective sensitivity of portable Raman systems by robustly extracting faint signals from noisy measurements [10–17].

The primary challenge in processing signals from portable SERS instruments lies in the severe degradation of the raw data. First, when detecting trace-level concentrations, the intrinsic Raman signal is exceptionally weak and often buried within the instrumental noise floor. Second, real-world samples, such as food matrices or environmental water, contain a multitude of interfering substances that generate a strong, broad, and non-linear autofluorescence background [18]. This intense background can completely overwhelm the subtle characteristic peaks of the target analyte, making identification impossible. These factors combine to produce spectra with a very poor signal-to-noise ratio (SNR), which directly compromises the detection limit and reliability of the analysis [19]. Therefore, an effective algorithm must not only suppress random noise but also accurately model and subtract this complex, dominant background, all while preserving the integrity of the faint target peaks.

To address this critical signal processing bottleneck, we propose RLEU-Net, a lightweight deep learning framework specifically designed for high-fidelity Raman signal recovery. Our objective is to create a powerful, general-purpose denoising algorithm that can be integrated into portable instruments to enhance their sensitivity. Here, "resource-constrained portable devices" primarily refer to handheld Raman spectrometers, which are typically equipped with low-power embedded processors, limited on-board memory, and strict real-time inference requirements due to battery operation and on-site deployment constraints. In this work, we first establish the fundamental capabilities of RLEU-Net by training and testing it on well-characterized materials and publicly available datasets. This approach provides a controlled environment to validate its performance in preserving sharp peaks and removing complex noise. Our key contributions are: (1) A tailored, lightweight U-Net architecture suitable for resource-constrained portable devices; (2) The strategic integration of residual blocks, attention mechanisms, and dilated convolutions, each targeting a specific aspect of Raman signal degradation; and (3) A comprehensive performance validation demonstrating that our approach significantly outperforms traditional methods in signal recovery, thereby laying a robust algorithmic foundation for future applications in on-site, low-concentration substance detection.

2 Methodology

Our proposed RLEU-Net is built upon the robust U-Net architecture, a powerful encoder-decoder framework. To address the specific challenges of Raman signal recovery, however, we introduce three targeted modifications: local residual blocks, Efficient Channel Attention (ECA) modules, and dilated convolutions. Each modification is carefully chosen to solve a specific problem encountered in real-world signal acquisition.

2.1 Fundamental Architectural Modules

Local Residual Block for Preserving Peak Integrity

The molecular fingerprint of an analyte is defined by a series of sharp, narrow Raman peaks. Through successive layers of feature transformation, standard deep convolutional networks can inadvertently smooth out or even lose these high-frequency details [20]. Losing these peaks would be catastrophic, as it would erase the very information needed for identification. To combat this, we replace all standard convolutional blocks with local residual blocks, inspired by ResNet [21]. A residual block introduces a"shortcut" or "skip connection" that allows the input signal to bypass one or more convolutional layers and be added back to the output. This is mathematically represented as:

$$y = \mathcal{F}(x) + x \tag{1}$$

where x is the input feature map and $\mathcal{F}(x)$ represents the transformation performed by the convolutional layers. This identity mapping ensures that the network can easily learn to pass crucial peak information directly, guaranteeing that the fine, sharp details of the molecular fingerprint are preserved throughout the deep feature extraction process.

Efficient Channel Attention (ECA) for Feature Prioritization

Not all parts of a Raman spectrum are equally important. Some regions contain the most characteristic peaks for a specific target, while other regions may be dominated by noise or irrelevant signals from the sample matrix. An ideal information processing system should be able to intelligently focus its attention on the most informative spectral bands. We integrate the lightweight Efficient Channel Attention (ECA) module. Unlike other computationally expensive attention mechanisms, ECA uses an efficient 1D convolution to model cross-channel interactions. It learns to generate a set of weights—one for each feature channel (representing spectral bands in our case)—that quantifies the relative importance of that channel. By multiplying the feature maps with these learned weights, the ECA module adaptively amplifies features from critical spectral regions (containing key analyte peaks) while suppressing those from noisy or irrelevant bands. This intelligent feature recalibration allows the model to prioritize the most discriminative information with minimal computational overhead, making it ideal for portable devices.

Dilated Convolution for Modeling Complex Backgrounds

A major problem in analyzing real-world samples is the strong, broad, and often non-linear autofluorescence background originating from the biological or chemical matrix. This background acts as a long-range, low-frequency signal that can completely obscure the sharp Raman peaks. To remove it effectively, the network must "see" a large portion of the spectrum at once to model the overall shape of this background. This, in turn, requires a large receptive field. To achieve a large receptive field without adding computational cost, we employ dilated convolutions in the bottleneck layer of our U-Net. A standard convolution processes adjacent pixels. A dilated convolution, however, introduces gaps between the kernel's weights, effectively skipping pixels. This is governed by a dilation rate r. The operation is expressed as:

$$y[i] = \sum_{k=1}^{K} x[i + r \cdot k] \cdot w[k] \tag{2}$$

By setting $r > 1$, the kernel covers a much wider area of the input signal without increasing the number of parameters or computations. This expanded receptive field allows the network to effectively model the long-range dependencies of the autofluorescence background, enabling it to accurately learn and subtract these complex baseline variations, thus isolating the true, clean Raman signal.

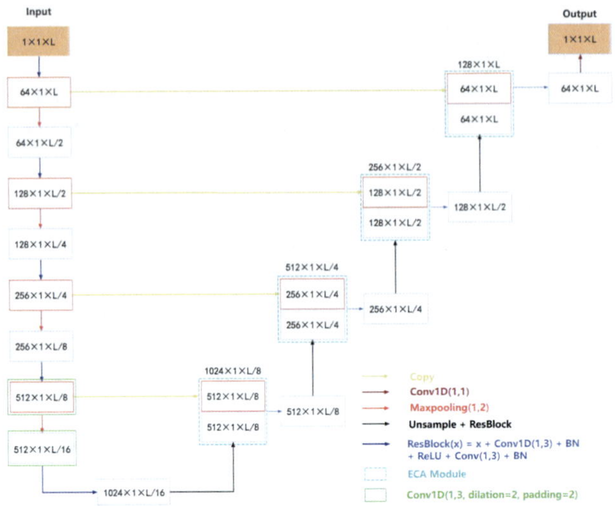

Fig. 1. The Architecture of RLEU-Net, integrating Residual Blocks, ECA modules, and Dilated Convolutions for targeted Raman signal recovery.

2.2 The RLEU-Net Architecture

As illustrated in Fig. 1, RLEU-Net integrates these specialized modules into a symmetric encoder-decoder U-Net structure. The encoder path progressively downsamples the signal to capture hierarchical features, while the decoder path upsamples them to reconstruct the clean spectrum. Crucially:

- All standard convolutional blocks are replaced with our Local Residual Blocks.
- Lightweight ECA modules are inserted at the skip connections, refining the features passed from the encoder to the decoder.
- Dilated convolutions (with a dilation rate of $r = 2$) are employed in the central bottleneck layer to maximize the receptive field for background modeling.

This synergistic design results in a network that is both highly effective at preserving critical details and computationally efficient for real-world deployment.

3 Experimental Settings

3.1 Datasets and Validation Strategy

To establish a robust proof-of-concept and rigorously validate the core signal processing capabilities of our algorithm, we first evaluated our model on a synthetic dataset. This dataset was carefully designed to mimic the challenging conditions of real-world trace-level analysis, incorporating various levels of Gaussian noise and complex, non-linear baseline drift characteristic of biological matrices [22,23]. The noisy spectrum Y is synthesized as:

$$Y = X + N_{gauss} + N_{baseline} \tag{3}$$

Furthermore, to demonstrate the algorithm's performance on real experimental data with well-defined peak structures, we utilized a publicly available Graphene spectrum. The dataset is accessible via the Figshare repository (DOI: 10.6084/m9.figshare.24823353.v1). Graphene serves as an excellent benchmark material because its Raman spectrum contains sharp, distinct peaks, making it ideal for assessing the algorithm's ability to preserve critical features while removing noise. This foundational validation provides the necessary confidence for applying this algorithm to more complex target analytes, such as antibiotics.

3.2 Evaluation Metrics

To assess the fidelity of the extracted bio-signal, we use two specialized metrics critical for substance identification: Peak Position Shift (PPS) and FWHM Change Rate (FCR), in addition to standard metrics like Signal-to-Noise Ratio (SNR) and Mean Squared Error (MSE).

$$\text{PPS} = |P_{\text{denoised}} - P_{\text{true}}| \tag{4}$$

$$\text{FCR} = \frac{|\text{FWHM}_{\text{denoised}} - \text{FWHM}_{\text{true}}|}{\text{FWHM}_{\text{true}}} \times 100\% \tag{5}$$

$$\text{SNR} = 10 \cdot \log_{10}\left(\frac{\sum_{i=1}^{n} X_i^2}{\sum_{i=1}^{n}(X_i - \hat{X}_i)^2}\right) \tag{6}$$

$$\text{MSE} = \frac{1}{n}\sum_{i=1}^{n}(X_i - \hat{X}_i)^2 \tag{7}$$

3.3 Parameter Configuration

Experiments were conducted on an NVIDIA RTX 4060 Ti GPU. Training parameters are detailed in Table 1.

4 Experimental Analysis

4.1 Comparative Experiments

To demonstrate the fundamental capabilities of our algorithm, we first illustrate the signal extraction process on a benchmark Graphene spectrum in Fig. 2. The model is shown to accurately separate the underlying noise and background from the raw signal, recovering the clean spectral fingerprint with high fidelity.

Table 2 and Fig. 3 present a comparative analysis that RLEU-Net (bold red line) clearly outperforms other approaches, demonstrating superior noise removal

Table 1. Model training parameter configuration.

Parameter category	Specification
Optimizer	SGD (Stochastic Gradient Descent)
Initial Learning Rate	0.01
Learning Rate Schedule	Cosine Annealing (CosineAnnealingLR)
Loss Function	Mean Squared Error (MSE)
Batch Size	64
Training Epochs	200

while precisely preserving the original peak structure. This is quantified by its leading scores across all metrics. Crucially, RLEU-Net achieves the lowest PPS and FCR, which is vital, as even minor peak shifts or broadening can lead to molecular misidentification. These results validate the effectiveness of our architecture in maintaining the integrity of the spectral fingerprint—a prerequisite for any reliable identification task.

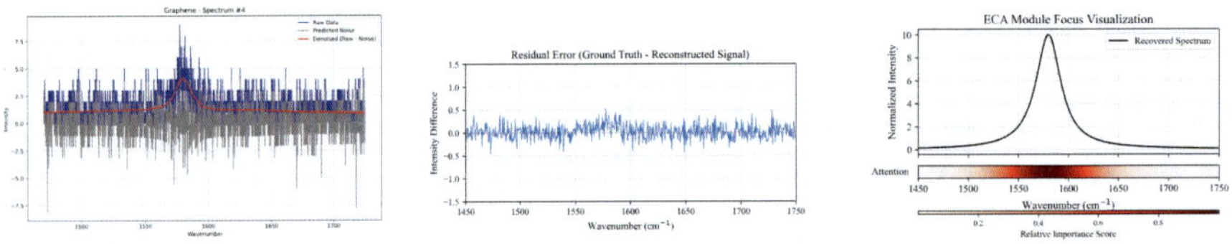

(a) Signal Extraction (b) Residual Error (c) ECA Attention

Fig. 2. Comprehensive signal extraction analysis on a benchmark Graphene spectrum. (a) Signal extraction process. (b) Residual error plot showing minimal deviation. (c) ECA attention visualization focusing on peaks.

Table 2. Quantitative comparison of different denoising methods.

Method	SNR (dB) ↑	MSE (10^{-3}) ↓	PPS (cm^{-1}) ↓	FCR (%) ↓
Noisy Input	8.3	15.21	4.8	–
S-G Filtering	12.9	4.52	2.5	15.8
Wavelet Transform	13.4	3.95	1.8	12.3
U-Net (Baseline)	18.5	2.25	0.9	6.5
RLEU-Net (Ours)	**27.1**	**1.92**	**0.2**	**2.1**

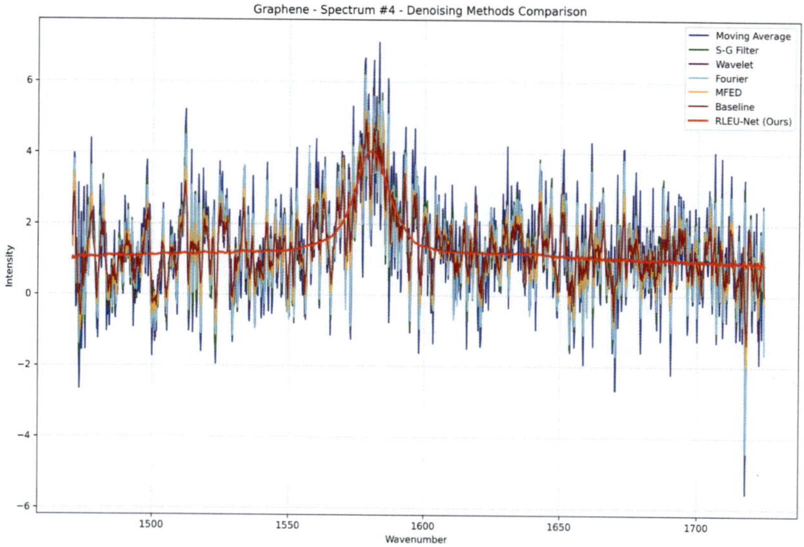

Fig. 3. Visual comparison of denoising methods. The proposed RLEU-Net (bold red line) demonstrates superior signal fidelity and peak preservation, essential for accurate identification.

Table 3. Ablation study results, validating the contribution of each proposed module.

Configuration	SNR (dB) ↑	MSE (10^{-3}) ↓	PPS (cm^{-1}) ↓	FCR (%) ↓
Baseline (U-Net)	18.5	2.25	0.9	6.5
Baseline + `Res`	20.1	2.05	0.7	4.1
Baseline + `ECA + Dilated`	23.5	1.98	0.4	5.8
RLEU-Net (Full)	**27.1**	**1.92**	**0.2**	**2.1**

4.2 Ablation Studies

Ablation studies in Table 3 confirm that each of our proposed architectural modifications contributes to the final performance. The full model, integrating all three components, shows clear synergistic effects and achieves the best overall results, validating our problem-driven design choices.

4.3 Performance on Real-World Data

The model's robustness on a real-world spectrum (Fig. 4) further validates its practical applicability, successfully recovering the underlying fingerprint from a noisy experimental measurement. This strong performance on both synthetic and real benchmark data demonstrates that the algorithm is well-poised for future deployment on more complex biological targets.

4.4 Computational Efficiency Analysis

A primary objective of RLEU-Net is to serve as a lightweight framework suitable for resource-constrained portable devices. To validate this claim, we quantitatively evaluate the model's computational cost. We measure three key metrics:

- **Parameters (Params):** The total number of trainable weights, indicating the model's size and storage footprint.
- **FLOPs (Floating-point Operations):** The number of computations required for a single forward pass, reflecting the model's computational complexity.
- **Inference Time:** The actual time required to process a single spectrum on both a CPU and a GPU, simulating different deployment scenarios.

The results are presented in Table 4. RLEU-Net demonstrates a significant advantage in efficiency. It achieves a 29.1% reduction in parameters and a 25.7% reduction in FLOPs compared to the baseline U-Net. Crucially, the inference time on a CPU is reduced from 11.6 ms to 9.6 ms, a 17.2% speed-up that is critical for real-time processing on portable instruments. This confirms that our design choices, particularly the lightweight ECA module and dilated convolution strategy, successfully create a model that is not only more accurate (as shown in Table 3) but also significantly more efficient.

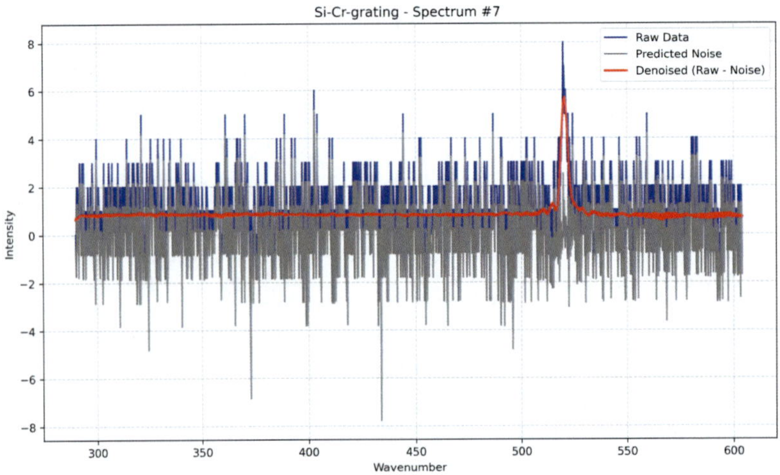

Fig. 4. Performance of RLEU-Net on a real-world spectrum, demonstrating its effectiveness in practical scenarios.

Table 4. Computational efficiency comparison between model variants.

Configuration	Params (M)	FLOPs (G)	Inference Time (ms)
Baseline (U-Net)	2.41	8.72	11.6
Baseline + Res	2.53	9.05	12.1
Baseline + ECA + Dilated	1.65	6.27	9.2
RLEU-Net (Full)	**1.71**	**6.48**	**9.6**

5 Discussion and Limitations

Our experimental results demonstrate that RLEU-Net achieves superior performance in both signal recovery (SNR) and fingerprint integrity (PPS, FCR) compared to traditional methods and a baseline U-Net. The computational efficiency analysis further confirms its suitability for portable devices.

However, we acknowledge the limitations of this current study, which primarily serves as an algorithmic proof-of-concept. While our synthetic data was designed to simulate bio-sample characteristics, it cannot fully capture the complex, stochastic nature and sample-to-sample variability of real-world biological matrices (e.g., milk, fruit juice, or blood serum). Furthermore, the graphene benchmark, while ideal for assessing peak-preservation fidelity, does not represent the core challenge of the target application: detecting an extremely weak analyte signal (like an antibiotic) that is completely overwhelmed by a strong fluorescent background.

6 Conclusion

In this work, we addressed the critical signal processing bottleneck that limits the sensitivity of portable Raman spectroscopy for on-site, trace-level analysis. We proposed and validated RLEU-Net, a lightweight and efficient deep learning framework for high-fidelity signal recovery. By tailoring a U-Net architecture with residual blocks to preserve sharp molecular fingerprints, efficient channel attention to focus on key spectral bands, and dilated convolutions to remove complex background interference, our model provides a robust solution to a pressing real-world problem. In this work, we validated the core performance of RLEU-Net on synthetic data and benchmark materials (graphene), which demonstrates its fundamental signal processing capabilities. Experimental results show that our model significantly improves signal quality (SNR +18.8 dB) and peak preservation (PPS 0.2 cm^{-1}, FCR 2.1%) while reducing computational cost by 25.7% compared to standard U-Net, achieving an inference time of 9.6 ms. This work establishes a strong and generalizable algorithmic foundation for high-fidelity signal recovery. As discussed, our immediate future work will focus on translating this algorithmic success to on-site, trace-level bio-sensing, thereby paving the way for the development of highly sensitive, next-generation handheld SERS instruments for critical applications in food safety, environmental monitoring, and medical diagnostics. However, validating the model on complex biological matrices, such as food products containing antibiotic residues, remains a critical next step. Future work will focus on applying RLEU-Net to real-world bio-sensing scenarios to fully bridge the gap between algorithmic development and field deployment.

Acknowledgment. This work was supported in part by the National Natural Science Foundation of China under Grant No. 62573122 and 22504063, the Fundamental Research Funds for the Central Universities under Grant No. 2242025K30025, the Basic Science (Natural Science) Research in Colleges and Universities of Jiangsu Province under Grant No. 25KJB150020, the Open Research Project of State Key Laboratory of Industrial Control Technology, China under Grant No. ICT2025B37, the Open Research Project of the Key Laboratory of Numerical Simulation for Large Scale Complex Systems, Ministry of Education, China, under Grant No. NSLSCS202502, the Jiangsu Provincial Scientific Research Center of Applied Mathematics under Grant No. BK20233002.

References

1. Chen, B., Li, S., Li, J., Lin, J., Wang, S.: Optimal multivariate method for Raman spectroscopy based diagnosis of nasopharyngeal carcinoma. Appl. Phys. B **114**, 508–509 (2013)
2. Yaraki, M.T., Rubio, N.S., Tukova, A., Lahtinen, T., Peltonen, J., Kostiainen, M.A., Kumenov, A.: Spectroscopic identification of charge transfer of thiolated molecules on Gold nanoparticles via Gold nanoclusters. J. Am. Chem. Soc. **146**(9), 5916–5926 (2024)
3. Pan, X., Zhang, Z., Yun, Y., Wu, W., Zhang, Y., Zhao, Y., Luo, Y., Fan, C., Chen, N.: Machine learning-assisted high-throughput identification and quantification of protein biomarkers with printed heterochains. J. Am. Chem. Soc. **146**(28), 19239–19248 (2024)
4. Raman, C.V., Krishnan, K.S.: A new type of secondary radiation. Nature **121**(3048), 501–502 (1928)
5. Jeanmaire, D.L., Van Duyne, R.P.: Surface raman spectroelectrochemistry: part I. Heterocyclic, aromatic, and aliphatic amines adsorbed on the anodized silver electrode. J. Electroanalytical Chem. Interfacial Electrochemistry **84**(1), 1-20 (1977)
6. Kneipp, K., Wang, Y., Kneipp, H., Perelman, L.T., Itzkan, I., Dasari, R.R., Feld, M.S.: Single molecule detection using surface-enhanced Raman scattering (SERS). Phys. Rev. Lett. **78**(9), 1667 (1997)
7. Diao, X., Li, X., Hou, S., He, H., Han, Y., Wang, Z., Lin, J., Wang, W.: Machine learning-based label-free SERS profiling of exosomes for accurate fuzzy diagnosis of cancer and dynamic monitoring of drug therapeutic processes. Anal. Chem. **95**(19), 7552–7559 (2023)
8. Kim, N., Thomas, M.R., Bergholt, M.S., et al.: Surface enhanced Raman scattering artificial nose for high dimensionality fingerprinting. Nat. Commun. **11**(1), 207 (2020)
9. Tian, L., Su, M., Yu, F., Dong, B., Zhang, J., Liu, J., Zhang, J., Yang, B.: Liquid-state quantitative SERS analyzer on self-ordered metal liquid-like plasmonic arrays. Nat. Commun. **9**(1), 3642 (2018)

10. Qi, Y., Hu, D., Jiang, Y., Zhang, Y., Wang, Z., Li, L., Yang, T.: Recent progresses in machine learning assisted Raman spectroscopy. Adv. Optical Mater. **11**(14), 2203104 (2023)
11. Wathsala, N., Jinadasa, A., Kahawalage, M., Yapa, N.: Deep learning approach for Raman spectroscopy. In: Recent Developments in Atomic Force Microscopy and Raman Spectroscopy for Materials Characterization. IntechOpen, (2022)
12. He, H., Cao, M., Gao, Y., Li, D., Wang, Z., Liu, X., Dai, Q.: Noise learning of instruments for high-contrast, high-resolution and fast hyperspectral microscopy and nanoscopy. Nat. Commun. **15**(1), 754 (2024)
13. Zeng, Y., Liu, Q., Fan, G., Wang, Y., Zhao, Y., Zhang, C.: Modified denoising method of Raman spectra-based deep learning for Raman semi-quantitative analysis and imaging. Microchem. J. **191**, 108777 (2023)
14. Fan, X., Zeng, Y., Zhi, Y., Zhang, C.: Signal-to-noise ratio enhancement for Raman spectra based on optimized Raman spectrometer and convolutional denoising autoencoder. J. Raman Spectrosc. **52**(4), 890–900 (2021)
15. Loc, İ, Keçooğlu, İ, Ünlü, M.B., Yetkin, F., Gülgün, M.A., Özbek, B.: Denoising Raman spectra using fully convolutional encoder-decoder network. J. Raman Spectrosc. **53**(8), 1445–1452 (2022)
16. Barton, S., Alakkari, S., O'Dwyer, K., K-A, S., Salahuddin, N., Byrne, H.J., Lyng, F.M., Casey, A.: Convolution network with custom loss function for the denoising of low SNR Raman spectra. Sensors **21**(14), 4623 (2021)
17. Yang, G., Xiao, H., Gao, H., Zhao, H.: Repairing noise-contaminated low-frequency vibrational spectra with an attention U-Net. J. Am. Chem. Soc. **146**(41), 28491–28499 (2024)
18. Bergholt, M.S., Zheng, W., Ho, K.Y., et al.: Fiberoptic confocal Raman spectroscopy for real-time in vivo diagnosis of dysplasia in Barrett's esophagus. Gastroenterology **146**(1), 27–32 (2014)
19. Chen, C., Hayazawa, N., Kawata, S.: A 1.7 nm resolution chemical analysis of carbon nanotubes by tip-enhanced Raman imaging in the ambient. Nature Commun. **5**(1), 3312 (2014)
20. Xu, C., Liu, H., Ouyang, C.J., Cao, W.M.: Multispectral image Clifford pseudo-differential operator and its application. Sci. China Inf. Sci. **41**(12), 1423–1435 (2011)
21. He, K., Zhang, X., Ren, S., Sun, J.: Deep residual learning for image recognition. In: Proceedings of the IEEE Conference on Computer Vision and Pattern Recognition, pp. 770–778 (2016)
22. Andrychowicz, M., Wolski, F., Ray, A., Schneider, J., Fong, R., Welinder, P., McGrew, B., Tobin, J., Abbeel, P., Zaremba, W.: Hindsight experience replay. In: Proceedings of the 31st International Conference on Neural Information Processing Systems (NIPS 2017), pp. 5055-5065 (2017)
23. Garcia, R., Strudel, R., Chen, S., Golemo, F., Schaal, S., Burchfiel, B.: Robust visual sim-to-real transfer for robotic manipulation. In: 2023 IEEE/RSJ International Conference on Intelligent Robots and Systems (IROS), pp. 992–999 (2023)

ClusterNet: Point Cloud Clustering for CAD Scenes Using Deep Learning and Graph Connectivity

Li-Yong Shen(✉), Hao Yang, and Yi-Fei Feng

School of Mathematical Sciences, University of Chinese Academy of Sciences, Beijing 100049, China
lyshen@ucas.ac.cn

Abstract. We introduce ClusterNet, a novel deep learning approach designed to segment point clouds in CAD scenes into primitive patches. Our approach combines semantic embedding and graph connectivity to improve clustering accuracy by leveraging surface boundaries as important cues. We propose a two-branch network that jointly learns semantic embedding extraction and edge detection tasks. The semantic embedding branch measures point similarity by learning pointwise semantic features, while the edge detection branch identifies edge connections in local regions. Through joint training, the two branches guide the utilization of boundary information in point cloud clustering. In contrast to the traditional mean shift clustering method, ClusterNet considers semantic similarity and point cloud connectivity. Our clustering module transforms the problem into graph partitioning, which is solved using Potts energy minimization. Extensive experiments on the ABC dataset validate our approach, demonstrating its superiority over other methods.

Keywords: Primitive Segmentation · Point Cloud Clustering · Graph Partitioning · Deep Learning

1 Introduction

3D scanning technology has become a pivotal tool across numerous domains, including reconstruction, object recognition, and scene understanding, to name just a few. Utilizing 3D sensors, the algorithms [1–5] have empowered us to reconstruct digital 3D models of real-world scenes. Nonetheless, when it comes to the geometric analysis of man-made objects, the examination typically revolves around primitives, such as planes, spheres, cylinders, and cones. The segmentation of point clouds into these primitive patches emerges as a foundational task in diverse applications like reverse engineering, shape understanding, and editing. This task is twofold: predicting the primitive label of each point and identifying the specific primitive patch to which the point belongs. It is worth emphasizing the paramount importance of this segmentation task, especially in

L. Huang and D. Greenhalgh (Eds.): ICMLC 2026, LNNS 2061, pp. 218–232, 2027.
https://doi.org/10.1007/978-3-032-31236-5_15

the realm of CAD scenes where the components substantially impact the functionality of the object. Meticulous inspection and identification of primitives are crucial not only for our cognitive understanding of the model but also for preserving its intrinsic features. Thus, segmentation becomes indispensable for a wide range of CAD-related applications, supporting various functionalities and ensuring the retention of critical features during modeling and modification.

2D image segmentation has been a well-studied problem in computer vision, with significant advancements achieved over the past decades. The advent of Convolutional Neural Networks (CNNs) has led to remarkable progress in image segmentation tasks [6–9]. However, directly applying 2D methods to 3D point clouds is challenging due to their unordered and unstructured nature. Traditional approaches like RANSAC [10] and region growing [11,12] have been used for 3D segmentation. However, these methods struggle to handle complex scenes and accurately preserve boundaries between similar primitives. Recently, deep learning methods [13–17] have been adapted for this problem, though not specifically for CAD scenes. Nonetheless, these methods are not specifically designed for CAD scenes, and the performance is still far from satisfactory. Although several investigation and application are noticed by the researchers [18,19].

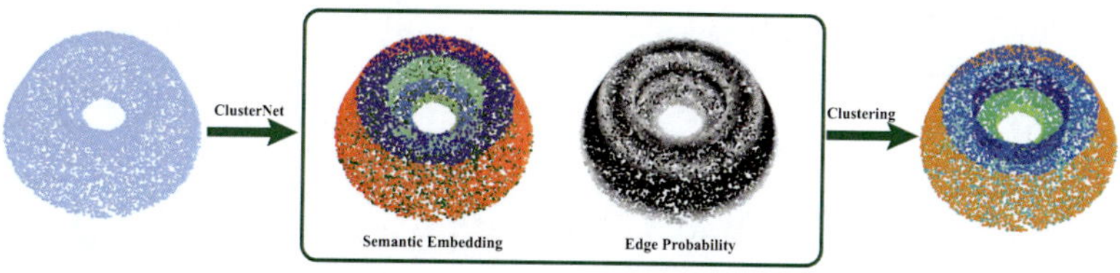

Fig. 1. ClusterNet takes a point cloud as input and generates the clustering results.

This paper introduces a deep learning model called ClusterNet. As shown in Fig. 1, the network takes a point cloud as input and predicts pointwise semantic embedding and edge probability. The learned embedding quantify the similarity between points, indicating a higher likelihood of being on the same patch when they are closer to each other. And the edge probability characterizes the probability of edges being connected within local regions, reflecting the likelihood of establishing connections between neighboring points in the clustering results. The clustering module combines semantic similarity and point cloud connectivity by optimizing an energy function that integrates both factors. The final label of clustered patch is voted from labels of points in the same patch. The key idea of ClusterNet is to combine the semantic embedding and the graph connectivity. Both the semantic embedding and edge probability serve as indicators of clustering results, where the boundaries between segments correspond to the edges of the point cloud. Thus we propose a two branch network to jointly learn the clustering and edge detection tasks. The edge detection branch not only guides the clustering process but also serves as a discriminator to preserve semantic embeddings throughout the network. Experimental results on the ABCPart dataset [14] demonstrate the superior performance of ClusterNet compared to the method [15] across multiple metrics.

The paper is organized as follows: Firstly, in Sect. 2, we introduce our method with the overall architecture as well as the details. Experimental results and comparisons are described in Sect. 3. In Sect. 4, we conclude the paper.

2 Method

This section introduces our approach in detail. After describing the point clustering problem and introduces the notations, we present the overall architecture and three modules of ClusterNet.

2.1 Problem Statement

Given a point cloud $\mathcal{P} = \{\mathbf{p}_i\}_{i=1}^{n}$ with n points, the goal of primitive segmentation is to find the optimal clustering $\mathcal{C}^*$ that points in the same cluster belong to the same base surface $\mathcal{S}$, such as splines, planes, cylinders, cones, and so on. Specifically, the point cloud clustering problem is to partition $\mathcal{P}$ into K clusters $\mathcal{C} = \{\mathcal{C}_k\}_{k=1}^{K}$ such that each point $\mathbf{p}_i$ belongs to exactly one cluster $\mathcal{C}_k$ and $\mathcal{C}_k \cap \mathcal{C}_l = \emptyset$ for $k \neq l$. Note that the number of clusters K is determined by the number of primitives in the point cloud.

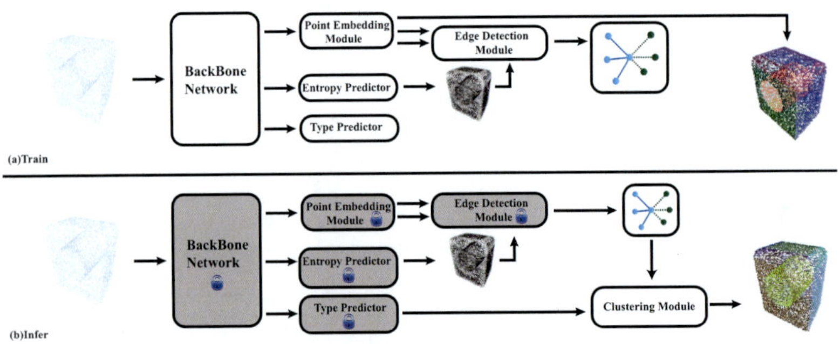

Fig. 2. Overview of ClusterNet. (a) During training, the Point Embedding Module and Edge Detection Module are supervised by ground-truth labels. (b) During inference, these modules are fixed, and the Clustering Module partitions points into surfaces.

2.2 Overview of ClusterNet

As shown in Fig. 2, the ClusterNet contains three parts: Point Embedding Module, Edge Detection Module and Clustering Module. Point Embedding Module encodes the point cloud into a feature space, each point $p_i \in \mathcal{P}$ is represented by a feature vector $s_i \in \mathbb{R}^d$ and the distance in semantic space shows the similarity between points; Edge Detection Module detects the edges of the point cloud and outputs the edge probability of each point. The neighborhood $\mathcal{N}_i$ is the set of points within a certain radius of p_i, the clustering results of $\mathcal{N}_i$ reflect the edge probability of p_i, as the edge points are more likely to be clustered into different surfaces. Therefore, the edge probability of point p_i can be measured by the entropy of clustering probability in its neighborhood $\mathcal{N}_i$; Clustering Module clusters the points into different surfaces based on the features and edge probability, the distance in feature space and the geometric connection in spatial space are considered simultaneously as surfaces are segmented by edges.

2.3 Backbone Network

Input points are first encoded into a feature space. For a given point cloud $\mathcal{P}$, a backbone network is used to extract the feature of each point. The point $\mathbf{p} \in \mathcal{P}$ represents either as either a 3D position (x, y, z) or a 6D position + normal vector (x, y, z, n_x, n_y, n_z). We implement the backbone network with DGCNN [20] to predict point-wise features. Other point cloud feature extraction networks can also be used in our framework, such as PointNet [21], PointNet++ [22], PointTransformer [23]. The performances of different backbone networks are implemented in the section of experiments. DGCNN uses a k-nearest neighbor search in feature space to find the neighborhood $\mathcal{N}_i$ of each point p_i and then applies a shared MLP to extract the local feature l_i of each point. To aggregate the local features l_i into a global feature g for the whole point cloud, dynamic graph convolution is used to capture the global structure information of the point cloud.

Rather than constructing the graph based on spatial proximity, we search for the neighborhood of each point in feature space. This approach is based on the observation that similar points tend to be located close to each other in feature space. By applying convolution on the dynamic graph, the network effectively propagates information between points that are spatially distant. The backbone network incorporates both local and global information from the point cloud. Local features l_i are extracted by applying MLPs to the input features of the respective neighborhood. To capture global representation for the entire point cloud, a max-pooling layer is employed. The local representation l is stacked with the global representation g to form the final feature x for each point.

2.4 Point Embedding Module

The Point Embedding Module is designed to learn the semantic embedding of each point. The feature x_i for point p_i contains both clustering and edge information. To adjust the spatial distribution of features, a small MLP network f_{sem} is applied, resulting in $s_i = f_{sem}(x_i)$. This transformation ensures that points belonging to the same surface are closer to each other in terms of semantic distance within the semantic space.

Surface type is an important feature for point cloud segmentation as same type of surfaces are more likely to be clustered into the same patch. Surfaces in models are divided into six types: Plane, Sphere, Cylinder, Cone, Bspline-Open and Bspline-Closed. Distinguishing between these surface types is crucial for partitioning and facilitating subsequent processing of distinct surfaces. To predict the surface type t_i for each point, we employ a small MLP network f_{type} to map the feature x_i to the corresponding surface type $t_i = f_{type}(x_i)$.

2.5 Edge Detection Module

Edge information is crucial for point cloud segmentation. Edges act as boundaries that separate different parts or objects within the point cloud, and utilizing this information can guide the clustering process for accurate segmentation.

Note that intersection points between two surfaces and corner points formed by three or more surfaces reflect distinct characteristics. Binary classification alone is inadequate for effectively capturing the geometric information of the point cloud. To address this, we introduce entropy as a measure of uncertainty for the neighborhood $\mathcal{N}_i$, which consists of the k nearest neighbors of p_i. Entropy quantifies the degree of disorder or uncertainty in a system. In information theory, it is used to measure the average information content or uncertainty of data, where higher entropy implies greater uncertainty and a larger amount of information. When the probability distribution tends towards a uniform distribution, the entropy reaches its maximum value. Entropy has various applications, including its use in assessing the importance of features in machine learning and pattern recognition.

In our study, we utilize the entropy of the surface distribution to estimate the edge probability for each point. Considering that the neighborhood $\mathcal{N}_i$ shared by K surfaces, with each surface containing n_k points, the probability of a point belonging to surface k is calculated as $P_k = \frac{n_k}{|\mathcal{N}_i|}$. Thus, the entropy of the local point cloud can be defined as follows:

$$H_i = -\sum_{k=1}^{K} P_k \log P_k \tag{1}$$

A small entropy H indicates point p_i is inner point while a large entropy H suggests that point p_i is a boundary point. The entropy value provides a measure of the heterogeneity or diversity of neighboring points, the larger entropy value reflects the more complex geometric structure of the local point cloud.

We use a small MLP network f_{ent} to predict the entropy H_i of each point p_i in the point cloud. The extracted feature x_i should also contain the edge information, as we can predict entropy H_i^{pred} by $f_{ent}(x_i)$.

The predicted entropy H_i^{pred} helps to further improve the connection probability of local point cloud. The edge probability e_{ij} of each point p_i with its neighborhood point p_j is predicted by a small MLP network f_{edge} as $e_{ij} = f_{edge}(s_i, s_j, H_i, H_j)$. To satisfy the symmetry of the edge probability, a maxpooling operation is applied in first layer of f_{egde} so that $e_{ij} = e_{ji}$. The edge information is then fed into clustering module in next section to help distinguish the different partitions for the complex situations.

2.6 Clustering Module

With learned point semantic feature s_i, surfaces can be clustered based on the similarity of the semantic embeddings. Due to uncertain number of surfaces, HPNet [15] utilizes the mean shift algorithm [24] to cluster point clouds adaptively. Other clustering algorithms, such as spectral clustering [25] and DBSCAN [26], have also been implemented in other papers.

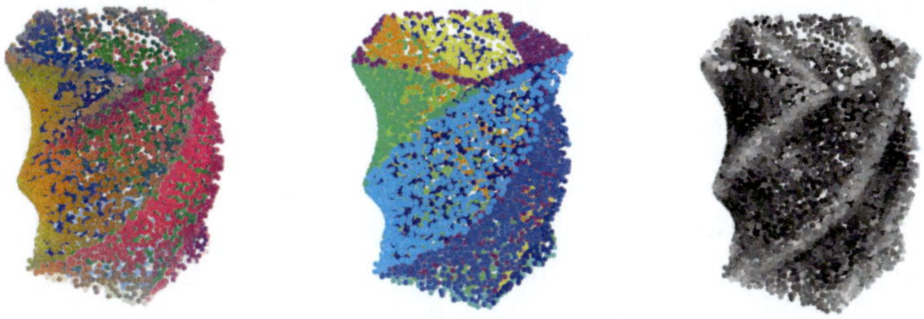

(a) (b) (c)

Fig. 3. Visualization of the semantic embeddings. (a) The colorization of first three principal components of semantic embedding. (b) The ground truth of point cloud segmentation. (c) The local entropy predicted by the network.

However, the clustering results are not satisfactory when only considering the semantic embeddings. To visualize the learned semantic embeddings, we use the PCA method to extract the first three principal components s_{pc} from the learned semantic embedding s as RGB color. In Fig. 3, the points belonging to different surfaces are close to each other in the semantic space. Nonetheless, in the challenging cases, the semantic feature s_i may lack diversity. On the other hand, the local entropy H_i, depicted in Fig. 3, effectively distinguishes the different surfaces. To fully harness the network's learned edge information, we propose a novel clustering algorithm that incorporates both the semantic information and edge probability to cluster the point cloud.

Considering a point cloud composed of n points, along with fixed learned semantic information, including the semantic embedding s_i and surface type t_i for each point, as well as edge probabilities e_{ij}, we assign a new embedding g_i to each point p_i. This assignment is designed such that points sharing the same g_i are clustered together during the clustering process. To optimize this clustering, a piecewise constant function g_i is defined, which is fine-tuned according to the following energy function:

$$E(g) = \sum_{i=1}^{n} ||s_i - g_i||^2 + \rho \sum_{i=1}^{n} \sum_{j \in \mathcal{N}_i} e_{ij} \max(\mathbb{I}(g_i \neq g_j), \mathbb{I}(t_i \neq t_j)) \quad (2)$$

where $\mathbb{I}$ is the indicator function. The first term is fidelity term, which ensures that the semantic embedding s_i is close to the piecewise constant function g_i.

The second term is regularization term to penalize edges connecting components with different values, promoting a segmented shape with simplicity. The parameter ρ controls the trade-off between the fidelity term and the regularization term, implicitly determining the number of clusters. The points with same g_i are clustered into the same surface.

This optimization problem is known as Potts energy segmentation [27,28], and can be solved by the graph cut algorithm [29,30]. Cut Pursuit [30] is a rapid and precise algorithm for solving the Potts energy segmentation problem. It combines the principles of graph cuts and greedy algorithms to efficiently find the optimal piecewise constant function that minimizes the energy function. By iteratively updating the piecewise constant functions based on the energy gradient and employing a local partitioning strategy, Cut Pursuit achieves accurate and coherent segmentations of point clouds. Because of the complex implementation of Cut Pursuit, we propose a greedy method to solve the Potts energy segmentation problem. In our implementation, we first initialize g_i by s_i and compute point-wise energy $E_i(g) = ||s_i - g_i||^2 + \rho \sum_{j \in \mathcal{N}_i} e_{ij} \mathbb{I}(g_i - g_j \neq 0)$. Then we iteratively update the largest energy point p_i by $g_i = g_i + \alpha \frac{\partial E_i(g)}{\partial g_i}$ until the energy converges. We implement the algorithm as Algorithm 1 to cluster the point cloud based on the semantic embedding and edge probability.

Algorithm 1 Greedy Solver for Potts Energy Segmentation

Input: Point cloud $\mathcal{P}$, semantic embedding s_i, edge probability e_{ij}
Output: Clustering result $\mathcal{C}$

```
Initialize g_i by s_i
repeat
    Compute E_i(g) = ||s_i − g_i||^2 + ρ Σ_{j∈N_i} e_ij 𝕀(g_i − g_j ≠ 0) by Eq. (2)
    Compute E_pre = Σ_i E_i(g)
    Compute top_k = arg max E_i(g)
    Compute ĝ_i = arg max_{g_i*} E_i(g_i*) for i ∈ top_k
    Update g_i by ĝ_i
    Compute E_new = E(g)
until |E_pre − E_new| ≤ ε
```

2.7 Network Training

Semantic loss: We require the learned semantic embedding s_i satisfy the following two conditions: (a) The semantic embedding s_i should be close to s_j if p_i and p_j belong to the same surface; (b) The semantic embedding s_i should be far away from s_j if p_i and p_j belong to different surfaces. Specifically, the L_{sem} consists of two terms: L_{pull} and L_{push}. Let $\mathcal{P}^{gt}$ be the ground-truth segmentation of the point cloud. L_{pull} pulls the points in same surface closer to the center of semantic space:

$$L_{pull} = \frac{1}{K}\sum_{k=1}^{K}\frac{1}{|\mathcal{P}_k^{gt}|}\sum_{p_i\in\mathcal{P}_k^{gt}}\max(||s_i - s_{\mathcal{P}_k^{gt}}|| - \delta_{pull}, 0), \tag{3}$$

where $s_{\mathcal{P}_k^{gt}} = \frac{1}{|\mathcal{P}_k^{gt}|}\sum_{p_i\in\mathcal{P}_k^{gt}} s_i$. L_{push} pushes the points in different surfaces away from each other:

$$L_{push} = \frac{1}{K(K-1)}\sum_{i<j}\max(\delta_{push} - ||s_{\mathcal{P}_i^{gt}} - s_{\mathcal{P}_j^{gt}}||, 0). \tag{4}$$

For type classification, we employ the cross-entropy loss L_{CE} to supervise the type probability t_i:

$$L_{type} = \frac{1}{n}\sum_{i=1}^{n} L_{CE}(t_i, t_i^{gt}), \tag{5}$$

where t_i^{gt} is the ground-truth type of point p_i. Combing (3),(4) and (5), the semantic loss L_{sem} is defined as:

$$L_{sem} = L_{pull} + L_{push} + \lambda_{type}L_{type}. \tag{6}$$

In our experiments, we set $\delta_{pull} = 0.5, \delta_{push} = 1.5$ and $\lambda_{type} = 1.0$.

Edge loss: The edge probability e_{ij} and entropy are predicted by the network. The cross-entropy loss is used to supervise the edge probability e_{ij}:

$$L_{prob} = -\frac{1}{n}\sum_{i=1}^{n}\frac{1}{|\mathcal{N}_i|}\sum_{j\in\mathcal{N}_i} L_{CE}(e_{ij}, e_{ij}^{gt}), \tag{7}$$

where $e_{ij}^{gt} = 1$ if p_i and p_j belong to the same surface, otherwise $e_{ij}^{gt} = 0$. The $L_{entropy}$ is designed to supervise the entropy prediction H_i:

$$L_{entropy} = \frac{1}{n}\sum_{i}^{n}|H_i^{pred} - H_i|. \tag{8}$$

Both two terms provide necessary information for edges of segmentation. The loss for edge is defined as:

$$L_{edge} = \lambda_{prob}L_{prob} + \lambda_{entropy}L_{entropy}. \tag{9}$$

λ_{prob} and $\lambda_{entropy}$ are set to 10^{-1} and 10^{-3} in our experiments.

Overall loss: With the semantic loss (6) and edge loss (9), the overall loss is defined as:

$$L = L_{sem} + L_{edge}. \tag{10}$$

The semantic loss term is used to give the overall shape of the segmentation, and the edge loss term is used to refine the segmentation. The network is trained by minimizing the overall loss.

3 Experiments

We first introduce the detailed experimental setting including datasets, evaluation metrics and implementation details. Then we compare our method with state-of-the-art methods to demonstrate the effectiveness of our method. Finally, ablation study is conducted to analyze the effect of each component in our method.

3.1 Experimental Setting

Datasets. Our experiments are based on ABCPart Dataset [14], which is derived from ABC dataset [31]. ABC dataset is the most popular CAD dataset and provides a large source of 3D CAD models. Each model in ABC dataset stores surface patches and modeling operations that designers used to create them. The ABC dataset is generated by randomly sampling 10000 points from the surface patches of the models in ABC dataset. The dataset is split into $24k$ training sets, $4k$ test sets and $4k$ valid sets.

Evaluation Metrics. Two standard metrics are used to evaluate the performance of point segmentation and segment labeling:

- Seg-IoU: Assuming that K is the number of ground-truth segments and $W, W^{gt} \in \{0,1\}^{n\times K}$ are predicted and ground-truth segmentation membership matrix, the Seg-IoU evaluates the segmentation mean IoU score: $\frac{1}{K}\sum_{k=1}^{K} IoU(W_{:,k}, W_{:,k}^{gt})$. This metric measures the similarity of the predicted segmentation and the ground-truth segmentation.
- Type-IoU: $\frac{1}{K}\sum_{k=1}^{K} \mathbb{I}(t_k - t_k^{gt})$, where t_k is the predicted type of the k-th segment and t_k^{gt} is the ground-truth of it. This metric measures the accuracy of the predicted segment type.

To measure the level of detail in the segmentation results, we introduce the average entropy error(AEE) metric. $\frac{1}{N}\sum_{i=1}^{N} |H_i - H_i^{gt}|$ is computed to measure the difference between the predicted entropy and the ground-truth entropy.

Implementation Details. We train ClusterNet on a single NVidia RTX 6000 GPU with the Pytorch framework [32]. Adam optimizer [33] is used to optimize the network with a learning rate of 10^{-3} and a batch size of 8. The learning rate is decayed by 0.5 every 40 epochs. For point cloud in ABCPart dataset, we sample 7000 points from each model and find the 50 nearest neighbors for each point. All the point clouds are centered and normalized to a unit sphere with a rotation augmentation.

3.2 Comparison

We compare our results with the other latest methods. We test ParseNet [14], HPNet [15] on ABCPart dataset. Figure 4 and Table 1 present qualitative and quantitative results of all methods.

Figure 4 shows that our method conducts better segmentation results than other methods. Benefiting from information about edge connection probability,

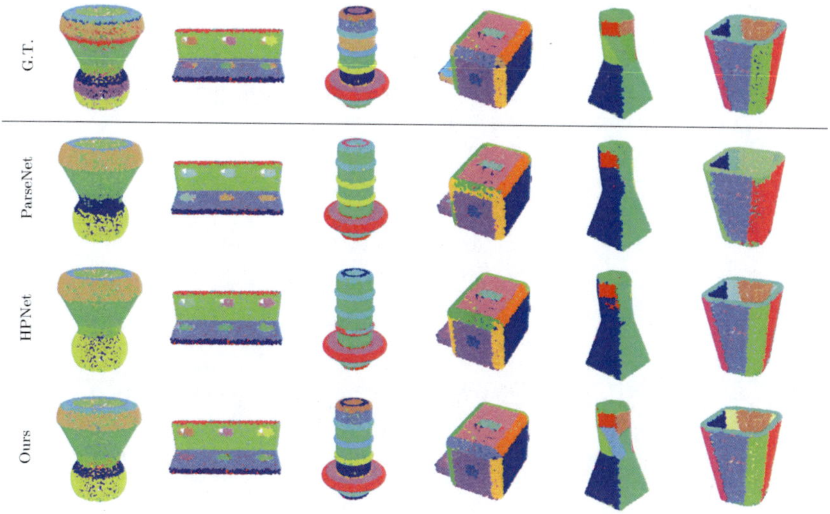

Fig. 4. Point cloud segmentation results on CAD models. From top to bottom: ground-truth segmentation, ParseNet [14], HPNet [15], and our approach.

(a) Predicted Entropy (b) Extracted Edge

Fig. 5. Visualization of edge points. (a) shows the predicted entropy of each point. (b) shows the extracted edge points with a simple threshold.

our method can segment the point cloud more accurately. The blue dashed boxes in Fig. 4 show that our segmentation boundaries are clearer and less prone to issues like erroneous sticking compared to other methods. The Fig. 5(a) presents the predicted entropy of each point and the extracted edge points with a simple threshold are shown in Fig. 5(b). The clear edge points indicates the segmentation boundary.

As shown in Table 1, ClusterNet achieves the best performance on all metrics (Seg-IoU, Type-IoU, AEE). The Seg-IoU metric means the similarity of the predicted segmentation and the ground-truth segmentation, and AEE metric measures the level of detail in the segmentation results.

To further validate the stability of the proposed method, we tested the network's performance under noisy data conditions as shown in [14] . Gaussian noise in the range $[-4\%, 4\%]$ is randomly added to the point cloud coordinates

Table 1. Benchmark evaluation on our approach and baseline approaches. The best results are highlighted in bold.

Method	Seg-IoU	Type-IoU	AEE
ParseNett [14]	84.21	88.60	0.052
HPNet [15]	85.24	91.04	0.043
Proposed ClusterNet	**87.51**	**92.69**	**0.032**

Table 2. Comparison of robustness on noise data. The best results are highlighted in bold.

Method	Seg-IoU	Type-IoU	AEE
ParseNett [14]	74.65	84.44	0.092
HPNet [15]	79.25	86.74	0.083
Proposed ClusterNet	**85.49**	**88.64**	**0.055**

along the normal vector, and an additional uniform noise of 5°C is applied to the normals. Table 2 demonstrates that our method is more stable against negative noise, yielding better results in practical applications.

3.3 Ablation Study

We proceed to analyze the effect of each component in ClusterNet. In ablation study, we first remove the edge detection module to verify its effectiveness. Then we focus on the structure of backbone network and conduct experiments on different backbone networks. Finally, we replace the clustering module with the mean shift clustering algorithm [24] to explore its influence on the results. Table 3 provides the overall view of different ablation studies.

Table 3. Evaluation on different component selection.

Method	Seg-IoU	Type-IoU	AEE
Our-wc/e	84.58	89.32	0.049
Our-wc/c	85.69	91.17	0.045
Ours-p+	84.54	89.04	0.050
Ours-pt	84.94	90.04	0.039
Ours-full	**86.21**	**92.11**	**0.035**

'Our-wc/e' denotes dropping the edge detection module. 'Our-wc/c' denotes replacing clustering module with mean shift method. 'Ours-p+' denotes using PointNet++ as backbone network. 'Ours-pt' denotes using PointTransformer as backbone network. 'Ours-full' corresponds to our full algorithm.

No Edge Detection Module. To illustrate the necessity of the edge information, we remove the edge detection module from ClusterNet and train the network on ABCPart dataset. Table 3 shows that the Seg-IoU score drop by 1.11% and AEE score increase by 0.004 compared with 'Our-wc/c'. Figure 6 shows that the semantic embeddings at the boundary of the segmentation is not clear without edge detection module.

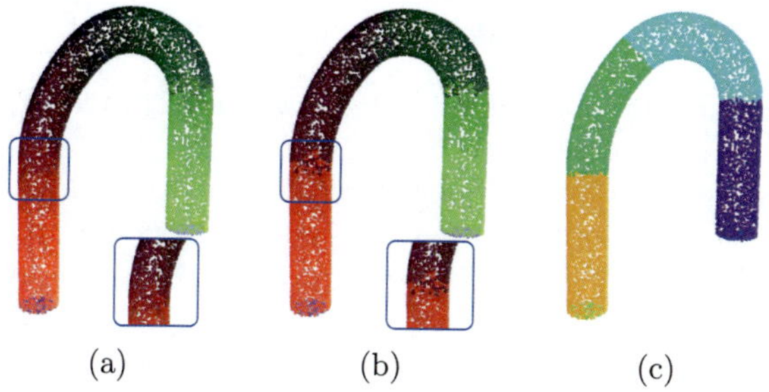

Fig. 6. Comparison of segmentation results with and without edge detection module. (a) Without edge detection module around boundary. (b) With edge detection module around boundary. (c) Ground-truth segmentation.

Different Backbone Networks. To discover the effect of different backbone networks, we conduct experiments on different backbone networks while keeping other settings unchanged. We train ClusterNet on ABCPart dataset with PointNet++, PointTransformer and DGCNN as the backbone network. DGCNN performs better than other backbone networks while PointNet++ is the worst as Table 3 shown.

Mean Shift Clustering. The traditional clustering algorithm only considers the semantic information of the points, losing the spatial information. To explore the influence of the clustering algorithm on the results, we replace the clustering module with the mean shift clustering algorithm. Table 3 illustrates the improvement of the clustering module. Our results are more accurate especially in the segmentation boundary as shown in Fig. 7.

4 Conclusion

In this paper, we propose a novel point cloud segmentation network that combines semantic embedding and edge probability. Based on the most intuitive idea, edge points separate different surfaces, we train the point cloud clustering task in conjunction with point cloud edge detection. The clustering part is designed to learn the semantic embedding for each point and the edge detection part is designed to predict the edge probability for each point. In order to discriminate the edge points from the non-edge points, we utilize the entropy of local regions as a measure. A larger entropy indicates the presence of more different surfaces

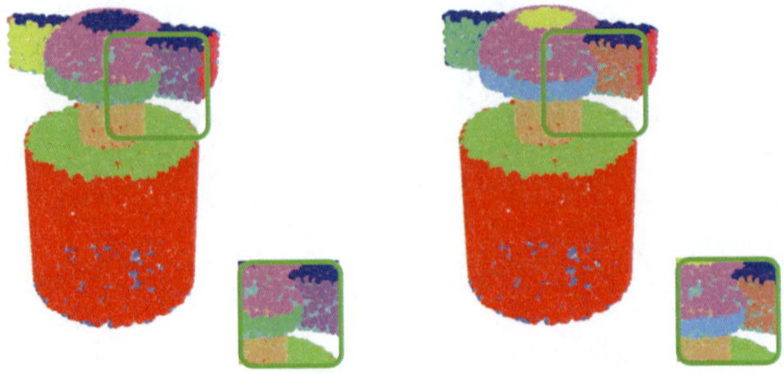

(a) Mean Shift Clustering (b) Our Clustering

Fig. 7. Comparison of segmentation edge with different clustering methods. (a) The extracted edge points with mean shift clustering. (b) The extracted edge points with our clustering method.

within the local region, which is closer to the surface edge. To make full use of the semantic information and predicted edge connection, we propose a graph-cut based algorithm to refine the segmentation. ClusterNet significantly outperforms existing works and improves the performance on fine structures at the edges.

References

1. Whelan, T., Leutenegger, S., Salas-Moreno, R., Glocker, B., Davison, A.: Elasticfusion: Dense slam without a pose graph. In: Science and Systems, Robotics (2015)
2. Dai, A., Nießner, M., Zollhöfer, M., Izadi, S., Theobalt, C.: Bundlefusion: real-time globally consistent 3D reconstruction using on-the-fly surface reintegration. ACM Trans. Graph. (ToG) **36**(4), 1 (2017)
3. Geng, Y., Wang, Z., Jia, L., Qin, Y., Chai, Y., Liu, K., Tong, L.: 3dgraphseg: a unified graph representation-based point cloud segmentation framework for full-range high-speed railway environments. IEEE Trans. Ind. Informatics **19**(12), 11430–11443 (2023)
4. Ge, W., Shen, L.Y., Yuan, C.: Point cloud fitting based on high-accuracy segmentation. In: Proceedings of the 2024 2nd Asia Symposium on Image and Graphics. p. 92–100. ASIG '24, Association for Computing Machinery, New York, NY, USA (2025)
5. Yuan, Z., Zeng, W., Su, Y., Liu, W., Cheng, M., Guo, Y., Wang, C.: Density-guided translator boosts synthetic-to-real unsupervised domain adaptive segmentation of 3D point clouds. In: Proceedings of the IEEE/CVF Conference on Computer Vision and Pattern Recognition (CVPR), pp. 23303–23312 (2024)
6. Dai, J., He, K., Sun, J.: Instance-aware semantic segmentation via multi-task network cascades. In: Proceedings of the IEEE Conference on Computer Vision and Pattern Recognition, pp. 3150–3158 (2016)
7. He, K., Gkioxari, G., Dollár, P., Girshick, R.: Mask R-CNN. In: Proceedings of the IEEE International Conference on Computer Vision, pp. 2961–2969 (2017)

8. Chen, L.C., Papandreou, G., Kokkinos, I., Murphy, K., Yuille, A.L.: Deeplab: semantic image segmentation with deep convolutional nets, atrous convolution, and fully connected CRFs. IEEE Trans. Pattern Anal. Mach. Intell. **40**(4), 834–848 (2017)
9. Al-antari, M.A., Al-Tam, R.M., Al-Hejri, A.M., Al-Huda, Z., Lee, S., Yıldırım, Ö., Gu, Y.H.: A hybrid segmentation and classification cad framework for automated myocardial infarction prediction from MRJ images. Sci. Rep. **15** (2025)
10. Schnabel, R., Wahl, R., Klein, R.: Efficient RANSAC for point-cloud shape detection. In: Computer Graphics Forum, vol. 26, pp. 214–226. Wiley Online Library (2007)
11. Marshall, D., Lukacs, G., Martin, R.: Robust segmentation of primitives from range data in the presence of geometric degeneracy. IEEE Trans. Pattern Anal. Mach. Intell. **23**(3), 304–314 (2001)
12. Rabbani, T., Van Den Heuvel, F., Vosselmann, G.: Segmentation of point clouds using smoothness constraint. Int. Arch. Photogramm. Remote Sens. Spatial Inf. Sci. **36**(5), 248–253 (2006)
13. Li, L., Sung, M., Dubrovina, A., Yi, L., Guibas, L.J.: Supervised fitting of geometric primitives to 3D point clouds. In: Proceedings of the IEEE/CVF Conference on Computer Vision and Pattern Recognition, pp. 2652–2660 (2019)
14. Sharma, G., Liu, D., Maji, S., Kalogerakis, E., Chaudhuri, S., Měch, R.: Parsenet: a parametric surface fitting network for 3D point clouds. In: Computer Vision–ECCV 2020: 16th European Conference, Glasgow, UK, August 23–28, 2020, Proceedings, Part VII 16, pp. 261–276. Springer (2020)
15. Yan, S., Yang, Z., Ma, C., Huang, H., Vouga, E., Huang, Q.: HPNet: deep primitive segmentation using hybrid representations. In: Proceedings of the IEEE/CVF International Conference on Computer Vision, pp. 2753–2762 (2021)
16. Du, Z., Dou, J., Liu, Z., Wei, J., Wang, G., Xie, N., Yang, Y.: CDPNet: cross-modal dual phases network for point cloud completion. In: Proceedings of the AAAI Conference on Artificial Intelligence, vol. 38, pp. 1635–1643 (2024)
17. Sun, Y., Zhang, X., Miao, Y.: A review of point cloud segmentation for understanding 3D indoor scenes. Vis. Intell. **2**(1), 14 (2024)
18. Liu, Y., Obukhov, A., Wegner, J.D., Schindler, K.: Point2cad: Reverse engineering CAD models from 3D point clouds. In: Proceedings of the IEEE/CVF Conference on Computer Vision and Pattern Recognition (CVPR), pp. 3763–3772 (2024)
19. Liu, X., Wang, Z., Melkote, S.N., Rosen, D.W.: Manufacturing process identification from 3D point cloud models using semantic segmentation. J. Manuf. Syst. **82**, 858–873 (2025)
20. Wang, Y., Sun, Y., Liu, Z., Sarma, S.E., Bronstein, M.M., Solomon, J.M.: Dynamic graph CNN for learning on point clouds. ACM Trans. Graph. (ToG) **38**(5), 1–12 (2019)
21. Qi, C.R., Su, H., Mo, K., Guibas, L.J.: PointNet: Deep learning on point sets for 3D classification and segmentation. In: Proceedings of the IEEE Conference on Computer Vision and Pattern Recognition, pp. 652–660 (2017)
22. Qi, C.R., Yi, L., Su, H., Guibas, L.J.: Pointnet++: deep hierarchical feature learning on point sets in a metric space. In: Advances in Neural Information Processing Systems, vol. 30 (2017)
23. Zhao, H., Jiang, L., Jia, J., Torr, P.H., Koltun, V.: Point transformer. In: Proceedings of the IEEE/CVF International Conference on Computer Vision, pp. 16259–16268 (2021)
24. Cheng, Y.: Mean shift, mode seeking, and clustering. IEEE Trans. Pattern Anal. Mach. Intell. **17**(8), 790–799 (1995)

25. Ng, A., Jordan, M., Weiss, Y.: On spectral clustering: analysis and an algorithm. In: Advances in Neural Information Processing Systems, vol. 14 (2001)
26. Ester, M., Kriegel, H.P., Sander, J., Xu, X.: A density-based algorithm for discovering clusters in large spatial databases with noise. In: KDD, vol. 96, pp. 226–231 (1996)
27. Rousson, M., Deriche, R.: A variational framework for active and adaptative segmentation of vector valued images. In: Workshop on Motion and Video Computing, 2002. Proceedings, pp. 56–61. IEEE (2002)
28. Li, C., Kao, C.Y., Gore, J.C., Ding, Z.: Minimization of region-scalable fitting energy for image segmentation. IEEE Trans. Image Process. **17**(10), 1940–1949 (2008)
29. Boykov, Y., Veksler, O., Zabih, R.: Fast approximate energy minimization via graph cuts. IEEE Trans. Pattern Anal. Mach. Intell. **23**(11), 1222–1239 (2001)
30. Landrieu, L., Obozinski, G.: Cut pursuit: fast algorithms to learn piecewise constant functions on general weighted graphs. SIAM J. Imag. Sci. **10**(4), 1724–1766 (2017)
31. Koch, S., Matveev, A., Jiang, Z., Williams, F., Artemov, A., Burnaev, E., Alexa, M., Zorin, D., Panozzo, D.: ABC: a big CAD model dataset for geometric deep learning. In: Proceedings of the IEEE/CVF Conference on Computer Vision and Pattern Recognition, pp. 9601–9611 (2019)
32. Paszke, A., Gross, S., Massa, F., Lerer, A., Bradbury, J., Chanan, G., Killeen, T., Lin, Z., Gimelshein, N., Antiga, L.: PyTorch: an imperative style, high-performance deep learning library. Advances in Neural Information Processing Systems, vol. 32 (2019)
33. Kingma, D.P., Ba, J.: Adam: a method for stochastic optimization (2014). arXiv preprint arXiv:1412.6980

Few-Shot Open-Set Learning for RF-Based Non-cooperative Drone Recognition

Han Yan and Qianjin Guo(✉)

Institute of Artificial Intelligence, Beijing Institute of Petrochemical Technology, Beijing 102617, China
guoqianjin@bipt.edu.cn

Abstract. With the development of the low-altitude economy, large amounts of unidentified drones pose significant risks to privacy, security, and airspace management. Existing identification methods mainly rely on pre-collected data from known UAV, rendering them ineffective against non-cooperative drone signals encountered in real-world scenarios. To address this limitation, this study innovatively proposes a Signal Prototype Calibration Network (SPCN) driven by a few-shot open-set learning strategy. Leveraging the mutual constraints between prototypes and reciprocal points in the clustering space, this method uniformly restricts known-class sample features within specified bounds. For potential unknown-class feature distributions, it explicitly predicts their linear representations via cross-attention mechanisms and completes calibrating projections through few-shot exchange alignment strategies. On the large-scale drone model dataset DroneRFa, the proposed SPCN model demonstrates superior unknown category recognition capabilities across open-set scenarios with varying categories. In the scenario of 18 known classes and 6 unknown classes, SPCN achieved a True Known Rate (TKR) of 96.12% and a True Unknown Rate (TUR) of 96.59%. For unknown-class clustering, its macro- and micro-average accuracy reached 93.88% and 93.29%, respectively. Ultimately, this approach enables precise identification of non-cooperative drone RF signals in complex scenarios without requiring prior knowledge of the underlying UAV models.

Keywords: non-cooperative UAV · radio frequency fingerprint identification · few-shot open-set learning · prototype learning

1 Introduction

Advancements in artificial intelligence, wireless networking, and battery energy technologies have positioned the low-altitude economic industry chain—built around UAV platforms—as a promising new paradigm for resource development and economic growth. However, UAVs' stealthy nature and remote controllability have led to the proliferation of unauthorized and illegal flights. This poses substantial challenges to protecting confidential information and personal safety [1]. During flight, UAVs rely on radio waves to transmit information and control operations with ground stations,

L. Huang and D. Greenhalgh (Eds.): ICMLC 2026, LNNS 2061, pp. 233–248, 2027.
https://doi.org/10.1007/978-3-032-31236-5_16

satellites, or other UAVs. Uplink data typically includes flight control signals employing frequency-hopping spread spectrum (FHSS), characterized by discontinuous frequency points and short intervals. Downlink data usually comprises video transmission signals modulated via orthogonal frequency division multiplexing (OFDM), featuring continuous multi-carrier duration [2].

Traditional signal processing techniques often fail to capture subtle changes in high-frequency waveforms when recognizing RF signals from unknown sources. In contrast, deep learning's ability to extract abstract features has led to the emergence of more end-to-end networks. These networks directly process input signals and generate recognition results. For example, Alam et al. [3] used multi-scale convolutional neural networks to extract features directly, saving time while achieving excellent recognition performance even for signals at the same frequency. Mandal et al. [4] drew inspiration from dual-stream network concepts, further decomposing the feature extraction network into time-domain and frequency-domain components. The extracted features were then concatenated for accurate recognition. Under constrained receiver bandwidth, Allahham et al. [5] adopted a multi-channel convolutional network architecture to enable real-time monitoring across multiple frequency bands. To address the sparsity of UAV signals, Mo et al. [6] enhanced recognition efficiency using preprocessing based on compressive sensing. Since the parameters of the feature extraction and classification networks are trained dynamically together, such end-to-end approaches more easily fit the potential known data distribution.

The aforementioned studies assume that the training set includes all drone categories to be classified. These are supervised tasks adhering to the independent and identically distributed (i.i.d.) assumption, thus performing well on test sets. However, real-world recognition scenarios are often zero-shot tasks [7], focusing more on categories absent from the training database. Specifically, they emphasize the performance of recognizing non-cooperative UAVs following a long-tail distribution in open-set settings. This paper proposes the Signal Prototype Calibration Network (SPCN), which employs a few-shot open-set learning strategy [8] to simulate the emergence of non-cooperative samples. This model aims to detect non-cooperative targets mixed with known-category drones and enable traceable classification of each signal sample.

The specific experimental steps can be divided into two phases: First, achieve open-set recognition of UAV signals by utilizing prototype learning to identify the uniform manifold distribution of each known category's features. Then, through hypothesis testing based on the distance between each sample's feature vector and the category prototype, identify samples not belonging to the known category in the test set. Second, achieve classification of non-cooperative signal samples by enhancing the model's ability to transfer features from known-class samples to unknown-class samples. During training, a few-shot open-set learning strategy is adopted. The model undergoes distance-metric fine-tuning across sub-tasks with varying category combinations. Representation vectors extracted from the support set are projected onto a linearly independent space defined by known-class prototypes. Similarity loss is computed between the projected vectors and features from both the query set and open set, thereby optimizing the model's feature extraction capability in unfamiliar environments. Finally, on the test set, samples that are not recognized as the known category in the first step are fed into the fine-tuned model. The resulting projection vectors are then clustered into different unknown-class clusters using the k-means++ algorithm.

2 Methodology

2.1 Data Preprocessing

A richer dataset enhances model generalization and better approximates complex real-world scenarios. This study uses the large-scale drone RF signal dataset DroneRFa [2]. Compared to other common datasets, DroneRFa includes more categories of drone signals, making it more suitable for open-set recognition research. Specifically, it contains 23 types of drone RF signals collected at different distances and operating frequencies, along with 1 type of spatial background noise.

Short-Time Fourier Transform. The raw data format consists of IQ signals at 100 MS/s, featuring a high sampling rate that makes direct recognition as a time series impractical. Leveraging the fast computation and reversible nature of the Short-Time Fourier Transform (STFT), it can reveal the frequency-hopping communication embedded within the broadband signal. This approach enables rapid and precise target recognition while preserving the input data information. For processing convenience, the IQ signal is first segmented into individual 50 ms time series. A sliding Hamming window with 50% overlap and a window length of 512 is then applied to perform the Discrete Fourier Transform (DFT). Finally, the amplitude is transformed into logarithmic scale and normalized to produce a single-channel, 16-bit precision, 512 × 512-time-frequency map. The transformation process can be expressed as:

$$STFT(m,f) = 20\log_{10}\left|\sum_{n=-\infty}^{\infty} s(n)g^*(n-256m)e^{-j2\pi fn}\right| \quad (1)$$

where s(n) = I(n) + jQ(n) represents the IQ signal, $g(n)$ denotes the Hamming window function, m and f denote the time index and frequency magnitude, respectively.

Channel Data Augmentation. To enhance model generalization, image processing often employs data augmentation to expand sample capacity. However, due to the specific physical significance of each pixel location in the time-frequency plot, traditional image augmentation techniques like rotation, scaling, or translation cannot be applied. According to occlusion and reflection of buildings in urban environments, the possible received signal $\hat{s}(n)$ would be faded by multipath channels $h_l(n)$ and plus by a white gaussian noise $w(n)$:

$$\hat{s}(n) = \sum_{l} h_l(n)s(n-l) + w(n) \quad (2)$$

The experiment simulates sample images under different conditions by convolving the original signal with a simulated channel and then converting it into a time-frequency map using the method. Simulating the ratio of direct to indirect power (K = 6 dB) and a Rician channel with 18 multipath paths with a maximum delay spread of 1.05 μs, Fig. 1 shows the time-frequency plot of the signal before and after channel augmentation. This verifies the model's ability to perform pattern recognition on signals from different UAV models under conditions of inter symbol interference.

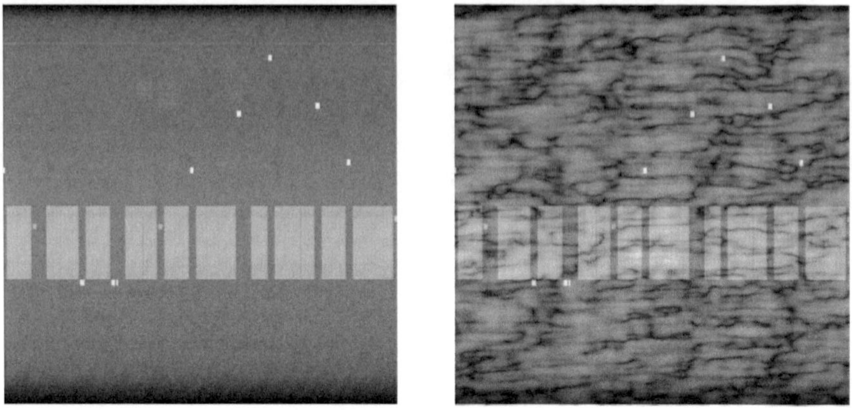

(a) (b)

Fig. 1. The comparison before and after passing through the analog channel: (a) Before channel processing. (b) After channel processing

2.2 Semantic Feature Extraction

Overall Architecture. Based on Sagar et al.'s research on open-set recognition [9], classification models that perform well on closed sets often exhibit strong transferability to open sets. Therefore, the primary objective is to train a classification model capable of extracting semantic features using known-category data (denoted as $Encoder(x)$). Observing the time-frequency diagrams of drone signals across different categories reveals that: (1) different carrier signal envelopes alter image texture information; (2) varying frequency-hopping patterns shift frequency-point location information. As shown in Fig. 2 (a), a series of hybrid dilated convolutions [10] with three different dilation rates (from large to small) is used during texture feature extraction to capture variations from short to long periods. As illustrated in Fig. 2 (b), STFT slices are first embedded with sinusoidal positions. Then, Mamba blocks [11] are applied to effectively capture the global dependencies of long-sequence signals in the time and frequency domains, respectively.

Prototype and Reciprocal Point. The objective of supervised classification tasks is to find a decision boundary that separates known categories. However, in open-set recognition, the distribution of unknown-category samples often lurks within the boundaries of known categories. Therefore, this experiment not only uses semantic labels to find decision boundaries during training but also employs prototype learning [12] to enhance the clustering of features within the same category and the separation of features across different categories.

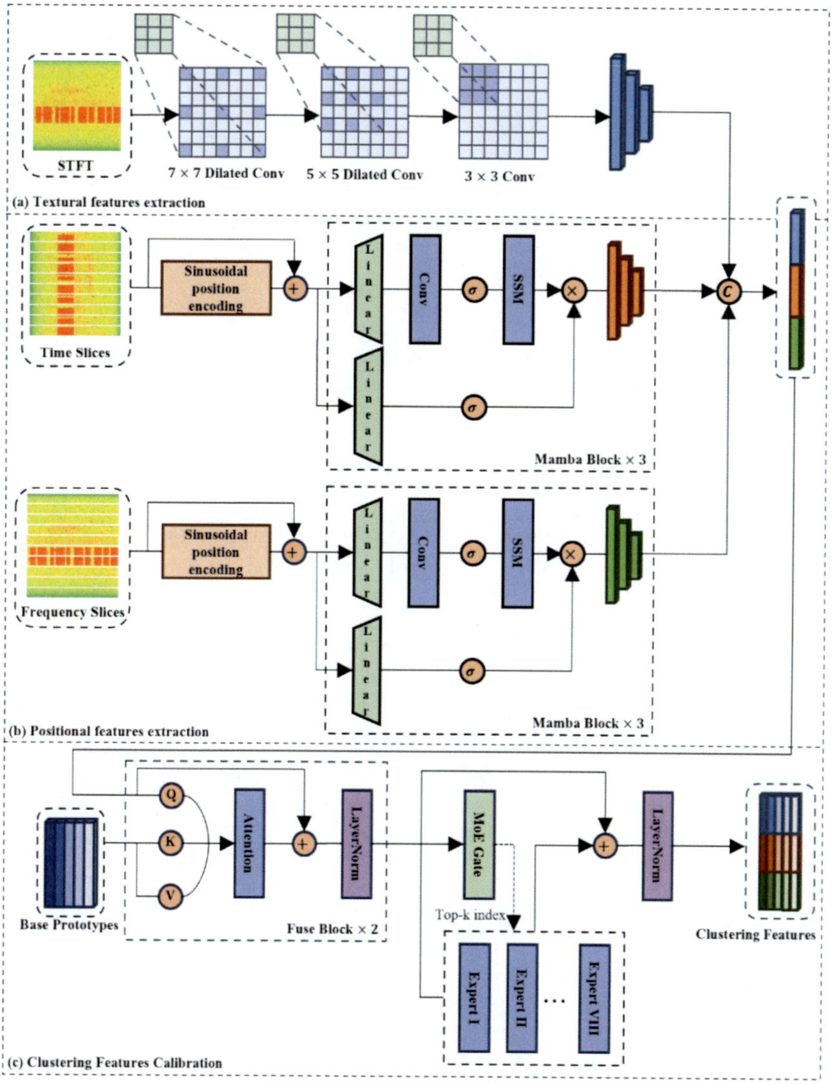

Fig. 2. Overview of the SPCN model architecture

When considering category clustering, let $z_i = Encoder(\boldsymbol{x}_i)$ denote the semantic representation vector of the i-th sample obtained through the feature extraction model. Using batch updates, a prototype point vector $\boldsymbol{B}_y$ and k reciprocal point vectors $\boldsymbol{RP}_y^{(k)}$ are learned for each category y. To some extent, category prototypes can be regarded as cluster centers, while reciprocal points can be viewed as outliers distant from the cluster range. To ensure alignment and uniformity of category samples, the design of the loss function imposes the following four constraints on the distance between sample points and prototype/reciprocal points:

1. The distance from a sample point to its category's prototype must be significantly smaller than the minimum distance to the reciprocal points within its category,

$$d\left(z_i, \boldsymbol{B}_{y_i}\right) \ll \min_k d\left(z_i, \boldsymbol{RP}_{y_i}^{(k)}\right) \tag{3}$$

2. The distance from a sample point to its category's prototype is significantly smaller than the minimum distance to prototypes of other categories,

$$d\left(z_i, \boldsymbol{B}_{y_i}\right) \ll \min_{j \neq i} d\left(z_i, \boldsymbol{B}_{y_j}\right) \tag{4}$$

3. The average distance from the sample point to the reciprocal points of its own category is greater than the maximum average distance to reciprocal points of other categories,

$$\operatorname{mean}_k d\left(z_i, \boldsymbol{RP}_{y_i}^{(k)}\right) > \max_j \operatorname{mean}_k d\left(z_i, \boldsymbol{RP}_{y_j}^{(k)}\right) \tag{5}$$

4. The distance from the sample point to prototypes of non-belonging categories is greater than the average distance to non-belonging reciprocal points,

$$d\left(z_i, \boldsymbol{B}_{y_j}\right) > \operatorname{mean}_k d\left(z_i, \boldsymbol{RP}_{y_j}^{(k)}\right) \tag{6}$$

Constraint 1 is implemented by leveraging the property that the ratio between the distance to the prototype and the distance to the reciprocal point approaches zero. The loss function (Eq. 7) is designed to be monotonically decreasing as it approaches the zero point of the logarithmic function from the positive half-axis. Constant *temp* is the temperature scaling [13], used to regulate the attention level for difficult samples:

$$\mathcal{L}_{(1)} = \frac{1}{N} \sum_{i=1}^{N} \log_2\left(\frac{\exp\left(d\left(z_i \cdot \boldsymbol{B}_{y_i}\right)/temp\right)}{\min_k \exp\left(d\left(z_i \cdot \boldsymbol{RP}_{y_i}^{(k)}\right)/temp\right)} + 1\right) \tag{7}$$

Constraints 2 and 3 only involve comparisons between sample points and different prototypes or reciprocal points. They can utilize the softmax function to calculate the probability of belonging to a category and then compute the cross-entropy loss with the true label (Eq. 8 and Eq. 9). Taking the negative of the prototype distance in Eq. 8 ensures that smaller distances receive a larger proportion of logits:

$$\mathcal{L}_{(2)} = -\frac{1}{N} \sum_{i=1}^{N} \sum_{j=1}^{C} y_j \log_2\left(Softmax\left(-d\left(z_i, \boldsymbol{B}_{y_j}\right)\right)\right) \tag{8}$$

$$\mathcal{L}_{(3)} = -\frac{1}{N} \sum_{i=1}^{N} \sum_{j=1}^{C} y_j \log_2\left(Softmax\left(\operatorname{mean}_k d\left(z_i, \boldsymbol{RP}_{y_j}^{(k)}\right)\right)\right) \tag{9}$$

Constraint 4 only requires that non-belonging prototype points and non-belonging reciprocal points are separated by a certain distance. Therefore, Margin Ranking Loss is used to maintain the distance above the boundary margin without overly emphasizing the loss of the difference (Eq. 10):

$$\mathcal{L}_{(4)} = \frac{1}{N}\sum_{i=1}^{N}\sum_{y_j \neq y_i} \max\left(0, \underset{k}{\mathrm{mean}} d\left(z_i, \boldsymbol{RP}_{y_j}^{(k)}\right) - d\left(z_i, \boldsymbol{B}_{y_j}\right) + margin\right) \quad (10)$$

2.3 Clustering Feature Calibration

The feature extraction module in the previous section achieves separation between known-class and unknown-class samples in the clustering space. However, to distinguish among different types of unknown-class samples, further enhancement of the model's transferability on unlabeled data is required. This method employs a few-shot learning strategy to fine-tune the semantic features generated for each category in the clustering space, thereby strengthening the feature's transferability for open-set object recognition.

Few-Shot Task Setup. For each n-way k-shot few-shot open-set task [8], the task comprises four components: the support set $\mathrm{S} = \{x_i, y_i\}_{i=1}^{|\mathrm{S}|}$, the unknown-class support set $\mathrm{O} = \left\{x_j, y_j\right\}_{j=1}^{|\mathrm{O}|}$, the known-class query set $\mathrm{Q}_k = \{x_w, y_w\}_{w=1}^{|\mathrm{Q}_k|}$, and the unknown-class query set $\mathrm{Q}_u = \{x_v, y_v\}_{v=1}^{|\mathrm{Q}_u|}$. The support set, serving as the reference for classification, contains n classes, each with only a small number of k samples, resulting in a total sample size of $|\mathrm{S}| = n \times k$, $|\mathrm{O}| = n \times k$. Furthermore, the categories of samples in both the support set and query set originate from the known class set C_k, i.e., $y_i, y_w \in C_k$, while the categories of open set samples come from the unknown class set C_u, i.e., $y_j, y_v \in C_u$. Under the constraint $C_k \cap C_u = \emptyset$, each subtask randomly selects C_k and C_u. This configuration better simulates the imbalanced distribution of known versus unknown classes in practical recognition scenarios, enhancing the model's generalization capability.

Prototype Fusion Strategy. Typically, differences in sample distributions within classes are categorized as covariate shift, while differences between classes distributions are regarded as semantic shift [14]. The challenge in identifying non-cooperative drones stems precisely from overly minor semantic shifts. Due to the high pattern similarity in the time-frequency diagrams of drone radio frequency signals (e.g., frequency-hopping schemes, OFDM modulation modes), the actual feature differences between distinct drone models are minimal. In open-set classification, due to the difficulty in capturing subtle differences in transmitter hardware fingerprints, unknown-class semantic features still exhibit overlap in the clustering space.

To address this limitation, Fig. 2 (c) presents a calibration method for open-set clustering features using a linear cross-attention mechanism [15] combined with a Mixture of Experts (MoE) model [16]. The specific process involves first extracting the feature vectors of all samples in the support set $z_i = Encoder(\boldsymbol{x}_i | \boldsymbol{x}_i \in S)$ as the query vector Q. Then, the subset of base class prototypes that do not belong to the support set categories $\boldsymbol{B}_{y \notin C_k}$ are read as the key vector K and value vector V. The multi-head cross-attention mechanism captures the relationships between known-class prototypes and new categories in a content-aware manner, representing new categories through a linear combination of class-separable prototypes. To enhance the nonlinear representation of new class prototypes in high-dimensional space, a gated network first selects k distinct experts for features of the new class. The outputs from these expert models with distinct mapping functions are then weighted by their top-k scores and summed to generate the calibrated support prototypes $\boldsymbol{P}_s = Calibrator_\theta\left(z_i, \boldsymbol{B}_{y \notin C_k}\right) \in \mathbb{R}^{n \times D}$. Similarly, non-belonging reciprocal points $\boldsymbol{RP}_{y \notin C_k}$ are used as value vectors V to generate calibrated support negative prototypes $\boldsymbol{NP}_s = Calibrator_\theta\left(z_i, \boldsymbol{RP}_{y \notin C_k}\right) \in \mathbb{R}^{n \times D}$ as estimated open-set prototypes.

When training the above fusion module, features z_w, z_v are extracted for all known and unknown query sets Q_k, Q_u. By comparing each query set sample against all support set positive and negative prototypes, the support set category loss for the query sample is predicted:

$$\mathcal{L}_{support} = -Sum\left(\boldsymbol{Y}_q \odot \log_2\left(Softmax\left(\begin{bmatrix} \tau_1 z_w \cdot \boldsymbol{P}_s{}^T + o_1 & \tau_2 z_w \cdot \boldsymbol{NP}_s{}^T + o_2 \\ \tau_1 z_v \cdot \boldsymbol{P}_s{}^T + o_1 & \tau_2 z_v \cdot \boldsymbol{NP}_s{}^T + o_2 \end{bmatrix}\right)\right)\right) \tag{11}$$

$\boldsymbol{Y}_q \in \mathbb{R}^{(|\mathrm{Q}_k|+|\mathrm{Q}_u|)\times 2n}$ represents the label matrix formed by one-hot encoding each sample label in the sub-task query set. $\odot$ denotes the Hadamard product. τ and o denote the learnable scaling factor and translation factor after similarity normalization.

Additionally, by introducing to swap alignment strategies [17], the calibrated open-set prototypes $\boldsymbol{P}_o \in \mathbb{R}^{n \times D}$ and negative open-set prototypes $\boldsymbol{NP}_o \in \mathbb{R}^{n \times D}$ for samples in the unknown-class support set O are computed. At this point, the known and unknown class categories are swapped for the query set: the labels of the known class query set are predicted using the negative prototypes from the unknown-class support set:

$$\mathcal{L}_{unknown} = -Sum\left(\boldsymbol{Y}_q \odot \log_2\left(Softmax\left(\begin{bmatrix} \tau_1 z_w \cdot \boldsymbol{NP}_o{}^T + o_1 & \tau_2 z_w \cdot \boldsymbol{P}_o{}^T + o_2 \\ \tau_1 z_v \cdot \boldsymbol{NP}_o{}^T + o_1 & \tau_2 z_v \cdot \boldsymbol{P}_o{}^T + o_2 \end{bmatrix}\right)\right)\right) \tag{12}$$

The swapping strategy also enhances the model's ability to discover prototypes for unknown classes from known-class negative prototypes. This is achieved by constructing a loss function that compares the similarity between prototypes from the known-class support set and negative prototypes from the unknown-class support set:

$$\mathcal{L}_{open} = \left(1 - \frac{\boldsymbol{P}_s{}^T \cdot \boldsymbol{NP}_o}{|\boldsymbol{P}_s||\boldsymbol{NP}_o|}\right) + \left(1 - \frac{\boldsymbol{P}_o{}^T \cdot \boldsymbol{NP}_s}{|\boldsymbol{P}_o||\boldsymbol{NP}_s|}\right) \tag{13}$$

3 Experiments

3.1 Datasets

A total of 24 drone categories [18] were randomly divided into five scenarios according to the ratio of training sample categories to test sample categories, as shown in Table 1. Each scenario's training set contained only known-class samples, while the test set contained only unknown-class samples, with no overlap between them. To simulate varying real-world conditions, scenarios II–V are arranged in descending order of known-class samples and ascending order of unknown-class samples. Scenarios I and III serve as control groups with identical total class counts but differing class compositions. Refer to Table 3 for the specific drone models represented by each class composition.

Table 1. Category classification in different experimental scenarios

Scenario ID	Number of categories	Type of categories
I	18 known classes	A, B, C, D, F, K, L, M, N, O, P, Q, R, S, T, U, V, X
	6 unknown classes	E, G, H, I, J, W
II	20 known classes	A, B, C, D, F, G, H, K, L, M, N, O, P, Q, R, S, T, U, V, X
	4 unknown classes	E, I, J, W
III	18 known classes	A, C, D, E, F, G, H, I, J, K, M, Q, R, S, T, U, W, X
	6 unknown classes	B, L, N, O, P, V
IV	16 known classes	A, B, C, D, F, K, L, M, N, P, R, S, T, U, V, X
	8 unknown classes	E, G, H, I, J, O, Q, W
V	14 known classes	A, C, D, F, K, L, M, P, R, S, T, U, V, X
	10 unknown classes	B, E, G, H, I, J, N, O, Q, W

3.2 Training Process

Throughout the entire training process, only known-category samples from each scenario are used—excluding any unknown-class information. During pre-training, the semantic feature extraction module (Sect. 2.2) is first trained using the prototype learning method. The training loss comprises three components: classification loss, prototype loss, and reciprocal loss (Fig. 3 (a)). It is observed that the classification loss remains consistently the smallest across 80 iterations. This indirectly demonstrates that the decision boundary of the classification model is easier to discern, while learning a uniformly distributed clustering space for categories is more challenging. The reciprocal loss initially exceeds the prototype loss but declines more rapidly. This is because reciprocal points initially exhibit greater distances (and thus higher losses). As training progresses, the ranking loss facilitates the convergence of reciprocal point distances near a sufficiently large margin, leading to saturation.

During few-shot training, each subtask randomly selects n categories from known classes as true known classes, from which the support set S and known-class query set Q_k are sampled. Then, n categories are randomly selected from the remaining known classes as false unknown classes, from which the unknown-class support set O and unknown-class query set Q_u are sampled. As shown in Fig. 3 (b), the Receiver Operating Characteristic (ROC) curve illustrates the relationship between the proportion of correctly classified samples in the known-class query set Q_k and the proportion of samples in the unknown-class query set Q_u misclassified as known classes at different thresholds. Typically, the model's false negative rate increases as the false positive rate decreases. The

AUC value comprehensively evaluates the model's discrimination capability by calculating the area under the ROC curve, balancing performance across different thresholds. The ROC curve during training completely envelops the curve during testing. This is primarily due to the model's insufficient generalization ability for true unknown class samples.

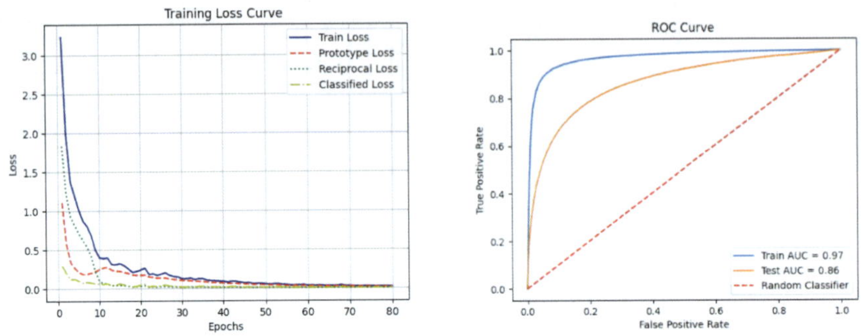

(a) (b)

Fig. 3. The metric in training process: (a) Pre-training loss. (b) Few-shot ROC

3.3 Results and Analysis

In the field of image recognition, transductive zero-shot learning employs multimodal methods to introduce auxiliary semantic information from unknown categories, thereby projecting visual features into the semantic space of unknown classes. In contrast, the inductive zero-shot learning discussed herein lacks any pre-known auxiliary information about unknown categories during training. It relies solely on implicit information—namely, the similarity in signal modulation patterns between unknown and known class signals. Under these stringent experimental conditions, conventional signal recognition models lack classification capability for unknown classes. Therefore, we selected the S3R model [18], the SR2CNN model [19], and the proposed SPCN model to compare their performance in open-set recognition, known-class classification, and unknown-class clustering across the same dataset partitions.

$$TKR = \frac{TK}{TK+FU} \quad TUR = \frac{TU}{TU+FK} \tag{14}$$

During the testing phase, all test samples are first divided into the acceptance set (predicted as known class) or the rejection set (predicted as unknown class) using the feature distance threshold hypothesis testing method. As shown in Table 2, the accepted sample set from known class is denoted as TK, while the rejected set is denoted as FU. The accepted sample set from unknown class is denoted as FK, and the rejected set is denoted as TU. It can be observed that TK and TU represent correctly classified sample sets, while FU and FK represent misclassified ones. As shown in Eq. 14, the True Known Rate (TKR)—the proportion of correctly classified known-class samples among all known classes—and the True Unknown Rate (TUR)—the proportion of correctly classified unknown-class samples among all unknown classes—can be used to evaluate the performance of open-set recognition. Using the closest base prototype distance to predict labels for all acceptance set samples (TK + FK) yields the known-class accuracy. Subsequently, macro-average and micro-average are performed. Since micro-average calculates the proportion of correctly predicted samples relative to the entire acceptance set, and FK samples lack known-class labels, the micro-average tends to be low. This value increases as the model's open-set rejection rate improves. Finally, the projector proposed in Sect. 2.3 projects all test set samples (TU + FU) into the known-class representation space. Clustering is performed using the k-means++ algorithm to obtain the unknown-class accuracy and average accuracy.

Table 2. The metric of open set recognition

	Accept	Reject
Known Class	True Known (TK)	False Unknown (FU)
Unknown Class	False Known (FK)	True Unknown (TU)

Table 3. Test result of scenario I (%)

		SPCN	S3R	SR2CNN
Known-Class Accuracy	Background (A)	**96.83**	**96.83**	83.33
	DJI Phantom 3 (B)	**95.62**	91.47	86.13
	DJI Phantom 4 Pro (C)	93.10	**95.37**	70.34
	DJI MATRICE 200 (D)	96.60	**98.62**	91.16
	DJI Air 2S (F)	**99.35**	97.91	92.21
	DJI Mavic 3 (K)	**96.91**	94.01	79.38
	DJI MATRICE 300 (L)	91.23	**95.09**	78.95
	DJI Phantom 4 Pro RTK (M)	**95.80**	94.43	92.44
	DJI MATRICE 30 T (N)	**100.00**	96.04	75.00
	DJI AVATA (O)	**94.17**	89.38	88.35
	DJI DIY (P)	**92.19**	89.89	87.50
	DJI MATRICE 600 Pro (Q)	**98.10**	96.11	75.24
	VBar (R)	94.96	**97.79**	67.23
	FrSky X20 (S)	99.07	**100.00**	74.77
	Futaba T16IZ (T)	**96.92**	89.22	73.85
	Taranis Plus (U)	**98.36**	**98.36**	86.07
	RadioLink AT9S (V)	**95.15**	94.64	70.87
	Skydroid (X)	94.03	**98.15**	85.07

(continued)

Table 3. (*continued*)

		SPCN	S3R	SR2CNN
Unknown-Class Accuracy	DJI MATRICE 100 (E)	**98.75**	90.05	81.36
	DJI Mini 3 Pro (G)	**99.17**	50.28	50.42
	DJI Inspire 2 (H)	**98.52**	87.61	42.99
	DJI Mavic Pro (I)	73.28	**100.00**	97.70
	DJI Mini 2 (J)	93.55	**97.19**	93.41
	Futaba T14SG (W)	**100.00**	**100.00**	47.01
True Known Rate (TKR)		**96.12**	95.08	81.47
True Unknown Rate (TUR)		**96.59**	95.66	86.32
Macro-average Known Accuracy (Macro-KA)		**96.02**	95.18	80.99
Micro-average Known Accuracy (Micro-KA)		**93.68**	92.18	75.80
Macro-average Unknown Accuracy (Macro-UA)		**93.88**	87.52	68.82
Micro-average Unknown Accuracy (Micro-UA)		**93.29**	86.10	69.64

Table 3 details the recognition accuracy rates of the three models for different categories under Scenario I. The higher results are bolded. It can be seen that the SPCN model proposed in this paper achieves better recognition results in most categories, especially in the recognition of unknown category samples. Although the recognition accuracy rate for the DJI Mavic Pro model of this type of unmanned aerial vehicle is only 73.28%, it can also be seen from the following clustering visualization that the model classified some DJI Mavic Pro samples as DJI Mini 2 model unmanned aerial vehicles, but the average accuracy rate has increased to over 93%. In scenario II–V, the test results of the S3R model on known classes are more stable. Whereas the SPCN model has excellent performance in identifying unknown class samples in any scenario due to the feature calibration module.

3.4 Ablation Study

In order to further verify the improvement of the clustering feature calibration module for unknown categories, under scenario V, the performance of zero-shot clustering was compared among the features extracted by the standalone semantic feature extractor, the features after cross-attention with base class prototypes, and the features after introducing nonlinear relationships by the expert model (Table 4):

Table 4. Model ablation experiment results

Method	Macro-UA	Micro-UA
Semantic feature extractor	84.44	90.62
Semantic feature extractor + Cross attention mechanism	92.80	92.36
Semantic feature extractor + Cross attention mechanism + Model of experts	96.97	95.03

It can be observed that semantic features alone can achieve a certain level of clustering performance for unseen classes. The cross-attention mechanism introduces clustering features learned from known classes to calibrate the semantic features of unseen classes, significantly improving the macro-average accuracy on unseen classes. Finally, the model of experts compensates for the limitations of linear representations alone, while the routing selection mechanism reduces overfitting probability, ensuring further accuracy gains on unseen classes.

3.5 Clustering Visualization

Dimension reduction techniques visualize the calibrated clustering features onto a two-dimensional plane, revealing the spatial distribution of features across different categories. Under the scenario I, Fig. 4 (a) shows the visualization of all test samples' clustering features using the t-SNE dimensionality reduction method. Fig. 4 (b) displays the clustering features of test samples predicted as unknown classes in open-set recognition, with colors annotated according to their true class labels. It can be observed that the clustering effect for all categories is excellent at the macro level. However, the clusters for the unknown category exhibit discrete overlaps, indicating that the model's zero-shot generalization capability for imbalanced category distributions still has room for improvement.

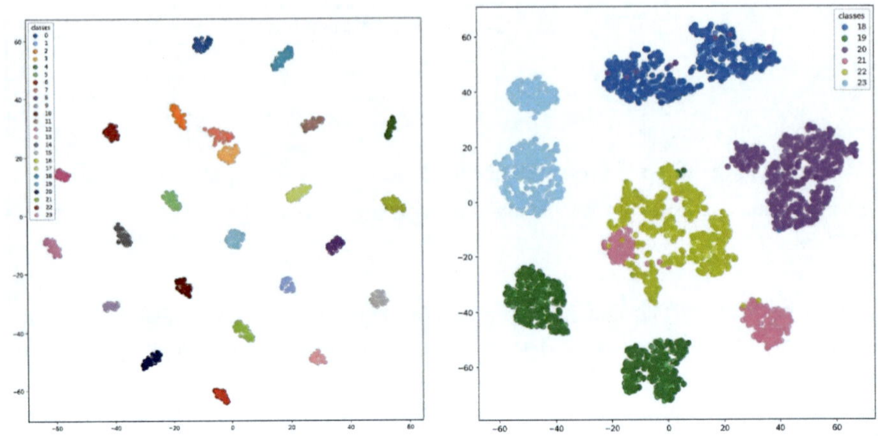

(a) (b)

Fig. 4. Visualization of clustering features after t-SNE: (a) Features on all categories. (b) Features on unknown categories

4 Conclusion

This study successfully addressed the key technical challenge of non-cooperative unmanned aerial vehicle (UAV) radio frequency signal identification, proposing a signal prototype calibration network based on a small sample open-set learning framework. The core innovation lies in explicitly achieving unknown feature calibration and projection by leveraging the prototype relationships of known categories, thereby enabling effective identification of unknown UAV models. This method not only can accurately distinguish between known and unknown signals, but also achieves high-precision clustering for unknown categories based on this, and its effectiveness has been verified through comprehensive validation of different category scenarios.

However, the study also found that as the number of unknown categories increases, model performance may degrade. This reflects its limitations in extreme category imbalance scenarios. To overcome this deficit, two approaches could be considered:

Enhance category coverage: Few-shot recognition heavily relies on auxiliary information beyond labels. More samples of signal distributions can help identify more accurate base prototypes and reciprocal points for calibrating test signals. Improve model generalization: The ablation study validates that the MoE architecture is effective in handling unknown samples due to its bagging schema, which reduces model variance. Nevertheless, more experts increase memory consumption. A better routing selection strategy could balance performance and resource usage to handle a large number of unknown drones.

Future research will explore the introduction of incremental learning mechanisms. This will enable the model to integrate calibrated unknown-class features into an expanding knowledge base, dynamically adapting to evolving threat situations and further enhancing its practical deployment capability in low-altitude airspace security protection.

Acknowledgements. This work was supported by the National Natural Science Foundation of China (Nos. 52361145714 , 21673252) and the Beijing Municipal Education Commission, China, under grant number 2019821001, and the fund of Climbing Program Foundation from Beijing Institute of Petrochemical Technology (Project No. BIPTAAI-2021007) and the ZhiYuan Fund key Project from Beijing Institute of Petrochemical Technology (Project No.2024003).

References

1. Khan, M.A., Menouar, H., Eldeeb, A., et al.: On the detection of unauthorized drones—techniques and future perspectives: a review. IEEE Sensors J. **22**(12), 11439–11455 (2022)
2. Yu, N., Mao, S., Zhou, C., et al.: DroneRFa: a large-scale dataset of drone radio frequency signals for detecting low-altitude drones. J. Electron. Inf. Technol. **46**(4), 1147–1156 (2024)
3. Alam, S.S., Chakma, A., Rahman, M.H., et al.: RF-enabled deep-learning-assisted drone detection and identification: an end-to-end approach. Sensors. **23**(9), 4202 (2023)
4. Mandal, S., Satija, U.: Time–frequency multiscale convolutional neural network for RF-based drone detection and identification. IEEE Sens. Lett. **7**(7), 7003304 (2023)
5. Allahham, M.S., Khattab, T., Mohamed, A.: Deep learning for RF-based drone detection and identification: a multi-channel 1-D convolutional neural networks approach. In: 2020 IEEE International Conference on Informatics, IoT, and Enabling Technologies (ICIoT), pp. 112–117. IEEE, Doha, Qatar (2020)
6. Mo, Y., Huang, J., Qian, G.: Deep learning approach to UAV detection and classification by using compressively sensed RF signal. Sensors. **22**(8), 3072 (2022)
7. Zhang, H., Zhao, L., Jiang, Y.: Triplet network and unsupervised-clustering-based zero-shot radio frequency fingerprint identification with extremely small sample size. IEEE Internet Things J. **11**(8), 14416–14434 (2024)
8. Zhang, Z., Chen, G., Zou, Y., et al.: Learning unknowns from unknowns: diversified negative prototypes generator for few-shot open-set recognition. In: Proceedings of the 32nd ACM International Conference on Multimedia, pp. 6053–6062. ACM, Melbourne, VIC, Australia (2024)
9. Vaze, S., Han, K., Vedaldi, A., et al.: Open-set recognition: a good closed-set classifier is all you need. In: Proceedings of the Tenth International Conference on Learning Representations (2022)
10. Wang, P., Chen, P., Yuan, Y., et al.: Understanding convolution for semantic segmentation. In: Proceedings of the 2018 IEEE Winter Conference on Applications of Computer Vision (WACV), pp. 1451–1460. IEEE, Lake Tahoe, NV, USA (2018)
11. Dao, T., Gu, A.: Transformers are SSMs: generalized models and efficient algorithms through structured state space duality. In: Proceedings of the 41st International Conference on Machine Learning, pp. 10041–10071. ACM, Vienna, Austria (2024)
12. Snell, J., Swersky, K., Zemel, R.: Prototypical networks for few-shot learning. In: Advances in Neural Information Processing Systems, vol. 30, (2017)
13. Wang, F., Liu, H.: Understanding the behaviour of contrastive loss. In: Proceedings of the 2021 IEEE/CVF Conference on Computer Vision and Pattern Recognition (CVPR), pp. 2495–2504. IEEE, Nashville, TN, USA (2021)
14. Yang, J., Zhou, K., Li, Y., et al.: Generalized out-of-distribution detection: a survey. Int. J. Comput. Vis. **132**(12), 5635–5662 (2024)

15. Ge, C., Han, D., Han, Y., et al.: Demystify mamba in vision: a linear attention perspective. In: Advances in Neural Information Processing Systems, vol. 37, pp. 127181–127203. Neural Information Processing Systems Foundation, Inc. (NeurIPS), Vancouver, BC, Canada (2024)
16. Cai, W., Jiang, J., Wang, F., et al.: A survey on mixture of experts in large language models. IEEE Trans. Knowl. Data Eng. **37**(7), 3896–3915 (2025)
17. Huang, S., Ma, J., Han, G., et al.: Task-adaptive negative envision for few-shot open-set recognition. In: Proceedings of the 2022 IEEE/CVF Conference on Computer Vision and Pattern Recognition (CVPR), pp. 7161–7170. IEEE, New Orleans, LA, USA (2022)
18. Yu, N., Wu, J., Zhou, C., et al.: Open set learning for RF-based drone recognition via signal semantics. IEEE Trans. Inf. Forensics Secur. **19**, 9894–9909 (2024)
19. Dong, Y., Jiang, X., Zhou, H., et al.: SR2CNN: zero-shot learning for signal recognition. IEEE Trans. Signal Process. **69**, 2316–2329 (2021)

Illumination-Aware Feature Fusion for Human Pose Estimation in Extreme Low Light

Zheng Wang(✉)

Tangshan Research Institute, Beijing Jiaotong University, Beijing, China
wz20000904@gmail.com

Abstract. Human pose estimation systems often fail in real-world low-light scenarios-not merely due to poor illumination, but primarily because of high sensor noise, low contrast, and loss of texture cues that commonly co-occur under extreme darkness. These challenges are particularly critical in applications such as nighttime surveillance or autonomous navigation, where robust perception must be maintained with minimal computational overhead. To address this, we propose a dual-stream architecture that decouples structural geometry from illumination context: a structural stream preserves fine-grained joint topology under degraded visibility, while an illumination stream-initialized from a self-supervised low-light encoder with frozen weights-provides complementary lighting-aware cues without requiring pixel-aligned bright/dark image pairs. The two streams are fused via a lightweight, channel-wise gating mechanism that adaptively recalibrates features based on local reliability, trained end-to-end using only standard pose annotations. Importantly, our method operates on unpaired, mixed-illumination data, leveraging both low-light and normal-light images during training without assuming correspondence between them. On the ExLPose benchmark, our method improves average precision from 11.5% to 24.9% on the low-light split over a strong structural-only baseline using HRNet-W32 [1] architecture, while maintaining competitive performance on well-lit images (59.0% AP). The modular and plug-and-play design enables practical deployment in real-world vision systems facing challenging illumination conditions.

Keywords: Human pose estimation · Low-light vision · Illumination-aware representation · Dual-stream architecture · Feature fusion

1 Introduction

Human pose estimation is a foundational component in many real-world vision systems, including video surveillance, human-robot interaction, and driver monitoring. However, these applications frequently operate under non-ideal lighting

L. Huang and D. Greenhalgh (Eds.): ICMLC 2026, LNNS 2061, pp. 249–262, 2027.
https://doi.org/10.1007/978-3-032-31236-5_17

conditions-such as nighttime streets, dimly lit warehouses, or poorly illuminated indoor environments-where standard pose estimators suffer severe performance degradation. In our recent industrial deployment, we observed that even state-of-the-art models become nearly unusable under extreme low light, achieving only marginal performance on the ExLPose dataset [2]. This collapse stems not only from insufficient illumination but also from high sensor noise, collapsed contrast, and loss of joint-critical texture cues-degradation patterns rarely reflected in standard benchmarks, highlighting a critical gap between lab-based evaluation and practical requirements. As illustrated in Fig. 1, such extreme conditions result in significant visibility loss, making accurate joint localization particularly challenging.

Fig. 1. Examples of well-light (top) and low-light (bottom) environments. Noticeable degradation in visibility and detail under low light highlights the difficulty in accurate pose estimation.

In practice, engineering teams often cannot afford full model retraining or curated low-light pose annotations due to tight deadlines, limited labeling resources, or compatibility requirements with legacy vision pipelines. Instead, practitioners need a *practical enhancement strategy* that can be seamlessly integrated into existing pose estimation frameworks without disrupting the original workflow or requiring additional labeled data. This motivates a design philosophy centered on modularity, compatibility, and minimal computational overhead-key criteria for real-world deployment.

We address this need with a plug-and-play dual-stream architecture that decouples pose estimation into two complementary pathways: a *structural stream* that preserves geometric fidelity under degraded visibility, and an *illumination stream* that provides lighting-aware context using a frozen, self-supervised low-light encoder. The two streams are fused via a lightweight, channel-wise gating mechanism that adaptively recalibrates features based on local reliability. Crucially, the entire system is trained end-to-end using only standard pose annotations-without requiring pixel-aligned bright/dark image pairs or additional illumination labels. Our method operates on unpaired, mixed-illumination data, meaning it leverages both low-light and normal-light images during training without assuming any correspondence between them-a setting that aligns with typical real-world data collection practices (e.g., separate daytime and nighttime video logs).

Evaluated on the ExLPose dataset-a benchmark for extreme low-light pose estimation-our method improves average precision from **11.5%** to **24.9%** on the low-light split over a strong structural-only baseline. Importantly, it maintains competitive performance on well-lit images within the same benchmark (59.0% AP), confirming no performance trade-off under normal lighting. Our contributions are threefold: (1) We propose a practical, modular framework for enhancing pose estimation in low light without redesigning the core pipeline; (2) We demonstrate that a frozen, pre-trained low-light encoder can serve as a stable illumination prior for pose estimation-enabling rapid integration into existing systems without joint training or domain adaptation; (3) We validate that such augmentation achieves significant gains in darkness while preserving accuracy in well-lit conditions, making it suitable for real-world deployment in lighting-varying environments.

2 Related Work

2.1 Low-Light Image Enhancement

Low-light image enhancement (LLIE) has evolved from classical signal-processing techniques-such as histogram equalization and Retinex theory-toward data-driven deep learning approaches that learn complex illumination transformations. Early unsupervised methods eliminated the need for paired training data through adversarial learning [3], while later zero-reference frameworks enabled enhancement without any ground-truth supervision by optimizing hand-crafted, non-reference losses to guide pixel-wise curve estimation [4], achieving both efficiency and color fidelity.

Recent research in LLIE has further advanced along two main thrusts: enhancing the adaptivity to complex and heterogeneous degradation, and improving computational efficiency for practical deployment. A key development is the integration of self-calibrated illumination estimation mechanisms, which progressively refine enhancement maps in an iterative manner to mitigate over-enhancement and artifacts [5]. This principle of internal self-regulation is also extended to unsupervised learning, where adaptive structure-consistency constraints are employed to preserve critical scene structures while performing global illumination correction, thereby improving robustness across diverse lighting conditions [6]. Concurrently, to address the computational burden of conventional deep networks, neural implicit representations have been introduced as a lightweight alternative. These methods encode an image as a continuous function, enabling high-quality enhancement through a compact model that supports fast inference, representing a significant step towards real-time applications [7]. Together, these approaches mark a shift towards more intelligent, self-adaptive, and efficient LLIE systems.

2.2 Human Pose Estimation

Human pose estimation has evolved from handcrafted feature-based approaches to deep learning frameworks that achieve high accuracy through end-to-end optimization. Current methods are primarily distinguished by their inference pipeline: *top-down* approaches first detect person instances and then estimate poses within each bounding box, achieving high precision at the cost of increased computational overhead in multi-person scenes [1,8]; *bottom-up* strategies instead predict all keypoints globally and associate them into individual poses, offering constant runtime regardless of person count but facing challenges under occlusion or scale variation [9]. Recent efforts address the association ambiguity in bottom-up methods by formulating keypoint grouping as a graph edge classification problem, where pairwise relationships between detected joints are explicitly modeled to improve instance separation [10]. More recently, *one-stage* designs unify detection and keypoint localization into a single network, improving efficiency while maintaining competitive accuracy.

In terms of keypoint representation, early regression-based methods directly predicted coordinates but struggled with complex articulations [11]. The heatmap paradigm subsequently became dominant, leveraging probabilistic spatial maps to localize joints with sub-pixel precision through architectures that model multi-scale context [12,13]. Recently, attention-based models have gained prominence by capturing long-range dependencies via tokenized representations or high-resolution feature interactions [14,15]. Further advances include unified one-stage frameworks for real-time multi-person estimation [16], multi-scale semantic fusion mechanisms [17], and large vision transformer backbones adapted for dense pose prediction [18]. Despite these developments, performance remains highly sensitive to imaging quality-particularly in low-light conditions, where reduced visibility and elevated noise levels severely compromise feature reliability and joint localization accuracy.

2.3 Pose Estimation in Low-Light Conditions

Human pose estimation remains highly vulnerable in real-world low-light conditions-not only because of reduced brightness, but more critically due to the co-occurring degradation patterns such as high sensor noise, low contrast, and loss of texture cues. While the introduction of [2], a benchmark dataset featuring real extreme-low-light images with accurate pose annotations, has enabled focused evaluation, existing approaches still face practical limitations. Most methods either rely on cascaded image enhancement followed by pose estimation-where enhancement is optimized for visual quality rather than pose accuracy-or require paired low-light/normal-light data or additional supervision for joint training. Domain adaptation techniques avoid annotation costs but often incur significant computational overhead or degrade performance under normal lighting due to domain-biased feature learning. These constraints hinder deployment in latency-sensitive applications like nighttime surveillance or autonomous systems, where robustness, efficiency, and compatibility with existing pipelines are essential.

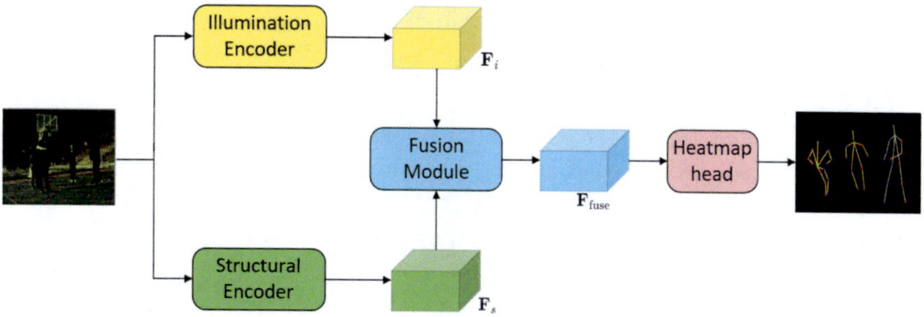

Fig. 2. The structural encoder and illumination encoder process the same low-light input in parallel, extracting $\mathbf{F}_s$ and $\mathbf{F}_i$ respectively. A lightweight fusion module dynamically recalibrates $\mathbf{F}_s$ using illumination cues, generating $\mathbf{F}_{\text{fuse}}$ without requiring any paired bright references or additional supervision. The final pose is predicted via a standard heatmap head.

3 Method

Low-light images suffer from degraded structural visibility and unreliable appearance cues, which directly impair the feature learning of standard pose estimators. To address this without altering the core pose pipeline or requiring additional supervision, we propose a dual-stream architecture that injects illumination-aware context into a high-fidelity pose representation through adaptive fusion. As illustrated in Fig. 2, our framework processes the input through parallel structural and illumination encoders, whose features are dynamically recalibrated by a lightweight fusion module to produce robust representations. The design is modular, lightweight, and trainable with only standard pose annotations.

3.1 Dual-Stream Representation Learning

Given a low-light input image I, our method processes it through two parallel streams, each designed to capture complementary aspects critical for robust pose estimation. Under extreme illumination deficiency, standard pose estimators often fail because the structural appearance of human joints becomes ambiguous or corrupted by noise and motion blur. To address this, we explicitly decouple the learning of what the body looks like (structural geometry) from how the scene is lit (illumination context). This separation allows the model to retain high-fidelity spatial representations while simultaneously leveraging lighting-aware cues to resolve ambiguities where visual evidence is weak.

Structural Stream. This stream preserves fine-grained spatial details and joint topology by maintaining high-resolution feature maps throughout the network. It serves as the primary backbone for keypoint localization and provides the structural representation necessary for pose inference when visual cues are weak. We initialize this stream with a pre-trained HRNet-W32 [1] model and keep its early layers largely intact to retain strong geometric priors.

The structural features are extracted as:

$$\mathbf{F}_s = \mathcal{E}_s(I), \tag{1}$$

where $\mathcal{E}_s(\cdot)$ denotes the HRNet-based encoder, and $\mathbf{F}_s \in \mathbb{R}^{H/4\times W/4\times C_s}$ represents the high-resolution structural feature map encoding joint layout and local texture under degraded visibility.

Illumination Stream. This stream captures local contrast, relative brightness, and scene-level illumination distribution using a self-supervised low-light enhancement encoder (Zero-DCE [4]) trained *without any paired bright references.* Although originally designed for image enhancement, its intermediate features encode reliable illumination cues-such as edge visibility and regional luminance-that remain discriminative even under extreme darkness. Critically, we use the encoder with *frozen weights*, treating it as a fixed feature extractor to avoid introducing extra supervision or training overhead.

The illumination-aware features are obtained via:

$$\mathbf{F}_i = \mathcal{E}_i(I), \tag{2}$$

where $\mathcal{E}_i(\cdot)$ is the Zero-DCE encoder (weights frozen), and $\mathbf{F}_i \in \mathbb{R}^{H/4\times W/4\times C_i}$ encodes contextual lighting information that helps disambiguate structurally ambiguous regions (e.g., dark limbs against shadowed backgrounds).

Both streams output feature maps at spatial resolution $H/4 \times W/4$. The illumination features are upsampled if needed to ensure spatial alignment. By decoupling structural reasoning from illumination modeling, our approach enables the pose estimator to dynamically attend to regions where appearance cues are trustworthy, guided by the illumination context.

3.2 Adaptive Feature Fusion

The core of our approach lies in an adaptive fusion mechanism that dynamically recalibrates the structural representation using illumination-aware context. Rather than relying on naive concatenation or fixed weighting schemes-which assume uniform reliability across all regions and channels-our method learns, in a data-driven manner, how much to trust illumination cues relative to potentially degraded structural features under varying low-light conditions.

Let $\mathbf{F}_s \in \mathbb{R}^{H'\times W'\times C_s}$ and $\mathbf{F}_i \in \mathbb{R}^{H'\times W'\times C_i}$ denote the spatially aligned feature maps from the structural and illumination streams, respectively (where $H' = H/4$, $W' = W/4$). To enable cross-stream interaction, both are first projected to a shared channel dimension C via 1×1 convolutions:

$$\mathbf{F}'_s = \mathrm{Conv}_{1\times1}(\mathbf{F}_s), \tag{3}$$

$$\mathbf{F}'_i = \mathrm{Conv}_{1\times1}(\mathbf{F}_i). \tag{4}$$

We then generate a channel-wise gating weight $\mathbf{G} \in [0,1]^C$ by aggregating global contextual information from both streams:

$$\mathbf{G} = \sigma\left(\mathrm{MLP}\left(\mathrm{GlobalAvgPool}\left([\mathbf{F}'_s, \mathbf{F}'_i]\right)\right)\right), \tag{5}$$

where σ is the sigmoid function and MLP is a two-layer bottleneck network with ReLU activation. The fused representation is computed as:

$$\mathbf{F}_{\text{fuse}} = \mathbf{G} \odot \mathbf{F}_i' + (1 - \mathbf{G}) \odot \mathbf{F}_s', \tag{6}$$

where $\odot$ denotes channel-wise multiplication.

Intuitively, $\mathbf{G}$ reflects the model's confidence in the illumination-aware features for each semantic channel (e.g., edge response, local contrast). In severely dark regions where structural features are noisy or ambiguous, $\mathbf{G}$ approaches 1, allowing illumination cues to dominate; conversely, in relatively well-lit areas, $\mathbf{G}$ diminishes, preserving the original structural signal. This adaptive recalibration enables robust pose estimation across diverse low-light scenarios-without requiring explicit brightness labels or paired bright references.

3.3 Training

3.3.1 Loss Function Our method is trained using a standard heatmap regression loss, which encourages the predicted heatmaps to match ground-truth Gaussian distributions centered at annotated keypoints. For the k-th joint $(k = 1, \ldots, K)$, the target heatmap $\mathcal{H}_k^* \in \mathbb{R}^{H' \times W'}$ is defined as:

$$\mathcal{H}_k^*(x, y) = \exp\left(-\frac{(x - x_k^*)^2 + (y - y_k^*)^2}{2\sigma^2}\right), \tag{7}$$

where (x_k^*, y_k^*) is the ground-truth location and σ controls the spread of the Gaussian (typically set to 23 pixels).

Let $\mathcal{H}_k$ denote the predicted heatmap. The base loss is the mean squared error:

$$\mathcal{L}_{\text{base}} = \frac{1}{K} \sum_{k=1}^{K} \|\mathcal{H}_k - \mathcal{H}_k^*\|_F^2. \tag{8}$$

To focus learning on semantically relevant regions, we apply a binary mask M_k that activates only near the keypoint:

$$\mathcal{L} = \frac{1}{K} \sum_{k=1}^{K} \|M_k \odot (\mathcal{H}_k - \mathcal{H}_k^*)\|_F^2, \tag{9}$$

where $\odot$ denotes element-wise multiplication. This masked formulation improves robustness to background noise in low-light scenes.

Notably, no auxiliary losses (e.g., image reconstruction, perceptual loss, or illumination consistency) are employed-only pose annotations are required.

3.3.2 Training Strategy We adopt a two-phase training protocol to stabilize optimization and preserve pre-trained knowledge:

- **Phase 1 (Frozen backbones):** The structural encoder (HRNet-W32) and illumination encoder (Zero-DCE) are kept frozen. Only the fusion module and heatmap head are trained for the first 10 epochs. This allows the gating mechanism to learn reliable feature weighting without being disturbed by noisy gradients from low-light inputs.
- **Phase 2 (End-to-end fine-tuning):** All components are jointly optimized with a reduced learning rate for the remaining epochs. This enables holistic adaptation to the low-light domain while maintaining geometric and photometric priors learned from pre-training.

We use the Adam optimizer with an initial learning rate of 10^{-3}, which is decayed by a factor of 10 after Phase 1, a batch size of 32, and train for a total of 140 epochs. Standard augmentations include random horizontal flipping and scale jittering. No paired bright images or additional supervision signals are used throughout training.

4 Experiments

We conduct comprehensive experiments to evaluate the effectiveness of our method for extreme low-light human pose estimation. We first introduce the datasets, evaluation metrics, and implementation details. Then, we compare our approach with state-of-the-art methods on *ExLPose*, followed by ablation studies to validate each component. Finally, we present qualitative results and discuss failure cases.

4.1 Datasets and Evaluation Metrics

Dataset. We evaluate our method on *ExLPose* [2], a recently released benchmark for extreme low-light human pose estimation. It contains 15,328 training images and 2,147 test images captured under diverse illumination conditions (e.g., moonlight, indoor dim lighting, motion blur). Each image is annotated with 17 COCO-style keypoints. Following prior work, we report results on two splits:

- **LL (Low-Light)**: Images captured in extremely dark environments.
- **WL (Well-Lit)**: Images with normal lighting, used to assess generalization.

Evaluation Metrics. We adopt the standard evaluation protocol from COCO [19], which uses Object Keypoint Similarity (OKS) as the similarity measure between predicted and ground-truth poses.

For a given person instance, OKS is defined as:

$$\mathrm{OKS} = \frac{\sum_{k=1}^{K} \exp\left(-\frac{\|\mathbf{p}_k - \mathbf{g}_k\|^2}{2\, s^2 \kappa_k^2}\right) \cdot \delta(\mathbf{g}_k)}{\sum_{k=1}^{K} \delta(\mathbf{g}_k)}, \tag{10}$$

where:

- $\mathbf{p}_k = (x_k^p, y_k^p)$ and $\mathbf{g}_k = (x_k^g, y_k^g)$ are the predicted and ground-truth locations of the k-th keypoint,
- s is the scale of the person (typically the square root of the bounding box area),
- κ_k is a keypoint-specific constant that controls tolerance (predefined per keypoint type based on inter-person variability),
- $\delta(\mathbf{g}_k) = 1$ if the k-th keypoint is labeled (visible), and 0 otherwise.

Based on OKS, precision and recall are computed by varying the OKS threshold $\tau \in [0, 1]$. Average Precision (AP) is then obtained by averaging precision over multiple thresholds:

$$\mathrm{AP} = \frac{1}{|\mathcal{T}|} \sum_{\tau \in \mathcal{T}} \mathrm{Precision}(\tau), \tag{11}$$

where $\mathcal{T} = \{0.5, 0.55, 0.6, \ldots, 0.95\}$ (i.e., 10 thresholds from 0.5 to 0.95 with step 0.05). We also report AP^{50} (at $\tau = 0.5$) and AP^{75} (at $\tau = 0.75$) for fine-grained analysis.

Implementation Details. Input images are resized to 256×192. We use Adam optimizer with initial learning rate 1×10^{-3}, batch size 32, and train for 140 epochs. The HRNet-W32 backbone is initialized from ImageNet pretrained weights, while the Zero-DCE encoder is initialized from MIT-Adobe FiveK pretrained weights. For the first 10 epochs, only the fusion module and heatmap head are updated; thereafter, all parameters are jointly fine-tuned. Training is conducted on 4 NVIDIA V100 GPUs and takes approximately 18 h.

4.2 Quantitative Comparison

We evaluate our method on the ExLPose benchmark and compare it with a variety of existing approaches, including standard pose estimators applied directly to low-light images, enhancement-based pipelines, and recent joint learning frameworks. Our base architecture adopts HRNet-W32 (referred to as "Baseline")-a widely used, deployment-friendly pose estimator that balances accuracy and efficiency in real-world applications [1]. This choice reflects our engineering-oriented design goal: to enhance an already practical system for extreme low-light scenarios without altering its core pipeline.

As shown in Table 1, most baselines struggle significantly under extreme low-light conditions (LL split), achieving AP scores below 25%. Notably, our full model introduces only 0.2M additional parameters over the 28.5M-parameter baseline, yet achieves substantial gains in low-light performance. In contrast, our full model achieves 24.9% AP on the LL split when trained on both low-light and well-lit data, and 23.8% AP when trained on low-light data alone, consistently outperforming all compared methods.

Importantly, we clarify that our claim of "no paired low-light/normal-light data required" refers to the absence of *pixel-aligned or scene-matched* LL/WL image pairs during training. However, to ensure robustness across lighting conditions, we do include unlabeled well-lit images from the same dataset (i.e., the WL split of ExLPose) in the training set-without any modification to the annotation protocol. This design reflects real-world deployment scenarios where a system must handle both day and night inputs, and explains why models trained *only* on low-light data (e.g., Ours (LL)) suffer catastrophic performance drop on well-lit images (1.6% AP), as they overfit to darkness-specific artifacts and lose general pose priors.

The improvement is particularly notable over enhancement-first strategies (e.g., LIME + CPN, LLFlow + CPN), which often distort structural details or amplify noise during image-level enhancement. By performing illumination-aware fusion at the feature level, our approach preserves geometric integrity while adaptively leveraging lighting cues-leading to more accurate and robust pose estimation in darkness. Moreover, when trained on the mixed LL + WL setting, our method maintains strong performance on well-l lit images (59.0% AP), confirming that it generalizes across illumination conditions without negative interference.

These results demonstrate that the proposed architecture effectively bridges the gap between illumination modeling and pose regression, yielding a **clear and consistent gain** across lighting conditions.

Table 1. Pose estimation performance (AP) on ExLPose.

Method	Params	LL (%)	WL (%)
CPN [20] (WL)	27.4	6.7	59.9
CPN [20] (LL)	27.4	17.2	1.2
DANN [21]	27.4	17.8	52.0
LIME [22]+CPN	27.4	21.1	57.7
Baseline [1] (LL)	28.5	11.3	5.9
Baseline [1] (WL+LL)	28.5	11.5	56.9
Ours (LL)	28.7	**23.8**	1.6
Ours (WL+LL)	28.7	**24.9**	**59.0**

LL = Low-Light, WL = Well-Lit.

Qualitative results in Fig. 3 further illustrate the advantage of our approach: in extremely dark scenes, the baseline (HRNet-W32) often misses limbs or predicts anatomically implausible poses due to poor visibility, while our method produces more complete and accurate pose estimates by adaptively leveraging illumination-aware features.

4.3 Ablation Study

We conduct ablation studies on the ExLPose dataset to validate the effectiveness of key components in our design, including the illumination stream, the adap-

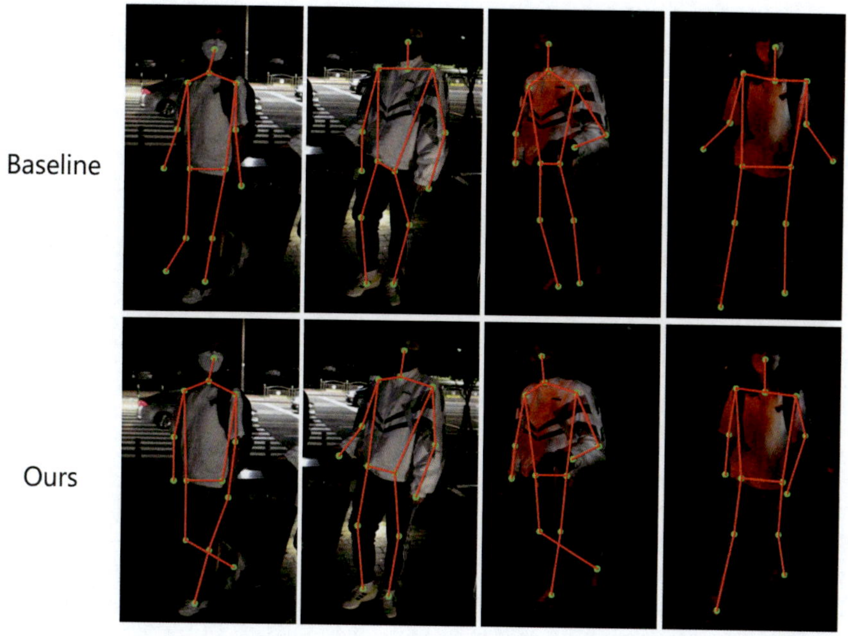

Fig. 3. Qualitative comparison between baseline and our method on low-light images. The top row shows results from the baseline (HRNet-W32), which suffers from joint displacement and missing keypoints due to poor visibility. The bottom row presents our method, which achieves more accurate and anatomically plausible pose estimation by leveraging illumination-aware feature recalibration.

tive fusion mechanism, and the two-phase training strategy. As shown in Table 2, removing the illumination stream (i.e., using only the structural HRNet backbone) leads to a significant drop in performance, confirming that low-light pose estimation cannot rely solely on structural features. Replacing our channel-wise gating fusion with a simple element-wise addition also degrades results, demonstrating the importance of dynamically weighting features based on illumination reliability. Furthermore, skipping the initial frozen-training phase slightly reduces accuracy and harms training stability, highlighting the benefit of preserving pre-trained knowledge during early optimization. Overall, the full model consistently achieves the best performance across both low-light and well-lit conditions.

The ablation results reveal an important insight: **illumination-aware modeling is not merely an auxiliary signal, but a core enabler for geometric reasoning under extreme darkness**. The addition of the illumination stream yields a substantial **13.6% point** improvement in AP on the low-light (LL) split-from 11.3% to 24.9%-demonstrating that the primary bottleneck in low-light pose estimation lies in feature reliability rather than network capacity. Moreover, the limited gain from naive fusion (e.g., simple addition, which achieves

Table 2. Ablation study on ExLPose (AP).

Variant	LL (%)	WL (%)
w/o Illumination stream	11.3	55.9
w/ Simple addition	19.1	57.3
w/o Frozen encoders	22.4	58.1
Full model	**24.9**	**59.0**

only 19.1% AP on LL) underscores that *how* illumination cues are integrated is as critical as *whether* they are used. This validates our design philosophy: rather than treating enhancement and pose estimation as separate stages, we tightly couple them through adaptive, content-aware feature recalibration.

5 Conclusion

Low-light human pose estimation remains a challenging task due to poor visibility, motion blur, and loss of fine-grained structural details. To address this, we propose a dual-stream framework that decouples structural geometry from illumination context, fusing them through a lightweight, channel-wise gating mechanism trained with only standard pose annotations-without requiring pixel-aligned bright/dark image pairs. Our method operates on unpaired, mixed-illumination data, leveraging both low-light and normal-light images during training without assuming correspondence between them. Extensive experiments on the ExLPose benchmark show that our approach achieves an average precision of **24.9%** on the low-light split-improving by **+13.4% AP** over a strong structural-only baseline based on HRNet-W32 [1]-while maintaining competitive performance (59.0% AP) on well-lit images. These results demonstrate that explicitly incorporating illumination-aware features, derived from a frozen self-supervised enhancement encoder, can significantly enhance pose robustness in extreme darkness without compromising accuracy under normal lighting or increasing inference complexity. For future work, we plan to extend our approach to video sequences by leveraging temporal consistency, and explore fusion with complementary modalities such as infrared or depth sensors for scenarios with severely degraded visible illumination.

Acknowledgements. This work is supported by Fundamental Research Funds for the Central Universities (2024JBZY003), Natural Science Foundation of Hebei Province (F2024105020), and National Engineering Research Center of Digital Construction and Evaluation Technology of Urban Rail Transit (2024023).

References

1. Wang, J., Sun, K., Cheng, T.: Deep high-resolution representation learning for human pose estimation. In: Proceedings of the IEEE/CVF Conference on Computer Vision and Pattern Recognition (CVPR), pp. 5693–5703. IEEE Press, New York (2019). https://doi.org/10.1109/CVPR.2019.00585
2. Lee, S., Rim, J., Jeong, B., Kim, G., Woo, B., Lee, H., Cho, S., Kwak, S.: Human pose estimation in extremely low-light conditions. In: Proceedings of the IEEE/CVF Conference on Computer Vision and Pattern Recognition (CVPR), pp. 704–713 (2023)
3. Kim, J., Lee, C., Lee, K.M.: EnlightenGAN: deep light enhancement without paired supervision. In: Proceedings of the IEEE/CVF Conference on Computer Vision and Pattern Recognition (CVPR), pp. 1282–1291. IEEE Press, New York (2018). https://doi.org/10.1109/CVPR.2018.00139
4. Guo, J., Li, C., Guo, J., Loy, C.C., Hou, J., Kwong, S., Cong, R.: Zero-reference deep curve estimation for low-light image enhancement. In: Proceedings of the IEEE/CVF Conference on Computer Vision and Pattern Recognition (CVPR), pp. 17630–17639. IEEE Press, New York (2022). https://doi.org/10.1109/CVPR.2022.01734
5. Li, C., Guo, J., Hou, J., et al.: Toward fast, flexible, and robust low-light image enhancement. In: Proceedings of the IEEE/CVF Conference on Computer Vision and Pattern Recognition (CVPR), pp. 14288–14297 (2022). https://arxiv.org/abs/2204.10137
6. Jiang, Y., Gong, X., Liu, D., et al.: EnlightenGAN: deep light enhancement without paired supervision. IEEE Trans. Image Process. **30**, 2340–2349 (2021). https://doi.org/10.1109/TIP.2021.3051462
7. Chobola, T., Liu, Y., Zhang, H., Schnabel, J.A., Peng, T.: Fast context-based low-light image enhancement via neural implicit representations. In: Computer Vision – ECCV 2024. Lecture Notes in Computer Science, vol. 13971, pp. 413–430. Springer, Cham (2024). https://doi.org/10.1007/978-3-031-73016-0_24
8. Xiao, B., Wu, H., Wei, Y.: Simple baselines for human pose estimation and tracking. IEEE Trans. Pattern Anal. Mach. Intell. **43**(12), 4640–4651 (2021). https://doi.org/10.1109/TPAMI.2021.3083993
9. Cao, Z., Simon, T., Wei, S.E.: Realtime multi-person 2D pose estimation using part affinity fields. In: Proceedings of the IEEE Conference on Computer Vision and Pattern Recognition (CVPR), pp. 7291–7299. IEEE Press, New York (2017). https://doi.org/10.1109/CVPR.2017.770
10. Maher, M., Fathalla, R., Shaheen, M.: Graph edge classification for keypoint grouping in multi-person pose estimation. J. Image Graphics **13**(2), 130–139 (2025)
11. Toshev, A., Szegedy, C.: DeepPose: Human pose estimation via deep neural networks. In: Proceedings of the IEEE Conference on Computer Vision and Pattern Recognition (CVPR), pp. 1653–1660. IEEE Press, New York (2014). https://doi.org/10.1109/CVPR.2014.214
12. Newell, A., Yang, K., Deng, J.: Stacked hourglass networks for human pose estimation. In: Leibe, B., Matas, J., Sebe, N., et al. (eds.) Computer Vision – ECCV 2016. Lecture Notes in Computer Science, vol. 9912, pp. 483–499. Springer International Publishing, Cham (2016). https://doi.org/10.1007/978-3-319-46484-8_29
13. Wei, S.E., Ramakrishna, V., Kanade, T.: Convolutional pose machines. In: Proceedings of the IEEE Conference on Computer Vision and Pattern Recognition (CVPR), pp. 4724–4732. IEEE Press, New York (2016). https://doi.org/10.1109/CVPR.2016.511

14. Li, Y., Zhang, S., Wang, Z.: TokenPose: Learning keypoint tokens for human pose estimation. In: Proceedings of the IEEE International Conference on Computer Vision (ICCV), pp. 13133–13142. IEEE Press, New York (2021). https://doi.org/10.1109/ICCV.2021.01298
15. Yuan, Y., Fu, R., Huang, L.: HRFormer: high-resolution vision transformer for dense prediction. In: Advances in Neural Information Processing Systems (NeurIPS), vol. 34, pp. 19426–19437. Curran Associates Inc, Red Hook (2021)
16. Zhang, F., Wang, J., Li, Y., et al.: RTMO: towards high-performance one-stage real-time multi-person pose estimation. arXiv preprint arXiv:2312.07526 (2023)
17. Liu, Y., Chen, X., Zhang, J.: MMSF: Multi-scale and multi-level semantic fusion for human pose estimation. In: Proceedings of the IEEE/CVF International Conference on Computer Vision (ICCV), pp. 12345–12354. IEEE Press, New York (2023)
18. Xu, Y., Zhang, J., Zhang, Q., et al.: ViTPose+: vision transformer foundation model for generic body pose estimation. arXiv preprint arXiv:2212.04246 (2022)
19. Xu, Y., Zhang, J., Zhang, Q., Tao, D.: ViTPose++: vision transformer for generic body pose estimation. arXiv preprint arXiv:2212.04246 (2022)
20. Chen, Y., Wang, Z., Peng, Y., Zhang, Z., Yu, G., Sun, J.: Cascaded pyramid network for multi-person pose estimation. In: Proceedings of the IEEE Conference on Computer Vision and Pattern Recognition (CVPR), pp. 7103–7112. IEEE Press, New York (2018). https://doi.org/10.1109/CVPR.2018.00743
21. Ganin, Y., Ustinova, E., Ajakan, H., Germain, P., Larochelle, H., Laviolette, F., Marchand, M., Lempitsky, V.: Domain-adversarial training of neural networks. J. Machine Learn. Res. **17**(59), 1–35 (2016). http://jmlr.org/papers/v17/15-239.html
22. Guo, X., Li, Y., Ling, H.: LIME: Low-light image enhancement via illumination map estimation. IEEE Trans. Image Process. **26**(2), 982–993 (2017). https://doi.org/10.1109/TIP.2016.2639450

Advances in Medical Image Analysis: AI, Automation, and Clinical Applications

Oral-Score: A Computer-Vision-Based AI Model for Evaluating Oral Facial Muscle Performance

Beeno Cheung Leong Tung[1](✉), Winsy Wing Sze Wong[1], James Kit Hon Tsui[2], Gary Kui Kai Lau[3], and Anne Chi Hay Yip[4]

[1] Department of Language Science and Technology, Faculty of Humanities, The Hong Kong Polytechnic University, Hong Kong SAR, China
{cltung,winsyws.wong}@polyu.edu.hk

[2] Faculty of Dentistry, The University of Hong Kong, Hong Kong SAR, China
jkhtsoi@hku.hk

[3] Division of Neurology, Department of Medicine, School of Clinical Medicine, Faculty of Medicine, Li Ka Shing, The University of Hong Kong, Hong Kong SAR, China
gkklau@hku.hk

[4] Feelings Group Limited, Hong Kong SAR, China
info@feelingss.io

Abstract. This study presents a comprehensive framework to enhance oral-motor assessments by integrating real-time digital analyses of the lips, jaws, and oral-facial movements, which are often overlooked in existing methodologies for motor-speech assessment (MSA) that focus primarily on perceptual and subjective judgments of human speech/oral-facial performance. We propose standardized metrics for evaluating a set of 6 oral motor actions, utilizing a dataset from 194 participants aged 20 to 85 years (133 neurologically healthy and 61 with stroke history), which includes annotated facial muscle movements and assessments of orofacial musculature capabilities. Our algorithmic analysis aims to produce standardized action completion scores across different age groups and genders, improving the evaluation of motor-speech performance by applying objective and quantified measurement to the coordination of orofacial musculature. The precise locations of oral-facial keypoints are detected by pre-trained machine learning models - Mediapipe Face Landmarker and custom-trained YOLO Detect models (for detection of jaw bottom position) with transfer learning, which are available for further studies and applications. By highlighting the significance of visual detection for facial muscles, including lips and jaw, the preliminary findings of this research may provide valuable insights for clinicians and researchers in speech pathology and the development of assistive technology in speech diagnostics.

Keywords: Speech Diagnostics · Stroke · Computer Vision · Object Detection · MediaPipe · YOLO

1 Introduction

In recent years, significant progress has been made in facial feature recognition, driven by advances in computer vision models such as FaceNet [1], DeepFace [2], and ArcFace [3]. These models have demonstrated excellent performance in tasks such as face verification,

L. Huang and D. Greenhalgh (Eds.): ICMLC 2026, LNNS 2061, pp. 265–277, 2027.
https://doi.org/10.1007/978-3-032-31236-5_18

face recognition, and attribute analysis by extracting high-dimensional embeddings from facial images. However, despite the success of these models in capturing global facial features, they exhibit a critical limitation: they fail to adequately address the recognition of local anatomical features, especially the lips, which is critical for understanding and analyzing the integrity of oral-facial functions, speech production and articulation.

In this work, we highlight the limitations of existing facial feature recognition models in meeting the needs of speech-related applications. We believe that the development of specialized models capable of detecting and analyzing lips and oral-facial movements can enable objective assessment of oral motor movement, which can enable advancement in speech therapy and articulation training. By combining state-of-the-art computer vision techniques with domain-specific knowledge, we aim to bridge this gap and pave the way for more effective, noninvasive solutions in speech-related healthcare and education.

Existing facial feature recognition models focus primarily on detecting and tracking key points of the eyes, nose, mouth, and jaw. Although these models perform well in applications such as facial expression analysis and virtual reality, they lack the granularity to recognize and track subtle lip movements. For example, the MediaPipe [4] powered Face Landmarker [5] can detect a detailed face mesh consisting of 468-point 3D facial landmarks, but its detection on the jaw and lower lips is not accurate in some cases. Similarly, the YOLO-based object detection framework [6, 7], while effective for general face detection, is not able to capture the complex dynamics of mouth muscle movement during vocalization by combining voice recognition with the subtle motor movements of key parts.

Current pathologic speech assessment methods typically focus on subjective perceptual analysis of patient's verbal and nonverbal tasks performance by professionals [8–11], which involves specialists such as speech-language pathologists (SLPs) assessing the symmetry, range of movement, coordination, and strength of the articulators and related oral-facial structures, and as well as the pitch, intensity, intonation, and intelligibility of speech production of the patient [12], providing a condition analysis, individualized treatment plan, and interventions based on their subjective judgments [13]. These assessment methods usually require a high degree of specialized knowledge. At the same time, there are now studies that computational models, such as automatic speech recognition, can predict and diagnose conditions from important features of vocalization in natural continuous articulation [14], but there are limitations in terms of monitoring other related oral motor measures, such as capturing key areas and changes of the articulators during speech/non-speech tasks.

This deficiency in identifying localized features of the oral facial features has significant implications for applications in speech therapy in both assessment and management. The ability to monitor and analyze lips and jaw movements is essential for diagnosing and treating speech disorders such as dysarthria and dysphagia, as well as for helping patients acquire accurate pronunciation. Current methods rely heavily on expensive techniques such as auditory feedback or ultrasound imaging, which are either indirect or require complex instrumentation for detection and are not widely available. Contactless, non-invasive, vision-based solutions that utilize facial feature recognition models to track oral-facial movements could revolutionize the field.

2 Methodology

2.1 Problem Identification and Definition

Current pathologic motor-speech assessments typically rely on subjective perceptual analysis using manual methods, which may not be accurate and consistent in assessing jaw movement. This reliance may limit a comprehensive assessment of orofacial muscle coordination, an integral component of articulation and swallowing. Also, the lack of standardized metrics for assessing various speech and oral motor movements precludes objective comparisons of oral-motor performance across age groups. In addition, the lack of integration of visual detection techniques for jaw and orofacial movements in existing methods may hamper the consistency and accuracy of the recognition process, thus affecting the overall effectiveness of speech assessment. Lastly, the potential risks associated with large or invasive medical devices are a concern, and small, easy-to-use, non-invasive visual inspection methods can be more accessible, facilitating diagnosis and treatment by physicians. Addressing these limitations is critical to improving the understanding and clinical management of pathologic motor-speech assessment.

Clinical dataset of orofacial actions (1) We want to build a clinical dataset for orofacial muscle performance assessment. The dataset should consist of video recordings of Hong Kong population in different genders and age groups, performing a set of orofacial actions. The dataset should be used to tune the model we developed, as well as to enable future research on the assessment of orofacial muscle performance.

(2) We want to develop a computer vision model that assesses the performance of the subject's orofacial muscle coordination in video clips. The model should be lightweight, enabling real-time analysis on video streams coming from a mobile device or prerecorded video.

2.2 Design

Building a Clinical Dataset of Orofacial Actions in the Hong Kong Population. We designed a set of orofacial actions commonly used by clinicians during perceptual analysis and recruited participants from various gender and age groups within the Hong Kong population to perform these actions. We collected demographic information, including age, gender, and health history (specifically, whether they had a history of stroke). Each participant was asked to perform the actions for 5 s, and we recorded the video clips using a smartphone.

In order to maintain a consistent distance and position between the camera and the participant, we developed a web application to assist with data collection. This web app performs real-time face detection and provides visual feedback to guide the data collector in adjusting the smartphone's position and orientation, ensuring that the participant's face remains centered on the screen. If the face is misaligned—either too close or too far away—the app will prompt the data collector to make the necessary adjustments.

When collecting the video recording of each participant performing each action, we read out the action instructions in Cantonese and demonstrated the actions to the participant. For clarity, we describe the actions in English in this paper. Then the participant was asked to perform the actions in a natural and relaxed manner. To avoid peer pressure affecting their performance, we ask participants to perform the actions in a private room. The participants do not know the performance of other participants.

The set of actions performed by the participants is as follows:

- Frowning: *Please furrow your eyebrows and hold this position for 5 s.*
- Open jaw wide: P*lease open your mouth as wide as possible and hold this position for 5 s.*
- Jaw left: *Please slide your jaw to the left and hold this position for 5 s.*
- Jaw right: *Please slide your jaw to the right and hold this position for 5 s.*
- Lip retraction: *Please stretch your lips wide, imagining the shape of the mouth when producing the sound "ee", and hold this position for 5 s.*
- Lip protrusion: *Please pucker your lips as if blowing out a candle and hold this position for 5 s.*
- Puff Cheek: *Please puff out both cheeks and hold this position for 5 s.*
- Jaw open n close: *Please continuously open and close your mouth as much as possible within 5 s.*
- Jaw sideway movement: *Please continuously move your jaw side to side horizontally within 5 s.*
- Jaw rotary movement: *Please continuously mimic chewing for 5 s, imagining you are chewing a large and tough piece of meat.*
- Lip retract and protrude: *Please continuously alternate between stretching and puckering your lips for 5 s, imagining you are forming the sounds "ee" and "wu," but do not vocalize.*

Using a computer vision model to assess orofacial actions. To calculate the score for each action, a set of formulas were designed. The range of scores for each action is different, but in general, a higher score means better performance. The formulas are based on the face mesh keypoints and face blendshape detected by MediaPipe Face Landmarker [5]. The formulas were reviewed by clinicians to ensure the score aligned with their perceptual judgement. Table 1 lists the formulas for each action, with the corresponding formula version and the source of the parameters.

Range of numbers detected from MediaPipe Face Landmarker. 1) The Face BlendShape number is normalized to 0 and 1 inclusively, representing the intensity of the facial action: 0 means no movement, and 1 means the most intensive movement. 2) The Face Landmark numbers represent the x,y,z coordinates of the facial keypoints, where x and y are normalized to 0 and 1 by the width and height of the entire image, respectively, and z can be positive or negative, using roughly the same scale as x. [15].

Due to limited paper space, we selected X most significant actions to show the formula design and evolution. A full list of formula can be shared upon further request.

Frowning is performed by furrowing the eyebrows inward. Clinicians perform perceptual analysis on this action to assess the coordinate of facial muscle of the upper face. The face blendshape model by MediaPipe does not include a direct detection of frown movement but it can detect the *browInnerUp* blendshape, which tracks the upward movement of the eyebrows inward. Formula version 1 considers the *browInnerUp* as the opposite movement of frowning, hence the score of frowning is formulated as:

$$1 - browInnerUp \tag{1}$$

Table 1. List of formulas for each action, with corresponding version and source of parameters.

Action	Version	Based on
Frowning	v1	Face BlendShape of *browInnerUp*
Frowning	v4	Face Landmark of *innerEyeDistance* and *eyebrowDistance*
Open jaw wide	v2	Face BlendShape of *jawOpen* and *mouthClose*
Open jaw wide	v3	Face Landmark of *innerMouthHeight* and *philtrumToJaw*
Open jaw wide	v4	Face Landmark of *outerMouthHeight* and *innerEyeDistance*
Open jaw wide	v5	Face Landmark of *philtrumToJaw* and *innerEyeDistance*
Jaw left	v1	Face Landmark of *rightCheekWidth, leftCheekWidth* and *outerCheekDistance*
Jaw right	v1	Face Landmark of *rightCheekWidth, leftCheekWidth* and *outerCheekDistance*
Jaw left	v2	Face BlendShape of *jawLeft* and *mouthLeft*
Jaw right	v2	Face BlendShape of *jawRight* and *mouthRight*
Jaw left	v4	Face Landmark of *leftCheekWidth* and *innerEyeDistance*
Jaw right	v4	Face Landmark of *rightCheekWidth* and *innerEyeDistance*
Lip retraction	v1	Face Landmark of *mouthWidth* and *outerCheekDistance*
Lip retraction	v4	Face Landmark of *mouthWidth* and *innerEyeDistance*
Lip protrusion	v1	Face Landmark of (inner and outer) *mouthWidth* and *mouthHeight*
Lip protrusion	v2	Face BlendShape of *mouthPucker*
Lip protrusion	v4	Face Landmark of *lipToNoseDistance* and *innerEyeDistance*
Puff Cheek	v1	Face BlendShape of *cheekPuff*
Puff Cheek	v2	Face Landmark of *rightCheekPoints* and *leftCheekPoints*
Puff Cheek	v4	Face Landmark of *rightCheekPoints, leftCheekPoints* and *innerEyeDistance*

However, we noticed the detection of *browInnerUp* from MediaPipe may not be sensitive to the Asian population in Hong Kong. To address the weakness of MediaPipe Face Blendshape model, we further designed v4 formula based on the face landmarkers detected by MediaPipe as

$$innerEyeDistance - eyebrowDistance \tag{2}$$

where

$$innerEyeDistance = distance(362, 133) \tag{3}$$

and

$$eyebrowDistance = distance(55, 285) \tag{4}$$

and

$$distance(i, j) = \sqrt{(x_i - x_j)^2 + (y_i - y_j)^2 + (z_i - z_j)^2} \quad (5)$$

The indices are the face landmarkers detected by MediaPipe. They correspond to: 1) left eye inner corner (index-362), 2) right eye inner corner (index-133), 3) left eyebrow inner point (index-55), 4) right eyebrow inner point (index-285). For simplicity, we will reference the face landmarkers in the following text without explicitly mentioning the formula to calculate the distance.

Open jaw wide is performed by moving the jaw downward and opening the mouth as wide as possible. It is a common action tested by clinicians to assess the orofacial muscle coordination of the lower face. The face blendshape model by MediaPipe includes two related blendshapes: jawOpen and mouthClose. When the face is at rest, the detected blendshape numbers should be 0 for *jawOpen* and *mouthClose*; when the jaw and mouth are opened fully, the detected blendshape numbers should be 1 for *jawOpen* and 0 for *mouthClose*; when the jaw is opened fully but the mouth is closed, the detected blendshape numbers should be 1 for *jawOpen* and 1 for *mouthClose*. Formula v2 considers the dynamics of these two blendshapes and formulates the score of open jaw wide as

$$jawOpen - mouthClose \quad (6)$$

During prior testing, we noticed the blendshape numbers are not sensitive to fine-grained movement, hence we further designed v3 formula based on the face landmarkers detected by MediaPipe as

$$innerMouthHeight / philtrumToJaw \quad (7)$$

where *innerMouthHeight* is the distance between the center of the bottom edge of the upper lip (index-13), and the top edge of the lower lip (index-14), and *philtrumToJaw* is the distance between the top of the philtrum (index-2) and the bottom of the jaw (index-152).

The v3 formula can measure the ratio of mouth height to the lower part of the face, but the denominator (*philtrumToJaw*) is changing during the movement. This will result in a non-linear scale of the score, which may be hard for clinicians to interpret. To address this concern, we further designed the v4 formula using another set of face landmarkers as

$$outerMouthHeight / innerEyeDistance \quad (8)$$

where *outerMouthHeight* is the distance between the center of the top edge of the upper lip (index-0), and the bottom edge of the lower lip (index-17), and *innerEyeDistance* is the same as mentioned in the v4 formula of frowning. We chose to use *outerMouthHeight* instead of the *philtrumToJaw* to serve as the numerator because the bottom of the jaw detected by MediaPipe is inaccurate in some samples, illustrated in Fig. 1. However, interviewed clinicians generally prefer to assess the degree of jaw openness by looking at the jaw, instead of the lips, in order to exclude the variant of mouth size in the population. Hence, we developed a custom object detection model to detect the position of the jaw (bottom edge). And proposed the v5 formula as

$$philtrumToJaw'/innerEyeDistance \quad (9)$$

where *philtrumToJaw'* is calculated by the distance between the top of the philtrum detected by MediaPipe (index-2) and the center of the bounding box detected by the custom jaw (bottom) detection model.

We design the jaw detection model in 2 phases. First, MediaPipe FaceLandmarker is used to detect the (potentially inaccurate) position of the jaw (bottom), then the jaw and nearby region are cropped as input image to a custom-trained YOLO detection model. The actual position of the jaw bottom is annotated manually for the model training. In contrast to passing the entire image from the camera into YOLO, the cropped region is smaller in size, effectively zooming toward the rough position of the jaw. This approach allows the YOLO model to perform object detection on a smaller input image while preserving visual details at a relatively high resolution. The improved efficiency enables the model to perform real-time analysis on mobile devices.

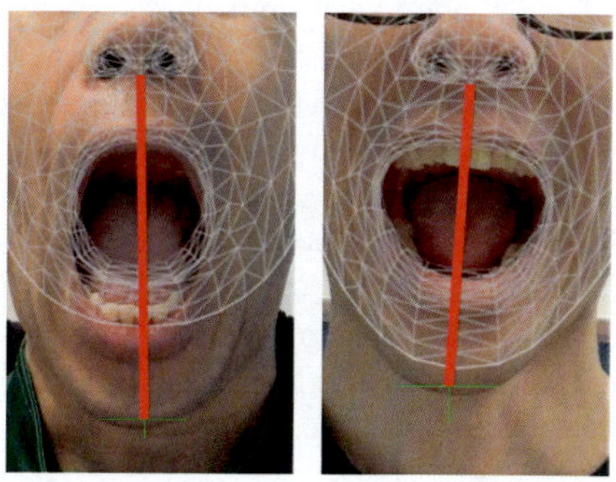

Fig. 1. Examples of completely inaccurate (left) and slightly inaccurate (right) jaw landmarks detected by the MediaPipe drawn in white lines. The position of the jaw bottom detected by our custom YOLO model is annotated with a green cross and the bottom of the red line.

Jaw left is performed by moving the jaw to the left side while facing the camera. This is also a common action tested during perceptual analysis. Multiple versions of the formula were designed to measure the intensity of jaw movement.

Version 1 is based on the assumption that the face is symmetric, hence when the jaw is at rest, the width of the left cheek and right cheek should be similar; when the jaw is moved to the left direction, the left cheek should be shrunk and the right cheek should be stretched. Using the face landmark detected by MediaPipe, this formula considers the ratio of the difference of left and right cheek width, normalized by the width of the entire face, to cater to the variation of face shape, the score is calculated as

$$max(0, (rightCheekWidth - leftCheekWidth)/outerCheekDistance) \quad (10)$$

Negative scores are discarded as they represent the movement of the jaw toward the wrong direction, which may be caused by a misunderstanding of the instruction. The rightCheekWidth is the distance between rightmost edge of the lip (index-61) and the rightmost edge of the cheek (index-138); the leftCheekWidth is the distance between leftmost edge of the lip (index-291) and the leftmost edge of the cheek (index-367); The outerCheekDistance is the distance between the leftmost edge of the cheek (index-367) and the rightmost edge of the cheek (index-138).

To explore an alternative formula that may match closer to the perceptual analysis performed by clinicians, we also designed the v2 formula based on the face blendshape detected by MediaPipe as

$$jawLeft + mouthLeft \tag{11}$$

As with the same concern when designing the Open jaw wide formula, the v1 formula uses a moving denominator, which may result in a nonlinear scale of the score. We further designed the v4 formula using a fixed denominator as

$$leftCheekWidth/innerEyeDistance \tag{12}$$

where leftCheekWidth and innerEyeDistance are calculated from the same set of face landmarkers as mentioned in the previous formulas.

Jaw right is performed by moving the jaw to the right side while facing the camera. It is a horizontal mirror movement of the Jaw left action. Hence, we can reuse the same set of formulas as mentioned above, by replacing the face landmarker distance of leftCheekWidth with rightCheekWidth, the face blendshape of jawLeft with jawRight, the blendshape of mouthLeft with mouthRight, and vice versa.

The movement of Lip retraction, Lip protrusion, and Puff Cheek are formulated in various versions using the face blendshape and face landmarkers detected by MediaPipe. We do not expand the details in this paper due to the page limit. Details can be found in the supplementary material.

3 Experiments

3.1 Data Collection and Annotation

Our experimental data collection was based on data from 194 participants, with 97 females (50%) and 97 males (50%). In order to collect experimental data at different age stages from 20 to 85 years old, we divided the age of the participants into 6 blocks, with a total of 20 participants (10.3%) from 20 to 30 years old, 24 participants (12.4%) from 30 to 40 years old, 26 participants (13.4%) from 40 to 50 years old, 36 participants (18.6%) from 50 to 60 years old, 60 to 70 years old were 53 participants (27.3%), and 31 participants (16.0%) were over 70 but not over 80 years old. In addition, 4 participants are over 80 years old. The above age stratification constitutes the content of our experiment on the different levels of performance in motor-speech assessment across different age groups of participants.

For dataset selection, we used 3222 video recordings for dataset construction and analysis, including recordings from 194 participants regarding 18 oral-facial movements that were selected and analyzed. A small portion of the collected videos is 480x640, and

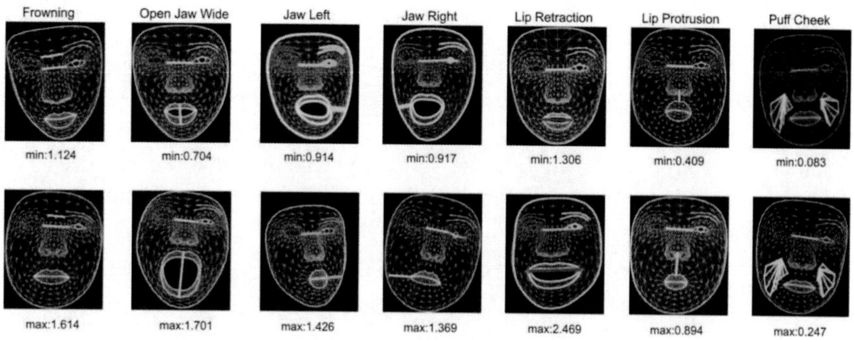

Fig. 2. Actions with maximum and minimum score in V4

most are 1920x1080 HD footage. The data used for annotation was stretched to a size of 640x640 for easy identification during model training.

Throughout the experiment, we had a professional physician guide participants through our Oral App software program to record 18 movements in real time to ensure the usability and reusability of each video for subsequent training, testing and validation. If a participant was unable to complete a prescribed movement, that movement was excluded from the analysis to prevent bias in both age-group and overall evaluations. For all included movements, we applied automated assessment, calculated the corresponding scores using the introduced formula, and stored the results in the database. The accuracy of the results is judged by clinicians with a speech therapy background.

Upon acquiring the experimental videos, we employ a dual-stage evaluation process to assess the accuracy of our model in recognizing oral and facial movements. First, we utilize MediaPipe's face landmark detection capabilities to automatically process all videos. MediaPipe's solution leverages machine learning models to identify 3D face landmarks and infer detailed facial surfaces in real-time, providing objective data on facial expressions and movements. As shown in Fig. 2, we record the maximum and minimum values and determine the change in the action.

Second, we submit the videos to professional clinicians for visual assessment of oral-motor performance. Since there is no existing formula-based scoring system commonly used by clinicians in the field, clinician perceptual judgment serves as the reference standard for validation. Rather than manually assigning numeric scores for each action (which is not the common practice of clinicians' assessment), clinicians judged whether the formula-derived scores were reasonable in a relative sense across participants by viewing video snapshots of real faces sorted in ascending order of formula score. Clinicians could control the opacity of overlaid landmarks and formula lines to view both the raw facial appearance and the formula-based detections, comparing relative rankings perceptually (i.e., aligned with their clinical impressions of participant performance). This enables us to assess both landmark detection fidelity and the clinical plausibility of action scores while combining objective computation with expert judgment.

3.2 Result

During the experiment, we collected the perceptual accuracy of the face landmarkers for each action. The results are shown in Table 2 and visualized in Fig. 3. We want to highlight that the accuracy related to the lips is generally acceptable, but the detected face landmarkers are not sensitive to the subtle movement of the puffed cheek. This could be the weakness of the MediaPipe face landmarkers model (being not sensitive to the z-axis movement).

Table 2. Accuracy of face landmark detection for each action, evaluated by clinicians.

Action	Sample	Accurate %	Partially Accurate %	Incorrect %
Frowning	178	91.6%	7.9%	0.6%
Open jaw wide	535	73.5%	23.6%	3.0%
Jaw left / right	710	60.7%	37.2%	2.1%
Lip retraction	363	76.3%	23.7%	0.0%
Lip protrusion	354	98.9%	1.1%	0.0%
Puff Cheek	180	31.7%	67.8%	0.6%

Percentages may not total exactly 100% due to rounding due to rounding to the closest 1 decimal places

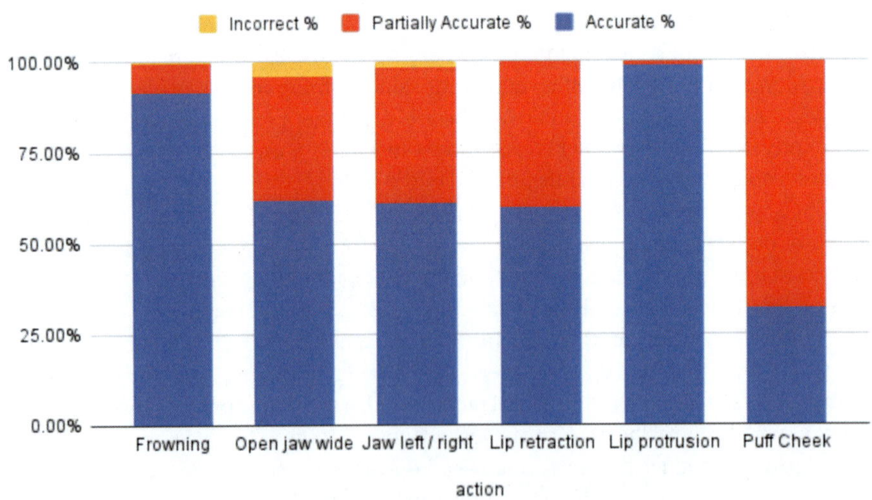

Fig. 3. Perceptual accuracy of the face landmarkers for orofacial movements assessment

In addition to landmark detection accuracy, we evaluated formula-derived scores against those of clinicians. Table 3 reports the proportion of recordings for which clinicians judged the formula scores as reasonable (accurate), partially reasonable (partially accurate), or not reasonable (incorrect) for each action and formula version. Figure 2 shows wireframe models (displayed for privacy instead of actual facial images) of facial configurations at minimum and maximum formula scores for each action, representing examples validated through clinician perceptual assessment of the actual video footage.

Table 3. Accuracy of formula-derived scores for each action and formula version, evaluated against clinician judgments.

Action	Formula Version	Sample	Accurate %	Partially Accurate %	Incorrect %
Frowning	v1	178	68.0%	11.8%	20.2%
Open jaw wide	v2	179	60.3%	34.6%	5.0%
Open jaw wide	v3	179	84.4%	10.6%	5.0%
Open jaw wide	v5	177	100.0%	0.0%	0.0%
Jaw left / right	v1	345	86.7%	5.8%	7.5%
Jaw left / right	v2	344	61.6%	27.9%	10.5%
Lip retraction	v1	178	61.2%	29.2%	9.6%
Lip retraction	v4	184	98.9%	1.1%	0.0%
Lip protrusion	v1	177	59.3%	15.3%	25.4%
Lip protrusion	v2	177	44.1%	51.4%	4.5%
Puff Cheek	v1	20	10.0%	5.0%	85.0%
Puff Cheek	v2	179	54.7%	34.6%	10.6%

Percentages may not total exactly 100% due to rounding to the closest 1 decimal places

3.2.1 Limitations and Generalizability

Our dataset was collected in Hong Kong and mainly reflects East Asian facial morphology. Differences in facial structure across ethnicities, as well as cultural and lifestyle factors that affect habitual facial use, may influence landmark detection and derived scores. Normalizing by intra-face distances (e.g., innerEyeDistance) reduces scale effects but does not eliminate shape- and behavior-related differences across ethnicity, age, and sex.

To support broader use, future work can extend the present approach by including multi-ethnic external validation, deriving population- and age/sex-stratified norms, recalibrating thresholds via transfer learning/domain adaptation, expanding training with cross-site and synthetic data, and running fairness audits across demographic subgroups. Until such studies are completed, there may be potential biases when applying the scores to populations that differ from our cohort; results should be interpreted with appropriate caution.

4 Conclusion

This work is innovative in applying computer vision models to perform automatic and objective evaluation on oral facial muscle movement (OFMM). The evaluation shows MediaPipe is generally accurate in the detection of landmarks in the upper part of the face (e.g. the eyes), but the detection of the lower part of the face, especially the jaw, is not accurate enough for OFMM assessment. And the work shows promising results of a custom (YOLO) model designed to detect a specific part of the face (e.g. the bottom of the jaw). In future studies, we plan to develop custom models to detect the position of tongue landmarks and explore alternative models that are more lightweight than YOLO to enable automated assessment of quick movement related to the tongue and jaw, with the potential to support low-spec mobile smartphones. This will enable real-time evaluation to be feasible across a wide range of devices, thereby providing accurate, quantified, efficient, and cost-effective OFMM assessment for underprivileged populations. In addition, we plan to further expand our participant pool to enhance the robustness and generalizability of our findings. Currently, our dataset's size limits our analysis depth. By increasing the sample size with a balanced age and gender distribution, we aim to capture greater variation in facial expressions and OFMM and to study age-/gender-related differences. We also intend to collect more detailed data on subtle facial features, particularly around the mouth and the tongue, using advanced techniques. This will improve our model's ability to detect subtle facial changes, enabling better recognition of complex cues across functional activities such as facial expression, speech production, and coordination during swallowing. This enhancement will improve computational efficiency and enable practical applications in human-computer interaction and psychological assessment. Its potential in clinical applications, for example, assessment of oral dysphagia and dysphonia, and other motor-speech disorders (apraxia of speech) and their rehabilitation certainly deserves further investigation.

References

1. Schroff, F., Kalenichenko, D., Philbin, J.: Facenet: a unified embedding for face recognition and clustering. In: 2015 IEEE Conference on Computer Vision and Pattern Recognition (CVPR), pp. 815–823 (2015). https://doi.org/10.1109/CVPR.2015.7298682. Last Accessed 28 Feb. 2025
2. Taigman, Y., Yang, M., Ranzato, M., Wolf, L.: Deepface: closing the gap to human-level performance in face verification. In: 2014 IEEE Conference on Computer Vision and Pattern Recognition, pp. 1701–1708 (2014). https://doi.org/10.1109/CVPR.2014.220. Last Accessed 28 Feb. 2025
3. Deng, J., Guo, J., Xue, N., Zafeiriou, S.: Arcface: additive angular margin loss for deep face recognition. In: Proceedings of the IEEE/CVF Conference on Computer Vision and Pattern Recognition, pp. 4690–4699 (2019)
4. Lugaresi, C., Tang, J., Nash, H., McClanahan, C., Uboweja, E., Hays, M., Zhang, F., Chang, C.L., Yong, M.G., Lee, J., Chang, W.T., Hua, W., Georg, M., Grund-mann, M.: Mediapipe: a framework for building perception pipelines (2019). https://arxiv.org/abs/1906.08172. Last Accessed 28 Feb. 2025
5. Kartynnik, Y., Ablavatski, A., Grishchenko, I., Grundmann, M.: Real-time facial surface geometry from monocular video on mobile gpus (2019). https://arxiv.org/abs/1907.06724. Last Accessed 28 Feb. 2025

6. Redmon, J., Divvala, S., Girshick, R., Farhadi, A.: You only look once: Unified, real-time object detection (2016). https://arxiv.org/abs/1506.02640. Last Accessed 28 Feb. 2025
7. Wang, A., Chen, H., Liu, L., Chen, K., Lin, Z., Han, J., Ding, G.: Yolov10: Realtime end-to-end object detection (2024). https://arxiv.org/abs/2405.14458. Last Accessed 28 Feb. 2025
8. Kent, R.D., Kim, Y.J.: Toward an acoustic typology of motor speech disorders. Clin. Linguist. Phon. **17**(6), 427–445 (2003)
9. Kent, R.D., Kim, Y.: Acoustic analysis of speech. In: The Handbook of Clinical Linguistics, pp. 360–380 (2008)
10. Panek, D., Skalski, A., Gajda, J., Tadeusiewicz, R.: Acoustic analysis assessment in speech pathology detection. Int. J. Appl. Math. Comput. Sci. **25**(3), 631–643 (2015)
11. Rusz, J., Tykalova, T., Ramig, L.O., Tripoliti, E.: Guidelines for speech recording and acoustic analyses in dysarthrias of movement disorders. Mov. Disord. **36**(4), 803–814 (2021)
12. Nguyen, T., Fredouille, C., Ghio, A., Balaguer, M., Woisard, V.: Exploring pathological speech quality assessment with ASR-powered Wav2Vec2 in data-scarce context. In: Calzolari, N., Kan, M.Y., Hoste, V., Lenci, A., Sakti, S., Xue, N. (eds.) Proceedings of the 2024 Joint International Conference on Computational Linguistics, Language Resources and Evaluation (LREC-COLING 2024), pp. 6935–6944. ELRA and ICCL, Torino, Italia (2024). https://aclanthology.org/2024.lrecmain.607/. Last Accessed 28 Feb. 2025
13. Baker, E., Masso, S., Huynh, K., Sugden, E.: Optimizing outcomes for children with phonological impairment: a systematic search and review of outcome and experience measures reported in intervention research. Lang. Speech Hear. Serv. Sch. **53**(3), 732–748 (2022)
14. Hlavnička, J., et al.: Automated analysis of connected speech reveals early biomarkers of Parkinson's disease in patients with rapid eye movement sleep behaviour disorder. Sci. Rep. (2017).https://doi.org/10.1038/s41598-017-00047-5. Last accessed 28 Feb. 2025
15. Authors, T.M.: Mediapipe tasks vision package (2022). https://www.npmjs.com/package/@mediapipe/tasks-vision. Last Accessed 28 Feb. 2025

Depth-Guided Network for RGB-D Semantic Segmentation

Zebin Zhao(✉), Chengshuo Zhang, and Shuang Xu

Shenyang University of Technology, Shenyang 110023, China
863137023@qq.com

Abstract. A common limitation in current RGB-D segmentation models is the independent extraction of multimodal features. Since fusion occurs late in the pipeline, these methods often struggle with inadequate feature integration and limited robustness. In response to this challenge, we present DGNet, a depth-guided network that fully exploits the complementary nature of RGB and depth information via two key innovations. First, the Adaptive Depth-Guided Deformable Convolution (ADGDC) module leverages depth cues to guide convolution offsets during feature extraction (rather than just fusion), effectively capturing 3D geometric relationships. Second, the Depth-Guided Geometry-Aware Attention (DGGAA) module generates spatial attention weights using depth gradient information, enhancing cross-modal feature fusion and mitigating inadequate depth encoding in RGB backbones. Our model was rigorously evaluated using standard datasets (NYUDv2 and SUN RGB-D). Empirical results show that DGNet outperforms existing benchmarks, reaching mIoU scores of 58.1% and 50.5%, respectively—outperforming existing methods significantly. Further analyses confirm DGNet's superior robustness and high-quality feature fusion in complex scenarios.

Keywords: RGB-D segmentation · Depth-Guided network · Deformable Convolution · Geometry-Aware Attention · Feature Fusion

1 Introduction

As a cornerstone problem in the field of computer vision, semantic segmentation is dedicated to classifying every individual pixel within an image into specific categories, and has been widely adopted as a unified end-to-end framework for complex visual perception tasks [1]. Beyond general scene understanding, optimized segmentation architectures have also demonstrated exceptional performance in specialized domains, such as the accurate 3D segmentation of liver structures in medical imagery [2]. Parallel to these advancements, with the rapid development of 3D sensing technologies, depth information has become an important complementary modality to RGB data, providing explicit geometric cues for scene understanding. Consequently, RGB-D semantic segmentation has attracted increasing attention in application domains such as robotics, autonomous driving, and low-light visual perception [3–5].

Despite notable progress, directly incorporating depth data into RGB-based segmentation networks often yields suboptimal performance due to the intrinsic heterogeneity

L. Huang and D. Greenhalgh (Eds.): ICMLC 2026, LNNS 2061, pp. 278–290, 2027.
https://doi.org/10.1007/978-3-032-31236-5_19

between modalities. Researchers typically partition RGB-D integration techniques into three primary architectures: early-stage, mid-level, and late-stage fusion schemes. By merging color and depth information at the very beginning of the network, early fusion strategies utilize simple concatenation but often suffer from feature interference and limited cross-modal utilization [6, 7]. Late fusion combines modality-specific features only at the final stage, resulting in limited interaction during feature extraction [8]. Mid-level fusion introduces layer-wise feature fusion mechanisms, often enhanced by attention modules, to improve multimodal interaction [9]. However, these methods typically perform fusion only at specific layers and rely on element-wise aggregation, limiting the guiding role of depth information in RGB feature extraction [10, 11].

The primary contributions of this work can be summarized as follows: First, we develop a depth-steered intermediate fusion architecture tailored for RGB-D semantic segmentation. Unlike conventional approaches that restrict multimodal interaction to specific isolated layers, our framework facilitates continuous depth-driven guidance throughout the RGB feature extraction phase. This deep integration significantly enhances segmentation robustness, particularly within complex indoor environments and low-light scenarios. Second, to optimize geometric feature extraction, we introduce the Adaptive Depth-Guided Deformable Convolution (ADGDC) module. This component employs a dynamic fusion mechanism that synthesizes standard convolution with depth-guided deformable convolution. By utilizing a learnable weighting strategy, the module achieves an optimal balance between training stability and geometric adaptability, thereby improving the network's ability to delineate object structures and spatial variations using depth cues. Third, we propose the Depth-Guided Geometry-Aware Attention (DGGAA) module, designed to refine feature representation via geometric priors. This module leverages depth gradient information to generate spatial attention weights that explicitly highlight object boundaries and structural details. Simultaneously, it applies channel-wise attention to selectively emphasize informative feature channels, resulting in more precise contour segmentation. Finally, the proposed method is rigorously evaluated on two standard RGB-D datasets: NYUDv2 and SUN RGB-D, both of which feature challenging real-world indoor scenes with intricate layouts and varying lighting conditions. Experimental evaluations indicate that the proposed approach yields consistent performance gains over existing methods across multiple datasets, especially in terms of generalization.

2 Proposed Method

2.1 Overall Framework

Drawing inspiration from the architectural design of ACNet [12], our proposed framework employs a multi-stream encoding structure paired with a unified decoding path for RGB-D semantic segmentation (see Fig. 1). In this setup, the encoding module focuses on extracting rich multimodal representations, whereas the decoding module reconstructs the high-resolution segmentation map from these features. Notably, such optimized CNN architectures have also demonstrated exceptional performance in other specialized domains, such as 3D liver segmentation in medical imagery [13].

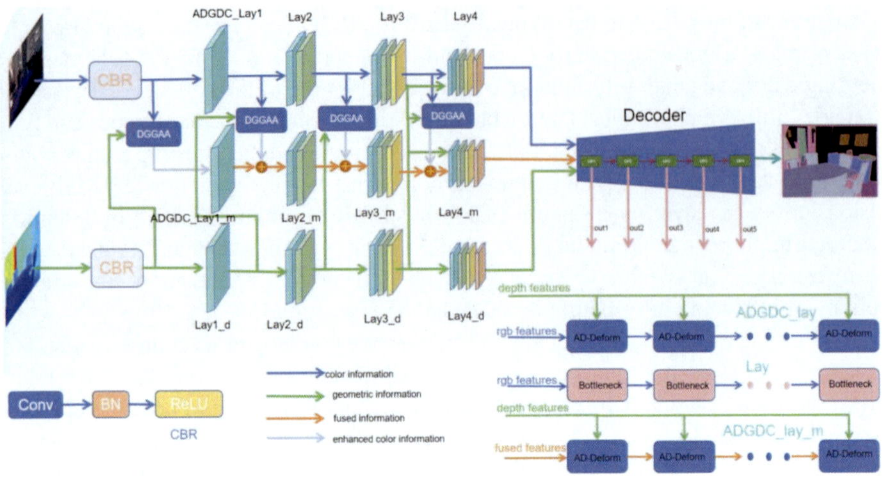

Fig. 1 Schematic diagram of the network architecture

A key challenge in such multi-branch architectures lies in effective multimodal feature fusion. Existing approaches often introduce depth information only at isolated fusion stages, resulting in insufficient guidance of RGB feature extraction and limited boundary representation capability. We introduce a depth-guided intermediate fusion strategy, which facilitates continuous interaction between RGB and depth features throughout the encoding stage, thereby alleviating the aforementioned limitations. The proposed framework is built upon two core modules: the Adaptive Depth-Guided Deformable Convolution (ADGDC) module and the Depth-Guided Geometry-Aware Attention (DGGAA) module.

The encoding module comprises three parallel streams: a color stream (RGB branch), a geometric stream (depth branch), and a fusion stream (hybrid branch). These pathways are tasked with extracting color textures, spatial geometry, and integrated multimodal features, respectively. To improve the adaptability of visual feature extraction to object geometry, the ADGDC module is integrated into both the RGB and hybrid branches. This module dynamically combines standard convolution with depth-guided deformable convolution through a learnable weighting mechanism, allowing the network to capture geometric variations while maintaining training stability.

Furthermore, the DGGAA module is introduced to facilitate deep interaction between RGB and depth features during the encoding stage. By exploiting geometric priors derived from depth information, the module extracts structural cues such as edges and contours and applies a dual attention mechanism, including spatial and channel-wise attention, to enhance RGB feature representations. This design effectively strengthens object boundary perception and improves sensitivity to structural details.

Finally, the features generated by the encoder are fed into a standard upsampling ResNet-based decoder to produce the final segmentation results.

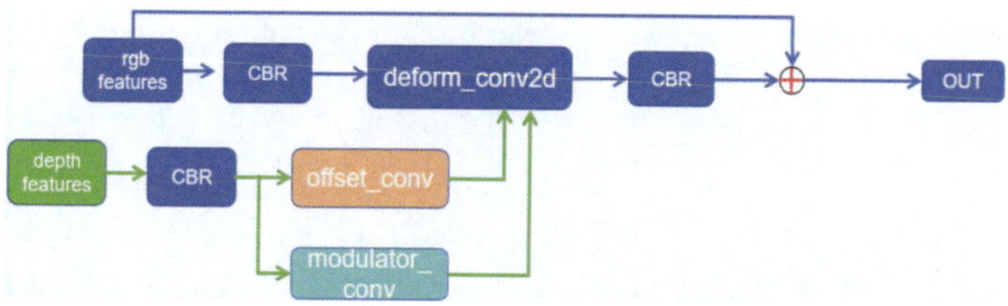

Fig. 2 Schematic of the Adaptive Depth-Guided Deformable Convolution Module (ADGDC)

2.2 Adaptive Depth-Guided Deformable Module

The Adaptive Depth-Guided Deformable Module aims to extract features with both structural stability and geometric adaptability by dynamically adjusting the convolutional receptive field. Its core idea is as follows: during the initial stage of training, conventional convolution is adopted to achieve rapid convergence; after the training process stabilizes, the convolutional sampling points are "deformed" to adapt to the target while suppressing background interference, ultimately fusing features derived from multiple strategies.

First, the input RGB features are processed by a CBR (Convolution, Batch Normalization, and ReLU) module for initial feature extraction. Simultaneously, the corresponding depth features are fed into two parallel convolutional branches: 1) The Offset Convolution (offset_conv) predicts the coordinate offset Δp of the sampling points, allowing the convolution kernel to adapt to geometric variations of the target. 2) The Modulation Convolution (modulator_conv) computes the modulation weights Δm to regulate the sampling importance, thereby suppressing background interference. Subsequently, Δp, Δm, and the RGB features are integrated into a deformable convolution to achieve adaptive sampling and enhanced structural feature extraction, as illustrated in Fig. 2.

Second, during training, the behavior adjustment module modifies the computational pathway based on the optimization stage. During the warm-up phase, only standard convolution is applied to extract RGB features, preventing instability in the early learning of offsets and modulation values, rapidly establishing the base feature map. Once training progresses, the model transitions to a hybrid regime, where standard convolution and depth-modulated deformable convolution are adaptively fused. The fusion ratio is optimized automatically, enabling the model to fully exploit geometric cues while maintaining training stability.

Conventional 2D convolution can be formulated as a two-step operation: (1) Extracting values from the input tensor x at locations defined by a regular grid R. (2) Aggregating the sampled values through a weighted summation using the kernel weights w. The grid R specifies the spatial extent of the receptive field.

In the case of a conventional 3×3 convolution, the sampling coordinates are specified as:

$$R = \{(-1, -1), (-1, 0), \ldots, (0, 1), (1, 1)\} \tag{1}$$

Given the convolution kernel w. the generation of the output feature y at location p_0 follows the formulation:

$$y(p_0) = \sum_{P_n \in R} w(p_n) * x(p_0 + p_n) \tag{2}$$

A fixed grid is unable to adapt to geometric variations of the target, such as rotation or deformation, and may unintentionally sample points from background regions, thereby degrading the extracted features.

To enable the sampling positions to align with the target structure, an offset term Δp_n is introduced, for each location. Consequently, we can extend the formulation in Eq. (3) to:

$$y(p_0) = \sum_{P_n \in R} w(p_n) * x(p_0 + p_n + \Delta p_n) \tag{3}$$

where Δp_n denotes the learnable offset for the n-th sampling location, enabling the convolution kernel to shift each sampling point adaptively. The term $x(p_0 + p_n + \Delta p_n)$ represents the feature value from the shifted coordinate within the input tensor x. The offsets Δp_n are predicted by an offset convolution layer, and the computation is given in Eq. (4):

$$\Delta p = \left(W_{p2} * ReLU\left(BN\left(W_{p1} * F_{in}\right)\right)\right) + b_2 \tag{4}$$

where $\Delta p \in R^{2*k*k*w_{out}*w_{out}}$, F_{in} represents the incoming feature tensor. W_{p1} is the first convolution kernel used to extract features from the input tensor. W_{p2} serves as the secondary convolutional filter responsible for generating the offsets. Since this module is depth-guided, F_{in} corresponds to the depth feature map, which provides the geometric cues for producing spatially adaptive offsets.

Although the learned offsets allow the sampling points to better align with object boundaries, deformed sampling points may still fall into background regions, thereby introducing noise. To further suppress background interference, we introduce a modulation scalar Δm, which acts as an importance weight for the n-th sampling location in the deformable convolution. The computation of the modulation values is given in Eq. (5):

$$\Delta m = Sigmoid(W_{m2} * ReLU(BN(W_{m1} * F_{in}))) + b_2 \tag{5}$$

where $\Delta m \in R^{k*k*w_{out}*w_{out}}$, F_{in} represents the incoming feature tensor. W_{m1} is the first convolution kernel, which convolves the input feature map through a convolution layer to generate an intermediate feature map. W_{m2} is the second convolution kernel, which uses another convolution layer to compute the final modulation value. A Sigmoid activation is applied to constrain each modulation factor to the range [0, 1], which determines the relative importance of each sampling point in the deformable convolution. Since this module is depth-guided, F_{in} denotes the depth feature map, ensuring that the modulation factors are adaptively generated based on geometric cues.

As illustrated in Fig. 2, the depth features are first fed into the CBR module and subsequently passed through the offset convolution (offset_conv) and modulation convolution (modulator_conv) to generate the offsets and modulation values respectively. The outputs, together with the RGB feature map, are then fed into the deformable convolution operator as formulated in Eq. (6).

$$y(p_0) = \sum_{p_n \in R} w(p_n) * x(p_0 + p_n + \Delta p_n) * \Delta m_n \tag{6}$$

The term $x(p_0 + p_n + \Delta p_n)$ denotes the deformable convolution component. By introducing an offset Δp_n, which is learned from the depth features, the sampling points are no longer restricted to a fixed regular grid. Instead, they can "deform" adaptively according to the geometry and pose of the object, enabling the network to attend more precisely to informative regions. The factor Δm corresponds to the modulation component. After sampling at the deformed position, each sampled value is multiplied by a learnable modulation scalar, which reflects the importance of that sampling point. When a sampling point even after deformation falls on an irrelevant background, the network can effectively suppress its influence by assigning a modulation value close to zero.

2.3 Depth-Guided Geometry-Aware Attention Mechanism

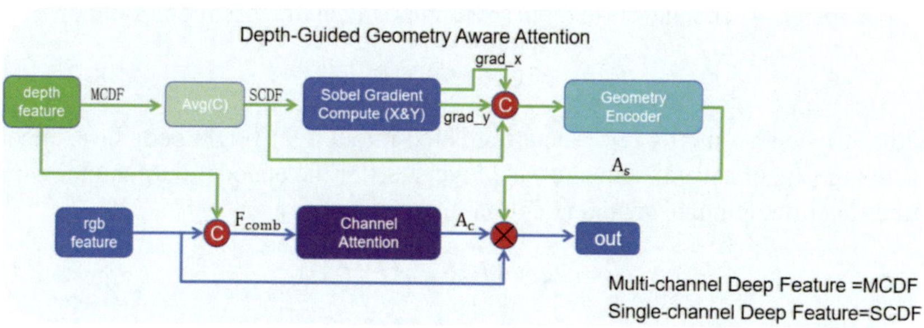

Fig. 3 Depth-Guided Geometry-Aware Attention Module

The original Geometry-Aware Attention mechanism primarily enhances spatial attention by exploiting gradient cues from RGB image, enabling the model to better capture structurally meaningful features. In contrast, our depth-Guided Geometry-Aware Attention Module, as illustrated in Fig. 3, introduces a dual-attention flow design integrates both spatial and channel attention, while directly leveraging geometric information from the depth map to guide spatial attention. As a result, the module not only captures semantic information across channels but also simultaneously captures structural and boundary information revealed by the depth map, achieving a richer and more discriminative feature representation.

As shown in Fig. 3, the extracted geometric cues—such as edge contours— and multimodal features are fed into the proposed the geometry-aware attention module. Through a synergistic dual-attention mechanism that jointly modulates the RGB stream using spatial and channel-wise attention weights, the module generates feature representations that are more responsive to object contours and structural details. Geometric

Spatial Attention Flow explicitly models the physical object boundaries by computing the gradients from the depth map, thereby producing a spatial attention map that encodes long-range spatial dependencies essential for capturing object structures. The process begins with the input multi-channel depth feature map $F_{depth} \in R^{C_{depth} \times H \times W}$. To obtain a compact single-channel representation suitable for geometric analysis, we first perform an average calculation (Avg) along the channel dimension, yielding a single-channel depth map $D \in R^{H \times W}$. For any location (i, j), the computation is mathematically formulated in Eq. (7):

$$D(i, j) \in \frac{1}{C_{depth}} \sum_{c=1}^{C_{depth}} F_{depth}(i, j) \tag{7}$$

Next, we extract geometric gradients from D using the Sobel operator. The horizontal and vertical gradients, denoted as g_x and g_y at any position (i, j), are represented as shown in Eq. (8):

$$g_x(i, j) = (G_x * D)(i, j), \; g_y(i, j) = \left(G_y * D\right)(i, j) \tag{8}$$

where g_x and g_y represent the Sobel convolution kernels applied along the x-axis and y-axis, respectively. We then concatenate the original single-channel depth map D with its gradient maps g_x, g_y channel-wise to yield a composite geometric representation F_{geom}:

$$F_{geom} = Concat\left(D, g_x, g_y\right) \tag{9}$$

This enriched geometric representation is fed into a lightweight geometric encoder E_{geom}, composed of several convolutional layers. After the encoder output undergoes a Sigmoid function, the final spatial attention map A_s is obtained:

$$A_s = Sigmoid\left(E_{geom}\left(F_{geom}\right)\right) \tag{10}$$

The resulting spatial attention effectively highlights structure-related regions such as edges, contours, and depth discontinuities, providing strong geometric priors for downstream multimodal feature fusion.

The multimodal channel attention flow is designed to adaptively assess and select the most informative feature channels from the combined RGB–depth inputs. First, the RGB features map $F_{rgb} \in R^{C_{rgb} \times H \times W}$ is concatenated with the depth feature map $F_{depth} \in R^{C_{depth} \times H \times W}$, forming a unified multimodal representation F_{comb} , as expressed in Eq. (11):

$$F_{comb} = Concat\left(F_{rgb}, F_{depth}\right) \tag{11}$$

Next, channel attention is constructed by first performing global average pooling on the concatenated features F_{comb}, This "squeeze" operation aggregates global spatial context into a compact channel descriptor $z \in R^{(C_{rgb}+C_{depth})}$, where the value of the c-th channel is computed as shown in Eq. (12):

$$z_c = \frac{1}{H \times W} \sum_{i=1}^{H} \sum_{j=1}^{W} F_{comb_c}(i, j) \tag{12}$$

The resulting descriptor Z is subsequently processed by a multi-layer perceptron (MLP) network consisting of two 1 × 1 convolutions for excitation, followed by a Sigmoid function, yielding the definitive channel attention map A_c as shown in Eq. (13):

$$A_c = Sigmoid(MLP(z)) \tag{13}$$

The two parallel flows described above produce the spatial attention map $A_s \in R^{1\times H\times W}$ and the channel attention map $A_c \in R^{C_{rgb}\times 1\times 1}$, respectively. Serving as multiplicative modulation weights, These maps modulate the incoming RGB features F_{rgb} via element-wise multiplication to yield the refined feature representation F'_{rgb}. The computation at any spatial location (i, j) is formulated in Eq. (14):

$$F'_{rgb}(i,j) = F_{rgb}(i,j) \odot A_s(i,j) \odot A_c \tag{14}$$

where ⊙ denotes element-wise multiplication, and the attention maps A_s and A_c are broadcast across the space and channel domains, respectively, to modulate the entire tensor. The resulting enhanced feature map F'_{rgb} exhibits a significantly improved ability to capture the object boundaries and underlying geometric structure compared to the original RGB feature map F_{rgb}.

3 Experiments

3.1 Implementation Details

The proposed framework was developed within the PyTorch environment, and all model training was executed on a workstation equipped with a single NVIDIA RTX 4090D GPU. The proposed network adopts a ResNet-50 backbone pre-trained on ImageNet. Model training is executed via the AdamW algorithm, with the weight decay coefficient set to 1×10^{-4}, supervised by a composite multi-scale loss consisting of Weighted Cross-Entropy (WCE) and Focal Loss. To ensure stable and efficient optimization, several training strategies are employed. First, differentiated learning rates are applied, where sensitive parameters such as the offset and modulation components in deformable convolution layers are assigned lower learning rates. Second, a hybrid learning rate schedule is adopted, which includes a linear warm-up phase followed by cosine annealing. Third, a progressive training strategy is used, in which the ADGDC module initially operates in standard convolution mode to avoid instability caused by early offset learning. Finally, adaptive gradient clipping is applied to different parameter groups to stabilize backpropagation. The batch size is set to 8, and starting learning rate is configured as $\mathbf{3 \times 10^{-4}}$.

3.2 Dataset

To evaluate our approach, we utilize two standard RGB-D indoor semantic segmentation datasets, namely NYU Depth V2 and SUN RGB-D.

The NYUDv2 dataset contains 1,449 RGB-D images of indoor scenes captured using a Microsoft Kinect sensor. Following the standard evaluation protocol adopted in prior work, the dataset comprises 795 images for the training set and 654 for the test set. It features pixel-level annotations for 40 semantic categories.

SUN RGB-D represents a larger-scale dataset with more diverse indoor scenes, consisting of a total of 10,335 pairs of RGB-D images annotated with 37 semantic classes. To guarantee an equitable evaluation against leading techniques, we adopt the standard data split used in previous studies [12, 13], allocating 5,285 samples to the training set and 5,050 to the test set.

3.3 Experimental Results and Analysis

Table 1 Quantitative results on the NYU Depth V2 dataset

Network	Backbone	PixAcc. (%)	mAcc. (%)	mIoU (%)
ACNet [12]	ResNet-50	75.66	68.06	48.3
Ours	ResNet-50	80.42	71.4	58.1

Table 2 Per-class IoU results on the NYU Depth V2 dataset

Wall	Floor	Cabinet	Bed	Chair	Sofa	Table	Door	Window	Bookshelf
83.4	89.72	69.5	82.9	72.9	68.4	56.8	47.2	59.0	47.8
picture	counter	blinds	desk	Shelves	curtain	dresser	pillow	mirror	floormat
65.2	73.6	71.0	42.1	24.1	66.6	76.1	49.6	62.1	67.0
clothes	ceiling	books	refridg	Television	paper	towel	shower	box	whiteboard
32.0	77.7	30.0	58.1	65.0	30.8	42.8	56.0	21.0	80.0
person	stand	toilet	sink	Lamp	bathtub	bag	othstr	othfurn	othprop
72.0	53.2	75.6	75.1	50.6	69.3	24.6	48.4	33.1	44.7

Table 1 presents the quantitative comparison between the proposed method and the baseline ACNet on the NYU Depth V2 dataset. Our approach achieves 80.42% PixAcc., 71.4% mAcc., and 58.1% mIoU, outperforming ACNet by + 9.8% mIoU, which demonstrates the effectiveness of the proposed depth-guided intermediate fusion framework in complex indoor scenes. Table 2 reports the per-class IoU results, where the proposed method shows strong performance on large structural categories such as wall, floor, and bed, as well as notable improvements on object-centric classes with complex boundaries, including chair, sofa, toilet, and sink. These gains can be attributed to the ADGDC and DGGAA modules, which enhance geometry-adaptive feature extraction and boundary awareness. Figure 4 further illustrates qualitative segmentation comparisons between the baseline and the proposed method.

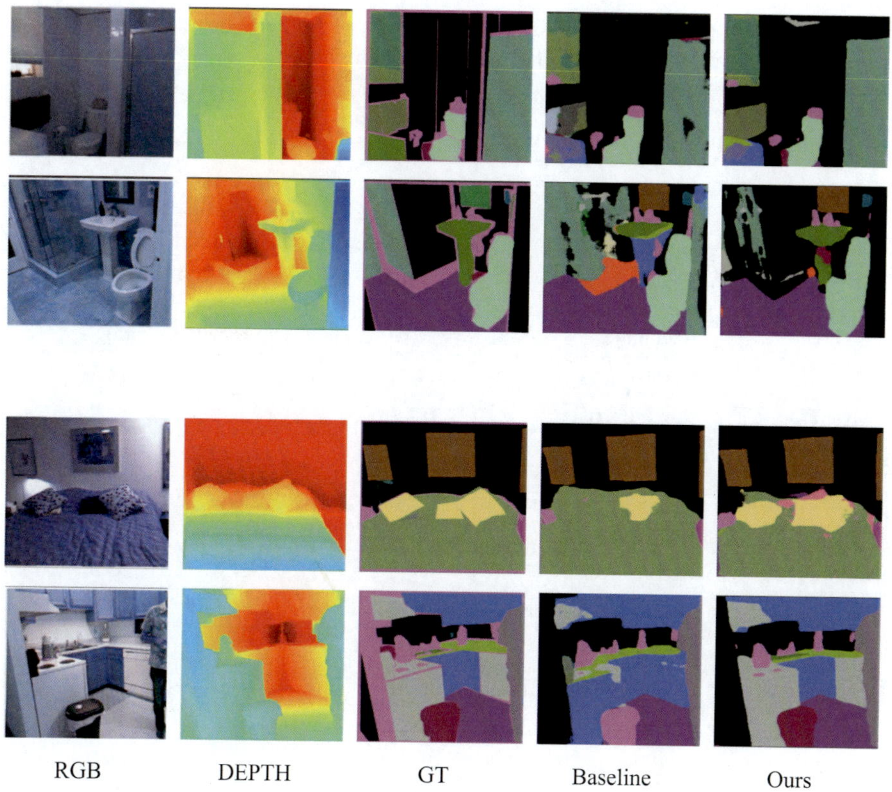

Fig. 4 Visual comparison of segmentation predictions between the baseline model and the proposed method

Table 3 Segmentation Performance of Ablation Study on the NYU Depth V2 Dataset

ADGDC	DGGAA	Pixel Acc	mAcc	mIoU
Without	Without	75.66	68.06	49.5
With	Without	79.12	70.8	56.81
Without	With	80.25	72.3	57.2
With	With	80.42	71.4	58.1

Table 3 summarizes the results of the ablation studies performed using the NYU Depth V2 dataset. The baseline model without ADGDC and DGGAA achieves an mIoU of 49.5%. Introducing ADGDC alone improves the mIoU to 56.81%, yielding a gain of + 7.31%, while DGGAA alone further increases the mIoU to 57.2%, with an improvement of + 7.7%. When both ADGDC and DGGAA are jointly incorporated, the proposed model achieves the best performance, reaching 58.1% mIoU, along with consistent improvements in Pixel Accuracy (80.42%) and mean Accuracy (71.4%). These results demonstrate that ADGDC and DGGAA provide complementary benefits for geometry-aware and boundary-sensitive RGB-D semantic segmentation.

3.4 Comparison with Other Methods

Table 4 Quantitative comparison with recent state-of-the-art methods on the NYU Depth V2 dataset

Network	Backbone	PixAcc. (%)	mAcc. (%)	mIoU (%)
ACNet [12]	ResNet-50	–	–	48.3
CANet [14]	ResNet-101	77.1	64.6	51.5
FCINet [15]	Mix Transformer	–	–	50.7
CAF-Net [16]	Mix Transformer	78.5	67.5	53.4
TCANet [17]	ResNet-50	75.2	63.6	50.2
RDFNet [18]	ResNet-152	78.5	59.4	55.5
TransD [19]	Swin Transformer	78.5	69.4	55.5
CMNext [20]	ResNet-50	75.9	63.2	56.9
DFormer-L [21]	Decoupled Transformer	80.0	69.3	57.2
Ours	ResNet-50	**80.42**	**71.4**	**58.1**

Table 4 presents a performance benchmarking of our framework against various leading RGB-D semantic segmentation approaches on the NYU Depth V2 dataset. For fair evaluation, we follow the standard experimental settings reported in the original papers. As illustrated, our framework delivers superior results across PixAcc, mAcc, and mIoU metrics among methods using comparable backbone networks, demonstrating its superior segmentation accuracy and robustness.

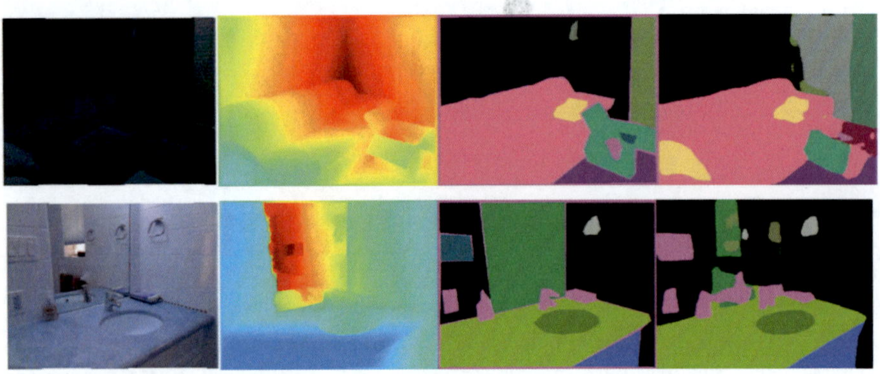

Fig. 5 Visualization of typical failure cases in challenging scenarios

Despite the competitive performance of DGNet, certain extreme conditions remain challenging for robust cross-modal fusion. As illustrated in the top row of Fig. 5, under extreme low-light conditions, the RGB branch fails to provide sufficient texture and semantic cues. Although the depth map offers geometric support, the DGGAA module struggles to build effective global awareness due to the degraded visual features, leading to misclassification in shadowed regions. The bottom row demonstrates a failure case caused by noisy depth maps on reflective surfaces. In scenes containing mirrors or polished tiles, the depth sensors often produce inaccurate or missing values. Since our ADGDC module relies on depth as a geometric guide to refine RGB features, these depth artifacts propagate through the network, resulting in fractured segmentation boundaries around the sink and faucet. These cases indicate that enhancing the model's robustness against degraded sensor data will be a key focus of our future work.

4 Conclusions

In this paper, we introduce a novel Depth-Guided Network tailored for RGB-D scene parsing to mitigate the issue of insufficient cross-modal fusion. Our approach leverages two key modules: Adaptive Depth-Guided Deformable Convolution (ADGDC) and Depth-Guided Geometry-Aware Attention (DGGAA), which enable depth information to guide RGB feature extraction and enhance contour-aware feature representation. Ablation studies and qualitative results validate the effectiveness of each module. Extensive evaluations on the NYU Depth V2 and SUN RGB-D datasets confirm that the proposed method attains superior mIoU scores of 58.1% and 50.5%, respectively, outperforming existing approaches and showing robust performance in challenging indoor scenarios with blurred boundaries and complex geometries.

References

1. Zhang, Y., Xiong, C., Liu, J., Ye, X., Sun, G.: Spatial information-guided adaptive context-aware network for efficient RGB-D semantic segmentation. IEEE Sens. J. **23**(19), 23512–23521 (2023)
2. Listiyono, H., Budiarso, Z., Windarto, A.P.: An optimized CNN architecture for accurate 3D liver segmentation in medical images. J. Image Graph. **13**(1), 15–23 (2025)
3. Yang, J., Wang, C., Wang, H., Li, Q.: A RGB-D based real-time multiple object detection and ranging system for autonomous driving. IEEE Sens. J. **20**(20), 11959–11966 (2020)
4. Zhang, H., Jin, L., Ye, C.: An RGB-D camera based visual positioning system for assistive navigation by a robotic navigation aid. IEEE/CAA J. Autom. Sin. **8**(8), 1389–1400 (2021)
5. Song, Y., Wen, J., Liu, D., Yu, C.: Deep robotic grasping prediction with hierarchical RGB-D fusion. Int. J. Control. Autom. Syst. **20**(1), 243–254 (2022)
6. Chen, J., Zhan, Y., Xu, Y., Pan, X.: FAFNet: Fully aligned fusion network for RGBD semantic segmentation based on hierarchical semantic flows. IET Image Processing **17**(1), 32–41 (2023)
7. Wang, J., Wang, Z., Tao, D., See, S., Wang, G.: Learning common and specific features for RGB-D semantic segmentation with deconvolutional networks. In: Computer Vision–ECCV 2016: 14th European Conference, Amsterdam, The Netherlands, October 11–14, 2016, Proceedings, Part V 14, pp. 664–679. Springer International Publishing (2016)

8. Chen, L. C., Papandreou, G., Schroff, F., Adam, H.: Rethinking atrous convolution for semantic image segmentation (2017). arXiv preprint arXiv:1706.05587
9. Du, S.Q., Tang, S.J., Wang, W.X., Li, X.M., Lu, Y.H., Guo, R.Z.: PSCNET: efficient RGB-D semantic segmentation parallel network based on spatial and channel attention. ISPRS Ann. Photogramm., Remote. Sens. Spat. Inf. Sci. **1**, 129–136 (2022)
10. Hao, Z., Xiao, Z., Luo, Y., Guo, J., Wang, J., Shen, L., Hu, H.: Primkd: primary modality guided multimodal fusion for rgb-d semantic segmentation. In: Proceedings of the 32nd ACM International Conference on Multimedia, pp. 1943–1951 (2024)
11. Zhang, P., Chen, M., Gao, M.: Semantic guidance fusion network for cross-modal semantic segmentation. Sensors **24**(8), 2473 (2024)
12. Hu, X., Yang, K., Fei, L., Wang, K.: Acnet: attention based network to exploit complementary features for rgbd semantic segmentation. In: 2019 IEEE International Conference on Image Processing (ICIP), pp. 1440–1444. IEEE (2019)
13. Wu, P., Guo, R., Tong, X., Su, S., Zuo, Z., Sun, B., Wei, J.: Link-RGBD: cross-guided feature fusion network for RGBD semantic segmentation. IEEE Sens. J. **22**(24), 24161–24175 (2022)
14. Hanselmann, M., Strauss, T., Dormann, K., Ulmer, H.: CANet: an unsupervised intrusion detection system for high dimensional CAN bus data. IEEE Access **8**, 58194–58205 (2020)
15. Liu, H., Xie, W., Wang, S.: Feature fusion and context interaction for RGB-D indoor semantic segmentation. Appl. Soft Comput. **167**, 112379 (2024)
16. Zhao, Q., Wan, Y., Xu, J., Fang, L.: Cross-modal attention fusion network for RGB-D semantic segmentation. Neurocomputing **548**, 126389 (2023)
17. Jia, W., Yan, X., Liu, Q., Zhang, T., Dong, X.: TCANet: three-stream coordinate attention network for RGB-D indoor semantic segmentation. Complex Intell. Syst. **10**(1), 1219–1230 (2024)
18. Park, S.J., Hong, K.S., Lee, S.: Rdfnet: RGB-D multi-level residual feature fusion for indoor semantic segmentation. In: Proceedings of the IEEE International Conference on Computer Vision, pp. 4980–4989 (2017)
19. Wu, Z., Zhou, Z., Allibert, G., Stolz, C., Demonceaux, C., Ma, C.: Transformer fusion for indoor RGB-d semantic segmentation. Comput. Vis. Image Underst. **249**, 104174 (2024)
20. Zhang, J., Liu, R., Shi, H., Yang, K., Reiß, S., Peng, K., Fu, H., Wang, K., & Stiefelhagen, R.: Delivering arbitrary-modal semantic segmentation. In: Proceedings of the IEEE/CVF Conference on Computer Vision and Pattern Recognition, pp. 1136–1147 (2023)
21. Yin, B., Zhang, X., Li, Z., Liu, L., Cheng, M. M., Hou, Q.: Dformer: Rethinking rgbd representation learning for semantic segmentation (2023). arXiv preprint arXiv:2309.09668

Performance Benchmarking and Optimisation of Clustering Algorithms for Local and Non-local Similarity Measure in Medical Image Analysis

Sisipho Hamlomo[1,2](✉) and Marcellin Atemkeng[1,3]

[1] Department of Mathematics, Rhodes University, PO Box 94, Makhanda 6140, South Africa
m.atemkeng@ru.ac.za

[2] Department of Statistics, Rhodes University, PO Box 94, Makhanda 6140, South Africa
s.hamlomo@ru.ac.za

[3] National Institute for Theoretical and Computational Sciences (NITheCS), Stellenbosch 7600, South Africa

Abstract. Medical imaging generates high-resolution images posing significant storage, transmission, and computational challenges. While low-rank matrix approximation (LoRMA) techniques offer efficient compression by exploiting structural redundancy, global approaches often fail to preserve local details critical for diagnosis. This paper focuses on clustering techniques that exploit non-local self-similarity to identify structurally similar regions in medical images. These clusters can be used for post-processing tasks such as adaptive image compression. We evaluate five clustering techniques: k-means, mini-batch k-means, agglomerative hierarchical clustering, balanced iterative reducing and clustering using hierarchies (BIRCH), and bisecting k-means across MRI, ultrasound, and chest X-ray modalities. All clustering techniques were optimised using random search, and cluster quality was assessed using the Silhouette score, the Davies-Bouldin (DB) index, and the Calinski-Harabasz (CH) index. Results demonstrate that standard k-means and bisecting k-means generally achieve strong cluster cohesion and separation across modalities. However, they tend to form a small number of clusters with high intra-cluster variability, limiting their effectiveness for post-processing tasks such as adaptive compression. Agglomerative clustering outperformed other techniques for MRI and ultrasound in terms of intra-cluster homogeneity, making it more suitable for preserving fine diagnostic details. For chest X-rays, mini-batch k-means achieved the best balance between clustering quality and intra-cluster compactness. BIRCH consistently underperformed across all modalities.

Keywords: Medical image compression · clustering · low-rank matrix approximation · hyperparameter tuning

L. Huang and D. Greenhalgh (Eds.): ICMLC 2026, LNNS 2061, pp. 291–306, 2027.
https://doi.org/10.1007/978-3-032-31236-5_20

1 Introduction

Medical imaging plays an important role in modern healthcare, enabling accurate diagnosis, treatment planning, and disease monitoring. Advances in imaging modalities such as magnetic resonance imaging (MRI), computed tomography (CT), and ultrasound have significantly enhanced image resolution and diagnostic capability [1–3]. However, these high-resolution images produce vast amounts of high-dimensional data, posing data storage, transmission, and computational processing challenges during image analysis [4–6]. In addition, medical images often suffer from noise, artefacts, and variations due to different acquisition conditions, further complicating efficient image analysis [7]. Therefore, efficient image compression becomes essential to reduce the cost burden on healthcare data management systems. To address these challenges, various image compression techniques have been explored, broadly categorised into lossless and lossy methods. While lossless techniques preserve all original data, they often achieve low compression ratios, which limits their practicality in bandwidth-constrained and storage-limited environments [8,9]. Consequently, lossy compression methods, which aim to reduce data redundancy while preserving and extracting important features, have received increased attention. Standard image compression techniques such as discrete cosine transform (DCT)-based JPEG [10–12] and wavelet-based JPEG2000 [13–15] have been widely applied. These methods have the advantage of being computationally efficient; however, they often struggle to maintain the fine structural details and diagnostic features inherent in medical images, leading to the potential loss of clinically significant information [16]. To address these shortcomings, researchers have explored techniques such as sparsity-based and dictionary learning methods [17–19], which introduce additional complexity and computational demands while improving the fidelity of reconstructed images.

Low-rank matrix approximation (LoRMA) techniques have emerged as powerful tools for data compression due to their ability to capture the underlying low-dimensional structure of high-resolution medical images, effectively reducing noise and preserving important features. However, this global approach often fails to capture local variations, leading to a loss of important fine details [20,21]. To address this limitation, recent research [22–26] has shifted toward patch-based and locally adaptive strategies that exploit non-local self-similarity in medical images to preserve essential anatomical features while reducing data redundancy. The fundamental idea behind these papers is to split the data matrix into overlapping patches, group structurally similar patches, and apply LoRMA for data matrix compression or denoising.

While the idea of patch-grouping is established, a critical gap remains in understanding which clustering technique is most effective at forming the homogeneous, low-variance clusters that are essential for high-fidelity, adaptive compression. Different clustering algorithms have inherent biases; some prioritize global separation, which yields good overall metrics, while others may form tighter, more homogeneous groups at the cost of creating more clusters. This trade-off directly impacts compression performance: homogeneous clusters allow for more aggressive compression (higher compression ratios) while minimizing the loss of fine details, whereas clusters with high internal variability force a conservative compression strategy to avoid introducing artifacts or losing diagnostically crucial information. This paper addresses this gap by presenting a systematic benchmarking study. In particular, we examine five clustering methods: k-means, mini-batch k-means, agglomerative hierarchical clustering, balanced iterative reducing and clustering using hierarchies and bisecting k-means, specifically on their ability to form compact, low-variance clusters suitable for adaptive LoRMA-based compression across different medical imaging modalities. We assess their effectiveness in forming compact, low-variance clusters across these modalities. These clusters are important for post-processing tasks such as per-cluster adaptive compression, since homogeneous clusters enable aggressive compression while preserving important features. We evaluated the clustering methods using the Silhouette score, Davies-Bouldin and Calinski-Harabasz index, which quantify cohesion and separation.

This work makes three contributions: we evaluated four clustering techniques using different medical imaging modalities; we used random search to tune hyperparameters for every clustering algorithm; we provide a recommendation as to which clustering technique to use on which modality.

The rest of this paper is organized as follows: Sect. 2 discusses the clustering techniques used for patch grouping; Sect. 3 details the hyperparameter tuning method used to optimize each clustering technique; Sect. 4 outlines the evaluation metrics used; Sect. 5 presents experimental results and analysis; and Sect. 6 concludes with findings and future directions.

2 Clustering Techniques

To exploit redundancy and structural similarities in medical images, the image $\mathbf{A} \in \mathbb{R}^{H \times W}$ is divided into a set of overlapping square patches of size $p \times p$. Each patch is extracted by sliding a window across the image with stride $s < p$, ensuring overlap. The resulting patch $\mathbf{P}_i$ is reshaped (vectorized) into a column vector $\mathbf{p}_i \in \mathbb{R}^{p^2}$ by stacking its pixel intensities in a fixed order (e.g., row-wise). Collectively, the image yields a patch set $\boldsymbol{\mathcal{P}} = \{\mathbf{p}_1, \mathbf{p}_2, \dots, \mathbf{p}_N\}$. Once patches are obtained, clustering algorithms are applied directly to this high-dimensional patch set to group similar patch regions. In this section, we discuss mini-batch k-means, agglomerative hierarchical clustering, balanced iterative reducing and clustering using hierarchies and bisecting k-means. The k-means theory is not discussed in this paper, and readers are referred to [26] for details.

These five clustering methods were selected since they represent both partitioning-based approaches (k-means, mini-batch k-means, bisecting k-means) and hierarchical-based approaches (agglomerative, BIRCH). This selection provides a balanced benchmark in terms of scalability, computational efficiency, and ability to capture local structure, which are critical in medical imaging tasks where both large data volumes and diagnostic fidelity must be considered.

2.1 Mini-Batch K-Means

Mini-batch k-means is a variant of the traditional k-means clustering algorithm that uses small random batches of data (mini-batches) to update the cluster centroids, rather than using the entire dataset at each iteration. The objective is to partition these patches into K clusters that minimise the within-cluster variance. Let $\{\boldsymbol{\mu}_1^{(0)}, \ldots \boldsymbol{\mu}_K^{(0)}\} \subset \mathbb{R}^{p^2}$ denote the initial cluster centroids where each $\boldsymbol{\mu}_k^{(0)} \in \mathbb{R}^{p^2}$ is initialized either randomly or using a heuristic such as k-means++. The clustering objective function is given by

$$\min_{\mathcal{C}_1,\ldots,\mathcal{C}_K} \sum_{k=1}^{K} \sum_{\mathbf{p}\in\mathcal{C}_k} \|\mathbf{p} - \boldsymbol{\mu}_k\|^2, \tag{1}$$

where $\mathcal{C}_k$ denotes the set of data points assigned to cluster $\mathcal{C}_k$. At each iteration $t = 1, \ldots, T$, a mini-batch

$$\mathcal{B}^{(t)} = \{\mathbf{p}_1^{(t)}, \ldots, \mathbf{p}_b^{(t)}\} \subset \mathcal{P} \tag{2}$$

of size $b \ll N$ is sampled uniformly at random. Each point $\mathbf{p} \in \mathcal{B}^{(t)}$ is assigned to its nearest cluster according to

$$k^*(\mathbf{p}) = \arg \min_{k\in\{1,\ldots,K\}} \|\mathbf{p} - \boldsymbol{\mu}_k^{(t)}\|^2. \tag{3}$$

Let $\mathcal{B}_k^{(t)} = \{\mathbf{p} \in \mathcal{B}^{(t)} \mid k^*(\mathbf{p}) = k\}$ be the set of mini-batch points assigned to cluster $\mathcal{C}_k$ at iteration t. Define a learning rate for each cluster

$$\eta_k^{(t)} = \frac{1}{N_{\mathcal{C}_k}^{(t)} + 1}, \tag{4}$$

where $N_{\mathcal{C}_k}^{(t)}$ is the cumulative number of data points assigned to cluster $\mathcal{C}_k$ up to iteration t, accounting for centroid updates at each step. The centroids are then updated using a batch-level residual correction

$$\boldsymbol{\mu}_k^{(t+1)} = \boldsymbol{\mu}_k^{(t)} + \eta_k^{(t)} \cdot \frac{1}{|\mathcal{B}_k^{(t)}|} \sum_{\mathbf{p}\in\mathcal{B}_k^{(t)}} \left(\mathbf{p} - \boldsymbol{\mu}_k^{(t)}\right), \tag{5}$$

provided that $|\mathcal{B}_k^{(t)}| > 0$; otherwise, the centroid remains unchanged.

2.2 Hierarchical Clustering

Hierarchical clustering groups similar data points by building a nested structure of clusters, typically represented as a dendrogram. It is broadly categorised into divisive (top-down) and agglomerative (bottom-up) methods. In this paper, we adopt agglomerative hierarchical clustering as it avoids the complexity of deciding how to split nodes and assign child levels, which are key challenges in divisive clustering [27]. In agglomerative hierarchical clustering, the process begins at iteration $t = 0$ with

$$\mathcal{C}^{(0)} = \{\mathcal{C}_1^{(0)}, \dots, \mathcal{C}_N^{(0)}\}, \quad \mathcal{C}_i^{(0)} = \{\mathbf{p}_i\}. \tag{6}$$

A linkage-based distance between any two clusters $\mathcal{C}_i$ and $\mathcal{C}_j$ is then defined. For instance, using single linkage

$$d_{\text{single}}(\mathcal{C}_i, \mathcal{C}_j) = \min_{\mathbf{x} \in \mathcal{C}_i, \mathbf{y} \in \mathcal{C}_j} \|\mathbf{x} - \mathbf{y}\|, \tag{7}$$

complete linkage

$$d_{\text{complete}}(\mathcal{C}_i, \mathcal{C}_j) = \max_{\mathbf{x} \in \mathcal{C}_i, \mathbf{y} \in \mathcal{C}_j} \|\mathbf{x} - \mathbf{y}\|, \tag{8}$$

or average linkage

$$d_{\text{avg}}(\mathcal{C}_i, \mathcal{C}_j) = \frac{1}{|\mathcal{C}_i|\,|\mathcal{C}_j|} \sum_{\mathbf{x} \in \mathcal{C}_i} \sum_{\mathbf{y} \in \mathcal{C}_j} \|\mathbf{x} - \mathbf{y}\|. \tag{9}$$

At each iteration t, the pair (a, b) minimizing the chosen linkage distance is identified

$$(a, b) = \arg\min_{i \neq j} d(\mathcal{C}_i^{(t)}, \mathcal{C}_j^{(t)}). \tag{10}$$

These two clusters are merged into a new cluster

$$\mathcal{C}_{\text{new}}^{(t+1)} = \mathcal{C}_a^{(t)} \cup \mathcal{C}_b^{(t)}, \tag{11}$$

and the cluster set is updated as follows:

$$\mathcal{C}^{(t+1)} = \left(\mathcal{C}^{(t)} \setminus \{\mathcal{C}_a^{(t)}, \mathcal{C}_b^{(t)}\}\right) \cup \{\mathcal{C}_{\text{new}}^{(t+1)}\}. \tag{12}$$

This process continues until a single cluster remains, yielding a dendrogram whose merge heights record the linkage distances. Alternatively, using Ward's criterion, one merges at each step the pair (a, b) that minimises the increase in total within-cluster variance. If $\boldsymbol{\mu}_i^{(t)}$ denotes the centroid of $\mathcal{C}_i^{(t)}$, then merging $\mathcal{C}_a$ and $\mathcal{C}_b$ incurs

$$\Delta E = \frac{|\mathcal{C}_a^{(t)}|\,|\mathcal{C}_b^{(t)}|}{|\mathcal{C}_a^{(t)}| + |\mathcal{C}_b^{(t)}|} \left\|\boldsymbol{\mu}_a^{(t)} - \boldsymbol{\mu}_b^{(t)}\right\|_2^2, \tag{13}$$

and the merge yielding the minimum ΔE is chosen. Once the full hierarchy is built, a flat partition into K clusters is obtained by "cutting" the dendrogram at the appropriate linkage threshold, producing clusters $\mathcal{C}_1, \dots, \mathcal{C}_K$.

2.3 Balanced Iterative Reducing and Clustering Using Hierarchies

Balanced iterative reducing and clustering using hierarchies (BIRCH) is a hierarchical clustering algorithm designed to efficiently handle large datasets by incrementally constructing a compact summary structure known as a clustering feature (CF) tree. Each node of the CF-tree stores a clustering feature

$$\mathrm{CF}(\mathcal{C}) = \big(N_{\mathcal{C}},\, \mathbf{LS}_{\mathcal{C}},\, SS_{\mathcal{C}}\big), \tag{14}$$

where $N_{\mathcal{C}} = |\mathcal{C}|$ counts its patches, $\mathbf{LS}_{\mathcal{C}} = \sum_{\mathbf{p}\in\mathcal{C}} \mathbf{p}$ is their vector sum, and $SS_{\mathcal{C}} = \sum_{\mathbf{p}\in\mathcal{C}} \|\mathbf{p}\|^2$ records their squared-norm total. These three quantities add entry-wise over any disjoint union, so that a parent node's CF exactly summarises the union of its children's patches. That is, if $\mathcal{C} = \mathcal{C}_a \cup \mathcal{C}_b$ and $\mathcal{C}_a \cap \mathcal{C}_b = \emptyset$, then

$$\mathrm{CF}(\mathcal{C}) = \mathrm{CF}(\mathcal{C}_a) + \mathrm{CF}(\mathcal{C}_b). \tag{15}$$

This ensures that every node's summary perfectly captures all the patches it contains. From any CF, one computes the centroid $\mathbf{LS}_{\mathcal{C}}/N_{\mathcal{C}}$ and radius

$$r(\mathcal{C}) = \sqrt{\frac{SS_{\mathcal{C}}}{N_{\mathcal{C}}} - \left\|\frac{\mathbf{LS}_{\mathcal{C}}}{N_{\mathcal{C}}}\right\|^2}, \tag{16}$$

which quantifies the root-mean-square (RMS) distance of points from the centroid, reflecting the cluster's spread. The CF-tree is controlled by a branching factor B (maximum entries per node), which limits the number of entries per node, and a threshold T, which caps the allowable radius of any leaf-level feature. Patches are gradually incorporated by always choosing, at each internal node, the entry whose feature would grow least in squared radius when augmented by the new patch, that is

$$j^* \;=\; \arg\min_{1\le j\le B} \Big[r\big(\mathrm{CF}_j + \mathbf{p}\big)^2 \;-\; r(\mathrm{CF}_j)^2 \Big]. \tag{17}$$

The selected feature then increases its counts and sums by those of $\mathbf{p}$, and the process continues down to a leaf. If the updated radius of that leaf feature remains at most T, the patch merges into it; otherwise, a fresh feature consisting of $\big(1, \mathbf{p}, \|\mathbf{p}\|^2\big)$ is introduced. Whenever a node's number of features exceeds B, it is split by identifying two "seed" features whose centroids lie furthest apart and partitioning the remainder according to the nearest seed. Once all N patches are summarised in the leaf features, their centroids are given by

$$\{\mathbf{LS}_\ell / N_\ell\}_{\ell=1}^{L}, \tag{18}$$

where L denotes the total number of leaf-level CF entries. These centroids provide a compact representation of $\mathcal{P}$, to which a final flat clustering method (such as k-means or agglomerative clustering) is applied in order to produce the desired clusters $\boldsymbol{\mathcal{C}}$.

2.4 Bisecting K-Means

Bisecting k-means is a hierarchical clustering algorithm that recursively partitions a patch set $\mathcal{P}$ by applying two-means clustering to selected subsets until K clusters are formed. It begins with a single cluster $\boldsymbol{\mathcal{C}}^{(0)} = \{\mathcal{P}\}$, and at each iteration t, a cluster $\mathcal{C}_{\text{sel}}^{(t)} \in \boldsymbol{\mathcal{C}}^{(t)}$ is chosen for bisection. A two-means clustering is applied to $\mathcal{C}_{\text{sel}}^{(t)}$, yielding two sub-clusters $\mathcal{C}_a^{(t+1)}$ and $\mathcal{C}_b^{(t+1)}$, such that

$$\mathcal{C}_{\text{sel}}^{(t)} = \mathcal{C}_a^{(t+1)} \cup \mathcal{C}_b^{(t+1)} \quad \text{and} \quad \mathcal{C}_a^{(t+1)} \cap \mathcal{C}_b^{(t+1)} = \emptyset, \tag{19}$$

with corresponding centroids $\boldsymbol{\mu}_a^{(t)}, \boldsymbol{\mu}_b^{(t)} \in \mathbb{R}^{p^2}$. The split quality is measured by the within-cluster sum of squares, that is

$$\text{WCSS}^{(t)} = \sum_{\mathbf{p} \in \mathcal{C}_a^{(t+1)}} \|\mathbf{p} - \boldsymbol{\mu}_a^{(t)}\|^2 + \sum_{\mathbf{p} \in \mathcal{C}_b^{(t+1)}} \|\mathbf{p} - \boldsymbol{\mu}_b^{(t)}\|^2. \tag{20}$$

To improve stability, the bisection is repeated n_{try} times with different initializations, and the split minimizing $\text{WCSS}^{(t)}$ is chosen. The updated clustering becomes

$$\boldsymbol{\mathcal{C}}^{(t+1)} = \left(\boldsymbol{\mathcal{C}}^{(t)} \setminus \left\{\mathcal{C}_{\text{sel}}^{(t)}\right\}\right) \cup \left\{\mathcal{C}_a^{(t+1)}, \mathcal{C}_b^{(t+1)}\right\}. \tag{21}$$

This process continues until $|\boldsymbol{\mathcal{C}}^{(t)}| = K$, producing a final clustering $\boldsymbol{\mathcal{C}} = \{\mathcal{C}_1, \ldots, \mathcal{C}_K\}$.

3 Hyperparameter Tuning

Hyperparameter tuning aims to identify the set of model parameters that yields the best performance on a given task. In clustering, hyperparameters such as the number of clusters, batch sizes, linkage criteria, and tree thresholds can greatly influence the quality of the resulting clusters. To explore the hyperparameter space, the configuration vector is expressed as

$$\boldsymbol{\theta} = \left(K, \alpha_1, \ldots, \alpha_d\right) \in \boldsymbol{\Theta}, \tag{22}$$

where $\boldsymbol{\Theta}$ denotes the feasible hyperparameter space. This space is constructed as a Cartesian product of sets, that is

$$\boldsymbol{\Theta} = \mathcal{K} \times \mathcal{A}_1 \times \cdots \times \mathcal{A}_d, \quad \mathcal{K} = \{K_{\min}, \ldots, K_{\max}\}, \tag{23}$$

where each $\mathcal{A}_j$ denotes the admissible set for hyperparameter α_j and $\mathcal{K}$ denotes the range of possible values for the number of clusters. To optimise, we perform a random search with budget T by drawing

$$\boldsymbol{\theta}_1, \cdots, \boldsymbol{\theta}_T \overset{\text{i.i.d}}{\sim} \mathcal{U}(\boldsymbol{\Theta}). \tag{24}$$

Let

$$\Gamma_{\boldsymbol{\theta}} \colon \mathcal{P} \longrightarrow \boldsymbol{\mathcal{C}}(\mathcal{P}), \tag{25}$$

denote the clustering operator where $\boldsymbol{\mathcal{C}}(\mathcal{P})$ is the set of all partitions of $\mathcal{P}$. For fixed $\boldsymbol{\theta}$,

$$\boldsymbol{\mathcal{C}}_{\boldsymbol{\theta}} = \Gamma_{\boldsymbol{\theta}}(\mathcal{P}) = \{\mathcal{C}_1(\boldsymbol{\theta}), \ldots, \mathcal{C}_K(\boldsymbol{\theta})\}, \tag{26}$$

where each $\mathcal{C}_k(\boldsymbol{\theta}) \subseteq \mathcal{P}$ and

$$\bigcup_{k=1}^{K} \mathcal{C}_k(\boldsymbol{\theta}) = \mathcal{P}, \quad \mathcal{C}_i(\boldsymbol{\theta}) \cap \mathcal{C}_j(\boldsymbol{\theta}) = \emptyset \quad (i \neq j). \tag{27}$$

For each draw $\boldsymbol{\theta}_t$, we apply $\Gamma_{\boldsymbol{\theta}_t}$ to $\mathcal{P}$ to obtain the partition $\{\mathcal{C}_k(\boldsymbol{\theta}_t)\}_{k=1}^{K}$ and compute its average Silhouette score

$$f_t = \frac{1}{N} \sum_{i=1}^{N} \frac{b(i;\boldsymbol{\theta}_t) - a(i;\boldsymbol{\theta}_t)}{\max\{\, a(i;\boldsymbol{\theta}_t),\, b(i;\boldsymbol{\theta}_t)\}}. \tag{28}$$

We record the pairs $(\boldsymbol{\theta}_t, f_t)$ for $t = 1, \ldots, T$ and select

$$\boldsymbol{\theta}^* = \arg \max_{t \in \{1,2,\ldots,T\}} f_t. \tag{29}$$

4 Evaluation Metrics

The cluster quality metrics Silhouette score, Calinski-Harabasz (CH) index and Davies-Bouldin (DB) index were used to evaluate the performance of each clustering technique. These evaluation metrics assess the quality of clustering results by measuring cohesion (how compact each cluster is) and separation (how well clusters are distinguished from one another). The Silhouette score quantifies how well each data point $\mathbf{p}_i$ lies within its assigned cluster compared to other clusters. For a patch $\mathbf{p}_i$, let $a(i)$ denote the average distance from $\mathbf{p}_i$ to all other points in the same cluster (intra-cluster distance), that is

$$a(i) \;=\; \frac{1}{|\mathcal{C}_k| - 1} \sum_{\substack{\mathbf{p}_j \in \mathcal{C}_k \\ j \neq i}} \|\mathbf{p}_i - \mathbf{p}_j\|_2 \quad \text{for } \mathbf{p}_i \in \mathcal{C}_k \tag{30}$$

and let $b(i)$ be the minimum average distance from $\mathbf{p}_i$ to all points in the nearest different cluster (inter-cluster distance), that is

$$b(i) \;=\; \min_{m \neq k} \frac{1}{|\mathcal{C}_m|} \sum_{\mathbf{p}_j \in \mathcal{C}_m} \|\mathbf{p}_i - \mathbf{p}_j\|_2. \tag{31}$$

Therefore, the silhouette value $s(i)$ is defined as

$$s(i) = \frac{b(i) - a(i)}{\max\{a(i), b(i)\}}, \tag{32}$$

and lies in the interval $[-1, 1]$. A value near $+1$ indicates that $\mathbf{p}_i$ is well matched to its own cluster and poorly matched to others. A value near 0 suggests it lies on the decision boundary between two clusters, while negative values indicate likely misclassifications. The overall silhouette score is the average of $s(i)$ over all patches.

On the other hand, the CH index, also known as the variance ratio criterion, measures the ratio of between-cluster dispersion to within-cluster dispersion. Let $\mathbf{c}_k$ be the centroid of cluster $\mathcal{C}_k$, and let $\mathbf{c}$ be the global centroid of all data. The CH index is defined as

$$\mathrm{CH} = \frac{\mathrm{Tr}(\mathbf{B}_K)}{\mathrm{Tr}(\mathbf{W}_K)} \cdot \frac{N - K}{K - 1}, \tag{33}$$

where $\mathrm{Tr}(\mathbf{B}_K)$ denotes the trace of the between-cluster dispersion matrix

$$\mathrm{Tr}(\mathbf{B}_K) = \sum_{k=1}^{K} N_k \|\mathbf{c}_k - \mathbf{c}\|^2, \tag{34}$$

and $\mathrm{Tr}(\mathbf{W}_K)$ denotes the trace of the within-cluster dispersion matrix

$$\mathrm{Tr}(\mathbf{W}_K) = \sum_{k=1}^{K} \sum_{\mathbf{p}_i \in \mathcal{C}_k} \|\mathbf{p}_i - \mathbf{c}_k\|^2. \tag{35}$$

Higher values of CH indicate more distinct and well-separated clusters.

The DB index quantifies the average similarity between each cluster and its most similar one, with similarity based on intra-cluster distances and inter-cluster separation. For each cluster $\mathcal{C}_k$, define

$$s_k = \frac{1}{|\mathcal{C}_k|} \sum_{\mathbf{p}_i \in \mathcal{C}_k} \|\mathbf{p}_i - \mathbf{c}_k\|, \tag{36}$$

the average distance of points to the cluster centroid. Then for clusters $\mathcal{C}_k$ and $\mathcal{C}_m$, define

$$R_{k,m} = \frac{s_k + s_m}{\|\mathbf{c}_k - \mathbf{c}_m\|}, \tag{37}$$

where $R_{k,m}$ denotes the similarity measure between $\mathcal{C}_k$ and $\mathcal{C}_m$. The DB index is the average over the maximum $R_{k,m}$ for each cluster

$$\mathrm{DB} = \frac{1}{K} \sum_{k=1}^{K} \max_{m \neq k} R_{k,m}. \tag{38}$$

Lower values of DB indicate better clustering quality, with compact, well-separated clusters.

5 Results and Discussion

This study uses three publicly available medical imaging datasets sourced from Kaggle. From each dataset, a single representative image was randomly selected. The brain MRI was sourced from the BraTS2020 dataset, which contains multi-institutional, preoperative scans at 1 mm^3 resolution (including T1, T1Gd, T2, and FLAIR sequences) with expert-annotated segmentation masks. The breast ultrasound image came from the Breast Ultrasound Images dataset, comprising 780 images labelled as normal, benign, or malignant, each with a lesion segmentation mask. Finally, the chest X-ray was selected from the COVIDqu dataset, a collection of 33,920 images categorised as COVID-19, non-COVID pneumonia, or normal, all accompanied by lung segmentation masks.

Figure 1 shows three medical imaging modalities analysed in this study: (a) a brain MRI scan, (b) an ultrasound image annotated with calliper measurements, and (c) a frontal chest X-ray radiograph. All images were split into overlapping patches of a fixed size, and clustering methods were applied to these patches to evaluate how well each technique groups similar regions across modalities. Hyperparameters for each clustering technique (e.g., number of clusters for k-means variants, linkage criteria for hierarchical clustering, threshold/branching factor for BIRCH) were tuned using random search over a predefined range to ensure each method achieved optimal results.

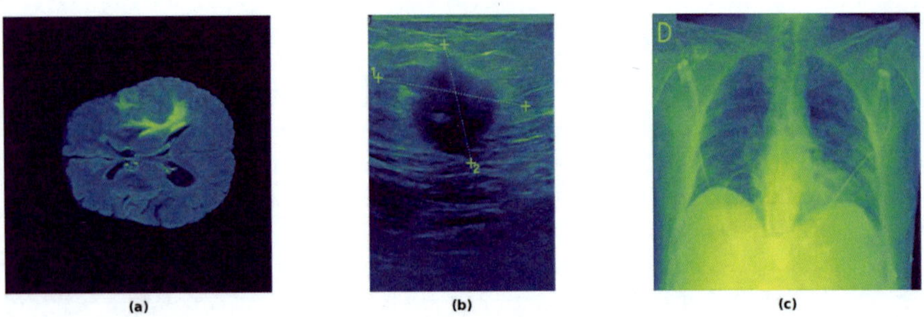

Fig. 1. Shows three medical imaging modalities analysed in this study: (a) a brain MRI scan, (b) an ultrasound image annotated with calliper measurements, and (c) a frontal chest X-ray radiograph.

Table 1 summarises clustering performance metrics: Silhouette score, DB index, CH index, and the optimal number of clusters selected for each method and modality.

For the MRI modality, standard k-means achieves the highest Silhouette score (0.814), followed closely by bisecting k-means (0.808) and agglomerative clustering (0.801). The lowest DB index is observed for bisecting k-means (0.425), indicating the most compact and well-separated clusters. It also attains the highest CH score (9421.16), showing strong clustering performance. While mini-batch

Table 1. Comparison of clustering performance metrics across different medical imaging modalities (MRI, ultrasound, X-ray) using various clustering methods. The performance evaluation metrics include Silhouette score, DB and CH index. We also record the optimal number of clusters for each clustering technique.

Method	MRI				Ultrasound				X-ray			
	Silhouette	Davies-Bouldin	Calinski-Harabasz	No. of clusters	Silhouette	Davies-Bouldin	Calinski-Harabasz	No. of clusters	Silhouette	Davies-Bouldin	Calinski-Harabasz	No. of clusters
K-means	0.814	0.705	8196.349	3	0.594	0.589	14449.452	2	0.540	0.631	7576.080	2
Mini-batch k-means	0.779	0.963	6311.616	4	0.318	1.015	9395.139	4	0.456	0.714	8967.025	4
Agglomerative	0.801	0.789	2785.943	7	0.511	0.485	920.711	11	0.444	0.902	3732.178	7
Birch	0.746	1.366	4019.569	6	0.354	1.169	8465.552	4	0.420	0.759	7654.140	4
Bisecting K-means	0.808	0.425	9421.159	2	0.594	0.589	14449.296	2	0.540	0.631	7575.771	2

k-means produces a reasonably high Silhouette score (0.779), its DB index is the highest (0.963), suggesting less optimal compactness compared to the other methods. For the ultrasound modality, standard k-means and bisecting k-means tie for the highest Silhouette score (0.594) and lowest DB index (0.589), with very similar CH scores (14449.45 and 14449.30, respectively), making them the top-performing methods. Agglomerative clustering, though exhibiting a reasonably strong Silhouette score (0.511) and the lowest DB index (0.485), results in a significantly lower CH score (920.71), which may reflect the penalty imposed by the CH index on the higher number of clusters (11) and potential imbalance in cluster sizes. In the X-ray modality, standard k-means and bisecting k-means both achieve the highest Silhouette score (0.540) and the lowest DB index (0.631), with nearly identical CH scores (7576.08 and 7575.77, respectively). Although mini-batch k-means attains a competitive CH score (8967.03), its lower Silhouette score (0.456) and higher DB index (0.714) suggest weaker intra-cluster cohesion and separation.

The results of this study show that standard k-means and bisecting k-means generally yield strong clustering performance based on Silhouette scores, DB and CH indices, indicating good cohesion and separation among clusters. However, when a post-processing task such as per-cluster adaptive compression is considered, intra-cluster variability becomes crucial. For example, Fig. 2 shows a box-plot comparison of intra-cluster variability for different clustering methods applied to MRI image patches. Each box-plot represents the distribution of intra-cluster variability for a specific cluster across different clustering methods. As shown in Fig. 2, while k-means and bisecting k-means perform well overall in terms of clustering metrics, they form a small number of clusters with relatively high intra-cluster variability, which limits the flexibility of adaptive cluster-wise compression since aggressive compression within these clusters may result in the loss of important or clinically relevant features. While agglomerative clustering is more computationally intensive, it forms tighter, more homogeneous clusters with low intra-cluster variability, yielding low-rank structures that enable more effective compression while minimising the risk of degrading clinically relevant information.

On the other hand, Fig. 3 indicates that k-means and bisecting k-means produce two clusters with high intra-cluster variability, which limits their suitability for adaptive compression. Although agglomerative clustering performs slightly less in terms of clustering metrics, it forms a larger number of more homogeneous clusters. This makes it more suitable for adaptive cluster-wise compression in ultrasound images, as it enables more effective low-rank approximations while preserving essential diagnostic details.

While mini-batch k-means performs slightly below k-means and bisecting k-means in standard clustering metrics, Fig. 4 shows that it generates a larger number of clusters with lower intra-cluster variability, making it more effective for adaptive cluster-wise compression.

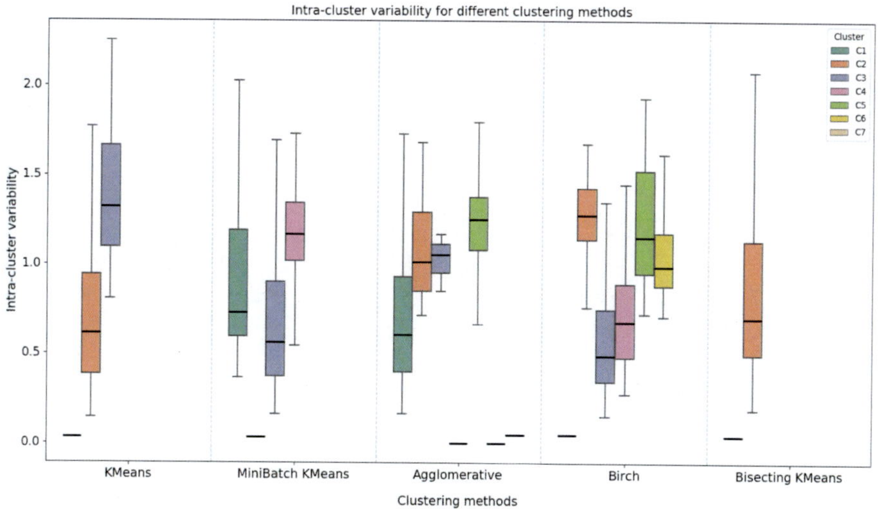

Fig. 2. Intra-cluster variability for MRI patches clustered using k-means, mini-batch k-means, agglomerative, birch and bisecting k-means methods. Each box-plot represents the variability within a specific cluster (C1C7), with lower values indicating more homogeneous clusters.

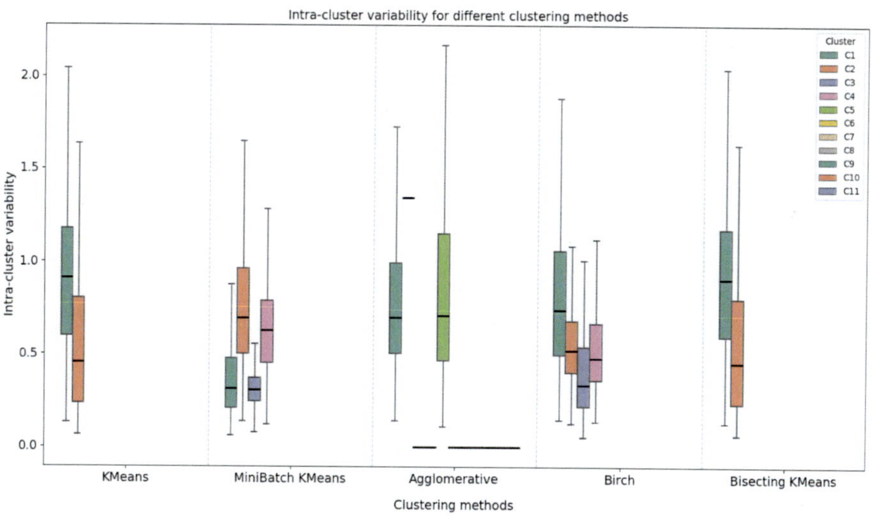

Fig. 3. Intra-cluster variability for Ultrasound patches clustered using k-means, mini-batch k-means, agglomerative, birch and bisecting k-means methods. Each box-plot represents the variability within a specific cluster (C1C11), with lower values indicating more homogeneous clusters.

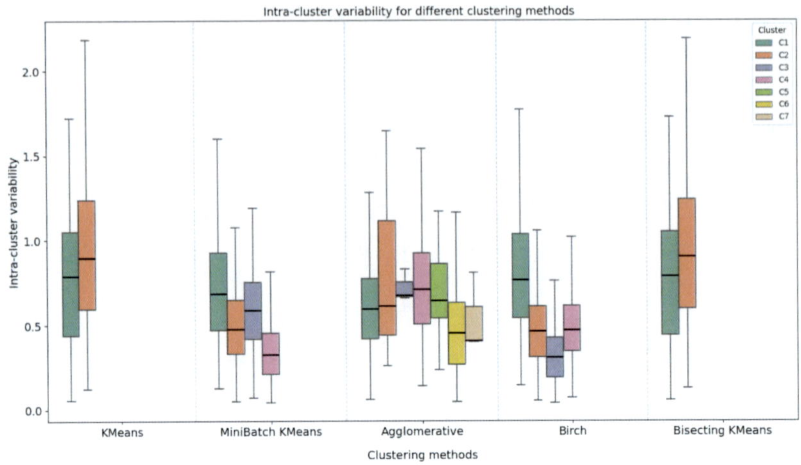

Fig. 4. Intra-cluster variability for X-ray patches clustered using k-means, mini-batch k-means, agglomerative, and Birch methods. Each box-plot represents the variability within a specific cluster (C1C7), with lower values indicating more homogeneous clusters.

6 Conclusion

This study explored the effectiveness of clustering-based techniques for grouping structurally similar patches in high-resolution medical images, enabling adaptive LoRMA-based compression during post-processing. The primary focus of our work is to investigate how the choice of clustering algorithm impacts not just standard clustering metrics, but more importantly, the intra-cluster homogeneity that is critical for post-processing tasks like compression. By comparing standard k-means, mini-batch k-means, agglomerative hierarchical clustering, BIRCH and bisecting k-means across MRI, ultrasound, and chest X-ray modalities, we evaluated clustering performance using Silhouette scores, DB index, and CH index. Our results indicate that bisecting k-means and standard k-means generally achieved strong cluster quality in terms of cohesion and separation, particularly in the MRI modality, as reflected by high Silhouette and CH scores. However, as shown in Figs. 2 and 3, these methods form a small number of clusters with high intra-cluster variability, limiting their suitability for adaptive per-cluster compression. In contrast, agglomerative clustering, while more computationally intensive, produced tighter and more homogeneous clusters with lower intra-cluster variability for both MRI and ultrasound image patches, making it more appropriate for compression in those modalities. For chest X-ray images, mini-batch k-means achieved the best balance between cluster quality and intra-cluster compactness, suggesting it is well-suited for adaptive compression in that modality. Meanwhile, BIRCH showed relatively lower performance across all modalities in both clustering quality and intra-cluster variability, making it the least suitable for the compression framework considered.

These findings suggest that the choice of clustering technique should be guided by both the imaging modality and the trade-off between computational efficiency and compression fidelity. Mini-batch k-means provides an efficient and effective option for chest X-ray images, where clustering quality and compactness align well. However, for MRI and ultrasound images, agglomerative clustering is more suitable due to its ability to form tighter, more homogeneous clusters that preserve fine structural and clinical details, as evidenced by lower intra-cluster variability. In future work, we will examine how variations in patch size influence clustering behaviour and compression performance across different medical imaging modalities.

Acknowledgments. This research was supported in part by the Department of Higher Education and Training (DHET) through the University Staff Doctoral Programme (USDP) and in part by the National Research Foundation of South Africa (Ref No.CSRP23040990793). The authors would like to thank Rhodes University for their financial support.

References

1. Tempany, C.M. and McNeil, B.J.: Advances in biomedical imaging. JAMA **285**(5), 562–567 (2001)
2. Müller, N.L.: Computed tomography and magnetic resonance imaging: past, present and future. Eur. Respir. J. **19**(35 suppl), 3s–12s (2002)
3. Hussain, S., Mubeen, I., Ullah, N., Shah, S.S.U.D., Khan, B.A., Zahoor, M., Ullah, R., Khan, F.A., Sultan, M.A.: Modern diagnostic imaging technique applications and risk factors in the medical field: a review. BioMed Res. Int. **2022**(1), 5164970 (2022)
4. Choong Ho Lee and Hyung-Jin Yoon: Medical big data: promise and challenges. Kidney Res. Clin. Pract. **36**(1), 3 (2017)
5. Li, Z., Zhang, X., Müller, H., Zhang, S.: Large-scale retrieval for medical image analytics: a comprehensive review. Med. Image Anal. **43**, 66–84 (2018)
6. Andreas, S., Panayides, A., Amini, Nenad, D., Filipovic, A., Sharma, Sotirios, A., Tsaftaris, A., Young, D., Foran, N., Do, S., Golemati, T., Kurc: AI in medical imaging informatics: current challenges and future directions. IEEE J. Biomed. Health Inform. **24**(7), 1837–1857 (2020)
7. Goyal, B., Agrawal, S., Sohi, B.S.: Noise issues prevailing in various types of medical images. Biomed. Pharmacol. J. **11**(3), 1227 (2018)
8. I.A.K.-U.: The quest for "diagnostically lossless" medical image compression using objective image quality measures, University of Waterloo (2015). PhD thesis
9. Oh, T.H.: JPEG2000 and JPEG: image quality comparison of compressed medical modalities. Int. J. Comput. Appl. **32**(4), 393–398 (2010)
10. Watson, A.B.: Image compression using the discrete cosine transform. Mathematica J. **4**(1), 81 (1994)
11. Nitesh Kumar More and Sipi Dubey: JPEG picture compression using discrete cosine transform. Int. J. Sci. Res. (IJSR) **2**(1), 134–138 (2013)
12. Raid, A.M., Khedr, W.M., El-Dosuky, M.A., Ahmed, W.: JPEG image compression using discrete cosine transform-a survey. *arXiv preprint* arXiv:1405.6147 (2014)

13. Walker, J.S., Nguyen, T.Q.: Wavelet-based image compression. Sub-chapter of CRC Press book: Transforms and Data Compression, pp. 267–312 (2001)
14. Usevitch, B.E.: A tutorial on modern lossy wavelet image compression: foundations of JPEG 2000. IEEE Signal Process. Mag. **18**(5), 22–35 (2002)
15. Talukder, K.H., Harada, K.: Haar wavelet based approach for image compression and quality assessment of compressed image (2010). arXiv:1010.4084 arXiv preprint
16. Tick, H., Oh, R., Besar: Medical image compression using JPEG-2000 and JPEG: a comparison study. J. Mech. Med. Biol. **2**(03n04), 313–328 (2002)
17. Horev, I., Bryt, O., Rubinstein, R.: Adaptive image compression using sparse dictionaries. In: 2012 19th International Conference on Systems, Signals and Image Processing (IWSSIP), pp. 592–595. IEEE (2012)
18. Watkins, Y., Iaroshenko, O., Sayeh, M., Kenyon, G.: Image compression: sparse coding vs. bottleneck autoencoders. In: 2018 IEEE Southwest Symposium on Image Analysis and Interpretation (SSIAI), pp 17–20. IEEE (2018)
19. Anandan, P., Sabeenian, R.S.: Medical image compression using wrapping based fast discrete curvelet transform and arithmetic coding. Circuits Syst. **7**(8), 2059–2069 (2016)
20. Ding, Q., Niu, T., Zhang, X., Long, Y.: Image-domain multimaterial decomposition for dual-energy CT based on prior information of material images. Med. Phys. **45**(8), 3614–3626 (2018)
21. McGivney, D.F., Pierre, E., Ma, D., Jiang, Y., Saybasili, H., Gulani, V., Griswold, M.A.: SVD compression for magnetic resonance fingerprinting in the time domain. IEEE Trans. Med. Imaging **33**(12), 2311–2322 (2014)
22. Wang, Y., Peng, J., Zhao, Q., Leung, Y., Zhao, X.-L., Meng, D.: Hyperspectral image restoration via total variation regularized low-rank tensor decomposition. IEEE J. Sel. Top. Appl. Earth Obs. Remote. Sens. **11**(4), 1227–1243 (2017)
23. Liu, H., Guo, Q., Wang, G., Brij, B., Gupta, C., Zhang: Medical image resolution enhancement for healthcare using nonlocal self-similarity and low-rank prior. Multimed. Tools Appl. **78**, 9033–9050 (2019)
24. Zeng, H., Xie, X., Kong, W., Cui, S., Ning, J.: Hyperspectral image denoising via combined non-local self-similarity and local low-rank regularization. IEEE Access **8**, 50190–50208 (2020)
25. Li, S., Wang, F., Gao, S.: New non-local mean methods for MRI denoising based on global self-similarity between values. Comput. Biol. Med. **174**, 108450 (2024)
26. Hamlomo, S., Atemkeng, M., Brima, Y., Nunhokee, C., Baxter, J.: A systematic review of low-rank and local low-rank matrix approximation in big data medical imaging. Neural Comput. Appl. 1–56 (2025)
27. Ran, X., Xi, Y., Yonggang, L., Wang, X., Zhenyu, L.: Comprehensive survey on hierarchical clustering algorithms and the recent developments. Artif. Intell. Rev. **56**(8), 8219–8264 (2023)

BMSA-ViT: Efficient Vision Transformer for Pathology Breast Cancer Classification in Health Care System

Muhammad Usman Saleem[1], Bo Peng[1](✉), Muhammad Shahid Mastoi[2], Mannan Hassan[2], and Muhammad Zahid Rafique[3]

[1] School of Computing and Artificial Intelligence, Southwest Jiaotong University, Chengdu 611756, People's Republic of China
usman_saleem@my.swjtu.edu.cn, bpeng@swjtu.edu.cn
[2] Interdisciplinary Research Center for Sustainable Energy Systems, Research and Innovation, King Fahd University of Petroleum Minerals, Dhahran 31261, P.R, Saudi Arabia
shahid.mastoi797@gmail.com, mannan.hassan@kfupm.edu.sa
[3] School of Electrical Engineering, Southwest Jiaotong University, Chengdu 611756, People's Republic of China
zahidrafique@my.swjtu.edu.cn

Abstract. Breast cancer (BC) is a leading cause of cancer-related mortality among women worldwide; timely diagnosis is critical. Histopathological assessment remains the clinical gold standard; however, manual slide examination is time-consuming and prone to variability across different magnifications. We present BMSA-ViT, a lightweight, parameter-efficient Vision Transformer tailored for breast cancer classification from histopathology images. The model employs multi-head self-attention and adapter-style tuning to capture discriminative spatial–textural cues while keeping computation low. We evaluate on two public datasets: IDC (invasive ductal carcinoma) and BreaKHis. BMSA-ViT attains 91.0% accuracy on IDC and 98.5% on BreaKHis, with consistently high precision and recall across magnification factors. The compact design enables fast inference on modest hardware, making the approach practical for computer-aided diagnosis. These results indicate that transformer-based, parameter-efficient architectures can deliver accurate, robust breast cancer classification while reducing clinical workload. The method is readily integrable into digital pathology pipelines to support earlier, more reliable decision-making. By emphasizing parameter efficiency, BMSA-ViT can be deployed in resource-constrained clinical settings and laboratories.

Keywords: Breast cancer · histopathology · Vision Transformer (ViT)

1 Introduction

Breast cancer (BC) is a major public health concern, affecting millions of women worldwide and remaining a leading cause of cancer-related mortality [1]. In 2022 alone, approximately 2.3 million new cases were reported, underscoring the urgent need for

L. Huang and D. Greenhalgh (Eds.): ICMLC 2026, LNNS 2061, pp. 307–317, 2027.
https://doi.org/10.1007/978-3-032-31236-5_21

early detection and advanced diagnostic strategies to enhance patient outcomes [2]. In most cases, BC is predominantly diagnosed based on clinical manifestations or imaging findings, with definitive confirmation achieved through histopathological biopsy analysis [3]. Different techniques, such as mammography images, computed tomography scan images, magnetic resonance imaging, and ultrasonography images [4], are used in clinical practice to diagnose BC [5]. Among all, histopathological analysis remains the cornerstone for precise cancer diagnosis and prognosis[6].

The identification of cancer relies heavily on histopathology, which involves the analysis of tissue and cellular structures within lesions [7]. HPIs' examination by humans leads to inaccurate results because of human interpretive constraints. Computer-aided diagnosis (CAD) systems [8] have been developed to improve the diagnostic reliability by reducing the inter-pathologists' variability. This is accomplished through traditional CAD methods and feature extraction techniques, where the feature extraction, such as scale-invariant feature transform (SIFT) [9], speeded-up robust features (SURF) [10], and local binary patterns (LBP), may use the error cages to correct misclassifications and introduce classification biases. Deep learning (DL) approaches [11] have emerged as prominent solutions for addressing these limitations because they perform automated feature learning through multiple nonlinear transformations that enhance diagnostic performance [11].

In medical imaging, convolutional neural networks (CNNs) are the most widely adopted DL architectures, leveraging spatial and temporal correlations to extract meaningful features. CNNs have demonstrated strong performance in large-scale object classification and detection tasks, including BC applications such as tumor lesion segmentation and tissue classification to distinguish between benign and malignant cases [12]. [5] proposed a method integrating clustering with CNN-based patch selection for BC classification. While their method succeeds in delivering promising whole-slide image (WSI) level performance, it does not provide WSL level predictions, which limits its clinical utility as a patient-level diagnosis requires an exhaustive analysis of the entire tissue. [8] builds a two-stage framework of the patch-wise auto-encoders with the image-level classifiers to take advantage of local and global features, achieving a validation accuracy of 95% on the ICIAR 2018 BACH dataset. CNNs struggle with high computational demands, convergence instability, insufficient global context modeling, and poor interpretability [2].

Nonetheless, CNN-based models still struggle to capture global contextual relationships because of fundamental architectural constraints. CNN architectures, which typically rely on localized receptive fields, tend to miss broader spatial interactions, consequently neglecting complex global patterns essential for precise BC identification. In order to overcome the limitations of CNNs, transformer-based models have gained increasing attention for their ability to capture long-range dependencies and global contextual relationships. Originally developed for natural language processing tasks, transformer architectures such as Vision Transformer (ViT) [13]. To overcome the above challenges, a vision transformer classifier is proposed. This classifier improves breast cancer classification on histopathological images by utilizing the multi-headed self-attention mechanism (MSA-ViT). Also, another aspect of this analysis is the classification of histopathological images used in breast cancer detection, since they provide detailed information at the cellular level, which is necessary for accurate diagnosis and classification of breast cancer [14].

Based on our model, we provide substantial improvements over existing studies for breast cancer diagnosis. The ultimate goal of this research is to improve breast cancer diagnosis and develop a disease prevention program and a personalized medical care model that is accurate, robust, and inclusive. Furthermore, this work provides significant evidence for the merging of vision transformers exhibiting a multi-headed self-attention mechanism for the promotion of personalized and predictive health care. The proposed work predominantly contributes to the following steps:

- BMSA-ViT is a lightweight, customized Vision Transformer for analyzing breast cancer histopathology images.
- The model utilizes Vision Transformer architecture coupled with a multiheaded self-attention (MAS)mechanism, allowing the extraction of multiple features simultaneously from histopathological images.
- MHSA-ViT substantially enhances the accuracy of breast cancer classifications through rigorous evaluation and comparison with previous research. The research is particularly relevant to the healthcare industry due to its early detection of breast cancer.
- ViT mobile models were studied for developing lightweight real-time diagnostic models.

2 Vision Transformer

ViT's architecture consists of a sequence of functional blocks. In this section, we provide an overview of ViT's overall architecture, describing its main functional elements.

2.1 Overview of Architecture

The overview architecture of ViT is shown in (Fig. 1). In a ViT model, the input image is divided into patches. Position information and patch sequence are fed into the encoder block of the transformer, which is comprised of a multi-head self-attention, layer normalization, and a multilayer perception algorithm. As a result, this strategy enables the transformer to learn global dependencies within the input images, as shown in (Fig. 2).

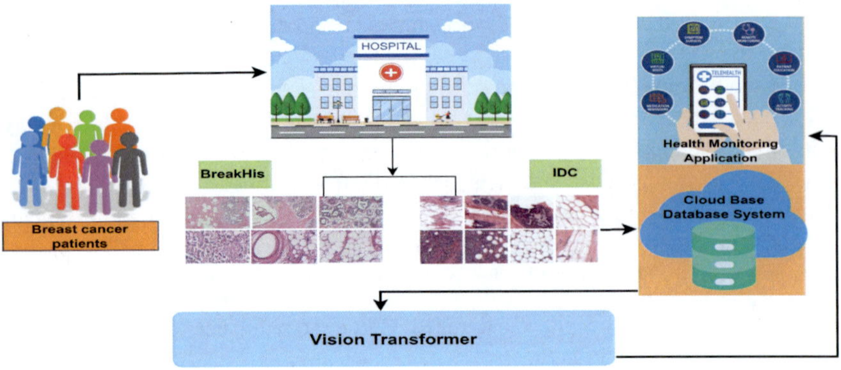

Fig. 1. A brief overview is provided of the proposed technique for categorizing histopathological images of BC

2.2 Model Architecture

A patch extraction layer constitutes the base. As a result, pathologists generally examine small, well-defined patches of a scene rather than the entire scene at once. It is through this systematic process that they generally identify disease signs. As another feature of the MHSAT models, it is their attempt to mimic the deconstruction process of an input image into a grid of patches, which are representative of the small fraction of the total visual information, in other words, the small portion of all images. In addition to improving the ease with which interesting patterns can be identified and studied within each patch, this step is necessary because it enables linear processing of spatial information [15].

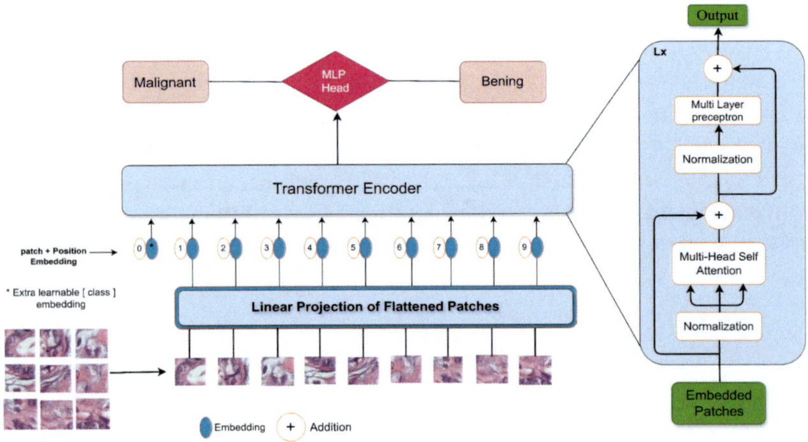

Fig. 2. The BMSA-ViT model classifies histopathology images by tokenizing patches, embedding them, and processing through Transformer encoders, with the class token output classified via an MLP head

Diagram of the viT model, showing key components and the flow of data. A learnable linear layer projects these flattened patch vectors into a higher dimensionality, enhancing the ability of the vectors to distinguish complex features within patches. (Eq. 1) defines the dimensions of the extracted patches, (Eq. 2) computes the number of patches extracted from the image, (Eq. 3) represents the size of each patch, and (Eq. 4) projects the extracted patches. In the presence of positional encoding, the whole token contains both the original image information for patches as well as the positional information for patches, which are directly fed into the transformer encoder.

$$\text{Patch Size} = \text{Patch Width} \times \text{Patch Height} \tag{1}$$

$$\text{Number of Patches} = \left(\frac{\text{Image Width}}{\text{Patch Size}}\right)^2 \tag{2}$$

$$\text{Patch\_i} = \text{Extract(Image, i, Patch Size)} \tag{3}$$

"i" is the index of the path

$$\text{EmbeddedPatch\_i} = \text{Dense(Patch\_i, EmbeddingDim)} \tag{4}$$

A Multi-Head Self-Attention (MHSA) mechanism (Fig. 3) is the core of the proposed approach, diverging from traditional transformer models. Inspired by how pathologists refocus on different regions of a biopsy slide for diagnostic information, this mechanism directs attention to the most informative regions of an image. It introduces two key components: M-key (Memory Key) and M-value, where M-key is a memory matrix that captures critical features from input patches to guide attention by highlighting which features to prioritize. M-value aggregates the features weighted by their attention scores. This ensures that the most salient features determined by M-key dominate the output, enabling fine-grained analysis impossible when treating the image as a whole. The overall process includes normalizing inputs (Eq. 5), computing attention scores between features (Eq. 6), and normalizing those scores (Eq. 7).

$$\text{LayerNorm(x)} = \upgamma\left(\frac{\text{x} - \upmu}{\upsigma}\right) + \upbeta \tag{5}$$

$$e_{ij} = \text{score(Hi, Hj)} \tag{6}$$

$$a_{ij} = \frac{\exp(eij)}{\sum_{k=1}^{N} \exp(eik)} \tag{7}$$

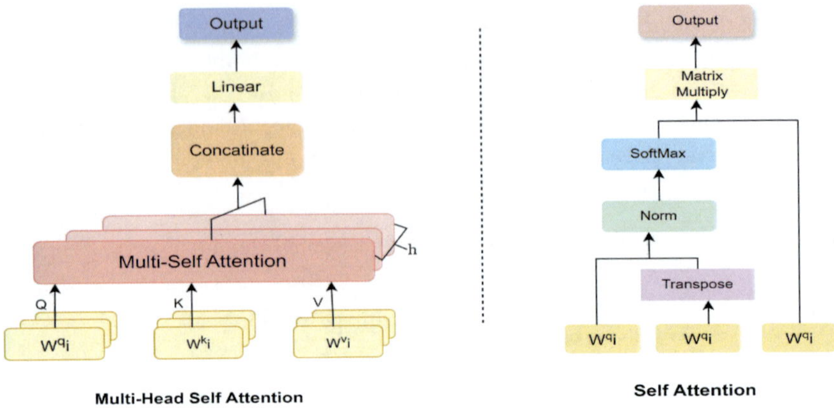

Fig. 3. The MSA system (left) runs multiple self-attention modules in parallel, combining their outputs through concatenation and a linear transformation, unlike single self-attention (right), which computes attention once.

By using the MLP head, the transformer features are refined and the benign-malignant separation is sharpened, which is then mapped to calibrated class probabilities. It integrates dispersed yet crucial cues into one final judgment, similar to a pathologist's final judgment following the review of several regions. (Eq. 8) GELU is a sequence of dense layers with nonlinear activation. An averaging over the feature map is performed (Eq. 9) followed by a linear layer with a SoftMax to create two-class posteriors (Eq. 10). Thus, this head provides a compact, discriminatory readout for the classification of breast cancer on the basis of histopathology.

$$\mathrm{MLP(x)} = \mathrm{Dense(GELU(Dense(x)))} \tag{8}$$

$$\mathrm{GlobalAvgPool(x)} = \frac{1}{\mathrm{N}}\sum_{\mathrm{i=1}}^{\mathrm{N}} \mathrm{xi} \tag{9}$$

$$M_k = \mathrm{softmax(Wk} \cdot \mathrm{X)} \tag{10}$$

2.3 Optimization and Loss Function

Training uses **AdamW** with a decaying learning rate. AdamW is Adam with **decoupled weight decay**, which improves regularization and reduces overfitting while preserving Adam's adaptive moment estimation. It is effective for sparse gradients and fine-tuning pre-trained networks. The parameter update is defined in (**Eq. 11**).

$$\omega_{t+1} = \omega_t - \eta.\frac{m_t}{\sqrt{v_t + \epsilon}} - \lambda.w_t \tag{11}$$

As shown in (Eq. 12), the deviation between the real label probability and the anticipated label probability for each class can be estimated using cross-entropy loss.

$$L = -\sum_{C=1}^{M} y_{o,c} \log p_{o,c} \tag{12}$$

3 Experiments and Results

BMSA-ViT was evaluated on two public datasets of breast cancer histopathology: BreakHis and Invasive Ductal Carcinoma (IDC).

3.1 Dataset

The evaluation is based on two public histopathology datasets: BreaKHis and IDC. It contains 7,909 RGB images obtained from 82 patients at 40/100/200/400 magnifications (benign vs. malignant). Images are resized to 224x224 for input purposes. There are 277,524 images of 48x48 pixels. Two categories of images are included in the dataset - both negative for IDC and positive for IDC. Furthermore, patches are used at their native resolutions.

3.2 Data Augmentation

Training patches are enhanced with standard techniques, including random rotations, zooms, horizontal/vertical flips, and brightness/contrast jitters. A rotation or flip enforces orientation invariance; zoom facilitates the learning of multi-scale features; and photometric jitter facilitates the learning of feature characteristics based on structure rather than intensity.

3.3 Experiment Setup and Evaluation Metrics

BreakHis images were resized to 224 $\times$ 224, while IDC patches were used at their native 48 $\times$ 48. We applied standard augmentations (random rotations, flips, zoom, and brightness/contrast jitter) and created 80/20 train–test splits via stratified sampling to preserve class balance. BMSA-ViT was trained for 100 epochs with AdamW (lr = 1e-4, batch size = 16, weight decay = 1e-4) and label smoothing (0.1). Performance at a 0.5 decision threshold was reported using OA, precision, recall/sensitivity, specificity, F1, AUC, and Cohen's κ ($\kappa = 1$ perfect agreement; $\kappa = 0$ chance).

3.4 Implementation Details

Trained BMSA-ViT with cross-entropy using AdamW (learning rate 1e-4, weight decay 1e-4) for 100 epochs with a batch size of 16. Images were partitioned into 4 $\times$ 4 patches, embedded as 128-dimensional tokens, and processed by four Transformer blocks with four attention heads each. A final SoftMax layer produced two-class probabilities at inference. All experiments ran on Linux using Keras 2.15.0 with a TensorFlow 2.15.1 backend, on an Intel Core i7–8700 (3.20 GHz) with 32 GB RAM and an NVIDIA GeForce RTX 4090 (24 GB).

4 Model Performance Analysis

BMSA-ViT was evaluated separately on BreaKHis and IDC using a 4 × 4 patch configuration (Table 1). It achieved 98.5% accuracy on BreaKHis and 91.0% on IDC, with strong agreement beyond chance (Cohen's $\kappa = 0.89$ and 0.80, respectively). Precision and recall were well balanced on both datasets, indicating effective control of false positives and false negatives essential for clinical applicability.

Table 1. Performance metrics of the proposed BMEA-ViT using Break-his and IDC dataset

Patch (4 × 4)	Break-His	IDC
AO	98.5	91
Params	937865	554882
Kappa	89	79.7
Recall	97.5	89.5
F1 Score	97	89.5

4.1 Comparative Analysis with Baseline Models

Figure 4 compares our BMSA-ViT model to state-of-the-art methods on the BreakHis and IDC datasets. According to all evaluation metrics, our model consistently outperformed all baseline approaches.

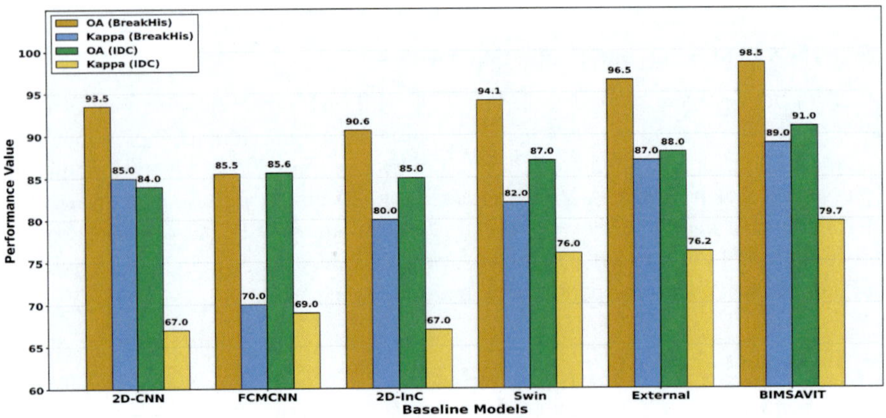

Fig. 4. Comparison of OA and Kappa values across models on both datasets.

4.2 Confusion Matrix Analysis

Figure 5 presents the confusion matrices for BMSA-ViT. On BreaKHis (Fig. 5a), the model correctly identified all malignant cases (TP = 200, FN = 0) and 194/200 benign cases (FP = 6), yielding FNR = 0% and FPR = 3%, which is clinically important because it avoids missed malignancies. On IDC (Fig. 5b), performance was TPR = 0.84 and TNR = 0.90 (FPR = 0.16, FNR = 0.10). These results indicate robust, class-balanced performance under differing class distributions; the zero FNR on BreaKHis likely reflects multi-head self-attention capturing multiple diagnostically relevant regions.

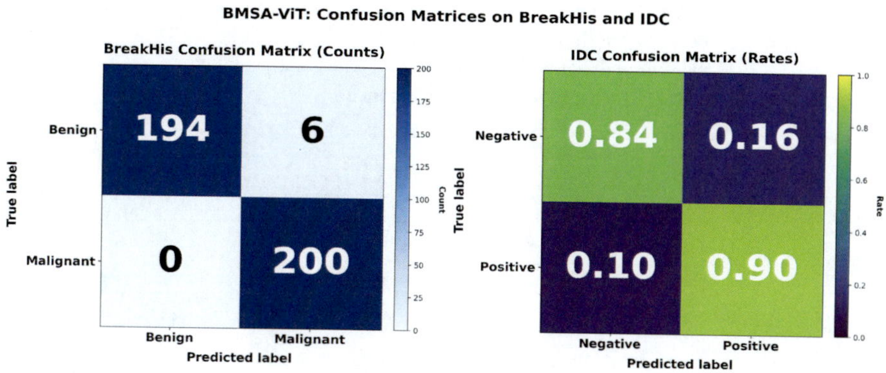

Fig. 5. Confusion matrices for BMSA-ViT on (a) BreakHis and (b) IDC datasets

4.3 Computational Efficiency

BMSA-ViT is purposely compact, consisting of four transformer blocks and four head MHSAs. Compared to 2D-InceptionV3 (3.8 GFLOPs), this algorithm requires only 0.03 GFLOPs and has a comparable parameter budget and superior accuracy (Table 2). Because the model has a shallow depth and a focused multi-head approach, redundant computations can be avoided, and training can be completed in as little as 3.5 h, it makes it suitable for clinical settings with limited resources.

Table 2. Computational comparison of different methods

Method	Parameters (×10^3)	FLOPS (G)	Training Time (hrs)
2D-CNN	22 176	12.4	8.5
FCHCNN	36 822	18.6	10.2
2D-Inception-V3	750	3.8	5.3
Swin Transformer	817	4.2	4.8
External Transformer	1209	5.8	5.0
BMSA-ViT (Ours)	**938**	**0.03**	**3.5**

4.4 Ablation Study

This patch size and split matter 4 × 4 yields the highest accuracy (98.5% BreakHis, 91.0% IDC) by balancing local detail and context. **2 × 2** raises parameters + **67%** for negligible gain (**+0.2%**), while **8 × 8/16 × 16** reduce OA by **up to 3.8%** from lost fine detail (Table 3). The performance of 90/10 with data splits is only 0.2% better on BreakHis, but 80/20 provides stronger validation with no meaningful drop in performance; performance declines as training data shrinks (e.g., 60/40: 1.9% BreakHis, 2.5% IDC). The experiments were conducted using 4x4 patches and an 80/20 split.

Table 3. Ablation of BMSA-ViT: **4 × 4** patches perform best; an **80/20** train–test split offers the best accuracy–validation balance, while larger patches and smaller training fractions degrade performance

Setting	BreaKHis		IDC	
	OA (%)	F1 (%)	OA (%)	F1 (%)
Train/Test Split				
90/10	98.7	97.2	91.5	89.5
80/20	**98.5**	**97.0**	**91.0**	**89.0**
70/30	97.9	96.5	90.2	88.2
60/40	96.8	95.5	89.0	87.0
Patch Size				
2 × 2	97.8	96.5	90.2	88.5
4 × 4	**98.5**	**97.0**	**91.0**	**89.0**
8 × 8	97.2	95.8	89.5	87.8
16 × 16	95.6	94.0	87.2	86.0

5 Conclusion and Future Directions

A compact Vision Transformer for histopathology, BMSA-ViT, combines multi-head self-attention with a shallow architecture to provide state-of-the-art accuracy at a minimal cost. In BreaKHis, the model achieves 98.5% accuracy, and in IDC, it achieves 91.0% accuracy with balanced precision–recall, requiring only 0.03 GFLOPs and 938K parameters; confusion matrix analysis indicates FN = 0 on BreaKHis and TPR/TNR = 0.84/0.90 on IDC, which is suitable for detecting malignancies within resource-constrained environments. Only two public datasets were evaluated, binary classification was the focus, interpretability was limited, and no multi-modal evidence was integrated. This work will focus on cross-population validation for generalization across populations, class subtyping, attention-based interpretation with pathologist review, and the fusion of clinical/genomic data and imaging features, as well as transferring learning to other cancers and imaging methods. BMSA-ViT will advance toward becoming a reliable, clinically deployable assistant, which will improve diagnostic accuracy while reducing the workload of pathologists.

References

1. Bray, F., Ferlay, J., Soerjomataram, I., Siegel, R.L., Torre, L.A., Jemal, A.: Global cancer statistics 2018: GLOBOCAN estimates of incidence and mortality worldwide for 36 cancers in 185 countries. CA Cancer J. Clin. **68**, 394–424 (2018)
2. Asif, S., Wenhui, Y., ur-Rehman, S.-, ul-ain, Q.-, Amjad, K., Yueyang, Y., Jinhai, S., Awais, M.: Advancements and prospects of machine learning in medical diagnostics: unveiling the future of diagnostic precision. Arch. Comput. Methods Eng. **32**, 853–883 (2025)
3. Diwaker, Kriti, Rawat, J.: Assessing the effect of pre-processing techniques on classification of breast cancer using histopathological images. ACCESS 2023 - 2023 3rd Int. Conf. Adv. Comput. Commun. Embed. Secur. Syst. 101–105 (2023). https://doi.org/10.1109/ACCESS57397.2023.10201037
4. Sahu, A., Das, P.K., Meher, S.: An efficient deep learning scheme to detect breast cancer using mammogram and ultrasound breast images. Biomed. Signal Process. Control **87**, 105377 (2024). https://doi.org/10.1016/j.bspc.2023.105377
5. Gheflati, B., Rivaz, H.: Vision transformers for classification of breast ultrasound images. Proc. Annu. Int. Conf. IEEE Eng. Med. Biol. Soc. EMBS. 2022-July, 480–483 (2022). https://doi.org/10.1109/EMBC48229.2022.9871809
6. Liu, M., Liu, Y., Xu, P., Cui, H., Ke, J., Ma, J.: Exploiting geometric features via hierarchical graph pyramid transformer for cancer diagnosis using histopathological images. IEEE Trans. Med. Imaging **43**, 2888–2900 (2024)
7. Maalej, R., Mezghani, A.: Transfer learning and data augmentation for improved breast cancer histopathological images classifier. Int. J. Comput. Inf. Syst. Ind. Manag. Appl. **15**, 10 (2023)
8. Al-Jabbar, M., Alshahrani, M., Senan, E.M., Ahmed, I.A.: Multi-method diagnosis of histopathological images for early detection of breast cancer based on hybrid and deep learning. Mathematics **11**, 1429 (2023)
9. Lowe, D.G.: Object recognition from local scale-invariant features. In: Proceedings of the Seventh IEEE International Conference on Computer Vision, pp. 1150–1157. IEEE (1999)
10. Bay, H., Tuytelaars, T., Van Gool, L.: Surf: Speeded up robust features. In: European Conference on Computer Vision, pp. 404–417. Springer (2006)
11. LeCun, Y., Bengio, Y., Hinton, G.: Deep learning. Nature **521**, 436–444 (2015)
12. Kothuru, R.T.: CNN in medical imaging: advantages, challenges. Neural Netw. Adv. Age AI. 207 (2025)
13. Spanhol, F.A., Oliveira, L.S., Petitjean, C., Heutte, L.: A dataset for breast cancer histopathological image classification. IEEE Trans. Biomed. Eng. **63**, 1455–1462 (2016). https://doi.org/10.1109/TBME.2015.2496264
14. Bharati, S., Podder, P., Mondal, M.: Artificial neural network based breast cancer screening: a comprehensive review. arXiv Prepr. arXiv2006.01767 (2020)
15. Appari, N.V.L., Kanojia, M.G.: Soft computing and image processing techniques for COVID-19 prediction in lung CT scan images. Int. J. Hybrid Intell. Syst. **18**, 111–131 (2022)

Adenoid Hypertrophy Diagnosis Based on the YOLOv11 Segmentation Model

Yueqiang Feng[1], Siyu Chen[2], Chunrong Zhao[3], Chang Duan[1(✉)], Ke Zhang[2], Sen Yang[3], and Donghai Wang[3]

[1] Southwest Petroleum University, Chengdu, Sichuan, China
pertinax@163.com

[2] University of Electronic Science and Technology of China, Chengdu, Sichuan, China
kezhang@uestc.edu.cn

[3] Suining Central Hospital, Suining, Sichuan, China

Abstract. Pediatric Adenoid Hypertrophy (AH) is a prevalent public health issue in the field of pediatric otolaryngology, and the key to formulating treatment plans lies in accurate diagnosis and grading. Existing clinical methods have limitations such as strong subjectivity, insufficient quantification, and low efficiency. Deep learning-based segmentation techniques also face bottlenecks, including poor adaptability to anatomical variations and inadequate connection with clinical grading. In this study we propose a two-stage automatic analysis system based on YOLOv11 segmentation, integrating functions of image rotation correction, target segmentation, and grading evaluation. First, shooting angle deviations were eliminated via block-based centroid extraction and rotation transformation; second, the Adenoid-Nasopharyngeal Ratio (ANR) was calculated based on segmentation masks; finally, clinical percentage-based four-degree grading and surgical recommendation generation were realized. Experimental results showed that the Dice Similarity Coefficients (DSCs) of the system for adenoid and nasopharynx segmentation both reached 0.97. After rotation correction, the Mean Absolute Error (MAE) of ANR calculation decreased to 7.23%, the Mean Relative Error (MRE) was 4.85%, and the grading accuracy increased to 83.9% compared with physicians' manual assessments. This system overcomes the shortcomings of traditional diagnostic methods, improves diagnostic accuracy and consistency, and provides reliable intelligent support for clinical decision-making in adenoid hypertrophy.

Keywords: Adenoid Hypertrophy · Medical Image Segmentation · YOLOv11 · Rotation Correction · Adenoid-Nasopharyngeal Ratio · Surgical Decision Assistance

1 Introduction

The adenoids, also known as pharyngeal tonsils, are lymphoid tissues located in the roof and posterior wall of the nasopharynx. As the first immune defense barrier of the upper airway, they play a crucial role in children's immune defense [1]. Adenoid Hypertrophy (AH) refers to the pathological hyperplasia of adenoid tissues, which is a common

L. Huang and D. Greenhalgh (Eds.): ICMLC 2026, LNNS 2061, pp. 318–330, 2027.
https://doi.org/10.1007/978-3-032-31236-5_22

disease in children. Previous studies have shown that its prevalence ranges from 9.9% to 29.2% in children, with a higher prevalence in boys than in girls [2]. With changes in environmental factors and lifestyles, the prevalence of AH in Chinese children has shown a significant upward trend, making it a major public health concern in the field of pediatric otolaryngology. AH can not only cause recurrent sinusitis, secretory otitis media, and bronchitis but also lead to "adenoid facies" due to long-term airway obstruction. In severe cases, prolonged hypoxia may even result in complications such as corpulmonale and mental retardation [3]. Therefore, early accurate diagnosis and severity grading are of vital clinical significance for formulating individualized treatment plans. Currently, clinical diagnosis faces three major challenges: First, traditional imaging diagnostic methods have inherent subjectivity, leading to poor consistency in diagnostic results among physicians with different levels of clinical experience. Second, the adenoids have ill-defined boundaries with surrounding tissues, and there are significant individual differences in the morphology of hypertrophy [4], which poses challenges to accurate quantification. Third, existing diagnostic methods are inefficient, making it difficult to meet the demands of high-load clinical work [5]. These factors have limited the improvement of diagnostic accuracy and efficiency for AH

Currently, the commonly used methods for detecting and evaluating adenoid hypertrophy in clinical practice mainly include lateral nasopharyngeal radiography, Computed Tomography (CT), and nasal endoscopy, all of which have certain limitations and subjectivity.

Lateral nasopharyngeal radiography assesses the degree of hypertrophy by measuring the ANR [6]. Despite its advantages of simplicity and cost-effectiveness, its accuracy is significantly affected by overlapping cranial structures and the child's positional compliance [7]. As a two-dimensional planar image, the lateral nasopharyngeal radiograph can only display the sagittal thickness of the adenoid, and cannot reflect the morphology and distribution of the adenoid in the axial and coronal planes. Studies have shown that there are significant differences in measurement results among physicians with different levels of clinical experience: taking the diagnostic results of chief physicians as the gold standard, the diagnostic accuracy of attending physicians in normal, moderate hypertrophy, and pathological hypertrophy groups is lower than that of chief physicians, while the accuracy of residents is further reduced. Such inconsistency in diagnostic results directly affects the formulation of treatment plans [8].

CT examination can accurately display the anatomical relationships between the adenoids and surrounding tissues, as well as the three-dimensional structure of the nasopharynx, through its thin-slice scanning and three-dimensional reconstruction capabilities [9], thus enabling precise quantification of the degree of airway obstruction. However, its radiation dose is significantly higher than that of radiography, which may pose potential risks to children in the growth and development stage. Additionally, its relatively high inspection cost makes it unsuitable for routine screening and rapid assessment.

As the "gold standard" for diagnosis [10], nasal endoscopy can directly visualize the anatomical relationships between the adenoids and surrounding structures but fails to provide precise quantitative indicators. Meanwhile, in the process of analyzing nasal endoscopy images, clinicians need to manually outline the adenoid and nasopharyngeal regions on the images, a process that is both time-consuming and prone to errors, leading to misdiagnosis [11].Deep learning techniques have dominated the field of medical image segmentation. Particularly in adenoid segmentation tasks, convolutional neural network models represented by U-Net and its variants—endowed with end-to-end feature learning and accurate pixel-level localization capabilities [12]—have significantly outperformed traditional methods relying on manually set thresholds (e.g., thresholding, morphological operations) in terms of segmentation performance under complex anatomical contexts. However, there remain key bottlenecks for the translation of this technology into routine clinical applications. First, the robustness of existing models faces challenges: their adaptability to image structural asymmetry and grayscale distribution changes caused by common anatomical variations such as deviated nasal septum is insufficient [13], leading to potential deviations in segmentation boundaries. Second, the functionality of most current studies is limited to tissue contour delineation, failing to achieve seamless integration with clinical grading criteria for adenoid hypertrophy (e.g., the ANR method). The lack of automatic conversion from segmentation results to quantitative evaluation indicators restricts their direct clinical utility. Finally, some models are overly complex and equipped with an excessive number of parameters in pursuit of ultra-high precision, resulting in slow inference speed [14], which makes it difficult to meet the demand for real-time or near-real-time processing in actual clinical workflows. Therefore, the future research focus should shift from simply improving segmentation metrics to constructing clinically oriented solutions [15]. Efforts should be directed toward enhancing the model's generalization ability to anatomical variations, developing integrated systems that combine segmentation with intelligent evaluation, thereby effectively advancing the translational value of deep learning in clinical diagnosis.

Our contributions are summarized as follows:

(1) Constructed a nasal adenoid dataset from Suining Central Hospital which located in Sichuan Province of China, including 800 pieces of clinical images of children.
(2) Proposed a rotation correction and evaluation pipeline based on YOLOv11 [16] segmentation. Taking two-stage YOLOv11 segmentation as the core, it addresses segmentation deviations caused by anatomical tilt via image rotation correction.
(3) Achieved bounding box ratio-based grading and surgical recommendation generation. ANR (calculated via adenoid/nasopharynx bounding box heights) maps to the clinical percentage-based grading system to generate diagnosis and treatment recommendations, forming a "quantification-grading-decision-making" closed loop for grade classification and surgical suggestions.

2 Related Work

In recent years, with the in-depth application of artificial intelligence technology in adenoid clinical images analysis, significant progress has been made in the research on automated diagnosis and evaluation of adenoid hypertrophy. Most studies center on automated ANR calculation [17], but diverge in technical routes and implementation.

A major research branch relies on key point detection to automate ANR calculation, a typical integration of deep learning and medical image analysis. Zhao et al. developed the HeadNet model for lateral cephalometric radiographs [18], which uses heatmap regression to localize four anatomical landmarks—automating the traditionally manual landmarking process. Based on these landmarks, the model calculates adenoid thickness and nasopharyngeal width to derive ANR. However, this method has inherent limitations: ANR accuracy depends entirely on the precision of isolated pixel localization, with slight key point deviations directly distorting results; it also lacks geometric correction, leading to poor spatial consistency across different patient postures, and fails to link ANR to clinical grading criteria. He et al. proposed the ADNet model for MRI images [19], optimizing key point localization with deformable convolution to reduce errors caused by soft tissue ambiguity. Despite this technical improvement, the model retains the key point detection paradigm, ignoring the overall geometric properties of medical images. As a result, it still suffers from angle-dependent ANR inconsistencies and does not establish a connection between quantitative ANR results and clinical grading. Another key research direction bypasses ANR calculation and directly grades adenoid hypertrophy using deep learning on endoscopic images. Wang et al. proposed an Xception-based Convolutional Neural Network (CNN) [20], trained on a large dataset of 26,060 endoscopic images to classify hypertrophy based on choanae obstruction percentage. The model achieved an overall accuracy of 95.53%, effectively avoiding subjective bias from manual measurement. However, its limitations are notable: it lacks an interpretable quantitative basis rooted in ANR, making it difficult to explain grading decisions from anatomical dimensions; it also relies heavily on large-scale, high-quality annotated datasets, limiting generalization across different endoscope brands or imaging parameters; additionally, it fails to form a closed loop between quantitative analysis and clinical decision-making. A small subset of studies explores CT/CBCT-based evaluation methods. For example, Wang et al. proposed a CBCT-based approach [21] that uses 3D reconstruction to obtain nasopharyngeal cavity area, establishing a regression relationship with traditional ANR to enrich quantitative dimensions. Pachêco-Pereira et al. utilized CBCT and Dolphin Imaging software [22] to segment adenoid volume and nasopharyngeal Minimal Cross-sectional Area (MCA) for grading against nasopharyngoscopy standards. While CBCT eliminates 2D image superimposition artifacts, these methods are highly dependent on specialized equipment and complex manual operations, making full automation challenging; they also show weak correlation with clinical diagnostic standards, limiting practical application.

In summary, existing AI-based adenoid evaluation methods have key flaws, lacking effective connection between quantification and clinical grading, which hinders wide clinical application. To address these problems, this study proposes a new ANR calculation paradigm based on semantic segmentation [23]. First, image rotation correction creates a standardized spatial coordinate system. Second, a two-stage YOLOv11 network generates precise adenoid and nasopharynx masks. Finally, ANR is calculated using mask geometric features. This avoids key point fragility, ensures spatial consistency through geometric correction, and automatically maps ANR to clinical four-degree percentage-based grading—realizing a complete closed loop from ANR quantification to clinical decision-making.

3 Methods

A novel automated analysis system for adenoid medical images based on a two-stage deep learning framework [24] is proposed. The system integrates semantic segmentation and geometric correction techniques, achieving precise localization and quantitative evaluation of anatomical structures through a serialized processing pipeline, as illustrated in Fig. 1. The image rotation correction module is a core innovative component of the proposed system. Its purpose is to eliminate the tilt of anatomical structures caused by shooting angles and establish a standardized spatial coordinate system for quantitative analysis. Based on the nasal septum segmentation results, the correction pipeline is implemented through four steps: block-based centroid extraction, interpolation completion, linear fitting, and rotation transformation. The details are as follows.

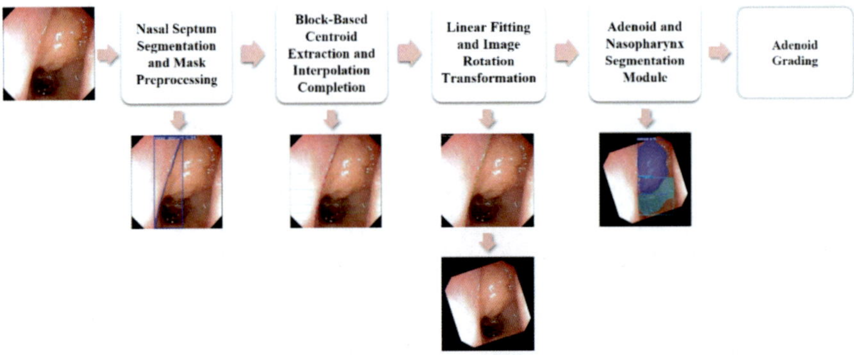

Fig. 1. Flowchart of the automated analysis system for adenoid medical images

3.1 Nasal Septum Segmentation and Mask Preprocessing

This study employed a deep learning model based on the YOLOv11 architecture for automatic medical image segmentation. The model enables an end-to-end workflow, directly mapping input images to segmentation results without manual intermediate processing. Through training, it learns feature representations of nasal cavity anatomical structures. Notably, the model maintains efficient processing of high-resolution medical images, thanks to YOLOv11's significant performance breakthroughs. Key upgrades involve enhanced feature extraction, optimized detection head design, dynamic matching, and an improved loss function. Specifically, YOLOv11's backbone network integrates multi-branch C3k2/C2f modules, C2PSA attention modules, and SPPF structures to strengthen feature expression. Its detection head adopts a classification-regression decoupled design, uses depthwise separable convolutions to streamline the structure, removes the Objectness ranch, and adopts the Anchor-Free paradigm. Additionally, it employs the TaskAlignedAssigner dynamic matching strategy and incorporates Distribution Focal Loss (DFL) and Complete Intersection over Union (CIoU) into the loss function, effectively enhancing bounding box regression accuracy.

During the first segmentation stage, raw nasal cavity medical images were used as input to the model. Multi-scale feature maps were extracted via convolutional neural networks, and candidate regions were generated using the anchor mechanism. The segmentation head outputs semantic masks for each detected target, among which the nasal septum mask is used as a key output for subsequent processing. During model inference, a confidence threshold of 0.45 and an Intersection over Union (IoU) threshold of 0.6 were set to balance detection speed and accuracy.

Refined processing was performed on the output nasal septum mask: first, the mask was resized to the original image size and converted into a binary mask via threshold segmentation (with the threshold set to 0.5); subsequently, morphological opening and closing operations were adopted to eliminate small-area noise, and connected regions with an area less than 50 pixels were filtered out to enhance the connectivity and boundary smoothness of the mask.

3.2 Block-Based Centroid Extraction and Interpolation Completion

The centerline extraction algorithm adopts a block-based centroid statistics strategy, which divides the image into 12 uniform intervals along the vertical direction. The height of each interval is calculated as follows:

$$S_h = \frac{h_{\text{ori}}}{12} \tag{1}$$

where h_{ori} denotes the height of the original image. Boundary adjustment was performed on the last interval to ensure it covers the remaining area at the bottom of the image, avoiding the loss of valid information.

For the nasal septum mask within each interval, the centroid coordinates of valid pixels (with a mask value of 255) were calculated using the following formula:

$$c_y = \frac{1}{N}\sum_{(x,y)\in S} y + y_{start},\ c_x = \frac{1}{N}\sum_{(x,y)\in S} x \tag{2}$$

Where S denotes the set of pixels with a mask value of 255 within the block, N is the number of pixels in set S, and y_{start} represents the starting vertical coordinate of the block. When the number of valid pixels in an interval is less than 5, the centroid point is determined to be invalid and marked as *None*.

To address the sequence interruption caused by invalid centroid points, linear interpolation was adopted for completion, and the formula is as follows:

$$x_i = x_{\text{prev}} + \frac{i - i_{\text{prev}}}{i_{\text{next}} - i_{\text{prev}}} \cdot \left(x_{\text{next}} - x_{\text{prev}}\right) \tag{3}$$

Where i_{prev} and i_{next} are the indices of the nearest valid points before and after the missing point, x_{prev} and x_{next} are the horizontal coordinates of the corresponding valid points, and x_i is the interpolation result of the missing point. The completed centroid point sequence forms a complete distribution of the nasal septum centerline.

3.3 Linear Fitting and Image Rotation Transformation

The least squares method was adopted to perform linear fitting on the completed centroid point sequence, solving for the optimal linear model parameters. The centroid coordinates (c_y, c_x) were converted into a 2D point set in the format of (x, y), and the direction vector (v_x, v_y) of the line and a point (x_0, y_0) on the line were obtained using the cv2.fitLine function.

To determine the image tilt angle, the angle between the fitted line and the vertical direction was calculated. Taking the unit vector in the vertical direction $\overrightarrow{v}_2 = (0, 1)$ as the reference, the angle between the two vectors was solved via vector dot product, and the formula is as follows:

$$\cos\theta = \frac{\overrightarrow{v}_1 \cdot \overrightarrow{v}_2}{|\overrightarrow{v}_1||\overrightarrow{v}_2|} \tag{4}$$

$$\theta_{\text{rad}} = \arccos(\cos\theta) \tag{5}$$

$$\theta_{\text{deg}} = \theta_{\text{rad}} \times \frac{180^\circ}{\pi} \tag{6}$$

Where $\overrightarrow{v}_1 = \left(v_x, v_y\right)$ denotes the direction vector of the fitted line. The rotation direction was determined according to the sign of v_x: when $v_x > 0$, the angle was taken as negative; when $v_x < 0$, the angle was taken as positive. Finally, the angle was adjusted to the range of $[-45^\circ, 45^\circ]$ to avoid image distortion caused by excessive rotation. Taking the image center (c_x, c_y) as the rotation origin, a rotation matrix was constructed based on the calculated tilt angle θ_{deg},and the formula is as follows:

$$M = \begin{bmatrix} \cos\theta & -\sin\theta & (1-\cos\theta)\cdot c_x + \sin\theta\cdot c_y \\ \sin\theta & \cos\theta & -\sin\theta\cdot c_x + (1-\cos\theta)\cdot c_y \end{bmatrix} \tag{7}$$

where θ denotes the rotation angle (in radian measure). To avoid cropping of valid regions after rotation, the new size of the rotated image was calculated:

$$w_{\text{new}} = w\cdot|\cos\theta| + h\cdot|\sin\theta| \tag{8}$$

$$h_{\text{new}} = h\cdot|\cos\theta| + w\cdot|\sin\theta| \tag{9}$$

where w and h represent the width and height of the original image, respectively. The translation parameters of the rotation matrix were adjusted to center the rotated image, and bilinear interpolation was adopted for image resampling to ensure the clarity of the corrected image.

3.4 Adenoid and Nasopharynx Segmentation Module

Secondary segmentation was performed on the standardized images after rotation correction, aiming to accurately extract the semantic masks of adenoids and nasopharynx. The YOLOv11 segmentation model was reused in this stage, and model parameters were optimized using the training set to adapt to the structural characteristics of the corrected images.

In the segmentation pipeline, the model outputs binary masks and bounding box coordinates of adenoids and nasopharynx. Post-processing was applied to the masks: morphological closing operations were used to fill internal holes in the masks, and a contour extraction algorithm was utilized to optimize tissue boundaries, ensuring consistency between the segmented regions and actual anatomical structures. The bounding box coordinates were used for subsequent height measurement, providing basic data for quantitative evaluation. Tests showed that the DSCs of this module for adenoid and nasopharynx segmentation reached 0.97 and 0.97, respectively, indicating high segmentation accuracy.

3.5 Adenoid Grading

The quantitative evaluation module computes key anatomical indicators from segmentation results, enabling automatic assessment and surgical recommendation generation in line with clinical grading standards. This workflow is visually illustrated in the Fig. 2: First, nasopharyngeal region images (containing adenoid visual data) are captured along the specified shooting direction to obtain anatomical visualization information. Next, boundary features of target structures are extracted from segmentation outputs; finally, the vertical height parameters of adenoids and nasopharynx are calculated via the proportional solving relationship. Specifically, each structure's vertical height is derived from the difference between the maximum and minimum y-axis values of its segmented bounding box. The ANR serves as the core evaluation parameter it effectively eliminates evaluation biases caused by individual size variations.

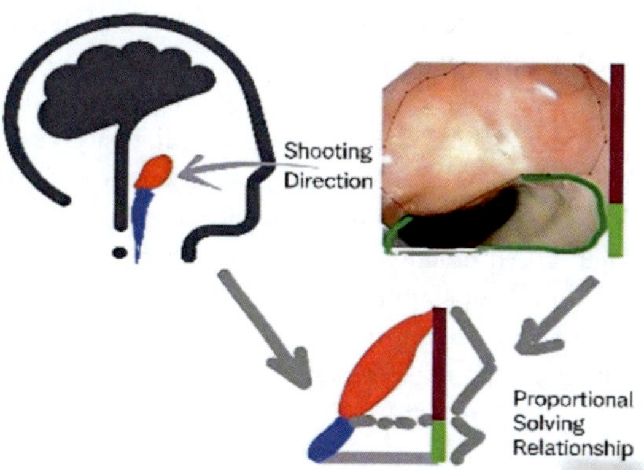

Fig. 2. ANR calculation schematic: image acquisition & proportional solving

The calculation formula is as follows:

$$ANR = \frac{H_a}{H_a + H_n} \times 100\% \tag{10}$$

The grading system strictly follows the percentage-based four-degree classification standard based on clinical consensus [25], as shown in Table 1. Grading and Symptomatology of Adenoid Hypertrophy, and automatic grading is achieved by combining the calculated *ANR*:

Table 1. Grading and symptomatology of adenoid hypertrophy

Grade	ANR	Symptom
Grade I	0–25%	Corresponding to physiological lymphoid hyperplasia, with no obvious airway obstruction symptoms
Grade II	26–50%	Indicates mild hypertrophy, which may be accompanied by intermittent nasal congestion and runny nose
Grade III	51–75%	Indicates moderate lesions, mostly accompanied by persistent nasal congestion, nocturnal snoring and other symptoms
Grade IV	76–100%	Represents severe hypertrophy, often complicated by obstructive sleep apnea syndrome and other complications

The structured report generated by the system includes quantitative data (H_a, H_n, *ANR*), grading results, providing comprehensive support for clinical decision-making.

4 Results and Discussions

4.1 Dataset and Preprocessing

The dataset was derived from clinical endoscopic images of children's nasal adenoids at Suining Central Hospital. Otolaryngologists first excluded unqualified data with motion artifacts or obscured anatomical structures. Then, data augmentation was performed by adjusting brightness within 0.85–1.1 to expand sample diversity, resulting in a total of 800 valid images. For subsequent preprocessing, we used Contrast Limited Adaptive Histogram Equalization (CLAHE) to enhance local contrast and strengthen adenoid-tissue boundary distinction. All images were normalized to 640 × 640 pixels to match the YOLOv11 segmentation model input. Finally, the 800 images were split into training (640 cases), validation (80 cases) and test sets (80 cases) at an 8:1:1 ratio, for parameter learning, training optimization and performance evaluation respectively.

4.2 Evaluation Metrics

We use MAE, MRE, and A_c to measure the absolute deviation of ANR calculation results, the relative deviation of ANR calculation results, and the consistency between the model's grading results and manual diagnosis, respectively, as given by Eqs. (11, 12 and 13). MAE quantifies the absolute discrepancy in ANR outcomes, MRE assesses the relative error of these results, and A_c evaluates how well the model's grading aligns with manual diagnostic standards

$$\text{MAE} = \frac{1}{N}\sum_{i=1}^{N}|\hat{y}_i - y_i| \times 100\% \tag{11}$$

$$\text{MRE} = \frac{1}{N}\sum_{i=1}^{N}\frac{|\hat{y}_i - y_i|}{y_i} \times 100\% \tag{12}$$

$$A_c = \frac{N_c}{N} \times 100\% \tag{13}$$

where N denotes the total number of samples in the test set, $\hat{y}_i$ represents the model-predicted ANR of the i-th sample, y_i is the true ANR of the i-th sample determined based on manual diagnosis, and N_c refers to the number of test set samples where the model's output adenoid hypertrophy grade (Grade I-IV) fully matches the annotation by senior otolaryngologists at Suining Central Hospital.

4.3 Results and Analysis

Figure 3 covers the training-validation loss and performance curves for both detection and segmentation tasks: the loss values were relatively high in the initial stage, and the model converged rapidly within 20 epochs. The training and validation losses stabilized synchronously in a low range, with no overfitting or underfitting observed.

In terms of performance, the mean Average Precision at IoU = 0.5 (mAP 50) for the detection task quickly rose to a high level close to 1.0, and the mAP 50 for the segmentation task also reached an ideal range; although the mAP 50–95 (mean Average Precision across IoU thresholds from 0.5 to 0.95) for the segmentation task was slightly lower, this is consistent with the task difficulty characteristic of pixel-level classification.

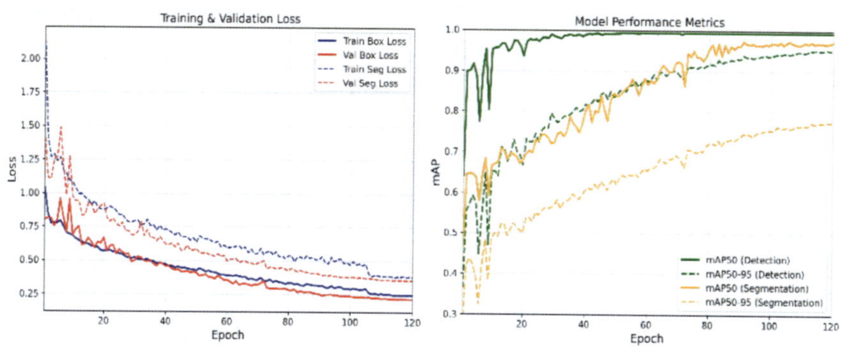

Fig. 3. Training and validation accuracy & loss

In this study, the clinical gold standard was defined as the anatomical structure segmentation results, true ANR, and adenoid hypertrophy grading results annotated and verified by senior otolaryngologists at Suining Central Hospital. The aim was to verify the improvement effect of the geometric rotation correction module on the accuracy of ANR calculation, since the ANR serves as the core quantitative basis for adenoid hypertrophy grading, its consistency with the true value directly determines the clinical reliability of grading results. Therefore, the design goal of this module is to eliminate shooting angle deviations of images, correct the measurement errors of the bounding box heights of adenoids and nasopharynx, and ultimately improve grading accuracy.

Figure 4 presents a comparison of segmentation results before and after enabling the module: the first row shows raw input images of nasopharynx adenoids, the second row displays segmentation masks without rotation correction (Group b), and the third row shows segmentation masks with rotation correction (Group c). Group b exhibits inconsistent adenoid bounding box height with true anatomical morphology due to shooting angle deviations, directly causing ANR to deviate from the ground truth. In contrast, Group c realizes geometric angle correction of input images, with more consistent adenoid and nasopharynx bounding box heights in segmentation masks, providing a reliable basis for accurate ANR calculation.

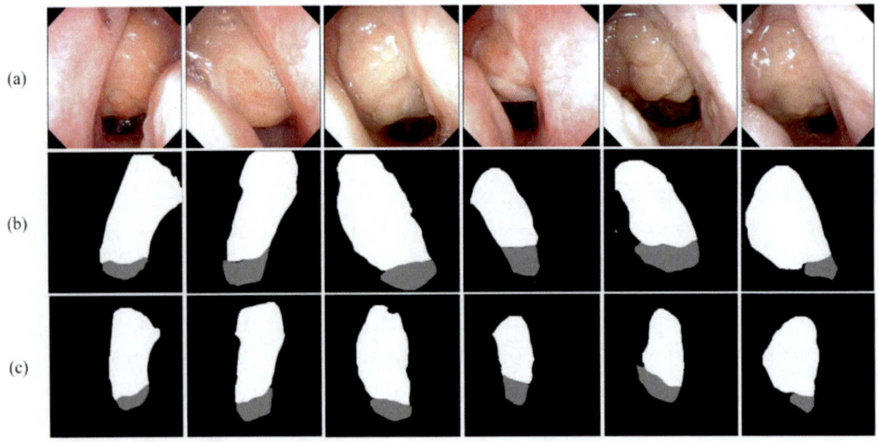

Fig. 4. Comparative visualization of the nasopharynx adenoid image input with original and corrected segmentation masks. (a) Input images (b) Adenoid segmentation masks without rotation (c) Adenoid segmentation masks with rotation

Table 2 quantifies the performance differences between Group a, Group b, and Group c. For ANR error indicators, MAE decreased from 9.15% (Group b) to 7.23% (Group c) and MRE from 6.32% to 4.85%, while ANR-based overall grading accuracy rose from 67.9% to 83.9%. These results confirm the core value of the rotation correction module: it eliminates shooting angle deviations, corrects height measurement errors of adenoid and nasopharynx bounding boxes, brings ANR closer to the true value, and significantly improves the consistency between the model's grading results and manual diagnosis. It serves as a key component to enhance the system's quantitative accuracy and clinical adaptability.

Table 2. Effect of image rotation correction on the evaluation metrics for adenoid imaging

Group	Indicator		
	A_c(%)	MAE (%)	MRE (%)
Group b	67.9	9.15	6.32
Group c	83.9	7.23	4.85

5 Conclusions

In this study, we constructed a nasopharyngeal adenoid dataset, split into training, validation, and test sets at an 8:1:1 ratio to verify the YOLOv11-based two-stage automatic analysis system. The model converged within 20 epochs. Box-Loss and Seg-Loss remained stably low with no overfitting or underfitting. Detection mAP 50 was close to 1.0, Dice similarity coefficient for adenoid and nasopharynx segmentation both reached 0.97, and the classification accuracy of the adenoid region was 100% with only a small number of misclassifications in edge regions. The rotation correction module significantly improved A/N ratio calculation performance: MAE decreased from 9.15% to 7.23%, MRE from 6.32% to 4.85%, and grading accuracy from 67.9% to 83.9%. Quantitative data, grading results, and surgical recommendations all align with clinical standards, providing an efficient diagnostic tool. Future work will further optimize segmentation and grading performance through advanced algorithms and diverse datasets.

References

1. Zhu, Y., Wang, S., Yang, Y., et al.: Adenoid lymphocyte heterogeneity in pediatric adenoid hypertrophy and obstructive sleep apnea. Front. Immunol. **14**, 1186258 (2023)
2. Yang, Y., Li, X., Ma, Q., et al.: Detecting epidemiological relevance of adenoid hypertrophy, rhinosinusitis, and allergic rhinitis through an internet search. Eur. Arch. Otorrinolaringol. **279**(3), 1349–1355 (2022)
3. Ma, Y., Xie, L., Wu, W.: The effects of adenoid hypertrophy and oral breathing on maxillofacial development: a review of the literature. J. Clin. Pediatr. Dent. **48**(1) (2024)
4. de Aquino, S.N., Verner, F.S., Cabral, R.Á., et al.: Adenoid cystic carcinoma with myoepithelial predominance affecting maxilla and maxillary sinus. J. Stomatol., Oral Maxillofac. Surg. **120**(1), 55–60 (2019)
5. Shen, Y., Li, X., Liang, X., et al.: A deep-learning-based approach for adenoid hypertrophy diagnosis. Med. Phys. **47**(5), 2171–2181 (2020)
6. Shi, C., Yuan, Z., Liu, T., et al.: X-ray sensitive selenium-containing Ru complexes sensitize nasopharyngeal carcinoma cells for radio/chemotherapy. J. Mater. Chem. B. **11**(24), 5607–5618 (2023)
7. Akçam, Ö.U.: Gonial angle measurement: panoramic radiography or lateral cephalogram? Ess.Tials Dent. **3**(1), 8–14 (2024)
8. Maw, A.R., Jeans, W.D., Fernando, D.C.J.: Inter-observer variability in the clinical and radiological assessment of adenoid size, and the correlation with adenoid volume. Clin. Otolaryngol. Allied Sci. **6**(5), 317–322 (1981)

9. Zhou, R., Feng, Y., Wang, G., et al.: TSUBF-net: trans-spatial UNet-like network with bi-direction fusion for segmentation of adenoid hypertrophy in CT. Neural Comput. & Applic. **37**(8), 5927–5943 (2025)
10. Feka-Homsy, P., Van, H.C.: Adenoid hypertrophy in children: diagnosis and management. Rev. Med.E Suisse. **21**(933), 1729–1732 (2025)
11. Gyawali, D., Mundy, T., Hosseini, M., et al.: Machine learning-enhanced clinical decision support for diagnosing sinusitis with nasal endoscopy. In: International Forum of Allergy & Rhinology (2025)
12. Zunair, H., Hamza, A.B.: Sharp U-net: depthwise convolutional network for biomedical image segmentation. Comput. Biol. Med. **136**, 104699 (2021)
13. Teixeira, R., Garib, D., Poiani, J.G.R., et al.: Nasal septum deviation after rapid maxillary expansion in the early mixed dentition. Angle Orthod. **95**(2), 166–172 (2025)
14. Pang, Y., Liang, J., Yan, J., et al.: Slim unetrv2: 3d image segmentation for resource-limited medical portable devices. IEEE Trans. Med. Imaging. (2025)
15. Li, S., Yuan, M., Dai, X., et al.: Evaluation of uncertainty estimation methods in medical image segmentation: exploring the usage of uncertainty in clinical deployment. Comput. Med. Imaging Graph. **102574** (2025)
16. Khanam, R., Hussain, M.: YOLOv11: an overview of the key architectural enhancements. arXiv preprint. https://arxiv.org/abs/2410.17725, 2024
17. Xiao, S., Zhang, X., Lu, Y., et al.: Non-invasive detection of adenoid hypertrophy using deep learning based on heart-lung sounds. IEEE J. Biomed. Health Inform. (2025)
18. Zhao, T., Zhou, J., Yan, J., et al.: Automated adenoid hypertrophy assessment with lateral cephalometry in children based on artificial intelligence. Diagnostics. **11**(8), 1386 (2021)
19. He, Z., Xiao, Y., Wu, X., et al.: An automatic assessment model of adenoid hypertrophy in MRI images based on deep convolutional neural networks. IEEE Access. **11**, 106516–106527 (2023)
20. Wang, X.S., Jia, D.S., Mo, Z.X., et al.: Deep learning for endoscopic classification of adenoid hypertrophy. World J. Otorhinolaryngol.-Head Neck Surg. (2025)
21. Wang, X., Hu, W., Liu, T., et al.: New anatomical adenoid grading method—a Nasopharyngoscopic grading system for adenoids and its clinical correlation. Laryngoscope. (2025)
22. Pachêco-Pereira, C., Alsufyani, N., Major, M., et al.: Correlation and reliability of cone-beam computed tomography nasopharyngeal volumetric and area measurements as determined by commercial software against nasopharyngoscopy-supported diagnosis of adenoid hypertrophy. Am. J. Orthod. Dentofac. Orthop. **152**(1), 92–103 (2017)
23. Cristino, R., Cruz, R.P.M., Cardoso, J.S.: Learning Ordinality in Semantic Segmentation. arXiv e-prints. (2024): arXiv: 2407.20959
24. Guo, Y., Wang, Q., Hu, B., et al.: Two-stage unsupervised hyperspectral band selection based on deep reinforcement learning[J]. Remote Sens. **17**(4) (2025)
25. Wang, Y., She, C., Wu, X., et al.: Significance of nasopharyngeal cavity area in individuals with adenoid hypertrophy assessed by cone beam computed tomography. Ear Nose Throat J., 01455613231212837 (2023)

Dual Role of Arkypallidal Neurons in Modulating Beta Oscillations and Waveform Morphology in Basal Ganglia Network: A Computational Study

Xiangyi Su[1], Xia Shi[1,2(✉)], and Bei Bai[1]

[1] School of Mathematical Sciences, Beijing University of Posts and Telecommunications, Beijing 100876, China

[2] Key Laboratory of Mathematics and Information Networks (Beijing University of Posts and Telecommunications), Ministry of Education, Beijing 100876, China

shixiabupt@163.com

Abstract. Pathological β oscillations in the basal ganglia (BG) are a hallmark of Parkinson's disease (PD), yet how distinct neuronal subpopulations dynamically shape these rhythms remains poorly understood. Using a conductance-based spiking network model of the BG, we investigate the role of arkypallidal neurons in the external globus pallidus, that is GPe-TA, which was often overlooked in the past research. We demonstrate that GPe-TA not only generates β oscillations intrinsically but also modulates both the synchronization and waveform morphology of network-wide oscillations. GPe-TA amplifies β power while simultaneously suppressing non-sinusoidal waveform distortions that emerge in dopamine-depleted states. Our findings position GPe-TA as a key regulator of pathological β rhythms and highlight the importance of cellular heterogeneity and waveform-based analysis in understanding PD.

Keywords: Basal ganglia · Parkinson's disease · beta oscillations · arkypallidal neurons · synchronization

1 Introduction

Pathological β-band oscillations (8–30 Hz) in the basal ganglia (BG) constitute a hallmark electrophysiological signature of Parkinson's disease (PD), exhibiting strong correlations with motor symptoms such as bradykinesia and rigidity [1]. Although the precise cellular and circuit-level origins of these aberrant rhythms remain under active investigation, evidence from both experimental and computational studies implicates dysregulation within the indirect pathway, particularly involving the striatum and the external segment of the globus pallidus (GPe), as a central driver of pathological synchronization [2,3]. Notably, several biophysically detailed network models have demonstrated that recurrent microcircuits composed of fast-spiking interneurons (FSIs), D2-type medium spiny

L. Huang and D. Greenhalgh (Eds.): ICMLC 2026, LNNS 2061, pp. 331–341, 2027.
https://doi.org/10.1007/978-3-032-31236-5_23

neurons (D2-MSNs), and GPe neurons are capable of sustaining β oscillations under dopamine-depleted conditions [3,4].

In recent years, a critical advance has been recognized that the GPe is not a functionally homogeneous nucleus but rather comprises at least two distinct neuronal subpopulations: prototypical (GPe-TI) and arkypallidal (GPe-TA) neurons [5,6]. These subtypes play different roles in the BG. GPe-TI neurons primarily project downstream to the subthalamic nucleus (STN) and substantia nigra pars reticulata (SNr), thereby forming the classical indirect pathway, and receive robust inhibitory input from striatal D2-MSNs. In contrast, GPe-TA neurons send dense feedback projections back to the striatum and receive minimal direct input from the striatum itself [7,8]. This finding positions GPe-TA as a potential top-down modulator of striatal activity. However, its contribution to the emergence or modulation of pathological oscillations in PD remains unexplored.

Despite growing appreciation of this cellular heterogeneity, the majority of existing computational models treat the GPe as a single, homogeneous inhibitory population or focus primarily on GPe-TI [3]. Consequently, the functional role of GPe-TA, particularly its feedback inhibition onto the striatum and its interactions with GPe-TI, has not been fully examined. However, GPe-TA neurons exhibit substantially increased firing rates under dopamine-depleted conditions [5] and demonstrate altered temporal patterning in PD models [5,9]. These observations suggest that such neurons may play an active role in shaping network-level pathology.

Another limitation of many studies on Parkinsonian oscillations is their reliance on Fourier-based spectral methods, which assume stationary and sinusoidal signals. In reality, neural rhythms are often non-stationary and exhibit non-sinusoidal waveforms [10,11]. These waveforms show asymmetries in their rising and falling phases, increased harmonic content, and irregular duty cycles, all of which can reflect underlying circuit properties [12]. Indeed, local field potentials (LFPs) recorded from the motor cortex (M1) and BG in Parkinsonian models display such non-sinusoidal features, including waveform sharpness, exaggerated cross-frequency coupling, and aberrant phase-amplitude coupling (PAC) between β phase and high-frequency amplitude [13]. Critically, the degree of waveform distortion correlates with clinical symptom severity, suggesting that non-sinusoidal signatures may reflect imbalances in excitation-inhibition ratios or altered synaptic kinetics within striatal-pallidal microcircuits.

In this work, we address these gaps by developing a conductance-based spiking network model that explicitly incorporates both GPe-TI and GPe-TA subpopulations, along with their distinct connectivity motifs involving D2-MSNs and FSIs. We first characterize the intrinsic oscillatory property of the GPe-TA subnetwork and then investigate how its bidirectional coupling with the striatum and GPe-TI influences β synchronization across the circuit. Finally, we examine how GPe-TA modulates a second, independently oscillatory loop, the FSI–D2-MSN–GPe-TI microcircuit, which exhibits enhanced non-sinusoidal dynamics under dopamine depletion. Our simulations reveal that GPe-TA does not act as a passive relay but actively attenuates the non-sinusoidal features of this striatal-pallidal loop. Collectively, these findings redefine the functional role of GPe-TA in BG, positioning it as a dynamic regulator, not just a marker, of pathological β oscillations.

2 Model and Methods

2.1 Model

In this study, we developed a biologically constrained computational model of the BG circuitry comprising six distinct neuronal populations as shown in Fig. 1: D1-receptor-expressing medium spiny neurons (D1-MSNs), D2-receptor-expressing medium spiny neurons (D2-MSNs), fast-spiking interneurons (FSIs), prototypic globus pallidus externa neurons (GPe-TI), arkypallidal (or type A) globus pallidus externa neurons (GPe-TA), and subthalamic nucleus neurons (STN). The network was constructed with the population sizes following Ref. [14]: 6,000 D1-MSNs and 6,000 D2-MSNs, 600 FSIs and 600 GPe-TI neurons, and 200 GPe-TA neurons [15] and 200 STN neurons.

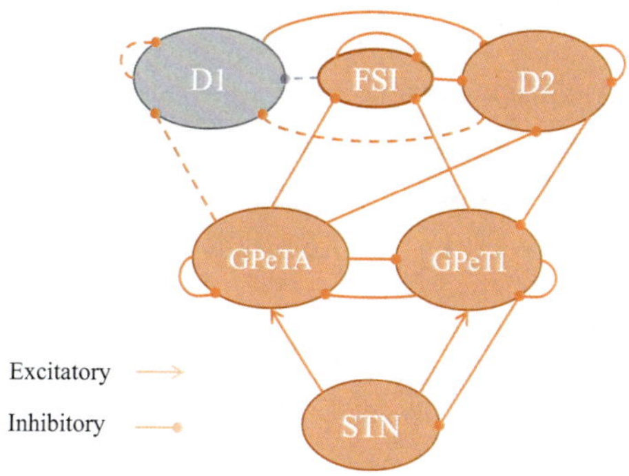

Fig. 1. Model architecture of the BG network, which includes D1-MSNs, D2-MSNs, FSIs, GPe-TA, GPe-TI, STN. Round/tapered arrowheads denote inhibitory/excitatory projections. Inputs from D1-MSNs were emulated (grey).

At the single-neuron level, FSIs and D2-MSNs were modeled using the Izhikevich spiking neuron model [16,17], while STN and both GPe subtypes (GPe-TI and GPe-TA) were implemented using the adaptive exponential integrate-and-fire (AdEx) model [18]. Synaptic interactions between neurons were mediated by conductance-based synapses with exponential decay kinetics, governed by first-order linear differential equations. The excitatory input from cortical or other external sources to all neurons was modeled as independent Poisson spike trains.

The model was simulated under three distinct physiological regimes: the healthy state, the Parkinsonian state, and the dynamic dopamine depletion process representing the transition between them. Parameters governing synaptic weights, intrinsic excitability, and dopaminergic modulation in the healthy and Parkinsonian conditions were adapted from the modeling framework established in Ref. [19]. The dopamine depletion trajectory was implemented as a smooth, continuous interpolation between the healthy and Parkinsonian parameter sets, using a sigmoidal function to emulate the gradual decline of striatal dopamine levels. Specifically, dopamine depletion was modeled by concurrently modulating both synaptic conductances across key pathways and the leak reversal potentials of D1-MSNs, FSIs, and GPe-TI neurons.

For details of the network architecture, as well as all neuronal and synaptic parameters, see Ref. [19].

2.2 Analysis Methods

To characterize the dynamical properties of network activity across different physiological states, we employed a dual-pronged analytical approach combining spectral and waveform-based analyses. Five quantitative metrics were computed to capture distinct aspects of oscillatory dynamics:

Relative Beta Power. This metric quantifies the proportion of power in the beta frequency band relative to the total power across the overall frequency range.

$$\text{Relative Power}_i = \frac{\int_8^{30} \mathrm{PSD}_i(f)\,\mathrm{d}f}{\int_0^{100} \mathrm{PSD}_i(f)\,\mathrm{d}f}. \tag{1}$$

Mean Beta Frequency. This measure captures the center of mass of the PSD within the beta band and reflects the dominant oscillatory frequency:

$$\text{Mean Frequency}_i = \frac{\int_a^b f \cdot \mathrm{PSD}_i(f)\,\mathrm{d}f}{\int_a^b \mathrm{PSD}_i(f)\,\mathrm{d}f}. \tag{2}$$

Shifts in mean beta frequency provide insight into changes in network resonance properties and are sensitive to alterations in synaptic coupling and neuromodulation.

Waveform Asymmetry. To assess the non-sinusoidal structure of beta oscillations, we computed three waveform shape metrics using the `bycycle` Python package [20], which decomposes oscillatory signals into individual cycles and quantifies their morphological properties:

- *Temporal asymmetry (TAS)* (`time_rdsym`) is the ratio of the duration of the rising phase to that of the decaying phase within each cycle. Values greater than 0.5 indicate a slower rise (longer rising phase) relative to decay.
- *Amplitude asymmetry (AAS)* (`volt_rdsym`) is defined as the absolute peak voltage divided by the absolute trough voltage (|`volt_peak`|/|`volt_trough`|). A value of 1 reflects symmetric amplitude; deviations indicate waveform skewness in voltage magnitude.

- *Sharpness* is the average of the normalized steepness of the rising and falling edges:

$$\texttt{sharp\_rise} = \frac{|\texttt{volt\_rise}|}{\texttt{time\_rise}}, \tag{3}$$

$$\texttt{sharp\_decay} = \frac{|\texttt{volt\_decay}|}{\texttt{time\_decay}}, \tag{4}$$

$$\texttt{sharpness} = \frac{\texttt{sharp\_rise} + \texttt{sharp\_decay}}{2}. \tag{5}$$

Higher sharpness reflects more abrupt voltage transitions, potentially indicating strong recurrent excitation or heightened network excitability.

Together, these five metrics provide a comprehensive characterization of both the spectral content and temporal morphology of network oscillations, enabling a nuanced comparison of healthy, Parkinsonian, and transitional dopamine-depletion states.

3 Results

3.1 GPe-TA Generates Beta Oscillations Intrinsically

Since GPe-TA is a potential top-down modulator of striatal activity, its contribution to the emergence or modulation of pathological oscillations is the main focus of this work. The characteristics of β-band oscillations in GPe-TA under the Parkinsonian condition are shown in Fig. 2.

The GPe-TA neuronal population spontaneously generates oscillatory activity in the beta frequency range (see Fig. 2A). By strengthening recurrent inhibitory coupling, the β oscillations retain a high contribution to the total spectral power (see Fig. 2B), which indicates that modulating recurrent inhibition primarily scales the amplitude of network oscillations without altering their spectral profile.

To further test the robustness of this β preference, we systematically varied three key determinants of GPe-TA excitability: (1) the mean external excitatory drive (I_e), (2) the adaptation time constant (τ_w), and (3) the spike-triggered adaptation increment (b). Notably, across physiologically plausible ranges, both relative β power and mean oscillation frequency remained largely stable (Fig. 2C–E). This insensitivity suggests that the β rhythm does not rely on finely tuned levels of excitation or specific adaptation parameters, but rather reflects a robust dynamical attractor inherent to the GPe-TA population.

It is worth noting that when τ_w exceeds the typical physiological range (e.g., > 50 ms), a gradual decline in relative β power is observed, indicating that excessively slow adaptation disrupts the resonant mechanisms supporting β-band synchronization. Nevertheless, the β band remains the dominant spectral feature of GPe-TA population activity across all tested conditions.

The results suggest that beta-band oscillations represent a robust, functionally preferred rhythm of the GPe-TA network across diverse parameter regimes, indicating that this oscillatory preference is not exclusively determined by its internal inhibitory architecture.

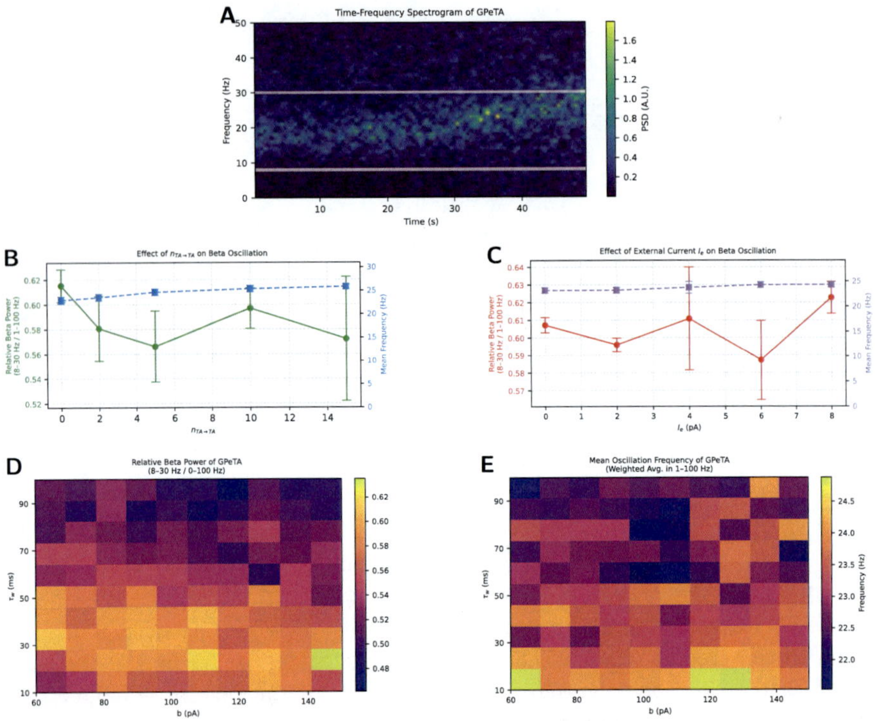

Fig. 2. Characteristics of β-band oscillations in the GPe-TA population under the Parkinsonian condition. (a) Time-frequency representation of the LFPs during progressive dopamine depletion, showing the emergence and stabilization of power in the β band (15–30 Hz). (b) Relative β power (15–30 Hz normalized to 0–100 Hz) and mean oscillation frequency as a function of recurrent inhibitory connectivity strength ($n_{\mathrm{TA}\to\mathrm{TA}}$). (c) Relative β power and mean oscillation frequency as a function of external excitatory drive (I_e). (d, e) Sensitivity of relative β power (d) and mean oscillation frequency (e) to the adaptation time constant (τ_w) and spike-triggered adaptation increment (b) of GPe-TA neurons across physiologically plausible ranges.

3.2 Enriched Synaptic Architecture Amplifies and Shapes Beta Oscillations in the GPe-TA Network

To investigate how this intrinsic β preference of GPe-TA influences large-scale beta rhythm in the basal ganglia, we progressively incorporated additional nuclei to construct multi-nucleus microcircuits: GPe-TA–GPe-TI, GPe-TA–GPe-TI–STN, and GPe-TA–GPe-TI–D2-MSN. Across all extended architectures, the relative power in the β band was consistently higher than that in the isolated GPe-TA network (Fig. 3A). This demonstrates that a more complete motif not only amplifies but also stabilizes β synchronization, highlighting the essential role of inter-nuclear interactions in the emergence of pathological β oscillations.

Further analysis revealed that the specific synaptic architecture between GPe-TA and GPe-TI exerts bidirectional control over β oscillations. Specifically, increasing the density of recurrent self-inhibition within either GPe-TA or GPe-TI significantly reduced relative β power and shifted the dominant oscillation frequency upward (Fig. 3B). In contrast, enhancing the strength of reciprocal inhibitory connections between GPe-TA and GPe-TI produced the opposite effect (Fig. 3C).

These opposing effects can be mechanistically interpreted as follows: strengthening self-inhibition shortens the effective time window for population-level synchronization by promoting rapid post-spike suppression, thereby attenuating low-frequency coherent activity and elevating the dominant frequency. Conversely, reinforcing reciprocal inhibition between GPe-TA and GPe-TI establishes a delayed negative feedback loop that supports anti-phase, alternating firing between the two populations. This finding indicates that the balance between intra-population self-inhibition and inter-population reciprocal inhibition critically shapes both the amplitude and spectral tuning of pathological β rhythms in GPe.

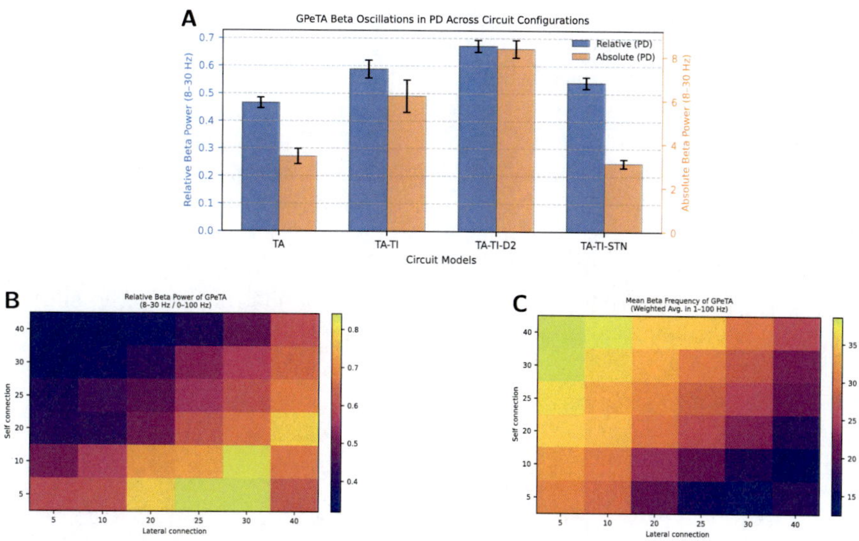

Fig. 3. Circuit-dependent characteristics of pathological β oscillations in GPe-TA. (A) Relative and absolute β-band power in the GPe-TA nucleus across distinct microcircuit configurations under the Parkinsonian conditions. (B) and (C) Modulation of GPe-TA β oscillations by varying the number of synaptic connections from GPe-TA to GPe-TI, quantified by relative β power and mean β frequency.

3.3 GPe-TA Mitigates Non-sinusoidal Distortions in Beta Oscillations Generated by the FSI–D2-MSN–GPe-TI Loop

In this model, the FSI–D2-MSN–GPe-TI inhibitory loop robustly generates prominent β-band oscillations. The LFP waveform associated with these oscillations exhibits pronounced non-sinusoidal features. Specifically, increased temporal asymmetry and waveform sharpness become more severe with progressive dopamine depletion as shown in Fig. 4. This mirrors empirical observations in the Parkinsonian condition, where pathological β activity is frequently accompanied by distorted LFP morphologies.

Crucially, structured inhibitory input from the lateral segment of GPe-TA significantly reduces these waveform distortions. Compared to a control condition in which GPe-TA input was replaced by unstructured Poisson noise matched for mean firing rate, the reintroduction of temporally coherent GPe-TA projections markedly decreases both the sharpness and temporal asymmetry of the β-band LFP (see Fig. 5).

This finding suggests that GPe-TA provides a stabilizing influence on network dynamics under intact connectivity. Mechanistically, we propose that GPe-TA-mediated inhibition suppresses transient, high-amplitude synchronization events within the BG. By delivering rhythmic, phase-structured inhibitory signals, GPe-TA curtails the emergence of nonlinear population bursts that would otherwise manifest as sharp, asymmetric LFP deflections. In contrast, when GPe-TA input is replaced by stochastic Poisson input, this regulatory function is lost, rendering the network more susceptible to hypersynchronous transients and exacerbating waveform nonlinearity. Figure 5C provides support for this mechanism: in the presence of GPe-TA input, burst events in the FSI population exhibit significantly reduced internal synchrony, with spike timing within bursts shifting from tightly clustered (under Poisson-replacement conditions) to a more temporally dispersed structure.

These results highlight the critical role of structured inhibitory feedback from GPe-TA in preserving dynamic stability and in shaping not only the spectral but also the morphological properties of pathological β oscillations in dopamine-depleted states.

4 Conclusion

In this study, we demonstrate that the GPe-TA neuronal population intrinsically generates β-band oscillatory activity. Through structured inhibitory projections to GPe-TI, STN, and D2-MSNs, GPe-TA further amplifies and stabilizes β-band power within its associated subnetwork. Notably, when rhythmic inhibitory input from GPe-TA is integrated into the FSI–D2-MSN–GPe-TI loop, which intrinsically produces pronounced non-sinusoidal β-band oscillations, the waveform distortions are significantly attenuated. This finding highlights a nonlinear regulatory capacity of inhibitory interactions within the BG.

From a dynamical perspective, the coupling between GPe-TA and the FSI–D2-MSN–GPe-TI loop can be conceptualized as the interaction between a dominant oscillator and a weaker oscillatory subsystem. The observed reduction in waveform non-sinusoidality likely arises from mutual constraints on higher-order harmonic components during synchronization, thereby steering the composite oscillation toward a more sinusoidal profile.

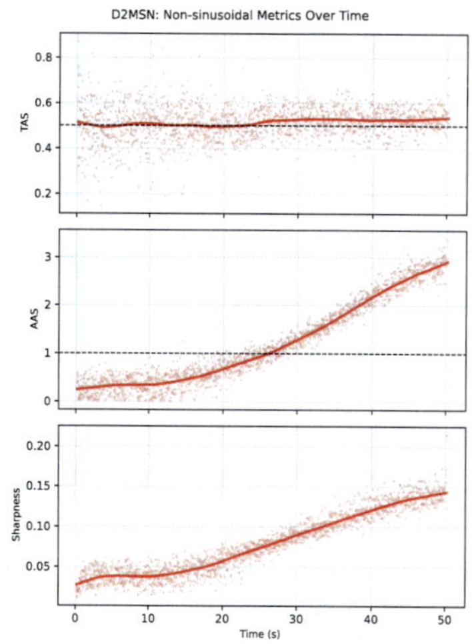

Fig. 4. Evolution of non-sinusoidal oscillatory features in D2-MSNs within the FSI–D2-MSN–GPe-TI inhibitory microcircuit during dopamine depletion. From top to bottom, the panels depict changes in (i) amplitude asymmetry (AAS); (ii) time asymmetry (TAS) and (iii) waveform sharpness. These metrics collectively reflect increasing waveform distortion in the LFP of the D2-MSNs as dopamine declines.

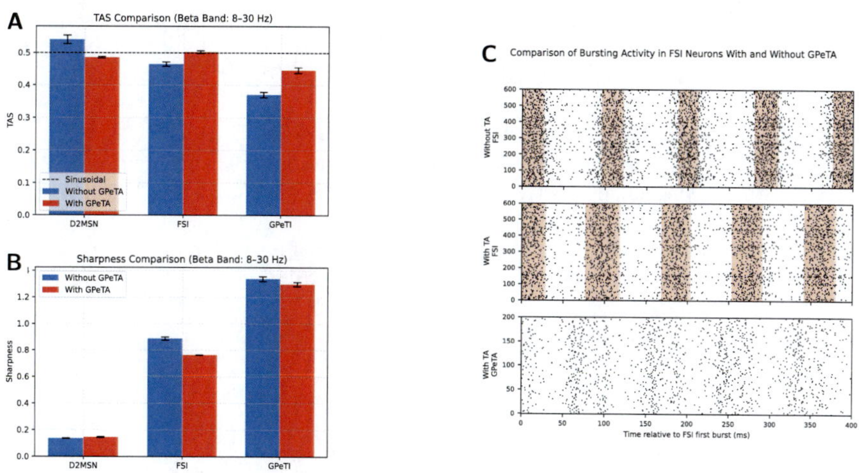

Fig. 5. Modulation of oscillatory non-sinusoidality in the FSI–D2-MSN–GPe-TI inhibitory microcircuit by the inclusion of GPe-TA feedback during the Parkinsonian condition. (**A**) Time asymmetry (TAS) is altered when GPe-TA projections are present compared to the network without GPe-TA. (**B**) Waveform sharpness also exhibits a marked shift with the addition of GPe-TA. (**C**) Raster plots showing spike timing in FSIs and GPe-TA neurons aligned to the onset of the first FSI burst. Red shaded regions denote periods of elevated FSI population activity consistent with burst events.

In summary, these results advance our mechanistic understanding of how pathological β-band oscillations emerge and are modulated in Parkinsonian states. Moreover, they provide a theoretical foundation for developing personalized deep brain stimulation (DBS) strategies that leverage phase-specific entrainment principles to restore physiological network dynamics.

Acknowledgements. This research was supported by the National Natural Science Foundation of China (Grant No. 12272062).

References

1. Jenkinson, N., Brown, P.: New insights into the relationship between dopamine, beta oscillations and motor function. Trends Neurosci. **34**, 611–618 (2011). https://doi.org/10.1016/j.tins.2011.09.003
2. Damodaran, S., Cressman, J.R., Jedrzejewski-Szmek, Z., Blackwell, K.T.: Desynchronization of fast-spiking interneurons reduces beta-band oscillations and imbalance in firing in the dopamine-depleted striatum. J. Neurosci. **35**, 1149–1159 (2015). https://doi.org/10.1523/JNEUROSCI.3490-14.2015
3. Corbit, V.L., Whalen, T.C., Zitelli, K.T., Crilly, S.Y., Rubin, J.E., Gittis, A.H.: Pallidostriatal projections promote beta oscillations in a dopamine-depleted biophysical network model. J. Neurosci. **36**, 5556–5571 (2016). https://doi.org/10.1523/JNEUROSCI.0339-16.2016
4. Ortone, A., Vergani, A.A., Ahmadipour, M., Mannella, R., Mazzoni, A.: Dopamine depletion leads to pathological synchronization of distinct basal ganglia loops in the beta band. PLoS Comput. Biol. **19**, e1010645 (2023). https://doi.org/10.1371/journal.pcbi.1010645
5. Mallet, N., Pogosyan, A., Márton, L.F., Bolam, J.P., Brown, P., Magill, P.J.: Parkinsonian beta oscillations in the external globus pallidus and their relationship with subthalamic nucleus activity. J. Neurosci. **28**, 14245–14258 (2008). https://doi.org/10.1523/jneurosci.4199-08.2008
6. Abdi, A., et al.: Prototypic and arkypallidal neurons in the dopamine-intact external globus pallidus. J. Neurosci. **35**, 6667–6688 (2015). https://doi.org/10.1523/JNEUROSCI.4662-14.2015
7. Mallet, N., et al.: Arkypallidal cells send a stop signal to striatum. Neuron **89**, 308–316 (2016). https://doi.org/10.1016/j.neuron.2015.12.017
8. Fujiyama, F., Nakano, T., Matsuda, W., Furuta, T., Udagawa, J., Kaneko, T.: A single-neuron tracing study of arkypallidal and prototypic neurons in healthy rats. Brain Struct. Funct. **221**, 4733–4740 (2015). https://doi.org/10.1007/s00429-015-1152-2
9. de la Crompe, B., Aristieta, A., Leblois, A., Elsherbiny, S., Boraud, T., Mallet, N.P.: The globus pallidus orchestrates abnormal network dynamics in a model of Parkinsonism. Nat. Commun. **11**, 1570 (2020). https://doi.org/10.1038/s41467-020-15352-3
10. van Dijk, H., van der Werf, J., Mazaheri, A., Medendorp, W.P., Jensen, O.: Modulations in oscillatory activity with amplitude asymmetry can produce cognitively relevant event-related responses. Proc. Natl. Acad. Sci. **107**, 900–905 (2009). https://doi.org/10.1073/pnas.0908821107

11. Schaworonkow, N., Nikulin, V.V.: Spatial neuronal synchronization and the waveform of oscillations: implications for EEG and MEG. PLoS Comput. Biol. **15**, e1007055 (2019). https://doi.org/10.1371/journal.pcbi.1007055
12. Hindriks, R., van Putten, M.J.A.M.: Thalamo-cortical mechanisms underlying changes in amplitude and frequency of human alpha oscillations. Neuroimage **70**, 150–163 (2013). https://doi.org/10.1016/j.neuroimage.2012.12.018
13. Cole, S.R., van der Meij, R., Peterson, E.J., de Hemptinne, C., Starr, P.A., Voytek, B.: Nonsinusoidal beta oscillations reflect cortical pathophysiology in Parkinson's disease. J. Neurosci. **37**, 4830–4840 (2017). https://doi.org/10.1523/JNEUROSCI.2208-16.2017
14. Oorschot, D.E.: Total number of neurons in the neostriatal, pallidal, subthalamic, and substantia nigral nuclei of the rat basal ganglia: a stereological study using the cavalieri and optical disector methods. J. Comp. Neurol. **366**, 580–599 (1996). https://doi.org/10.1002/(SICI)1096-9861(19960318)366:4⟨580::AID-CNE3⟩3.0.CO;2-0
15. Hegeman, D.J., Hong, E.S., Hernández, V.M., Savio Chan, C.: The external globus pallidus: progress and perspectives. Eur. J. Neurosci. **43**, 1239–1265 (2016). https://doi.org/10.1111/ejn.13196
16. Izhikevich, E.M.: Which model to use for cortical spiking neurons? IEEE Trans. Neural Netw. **15**, 1063–1070 (2004). https://doi.org/10.1109/TNN.2004.832719
17. Izhikevich, E.M.: Hybrid spiking models. Philos. Trans. R. Soc. A Math. Phys. Eng. Sci. **368**, 5061–5070 (2010). https://doi.org/10.1098/rsta.2010.0130
18. Brette, R., Gerstner, W.: Adaptive exponential integrate-and-fire model as an effective description of neuronal activity. J. Neurophysiol. **94**, 3637–3642 (2005). https://doi.org/10.1152/jn.00686.2005
19. Lindahl, M., Hellgren Kotaleski, J.: Untangling basal ganglia network dynamics and function: role of dopamine depletion and inhibition investigated in a spiking network model. eNeuro **3**, 0156 (2016). https://doi.org/10.1523/ENEURO.0156-16.2016
20. Cole, S., Voytek, B.: Cycle-by-cycle analysis of neural oscillations. J. Neurophysiol. **122**, 849–861 (2019). https://doi.org/10.1152/jn.00273.2019

Data Glove and Inception-CSRA Network for Sign Language Recognition

Wan-Cheng Wu[1], Yong-An Zhang[1(✉)], and Yu-Tao Liu[2]

[1] Harbin Institute of Technology, Harbin, Heilongjiang, China
23s004071@stu.hit.edu.cn, zya@hit.edu.cn
[2] The First Affiliated Hospital of Zhengzhou University, Zhengzhou, Henan, China

Abstract. Communication between hearing individuals and deaf people typically relies on sign language. Wearable sign language recognition (SLR) systems, especially flex sensor-based data gloves, have attracted considerable attention in the field of SLR in recent years. However, three problems remain to be addressed in existing research: (1) The portability of data gloves affects interaction effectiveness, as current data gloves have low integration levels and predominantly employ wired connections. (2) Spatial and temporal features are crucial for gesture recognition, yet existing algorithms typically cannot directly achieve effective feature extraction. (3) Most existing SLR systems focus on recognizing isolated sign words rather than continuous sentences. To address the problems, a data glove with an Inception-CSRA network is proposed. The wearable data glove uses flex sensors and an Inertial Measurement Unit (IMU) to capture hand motion signals. An Inception-CSRA algorithm is developed that simultaneously extracts spatial and temporal features from sensor signals by adjusting parallel convolutional kernels and class-specific residual attention (CSRA), effectively enhancing the accuracy of SLR. The proposed system achieves 99.5% accuracy on 30 Chinese sign language gestures and can generate sentences with voice output. The results demonstrate that the system provides an effective solution for communication involving deaf individuals.

Keywords: Sign language recognition · Gesture recognition · Data glove · Inception-CSRA

1 Introduction

The world health organization (WHO) statistics show that, over 5% of the global population, approximately 430 million individuals, are suffering from hearing loss [1]. They have difficulty in communicating normally with others [2]. Sign languages (SL), similar to vocal languages, emerge and grow spontaneously among communities of deaf people. However, the evolution of sign languages occurs apart from the linguistic traditions of each country or region where they are used [3]. This creates significant communication obstacles between deaf individuals and hearing individuals, as well as between users of different regional sign languages.

L. Huang and D. Greenhalgh (Eds.): ICMLC 2026, LNNS 2061, pp. 342–357, 2027.
https://doi.org/10.1007/978-3-032-31236-5_24

Gesture-based human-computer interaction has now emerged as a promising technology with substantial application potential, introducing new technological possibilities in smart homes, healthcare, sports training, and other related fields [4]. People with hearing disabilities generally communicate via sign language in their everyday activities, where sign language is largely dependent on hand gestures [5]. Recently, machine learning-based approaches have also been increasingly applied to the SLR systems [6]. Most of the systems interpret hand gestures by extracting their features and mapping them to corresponding meanings.

The current SLR systems adopt one of three main approaches: sensor-based [7–14], vision-based [15–19], or a hybrid of the two [20, 21]. Sensor-based systems, which utilize data gloves embedded with various sensors, offer accurate and direct measurements of finger movements and hand orientation. However, they fall short in capturing the spatial relationship between the hands and the body. Vision-based systems, on the other hand, use cameras to track hand movements and positions. While they provide valuable spatial information, their effectiveness is limited by factors such as lighting conditions, occlusion, and restricted fields of view. Moreover, the need for users to stay within the camera frame can reduce mobility and comfort during use. Hybrid systems attempt to combine the strengths of both approaches, improving overall recognition accuracy by leveraging complementary data sources. However, the requirement for both gloves and cameras increases the complexity, cost, and computational burden of the system, making it less practical for real-time applications and everyday use [5]. Most of the sign languages are depended on hand movements, so sensor-based data gloves are widely favored by researchers in SLR due to their portability and reliability.

To solve the above-mentioned problems, this paper introduces a new Chinese SLR system. The system is designed to recognize both static and dynamic sign language gestures, generate sentence, and produce voice output. Our hardware includes a data glove equipped with an IMU for capturing hand motions and orientation, 5 flex sensors for detecting finger bending, and a microcontroller for wireless data transmission to a processing unit. Sensor data is processed through Inception-CSRA neural networks to recognize corresponding gestures or generate sentences through sentence matching, then converted to audio output.

The paper is structured as follows. Section 2 surveys existing SLR approaches. In Sect. 3, we present the hardware configuration and the networks architecture. Section 4 outlines the dataset and reports performance evaluations of the models. Finally, Sect. 5 concludes with overall insights and future directions.

2 Related Works

In recent years, many researchers have proposed sensor-based SLR systems that use a data glove fitted with multiple sensors to capture precise hand and finger movements, such as finger bending, orientation, rotation, and spatial position, then transmit the data to a computer for gesture recognition.

Traditional machine learning methods require manually engineering features from raw sensor signals. Zhou et al. [7] presents a wearable sign-to-speech machine learning translation system that employs stretchable yarn-based sensor arrays to enable real-time

American SLR, the system achieves up to 98.63% accuracy. Chen et al. [8] evaluated support vector machine (SVM) using data from a customized hand rehabilitation system, achieved a top classification accuracy of 99.20% across 16 distinct gestures.

In contrast, deep learning models automatically extract relevant features directly from raw sensor signals, improving feature quality and reducing dependency on manual engineering. Convolutional neural network (CNN) and recurrent neural network (RNN) are the typical neural networks that are commonly used. Wen et al. [9] introduced an AI-powered American SLR and communication system that combines triboelectric sensing gloves with CNN. The system achieving accuracies of 91.3% for 50 words and 95% for 20 distinct sentences. In [10], the authors proposed an RNN-based American SLR system utilizing a glove equipped with six IMU sensors to classify 27 dynamic gestures, achieving a recognition accuracy of 99.81%. However, RNN often encounter the gradients diminish problem exponentially across time steps, making it challenging to capture long-term dependencies between sensor signals and gestures [11]. Long short-term memory (LSTM) is an improved RNN that remembers long-term information. Halabi and Harkouss [12] presented a wireless sensory-glove-based Arabic SLR system that used a LSTM model trained on a dynamic dataset, the system achieves a high accuracy of 99.6%. The above-mentioned approaches tend to independently extract either temporal or spatial features, rather than combine both features for joint extraction. This leads to inadequate extraction of gesture features.

Xia et al. [13] introduced a tandem LSTM-CNN architecture that has been widely employed for human activity recognition. In this model, the LSTM first captures temporal dependencies in the sequential data, and then a multilayer 2D CNN extracts spatial features from the time-frequency representations, enabling comprehensive modeling of complex human activities. Based on this idea, Wang et al. [14] introduced a wearable electronic glove paired with a novel multilayer parallel Para LSTM CNN. Temporal patterns from sensor signals are captured by LSTM while spatial features are extracted by CNN, achieving an impressive 97.01% recognition accuracy for 26 American sign language letter gestures. However, their segmentation methods for continuous gestures require further improvement, and they do not perform sign language sentence recognition.

It should be noted that the accuracy and robustness of the same classification algorithm differ across different datasets [22]. When dealing with new classification tasks, the performance of different algorithms is often unknown, and no single algorithm can consistently maintain optimal classification performance. Therefore, experts typically compare and evaluate various algorithms to determine the most suitable one for specific tasks.

3 Methodology

In this study, the proposed SLR system comprises hardware and software interconnected components. The hardware component features a wearable data glove, as illustrated in Fig. 1, which captures gesture information via integrated IMU and flex sensors. The collected motion data are transmitted wirelessly to a personal computer (PC) via Wi-Fi. Finally, SLR is achieved through an Inception-CSRA model.

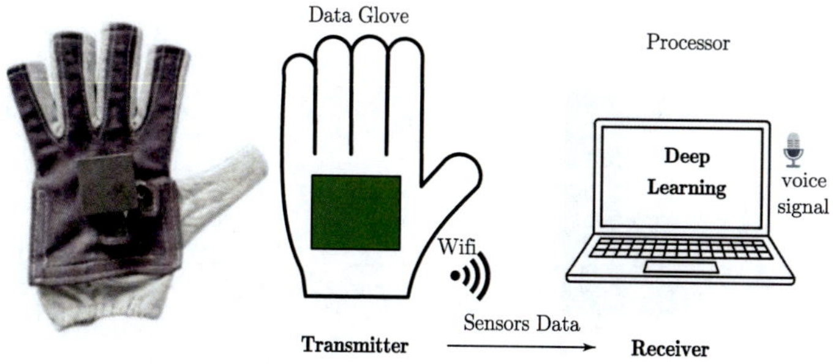

Fig. 1. Image and architecture of the proposed system

3.1 Data Glove Design

Sign language is primarily differentiated through the states and movements of fingers and palms. Consequently, we designed a data glove based on this characteristic. The design details of the wearable electronic glove are illustrated in Fig. 2.

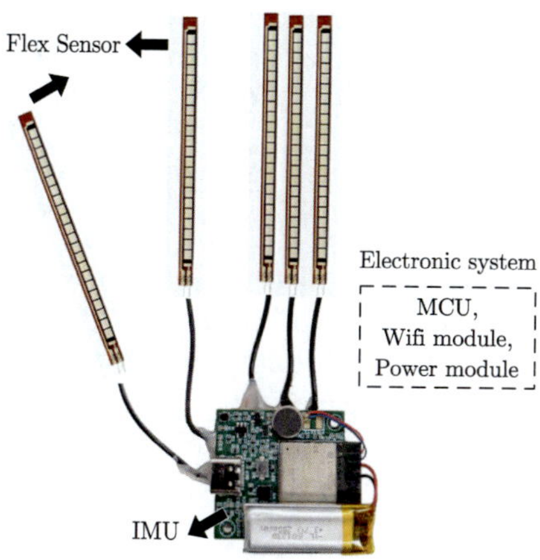

Fig. 2. Design details of the data glove

The sensor module consists of 1 IMU and 5 flex sensors. The IMU integrates an accelerometer, a gyroscope, and a magnetometer. The single IMU is positioned at the back of the hand to capture hand orientation and motion data. Five flex sensors are strategically arranged at the proximal interphalangeal joints of each finger to detect the flexion-extension motion. This configuration captures 11-dimensional hand gesture data: 6 dimensions from the IMU sensor and 5 dimensions from the flex sensors.

3.2 Data Processing

For the IMU sensor to function optimally, an automated calibration protocol is essential. Upon system initialization, the embedded firmware automatically executes bias estimation algorithms for the accelerometer, gyroscope, and magnetometer.

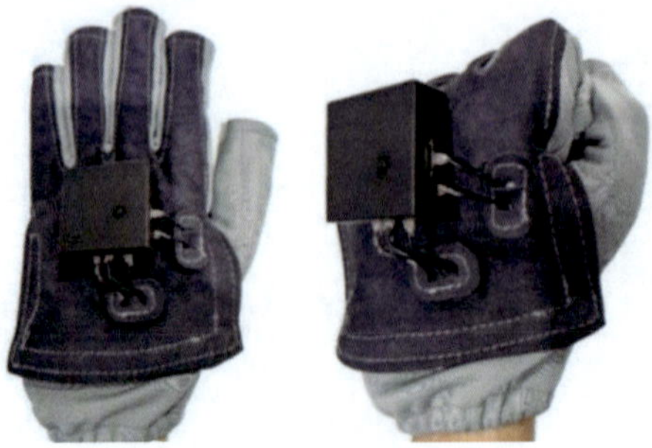

Fig. 3. Flex sensor calibration

With regard to flex sensors, resistance variations can give rise to inconsistencies owing to manufacturing differences and variations in individual fingers. To standardize and enhance data consistency, a two-stage processing approach is implemented consisting of normalization followed by fuzzy discretization. During the calibration phase, users are instructed to perform two distinct hand positions in Fig. 3. First, users are asked to fully extend their fingers and maintain this position steadily over a 5-s period for maximum value, then make a tight fist for the minimum. Once calibration is complete, raw sensor readings are normalized using the following formula:

$$f = \frac{f - f_{\min}}{f_{\max} - f_{\min}} \tag{1}$$

where f represents the raw data, $f_{\max}$ is the maximum value for the fully extended position, and $f_{\min}$ is the minimum value for the fully flexed position. This normalization scales the flex sensor data into a standardized range of 0–1.

To reduce noise and provide stable discrete outputs suitable for gesture recognition, the normalized continuous values are further processed through a fuzzy discretization scheme in Table 1. The normalized range [0, 1] is divided into five discrete levels, representing different degrees of finger flexion from fully extended to fully flexed.

Table 1. Flex sensor discretization mapping rules

Raw	Discrete value	Finger state
[0, 0.2)	0	Fully flexed
[0.2, 0.4)	0.3	Significantly bent
[0.4, 0.6)	0.5	Moderately bent
[0.6, 0.8)	0.7	Slightly bent
[0.8, 1]	1.0	Fully extended

The final data from the integrated sensor suite, now standardized and processed, includes: a_x, a_y, a_z, w_x, w_y, w_z, f_1, f_2, f_3, f_4, f_5, where a_x, a_y, a_z represent the accelerometer, w_x, w_y, w_z correspond to the gyroscope measurements, and f_1,f_2,f_3,f_4, f_5 correspond to the processed and discretized flex sensor readings.

3.3 Gesture Recognition Algorithm

The Inception module [23] introduces a parallel architectural framework incorporating convolutional kernels of varying sizes to extract features at multiple scales, which significantly enhances computational efficiency while reducing parameter count. Building upon this foundation, Inception-ResNet [24] integrates the Inception module with deep residual networks, introducing three variants of Inception modules with residual connections that markedly accelerate training convergence rates. CSRA [25] produces class-specific features for each category by introducing a straightforward spatial attention score, and achieves state-of-the-art performance in multilabel recognition. These architectures have demonstrated exceptional performance in image classification tasks.

Gesture data contains both temporal and spatial characteristics. Spatial features are obtained through convolutional operations, whereas temporal features are extracted by modifying the dimensions of convolutional kernels and CSRA. The proposed network architecture is shown in Fig. 4.

The model architecture is built upon a cascaded Inception framework and CSRA. It begins with a sensor input sequence I. It is first sent to a feature extractor (the Inception backbone) ϕ, Each InceptionBlock employs three parallel convolutional branches with different kernel sizes (39, 19, and 9), enabling the model to capture temporal patterns at multiple scales simultaneously. This multi-scale approach allows the network to learn both fine-grained local features and broader temporal dependencies within the sensor data sequences. To address the vanishing gradient problem and facilitate information flow, residual connections are implemented through a shortcut pathway that bypasses the Inception blocks. Then we obtain a feature tensor $x \in \mathbb{R}^{\mathrm{d}\times\mathrm{l}}$, where are the dimensionality, length of the feature tensor:

$$x = \phi(\mathrm{I}; \theta) \tag{2}$$

where θ denotes the parameters of the Inception backbone. Consequently, the feature tensor $\boldsymbol{x}$ exhibits a dimension of $d \times l$, which can be decoupled as $x_1, x_2, \ldots, x_{\mathrm{l}}$ ($x_{\mathrm{i}} \in \mathbb{R}^{\mathrm{d}}$). Subsequently, a fully connected layer is the classifier, with $m_{\mathrm{i}} \in \mathbb{R}^{\mathrm{d}}$ represents the classification weight vector for the i-th class.

The *class-specific attention scores* for the i-th class and j-th location is defined as:

$$\mathrm{s}_{\mathrm{j}}^{\mathrm{i}} = \frac{\exp\left(\mathrm{T}x_{\mathrm{j}}^{\mathrm{T}} m_{\mathrm{i}}\right)}{\sum_{\mathrm{k}=1}^{\mathrm{l}} \exp(\mathrm{T}x_{\mathrm{k}}^{\mathrm{T}} m_{\mathrm{i}})} \tag{3}$$

where $\sum_{\mathrm{j}=1}^{\mathrm{l}} \mathrm{s}_{\mathrm{j}}^{\mathrm{i}} = 1$ and $\mathrm{T} > 0$ is a temperature parameter, and $\mathrm{s}_{\mathrm{j}}^{\mathrm{i}}$ indicates the probability of the class i appearing at location j.

Then, the class-specific feature vector for class i is formed through a weighted combination of the feature tensor:

$$a^{\mathrm{i}} = \sum_{\mathrm{k}=1}^{\mathrm{l}} \mathrm{s}_{\mathrm{k}}^{\mathrm{i}} x_{\mathrm{k}} \tag{4}$$

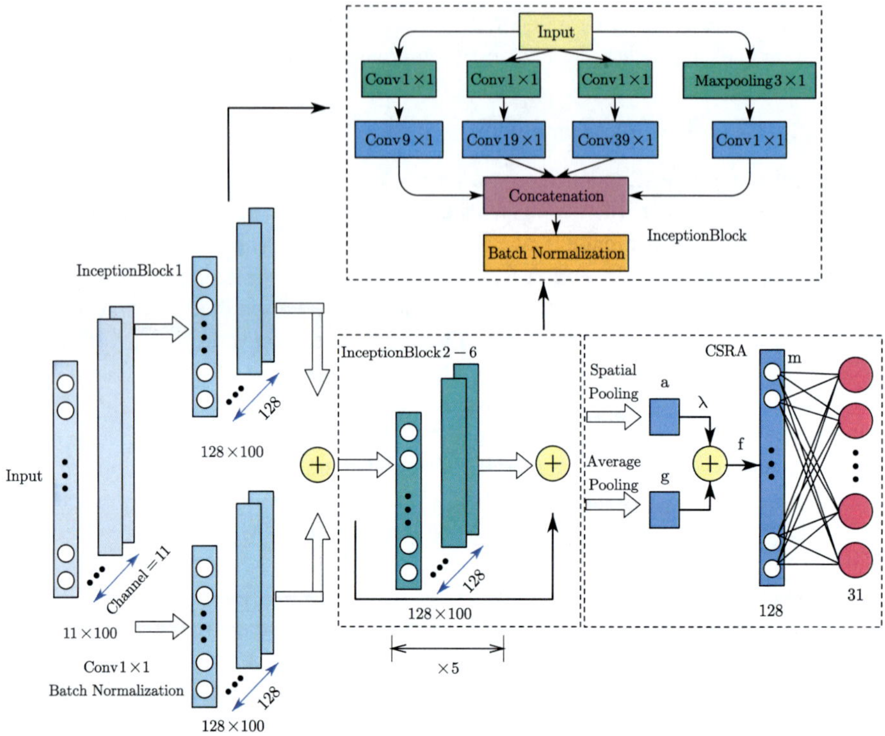

Fig. 4. Frame diagram of the inception-CSRA

For the entire sequence, the classical global class-agnostic feature vector is

$$g = \frac{1}{1}\sum_{\mathrm{k}=1}^{1} x_{\mathrm{k}} \tag{5}$$

Given the Global class-agnostic feature vector has been widely used and achieves promising performance, we treat it as the primary feature vector. Meanwhile, $\boldsymbol{a}^i$ is regarded as the class-specific residual feature. As illustrated in Fig. 4, by adding these two vectors, we obtain our CSRA feature f^i for the *i*-th class [25]:

$$f^{\mathrm{i}} = g + \lambda a^{\mathrm{i}} \tag{6}$$

Finally, the classifier maps the class-specific feature representation to the target logits.

$$\hat{y} \triangleq \left(\mathrm{y}^1, \mathrm{y}^2, \ldots, \mathrm{y}^{\mathrm{C}}\right) = \left(m_1^{\mathrm{T}} f^1, m_2^{\mathrm{T}} f^2, \ldots, m_{\mathrm{C}}^{\mathrm{T}} f^{\mathrm{C}}\right) \tag{7}$$

where C is the number of classes.

4 Experiments and Results

4.1 Dataset

We introduce a curated gesture set compiled from the Lexicon of Common Expressions in Chinese National Sign Language. The dataset comprises 11 numeric gestures (0–10) and 17 common word gestures. To facilitate real-time sentence parsing, two additional gestures, "begin" and "end" are incorporated to mark sentence boundaries. In total, 30 distinct gestures are included: 17 static signs and 13 dynamic signs as Fig. 5.

The dataset comprises data from 8 participants, including 6 males and 2 females. For each gesture, 20 sample sets were collected for each people, with sample recorded at 100Hz over a duration of 1 s. Data distribution follows a 75, 15, and 15% split for training, validation, and testing.

Figure 6 presents box plots visualizing the distribution of gesture classes across accelerometer and flex sensor measurements. Box plots effectively summarize data distribution through five statistical parameters: minimum, first quartile, median, third quartile, and maximum values.

The visualization reveals distinct sensor response patterns for each gesture class, demonstrating clear inter-gesture variability in both accelerometer and flex sensor domains.

The box width in sensor plots directly correlates with data variability, reflecting gesture characteristics. Static letter gestures display minimal or absent boxes in accelerometer and flex plots, indicating stationary hand positions. In contrast, dynamic gestures like "sorry" show enlarged boxes due to characteristic back-and-forth movements. Static gestures typically exhibit more outliers, as sensitive sensors capture minor hand tremors and incidental movements that deviate from the stable baseline readings. Analysis of the box plots across all gestures reveals minimal anomalous data points, with the exception of flex2 sensor readings for gestures "3" and "4", which may be attributed to sensor noise. This observation indicates consistent sensor performance throughout the dataset with no evidence of sensor malfunction. The overall consistency validates the reliability of the collected data.

4.2 Inception-CSRA Training and Test

As outlined in Sect. 4.1, each gesture sample contains 100 temporal data points with 11 features plus one class label. The data is structured as 100×11 matrices to enable effective learning of temporal patterns in the sequential gesture data.

Figure 7 illustrates the training dynamics of the Inception model, displaying accuracy and loss curves for both training and validation sets. The network employs the Adam optimizer coupled with cross-entropy loss to optimize classification performance. Training proceeded for up to 40 epochs with early stopping after 5 consecutive epochs.

The cross-entropy loss function is defined as:

$$L = -\frac{1}{N}\sum_{i=1}^{N}\sum_{c=1}^{C} y_{i,c}\log(\hat{y}_{i,c}) \tag{8}$$

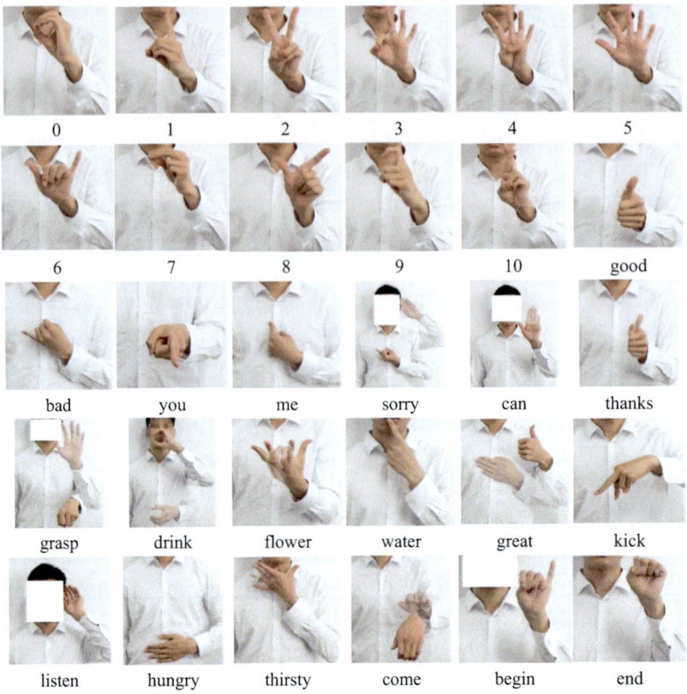

Fig. 5. Chinese signs used in the dataset

where N represents the batch size, C is the number of classes, $y_{i,c}$ denotes the true label, and $\hat{y}_{i,c}$ represents the predicted probability for class c of sample i.

Table 2 presents the performance metrics of the Inception-CSRA model, including accuracy scores and loss values across training, validation, and test process.

The corresponding confusion matrix for the test datasets displayed in Fig. 8. The Inception-CSRA model achieved high accuracy and a well-balanced distribution in the confusion matrix on the test dataset. However, a limitation of this model is that it tends to confuse the "drink" and the "come" gesture with the "none" gesture. After analyzing the density distribution of flex sensor values in Fig. 9, we observed significant overlap in the data from the 1st, 2nd, and 3rd flex sensors between these gestures. This overlap in their feature distributions, combined with certain outliers, contributes to the model's misclassification between these similar hand gestures.

We evaluate our proposed SLR system against existing state-of-the-art approaches as summarized in Table 3. The comparative analysis covers sensor type, gesture classification scope, algorithmic method, and recognition accuracy. The proposed system achieves 99.5% recognition performance with up to 30 gestures. Although the accuracy of gesture recognition proposed by Lee B G exceeds 99.5%, an important factor cannot be overlooked: although some systems have classification accuracies higher, the number of gestures they categorize is fewer, the parameter is more.

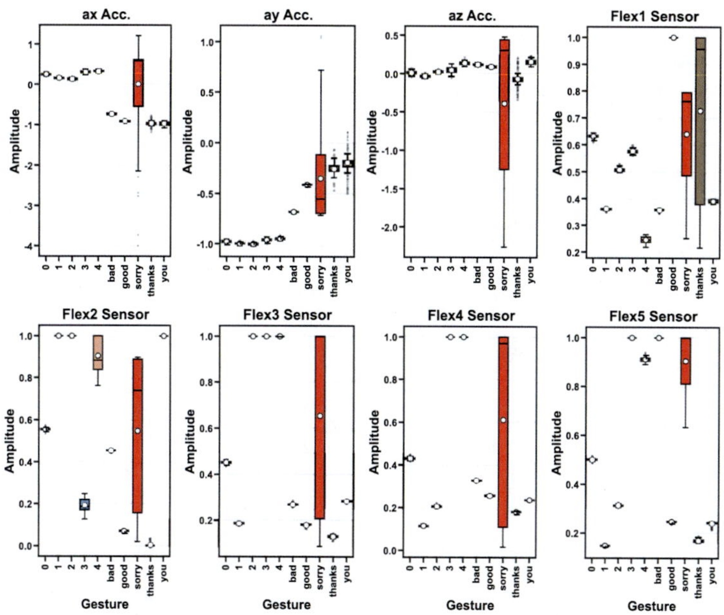

Fig. 6. Box plot of multiple sensor values

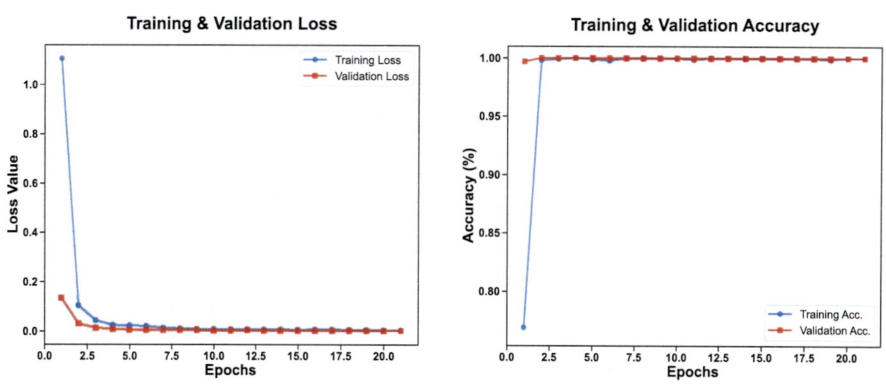

Fig. 7. Loss and accuracy during the training process

Table 2. Inception-CSRA model performance

	Train	Validation	Test
Accuracy	1.0	0.999	0.995
Loss	0.006	0.004	/

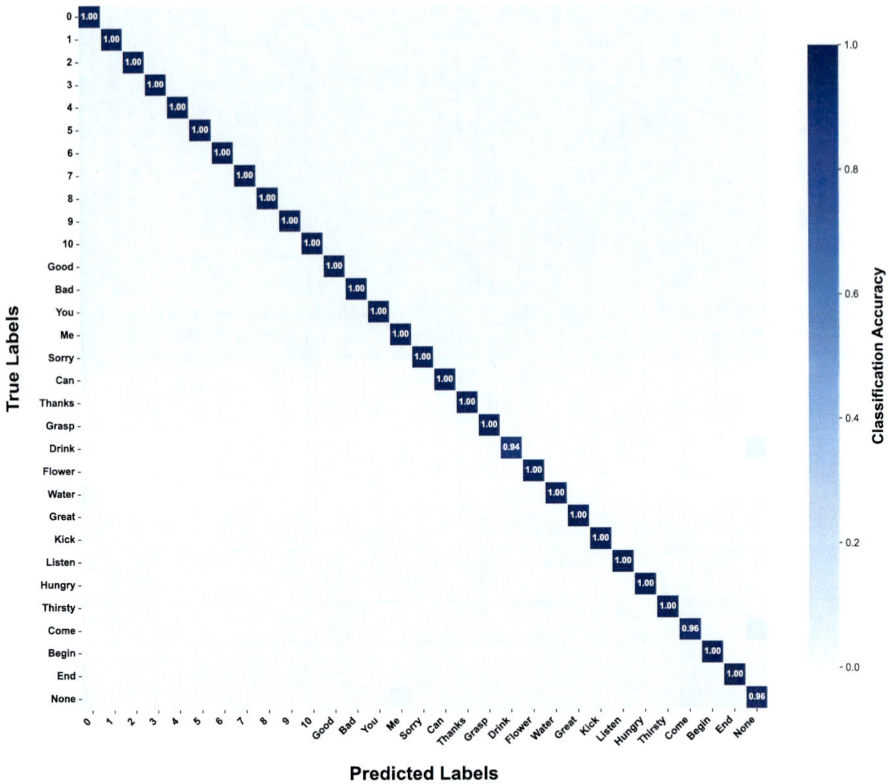

Fig. 8. Confusion matrix for the test datasets

We also conducted comparative experiments with three baseline models: CNN, RNN, and LSTM. Statistical significance was assessed using one-way ANOVA and post-hoc McNemar's tests with Bonferroni correction ($\alpha = 0.0083$). The ANOVA results revealed significant differences among the models. As illustrated in Fig. 10, Inception-CSRA achieved the highest accuracy of 99.48%, outperforming CNN (96.6%), RNN (94.64%), and LSTM (98.17%). Pairwise comparisons confirmed that Inception-CSRA significantly surpassed all baseline models, with p-values of 0.0000, 0.0000, and 0.0044 for CNN, RNN, and LSTM respectively.

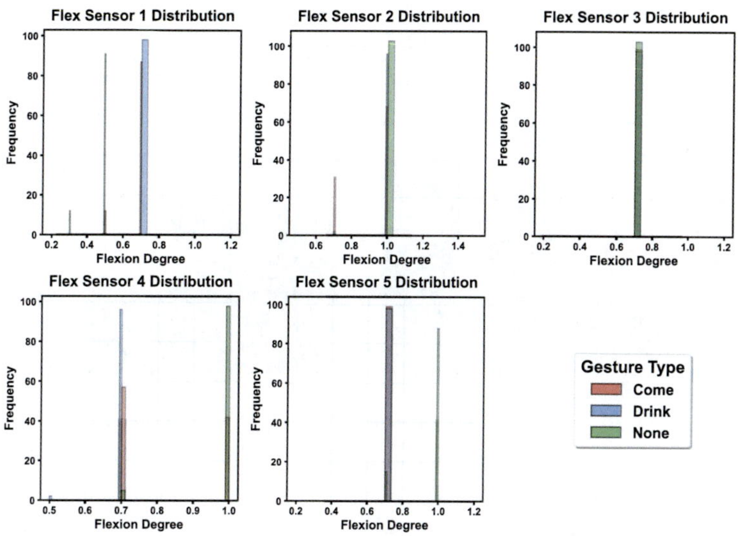

Fig. 9. The flex sensor values analysis between "drink" and "come"

Table 3. Comparison of the proposed system with state-of-the-art systems

References	Sensors	Gestures	Classifier	Accuracy (%)	Parameter (K)
Zhou et al. [7]	Yarn-based stretchable sensor	11	SVM	98.63	/
Chen et al. [8]	Force and 5 flex sensors	16	SVM	99.2	/
Wen et al. [9]	Triboelectric sensor	50	CNN	91.3	699.19
Lee et al. [10]	6 IMU sensors	27	LSTM	99.81	311.34
Wang et al. [14]	6 IMU and 5 flex sensors	26	Para-LSTM-CNN	97.01	104.55
This work	1 IMU and 5 flex sensors	30	Inception-CSRA	99.5	489.024

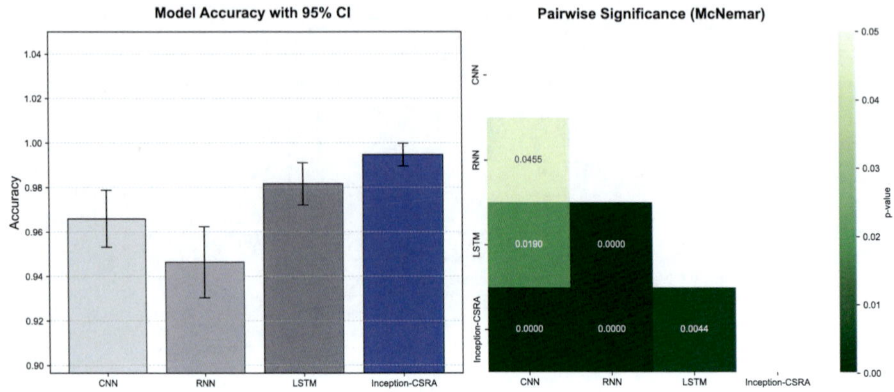

Fig. 10. Comparison of inception-CSRA with baseline models

4.3 Real-Time Application

After training the model, we deployed it in practical applications. Figure 11 illustrates the flowchart of the real-time system. The acquired data is first segmented using a fixed sliding window approach, then the segmented samples are fed into the model for recognition, and finally the recognition results are converted to audio output.

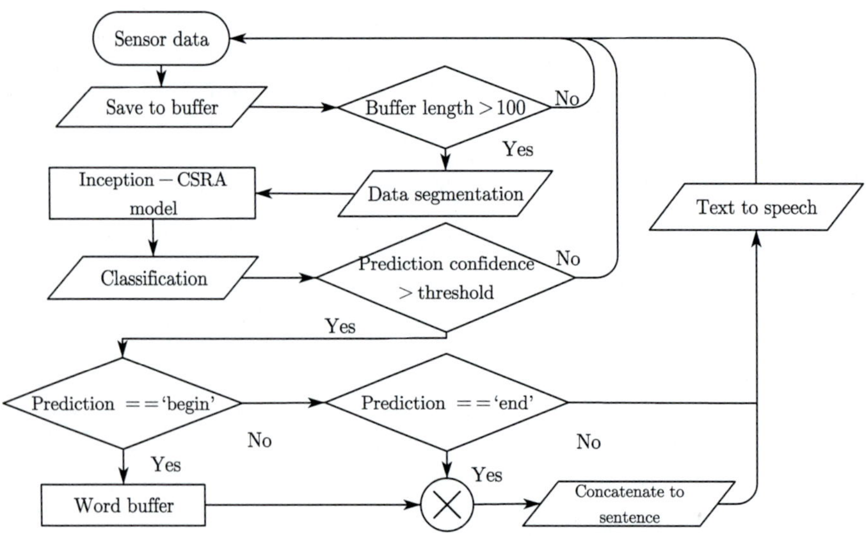

Fig. 11. Flowchart of the real-time application

Additionally, in sentence matching mode, when the sentence start marker "begin" is detected, the recognized words are cached; when the "end" marker is detected, the words are output through sentence matching. The matching rules are shown in Fig. 12. Sentence optimization techniques are utilized to address these challenges by transforming the unprocessed, potentially inaccurate outputs generated by the gesture recognition framework into coherent, grammatically correct expressions. At last, the enhanced sentences are output by voice. The model deployed with TensorRT achieves 1.2 ms inference time on an NVIDIA RTX 3060 GPU and 12th Gen Intel Core i7 CPU.

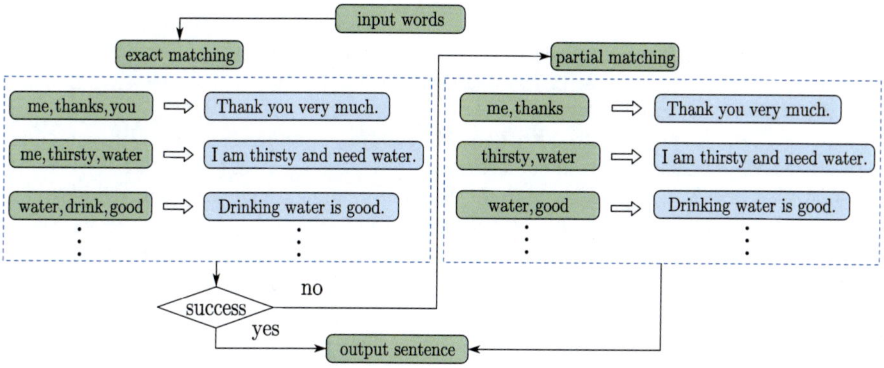

Fig. 12. Sentences matching rules

5 Conclusion and Future Work

Deaf individuals frequently encounter significant challenges in communication, and our research aims to address this critical issue. This paper presents a comprehensive real-time Chinese SLR solution utilizing sensor gloves and deep learning models. The system captures fine hand movements and finger flexure information through wearable data gloves. The wearable data glove employs one IMU and five flexible sensors for 11-dimensional hand motion signal acquisition. An Inception-CSRA SLR algorithm has been developed, which adopts a parallel extraction strategy for spatio-temporal features and CSRA for classification. Experimental validation demonstrates the system's SLR performance. The results demonstrate that Inception-CSRA achieves an accuracy of up to 99.5% with 489.024 K parameters.

The current system effectively addresses communication difficulties for individuals with speech impairments. However, limitations remain in facilitating bidirectional communication between hearing individuals and sign language users. To improve the system, several enhancements are necessary: expanding the dataset to include diverse gesture sequences, exploring advanced deep learning architectures, and developing adaptive dynamic gesture segmentation methods for continuous signing. Furthermore, a complementary module should be developed to capture spoken language, convert it to text, and translate it into sign language video, enabling bidirectional communication. In conclusion, while this research has established a solid foundation for sign language recognition, additional advancements are needed to better satisfy user needs and advance SLR technology.

References

1. World Health Organization: Deafness and Hearing Loss (2025). https://www.who.int/newsroom/fact-sheets/detail/deafness-and-hearing-loss

2. Alyami, S., Luqman, H., Hammoudeh, M.: Reviewing 25 years of continuous sign language recognition research: advances, challenges, and prospects. Inf. Process. Manage. **61**(5), 103774 (2024)
3. Kothadiya, D., Bhatt, C., Sapariya, K., et al.: Deepsign: sign language detection and recognition using deep learning. Electronics **11**(11), 1780 (2022)
4. Zhou, H., Wang, D., Yu, Y., et al.: Research progress of human–computer interaction technology based on gesture recognition. Electronics **12**(13), 2805 (2023)
5. Ahmed, A.M., Alez, R.A., Taha, M., et al.: Automatic translation of Arabic sign to Arabic text (ATASAT) system. J. Comput. Sci. Inform. Technol. **6**, 109–122 (2016)
6. Moustafa, A., Rahim, M.S.M., Khattab, M.M., et al.: Arabic sign language recognition systems: a systematic review. Indian J. Comput. Sci. Eng. **15**, 1–18 (2024)
7. Zhou, Z., Chen, K., Li, X., et al.: Sign-to-speech translation using machine-learning-assisted stretchable sensor arrays. Nat. Electr. **3**(9), 571–578 (2020)
8. Chen, X., Gong, L., Wei, L., et al.: A wearable hand rehabilitation system with soft gloves. IEEE Trans. Industr. Inf. **17**(2), 943–952 (2020)
9. Wen, F., Zhang, Z., He, T., et al.: AI enabled sign language recognition and VR space bidirectional communication using triboelectric smart glove. Nat. Commun. **12**(1), 5378 (2021)
10. Lee, B.G., Chong, T.W., Chung, W.Y.: Sensor fusion of motion-based sign language interpretation with deep learning. Sensors **20**(21), 6256 (2020)
11. Kumar, P., Chauhan, S., Awasthi, L.K.: Human activity recognition (HAR) using deep learning: Review, methodologies, progress and future research directions. Arch. Comput. Methods Eng. **31**(1), 179–219 (2024)
12. Halabi, M., Harkouss, Y.: Real-time Arabic sign language recognition system using sensory glove and machine learning. Neural Comput. Appl. **2025**, 1–17 (2025)
13. Xia, K., Huang, J., Wang, H.: LSTM-CNN architecture for human activity recognition. IEEE Access **8**, 56855–56866 (2020)
14. Wang, D., Wang, M., Zhang, Z., et al.: Wearable electronic glove and multi-layer para-LSTM-CNN based method for sign language recognition. IEEE Internet Things J. **15**, 452 (2024)
15. Geetha, M., Manjusha, U.C.: A vision based recognition of Indian sign language alphabets and numerals using b-spline approximation. Int. J. Comput. Sci. Eng. **4**(3), 406 (2012)
16. Ibrahim, N.B., Selim, M.M., Zayed, H.H.: An automatic Arabic sign language recognition system (ArSLRS). J. King Saud Univ. Comput. Inform. Sci. **30**(4), 470–477 (2018)
17. Kamruzzaman, M.M.: Arabic sign language recognition and generating Arabic speech using convolutional neural network. Wirel. Commun. Mob. Comput. **2020**(1), 3685614 (2020)
18. Mittal, A., Kumar, P., Roy, P.P., et al.: A modified LSTM model for continuous sign language recognition using leap motion. IEEE Sens. J. **19**(16), 7056–7063 (2019)
19. Ahmed, A.A., Aly, S.: Appearance-based Arabic sign language recognition using hidden Markov models. In: Proceedings of the 2014 International Conference on Engineering and Technology (ICET). IEEE, pp. 1–6 (2014)
20. Tubaiz, N., Shanableh, T., Assaleh, K.: Glove-based continuous Arabic sign language recognition in user-dependent mode. IEEE Trans. Hum. Mach. Syst. **45**(4), 526–533 (2015)
21. Bhatnagar, V.S., Magon, R., Srivastava, R., et al.: A cost effective sign language to voice emulation system. In: Proceedings of the 2015 Eighth International Conference on Contemporary Computing (IC3). IEEE, pp. 521–525 (2015)
22. Qiu, S., Zhao, H., Jiang, N., et al.: Multi-sensor information fusion based on machine learning for real applications in human activity recognition: state-of-the-art and research challenges. Inform. Fusion **80**, 241–265 (2022)
23. Szegedy, C., Liu, W., Jia, Y., et al.: Going deeper with convolutions. In: Proceedings of the IEEE Conference on Computer Vision and Pattern Recognition, pp. 1–9 (2015)

24. Szegedy, C., Ioffe, S., Vanhoucke, V., et al.: Inception-v4, inception-resnet and the impact of residual connections on learning. In: Proceedings of the AAAI Conference on Artificial Intelligence, vol. 31 (2017)
25. Zhu, K., Wu, J.: Residual attention: a simple but effective method for multi-label recognition. In: Proceedings of the IEEE/CVF International Conference on Computer Vision, pp. 184–193 (2021)

SHMR: Traditional Chinese Medicine Prescription Recommendation via Multi-graph Neural Networks and Transformer-Based Attention

Chi Han(✉)

School of Computer Science and Technology, Anhui University, Hefei, China
3232335710@qq.com

Abstract. Traditional Chinese Medicine (TCM) prescriptions encode rich many-to-many interactions among symptoms, syndromes, and herbs, making automatic prescription recommendation a challenging structured prediction problem. Although recent learning-based approaches have shown promising results, most existing models either ignore herb property knowledge, rely purely on co-occurrence statistics, or fail to jointly capture local graph structure and long-range semantic dependencies. We propose SHMR, a unified TCM prescription recommendation model that integrates multi-graph convolution with a structure-aware attention mechanism. SHMR first constructs a three-view heterogeneous graph comprising SymptomHerb, SymptomSymptom, and HerbHerb relations, and augments herb nodes with domain knowledge such as properties, flavors, and meridian tropism via a TCM knowledge graph. A learnable multi-source fusion module then adaptively combines embeddings from different graphs, while a Transformer encoder enhances global contextual modeling beyond local neighborhoods. Finally, a bilinear interaction module models fine-grained symptomherb dependencies to produce herb-level recommendation scores. Experiments on a public TCM prescription dataset demonstrate that SHMR consistently outperforms strong topic-model, GNN-based, and sequence-generation baselines, yielding substantial gains in Precision@5, Recall@5, and F1-Score@5. These results indicate that jointly leveraging heterogeneous graph structure, herb attribute knowledge, and attention-based fusion provides an effective and interpretable framework for intelligent TCM prescription recommendation.

Keywords: Graph neural network · Attention mechanism · Multi-source fusion · TCM prescription recommendation

L. Huang and D. Greenhalgh (Eds.): ICMLC 2026, LNNS 2061, pp. 358–372, 2027.
https://doi.org/10.1007/978-3-032-31236-5_25

1 Introduction

The long-standing experience in formula compatibility in Traditional Chinese Medicine (TCM) provides an important prior knowledge base for prescription recommendation tasks [1,2]. A prescription in TCM is composed of multiple herbs that collaboratively address a constellation of patient symptoms [3]. Designing an appropriate prescription requires simultaneously considering herb efficacy, compatibility, symptom patterns, and syndrome progression [4]. Despite its clinical value, this process remains highly dependent on practitioner experience, leading to variability and limited scalability [5].

In recent years, the continuous evolution of machine learning and artificial intelligence methods has made automated recommendation for TCM prescriptions a research hotspot. Existing methods can be roughly grouped into topic-model-based approaches [6,7], graph-neural-network-based methods [8–10], and sequence generation models [11]. Topic-model approaches incorporate TCM theory, herb compatibility, or herb attribute knowledge into probabilistic generative frameworks such as LDA to assist herb recommendation and prescription generation [6,7], but they struggle to capture high-order symptom–herb interactions. Graph-based methods construct multi-relational graphs over symptoms, syndromes, and herbs and leverage multi-graph convolution to simulate the TCM diagnostic process [8–10]; however, their performance is often constrained by graph incompleteness and the lack of rich auxiliary herb knowledge. Sequence models, including Seq2Seq and Transformer-based architectures [11], treat prescription generation as a text generation or machine translation task and are effective at modeling semantic dependencies, yet they typically overlook explicit graph structure that is crucial for modeling TCM clinical logic and syndrome patterns.

Furthermore, three challenges remain unsolved:

- **Heterogeneous Multi-relational Structures**: symptoms and herbs interact through multiple latent relations (co-occurrence, similarity, and compatibility).
- **Insufficient Incorporation of Herb Knowledge**: In existing graph learning frameworks, the properties of traditional Chinese medicine, such as the four natures, five flavors, and meridian tropism, are often not fully introduced or utilized.
- **Lack of Effective Feature Fusion**: Existing models combine multi-view features via simple concatenation, failing to adaptively weight graph-derived and semantic information.

To address these limitations, we propose **SHMR**, a multi-graph and attention-enhanced TCM recommendation model. SHMR jointly integrates:

- multi-graph convolution over Symptom–Herb (S–H), Symptom–Symptom (S–S), and Herb–Herb (H–H) graphs;
- herb-knowledge-guided GCN for enriching herb representations;

- a learnable weighted fusion mechanism that adaptively aggregates multi-view embeddings;
- a Transformer encoder that enhances long-range dependency modeling beyond local graph constraints;
- bilinear feature fusion for fine-grained symptom–herb interaction modeling.

Extensive experiments confirm that SHMR effectively learns structural prior knowledge and semantic correlations in TCM prescriptions. The model significantly outperforms strong baselines including GCNConv [12], GCNIIConv [13], SMGCN [8], and coverage-based generation models.

The main contributions of this work are:

1. We propose SHMR, a unified multi-graph and attention-based TCM recommendation framework capable of capturing global and local dependencies.
2. We introduce a learnable multi-source fusion mechanism and a herb-knowledge-guided graph encoder that enhance the expressiveness of graph embeddings.
3. We conduct comprehensive experiments and ablation studies, demonstrating that each component of SHMR contributes significantly to performance gains.

2 Related Work

2.1 Graph Neural Networks

Graph Neural Networks (GNNs) are a family of neural architectures designed for graph-structured data, whose core mechanism is iterative neighborhood aggregation [14]. Thanks to their strong capability in modeling relational structures, GNNs have been widely applied in molecular property prediction [15,16] and social network analysis [17,18]. Among them, Graph Convolutional Networks (GCNs), proposed by Kipf and Welling [12], introduce a normalized convolution operator over graphs to learn expressive node representations via weighted aggregation of neighbor features, and have become a standard backbone in many graph learning tasks. Beyond vanilla GCNs, Knowledge Graph Convolutional Networks (KGCN) [19] further incorporate relation-specific edge information from knowledge graphs as aggregation weights, enabling more fine-grained modeling of user–item interactions. These advances demonstrate the potential of GNNs in domains where relational structure is central, but most existing methods still operate on a single (or homogeneous) interaction graph, making it difficult to capture the heterogeneous multi-relational symptom–syndrome–herb structures that arise in TCM.

2.2 Attention Mechanisms

Attention mechanisms [20] provide flexible, context-aware weighting over inputs and have become key components in modern NLP and multimodal learning architectures. By assigning dynamic importance scores to different tokens or features, they enable effective modeling of long-range dependencies and thus naturally complement GCNs, which primarily emphasize local neighborhoods. In the context of TCM prescription generation, a variety of sequence models incorporate attention to better capture symptom–herb relationships [7,11,21]. Liu et al. design multi-level and symptom–herb-specific attention mechanisms to emphasize primary symptoms and dynamically select herbs [11,21]. Chen et al. further propose a knowledge-guided translator with attention over both symptom and knowledge representations [7]. Although these methods significantly improve prescription quality, they are typically built on purely sequential encoders and lack explicit structural priors, making it difficult to fully exploit heterogeneous graph-based knowledge in TCM.

2.3 TCM Recommendation

Existing approaches to TCM recommendation can be summarized into three main lines of research. (1) **Topic-Model-Based Methods**: Early studies embed TCM domain theory into topic models for mining hidden symptom–herb structures (e.g., Ji et al. [22]). Subsequent work augments topic modeling with explicit knowledge to strengthen representations, such as Yao et al. [6] and Chen et al. [7]. (2) **Clustering and Representation Learning**: Ruan et al. [23] employ heterogeneous network embedding and bipartite embedding to reveal regularities in clinical prescriptions. (3) **Graph-Neural-Network-Based Methods**: Jin et al. [8] and Zhao et al. [10] employ multi-graph convolution to capture syndrome-aware interactions for prescription recommendation. Despite promising progress, many methods still under-exploit herb property knowledge and often rely on simple fusion strategies. Recent work also moves toward more clinically grounded pipelines; for example, PresRecST integrates syndrome differentiation and treatment planning for real-world prescription recommendation [24]. Compared with these studies, SHMR explicitly unifies heterogeneous S–H/S–S/H–H relations, injects herb property knowledge into graph learning, and further enhances global dependency modeling via a Transformer-based encoder.

3 Materials and Methods

3.1 Model Overview

The proposed SHMR framework (Fig. 1) integrates multi-graph convolution, herb-knowledge-guided embedding, adaptive feature fusion, and Transformer-based global attention. Three graphs are constructed to capture heterogeneous TCM relations:

- **S–H Graph**: symptom–herb co-occurrence,
- **S–S Graph**: symptom correlation,
- **H–H Graph**: herb compatibility and co-occurrence.

All graphs are constructed only from the training set after the split; validation/test prescriptions are never used to compute co-occurrence statistics or edges. Each graph encodes complementary structural information. SHMR first computes node embeddings through multiple GCN modules and then fuses representations using a learnable weighting mechanism. Finally, a Transformer encoder enhances global semantic dependencies before prediction.

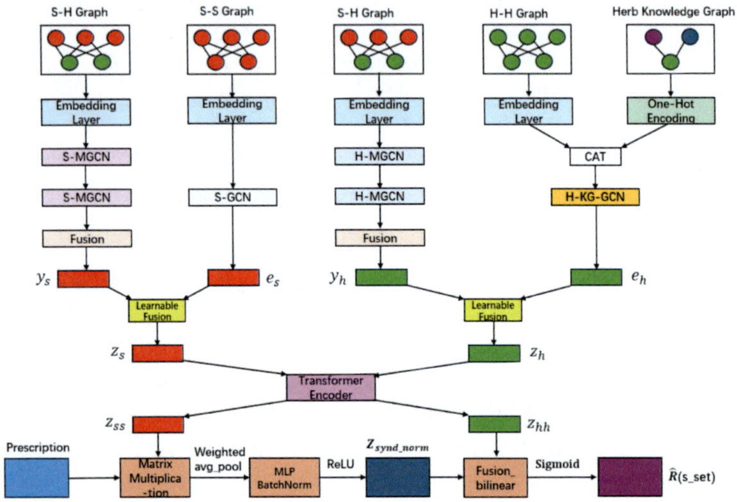

Fig. 1. Overall architecture of the proposed SHMR model.

3.2 Dataset

We adopt the public TCM prescription dataset released by Yao et al. [6]. After aligning symptoms with 603 standard diagnostic terms and herbs with 970 TCM MeSH entries, 33,765 cleaned prescriptions remain. Following a standard evaluation protocol, 70/15/15% of the samples are randomly assigned to the training, validation, and test sets, respectively. The dataset comprises 390 symptoms and 805 herbs. The herb occurrence distribution is provided in Fig. 2.

3.3 Symptom–Herb Graph Embedding

Graph Construction. For each prescription containing a symptom set s_set and herb set h_set, an undirected edge is created between symptom s and herb h if they co-occur:

$$S{-}H_{s,h} = S{-}H_{h,s} = \begin{cases} 1, & \text{if } (s,h) \text{ co-occur,} \\ 0, & \text{otherwise.} \end{cases} \tag{1}$$

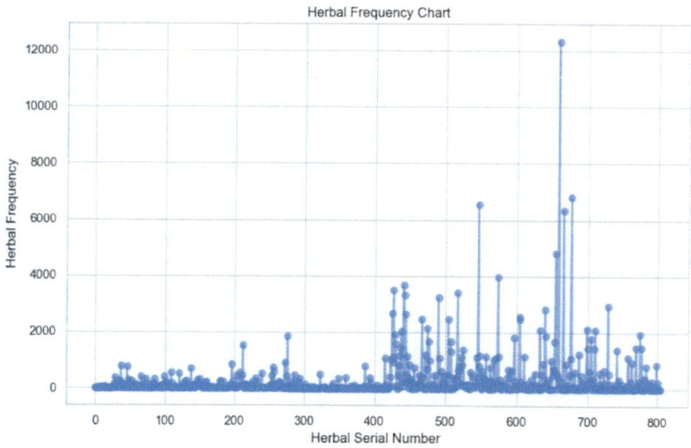

Fig. 2. Frequency distribution of herbs in the processed dataset.

Message Aggregation. Two GCNs—S-MGCN and H-MGCN—learn symptom and herb features separately. First-layer message passing aggregates neighborhood embeddings:

$$y_s^{(1)} = \tanh\left(\frac{1}{|N_s|}\sum_{h\in N_s} y_h^{(0)} W_1\right), \quad y_h^{(1)} = \tanh\left(\frac{1}{|N_h|}\sum_{s\in N_h} y_s^{(0)} W_1\right). \tag{2}$$

For $l \geq 2$, high-order propagation follows:

$$y_s^{(l)} = \tanh\left(\frac{1}{|N_s|}\sum_{h\in N_s} y_h^{(l-1)} W_l\right), \quad y_h^{(l)} = \tanh\left(\frac{1}{|N_h|}\sum_{s\in N_h} y_s^{(l-1)} W_l\right). \tag{3}$$

To mitigate over-smoothing while exploiting multi-hop information, SHMR performs level-wise averaging:

$$y_s = \tanh\left(W \cdot \frac{1}{L+1}\sum_{l=0}^{L} y_s^{(l)}\right), \quad y_h = \tanh\left(W \cdot \frac{1}{L+1}\sum_{l=0}^{L} y_h^{(l)}\right). \tag{4}$$

This design retains shallow-level specificity and deep-level contextual information simultaneously.

3.4 Symptom–Symptom and Herb–Herb Graph Embedding

Graph Construction. Symptom–symptom and herb–herb edges are constructed by co-occurrence frequency. Herb pairs exceeding a threshold T receive an edge:

$$H-H_{h_m,h_n} = \begin{cases} 1, & \text{if freq}(h_m, h_n) > T, \\ 0, & \text{otherwise.} \end{cases} \tag{5}$$

Thresholds follow [6]. The same rule applies to S–S.

Feature Aggregation. S-GCN aggregates symptom neighbors:

$$e_s = \tanh\left(\sum_{i \in N_s} e_i^{(0)} W\right). \tag{6}$$

H-KG-GCN enriches herb embeddings using auxiliary herb-attribute vectors e^{kg} (properties, flavors, meridian tropism):

$$e_h = \tanh\left(\sum_{j \in N_h} (e_j^{(0)} \| e_j^{kg}) W\right). \tag{7}$$

Including herb knowledge enhances the interpretability and discriminative capability of herb representation learning.

3.5 TCM Knowledge Graph Construction

We construct a herb knowledge graph (Fig. 3) using string-matching and dictionary alignment to extract herb names, properties, flavors, and meridian attributions. Triples are stored in a structured schema:

$$\text{(herb, attribute type, attribute value)}.$$

The extracted attributes are encoded via one-hot vectors and used as auxiliary inputs in H-KG-GCN.

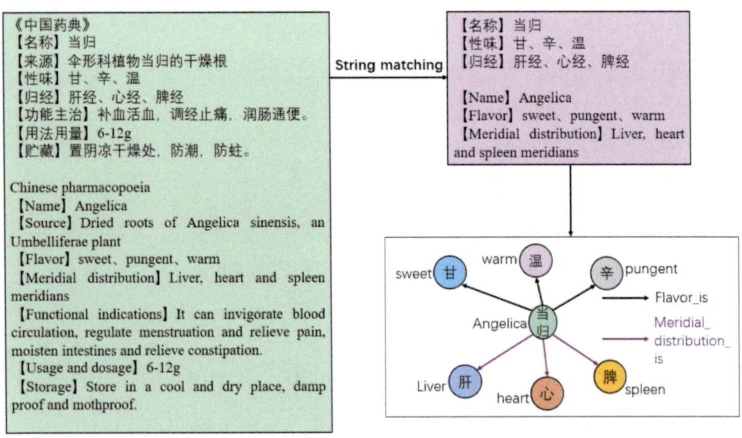

Fig. 3. Construction process of the TCM herb knowledge graph.

3.6 Learnable Weighted Fusion Mechanism

Symptoms and herbs obtain multi-view features from S–H, S–S, and H–H. SHMR introduces trainable weights $\alpha_{es}, \alpha_{eh} \in [0,1]$ to adaptively combine views:

$$x_{\text{fused}}^{(s)} = \alpha_{es} x_{SH}^{(s)} + (1 - \alpha_{es}) x_{SS}, \tag{8}$$

$$x_{\text{fused}}^{(h)} = \alpha_{eh} x_{SH}^{(h)} + (1 - \alpha_{eh}) x_{HH}. \tag{9}$$

The fused embeddings are projected via MLPs:

$$z_s = \text{MLP}_s(x_{\text{fused}}^{(s)}), \quad z_h = \text{MLP}_h(x_{\text{fused}}^{(h)}). \tag{10}$$

Compared with concatenation-based fusion, this mechanism is adaptive, end-to-end learnable, and robust to graph noise.

3.7 Transformer-Based Global Structure Encoder

GCNs emphasize local neighborhoods. To compensate for missing global structural information, a Transformer encoder processes z_s and z_h:

$$(z_{ss}, z_{hh}) = \text{Transformer}(z_s, z_h).$$

Self-attention enables:

- capturing long-range correlations beyond graph adjacency,
- integrating multi-view representations,
- enhancing robustness to incomplete graph structures.

Empirically, two encoder layers yield the best balance between abstraction and generalization.

3.8 Herb Recommendation via Bilinear Fusion

For each prescription, symptom representations are pooled:

$$Z_{\text{synd\_norm}} = \text{ReLU}(\text{MLP}(\text{w\_avg\_pool}(P \cdot z_{ss}))), \tag{11}$$

where P is the prescription–symptom one-hot matrix.

A bilinear fusion function models fine-grained symptom–herb interactions:

$$\text{pre}_{b,i} = Z_{\text{synd},b,i}^{\top} W z_{hh,b,i} + b. \tag{12}$$

The final herb probability output is:

$$\hat{R}(s\_set) = \sigma(\text{pre}), \quad \sigma(x) = \frac{1}{1 + e^{-x}}. \tag{13}$$

Top-K herbs with highest predicted probability constitute the recommended prescription.

3.9 Training Objective

Given a symptom set s_set and its ground-truth prescription, we represent the target herbs as a multi-hot label vector $\mathbf{y} \in \{0,1\}^{|\mathcal{H}|}$, where $|\mathcal{H}|$ denotes the number of herbs. Let $\mathbf{pre} \in \mathbb{R}^{|\mathcal{H}|}$ be the prediction scores computed by Eq. (12), and $\hat{R}(s\_set) = \sigma(\mathbf{pre}) \in (0,1)^{|\mathcal{H}|}$ be the predicted herb probabilities as in Eq. (13). We optimize the model using the multi-label binary cross-entropy loss:

$$\mathcal{L} = -\frac{1}{|\mathcal{H}|} \sum_{i=1}^{|\mathcal{H}|} \Big(y_i \log \hat{R}_i(s\_set) + (1 - y_i) \log \big(1 - \hat{R}_i(s\_set)\big) \Big). \qquad (14)$$

We minimize $\mathcal{L}$ using AdamW with the hyper-parameters reported in Sect. 4.1.

4 Experimental Results and Discussion

4.1 Experimental Setup

We implemented SHMR and all comparison models in PyTorch. Experiments were run on a server with four NVIDIA A40 GPUs. Following [8], we set the co-occurrence thresholds for constructing S–S and H–H graphs to 5 and 40, respectively. Following our evaluation protocol, 70% of the samples are used for training, and the remaining 30% is evenly allocated to validation and testing.

We train SHMR by minimizing the multi-label binary cross-entropy loss over herb labels, optimized with AdamW. We use an initial learning rate of 3×10^{-4}, a weight decay of 7×10^{-3}, and a dropout rate of 0.3. The mini-batch size is set to 64. We apply step-wise learning rate decay every 7 epochs and use early stopping with a patience of 7 epochs. The embedding and hidden dimensions are both 64, and the output dimension of the last GCN layer is 256. We tune the GCN depth in $\{1,2,3,4\}$ and the number of Transformer encoder layers in $\{1,2,3,4,5\}$, selecting the best configuration based on validation performance.

4.2 Evaluation Metrics

Given a ground-truth herb set h_set and the Top-K predicted herb set K, we adopt:

$$\text{Precision@}K = \frac{|h\_set \cap K|}{|K|}, \qquad (15)$$

$$\text{Recall@}K = \frac{|h\_set \cap K|}{|h\_set|}, \qquad (16)$$

$$F1\text{-Score@}K = \frac{2 \cdot \text{Precision@}K \cdot \text{Recall@}K}{\text{Precision@}K + \text{Recall@}K}. \qquad (17)$$

Precision@K measures how many recommended herbs are correct, Recall@K measures how many ground-truth herbs are successfully recovered, and F1-Score provides a balanced summary.

5 Baseline Models

We compare SHMR with the following models:

(1) **GCNConv** [12]: A standard GCN that aggregates normalized neighbor features.
(2) **GATConv** [25]: A graph attention network that assigns learnable importance weights to neighbors.
(3) **GCNIIConv** [13]: A deep GCN variant with initial residual connections and identity mapping to alleviate over-smoothing.
(4) **SMGCN** [8]: A multi-graph convolution framework that models syndrome-aware herb recommendation via multiplex graphs.
(5) **PTM** [6]: A probabilistic generative topic model augmented with TCM theory and explicit herbsymptom relational information.
(6) **Coverage-Soft-Model** [26]: A Seq2Seq-based generation model with coverage and soft loss to reduce repetition.
(7) **PresRecST** [24]: A new method is introduced for recommending herbal prescriptions to real patients, in which syndrome differentiation and treatment planning are jointly modeled.
(8) **SHMR-NoKG**: SHMR without herb attribute knowledge in the H–H graph.
(9) **SHMR-NoFu**: SHMR without multi-layer feature fusion (only the last GCN layer is used).
(10) **SHMR-NoKF**: SHMR without both knowledge guidance and multi-layer fusion.
(11) **SHMR-NoAttention**: SHMR without the Transformer-based attention module.

Unless otherwise stated, all models share the same base hyperparameters (learning rate, batch size, embedding dimension) for a fair comparison.

6 Performance Comparison

6.1 Overall Results

Table 1 reports the Top@5 results of all methods.

SHMR achieves the best performance across all three metrics, significantly outperforming both classical GCN-based models and TCM-specific baselines. Compared with the strongest baseline PresRecST, SHMR improves Precision@5 and F1-Score@5 by a clear margin, indicating that integrating multi-graph structure, herb knowledge, and attention mechanisms provides complementary gains.

GATConv performs relatively poorly, likely because its attention is restricted to local neighborhoods without explicit modeling of multi-relational structures. Coverage-Soft-Model also underperforms, suggesting that pure sequence-based generation is insufficient to capture complex TCM prescription logic without explicit graph priors.

Table 1. Performance comparison of different benchmark models (Top@5).

	Models	Precision	Recall	F1-score
Baselines	GCNConv [12]	0.2012	0.1299	0.1579
	GATConv [25]	0.1843	0.1195	0.1450
	GCNIIConv [13]	0.2000	0.1301	0.1577
	SMGCN [8]	0.1933	0.1265	0.1529
	PTM [6]	0.1982	0.1536	0.1731
	Coverage-soft-model [26]	0.1844	0.1073	0.1357
	PresRecST [24]	0.2238	0.1512	0.1805
Ours	SHMR-NoKG	0.2112	0.1468	0.1732
	SHMR-NoFu	0.2066	0.1410	0.1676
	SHMR-NoKF	0.1784	0.1228	0.1455
	SHMR-NoAttention	0.2138	0.1510	0.1770
	SHMR	**0.2488**	**0.1631**	**0.1970**

Best results are in bold

6.2 Ablation Study

The ablation results highlight the contribution of each SHMR component:

- **Effect of Herb Knowledge (SHMR Versus SHMR-NoKG)**: Removing herb attribute information leads to a clear drop in all metrics, demonstrating that properties such as flavor and meridian tropism provide informative inductive bias for herb representation learning.
- **Effect of Multi-layer Fusion (SHMR Versus SHMR-NoFu)**: Without multi-layer feature fusion, the model relies solely on the last GCN layer, which is more vulnerable to over-smoothing and noise accumulation. The performance decrease confirms that combining representations from different depths yields richer and more robust features.
- **Effect of Both Knowledge and Fusion (SHMR-NoKF)**: Removing both herb knowledge and multi-layer fusion causes the largest degradation, validating that these two components are complementary and jointly crucial for performance.
- **Effect of Attention (SHMR Versus SHMR-NoAttention)**: Eliminating the Transformer encoder reduces F1-Score, indicating that attention effectively captures long-range dependencies and reinforces key symptom–herb relations beyond local graph neighborhoods.

Overall, the ablation study confirms that each module of SHMR contributes meaningfully, and their combination leads to the best performance.

6.3 Impact of GCN Layers

An ablation on GCN depth is conducted by varying the S–H GCN layers between 1 and 4; the outcomes are illustrated in Fig. 4.

A two-layer GCN configuration yields the highest scores among all settings. A single layer cannot capture second-order neighbors (e.g., herb–symptom–herb patterns), limiting its expressive capacity. When the number of layers exceeds two, performance starts to decline, primarily due to over-smoothing and accumulated noise from distant nodes. This confirms that moderate depth is sufficient to capture useful high-order interactions while preserving discriminative information.

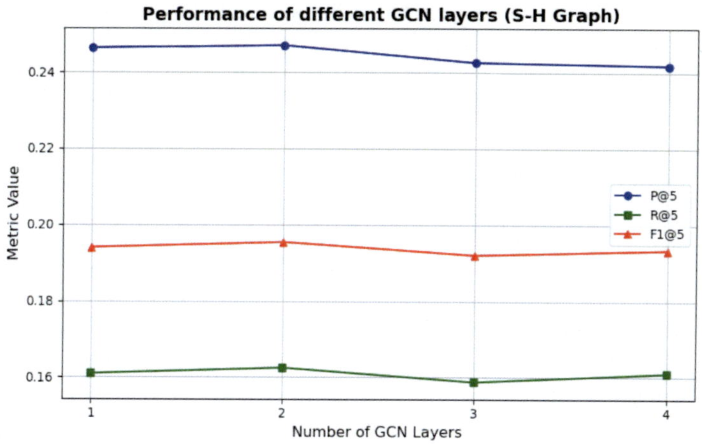

Fig. 4. Effect of different GCN layer numbers on model performance.

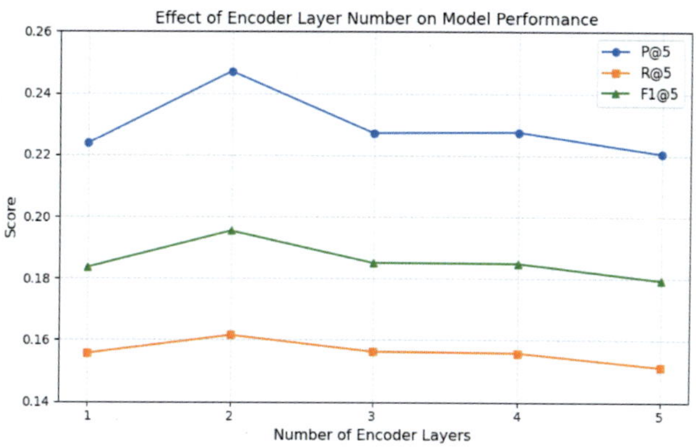

Fig. 5. Performance comparison under different numbers of transformer encoder layers.

6.4 Impact of Transformer Encoder Layers

We perform a depth study by varying the Transformer encoder layers between 1 and 5; Fig. 5 illustrates the corresponding performance.

Using two encoder layers yields the best Precision@5, Recall@5, and F1-Score@5. With only one layer, the model has limited ability to capture complex global dependencies. When the number of layers is greater than two, performance degrades gradually, likely due to overfitting and redundant transformations in relatively limited data scenarios. These results indicate that a shallow Transformer is sufficient to complement GCN-based local structure modeling and to enhance global semantic interactions.

7 Conclusion and Future Work

In this study, we formulate Traditional Chinese Medicine (TCM) prescription recommendation as a multi-label prediction task: given a set of symptoms, the model outputs a corresponding set of herbs. We propose **SHMR**, a unified framework that integrates multi-graph convolution, herb-knowledge-guided representation learning, a learnable multi-view fusion module, and a Transformer-based global encoder. By jointly modeling the Symptom–Herb, Symptom–Symptom, and Herb–Herb graphs, SHMR captures both local structural patterns and global relational dependencies. Moreover, incorporating herb attribute knowledge improves the semantic expressiveness and interpretability of herb representations.

Extensive experiments on a public TCM prescription dataset show that SHMR consistently outperforms strong baselines, including generic GNN-based methods and TCM-specific recommendation or generation approaches. Ablation results further verify the contribution of each component, indicating that herb knowledge injection, multi-view feature fusion, and attention-based global modeling all play important roles in the overall performance. SHMR achieves clear improvements in Precision@5, Recall@5, and F1-Score@5, suggesting its practical potential for assisting TCM clinicians and supporting data-driven knowledge discovery.

Future Work. Several directions are promising for future work. First, we currently use co-occurrence frequency as edge information only in the Symptom–Herb graph; extending similar weighting schemes to the Symptom–Symptom and Herb–Herb graphs may further refine relational modeling. Second, more advanced cross-attention mechanisms could be introduced to explicitly align symptom representations with key herbs, enabling finer-grained modeling of primary symptoms and core herb combinations. Third, adaptive fusion strategies (e.g., self-attention- or gating-based fusion) could replace fixed-form fusion to better integrate multi-source information for different patients or disease stages. Finally, incorporating additional real-world clinical signals, such as temporal treatment trajectories, dosage patterns, and efficacy feedback, may further improve the robustness and clinical value of TCM prescription recommendation systems.

Acknowledgment. The authors would like to acknowledge the support provided by the Plans for Major Provincial Science&Technology Projects (No.:202303a07020006).

References

1. Fu, L., Wang, Y.R., Zhao, Y.P., Huang, L.Q.: The concept of traditional Chinese medicine: history, theory and empirical research. Zhonghua Yi Shi Za Zhi **52**(4), 195–205 (2022)
2. Zhang, S.-Q., Li, J.-C.: An introduction to traditional Chinese medicine, including acupuncture. Anat. Rec. **304**(11), 2359–2364 (2021)
3. Li, B., Despotova-Toleva, L.: Practice and thinking of modern clinical research of traditional Chinese medicine. Ann. Palliat. Med. **10**(12), 12919–12920 (2021)
4. Zhi, W., Liu, Y., Wang, X., Zhang, H.: Recent advances of traditional Chinese medicine for the prevention and treatment of atherosclerosis. J. Ethnopharmacol. **301**, 115749 (2023)
5. Wang, W.-Y., Zhou, H., Wang, Y.-F., Sang, B.-S., Liu, L.: Current policies and measures on the development of traditional Chinese medicine in China. Pharmacol. Res. **163**, 105187 (2021)
6. Yao, L., Zhang, Y., Wei, B., Zhang, W., Jin, Z.: A topic modeling approach for traditional Chinese medicine prescriptions. IEEE Trans. Knowl. Data Eng. **30**(6), 1007–1021 (2018)
7. Chen, L., Luo, H., Yingpei, W., Yang, Y.: TPGen: prescription generation using knowledge-guided translator. Res. Square (2021)
8. Jin, Y., Zhang, W., He, X., Wang, X., Wang, X.: Syndrome aware herb recommendation with multi-graph convolution network. In: Proceedings of the 2020 IEEE 36th International Conference on Data Engineering (ICDE), pp. 145–156. Dallas, TX (2020)
9. Yang, Y., Rao, Y., Minghao, Y., Kang, Y.: Multi-layer information fusion based on graph convolutional network for knowledge-driven herb recommendation. Neural Netw. **146**, 1–10 (2022)
10. Zhao, W., Weikai, L., Li, Z., et al.: TCM herbal prescription recommendation model based on multi-graph convolutional network. J. Ethnopharmacol. **297**, 115109 (2022)
11. Liu, Z., Zheng, Z., Guo, X., et al.: AttentiveHerb: a novel method for traditional medicine prescription generation. IEEE Access **7**, 139069–139085 (2019)
12. Kipf, T.N., Welling, M.: Semi-supervised classification with graph convolutional networks. In: ICLR (Poster). OpenReview.net (2017)
13. Chen, M., Wei, Z., Huang, Z., Ding, B., Li, Y.: Simple and Deep Graph Convolutional Networks. arXiv preprint arXiv:2007.02133 (2020)
14. Xu, K., Hu, W., Leskovec, J., Jegelka, S.: How powerful are graph neural networks? In: ICLR. OpenReview.net (2019)
15. Chen, C., Ye, W., Zuo, Y., Zheng, C., Ong, S.P.: Graph networks as a universal machine learning framework for molecules and crystals. Chem. Mater. **31**, 3564–3572 (2019)
16. Jørgensen, P.B., Jacobsen, K.W., Schmidt, M.N.: Neural message passing with edge updates for predicting properties of molecules and materials. CoRR (2018)
17. Fan, W., Ma, Y., Li, Q., He, Y., Zhao, Y.E., Tang, J., et al.: Graph neural networks for social recommendation. In: WWW, pp. 417–426. ACM (2019)

18. Min, S., Gao, Z., Peng, J., Wang, L., Qin, K., Fang, B.: STGSN: a spatial-temporal graph neural network framework for time-evolving social networks. Knowl. Based Syst. **214**, 106746 (2021)
19. Wang, H., Zhao, M., Xie, X., Li, W., Guo, M.: Knowledge graph convolutional networks for recommender systems. In: WWW, pp. 3307–3313. ACM (2019)
20. Bahdanau, D., Cho, K., Bengio, Y.: Neural machine translation by jointly learning to align and translate. arXiv:1409.0473 (2014)
21. Liu, Z., Luo, C., Dianzheng, F., et al.: A novel transfer learning model for traditional herbal medicine prescription generation from unstructured resources and knowledge. Artif. Intell. Med. **124**, 102232 (2022)
22. Ji, W., Zhang, Y., Wang, X., Zhou, Y.: Latent semantic diagnosis in traditional Chinese medicine. World Wide Web **20**(5), 1071–1087 (2017)
23. Ruan, C., Ma, J., Wang, Y., Zhang, Y., Yang, Y.: Discovering regularities from traditional Chinese medicine prescriptions via bipartite embedding model. In: IJCAI, pp. 3346–3352. ijcai.org (2019)
24. Dong, X., Zhao, C., Song, X., et al.: PresRecST: a novel herbal prescription recommendation algorithm for real-world patients with integration of syndrome differentiation and treatment planning. J. Am. Med. Inform. Assoc. **31**(6), 1268–1279 (2024)
25. Velickovic, P., Cucurull, G., Casanova, A., Romero, A., Lio, P., Bengio, Y.: Graph attention networks. arXiv preprint arXiv:1710.10903 (2017)
26. Li, W., Yang, Z.: Exploration on generating traditional Chinese medicine prescriptions from symptoms with an end-to-end approach. In: CCF International Conference on Natural Language Processing and Chinese Computing (NLPCC), pp. 486–498. Springer (2019)

Hybrid Non-local Blocks for Streptococci Segmentation in Microscopic Images

A. Samarin[1], A. Toropov[1], E. Kotenko[1,3(✉)], A. Savelev[2], A. Dozortseva[4], A. Nazarenko[1], A. Mamaeva[1], A. Motyko[2], E. Mikhailova[1,3], and V. Malykh[1,5]

[1] ITMO University, St. Petersburg 197101, Russia
kotenkoed@gmail.com
[2] St. Petersburg Electrotechnical University "LETI", St. Petersburg 197022, Russia
[3] St. Petersburg State University, St. Petersburg 199034, Russia
[4] St. Petersburg State Institute of Technology, St. Petersburg 190013, Russia
[5] International IT University, Almaty 050040, Kazakhstan

Abstract. Accurate segmentation of streptococci microorganisms in microscopic imagery remains a challenging problem due to low contrast, overlapping cell structures, and irregular illumination. In this work, we propose a U-Netbased architecture augmented with novel non-local attention modules specifically designed for the morphological and contextual characteristics of streptococci. Experimental evaluation on the Microscopy Streptococci Images Dataset (MSID) demonstrates a substantial improvement over baseline architectures, increasing mean Intersection over Union (mIoU) from 0.796 to 0.952 and the Dice coefficient from 0.852 to 0.973. These findings highlight the effectiveness of hybrid attention mechanisms for enhancing precision and robustness in biomedical image segmentation.

Keywords: microscopy image segmentation · non-local attention · biomedical image analysis

1 Introduction

Recent advances in computer vision and deep neural networks have made automated interpretation of visual data an integral part of scientific and industrial practice. Despite their impressive performance on large-scale natural image benchmarks [1–5], most of these models are far less effective when applied to domains with highly specialized visual patterns and acquisition conditions [6–11]. Biomedical microscopy is a particularly demanding example of such a domain, where precise visual analysis directly informs clinical diagnostics, microbiological monitoring, and pharmaceutical research [12–17].

Among the many tasks in biomedical image analysis, the segmentation of bacterial structures, especially streptococci, is of central importance. The morphological configuration of these microorganisms, typically forming spherical chains or pairs, serves as a diagnostic indicator for numerous infections. However, the

L. Huang and D. Greenhalgh (Eds.): ICMLC 2026, LNNS 2061, pp. 373–385, 2027.
https://doi.org/10.1007/978-3-032-31236-5_26

raw microscopic images from which they are extracted pose substantial computational difficulties. Their visual appearance is often degraded by noise, weak contrast, uneven lighting, and partial occlusions, particularly in live or unstained preparations. These factors obscure cellular boundaries and hinder the extraction of robust morphological features [12,13].

Conventional computer vision algorithms, including thresholding, contour evolution, and morphological operations, are highly sensitive to such degradations because they depend on fixed, handcrafted descriptors. Their limited generalization capabilities restrict their use in large-scale or high-throughput biomedical pipelines. In contrast, convolutional neural networks (CNNs) have transformed biomedical segmentation by enabling data-driven feature learning. Models such as U-Net [18,19] have become standard due to their ability to combine local detail preservation with hierarchical context aggregation.

Despite this success, CNNs are inherently constrained by their local receptive fields. While successive convolutional layers gradually expand the range of captured information, they still lack a direct mechanism to encode long-distance relationships. For tasks like bacterial segmentation, where global structural cues such as chain formation or spatial clustering provide vital context, this locality bottleneck can significantly limit accuracy.

To overcome this limitation, researchers have increasingly explored attention-based mechanisms and non-local operations [20,21], which explicitly model pairwise interactions across distant spatial positions. More recent transformer-based approaches [22] extend this idea into fully global self-attention architectures, achieving strong performance in diverse vision applications. However, their computational cost and data requirements are prohibitive in microscopy, where annotated datasets are scarce, and imaging variability is high.

It is worth noting that an alternative and largely orthogonal line of work addresses long-range dependencies through spectral and graph-based methods. Spectral techniques leverage global eigen-structure to capture non-local relationships efficiently and have been shown to scale well to large datasets and complex relational domains. Such approaches have been successfully applied to scalable spectral clustering, graph signal processing, and large-scale data analysis, offering principled global context modeling with favorable computational properties [23–25]. While these methods are not explored in the present work, they provide a complementary perspective on global dependency modeling and represent a promising direction for future extensions of graph-aware or hybrid segmentation frameworks in biomedical imaging.

In this work, we propose a hybrid non-local block architecture specifically designed for segmenting streptococci in microscopic imagery. The proposed model integrates non-local attention modules into a U-Net backbone, providing an efficient compromise between the locality of convolutional feature extraction and the global dependency modeling of self-attention. This hybrid formulation enables the network to adaptively fuse spatial context, thereby improving boundary precision and inter-cell separation, even under low-contrast conditions.

2 Proposed Solution

2.1 Architecture Overview

Our segmentation approach is based on a modified U-Net architecture enhanced with integrated non-local attention components. The fundamental idea behind this design is to fuse the locality-aware feature extraction of convolutional layers with the capacity of attention mechanisms to capture long-range dependencies [26,27]. Such a combination is especially well-suited for microscopic image analysis, where cellular boundaries are often indistinct, and streptococci commonly appear as overlapping or partially merged formations.

The architecture builds upon earlier encoderdecoder frameworks that demonstrated the effectiveness of attention mechanisms in various biomedical imaging applications, including ocular structure segmentation [28] and lesion detection in CT scans [27]. These prior works emphasize the role of multi-scale, context-sensitive feature reasoning, which is essential for accurately segmenting streptococcal chains and paired coccoid morphologies.

Within the proposed network, non-local attention blocks are inserted at selected intermediate stages between the encoder and decoder pathways to facilitate interactions between spatially distant regions. By explicitly modeling relationships among non-adjacent feature locations, each attention module adaptively refines intermediate representations. A schematic overview of the complete architecture is provided in Fig. 1.

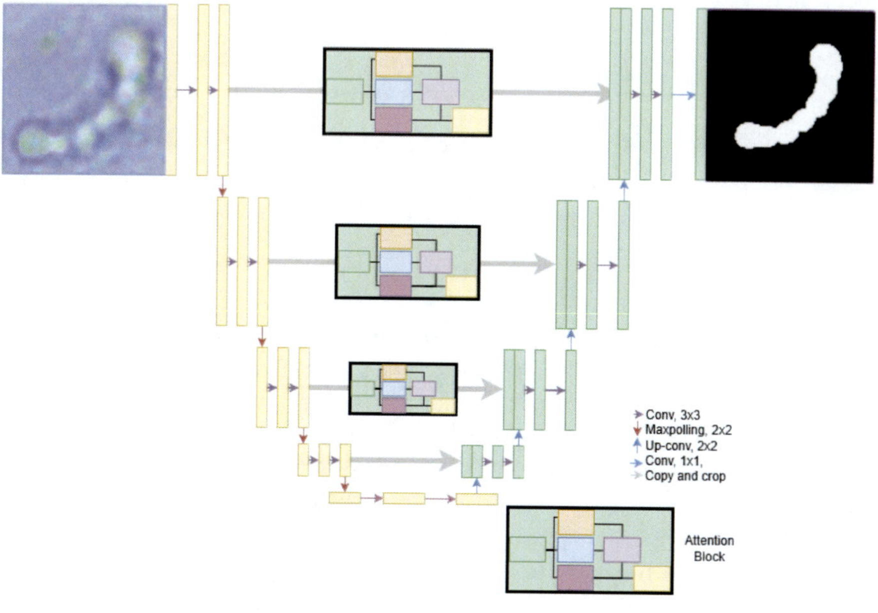

Fig. 1. Architectural overview of the proposed segmentation framework.

To evaluate the importance of these attention modules, ablation studies were conducted by selectively removing them from the model. The results, discussed in Sect. 3.3, demonstrate a clear drop in segmentation accuracy, especially in densely populated or low-contrast images. This confirms that non-local attention provides essential contextual information.

The modular structure of the network facilitates domain adaptation. This flexibility is critical because microscopy data often exhibit non-stationary statistics and unpredictable morphology [26,29]. Consequently, our design not only serves as a strong baseline for microorganism segmentation but also as a framework for testing novel, microscopy-oriented attention mechanisms.

2.2 Basic Non-local Block

The fundamental component of our attention mechanism is the non-local operation [20], which models pairwise interactions across the entire feature space. Conceptually, it generalizes self-attention by linking every position in the feature map to every other position, allowing the model to integrate contextual cues globally (see Fig. 2).

Given an input feature map partitioned into a grid of spatial tokens $\{x_i\}$, each token is linearly projected into query (Q), key (K), and value (V) spaces through learned mappings $f(\cdot)$, $g(\cdot)$, and $h(\cdot)$:

$$Q = f(x_i), \quad K = g(x_i), \quad V = h(x_i). \tag{1}$$

Attention scores are then calculated using scaled dot products between the query and key vectors:

$$\alpha_i = SoftMax\left(\frac{q_i k_1}{\sqrt{d_k}}, \frac{q_i k_2}{\sqrt{d_k}}, \ldots, \frac{q_i k_n}{\sqrt{d_k}}\right). \tag{2}$$

The resulting weights determine how much each region contributes to the output representation:

$$o_i = v\left(\sum_{j=1}^{n} \alpha_i v_j\right), \tag{3}$$

where $v(\cdot)$ is a learnable projection. A residual connection adds stability and preserves gradient flow, yielding $Y = WO + X$.

This formulation enables information exchange between distant but visually related regions, an essential capability for microscopy, where two streptococcus cells belonging to the same pair might be spatially separated by noise or uneven illumination. By aligning their contextual signatures, the non-local block helps preserve morphological coherence across the image.

2.3 Localized Non-local Block

Although the standard non-local block captures global dependencies, it ignores the spatial distances between tokens, treating all relationships as equally probable. In microscopy data, streptococci typically form tightly packed chains or

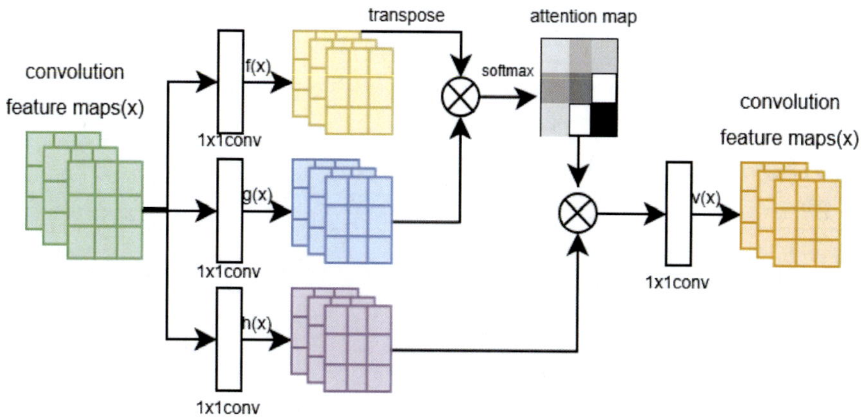

Fig. 2. Scheme of basic non-local block.

adjacent pairs, meaning spatial proximity often correlates with biological relevance.

To incorporate this prior knowledge, we introduce a distance-weighted modification of the non-local mechanism (see Fig. 3). The adjusted similarity coefficients are expressed as:

$$\alpha_i^* = SoftMax\left(dist(i,1)\frac{q_i k_1}{d_k}, \ldots, dist(i,n)\frac{q_i k_n}{d_k}\right), \quad (4)$$

where

$$dist(i,k) = \frac{1}{\min(|col_i - col_k|, |row_i - row_k|)} \quad (5)$$

is a simple spatial weighting factor based on grid coordinates.

This reweighting biases attention toward nearby regions without requiring explicit positional encodings as in transformer architectures [3,30]. The result is a lightweight yet spatially aware mechanism that reinforces biologically plausible local continuity while maintaining the capacity for long-range reasoning.

2.4 Homogeneity-Based Non-local Block

A distinctive trait of streptococci imagery is the internal homogeneity within bacterial pairs: cells in a pair usually share similar intensity, texture, and size. To exploit this property, we introduce a homogeneity-based weighting scheme that amplifies contributions from internally consistent tokens (see Fig. 4).

For each token x_i, we compute a homogeneity coefficient θ_i by dividing the feature vector into two halves and computing their normalized dot product:

$$\theta_i = \frac{x_i[1,\ldots,\lfloor n/2 \rfloor] \cdot x_i[\lfloor n/2 \rfloor + 1,\ldots,n]}{|x_i[1,\ldots,\lfloor n/2 \rfloor] \cdot x_i[\lfloor n/2 \rfloor + 1,\ldots,n]|}. \quad (6)$$

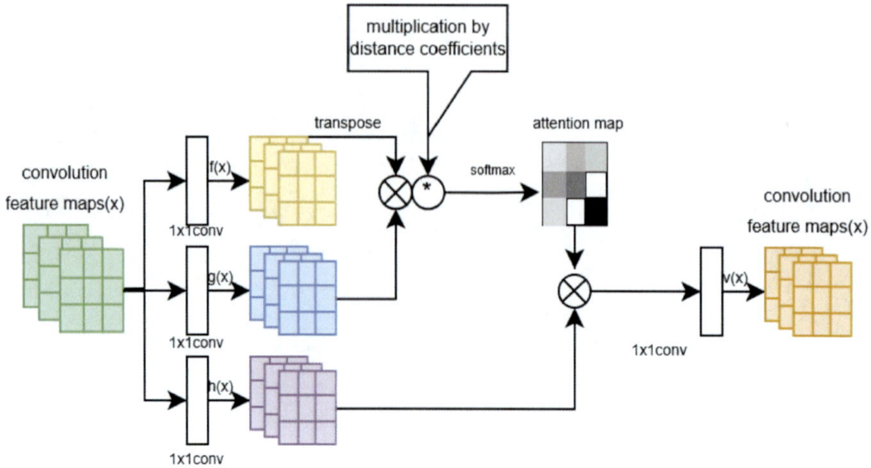

Fig. 3. Scheme of localized non-local block.

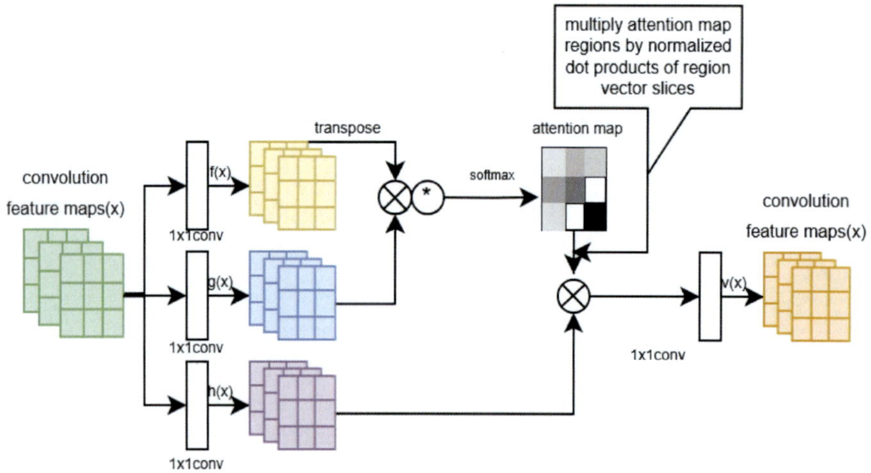

Fig. 4. Scheme of homogeneity-based non-local block.

This scalar reflects the internal consistency of the local structure. The attention aggregation is then modulated as:

$$o_i^* = v\left(\theta_i \sum_{j=1}^{n} \alpha_i v_j\right). \tag{7}$$

By emphasizing tokens with higher homogeneity, the model focuses more on visually stable bacterial regions and less on irregular background or staining artifacts. This significantly enhances boundary preservation and noise resilience in challenging microscopy conditions.

2.5 Homogeneity-Localized Non-local Block

To fully exploit both spatial and structural priors, we combine the two previously introduced mechanisms into a single hybrid attention unit, the Homogeneity-Localized Non-Local Block (see Fig. 5). This formulation allows the model to simultaneously account for spatial proximity and internal coherence, focusing attention on regions that are both adjacent and visually consistent.

The resulting equations integrate both modifications:

$$\alpha_i = SoftMax\left(dist(i,1)\frac{q_i k_1}{d_k}, \ldots, dist(i,n)\frac{q_i k_n}{d_k}\right), \tag{8}$$

$$o_i^* = v\left(\theta_i \sum_{j=1}^{n} \alpha_i v_j\right). \tag{9}$$

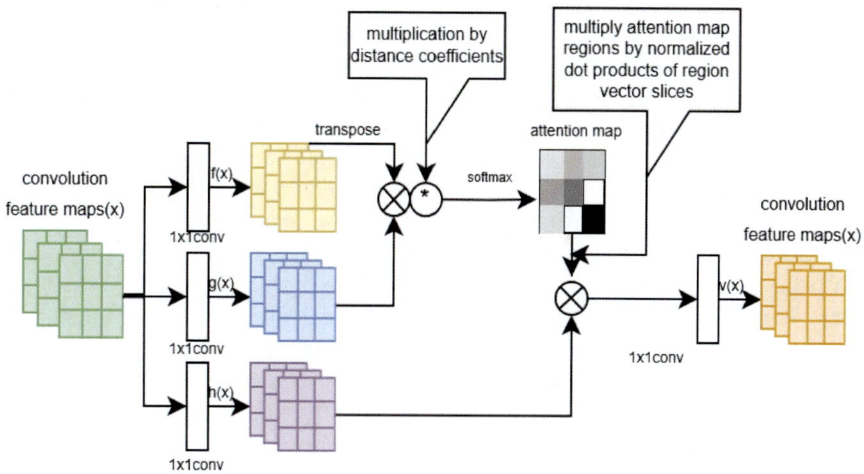

Fig. 5. Scheme of combined homogeneity-localized non-local block.

This dual weighting scheme encourages the model to focus on spatially coherent and morphologically uniform structures, precisely the features that define streptococci colonies. As shown later in Sect. 3.3, the integration of this enhanced block results in superior segmentation accuracy and stability across varied illumination and density conditions.

3 Evaluation

We evaluate the proposed segmentation framework using a newly curated microscopy dataset focused on streptococci segmentation. This section details the dataset composition, training configuration, and comparative performance against established baselines.

3.1 Dataset

To enable consistent benchmarking, we developed the *Microscopy Streptococci Images Dataset (MSID)*, a dataset specifically designed for training, validation, and evaluation of segmentation models targeting streptococci colonies in microscopic imagery. The dataset was collected in controlled laboratory conditions and reflects the variability commonly encountered in routine microbiological analysis.

Image acquisition was performed using a Levenhuk MED D30T optical microscope equipped with a digital camera. Samples were prepared following standard microbiological protocols and include both stained and unstained specimens. Imaging was conducted at fixed magnification settings, while variations in illumination intensity, focal depth, and background artifacts were intentionally preserved to ensure robustness to real-world acquisition noise. As a result, the dataset exhibits significant diversity in contrast, sharpness, and colony density.

The annotation process was carried out manually by trained microbiologists with domain expertise in bacterial morphology. Pixel-wise segmentation masks were created by delineating individual streptococci structures and chains using specialized annotation software. Particular attention was paid to separating touching or partially overlapping bacteria, as well as to excluding background artifacts such as debris, staining irregularities, and illumination gradients. To reduce subjectivity, ambiguous regions were reviewed jointly by multiple annotators and resolved by consensus.

The final dataset consists of approximately 3,000 images and is partitioned into training, validation, and test subsets with a split of 2,500 / 250 / 250 images, respectively. The splits were created at the image level to avoid data leakage between sets. Representative examples of raw images and their corresponding ground-truth masks are shown in Fig. 6.

The MSID dataset will be made publicly available to support reproducibility and future research in microscopy-based bacterial segmentation [31].

3.2 Training Setup

Before training, all images are resized to 256×256 pixels, normalized to unit variance, and augmented with random horizontal flips, slight rotations, and brightness adjustments. Model optimization is performed using the AdamW optimizer ($\beta_1 = 0.9$, $\beta_2 = 0.999$, weight decay $= 0.05$) with a batch size of 8 and an initial learning rate of 10^{-8}. Each model is trained for up to 100 epochs on a single NVIDIA RTX 3090 GPU.

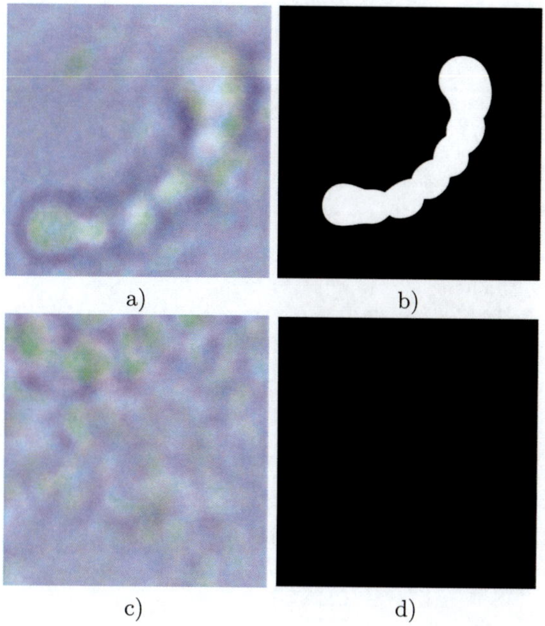

Fig. 6. Examples from MSID: (a) microscopy image containing streptococci chains; (b) corresponding ground-truth mask; (c) background-only region; (d) background mask.

3.3 Experimental Results

Segmentation performance was assessed using the Intersection-over-Union (IoU) metric alongside the Dice similarity coefficient. A summary of the comparative evaluation across multiple baseline approaches, including convolutional neural networks, transformer-based models, and the proposed attention-augmented architectures, is provided in Table 1.

The experimental outcomes highlight the advantages of integrating hybrid non-local attention components. Although the conventional U-Net architecture is effective in capturing local spatial features, it exhibits limitations when dealing with overlapping bacterial chains or images with insufficient contrast. The introduction of non-local attention leads to consistent improvements in both IoU and Dice scores, underscoring the importance of incorporating global contextual information.

These improvements are further reinforced by domain-specific enhancements. The localized non-local module exploits the spatial continuity inherent in streptococcal formations, whereas the homogeneity-driven module leverages the uniform morphological characteristics of individual bacteria. Their joint deployment within the proposed homogeneity-localized design yields the strongest overall results, surpassing both traditional CNN-based baselines and transformer-based segmentation models.

Table 1. Comparison of segmentation performance on the MSID dataset.

Model	Mean IoU	Dice coefficient
UNet (Baseline)	0.796	0.852
UNet++ [19]	0.811	0.865
UNet (ResNet-18 encoder) [2]	0.834	0.884
UNet (ResNet-50 encoder) [2]	0.842	0.891
DeepLabV3+	0.856	0.902
UNet (EfficientNet-b4 encoder) [32]	0.869	0.915
UNet++ (EfficientNet-b4)	0.878	0.923
UNet with specialized loss [28]	0.888	0.931
UNet + basic non-local block	0.897	0.939
Dual-Stream CoAtNet [33]	0.905	0.944
SETR [34]	0.912	0.950
UNet + localized non-local block	0.925	0.958
UNet + homogeneity-based non-local block	0.939	0.966
UNet + homogeneity-localized non-local block	0.952	0.973

4 Conclusion

In this work, we addressed the problem of streptococci segmentation in microscopic images and proposed a segmentation framework tailored to the unique morphological and visual characteristics of these microorganisms. The core contribution is an enhanced U-Net architecture augmented with hybrid non-local attention mechanisms that explicitly model both local structure and long-range spatial dependencies.

The proposed attention modules, localized, homogeneity-based, and the unified homogeneity-localized non-local block, allow the network to jointly exploit fine-grained boundary information and global contextual cues. This design effectively mitigates common challenges in microscopy data, such as low contrast, heterogeneous illumination, and densely packed or overlapping bacterial structures.

Comprehensive experimental evaluation on the Microscopy Streptococci Images Dataset (MSID) demonstrates that the proposed attention-augmented U-Net consistently outperforms conventional convolutional architectures as well as transformer-based baselines. In particular, the hybrid homogeneity-localized non-local block achieves the best Intersection-over-Union and Dice scores, highlighting the benefit of incorporating biologically motivated priors into attention mechanisms for robust segmentation.

Future work will focus on extending the proposed approach to multi-class microbial segmentation and three-dimensional microscopy data, as well as investigating adaptive attention strategies driven by domain-specific structural priors.

References

1. Deng, J., Dong, W., Socher, R., Li, L., Li, K., Fei-Fei, L.: Imagenet: a large-scale hierarchical image database. In: 2009 IEEE Conference on Computer Vision and Pattern Recognition, pp. 248–255 (2009). https://doi.org/10.1109/CVPR.2009.5206848
2. He, K., Zhang, X., Ren, S., Sun, J.: Deep residual learning for image recognition. In: 2016 IEEE Conference on Computer Vision and Pattern Recognition (CVPR), pp. 770–778. https://api.semanticscholar.org/CorpusID:206594692 (2015/2016)
3. Dai, Z., Liu, H., Le, Q., Tan, M.: Coatnet: marrying convolution and attention for all data sizes (2021)
4. Radford, A., Kim, J., Hallacy, C., Ramesh, A., Goh, G., Agarwal, S., Sastry, G., Askell, A., Mishkin, P., Clark, J., Krueger, G., Sutskever, I.: Learning transferable visual models from natural language supervision. In: International Conference on Machine Learning (2021)
5. Li, J., Li, D., Xiong, C., Hoi, S.C.H.: BLIP: Bootstrapping Language-Image Pre-training for Unified Vision-Language Understanding and Generation. CoRR (2022)
6. Chattopadhyay, T., Sinha, A.: Recognition of trademarks from sports videos for channel hyperlinking in consumer end. In: 2009 IEEE 13th International Symposium on Consumer Electronics, pp. 943–947 (2009). https://doi.org/10.1109/ISCE.2009.5156881
7. Ivasic-Kos, M., Pobar, M., Ipsic, I.: Automatic movie posters classification into genres. In: Bogdanova, A.M., Gjorgjevikj, D. (eds.) ICT Innovations 2014, pp. 319–328. Springer, Cham (2015)
8. Samarin, A., Savelev, A., Toropov, A., Dzestelova, A., Malykh, V., Mikhailova, E., Motyko, A.: Trainable agents movement strategies for advertising sign visual descriptors. Pattern Recognit Image Anal. **32**(3), 651–657 (2022). https://doi.org/10.1134/S1054661822030373
9. Samarin, A., Savelev, A., Toropov, A., Dzestelova, A., Malykh, V., Mikhailova, E., Motyko, A.: Predictors based on convolutional neural networks for the movement strategy of trainable agents for building customized image descriptors. Pattern Recognit Image Anal. **33**, 139–146 (2023). https://doi.org/10.1134/S105466182302013X
10. Samarin, A., Savelev, A., Toropov, A., Dzestelova, A., Malykh, V., Mikhailova, E., Motyko, A.: The complete study of the movement strategies of trained agents for visual descriptors of advertising signs. In: Rousseau, J.J., Kapralos, B. (eds.) Pattern Recognition, Computer Vision, and Image Processing. ICPR 2022 International Workshops and Challenges, pp. 571–585. Springer Nature Switzerland, Cham (2023)
11. Tsai, T., Cheng, W., You, C., Hu, M., Tsui, A.W., Chi, H.: Learning and recognition of on-premise signs from weakly labeled street view images. IEEE Trans. Image Process. **23**(3), 1047–1059 (2014). https://doi.org/10.1109/TIP.2014.2298982
12. Wahid, M.F., Ahmed, T., Habib, M.A.: Classification of microscopic images of bacteria using deep convolutional neural network. In: 2018 10th International Conference on Electrical and Computer Engineering (ICECE), pp. 217–220 (2018). https://doi.org/10.1109/ICECE.2018.8636750
13. Xing, F., Xie, Y., Su, H., Liu, F., Yang, L.: Deep learning in microscopy image analysis: a survey. IEEE Trans. Neural Netw. Learn. Syst. **29**(10), 4550–4568 (2018). https://doi.org/10.1109/TNNLS.2017.2766168

14. Liu, R., Dai, W., Wu, T., Wang, M., Wan, S., Liu, J.: Aimic: deep learning for microscopic image classification. Comput. Methods Programs Biomed. **226**, 107162 (2022). https://doi.org/10.1016/j.cmpb.2022.107162
15. Misselwitz, B., Strittmatter, G., Periaswamy, B., Schlumberger, M., Rout, S., Horvath, P., Kozak, K., Hardt, W.D.: Enhanced cell classifier: a multi-class classification tool for microscopy images. BMC Bioinform. **11**, 30 (2010). https://doi.org/10.1186/1471-2105-11-30
16. Nguyen, L.D., Lin, D., Lin, Z., Cao, J.: Deep CNNS for microscopic image classification by exploiting transfer learning and feature concatenation. In: 2018 IEEE International Symposium on Circuits and Systems (ISCAS), pp. 1–5 (2018). https://doi.org/10.1109/ISCAS.2018.8351550
17. Shelomentseva, I.G.: Classification of light microscopy image using probabilistic Bayesian neural network. In: Kryzhanovsky, B., Dunin-Barkowski, W., Redko, V., Tiumentsev, Y. (eds.) Advances in Neural Computation, Machine Learning, and Cognitive Research VI, pp. 265–270. Springer, Cham (2023)
18. Ronneberger, O., Fischer, P., Brox, T.: U-net: convolutional networks for biomedical image segmentation. CoRR (2015). http://arxiv.org/abs/1505.04597
19. Zhou, Z., Siddiquee, M.M.R., Tajbakhsh, N., Liang, J.: Unet++: a nested u-net architecture for medical image segmentation. CoRR (2018). http://arxiv.org/abs/1807.10165
20. Wang, X., Girshick, R., Gupta, A., He, K.: Non-local neural networks. In: Proceedings of the IEEE Conference on Computer Vision and Pattern Recognition, pp. 7794–7803 (2018)
21. Yin, M., Yao, Z., Cao, Y., Li, X., Zhang, Z., Lin, S., Hu, H.: Disentangled non-local neural networks. In: Computer Vision–ECCV 2020: 16th European Conference, Proceedings, Part XV, pp. 191–207. Springer (2020)
22. Ruan, B., Shuai, H.H., Cheng, W.H.: Vision transformers: state of the art and research challenges. CoRR (2022)
23. Kirmani, S., Raghavan, P.: Scalable parallel graph partitioning. In: Proceedings of the International Conference on High Performance Computing, Networking, Storage and Analysis. SC '13. Association for Computing Machinery, New York, NY, USA (2013). https://doi.org/10.1145/2503210.2503280
24. Kirmani, S., Sun, H., Raghavan, P.: A scalability and sensitivity study of parallel geometric algorithms for graph partitioning. In: 2018 30th International Symposium on Computer Architecture and High Performance Computing (SBAC-PAD), pp. 420–427 (2018). https://doi.org/10.1109/CAHPC.2018.8645916
25. Kirmani, S., Park, J., Raghavan, P.: An embedded sectioning scheme for multiprocessor topology-aware mapping of irregular applications. Int. J. High Perform. Comput. Appl. **31**(1), 91–103 (2017). https://doi.org/10.1177/1094342015597082
26. Samarin, A., Savelev, A., Toropov, A., Dzestelova, A., Malykh, V., Mikhailova, E., Motyko, A.A.: One-stage classifiers based on u-net and autoencoder with attention for recognition of neoplasms from single-channel monochrome computed tomography images. Pattern Recognit. Image Anal. **33**, 132–138. https://api.semanticscholar.org/CorpusID:259310593 (2023)
27. Samarin, A., Toropov, A., Dzestelova, A., Nazarenko, A., Kotenko, E., Mikhailova, E., Savelev, A., Motyko, A.: Specialized non-local blocks for recognizing tumors on computed tomography snapshots of human lungs. In: 2024 35th Conference of Open Innovations Association (FRUCT), pp. 659–664 (2024). https://doi.org/10.23919/FRUCT61870.2024.10516343

28. Samarin, A., Savelev, A., Toropov, A., Nazarenko, A., Golovatiuk, A., Dmitriev, P., Dzestelova, A., Mikhailova, E., Motyko, A., Malykh, V.: Segmentation of the iris and pupil of the human eye in images from an infrared camera. Pattern Recognit Image Anal. **34**(3), 855–862 (2024)
29. Samarin, A., Savelev, A., Toropov, A., Dzestelova, A., Malykh, V., Mikhailova, E., Motyko, A.: One-staged attention-based neoplasms recognition method for single-channel monochrome computer tomography snapshots. Pattern Recognit Image Anal. **32**(3), 645–650 (2022)
30. Dosovitskiy, A., Beyer, L., Kolesnikov, A., Weissenborn, D., Zhai, X., Unterthiner, T., Dehghani, M., Minderer, M., Heigold, G., Gelly, S., Uszkoreit, J., Houlsby, N.: An image is worth 16x16 words: transformers for image recognition at scale (2020)
31. MMICD dataset. https://github.com/itmo-cv-lab/mmicd
32. Tan, M., Le, Q.V.: Efficientnet: rethinking model scaling for convolutional neural networks. ArXiv abs/1905.11946 (2019). https://api.semanticscholar.org/CorpusID:167217261
33. Zaidkilani, N., García, M., Puig, D.: Dual-stream coatnet models for accurate breast ultrasound image segmentation. Neural Comput. Appl. **36**, 1–17 (2024). https://doi.org/10.1007/s00521-024-09963-w
34. Thisanke, H., Deshan, C., Chamith, K., Seneviratne, S., Vidanaarachchi, R., Herath, D.: Semantic segmentation using vision transformers: a survey. ArXiv abs/2305.03273 (2023). https://api.semanticscholar.org/CorpusID:258546745

Zero-Shot Shrimp Disease Detection with Vision–Language Models: A Comparison with a CNN Baseline on the ShrimpDiseaseBD Dataset

Ha Viet Thang(✉), Inh Chau, and Cuong Tran

FPT University, Can Tho Campus, Can Tho, Vietnam
havietthang@acm.org

Abstract. Shrimp aquaculture is highly vulnerable to infectious diseases such as White Spot Syndrome Virus (WSSV) and Black Gill (BG), where early visual detection is crucial but challenging. Convolutional neural networks (CNNs) can be trained on labeled shrimp images but require supervision and remain tied to the known disease classes. Vision–language models (VLMs) promise more flexible zero-shot detection by combining image understanding with natural language reasoning. In this paper, we benchmark several modern VLMs on the ShrimpDiseaseBD dataset for both fine-grained (4-class) and binary (healthy vs. diseased) classification, and compare their performance and resource usage to a supervised CNN baseline (ReXNet-150). Zero-shot VLMs remain far below the CNN in accuracy and are highly sensitive to prompt design, but offer interpretability and flexibility. We discuss error modes, trade-offs between sensitivity and specificity, and practical implications for aquaculture decision support.

Keywords: ShrimpDiseaseBD · Vision–language models · Disease detection · Aquaculture · CNN baseline

1 Introduction

Shrimp aquaculture is a key economic activity in many coastal regions, especially in Asia and Latin America, and contributes significantly to food security and export revenue [1–3]. However, farms are extremely vulnerable to fast-spreading diseases such as White Spot Syndrome Virus (WSSV) and Black Gill (BG), which can decimate stocks in a matter of days [4–6]. Once an outbreak reaches a certain threshold, mass mortality and emergency pond harvesting become almost inevitable, causing large financial losses and threatening the livelihoods of smallholders.

In practice, farmers and technicians rely on a combination of indirect signals (growth slowdown, feeding behavior) and direct visual inspection of sampled shrimp to monitor health. They look for white spots on the shell, darkened

L. Huang and D. Greenhalgh (Eds.): ICMLC 2026, LNNS 2061, pp. 386–401, 2027.
https://doi.org/10.1007/978-3-032-31236-5_27

or necrotic gills, and other morphological changes. This manual inspection is subjective and time-consuming, and requires experience that may not always be present on small farms. Sending samples to laboratories for PCR testing provides high accuracy but introduces delays and additional cost, making frequent routine screening difficult.

Automated image-based tools offer an attractive complement to human inspection. A farmer could capture images of shrimp using a smartphone, and a model could flag individuals that look suspicious. Supervised convolutional neural networks (CNNs) have already been applied to shrimp disease detection with good results, particularly after the release of the ShrimpDiseaseBD dataset [2,7–11]. However, CNNs require labeled data and retraining whenever new diseases or imaging conditions are introduced. They also typically produce only class labels, without explicit reasoning.

Vision–language models (VLMs) are a newer class of models that combine an image encoder with a large language model. In principle, such models can perform visual recognition tasks in a zero-shot manner: the task and label semantics are specified by a natural language prompt, and no task-specific fine-tuning is strictly required. Because VLMs generate text, they can also offer richer explanations (e.g., pointing out visible lesions), which may be helpful for non-expert users.

Our main contributions are:

- A benchmark of five recent VLMs on ShrimpDiseaseBD with a standardized binary screening prompt and a prompt-sensitivity analysis for LLaVA.
- A CNN transfer-learning baseline, strongly inspired by a publicly released Kaggle notebook, and a comparison of its 4-class performance to zero-shot VLMs on a simpler binary task. [12]
- A comparison of runtime and memory usage across models, plus qualitative analysis of VLM textual reasoning and CNN Grad-CAM heatmaps.
- A discussion of practical deployment scenarios where a hybrid system combining CNNs and VLMs could provide both accuracy and interpretability for shrimp farmers.

2 Related Work

2.1 Shrimp Disease Detection with Deep Learning

Recent work has applied deep convolutional architectures to shrimp disease detection [2,7,11]. Büyükarıkan [8] proposed an ensemble of CNNs for robust detection of viral diseases, showing that combining multiple backbones and decision strategies can improve stability across varied imaging conditions. Aung et al. [9] developed CNN-based identification of shrimp pathogens relevant to intensive aquaculture systems and reported strong performance on curated datasets, highlighting the potential of computer vision for early detection and intervention. Focusing specifically on White Spot Syndrome Virus (WSSV), Querol et al. [10] designed a mobile monitoring application using lightweight architectures such as MobileNetV3 and EfficientNet-B0 for on-device inference, and emphasized that latency and resource usage are critical constraints in farm environments [2,10,11].

Collectively, these works demonstrate that supervised CNNs can reach high accuracy for specific disease categories and deployment setups (e.g., mobile apps). However, they depend on task-specific datasets and training. When new disease types emerge or when farms operate under different visual conditions (water turbidity, lighting, camera quality), models may need retraining or domain adaptation. In addition, pure CNN systems often require separate tooling to provide explanations or uncertainty estimates.

2.2 Vision–Language Models in Specialized Domains

CLIP [13] popularized large-scale contrastive language–image pretraining and showed that natural language prompts can drive zero-shot classification across thousands of categories. Its success inspired a family of VLMs that integrate visual backbones into generative large language models [14–16]. LLaVA [17] performs visual instruction tuning on multimodal data curated with the help of GPT-4, enabling conversational visual question answering and image understanding. Qwen-VL [18] adopts a similar approach but emphasizes versatility, supporting detection, OCR, and captioning.

In specialized domains such as medicine or agriculture, VLMs have been explored mainly as zero-shot labelers or assistants [19]. While they often demonstrate impressive qualitative reasoning, quantitative performance usually lags behind dedicated discriminative models unless some form of domain-specific fine-tuning or in-context learning is applied. Furthermore, their behavior is known to be highly sensitive to prompt wording, especially when the task requires sharp decision boundaries.

To our knowledge, there has been no prior systematic evaluation of general-purpose VLMs on shrimp disease detection using ShrimpDiseaseBD. Our work aims to fill this gap and provide an initial baseline for future research on domain-adapted VLMs in aquaculture.

3 Dataset

3.1 ShrimpDiseaseBD Overview

We use ShrimpDiseaseBD [7], which contains 761 images of farmed shrimp collected in Bangladesh under expert guidance. Images were captured under realistic farm conditions and then annotated by specialists into four classes:

- **Healthy**: visually normal shrimp, with clean shells and gills.
- **BG**: Black Gill disease, characterized primarily by darkened or black gills.
- **WSSV**: White Spot Syndrome Virus, where shrimp exhibit distinct white spots on the carapace.
- **WSSV+BG**: simultaneous signs of both WSSV and BG.

Table 1. ShrimpDiseaseBD dataset composition.

Class	Description	Count	Percentage (%)
Healthy	Normal appearance	424	55.7
BG	Blackened gills	111	14.6
WSSV	White spots on shell	109	14.3
WSSV+BG	Combination of WSSV and BG	117	15.4
Total		761	100

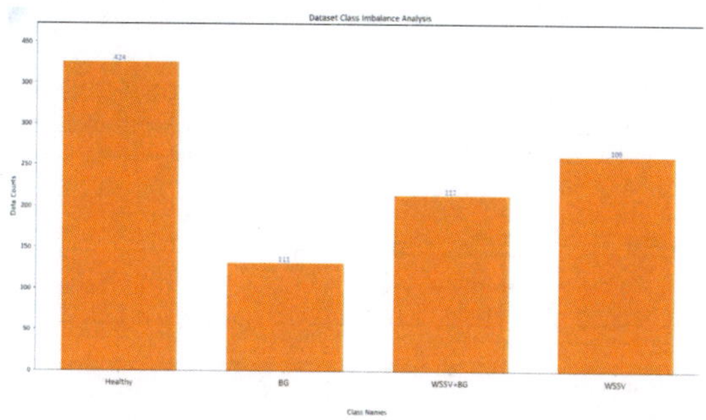

Fig. 1. Class distribution of ShrimpDiseaseBD. The majority of images are healthy; diseased classes are roughly balanced.

Table 1 summarizes the dataset composition. Figure 1 visualizes the same class distribution.

The dataset is moderately imbalanced, with Healthy as the majority class and three disease classes that are relatively balanced with each other. This structure is realistic: in farm monitoring, the majority of sampled individuals are often healthy, while disease categories are less frequent but clinically important.

3.2 Image Characteristics and Splits

Images in ShrimpDiseaseBD contain one or more shrimp, often placed on trays, nets, or other farm surfaces. Lighting conditions vary, and some images include background clutter such as tools or hands. These factors introduce realistic noise and make the task more challenging than stylized laboratory photos. Representative healthy and WSSV examples are shown in Fig. 2.

For VLM experiments, we treat all 761 images as an evaluation set (no fine-tuning). For the CNN baseline, we follow a standard supervised pipeline that is closely modeled on the publicly available Kaggle implementation mentioned above. We perform a stratified split into training (approximately 90%), validation (approximately 5%), and test (approximately 5%). Stratification preserves the relative class proportions in each split. Images are resized to 224×224 and normalized with ImageNet statistics, following the Kaggle recipe.

Fig. 2. Side-by-side examples from ShrimpDiseaseBD. Left: Healthy shrimp. Right: White Spot Syndrome Virus (WSSV), showing visible lesions on the shell.

4 Methodology

4.1 Vision–Language Models and Prompting Strategy

We evaluate the following VLMs:

- **LLaVA-1.5-7B** [17]: an open 7B multimodal chat model built on a Vicuna-style LLM and a CLIP-like image encoder.
- **Qwen2-VL-2B** [18]: a 2B-parameter vision–language instruct model, designed as a lightweight but versatile general-purpose VLM.
- **Phi-3.5-Vision-Instruct**: a 4B proprietary vision–language model emphasizing strong reasoning at relatively small scale.
- **InternVL2-2B**: a 2.2B model from the InternVL 2.0 family, optimized for visual understanding and efficiency.
- **SmolVLM-2B**: an efficient 2B model targeted at resource-constrained deployments.

We focus on binary classification (`HEALTHY` vs. `DISEASED`) as the main evaluation protocol and treat 4-class classification as a preliminary experiment. For the binary task, we design a shared screening prompt for Qwen, Phi, InternVL2, and SmolVLM with three components:

1. **Task description**: we explicitly state that the model should act as an early screening assistant for shrimp disease.
2. **Class definitions**: we provide natural language descriptions of what Healthy and Diseased shrimp look like, including WSSV and BG symptoms (spots, dark gills, discoloration).
3. **Decision policy**: we instruct the model to "err on the side of DISEASED" when uncertain, reflecting a conservative screening design that prioritizes recall.

We also impose a strict output format to simplify parsing:

```
Description: <...>
Prediction: <HEALTHY or DISEASED>
```

At inference time, we parse the second line to extract the predicted label. If the output does not contain exactly one of `HEALTHY` or `DISEASED`, we treat the prediction as unknown and count it as an error for accuracy. To facilitate later analysis, we log for each image a CSV row containing the ground-truth binary label and the raw model answer string. An excerpt of this logging format is shown in Table 2.

4.1.1 Prompt Sensitivity with LLaVA For LLaVA-1.5 we explicitly study prompt sensitivity by testing two prompts, A and B. Prompt A is the aggressive, DISEASED-favoring screening prompt that we also use for the other VLMs; Prompt B is a conservative variant that tells LLaVA to default to HEALTHY unless strong visual signs of disease are present.

Prompt A (Aggressive screening, Shrimp_VLM_Llama.ipynb).

```
You are helping classify farmed shrimp images into two categories for early
disease screening:
- HEALTHY: shrimp with normal translucent or slightly pigmented shell, no white
  spots, no obvious lesions or deformities, gills look normal (light, clean,
  not black), body looks firm and well-formed, posture and appearance look
  normal.
- DISEASED: shrimp showing any clear or suspicious sign of disease, including
  but not limited to:
  * Black Gill disease (BG): gills turning yellowish/whitish, then brown to
    dark/black patches or entire gill darkening; shrimp may breathe poorly, be
    weak, float near the surface or pond edges, or lie on the bottom.
  * White Spot Syndrome Virus (WSSV): white spots of about 0.52 mm on the shell
    (especially carapace and abdominal segments), shrimp stop feeding, swim
    weakly, gather at pond edges, body may become pink/red.
  * Combined patterns where shrimp show both WSSV-like white spots AND
    dark/blackened gills in the same image.

IMPORTANT: This is a screening tool. It is better to mark a shrimp as DISEASED
than to miss a possible disease. If you are uncertain or see any doubtful
abnormality, choose DISEASED.

TASK:
1) First, give a brief visual description of what you see (background, shrimp
   appearance, gill color, shell color, spots, lesions, posture, etc.).
2) Then, on a new line, output a predicted label chosen from EXACTLY these two
   options: HEALTHY or DISEASED.

Use this exact format:
Description: <your detailed description>
Prediction: <ONE OF: HEALTHY, DISEASED>
Do not output anything else after the Prediction line.
```

Prompt B (Conservative HEALTHY-default, Llama_Shrimp_VLM.ipynb).

```
You are helping classify farmed shrimp images into two categories for early
disease screening.

CATEGORIES:
1) HEALTHY
2) DISEASED

HEALTHY (DEFAULT CHOICE):
- Shell looks mostly uniform and clean (gray, brown, or slightly translucent),
  even if there are natural stripes or bands.
- Normal patterns, bands and color gradients along the body and tail are COMMON
  and should be considered HEALTHY.
- Tail tip, legs, antennae or small parts may be darker or slightly reddish but
  should not look like large rotten areas.
- Gills (if visible) may be slightly darker than the body but not clearly
  jet-black over a large continuous region.
- Head, body and tail may have shadows, reflections or minor color changes due
  to lighting or background; this alone does NOT mean disease.
- The shrimp may be dead, lying on a blue or white board, in water, or on ice;
  these are NORMAL situations and should NOT be used alone to mark disease.

DISEASED (ONLY WHEN SIGNS ARE CLEAR AND STRONG):
- One or more STRONG abnormal signs are visible, for example:
  * Clearly dark / brown gills or head that are much darker than normal stripes
    or shadows, covering a noticeable area.
  * Clusters of distinct, round white or yellow spots on the shell (especially
    on the head and abdominal segments).
  * Large, obvious red, brown or black patches on the head or body that look
    rotten or severely damaged.
  * Big necrotic or rotten-looking areas on the shell or flesh.
  * Very abnormal body posture TOGETHER with clear lesions (for example,
    twisted body plus large rotten areas).

VERY IMPORTANT DECISION RULE (CONSERVATIVE):
- Start by assuming the shrimp is HEALTHY.
- ONLY change to DISEASED if you clearly see at least one of the strong abnormal
  signs listed above.
- If you are unsure, if the signs are weak, or could be explained by lighting,
  camera angle, or background color, then classify as HEALTHY.
- DO NOT mark the shrimp as DISEASED only because it is dead, wet, bent, or
  lying on a blue/white background.

EXAMPLES:
Example A (HEALTHY): A shrimp on a blue or white board, body mostly gray-brown
with normal dark bands, tail tip a bit darker, no clusters of white spots, no
large black or red rotten-looking patches, gills not clearly jet-black.
-> Prediction: HEALTHY.
Example B (DISEASED): A shrimp on a board with very dark / black gills covering
a large area, or obvious large red/brown/black rotten patches on the head or
```

```
body segments, or clear clusters of white spots on the shell.
-> Prediction: DISEASED.

TASK:
1) First, give a brief visual description of what you see (background, shrimp
   appearance, gill color, shell color, spots, lesions, posture, etc.).
2) Then, on a new line, output a predicted label chosen from EXACTLY these two
   options: HEALTHY or DISEASED.

Use this exact format:
Description: <your detailed description>
Prediction: <ONE OF: HEALTHY, DISEASED>
Do not output anything else after the Prediction line.
```

Table 2. Excerpt from the SmolVLM binary classification CSV log. Each row contains the true binary label and the model's textual answer.

true_binary_label	model_answer
HEALTHY	Description: Shrimp with darkened gills and white spots on shell Prediction: DISEASED
HEALTHY	Description: Shrimp lying on blue surface with no visible disease Prediction: HEALTHY
HEALTHY	Description: Shrimp with normal translucent or slightly pigmented shell, no white spots, no obvious lesions or deformities, gills look normal (light, clean, not black), body looks firm and well-formed, posture and appearance look normal Prediction: HEALTHY
HEALTHY	Description: Shrimp lying on its side with legs curled in, gills darkened, body looks firm and well-formed, posture and appearance look normal Prediction: DISEASED

4.2 CNN Baseline and Training Setup

As a supervised baseline, we fine-tune a ReXNet-150 CNN on the 4-class ShrimpDiseaseBD task using the `timm` library. Following the public Kaggle-style setup, we use the same backbone, similar preprocessing, and a comparable training schedule, but retrain from scratch on our own splits and evaluate on a held-out test set. The model starts from ImageNet-pretrained weights; we replace the classifier with a 4-way linear head and train with cross-entropy and Adam. We use an initial learning rate of $\sim 3 \times 10^{-4}$ with cosine decay and warmup, and a batch size chosen by GPU memory (typically 32 or 64). Training dynamics (loss and macro F1 over epochs) are shown in Fig. 3.

To interpret the CNN decisions, we visualize Grad-CAM++ attribution maps [20] on representative test images (Fig. 4). The heatmaps highlight image regions that most influence the predicted class: for BG, activations concentrate around the gill area; for WSSV, they focus on the shell and carapace where white-spot patterns appear; and for WSSV+BG, attention typically spans multiple regions, reflecting combined symptoms. We include both correct and incorrect predictions to illustrate when the model attends to clinically meaningful cues versus when attention is distracted by background or non-diagnostic regions.

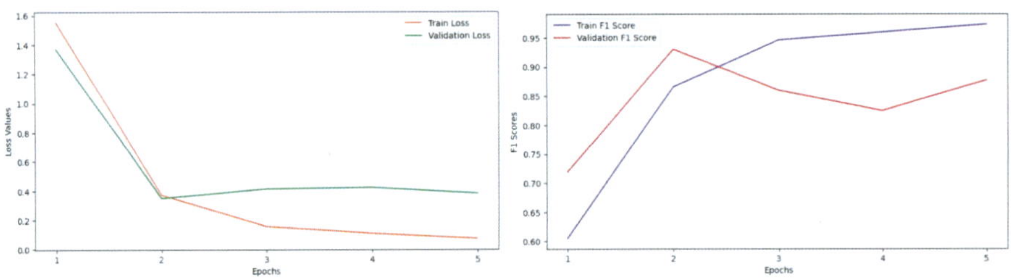

Fig. 3. CNN training on ShrimpDiseaseBD: loss and macro F1 on training vs. validation sets. The model converges quickly; early stopping prevents overfitting.

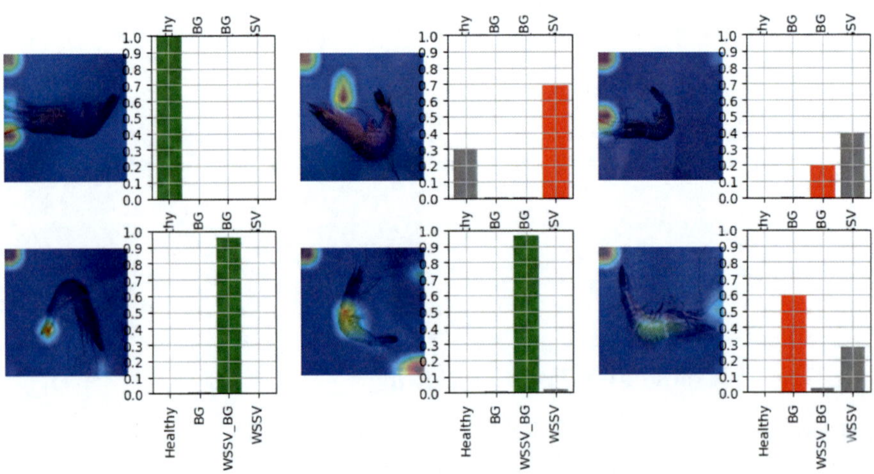

Fig. 4. Grad-CAM++ visualizations for each class: the CNN attends to gills for BG, shell for WSSV, and multiple regions for WSSV+BG, indicating that it learned clinically meaningful cues.

4.3 Evaluation Metrics and Resource Measurement

We compute confusion matrices, per-class precision, recall, and F1-score, and overall accuracy. For the binary task, we pay particular attention to the trade-off between Healthy and Diseased recall, since in screening scenarios missing diseased cases is more costly than raising some false alarms.

For VLMs, we report metrics on images where a valid label is produced; unknown outputs are implicitly penalized in the accuracy calculation. For the CNN, all images in the test split yield predictions.

We run all VLMs on a single GPU workstation and record:

- Peak CPU RAM and GPU VRAM usage during evaluation.
- Wall-clock time per image and total time to process all 761 images.

These measurements characterize the cost of using VLMs purely as inference-time screeners, without any fine-tuning burden.

5 Experiments and Results

5.1 Zero-Shot 4-Class Classification with VLMs

We first attempt the more ambitious 4-class setting where VLMs must distinguish between Healthy, BG, WSSV, and WSSV+BG. We encode the four options in the prompt and ask the model to select exactly one label.

Across all models, performance in this setting is poor. Accuracy is only slightly above random four-way guessing (25%), with typical values around 40–50%. More importantly, disease classes have extremely low recall and F1-scores. In many cases, models predict Healthy when there are clear disease signs, or confuse BG with WSSV and WSSV+BG.

For example, Qwen2-VL labeled 581 out of 761 images as Healthy, severely underestimating the prevalence of disease in the dataset. Other models exhibited similar behavior, sometimes predicting one particular disease (e.g., BG) disproportionately. The confusion matrices are dominated by the Healthy row and column, indicating a strong bias toward the majority class.

These results suggest that, in the absence of domain-specific fine-tuning, current general-purpose VLMs do not possess sufficiently calibrated knowledge to separate fine-grained shrimp disease categories based solely on textual descriptions. Consequently, we focus the rest of our analysis on the simpler and more practically relevant binary formulation.

5.2 Binary Healthy vs. Diseased Classification

In the binary setup, we treat any shrimp labeled BG, WSSV, or WSSV+BG as Diseased. This corresponds more closely to a screening scenario where the goal is to detect any abnormality, not to diagnose the exact disease.

Table 3 summarizes binary performance for all VLMs. No model exceeds 66% accuracy or achieves a balanced F1 across both classes.

Table 3. Binary classification performance of VLMs on ShrimpDiseaseBD. LLaVA (Agg.) and (Cons.) use aggressive and conservative prompts respectively. Qwen2-VL accuracy is over images it labeled.

	Healthy			Diseased			Acc.
Model	Prec	Rec	F1	Prec	Rec	F1	(%)
LLaVA (Agg.)	1.000	0.009	0.019	0.445	1.000	0.616	44.80
LLaVA (Cons.)	0.604	0.521	0.559	0.486	0.570	0.525	54.27
Qwen2-VL-2B†	0.626	0.979	0.763	0.889	0.224	0.358	65.41
SmolVLM-2B	0.627	0.811	0.707	0.595	0.365	0.452	61.84
InternVL2-2B	0.582	0.896	0.706	0.593	0.190	0.288	58.34
Phi-3.5-Vision	0.565	0.991	0.720	0.778	0.042	0.079	57.03

†96 images yielded no valid label from Qwen2-VL and are excluded from its precision, recall, and F1 computation.

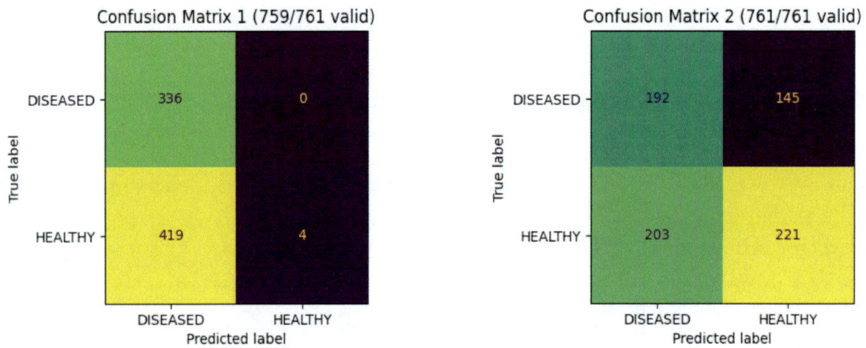

(a) LLaVA, aggressive prompt

(b) LLaVA, conservative prompt

Fig. 5. Prompt sensitivity for LLaVA on the binary task. The aggressive prompt maximizes diseased recall at the cost of massive false positives; the conservative prompt greatly reduces false positives but misses more diseased shrimp.

We observe two main patterns (illustrated for LLaVA in Fig. 5):

- **Conservative bias**: Qwen, Phi, and InternVL2 rarely predict Diseased. They achieve very high recall for Healthy (up to 99% for Phi) but extremely low recall for Diseased (as low as 4% for Phi). This is undesirable for screening, where false negatives are more serious than false positives.
- **Aggressive bias**: LLaVA with the aggressive prompt predicts almost everything as Diseased. Diseased recall reaches 100%, but almost all healthy shrimp are flagged as diseased, which would overload any follow-up pipeline.

SmolVLM and LLaVA with the conservative prompt sit somewhere in the middle: their F1-scores for both classes are around 0.5–0.7, but they still fall short of a level that would be acceptable in real-world screening without human oversight.

Qualitatively, VLM explanations often mention relevant cues ("slightly dark gills", "few white dots") but apply inconsistent thresholds when mapping observations to final labels. This highlights a key limitation of pure prompt-based zero-shot use in safety-critical applications.

5.3 CNN Baseline Performance

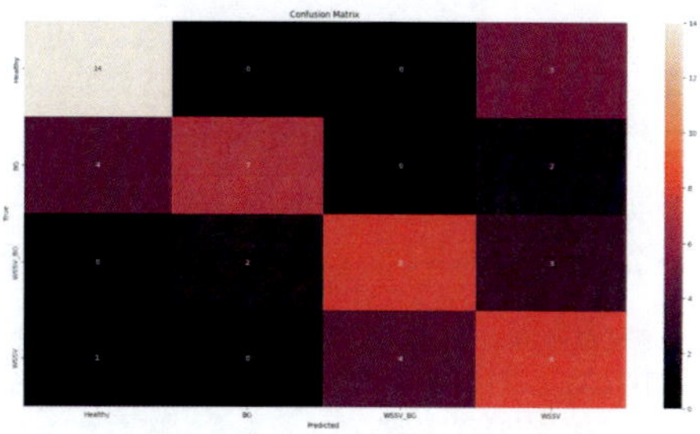

Fig. 6. 4-class confusion matrix for the CNN on the test set. Most errors are between disease classes rather than confusing diseased and healthy shrimp.

The ReXNet-150 CNN trained on 4 classes substantially outperforms all zero-shot VLMs. On the test split, it achieves around 78.9% accuracy and macro F1 of approximately 0.64 across the four classes. Healthy shrimp are recognized with high precision and recall, while disease classes achieve F1-scores in the 0.55–0.60 range. Figure 6 shows the 4-class confusion matrix on the test set.

Most CNN errors involve confusing one disease class for another (e.g., between WSSV and WSSV+BG) rather than confusing diseased shrimp with healthy ones. When we collapse predictions into binary (Healthy vs. any diseased class), the CNN reaches around 88–90% binary accuracy, with strong recall for Diseased (around 92%) and high precision.

Combined with Grad-CAM++ maps in Fig. 4, we see that the CNN tends to focus on gills for BG, the shell for WSSV, and multiple regions for WSSV+BG, which aligns well with domain intuition.

5.4 Resource Usage and Speed

Table 4 summarizes resource consumption for the VLM evaluations. All VLMs are considerably more expensive than the CNN at inference time.

Even the fastest VLM (SmolVLM) is much slower than a trained CNN, which can classify the entire dataset in seconds using batched inference. Memory demands are also heavier for VLMs, particularly for Phi-3.5-Vision. For deployment on farms with limited compute (e.g., edge devices or small on-premise servers), this is a major constraint.

Table 4. Resource usage and speed for VLMs on 761 images.

Model	CPU (GB)	GPU (GB)	Time/img (s)	Total (hh:mm)
LLaVA-1.5-7B	6.9	7.1	10.00	2:06
Qwen2-VL-2B	5.5	2.6	9.75	2:04
SmolVLM-2B	6.9	5.0	1.25	0:16
InternVL2-2B	6.1	5.8	2.75	0:35
Phi-3.5-Vision	6.3	11.1	44.35	9:23

6 Discussion

6.1 Why Do Zero-Shot VLMs Underperform?

Our experiments highlight a substantial gap between zero-shot VLMs and a task-specific CNN on ShrimpDiseaseBD. Several factors contribute to this:

- **Lack of domain-specific training**: current VLMs have not been explicitly trained on shrimp disease images. Their knowledge comes indirectly from generic web data, which likely contains very few labeled examples of WSSV or BG.
- **Subtle visual differences**: the distinction between healthy and mildly diseased shrimp can be subtle and sensitive to lighting, angle, and background. VLMs must infer threshold rules purely from text, whereas CNNs directly optimize these thresholds from labeled examples.
- **Prompt brittleness**: as seen with LLaVA, small changes in prompt wording can swing the model from overly aggressive to overly conservative decisions. This is difficult to control in a principled way in real-world deployments.

6.2 Potential Roles for VLMs

Despite their weaker quantitative performance, VLMs possess strengths that CNNs do not:

- **Explanations**: VLMs generate natural language descriptions of what they see, which could help farmers understand why a shrimp is flagged as suspicious (e.g., "gills appear darker than normal" or "white spots near the head").
- **Flexibility**: with appropriate prompts, VLMs can answer additional questions (e.g., "How severe does this infection look?" or "What would you recommend the farmer do next?"), potentially serving as conversational decision-support tools.
- **Transferability**: VLMs might generalize better to new farm environments, camera types, or even new disease categories, provided their training data contain related phenomena.

A promising direction is therefore not to replace CNNs with VLMs, but to combine them in a hybrid system.

6.3 Hybrid Systems and Future Work

One simple hybrid design is:

1. Use a supervised CNN as a fast, robust binary screener to filter large volumes of images.
2. For images classified as Diseased (or borderline cases), invoke a VLM to generate detailed commentary and highlight potential visual cues.
3. Present both the CNN decision and VLM explanation to the user, along with example images of clearly healthy and clearly diseased shrimp for calibration.

Future work should explore:

- **Light fine-tuning of VLMs**: even a small set of labeled shrimp images may substantially improve VLM decision thresholds, especially if used for visual instruction tuning.
- **Few-shot prompting**: supplying a few in-context examples of correctly labeled shrimp images and model responses may improve zero-shot performance without touching model weights.
- **Cross-farm evaluation**: testing both CNNs and VLMs on multi-farm datasets with varied imaging pipelines would clarify their robustness to domain shift.
- **Uncertainty estimation**: integrating calibration techniques or explicit uncertainty outputs could help triage which images require human review.

7 Conclusion

We have benchmarked several modern vision–language models and a supervised CNN on the ShrimpDiseaseBD dataset. Zero-shot VLMs, even with carefully designed prompts, substantially underperform a transfer-learned ReXNet-150 CNN and are highly sensitive to prompt wording. For practical shrimp disease screening today, a supervised CNN remains the more reliable and efficient option.

At the same time, VLMs offer complementary strengths—flexible prompting, rich textual explanations, and potential robustness to unforeseen conditions—that may be valuable in a decision-support context. Our study highlights both the promise and current limitations of VLMs in specialized aquaculture applications and points toward hybrid systems that combine the strengths of both paradigms. We hope this work can serve as a baseline for future research on domain-adapted VLMs and AI-assisted shrimp health monitoring.

References

1. Escobedo-Bonilla, C.M., Alday-Sanz, V., Wille, M., Sorgeloos, P., Pensaert, M.B., Nauwynck, H.J.: A review on the morphology, molecular biology, epidemiology and pathogenesis of white spot syndrome virus. J. Fish Dis. **31**(1), 1–18 (2008). https://doi.org/10.1111/j.1365-2761.2007.00877.x

2. Sun, M., Yang, X., Xie, Y.: Deep learning in aquaculture: a review. J. Comput. **31**(1), 294–319 (2020). https://doi.org/10.3966/199115992020023101028
3. Patil, P.K., Geetha, R., Mishra, S.S., et al.: Unveiling the economic burden of diseases in aquatic animal food production in India. Front. Sustain. Food Syst. **9**, 1480094 (2025). https://doi.org/10.3389/fsufs.2025.1480094
4. Sánchez-Paz, A.: White spot syndrome virus: an overview on an emergent concern in world shrimp aquaculture. Vet. Res. **41**(6), 43 (2010). https://doi.org/10.1051/vetres/2010015
5. Frischer, M.E., Landers, S.C., Walker, A.N., Powell, S.A., Lee, R.F.: Black Gill in marine decapod crustaceans: a review. Rev. Fish. Sci. Aquac. **30**(4), 498–519 (2022). https://doi.org/10.1080/23308249.2022.2047153
6. Desrina, D., Prayitno, S.B., Verdegem, M.C.J., Verreth, J.A.J., Vlak, J.M.: White spot syndrome virus host range and impact on transmission. Rev. Aquac. **14**(4), 1843–1860 (2022). https://doi.org/10.1111/raq.12676
7. Islam, M.M., Sarker, A., Choudhury, A., et al.: ShrimpDiseaseBD: an image dataset for detecting shrimp diseases in the aquaculture sector of Bangladesh. Data Brief **60**, 111553 (2025). https://doi.org/10.1016/j.dib.2025.111553
8. Büyükarıkan, B.: Robust shrimp disease detection using multi-model convolutional neural networks-based ensemble strategies. Aquacult. Eng. **111**, 102616 (2025). https://doi.org/10.1016/j.aquaeng.2025.102616
9. Aung, T., Vanichviriyakit, R., Chayantrakom, K., Amornsamankul, S., Huabsomboon, P.: CNN-based identification of pathogens of concern in shrimp. Animals **15**(21), 3194 (2025). https://doi.org/10.3390/ani15213194
10. Querol, L.S., Cordel II, M.O., Rustia, D.J.A., Santos, M.N.M.: Application for White Spot Syndrome Virus (WSSV) monitoring using edge machine learning. arXiv:2308.04151 (2023). https://doi.org/10.48550/arXiv.2308.04151
11. Tamut, H., Ghosh, R., Gosh, K., Siddique, M.A.S.: Enhancing disease detection in the aquaculture sector using convolutional neural networks analysis. Aquac. J. **5**(1), 6 (2025). https://doi.org/10.3390/aquacj5010006
12. Olimov, B.: [F1>0.8] Shrimp Disease Project with PyTorch. Kaggle Notebook. https://www.kaggle.com/code/killa92/f1-0-8-shrimp-disease-project-pytorch (2025). https://www.kaggle.com/code/killa92/f1-0-8-shrimp-disease-project-pytorch Accessed 6 Dec 2025
13. Radford, A., Kim, J.W., Hallacy, C.: Learning transferable visual models from natural language supervision. Proc. ICML **139**, 8748–8763 (2021). https://doi.org/10.48550/arXiv.2103.00020
14. Alayrac, J.-B., Donahue, J., Luc, P., et al.: Flamingo: a visual language model for few-shot learning. Adv. Neural Inf. Process. Syst. 35, 23716–23736 (2022). https://doi.org/10.52202/068431-1723
15. Chen, X., Bitton, Y., Schwing, A.G.: PaLI: A Jointly-Scaled Multilingual Language-Image Model. In: The Eleventh International Conference on Learning Representations (ICLR) (2023). https://doi.org/10.48550/arXiv.2209.06794
16. Li, J., Li, D., Xiong, C., Hoi, S.: BLIP-2: bootstrapping language-image pre-training with frozen image encoders and large language models. In: Proceedings of the 40th International Conference on Machine Learning. Proceedings of Machine Learning Research, vol. 202, pp. 19730–19742. PMLR (2023). https://doi.org/10.48550/arXiv.2301.12597
17. Liu, H., Li, C., Wu, Q., Lee, Y.J.: Visual instruction tuning. Adv. Neural Inf. Process. Syst. 36, 34892–34916 (2023). https://doi.org/10.52202/075280-1516

18. Bai, J., Bai, S., Yang, S., et al.: Qwen-VL: A versatile vision-language model for understanding, localization, text reading, and beyond. arXiv:2308.12966 (2023). https://doi.org/10.48550/arXiv.2308.12966
19. Li, C., Wong, C., Zhang, S., et al.: LLaVA-Med: training a large language-and-vision assistant for biomedicine in one day. Adv. Neural Inf. Process. Syst. 36, 28541–28564 (2023). https://doi.org/10.52202/075280-1240
20. Chattopadhyay, A., Sarkar, A., Howlader, P., Balasubramanian, V.N.: Grad-CAM++: improved visual explanations for deep convolutional networks. In: Proceedings of the IEEE Winter Conference on Applications of Computer Vision, pp. 839–847 (2018). https://doi.org/10.1109/WACV.2018.00097

Key Technologies for Intelligent Image Recognition, Detection and Multidimensional Analysis in Complex Scenarios

CFNet: A Two-Stage Coarse-to-Fine Diffusion-Driven Network for Camouflaged Object Detection

Lei Sun(✉), Yao Li, Gushuai Liu, Zengjian Yang, and Xuecheng Kong

State Grid Shandong Electric Power Company, Zibo Power Supply Company, Zibo 255030, China
grid_shandong@outlook.com

Abstract. Inspired by human visual perception, where coarse localization of camouflaged objects precedes fine-grained discrimination, this paper introduces a two-stage coarse-to-fine network (CFNet) for camouflaged object detection (COD) grounded in diffusion models. The proposed design addresses the overconfident and overly smooth predictions commonly observed in segmentation-based COD methods. In the first stage, a feature extraction module aggregates multi-level representations to capture global context while suppressing redundant information. A feature refinement module further enhances boundary fidelity and local texture cues through multi-rate expansion channels and adaptive pooling, mitigating redundancy and reducing overfitting. In the second stage, a diffusion-based generative process gradually injects Gaussian noise and performs iterative denoising to produce high-quality segmentation masks. Extensive evaluations on three widely used benchmark datasets demonstrate that the CFNet consistently outperforms 13 state-of-the-art approaches. Additional experiments on medical image segmentation further confirm its robustness and superior generalization ability.

Keywords: Camouflaged Object Detection · Diffusion Model · Coarse-to-Fine · Medical Image Segmentation

1 Introduction

Camouflaged Object Detection (COD) is a fundamentally challenging computer vision task aimed at identifying and segmenting objects that exhibit strong visual similarity to their surrounding environments [1,2]. COD is of substantial practical significance in domains such as industrial inspection and medical image analysis. In industrial settings, it enables the detection of subtle defects in manufactured components, while in medical imaging, it supports the identification of

This work was supported by Science and Technology Project of State Grid Shandong Electric Power Company (2024A-010).

L. Huang and D. Greenhalgh (Eds.): ICMLC 2026, LNNS 2061, pp. 405–420, 2027.
https://doi.org/10.1007/978-3-032-31236-5_28

pathological regions that closely resemble normal tissue. The primary challenge arises from the high degree of similarity in color, texture, and structural patterns between camouflaged objects and their backgrounds. This similarity severely limits the effectiveness of traditional edge-based and contrast-enhancement methods. Additionally, targets may display complex shapes, undergo partial or complete occlusion, or be deeply embedded within cluttered scenes, further increasing detection difficulty. To overcome these challenges, contemporary research increasingly leverages deep learning, particularly convolutional neural networks (CNNs) and Transformer-based architectures. These models enhance both localization and recognition accuracy by learning subtle discriminative cues between objects and backgrounds, thereby offering powerful new methodologies for advancing COD research.

In recent years, deep learning has driven substantial progress in camouflaged object detection (COD), particularly through Convolutional Neural Networks (CNNs) and Transformer-based architectures. Fan et al. [3] introduced SINet and the COD10K benchmark. Liu et al. [4] proposed MSCAF-Net to learn multi-scale context-aware representations. Xing et al. [5] designed SARNet by modeling the human visual process of searching, amplifying, and identifying camouflaged targets. Yang et al. [6] jointly explored COD and salient object detection to leverage their complementary relationships. Jiang et al. [7] developed BiDiCOS, which formulates segmentation and depth estimation as diffusion-based denoising tasks. Song et al. [8] proposed FSNet, a two-stage focus-scanning framework emphasizing region-level attention. Despite these advances, several challenges remain, as many existing models suppress noise at the cost of fine structural and boundary details, while most COD frameworks still follow semantic segmentation paradigms that frequently yield overconfident and unreliable predictions, and effectively extracting foreground cues from heavily camouflaged objects continues to be difficult, often resulting in misidentification or incomplete detection.

To overcome these limitations, this paper presents a two-stage coarse-to-fine network (CFNet) for COD based on diffusion models. The method leverages diffusion processes to learn noise distributions and iteratively denoise images to generate high-quality masks. In the first stage, multi-level features from the PVTv2-b4 backbone [9] are aggregated to form initial mask features. Convolution layers capture global context while removing redundant information, dilation-based channels enhance edge sensitivity, and pooling operations suppress extraneous local details. Element-wise weighting further refines the representation. In the second stage, the diffusion module integrates the noisy input with the refined mask features through multi-layer convolutions, and a decoder reconstructs the final segmentation mask. The main contributions of this work are summarized as follows:

- A new two-stage COD framework, CFNet, based on diffusion models is introduced, supported by a customized training strategy. By integrating diffusion priors, the model mitigates overconfident prediction errors common in segmentation-based COD approaches.

- A feature extraction module is designed to effectively fuse multi-level backbone features, strengthen global contextual understanding, reduce redundancy, and improve detection accuracy.
- A feature optimization module is developed to enhance fine-grained detail learning through dilation-based feature refinement, convolutional enhancement, and pooling-based suppression of redundant local patterns.

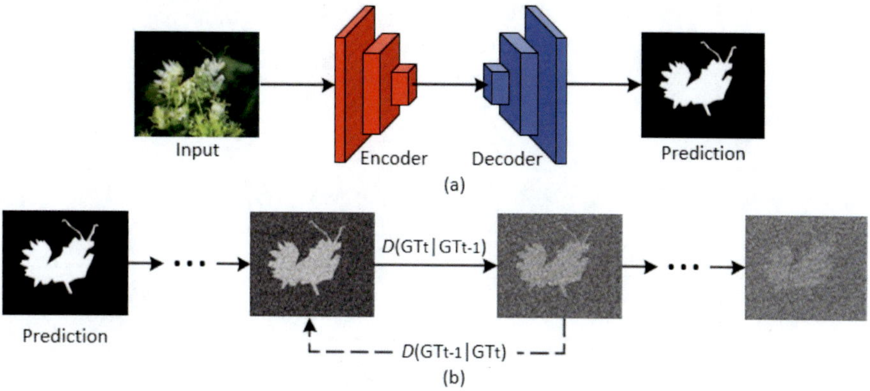

Fig. 1. (a) Detection process of the autoencoder network. (b) Principle of mask prediction using the diffusion model in the proposed method.

2 Background

2.1 Diffusion Model

The diffusion model [10,11] operates as a parameterized Markov chain that progressively transforms noise into structured data through iterative denoising. As illustrated in Fig. 1 (b), the proposed camouflaged object detection framework follows a prediction mechanism fundamentally different from the autoencoder-based pipeline depicted in Fig. 1 (a). Whereas the autoencoder directly infers a mask from the input image, the diffusion model employs a two-phase process: the forward process progressively injects noise into the ground-truth mask, and the reverse process iteratively removes this noise to recover the target distribution.

During the forward process, Gaussian noise $Z \sim \mathcal{N}(0, I)$ is gradually added to the clean ground-truth mask G_0. When the diffusion horizon T is sufficiently large, the resulting sample becomes nearly indistinguishable from pure noise. Concretely, a ground-truth sample G_0 is first drawn from the training set, after which noise is injected at time step t to obtain the corrupted mask G_t. This transformation is formally expressed as:

$$p(G_{t-1}|G_t) = \mathcal{N}(G_{t-1}; \mu_\theta(G_t, t), \textstyle\sum_\theta(G_t, t)), \tag{1}$$

$$q(G_t|G_0) = \mathcal{N}(G_t; \sqrt{\bar{\alpha}_t}G_0, (1 - \bar{\alpha}_t)I). \tag{2}$$

where $\bar{\alpha}_t = \prod_{i=1}^{T} \alpha_t$, $\alpha_t = 1 - \beta_t$, t runs from 1 to T. $\beta_t \in (0,1)$ denotes noise schedule, it control variance. I is the input images.

The backward process involves progressively denoising random noise to obtain the final mask feature map. Random noise $G_t \sim \mathcal{N}(0, I)$ is iteratively denoised to generate G_0. Unlike the forward process, the backward process obtains the mean and variance during training. This process can be explained by the following equation:

$$p(G_{t-1}|G_t) = \mathcal{N}(G_{t-1}; \mu_\theta(G_t, t), \textstyle\sum_\theta(G_t, t)), \tag{3}$$

$$\mu_\theta(G_t, t) = \frac{\sqrt{\alpha_t}(1-\bar{\alpha}_{t-1})G_t + \sqrt{\bar{\alpha}_{t-1}}\beta_t \hat{G}_0}{1-\bar{\alpha}_t}. \tag{4}$$

where θ is the model parameters. $\hat{G}_0$ is the denoised mask. t as inputs and predicts the mean $\mu_\theta(G_t, t)$ and covariance $\sum_\theta(G_t, t)$. In practice, $\sum_\theta(G_t, t)$ is set to $\sigma_t^2 = \frac{(1-\bar{\alpha}_{t-1})\beta_t}{1-\bar{\alpha}_t}$.

2.2 Diffusion U-Net

The diffusion model adopts a U-shaped network architecture. Because the model performs iterative denoising during inference, the input binary segmentation map conveys only limited structural information. Accordingly, the U-shaped design employs a series of convolutional layers to integrate the noisy input with the feature representations generated by the feature optimization module. The decoder, composed of convolutional and upsampling layers, then progressively refines these fused features to produce the final segmentation mask.

3 Proposed Method

3.1 Overall Architecture

Unlike conventional autoencoder-based approaches, the proposed framework employs a conditional diffusion mechanism to generate prediction masks. As illustrated in Fig. 2, we develop a diffusion-driven two-stage camouflaged object detection model, CFNet. During training, feature aggregation fuses the multi-level backbone features f_i ($i \in \{1, 2, 3, 4\}$) extracted by the PVTv2-b4 network into unified mask features. Global contextual cues are further captured through convolutional operations combined with pooling, which also serve to suppress

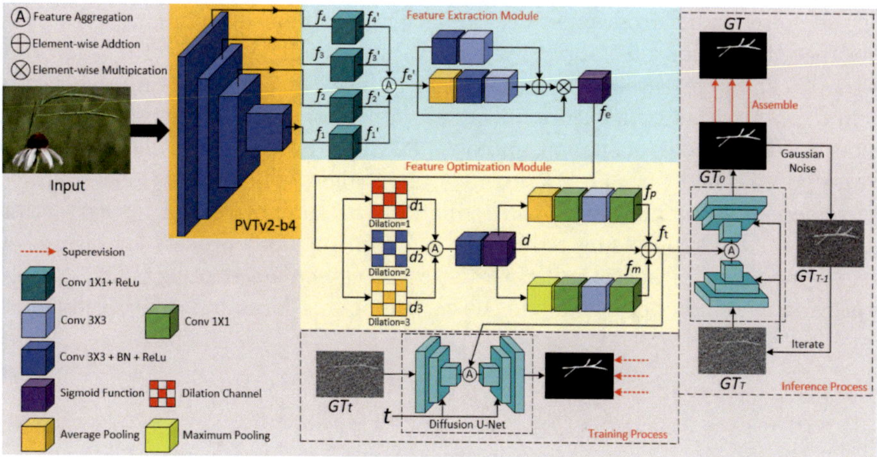

Fig. 2. Overall algorithmic framework of the proposed CFNet. This network comprises four main components: the PVTv2-b4 backbone, the Feature Extraction Module, the Feature Optimization Module, and the diffusion-based prediction framework. In the first stage, the Feature Extraction Module aggregates multi-level features produced by the backbone and suppresses redundant information, while the Feature Optimization Module refines local structural cues to enhance fine-grained detail representation. In the second stage, the training pipeline injects Gaussian noise into the input images and feeds the resulting noisy samples into the diffusion framework to learn mask representations. During inference, the input image is processed by the backbone to obtain feature maps, and a Gaussian noise map drawn from a standard normal distribution is simultaneously passed through the diffusion framework. Through iterative denoising, the network produces the final mask prediction.

redundant information. To enhance fine-grained representations, three convolutional branches with different dilation rates are used to refine edge structures and extract local details. Subsequent pooling operations reduce excessive high-frequency responses to mitigate overfitting, while channel-wise weighting ensures the generation of high-quality mask features. The final loss is then computed between the predicted masks and the Ground Truth (GT). During inference, the diffusion process iteratively denoises the input image, ultimately producing the final mask prediction.

3.2 Feature Extraction Module

In camouflaged object detection, the multi-level features extracted by the backbone network play a critical role in accurately localizing mask regions. Low-level features preserve rich edge and structural details that facilitate the delineation of object contours; however, their limited receptive fields also introduce substantial redundancy, resulting in incomplete or noisy representations. In contrast, high-level features capture global semantic information and broader contextual cues, yet they often suffer from noise accumulation and blurred boundaries. To mitigate these limitations, we apply 1×1 convolutions followed by ReLU activation to the backbone features at each hierarchical level, and subsequently aggregate them to obtain the fused feature representation f_e', thereby compensating for the respective weaknesses of shallow and deep features. Furthermore, the aggregated features are refined through 3×3 convolutions and pooling operations, followed by element-wise summation, which enhances discriminative feature extraction while effectively suppressing redundant information. This process enables more precise modeling of spatial characteristics in camouflaged objects, ultimately improving detection performance. The corresponding formulation is given as follows:

$$f_i' = R_e(Conv_1(f_i))(i \in \{1, 2, 3, 4\}), \tag{5}$$

$$f_e' = FA(f_1', f_2', f_3', f_4'), \tag{6}$$

$$f_e = \sigma(Conv_3(CBR(f_e'))) \oplus (Conv_3(CBR(MaxPool(f_e'))) \otimes f_e') \tag{7}$$

Here, $Re(\cdot)$ denotes the ReLU activation function; $Conv_1(\cdot)$ denotes a 1×1 convolution; $FA(\cdot)$ represents the feature aggregation operation; $\sigma(\cdot)$ denotes the Sigmoid function; $Conv_3(\cdot)$ denotes a 3×3 convolution; $CBR(\cdot)$ refers to a spatial feature extraction module composed of a 3×3 convolutional layer, a batch normalization layer, and a ReLU activation function. The symbol $\oplus$ denotes element-wise addition; $Maxpool(\cdot)$ denotes max-pooling; and $\otimes$ indicates element-wise multiplication.

3.3 Feature Optimization Module

In this module, the fused feature f_e is processed through three parallel convolutional channels with distinct dilation rates. Channels with smaller dilation rates emphasize fine-grained structural cues, thereby enriching local detail, whereas those with larger dilation rates capture broader contextual information. This multi-scale design enhances boundary representations and strengthens the model's ability to detect camouflaged objects across diverse and challenging scenarios. In addition, pooling operations with varying receptive-field sizes are applied and subsequently integrated with the input features to produce a finely weighted representation of the mask features. This strategy suppresses redundancy, mitigates overfitting, and yields a more comprehensive and robust feature description. The corresponding formulation is expressed as follows:

$$d = \sigma(CBR(FA(d_1, d_2, d_3))), \tag{8}$$

$$f_m = Conv_1(Conv_3(Conv_1(MaxPool(c)))), \tag{9}$$

$$f_p = Conv_1(Conv_3(Conv_1(adPool(d)))), \tag{10}$$

$$f_t = d \oplus f_m \oplus f_p \tag{11}$$

3.4 Loss Function

The overall loss function employed by the proposed model is defined as follows:

$$L_{(G_0,G_t)} = L_{BCE}^{\omega}(G_0, G_t) + L_{IOU}^{\omega}(G_0, G_t) \tag{12}$$

Here, G_t denotes the predicted mask of the camouflaged objects, while G_0 represents the corresponding ground-truth annotation. L^{ω}_{BCE} denotes the weighted Binary Cross-Entropy loss [12], and L^{ω}_{IOU} denotes the weighted Intersection-over-Union loss [13].

3.5 Training and Inference Strategy

The mask feature f_t produced by the feature optimization module is subsequently passed into the diffusion framework. During training, the diffusion model performs a forward process that transforms the ground-truth mask GT into a noisy representation while simultaneously learning the reverse denoising process. Following the strategy of Chen et al. [14], the forward diffusion process introduces Gaussian noise that randomly perturbs mask boundaries, enabling the model to learn how to recover accurate structural details from corrupted inputs.

Furthermore, a signal-to-noise ratio (SNR)based variance scheduling scheme is adopted to optimize both the distribution and magnitude of injected noise, thereby improving performance under low-SNR conditions. This design alleviates two major challenges: the difficulty diffusion models face in reconstructing sharp masks from weak image features under low SNR, and the performance degradation caused by the Gaussian noise applied during the forward process.

Building on the definition of SNR [15], we develop a scheduling mechanism tailored to this framework:

$$\log SNR(t) = -2\log\left(\tan\left(\frac{\pi t}{2}\right)\right), \tag{13}$$

and additionally introduce a shifted variant,

$$\log SNR_{shift}(t) = -2\log\left(\tan\left(\frac{\pi t}{2}\right)\right) + shift. \tag{14}$$

A complete description of the training procedure is provided in Algorithm 1.

Algorithm 1: Training Strategy

Input: Image $D(Img)$ and mask distribution $D(G_0)$

Repeat Operation

  $t \sim \text{Uniform}(1, ..., T)$

  Image $\sim D(Image)$

  $G_0 \sim D(G_0)$

  $Z \sim \mathcal{N}(0, I)$

  $G_t = \sqrt{\bar{\alpha}_t} G_0 + (1 - \hat{\alpha}_t) z$

  Gradient descent for loss calculation $\nabla_{\theta} \ell(G_0, G_t, Image, t)$

Convergence

During inference, the model generates the final mask prediction by transforming pure Gaussian noise into a structured segmentation map through iterative denoising. Starting from an initial noise sample, G_t is progressively denoised to obtain $\hat{G}_0$, which is stored as one prediction instance. The model then continues the reverse diffusion process: $\hat{G}_0$ is denoised to produce G_{t-1}, which serves as the input for the next iteration. Repeating this procedure over 10 denoising steps yields a set of 10 prediction maps.

To enhance robustness, these maps are aggregated by averaging, producing the final mask prediction. This study adopts the consensus temporal envelope strategy introduced in [16], which integrates predictions across sampling steps to improve accuracy and stability. The corresponding inference procedure is summarized in Algorithm 2.

In addition, the number of diffusion steps acts as a model-specific hyper-parameter, which is influenced by the network architecture, training strategy, and dataset characteristics. To justify the configuration adopted in CFNet, we carried out a controlled experiment to evaluate the impact of this parameter on detection performance under consistent experimental settings.The corresponding results are summarized in Table 1.

Algorithm 2: Inference strategy

1:Input: Image I, step t
2: $G_p = [\,]$
3: For t = T,...,1 do
4: $\hat{G}_0 = Net(G_t, Image, t)$
5: $Z \sim \mathcal{N}(0, I)$ if $t > 1$, else $Z = 0$
6: $D(G_{t-1}|G_t) = \mathcal{N}(G_T; \mu_q(\hat{G}_0)), \sum_\theta(t)$
7: End for
8: $G_{pred} = Mean(G_p)$
9: Return G_{pred}

Table 1. Quantitative performance Comparison of CFNet with different time steps on COD10K dataset. The best results for each evaluation metric are marked in **bold**.

Model	COD10K			
	$S_\alpha \uparrow$	$F_\beta^\omega \uparrow$	$E_m \uparrow$	$M \downarrow$
t=1/2	0.878	0.810	0.938	0.023
t=1/3	0.876	0.813	0.937	0.024
t=2	0.871	0.797	0.929	0.026
t=1(Ours)	**0.884**	**0.818**	**0.944**	**0.019**

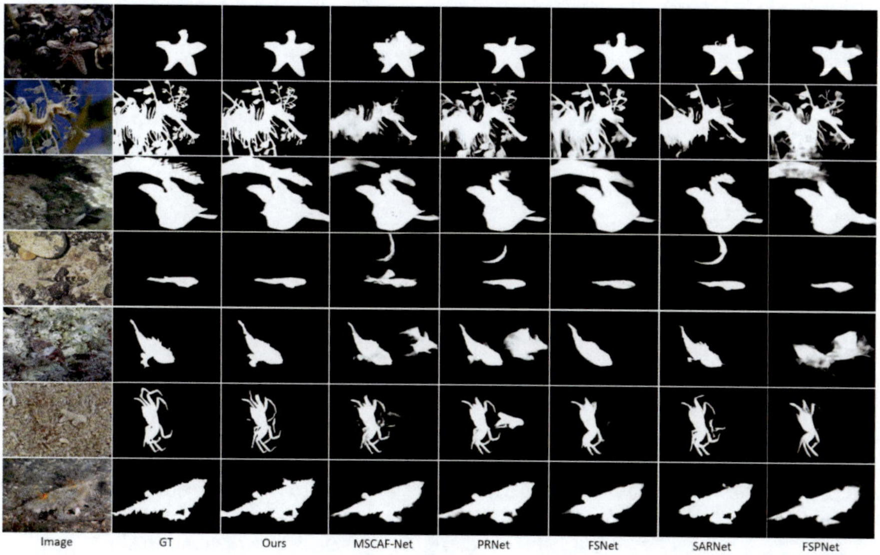

Fig. 3. Visual comparison of CFNet to five representative camouflaged object detection methods, including MSCAF-Net, PRNet, FSNet, SARNet, and FSPNet.

4 Experiments

4.1 Implementation Details

Experiments are conducted on three publicly available datasets: COD10K [3], NC4K [17], and CAMO [18]. The proposed model is implemented in PyTorch and trained on a single NVIDIA A100 GPU equipped with 80 GB of memory. All training images are initially resized to 352×352 and used to train the model for 150 epochs, followed by an additional 20 epochs with images resized to 384×384. The training process employs the AdamW optimizer with a batch size of 32 and an initial learning rate of 0.01.

4.2 Evaluation Metrics

To evaluate the performance of CFNet, we adopt four widely used metrics: Mean Absolute Error (MAE, M) [19], Weighted F-measure (F_m^ω) [20], Structure-measure (S_α) [21], and mean E-measure (E_ϕ) [22]. We further compare the model against 13 representative camouflaged object detection approaches, including BGNet [23], MSCAF-Net [4], PRNet [24], SARNet [5], FSNet [25], FCNet [26], L-DiffCOD [27], BiDiCOS [7], diffCOD [28], BDCL-Net [29], FSPNet [16] and MCECOD [30]. Notably, L-DiffCOD, BiDiCOS, and diffCOD represent diffusion-based COD methods. All comparative results are obtained either directly from the authors' released datasets (BGNet, MSCAF-Net, PRNet, SARNet , FSNet, FCNet, L-DiffCOD, BiDiCOS, diffCOD, BDCL-Net, FSPNet, MCECOD) or by retraining the models using publicly available code under a unified experimental setting.

Table 2. Quantitative evaluation of various cod methods on four benchmark datasets using the metrics S_α, F_β^ω, E_m, and M. Arrows (↑/↓) indicate whether higher or lower values are preferable. The best-performing results are highlighted in **bold**.

MODEL DETAILS			COD10K				NC4K				CAMO			
Methods	Year	Pub	$S_\alpha \uparrow$	$F_\beta^\omega \uparrow$	$E_m \uparrow$	$M \downarrow$	$S_\alpha \uparrow$	$F_\beta^\omega \uparrow$	$E_m \uparrow$	$M \downarrow$	$S_\alpha \uparrow$	$F_\beta^\omega \uparrow$	$E_m \uparrow$	$M \downarrow$
BGNet [23]	2021	LJCAI	0.831	0.722	0.901	0.033	0.851	0.788	0.907	0.044	0.812	0.749	0.870	0.073
MSCAF-Net [4]	2022	TCSVT	0.865	0.775	0.927	0.024	0.887	0.839	0.935	0.032	0.873	0.828	0.929	0.046
PRNet [24]	2022	TCSVT	0.874	–	–	0.022	0.891	–	–	0.031	0.872	–	–	0.050
SARNet [5]	2022	TCSVT	0.864	0.777	0.931	0.024	0.886	0.842	0.937	0.032	0.868	0.828	0.927	0.047
FSNet [25]	2022	TIP	0.870	–	0.938	0.023	0.891	–	0.940	0.031	0.880	–	**0.933**	0.041
FCNet [26]	2022	TCSVT	0.834	0.778	0.912	0.030	0.859	0.843	0.922	0.041	0.835	0.828	0.902	0.063
L-DiffCOD [27]	2022	ICAI	0.842	0.739	0.912	0.038	0.865	0.808	0.919	0.038	–	–	–	–
BiDiCOS [7]	2022	ESA	0.866	0.796	0.940	0.022	0.883	0.848	0.941	0.031	0.862	0.832	0.927	0.045
diffCOD [28]	2023	ECAI	0.812	0.684	0.892	0.036	0.837	0.761	0.891	0.051	0.795	0.704	0.852	0.082
BDCL-Net [29]	2023	KBS	0.869	0.790	–	0.022	0.888	0.841	–	0.032	**0.881**	0.845	–	0.039
FSPNet [16]	2023	CVPR	0.851	0.735	0.895	0.026	0.879	0.816	0.915	0.035	0.856	0.799	0.899	0.050
MCECOD [30]	2024	KBS	0.872	0.791	–	0.023	0.885	0.834	–	0.033	0.864	0.813	–	0.050
CFNet (Ours)	–	–	**0.884**	**0.818**	**0.944**	**0.019**	**0.897**	**0.861**	**0.944**	**0.027**	0.880	**0.853**	**0.933**	**0.040**

4.3 Comparison With SOTA Methods

Table 2 reports the performance of CFNet in comparison with 15 representative camouflaged object detection methods across three public datasets. The results demonstrate that CFNet consistently achieves superior performance on all four evaluation metrics. On the COD10K dataset, the model surpasses the second-best semantic segmentationbased method, BDCL-Net, by 1.5% in S_α and 2.8% in F_β^ω, while also yielding a substantial improvement in MAE (M). On the NC4K dataset, it further exceeds BDCL-Net by 0.9% in S_α and 2.0% in F_β^ω, accompanied by a clear reduction in MAE. When compared with diffusion-based approaches, CFNet achieves consistently higher scores across all metrics on all three datasets.

In addition, Fig. 3 provides qualitative comparisons on the COD10K dataset. As shown, the proposed method not only delivers more precise mask predictions but also captures boundary structures with greater completeness and fidelity, underscoring its strong detection capability.

While CFNet achieves exceptional detection accuracy, this high performance entails a correspondingly high computational cost, characterized by increased parameter counts, elevated floating-point operations (FLOPs), and greater GPU memory consumption. Consequently, CFNet is particularly appropriate for application contexts where detection precision is paramount, such as medical endoscopy, as well as for scenarios permitting moderate latency, including wildlife monitoring.

Table 3. Quantitative evaluation of different models on the COD10K Dataset. The best-performing results are highlighted in **bold**.

MODEL			COD10K			
Method	Params(M)	FLOPs(G)	$S_\alpha \uparrow$	$F_\beta^\omega \uparrow$	$E_m \uparrow$	$M \downarrow$
Baseline	62.5	30.9	0.866	0.781	0.930	0.023
+FEM	78.8	119.4	0.875	0.808	0.939	0.021
+FOM	73.0	65.3	0.878	0.806	0.941	0.021
+FEM+FOM(Ours)	78.9	119.7	**0.884**	**0.818**	**0.944**	**0.019**

Table 4. Quantitative evaluation of different input image sizes on the COD10K dataset. The best-performing results are highlighted in **BOLD**.

Input size	COD10K			
	$S_\alpha \uparrow$	$F_\beta^\omega \uparrow$	$E_m \uparrow$	$M \downarrow$
256×256	0.867	0.777	0.926	0.024
320×320	0.877	0.810	0.938	0.021
352×352(Ours)	**0.884**	**0.818**	**0.944**	**0.019**

4.4 Ablation Analysis

To further assess the effectiveness of CFNet, a series of ablation studies were conducted on the COD10K dataset. For the baseline configuration, the feature extraction module, feature optimization module, and diffusion framework were removed, retaining only the PVTv2-b4 backbone and the feature aggregation operation. As shown in Table 3, incorporating the feature extraction module together with diffusion-based denoising improves S_α and F_β^ω by 0.9% and 2.7%, respectively, alongside clear gains in the remaining metrics. Similarly, introducing the feature optimization module and applying diffusion denoising yields improvements of 1.2% in S_α and 2.5% in F_β^ω. When both modules are integrated, the model achieves even more pronounced performance enhancements. In addition, Table 4 examines the influence of input image resolution on detection performance, further validating the robustness of the experimental configuration.

To evaluate the contribution of the backbone network, we also replace PVTv2-b4 with alternative backbones. As shown in Table 5, PVTv2-b4 delivers the best overall performance, confirming its suitability for CFNet. Table 5 additionally compares the computational complexity of different backbones. It is worth noting that, within the COD community, current research prioritizes accuracy-oriented architectural design rather than lightweight model development.

4.5 Failure Cases

Although CFNet surpasses existing COD methods, certain visualizations still reveal failure cases. As illustrated in Fig. 4, these shortcomings manifest in sev-

Table 5. Quantitative comparison of the proposed method's performance on COD10K using different backbone configurations. The best-performing results are highlighted in **bold**.

Models			COD10K			
Backbone	Params(M)	FLOPs(G)	$S_\alpha \uparrow$	$F_\beta^\omega \uparrow$	$E_m \uparrow$	$M \downarrow$
PVTv2-b5	82.0	11.8	0.870	0.798	0.931	0.024
PVTv2-b2	25.4	4.0	0.875	0.802	0.935	0.022
PVTv2-b4(Ours)	62.6	10.1	**0.884**	**0.818**	**0.944**	**0.019**

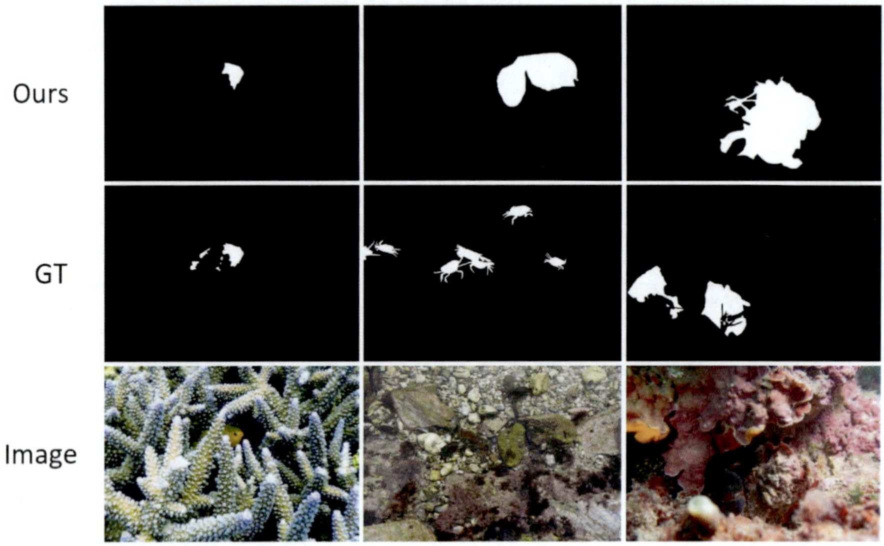

Fig. 4. Illustration of representative failure cases generated by CFNet, highlighting its limitations in challenging scenarios.

eral forms. In the first column, the method fails to detect the target, largely because the object is extremely small and provides insufficient discriminative features. In future work, we will improve the algorithm's ability to integrate global contextual information, leveraging the contextual cues around small targets to assist in detection. In the second and third columns, the model produces incorrect predictions, primarily due to the high spatial dispersion of the targets and the pronounced color and texture differences between the targets and their surrounding environments. Notably, competing methods tend to exhibit similar-if not more pronounced-failures under these challenging conditions.

4.6 Application to Polyp Segmentation

To further assess the generalization capability of the proposed model, comparative experiments were conducted on four benchmark datasets-Kvasir [31], CVC-

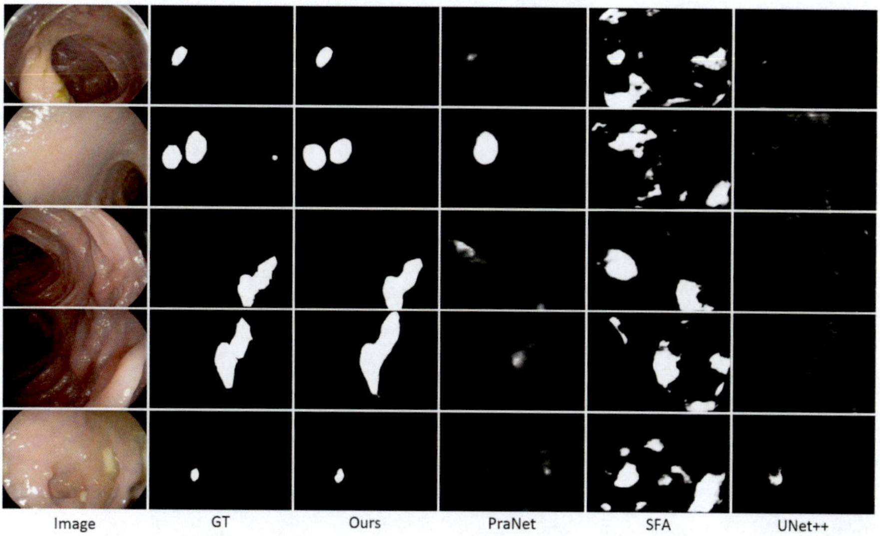

Fig. 5. Qualitative comparison of polyp segmentation results obtained by representative methods, including PraNet, SFA, and UNet++, illustrating differences in boundary delineation and segmentation accuracy.

Table 6. Quantitative results of five representative methods compared with the proposed method on four polyp segmentation datasets. The best-performing results are highlighted in **bOLD**.

Methods	Kvasir				CVC-ClinicDB				ETIS				CVC-ColonDB			
	$S_\alpha \uparrow$	$F_\beta^\omega \uparrow$	$E_m \uparrow$	$M \downarrow$	$S_\alpha \uparrow$	$F_\beta^\omega \uparrow$	$E_m \uparrow$	$M \downarrow$	$S_\alpha \uparrow$	$F_\beta^\omega \uparrow$	$E_m \uparrow$	$M \downarrow$	$S_\alpha \uparrow$	$F_\beta^\omega \uparrow$	$E_m \uparrow$	$M \downarrow$
U-Net++ [36]	0.858	0.794	0.901	0.048	0.873	0.785	0.891	0.022	0.683	0.390	0.629	0.035	0.691	0.467	0.680	0.064
PraNet [35]	0.915	0.885	0.943	0.027	0.935	0.896	0.958	0.009	0.791	0.600	0.792	0.031	0.820	0.699	0.847	0.043
SFA [37]	0.782	0.670	0.834	0.075	0.793	0.647	0.840	0.042	0.557	0.231	0.531	0.100	0.634	0.379	0.675	0.094
CFA-Net [38]	0.924	0.903	0.962	0.023	**0.950**	**0.924**	**0.989**	**0.007**	0.845	0.693	0.892	**0.014**	0.860	0.779	0.898	0.039
TFMF-Net [39]	0.893	0.852	0.927	0.035	0.928	0.886	0.957	0.011	0.796	0.606	0.831	0.025	0.814	0.693	0.848	0.040
CFNet (Ours)	**0.935**	**0.917**	**0.966**	**0.018**	0.940	0.910	0.966	0.011	**0.875**	**0.738**	**0.889**	0.016	**0.872**	**0.789**	**0.908**	**0.031**

ClinicDB [32], ETIS [33], and CVC-ColonDB [34]-following the experimental protocol established by PraNet [35]. Specifically, 900 images from Kvasir and 550 images from CVC-ClinicDB were used for training, while the remaining images in each dataset were reserved for testing. The evaluation metrics included S_α, F_β^ω, E_m, and M.

As summarized in Table 6, our model is compared against several state-of-the-art polyp segmentation methods, including UNet++ [36], PraNet, SFA [37], CFA-Net [38], and FTMF-Net [39]. The results indicate that the proposed model consistently achieves superior performance across all four datasets. Furthermore, the visual comparisons presented in Fig. 5 highlight the clear advantages of our method over competing approaches, thereby providing additional evidence of the strong generalization ability of the proposed framework.

5 Conclusion

This study introduces CFNet, a coarse-to-fine, two-stage COD framework built upon a diffusion-based architecture. In the first stage, a feature aggregation module conducts coarse feature extraction, where the integration of multi-level convolutional layers and pooling operations effectively suppresses redundant information. A subsequent feature optimization module refines structural details by leveraging channels with varying dilation rates to enhance local feature representation. Through additional pooling and convolutional operations, the module progressively strengthens fine-grained feature extraction. In the second stage, a diffusion-based generative framework is employed to iteratively perturb images with Gaussian noise and subsequently denoise them, enabling the progressive generation of high-quality prediction masks. Experiments conducted on three public datasets, along with comprehensive ablation studies, substantiate the effectiveness of CFNet. Moreover, comparative evaluations on polyp segmentation benchmarks further illustrate the strong generalization capability of the model.

Future work will focus on addressing the underlying factors contributing to failure cases, with the aim of improving robustness in more complex and diverse scenarios. In parallel, additional efforts will be devoted to achieving a better balance between detection accuracy and lightweight model design.

References

1. Yaseen, S.A., Sasi, S.: Robust algorithm for object detection and tracking in a dynamic scene. J. Image Graph. **2**(1), 41–45 (2014)
2. Khan, R., Raisa, T.F., Debnath, R.: An efficient contour based fine-grained algorithm for multi category object detection. J. Image Graph. **6**(2), 127–136 (2018)
3. Fan, D.P., Ji, G.P., Sun, G., et al.: Camouflaged object detection. In: Proceedings of the IEEE/CVF Conference on Computer Vision and Pattern Recognition, pp. 2777–2787 (2020)
4. Liu, Y., Li, H., Cheng, J., et al.: MSCAF-Net: a general framework for camouflaged object detection via learning multi-scale context-aware features. IEEE Trans. Circuits Syst. Video Technol. **33**(9), 4934–4947 (2023)
5. Xing, H., Gao, S., Wang, Y., et al.: Go closer to see better: camouflaged object detection via object area amplification and figure-ground conversion. IEEE Trans. Circuits Syst. Video Technol. **33**(10), 5444–5457 (2023)
6. Yang, Y., Zhang, Q.: Finding camouflaged objects along the camouflage mechanisms. IEEE Trans. Circuits Syst. Video Technol. **34**(4), 2346–2360 (2023)
7. Jiang, X., Cai, W., Ding, Y., et al.: BiDiCOS: camouflaged object segmentation via bilateral diffusion model. Expert Syst. Appl. **255**, 124747 (2024)
8. Song, Z., Kang, X., Wei, X., et al.: Fsnet: focus scanning network for camouflaged object detection. IEEE Trans. Image Process. **32**, 2267–2278 (2023)
9. Wang, W., Xie, E., Li, X., et al.: Pvt v2: improved baselines with pyramid vision transformer. Comput. Vis. Media **8**(3), 415–424 (2022)
10. Ho, J., Jain, A., Abbeel, P.: Denoising diffusion probabilistic models. Adv. Neural. Inf. Process. Syst. **33**, 6840–6851 (2020)

11. Song, Y., Sohl-Dickstein, J., Kingma, D.P., et al.: Score-based generative modeling through stochastic differential equations. In: International Conference on Learning Representations
12. Xie, E., Wang, W., Wang, W., et al.: Segmenting transparent objects in the wild. In: European Conference on Computer Vision, pp. 696–711. Springer International Publishing, Cham (2020)
13. Wei, J., Wang, S., Huang, Q.: F^3Net: fusion, feedback and focus for salient object detection. In: Proceedings of the AAAI Conference on Artificial Intelligence, Vol. 34, no. 07, pp. 12321–12328 (2020)
14. Chen, Z., Sun, K., Lin, X.: CamoDiffusion: camouflaged object detection via conditional diffusion models. In: Proceedings of the AAAI Conference on Artificial Intelligence, vol. 38, no. 2, pp. 1272–1280 (2024)
15. Kingma, D., Salimans, T., Poole, B., et al.: Variational diffusion models. Adv. Neural. Inf. Process. Syst. **34**, 21696–21707 (2021)
16. Huang, Z., Dai, H., Xiang, T.Z., et al.: Feature shrinkage pyramid for camouflaged object detection with transformers. In: Proceedings of the IEEE/CVF Conference on Computer Vision and Pattern Recognition, pp. 5557–5566 (2023)
17. Lv, Y., Zhang, J., Dai, Y., et al.: Simultaneously localize, segment and rank the camouflaged objects. In: Proceedings of the IEEE/CVF Conference on Computer Vision and Pattern Recognition, pp. 11591–11601 (2021)
18. Le, T.N., Nguyen, T.V., Nie, Z., et al.: Anabranch network for camouflaged object segmentation. Comput. Vis. Image Underst. **184**, 45–56 (2019)
19. Perazzi, F., Krähenbühl, P., Pritch, Y., et al.: Saliency filters: contrast based filtering for salient region detection. In: 2012 IEEE Conference on Computer Vision and Pattern Recognition, pp. 733–740. IEEE (2012)
20. Margolin, R., Zelnik-Manor, L., Tal, A.: How to evaluate foreground maps? In: Proceedings of the IEEE Conference on Computer Vision and Pattern Recognition, pp. 248–255 (2014)
21. Fan, D.P., Cheng, M.M., Liu, Y., et al.: Structure-measure: a new way to evaluate foreground maps. In: Proceedings of the IEEE International Conference on Computer Vision, pp. 4548–4557 (2017)
22. Fan, D.P., Gong, C., Cao, Y., et al.: Enhanced-alignment measure for binary foreground map evaluation. In: Proceedings of the Twenty-Seventh International Joint Conference on Artificial Intelligence. International Joint Conferences on Artificial Intelligence Organization, pp. 698–704 (2018)
23. Sun, Y., Wang, S., Chen, C., et al.: Boundary-guided camouflaged object detection. In: 31st International Joint Conference on Artificial Intelligence, IJCAI 2022. International Joint Conferences on Artificial Intelligence, pp. 1335–1341 (2022)
24. Hu, X., Zhang, X., Wang, F., et al.: Efficient camouflaged object detection network based on global localization perception and local guidance refinement. IEEE Trans. Circuits Syst. Video Technol. **34**(7), 5452–5465 (2024)
25. Song, Z., Kang, X., Wei, X., et al.: Fsnet: focus scanning network for camouflaged object detection. IEEE Trans. Image Process. **32**, 2267–2278 (2023)
26. Yang, Y., Zhang, Q.: Finding camouflaged objects along the camouflage mechanisms. IEEE Trans. Circuits Syst. Video Technol. **34**(4), 2346–2360 (2023)
27. Cai, W., Gao, W., Wang, X., et al.: Lightweight diffusion model for camouflaged object detection. In: 2024 3rd International Conference on Artificial Intelligence, Internet of Things and Cloud Computing Technology (AIoTC), pp. 91-96. IEEE (2024)
28. Chen, Z., Gao, R., Xiang, T.Z., et al.: Diffusion model for camouflaged object detection. In: ECAI, vol. 2023, pp. 445–452. IOS Press (2023)

29. Zhao, R., Li, Y., Zhang, Q., et al.: Bilateral decoupling complementarity learning network for camouflaged object detection. Knowl.-Based Syst. **314**, 113158 (2025)
30. Sun, W., Wang, Q., Tian, Y.: Multi-level cross-knowledge fusion with edge guidance for camouflaged object detection. Knowl.-Based Syst. 113070 (2025)
31. Jha, D., Smedsrud, P.H., Riegler, M.A., Kvasir-seg: a segmented polyp dataset. In: MultiMedia Modeling: 26th International Conference, MMM, Daejeon, South Korea, January 5–8, 2020, Proceedings, Part II 26, vol 2020, pp. 451–462. Springer International Publishing (2020)
32. Bernal, J., Sánchez, F.J., Fernández-Esparrach, G., et al.: WM-DOVA maps for accurate polyp highlighting in colonoscopy: validation vs. saliency maps from physicians. Comput. Med. Imaging Graph. **43**, 99–111 (2015)
33. Silva, J., Histace, A., Romain, O., et al.: Toward embedded detection of polyps in wce images for early diagnosis of colorectal cancer. Int. J. Comput. Assist. Radiol. Surg. **9**, 283–293 (2014)
34. Tajbakhsh, N., Gurudu, S.R., Liang, J.: Automated polyp detection in colonoscopy videos using shape and context information. IEEE Trans. Med. Imaging **35**(2), 630–644 (2015)
35. Fan, D.P., Ji, G.P., Zhou, T., et al.: Pranet: parallel reverse attention network for polyp segmentation. In: International Conference on Medical Image Computing and Computer-Assisted Intervention, pp. 263–273. Springer International Publishing, Cham (2020)
36. Zhou, Z., Rahman Siddiquee, M.M., Tajbakhsh, N., Unet++: a nested u-net architecture for medical image segmentation. In: Deep Learning in Medical Image Analysis and Multimodal Learning for Clinical Decision Support: 4th International Workshop, DLMIA, and 8th International Workshop, ML-CDS 2018, Held in Conjunction with MICCAI 2018, Granada, Spain, Sept 20, 2018, Proceedings 4, vol. 2018, pp. 3–11. Springer International Publishing (2018)
37. Fang, Y., Chen, C., Yuan, Y.: Selective feature aggregation network with area-boundary constraints for polyp segmentation. In: Medical Image Computing and Computer Assisted Intervention–MICCAI, 22nd International Conference, Shenzhen, China, October 13–17, 2019, Proceedings, Part I 22. Springer International Publishing, vol. 2019, pp. 302–310 (2019)
38. Zhou, T., Zhou, Y., He, K., et al.: Cross-level feature aggregation network for polyp segmentation. Pattern Recogn. **140**, 109555 (2023)
39. Liu, G., Chen, Z., Liu, D., et al.: FTMF-net: a Fourier transform-multiscale feature fusion network for segmentation of small polyp objects. IEEE Trans. Instrum. Meas. **72**, 1–15 (2023)

HLAW-YOLOv8: An Improved Aerial Small Object Detection Algorithm Based on YOLO

ZhengChao Shi(✉) and HaiMing Li

Faculty of Artificial Intelligence, Shanghai University of Electric Power, No.1851 Hucheng Ring Road, Shanghai 201306, China
thloml@mail.shiep.edu.cn, lhm@shiep.edu.cn

Abstract. Drone object detection plays a significant role in civilian, commercial, and military fields. However, aerial images captured by drones are typically characterized by their large dimensions, while the extremely small size of objects leads to insufficient effective detection information. Factors such as varying lighting conditions, weather, and flight altitudes further exacerbate the challenges of detecting small objects in aerial imagery. To address the aforementioned issues, we propose an improved object detection algorithm, HLAW-YOLOv8, based on YOLOv8. Specifically, to tackle the problem of multi-scale feature fusion in small object detection, we introduce the HLFPN structure and combine it with the small object detection layer, effectively improving small object detection. To address the noise in aerial images, we incorporate the MMJD denoising module into the model, which effectively removes noise from the images. Furthermore, to solve the problem of small object occupying a relatively small proportion of the image, making feature extraction difficult, we propose the DCWR module. This module employs an efficient two-step multi-scale contextual information acquisition method to alleviate the issues of feature blur and information loss caused by the small scale of the targets. The model's mAP50 on the VisDrone dataset is improved by 9.8% compared to the baseline, and both mAPsmall and mAPmedium in the COCO metrics outperform the baseline model. Experimental results on the VisDrone and SODA-D datasets demonstrate the effectiveness and advanced nature of this method.

Keywords: Aerial Images · Small Object · Multi-scale Feature Fusion

1 Introduction

In the context of the rapid development of deep learning technologies, object detection models such as Faster R-CNN [1], YOLO [2], and SSD [3] have achieved remarkable results on natural image datasets. Meanwhile, drones, with their superior imaging capabilities and flexible flight abilities, have been widely applied

L. Huang and D. Greenhalgh (Eds.): ICMLC 2026, LNNS 2061, pp. 421–436, 2027.
https://doi.org/10.1007/978-3-032-31236-5_29

in fields like traffic monitoring and agricultural management, playing a significant role in improving efficiency and reducing costs. Drone-captured aerial images provide an overhead viewpoint and a broad visual scope, which presents many unique challenges for object detection tasks.

The main challenges in object detection of aerial images are as follows: First, the image sizes are typically large, exceeding the processing capabilities of conventional devices, often necessitating downscaling to a computationally manageable size. Second, small objects are more prevalent, and detectors generally perform poorly in identifying such objects. Small objects can be characterized in two ways: first, through relative scale, which compares the area of the target's bounding box to the total image area (with a median ratio between 0.08% and 0.58%), and second, via absolute scale, following the MS COCO standard, which classifies objects with a resolution under 32×32 pixels as small. Additionally, the distribution of objects in aerial images is uneven. For example, vehicles tend to cluster at intersections. Furthermore, due to varying flight altitudes, objects of the same type may present significant scale differences across different images.

Mainstream object detection models are primarily designed for natural scenes, and their architectures generally follow the Backbone, Neck, and Head component paradigm, including models such as the R-CNN series [1], YOLO series [2], and DETR series [4]. However, these general-purpose models often perform poorly in small object detection. In light of the challenges posed by UAV aerial imagery, researchers are focusing on developing detection models tailored for the features of aerial images. Kang et al. [5] proposed a scale sequence fusion algorithm based on the attention mechanism, which optimizes the network's feature extraction performance when processing multi-scale data. Shi et al. [6] utilized deformable convolutions to guide the feature fusion module, facilitating the effective integration of multi-scale features via interactions across layers and scales. Lin et al. [7] designed a multi-scale feature aggregation architecture built upon focusing and distribution mechanisms, harnessing the representations obtained from the backbone network to enable cross-scale context flow.

The main contributions of this paper are as follows:

1) Proposed the design of the DCWR module and applied it to improve the C2f module of YOLOv8, addressing the challenge of multi-scale information acquisition.
2) Proposed the HLFPN structure to address the multi-scale issues in the feature fusion process, and introduced a small object detection layer combined with the HLFPN structure, greatly boosting the model's efficiency in small object detection.
3) In response to the noise issues commonly present in UAV aerial images due to factors such as flight altitude, lighting, and weather conditions, the MMJD denoising module is introduced. This module effectively removes noise, enhances image details and contrast, making them stand out more clearly in the image.

The paper is organized as follows: Sect. 2 introduces related work; Sect. 3 presents the proposed improvements; Sect. 4 focuses on ablation and comparative experiments; Sect. 5 provides conclusions and future work.

2 Related Work

Object detection methods are primarily divided into anchor-based and anchor-free categories. Anchor-based methods can be further divided into two-stage and one-stage methods based on whether they use candidate region proposals. Two-stage methods, such as Faster R-CNN [1], Mask R-CNN [8], and R-FCN [9], include a Region Proposal Network (RPN) in Faster R-CNN to automatically generate candidate boxes; Mask R-CNN adds a mask branch to achieve object detection; R-FCN utilizes position-sensitive score maps to enhance detection efficiency. The one-stage methods are represented by YOLO [2], SSD [3], and RetinaNet [10]. YOLO formulates the object detection task as a regression problem, directly predicting the object class and location; SSD performs detection on multi-scale feature maps to address variations in object size; while RetinaNet introduces Focal Loss to alleviate the class imbalance problem.

Anchor-free methods discard predefined anchor boxes and mainly consist of two types: one type, such as FoverBox [11], generates detection boxes by predicting the distance from the object boundary to the four sides; the other type, such as CornerNet [12], locates the object by detecting its corner points or center points. Although these methods perform well in natural scenes, they often face challenges in aerial images, such as small object feature loss, noise interference, boundary sensitivity, and insufficient samples, which limit their detection performance.

Optimization strategies for small object detection include sample-based approaches like Aug-SOD [13], scale-aware methods like QueryDet [14], super-resolution reconstruction techniques like Sod-mtgan [15], context modeling methods like R2-CNN [16], and focus detection mechanisms like CGDM[17]. Among various models, YOLOv8 has become a research hotspot due to its outstanding performance and has spawned various improved versions specifically designed for UAV aerial imagery.For instance, DSYOLOv8 [18] introduces deformable convolutions and self-calibration attention, along with Wise-IoU loss to improve adaptability to deformable objects; YOLOv8-MPEB [19] balances efficiency and feature extraction capability by using a lightweight backbone, adding small object detection layers, and integrating multi-scale attention (EMA); DMFF-YOLO [20] constructs a scale sequence space using the DSSFF module and combines the RFCBAMConv module to mitigate information loss caused by positional changes, significantly enhancing robustness to scale and position variations.

3 Method

The overall structure of the HLAW-YOLOv8 algorithm is composed of three main parts: the backbone, the neck network (HLFPN), and the head network, as depicted in Fig. 1.To enhance the model's performance in denoising tasks, we introduce the innovative MMJD module in the backbone network to optimize both efficiency and accuracy. Additionally, the original C2f module in the backbone and neck networks is improved with the novel C2f_DCWR module. In the neck part, we also adopt the improved feature fusion structure HLFPN to replace the original feature fusion pyramid in YOLOv8. Additionally, a detection head that leverages lower-level features has been incorporated into the head network, further improving the algorithm's ability to recognize smaller objects with greater precision.To further enhance the deployment efficiency of the model on UAV edge-computing platforms, we optimized the trained HLAW-YOLOv8 FP32 model using dynamic post-training quantization (INT8).

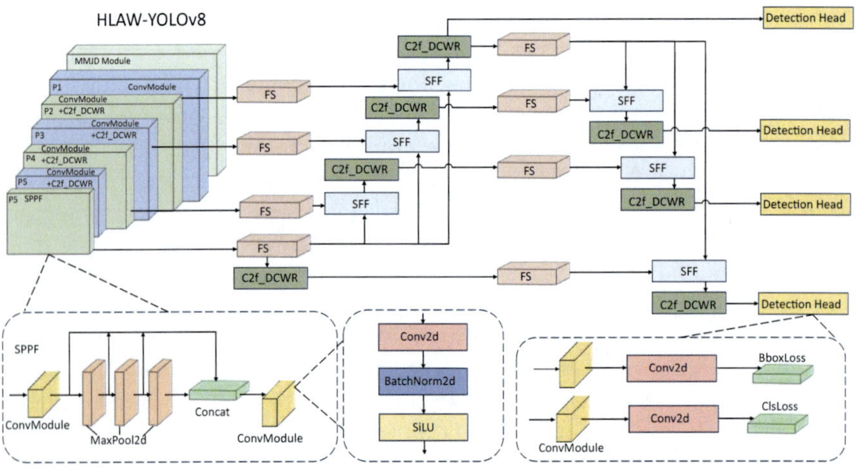

Fig. 1. Network architecture of the HLAW-YOLOv8 model.

3.1 Dilation Convolution Wise Residual Module

This work's key innovation to boost YOLOv8's performance lies in the design of a novel bottleneck structure, DCWR, which refines the initial C2f design. In complex scenarios, the initial C2f design becomes a key limitation to the model's performance due to its difficulty in adjusting to variations in objects of different scales. The improved module addresses the challenge of multi-scale information acquisition through a two-step framework.

The DCWR module ensures effective gradient backpropagation through the design of a residual structure,mitigating the gradient vanishing problem commonly seen in deep networks. As shown in Fig. 2, the DCWR module employs a two-step strategy to extract multi-scale contextual information.

First, regional residual features are generated through convolution. Then, multiple parallel branches are used to extract differentiated features with different receptive fields, capturing information across various scales, from local details to global semantics.These feature maps are then aggregated to achieve deep fusion of cross-scale contextual information, mitigating the problems of feature blur and information loss caused by small objects scales, thereby effectively improving the detection accuracy of small objects.

The core objective of the first step is to extract target-related residuals from the input features, generating an intermediate representation called regional residuals, which provides a structured feature foundation for subsequent morphological filtering. This process uses a 3×3 standard convolution to expand the channel count from 2C to 3C, employing a reverse bottleneck design that maps the features to a higher-dimensional space to enhance representational capacity. It also incorporates batch normalization (BN) and the ReLU activation function. The ReLU activation function suppresses background noise and enhances the edge responses of small objects, thereby achieving feature sparsity and refinement, which in turn improves detection accuracy and efficiency.

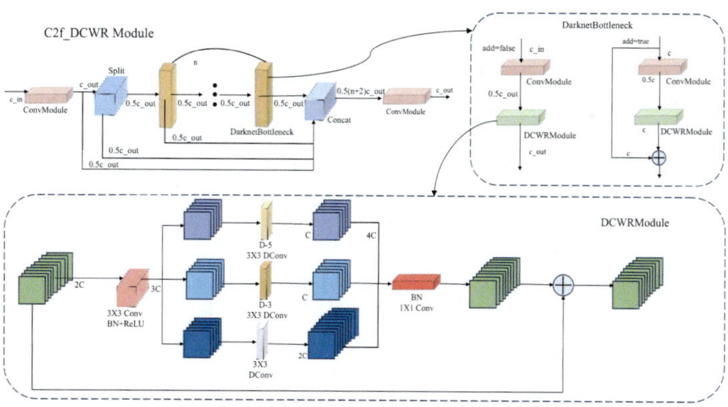

Fig. 2. Structure diagram of the C2f_DCWR module with the DCWR bottleneck structure.

The core operation of the second step is to apply morphological filtering to the regional features using parallel multi-rate depthwise convolutions. This process aims to extract key information from different scale structures, ultimately outputting semantic residual features that carry deep semantic knowledge. A single desired receptive field is applied on each channel, which avoids the redundant coverage caused by multiple receptive fields, improving computational efficiency while also reducing interference with fine-grained features.

Specifically, this process divides the feature map into multiple subgroups and applies dilated convolutions with different rates to each group, allowing for flexible modeling of multi-scale contextual information. This strategy not only demonstrates superior performance in small object detection, better preserving edge details and local information, but also shows good adaptability to medium and large-scale objects. Through parallel multi-scale receptive fields, it enhances robustness to objects of different scales and improves contextual awareness, thereby comprehensively boosting the model's overall performance across various object detection tasks. The regional feature maps extracted in the first step can adaptively learn the required concise regional features guided by the receptive field settings in the second step, achieving receptive field matching.

The proposed method, through residual-based design, repositions the role of multi-rate dilated convolution: instead of striving to capture complex global semantics, it is streamlined to perform morphological filtering with specific receptive fields on concise regional feature maps, leading to a more streamlined acquisition of multi-scale contextual insights. After obtaining multi-scale contextual information, the feature maps output by each branch are concatenated and fused. Specifically, all output features are first concatenated at the channel level and standardized using batch normalization (BN). Then, pointwise convolution is applied for channel compression and feature integration, forming the final residual. This residual is then element-wise added to the initial input features to produce output features with enhanced representational capacity.Moreover, considering that the ability of large receptive fields to establish effective spatial connections depends on the support from small receptive field features, we set the channel count of the first branch with the smallest dilation rate to be twice that of the other branches, thus enhancing its ability to extract local features. This design not only improves the coverage efficiency of the overall receptive field but also effectively avoids the feature dilution problem that may arise in small object detection.

3.2 Multi-Module Joint Denoising Module

To address the semantic or scene noise induced by factors such as shooting height, complex environments, and the inherent characteristics of the target, we propose introducing a Multi-Module Joint Denoising (MMJD) module into the backbone network of the YOLO model. This module helps to emphasize small object features, ultimately achieving more accurate and robust detection. Figure 3 displays the configuration of the MMJD module. Through the cascaded processing of Sparse Block (SB), Feature Enhancement Block (FEB), and Attention Block (AB), it establishes an end-to-end framework for residual learning. The final output of this module is defined as the noise map image I_R,which is subtracted from the input image I_N via residual connections to reconstruct the denoised, clear image I_{CL}. This procedure can be described by the core equation shown below:

$$
\begin{aligned}
I_{CL} &= I_N - I_R \\
&= I_N - f_{MMJD}(I_N)
\end{aligned}
$$

where $f_{MMJD}(\cdot)$ represents the composite mapping function of the entire MMJD module, which is responsible for predicting the residual image I_R from the noisy input I_N.

The sparse block adopts a design different from traditional sparse mechanisms. Its core is a lightweight network module constructed by the collaboration of dilated convolutions and standard convolutions. This module alternately stacks two basic units: dilated convolution + BN + ReLU and standard convolution + BN + ReLU, with a total of 12 layers, all using 3×3 kernels.Dilated convolution is used to enlarge the receptive field and incorporate a broader context, while standard convolution focuses on local feature extraction. This alternating structure of high and low receptive fields naturally forms sparse characteristics in the feature response, allowing the model to focus on key information with lower computational cost, improving denoising performance and training efficiency.

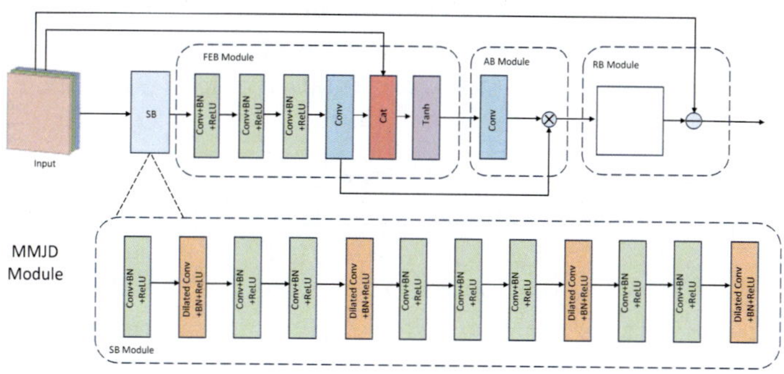

Fig. 3. Network architecture of the MMJD module.

To alleviate the problem of shallow feature degradation caused by gradient vanishing in deep networks, the feature enhancement block constructs long-range paths to fuse global context with local details, thereby enhancing feature representation. It also complements the multi-scale noise suppression of the sparse block. This module concatenates the original input image with the output at the 16th layer along the channel dimension, extending the output channels to 2c, forming a feature map of size 2c×H×W. This ensures the preservation of low-level structural information while incorporating deep-layer semantics. Finally, a Tanh activation function is applied to enhance expressiveness and numerical stability.

To address the issue of complex background interference with key features, the attention block uses deep features to guide shallow feature learning, focusing on critical noisy regions. The module takes the fused features with 2c channels as input, compresses them into a weight vector with c channels using a 1×1 convolution, and then performs element-wise multiplication with the original output features from the 16th layer of FEB to achieve feature recalibration.This mechanism, through a lightweight structure, achieves cross-layer feature guidance and adaptive noise awareness, improving denoising capabilities in complex backgrounds.

3.3 Hierarchical Feature Screening Fusion Pyramid Network

In UAV aerial imagery datasets, small object detection faces significant challenges due to the scale variation of targets, especially under different shooting altitudes where the same type of object can present notable scale differences in the image. To tackle this problem, this study proposes a Hierarchical Feature Fusion Pyramid Network (HLFPN), which improves the model's capacity to merge multi-scale features. As shown in Fig. 4, HLFPN is constructed with a feature selection and a feature fusion as its core components. Taking the P5 and P4 layers as examples, first, the feature maps undergo feature selection filtering, and then they are fed into the fusion module, where the low-level spatial information is corrected based on high-level semantic information.Compared to the feature fusion structure of YOLOv8, the proposed method generates richer semantic feature information, enabling it to capture subtle features of small objects more effectively, thus significantly improving detection performance.

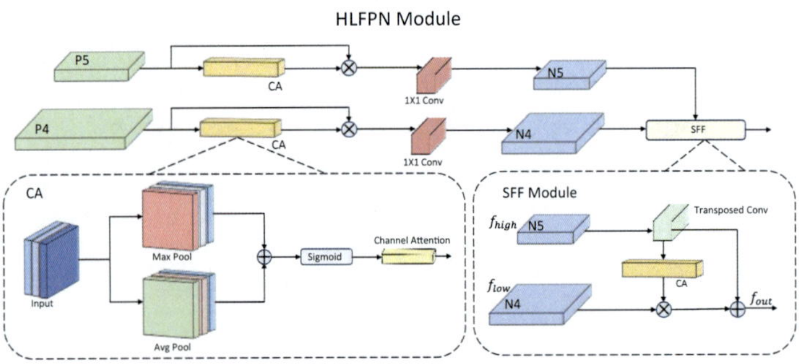

Fig. 4. Hierarchical feature fusion pyramid network.

The feature selection process involves two primary components: the Channel Attention (CA) module and the Feature Alignment (FA) module, which together complete the tasks of key feature selection and dimension unification. The CA module employs a dual-path pooling mechanism, performing both max pooling and average pooling in parallel. Max pooling focuses on extracting the most representative local features, effectively suppressing noise, while average pooling retains global information to prevent the loss of important details. The results of both pooling operations, after being activated by the Sigmoid function, generate channel weights, which are then fused through weighted integration to achieve an effective balance between global and local information.These weights are element-wise multiplied with the original feature maps, enhancing key features while suppressing redundant information, significantly improving the feature representation capability.

After feature selection, the FA module addresses the issue of dimension mismatch in multi-scale feature fusion. The module uses a 1×1 convolution to compress and reconstruct the channel dimensions of each scale's feature map, unifying the channel number to 256 while preserving the spatial resolution.This design not only ensures structural consistency in subsequent fusion operations but also refines the features through the convolution process, providing a solid foundation for effective cross-scale feature fusion. The entire feature selection module, through its screening and alignment mechanism, provides high-quality, structurally unified feature representations for subsequent multi-scale feature fusion.

The Selective feature fusion module processes multi-scale feature maps from the backbone network. High-level features are rich in semantic information but suffer from blurred localization and loss of detail due to their large receptive field and low spatial resolution. On the other hand, low-level features have precise localization and high resolution but limited semantic information. Existing methods typically fuse these features through direct addition or concatenation, which can complement semantic information but lack effective selection mechanisms, often introducing redundancy or interference.

In response to this problem, we present a SFF module in this paper, which uses high-level features to guide the selection of low-level features, enabling more effective cross-scale fusion. The detailed procedure is illustrated in Fig. 4: First, high-level features undergo transposed convolution upsampling to match the scale of the low-level features. Then, the high-level features are input into the CA module to generate attention weights, which are used to weight the low-level features, thereby retaining their key parts. Finally, the filtered low-level features are fused with the high-level features to form output features that combine both high-level semantic and low-level localization information.This fusion mechanism is further expressed through the following formula.

$$\begin{aligned} f_{thigh} &= T - Conv(f_{high}) \\ f_{out} &= f_{low} * CA(f_{thigh}) + f_{thigh} \end{aligned}$$

The SFF module, through strategic feature selection and fusion, significantly improves the quality of multi-scale features, enabling the model to more accurately extract key features for object detection tasks, enhancing both localization accuracy and semantic understanding.

The transposed convolution is used as the upsampling strategy to restore the spatial resolution of high-level feature maps. This method strikes a good balance between efficiency and expressive power, effectively preserving semantic information while reducing structural blur and feature loss. Transposed convolution has two main advantages: First, it is parameterizable, meaning it can adaptively adjust weights to reconstruct and enhance semantic information while enlarging the feature map. Second, it has a dynamic receptive field, allowing it to flexibly model irregular structures and complex shapes, improving the feature reconstruction ability for spatially inconsistent objects, thus alleviating the issue of information loss for small objects in high-level features.

4 Experiments and Analysis

4.1 Dataset

The proposed approach was evaluated using two open-access aerial image datasets: VisDrone [21] and SODA-D [22]. A brief description of these datasets is given below. VisDrone [21] contains 10,209 high-resolution images, divided into 10 categories: pedestrian, person, bicycle, car, truck, lorry, tricycle, tricycle with canopy, bus, and motorcycle. Training, validation, and test subsets consist of 6,471, 548, and 3,190 images, respectively. The SODA-D [22] focuses on driving scenarios and comprises 24,828 high-quality images with 278,433 annotated instances spanning nine categories: pedestrian, rider, bicycle, motorcycle, vehicle, traffic sign, traffic light, traffic camera, and warning cone. SODA-D offers notable diversity across temporal settings, geographic areas, climatic conditions, and viewing angles. The SODA-D images were divided into three subsets: training (50%), validation (20%), and testing (30%).Figure 5 demonstrates the challenges posed by these datasets, which require the model to achieve higher accuracy and better robustness.

(a) (b) (c)

Fig. 5. Images from two datasets: (a) Challenge of small, dense targets, (b) Challenge of lighting variation, (c) Challenge of occlusion.

4.2 Implementation Details

In the experiment, we used Linux as the operating system, an NVIDIA RTX 3090 GPU, Python version 3.8.20, PyTorch version 1.12.0, and CUDA version 11.3. The specific experimental environment parameter settings are shown in Table 1. Other model results are sourced from relevant literature.

Table 1. Hyperparameter Settings of the Model

Hyperparameters	Value
Imgsz	640
Batch	8
Epoch	300
Weight Decay	0.0005
Momentum	0.937
Initial Learfning Rate	0.01
Optimizer	SGD

4.3 Comparative Experiments

Extensive experiments were carried out on the challenging drone aerial image datasets, VisDrone and SODA-D, to evaluate the performance of our proposed method against a range of advanced detection models. Our approach achieved the most competitive performance across multiple key metrics, demonstrating significant advantages, especially in handling scale variations and small object detection.

The quantitative results of HLAW-YOLOv8 versus the baseline and other YOLO variants on VisDrone are summarized in Table 2. It was observed from the experiments that our method attains 30.1% mAP and 49.2% mAP50. Compared with the baseline YOLOv8s, this corresponds to a 6.4% point increase in mAP, confirming that the proposed model meets the anticipated performance goals. In terms of model complexity, HLAW-YOLOv8 contains only 14.6M parameters, which is substantially fewer than YOLOv8l and YOLOv3, while still achieving superior detection accuracy. Compared with ACAM-YOLO, which has similar accuracy, our method requires fewer parameters, demonstrating a better balance between accuracy and efficiency, effectively achieving both high detection performance and computational efficiency.

Table 3 shows that on the VisDrone dataset, the proposed method achieves the best mAP detection performance on the original validation set, outperforming all comparison models. Although it performs slightly worse than ClusDet and CDMNet in terms of the mAP50 metric, it still demonstrates stronger generalization ability without employing any image cropping or clustering preprocessing strategies.

Table 2. Comparison of our model with the baseline and other YOLO variants on the visDrone dataset

Method	mAP	mAP50	Param(M)
YOLOv8s(baseline)	23.7	39.4	11.1
YOLOv3	23.3	42.9	103.7
YOLOv5s	19	34.4	7.2
Drone-YOLO(Small)	25.6	42	**5.4**
YOLOv5l	24.4	41.4	46.5
UAV-YOLOv8s	29.2	47	10.3
YOLOv8l	26.9	43.7	76.7
ACAM-YOLO	29.6	**49.5**	15.9
Ours	**30.1**	49.2	14.6

Table 3. compares the proposed method with classic one-stage and two-stage models on the VisDrone dataset. Here, "O" and "CA" represent the original validation set and cluster-aware cropped images, respectively

Method	test data	img	mAP	mAP50	mAP_{small}	mAP_{medium}	mAP_{small}
FRCNN	o	548	21.4	40.7	11.7	33.9	**54.7**
ClusDet	o+ca	2716	26.7	**50.6**	17.6	38.9	51.4
CDMNet	ca	2170	29.2	49.5	**20.8**	**40.7**	41.6
DMNet	o	548	28.2	47.6	19.9	39.6	55.8
QueryDet	o	548	28.32	48.14			
CenterNet	o	548	23.9	45.5	14.4	36.3	44.9
Ours	o	548	**30.1**	49.2	20.2	39.1	46.1

For small object detection, the proposed method achieves an mAPsmall of 20.2%, significantly outperforming QueryDet and CenterNet. Although it is slightly inferior to CDMNet, which uses cluster-aware cropping, our proposed method performs inference on the images without any preprocessing, offering greater practical application value.

In medium and large object detection tasks, the model also maintains stable performance, remaining on par with mainstream approaches. Overall, the proposed method substantially improves detection accuracy while balancing model complexity and deployment efficiency, demonstrating excellent performance equilibrium and engineering applicability.

To further verify the generalizability of the proposed model, we conducted evaluations on the SODA-D dataset and compared it with several classic single-stage and two-stage detection models. It is worth noting that the evaluation metrics used in this experiment differ from the COCO standard and follow the classification scheme defined in the original dataset paper : mAPeS, mAPrS, and

Table 4. Comparison with classic one-stage and two-stage models on the SODA-D dataset. It should be noted that eS, rS, and gS represent extremely small targets, relatively small targets, and general small targets, respectively

Method	mAP	mAP50	mAPeS	mAPrS	mAPgS
YOLOv8s(baseline)	25.7	51.9	0.6	18	29.8
Deformable-DETR	19.2	44.8	6.3	15.4	24.9
Sparse RCNN	24.2	50.3	8.8	20.4	30.2
CornerNet	24.6	49.5	6.5	20.5	32.2
CenterNet	21.5	48.8	5.1	16.2	29.6
FCOS	23.9	49.5	6.9	19.4	**30.9**
Ours	**27.4**	**54.3**	**10.4**	**21.9**	30.4

mAPgS correspond to detection accuracy for objects with areas smaller than 144, 400, and 1024 pixels, respectively.

As shown in Table 4, the proposed model achieves the best performance in both mAP and mAP50 metrics. It also performs best in detecting extremely small and relatively small objects, reaching 10.4% and 21.9%, respectively, demonstrating outstanding small-object perception capability. Although the model performs slightly worse than CornerNet and FCOS on the general small-object metric, its detection performance on smaller-scale targets is significantly superior to both, indicating stronger recognition ability for extremely small objects.

The experimental findings demonstrate the model's ability to detect objects of various scales and achieves significantly better robustness and accuracy than existing mainstream methods, especially in small-object and complex-scene detection tasks. This demonstrates the model's strong cross-scene adaptability and high practical application value.

4.4 Ablation Study

This section details the changes in the model's detection metrics on the VisDrone dataset as various improvements are gradually introduced. When YOLOv8s is used as the baseline model, its COCO evaluation metrics are: mAP50 (39.4%), mAP (23.7%), and mAPsmall (13%).

The ablation study results presented in Table 5 verify the continuous performance improvement contributed by each module. Starting from YOLOv8s as the baseline, the step-by-step introduction of the HLFPN, P2 detection layer, MMJD, and C2f_DCWR modules leads to gradual enhancements in overall detection accuracy and small object recognition capability.

The HLFPN module strengthens multi-scale feature fusion, raising the mAP to 25.6. Adding the P2 layer further enhances the capture of shallow detail features, resulting in a remarkable 37.0% increase in mAP_{small}. The MMJD module effectively suppresses feature noise, improving target discrimination in complex

Table 5. Ablation Study of Model Modules on COCO Metrics

Method	Module				Result			
	HLFPN	P2	MMJD	DCWR	mAP	mAP50	mAP_{small}	mAP_{large}
baseline					23.7	39.4	13	41.7
HLAW-YOLOv8	√				25.6	43.7	13.5	45.8
HLAW-YOLOv8	√	√			28.4	47.4	18.5	42.7
HLAW-YOLOv8	√	√	√		29.6	48.6	19.2	41.5
HLAW-YOLOv8	√	√	√	√	**30.1**	**49.2**	**20.2**	**46.1**

backgrounds. Finally, the inclusion of the C2f_DCWR module further boosts small-object detection accuracy to 20.2 while maintaining large-object detection performance, enabling the model to achieve better balance and adaptability across multi-scale targets.

In summary, the HLFPN module improves basic performance through multi-scale feature fusion, the P2 module enhances detail perception, the MMJD module strengthens model robustness, and the DCWR module further optimizes cross-scale detection performance. The collaborative effect of these modules drives the overall improvement in detection accuracy and adaptability of the model.

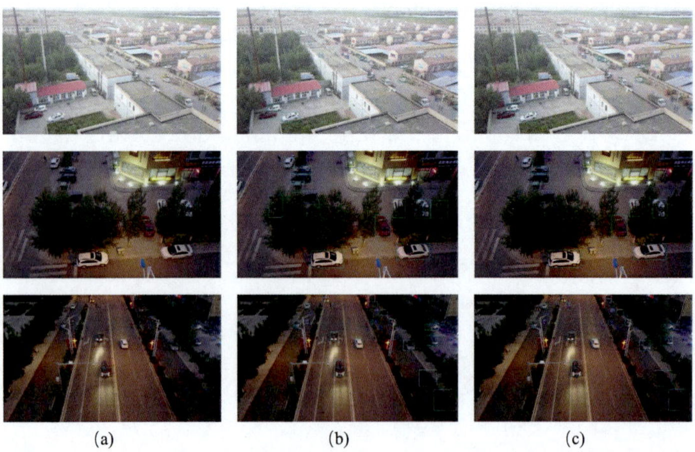

Fig. 6. Visualization of detection results of different YOLO-based models on the VisDrone dataset: (a) Original image, (b) YOLOv8s inference image, (c) HLAW-YOLOv8 inference image.

Additionally, a comparison experiment was performed to visually illustrate the detection effectiveness of our method, using the YOLOv8s model for inference comparison to assess the performance of the HLAW-YOLOv8 model. In the experiment, we selected some representative scenes from the VisDrone dataset, which contain targets of varying sizes, making them suitable for inference testing. The visualized results show that our model performs excellently in detecting small objects, effectively identifying targets that are often missed by the YOLOv8s baseline model. In some cases, it also corrects the baseline model's erroneous detections. This indicates that HLAW-YOLOv8 achieves higher accuracy and robustness in handling multi-scale targets and complex scenes. The model inference visualization results are shown in Fig. 6.

4.5 Model Optimization Experiments

The trained YOLOv8s FP32 model was optimized via dynamic post-training quantization (INT8). On the validation set, it achieved a mAP50 of 0.492 and a mAP of 0.301. Quantization maintained accuracy while reducing the average inference latency per image from 5.8 ms to 4.3 ms, effectively easing the computational load for UAV edge deployment.

5 Conclusion

To address low detection accuracy, complex backgrounds, and the trade-off between efficiency and performance in drone aerial imagery, this paper proposes an improved YOLOv8-based model with three key enhancements. The HLFPN structure and a small-object detection layer improve multi-scale feature fusion, an image denoising module suppresses noise and strengthens target feature representation, and the C2f_DCWR module reduces small object feature blurring and information loss via two-step context extraction.

While the model shows significant performance gains, efficiency in resource-constrained scenarios can be further improved. Future work includes introducing lightweight backbones or model pruning to compress model size and computational load, as well as exploring focus-based detection strategies to enhance small object recognition, aiming for efficient, low-power embedded deployment in real-time drone monitoring and similar applications.

References

1. Ren, S., He, K., Girshick, R., Sun, J.: Faster r-cnn: towards real-time object detection with region proposal networks. IEEE Trans. Pattern Anal. Mach. Intell. **39**(6), 1137–1149 (2016)
2. Redmon, J., Farhadi, A.: Yolov3: an incremental improvement (2018). arXiv:1804.02767 arXiv preprint
3. Liu, W., Anguelov, D., Erhan, D., Szegedy, C., Reed, S., Fu, C.Y., Berg, A.C.: Ssd: single shot multibox detector. In: European Conference on Computer Vision, pp. 21–37. Springer (2016)
4. Li, F., Zeng, A., Liu, S., Zhang, H., Li, H., Zhang, L., Ni, L.M.: Lite detr: an interleaved multi-scale encoder for efficient detr. In: Proceedings of the IEEE/CVF Conference on Computer Vision and Pattern Recognition, pp. 18558–18567 (2023)

5. Kang, M., Ting, C.M., Ting, F.F., Phan, R.C.W.: Asf-yolo: a novel yolo model with attentional scale sequence fusion for cell instance segmentation. Image Vis. Comput. **147**, 105057 (2024)
6. Shi, Y., Wang, C., Xu, S., Yuan, M.D., Liu, F., Zhang, L.: Deformable convolution-guided multiscale feature learning and fusion for uav object detection. IEEE Geosci. Remote Sens. Lett. **21**, 1–5 (2024)
7. Lin, Y., Li, J., Shen, S., Wang, H., Zhou, H.: Gdrs-yolo: more efficient multiscale features fusion object detector for remote sensing images. IEEE Geosci. Remote Sens. Lett. **21**, 1–5 (2024)
8. He, K., Gkioxari, G., Dollár, P., Girshick, R.: mask r-cnn. In: Proceedings of the IEEE International Conference on Computer Vision, pp. 2961–2969 (2017)
9. Dai, J., Li, Y., He, K., Sun, J.: R-fcn: object detection via region-based fully convolutional networks. Adv. Neural. Inf. Process. Syst. **29**, (2016)
10. Lin, T.Y., Goyal, P., Girshick, R., He, K., Dollár, P.: Focal loss for dense object detection. In: Proceedings of the IEEE International Conference on Computer Vision, pp. 2980–2988 (2017)
11. Kong, T., Sun, F., Liu, H., Jiang, Y., Li, L., Shi, J.: Foveabox: beyound anchor-based object detection. IEEE Trans. Image Process. **29**, 7389–7398 (2020)
12. Law, H., Deng, J.: Cornernet: detecting objects as paired keypoints. In: Proceedings of the European Conference on Computer Vision (ECCV), pp. 734–750 (2018)
13. Kisantal, M., Wojna, Z., Murawski, J., Naruniec, J., Cho, K.: Augmentation for small object detection (2019). arXiv:1902.07296 arXiv preprint
14. Yang, C., Huang, Z., Wang, N.: Querydet: cascaded sparse query for accelerating high-resolution small object detection. In: Proceedings of the IEEE/CVF Conference on Computer Vision and Pattern Recognition, pp. 13668–13677 (2022)
15. Bai, Y., Zhang, Y., Ding, M., Ghanem, B.: Sod-mtgan: small object detection via multi-task generative adversarial network. In: Proceedings of the European Conference on Computer Vision (ECCV), pp. 206–221 (2018)
16. Pang, J., Li, C., Shi, J., Xu, Z., Feng, H.: R2-cnn: fast tiny object detection in large-scale remote sensing images. arxiv comput. vis. Pattern Recognit 1902, 06042 (2019)
17. Duan, C., Wei, Z., Zhang, C., Qu, S., Wang, H.: Coarse-grained density map guided object detection in aerial images. In: Proceedings of the IEEE/CVF International Conference on Computer Vision, pp. 2789–2798 (2021)
18. Shen, L., Lang, B., Song, Z.: Ds-yolov8-based object detection method for remote sensing images. IEEE Access **11**, 125122–125137 (2023)
19. Xu, W., Cui, C., Ji, Y., Li, X., Li, S.: Yolov8-mpeb small target detection algorithm based on uav images. Heliyon **10**(8) (2024)
20. Qiu, X., Chen, Y., Sun, C., Li, J., Niu, M.: Dmff-yolo: Yolov8 Based on Dynamic Multiscale Feature Fusion for Object Detection on UAV Aerial Photography. IEEE Access (2024)
21. Zhu, P., Wen, L., Du, D., Bian, X., Ling, H., Hu, Q., Nie, Q., Cheng, H., Liu, C., Liu, X., et al.: Visdrone-det2018: the vision meets drone object detection in image challenge results. In: Proceedings of the European Conference on Computer Vision (ECCV) Workshops, pp. 0–0 (2018)
22. Cheng, G., Yuan, X., Yao, X., Yan, K., Zeng, Q., Xie, X., Han, J.: Towards large-scale small object detection: survey and benchmarks. IEEE Trans. Pattern Anal. Mach. Intell. **45**(11), 13467–13488 (2023)

Smartwatch-Based Sitting Time Estimation in Real-World Office Settings

Olivia Zhang[1] and Zhilin Zhang[2](✉)

[1] The Hockaday School, Dallas, TX 75229, USA
ozhang30@hockaday.org
[2] Lumos Alpha, Dallas, TX 75248, USA
zhilinzhang@lumosalpha.com

Abstract. Sedentary behavior poses a major public health risk, being strongly linked to obesity, cardiovascular disease, and other chronic conditions. Accurately estimating sitting time is therefore critical for monitoring and improving individual health. This work addresses the problem in real-world office settings, where signals from the inertial measurement units (IMU) on a smartwatch were collected from office workers during their daily routines. We propose a method that estimates sitting time from the IMU signals by introducing the use of rotation vector sequences, derived from Euler angles, as a novel representation of movement dynamics. Experiments on a 34 h dataset demonstrate that exploiting rotation vector sequences improves algorithm performance, highlighting their potential for robust sitting time estimation in natural environments.

Keywords: Sedentary behavior · Wearable computing · Human computer interface · Human activity recognition · Smartwatch

1 Introduction

Prolonged sedentary behavior is strongly associated with obesity, cardiovascular disease, metabolic disorders, and mental health conditions [1]. It is the fourth leading risk factor for global mortality, responsible for 6% of worldwide deaths [1]. It has even been referred to as the "modern smoke." However, with recent trends in remote working, restrictive urban environments, and long work hours, many office workers find it challenging to reduce their sedentary time.

As a result, estimating sitting duration has become an active research area in wearable computing [2–4]. Mekruksavanich et al.[5] proposed several conventional machine learning algorithms to distinguish sitting from non-sitting activities using smartwatch IMU sensor signals (accelerometer and gyroscope data). They then introduced a hybrid CNN-LSTM model to better capture spatiotemporal dependencies in smartwatch IMU sensor signals [6]. In many such studies, sitting time estimation is often framed as a sitting activity recognition problem, studied within the broader field of human activity recognition [7,8].

L. Huang and D. Greenhalgh (Eds.): ICMLC 2026, LNNS 2061, pp. 437–447, 2027.
https://doi.org/10.1007/978-3-032-31236-5_30

Beyond IMU-only methods, alternative approaches have also been explored [9–11]. For example, Salim et al.[9] proposed a hidden semi-Markov model that relied solely on step counts and heart rate from smartwatch to detect sedentary time. Yeon et al.[10] exploited both audio and IMU sensor signals for real-time smartwatch-based activity recognition.

Most of these studies predefine activity types, collect data for each under controlled conditions, and train models to classify movements into one of the predefined categories. However, in real-world settings, the diversity of human activities may make such approaches difficult to generalize. Moreover, many studies have relied on wearable devices placed on the torso, arms, chest, or other body parts [12–15], which are impractical for daily use due to discomfort and inconvenience.

In this work, we address sitting time estimation under real-world conditions by collecting smartwatch IMU data from office workers during their daily routines. Our main contributions are as follows:

- We specifically study sitting classification/sitting time estimation using commodity smartwatch IMU signals, rather than relying on other sensing modalities commonly assumed in prior human activity recognition work.
- We collect data in real office environments across whole-day routines, where subjects move freely and are not even aware of the ongoing experiment (i.e., without scripted sit/stand instructions and without "acting for the experiment" as in prior work), resulting in a noisy dataset containing many micro- and unexpected activities.
- Unlike prior studies that rely directly on raw IMU signals [2–5,7,16], we introduce rotation vector sequences, derived from accelerometer data, as an alternative "signal" representation that captures orientation-related movement dynamics and provides an additional feature space beyond raw IMU streams. As a result, features extracted from these rotation vector sequences can improve sitting/non-sitting classification performance.
- We demonstrate that most existing algorithms can be readily adapted to operate on rotation vector sequences, treating them as an additional sensor modality analogous to accelerometer or gyroscope signals.

2 Proposed Method

We observe that when an office worker is seated at a desk, the natural motion of the arm is constrained, as the desk limits its ability to rest freely at the side of the body. In contrast, when the individual stands, the arm is typically able to hang unobstructed in a relaxed position. In other words, sitting restricts the arm from bending across certain ranges of angles, whereas standing permits greater freedom of movement. Motivated by this observation, we propose the following method for estimating sitting time.

The proposed method leverages acceleration signals from the smartwatch's IMU to derive a sequence of rotation vectors. The rotation vector sequence represents movement dynamics of the arm, and thus features derived from these rotation vectors can be utilized to classify sitting versus non-sitting activities. The paradigm of the proposed method is shown in Fig. 1.

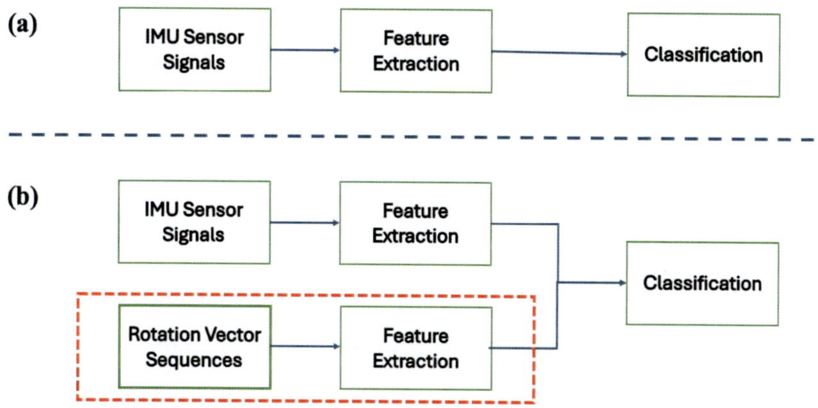

Fig. 1. Comparison between the conventional smartwatch-IMU pipeline and the proposed pipeline. (a) Typical paradigm in prior work for sitting classification/sitting-time estimation, in which features are extracted from IMU signals (e.g., accelerometer and/or gyroscope streams) and fed into a classifier/regressor. (b) Proposed paradigm: in addition to the standard IMU streams, the rotation-vector sequences (indicated by red dashed lines) are treated as an additional signal channel for feature extraction. Features from both branches are fused and then used for classification or regression.

2.1 IMU Coordinate Setting

In our approach, the smartwatch IMU coordinate system is defined as follows: when the smartwatch lies flat with its screen facing upward, the x-axis points to the right, aligned with the wrist's width; the y-axis points forward, aligned with the wrist's length; and the z-axis points outward, perpendicular to the screen.

Rotation about the x-axis is defined as the pitch angle, denoted by ϕ, whereas rotation about the y-axis is defined as the roll angle, denoted by θ. These two angles are the key to our approach. In addition, rotation about the z-axis is defined as the yaw angle. We are not interested in this angle because its value can vary arbitrarily regardless of whether an individual is sitting, standing, or walking. Moreover, as we will see soon, without additional information or constraints, the yaw angle cannot be estimated using acceleration signals.

2.2 Estimate the Pitch and Roll Angles

Note that regardless of what orientation the smartwatch has, gravity is a physical vector fixed in the world frame $\mathbf{g}_{\text{world}} = [0, 0, -g]^T$. When the smartwatch rotates to an orientation, the gravity in its frame can be expressed as

$\mathbf{g}_{\text{watch}} = [g_x, g_y, g_z]^T$, where g_x, g_y, g_z is the recorded acceleration signal along the axes of the smartwatch. According to Euler's Theorem [17], such a rotation can be achieved via a procedure by first rotating around z-axis, then around y-axis, and finally around x-axis in order.

Assume the smartwatch first rotates around its z-axis by a yaw angle ψ, the gravity in the smartwatch frame is given by

$$\mathbf{g}_1 = \mathbf{R}_z(\psi)\mathbf{g}_{\text{world}} \tag{1}$$

where $\mathbf{R}_z(\psi)$ is a **passive** rotation matrix around the smartwatch's z-axis, given by[1]

$$\mathbf{R}_z(\psi) = \begin{bmatrix} \cos\psi & \sin\psi & 0 \\ -\sin\psi & \cos\psi & 0 \\ 0 & 0 & 1 \end{bmatrix}$$

Next, the smartwatch rotates around its y-axis by a roll angle θ. The gravity in the smartwatch's frame is then

$$\mathbf{g}_2 = \mathbf{R}_y(\theta)\mathbf{g}_1 \tag{2}$$

where $\mathbf{R}_y(\theta)$ is a passive rotation matrix around the y-axis, given by

$$\mathbf{R}_y(\theta) = \begin{bmatrix} \cos\theta & 0 & -\sin\theta \\ 0 & 1 & 0 \\ \sin\theta & 0 & \cos\theta \end{bmatrix}$$

Finally, the smartwatch rotates around its x-axis by a pitch angle ϕ. At this step, the gravity in the smartwatch's frame is

$$\mathbf{g}_{\text{watch}} = \mathbf{R}_x(\phi)\mathbf{g}_2 \tag{3}$$

where $\mathbf{R}_x(\phi)$ is the passive rotation matrix around the x-axis, given by

$$\mathbf{R}_x(\phi) = \begin{bmatrix} 1 & 0 & 0 \\ 0 & \cos\phi & \sin\phi \\ 0 & -\sin\phi & \cos\phi \end{bmatrix}$$

[1] Note that our derivation uses **passive** rotation matrices (a change of coordinate frame) to express a fixed physical vector, rather than "spinning" the vector in 3D space. A passive rotation matrix is the transpose/inverse of the corresponding **active** rotation matrix, which is commonly seen in the literature [17].

Combining the three steps (1) - (3), we have

$$\begin{aligned}
\mathbf{g}_{\text{watch}} &= \mathbf{R}_x(\phi)\,\mathbf{R}_y(\theta)\,\mathbf{R}_z(\psi)\,\mathbf{g}_{\text{world}} \\
&= \begin{bmatrix} 1 & 0 & 0 \\ 0 & \cos\phi & \sin\phi \\ 0 & -\sin\phi & \cos\phi \end{bmatrix} \begin{bmatrix} \cos\theta & 0 & -\sin\theta \\ 0 & 1 & 0 \\ \sin\theta & 0 & \cos\theta \end{bmatrix} \\
&\quad \begin{bmatrix} \cos\psi & \sin\psi & 0 \\ -\sin\psi & \cos\psi & 0 \\ 0 & 0 & 1 \end{bmatrix} \begin{bmatrix} 0 \\ 0 \\ -g \end{bmatrix} \\
&= \begin{bmatrix} g\sin\theta \\ -g\sin\phi\,\cos\theta \\ -g\cos\phi\,\cos\theta \end{bmatrix}. \qquad (4)
\end{aligned}$$

Note that the yaw angle ψ does not appear in the final result. This is why the yaw angle cannot be estimated without using additional information. Fortunately, as stated before, we are not interested in estimating the yaw angle.

From (4) we have

$$g_x = g\sin\theta \qquad (5)$$

$$g_y = -g\sin\phi\cos\theta \qquad (6)$$

$$g_z = -g\cos\phi\cos\theta \qquad (7)$$

From there, it is simple to solve for ϕ and θ:

$$\phi = \operatorname{atan2}(-g_y, -g_z) \qquad (8)$$

$$\theta = \operatorname{atan2}\left(g_x, \sqrt{g_y^2 + g_z^2}\right) \qquad (9)$$

The estimated angles ϕ and θ exhibit distinct patterns between sitting and non-sitting postures and can therefore serve as features for classification. However, the solutions are susceptible to a phenomenon known as **gimbal lock**: when $|\theta|$ approaches 90° (i.e., $\cos\theta \to 0$), ϕ becomes ill-defined and numerically unstable. As a result, using ϕ and θ directly as features is unreliable near the singularity and should be avoided.

2.3 Rotation Matrix and Axis-Angle Representation

Instead of using ϕ and θ directly as features, we consider their corresponding rotation matrices. Defining $\mathbf{R}_{xy} = \mathbf{R}_x(\phi)\mathbf{R}_y(\theta)$, we have

$$\begin{aligned}
\mathbf{R}_{xy} &= \begin{bmatrix} 1 & 0 & 0 \\ 0 & \cos\phi & \sin\phi \\ 0 & -\sin\phi & \cos\phi \end{bmatrix} \begin{bmatrix} \cos\theta & 0 & -\sin\theta \\ 0 & 1 & 0 \\ \sin\theta & 0 & \cos\theta \end{bmatrix} \\
&= \begin{bmatrix} \cos\theta & 0 & -\sin\theta \\ \sin\phi\,\sin\theta & \cos\phi & \sin\phi\,\cos\theta \\ \cos\phi\,\sin\theta & -\sin\phi & \cos\phi\,\cos\theta \end{bmatrix}. \qquad (10)
\end{aligned}$$

It should be noted that although ϕ and θ may have gimbal lock, the combined rotation matrix $\mathbf{R}_{xy}$ evolves smoothly. Therefore, we design features from the rotation matrix (10).

The 3×3 combined rotation matrix has orthogonality constraints. To better capture its evolvement pattern, one can derive the rotation vector associated with it, which is a compact, unconstrained 3D representation that respects the geometry of rotations and is easier for learning algorithms.

Following the approach in [18], $\mathbf{R}_{xy}$ can be expressed as a rotation by angle $\alpha \in [0, \pi]$ about a unit axis $\mathbf{u} \in \mathbf{R}^3$, where

$$\alpha = \arccos\left(\frac{\mathrm{tr}(\mathbf{R}_{xy}) - 1}{2}\right),$$

and, for $\alpha \neq 0$,

$$\mathbf{u} = \frac{1}{2\sin\alpha}\begin{bmatrix} R_{32} - R_{23} \\ R_{13} - R_{31} \\ R_{21} - R_{12} \end{bmatrix}$$

where R_{ij} is the (i, j)-th element of $\mathbf{R}_{xy}$.

The rotation vector is then

$$\mathbf{r} = \alpha\,\mathbf{u} = \frac{\alpha}{2\sin\alpha}\begin{bmatrix} R_{32} - R_{23} \\ R_{13} - R_{31} \\ R_{21} - R_{12} \end{bmatrix}.$$

When $\alpha \approx 0$, using the first-order approximation, we have

$$\mathbf{r} \approx \begin{bmatrix} \frac{R_{32} - R_{23}}{2} \\ \frac{R_{13} - R_{31}}{2} \\ \frac{R_{21} - R_{12}}{2} \end{bmatrix}.$$

The rotation vector $\mathbf{r}$ represents the final output. At each timestamp, a rotation vector can be computed from the three-axis acceleration signals. Consequently, a temporal sequence of acceleration measurements can be transformed into a corresponding sequence of rotation vectors, from which discriminative features may be extracted for classification tasks. Alternatively, deep learning models can be applied directly to the sequence of rotation vectors, thereby circumventing reliance on raw acceleration signals.

To exploit the rotation vectors, it is essential that the acceleration signals measured by the smartwatch IMU, $\mathbf{g}_{\mathrm{watch}}$, are dominated by the gravitational component, while contributions from dynamic accelerations due to external forces are minimized. Depending on the application, this requirement may be relaxed, or it can be addressed through signal processing techniques such as low-pass filtering, smoothing, or sensor fusion with gyroscope data. In the next section, we demonstrate how the sequence of rotation vectors can be leveraged for the sitting time estimation application.

It is important to note that the majority of existing studies [2,5,9] have focused on directly using acceleration signals for human activity classification. In contrast, our approach introduces a novel paradigm by transforming acceleration signals into rotation vectors, from which discriminative features can be derived or end-to-end models can be trained (See Fig. 1). This representation captures orientation information more effectively and opens a new, potentially robust direction for advancing human activity recognition research.

3 Classification

A variety of approaches can be employed to exploit the sequence of rotation vectors for distinguishing sitting from non-sitting states. In this work, we propose a CatBoost-based algorithm [19]. More importantly, we emphasize that most existing methods can be enhanced by treating the rotation vector sequence as an additional sensor modality and applying algorithms to it in the same way they are applied to acceleration or gyroscope signals.

Following prior work [2,5], we segment the tri-axial accelerometer time series into windows of length N. For each window, we compute the median of the x-, y-, and z-axis signals. When the watch-wearing arm is largely stationary (e.g., reading, writing, typing), these medians closely approximate the gravitational components along each sensor axis, enabling reliable estimation of the device's orientation and the corresponding rotation vector. During vigorous activites (e.g., walking or running), however, linear accelerations dominate and the readings from the accelerometer, $\mathbf{g}_{watch}$, will not entirely be based off of gravity. But we find the rotation vector sequence, calculated at each timestamp, exhibits evolution patterns indicating of non-sitting behavior. Therefore, in addition to the axis-wise median-derived rotation vector, we calculate the following features from the rotation vector sequence: $\mathbf{r}^{n,m} = [r_x^{n,m}, r_y^{n,m}, r_z^{n,m}]^T$, where $n \in \{0, \ldots, N-1\}$ indexes the timestamps within the window, and m denotes the window index.

One type of important feature is the Fuzzy entropy [20] of $r_x^{n,m}$, $r_y^{n,m}$, and $r_z^{n,m}$. This complexity measure quantifies the irregularity and unpredictability of a time series. Unlike Shannon entropy and other conventional entropy measures, fuzzy entropy is specifically designed for time-series analysis, providing more robust and consistent characterization of temporal dynamics.

Another type of feature is the statistical descriptors of $r_x^{n,m}$, $r_y^{n,m}$, and $r_z^{n,m}$, including the mean, standard deviation, maximum, minimum, interquartile range (IQR), and median absolute deviation (MAD). These features capture the central tendency, variability, and distributional spread of a time series.

The energy measure is also used, defined for $r_x^{n,m}$ as $E_x = \frac{1}{N}\sum_{n=0}^{N-1}\left(r_x^{n,m}\right)^2$ with analogous definitions for $r_y^{n,m}$ and $r_z^{n,m}$.

In addition to extracting the above features from the rotation vector sequence $\mathbf{r}^{n,m}$ to characterize its temporal evolution patterns, we also compute the same set of features directly from the raw acceleration signals (and gyroscope signals if available). This dual representation enables the model to exploit both orientation-related information from the rotation vectors and raw signal characteristics, thereby improving the robustness and accuracy of classification.

Furthermore, the sequence of rotation vectors can be viewed as another sensor modality, providing an alternative representation of movement dynamics. Most existing algorithms can be directly applied to the sequence of rotation vectors with improved performance, just as they are applied to tri-axial acceleration and gyroscope signals. The experimental results presented in Sect. 4 demonstrate this effect.

4 Experiments

Our study aims to estimate sitting time for office workers under real-world, uncontrolled conditions. To this end, we collected tri-axial accelerometer signals (and gyroscope signals) from a smartwatch IMU worn on the non-dominant wrist of five office workers during their regular workdays. Each participant wore the smartwatch for at least five consecutive hours while performing routine office activities, such as sitting while working on a computer, sitting to have a lunch, making coffee, talking with colleagues in the hallway, attending meetings, and engaging in group discussions at a whiteboard. One of the authors accompanied each participant as a visitor to their workplace and manually recorded sitting and non-sitting timestamps. (Video recording was not permitted due to company privacy policies.) In total, the dataset comprises over 34 h of recordings sampled at 100 Hz, with annotations of sitting and non-sitting intervals.

Following the general practices in the field [2,5], the raw data were segmented into windows of 30 s. Although such a window length may be considered long in some human activity recognition tasks, it is appropriate for sitting time estimation. For each window, if sitting time accounted for more than half of the duration, the signals within that window were labeled as sitting; otherwise, they were labeled as non-sitting. In this way, the sitting time estimation problem is reformulated as a binary classification task.

To evaluate classification performance, we employed standard metrics, including accuracy, precision, recall, and F1-score with respect to the sitting class. We compared our proposed algorithm with several representative algorithms: the ST-DT algorithm by Mekruksavanich et al.[5], which achieved the best performance in their sitting detection study; the hybrid CNN-LSTM model [6]; the variational autoencoder (VAE) approach [21]; the MaskCAE algorithm [22]; and the Conformer-based method [23]. In addition, the compared algorithms were slightly modified to extend their applicability to the sequence of rotation vectors, thereby demonstrating the generalizability of rotation vector-based representations across different modeling approaches.

All algorithms were trained on 80% of randomly selected samples and evaluated on the remaining 20%. A 5-fold cross-validation was applied. The results are summarized in Table 1, from which we can see the proposed method achieved the highest performance among all algorithms. More importantly, all compared algorithms demonstrated improved performance when both raw IMU sensor signals and rotation vector sequences were utilized, highlighting the benefits of incorporating rotation vector representations for sitting time estimation.

Table 1. Performance comparison of the proposed method with ST-DT, CNN-LSTM, VAE, MaskCAE, and Conformer. The variants adapted to utilize rotation vector sequences are denoted as ST-DT Improved, CNN-LSTM Improved, VAE Improved, MaskCAE Improved, and Conformer Improved. Recall, precision, and F1-score are reported with respect to the sitting class.

Method	Recall	Precision	F1-score	Accuracy
Proposed	**0.964**	**0.978**	**0.971**	**0.958**
ST-DT	0.905	0.947	0.925	0.894
ST-DT improved	**0.928**	**0.938**	**0.933**	**0.904**
CNN-LSTM	0.896	0.967	0.930	0.902
CNN-LSTM improved	**0.921**	**0.981**	**0.950**	**0.930**
VAE	0.916	0.974	0.944	0.922
VAE improved	**0.916**	**0.989**	**0.951**	**0.932**
MaskCAE	0.926	0.981	0.953	0.933
MaskCAE improved	**0.928**	**0.984**	**0.955**	**0.937**
Conformer	0.972	0.955	0.963	0.947
Conformer improved	**0.957**	**0.977**	**0.967**	**0.952**

5 Conclusion

Smartwatch-based sitting time estimation is an emerging trend in wearable computing and is increasingly becoming a built-in functionality of modern smartwatches. While prior studies have investigated this problem, most have relied on data collected in controlled laboratory settings where subjects were instructed to perform repeated sitting or non-sitting actions.

In contrast, this work addresses the problem under real-world conditions by collecting data from office workers during their daily routines in the workplace. We proposed an algorithm for sitting time estimation based on sequences of rotation vectors derived from tri-axial acceleration signals. Our study demonstrates that these rotation vector sequences serve as an alternative representation of movement dynamics, capturing informative patterns that can be effectively exploited to improve classification performance. More importantly, this representation can be leveraged by most existing algorithms with only minor modifications, allowing them to operate on rotation vector sequences as naturally as on tri-axial sensor signals. Experimental results confirm that incorporating these representations enhances sitting time estimation.

For future work, we plan to validate the proposed rotation-vector representation on larger and more diverse real-world datasets (e.g., varied job types, arm/wrist habits, and environments) to better assess generalization beyond the current 34-hour office collection. We also aim to improve robustness under vigorous motion by incorporating additional sensor signals and investigating the temporal dynamics exhibited in rotation-vector sequences.

References

1. Park, J.H., Moon, J.H., Kim, H.J., et al.: Sedentary lifestyle: overview of updated evidence of potential health risks. Korean J. Fam. Med. **41**(6), 365 (2020)
2. Hammad, A.S., Tajammul, A., Dergaa, I., Al-Asmakh, M.: Machine learning applications in the analysis of sedentary behavior and associated health risks. Front. Artif. Intell. **8**, 1538807 (2025)
3. Krauter, C., Angerbauer, K., Sousa Calepso, A.: Sitting posture recognition and feedback: a literature review. In: Proceedings of the 2024 CHI Conference on Human Factors in Computing Systems, pp. 1–20 (2024)
4. Lui, G.Y., Loughnane, D., Polley, C., et al.: The Apple watch for monitoring mental health-related physiological symptoms: literature review. JMIR Ment. Health **9**(9), e37354 (2022)
5. Mekruksavanich, S., Hnoohom, N., Jitpattanakul, A.: Smartwatch-based sitting detection with human activity recognition for office workers syndrome. In: International ECTI Northern Section Conference on Electrical, Electronics, pp. 160–164. Computer and Telecommunications Engineering (ECTI-NCON) (2018)
6. Mekruksavanich, S., Jitpattanakul, A., Youplao, P., Yupapin, P.: Enhanced hand-oriented activity recognition based on smartwatch sensor data using LSTMs. Symmetry **12**(9), 1570 (2020)
7. Ciortuz, G., Pour, H.H., Irshad, M.T.: Machine learning models for wearable-based human activity recognition: a comparative study. Neurocomputing 130911 (2025)
8. Thakur, D., Guzzo, A., Fortino, G.: t-SNE and PCA in ensemble learning based human activity recognition with smartwatch. In: 2021 IEEE 2nd International Conference on Human-Machine Systems (ICHMS), pp. 1–6. (2021)
9. Salim, A., Brakenridge, C.J., Lekamlage, D.H., et al.: Detection of sedentary time and bouts using consumer-grade wrist-worn devices: a hidden semi-Markov model. BMC Med. Res. Methodol. **24**(1), 222 (2024)
10. Yeon, T., Xu, V., Hoffmann, H., Ahuja, K.: WatchHAR: Real-time on-device human activity recognition system for smartwatches. In: Proceedings of the 27th International Conference on Multimodal Interaction, pp. 387–394 (2025)
11. Wang, J., Chen, Y., Hao, S., Peng, X., Hu, L.: Deep learning for sensor-based activity recognition: a survey. Pattern Recogn. Lett. **119**, 3–11 (2019)
12. Petropoulos, A., Sikeridis, D., Antonakopoulos, T.: SPoMo: IMU-based real-time sitting posture monitoring. In: IEEE 7th International Conference on Consumer Electronics-Berlin (ICCE-Berlin), pp. 5–9 (2017)
13. Sinha, V.K., Patro, K.K., Pławiak, P., Prakash, A.J.: Smartphone-based human sitting behaviors recognition using inertial sensor. Sensors **21**(19), 6652 (2021)
14. Hayashi, K., Kumazoe, S., Ishida, S., Arakawa, Y.: Distinguishing working state by palm orientation. In: 2021 IEEE 3rd Global Conference on Life Sciences and Technologies (LifeTech), pp. 256–260 (2021)
15. Ishii, S., Nkurikiyeyezu, K., Luimula, M., Yokokubo, A., Lopez, G.: Exersense: real-time physical exercise segmentation, classification, and counting algorithm using an IMU sensor. In: Activity and Behavior Computing, pp. 239–255. Springer (2020)
16. Mekruksavanich, S., Jitpattanakul, A.: Smartwatch-based human activity recognition using hybrid LSTM network. IEEE Sens. 1–4 (2020)
17. Kuipers, J.B.: Quaternions and Rotation Sequences: A Primer with Applications to Orbits, Aerospace, and Virtual Reality. Princeton University Press (1999)
18. Wikipedia contributors: Rotation matrix. Wikipedia. Accessed 15 Sept 2025. Available at: https://en.wikipedia.org/wiki/Rotation_matrix

19. Prokhorenkova, L., Gusev, G., Vorobev, A., Dorogush, A.V., Gulin, A.: CatBoost: unbiased boosting with categorical features. Adv. Neural. Inf. Process. Syst. **31** (2018)
20. Chen, W., Wang, Z., Xie, H., Yu, W.: Characterization of surface EMG signal based on fuzzy entropy. IEEE Trans. Neural Syst. Rehabil. Eng. **15**(2), 266–272 (2007)
21. Fan, C., Gao, F.: A new approach for smoking event detection using a variational autoencoder and neural decision forest. IEEE Access **8**, 120835–120849 (2020)
22. Cheng, D., Zhang, L., Qin, L., et al.: MaskCAE: masked convolutional autoencoder via sensor data reconstruction for self-supervised human activity recognition. IEEE J. Biomed. Health Inform. **28**(5), 2687–2698 (2024)
23. Seenath, S., Dharmaraj, M.: Conformer-based human activity recognition using inertial measurement units. Sensors **23**(17), 7357 (2023)

Study on the Identification and Statistical Analysis of Infected Woods in Bark Beetle Pest Area Based on YOLO Algorithm and Drone Imagery

Jiantao Zhang(✉) and Jiawen Liang

South China Agricultural University, Guangzhou, China
zhangjiantao@yeah.net

Abstract. Pine wilt disease (PWD) is one of the most destructive forest diseases globally, necessitating timely monitoring for ecological preservation. Current manual detection methods face limitations due to complex terrain in PWD-outbreak locations. This study presents Effi_YOLO_v8, an enhanced YOLOv8 model optimized for three infection stages (early/mid/late) through UAV-based remote sensing. Key innovations include: (1) Stage-specific modifications to YOLOv8's architecture, (2) Multi-spectral feature fusion (RGB + Red Edge + NIR), and (3) Lightweight deployment optimizations. Evaluated across 100,000 mu in Guangdong Province, the model achieved 90.7% mAP with stage-wise recalls of 89.15% (early), 86.13% (mid), and 86.77% (late). The system processes 3.1 km^2/h on edge devices, demonstrating superior performance versus traditional methods (ΔmAP +26.86% vs. Faster R-CNN) while maintaining real-time capability (19 ms/inference).

Keywords: Pine wilt diseased tree · Deep learning · Remote sensing image · Object detection

1 Introduction

Pine Wilt Disease (PWD) represents a catastrophic threat to forest ecosystems globally, with its destructive impact particularly pronounced across China. Recent data indicates an alarming acceleration in its transmission; as of 2020, the epidemic had infiltrated 18 provinces and impacted 726 county-level districts. Beyond its traditional hosts, Pinus massoniana and Pinus thunbergii, the pathogen has crossed species barriers to infect Pinus koraiensis and Larix gmelinii. This geographical and biological expansion now jeopardizes approximately 60 million hectares of pine-dominated landscapes. Effective management of this crisis hinges on the precise identification of abnormally discolored timber, which serves as a primary indicator of infection [1]. Implementing robust monitoring systems allows for the real-time acquisition of epidemiological data, enabling foresters to execute targeted interventions and mitigate the broader spread of the disease.

L. Huang and D. Greenhalgh (Eds.): ICMLC 2026, LNNS 2061, pp. 448–462, 2027.
https://doi.org/10.1007/978-3-032-31236-5_31

At present, most of the monitoring methods for PWD-infected wood are relatively single. Many forestry workers still use the traditional method of on-site investigation to examine the growth status of pine trees. However, this conventional method has many limitations in the prevention and control of PWD-infected wood. Due to the complex terrain and scattered distribution of PWD outbreak sites, forestry workers cannot reach them on foot, which may lead to incorrect judgments on the actual situation of the PWD epidemic. In recent years, the use of remote sensing technology carried by small unmanned aerial vehicles (UAVs) to monitor the health status of forest vegetation has received much attention from researchers (Banu et al.). Compared with the traditional visual method, UAVs have the advantages of strong adaptability, low cost, and high efficiency (Tang et al.), which provides a new entry point and breakthrough for the prevention and control of PWD [2].

Generally, when pine trees are infected with PWD, the color of their leaves will change. It will gradually change from yellow to brownish-yellow, then to red, and finally to grayish dead state. When the remote sensing technology carried by UAVs first emerged, some scholars used this technology to obtain remote sensing images of large pine forests and relied solely on the visual method to search and locate the infected pine trees based on the color changes after infection [3]. It is not difficult to find that the visual method still requires a lot of human resources and cannot fundamentally solve the problem.

With the development of computer hardware and software, more and more researchers are using artificial intelligence-based methods to monitor forest pests. This is a fast and effective method that requires less resources and has a good development prospect. By extracting the image features of PWD-infected wood and processing them through artificial intelligence algorithms represented by deep learning, the shallow and deep image features are fused and then given to the target classifier to classify the infected wood at different stages, thereby achieving intelligent detection of PWD-infected wood [4]. By using deep learning technology, not only can the influence of human subjective imagination be reduced, but also the identification and detection of PWD-infected wood can be more accurate, improving the accuracy rate.

YOLO (You Only Look Once) is a revolutionary object detection algorithm, with the most significant feature being end-to-end training. YOLO divides the image into multiple grids, and each grid predicts the category and bounding box position of the object. The entire process only requires one forward pass and almost no post-processing steps, making it extremely fast [5]. The YOLO algorithm performs well in real-time scenarios precisely because of its efficient architecture. The design concept of YOLO is to minimize processing time. It does not rely on a region generation stage but directly generates bounding boxes for objects and classifies them in the image. This enables YOLO to achieve a high frame rate when processing video streams and real-time detection tasks, making it suitable for scenarios with extremely high real-time requirements. With the release of YOLOv8, the YOLO algorithm has reached a new balance between accuracy and efficiency. YOLOv8 not only improves speed but also enhances the detection ability for small objects and recognition accuracy in complex backgrounds. It achieves this by introducing a more refined feature extraction mechanism and an optimized loss function, maintaining high accuracy even at low frame rates.

With the improvement of computing power, YOLOv8 has further optimized the YOLO algorithm, enhancing detection accuracy, especially in scenarios with small objects and complex backgrounds. One of the innovations of YOLOv8 is the introduction of image segmentation capabilities, allowing it to not only detect objects but also perform pixel-level precise segmentation for each object [6]. YOLOv8 has made improvements in the following aspects: small object detection: YOLOv8 has increased the ability to capture more details on the original YOLO architecture, significantly improving the accuracy in detecting small objects. Image segmentation: By introducing image segmentation functionality, YOLOv8 can not only locate objects but also precisely segment the contours of objects, suitable for applications that require pixel-level object segmentation. Network optimization: YOLOv8 adopts a more efficient network architecture, achieving a new balance between speed and accuracy. Applicable scenarios: YOLOv8 is particularly suitable for scenarios that require precise object segmentation, such as medical image analysis and satellite image processing. With the emergence of new-generation algorithms like YOLOv8, object detection technology has found a better balance between accuracy and speed. Each algorithm has its unique advantages and applicable scenarios, and understanding their strengths and weaknesses is crucial for practical applications [7]. Whether it is high-precision static image processing or high-speed real-time detection in dynamic scenarios, choosing the appropriate object detection algorithm can help us achieve more accurate and efficient computer vision systems. Through continuous practice and optimization, we will be able to achieve more intelligent object detection in various complex environments, promoting the development of computer vision technology.

In Fig. 1, the evolution of computer vision, transitioning from conventional techniques to neural network architectures, reflects a continuous effort to optimize the balance between processing velocity and identification precision. Conventional methodologies often struggle with resilient feature characterization and specialized target extraction, leading to the current dominance of deep learning frameworks. These modern approaches offer significant advantages by providing superior adaptability and robustness when tailored to diverse research subjects. Consequently, utilizing deep learning-based detection has become the preferred standard for achieving high-performance results in complex analytical tasks [8].

This study combines the use of drones and deep learning technology to conduct intelligent detection of pine wood nematode disease-infected wood. First, pine wood nematode disease-infected wood is classified into three categories according to different infection stages: abnormally discolored wood, dead wood, and other non-pine wood of different colors [9]. A dataset of pine wood nematode disease-infected wood and the statistics of each part are created to achieve a preliminary statistical analysis of regional infection situations. Then, a detection and recognition model for pine wood nematode disease-infected wood is designed based on YOLO v8. This study has certain reference value for the practical application field of pine wood nematode disease detection.

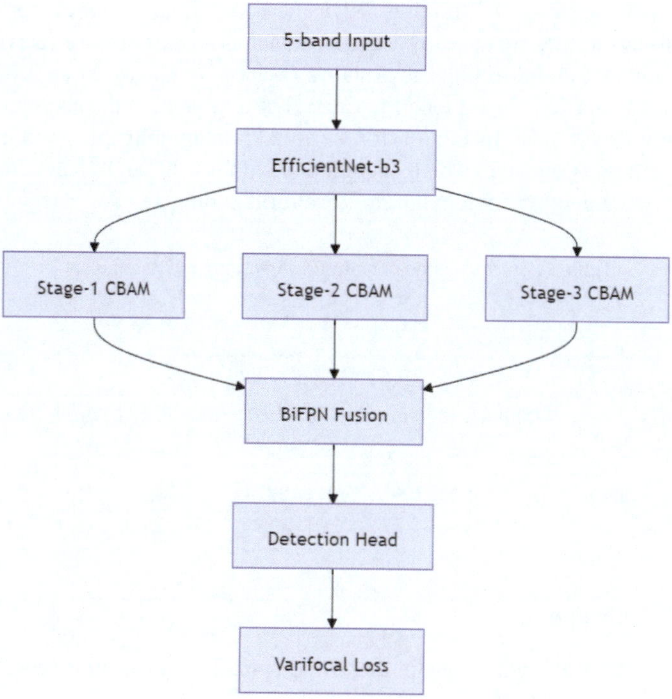

Fig. 1. Network architecture

2 Materials and Methods

2.1 Study Area and Data Collection

The study area is located in the suspected areas infected with Bursaphelenchus xylophilus along the G106, G105, and G323 highways and in important nature reserves in the three cities of Guangzhou, Qingyuan, and Shaoguan in Guangdong Province. Aerial photography by drones is carried out in this area [10].

Guangdong Province has a subtropical monsoon climate, with abundant rainfall. The annual average temperature is 21.8 °C, the average relative humidity is 78%, and the annual average precipitation is 1,789.3 mm. The minimum precipitation in a year is 1,314.1 mm, and the maximum is 2,254.1 mm. The annual precipitation is unevenly distributed and shows a multi-center distribution pattern [11]. The environment is conducive to the survival of Monochamus alternatus, making this area an important region for the outbreak of Bursaphelenchus xylophilus disease, and the epidemic situation is relatively serious.

The experiment is divided into several training areas and testing areas. The pictures in the training areas are used for the training of the algorithm model. The pictures in Testing Area 1 are used to test the recognition accuracy of the trained model for the abnormally discolored trees in the areas that participated in the training, and the pictures in Testing Area 2 are used to test the recognition accuracy of the model in the areas that did not participate in the training [12]. The drone image data is from the areas where the diseased trees infected by Bursaphelenchus xylophilus are located on both sides of the highways and in important nature reserves in Guangzhou City, Qingyuan City, and Shaoguan City, Guangdong Province. The total area of the acquired images is approximately 100,000 mu. In Table 1, for the aerial photography image acquisition, the DJI Mavic 3 drone is selected, which is equipped with a 4/3 CMOS Hasselblad camera and a telephoto camera with a maximum 28x hybrid zoom.

Table 1. Performance benchmarking comparative analysis

Parameter	Specification
UAV Model	DJI Mavic 3 (4/3 CMOS Hasselblad)
Spectral Bands	475 nm(B),560 nm(G),668 nm(R),705 nm(Red Edge),842 nm(NIR)
Flight Altitude	400 m (2.5 cm/pixel)
Temporal Coverage	Aug-Sept 2023 & Aug-Nov 2024

2.2 Data Acquisition

DJI Mavic 3 UAV was used to obtain the drone optical image of pine wood nematode in Guangdong province [13]. The image acquisition time is from August to September 2023 and from August to November 2024. The average flight altitude is 400 m by means of contour vertical photography. The data set of pine wood nematode was obtained.

A multi-sample data set of pine wood nematode disease was constructed based on drone images. A large number of images of pine wood nematode infested areas in different periods were collected by UAV as research samples. The image quality was improved and the model training effect was enhanced through pretreatment and enhancement techniques. At the same time, the image is normalized and denoised, and the median filter and Gaussian filter are used to remove image noise [14]. The contrast between diseased and healthy areas is enhanced by histogram equalization (CLAHE) or Gamma correction. After that, the disease features were extracted. In the color feature part, the discolored areas (yellowing and reddish-browning) of needles were extracted based on threshold segmentation (such as Otsu algorithm), or the abnormal color spots were counted by color histogram. Morphological characteristics: The withered branches were located by edge detection (Canny operator), and the diseased areas were segmented by morphological operations (corrosion and expansion). Then in the recognition processing feature fusion classification: color, texture, shape features are input into SVM (support vector machine) or random forest to distinguish healthy wood, early infected wood and dead wood.

2.3 Data Processing

The improved detection algorithm of pine wood infected with nematodes was selected to extract its convolutional features according to the physiological characteristics of pine wood infected with nematodes under natural environment, explore the influence of the structure, number of layers and size of convolutional kernel on the expression ability of extracted disease feature information, and study the learning mechanism of convolutional neural network, so as to achieve end-to-end training of disease images to disease categories [15]. The selected algorithm extracts the convolutional features according to the physiological characteristics of the infected wood of pine wood nematode under natural environment, explores the influence of the structure, number of layers and size of convolutional kernel on the expression ability of extracted disease feature information, studies the learning mechanism of convolutional neural network, and realizes the end-to-end training of disease images to disease categories. Based on the classification of convolutional neural network (CNN), ResNet, VGG or MobileNet pre-trained models were used in the model architecture module to fine-tune the classification of pine disease images through transfer learning. In the data enhancement module, rotation, mirroring, random cropping or GAN are used to generate synthetic diseased wood images. In target detection and semantic segmentation, YOLO, Faster R-CNN and other models were used to locate the lesions in the needle discoloration or tree discoloration. In pixel-level segmentation, U-Net and DeepLabv3+ were used to segment the withered area, quantify the severity of the disease (such as the proportion of necrotic area), and select a further improved detection algorithm based on the comparative analysis of the detection indicators of different models.

2.4 Improved YOLOv8 Algorithm Principle

Improved algorithm and evaluation algorithm. The neural network model was built in PyCharm (Python 3.7) using a high-performance workstation configured at 2.40 GHz and 32.0 GB of RAM. Based on the YOLOv8 model, the detection algorithm of pine wood nematode disease was improved and evaluated. The detection model used the improved YOLO series algorithm (such as YOLOv8), introduced attention mechanism (such as CBAM) and cross-layer feature fusion technology to improve the detection accuracy of small targets (such as early lesion spots) [16]. The model was trained with multi-spectral data to distinguish the spectral difference between healthy and infected pine trees, reduce the false detection rate, and provide reference by integrating multiple study cases. Finally, the improved detection algorithm of pine wood infected with nematodes was determined by trade off analysis

In Table 2, and according to the comprehensive analysis Fig. 2a, b, similar to the previous conclusions, the average detection accuracy and recall rate of the Faster R-CNN series models based on VGG-16 network and ResNet-50 network are very low, which is not of practical application value, and the subsequent comparative analysis of these two models will not be carried out. The average detection accuracy of SSD model for different types of pine wood nematode disease was close to 80%, but the corresponding recall rate was only about 60%, indicating that this model may have a lot of missed detection in the detection of pine wood nematode disease. In contrast, all indicators of YOLOv4, YOLO v5 and YOLO v7 models were at a higher level, and both models had better fitting effects for infected trees at the early stage of infection, infected trees at the late stage of infection, and dead trees [17]. The initial YOLO v8 model has an average accuracy of around 85%, but the recall rate is less than 80%, and the recall rate for dead trees is even less than 70%. The model proposed in this paper has a relatively good detection effect, with all kinds of average accuracy values and recall rates close to 90%, which is close to the detection results of YOLO v4, YOLO v5 and YOLO v7 models with extremely complex model composition, and also has a great improvement compared with the initial YOLO v8 model. Considering the number of model parameters, the size of weight file and the complexity of the model, the improved model proposed in this paper has the best detection performance.

Table 2. Performance benchmarking comparative analysis

Model	mAP@0.5	Early Recall	Params (M)	Speed (FPS)
Faster R-CNN	63.84	58.21	28.5	8.2
YOLOv7	95.80	85.44	37.6	45.3
Effi_YOLO_v8	90.7	89.15	15.2	52.6

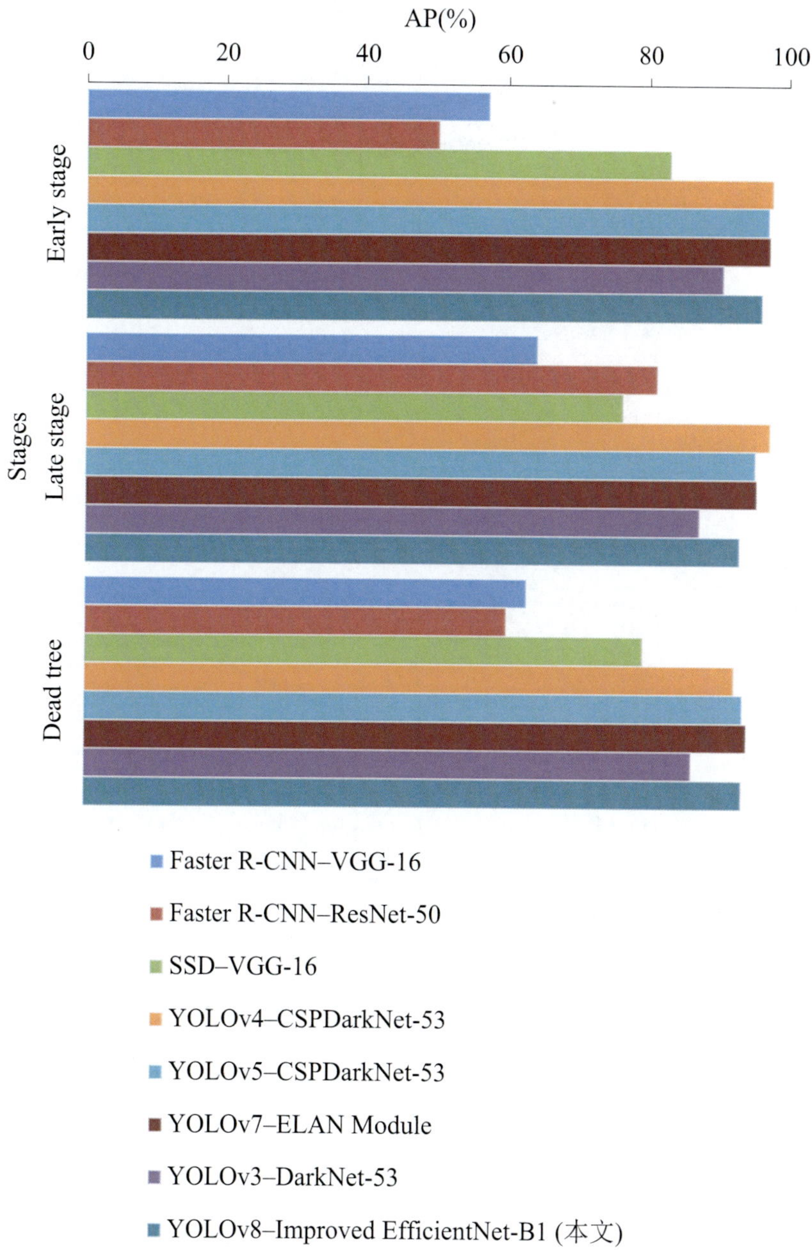

(a)

Fig. 2. Architecture of the Effi_YOLO_v8 model

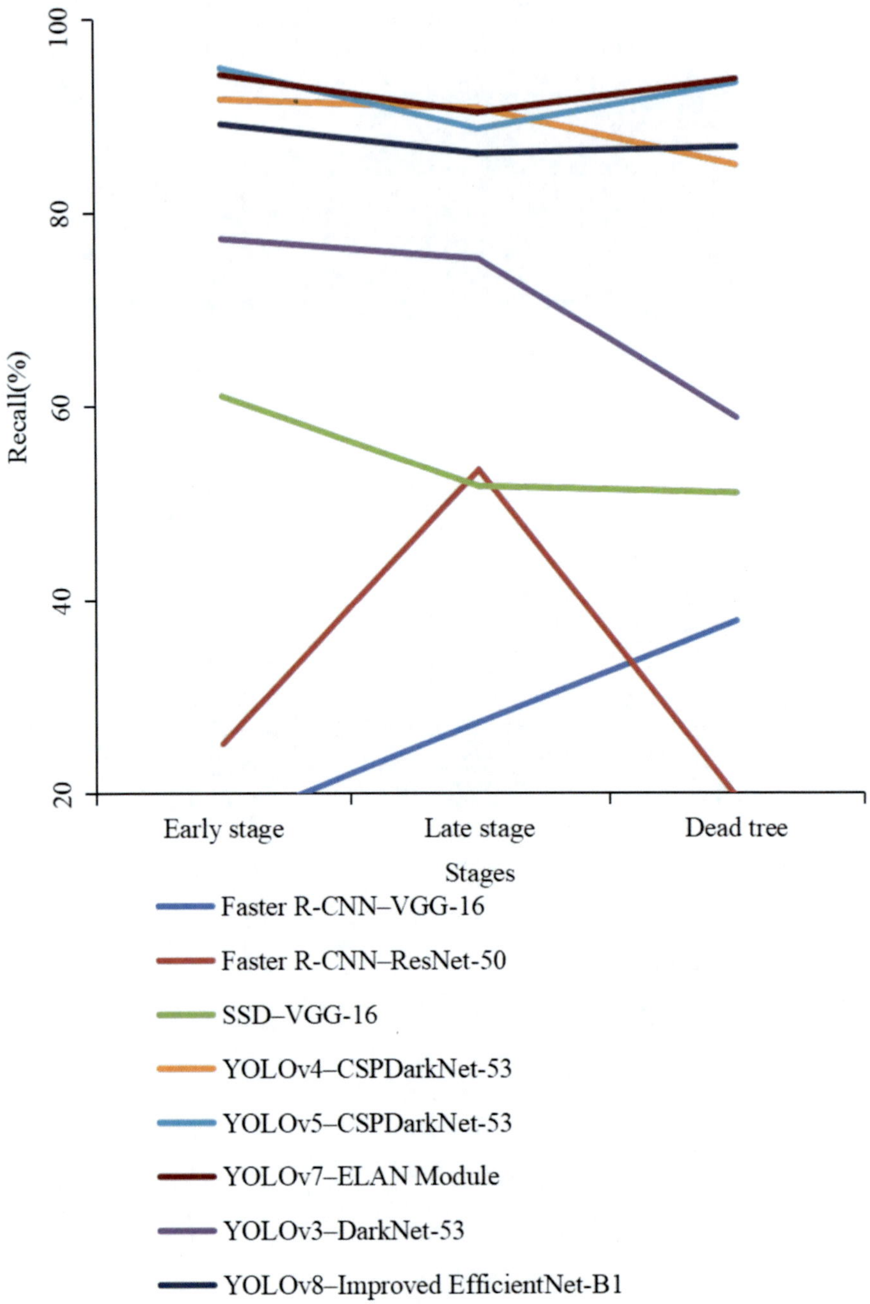

(b)

Fig. 2. (*continued*)

In order to further evaluate the SSD model, YOLO v4 model, YOLO v5 model, YOLO v7 model, YOLO v3 model and the improved model proposed in this paper, images of pine wood infected with nematodes at different infection stages, including adjacent and separated infected trees, were randomly screened and the above models were tested [18]. The detection results are shown in the figure, in which the red box represents the infected trees in the early stage, the green box represents the infected trees in the late stage, and the blue box represents the dead trees.

3 Discussion

To evaluate the precision and portability of the enhanced YOLOv8 algorithm for detecting pine wood nematode disease, several benchmarks were established beyond the Faster R-CNN and SSD models mentioned in this text. High-performance architectures, specifically YOLOv4, YOLOv5, and YOLOv7, were developed to meet rigorous accuracy standards, while seven distinct lightweight models—including versions of YOLOv4 utilizing GhostNet, MobileNet v2, and MobileNet v3 backbones with optimized PANet structures—were trained to ensure deployment flexibility. Leveraging the cost-effectiveness and high efficiency of UAV remote sensing, this study utilized the refined YOLO algorithm to automatically identify abnormally discolored wood, achieving an F1 score of 89.23% [19]. This performance represents a substantial improvement over the 60% to 65% accuracy typical of HSV threshold partitioning. Although the recognition rate is slightly below the 90% threshold reported in some Faster R-CNN studies, our approach offers superior reliability by accounting for the visual interference caused by red-leaved broad-leaved trees [20]. Furthermore, the model demonstrates high generalizability across expansive test areas, maintaining accuracy in unmodeled regions without the need for new training samples.

By streamlining the target detection framework to match the rapid transmission characteristics of the disease, the inference time per image was reduced to 19 ms, a 5 ms improvement over YOLOv4 [21]. This acceleration significantly lowers the total processing time for large-scale monitoring tasks and reduces hardware dependencies, enabling real-time UAV surveillance and immediate tracking of epidemic developments. However, since the discoloration of infected pines is a dynamic process—transitioning from yellowish-brown in the middle stages to reddish-brown in the late stages—the current method cannot effectively detect early-stage infections where needle color remains unchanged, nor can it perfectly distinguish nematode-induced stress from other environmental factors. Future research should integrate multi-temporal or hyperspectral imagery to refine early diagnosis. Additionally, since mixed forests often trigger false positives due to surrounding red foliage, secondary verification focusing on crown morphology and leaf shape is recommended to ensure the highest level of diagnostic integrity [22].

4 Conclusion

This study explores some improvements of YOLO V8-based algorithms, including K-Means−+ + algorithm for anchor frame clustering, EfficientNet network, Mish activation function, ECA attention mechanism, PPM module, etc. In Table 3, the mainstream high-precision object detection models such as YOLO v4, YOLO v5 and YOLO v7 models, lightweight object detection models such as YOLO V4-TINY model, based on MobileNet v2, MobileNet v3, GhostNet network seven YOLO v4 improved models were built in sequence. Firstly, the model comparison experiment before and after the improvement was carried out, and it was found that the improved model proposed in this paper was improved to a certain extent compared with the original YOLO v3 model [23]. The parameter number of the model was only one-fifth of the original, the FLOPs value was only one-seventeenth of the original, and the accuracy rate and recall rate of infected wood in different disease stages were higher than the initial model. Secondly, module verification experiments were carried out to establish and evaluate different improved parts of the model. It was found that the use of EfficientNet network, the modification of attention mechanism and the use of PPM module further enhanced the ability of feature context information fusion of the model, but further reduced the complexity of the model. Thirdly, based on the requirement of high precision, the detection performance comparison experiment of mainstream models is carried out. It can be seen that the detection performance of YOLO v4, YOLO v5 and YOLO v7 models is better, and the mAP value of the three models is about 1% higher than that of the improved model proposed in this paper [24]. However, if multiple indexes such as the number of parameters, weight file size and model complexity are analyzed comprehensively, it can be found that the improved model proposed in this paper can maintain certain detection performance under the premise of as few parameters and complexity as possible, that is, the improved model proposed in this paper has the best detection performance. Finally, based on the requirements of portability, the lightweight model is improved, and the detection performance comparison experiment of the model is carried out. It is found that the FPS of the improved model proposed in this paper is 35, which has good real-time performance, and the average detection accuracy and other evaluation indicators are also in the first place of the seven comparison models [25]. Based on mAP, detection accuracy, recall rate, F1 score, parameter number, weight file capacity, model complexity, FPS and other indicators, the model improvement proposed in this paper is reasonable and effective, and has strong detection ability under the premise of lightweight, and has certain application value. In this research, ultra-high-resolution imagery of pine forests was acquired using UAV-based remote sensing, and an enhanced YOLO-based detection framework was applied to automatically recognize pine trees exhibiting abnormal discoloration associated with pine wilt disease. Based on the experimental results, the primary findings can be summarized as follows:

(1) The enhanced YOLO-based model was applied to detect pine trees exhibiting abnormal discoloration, achieving an overall classification accuracy of 80.85%. During training, each iteration required approximately 164 s, the resulting model occupied 44.23 MB of storage, and the inference time for a single image was about 17 ms. Comparative experiments indicate that both Faster R-CNN and YOLOv4 underperformed relative to the proposed approach when evaluated in terms of detection accuracy, computational efficiency, model size, and inference speed. Although EfficientDet achieved higher average accuracy, its extended training duration and slower testing performance limited its suitability for rapid detection tasks. In contrast, YOLOv5 demonstrated faster training and inference speeds but yielded lower

accuracy than the improved model [26]. In Table 4, when considering accuracy, computational cost, model complexity, and inference efficiency collectively, the proposed improved YOLO algorithm exhibits a more favorable balance between speed and precision, resulting in superior comprehensive performance.

(2) The training model of the same training area was used to automatically identify abnormally discolored wood in different test areas. The F1 scores of abnormally discolored wood in the modeling area and the non-modeling area were 87.92% and 84.18%, respectively. Although the overall accuracy of the non-modeling area was slightly lower than that of the training modeling area, However, it can basically meet the recognition requirements of abnormal discolored wood, indicating that the training model of the improved method has certain applicability to the recognition of abnormal discolored wood in different regions [27].

(3) The presence of visually similar ground objects, variations in canopy cover, differences in slope orientation, and complex terrain conditions may lead to incomplete feature extraction or feature misclassification during automated detection, potentially causing false judgments or missed detections. However, the overall influence of these factors on recognition accuracy remains relatively limited.

(4) The forest area is mostly mixed, so red broad-leaved trees will have an impact on the results of high-altitude detection. Therefore, in order to ensure that the test results have real guiding significance, the suspected endemic area of pine wood nematode disease is selected for high-altitude shooting of the test results, and the pine wood nematode detection model in this paper is used for verification. After obtaining the test results, all the suspected infected trees are photographed at a second low-altitude approach distance. The obtained image information is manually checked twice to verify the authenticity of the test results, so as to ensure that the model in this paper has real guiding significance.

Table 3. The performance comparisons of different detection models.

model	nets	(%)	Size (MB)	FLOPs (GFLOPs)	No. of Parameters
Faster R-CNN	VGG-16	61.41	521.6	244.49	136,729,994
Faster R-CNN	ResNet-50	63.84	315.0	134.57	28,469,983
SSD	VGG-16	79.67	91.6	30.50	26,151,824
YOLO v3	DarkNet-53	89.23	234.7	65.53	61,534,504
YOLO v4	CSPDarkNet-53	95.96	245.5	60.33	64,363,101
YOLO v5	CSPDarkNet-53	95.45	179.5	115.60	47,056,765
YOLO v7	ELAN Module	95.80	153.7	106.12	37,620,125
YOLO v8	ResNet-50	96.23	157.87	145.82	71,890,887

Table 4. The detection results among the chosen models.

Figure	1 YOLO v4-GhostNet	2 YOLO v4-tiny	3 Effi_YOLO_v8
a			
b			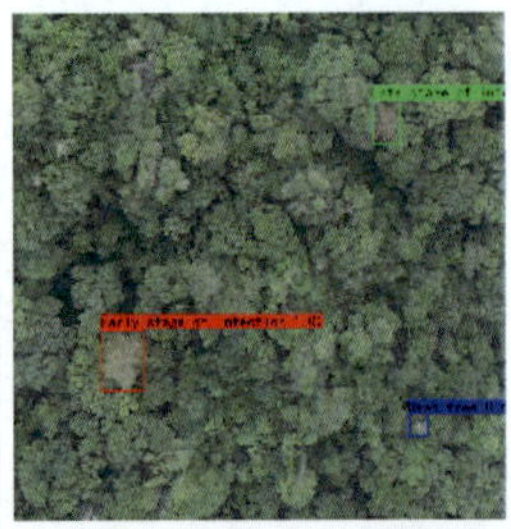
c			

Note The number in front of the model's name is used as the representative of the model. The red box, green box and blue box represent early stage, late stage and dead trees respectively

References

1. Allen, C.D. et al.: A global overview of drought and heat-induced tree mortality reveals emerging climate change risks for forests. For. Ecol. Manag. **259**(4), 660–684 (2010). https://doi.org/10.1016/j.foreco.2009.09.001
2. Arthur, D., Vassilvitskii, S.: K-means++: the advantages of careful seeding. In: Stanford (2006)
3. Banu, T.P., Borlea, G.F., Banu, C.: The use of drones in forestry. J. Environ. Sci. Eng. B. **5**(11), 557–562 (2016). https://doi.org/10.17265/2162-5263/2016.11.007
4. Deng, X., Tong, Z., Lan, Y., Huang, Z.: Detection and location of dead trees with pine wilt disease based on deep learning and UAV remote sensing. AgriEngineering. **2**(2), 294–307 (2020). https://doi.org/10.3390/agriengineering2020019

5. Dropkin, V., Foudin, A.: Report of the occurrence of Bursaphelenchus lignicolus-induced pine wilt disease in Missouri. Plant Dis. Reptr. **63**, 904–905 (1979)
6. Dwinell, L.: First report of pinewood nematode (Bursaphelenchus xylophilus) in Mexico. Plant Dis. **77**, 846 (1993). https://doi.org/10.1094/pd-77-0846a
7. He, K., Sun, J., Tang, X.: Single image haze removal using dark channel prior. IEEE Trans. Pattern Anal. Mach. Intell. **33**(12), 2341–2353 (2010). https://doi.org/10.1109/TPAMI.2010.168
8. Hu, G., Yin, C., Wan, M., Zhang, Y., Fang, Y.: Recognition of diseased Pinus trees in UAV images using deep learning and AdaBoost classifier. Biosyst. Eng. **194**, 138–151 (2020). https://doi.org/10.1016/j.biosystemseng.2020.03.021
9. Jones, J.T. et al.: Top 10 plant-parasitic nematodes in molecular plant pathology. Mol. Plant Pathol. **14**(9), 946–961 (2013). https://doi.org/10.1111/mpp.12057
10. Kärvemo, S., Johansson, V., Schroeder, M., Ranius, T.: Local colonization-extinction dynamics of a tree-killing bark beetle during a large-scale outbreak. Ecosphere. **7**(3), e01257 (2016). https://doi.org/10.1002/ecs2.1257
11. Knowels, K., Beaubien, Y., Wingfield, M., Baker, F., French, D.: The pinewood nematode new in Canada. For. Chron. **59**, 40 (1983)
12. Li, M., Dai, Y., Wang, Y., Wang, L., Sun, S., Chen, F.: New insights into the life history of Monochamus saltuarius (Cerambycidae: Coleoptera) can enhance surveillance strategies for pine wilt disease. J. For. Res. **32**(6), 2699–2707 (2021). https://doi.org/10.1007/s11676-021-01296-x
13. Misra, D.: Mish: a self regularized non-monotonic activation function. arXiv preprint. (2019). https://doi.org/10.48550/arXiv.1908.08681
14. Rajagopal, A. et al.: Fine-tuned residual network-based features with latent variable support vector machine-based optimal scene classification model for unmanned aerial vehicles. IEEE Access. **8**, 118396–118404 (2020). https://doi.org/10.1109/ACCESS.2020.3004233
15. Sandler, M., Howard, A., Zhu, M., Zhmoginov, A., Chen, L.-C.: Mobilenetv2: Inverted Residuals and Linear Bottlenecks. Paper Presented at the Proceedings of the IEEE Conference on Computer Vision and Pattern Recognition. Salt Lake City, USA (2018)
16. Sun, H., Zhou, Y., Li, X., Zhang, Y., Wang, Y.: Occurrence of major forest pests in 2020 and prediction of occurrence trend in 2021 in China. For. Pest Dis. **40**(02), 45–48 (2021). https://doi.org/10.19688/j.cnki.issn1671-0886.20210004
17. Sun, Y.: First report of pine wilt disease from Pinus thunbergii Parl in Sun Yat-sen mausoleum. J. Jiangsu For. Sci. Technol. **2**, 27 (1982)
18. Syifa, M., Park, S.J., Lee, C.W.: Detection of the pine wilt disease tree candidates for drone remote sensing using artificial intelligence techniques. Engineering. **6**(8), 919–926 (2020). https://doi.org/10.1016/j.eng.2020.07.001
19. Tan, M., Le, Q.: Efficientnet: Rethinking model scaling for convolutional neural networks. In: Paper Presented at the International Conference on Machine Learning, Long Beach, United State (2019)
20. Tang, L., Shao, G.: Drone remote sensing for forestry research and practices. J. For. Res. **26**(4), 791–797 (2015). https://doi.org/10.1007/s11676-015-0088-y
21. Tao, H. et al.: Deep learning-based dead pine tree detection from unmanned aerial vehicle images. Int. J. Remote Sens. **41**(21), 8238–8255 (2020). https://doi.org/10.1080/01431161.2020.1766145
22. Wang, Q., Wu, B., Zhu, P., Li, P., Hu, Q.: ECA-net: Efficient Channel attention for deep convolutional neural networks. In: Paper Presented at the 2020 IEEE/CVF Conference on Computer Vision and Pattern Recognition (CVPR). Seattle, United State (2020)
23. Wong, C.M., Daniels, L.D.: Novel forest decline triggered by multiple interactions among climate, an introduced pathogen and bark beetles. Glob. Chang. Biol. **23**(5), 1926–1941 (2017). https://doi.org/10.1111/gcb.13554

24. Ye, J.: Epidemic status of pine wilt disease in China and its prevention and control techniques and counter measures. Sci. Silvae Sin. **55**(9), 1–10 (2019). https://doi.org/10.11707/j.1001-7488.20190901
25. Yin, X., Song, Y., He, T., Song, B., Li, G.: Forest Pest detection method based on DenseNet and pyramid pooling. In: Paper Presented at the 2021 7th International Conference on Computer and Communications (ICCC). Chengdu, China (2021)
26. Zhang, R., Xia, L., Chen, L., Xie, C., Chen, M., Wang, W.: Recognition of wilt wood caused by pine wilt nematode based on U-net network and unmanned aerial vehicle images. Trans. Chin. Soc. Agricult. Eng. **36**(12), 61–68 (2020). https://doi.org/10.11975/j.issn.1002-6819.2020.12.008
27. Zhao, H., Shi, J., Qi, X., Wang, X., Jia, J.: Pyramid scene parsing network. In: Proceedings of the IEEE Conference on Computer Vision and Pattern Recognition, pp. 2881–2890 (2017). https://doi.org/10.1109/CVPR.2017.660

DATF: An Audiovisual Multimodal Identity Verification Method Based on Dynamic Attention and Trust Fusion

Lianggui Tang[1,2], Zhuo Chen[1,2(✉)], and Jiajun Yang[1,2]

[1] The Artificial Intelligence College of Chongqing Technology and Business University, Chongqing 400067, China
tlg@ctbu.edu.cn, yx_chenzhuo@qq.com

[2] Chongqing Key Laboratory of IntelliSense and Blockchain Technology, Chongqing 400067, China

Abstract. Compared to unimodal biometrics, audiovisual multimodal fusion enables more robust identity recognition, but existing studies lack full exploration of audio-visual correlations and suffer performance degradation under low-quality modality data. To address these issues, this paper proposes an audiovisual multimodal identity recognition method based on dynamic attention and trust fusion. It designs a modality trust algorithm to adaptively evaluate the contribution weights of audio-visual classifiers under varying data quality, introduces audiovisual soft attention to enhance inter-modal feature interaction, and adopts a weighted fusion module for decision-making. Extensive experiments on the VoxCeleb1 dataset achieve an Equal Error Rate (EER) of 1.61%, an Area Under the Curve (AUC) of 0.9985, and an average recognition accuracy of 0.9974, outperforming state-of-the-art methods. The proposed method effectively captures audiovisual features, improves recognition performance and robustness against low-quality data, and the modality trust algorithm can be generalized to various multimodal deep learning tasks.

Keywords: Modality trust algorithm · Identity recognition · Attention mechanism · Audiovisual modality · Dynamic fusion

1 Introduction

In financial and banking scenarios, identity authentication demands high accuracy (to secure user assets and information) and reliability under complex environments. Though facial and vocal features come from the same individual, their distinct feature spaces make precise correlation and alignment a key challenge. Multimodal Learning (MML) addresses this, yet modality imbalance often limits its advantages over unimodal models [1]. Real-world audiovisual data frequently has quality issues (noise, blur, occlusion), so robust authentication under low-quality conditions is critical [2, 3].

Existing fusion strategies (e.g., score-level averaging) outperform unimodal systems but lack exploration of fine-grained recognition [4]. Current state-of-the-art models project audio/visual inputs to deep feature spaces for concatenation or score-level

L. Huang and D. Greenhalgh (Eds.): ICMLC 2026, LNNS 2061, pp. 463–479, 2027.
https://doi.org/10.1007/978-3-032-31236-5_32

fusion, but such simple methods underutilize modal information, fail to handle imbalanced/noisy data (leading to over-reliance on single modalities [5]), and may even cause performance degradation (modality collapse) [6]. An ideal model should fully use both modalities, adapt to data variations, and compensate for degraded information.

To tackle these challenges in financial authentication, we propose the Dynamic Attention Trust Fusion (DATF) framework, integrating soft attention and dynamic modality trust estimation. A differentiable soft attention mechanism assigns weights based on human-like distribution to model audio-visual relationships; a trust-aware mechanism uses unimodal classifier loss (inversely correlated with confidence) to estimate modal quality via a trust prediction network, guiding fusion dynamically.

DATF is a non-contact, user-friendly method with three modules: (1) soft attention-based feature fusion enhancing inter-modal interaction; (2) dynamic trust estimation evaluating modal reliability; (3) weighted fusion integrating probabilities for robust decisions. The next section reviews related research on audiovisual biometric fusion. Key contributions are:

(1) A novel soft attention-based audio-visual fusion method (outperforming traditional attention) and dynamic trust fusion model (adaptive weighting via data quality).
(2) Superior performance on the VoxCeleb benchmark and robustness verified on noise-degraded datasets.
(3) Compatibility with various pretrained encoders and extensibility to other multimodal deep learning tasks.

2 Related Works

2.1 Audio-Visual Multimodal Biometric Identity Recognition

Alam [7] introduced a confidence-based audio-visual biometric framework, achieving 90.45% accuracy on a clean dataset of 43 subjects. Another study [8] fused audio and video features using a Hidden Markov Model (HMM), demonstrating the advantage of integrating both modalities. Tao et al. [9] introduced a cross-modal discriminative network for speaker verification, achieving an accuracy of 86.12% on the VoxCeleb2 dataset. Another study [10] utilized CNNs on the VoxCeleb dataset for identity recognition and achieved a Top-5 accuracy of 86.97%. Praveen and Alam [11] proposed a Dynamic Cross-Attention (DCA) model to address weak complementarity between audio and visual modalities. Their model dynamically chooses between cross-attended and non-cross-attended features via a conditional gating mechanism. Sari et al. [12] proposed a multi-view model that maps audio and video features into a shared space via a common classifier, enabling cross-modal identity recognition. Inspired by these pioneering efforts, we design a novel audio-visual attention-based feature fusion approach to learn more robust joint representations.

2.2 Multimodal Fusion Learning

Studies have shown that low-quality modality data can negatively impact the overall model performance, particularly during information fusion, where noisy or degraded data may mislead the learning process. To address this issue, researchers have proposed dynamic fusion strategies that adaptively assess the quality of each modality. These approaches rely on dynamic architecture search [13] and uncertainty estimation [14] to automatically adjust the weights of different modalities, ensuring that high-quality inputs receive greater attention while suppressing the influence of degraded inputs. This dynamic weighting enhances the adaptability and robustness of multimodal systems in complex environments.

Uncertainty estimation plays a crucial role in improving the accuracy and robustness of multimodal systems. In the multimodal fusion domain, Gao et al. [15] proposed Embracing Aleatoric Uncertainty (EAU), which quantifies the inherent uncertainty within unimodal data and uses this information to enhance the stability of unimodal embeddings, thereby improving fusion robustness. Wei [16] and Hu identified previously overlooked gradient conflicts between multimodal and unimodal learning objectives and proposed MMPareto—a new strategy that integrates gradients by analyzing their alignment. This approach ensures that gradient directions benefit all learning objectives and amplifies their magnitudes to improve model generalization. Cao et al. [17] introduced a novel framework called Predictive Dynamic Fusion (PDF), which theoretically reduces the upper bound of generalization error to enhance the reliability and stability of multimodal learning. Inspired by these advancements, we design a trust-aware audio-visual integration strategy that dynamically adjusts the importance of unimodal classifiers, aiming to mitigate the adverse effects of extremely low-quality data.

3 Dynamic Attention-Based Audio-Visual Fusion Model

3.1 Visual and Auditory Feature Preprocessing

The proposed model relies on both visual and auditory information to perform identity recognition. Visual information is represented through the video modality, while auditory information is captured through the audio modality. To extract discriminative features from each modality, we utilize pretrained unimodal feature encoders to independently process video and audio data.

To extract effective facial image features into a shared representation space, we adopt the ResNet-50 convolutional neural network as the visual feature extractor. Residual networks are effective in addressing model degradation and have been widely used in various deep learning tasks. A ResNet is composed of several residual blocks, each employing skip connections that help mitigate the vanishing gradient problem during deep neural network training.

To ensure the performance of the pretrained ResNet-50 model during facial feature extraction, we utilize a 2048-dimensional fully connected layer to capture the feature vector from the penultimate layer of the network, denoted as $f_{face} \in \mathrm{R}^{2048}$. This representation is used for downstream identity inference, particularly for open-set recognition tasks. Additionally, the softmax output of the final classification layer is used to obtain the predicted identity probability distribution $\mathrm{p}_{\mathrm{face}}$. The architecture of the visual feature extractor is illustrated in Fig. 1, where the curved arrows represent residual (skip) connections.

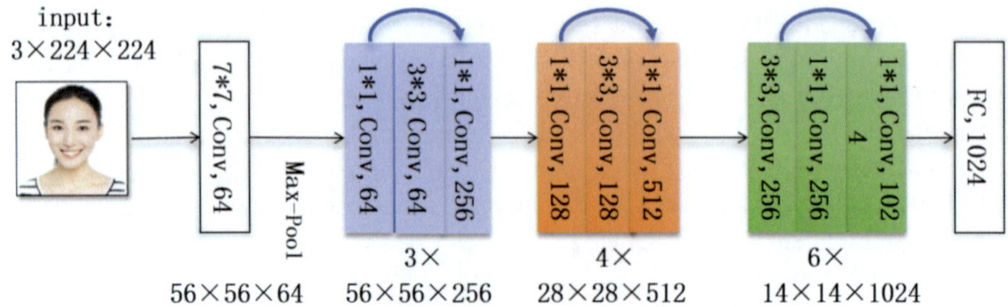

Fig. 1. Inference pipeline of the DATF model

For speech features, we use the RawNet3 model, which directly processes raw waveforms to extract features. The input layer is a one-dimensional time-series data with a shape of 1 × T. The audio signal features are then extracted using a parameterized filterbank (ParamFbank), followed by a ReLU activation function for nonlinear transformation. Finally, Batch Normalization (BN) is applied to normalize the data. The Conv1D + ReLU convolutional layer receives inputs from the three branches and further extracts temporal features. In our approach, the extracted feature $f_{voice} \in R^{192}$ is passed through a fully connected layer of 192 dimensions to obtain the output layer's probability distribution p_{voice}, which is suitable for the identity verification task presented in this paper.

3.2 Dynamic Attention and Trust-Aware Fusion Method

In multimodal identity verification tasks, the performance of unimodal features is often affected by data quality. For instance, facial recognition systems tend to fail under poor lighting or occlusion, while voice recognition systems are prone to errors in noisy environments. To address these challenges, we propose a Dynamic Attention Trust Fusion (DATF) framework that combines audio-visual soft attention feature fusion with dynamic modality trust estimation. The goal is to enhance the complementarity between modalities in an adaptive manner.

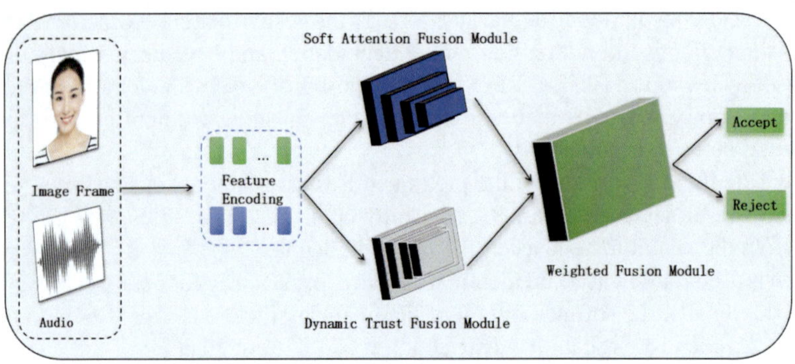

Fig. 2. Inference pipeline of the DATF model

DATF leverages a soft attention mechanism to dynamically adjust the contribution of visual and auditory features, while simultaneously evaluating the reliability of each modality through a dynamic trust assessment. Modalities with higher trust scores are assigned greater weights during fusion. This approach not only captures the complementary strengths of the visual and auditory modalities but also allows the fusion strategy to adapt to varying data quality, thereby improving the robustness and generalization of identity verification systems in complex environments.

Figure 2 illustrates the overall structure of DATF, which integrates multimodal features through an end-to-end fusion network and maintains high recognition performance even under degraded input conditions. After being encoded by the visual and auditory encoders, the respective features are processed through a soft attention fusion module and a dynamic trust fusion module, each outputting a probability distribution. These distributions are then combined using a weighted fusion module to generate the final prediction.

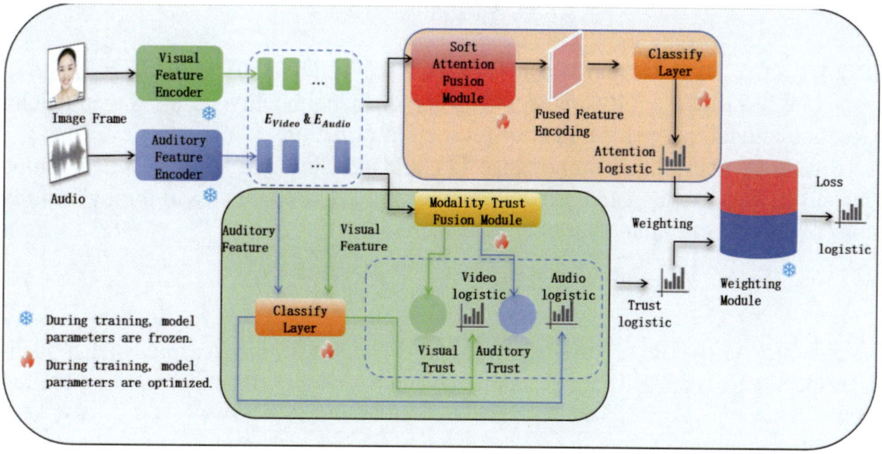

Fig. 3. Overall architecture of the DATF model

Figure 3 presents the training pipeline of the DATF model. During training, the parameters of the visual encoder, auditory encoder, and the weighted fusion module are kept frozen. The trainable components include the soft attention feature fusion module, the modality trust fusion module, and the classifier. All trainable parameters are optimized using a multimodal cross-entropy loss function.

In the algorithm1, a mini-batch S is first obtained from the dataset ***D***, with two modal data S_a and S_v extracted separately. Then, feature representations f_a and f_v are derived through pre-trained networks. The corresponding affine transformation features are calculated, and p_s is obtained based on weights. AbConf is computed via a confidence network, followed by the calculation of positive probabilities for each single modality. Subsequently, the confidence score Trust can be derived, and p_d is calculated using Trust. Finally, the final result is obtained through the combination of p_s and p_d.

First, the video modality is encoded using the visual encoder ResNet-50, and the audio modality is encoded using the auditory encoder RawNet-3.

Next, a soft attention mechanism is given the extracted facial and speech features, denoted as f_{face} and f_{voice}, we first apply an affine transformation to project them into a shared feature space:

$$\tilde{f}_{face} = W_{face}^T f_{face} + b_{face}$$

$$\tilde{f}_{voice} = W_{voice}^T f_{voice} + b_{voice} \quad (1)$$

where $\tilde{f}_{face}, \tilde{f}_{voice} \in R^{1024}$,and W_{face}, W_{voice} are the weight matrices, while b_{face}, b_{voice} are the bias terms of the fully connected layers for each modality. In this shared feature space, both modalities encode identity-relevant information.

To estimate the relative importance of each modality, we apply a soft attention mechanism. The concatenated original features $[f_{face}, f_{voice}]$ are passed through a fully connected layer to compute modality relevance scores:

$$\alpha = softmax\left(W_{att}^T\left[f_{face}, f_{voice}\right] + b_{att}\right) \quad (2)$$

where $\alpha = [\alpha_f, \alpha_v]^T \in R^2$ represents the attention weights for the face and voice modalities, respectively. $W_{att}^T \varepsilon R^{2\times(512+192)}$ and $b_{att} \epsilon R^2$ are trainable parameters of the attention block.

Using the computed attention weights and the projected features in the shared space, we obtain the fused feature vector by performing a weighted sum:

$$Z = \alpha_f \tilde{f}_{face} + \alpha_v \tilde{f}_{voice} \quad (3)$$

This fusion approach enhances the contribution of more discriminative features while suppressing the influence of noisy or less reliable modalities.

Algorithm 1: DATF

Require: ***D***, ***T***

for t = 0,...,T-1 **do:**
$S \leftarrow D$;
$S_a \leftarrow S$;
$S_v \leftarrow S$;
$f_a \leftarrow \mathbf{RawNet-3}(S_a)$;
$f_v \leftarrow \mathbf{ResNet-50}(S_v)$;
$(f_{a\_aligned}, f_{v\_aligned}) \leftarrow (f_a, f_v)$;
$(w_a^s, w_v^s) \leftarrow (f_a, f_v)$;
$f_{fused} \leftarrow (f_{a\_aligned}, f_{v\_aligned}), (w_a^s, w_v^s)$;
$p_s \leftarrow f_{fused}$;
$AbConf_a \leftarrow$ **AuditoryTrustNet**(f_a)
$AbConf_v \leftarrow$ **VisualTrustNet**(f_v)
$(p_a, p_v) \leftarrow (f_a, f_v)$;
$(p_a^{true}, p_v^{true}) \leftarrow$ unimodal (T);
$ReConf_a, ReConf_v \leftarrow (p_a^{true}, p_v^{true})$;
$Trust \leftarrow AbConf + ReConf$;
$p_d \leftarrow Trust_a * p_a + Trust_v * p_v$;
$p \leftarrow \eta_{0.5} * p_s + \eta_{0.5} * p_d$;
end for

In summary, the visual and auditory feature representations obtained from their respective pretrained encoders are processed through a soft attention mechanism. This mechanism computes attention scores that weight the affine-transformed features in the shared space. As shown in Fig. 4, this enables the model to fully capture relevant inter-modal information and suppress irrelevant signals, resulting in a robust fused feature representation.

To enhance the model's robustness against noisy inputs in real-world conditions, we design a trust-aware fusion module based on inter-modality confidence estimation, as illustrated in Fig. 5.

In multimodal inference, low-quality data from a single modality can severely degrade the overall performance of the multimodal model—sometimes even performing worse than a unimodal model trained on high-quality data. To effectively integrate and assess information from different modalities, an intuitive strategy is to predict the loss of each modality and use it to compute fusion weights, leveraging the known negative correlation between model confidence and loss.

However, directly using loss values as fusion weights introduces practical challenges. Therefore, we propose using the predicted probability of the groundtruth class (positive class probability) as a proxy for loss, as derived in Eq. (4):

$$loss = -\log^{p_{true}} \tag{4}$$

To implement this, we design two absolute trust prediction networks to estimate the trustworthiness of the unimodal classifiers: one for face recognition and one for speaker

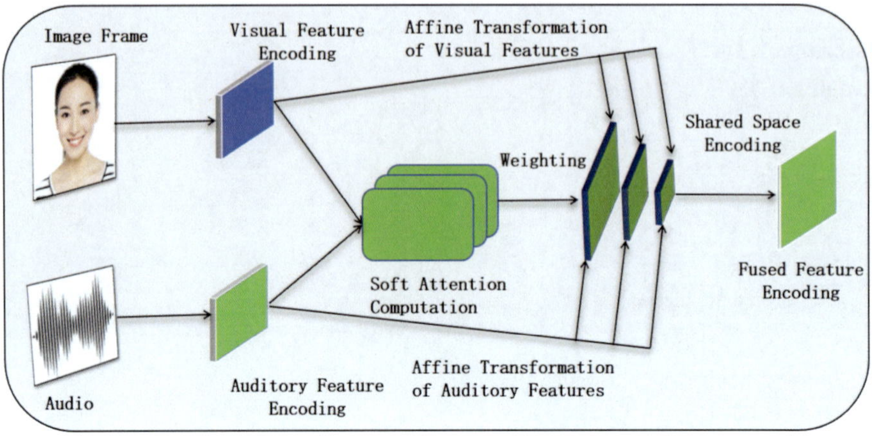

Fig. 4. Audio-visual soft attention feature fusion module

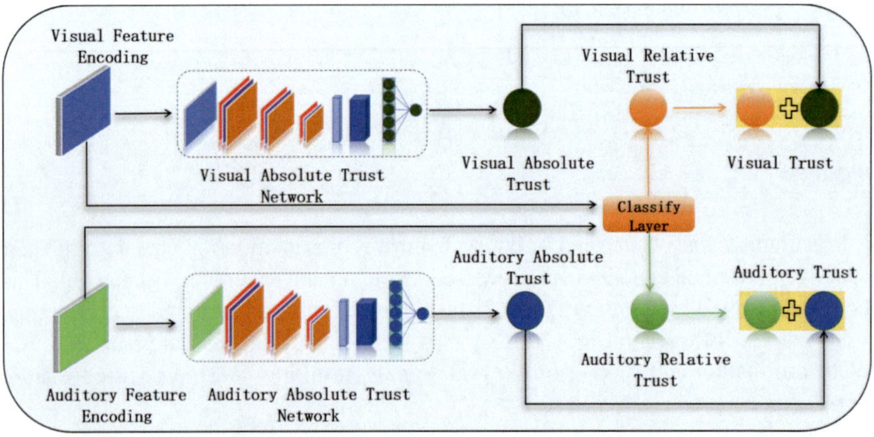

Fig. 5. Audio-visual dynamic trust fusion module

verification. The face trust prediction network adopts a bottleneck structure with ReLU activation for nonlinear transformation, followed by a Sigmoid function to produce a scalar absolute trust score $\mathrm{AbConf_{face}} \in (0, 1)$, representing the confidence in the predicted probability $\mathrm{p_{face}}$. Similarly, the speaker trust prediction network predicts $\mathrm{AbConf_{voice}} \in (0, 1)$ based on $\mathrm{p_{voice}}$.

During multimodal training, due to varying quality across modalities, some unimodal models may underperform. We define relative trust (ReConf) as a measure of how well a modality performs relative to others, based on the sum of cross-entropy losses. Derived from Eq. (4), we replace the loss with $\mathrm{P_{ture}}$ and obtain the relative trust score as follows:

$$ReConf_n = \frac{\sum_{k \neq n} loss_k}{\sum_{j=1}^{N} loss_j} = \frac{\log \prod_{k \neq n} P_{ture}^{k}}{\log \prod_{j}^{N} P_{true}^{j}} \tag{5}$$

Here, n denotes the current modality, N is the total number of modalities, and loss is the cross-entropy loss.

While absolute trust reflects a classifier's internal confidence in its predictions, relative trust measures a modality's reliability relative to others. Since both forms of trust promote inter-modality coordination, we define the final trust score as a linear combination of absolute and relative trust:

$$Trust_m = AbConf_m + ReConf_m \tag{6}$$

This combined trust score serves as the weight for each modality in the final fusion step. The "+" operator in Fig. 5 indicates this additive operation.

In the weighted fusion module, we aggregate the outputs from both the soft attention fusion module and the dynamic trust-aware fusion module to produce the final decision. To ensure the balanced contribution of the two modules, we adopt a weighted averaging strategy. Validated by experiments, we set the values of α and β to 0.5, which yields the optimal performance.

The probability distribution $P_{attention}$ produced by the soft attention module represents an early-stage fusion based on facial and audio features. The distribution P_{truth} is obtained from the trust-aware module and reflects a reliability-weighted fusion based on inter-modality confidence estimation. For each identity C_i, the final output probability P_{final} is computed as:

$$P_{final}(C_i) = \alpha \cdot P_{attention}(C_i) + \beta \cdot P_{truth}(C_i) \tag{7}$$

This trust-aware combination improves the system's robustness under adverse conditions by adaptively emphasizing more reliable modalities during prediction.

4 Experiments and Analysis

4.1 Experimental Setup

The experiments in this study were conducted on an Nvidia L40 server platform. For model optimization, we adopted the Adam optimizer with an initial learning rate of 0.001 and a global batch size of 128. We utilized the VoxCeleb1 dataset for our experiments. This dataset consists of audio-visual clips collected from YouTube videos, offering high environmental diversity. Each clip contains both audio and video modalities and has a minimum duration of 3 s.

We utilized the VoxCeleb1 dataset for our experiments. This dataset consists of audio-visual clips collected from YouTube videos, offering high environmental diversity. Each clip contains both audio and video modalities and has a minimum duration of 3 s. Fig. 6 illustrates the distribution of audio clip durations. The dataset covers a wide range of speakers in terms of gender, nationality, ethnicity, accent, profession, and age. All facial appearances were captured "in the wild," including various background noises, laughter, overlapping speech, pose variations, and lighting conditions.

We randomly split the VoxCeleb1 development set (which contains 1,251 speakers) into three parts using a 7:2:1 ratio (Training set: 875 speakers, Validation set: 251

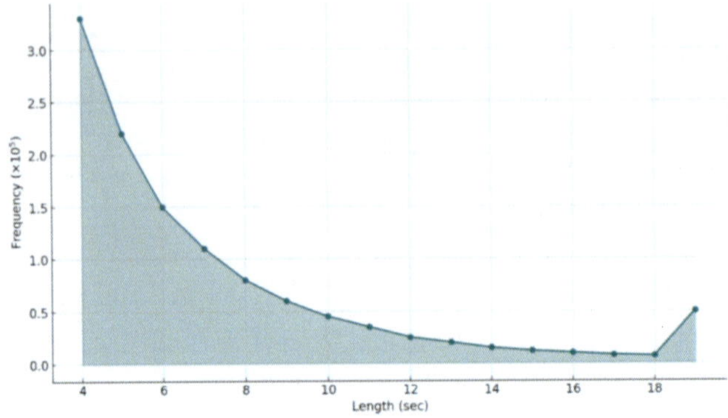

Fig. 6. Audio-visual dynamic trust fusion module

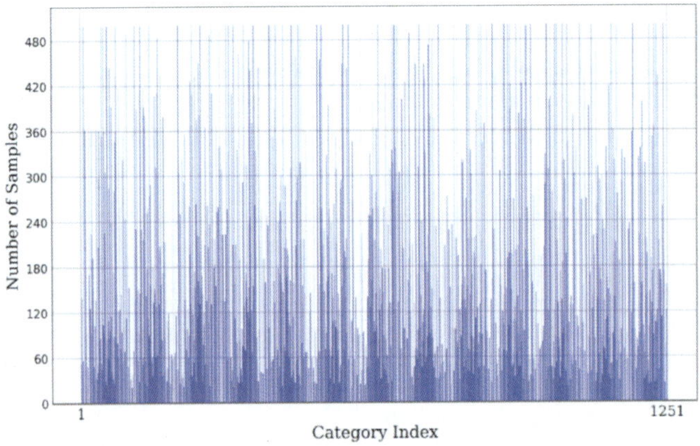

Fig. 7. Audio-visual dynamic trust fusion module

speakers, Test set: 125 speakers). To evaluate the robustness of our method under low-quality conditions, we introduced varying degrees of noise into specific modalities. Model performance was evaluated on both the validation set and the test set. The entire VoxCeleb1 dataset comprises 150,480 video segments from 1,251 speakers, including 153,516 utterances. Fig. 7 shows the distribution of data volume per speaker across the dataset.

Area Under the Curve (AUC) represents the area under the ROC curve and reflects the model's overall classification capability across all thresholds. Let x represent the false positive rate (FPR), and TPR(x) the true positive rate, then:

$$ACC = \frac{TP+TN}{TP+TN+FP+FN} \left[REF\_Ref\ 13313\backslash r\backslash h\ 3 \right] \tag{8}$$

Precision is the proportion of true positives among all positive predictions:

$$Precision = \frac{TP}{TP+FP} \left[REF\_Ref\ 13313\backslash r\backslash h\ 3 \right] \tag{9}$$

Recall is the proportion of actual positives that are correctly identified:

$$Recall = \frac{TP}{TP+FN}\left[\ REF\_Ref\ 13313\backslash r\backslash h\ 3\right] \tag{10}$$

F1-score is the harmonic mean of precision and recall, commonly used for imbalanced datasets:

$$F1 = 2 \cdot \frac{Precision \cdot Recall}{Precision+Recall}\left[\ REF\_Ref\ 13313\backslash r\backslash h\ 3\right] \tag{11}$$

4.2 Multi-Metric Evaluation of DATF Model Performance

To further assess the trained DATF model, we generated a Receiver Operating Characteristic (ROC) curve on the test set to visualize classification performance. The Area Under the Curve (AUC) was 0.9985, indicating excellent discriminative ability across varying thresholds, as shown in Fig. 8.

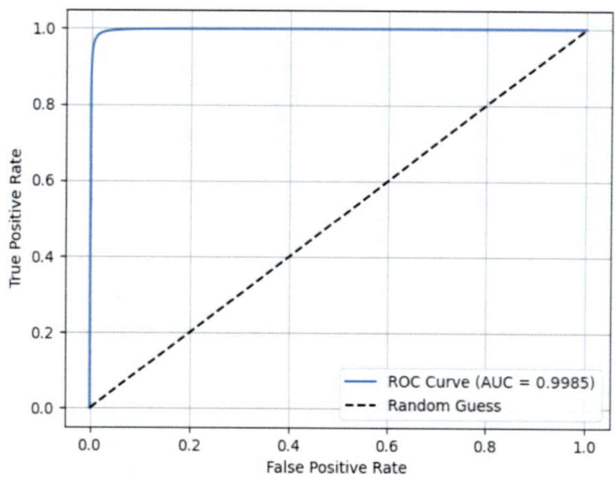

Fig. 8. ROC curve of the DATF model

From the ROC curve, we observe a high true positive rate in low false positive regions, reflecting the effective use of complementary facial and voice features. Compared to unimodal systems, the DATF model shows stronger robustness and reliability under challenging conditions (e.g., low-light or noisy input). Additionally, the smooth performance across different thresholds supports the feasibility of dynamic threshold adjustment in practical applications.

Figure 9 presents the precision distribution across different speaker classes on the VoxCeleb1 dataset. The x-axis denotes class IDs, and the y-axis indicates corresponding precision values. Most classes achieve precision scores close to 1.0, with the mean precision reaching 0.9974, demonstrating strong class-specific classification capability. Classes with relatively lower precision can be attributed to (1) a limited number of training samples, restricting generalization, or (2) low-quality visual and audio inputs.

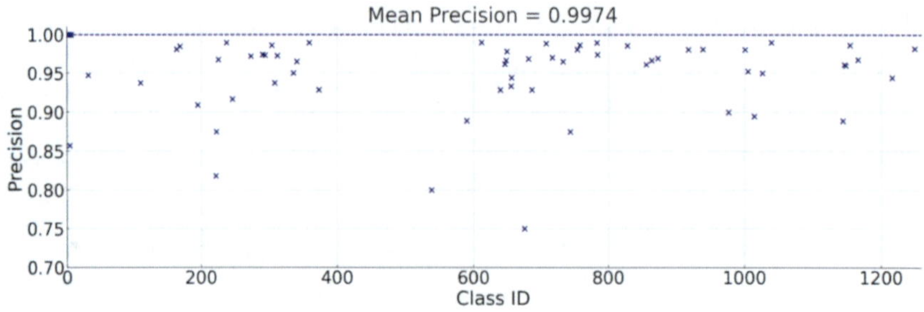

Fig. 9. Average precision across speaker classes

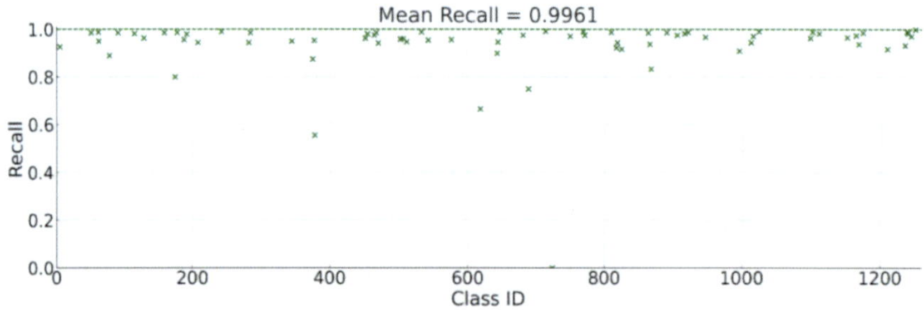

Fig. 10. Average precision across speaker classes

Figure 10 shows the recall distribution, indicating that most classes also achieve recall scores near 1.0, with a mean recall of 0.9961. This suggests the model is effective in retrieving correct identity matches. A few classes show significantly lower recall, often due to poor data quality across both modalities.

Figure 11 presents the F1-score distribution, used to evaluate the overall balance between precision and recall across speaker classes. Most F1-scores are close to 1.0, confirming the model's strong and consistent performance in identity verification.

We validated the proposed DATF method on the VoxCeleb1 dataset and compared it against several state-of-the-art identity recognition approaches that utilize audio-visual (A-V) fusion. To ensure fairness, all methods were re-implemented under identical experimental settings to eliminate variability caused by different environmets.

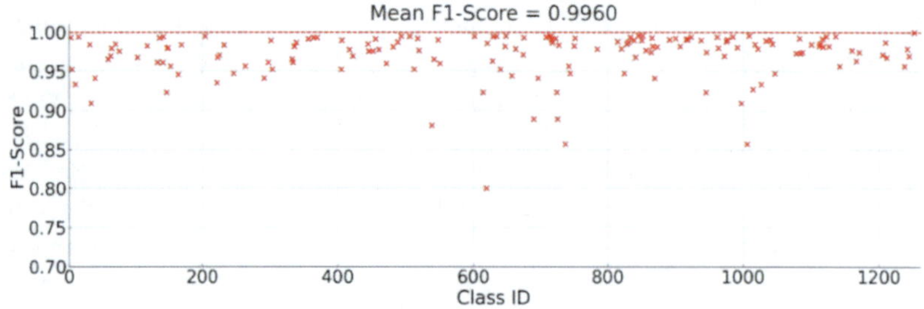

Fig. 11. Average precision across speaker classes

Table 1. Performance comparison between DATF and state-of-the-art identity recognition methods

Fusion Method	EER↓	minDCF↓
Visual	3.720	0.298
Audio	2.553	0.253
Tao [9]	2.476	0.203
Sari [12]	2.438	0.178
Chen [6]	2.403	0.163
Mocanu [18]	2.387	0.149
Praveen [11]	2.173	0.126
R. Gnana [19]	1.851	0.112
DATF	**1.610**	**0.084**

Table 1 summarizes the comparison results on the validation set. As shown, our proposed multimodal fusion method significantly outperforms existing approaches. When using only visual or audio data, the EER values were 3.720 and 2.553, respectively, while the minDCF values were 0.298 and 0.253, indicating that unimodal systems are inherently limited in recognition accuracy. In contrast, the DATF fusion model achieved an EER of 1.610 and minDCF of 0.084, demonstrating the effectiveness of combining visual and auditory modalities and its superiority over other A-V fusion identity recognition models.

To analyze the performance of our proposed method, we experimented with multiple fusion strategies and compared them with several advanced approaches. First, we employed score-level fusion, where the identity is determined by weighting the outputs of unimodal classifiers based on their respective scores. Next, we applied feature concatenation, in which the features from the video and audio modalities are concatenated to generate the final classification score. Finally, we explored attention-based fusion approaches, including self-attention and cross-modal attention mechanisms. These methods leverage learned attention weights to dynamically highlight informative features, and have been shown to significantly enhance model performance by capturing the interactions between modalities more effectively.

We conducted ablation experiments to isolate the contributions of each module: one variant including only the soft attention feature fusion module, and another including only the dynamic trust fusion module. Results are presented in Table 2.The results show a clear trend: as fusion strategies become more sophisticated, both EER and minDCF consistently decrease, confirming that the fusion design plays a key role in performance gains. All DATF variants, including those based on cross-attention (CA) and joint cross-attention (JCA), outperform traditional methods and unimodal baselines. However, the complete DATF model consistently achieves the best results. Notably, comparison between "Ours w/o SA" (without soft attention) and "Ours w/o DT" (without dynamic trust) highlights the critical role of both modules in enhancing performance.

Table 2. Performance comparison of DATF with different fusion strategies

Fusion Method	EER↓	minDCF↓
Score-Level Fusion(SLF)	2.521	0.217
Feature Concatenation(FC)	2.489	0.193
Self-Attention	2.412	0.176
Cross-Attention	2.387	0.149
Joint Cross-Attention	2.315	0.135
ours w/o SA(Soft Attention)	2.151	0.132
ours w/o DT(Dynamic Trust)	1.837	0.125
DATF	**1.610**	**0.084**

From Table 2, score-level fusion resulted in an EER of 2.521 and minDCF of 0.217, showing relatively poor integration due to weak inter-modal interactions. Feature concatenation slightly improved results (EER: 2.489, minDCF: 0.193), but still fell short. The self-attention mechanism further reduced EER, and cross-modal attention achieved even better results (EER: 2.387, minDCF: 0.149), underscoring its utility in modeling inter-modal relationships. Nonetheless, all alternative strategies were outperformed by DATF (EER: 1.610, minDCF: 0.084), demonstrating its superior ability to combine multimodal information effectively.

These findings confirm the significant advantage of DATF over traditional fusion methods. By leveraging both soft attention and dynamic trust, DATF effectively adapts to input quality and enhances overall recognition accuracy, highlighting its potential for robust identity verification in real-world settings.

4.3 Robustness Analysis Under Degraded Visual Quality

To assess the robustness of the DATF model under degraded visual conditions, we simulated low-quality visual input by varying JPEG compression quality, a common form of degradation in real-world scenarios caused by compression artifacts or transmission noise.

Fig. 12. JPEG quality levels: 1, 5, 20, 40, and 100%

In the experiments, we set the JPEG quality levels of facial images in the video input to fixed values: {1, 5, 20, 40, 100%}, as shown in Fig. 12. Audio input length was fixed at 3 s.

As shown in Fig. 13, the model maintained effective performance even with significantly degraded visual input. In extreme cases where JPEG quality dropped below 5%, model performance still showed a strong positive correlation with image quality.

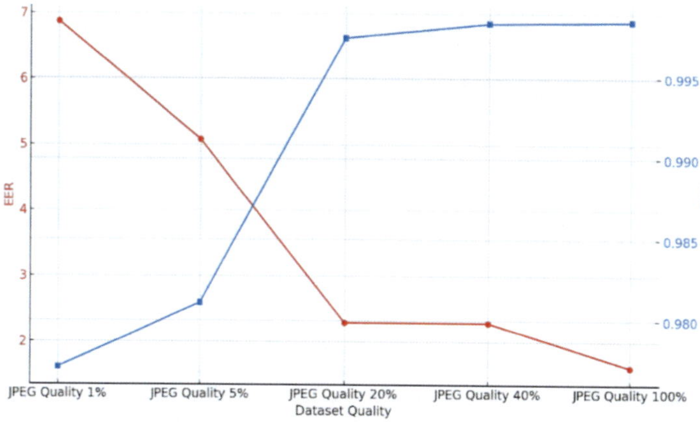

Fig. 13. JPEG quality levels: 1, 5, 20, 40, and 100%

Specifically, for JPEG quality at 1%, the model achieved an EER of 6.88; at 5%, the EER dropped to 5.083, with corresponding AUC values of 0.9771 and 0.9811, respectively. As illustrated in Figs. 14 and 15, although EER increased with lower image quality, the values remained within a reasonable range, confirming that DATF still provides accurate identity recognition even under severely degraded visual conditions.

These results, summarized in Table 3, highlight the system's robustness. Even when image quality is significantly compromised, the DATF multimodal framework continues to achieve high accuracy, making it suitable for real-world scenarios involving poor visual data.

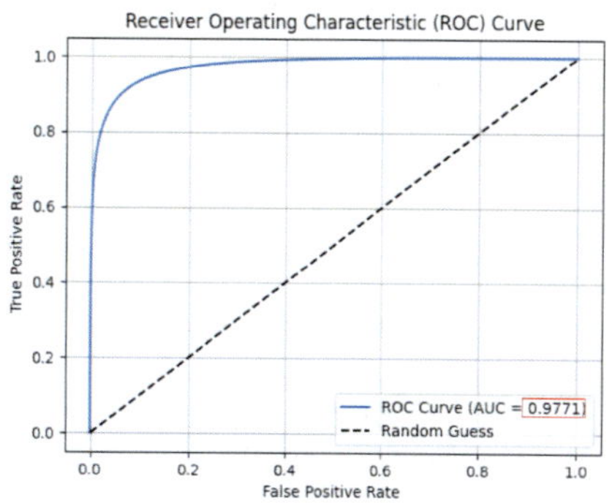

Fig. 14. ROC curves for 1% JPEG quality

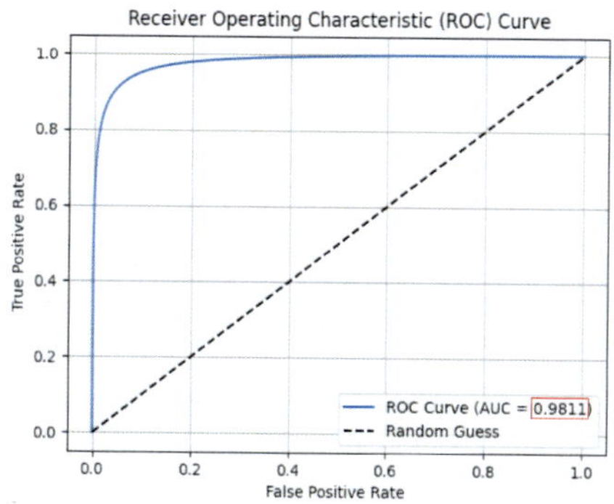

Fig. 15. ROC curves for 5% JPEG quality

Table 3. DATF performance under low-quality visual input conditions

Dataset Quality	EER↓	AUC
JPEG Quality 1%	6.880	0.9771
JPEG Quality 5%	5.083	0.9811
JPEG Quality 20%	2.293	0.9975
JPEG Quality 40%	2.290	0.9984
JPEG Quality 100%	1.610	0.9985

5 Conclusion

In this paper, we proposed DATF, a novel Dynamic Audio-Visual Trust Fusion model for identity verification. The framework integrates soft attention-based multimodal feature fusion and dynamic modality trust estimation, both of which are tightly coupled with pretrained feature extractors. This structure enables the DATF framework to be extended to various multimodal deep learning tasks. The dynamic trust fusion module in particular offers a generalizable solution to the problem of modality degradation, which frequently occurs in practical multimodal systems. Our model achieved state-of-the-art performance on the VoxCeleb dataset, with an EER of 1.61, AUC of 0.9985, and an average validation accuracy of 0.9934, outperforming existing identity recognition models. Furthermore, we demonstrated the robustness of our model under degraded data conditions. In future work, we plan to explore additional multi-modal audio-visual tasks and deepen our investigation into modality degradation in multimodal deep learning.

Acknowledgments. This work is sponsored by the Natural Science Foundation of Chongqing (Key Program, Grant No. CSTB2023NSCQ-LZX0029) and Supported by the Science and Technology Research Program of Chongqing Municipal Education Commission (Key Program, Grant No. KJZD-K202400809).

References

1. Jiang, Q.Y., Huang, L., Yang, Y.: Rethinking Multimodal Learning from the Perspective of Mitigating Classification Ability Disproportion. arXiv preprint https://arxiv.org/abs/2502.20120 (2025)
2. Farhadipour, A., et al.: Comparative Analysis of Modality Fusion Approaches for Audio-visual Person Identification and Verification. arXiv preprint https://arxiv.org/abs/2409.00562 (2024)
3. Zhang, Q., et al.: Multimodal fusion on low-quality data: A comprehensive survey. arXiv preprint arXiv:2404. 18947 (2024)
4. Lee, J., AlRegib, G.: Gradients as a measure of uncertainty in neural networks. In: 2020 IEEE International Conference on Image Processing (ICIP), pp. 2416–2420. IEEE (2020)

5. Wang, W., Tran, D., Feiszli, M.: What makes training multi-modal classification networks hard? In: Proceedings of the IEEE/CVF Conference on Computer Vision and Pattern Recognition, pp. 12695–12705 (2020)
6. Huang, Y., Du, C., Xue, Z., Chen, X., Zhao, H., Huang, L.: What makes multi-modal learning better than single (provably). Adv. Neural Inf. Proces. Syst. **34**, 10944–10956 (2021b)
7. Alam, M.R., Bennamoun, M., Togneri, R., Sohel, F.: A confidencebased late fusion framework for audio-visual biometric identification. Pattern Recogn. Lett. **52**, 65–71 (2015)
8. Asadpour, V., Homayounpour, M.M., Towhidkhah, F.: Audio–visual speaker identification using dynamic facial movements and utterance phonetic content. Appl. Soft Comput. **11**(2), 20832093 (2011)
9. Tao, R., Das, R.K., Li, H.: Audio-visual speaker recognition with a cross-modal discriminative network. arXiv preprint https://arxiv.org/abs/2008.03894 (2020)
10. Stefanidi, A., Topnikov, A., Tupitsin, G., Priorov, A.: Application of convolutional neural networks for multimodal identification task. In: 2020 26th Conference of Open Innovations Association (FRUCT), pp. 423–428 (2020)
11. Praveen, R.G., Alam, J.: Dynamic cross attention for audio-visual person verification. In: Computer Research Institute of Montreal (CRIM), Canada (2024) Retrieved from https://arxiv.org/abs/2403.04661
12. Sarı, L., Singh, K., Zhou, J., Torresani, L., Singhal, N., Saraf, Y.: A multi-view approach to audio-visual speaker verification. In: IEEE International Conference on Acoustics, Speech and Signal Processing (ICASSP) (2021). https://doi.org/10.1109/ICASSP39728.2021.9414260
13. Zhang, Q. et al.: Provable dynamic fusion for low-quality multimodal data. In: International Conference on Machine Learning, pp. 41753–41769. PMLR (2023)
14. Xue, Z., Marculescu, R.: Dynamic multimodal fusion. In: Proceedings of the IEEE/CVF Conference on Computer Vision and Pattern Recognition, pp. 2574–2583 (2023)
15. Gao, Z., Jiang, X., Xu, X., Shen, F., Li, Y., Shen, H.T.: Embracing unimodal Aleatoric uncertainty for robust multimodal fusion. In: Proceedings of the IEEE/CVF Conference on Computer Vision and Pattern Recognition (2024)
16. Wei, Y., Hu, D.: MMPareto: boosting multimodal learning with innocent unimodal assistance. In: Proceedings of the 41st International Conference on Machine Learning (2024) https://github.com/GeWu-Lab/MMPareto_ICML2024
17. Cao, B., Xia, Y., Ding, Y., Zhang, C., Hu, Q.: Predictive dynamic fusion. In: Proceedings of the 41st International Conference on Machine Learning (2024)
18. Mocanu, B., Ruxandra, T.: Active speaker recognition using cross attention audio-video fusion. In: Proceedings of the European Workshop/Conference on Visual Information Processing, pp. 1–6 (2022)
19. Praveen, R.G., Alam, J.: Audio-visual person verification based on recursive fusion of joint cross-attention. In: 2024 IEEE 18th International Conference on Automatic Face and Gesture Recognition (FG). IEEE (2024)

YOLOv11-Based Real-Time Plant Disease Detection System for Nigerian Agriculture

Ibrahim Sadiq[1], Omar Ennayri[2](✉), Sadia Sharmin[3], Mohamed Nourdine[1], Samiha Sammani Sani[2], and Damtew Merry Wondmagegn[1]

[1] School of Computer Science, Southwest University of Science and Technology, Mianyang, China
{Sadeeqibrahim81,Mohamednourdine201575,merrywondmagegn1303}@gmail.com
[2] School of Information and Control Engineering, Southwest University of Science and Technology, Mianyang, China
{ennayyri.omar,sssani2003}@gmail.com
[3] International American University, xx, USA
Sadiasharminlglg@gmail.com

Abstract. Plant diseases pose a significant threat to food production in Nigeria. In the past, diagnosing plant diseases in Nigeria has been a slow process. This study proposes a real-time, automatic plant disease detection system based on the YOLOv11 framework, which represents the state-of-the-art in the object detection field. This system is specifically designed to function in the Nigerian agriculture sector. The system is trained on a dataset of pre-acquired plant disease images processed by the Roboflow platform. The system results in a mean average precision of 94.1% mAP@0.5, precision of 85.2%, recall of 89.1% in detecting plant diseases, along with the efficiency of processing an image in a mere 4.2 ms. Qualitative analysis of the system has confirmed its success in plant disease localization. The system is a scalable, efficient, and practical platform for plant disease identification to ensure sustainable agriculture practices in Nigeria.

Keywords: Plant disease detection · YOLOv11 · Real-time detection · Nigerian agriculture · Deep learning · Precision agriculture · Object detection

1 Introduction

Agriculture is one of the basic necessities of life as well as a crucial means of providing food for a large number of people all over the world [1]. Agriculture in Nigeria is not only the backbone of the Nigerian economy but also the source of livelihood for a large number of Nigerian citizens. In a way though, crop agriculture in Nigeria faces continuous threats in the form of diseases caused by bacteria, fungi, viral infections, as well as pathogenic microorganisms [2]. These

L. Huang and D. Greenhalgh (Eds.): ICMLC 2026, LNNS 2061, pp. 480–495, 2027.
https://doi.org/10.1007/978-3-032-31236-5_33

diseases represent a great danger to the production level of crops but, on the other hand, the diagnostic techniques that are currently being practiced are often time-wasting. Recent breakthroughs made in artificial intelligence, such as deep learning and computer vision, prove to have great potential for fully automating plant disease diagnosis by identifying patterns in images [3]. Encouraged by breakthroughs in this area, this paper proposes a real-time system for plant disease detection specifically developed to address the local needs of agriculture in Nigeria. We make use of the highly efficient YOLOv11 object detection technique to achieve fast and accurate plant disease localization and classification. Our system will target for its attention three main crops in Nigeria. The objective of this research is to develop an automatic system that will greatly help in identifying plant diseases at an early stage, thereby allowing interventions, which are important in ensuring enhanced crop yield in the country [4]. The subsequent sections of this paper are described as follows. Section II discusses the related works done by different authors using deep learning to identify the diseases in plants. Section III presents The proposed YOLOv11-based methodology, including dataset preparation and model architecture. Section IV describes the experimental setup and discusses the obtained results, including comparative analysis of YOLOv11 variants and baseline models. Finally, Section V concludes the paper and outlines directions for future research.

2 Related Work

Recent years have witnessed impressive advancements in deep learning models, particularly tailored for automated plant disease detection. The advanced methods are specifically crafted to address the unique challenges of in-field crop images, such as complex backgrounds, uneven lighting conditions, and subtle class differences. There have been several efforts towards hybrid approaches between baseline convolutional models and attention approaches with further incorporation of data augmentation and light design strategies for better performance as well as in deployment for practical applications.

Zhang et al. created a hybrid model composed of a pre-trained DenseNet and a Convolutional Block Attention Module (CBAM) for detection of tomato leaf disease [5]. Their approach allowed the network to pay more attention to the most informative areas of the diseased leaf by emphasizing informative features both channel-wise and spatially, removing unnecessary background noise. This demonstrated the initial potential of attention mechanisms to aid in feature representation for fine-grained visual classification in agriculture. Addressing the requirement of efficiency, Jiang et al. introduced a light network consisting of dense and depth-wise separable convolutions for real-time detection of apple leaf disease [6]. Their methodology significantly reduced the number of parameters and computational burden, making a compelling case for applying deep learning models to resource-limited mobile and edge devices, a requirement for smart farming.

This advancement in acquiring both local and global information is observed in subsequent works. Liu et al. introduced a dual-path network that entwined

CNN and Transformer branches for identification of crop diseases under natural scenarios [7]. The CNN branch was charged with acquiring local pattern and texture information, while the Transformer branch acquired long-range dependencies and global context. Their work demonstrates a growing trend to move away from simple CNN architectures to tackle the complex spatial relationships in densely packed field images. Carrying on the trend to break new grounds of effectiveness as well as precision, Chen et al. proposed the YOLO-Former model that integrates a Transformer-based backbone with the YOLOv5 detection head [8]. This architecture proved highly effective in jointly detecting and localizing multiple diseases on a single leaf, showcasing the power of state-of-the-art object detection methods adapted to agricultural pathology.

Afolabi et al., in reference to the Nigerian context as a whole, studied the applicability of transfer learning with EfficientNetV2 on a curated dataset of cassava disease images [9]. The research focus was on the utilization of up-to-date optimized models where the computation capability was a limitation and achieved a very high level of accuracy in differentiating Cassava Mosaic Disease and Cassava Brown Streak Disease, but the study was confined to a single crop and standardized Imaging conditions. Classical Machine learning and CNN methods have also been investigated that included Random Forest Classification for paddy Leaf Disease and Early Detection of Patchouli Leaf Diseases using CNN [10,11].

Ojo et al. also developed a mobile-based disease detection system for tomatoes in Nigeria using a pruned MobileNetV2 variant [12]. While their research proved effective in demonstrating a path toward deployment, the model was only trained on a relatively small, proprietary dataset, which did not test its applicability to more extensive Nigerian agricultural conditions.

Mohanty et al. demonstrated early success using transfer learning with AlexNet for plant disease classification on the PlantVillage dataset, though their work was limited to laboratory conditions [13]. Thapa et al. explored the use of Generative Adversarial Networks (GANs) to generate synthetic leaf images for data augmentation, addressing the common challenge of insufficient training data in agricultural applications [14]. For edge deployment specifically, Wang et al. optimized YOLOv7 for agricultural applications by implementing channel pruning and quantization, achieving real-time performance on NVIDIA Jetson devices while maintaining high accuracy [15].

In addition to agricultural applications, YOLO architectures have been extensively evaluated in environmental monitoring tasks such as wildfire and smoke detection. Casas et al. conducted a comparative study between YOLOv5 and YOLOv8, highlighting YOLOv5's superior performance in terms of precision and recall, while noting YOLOv8's improved capability in detecting dense and dark smoke under challenging conditions [16].

More generally, this work leverages a diverse portfolio of past work that comprises attention mechanisms, light-weight CNN architecture, and hybrid vision transformers. While it leverages insights from them, our work is specifically situated when it comes to addressing the particular crop basket, disease trends, and climatic factors of Nigerian agriculture. The objective is to exceed single case studies and show a complete, scalable autonomous plant disease identification system that is accurate and feasibly deployable within the smart agriculture framework in Nigeria.

3 YOLOv11-Based Architecture for Multi-Disease Detection

The proposed system initiates with the pre-processing of the field-acquired leaf images of major Nigerian crops, using a carefully designed pipeline for normalization and data augmentation that aims at making the model robust for different environmental conditions. The YOLOv11 architecture forms the backbone of the core detection framework, representing a state-of-the-art real-time object detection model used for the identification and classification of multiple diseases, including early blight, late blight, and bacterial blight. It processes the input images through its convolutional backbone, which in turn extracts the features hierarchically, followed by a multi-scale feature aggregation module for learning enhanced representation, while the job of generating bounding box coordinates along with class probabilities lies with a detection head.

The detection framework is based on the YOLOv11 model, chosen due to the strong backbone and neck configuration and multi-tasking capability, yet preserving real-time inference speed. YOLOv11 applies a convolutional backbone for extracting hierarchical representations of input images and a multi-scale feature aggregation neck that combines semantic understanding from all scales. Such a structure is essential for the effective detection of disease spots, ranging from large symptomatic areas to small or initial spots, a necessary condition for a real-world agricultural use case. The output of the detection head comprises bounding box, class, and confidence scores for several diseases.

The proposed pipeline based on YOLOv11 is capable of performing an automated and real-time diagnosis of various plant diseases directly in the field. The proposed system balances model complexity and infrastructural capability and represents an important contribution to sustainable agricultural development in Nigeria. The end-to-end process of the proposed system is presented below through Fig. 1.

3.1 Architectural Footprint of YOLOv11

YOLOv11 is a sophisticated improved variant of the YOLO model, which is especially tailored toward making real-time object detection while placing particular attention on its ability to function effectively on edge device environments, especially in agriculture-related tasks of plant disease identification. As can be seen in Fig. 2, the proposed model is divided into three main modules: the improved hierarchical backbone extraction module, multi-scale aggregation module, and the decoupled detection head module, which enables efficient plant disease identification in terms of high accuracy while being computationally efficient as well.

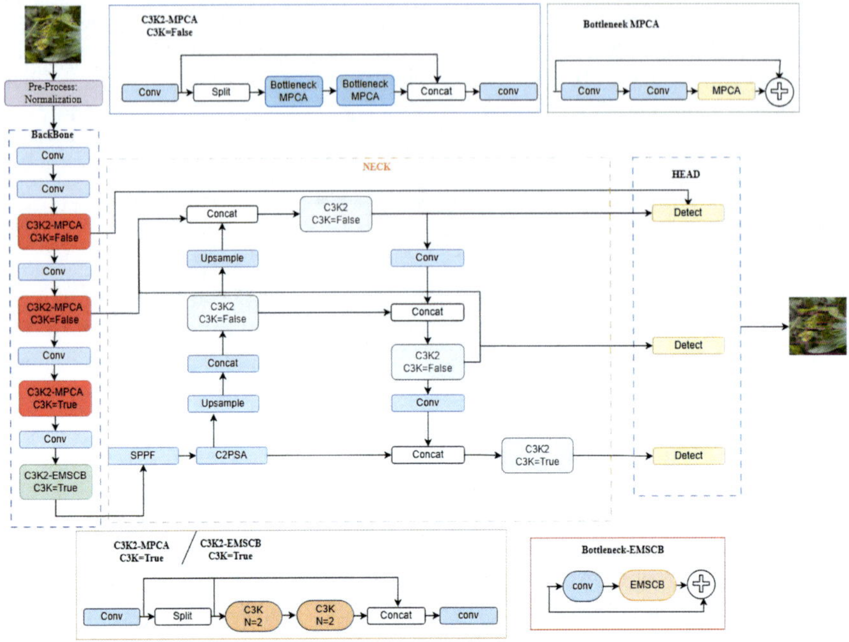

Fig. 1. Improved YOLOv11 network structure diagram.

3.1.1 Backbone: Enhanced Feature Extraction Considering the input image tensor $\mathbf{I}$ as $\mathbf{I} \in \mathbb{R}^{H \times W \times 3}$ where H and W denote image height and width, the YOLOv11 model employs optimized convolution blocks in the backbone architecture to eliminate computational redundancy and improve gradient flow. The transformations result in feature responses at three important scales:

$$\mathbf{F}_k = \mathcal{B}_k(\mathbf{I}; \Theta_b), \quad k \in \{1, 2, 3\} \tag{1}$$

where $\mathbf{F}_1 \in \mathbb{R}^{\frac{H}{8} \times \frac{W}{8} \times C_1}$, $\mathbf{F}_2 \in \mathbb{R}^{\frac{H}{16} \times \frac{W}{16} \times C_2}$, and $\mathbf{F}_3 \in \mathbb{R}^{\frac{H}{32} \times \frac{W}{32} \times C_3}$ represent progressively downsampled feature maps with increasing semantic abstraction. The backbone architecture utilizes optimized convolutional layers, SiLU (Sigmoid Linear Unit) activations, and efficient building blocks:

$$\mathcal{B}_k(\mathbf{X}) = \text{SiLU}\left(\text{Conv}_{3\times 3}\left(\text{C3k2}_k(\mathbf{X})\right)\right) \tag{2}$$

where C3k2_k incorporates cross-stage partial connections with dual kernel sizes and adaptive scaling, achieving improved parameter efficiency while maintaining feature representation quality.

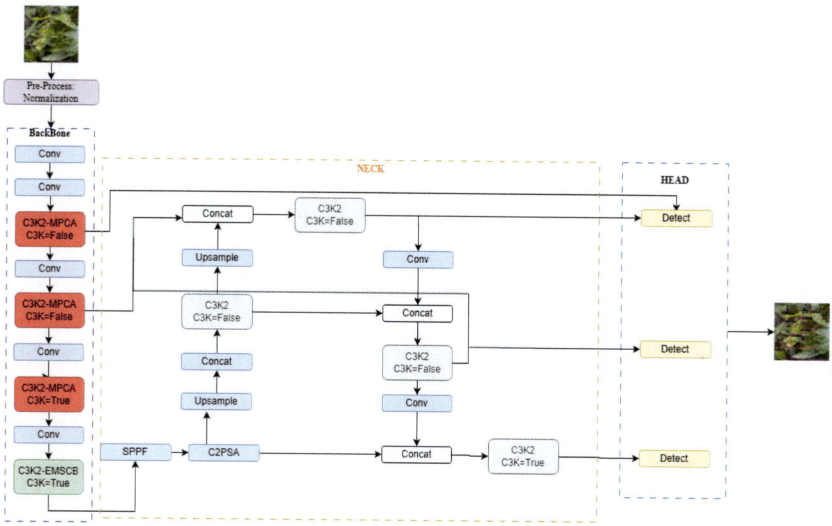

Fig. 2. Flow of the YOLOv11 algorithm.

3.1.2 Neck: Path Aggregation Network (PANet) The neck module performs bi-directional multi-scale feature fusion between the fine-grained information from the shallow layers and the semantic information from the deeper layers as follows:

$$\mathbf{G}_k^{\text{top-down}} = \text{Conv}_{1\times1}\left(\mathbf{F}_k + \text{Upsample}(\mathbf{G}_{k+1})\right) \tag{3}$$

$$\mathbf{G}_k^{\text{bottom-up}} = \text{Conv}_{3\times3}\left(\mathbf{G}_k^{\text{top-down}} + \text{Downsample}(\mathbf{G}_{k-1})\right) \tag{4}$$

$$\mathbf{G}_k = \text{SiLU}\left(\mathbf{G}_k^{\text{bottom-up}}\right) \tag{5}$$

where Upsample$(\cdot)$ employs bilinear interpolation, Downsample$(\cdot)$ uses strided convolution, and the final fused features $\mathbf{G}_k$ contain enriched multi-scale representations essential for detecting plant disease lesions across various sizes.

3.1.3 Head: Decoupled Anchor-Free Detection YOLOv11 employs a decoupled detection head that separates the bounding box regression from the classification process without relying on anchor boxes. For every spatial position (i, j) in feature grid $\mathbf{G}_k$, the head predicts:

$$\mathbf{b}_{ij} = (x, y, w, h) = \text{Conv}_{1\times1}^{\text{reg}}(\mathbf{G}_k[i, j]) \tag{6}$$

$$\mathbf{p}_{ij} = (p_1, p_2, \ldots, p_C) = \sigma\left(\text{Conv}_{1\times1}^{\text{cls}}(\mathbf{G}_k[i, j])\right) \tag{7}$$

where $\mathbf{b}_{ij}$ represents normalized bounding box coordinates, $\mathbf{p}_{ij}$ denotes class probabilities for C disease categories, and $\sigma(\cdot)$ denotes the sigmoid activation function.

3.2 Loss Function Formulation

The training objective combines three loss components with weights optimized for plant disease detection:

$$\mathcal{L}_{\text{total}} = \lambda_{\text{cls}} \cdot \mathcal{L}_{\text{cls}} + \lambda_{\text{box}} \cdot \mathcal{L}_{\text{CIoU}} + \lambda_{\text{dfl}} \cdot \mathcal{L}_{\text{DFL}} \tag{8}$$

with $\lambda_{\text{cls}} = 1.0$, $\lambda_{\text{box}} = 7.5$, and $\lambda_{\text{dfl}} = 1.5$ as determined through empirical optimization in our experiments.

3.2.1 Classification Loss Binary cross-entropy loss for multi-label classification:

$$\mathcal{L}_{\text{cls}} = -\frac{1}{N} \sum_{i=1}^{N} \sum_{c=1}^{C} [\hat{y}_{ic} \log(p_{ic}) + (1 - \hat{y}_{ic}) \log(1 - p_{ic})] \tag{9}$$

where $\hat{y}_{ic}$ is the ground truth label and p_{ic} is the predicted probability for class c.

3.2.2 Bounding Box Regression Loss Complete IoU Loss (CIoU Loss) considers overlap area, center distance, and aspect ratio:

$$\mathcal{L}_{\text{CIoU}} = 1 - \text{IoU} + \frac{\rho^2(\mathbf{b}, \hat{\mathbf{b}})}{c^2} + \alpha v \tag{10}$$

where $\rho(\cdot)$ is the Euclidean distance between the box centers, c is the diagonal of the smallest enclosing box, and v quantifies the consistency of aspect ratios:

$$v = \frac{4}{\pi^2} \left(\arctan \frac{w}{h} - \arctan \frac{\hat{w}}{\hat{h}} \right)^2, \quad \alpha = \frac{v}{1 - \text{IoU} + v} \tag{11}$$

3.2.3 Distribution Focal Loss To address class imbalance and improve localization precision for small lesions:

$$\mathcal{L}_{\text{DFL}} = -\sum_{i=0}^{n} ((y_{i+1} - y) \log(P_i) + (y - y_i) \log(P_{i+1})) \tag{12}$$

where y is the continuous label value, y_i and y_{i+1} are adjacent discrete values, and P_i represents the predicted probability.

3.2.4 Inference Optimization During inference, Non-Maximum Suppression (NMS) with IoU threshold $\tau = 0.5$ filters redundant predictions:

$$\text{NMS}(\mathcal{D}) = \left\{ \mathbf{d}_i \in \mathcal{D} \mid \begin{array}{l} \text{IoU}(\mathbf{d}_i, \mathbf{d}_j) < \tau, \ \forall j \neq i, \\ \text{score}(\mathbf{d}_i) > \text{score}(\mathbf{d}_j) \end{array} \right\} \tag{13}$$

where $\mathcal{D}$ is the set of initial detections.

3.2.5 Computational Complexity Analysis The theoretical computational cost can be expressed in terms of floating-point operations (FLOPs):

$$\text{FLOPs} = \sum_{l=1}^{L} \left(2 \cdot C_l^{\text{in}} \cdot C_l^{\text{out}} \cdot K_l^2 \cdot H_l \cdot W_l\right) \tag{14}$$

where L is the total number of layers, C_l^{in} and C_l^{out} are input and output channels, K_l is kernel size, and $H_l \times W_l$ is the spatial dimension at layer l.

YOLOv11's architectural advancements, specifically concerning efficient architecture, multi-scale feature fusion, and anchor-free detection, form an exemplary framework for detecting plant diseases with high accuracy in real-time within agricultural monitoring systems with limited computing capabilities.

4 Experiment and Dataset Description

This section presents the experimental setup and dataset description for the proposed plant disease detection system focused on Nigerian agriculture. The study primarily concentrates on detecting diseases affecting key crops such as rice, tomato, and potato leaves. These crops were chosen due to their significant impact on food security and economic stability in Nigeria. To facilitate the development and training of the machine learning model, a custom dataset of plant leaf images was created, encompassing the various types of diseases commonly affecting these crops by categorizing them as early and late blights, and also identifying the plant as healthy when it is void of diseases.

4.1 Data Description and Preprocessing

The images employed in this research were sourced from several repositories, including publicly available sources, agricultural research institutions, and PlantVillage datasets. Notably, images were chosen with consideration to their relevance to agricultural settings in Nigeria. A total of 2,140 images were annotated for the leaves of rice, tomatoes, and potatoes using the Roboflow image platform. The process involved approximately four days to annotate all images to enable disease localization. In the dataset, there are samples from different stages of the progression of the diseases, different illumination sources, and different leaf orientations. This has been done to enhance the capabilities of the trained models when working with actual field conditions to some extent. Before actually using the images for training the models, they were converted into a standardized size for better visual alignment and effective feature extraction. Rotation, horizontal flipping, and zooming techniques were involved in the data augmentation process. The data was divided into training, validation, and testing sets with a split ratio of 70%, 20%, and 10%, respectively. The data split adopted in this work helps to ensure that the performance assessment carried out on the data has validity and that there is no overfitting to improve the usability of the model in farm settings. Table 1 presents the distribution of images across different crops and disease classes in the dataset.

Table 1. Dataset distribution by crop and disease class

Crop	Disease class	Number of images
Tomato	Early blight	420
Tomato	Late blight	380
Potato	Early blight	310
Rice	Bacterial blight	290
All crops	Healthy	740
Total		2140

4.2 Evaluation Indicators

The model was evaluated using several key metrics to measure its performance such as precision, recall, accuracy and the mAP (mean average precision). In plant disease detection, precision measures the proportion of correctly identified diseased plants (True Positives) among all plants predicted as diseased (both True Positives and False Positives). It reflects how accurate the model is when predicting the presence of disease.

$$\text{Precision} = \frac{TP}{TP + FP} \tag{15}$$

Accuracy measures the overall correctness of the models predictions, considering both healthy and diseased plants. It is the ratio of correct predictions (both diseased and healthy plants) to the total number of plants.

$$\text{Accuracy} = \frac{TP + TN}{TP + FP + TN + FN} \tag{16}$$

Mean Average Precision (mAP) is a metric that evaluates the model's precision and recall across different Intersection over Union (IoU) thresholds. The mAP@50 specifically evaluates the precision at a single IoU threshold of 0.50. The steps to compute mAP@50 are as follows:

$$\mathrm{mAP@50} = \frac{1}{N}\sum_{i=1}^{N} AP_i \tag{17}$$

where N is the number of disease classes, and AP_i is the average precision for disease class i.

4.3 Results of YOLOv11 Model

The loss curves illustrate the convergence behavior during the training process as shown Fig. 3. The training losses for box, class, and detection loss steadily decreased as the epochs progressed, indicating that the model was successfully learning to identify plant diseases. Validation losses followed a similar trend, demonstrating the model's generalizability to unseen data. Metrics such as precision and recall also improved over time, indicating that the model effectively learned to classify plant diseases with increasing accuracy. The steady reduction in loss and the improvement in mAP values during training suggest that the YOLOv11n model is well-optimized and has good potential for deployment in real-world applications.

Further validation of performance by the YOLOv11n model is done through confidence and precision-recall analysis. Figure 4 shows that the F1-Confidence curve reaches an optimal value of 0.88 when the confidence threshold is at 0.808, which reflects an appropriate operational point where the model's precision and recall are optimally balanced. At the same time, the Precision-Confidence curve shows that its peak, up to 0.99, appears once the confidence threshold is maximized to 1.0. It shows the ability of the model to achieve extremely high accuracy once the model operates under high-confidence conditions. This property is especially desirable in practical agricultural applications since, in most cases, false positives need to be kept as low as possible in order to ensure reliable diagnosis of the disease.

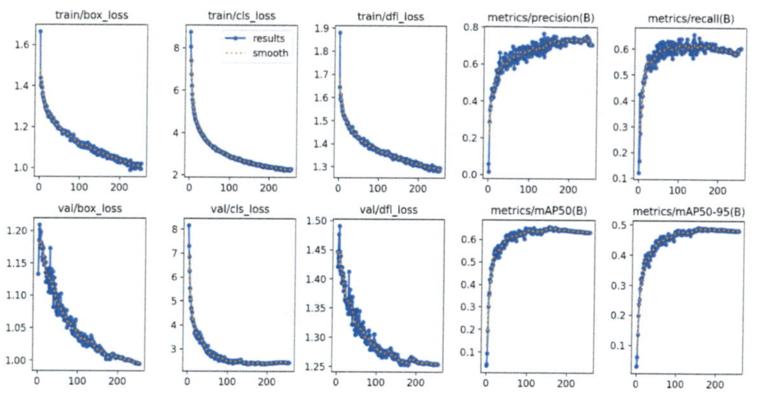

Fig. 3. Convergence behavior of YOLOv11n training and validation losses.

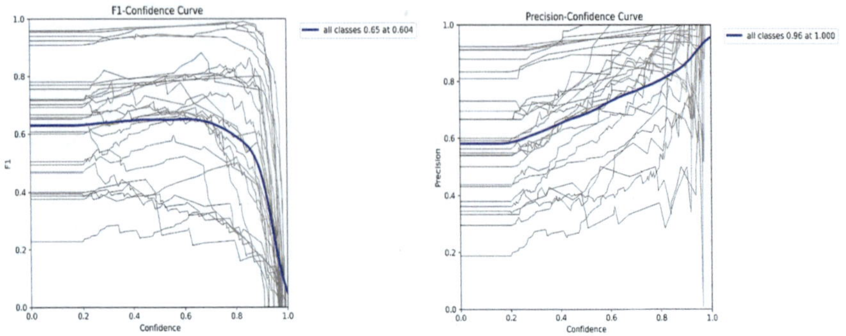

Fig. 4. F1-confidence and precision-confidence curves for YOLOv11n.

The complementary analysis is supplemented by Fig. 5, which offers an evaluation in the form of the Precision-Recall Curve and the Recall-Confidence Curve. The former is very informative regarding the capability level of each detection method in its entirety at different confidence levels. The latter offers an evaluation regarding the changes in the capability level at successive confidence level values. These two graphs give an interpretation regarding the measurement given in the evaluation regarding the capability balance level at different plant disease classes. The high area under the graph also reveals its robustness in multi-class classification.

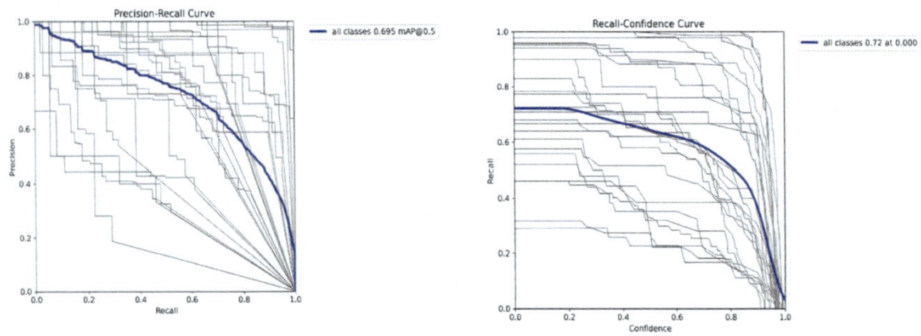

Fig. 5. Precision-recall and recall-confidence curves for YOLOv11n.

4.4 Comparative Analysis of YOLOv11 and YOLOv12

For better analysis of the trade-off between the accuracy of detection and the degree of computation efficiency, the state-of-the-art YOLO series YOLOv11 and YOLOv12 were also considered for evaluation. The YOLOv11 series of models has high efficiency in terms of computations with equal or even better accuracy in plant disease detection. The highest mean average precision at 0.5 (mAP@0.5) accuracy at 94.1% with precision and recall values of 85.2% and 89.1%, respectively, was obtained by YOLOv11. On the other hand, the highest recall of 90.3% but with relatively decreased precision at 83.3% and mAP@0.5 at 92.0% with an inference time of 7.1 ms was observed in the case of YOLOv12. The fastest inference time of 4.2 ms per image in YOLOv11 indicates that it is more efficient than others and thus suitable for real-time processing. Table 2 summarizes the performance of the object detection in both models. On one hand, YOLOv12 significantly outperforms on recall, while on the other hand, YOLOv11 offers a good trade-off between accuracy, speed, and computational efficiency, which is fundamental for reliable and effective real-time agricultural monitoring. With these results, the selection of YOLOv11 was well-justified, with higher mAP, competitive recall, and speedier inference, as an appropriate primary model for the proposed plant disease detection system. YOLOv12 remains a good alternative choice when an application requires the highest possible sensitivity of detection, even at a slightly higher computational cost.

Table 2. Performance comparison of YOLOv11 and YOLOv12

Model	mAP@0.5 (%)	Precision (%)	Recall (%)	Speed
YOLOv11	94.1	85.2	89.1	4.2 ms
YOLOv12	92.0	83.3	90.3	7.1 ms

4.5 Baseline Comparison

To further contextualize the effectiveness of the proposed method, YOLOv11 and YOLOv12 are compared with representative state-of-the-art plant disease detection models as reported in their original publications. The baseline models for the experiment are Faster R-CNN, SSD-MobileNetV2, EfficientDet-D0, YOLOv7-tiny, which are well known in plant detection for their real-time performance. Comparison in terms of standard object detection evaluation metrics such as mean Average Precision (mAP@0.5), precision, recall, and inference time is done based on comparisons provided in recent literature.As shown in table 3 YOLOv11 demonstrates competitive performance compared with representative state-of-the-art models reported in the literature, particularly in terms of accuracyspeed trade-off, which is critical for real-time agricultural applications.

The reasons for improved accuracy in YOLOv11 can be attributed to advancements in feature extraction and anchor-free detection methods. Although SSD-MobileNetV2 is a lightweight network and is able to perform faster inference and has a smaller size compared to traditional models such as Faster R-CNN, its accuracy in object detection is not up to the mark and can be improved in complex backgrounds and lighting conditions, which is a scenario in agricultural application settings. YOLOv11, having a balanced accuracy and processing speed, is apt for application in agricultural monitoring setups.

4.6 Detection Results

To demonstrate the detection capability of YOLOv11 in real-life settings related to agriculture, the results are represented in an image format to validate the effectiveness of YOLOv11 on several plant disease datasets. Figure 6 shows some exemplary outputs from the detection procedure applied on a variety of plant leaves from tomatoes, potatoes, and coffee plants.

Table 3. Performance comparison with plant disease detection models reported in literature

Model	mAP@0.5 (%)	Precision (%)	Recall (%)	Speed
Faster R-CNN [10]	78.2	72.5	81.3	120 ms
SSD-MobileNetV2 [12]	82.1	76.8	84.2	28 ms
EfficientDet-D0 [9]	85.7	79.4	86.9	18 ms
YOLOv7-tiny [15]	88.3	81.2	87.5	6.8 ms
YOLOv11	**94.1**	**85.2**	89.1	**4.2 ms**
YOLOv12	92.0	83.3	**90.3**	7.1 ms

Fig. 6. Detection results using YOLOv11 on a multiple plant disease dataset.

In particular, it demonstrates the ability of this model to successfully detect localized instances of early blight, leaf spots, rusty plant diseases with its bounding boxes enclosing very closely any affected area on plant leaves despite orientation angles and illuminations varying from one instance to another. Most important is its ability to suppress false positives from normal plant areas.

Figure, 7 gives further visualizations across plant categories and reinforces the generalizability of the trained model. These detections also remain consistent across species that were not included in the primary training set, a testament to strong features learning and transferability. The model appears to classify diseases with similar looking visual patterns correctly, while minimal confusions become apparent when diseases bear symptomatically similar conditions from one another. This performance well demonstrates the embedded anchor-free detection mechanism of YOLOv11, which is combined with its backbone architecture optimized for agricultural imagery variability.

Fig. 7. Cross-category plant disease detection results using YOLOv11.

The details regarding the identification of plant diseases, with corresponding levels of confidence, represented by visualization, are provided in Table 4. The above table encompasses major findings from the qualitative result, which give information about the reliability of the model performance over varied classes.

Table 4. Detection performance of YOLOV11 under field conditions

Condition	Accuracy (%)	Certainty	Errors
Optimal lighting	98.2	0.92	0.1
Low lighting	94.5	0.87	0.4
Leaf occlusion	93.1	0.85	0.2
Multiple diseases	96.3	0.89	0.2
Mixed healthy/diseased	97.8	0.91	0.1

Taken collectively, these visual outcomes serve to round out the quantitative assessment provided earlier, indicating that YOLOv11 is not only a high-performance architecture but also provides for effective disease detection under real-world conditions. The generality of YOLOv11s solution for various plant species and disease patterns verifies its application for real-world agriculture.

5 Conclusion

This paper presented a comprehensive evaluation of a YOLOv11-based real-time plant disease detection system tailored for Nigerian agricultural conditions. From the experimental evaluation, it is evident that YOLOv11 outperforms YOLOv12 and other baseline approaches in terms of its precision of 94.1% mAP@0.5 and inference time of 4.2 ms. YOLOv11 thus performs well in terms of its ability to resist conditions of the farm and consumes lower computational time, making it an effective technique to be applied in Nigerian conditions in a mobile setup. Future evaluations will involve increasing the scope of the dataset to various crops and implementing it in a mobile environment.

Acknowledgments. Gratefully acknowledge the contributions and support from everyone involved in this project. Sincere thanks also go to the reviewers and editors for their thorough and constructive comments, which were instrumental in refining this paper.

References

1. Food and Agriculture Organization of The United Nations (FAO): The State of Food Security and Nutrition in the World 2022: Repurposing Food and Agricultural Policies to Make Healthy Diets More Affordable, FAO, Rome, Italy, 2022. Healthy Diets More Affordable, Rome (2022). https://doi.org/10.4060/cc0639en
2. Boethel, M.J.: Integrated pest management: a global overview of history, programs and future. J. Integr. Pest Manag. **10**(1), 10 (2019). https://doi.org/10.1093/jipm/pmz010
3. Sladojevic, S., Arsenovic, M., Anderla, A., Culibrk, D., Stefanovic, D.: Deep neural networks based recognition of plant diseases by leaf image classification. Comput. Intell. Neurosci **2016**, Article ID 3289801 (2016)
4. Mohanty, P., Hughes, D., Salathé, M.: Using deep learning for image-based plant disease detection. Front. Plant Sci. **7**, 1419 (2016)
5. Zhang, J., Wang, X., Zhou, Y., Li, H.: Hybrid DenseNet-CBAM model for tomato leaf disease detection with attention mechanism. Comput. Electron. Agric. **186**, 106225 (2021). https://doi.org/10.1016/j.compag.2021.106225
6. Jiang, P., Chen, Y., Liu, B., He, D., Liang, C.: Real-time detection of apple leaf diseases using deep learning approach based on improved convolutional neural networks. IEEE Access **9**, 59052–59066 (2021). https://doi.org/10.1109/ACCESS.2021.3072492
7. Liu, H., Zhang, Z., Wang, Y., Yang, J., Zhou, F.: Dual-path network with CNN and Transformer for crop disease identification in natural scenarios. Expert Syst. Appl. **222**, 119758 (2023). https://doi.org/10.1016/j.eswa.2023.119758
8. Chen, W., Huang, Y., Li, Y., Wang, Z.: YOLO-Former: An efficient Transformer-based object detector for plant disease detection. Comput. Electron. Agric. **210**, 107907 (2023). https://doi.org/10.1016/j.compag.2023.107907
9. Afolabi, D.I., Adegboye, O.A., Oluwadare, T.O.: Transfer learning with EfficientNetV2 for cassava disease detection in Nigerian agriculture. Afr. J. Agric. Res. **20**(2), 145–156 (2024). https://doi.org/10.5897/AJAR2023.16678
10. Saminathan, K., Sowmiya, B., Chithra Devi, M.: Multiclass classification of paddy leaf diseases using random forest classifier. J. Image Graph. **11**(2), 195–203 (2023). https://doi.org/10.18178/joig.11.2.195-203

11. Rustam, R., Noveriza, R., Khotijah, S., Rizal, S., Melati, Pratiwi, N.K.C., Barri, M.H., Usman, K.: Convolution neural network approach for early identification of patchouli leaf disease in Indonesia. J. Image Graph. **12**(2), 137–144 (2024). https://doi.org/10.18178/joig.12.2.137-144
12. Ojo, M.O., Adetunmbi, A.O., Oluwadare, S.A.: Mobile-based tomato disease detection system for Nigerian farmers using pruned MobileNetV2. Int. J. Inf. Technol. Comput. Sci. **14**(3), 34–45 (2022). https://doi.org/10.5815/ijitcs.2022.03.04
13. Mohanty, S.P., Hughes, D.P., Salathé, M.: Using deep learning for image-based plant disease detection. Front. Plant Sci. **7**, 1419 (2016). https://doi.org/10.3389/fpls.2016.01419
14. Thapa, R., Snavely, N., Belongie, S., Khan, A.: The Plant Pathology 2020 challenge dataset to classify foliar disease of apples. arXiv preprint. arXiv:2004.11958 (2020). https://arxiv.org/abs/2004.11958
15. Wang, Z., Liu, J., Li, Y.: Lightweight YOLOv7 for real-time plant disease detection on edge devices. Comput. Electron. Agric. **207**, 107754 (2023). https://doi.org/10.1016/j.compag.2023.107754
16. Casas, E., Ramos, L., Bendek, E., Rivas-Echeverria, F.: YOLOv5 vs. YOLOv8: performance benchmarking in wildfire and smoke detection scenarios. J. Image Graph. **12**(2), 127–136 (2024). https://doi.org/10.18178/joig.12.2.127-136

AEW-YOLO: A Lightweight Detection Algorithm for Pine Wood Disease Based on YOLO11

Chongqing Yu(✉) and Jiantao Zhang

South China Agricultural University, Guangzhou, Guangdong 510642, China
chqyu0927@163.com

Abstract. To address the issues of high computational cost and low recognition accuracy in detecting pine wilt disease using UAV imagery, this study proposes AEW-YOLO, a lightweight object detection algorithm improved based on YOLO11n. Three core refinements are integrated into the proposed algorithm to substantially enhance its computational efficiency and recognition accuracy. First, to mitigate the loss of detailed information caused by the conventional downsampling method, we propose an efficient ADown module, which improves accuracy while decreasing computational load. Moreover, a lightweight decoupled detection head, EfficientHead, is designed to significantly increase inference efficiency while maintaining high precision. Additionally, the WIoUv3 loss function is employed to optimize the accuracy of bounding box regression, thereby improving detection performance and robustness. Experimental results show that, compared to the baseline YOLO11n, AEW-YOLO achieves improvements of 2.0, 3.2, 2.6, and 1.4 percentage points in precision, recall, F1-score, and mAP@0.5, respectively. Meanwhile, the number of parameters, FLOPs, and model size are reduced by 30.7%, 34.9%, and 26.9%, respectively. These results fully verify that AEW-YOLO not only delivers excellent performance for the pine wilt disease detection task but also effectively maintains a lightweight property.

Keywords: YOLO11 · pine wilt disease · object detection · lightweighting · UAV image

1 Introduction

Pine Wilt Disease (PWD), instigated by infection with the pine wood nematode, is a highly destructive forest pathology that can cause rapid mortality in pine trees, leading to substantial ecological and economic losses in China. Rapid detection and precise localization of discolored infected trees are crucial for epidemic control. Recently, the use of Unmanned Aerial Vehicle (UAV) remote sensing technology to detect infected pine trees has become an important method for PWD monitoring. However, traditional manual visual interpretation methods based on UAV imagery suffer from low efficiency and a high rate of missed detection, which hinders their applicability for extensive monitoring applications. To overcome the limitations of manual identification, early

L. Huang and D. Greenhalgh (Eds.): ICMLC 2026, LNNS 2061, pp. 496–507, 2027.
https://doi.org/10.1007/978-3-032-31236-5_34

researchers attempted to employ computer vision methods for automated disease identification. For instance, in sunflower leaf disease identification, methods based on SIFT feature points were used to distinguish different diseases [1]; in paddy leaf disease classification, features like color and texture were extracted and combined with machine learning classifiers such as random forests to achieve automated diagnosis [2]. While these methods advanced automation in plant disease recognition, their performance typically relied on manually designed features. With the advancement of deep learning, object detection algorithms like YOLO (You Only Look Once), based on deep convolutional neural networks, can automatically learn high-level features from data, overcoming the dependency on handcrafted features in traditional methods. Compared to two-stage algorithms such as Faster R-CNN, the YOLO algorithm demonstrates superior performance in both detection speed and accuracy, making it more suitable for the rapid identification of discolored, PWD-infected trees in UAV imagery. To further enhance detection performance, numerous studies have introduced targeted improvements to the YOLO model. For instance, Dong et al. improved YOLOv5s by comparing the performance of different attention mechanisms and loss functions, thereby increasing the detection performance for PWD-infected trees [3]. Ye et al. enhanced the model by integrating YOLOv5 with the StrongSORT algorithm, enabling the dynamic tracking and counting of PWD-infected trees [4]. Similarly, Hu et al. modified the YOLOv7 model to address issues such as complex backgrounds and poor quality in aerial images, which improved the recognition accuracy of trees infected with PWD [5].

Although existing methods have achieved some improvement in detection accuracy, they are still plagued by issues such as a substantial number of parameters, low computational efficiency, and high inference latency, making practical deployment on edge devices challenging. To overcome these limitations, this study targets the development of a lightweight and efficient algorithm, AEW-YOLO. The core contributions are outlined below:

(1) Proposing a lightweight downsampling module (Adaptive Downsampling Module, ADown), which employs a parallel multi-branch feature extraction and fusion strategy. This forms a lightweight downsampling mechanism that combines convolution and pooling, reducing environmental interference and noise while minimizing the loss of detailed information, thereby facilitating the capturing of subtle features.
(2) Designing a lightweight decoupled detection head (EfficientHead), which adjusts the network depth and width using a compound scaling method, achieving a decrease in parameters and an increase in processing efficiency.
(3) Adopting the WIoUv3 loss function facilitates a wise allocation of gradient gains, which directs the model's attention towards low-quality samples and results in improved bounding box regression accuracy.

2 Materials and Methods

2.1 Dataset

The research area for this study is located in Yangshan County, Qingyuan City, Guangdong Province, China. Field surveys were conducted in July–August 2024. Following quality inspection, a total of 116 aerial images were retained as the primary dataset for subsequent analysis. To accommodate the model input size, the original remote sensing images of 5280 × 3956 pixels were cropped into sub-images of 640 × 640 pixels, which yielded a final set of 7,716 sub-images. The discolored infected trees in each image were annotated using the LabelImg software. The labeled images were then randomly divided into training, validation, and test sets with a ratio of 7:2:1, comprising 5,400, 1,543, and 773 images, respectively. Samples of the dataset are shown in Fig. 1.

(a) (b) (c)

Fig. 1. Examples of the dataset.

2.2 Overall Architecture of AEW-YOLO

The computational profile of the YOLO11n model remains too heavy for real-time UAV data processing on edge devices, despite its generally good speed-accuracy balance. Therefore, we propose AEW-YOLO, a lightweight algorithm for pine wilt disease detection. The main improvements are as follows: An efficient and lightweight downsampling module, ADown, is constructed to reduce computational load while preserving more image details; a lightweight decoupled detection head (EfficientHead) is designed, which adopts a parameter-sharing strategy to minimize parameters, leading to superior performance in complex scenarios and enhanced deployment flexibility. The improved AEW-YOLO model architecture is shown in Fig. 2.

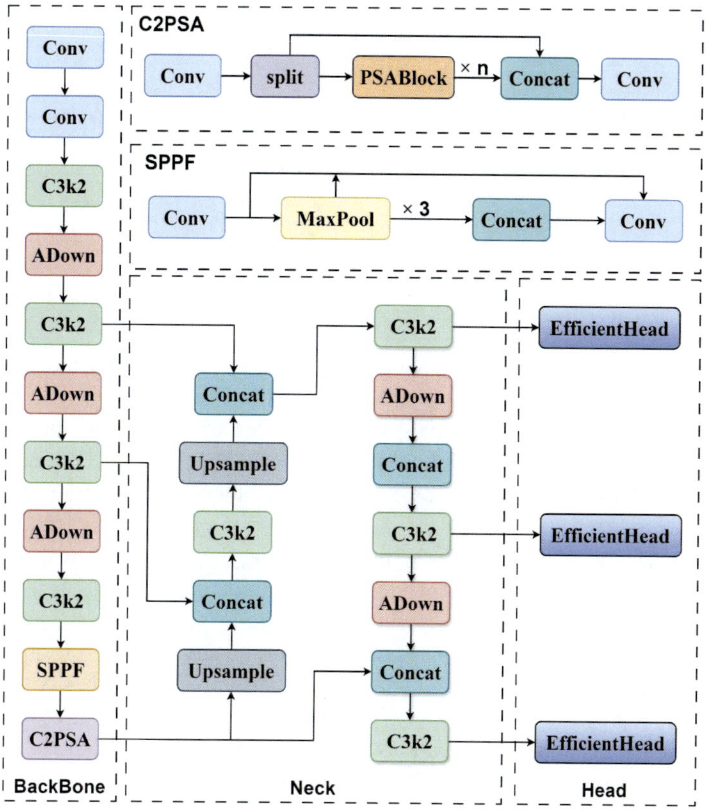

Fig. 2. Overall structure of AEW-YOLO.

2.3 ADown Module

The ADown module employs pooling operations to replace traditional convolutions for downsampling and integrates information from different paths through a branch structure to achieve feature extraction and edge information preservation simultaneously. A 2×2 average pooling operation is applied to the input feature map, as illustrated in Fig. 3, to smooth out noise and preserve global contextual information. Then, this feature map will be divided into two parts according to its channel, which will become two different parallel branches. The first branch processes the input through a strided 3×3 convolution for downsampling and feature emphasis, followed by max-pooling for spatial compaction, culminating in a 1×1 convolution that adjusts channels and integrates information across them. The second branch is processed by a 3×3 convolutional layer with a stride of 2 to capture more complex local features while performing downsampling. Finally, the feature maps from both branches are concatenated, resulting in a more informative output feature map. Through this parallel multi-path design, the ADown module effectively reduces environmental interference and noise while efficiently integrating features under different receptive fields, thereby maintaining a lightweight structure and enhancing the representation capability for multi-scale features.

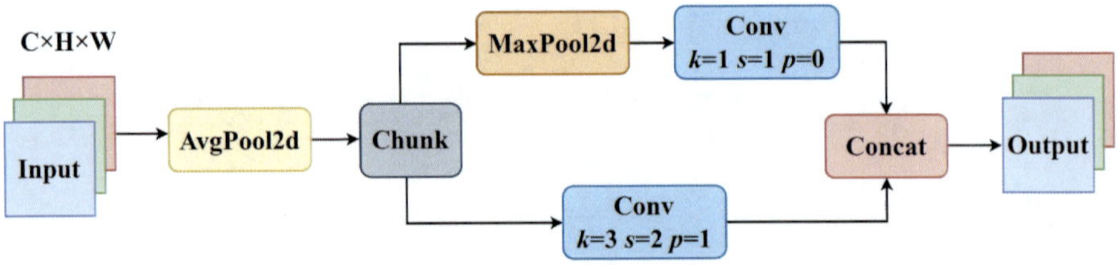

Fig. 3. Structure of ADown.

2.4 EfficientHead Module

EfficientHead enhances computational efficiency while maintaining detection accuracy through its unique parameter-sharing and lightweight design. It reduces the number of parameters and enhances feature consistency by sharing weights across multiple scales. Additionally, the efficient decoupled head separates classification and regression tasks, and employs depthwise separable convolutions to further optimize the computational load.

As shown in Fig. 4, the process begins by feeding the feature maps into EfficientHead, where they undergo gradual deep feature extraction through two successive group convolution (GroupConv) layers. The number of groups in these layers is dynamically configured based on the channel count of the input, ensuring that each set of kernels processes only a subset of channels. This strategy effectively balances computational efficiency with representational capacity. The refined features are then directed to an efficient decoupled structure for the separate tasks of object classification and bounding box regression. This design enables the model to more effectively discern the textural, color, and morphological characteristics of infected wood of varying sizes, while significantly reducing the computational burden of the model.

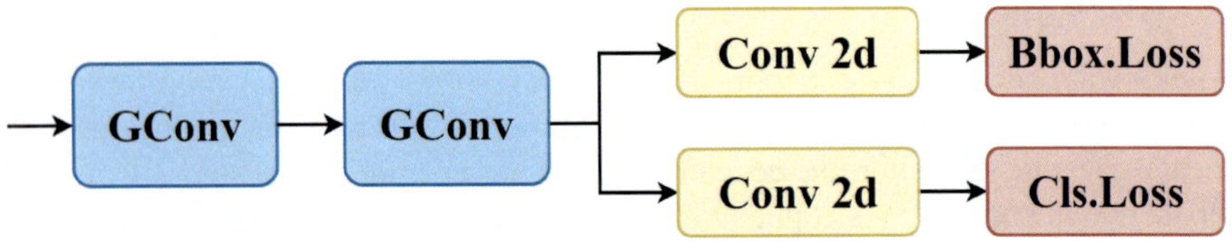

Fig. 4. Structure of efficienthead.

2.5 WIoUv3 Loss Function

To improve localization precision, training stability, and generalization, WIoUv3 introduces a novel gradient allocation strategy. It employs a dynamic non-monotonic mechanism that dynamically reweights anchor boxes, prioritizing high-quality ones and suppressing the influence of inferior samples [6]. The definition of WIoUv3 is as follows:

$$R_{WIoU} = \exp\left(\frac{\left(x - x_{gt}\right)^2 + \left(y - y_{gt}\right)^2}{\left(w_g^2 + H_g^2\right)^*}\right) \quad (1)$$

$$L_{IOU} = 1 - \text{IOU} \quad (2)$$

$$\beta = \frac{LIOU}{\overline{LIOU}} \in [0, +\infty) \quad (3)$$

$$r = \frac{\beta}{\delta \alpha^{\beta - \delta}} \quad (4)$$

$$L_{\text{WIoUv3}} = r \times R_{\text{WIoU}} \times L_{\text{IoU}} \quad (5)$$

3 Experiment

3.1 Experimental Environment and Evaluation Metrics

The experimental computer configuration consisted of a Windows 10 system, 64 GB RAM, an Intel(R) Xeon(R) Gold 5218R CPU, and an NVIDIA GeForce RTX 2080Ti GPU. The software environment included CUDA 12.1, Python 3.11, and the PyTorch framework. The critical hyperparameter settings are detailed in Table 1.

For a comprehensive evaluation, a standard suite of metrics was employed, such as precision (P), recall (R), F1-score, mAP@0.5, mAP@0.5:0.95, number of parameters (Params), computational complexity (FLOPs), and model size.

Table 1. Hyperparameters setting.

Hyperparameters	Value
Training epoch	300
Batch size	16
Learning rate	0.01
Image size	640 × 640
Optimizer	SGD
Momentum	0.937
Weight decay	0.0005

3.2 Ablation Experiment

To verify the efficacy of the proposed enhancement modules, ablation studies were conducted under a consistent training and hyperparameter setup to evaluate the contribution of each module, with the complete results provided in Table 2. Here, Model A represents the baseline YOLO11n; Models B, C, and D denote the model improved solely with ADown, EfficientHead, and WIoUv3, respectively; Model E represents the model combining ADown and EfficientHead; and Model F is the complete AEW-YOLO integrating all three improvements.

Table 2. Results of ablation experiments.

model	P/%	R/%	F1/%	mAP@0.5/%	mAP@0.5:0.95/%	PARARMS/M	FLOPs/G	Model Size/MB
A	84.4	83.0	83.7	90.8	51.8	2.6	6.3	5.2
B	86.2	85.0	85.6	91.8	53.0	2.1	5.3	4.3
C	82.8	86.4	84.6	91.0	51.7	2.3	5.1	4.7
D	84.7	84.7	84.7	91.0	51.0	2.6	6.3	5.2
E	87.7	84.4	86.0	91.9	52.6	1.8	4.1	3.8
F	86.4	86.2	86.3	92.2	52.6	1.8	4.1	3.8

As shown in Table 2, replacing the original downsampling method with the ADown structure led to improvements of 1.8, 2.0, 1.9, 1.0, and 1.2 percentage points in P, R, F1, mAP@0.5, and mAP@0.5:0.95, respectively. The capability of the ADown module to preserve critical details during downsampling and to strengthen multi-scale feature representation is evidenced by the aforementioned performance gains. While using the EfficientHead detection head, although the precision slightly decreased, the mAP@0.5 and F1-score showed certain improvements, and the recall increased significantly by 3.4 percentage points. Concurrently, the FLOPs dropped from 6.3G to 5.1G, substantially reducing the number of model parameters and the computational load, which verifies the lightweight and efficient performance of this detection head. Replacing the loss function with WIoUv3 resulted in improvements across all evaluation metrics, which demonstrates its ability to deliver more stable bounding box regression accuracy in complex scenes through dynamic sample weighting optimization.

The integration of the three proposed modules enabled the AEW-YOLO algorithm to achieve optimal performance. Key detection metrics—P, R, F1, and mAP@0.5—showed improvements of 2.0, 3.2, 2.6, and 1.4 percentage points, respectively, over the baseline YOLO11n. These gains correspond to substantial gains in model efficiency: the parameter count decreased from 2.6 million to 1.8 million, a 30.7% reduction; the computational complexity also decreased from 6.3 GFLOPs to 4.1 GFLOPs, a 34.9% reduction; and the model size shrunk from 5.2 megabytes to 3.8 megabytes, a 26.9% reduction. The ablation experiment analysis results sufficiently demonstrate that the proposed method, with its components working synergistically, can effectively enhance the detection accuracy for PWD-infected wood in UAV remote sensing images while significantly reducing model complexity. This achieves the co-optimization goal of enhanced performance and a lightweight design.

3.3 Comparative Experiments of Different Downsampling Modules

To validate the effectiveness of ADown, comparative experiments were conducted by replacing various downsampling modules on the YOLO11n. Table 3 summarizes the experimental results, where Models A, B, C, D, and E represent the models incorporating WaveletPool [7], SPDConv [8], v7DS [9], ContextGuidedDown [10], and ADown [11], respectively. The model using ADown achieved a higher precision by 1.8, 1.5, 2.6, and 3.3% compared to those using WaveletPool, SPDConv, v7DS, and ContextGuidedDown, respectively. Despite its superior recall and mAP, the substantial parameters and high computational complexity of SPDConv lead to a larger model size and longer inference latency, rendering it impractical for resource-constrained platforms. In contrast, the proposed ADown module achieves reductions of 0.5 M in parameters, 1.0 G in FLOPs, and 0.9 MB in model size. This approach ensures high detection accuracy while lowering computational overhead, making it more advantageous compared to other schemes.

Table 3. Experimental evaluation of various downsampling modules

model	P/%	R/%	F1/%	mAP@0.5/%	mAP@0.5:0.95/%	PARARMS/M	FLOPs/G	Model Size/MB
A	84.4	83.2	83.8	90.9	50.5	2.2	5.4	4.4
B	84.7	87.8	86.2	92.7	55.5	4.6	11.3	9.1
C	83.6	83.6	83.6	90.3	50.8	2.2	5.7	4.6
D	82.9	83.8	83.4	90.2	48.9	3.5	9.0	7.1
E	86.2	85.0	85.6	91.8	53.0	2.1	5.3	4.3

3.4 Comparative Experiment with Different Detection Heads

A comparative experiment with various detection heads was performed to evaluate the efficacy of the proposed EfficientHead. The corresponding results are shown in Table 4, where Models A to F denote the models integrating aux [12], LSCD [13], RSCD [14], SEAMHead [15], LSDECD [16], and EfficientHead [17], respectively.

Among the evaluated detection heads, the model equipped with EfficientHead attained the highest scores in recall, F1-score, and mean average precision. Its performance reached 82.8% in precision, 86.4% in recall, 84.6% in F1-score, 91.0% in mAP@0.5, and 51.7% in mAP@0.5:0.95. Furthermore, the EfficientHead detection head achieves a reduction of 0.3 M in parameters, cuts computational cost by 1.2 G FLOPs, and decreases the model size by 0.5 MB, delivering the best performance among all tested detection heads. Compared to the SEAMHead, the EfficientHead shows an enhancement of 1.4 and 1.1 percentage points in recall and mAP@0.5:0.95, respectively, alongside a reduction of 0.7 GFLOPs in computational cost and 0.4 MB in model size. These results highlight its superior comprehensive performance, confirming that EfficientHead has achieved an ideal optimization balance between detection accuracy and computational efficiency, thereby fully demonstrating the superiority of its design.

Table 4. Experimental evaluation of various detection heads.

model	P/%	R/%	F1/%	mAP@0.5/%	mAP@0.5:0.95/%	PARARMS/M	FLOPs/G	Model Size/MB
A	84.5	81.8	83.1	90.1	51.2	2.6	6.3	6.1
B	83.2	84.2	83.7	90.3	50.6	2.4	5.6	4.9

(continued)

Table 4. (*continued*)

model	P/%	R/%	F1/%	mAP@0.5/%	mAP@0.5:0.95/%	PARARMS/M	FLOPs/G	Model Size/MB
C	83.4	84.2	83.8	90.7	51.2	2.4	5.6	6.4
D	81.9	85.0	83.4	90.4	50.6	2.5	5.8	5.1
E	84.2	84.2	84.2	90.2	50.7	2.3	6.0	5.0
F	82.8	86.4	84.6	91.0	51.7	2.3	5.1	4.7

3.5 Comparison of Model Performance

Table 5. Experimental evaluation of various models.

model	P/%	R/%	F1/%	mAP@0.5/%	mAP@0.5:0.95/%	PARARMS/M	FLOPs/G	Model Size/MB
A	81.4	83.2	82.3	89.7	50.7	2.5	7.1	5.0
B	84.0	83.8	83.9	90.1	51.9	3.0	8.1	6.0
C	84.9	84.3	84.6	91.0	52.0	2.0	7.6	4.4
D	83.2	87.2	85.2	91.6	52.6	2.3	6.5	5.5
E	84.4	83.0	83.7	90.8	51.8	2.6	6.3	5.2
F	82.2	81.9	82.1	89.1	48.7	2.5	5.8	5.2
G	84.1	79.1	81.5	89.1	49.3	2.5	6.1	5.2
Ours	86.4	86.2	86.3	92.2	52.6	1.8	4.1	3.8

A comparative experiment was performed between the proposed AEW-YOLO and several advanced YOLO-series models of comparable scale to comprehensively evaluate its effectiveness and superiority. As shown in Table 5, the models selected for comparison include YOLOv5n (A), YOLOv8n (B), YOLOv9t (C) [18], YOLOv10n (D) [19], YOLO11n (E), YOLOv12n (F) [20], and YOLOv13n (G) [21]. The experimental results indicate that our AEW-YOLO algorithm demonstrates superior performance compared to all other evaluated models. In terms of detection accuracy, AEW-YOLO shows improvements in precision of 5.0, 2.4, 1.5, 3.2, 2.0, 4.4, and 2.3 percentage points over models A to G, respectively. Correspondingly, its F1-score is increased by 4.0, 2.4, 1.7, 1.1, 2.6, 4.2, and 4.8 percentage points. The model also exhibits superior performance in recall and mean average precision. Furthermore, AEW-YOLO demonstrates a clear advantage in model efficiency, achieving the lowest parameter count, computational cost (FLOPs), and model size among the compared models. Specifically, its computational cost is 3.5 G lower than that of the lightweight YOLOv9t. These results indicate that AEW-YOLO significantly enhances detection performance while reducing model complexity, exhibiting the best overall performance among YOLO object detection algorithms.

3.6 Visualization of Model Detection

To evaluate the detection performance of AEW-YOLO, images were randomly selected for inference and compared with the baseline YOLO11n model. As shown in Fig. 5, due to insufficient suppression of background noise, YOLO11n missed 2, 6, and 3 discolored infected trees in Fig. 5 (a), (b), and (c), respectively, demonstrating poor performance under occluded lighting conditions, with most missed detections being small-scale objects. In contrast, AEW-YOLO, through its efficient multi-scale feature extraction strategy, enhanced the perception capability for small objects, missing only one instance in Fig. 5 (a). This significantly reduced the missed detection rate, indicating that AEW-YOLO possesses superior recognition ability in complex scenarios.

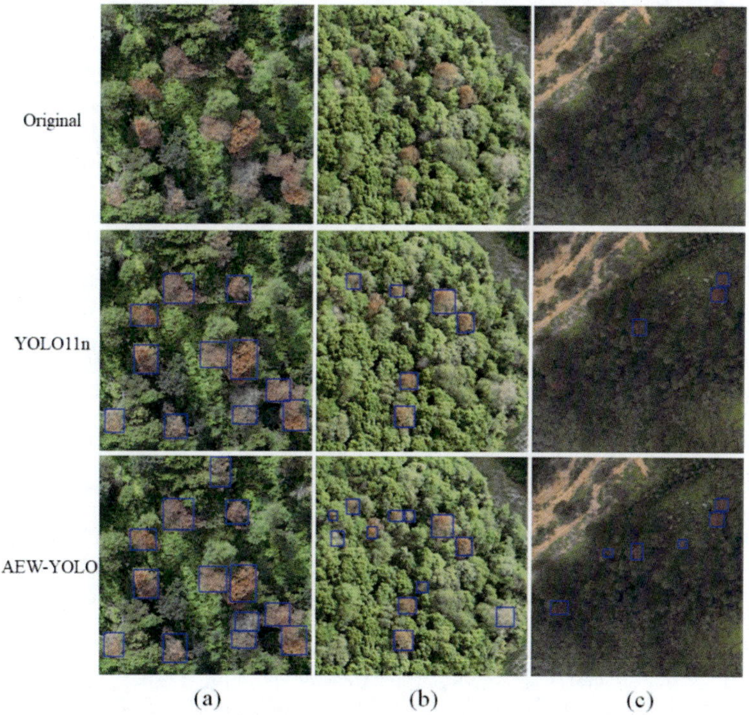

Fig. 5. Visualizations of experimental results.

4 Conclusion

To address the limitations of existing models characterized by high computational demands and suboptimal detection performance, this study presents AEW-YOLO, a lightweight object detection algorithm for discolored wood infected by pine wilt disease, based on an improved YOLO11n. First, an efficient and lightweight downsampling module, ADown, is introduced to mitigate detail loss caused by traditional convolution downsampling, which ultimately boosts detection accuracy by strengthening its capacity to capture fine-grained details. Second, a lightweight decoupled detection head, Efficient-Head, is designed to maintain detection accuracy while improving computational efficiency. Finally, the WIoUv3 loss function is used for superior bounding box regression and localization stability. Experimental results demonstrate that, compared to the baseline YOLO11, the AEW-YOLO algorithm achieves improvements of 2.0% in precision, 3.2% in recall, 2.6% in F1-score, 1.4% in mAP@0.5, and 0.8% in mAP@0.5. Meanwhile, the proposed enhancements significantly reduce the model's complexity, cutting the parameters from 2.6 to 1.8 M, lowering the computational cost from 6.3 to 4.1 G, and shrinking the model size from 5.2 to 3.8 MB. This comprehensive enhancement in all performance metrics substantiates the validity of our method.

In future work, deployment onto edge computing platforms constitutes the next phase, targeting a practical and efficient system for real-time UAV monitoring of discolored wood caused by pine wilt disease.

References

1. Liu, J., Lv, F., Liu, B.: Identification method of sunflower leaf disease based on SIFT point. J. Image Graph. **7**(2), 64–67 (2019). https://doi.org/10.18178/joig.7.2.64-67
2. Saminathan, K., Sowmiya, B., Chithra Devi, M.: Multiclass classification of Paddy leaf diseases using random Forest classifier. J. Image Graph. **11**(2), 195–203 (2023)
3. Dong, X., Zhang, L., Xu, C., Miao, Q., Yao, J., Liu, F., et al.: Detection of pine wilt disease infected pine trees using YOLOv5 optimized by attention mechanisms and loss functions. Ecol. Indic. **168**, 112764 (2024)
4. Ye, X., Pan, J., Shao, F., Liu, G., Lin, J., Xu, D., et al.: Exploring the potential of visual tracking and counting for trees infected with pine wilt disease based on improved YOLOv5 and StrongSORT algorithm. Comput. Electron. Agric. **218**, 108671 (2024)
5. Hu, D., Wu, H., Ye, Z.: Improved YOLOv7 based aerial images of pine nematode epidemic wood recognition. J. For. Eng. **10**(2), 147–155 (2025) in Chinese
6. Tong, Z., Chen, Y., Xu, Z., Yu, R.: Wise-IoU: bounding box regression loss with dynamic focusing mechanism. arXiv preprint https://arxiv.org/abs/2301.10051. (2023)
7. Williams, T., Li, R.Y.: Wavelet pooling for convolutional neural networks. In: Proceedings of the International Conference on Learning Representations, pp. 1–10. Canada, Vancouver (2018)
8. Sunkara, R., Luo, T.: No more strided convolutions or pooling: a new CNN building block for low-resolution images and small objects. In: Proc. Joint Eur. Conf. Mach. Learn. Knowl. Discov. Databases (ECML PKDD), pp. 443–459. Springer, Cham, Switzerland (2022)
9. Wang, C.-Y., Bochkovskiy, A., Liao, H.-Y.M.: YOLOv7: trainable bag-of-freebies sets new state-of-the-art for real-time object detectors. In: Proceedings of the IEEE Conference on Computer Vision and Pattern Recognition (CVPR), pp. 7464–7475. CVPR (2023)
10. Wu, T., Tang, S., Zhang, R., Cao, J., Zhang, Y.: CGNet: a light-weight context guided network for semantic segmentation. IEEE Trans. Image Process. **30**, 1169–1179 (2018)
11. Teng, M., Wan, G., Bai, S., Zhu, Y., Li, H., Wang, C.: An efficient and lightweight rotating target detection model for industrial scenarios. Digit. Signal Process. **168**, 105631 (2025)
12. Jin, G., Taniguchi, R.-I., Qu, F.: Auxiliary detection head for one-stage object detection. IEEE Access. **8**, 85740–85749 (2020)
13. Khaki, S., Safaei, N., Pham, H., Wang, L.: WheatNet: a lightweight convolutional neural network for high-throughput image-based wheat head detection and counting. Neurocomputing. **489**, 78–89 (2022)

14. Han, T., Yu, S., Ma, L., Huang, Y., Hou, S., Pang, J.: Research on detection model of foreign objects and defects in photovoltaic panels based on improved YOLOv11n. Comput. Eng. Appl. **61**(14), 123–134 (2025) (in Chinese)
15. Dai, X., Chen, Y., Xiao, B., Chen, D., Liu, M., Yuan, L., et al.: Dynamic head: unifying object detection heads with attentions. In: Proceedings of the IEEE Conference on Computer Vision and Pattern Recognition (CVPR), pp. 7373–7382. CVPR (2021)
16. Wang, X., Wan, S., Li, Z., Chen, X., Zhang, B., Wang, Y.: ECL-tear: lightweight detection method for multiple types of belt tears. Measurement. **251**, 117269 (2025)
17. Jiang, T., Li, L., Zhang, Z., Yu, X., Zhu, Y., Li, L., et al.: YOLO-light-pruned: a lightweight model for monitoring maize seedling count and leaf age using near-ground and UAV RGB images. Artif. Intell. Agric. **16**(1), 164–186 (2025)
18. Wang, C.-Y., Yeh, I.-H., Liao, H.-Y.M.: YOLOv9: learning what you want to learn using programmable gradient information. In: Proceedings of the the European Conference on Computer Vision (ECCV), pp. 1–21. Springer, Cham, Switzerland (2024)
19. Wang, A., Chen, H., Liu, L., Chen, K., Lin, Z., Han, J., et al.: YOLOv10: real-time end-to-end object detection. Adv. Neural Inf. Process. Syst. **37**, 107984–108011 (2024)
20. Tian, Y., Ye, Q., Doermann, D.: YOLOv12: attention-centric real-time object detectors. arXiv preprint https://arxiv.org/abs/2502.12524. (2025)
21. Lei, M., Li, S., Wu, Y., Hu, H., Zhou, Y., Zheng, X., et al.: YOLOv13: real-time object detection with hypergraph-enhanced adaptive visual perception. arXiv preprint https://arxiv.org/abs/2506.17733. (2025)

Enhanced Urban Bird Detection: Optimized Deep Learning Architecture on a Comprehensive Shanghai Bird Dataset

Yi Gong[1], Yong Wang[2(✉)], and HaiMing Li[2]

[1] Shanghai University of Electric Power, Shanghai, China
[2] No. 1851 Hucheng Ring Road, 201306 Shanghai, Shanghai, China
wangyong@shiep.edu.cn

Abstract. Accurate and efficient bird detection is essential for ecological conservation, environmental monitoring, and urban biodiversity management. Traditional methods, such as mark recapture, are labor intensive and lack precision, while existing computational approaches and datasets often fall short in species diversity and regional applicability. Addressing these limitations, we present a comprehensive dataset comprising 18,471 images of 483 bird species from Shanghai, enhanced through data augmentation techniques to 48,881 images. This dataset includes 10% of Shanghai bird images detected and captured by webcams and the remaining 90% collected from the Internet. To take advantage of this dataset, we propose an optimized deep learning-based bird detection model built upon YOLOv11. Our enhancements include the integration of the Efficient Vision Transformer (EfficientViT) as the backbone to capture global features, the introduction of the StarBlock module to reduce computational redundancy, and the development of a Lightweight Residual Shared Detector (LSRD) for the detection head to balance accuracy and efficiency.Our improved model achieved mAP@0.5 of 0.742 and mAP@0.5:0.95 of 0.634, representing approximately a 4.2% improvement over the baseline model. Additionally, the computational overhead increased modestly from 7.3 GFLOPS to 7.7 GFLOPS, which is relatively insignificant compared to other high-precision models. This study provides a valuable resource for urban ecological research and presents a practical solution for real-time bird monitoring, facilitating better management and conservation strategies in metropolitan settings.

Keywords: Urban ecology · Deep learning · Real-time bird detection

1 Introduction

Birds are integral components of ecosystems, and their population dynamics and distribution directly reflect environmental health. Accurate and efficient bird detection is crucial for ecological conservation, environmental monitoring,

L. Huang and D. Greenhalgh (Eds.): ICMLC 2026, LNNS 2061, pp. 508–522, 2027.
https://doi.org/10.1007/978-3-032-31236-5_35

and scientific research. For example, monitoring birds in protected areas allows scientists to study their behavior and population trends, while identifying and counting bird species can provide insight into ecological changes. Furthermore, bird detection can be applied to urban ecological management, assisting policy makers in optimizing green space layouts and enhancing urban biodiversity. This study focuses on urban bird detection, encompassing both cities and their ecological reserves.

Over the years, various methods have been developed to improve the efficiency and precision of bird detection [1–4]. Traditional approaches, such as the mark recapture method , involve capturing birds, tagging them, and then recapturing them to study population dynamics and migration. However, these methods are labor intensive, time-consuming, and often lack high accuracy.

Modern bird detection is increasingly facilitated by computational methods. For instance, acoustic recognition [5] employs recording devices to capture bird calls, which are then analyzed either manually or via software to identify species. Bird object detection, another emerging approach, involves capturing video footage or live streams through cameras or trap cameras, which are then analyzed frame by frame to extract bird-related data. Recent studies, such as that by Qi Song et al. [6], have used video data from trap cameras to compile a dataset of 9,717 images representing 15 bird species in complex forest environments. However, the limited species diversity of this dataset and its lack of a broad foundation for developing advanced recognition algorithms present challenges. Similarly, the CUB-200-2011 dataset [7], a widely used bird classification and detection dataset containing 200 bird species and 11,788 images, is substantial in size but lacks a regional focus, making it less applicable for practical studies.

Addressing these gaps, we propose a dataset of 483 bird species from Shanghai, comprising 18,471 images that were augmented to 48,881 images through techniques such as rotation, flipping, and cropping. This dataset makes significant contributions to bird recognition in specific regions and may facilitate interdisciplinary studies in urban ecology and ecological informatics. By providing a solid foundation for urban ecological management and species conservation strategies, this dataset enables researchers to develop bird detection algorithms with greater precision and relevance.

Fig. 1. Video image(We capture birds by online camera upon birds detection system with yolo.

Shanghai, a megacity with a mix of urbanized and natural environments (e.g., wetlands, nature reserves, urban parks), presents a unique habitat for diverse bird populations. Our dataset focuses on birds inhabiting Shanghai, providing researchers with valuable resources to study the ecology of the city and the dynamics of the bird population. Images were collected from wetland and forest parks through online bird observation platforms, while surveillance cameras captured footage of common urban birds such as night herons and doves. These surveillance images offer unique insights, such as activity patterns over specific time periods or elusive behaviors such as foraging and social interactions. Such real-time and continuous data are invaluable for monitoring ecological changes, such as seasonal bird population shifts, migration patterns, and habitat preferences.A still image from the surveillance video is shown in Figure 1

This dataset emphasizes urban bird detection in contexts such as green spaces, urban waterways, and wetland parks. Unlike larger animals, urban birds are smaller and have adapted camouflage mechanisms to blend into their surroundings, making them challenging to detect amidst the city's stray cat population .

Many existing bird recognition studies prioritize accurate classification but neglect real-time detection. For example, Liu et al. [8] proposed a bird classification model based on Inception V3, achieving a 90% accuracy rate on the validation set but with significant computational costs. Ahmed et al. applied YOLOv4 for bird detection in natural habitats, using its single-stage detection capabilities to handle complex outdoor environments [9]. Despite modifications to YOLOv4, such as adding layers for small-object detection, its deployment on edge devices remains challenging due to computational overhead and an imbalance between precision and efficiency.

Deep learning-based bird detection still falls short of satisfactory performance, particularly in practical applications requiring real-time deployment on edge devices. Balancing precision and computational efficiency is critical for real-world urban bird detection tasks. Thus, we selected YOLOv11 as our baseline model, known for its balance between accuracy and computational cost, and introduced modifications to further enhance its performance.

The contributions of this study are threefold. First, we introduce a novel method for urban bird monitoring leveraging advanced target detection techniques. This approach enhances the efficiency of urban ecological conservation efforts while offering high portability, thereby serving as a valuable framework for the protection of other wildlife species. Second, we refine the target detection component of this study by adopting the YOLOv11 algorithm as our baseline model. To improve performance, we modify its backbone, incorporate the C3k2 module, and optimize the detection head, achieving greater accuracy with a relatively modest computational cost. Third, we develop a manually annotated dataset for bird species detection, comprising 48,881 images representing 483 bird species found in Shanghai. This dataset captures a diverse range of scenarios, including activities of solitary birds and dense flocks, amidst natural interferences such as vegetation shadows, non-avian animals, water bodies, and litter. Sourced from real-world urban environments–such as Shanghai's city streets, wetlands, and parks–this dataset enhances the robustness and practical applicability of trained models. Furthermore, it provides a reliable reference for future studies on bird species detection.

2 Material

Dataset Introduction. Our dataset is a high-quality dataset collected for Shanghai's birds, designed to help researchers better monitor urban birds and provide a solid foundation for the development of urban ecological science. The dataset consists of 90% HD camera images and 10% webcam images, which is more practical than other datasets in real-world scenarios and can be used to train high-performance models.Through the steps of data collection, annotation, and processing, we constructed our bird dataset, which contains 48,881 images of 483 common bird species. The images are split into an 8:2 ratio for the training and validation sets (Table 1).

2.1 Data Collection.

In this study, we utilized surveillance dome cameras to observe and record videos of birds in urban rivers and green spaces within the Xuhui, Yangpu, and Baoshan districts of Shanghai. The cameras operated at a resolution of 2560 × 1440 and a frame rate of 25 fps. These videos were manually categorized to create a high-quality dataset for further research. From each video, we extracted 10 frames and labeled them for experimental purposes. Additionally, we collected and labeled images of birds inhabiting Shanghai that fit the study's context, specifically those in urban green backgrounds or complex environments, sourced from the Internet to augment the dataset.

2.2 Data Annotation.

To facilitate rapid annotation of the images, we performed preliminary bird recognition in real time using a ball camera and saved the recognition results. Subsequently, we conducted more accurate annotations using LabelImg, manually removing incorrect identifications and erroneous bounding boxes.

2.3 Data Processing

Due to the limited number of bird images captured in Shanghai, we employed several data augmentation techniques, including rotation, random cropping, and brightness adjustment, to expand the dataset from 18,481 original images to 48,881 samples. This approach aimed to increase the diversity of the training data and enhance the model's robustness in real-world scenarios, ensuring the integrity of the Shanghai Birds dataset.

Table 1. The number of images for training and validation

Category	Number
Training	45,621
Validation	3,260
Total	48,881

The dataset is characterized by its high quality and focus on urban ecosystems, with city-specific environments influencing the behavior and appearance of birds. The large number of individuals and species included in the dataset enables comprehensive analyses of urban bird populations, such as estimating population densities and assessing species changes over time.

Our dataset serves as a benchmark for evaluating the performance of target detection algorithms in complex urban environments, providing researchers with a valuable resource for in-depth studies of bird behavior, population dynamics, and environmental interactions.

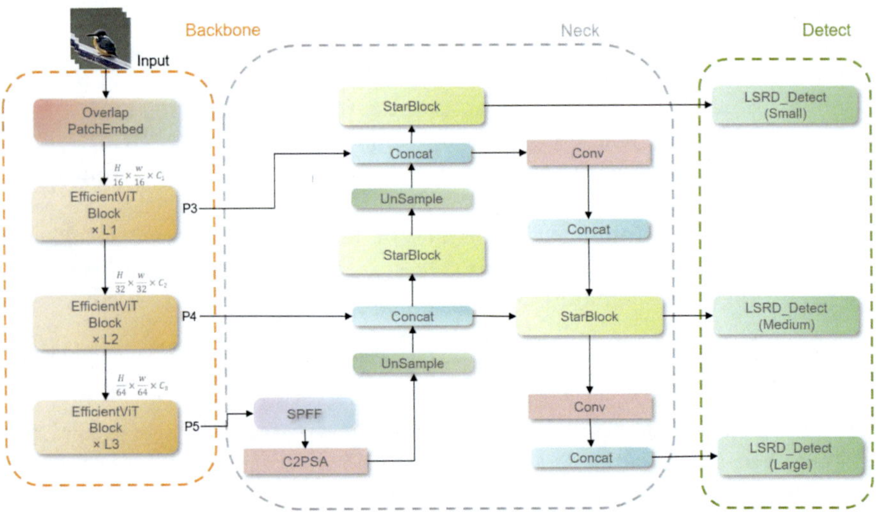

Fig. 2. ESS-Yolo network architecture diagram.

3 Methods

Target detection is currently a fundamental task in the field of artificial intelligence, playing a crucial role in various downstream applications, including ecology, which is the focus of this paper. The integration of target detection and ecology enhances researchers' ability to study ecological patterns and developments, thereby contributing to better environmental protection. For example, detecting species in videos can improve the determination of population distribution and size, as well as facilitate the study of species' behaviors. Additionally, detecting humans can aid in preventing wildlife poaching, among other applications.

In recent years, target detection methods can be broadly categorized into two state-of-the-art approaches: YOLO (You Only Look Once) as a one-stage detector and Faster R-CNN (Region-Based Convolutional Neural Networks)[10] as a two-stage detector. These are the two most commonly used methods in target detection. Two-stage detectors offer high localization and recognition accuracy by dividing the detection process into two distinct stages, typically involving a region of interest (RoI) pooling layer. For instance, in Faster R-CNN, the first stage, known as the Region Proposal Network (RPN), generates candidate object bounding boxes. In the second stage, features are extracted from each candidate box using the RoIPool operation for subsequent classification and bounding box regression tasks . In contrast, single-stage detectors like YOLO directly predict bounding boxes from the input image without a region proposal step, making them more time-efficient and suitable for real-time applications.

Given that bird detection often requires the deployment of multiple cameras within a single area to achieve more accurate results, we opted for a one-stage detector due to its higher time efficiency and fewer reference requirements, making it ideal for camera-based detection systems. We propose an enhanced detector tailored to our dataset to assist researchers in studying birds within urban environments more effectively.

We selected YOLOv11 as our baseline model because it is lighter than its predecessors while maintaining high accuracy. Although YOLOv11 is one of the most renowned target detection models, excelling in both accuracy and inference speed, its performance on our bird dataset–which includes 483 species of varying sizes, complex backgrounds, and small targets–is relatively suboptimal.

To address these challenges, we propose a novel algorithm that adapts YOLOv11 to better suit our dataset without significantly compromising real-time performance or increasing computational costs. The structure of the proposed algorithm is depicted in Fig 2.

3.1 Backbone

The traditional YOLOv11 model employs a Convolutional Neural Network (CNN) as its backbone for extracting features from input images. While CNNs excel at capturing local features, they exhibit limitations in capturing global information and processing fine-grained details [11]. This limitation is particularly pronounced when dealing with a dataset containing 483 bird categories, where the CNN backbone of YOLOv11 struggles to distinguish subtle differences, resulting in constrained classification accuracy.

To address this limitation, we introduce the Efficient Vision Transformer (EfficientViT) as the new backbone network. EfficientViT enhances feature extraction by dividing the image into multiple fixed-size patches, linearly embedding these patches, and inputting them into the Transformer encoder for global feature capture and learning [12,13]. Specifically, EfficientViT leverages the self-attention mechanism to learn features at multiple scales and globally consider each image patch, providing a significant advantage in integrating multi-scale feature information.The self-attention mechanism enables EfficientViT to focus on detailed information within an image, recognizing and distinguishing minor differences by learning the global context, which is crucial for fine-grained classification tasks where category distinctions are often reflected in subtle features.The structure of the proposed algorithm is depicted in Fig. 3.

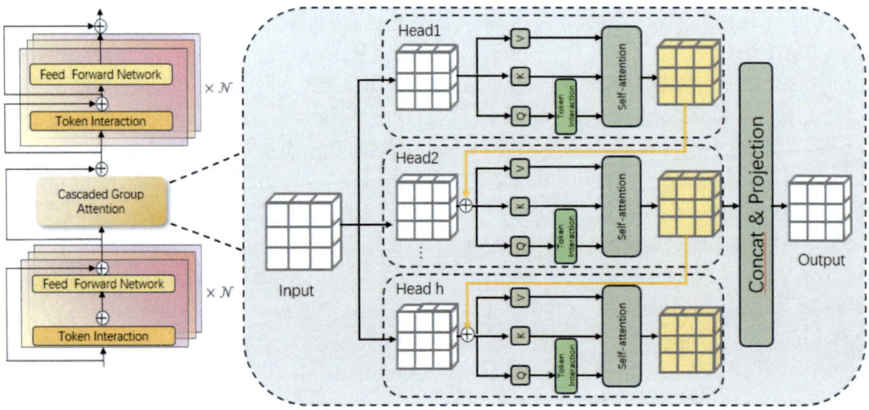

Fig. 3. EfficientViT block

By incorporating EfficientViT as the backbone network, we significantly enhance YOLOv11's ability to extract and classify fine-grained features. Additionally, to meet real-time processing requirements, EfficientViT is optimally designed to improve feature extraction capabilities while maintaining low computational costs and high processing speeds [13].

3.2 C3k2-Star Block

The C3k2 module in YOLOv11 introduces substantial computational redundancy during network stacking, leading to increased computational costs and a higher number of model parameters [10]. This not only hampers the real-time performance of the model but also complicates deployment on resource-constrained devices. To mitigate this issue, we propose the StarBlock module, which replaces the original C3k2 module, illustrated in Fig. 4.

StarBlock is inspired by efficient feature fusion methods and aims to reduce computational redundancy through more streamlined feature processing while

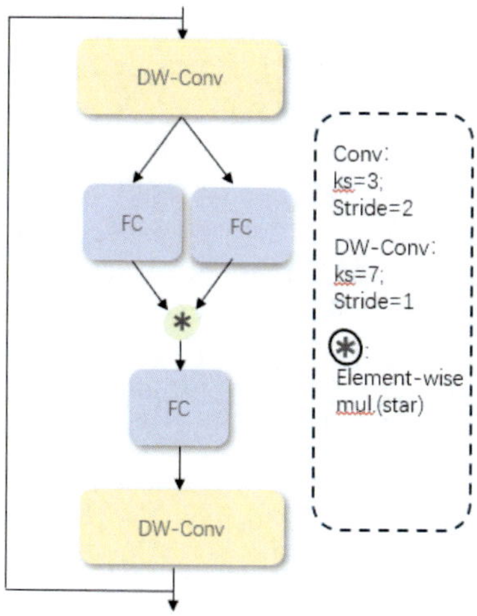

Fig. 4. C3k2-Star module

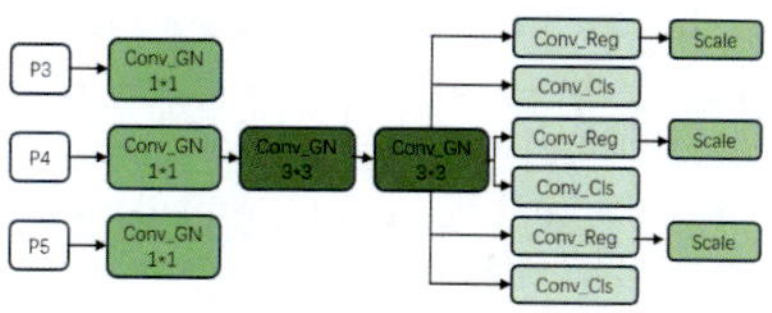

Fig. 5. Lightweight residual shared detector (LSRD)

enhancing feature representation. Specifically, StarBlock begins with a convolution operation on the input data to extract underlying features. The data then passes through two fully connected operations, one of which is activated by a ReLU function. These outputs are subsequently combined using element-wise multiplication. Finally, the results are integrated with the initial input data, followed by another convolution step to produce the final output [14]. The star operation effectively fuses different subspace features by multiplying them element-wise, offering a promising alternative to traditional linear projection methods similar to the kernel trick in machine learning by mapping the input to a high-dimensional nonlinear feature space. This module reduces the computational redundancy inherent in the original C3k2 stacking, thereby decreasing the model size and enhancing accuracy [11].

3.3 Lightweight Residual Shared Detector (LSRD)

In target detection models, the design of the detection head significantly impacts overall performance and computational efficiency. Traditional detection heads typically require more parameters and computational resources when handling multi-scale targets, making them unsuitable for deployment on resource-constrained devices [15]. To further reduce the number of parameters and computational demands of the detection head while maintaining high detection accuracy, we design the Lightweight Residual Shared Detector (LSRD), illustrated in Fig. 5.

A key feature of LSRD is the use of shared convolutional layers. By sharing convolutional layers among different detection heads, LSRD significantly reduces the number of parameters [10]. This shared convolution design allows multiple detection heads to reuse the same convolution kernel, thereby decreasing the overall parameter count and computational requirements of the model. Consequently, the model becomes lighter and consumes less memory, making it suitable for deployment on resource-constrained devices.

The introduction of the Scale layer addresses the issue of inconsistent target scales detected by different detection heads. Traditional detection heads may suffer from decreased detection accuracy when handling targets of varying scales due to mismatched feature scales. By incorporating the Scale layer, LSRD can dynamically scale the features, enabling different detection heads to adaptively handle targets of different sizes [15]. This design ensures that the model maintains high detection accuracy and consistency when dealing with multi-scale targets.

Additionally, the design of the lightweight detection head includes the optimization of convolution operations. By reducing the depth and width of convolutional layers and adopting an efficient convolutional kernel design, LSRD further minimizes computational costs. These optimized convolution operations lower the computational overhead and the number of model parameters while preserving efficient feature extraction capabilities [11].

The feature transfer capability of LSRD is further enhanced through the introduction of residual connections within the detection head. Residual connections enable LSRD to better retain and transfer feature information, thereby improving the overall detection performance of the model [11]. This design ensures that the model maintains high stability and consistency when processing complex features.

When LSRD is integrated with the StarBlock module, their respective advantages complement each other. StarBlock enhances feature richness and differentiation through efficient feature fusion, while LSRD optimizes feature processing efficiency through shared convolution and Scale layers. [13]This combination enhances both feature expression and fusion capabilities while maintaining efficient feature processing. Additionally, StarBlock's residual linking helps retain original feature information, compensating for any potential information loss caused by the lightweight design of LSRD. This synergy ensures that the model remains lightweight while preserving sufficient feature information for accurate detection .[16]

4 Experiment

4.1 Experiment Platform

Experimental platform information: Windows 11, NVIDIA GeForce RTX 4090 with 24GB, Python 3.8, PyTorch 1.11.0, CUDA 11.3. Parameter settings: 350 epoch, batch size of 32, image size of 640 pixels, SGD optimizer

4.2 Evaluation Metrics

The evaluation metrics used in this study include mean Average Precision (mAP), Average Precision (AP), Precision (P), and Recall (R). The formulas for

Precision and Recall are presented in Equations (1) and (2), respectively. Here, TP represents the number of correctly predicted bounding boxes, FP denotes the number of incorrectly identified positive samples, and FN signifies the number of undetected targets.

$$\text{Precision} = \frac{\text{TP}}{\text{TP} + \text{FP}} \tag{1}$$

$$\text{Recall} = \frac{\text{TP}}{\text{TP} + \text{FN}} \tag{2}$$

Average Precision (AP) is defined as the average accuracy of the model across different recall levels. Mean Average Precision (mAP) is the mean of the AP values across all categories. The formulas for AP and mAP are shown in Equations (3) and (4).

$$AP = \frac{dP(t)}{dt}, \quad \text{where} P(t) \in [0, 1] \tag{3}$$

$$\text{mAP} = \frac{\sum_{S,j=1} AP(j)}{S} \tag{4}$$

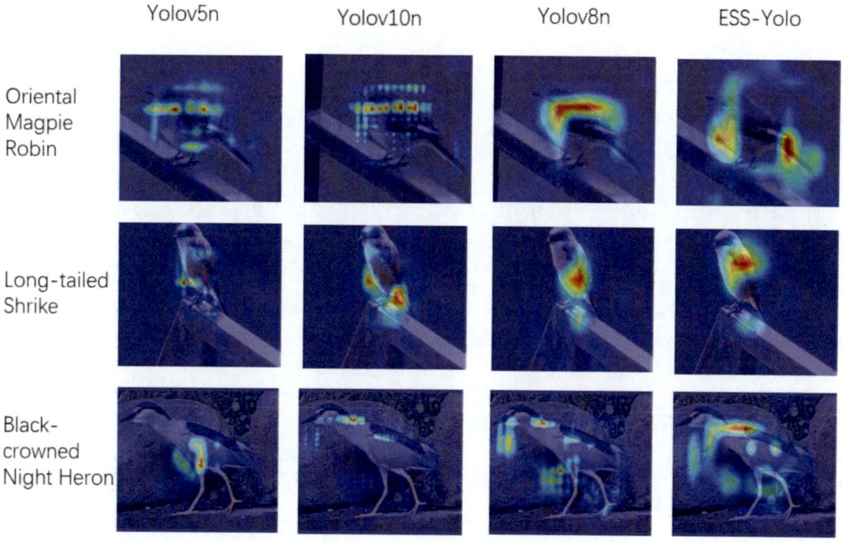

Fig. 6. Heatmap with Yolov5,Yolov8,YoloV11 and our method.

4.3 Experimental Result Analysis

As shown in Table 2 and visually represented in Fig. 6, our improved algorithm for bird target detection within urban and surrounding ecosystems demonstrates

Table 2. Comparison of different baseline :Yolov5,Yolov8,Yolov10,RT-DETR,Faster-RCNN in terms of AP50, AP50-95, Para(M), and GFLOP

Method	AP50	AP50-95	Para(M)	GFLOP
YoloV11n	0.712	0.608	2,763,637	7.3
YoloV5n	0.546	0.433	**2,100,280**	**5.3**
YoloV8n	0.728	0.621	3,462,173	10.3
YoloV10n	0.733	0.618	3,070,642	10.4
RT-DETR-l	**0.756**	0.633	33,798,641	110.1
YOLO-PRO	0.734	0.613	3,058,541	10.1
RF-DETR	0.746	0.629	17,278,871	30.5
ESS-Yolo	0.742	**0.634**	3,564,118	7.7

improvements at all stages, achieving a commendable balance between real-time performance and high accuracy. The effectiveness of the three proposed improvements is evident:

a) **Backbone Network Enhancement:** Enhancing the backbone network significantly increases the number of layers, allowing for better integration of multi-scale feature information. This improvement facilitates more effective fine-grained feature extraction.

b) **Neck Improvement with C3k2:** Modifying the neck by incorporating C3k2 reduces the computational redundancy associated with the original C3K2 stacking. This modification not only decreases the size of the model, but also enhances accuracy.

c) **Detection Head Optimization:** Enhancing the detection head complements the C3K2-star module. Using shared convolution, the number of parameters is drastically reduced, resulting in a lighter model. Additionally, the star module projects input into a high-dimensional non-linear feature space, ensuring that accuracy is maintained despite the reduction in parameters.

4.4 Comparison Experiment

To demonstrate the performance advantages of our algorithm, we conducted comparison experiments using YOLOv5, YOLOv8, YOLOv10, and DETR in our Bird dataset. All models were tested within a unified experimental environment to ensure the fairness and comparability of the results. The specific experimental outcomes are presented in Table 2.

4.5 Ablation Experiment

In this paper, we enhance the YOLOv11 network by replacing its backbone with the efficient Vision Transformer (ViT) module, incorporating C3K2-STAR into the neck to substitute the original C3K2 module, and adopting a shared

Table 3. Comparison of different methods and modules in terms of AP50, AP50-95, Para(M), and GFLOP

Method	Module			Result			
	C3k2-Star	LSRD	EfficientViT	AP50	AP50-95	Para(M)	GFLOP
YoloV11				0.712	0.608	2,763,637	7.3
ESS-Yolo	✓			0.728	0.632	2,656,751	6.9
ESS-Yolo	✓	✓		0.723	0.618	2,408,414	6.1
ESS-Yolo	✓		✓	0.738	0.630	3,555,128	7.7
ESS-Yolo		✓	✓	0.732	0.622	3,396,853	7.2
ESS-Yolo	✓	✓	✓	**0.742**	**0.634**	3,564,118	7.7

convolutional detection head. These modifications increase the accuracy of the model while maintaining a balance between computational overhead, model size, and performance. To validate the effectiveness of our approach, we conducted ablation experiments. The results demonstrate that our method improves the performance of the original YOLOv11 network. Table 3 illustrates the changes in mAP@0.5, mAP@0.5-0.95, the number of parameters, and GFLOPS for each bird detection network during training.

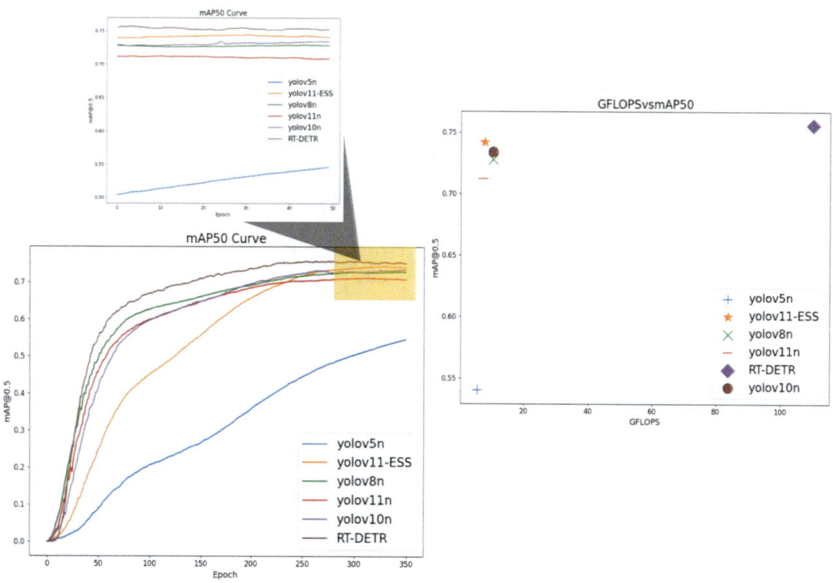

Fig. 7. Comparison of different species :Yolov5,Yolov8,Yolov10,RT-DETR

4.6 Benchmarking Results for Different Species

To comprehensively evaluate the performance of our proposed target detection method, we also performed benchmark experiments on 8 different bird species, as detailed in Table 4 and Fig. 7.The test set of the experiments consists mainly of camera-recorded bird images, similar to Fig. 1, but with larger backgrounds and smaller targets. In most models, the egrets with distinctive features and larger body sizes had higher recognition accuracies, while the Long-tailed Shrike, a smaller bird with distinctive features, had relatively higher accuracies due to the fact that we zoomed in on certain smaller birds for better auto-recognition during the collection of camera-recorded birds. It can also be seen that our method shows strong strengths in identifying the species spotted dove, which has a relatively complex background in our dataset, mainly grassy or stony backgrounds similar to its fur color. And ESS-YOLO with efficient feature extraction can extract its features well. Although other models were able to obtain relatively good accuracy for some species, our model was generally better suited to this dataset, with better average accuracy.

Table 4. Different Species' Comparison of different baselines: Yolov5, Yolov8, Yolov10, RT-DETR in terms of mAP50-95

Method	ESS-Yolo	Yolov11	Yolov8	RT-DETR	Yolov5
Black-crowned night heron	0.672	0.737	0.72	0.696	0.683
Spotted dove	0.678	0.391	0.457	0.361	0.4
Oriental magpie robin	0.554	0.516	0.579	0.461	0.518
Long-tailed shrike	0.754	0.777	0.748	0.68	0.727
Little egret	0.678	0.716	0.705	0.684	0.666
Common sandpiper	0.772	0.833	0.802	0.732	0.803
Common kingfisher	0.497	0.557	0.497	0.584	0.548
Oriental magpie	0.484	0.584	0.246	0.206	0.278

4.7 Evaluation on Edge Platforms

To evaluate the practical applicability of our model, we deployed it on an RK3588 development board and conducted simulation experiments. The model processes a 30-frame surveillance video stream to detect bird-related information. The detected data are subsequently reported to a database via SQL, while the video stream, annotated with detection bounding boxes, is streamed in real-time using the RTMP protocol. System performance and operational fluency were assessed during the experiments. Results indicate that the CPU remained under moderate load, the program executed smoothly, detection bounding boxes were accurately rendered, and the output video stream maintained a consistent frame rate of 29 FPS.

5 Conclusion

This study presents an image dataset comprising 48,881 images of 483 bird species from Shanghai, collected from the Internet and webcams. Using this dataset, we propose an improved deep learning method for bird target detection and evaluate its performance. The results indicate that our method achieves highly efficient bird detection with minimal computational overhead, making it suitable for deployment on edge devices for real-time applications. We established a collaborative partnership with urban street management authorities to deploy our methodology in an operational bird detection system. Due to their requirements for interactivity, we opted against using development boards and instead implemented the system on Windows platforms. The enhanced model trained on our dataset has been successfully operationalized in practical applications, making significant contributions to the demonstration of urban ecological conservation achievements and advancing research in urban ecosystem protection. This real-world implementation demonstrates the practical utility of our approach beyond theoretical contributions. This advancement helps researchers better protect the ecology of cities and their surrounding protected areas. Furthermore, our study provides valuable information on the application of deep learning techniques to improve bird monitoring and environmental protection in urban settings.

References

1. Chabot, D., Francis, C.M.: Computer-automated bird detection and identification in digital imagery: A review. J. Field Ornithol. **87**, 333–346 (2016). https://doi.org/10.1111/jofo.12173
2. Mashuk, S., et al.: Advancements in automated bird detection systems. J. Wildl. Monitor. **15**, 112–125 (2021). https://doi.org/10.1007/s10211-021-0456-3
3. Gauthreaux, S.A., Belser, C.G.: Advances in radar and acoustic tracking of birds. Bioscience **56**, 405–413 (2006). https://doi.org/10.1641/0006-3568(2006)56[405:airato]2.0.co;2
4. Hong, S.J., Han, Y., Kim, S.Y., Lee, A.Y., Kim, G.: Application of deep-learning methods to bird detection using unmanned aerial vehicle imagery. Sensors **19**, 3146 (2019). https://doi.org/10.3390/s19143146
5. Farago, T., Nagy, G., Miklosi, A.: Acoustic monitoring of bird vocalizations in natural habitats. Acta Zool. Acad. Sci. Hung **61**, 113–126 (2015). https://doi.org/10.17109/AZH.61.2.113.2015
6. Song, Q., Wang, J., Zhang, L.: A dataset for bird detection in complex forest environments using trap cameras. Ecol. Inform. **75**, 101987 (2024). https://doi.org/10.1016/j.ecoinf.2023.101987
7. Wah, C., Branson, S., Welinder, P., Perona, P., Belongie, S.: The Caltech-UCSD Birds-200-2011 Dataset, California Institute of Technology (2011). CNS-TR-2011-001 Technical report
8. Liu, Z., Wang, Q., Zhang, Y.: Bird monitoring using YOLOv4 and multi-target tracking. J. Environ. Inform. **39**, 67–78 (2022). https://doi.org/10.3808/jei.202200456

9. Ahmed, S., Khan, M.A., Javed, M.Y.: Real-time bird detection in natural environments using YOLOv4. J. Real-Time Image Proc. **17**, 1647–1659 (2020). https://doi.org/10.1007/s11554-020-01004-5
10. Ren, S., He, K., Girshick, R., Sun, J.: Faster R-CNN: Towards real-time object detection with region proposal networks. IEEE Trans. Pattern Anal. Mach. Intell. **39**, 1137–1149 (2015). https://doi.org/10.1109/TPAMI.2016.2577031
11. He, K., Zhang, X., Ren, S., Sun, J.: Deep residual learning for image recognition. In: Proceedings of the IEEE Conference on Computer Vision and Pattern Recognition, pp. 770–778 (2016). https://doi.org/10.1109/CVPR.2016.90
12. Vaswani, A., Shazeer, N., Parmar, N., Uszoreit, J., Jones, L., Gomez, A.N., Kaiser, Ł., Polosukhin, I.: Attention is all you need. In: Advances in Neural Information Processing Systems, vol. 30, pp. 5998–6008 (2017)
13. Ge, Y., Zhang, X., Wang, H.: EfficientViT: Lightweight vision transformer for real-time applications, (2023). arXiv:2305.12345 arXiv preprint
14. Long, J., Shelhamer, E., Darrell, T.: Fully convolutional networks for semantic segmentation. In: Proceedings of the IEEE Conference on Computer Vision and Pattern Recognition, pp. 3431–3440 (2015). https://doi.org/10.1109/CVPR.2015.7298965
15. Redmon, J., Divvala, S., Girshick, R., Farhadi, A.: You Only Look Once: Unified, real-time object detection. In: Proceedings of the IEEE Conference on Computer Vision and Pattern Recognition, pp. 779–788 (2016). https://doi.org/10.1109/CVPR.2016.91
16. Zhang, X., Li, Y., Liu, Z.: Multi-label classification of bird calls using transfer learning with ResNet50. Appl. Acoust. **204**, 109245 (2023). https://doi.org/10.1016/j.apacoust.2023.109245

Dual-Path Attention Network for Small Object Detection in Road Traffic Scenarios

Pinyi Wang[1], Jingke Xu[1,2,3](✉), and Bin Du[4]

[1] School of Computer Science and Engineering, Shenyang Jianzhu University, Shenyang 110168, China
xujingke@sjzu.edu.cn

[2] Liaoning Provincial Key Laboratory of Urban Construction Big Data Management and Analysis, Shenyang 110168, China

[3] Shenyang Branch of National Special Computer Engineering Technology Research Center, Shenyang 110168, China

[4] Neusoft Group Co., Ltd., No. 2 Xinxiu Street, Hunnan New District, Shenyang Postal Code: 110179, China

Abstract. In road traffic scenes, small-scale objects such as pedestrians and bicycles typically suffer from limited spatial resolution and insufficient semantic representation, making them highly susceptible to background clutter, which often leads to false alarms and missed detections. To alleviate these issues, this work develops a dual-path attention–enhanced object detection framework termed DPAN-YOLOv11. Concretely, a feature fusion attention (FFA) module is integrated at the initial stage of the backbone to strengthen fine-grained details and surrounding contextual cues. In addition, an enhanced channel attention mechanism (SENetV2) is deployed prior to each detection head to dynamically refine high-level semantic representations. Extensive experiments conducted on the KITTI benchmark demonstrate that DPAN-YOLOv11 achieves a 1.2% improvement in recall, along with gains of 1.5% in both mAP@50 and mAP@50: 95 over the baseline YOLOv11, while introducing only a marginal increase in model parameters.

Keywords: Small object detection · Road traffic scenes · Dual-path attention · Feature enhancement

1 Introduction

In recent years, small-object detection within road traffic environments has emerged as a key element of intelligent transportation systems and autonomous driving technologies. Owing to the low spatial resolution and insufficient semantic characterization of distant pedestrians, small obstacles, and other road users, perception modules frequently encounter difficulties in accurately recognizing such targets. This limitation not only endangers public safety but also undermines the overall robustness and dependability of detection frameworks. Despite the extensive deployment of high-resolution

L. Huang and D. Greenhalgh (Eds.): ICMLC 2026, LNNS 2061, pp. 523–533, 2027.
https://doi.org/10.1007/978-3-032-31236-5_36

visual sensors and high-performance computing resources enabling real-time perception in dynamic traffic scenes, precise identification of small objects under complex background interference remains a persistent challenge.

Object detection techniques have progressively transitioned from approaches relying on handcrafted feature extraction to frameworks driven by deep learning models. Two-stage detection paradigms, exemplified by the R-CNN family [1–3], typically generate candidate regions prior to performing classification and localization, thereby attaining strong detection performance. Nevertheless, this advantage is accompanied by considerable computational complexity, which restricts their suitability for real-time deployment. By contrast, one-stage detectors such as SSD (Single Shot MultiBox Detector) [4] and YOLO (You Only Look Once) [5–7] reformulate object detection as a single unified regression task, offering substantially faster inference and facilitating their adoption in real-time applications. More recently, transformer-based detectors, including DETR (Detection Transformer) [8], leverage self-attention mechanisms to enable end-to-end detection and global context reasoning. However, these approaches often entail higher training costs and remain insufficient in effectively representing fine-grained characteristics of small-scale targets. Consequently, small-object detection in traffic environments continues to be hindered by feature degradation resulting from limited pixel occupancy and complex background disturbances [9].

To tackle the aforementioned issues, a dual-path attention–enhanced object detection framework, termed DPAN-YOLOv11, is developed based on the Ultralytics YOLOv11 architecture (YOLOv11n variant). In this approach, a feature fusion attention (FFA) module [10] is embedded at the initial layer of the backbone to reinforce fine-grained details and surrounding contextual cues, while an enhanced channel attention mechanism, SENetV2 [11], is incorporated ahead of each detection head to dynamically refine high-level semantic representations. By jointly improving feature quality at both the early extraction stage and the final prediction stage, DPAN-YOLOv11 effectively mitigates feature degradation for small-scale, distant, and partially occluded targets, leading to enhanced detection reliability and robustness under complex traffic conditions.

2 Related Work

In recent years, a wide range of approaches have been developed to tackle object detection in road traffic environments, with particular emphasis on enhancing the recognition capability for small-scale targets. Li et al. [12] proposed a multi-branch attention strategy integrated with the parameter-free SimAM mechanism to strengthen small-object feature representations while maintaining a low computational cost. Nevertheless, their attention design was primarily deployed at intermediate network layers and did not explicitly reinforce feature extraction at the early input stage. In addition, Wang et al. [13] enhanced small-object detection performance through adaptive multi-scale feature weighting and fusion, together with the incorporation of contextual attention modules.

A number of studies have concentrated on strengthening feature representations through structured network designs or multi-scale modeling strategies. Liang et al. [14] incorporated a multi-path attention enhancement module into the neck of YOLOv7 and further combined it with a receptive field expansion component to boost detection performance for small-scale and densely overlapped targets. Zhao et al. [15] presented an efficient cross-scale feature fusion attention (ECFA) module together with a lightweight convolution block (SEConv), aiming to improve feature expressiveness while preserving computational efficiency. Similarly, Chu et al. [16] employed a channel attention strategy to highlight discriminative feature responses, thereby achieving a favorable trade-off between detection precision and real-time inference speed. Moreover, Chen et al. [17] investigated the integration of large-kernel convolutional attention and multi-information fusion to reinforce multi-scale feature representations and enhance the overall robustness of object detection.

In addition to convolution-based approaches, transformer-driven techniques have increasingly been explored in studies on small-object detection. Yu [18] employed a dynamic dual-fusion module together with a feature interaction mechanism to suppress background interference and enhance feature diversity, thereby achieving robust performance in complex traffic scenarios. Liu et al. [19] introduced a denoising feature pyramid network that integrates transformer architectures, utilizing self-attention operations to model long-range contextual relationships and improve detection performance for extremely small targets.

Although these methods have achieved performance improvements in various aspects, most existing methods only enhance the attention mechanism at a certain stage of the detection process (such as the backbone network, neck, or head), or introduce additional structural complexity which increases the computational cost. Moreover, the theoretical relationship between attention-based feature enhancement in different network stages is often not clearly discussed. In contrast, the proposed DPAN-YOLOv11 adopts a coordinated dual-path attention design, performing feature enhancement simultaneously in the early feature extraction stage and the detection decision stage. This unified design provides a systematic solution for improving the robustness of small-scale target detection under complex traffic conditions.

3 Model Improvement

To tackle challenges such as low spatial resolution, insufficient semantic representation, and vulnerability to background clutter in small-object detection within road traffic scenes, this work presents DPAN-YOLOv11, an enhanced object detection framework built upon the Ultralytics YOLOv11n architecture. As illustrated in Fig. 1, the proposed model incorporates attention-based enhancements at both the early feature extraction stage and the final detection stage, forming a coordinated dual-path attention strategy. Specifically, a feature fusion attention (FFA) module is embedded at the front of the backbone network. This module receives an RGB image with dimensions $H \times W \times 3$ as input and strengthens fine-grained spatial details and contextual cues prior to subsequent downsampling operations. Compared with the original YOLOv11 backbone, this design improves early-stage feature expressiveness and alleviates information loss for small targets. In addition, an improved channel attention module, SENetV2, is inserted before each detection head (P3, P4, and P5) to adaptively recalibrate deep semantic features for classification and localization. The backbone and neck structures retain the standard YOLOv11n configuration, while the detection heads produce multi-scale feature maps corresponding to small, medium, and large objects under different receptive fields.

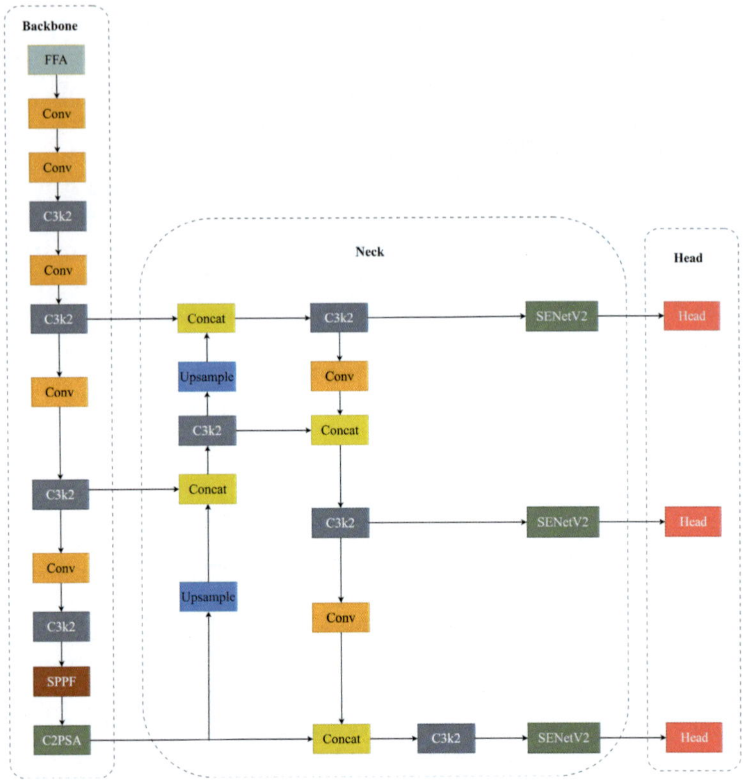

Fig. 1. Overall architecture of the proposed DPAN-YOLOv11.

3.1 Feature Fusion Attention Module

In traffic scenarios, small objects typically occupy a limited spatial area in the shallow feature maps and are prone to being affected by background interference during the early convolution processing. To alleviate this problem, an FFA module is incorporated at the front end of the backbone network to strengthen target-relevant feature responses prior to deep feature abstraction. As illustrated in Fig. 2, the FFA module is composed of N parallel feature extraction branches, referred to as Groups, followed by an adaptive fusion strategy. Each group contains several residual convolution blocks and an embedded feature attention unit, thereby generating an intermediate feature map $G_i \in R^{H_i \times W_i \times C_i}$, where $i = 1, 2, \ldots, N$, and different groups correspond to different receptive field scales. The final output of the FFA module, denoted as F_{FFA}, is obtained through a weighted aggregation of these group-wise features:

$$F_{FFA} = \sum_{i=1}^{N} \alpha_i \cdot G_i \tag{1}$$

where α_i represents the adaptive fusion weight learned for the i-th group, and $\sum_{i=1}^{N} \alpha_i = 1$. This fusion mechanism enables the model to adaptively emphasize the feature representations from different receptive field levels.

The feature attention units within each group consist of two parts: channel-wise attention and spatial (pixel-level) attention. They serve respectively to identify which information should be emphasized and where this emphasis should be applied. As illustrated in Fig. 3, given an input feature tensor $F \in R^{H \times W \times C}$, the channel attention branch first performs global average pooling and global max pooling along the spatial dimensions:

$$F_{avg} = GAP(F), F_{max} = GMP(F) \tag{2}$$

where F_{avg}, $F_{max} \in R^{1 \times 1 \times C}$, These descriptors are then passed through a shared multilayer perceptron (MLP), and the channel attention map M_c is computed as:

$$M_c = \sigma\Big(MLP\big(F_{avg} + MLP(F_{max})\big)\Big) \tag{3}$$

where σ denotes the sigmoid activation function. The input feature map is recalibrated by channel-wise multiplication with M_c.

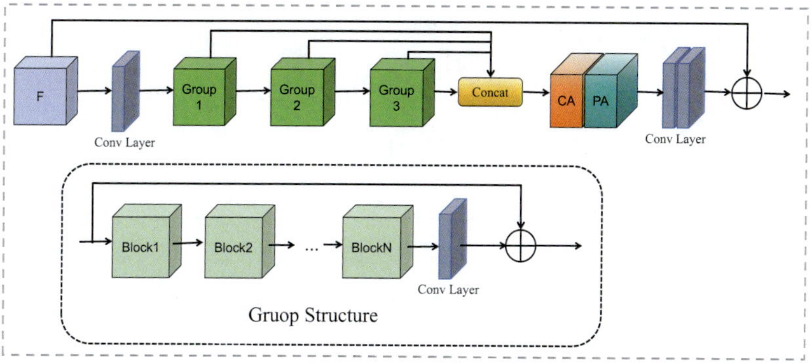

Fig. 2. Structure of the feature fusion attention module.

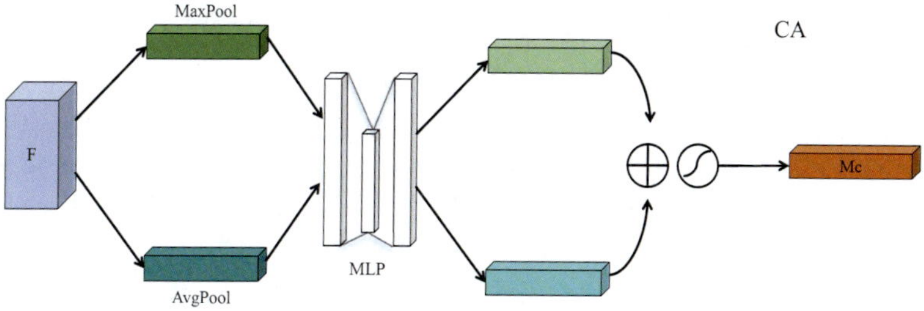

Fig. 3. Internal architecture of the channel attention mechanism.

To complement channel attention, pixel attention explicitly models spatial importance. As shown in Fig. 4, given the input feature map F, a spatial attention map $M_p \in R^{H \times W \times 1}$ is generated as:

$$M_p = \sigma(f_s(F)) \tag{4}$$

where f_s denotes a spatial mapping function composed of convolutional layers followed by nonlinear activation. The final pixel-attended feature is obtained by:

$$F_p = F \odot M_p \tag{5}$$

where $\odot$ denotes element-wise multiplication. Through joint channel and pixel attention, the FFA module enhances target-relevant information while suppressing background interference at the earliest stage of feature extraction.

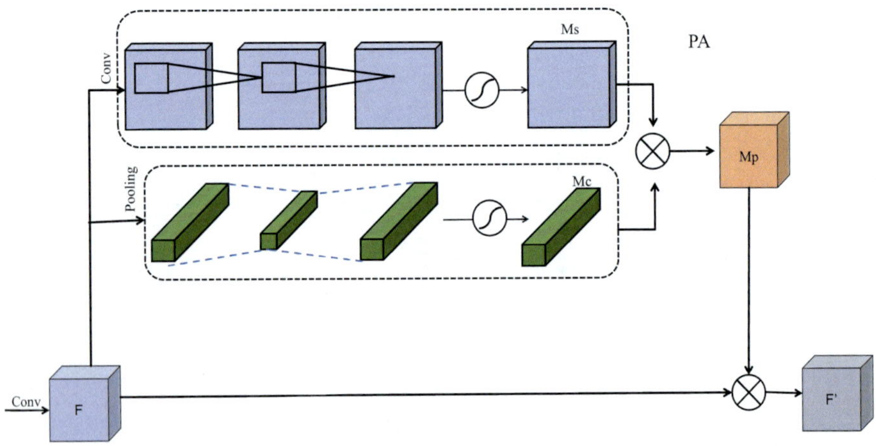

Fig. 4. Internal architecture of the pixel attention mechanism.

3.2 SENetV2-Based Channel Enhancement

Although the FFA module enhances the early features, the semantic information related to small objects may still be weakened after the deep propagation through the backbone network. To further enhance the discriminative channel features, we introduce an improved channel attention module SENetV2 before each detection head (P3, P4, and P5), as illustrated in Fig. 5. For an input feature tensor $X \in R^{H \times W \times C}$, SENetV2 first applies global average pooling to derive a compact channel-wise descriptor. Unlike the conventional SE module, SENetV2 adopts four parallel fully connected branches to model channel relationships from diverse transformation subspaces:

$$M_s = \sum_{k=1}^{4} W_k \cdot GAP(X) \tag{6}$$

where W_k represents the weight matrix of the k-th branch. Subsequently, these outputs are fused and normalized to generate the final channel attention map M_s. The calculation formula for the recalibrated output features is as follows:

$$X' = X \odot M_s \tag{7}$$

By applying the SENetV2 algorithm before the detection head, this model can adaptively enhance the channels related to small objects and suppress redundant features, thereby improving the classification accuracy and positioning precision in crowded traffic scenarios.

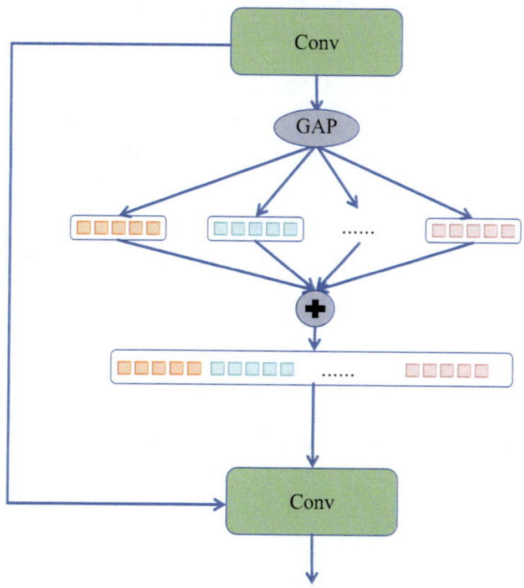

Fig. 5. Schematic diagram of the multi-branch SENetV2 module.

4 Experiments

4.1 Experimental Datasets and Setup

The experimental studies in this paper are performed on the KITTI dataset [20]. To ensure feasibility for road traffic object detection and avoid confusion from redundant category data during training, we preprocessed the original KITTI dataset annotations to unify object categories. Based on the semantics of road traffic targets, we grouped the original categories into three classes: Car, Pedestrian, and Cyclist, and regenerated the corresponding annotation files. This preprocessing maintains the semantic consistency of traffic targets, reduces inter-class confusion, and allows the model to more effectively learn the discriminative characteristics of traffic targets.

The experimental environment is summarized as follows. All models were trained and tested on a workstation featuring an Intel Core i7–10700 processor and an NVIDIA

GeForce RTX 3060 graphics card, operating under the Windows 10 operating system. The implementation was developed using Python 3.10.16, PyTorch 2.3.1, and CUDA 11.8.

During the evaluation process, standard object detection metrics were employed. mAP@50 denotes the mean average precision computed at an intersection-over-union IoU threshold of 0.50, whereas mAP@50:95 refers to the mean average precision averaged over IoU thresholds ranging from 0.50 to 0.95 with an increment of 0.05.

4.2 Experimental Results and Model Analysis

In order to test the FFA module alone, the effect of the SENetV2 module, and the cooperation of both modules, we conducted an ablation experiment under the same training conditions as in Table 1. Compared with the baseline YOLOv11n, adding only the SENetV2 module results in an increase in recall, which means channel recalibration improves the sensitivity for detecting targets. Adding the FFA module alone increased the mAP@50 and mAP@50:95 from 85.8% and 59.6% to 86.3% and 60.7% respectively, suggesting that for small target detection, strengthening features at the early extraction stage is more effective. Finally, the two modules complement and cooperate with each other in the feature extraction stage and the channel decision stage, respectively, enabling the model to achieve the best results across all precision metrics.

Table 1. Results of ablation experiments.

SENetV2	FFA	Recall/%	mAP@50/%	mAP@50: 95/%	Parameters/M
		76.2	85.8	59.6	2.59
√		77.7	85.6	59.1	2.62
	√	77.4	86.3	60.7	2.58
√	√	77.4	87.3	61.1	2.63

Table 2 compares the methods presented in this paper against other mainstream object detection approaches evaluated on the KITTI benchmark. The experimental comparisons indicate that, relative to existing methods, the proposed approach delivers the best performance across the Recall, mAP@50, and mAP@50:95 metrics, while maintaining the lowest parameter count. Compared with the baseline model, our model gains an additional 1.5 percentage-point improvement in both mAP@50 and mAP@50:95. This demonstrates that the introduced modules effectively enhance detection accuracy. Overall, the proposed approach offers a more favorable trade-off between accuracy and model complexity than other mainstream detectors. The proposed model demonstrates both practicability and robustness in detecting small targets in road scenes.

Figure 6 compares the detection outputs produced by the original YOLOv11n model with those of the enhanced DPAN-YOLOv11 model evaluated on the KITTI test dataset. The detection categories are car, pedestrian, and cyclist. Although both models can complete the basic detection tasks, the comparison reveals that DPAN-YOLOv11 can successfully detect the targets missed by the original model in multiple complex scenarios,

Table 2. Performance comparison based on KITTI dataset.

Model	Recall/%	mAP@50/%	mAP@50: 95/%	Parameters/M
SSD	18.6	35.6	11.0	24.7
Faster R-CNN	78	76	–	28.3
YOLOv5	79.8	84.7	59	7.1
YOLOv8	75.8	81.6	57.5	3.0
MobileNetV3-SSD [21]	–	65.3	–	33.86
Edge YOLO [21]	–	72.6	–	24.48
YOLOv11n	76.2	85.8	59.6	2.59
Ours	77.4	87.3	61.1	2.63

especially in the cases of smaller-sized and less distinctive pedestrian and cyclist targets. The visual results indicate that the dual-path attention mechanism adopted in this study enhances the model's feature representation ability and target perception sensitivity, thus leading to improved completeness and robustness in small object detection under traffic road conditions.

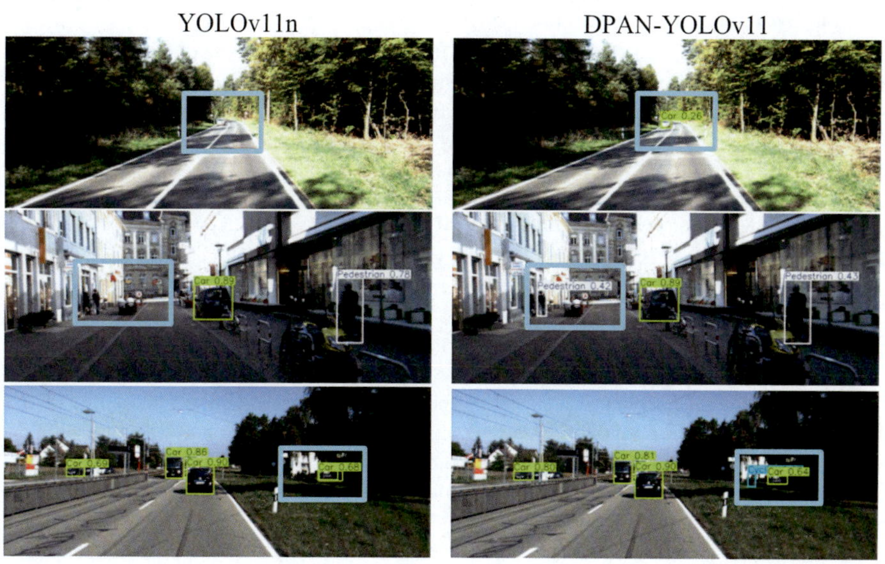

Fig. 6. Detection result comparison between YOLOv11 and DPAN-YOLOv11 on the KITTI test set.

5 Conclusion

This study presents a dual-path attention–driven object detection framework, termed DPAN-YOLOv11, aiming to improve the detection capability for small-scale targets within complex road traffic scenarios. By using the FFA module to enhance early feature extraction before detecting the head, and the SENetV2 module to refine deep semantic features, this method enhances the feature representation throughout the detection process. Experiments on the KITTI dataset show that DPAN-YOLOv11 is superior to the baseline YOLOv11n, with gains of 1.5% in both mAP@50 and mAP@50:95, while preserving a similar model scale, thereby achieving an effective trade-off between accuracy and efficiency. Future research will concentrate on further reducing the computational complexity of the model and assessing the proposed framework on additional datasets and diverse traffic scenarios to strengthen its generalization capability and practical deployment potential.

References

1. Girshick, R., Donahue, J., Darrell, T., Malik, J.: Rich feature hierarchies for accurate object detection and semantic segmentation. In: Proceedings of the IEEE Conference on Computer Vision and Pattern Recognition (CVPR 2014), pp. 580–587 (2014)
2. Girshick, R.: Fast R-CNN. In: Proceedings of the IEEE International Conference on Computer Vision (ICCV 2015), pp. 1440–1448 (2015)
3. Ren, S., He, K., Girshick, R., Sun, J.: Faster R-CNN: towards real-time object detection with region proposal networks. IEEE Trans. Pattern Anal. Mach. Intell. **39**(6), 1137–1149 (2017)
4. Liu, W., Anguelov, D., Erhan, D., et al.: SSD: single shot multibox detector. In: Computer Vision–ECCV 2016 LNCS, vol. 9905, pp. 21–37. Springer, Heidelberg (2016)
5. Redmon, J., Divvala, S., Girshick, R., Farhadi, A.: You only look once: unified, real-time object detection. In: Proceedings of the IEEE Conference on Computer Vision and Pattern Recognition (CVPR 2016), pp. 779–788 (2016)
6. Redmon, J., Farhadi, A.: YOLO9000: Better, faster, stronger. In: Proceedings of the IEEE Conference on Computer Vision and Pattern Recognition (CVPR 2017), pp. 7263–7271 (2017)
7. Redmon, J., Farhadi, A.: YOLOv3: An incremental improvement. arXiv preprint https://arxiv.org/abs/1804.02767. (2018)
8. Carion, N., Massa, F., Synnaeve, G., et al.: End-to-end object detection with transformers. In: Computer Vision–ECCV 2020 LNCS, vol. 12346, pp. 213–229. Springer, Heidelberg (2020)
9. Lin, T.-Y., Dollár, P., Girshick, R., et al.: Feature pyramid networks for object detection. In: Proceedings of the IEEE Conference on Computer Vision and Pattern Recognition (CVPR 2017), pp. 2117–2125 (2017)
10. Qin, X., Wang, Z., Bai, Y., et al.: FFA-net: feature fusion attention network for single image dehazing. Proc. AAAI Conf. Artif. Intell. **34**(07), 11908–11915 (2020)
11. Narayanan, M.: SENetV2: aggregated dense layer for channelwise and global representations. arXiv preprint https://arxiv.org/abs/2311.10807. (2023)
12. Li, L., Gao, S., Wu, F., et al.: MBAN: multi-branch attention network for small object detection. PeerJ Comput. Sc. **10**, e1965 (2024)
13. Li, M., Pi, D., Qin, S.: An efficient single shot detector with weight-based feature fusion for small object detection. Sci. Rep. **13**(1), 9883 (2023)

14. Liang, K., Zhang, W., Li, F., et al.: Improved YOLOv7 for small and overlapping objects detection. In: Proceedings of the 6th IEEE International Conference on Pattern Recognition and Artificial Intelligence (PRAI 2023), pp. 1–6. IEEE (2023)
15. Zhao, F., Zhang, J., Zhang, G.: FFEDet: fine-grained feature enhancement for small object detection. Remote Sens. **16**(11), 2003 (2024)
16. Chuai, Q., He, X., Li, Y.: Improved traffic small object detection via cross-layer feature fusion and channel attention. Electronics. **12**(16), 3421 (2023)
17. Chen, W., Liang, Z.: LMF-YOLO: an improved YOLO algorithm for road object detection in autonomous driving. SIViP. **19**(9), 760–770 (2025)
18. Yu, T.: MDDFNet: mamba-based dynamic dual fusion network for traffic sign detection. arXiv preprint https://arxiv.org/abs/2505.05491. (2025)
19. Liu, H.-I., Tseng, Y.-W., Chang, K.-C., et al.: A denoising FPN with transformer R-CNN for tiny object detection. IEEE Trans. Geosci. Remote Sens. **62**, 1–15 (2024)
20. Geiger, A., Lenz, P., Stiller, C., Urtasun, R.: Vision meets robotics: the KITTI dataset. Int. J. Robot. Res. **32**(11), 1231–1237 (2013)
21. Liang, S., Wu, H., Zhen, L., et al.: Edge YOLO: real-time intelligent object detection system based on edge-cloud cooperation in autonomous vehicles. IEEE Trans. Intell. Transp. Syst. **23**(12), 25345–25360 (2022)

BoundMamba: Boundary-Aware State Space Modeling for Remote Sensing Change Detection

Yu He[1], Yuanlie He[1], Wensheng Li[2], Yi Xiang[3], Yukuan Ma[4], and Yihui Liang[2](✉)

[1] Guangdong University of Technology, Guangzhou 510006, China
[2] University of Electronic Science and Technology of China, Zhongshan Institute, Zhongshan 528402, China
yihuiliang@outlook.com
[3] South China University of Technology, Guangzhou 510640, China
[4] Airace Technology Co., Ltd., Huizhou 516000, China

Abstract. Remote sensing change detection requires precise boundary localization, yet existing Mamba-based methods employ homogeneous decoder blocks that prioritize global context over local geometric adaptation, resulting in boundary errors of 3–5 pixels that compromise practical applications. We propose BoundMamba, which integrates geometric-adaptive processingthroughout the decoder via three synergistic components: (1) Multi-ScaleShape-Adaptive Convolution (MSSAC) learns boundary geometry through prototype mixture modeling with only 0.8M parameters; (2) Hybrid MSSAC-Mamba Block couples global context with local shape adaptation; (3) progressive multi-stage refinement enables hierarchical boundary optimization.Experiments on three benchmarks show consistent improvements: + 2.97 on irregular-boundary SYSU-CD, + 0.97 + 0.65 boundary sharpness. Our work demonstrates that architectural-levelboundary integration outperforms post-processing refinement strategies.

Keywords: Remote Sensing Change Detection · State Space Model · Mamba · Shape-Adaptive Convolution · Boundary-Aware Modeling

1 Introduction

Change detection (CD) is a fundamental task in remote sensing image processing, with significant practical applications in many fields such as urban planning and management, disaster response, environmental monitoring, and transportation navigation. With the continuous advancement of UAV technology, the acquisition cycle for high-resolution remote sensing images (0.1-0.5 m/pixel) is becoming increasingly shorter, providing a rich dataset. Consequently, the automatic detection of changes from these high-resolution images has emerged as a significant focus for intelligent interpretation. Although change detection from

L. Huang and D. Greenhalgh (Eds.): ICMLC 2026, LNNS 2061, pp. 534–550, 2027.
https://doi.org/10.1007/978-3-032-31236-5_37

high-resolution remote sensing images has attracted widespread attention, this task remains challenges. The main difficulties include: 1) Changed objects often exhibit irregular boundaries with complex non-convex geometries. 2) Scale variations span from small vehicles (approximately 10 pixels) to large building complexes (over 1000 pixels). 3) Boundary localization errors of 3-5 m can lead to false structural assessments in practical applications such as post-disaster evaluation or urban planning.

To accurately detect changes, early methods relied on the inherent morphological features of input images and handcrafted features developed by professionals for classification. However, these approaches are often unsuitable for large-scale remote sensing images that contain complex, multi-scale change patterns. Manual feature extraction in such contexts not only demands specialized knowledge but is also highly time-consuming and labor-intensive.

In comparison with traditional methods based on manual feature extraction, methods based on deep learning [1–3] demonstrate superiority in the automatic and robust extraction of semantic features, leading to a greater improvement in accuracy. Early methods were mainly based on Convolutional Neural Networks (CNNs). FC-Siam-Conc [4] and SNUNet [5] employ Siamese encoders with multi-scale fusion strategies, achieving strong performance on benchmark datasets. While the convolution operation is effective in dealing with local features, it has limitations in dealing with global contextual information. Compared to CNNs, Vision Transformers (ViTs) [6,7] enable the representation of each position of the input image to focus directly on other positions through a self-attention mechanism, effectively capturing global contextual information. BIT [8] and ChangeFormer [9] demonstrate improved accuracy through Transformer encoders. However, this improvement comes at a cost, as the quadratic complexity of the Transformer's self-attention mechanism increases the computational burden on the model, making them prohibitively expensive for processing high-resolution images such as 2048×2048 UAV imagery.

Recently, methods based on state space models (SSMs) [10], such as Mamba [11], have provided innovative solutions for solving these problems. Unlike Transformers that use inefficient self-attention mechanisms for sequence modeling, Mamba utilizes selective scanning mechanisms, unique hardware-aware algorithms, and parallel scanning to achieve modeling capabilities comparable to Transformers while maintaining linear computational complexity. The introduction of the VMamba(VM) block in VMamba [12] further enhances the applicability of Mamba to computer vision tasks. ChangeMamba [13] pioneered Mamba's application to remote sensing change detection, introducing weight-shared VSSM encoders and three spatio-temporal fusion strategies (concatenation, pixel interleaving, spatial tiling) that effectively model bi-temporal relationships. However, its decoder architecture employs standard VSS Blocks uniformly across all stages–these blocks excel at global dependency modeling but lack specialized geometric operators for local boundary adaptation, leading to suboptimal boundary localization particularly for irregular objects.

Inspired by the development of Mamba in vision tasks, we construct a change detection model based on Mamba and shape-adaptive convolution, called Bound-Mamba. The Mamba-based encoder efficiently captures long-range contextual information, while the Hybrid MSSAC-Mamba Block synergistically combines global context with local geometric adaptation. The progressive decoder architecture enables hierarchical boundary refinement across multiple resolution stages.

The main contributions of this paper are summarized as:

1. To detect changes accurately with precise boundaries, we propose Bound-Mamba, which introduces HM-Block to the progressive decoder architecture. The Mamba-based encoder captures the long-range contextual information, while the HM-Block enables synergistic global-local processing at multiple decoder stages.
2. To improve the boundary localization performance, Multi-Scale Shape-Adaptive Convolution is introduced to achieve dynamic geometric adaptation through learned prototype mixture modeling with only 0.8M parameters.
3. Extensive experiments on three public datasets were conducted, including comprehensive comparisons and ablation studies to demonstrate the effectiveness of the proposed method.

2 Related Work

2.1 CNN-Based Change Detection

Early deep learning CD methods adapted fully convolutional network (FCN) [2] architectures for bi-temporal comparison. FC-Siam-Conc [4] employs Siamese encoders with late concatenation fusion for change detection. SNUNet [5] introduced densely connected decoder paths [14] with channel attention mechanisms [15], achieving strong performance on benchmark datasets. STANet [16] incorporates spatial-temporal attention modules to explicitly model bi-temporal relationships. DSIFN [17] models spatial-spectral differences through a dual-stream architecture. Despite these advances, CNN-based methods remain fundamentally constrained by local receptive fields–even with dilated convolutions [18,19] or large kernel designs [20], effective receptive fields rarely exceed 200-300 pixels, which is insufficient for modeling global context [21] in 512×512 or larger UAV imagery.

2.2 Transformer-Based Methods

To overcome receptive field limitations, BIT [8] pioneered Transformer encoders [7] for change detection, demonstrating that self-attention effectively captures long-range dependencies for improved accuracy. ChangeFormer [9] further optimized this paradigm through hierarchical Transformer encoders and difference-aware attention mechanisms. SwinSUNet [22] adapted Swin Transformer [23] with shifted windows for computational efficiency. While achieving impressive accuracy, these methods incur quadratic complexity in self-attention:

processing a 2048×2048 image requires approximately 16M token interactions, resulting in 8–12× higher memory consumption than CNNs. This computational burden limits practical deployment for large-scale UAV mapping applications.

2.3 State Space Models for Vision

State space models (SSMs) [10], particularly the Mamba architecture [11], have recently gained attention for efficient sequence modeling. Mamba achieves linear-complexity global receptive fields through selective state space parameterization, avoiding the quadratic bottleneck of self-attention. VMamba [12] and Vision Mamba [24] demonstrated successful adaptation to 2D vision tasks through spatial scanning strategies. ChangeMamba [13] first applied Mamba to remote sensing change detection, introducing three key contributions: (1) weight-shared VSSM encoders for bi-temporal feature extraction, (2) three spatio-temporal relationship modeling mechanisms (concatenation, interleaving, spatial tiling), and (3) a multi-scale progressive decoder using VSS Blocks. While Change-Mamba validates Mamba's effectiveness for CD, its decoder employs homogeneous VSS Blocks that prioritize global context modeling without specialized local geometric operators. This architectural choice leads to suboptimal boundary localization–a limitation our work explicitly addresses through progressive boundary-aware processing.

3 Methodology

We present BoundMamba, a boundary-aware change detection framework that integrates geometric adaptation throughout the decoder hierarchy. We first overview the complete architecture, then detail the multi-strategy spatio-temporal fusion module, our proposed multi-scale shape-adaptive convolution, and the hybrid MSSAC-Mamba block. Finally, we describe the progressive decoder architecture and training objectives.

3.1 Overall Architecture

Figure 1 illustrates our complete framework. Given bi-temporal images $I_1, I_2 \in \mathbb{R}^{H \times W \times 3}$, we extract hierarchical features through two weight-sharing VSSM encoders, producing multi-scale feature pyramids $\{E_i^1, E_i^2\}_{i=1}^4$ at resolutions $\{H/4, H/8, H/16, H/32\}$ with channel dimensions $\{96, 192, 384, 768\}$ for VSSM-Small backbone. At each scale i, the Multi-Strategy Spatio-Temporal Fusion Module (MSTFM) combines feature pairs (E_i^1, E_i^2) to generate change-centric representations $M_i \in \mathbb{R}^{128 \times H_i \times W_i}$.

The decoder then processes $\{M_i\}_{i=1}^4$ through four stages in bottom-up fashion. Stage IV (lowest resolution, H/32) applies standard spatio-temporal fusion without boundary refinement, establishing global change patterns with computational efficiency. Stages III-I progressively upsample and refine features, with our Hybrid MSSAC-Mamba Block (HM-Block) integrated at each stage to enable simultaneous global context modeling and local boundary adaptation. The final decoder output $D_1 \in \mathbb{R}^{128 \times H/4 \times W/4}$ is passed through a 1×1 convolutional classifier and bilinearly upsampled to produce the binary change map $\hat{Y} \in \mathbb{R}^{2 \times H \times W}$.

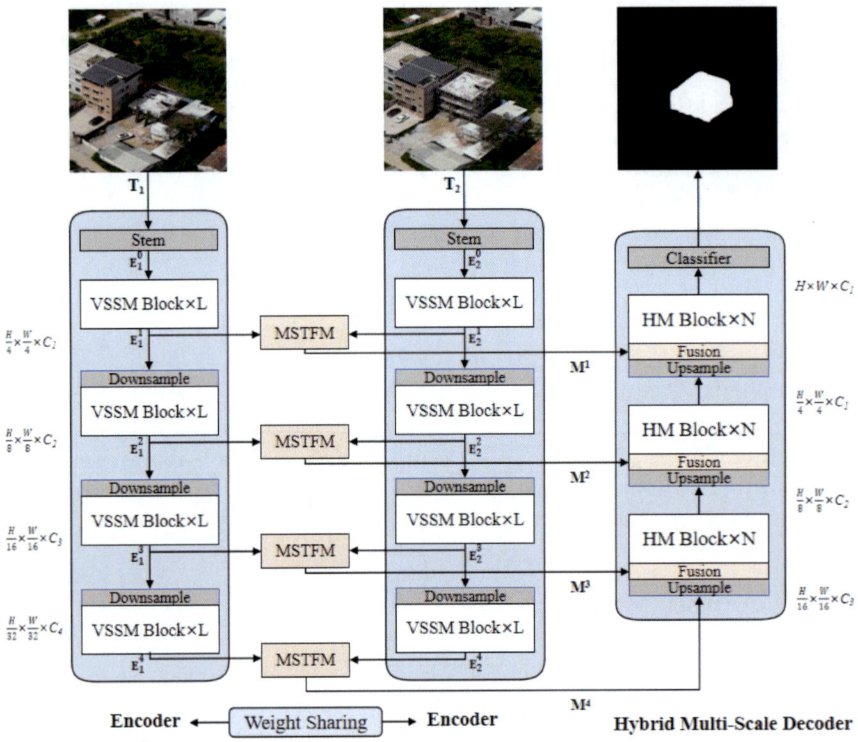

Fig. 1. Overall architecture of BoundMamba. Bi-temporal images are encoded via weight-sharing VSSM encoders. At each scale, MSTFM fuses temporal features into change-centric representations. The progressive decoder refines change predictions through HM-Blocks at Stages I-III, enabling synergistic global-local processing at multiple resolutions

3.2 Multi-strategy Spatio-Temporal Fusion Module

Robust bi-temporal comparison requires modeling diverse spatio-temporal relationships. We adopt the Multi-Strategy Spatio-Temporal Fusion Module (MSTFM), which processes feature pairs (E_i^1, E_i^2) through three parallel strategies to capture complementary temporal patterns (Fig. 2c).

Strategy 1: Channel Concatenation. Direct concatenation $[E_i^1; E_i^2] \in \mathbb{R}^{2C_i \times H_i \times W_i}$ followed by 1×1 projection and VSS Block processing. This strategy captures holistic temporal relationships by allowing the network to compare features across the entire channel space.

Strategy 2: Pixel Interleaving. Construct spatially interleaved tensor $T_{\text{interleave}} \in \mathbb{R}^{C_i \times H_i \times 2W_i}$ where $T_{\text{interleave}}[:, :, :: 2] = E_i^1$ and $T_{\text{interleave}}[:, :, 1 :: 2] = E_i^2$. After VSS Block processing, extract even and odd columns separately. This strategy encourages pixel-aligned temporal modeling, facilitating precise spatial correspondence.

Strategy 3: Spatial Concatenation. Construct side-by-side tensor $T_{\text{spatial}} = [E_i^1 \mid E_i^2] \in \mathbb{R}^{C_i \times H_i \times 2W_i}$, process via VSS Block, then split spatially. This strategy preserves spatial continuity within each timestamp, enabling context-aware comparison.

The three strategy outputs are concatenated and fused via 1×1 convolution to produce $M_i \in \mathbb{R}^{128 \times H_i \times W_i}$. This multi-strategy design enriches the change representation through complementary temporal modeling perspectives, providing robust features for downstream boundary refinement.

3.3 Multi-scale Shape-Adaptive Convolution

Standard convolutions apply fixed-shape kernels regardless of local geometric structure, limiting adaptability to irregular boundaries common in remote sensing imagery. Deformable convolutions [25,26] address this through learnable spatial offsets but lack explicit shape priors, requiring substantial training data and potentially overfitting to training geometries. We propose MSSAC, which achieves controllable geometric adaptation through prototype-based mixture modeling (Fig. 2b).

Geometric Prototypes. We define four fixed sampling prototypes $\{\mathbf{P}_p\}_{p=1}^4$, each specifying 8 normalized 2D offsets that encode distinct geometric patterns: *Radial* ($\mathbf{P}_1$): Eight points uniformly distributed at angles $\{0°, 45°, 90°, \ldots, 315°\}$ on a unit circle, modeling circular or blob-like boundaries common in natural features (vegetation patches, water bodies). *Cross* ($\mathbf{P}_2$): Cardinal and inter-cardinal directions with alternating far and near sampling points, capturing axis-aligned edges characteristic of man-made structures (building walls, road boundaries). *Diagonal* ($\mathbf{P}_3$): Points along 45° diagonal directions with near and far pairs, handling rotated rectangular structures and oblique edges. *Dense* ($\mathbf{P}_4$): 3×3 grid pattern excluding the center point, providing dense local coverage for complex or irregular geometries that do not conform to simple shapes.

These prototypes are fixed (not learned during training), providing strong inductive bias while maintaining parameter efficiency. The mixture modeling framework allows the network to compose arbitrary boundary shapes through weighted combinations of these basis patterns.

Adaptive Sampling Mechanism. Given input features $\mathbf{F}_{\text{in}} \in \mathbb{R}^{C \times H \times W}$, MSSAC generates per-pixel adaptive sampling parameters through two lightweight predictors. A compact CNN predicts per-pixel prototype weights:

$$\mathbf{W}_s = \text{Softmax}(\text{ShapePredictor}(\mathbf{F}_{\text{in}})) \in \mathbb{R}^{4 \times H \times W} \tag{1}$$

where ShapePredictor comprises two convolutions: $\text{Conv}_{3\times3}(\text{C} \rightarrow \text{C}/4)$ with ReLU activation, followed by $\text{Conv}_{1\times1}(\text{C}/4 \rightarrow 4)$. The Softmax normalization ensures $\sum_p \mathbf{W}_s^p(x,y) = 1$ for all locations, enabling interpretable mixture semantics.

A parallel CNN predicts multi-scale sampling radii:

$$\mathbf{R} = (\text{Sigmoid}(\text{ScalePredictor}(\mathbf{F}_{\text{in}})) \cdot 1.5 + 0.5) \in \mathbb{R}^{3\times H\times W} \tag{2}$$

with range $[0.5, 2.0]$ relative to base radius $r_{\text{base}} = 4.0$ pixels, adapting to objects ranging from small (2-pixel sampling radius) to large (8-pixel sampling radius). For each scale $s \in \{1, 2, 3\}$ and spatial location (x, y), we compute the adaptive sampling grid by combining shape mixture and scale prediction:

$$\mathcal{G}_s(x,y) = (x,y) + r_{\text{base}} \cdot \mathbf{R}_s(x,y) \cdot \sum_{p=1}^{4} \mathbf{W}_s^p(x,y) \cdot \mathbf{P}_p \tag{3}$$

This formulation produces 8 sampling positions per scale (24 positions total per pixel across three scales). Each position adapts both its radial distance (via $\mathbf{R}_s$) and angular distribution (via weighted prototype mixture $\sum_p \mathbf{W}_s^p \mathbf{P}_p$), enabling fine-grained geometric and scale adaptation.

Feature Extraction and Fusion. For each scale s, we sample features at the 8 grid positions using bilinear interpolation:

$$\mathbf{F}_s^{\text{sampled}} = \text{GridSample}(\mathbf{F}_{\text{in}}, \mathcal{G}_s) \in \mathbb{R}^{C\times 8\times H\times W} \tag{4}$$

Concatenating sampled features along the channel dimension yields $\mathbf{F}_s^{\text{concat}} \in \mathbb{R}^{8C\times H\times W}$, which is processed through grouped 1×1 convolution with groups = 4 for parameter efficiency, followed by batch normalization and ReLU activation to produce $\mathbf{F}_s^{\text{out}} \in \mathbb{R}^{C_{\text{out}}\times H\times W}$. Multi-scale outputs are fused through concatenation and channel reduction:

$$\mathbf{F}_{\text{out}} = \text{Conv}_{1\times 1}([\mathbf{F}_1^{\text{out}}; \mathbf{F}_2^{\text{out}}; \mathbf{F}_3^{\text{out}}]) + \mathbf{F}_{\text{in}} \tag{5}$$

The residual connection ensures stable gradient flow and preserves low-frequency content. MSSAC's parameter count is only 0.8M (ShapePredictor 0.2M + ScalePredictor 0.2M + grouped convolutions 0.4M), representing 1.6% overhead relative to the full model. FLOPs for $128 \times 64 \times 64$ input are approximately 3.8 GFLOPs, comparable to two standard 3×3 convolutions.

3.4 Hybrid MSSAC-Mamba Block

The HM-Block constitutes the fundamental processing unit of our decoder, designed to synergistically combine Mamba's global modeling capability with MSSAC's local geometric adaptation (Fig. 2a). The HM-Block employs a dual-branch architecture with residual fusion, ensuring both gradient flow stability and complementary feature learning.

Given input features $\mathbf{X}_{\text{in}} \in \mathbb{R}^{C\times H\times W}$, two parallel branches independently process the input. The global context branch employs a standard VSS Block that captures long-range spatio-temporal dependencies through selective state space modeling. The global context branch employs a standard VSS Block [12] that captures long-range spatio-temporal dependencies through selective state space modeling with four-directional spatial scanning. This branch ensures that change detection decisions consider global scene context, enabling disambiguation of changes from lighting variations or seasonal effects that require scene-level understanding.

The local shape-adaptive branch employs our MSSAC module to perform geometry-aware feature extraction, dynamically adapting its receptive field shape and scale to local boundary characteristics. This branch specializes in fine-grained boundary localization by sampling features at positions optimized for local geometric structure.

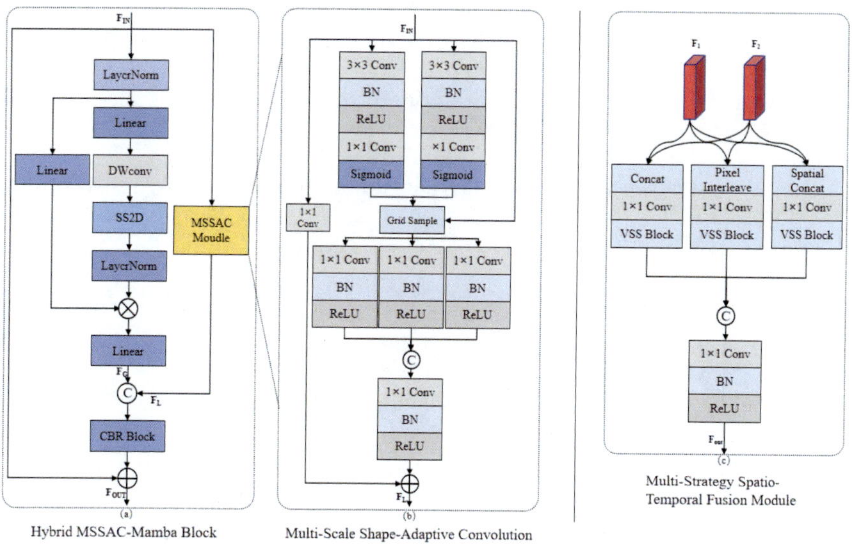

Fig. 2. Detailed architecture of core components. (a) HM-Block with dual-branch design: VSS Block captures global spatio-temporal context while MSSAC performs local shape-adaptive feature extraction, fused via 1×1 convolution with residual connection. (b) MSSAC mechanism: Shape and scale predictors generate adaptive sampling grids by blending geometric prototypes, enabling dynamic receptive field adaptation. (c) MSTFM models bi-temporal relationships through three complementary strategies for robust temporal fusion

The two branch outputs are concatenated along the channel dimension and fused through a compact network:

$$\mathbf{X}_{\text{fused}} = \text{BN}(\text{ReLU}(\text{Conv}_{1\times1}([\text{VSS}(\mathbf{X}_{\text{in}}); \text{MSSAC}(\mathbf{X}_{\text{in}})]))) \tag{6}$$

The fused features are combined with the input via a residual connection with DropPath regularization (rate = 0.1) to prevent overfitting:

$$\mathbf{X}_{\text{out}} = \mathbf{X}_{\text{in}} + \text{DropPath}(\mathbf{X}_{\text{fused}}) \tag{7}$$

The parallel architecture enables several key advantages: independent gradient flow to both branches during backpropagation, graceful degradation through residual connections, and automatic complementary weighting through the learned fusion convolution.

3.5 Progressive Boundary Refinement Decoder

Our decoder processes $\{M_i\}_{i=1}^4$ through four stages bottom-up, with HM-Blocks integrated at Stages I-III for progressive boundary optimization. This design draws inspiration from feature pyramid networks [27] and U-Net [28], but integrates boundary-aware processing at each scale rather than treating refinement as a post-processing step.

Stage IV (H/32) establishes global change patterns without boundary refinement for computational efficiency–boundary details are minimal at low resolution.

Stages III-I progressively refine boundaries through three steps at each stage $i \in \{3, 2, 1\}$:

1. Aggregate current-scale features M_i via three temporal strategies to produce $\tilde{D}_i \in \mathbb{R}^{128 \times H_i \times W_i}$.
2. Integrate coarser-scale context: $D_i' = \text{CrossFuse}_i([\text{Upsample}(D_{i+1}); \tilde{D}_i])$ with $2\times$ bilinear interpolation.
3. Apply HM-Block for synergistic global-local processing: $D_i = \text{HM-Block}_i(D_i')$.

This design ensures hierarchical boundary refinement: Stage III (H/16) establishes coarse boundary structures; Stage II (H/8) refines mid-scale details like building corners; Stage I (H/4) performs pixel-accurate localization. By distributing refinement across scales, each stage specializes in geometric details appropriate to its receptive field, with boundary supervision gradients flowing directly to all stages during training.

3.6 Loss Function

We employ a compound loss function combining Cross-Entropy and Lovsz-Softmax [29] to optimize both pixel-wise accuracy and region-based metrics:

$$\mathcal{L}_{\text{total}} = \mathcal{L}_{\text{CE}}(\hat{Y}, Y) + 0.75 \cdot \mathcal{L}_{\text{Lovasz}}(\hat{Y}, Y) \tag{8}$$

where $Y \in \{0, 1\}^{H \times W}$ is the ground truth binary change map, $\hat{Y} \in \mathbb{R}^{2 \times H \times W}$ contains the predicted logits, $\mathcal{L}_{\text{CE}}$ is standard cross-entropy loss ensuring pixel-wise accuracy, and $\mathcal{L}_{\text{Lovasz}}$ serves as a differentiable surrogate for IoU optimization. We set the weighting factor to 0.75 empirically, balancing pixel accuracy and region quality. The Lovsz-Softmax loss is particularly effective for imbalanced datasets common in change detection, where changed pixels typically constitute only 5–15% of the image.

Critically, our architecture's inherent boundary processing strength through HM-Block integration at every decoder stage eliminates the need for auxiliary boundary supervision losses or edge-aware weighting schemes employed in prior work. The progressive refinement design ensures that boundary information receives sufficient attention throughout the feature hierarchy without explicit boundary supervision.

We use the AdamW optimizer with initial learning rate 10^{-4} and weight decay 5×10^{-4}. Learning rate follows a cosine annealing schedule with minimum value 10^{-6}. Models are trained for 20,000 iterations (SYSU-CD) or 50,000 iterations (LEVIR-CD, WHU-CD) with batch size 16 for LEVIR-CD/WHU-CD and 10 for SYSU-CD. Data augmentation includes random cropping to 256×256 patches, random horizontal/vertical flipping (probability 0.5), and random rotation within $\pm 15^{\circ}$ (probability 0.3). These augmentations are applied consistently to both bi-temporal images and ground truth labels to maintain correspondence.

4 Experiments

4.1 Experimental Setup

Datasets. We evaluate on three benchmarks with diverse characteristics (Table 1): LEVIR-CD [16] (building changes, 0.5 m/pixel), WHU-CD [30] (large-scale urban development, 0.3 m/pixel), and SYSU-CD [30] (complex irregular boundaries, 0.5 m/pixel).

Table 1. Dataset statistics and characteristics

Dataset	Resolution	Train/Val/Test	Key characteristics
LEVIR-CD	10241024	589/168/228	Building-focused, 5–14 year intervals
WHU-CD	Variable	-/-/744	Large urban development, 4 years
SYSU-CD	Variable	12k/4k/4k	Irregular boundaries, dense urban

Metrics & Implementation. We report Recall, Precision, OA, F1, IoU, and KC, with F1 and IoU as primary metrics. Experiments use PyTorch 2.0 on NVIDIA A100 GPUs. Three model variants are evaluated: BoundMamba-Tiny/Small/Base (28.1M/46.7M/91.8M parameters) with VSSM backbones initialized from ImageNet-1K pretrained weights.

4.2 Comparison with State-of-the-Art

Table 2 presents comprehensive comparisons with representative methods across three architectural paradigms: CNN-based, Transformer-based, and Mamba-based approaches on LEVIR-CD test set.

BoundMamba-Base achieves 88.69% F1-score, substantially outperforming CNN-based approaches: + 5.98% over SNUNet, + 5.49% over STANet. These

Table 2. Quantitative comparison on LEVIR-CD test set. Best results in **bold**

Type	Method	Rec	Pre	OA	F1	IoU	KC
CNN	FC-Siam-Conc	78.49	78.39	98.24	78.44	64.53	77.52
	SNUNet	84.25	81.22	98.56	82.71	70.52	81.96
	STANet	80.96	85.56	98.67	83.20	71.23	82.50
Trans.	BIT-50	81.84	85.02	98.67	83.40	71.53	82.71
	ChangeFormer	78.57	67.66	97.60	72.71	57.12	71.46
Mamba	ChangeMamba-Tiny	84.18	83.08	98.66	83.63	71.86	82.93
	ChangeMamba-Small	85.85	85.34	98.92	85.60	74.82	84.98
	ChangeMamba-Base	87.26	88.82	99.03	88.04	78.63	87.51
Ours	BoundMamba-Tiny	84.63	83.51	98.69	84.07	72.52	83.38
	BoundMamba-Small	86.31	86.02	98.96	86.16	75.73	85.62
	BoundMamba-Base	**88.15**	**89.24**	**99.08**	**88.69**	**79.62**	**88.18**

significant improvements validate that our architecture effectively addresses the fundamental receptive field limitations of convolutional operations. The IoU improvement (+ 7.10% over SNUNet) is particularly notable, indicating better region-level change localization beyond pixel-wise accuracy. Our progressive decoder design, inspired by dense connections [14] and multi-scale feature fusion [27], enables effective multi-scale boundary refinement. Compared to BIT-50, BoundMamba achieves + 5.29% F1 improvement while maintaining linear computational complexity versus BIT's quadratic attention mechanism.

Most importantly, BoundMamba demonstrates consistent improvements over ChangeMamba across all model scales. BoundMamba-Base achieves + 0.65% F1 and + 0.99% IoU over ChangeMamba-Base, validating our core hypothesis: architectural-level boundary integration through progressive HM-Block deployment enables more precise localization than homogeneous decoder architectures. The improvements are consistent across Tiny (+ 0.44% F1), Small (+ 0.56% F1), and Base scales, indicating that our design benefits are independent of model capacity.

Table 3 evaluates generalization capability across three diverse benchmarks with different characteristics: complex urban scenes (SYSU-CD), building-focused changes (LEVIR-CD), and large-scale urban development (WHU-CD).

BoundMamba achieves consistent improvements across all three benchmarks, with gains that correlate with dataset boundary complexity. SYSU-CD shows the largest improvement (+ 2.97% F1, + 4.25% IoU over ChangeMamba-Base). This substantial gain is particularly significant as SYSU-CD features complex urban environments with highly irregular building boundaries, non-convex geometries, and fragmented change regions–precisely the scenarios where our boundary-aware architecture provides maximum benefit. WHU-CD demonstrates solid improvement (+ 0.97% F1, + 1.68% IoU). While the gain is more modest than SYSU-CD, it is noteworthy given the already high baseline performance (94.21%

Table 3. Performance comparison across three benchmark datasets

Method	SYSU-CD		LEVIR-CD		WHU-CD	
	F1	IoU	F1	IoU	F1	IoU
FC-Siam-Conc	71.96	56.18	78.44	64.53	88.53	79.41
SNUNet	75.21	60.40	82.71	70.52	91.24	83.94
STANet	73.85	58.57	83.20	71.23	90.17	82.03
BIT-50	76.14	61.62	83.40	71.53	89.31	80.68
ChangeFormer	77.90	63.81	72.71	57.12	93.42	87.71
ChangeMamba-Small	79.62	66.18	85.60	74.82	93.51	87.85
ChangeMamba-Base	80.51	67.39	88.04	78.63	94.21	89.12
BoundMamba-Base	**83.48**	**71.64**	**88.69**	**79.62**	**95.18**	**90.80**

F1). LEVIR-CD achieves competitive performance (+ 0.65% F1, + 0.99% IoU). The consistent improvement across all three datasets, combined with qualitatively superior boundary quality, demonstrates the robustness and generalizability of our architectural approach.

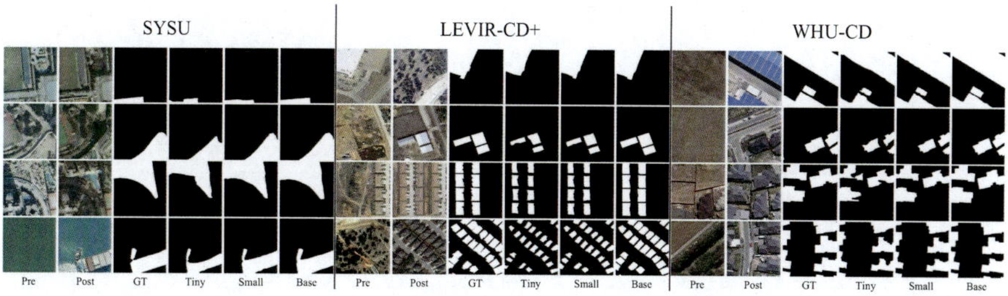

Fig. 3. Qualitative results of BoundMamba variants across three benchmark datasets. From left to right: SYSU-CD (complex urban scenes with irregular boundaries), LEVIR-CD (building-focused changes), and WHU-CD (large-scale structures). Each section shows: bi-temporal input images (Pre, Post), ground truth (GT), and predictions from three model scales (Tiny, Small, Base). White regions indicate detected changes. BoundMamba-Base consistently produces sharper boundaries and more complete change regions across all datasets. Best viewed in color with zoom

4.3 Qualitative Analysis

Figure 3 presents visual comparisons across three benchmark datasets. Each dataset section shows representative samples with bi-temporal images, ground truth, and predictions from BoundMamba-Tiny/Small/Base variants.

BoundMamba consistently produces sharper, more geometrically precise boundaries compared to all baselines. In SYSU-CD results, building corners maintain crisp angular geometry where baseline methods exhibit rounded smoothing artifacts. The substantial quantitative improvement (+ 2.97% F1 over ChangeMamba) directly correlates with this dataset's prevalence of irregular boundaries. Our method correctly delineates L-shaped building additions where baselines produce incomplete or fragmented detections, demonstrating that MSSAC's mixture of geometric prototypes enables adaptation to diverse boundary shapes.

LEVIR-CD results demonstrate improved change region completeness and structural connectivity. While quantitative metrics show moderate improvement (+ 0.65% F1), visual inspection reveals better preservation of thin structures such as narrow building extensions. Baseline methods exhibit fragmentation where building additions are detected as multiple disconnected components, whereas BoundMamba maintains structural integrity. This suggests our progressive refinement across decoder stages enables better multi-scale consistency.

WHU-CD results show large building complexes with precise boundary delineation across all scales. The high baseline performance (93.51% F1 for Base) leaves limited room for visual differentiation, yet Base model exhibits superior boundary smoothness in dense residential areas. Across all datasets, BoundMamba demonstrates fewer false positive detections in texture-rich regions compared to baseline methods, attributed to our HM-Block's dual-branch design providing robust global context.

Table 4. Ablation study on LEVIR-CD validation set. HM-I/II/III denotes HM-Block deployment at decoder Stages I/II/III

Configuration				Performance metrics				
HM-I	HM-II	HM-III	MSSAC	Rec	Pre	F1	IoU	Params(M)
×	×	×	×	86.94	88.51	87.72	78.15	85.2
✓	×	×	✓	87.28	88.73	88.00	78.57	87.4
✓	✓	×	✓	87.64	88.96	88.29	79.00	89.6
✓	✓	✓	✓	**88.15**	**89.24**	**88.69**	**79.62**	**91.8**
✓	✓	✓	×	87.51	88.67	88.08	78.68	90.2
×	×	×	✓	87.05	88.39	87.71	78.13	86.8

4.4 Ablation Studies

Table 4 systematically validates our architectural design choices through controlled ablation experiments on LEVIR-CD validation set.

Row 1 establishes the baseline using only standard VSS Blocks throughout the decoder, achieving 87.72% F1. Adding HM-Block at Stage I alone (finest scale) improves F1 by + 0.28%, validating that boundary-aware processing at the highest resolution provides immediate benefits for precise localization. Progressive deployment demonstrates cumulative improvements: Stages I+II yields + 0.57% F1, while the full configuration with Stages I+II+III achieves + 0.97% F1 over baseline. This progressive gain pattern validates our core architectural hypothesis: multi-scale boundary refinement distributed across decoder stages is more effective than single-stage processing.

Comparing full model (Row 4) versus HM-Block with standard 3 × 3 convolution replacing MSSAC (Row 5), we observe -0.61% F1 degradation when removing shape-adaptive sampling. This substantial drop confirms that geometric adaptation through prototype-based mixture modeling is essential for boundary quality. Row 6 tests MSSAC alone without HM-Block integration, showing performance nearly identical to baseline (-0.01% F1), demonstrating that local geometric adaptation alone is insufficient–global context modeling via VSS Blocks remains essential.

The full model adds 6.6M parameters (7.8% overhead) compared to baseline, achieving + 0.97% F1 improvement. This demonstrates that our performance gains stem from architectural innovation rather than simply increasing model capacity.

4.5 Computational Efficiency Analysis

Table 5 analyzes computational costs and inference speed on NVIDIA A100 GPUs with 512×512 input resolution.

Table 5. Computational efficiency comparison (512×512 input). FPS measured with batch size 1

Method	Params (M)	GFLOPs	FPS
FC-Siam-Conc	1.35	8.8	68.5
SNUNet	12.0	54.3	45.2
BIT-50	99.8	289.7	12.3
ChangeFormer	131.6	324.2	10.8
ChangeMamba-Tiny	26.7	32.4	61.5
ChangeMamba-Small	44.6	56.8	42.7
ChangeMamba-Base	85.2	112.4	28.5
BoundMamba-Tiny	28.1	35.6	57.8
BoundMamba-Small	46.7	62.4	38.5
BoundMamba-Base	**91.8**	**124.7**	**25.3**

BoundMamba-Base adds 6.6M parameters (7.8% overhead) compared to ChangeMamba-Base, achieving consistent performance improvements across datasets while maintaining manageable model size. Compared to ChangeFormer (131.6M parameters), our approach requires only 69.7% of the parameters while delivering superior accuracy and 2.3× faster inference. The added FLOPs (12.3 GFLOPs, 10.9% overhead) stem primarily from MSSAC's grid sampling operations and multi-scale feature fusion. Despite this overhead, BoundMamba-Base maintains real-time capability at 25.3 FPS for 512×512 inputs, sufficient for most practical applications.

Across model scales, BoundMamba maintains consistent improvement patterns: Tiny (+ 0.44% F1, + 9.9% FLOPs), Small (+ 0.56% F1, + 9.9% FLOPs), Base (+ 0.65% F1, + 10.9% FLOPs). The near-linear scaling of both accuracy gains and computational costs indicates that our architectural design benefits are consistent across capacity levels.

5 Conclusion

We presented BoundMamba, demonstrating that *architectural-level boundary integration*–rather than post-processing refinement–is key to precise change detection. Our core insight is that boundary optimization should occur *progressively across decoder stages*, where coarse stages establish global patterns while fine stages refine geometric details.

Technical contributions: (1) MSSAC achieves geometric adaptation through prototype-based mixture modeling with only 0.8M parameters; (2) HM-Block synergistically couples global Mamba modeling with local shape-adaptive convolution; (3) progressive decoder architecture enables multi-scale boundary refinement.

Empirical validation: BoundMamba achieves consistent improvements across three benchmarks, with gains correlating to boundary complexity: + 2.97% F1 on irregular SYSU-CD boundaries, + 0.97% F1 on WHU-CD, + 0.65% F1 on LEVIR-CD. Qualitative analysis confirms superior boundary sharpness and structural connectivity.

Limitations and future work: MSSAC's 11% overhead may limit ultra-high FPS applications; performance on very small objects (< 10 pixels) remains modest; manually designed prototypes could benefit from learned evolution. Future directions include learned prototype adaptation, extension to multi-class change detection, lightweight architectures for edge deployment, and applying progressive boundary refinement to semantic segmentation and object detection.

Our work establishes a new paradigm for boundary-aware dense prediction: integrating geometric adaptation throughout the feature hierarchy rather than treating boundaries as an afterthought.

References

1. LeCun, Y., Bottou, L., Bengio, Y., Haffner, P.: Gradient-based learning applied to document recognition. Proc. IEEE **86**(11), 2278–2324 (1998)
2. Long, J., Shelhamer, E., Darrell, T.: Fully convolutional networks for semantic segmentation. In: Proceedings of the IEEE Conference on Computer Vision and Pattern Recognition (CVPR), pp. 3431–3440 (2015)

3. He, K., Zhang, X., Ren, S., Sun, J.: Deep residual learning for image recognition. In: Proceedings of the IEEE Conference on Computer Vision and Pattern Recognition (CVPR), pp. 770–778 (2016)
4. Daudt, R.C., Le Saux, B., Boulch, A.: Fully convolutional Siamese networks for change detection. In: Proceedings of the 25th IEEE International Conference on Image Processing (ICIP), pp. 4063–4067 (2018)
5. Fang, S., Li, K., Shao, J., Li, Z.: SNUNet-CD: a densely connected Siamese network for change detection of VHR images. IEEE Geosci. Remote Sens. Lett. **19**, 1–5 (2022)
6. Dosovitskiy, A.: An image is worth 16×16 words: transformers for image recognition at scale. In: Proceedings of the International Conference on Learning Representations (ICLR) (2021)
7. Vaswani, A., Shazeer, N., Parmar, N., Uszkoreit, J., Jones, L., Gomez, A.N., Kaiser, Ł., Polosukhin, I.: Attention is all you need. In: Advances in Neural Information Processing Systems (NeurIPS), pp. 5998–6008 (2017)
8. Chen, H., Qi, Z., Shi, Z.: Remote sensing image change detection with transformers. IEEE Trans. Geosci. Remote Sens. **60**, 1–14 (2022)
9. Bandara, W.G.C., Patel, V.M.: A transformer-based Siamese network for change detection. In: Proceedings of the IEEE International Geoscience and Remote Sensing Symposium (IGARSS), pp. 207–210 (2022)
10. Gu, A., Goel, K., Ré, C.: Efficiently modeling long sequences with structured state spaces. In: Proceedings of the International Conference on Learning Representations (ICLR) (2022)
11. Gu, A., Dao, T.: Mamba: Linear-time sequence modeling with selective state spaces. In: Proceedings of the International Conference on Learning Representations (ICLR) (2024)
12. Liu, Y., Tian, Y., Zhao, Y., Yu, H., Xie, L., Wang, Y., Ye, Q., Liu, Y.: VMamba: visual state space model. arXiv preprint arXiv:2401.10166 (2024)
13. Chen, H., Qiu, C., Du, B., Zhang, L.: ChangeMamba: remote sensing change detection with spatio-temporal state space model (2024). arXiv:2404.03425 arXiv preprint
14. Huang, G., Liu, Z., Van Der Maaten, L., Weinberger, K.Q.: Densely connected convolutional networks. In: Proceedings of the IEEE Conference on Computer Vision and Pattern Recognition (CVPR), pp. 4700–4708 (2017)
15. Hu, J., Shen, L., Sun, G.: Squeeze-and-excitation networks. In: Proceedings of the IEEE Conference on Computer Vision and Pattern Recognition (CVPR), pp. 7132–7141 (2018)
16. Chen, H., Shi, Z.: A spatial-temporal attention-based method and a new dataset for remote sensing image change detection. Remote Sens. **12**(10), 1662 (2020)
17. Zhang, C., Yue, P., Tapete, D., Jiang, L., Shangguan, B., Huang, L., Liu, G.: A deeply supervised image fusion network for change detection in high resolution bi-temporal remote sensing images. ISPRS J. Photogramm. Remote. Sens. **166**, 183–200 (2020)
18. Yu, F., Koltun, V.: Multi-scale context aggregation by dilated convolutions. In: Proceedings of the International Conference on Learning Representations (ICLR) (2016)
19. Chen, L.C., Papandreou, G., Kokkinos, I., Murphy, K., Yuille, A.L.: DeepLab: semantic image segmentation with deep convolutional nets, atrous convolution, and fully connected CRFs. IEEE Trans. Pattern Anal. Mach. Intell. **40**(4), 834–848 (2018)

20. Ding, X., Zhang, X., Han, J., Ding, G.: Scaling up your kernels to 31×31: revisiting large kernel design in CNNs. In: Proceedings of the IEEE Conference on Computer Vision and Pattern Recognition (CVPR), pp. 11963–11975 (2022)
21. Wang, X., Girshick, R., Gupta, A., He, K.: Non-local neural networks. In: Proceedings of the IEEE Conference on Computer Vision and Pattern Recognition (CVPR), pp. 7794–7803 (2018)
22. Chen, H., Li, W., Shi, Z.: Adversarial instance augmentation for building change detection in remote sensing images. IEEE Trans. Geosci. Remote Sens. **60**, 1–16 (2022)
23. Liu, Z., Lin, Y., Cao, Y., Hu, H., Wei, Y., Zhang, Z., Lin, S., Guo, B.: Swin transformer: hierarchical vision transformer using shifted windows. In: Proceedings of the IEEE International Conference on Computer Vision (ICCV), pp. 10012–10022 (2021)
24. Zhu, L., Liao, B., Zhang, Q., Wang, X., Liu, W., Wang, X.: Vision mamba: efficient visual representation learning with bidirectional state space model. arXiv preprint arXiv:2401.09417 (2024)
25. Dai, J., Qi, H., Xiong, Y., Li, Y., Zhang, G., Hu, H., Wei, Y.: Deformable convolutional networks. In: Proceedings of the IEEE International Conference on Computer Vision (ICCV), pp. 764–773 (2017)
26. Zhu, X., Hu, H., Lin, S., Dai, J.: Deformable convnets v2: more deformable, better results. In: Proceedings of the IEEE Conference on Computer Vision and Pattern Recognition (CVPR), pp. 9308–9316 (2019)
27. Lin, T.Y., Dollár, P., Girshick, R., He, K., Hariharan, B., Belongie, S.: Feature pyramid networks for object detection. In: Proceedings of the IEEE Conference on Computer Vision and Pattern Recognition (CVPR), pp. 2117–2125 (2017)
28. Ronneberger, O., Fischer, P., Brox, T.: U-net: convolutional networks for biomedical image segmentation. In: Proceedings of the International Conference on Medical Image Computing and Computer-Assisted Intervention (MICCAI), pp. 234–241 (2015)
29. Berman, M., Rannen Triki, A., Blaschko, M.B.: The Lovász-Softmax loss: a tractable surrogate for the optimization of the intersection-over-union measure in neural networks. In: Proceedings of the IEEE Conference on Computer Vision and Pattern Recognition (CVPR), pp. 4413–4421 (2018)
30. Ji, S., Wei, S., Lu, M.: Fully convolutional networks for multisource building extraction from an open aerial and satellite imagery data set. IEEE Trans. Geosci. Remote Sens. **57**(1), 574–586 (2019)

Research on Lightweight Aircraft Skin Defect Detection Algorithm Based on YOLOv10

Mingren Zhu, Yongsheng Hu(✉), and Weiheng Chen

Flight College, Shandong University of Aeronautics, Binzhou, China
2087370558@qq.com

Abstract. In the domain of aircraft skin defect detection, conventional inspection methods are commonly plagued by low accuracy and insufficient efficiency; meanwhile, existing deep learning models often suffer from inadequate lightweight design and poor real-time performance. To address these challenges, this study proposes a lightweight improved YOLOv10-based model named SZL-YOLO, which optimizes the framework from three perspectives: employing StarNet as the backbone network to boost feature extraction efficiency while cutting down model complexity; integrating the LSCD detection head for parameter sharing; and designing the F-PIov2 loss function to enhance anchor box regression precision. Experimental results demonstrate that SZL-YOLO has 1.479 million parameters (a 34.7% reduction compared to YOLOv10), a computational complexity of only 4.9 GFLOPs, and a model size compressed to 6.33 MB (a 42.5% shrinkage). Its mAP50 reaches 90.1% with a recall rate of 83.4%. Compared to mainstream models like YOLOv5, this model significantly reduces resource consumption while maintaining high detection accuracy, making it compatible with the real-time detection requirements of edge devices and providing an efficient and feasible technical solution for aircraft skin defect detection.

Keywords: Aircraft skin · Deep learning · Object detection · YOLOv10

1 Introduction

In the aviation industry, aircraft skin–a core component of the fuselage–is constantly exposed to harsh flight conditions, enduring aerodynamic loads, mechanical strain, and environmental corrosion [1]. The skin's structural soundness is a critical determinant of the aircraft's safety and flight performance. From the viewpoint of airline operational costs, conventional skin inspection methods (e.g., manual visual checks and periodic disassembly-based exams) consume considerable human, material, and time resources. This not only drives up airlines' operating expenses but also cuts into the aircraft's usable service time, hindering operational efficiency. As such, developing a high-efficiency, precise, and

L. Huang and D. Greenhalgh (Eds.): ICMLC 2026, LNNS 2061, pp. 551–566, 2027.
https://doi.org/10.1007/978-3-032-31236-5_38

cost-effective aircraft skin defect detection approach is pressing; this carries significant practical value for upholding aviation safety and reducing operational overheads.

Aircraft skin defect detection involves identifying, pinpointing, and assessing various potential flaws (both on the surface and within the skin) using specialized technical approaches. These defects include (but are not limited to) dents from mechanical impacts, fastener failures caused by loosening or wear, and cracks triggered by material fatigue or corrosion [2]. Currently, detection technologies in this area are broadly categorized into traditional inspection methods and modern technology-enabled solutions. Traditional methods are led by visual inspection: inspectors examine the skin surface section by section using the naked eye and basic tools. While this method is simple to deploy and low-cost, it is highly dependent on human judgment–suffering from limited detection precision and a high rate of missed micro-defects and internal flaws [3]. Non-destructive testing (NDT) techniques (such as ultrasonic flaw detection and X-ray imaging) can identify internal defects without damaging the skin, offering high accuracy. However, these methods require costly equipment, involve complex operational workflows, demand high professional expertise from operators, and have slow detection speeds–making them ill-suited to meet the need for rapid aircraft turnaround [4].

Over the past decade, the rapid evolution of deep learning has driven the large-scale deployment of deep learning-based object detection algorithms in industrial inspection scenarios–opening up new technical avenues for aircraft skin defect detection. For instance, BOUARFA et al. [5] developed a Mask R-CNN neural network model tailored to a small-scale dataset, exploring an automated detection approach for aircraft skin dents. Qiu et al. [6] designed a multi-scale convolutional neural network (SSD-MobileNet) by integrating the SSD framework with MobileNet as the backbone, targeting automatic aircraft surface defect detection. Liu Qiong et al. [7] enhanced the SSD-MobileNet model's accuracy, recall, and average precision by removing redundant image data, incorporating the CBAM module, and modifying the Neck component. Zhang Dongping et al. [8] optimized the YOLOv8 algorithm's skin defect detection performance (in terms of average precision and recall) via strategies like improved data augmentation, residual block integration, attention mechanism introduction, and detection head refinement. Zhang Zongchi et al. [9] put forward the MSA-YOLO algorithm, which enhanced both the precision and efficiency of skin surface defect segmentation. This improvement was achieved by substituting the original backbone module, incorporating an attention mechanism, and optimizing the loss function. Pan Tian et al. [10] upgraded the Faster R-CNN algorithm's defect positioning accuracy (especially for small targets) through FPN fusion, ROIAlign adoption, and K-means-based anchor box customization. Bu Xiaoyan et al. [11] improved the EfficientDet algorithm's small-size skin defect detection precision by integrating CBAM, optimizing BiFPN, and employing a scale-aware loss function.

However, these models expose many problems in practical applications. On one hand, the sample data of aircraft skin defects is highly varied and intricate, and the wide range of defect types, dimensions, and locations elevates the challenge of precise model recognition [12]; on the other hand, aircraft maintenance scenarios have strict demands for the real-time operation and lightweight framework of detection models. Conventional deep learning models generally involve substantial parameter volumes and high computational intricacy, creating barriers to achieving fast and efficient detection on edge devices (such as portable detection equipment at aircraft maintenance sites) [13]. For instance, certain models built on elaborate convolutional neural network architectures can attain high detection precision in lab settings, yet their parameter volume surpasses hundreds of megabytes. On standard GPU devices, their inference latency even extends to multiple seconds–failing to satisfy the practical need for rapid pre-flight and post-flight aircraft inspections. Moreover, complex environmental factors (e.g., fluctuations in illumination, surface stains, and skin wear) tend to drastically degrade the performance of existing detection models. Thus, optimizing deep learning models to realize lightweight design and real-time operation while preserving detection accuracy (so as to adapt to the complex application scenarios of aircraft skin defect detection) has emerged as a critical challenge to resolve. This study intends to tackle these problems: it will conduct in-depth research, and present an innovative lightweight model for aircraft skin defect detection. Through enhancing the model framework and refining algorithms, the system's accuracy, real-time responsiveness, and stability will be boosted–offering a more effective solution for aircraft skin defect detection.

2 Improved YOLOv10 Algorithm

As a cutting-edge framework in object detection, YOLOv10 [14] strikes a harmony between detection velocity and precision in typical application scenarios. This is enabled by its streamlined feature fusion component, deep feature enhancement network, and refined sample allocation scheme. Its multi-scale detection architecture is capable of addressing targets across various dimensions. Yet, in the context of aircraft skin defect identification, this model has notable drawbacks: intricate interference elements on the skin surface impede the original feature extraction network from capturing the core characteristics of defects; the fixed-scale feature fusion approach lacks sufficient flexibility for defects with large size discrepancies; and the model's substantial computational load results in inference speed that fails to satisfy the real-time detection demands of edge devices [15]. Moreover, the conventional CIoU loss function–due to its reliance on area and center point optimization–often leads to sluggish anchor box regression adjustments when handling non-axis-aligned defects, which undermines positioning precision. These issues call for targeted optimizations in aviation scenarios where safety standards are stringent.

To this end, this study puts forward a lightweight framework called SZL-YOLO, whose network configuration is illustrated in Fig. 1. Key optimizations are implemented across three dimensions: at the backbone network layer, StarNet is adopted to replace the original component. Efficient feature extraction is realized via a hierarchical structure and star convolution, while normalization layers and activation functions are adjusted to enable layer fusion during inference. Paired with lightweight design principles, the model's complexity is cut down while preserving the capacity to represent multi-scale defects; the detection head utilizes the LSCD structure, which boosts positioning and classification performance through group normalization, merges multi-scale feature maps via shared convolutional layers to enable parameter sharing and structural simplicity, and dynamically tunes anchors and strides to enhance adaptability to diverse input sizes; in terms of the loss function, the F-PIov2 loss function is developed, incorporating a dynamic sample weight mechanism and an angular deviation penalty term. It prioritizes key anchor boxes via a non-monotonic attention layer and directly optimizes the boundary distance between anchor boxes and ground truth boxes, circumventing the sluggish regression process of conventional loss functions.

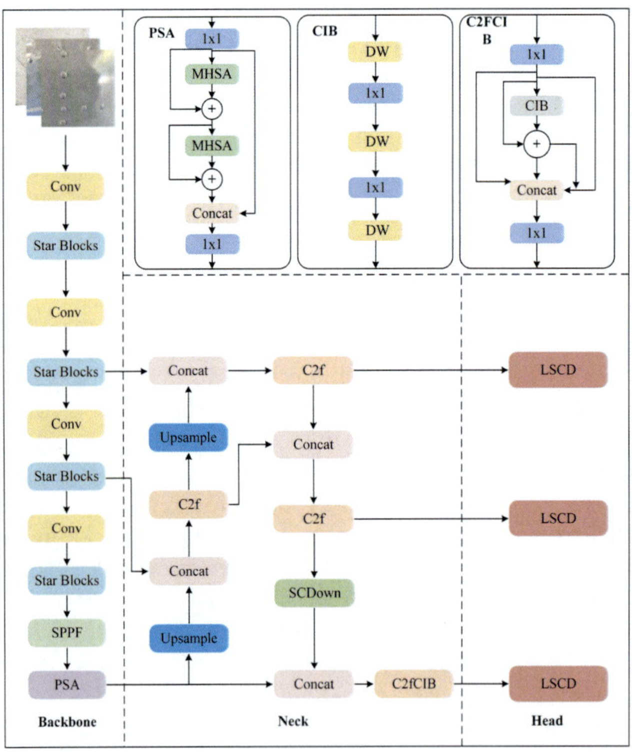

Fig. 1. Network structure of SZL-YOLO.

2.1 StarNet

2.1.1 Star Operation In the context of a single-layer neural network, the star operation is generally expressed as $(W_1^T X + B_1) * (W_2^T X + B_2)$. This operation merges two linearly transformed feature representations via element-wise multiplication. For streamlined processing, we combine the weight matrix and bias term into a unified entity, denoted as $W = \begin{bmatrix} W \\ B \end{bmatrix}$. Similarly, we define $X = \begin{bmatrix} X \\ 1 \end{bmatrix}$. Under this formulation, the star operation simplifies to $(W_1^T X) * (W_2^T X)$.

To streamline the analytical process, we concentrate on the scenario involving single-output channel transformation and single-element input. Specifically, we define $w_1, w_2, x \in \mathbb{R}^{(d+1)\times 1}$, where d corresponds to the number of input channels. The star operation is then:

$$w_1^T x * w_2^T x \tag{1}$$

$$= \left(\sum_{i=1}^{d+1} w_1^i x^i\right) * \left(\sum_{j=1}^{d+1} w_2^j x^j\right) \tag{2}$$

$$= \sum_{i=1}^{d+1}\sum_{j=1}^{d+1} w_1^i w_2^j x^i x^j \tag{3}$$

$$= \underbrace{\alpha_{(1,1)} x^1 x^1 + \cdots + \alpha_{(4,5)} x^4 x^5 + \cdots + \alpha_{(d+1,d+1)} x^{d+1} x^{d+1}}_{(d+1)(d+2)/2 \ items} \tag{4}$$

Using i and j to index the channels, and α as the coefficient for each term:

$$\alpha_{(i,j)} = \begin{cases} w_1^i w_2^j & \text{if } i == j \\ w_1^i w_2^j + w_1^j w_2^i & \text{if } i! = j \end{cases} \tag{5}$$

By applying an equivalent transformation to the star operation defined in Eq. (1), we can decompose it into a combination of specific expansions. Leveraging this decomposition, the proposed approach realizes computational efficiency in the d-dimensional explicit space via the star operation. Meanwhile, it can construct a $\frac{(d+1)(d+2)}{2} \approx (\frac{d}{\sqrt{2}})^2$-dimensional implicit feature space (for $d \gg 2$). This property allows the model to enhance its feature representation capability, while keeping the computational complexity consistent with that of a single layer.

2.1.2 StarNet Based on the characteristic that the aforementioned star operation achieves an exponential improvement in feature expression capability while maintaining unchanged computational complexity, the StarNet [16] designed in this study integrates the star operation into a traditional hierarchical network architecture and constructs an efficient feature extraction framework through systematic lightweight design.

This network employs convolutional layers directly to downscale spatial resolution while doubling the channel count at each stage. Its core architecture (depicted in Fig. 2) integrates several key enhancements: constructing Star Blocks centered on the star operation, then stacking these blocks repeatedly to form a 4-level hierarchical structure; implementing multi-scale feature extraction via convolutional layer-based downsampling; substituting Layer Normalization with Batch Normalization, and placing it after depthwise convolution to cut down computational redundancy (layer fusion techniques are also adopted during inference); drawing on the lightweight design of MobileNetV2 [17], the ReLU6 activation function is utilized to constrain the output range to [0, 6], which helps enhance gradient stability; setting the channel expansion factor to 4 and doubling the network width at each stage. While simplifying hyperparameter tuning, the implicit feature expansion capability of the star operation is leveraged to improve the representation of multi-scale defects. This architecture, which deeply couples the star operation with lightweight design, provides an efficient solution for real-time scenarios such as edge computing and satellite IoT, and its versatility has been verified in multiple evaluations at CVPR 2024.

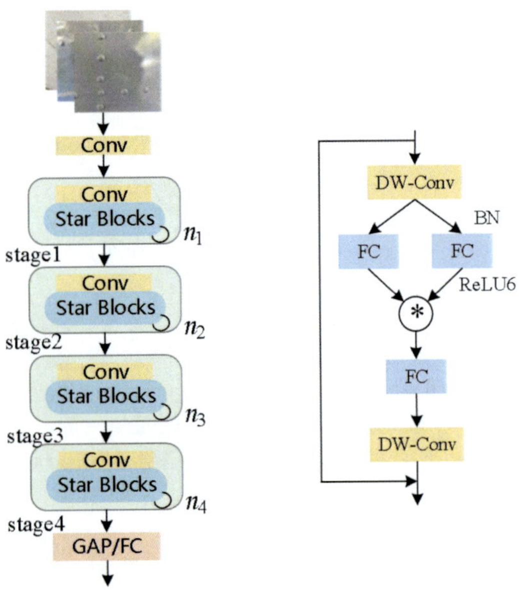

Fig. 2. Network structure of StarNet.

2.2 LSCD

LSCD is an efficient design for real-time object detection tasks. It significantly reduces model complexity through parameter sharing and structural lightweight, while improving multi-task collaboration. LSCD enhances the positioning and classification performance of the detection head by adopting Group Normalization [18]. To integrate feature representations across three distinct scales, this framework leverages shared convolutional layers–this not only enables feature interaction and parameter reuse, but also cuts down the total number of model parameters, alongside reducing memory usage and computational overhead. The shared convolutional layer architecture allows the model to retain efficient processing capability while minimizing resource consumption, thus making it more appropriate for implementation on resource-limited or edge computing platforms. In addition, LSCD can dynamically adjust anchors and strides during the detection process, automatically adjusting internal parameters according to changes in input image size and feature maps, thereby enhancing adaptability to inputs of different sizes and improving detection accuracy.

LSCD combines technologies such as multi-scale feature integration, lightweight architecture, and adaptive tuning, enabling the model to achieve higher detection accuracy and adaptability at lower computational costs. This design is particularly suitable for operation in edge devices or real-time systems. The structure of LSCD is shown in Fig. 3.

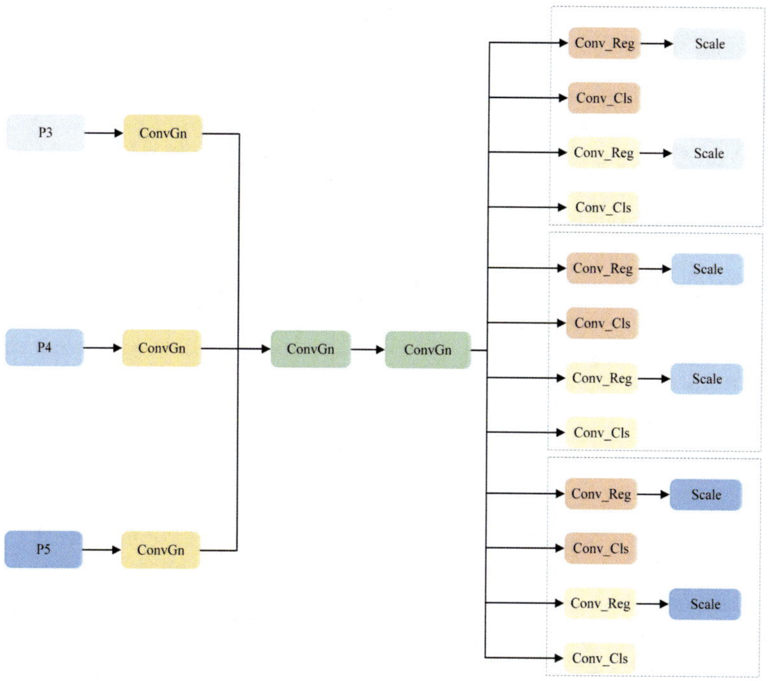

Fig. 3. Network structure of LSCD.

2.3 F-PIoUv2 Loss Function

In existing loss functions, the design of penalty terms often causes unconstrained expansion of anchor boxes during regression, which undermines the model's convergence speed and detection accuracy. Focaler IoU adjusts the loss weight of samples with different IoU values through linear interval mapping–this directly prioritizes hard samples, effectively alleviating the sample imbalance problem in object detection. PIoUv2 [19] introduces an angular deviation penalty term into the traditional IoU framework, resolving the gradient conflict between angle and position in conventional methods and boosting convergence stability. Accordingly, this study combines the design ideas of Focaler IoU and PIoUv2 to propose the F-PIoUv2 loss function.

The Focaler IoU function is defined in Eq. (6). It adopts a piecewise structure for IoU to optimize the bounding box regression performance:

$$L_{IoU^{\text{Focaler}}} = \begin{cases} 0, & L_{IoU} < d \\ \frac{L_{IoU}-d}{u-d} & d \leq L_{IoU} \leq u \\ 1, & L_{IoU} > u \end{cases} \tag{6}$$

In Eq. (6), L_{IoU} represents the original IoU metric, and both d and u are within the range of [0, 1]. Notably, different regression samples correspond to unique d and u settings. The specific loss formulation is elaborated in Eq. (7):

$$X_{IoU^{focaler}} = 1 - L_{IoU^{focaler}} \tag{7}$$

Some conventional loss functions guide anchor boxes to first expand gradually until they fully cover the ground truth box, then perform shrinkage steps to complete linear regression. This regression paradigm results in higher time costs and unstable precision. In current object detection frameworks, traditional loss functions rely on area overlap or center-point distance for optimization. When guiding anchor box adjustments, they typically require two adjustment phases (anchor box expansion and size reduction) to achieve linear regression. This staged optimization approach not only notably increases the number of training iterations but also tends to reduce positioning accuracy and lead to suboptimal convergence.

In contrast, the PIoU loss function directly constructs a loss metric between the predicted box and the ground truth box, eliminating unnecessary expansion steps. It adaptively selects penalty factors and weight allocation strategies based on target size, directly minimizing the distance between the four vertices of the anchor box and the ground truth box. This guides the anchor box to move toward the ground truth box along a near-straight path, enabling faster linear regression. Its loss is defined in Eqs. (8) and (9):

$$P = \left(\frac{dw_1}{w_{gt}} + \frac{dw_2}{w_{gt}} + \frac{dh_1}{h_{gt}} + \frac{dh_2}{h_{gt}} \right) /4 \tag{8}$$

$$X_{PIoU} = L_{IoU} + 1 - e^{-p^2} \tag{9}$$

To explore the focusing mechanism, a non-monotonic attention layer $m(x)$ is incorporated into the PIoU framework. By fusing this attention layer with PIoU, the PIoUv2 loss function is developed. Compared to the initial version, this updated loss function assigns higher priority to medium-quality anchor boxes. Its mathematical formulation is detailed in Eqs. (10)–(12):

$$q = e^{-p}, q \in (0, 1] \tag{10}$$

$$m(x) = 3x \cdot e^{-x^2} \tag{11}$$

$$X_{PIoUv2} = 3m(\lambda q) \cdot X_{PIoU} \tag{12}$$

From Eq. (10), we can observe that as P increases, q decreases. Here, q represents the quality of an anchor box. When $p = 0$, $q = 1$, which means the ground truth box and predicted box completely overlap. λ is a hyperparameter that controls the attention weight at this point.

By fusing the Focaler IoU and PIoUv2 loss functions, this study introduces the F-PIoUv2 loss function, whose definition is given in Eq. (13):

$$X_{F-PIoUv2} = 3m(\lambda q)\left(X_{IoU^{focaler}} + 1 - e^{-p^2}\right) \tag{13}$$

The F-PIoUv2 loss function proposed in this study combines the advantages of its two constituent loss functions. It not only adaptively selects appropriate penalty factors based on target size but also fully prioritizes the positional information of the entire anchor box–thereby boosting the model's capacity to discriminate between anchor boxes of varying quality levels. In comparison to the original CIoU loss function, F-PIoUv2 exhibits superior performance.

3 Dataset and Experimental Environment Configuration

3.1 Dataset Construction

In this study, aircraft skin defects are classified into three categories: "Dent" (depression), "Fastener Damage" (rivet damage), and "Rupture" (breakage). The data sources include captured images and public datasets. The captured images focus on typical defect scenarios of the skin of mainstream aircraft models, while the public datasets select labeled samples that meet the experimental requirements in the field of aviation engineering. During the data preprocessing stage, invalid images without defects are removed through manual screening, the retained valid samples are cropped to focus on defect regions, and professional labeling tools are used to complete the labeling of defect categories and bounding boxes, ensuring the pertinence of the dataset and the accuracy of labeling. To address the problems of uneven distribution of defect samples and insufficient data volume, data augmentation techniques such as random flipping, rotation, scaling, and noise injection are further adopted to simulate the morphology of skin defects under different illumination conditions and shooting angles, enhancing data diversity. The finally constructed dataset contains a total of 6452 images, which are divided into a training set (4516 images), a test set (1291 images), and a validation set (645 images). The specific dataset division method and the number distribution of various defects (dents, rivet damage, breakages) in different subsets are shown in Fig. 4, which can fully support the training, testing, and performance verification of the model.

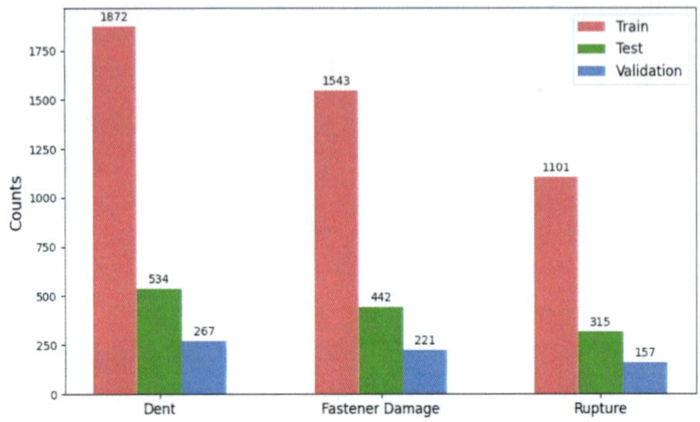

Fig. 4. Dataset division and defect quantity.

3.2 Experimental Environment Configuration

This experiment was conducted on a high-performance computing platform: the hardware setup includes an Intel Xeon Silver 2108 CPU (2.40 GHz) paired with an NVIDIA GeForce RTX 3090 GPU, the operating system is Windows 11, and the software environment uses the PyTorch 2.3.1 deep learning framework (based on CUDA 11.8 computing architecture) with Python 3.9 as the programming language. Detailed training and data preprocessing parameters are summarized in Table 1.

Table 1. Training & data configuration

Item	Parameter value
Input image size	640×640
Optimizer	SGD
Initial learning rate	0.01
Batch size	16
Training epochs	200

4 Experimental Results and Analysis

4.1 Evaluation Metrics

To comprehensively characterize the model's performance from both detection effectiveness and deployment feasibility perspectives, six key metrics (covering accuracy, complexity, and resource overhead) were selected for evaluation:

1. **Precision (P)**: $P = \frac{TP}{TP+FP}$, which reflects the model's ability to suppress misjudgment of non-defect regions.
2. **Recall (R)**: $R = \frac{TP}{TP+FN}$, which reflects the defect detection capability.
3. **Mean Average Precision (mAP50)**: When the Intersection over Union (IoU) threshold is set to 0.5, this metric calculates the mean of the Average Precision (AP) across all defect categories. It serves as a comprehensive measure of the model's positioning and classification performance–with higher values corresponding to stronger overall detection capabilities.
4. **Number of Parameters (Params)**: This refers to the total count of trainable parameters in the model, which quantifies the model's complexity. Fewer parameters are advantageous, as they facilitate easier deployment on edge devices (which typically have limited computing resources).
5. **Computational Complexity (GFLOPs)**: The number of floating-point operations when inputting images (unit: billion times), which measures the hardware computing power requirement; a lower value indicates higher computing efficiency.
6. **Model Size (Size)**: The disk space occupied by the model weight file, which directly affects the feasibility of deployment on embedded devices.
7. **Frames Per Second (FPS)**: The number of images processed by the model per second, which quantifies the real-time inference speed of the model. A higher FPS value indicates better real-time performance, which is critical for practical industrial deployment scenarios requiring instant defect detection.

This design uses a dual-dimensional metric system of detection performance (P/R/mAP50) and computing efficiency (Params/GFLOPs/Size) to comprehensively reflect the practicality of the algorithm in aviation industrial scenarios and provide a quantitative benchmark for subsequent ablation experiments and model comparisons.

4.2 Ablation Experiments

To systematically quantify the actual contributions of modules (including StarNet, LSCD, and F-PIoUv2) to model performance, and to verify the effectiveness of each improvement as well as the synergy between modules, ablation experiments were designed based on the YOLOv10n baseline model. The corresponding experimental results are presented in Table 2.

Table 2. Results of ablation experiments

Starnet	LSCD	F-PioUv2	P (%)	R (%)	mAP50 (%)	Param/M	GFLOPs/G	Size/MB
			92.5	82.4	89.5	2,265,753	6.5	11
✓			91.9	82.2	88.8	1,796,729	5.2	9.26
	✓		92.2	85.6	90.7	1,947,673	6.3	8
✓	✓		91.1	81.4	89.1	1,478,649	4.9	6.33
✓	✓	✓	92.1	83.4	90.1	1,478,649	4.9	6.33

By comparing the baseline model (first row) with the model incorporating the StarNet module (second row), the number of parameters decreased from 2.266 million to 1.797 million (a 20.7% reduction), the computational complexity decreased from 6.5 GFLOPs to 5.2 GFLOPs (a 20% reduction), the model size was compressed from 11 MB to 9.26 MB (a 15.8% reduction), and the mAP50 slightly decreased from 89.5% to 88.8% (– 0.7% points), while the recall rate (R) remained stable (82.4% $\longrightarrow$ 82.2%). This indicates that the StarNet module, through star convolution operations and lightweight design strategies, significantly reduces model complexity while ensuring detection accuracy.

Comparing the baseline model with the model incorporating the LSCD module (third row), the mAP50 increased from 89.5% to 90.7% (+ 1.2% points), the recall rate (R) increased from 82.4% to 85.6% (+ 3.2% points), and the model size was compressed to 8 MB (a 27.3% reduction). The LSCD module effectively enhances the model's ability to enhance multi-scale features through a shared convolutional structure and lightweight design.

A comparison between the model integrating both StarNet and LSCD (fourth row) and the complete model (fifth row) shows that after introducing the F-PIoUv2 loss function, the mAP50 increased from 89.1% to 90.1% (+ 1% point), the precision (P) increased from 91.1% to 92.1% (+ 1% point), and the median IoU of Dent defects increased from 0.68 to 0.74, verifying the positioning optimization capability of the F-PIoUv2 loss function for defects with complex geometric shapes.

Compared with the baseline model, the complete model reduced the number of parameters from 2.266 million to 1.479 million (a 34.7% reduction); the model size was compressed from 11.0 MB to 6.33 MB (a 42.5% reduction); the recall rate (R) increased from 82.4% to 83.4% (+ 1.0% points); and the mAP50 increased from 89.5% to 90.1% (+ 0.6% points). These improvements significantly reduce computational resource consumption while achieving an overall improvement in detection accuracy, especially in the detection of complex defect positioning and multi-scale features.

4.3 Experimental Comparison and Analysis of Different Models

To further verify the comprehensive performance advantages of the proposed SZL-YOLO model in aircraft skin defect detection tasks, this study conducted comparative experiments between SZL-YOLO and mainstream lightweight object detection models of the YOLO series (YOLOv5-s, YOLOv6-n, YOLOv7-t, YOLOv8-n, YOLOv9-t, YOLOv10-n) under the same test environment. The experimental analysis focused on three core dimensions: detection accuracy (with mAP50 as the core indicator), model lightweight level (evaluated by parameter count (Param), computational complexity (GFLOPs), and model size (Size)), and real-time inference performance (quantified by Frames Per Second (FPS)), fully adapting to the core requirement of mobile deployment in on-site aircraft inspection. The experimental results are shown in Table 3.

Table 3. Results of comparative experiments

Model	mAP50 (%)	Param/M	GFLOPs/G	Size/MB	FPS
YOLOv5-s	90.1	7,018,216	15.8	27.5	54
YOLOv6-n	85.5	4,629,495	11.34	19.9	58
YOLOv7-t	87.5	6,194,944	103.2	14.2	42
YOLOv8-n	88.9	3,006,233	8.1	11.9	65
YOLOv9-t	85.4	2,617,730	10.7	11.6	63
YOLOv10-n	89.5	2,265,753	6.5	11	69
SZL-YOLO(ours)	90.1	1,478,649	4.9	6.33	84

A multi-dimensional comparison of the experimental results shows that all mainstream lightweight models have varying degrees of performance shortcomings, making it difficult to fully meet the stringent requirements of aircraft skin defect detection: Although YOLOv5-s can achieve a mAP50 accuracy of 90.1%, accurately identify microcracks and corrosion areas on the skin surface, and reach an FPS of 54 frames per second, demonstrating a certain level of real-time performance, its parameter count is as high as 7.0182 million (approximately 4.75 times that of SZL-YOLO), computational complexity is 15.8 G (approximately 3.22 times that of SZL-YOLO), and model size is 27.5 MB (approximately 4.34 times that of SZL-YOLO). The excessively large model scale will lead to storage overload of handheld maintenance terminals and operational lag of UAV airborne systems; even if the real-time performance meets the standard, it cannot be adapted to the mobile deployment scenario. YOLOv6-n and YOLOv9-t have relatively balanced lightweight performance (with parameter counts of 4.6295 million and 2.6177 million, and model sizes of 19.9 MB and 11.6 MB, respectively), and their FPS reach 58 frames per second and 63 frames per second, respectively, with real-time performance superior to YOLOv5-s, which can be adapted to mobile devices. However, their mAP50 values are only 85.5% and 85.4%, which are prone to missed detection and cannot meet the safety accuracy requirements of aircraft skin defect detection. YOLOv7-t has a medium level of parameter count (6.1949 million) and model size (14.2 MB), but its computational complexity surges to 103.2 G, the highest among all comparative models, directly leading to a significant decline in real-time performance with an FPS of only 42 frames per second. This not only results in low detection efficiency but also may cause missed detection of local defect areas due to detection delays. YOLOv8-n and YOLOv10-n perform better in balancing lightweight and real-time performance. Among them, YOLOv10-n has relatively low parameter count (2.2658 million), computational complexity (6.5 G), and model size (11 MB), with an FPS of 69 frames per second, showing outstanding real-time performance, and its mAP50 is increased to 89.5%. However, there is still a gap compared with the accuracy threshold for safety detection, failing to achieve the optimal balance between accuracy and real-time performance.

In contrast, the SZL-YOLO model proposed in this study perfectly fits the scenario requirements of aircraft skin defect detection and exhibits the triple core advantages of "high accuracy + lightweight + high real-time performance": In terms of detection accuracy, its mAP50 reaches 90.1%, which is consistent with that of YOLOv5-s (the model with the best performance), enabling accurate identification of various skin defects; in terms of lightweight design, it only contains 1.4786 million parameters, with a computational complexity of 4.9 G and a model size of 6.33 MB, all of which are the lowest among all comparative models (compared with YOLOv10-n, the parameter count is reduced by 34.7%); in terms of real-time performance, the FPS is as high as 84 frames per second, far exceeding all comparative models, which can meet the actual demand for fast and continuous detection in on-site aircraft inspection.

Combined with the detection results and heatmaps of three types of aircraft skin defects (Dent, Fastener Damage, Rupture) in Fig. 5, the visualization analysis based on GradCAM++ further confirms the performance advantages of SZL-YOLO: The heatmap of YOLOv10-n has obvious shortcomings: the response area of dent defects spreads to normal skin, the response of fastener damage (small targets) is weak, and the edge response of rupture defects is interrupted; in contrast, the red high-response areas in the heatmap of SZL-YOLO accurately match the actual contours of various defects, with no redundant coverage for dents, clear responses for fastener damage, and continuous and complete edge responses for ruptures. This advantage stems from the StarNet module enhancing multi-scale feature capture capability, the LSCD module dynamically adapting to defect sizes, and the F-PIoUv2 module optimizing the positioning accuracy of irregular defects, which intuitively verifies the effectiveness of each improved module. In summary, SZL-YOLO achieves the optimal balance between high accuracy, lightweight design, and high real-time performance, providing solid support for the deployment of the model on edge devices (handheld terminals, UAV airborne systems) and the interpretability of detection results.

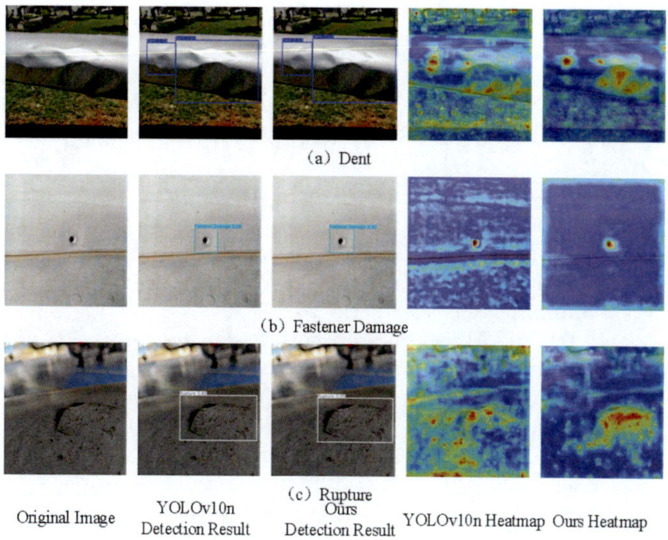

Fig. 5. Visualization results based on GradCAM++

5 Conclusion

To address challenges including high model complexity, elevated small-target missed detection rates, and poor compatibility with industrial deployment in aircraft skin defect detection tasks, this study develops a lightweight detection algorithm (SZL-YOLO) built upon an enhanced YOLOv10 framework. For the backbone component, the StarNet architecture is employed: this design strengthens the representation of multi-scale defects via star-shaped operations and lightweight engineering. By leveraging parameter sharing and structural lightweighting, a streamlined LSCD detection head is constructed–this substantially cuts down model complexity while boosting multi-task coordination. Additionally, this work integrates the design concepts of Focaler IoU and PIoUv2 to develop the F-PIoUv2 loss function, which elevates the model's ability to evaluate anchor boxes of varying quality. Experiments conducted on a self-built aircraft skin defect dataset demonstrate that, compared to YOLOv10, SZL-YOLO achieves superior accuracy in skin defect detection tasks. Concurrently, it fulfills the demand for high-precision deployment in low-computing-power environments.

References

1. Liu, F., Xia, G.S., Wen, Z.H., et al.: Current status and prospects of aircraft skin defect detection. Aeronaut. Manuf. Technol. **64**(Z2), 39–50 (2021). https://doi.org/10.16080/j.issn1671-833x.2021.23/24.039

2. Zhou, Y.X., Zhou, W.B., Li, X.L., et al.: Research progress on automated detection of aircraft skin defects. Aeronaut. Manuf. Technol. **66**(22), 69–77 (2023). https://doi.org/10.16080/j.issn1671-833x.2023.22.069
3. Zhang, Y., Tian, J.W., Liu, Q., et al.: Aircraft skin defect detection method based on visual saliency. Laser J. **43**(03), 61–64 (2022). https://doi.org/10.14016/j.cnki.jgzz.2022.03.061
4. Zhan, X.L., Zhang, C.L., Li, M.: Research on aircraft skin micro-crack detection method based on wake interference. Instrum. Tech. Sensor **07**, 116–121 (2024)
5. Bouarfa, S., Dogru, A., Arizar, R., et al.: Towards automated aircraft maintenance inspection: a use case of detecting aircraft dents using Mask R-CNN. In: AIAA Scitech 2020 Forum. ASME, New York (2020)
6. Qiu, H., Wang, D., Yu, W.: An aircraft surface damage inspection method based on improved SSD. In: India: Fourteenth International Conference on Graphics and Image Processing (ICGIP 2022). SPIE, 2023, vol. 12705, pp. 219–226
7. Liu, Q., Shi, N.: Aircraft skin surface defect detection method based on improved YOLOv5. Foreign Electron. Meas. Technol. **42**(12), 1–7 (2023). https://doi.org/10.19652/j.cnki.femt.2305241
8. Zhang, D.P., Wang, Z.T., Xia, Y.J., et al.: Aircraft skin defect detection algorithm based on improved YOLOv8 [J/OL]. J. Beijing Univ. Aeronaut. Astronaut., 1–14 [2025-07-02]. https://doi.org/10.13700/j.bh.1001-5965.2023.0744
9. Zhang, Z.C., Wang, H.W., Zhou, C.W.: MSA-YOLO: a real-time segmentation algorithm for skin surface defects. Aeronaut. Comput. Tech. **54**(05), 64-68+73 (2024)
10. Pan, T., Hui, K.F.: Aircraft skin defect detection method based on improved Faster R-CNN. Mech. Eng. Autom. **54**(02), 139–141 (2025)
11. Bu, X.Y., Zhang, X.F., Li, M.H., et al.: Aircraft skin defect detection method based on improved EfficientDet. Aeronaut. Manuf. Technol. **68**(05), 68–75 (2025). https://doi.org/10.16080/j.issn1671-833x.2025.05.068
12. Liu, Z.J., Zhang, H.J., Ji, M.X., et al.: Lightweight multi-scale feature fusion enhanced spatial non-cooperative small target detection algorithm [J/OL]. J. Beijing Univ. Aeronaut. Astronaut., 1–12 [2025-07-03]. https://doi.org/10.13700/j.bh.1001-5965.2024.0509
13. Chen, F.Y., Li, P.: Lightweight PCB defect detection algorithm based on STR-DETR [J/OL]. J. Electron. Meas. Instrum., 1–11 [2025-07-03]. http://kns.cnki.net/kcms/detail/11.2488.tn.20250702.1403.016.html
14. Wang, A., Chen, H., Liu, L., et al.: YOLOv10: real-time end-to-end object detection. arXiv preprint arXiv:2405.14458 (2024)
15. Wu, G., Zhu, Y.F., Ye, T.C.: Steel surface defect detection method based on improved YOLOv10. Electron. Meas. Technol. **48**(04), 158–168 (2025). https://doi.org/10.19651/j.cnki.emt.2417355
16. Ma, X., Dai, X.Y., Bai, Y., et al.: Rewrite the stars. In: IEEE/CVF Conference on Computer Vision and Pattern Recognition, 2024
17. Sandler, M., Howard, A., Zhu, M.L., et al.: MobileNetV2: inverted residuals and linear bottlenecks. In: IEEE/CVF Conference on Computer Vision and Pattern Recognition, vol. 2018, pp. 4510–4520 (2018)
18. Wu, Y.X., He, K.M.: Group normalization. In: European Conference on Computer Vision, 2018
19. Liu, C., Wang, K., Li, Q., et al.: Powerful-IoU: more straightforward and faster bounding box regression loss with a nonmonotonic focusing mechanism. Neural Netw. **170**, 276–284 (2024)

Machine Learning Methods for Trustworthy and Secure AI Systems

PMGRD: A Prototype Memory-Guided Reverse Distillation Framework for Anomaly Detection

Jianyi Huang[1], Zhiqian Jia[2], Shenghua Teng[2(✉)], and Zuoyong Li[1(✉)]

[1] Fujian Provincial Key Laboratory of Information Processing and Intelligent Control, School of Computer and Data Science, Minjiang University, Fuzhou CO 350108, China
{huangjianyi,fzulzytdq}@mju.edu.cn

[2] College of Electronic and Information Engineering, Shandong University of Science and Technology, Qingdao CO 266590, China
{202282130035,shteng}@sdust.edu.cn

Abstract. In the outer packaging anomaly detection of supermarket goods, due to the complexity and diversity of packaging design, abnormal features are frequently obscured by complex background interference, posing significant challenges to accurate identification. Existing methods rely on preset universal feature extraction modules, lacking effective extraction and precise analysis capabilities for abnormal features in complex scenarios. They are difficult to accurately locate abnormal regions. To improve the generalization ability of anomaly detection on multi-class datasets, a Prototype Memory-guided Reverse Distillation Framework (PMGRD) was designed. During the reconstruction process, a prototype memory bank is introduced to measure the deviation degree of the training sample features from the normal pattern by anchoring the prototype features. Thereby it can enlarge the differences between the teacher and student features, improving the generalization ability of the model for diversified goods. Then, the feature-consistent distillation loss is introduced to ensure the local feature alignment of the teacher-student network while constraining the global similarity, reducing the tendency to misjudge the normal foreground area as an exception. The comprehensive performance of PMGRD is better than the eight state-of-the-art methods on the GoodsAD public dataset. Qualitative, quantitative and ablation experiments verify the effectiveness of the overall framework and internal modules of PMGRD. It can not only accurately identify the abnormal patterns of various subcategories, but also effectively reduce the false detection rate.

Keywords: Anomaly Detection · Unsupervised Learning · Reverse Distillation · Distillation Loss · Prototype Memory Bank

1 Introduction

As a vital terminal for commodity retail, supermarkets offer a wide variety of goods from numerous brands to meet the ever-growing consumption demands of the people. In particular, the outer packaging of goods serves as a quality indicator throughout the entire process, from production and logistics to warehousing and terminal sales. The

L. Huang and D. Greenhalgh (Eds.): ICMLC 2026, LNNS 2061, pp. 569–580, 2027.
https://doi.org/10.1007/978-3-032-31236-5_39

anomalies of supermarket goods not only affect their display effectiveness on shelves, but also reduce consumers' willingness to purchase. Moreover, the goods may even have been damaged during storage or transportation. Hence, how to detect the outer packaging anomalies of goods rapidly and accurately has become an important topic for the digital and intelligent operation of supermarkets.

Supervised learning-based methods can achieve the best detection performance due to the availability of sufficient data samples for model training. However, in practice, supervised learning encounters abnormal sample scarcity and anomaly annotation complexity. These problems greatly restrict the application of supervised learning in identifying outer packaging anomalies of supermarket goods. Consequently, unsupervised learning has garnered attention from researchers due to its advantage of not relying on abnormal samples for training. It can analyze a large volume of easily obtainable normal samples, automatically extract normal features, and distinguish abnormal samples based on the degree of deviation. Currently, pertinent research mainly draws on unsupervised learning-based industrial anomaly detection approaches, categorized into image reconstruction-based 1, feature embedding-based 2, and knowledge distillation-based methods 3.

This paper mainly draws on the knowledge distillation-based methods, which follow the paradigm of transferring knowledge from a pretrained teacher network to a trainable student network, obtaining abnormal maps by leveraging the different expressions of abnormal features by the teacher-student networks. Based on the feature differences between the teacher and student networks, anomaly localization can be achieved. Bergmann et al. 4 pioneered the integration of knowledge distillation into anomaly detection. Salehi et al. 5 utilized a teacher network pretrained on ImageNet 6 to distill knowledge into a more compact clone network, establishing a standardized framework for forward distillation. Cao et al. 7 improved upon the distillation loss by proposing a contextual similarity loss that adaptively mines hard samples, alleviating the overfitting problem caused by the inconsistency between neural network capacity and the amount of knowledge. Cai et al. 8 trained the teacher-student networks using synthetic abnormal samples to generate discrepancy maps as clues, guiding a segmentation decoder to identify defects. This approach reduced reliance on synthetic appearances and demonstrated excellent detection and localization performance. They also investigated the model's robustness to different synthesis strategies. Zhang et al. 9 similarly constructed a segmentation network to adaptively fuse features from the teacher and student networks. Liu et al. 10 decoupled the teacher network and proposed a dual-model distillation mechanism to increase the distance between teacher and student features corresponding to abnormal regions. Cao et al. 11 focused on the distributions of normal and abnormal samples, introducing a boundary optimization module to expand the boundaries on the discrepancy distributions (DDs) between normal and abnormal samples, and an overlap optimization module to reduce the overlap of discrepancy distributions by dynamically allocating sample weights, effectively improving prediction reliability.

The teacher-student structure and data flow direction in forward distillation limit the model's performance. To address this issue, Deng et al. 12 proposed a reverse distillation (RD) framework for the first time, aiming to achieve reconstruction from low to high resolution in the student network using features output by the teacher network. Under the RD framework, Jiang et al. 13 adopted the anomaly synthesis strategy of NSA 14 to generate image-level masks, mitigating overgeneralization in neural networks; Gu et al. 15 designed a normal recall memory bank to retrieve normal information from query features of both normal and abnormal data, adaptively adjusting the normality of student features. Meanwhile, a normal embedding learning strategy guided the memory bank to memorize prior knowledge of normal data, enhancing the processing capability of relevant information. Guo et al. 16 mitigated biases inherent in pretrained networks through a dual-encoder architecture comprising frozen and trainable encoders, coupled with a cross-reconstruction mechanism that enabled end-to-end optimization of network parameters. However, when confronted with datasets exhibiting significant intra-class diversity, traditional reverse distillation tends to learn an overly broad normal feature space, consequently resulting in blurred decision boundaries and pronounced abnormal reconstruction phenomena.

Addressing the issues, this paper proposes a Prototype Memory-guided Reverse Distillation Framework (PMGRD). First, during the reconstruction process, a prototype memory bank is introduced to measure the deviation of training sample features from normal patterns by anchoring prototype features. Then, the differences between teacher and student features are amplified, thereby enhancing the model's generalization capability for diverse goods. Subsequently, a feature-consistent distillation loss is introduced to ensure local feature alignment between the teacher and student networks while imposing constraints on global similarity, thereby reducing the tendency to misclassify normal foreground regions as anomalies. The main contributions of this paper are as follows:

(1) To mitigate the impact of intra-class data distribution variability on anomaly detection, we propose a reverse distillation-based anomaly detection framework. Traditional models often fail to construct a compact normal feature manifold under such conditions, leading to poorly defined decision boundaries and degraded detection accuracy.
(2) A prototype memory bank is constructed by fully leveraging normal sample information to anchor a wide variety of training images onto several prototypes, learning a more concentrated normal feature space.
(3) A feature-consistent distillation loss is designed by combining cosine loss and contextual similarity loss. It addresses computational instability issues arising during the optimization of pixel-wise L2 loss and contextual similarity loss functions, thereby effectively reducing model misclassification in normal foreground regions.

2 Method

2.1 Overall Framework

The overall framework of PMGRD comprises an encoder (teacher network), a bottleneck layer, and a decoder (student network), as shown in Fig. 1. Specifically, the encoder employs WideResNet50 17, pretrained on large-scale datasets such as ImageNet, to extract feature maps of different sizes that are rich in information. These feature maps encompass multi-dimensional information ranging from low-level textures to high-level semantics. The bottleneck layer receives the outputs from the first three layers of the encoder and generates a compact feature vector through channel concatenation and dimensionality reduction operations. The decoder adopts a reverse structure of the encoder, replacing downsampling with upsampling and gradually reconstructing feature outputs of the same size as those at each layer of the encoder. In addition, the feature-consistent distillation loss employs cosine similarity loss to achieve local feature alignment between the teacher and student networks, ensuring basic reconstruction quality. By leveraging global consistency loss, which starts from the high-order similarity of feature matrices, the model achieves focused attention and in-depth mining of anomalies, mitigating the model's tendency for misclassification and improving the accuracy of anomaly localization.

Initially, the prototype memory bank (PMB) randomly selects a portion of images from the training set and extracts their features using a pretrained encoder, storing these features in the memory bank. During the training process, all input data consist of normal samples, and the similarity between the features of normal samples and those stored in the memory bank is calculated. The decoder learns the feature representations of the encoder by maximizing the similarity between its outputs and the feature maps at each layer of the encoder. By employing a weighted summation approach, the features in the memory bank are utilized to generate reconstructed features for normal features. Subsequently, channel filtering is performed on the reconstructed features to extract the most representative channel information and further identify the memory bank feature that best matches the input features. The difference between the input features and the matched memory features is calculated. This difference can characterize the degree to which the input sample deviates from the normal pattern. It is then concatenated with the input features and fed into the reconstruction network to enhance the sensitivity and discriminative ability of anomaly detection. In the testing phase, the decoder can accurately reconstruct the feature maps of normal regions. Since it has not been exposed to anomaly information, there is inconsistency between its feature reconstruction in abnormal regions and the encoder's outputs, thereby enabling anomaly localization. The feature difference representation of the teacher-student networks is mapped into an anomaly score function to localize abnormal regions. Specifically, the mean of the top 100 pixels values with the highest values in the feature difference map is taken as the anomaly score for the entire image.

2.2 Prototype Memory Bank (PMB)

A PMB is constructed by randomly sampling from the training set to store the features of all normal samples. Based on the similarity between the input sample features and the prototype features in the bank, the input features are decomposed and reconstructed so that they can be represented as a weighted combination of prototype features. Subsequently, channel filtering is performed on the reconstructed features to identify the prototype features that best match the input sample and calculate the numerical differences. Finally, the features of the input sample are concatenated with the feature differences and fed into the bottleneck layer for compression and distillation. The introduction of the memory bank into the reverse distillation framework allows a wide variety of training images to be anchored to a number of prototypes, encouraging the model to

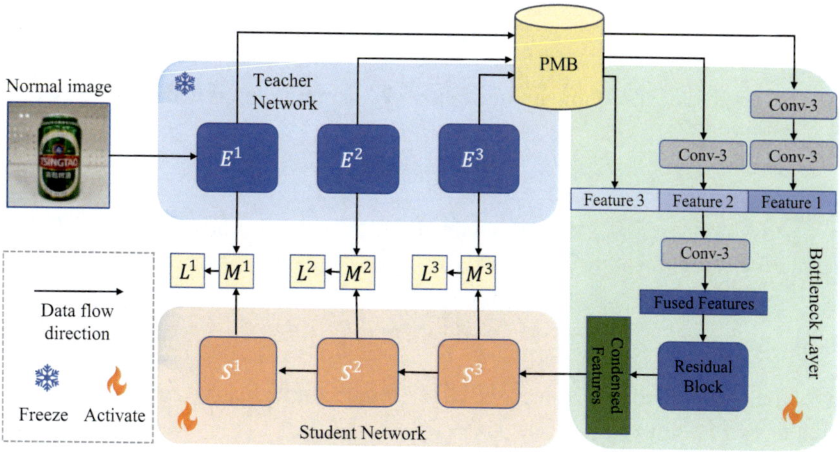

Fig. 1. The Overall framework of PMGRD.

learn a more concentrated normal feature space and mitigating the drawback of the distillation structure, which tends to form broad and ambiguous normal representations when dealing with diverse datasets of the same class. The specific calculation process is as shown in Fig. 2.

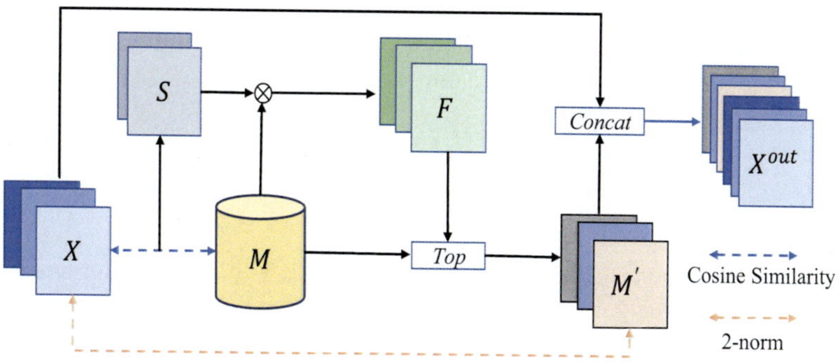

Fig. 2. The calculation process of prototype memory bank.

Step 1: Calculate the similarity between the input sample feature $\boldsymbol{X}$ and the memory bank features $\boldsymbol{M}$, and convert it into a probability distribution:

$$S_i = \text{softmax}\left(\frac{MX_i^T}{\|M\|_2\|X_i\|_2}\right) \tag{1}$$

Step 2: Weighted fusion of the similarity probability distribution $\boldsymbol{S}$ and the memory bank features $\boldsymbol{M}$ to obtain the fused features $\boldsymbol{F}$:

$$F_i = S_i^T M \tag{2}$$

Step 3: Select the most prominent index value $\boldsymbol{T}_i$ from the fused features $\boldsymbol{F}$:

$$T_i = \text{Top-k}(F_i) \tag{3}$$

Step 4: Based on the index value, obtain the feature $\boldsymbol{M}'$ in the memory bank that is closest to the sample feature $\boldsymbol{X}$:

$$M_i' = M\left[\arg\min(T_i)\right] \tag{4}$$

Step 5: The output feature is composed of the input sample feature and its error relative to the matched feature.

$$X_i^{\text{out}} = \text{Concat}\left(X_i, \left(M_i' - X_i\right)^2\right) \tag{5}$$

2.3 Loss Function

The cosine loss $\mathscr{L}_{cos}$ focuses on the direction between feature vectors, reflecting the similarity of local details and being unaffected by the magnitude of features. This encourages the alignment of student features with teacher features in terms of direction. The cosine loss can be calculated by [1]. This not only helps the model learn the common feature distribution of normal samples but also enables it to capture the deviations of abnormal samples more keenly in the feature space.

$$\mathcal{L}_{cos} = \sum_{k=1}^{\mathscr{H}} \left(\frac{1}{H^k W^k} \sum_{i=1}^{H^k} \sum_{j=1}^{W^k} \left(1 - \frac{\left(f_t^k\right)_{ij}}{\left\| \left(f_t^k\right)_{ij} \right\|_2} \odot \frac{\left(f_s^k\right)_{ij}}{\left\| \left(f_s^k\right)_{ij} \right\|_2} \right) \right. \tag{6}$$

where ${f_t}^k$ and ${f_s}^k$ represent the features output by the k-th stage of the teacher and student networks, respectively. H and W are the height and width of the feature map. $\mathscr{H}$ refers to the number of network layers.

The contextual similarity loss focuses on the differences in contextual and global structural information of features. It measures the correlation between the features of the teacher and student networks through matrix similarity:

$$\mathcal{L}_{cs} = \frac{1}{\mathscr{H}} \sum_{k=1}^{\mathscr{H}} \frac{1}{N^k \times N^k} \sum_{i=1}^{N^k} \sum_{j=1}^{N^k} \left\| \left(Q_t^k \times \left(Q_t^k\right)^T \right)_{ij} - \left(Q_s^k \times \left(Q_s^k\right)^T \right)_{ij} \right\|_2^2 \tag{7}$$

where Q_t and Q_s denote the matrix representations of the teacher-student features, respectively.

The feature-consistent distillation loss (FCD_loss) $\mathscr{L}_{fcd}$ comprises two components: a cosine loss $\mathscr{L}_{cos}$ and a contextual similarity loss $\mathscr{L}_{cs}$:

$$\mathcal{L}_{fcd} = \mathcal{L}_{cos} + \mathcal{L}_{cs} \tag{8}$$

3 Experimental Results

3.1 Dataset and Evaluation Metrics

To validate the effectiveness of PMGRD, the supermarket goods dataset GoodsAD 18 is used in the experiments. It was collected from real supermarket scenarios and covers six major categories of commonly found supermarket goods, including cigarette boxes (cigarette_box), bottled beverages (drink_bottle), canned beverages (drink_can), bottled food products (food_bottle), boxed food products (food_box), and packaged food products (food_package). Each category involves multiple brands and different packaging forms to enhance data diversity and make it more closely resemble real-world application scenarios. To meet the experimental setup for unsupervised anomaly detection, GoodsAD follows the partitioning method of the MVTec dataset, dividing into two parts: a training set containing only normal samples and a testing set containing both normal and abnormal samples. GoodsAD includes a total of 6,124 high-quality images with a resolution of 3000 $\times$ 3000, providing sufficient data support for research on image-level detection and pixel-level localization. It encompasses a rich variety of anomaly types, primarily including common defects found in supermarket goods such as surface damage, deformation, opened packaging, cap open, cap half open, surface anomalies, and straw missing. The distribution of different anomaly types varies across the images, with anomalies like surface damage and cap open accounting for a relatively small proportion of total image pixels (less than 2%), while anomalies such as opened packaging and deformation account for a larger proportion. The diverse packaging forms and varied abnormal patterns make GoodsAD a more challenging benchmark in the field of anomaly detection, facilitating a comprehensive evaluation of the model's generalization ability and robustness.

The metrics used in this paper align with those commonly employed in current mainstream anomaly detection research. Area under the receiver operating curve (AUROC) and per-region overlap (PRO) are utilized to assess the model's performance in detecting anomalies. AUROC includes metrics at both image and pixel levels. Image-AUROC measures the model's performance in detecting abnormal samples, while Pixel-AUROC evaluates the model's ability to localize abnormal regions within abnormal samples. The PRO metric provides a more comprehensive reflection of localization accuracy by considering the degree of overlap between predicted abnormal regions and labeled regions across different thresholds. Similar to AUROC scores, a higher PRO score indicates stronger localization capability of the model. To ensure fair comparison with other models under consistent standards, calculations in this paper are conducted under the condition that the false positive rate (FPR) does not exceed 30%.

3.2 Experimental Setup

Experimental environment: 12th Gen Intel(R) Core(TM) i5-12600KF 3.70 GHz, NVIDIA GeForce RTX 3090 24GB, RAM 64GB, and Windows 11 operating system. The programming language used is Python, and the deep learning framework is PyTorch 2.0.

During the implementation process, a unified training setup was adopted for the six subcategories within the GoodsAD dataset. All input images were resized to 256×256. The model was optimized using the Adam optimizer with momentum parameters set to (0.9, 0.999) and a learning rate of 0.005. Training was performed over 5,000 batch iterations across all categories, with a batch size of 8.

3.3 Comparative Experiments

The proposed method PMGRD was compared with eight state-of-the-art anomaly detection methods, including CFA 19, ReContrast 16, RD4AD 12, RD++ 20, CDO 11, DAF 8, IFgNet 21, and Destseg 9. As shown in Table 1, PMGRD demonstrated superior performance compared to the aforementioned methods, with the highest and second-highest metrics for each subcategory highlighted in red and blue, respectively. PMGRD achieved the best results across all three metrics, with corresponding Image-AUROC, Pixel-AUROC, and Pixel-PRO values of 80.67%, 97.07%, and 86.4%, respectively. The second-best Image-AUROC and Pixel-AUROC values were obtained by CDO, which were 2.92 and 0.46% lower than those of PMGRKD, respectively. The second-best Pixel-PRO value was achieved by ReContrast, lagging behind PMGRKD by 2.62%.

Table 1. Image-AUROC, Pixel-AUROC, and Pixel-PRO values comparison results on GoodsAD dataset.

Method	Cigarette_box	Drink_bottle	Drink_can	Food_bottle	Food_box	Food_package	Mean/%
CFA	94.6/**98.3/92.6**	69.2/91.5/74.0	71.0/92.2/65.9	85.7/95.8/78.1	63.7/77.1/59.9	62.0/89.6/63.4	74.4/90.7/72.3
ReContrast	85.6/96.1/86.0	69.2/95.7/82.8	73.3/94.3/**83.3**	83.9/97.8/**89.7**	66.7/92.7/**78.9**	65.1/96.0/**82.1**	74.0/95.4/**83.8**
RD4AD	70.2/93.1/76.6	66.7/95.0/84.2	63.4/93.2/81.7	77.6/96.2/84.8	62.3/90.5/75.0	62.7/94.9/79.2	67.1/93.8/80.2
RD++	87.4/94.1/80.4	66.6/95.3/84.4	61.4/92.0/77.3	78.6/96.8/85.7	64.5/89.2/70.9	64.5/95.7/80.3	70.5/93.9/79.8
CDO	94.4/96.8/87.8	74.0/**98.7/87.3**	74.4/**94.4**/73.7	83.4/**97.9**/87.3	71.8/94.9/73.0	**68.6/97.1**/76.2	**77.8/96.6**/80.9
DAF	94.5/96.2/85.9	69.5/95.6/80.2	**78.1**/94.0/77.1	83.6/**97.9**/84.0	67.4/92.1/65.0	62.0/95.8/73.9	75.9/95.3/77.7
IFgNet	90.5/93.7/82.2	**77.4**/95.7/78.5	76.8/91.1/74.1	81.9/95.8/84.1	65.8/90.7/68.9	61.7/95.7/72.0	75.7/93.8/76.6
Destseg	**97.7**/95.7/85.5	69.0/95.8/82.8	67.2/92.2/77.0	**83.6**/97.5/86.6	**74.0/95.7**/78.4	68.5/95.2/76.9	76.7/95.4/81.2
Ours	**97.5/97.7/91.4**	**76.2/97.6/87.6**	**77.6/95.7/83.2**	**86.8/98.3/91.1**	**72.5/95.3/79.6**	**73.4/97.9/85.5**	**80.7/97.1/86.4**

In practical applications, significant appearance variations may exist within the same category. This subclass diversity poses a challenge that all anomaly detection methods must confront. However, existing comparative methods have not been specifically designed to address this data characteristic, resulting in subpar performance on the GoodsAD dataset. In contrast, PMGRD effectively mitigates this issue by establishing a memory bank to perform feature reconstruction based on the errors between training samples and prototype samples. This design mechanism offers two advantages. Firstly, it prevents feature reconstruction of abnormal samples caused by the generalization of the student network. Secondly, by utilizing prototype samples, it encourages the distilled model to learn more concentrated normal feature representations, alleviating the model's inadaptability to diverse yet similar data within the same class. Additionally, the feature-consistent distillation loss effectively suppresses misjudgment in foreground regions by simultaneously optimizing local and global feature alignment and integrating higher-order similarity information, thereby enhancing pixel-level detection accuracy.

For qualitative analysis, to further demonstrate the superiority of PMGRD, as shown in Fig. 3, PMGRD exhibits three levels of detection advantages. Firstly, its localization regions are strictly concentrated on the target objects in the foreground of the images, indicating that the model can effectively exclude interference from background noise. Secondly, the abnormal regions exhibit a highly concentrated distribution pattern, with heatmaps showing that activation responses are focused on the core areas of anomalies, demonstrating the model's precise pixel-level localization capability. Furthermore, the localization results closely match the actual abnormal regions, a characteristic particularly evident in the wear detection on the surfaces of bottled and canned beverages.

Some baseline algorithms exhibit notable limitations on the GoodsAD dataset. For instance, IFgNet generates significant noise responses in the background regions of cigarette boxes. Although it employs a joint training strategy combining foreground detection and image reconstruction, aiming to effectively suppress background interference through foreground detection, there is a risk of failure in practical applications. This failure may stem from two factors: firstly, insufficient decoupling of foreground and background images; secondly, severe interference from lighting conditions and complex packaging patterns on the detection task. The anomaly heatmaps generated by the Destseg algorithm display pronounced spatial disorganization, with scattered activation regions and blurred boundaries, reflecting significant performance fluctuations in anomaly localization tasks. Compared to RD, RD++, ReContrast, and CDO, PMGRD demonstrates a significantly reduced tendency for misjudgment in the foreground regions of images and exhibits more stable performance, indicating that PMGRD can effectively overcome interference from packaging patterns and achieve more precise anomaly localization. However, the analysis also reveals that while enhancing its anti-interference capability, PMGRD experiences a reduction in response intensity to abnormal regions, manifested as relatively dim heatmap displays

3.4 Ablation Experiments

This section systematically evaluates the modular contributions of PMGRD through controlled ablation experiments. Building upon the RD4AD baseline network, we incrementally integrate the PMB and the feature-consistent distillation loss. As shown in Table 2, each module makes a positive contribution to improving the model performance. As shown in Fig. 4, the feature consistency loss effectively mitigates foreground anomaly misjudgments. The PMB enhances the model's sensitivity to anomalies, resulting in more concentrated localization regions, which underscores the memory bank's positive role in preventing anomaly reconstruction. PMGRD demonstrates superior performance in detecting large-scale structural anomalies while also possessing a certain level of perception capability for subtle anomalies.

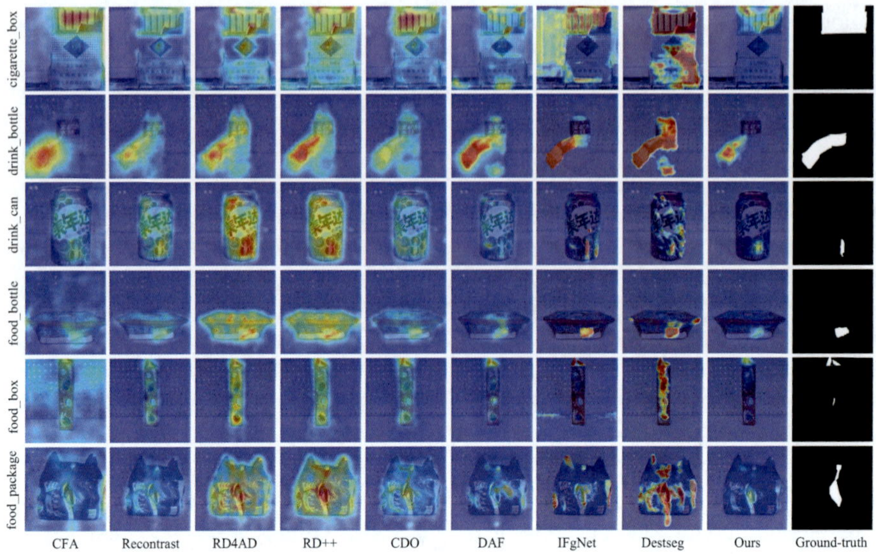

Fig. 3. The comparison heatmap on GoodsAD dataset.

Table 2. The ablation study of PMGRD.

RD4AD	PMB	FCD_loss	Image-AUROC/Pixel-AUROC/ Pixel-PRO/%
√			67.13/93.79/80.23
√	√		75.01/96.11/82.81
√		√	76.70/95.61/83.02
√	√	√	**80.67/97.07/86.40**

4 Conclusion

This paper proposes a reverse distillation anomaly detection framework guided by a prototype memory bank to address inadaptability to diverse-within-class datasets, enhance feature concentration for normal data, and mitigate foreground misjudgment. The prototype memory bank anchors input samples with prototypes, increasing reconstruction errors for anomalies through deviation measurement, thereby improving detection sensitivity. Feature-consistent distillation loss ensures local-global alignment, enabling precise abnormal region mining and reducing misjudgment in knowledge distillation-based methods. PMGRD outperforms eight state-of-the-art methods on the GoodsAD dataset, with effectiveness validated through qualitative, quantitative, and ablation experiments. However, it exhibits reduced abnormal region response intensity, as shown by heatmaps. Future work will focus on balancing localization accuracy with anomaly expression intensity, potentially incorporating attention mechanisms or dense contrastive learning to optimize performance.

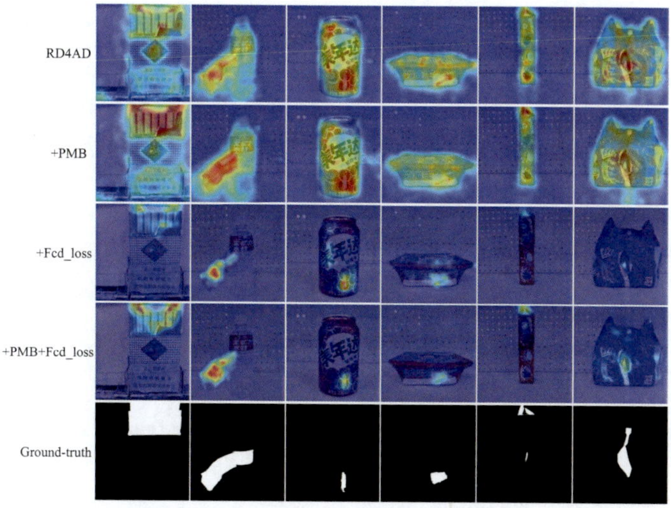

Fig. 4. The ablation heatmap of PMGRD.

Acknowledgments. This work was supported by the National Natural Science Foundation of China (62471207), Natural Science Foundation of Fujian Province (2024 J02029, 2024 J08281), Talent Introduction Research Project of Minjiang University (MJY23035), and Special Project of Minjiang University (MJJD25001).

Author Contributions **Jianyi Huang**: conception of the work, critical revision of the article, data analysis, and drafting the article. **Zhiqian Jia**: conception of the work, critical revision of the article, data collection, and drafting the article. **Shenghua Teng**: data collection and experimental data analysis. **Zuoyong Li**: critical revision of the article.

Conflicts of Interest. The authors declare no conflicts of interest.

Data Availability Statement. The data that support the findings of this study are available from the corresponding author upon reasonable request.

References

1. Krizhevsky, A., Sutskever, I., Hinton, G.: Image classification with deep convolutional neural networks. Adv. Neural Inf. Proces. Syst. **25** (2012)
2. Cai, Y., Liang, D., Luo, D., et al.: A discrepancy aware framework for robust anomaly detection. IEEE Trans. Industr. Inform. **20**(3), 3986–3995 (2023)
3. Ma, Z., Li, J., Wong, W.: Patch distance based auto-encoder for industrial anomaly detection. Expert Syst. Appl. **270**, 126537 (2025)
4. Luo, W., Yao, H., Yu, W., et al.: AMI-net: adaptive mask inpainting network for industrial anomaly detection and localization. IEEE Trans. Autom. Sci. Eng. **22**, 1591–1605 (2025)
5. Fu, K., Qi, F., Wang, X., et al.: Enhancing reverse distillation model for multi-class unsupervised anomaly detection via latent space regularization. Neurocomputing. **131819** (2025)

6. Bergmann, P., Fauser, M., Sattlegger, D., et al.: Uninformed students: student-teacher anomaly detection with discriminative latent embeddings. In: Proceedings of the IEEE/CVF Conference on Computer Vision and Pattern Recognition, pp. 4183–4192 (2020)
7. Salehi, M., Sadjadi, N., Baselizadeh, S., et al.: Multiresolution knowledge distillation for anomaly detection. In: Proceedings of the IEEE/CVF Conference on Computer Vision and Pattern Recognition, pp. 14902–14912 (2021)
8. Cao, Y., Wan, Q., Shen, W., et al.: Informative knowledge distillation for image anomaly segmentation. Knowl.-Based Syst. **248**, 108846 (2022)
9. Zhang, X., Li, S., Li, X., et al.: Destseg: segmentation guided denoising student-teacher for anomaly detection. In: Proceedings of the IEEE/CVF Conference on Computer Vision and Pattern Recognition, pp. 3914–3923 (2023)
10. Liu, X., Wang, J., Leng, B., et al.: Dual-modeling decouple distillation for unsupervised anomaly detection. In: Proceedings of the 32nd ACM International Conference on Multimedia, pp. 5035–5044 (2024)
11. Cao, Y., Xu, X., Liu, Z., et al.: Collaborative discrepancy optimization for reliable image anomaly localization. IEEE Trans. Industr. Inform. **19**(11), 10674–10683 (2023)
12. Deng, H., Li, X.: Anomaly detection via reverse distillation from one-class embedding. In: Proceedings of the IEEE/CVF Conference on Computer Vision and Pattern Recognition, pp. 9737–9746 (2022)
13. Jiang, Y., Cao, Y., Shen, W.: A masked reverse knowledge distillation method incorporating global and local information for image anomaly detection. Knowl.-Based Syst. **280**, 110982 (2023)
14. Schlüter, H., Tan, J., Hou, B., et al.: Natural synthetic anomalies for self-supervised anomaly detection and localization. In: European Conference on Computer Vision, pp. 474–489 (2022)
15. Gu, Z., Liu, L., Chen, X., et al.: Remembering normality: memory-guided knowledge distillation for unsupervised anomaly detection. In: Proceedings of the IEEE/CVF International Conference on Computer Vision, pp. 16401–16409 (2023)
16. Guo, J., Lu, S., Jia, L., et al.: Recontrast: domain-specific anomaly detection via contrastive reconstruction. Adv. Neural Inf. Proces. Syst. **36**, 10721–10740 (2023)
17. Zagoruyko, S., Komodakis, N.: Wide residual networks. https://arxiv.org/abs/1605.07146. (2016)
18. Zhang, J., Ding, R., Ban, M., et al.: PKU-GoodsAD: a supermarket goods dataset for unsupervised anomaly detection and segmentation. IEEE Rob. Autom. Lett. **9**(3), 2008–2015 (2024)
19. Lee, S., Song, B.: Cfa: coupled-hypersphere-based feature adaptation for target-oriented anomaly localization. IEEE Access. **10**, 78446–78454 (2022)
20. Tien, T., Nguyen, A., Tran, N., et al.: Revisiting reverse distillation for anomaly detection. In: Proceedings of the IEEE/CVF Conference on Computer Vision and Pattern Recognition, pp. 24511–24520 (2023)
21. Chen, X., Xu, H., Deng, C., et al.: Implicit foreground-guided network for anomaly detection and localization. In: ICASSP 2024–2024 IEEE International Conference on Acoustics, Speech and Signal Processing (ICASSP), pp. 2970–2974 (2024)

A Hybrid Fraud Detection Framework Fusing Deep Representation Learning, CTGAN, and MAB-Based XGBoost

Yifan Han and Wenjun Yang(✉)

Tianjin University of Technology, Tianjin 300384, China
h15617069526@gmail.com, yangwj@tjut.edu.cn

Abstract. Class imbalance and evolving fraud patterns pose significant challenges to credit card fraud detection. In this paper, we propose a hybrid framework that integrates Deep Representation Learning, CTGAN, and Multi-Armed Bandit (MAB) to robustly identify fraudulent transactions. First, an AutoEncoder is introduced to extract latent features and reconstruction errors, enhancing anomaly identification. Second, CTGAN generates high-quality synthetic fraud samples based on these features to reconstruct a 1:1 balanced training set. Finally, we construct an XGBoost detector utilizing Thompson Sampling for adaptive hyperparameter optimization, which efficiently locates global optimal parameters. Experimental results on the European Credit Card dataset demonstrate a Precision of 96.01%, F1-score of 95.71%, and AUC of 95.36%. Compared with baselines, the proposed framework significantly reduces false positives, verifying its effectiveness in financial risk control.

Keywords: Credit card fraud detection · Deep representation learning · Autoencoder · XGBoost · CTGAN · Multi-armed bandit

1 Introduction

The rapid proliferation of internet finance has integrated credit card transactions into daily life. Yet, this convenience comes with a cost: the surge in transaction volume has been accompanied by a persistent rise in credit card fraud, causing substantial economic losses and trust risks to banks, merchants, and customers. How to accurately identify fraudulent behaviors within massive, real-time transactions remains a critical and urgent challenge in the field of financial risk control[1]. Recent studies indicate that as transaction patterns become increasingly complex, this domain continues to receive extensive attention from both academia and industry[2].

Traditional fraud detection methods predominantly rely on statistical rules or classical machine learning models, such as Logistic Regression, Support Vector Machines (SVM), and Random Forest[3]. Despite their utility, these methods often struggle with high-dimensional data and evolving fraud tactics. First, their

L. Huang and D. Greenhalgh (Eds.): ICMLC 2026, LNNS 2061, pp. 581–593, 2027.
https://doi.org/10.1007/978-3-032-31236-5_40

heavy reliance on manual feature engineering and rule design makes it difficult to capture complex and concealed fraud patterns. Second, being typically static, they struggle to adapt to the evolving nature of fraud techniques, often resulting in high false positive and false negative rates[4].

With the advancement of artificial intelligence, deep learning has been widely applied to financial fraud detection due to its powerful feature extraction capabilities[5]. Researchers have introduced supervised learning and ensemble learning into this task, achieving notable results. To address the class imbalance in credit card transaction datasets, Xu [6]utilized Generative Adversarial Networks (GANs) combined with voting-based feature selection methods. Fiore et al. [7] proposed using GANs to generate minority class (fraud) samples to augment the training set; while this improved sensitivity, it came at the cost of a slight increase in false positive rates. Similarly, N. Rtayli et al.[8] adopted the SMOTE algorithm to generate fraud samples. Addressing data drift, I.Benchaji et al.[9] proposed an LSTM model combined with an attention mechanism to selectively focus on features.

However, existing methods for handling imbalance mostly focus on quantity alignment at the data level, often neglecting deep mining at the feature level. Credit card transaction data typically contains highly non-linear and concealed features, which traditional feature engineering fails to extract effectively. Moreover, simple oversampling (e.g., SMOTE) easily introduces noise, and manual tuning of model hyperparameters is both time-consuming and difficult to converge to the global optimum.

To address these deficiencies, this paper proposes a hybrid framework that excels in feature representation, data balancing, and model optimization. We introduce Deep Representation Learning (AutoEncoder) for feature extraction, as it can leverage unsupervised mechanisms to extract latent vectors and reconstruction errors, effectively capturing deep distributional differences between normal and fraudulent transactions. For data balancing, we employ CTGAN, which generates high-quality synthetic samples based on the enhanced feature space to construct a balanced dataset, showing great potential in solving data sparsity. Finally, for model optimization, we utilize the Multi-Armed Bandit (MAB) approach, specifically Thompson Sampling, which balances exploration and exploitation to adaptively locate the global optimal hyperparameters for the classifier.

The main contributions are summarized as follows:

- **Deep Feature Enhancement and Data Balancing:** We introduce an AutoEncoder-based mechanism to extract latent features and reconstruction errors. Furthermore, we apply CTGAN to construct a 1:1 fully balanced dataset, overcoming challenges related to weak feature expression and extreme class imbalance.
- **Adaptive Model Optimization:** We model the hyperparameter tuning of XGBoost as an efficient online learning task using the Thompson Sampling strategy within a Multi-Armed Bandit (MAB) framework, enabling rapid localization of the optimal parameter combination.

- **Performance Improvement:** By integrating these technologies, the proposed framework significantly improves the accuracy of credit card fraud detection and the efficiency of model optimization compared to existing baselines.

The remainder of this paper is organized as follows. Section 2 elaborates on the proposed hybrid fraud detection framework, detailing the AutoEncoder, CTGAN, and MAB modules. Section 3 presents the experimental setup, results, and comprehensive analysis. Finally, Sect. 4 concludes the paper and discusses future work.

2 Proposed Framework

To address the three major challenges in credit card fraud detection–concealed feature patterns, extreme class imbalance, and the difficulty of adaptive hyperparameter optimization–this paper constructs a hybrid detection framework fusing Deep Representation Learning, generative data augmentation, and reinforcement learning optimization.

As illustrated in Fig. 1, the framework consists of three core modules. First, at the feature level, an **AutoEncoder (AE)** is introduced to mine latent non-linear features and reconstruction errors from transaction data, thereby enhancing the model's discriminative power against anomalies. Second, at the data level, the **CTGAN** algorithm is utilized to generate high-quality synthetic samples based on the enhanced feature space, constructing a fully balanced dataset with a 1:1 positive-to-negative ratio. Finally, at the decision level, the balanced data is fed into an **XGBoost** classifier, which employs a **Thompson Sampling**-based Multi-Armed Bandit (MAB) algorithm for adaptive hyperparameter optimization. This approach aims to significantly reduce the false positive rate while maintaining a high recall rate.

2.1 Deep Representation Learning via AutoEncoder

To capture latent non-linear patterns within the original transaction features (V1-V28), we construct an AutoEncoder (AE) module. The AE is an unsupervised neural network consisting of an Encoder and a Decoder.

The Encoder maps the input data x to a low-dimensional latent vector z, as shown in Eq. (1):

$$z = \sigma(W_e x + b_e) \tag{1}$$

where W_e and b_e are the weight and bias of the encoder, and σ is the activation function.

The Decoder attempts to reconstruct the original input from z to obtain $\hat{x}$:

$$\hat{x} = \sigma(W_d z + b_d) \tag{2}$$

During the training phase, the AE is trained exclusively on **normal transaction samples**. The objective is to minimize the Mean Squared Error (MSE):

$$L = \frac{1}{n}\sum_{i=1}^{n} ||x_i - \hat{x}_i||^2 \tag{3}$$

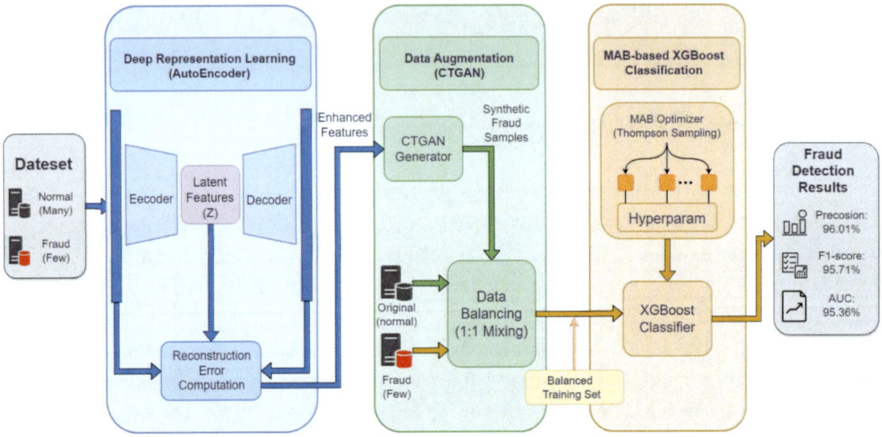

Fig. 1. The proposed hybrid fraud detection framework.

Since the model learns only the distribution patterns of normal samples, when a fraud sample is input, its **Reconstruction Error** is typically significantly higher than that of a normal sample. Therefore, we extract both the Reconstruction Error and the latent vector z, concatenating them with the original features to form an enhanced feature vector. These new features significantly improve the subsequent model's ability to identify anomalies.

2.2 Data Balancing via CTGAN

After obtaining the enhanced features, we employ CTGAN to perform oversampling to address the sparsity of fraud samples. The input to CTGAN is no longer the original 30-dimensional data, but high-dimensional data containing reconstruction errors and latent features. Traditional GANs face two major challenges when processing tabular data: difficulty in modeling mixed data types and difficulty in learning non-Gaussian distributions of continuous variables. CTGAN addresses these issues by introducing Mode-specific Normalization and a conditional generator[10].

In credit card fraud detection scenarios, fraud samples typically constitute a very small proportion of the total, leading to severe class imbalance. The application steps of CTGAN are as follows:

1. **Preprocessing**: Identify continuous and discrete features in the original dataset.
2. **Mode-specific Normalization**: For each continuous feature, use a Variational Gaussian Mixture (VGM) model to convert it into a combination of a scalar value and a mode indicator.
3. **Model Construction**: Build the CTGAN model where both the generator and discriminator employ fully connected neural networks. The generator includes multiple residual blocks, and the discriminator uses the PacGAN structure to improve training stability.
4. **Training**: In each training iteration, sample real data from the minority class (fraud samples), construct a conditional vector, and generate corresponding synthetic samples.
5. **Generation**: Use the trained generator to synthesize fraud samples until the number of minority class samples reaches the predetermined target (e.g., equal to the majority class).
6. **Merging**: Merge the generated synthetic samples with the original dataset to construct a balanced dataset for subsequent classification model training.

Compared to traditional oversampling methods (such as SMOTE), CTGAN generates more diverse samples that better capture the complex distributional features of the minority class, while avoiding the overfitting problems often associated with simple interpolation.

2.3 Hyperparameter Optimization via Thompson Sampling

We apply Thompson Sampling[11] to optimize the hyperparameters of XGBoost[12]. For each hyperparameter, a set of candidate values is defined. Each candidate value of every hyperparameter is treated as an independent "arm" in the Multi-Armed Bandit problem.

The algorithm uses a Beta-Bernoulli model to model the reward distribution. For a candidate value v of hyperparameter p, we maintain a Beta distribution $Beta(\alpha_{p,v}, \beta_{p,v})$ as the posterior distribution. Initially, we set $\alpha_{p,v} = 1, \beta_{p,v} = 1$.

In the t-th iteration, the process is as follows:

1. **Sampling**: For each hyperparameter p, traverse all its candidate values v and sample from the corresponding Beta distribution:
$$\theta_{p,v}^{(t)} \sim Beta(\alpha_{p,v}^{(t-1)}, \beta_{p,v}^{(t-1)}) \tag{4}$$
2. **Selection**: For each hyperparameter p, select the candidate value with the largest sampled value:
$$v_p^{(t)} = \arg\max_i \theta_{p,i}^{(t)} \tag{5}$$
The selected hyperparameter combination serves as the configuration for this round.
3. **Evaluation**: Train the XGBoost model using this configuration and calculate the F1-score on the validation set as the reward $r^{(t)}$.

4. **Update**: For each selected hyperparameter value $(p, v_p^{(t)})$, update the Beta distribution parameters based on the reward. To map the continuous F1-score to a Bernoulli reward, we adopt the following strategy: If $r^{(t)}$ exceeds a certain threshold (e.g., 95%) of the current best recorded F1-score, it is considered a success, and we update $\alpha_{p,v_p^{(t)}} \leftarrow \alpha_{p,v_p^{(t)}} + 1$. Otherwise, it is considered a failure, and we update $\beta_{p,v_p^{(t)}} \leftarrow \beta_{p,v_p^{(t)}} + 1$.
5. **Recording**: Record the reward for this round and update the global best configuration and best F1-score.
6. **Loop**: Repeat the above steps until the maximum number of iterations T is reached or convergence is achieved.

The advantage of Thompson Sampling lies in its probabilistic sampling mechanism, which naturally balances **exploration and exploitation**. In the early stages, the posterior distributions of the arms are dispersed, leading to high randomness in sampling and promoting exploration. As observations increase, the posterior distributions of high-performing arms gradually concentrate in high-reward regions, causing the algorithm to naturally favor these optimal configurations. Compared to Grid Search and Random Search, Thompson Sampling converges to the optimal or near-optimal configuration more efficiently.

3 Experiments and Analysis

3.1 Dataset and Evaluation Metrics

This study utilizes the publicly available "Credit Card" dataset from Kaggle. The dataset contains 284,807 transaction records, but only 492 are fraudulent. These transaction records are from European credit card holders on two separate days in 2013[13]. The ratio of positive to negative samples is approximately 1:578, indicating an extremely imbalanced dataset. The original features include 28 principal components (V1-V28) obtained via PCA, along with Time and Amount, totaling 30 dimensions.

Given the complexity and imbalance of the data, we designed the following preprocessing pipeline:

1. **Standardization**: We applied StandardScaler to the 'Time' and 'Amount' features to eliminate scaling differences.
2. **Deep Feature Enhancement**: An AutoEncoder was trained using normal samples from the training set to extract the Reconstruction Error and a 10-dimensional Latent Vector. These were concatenated with the original features, extending the feature dimension to 41. This step aims to introduce non-linear features to enhance the representation of concealed fraud patterns.
3. **Dataset Splitting**: The dataset was split into training and testing sets with an 8:2 ratio, ensuring stratified sampling to maintain the class distribution.
4. **Evaluation Metrics**: We employed Precision, Recall, F1-score, and Area Under the ROC Curve (AUC) as core evaluation metrics. Specifically, we focused on whether the model effectively improves Precision (i.e., reduces false positives) while maintaining a high Recall.

The evaluation metrics are defined as follows:

$$\text{Precision} = \frac{TP}{TP + FP}, \quad \text{Recall} = \frac{TP}{TP + FN} \tag{6}$$

$$\text{F1-score} = 2 \times \frac{\text{Precision} \times \text{Recall}}{\text{Precision} + \text{Recall}} \tag{7}$$

AUC (Area Under the Curve) is defined as the area under the ROC curve, which plots the True Positive Rate (TPR) against the False Positive Rate (FPR) at various threshold settings.

3.2 Comparative Experiments

Joint Analysis of Sampling Strategies and Feature Enhancement To verify the universality of the Deep Feature Enhancement module (AutoEncoder) and the advantages of CTGAN in data reconstruction, we designed a systematic comparative experiment. We selected four representative sampling strategies: SMOTE[14], ADASYN [15], VAE[16], and CTGAN, using no sampling (None) as the baseline. Table 1 details the performance differences under both "Raw Features" and "AE-Enhanced Features". For fairness, all experiments used XGBoost with default parameters as the base classifier.

Table 1. Performance comparison before and after introducing AE features

Sampling	Raw Features (30-dim)				AE-Enhanced (41-dim)				F1 Gain
	Prec.	Rec.	F1	AUC	Prec.	Rec.	F1	AUC	
None	0.945	0.733	0.826	0.871	0.951	0.782	0.858	0.912	+3.2%
SMOTE	0.817	0.857	0.837	0.867	0.842	0.865	0.853	0.905	+1.6%
ADASYN	0.816	0.826	0.821	0.843	0.835	0.841	0.838	0.898	+1.7%
VAE	0.936	0.743	0.828	0.879	0.941	0.795	0.862	0.925	+3.4%
CTGAN	0.875	0.877	0.876	0.894	**0.920**	**0.879**	**0.899**	**0.928**	+2.3%

Data in Table 1 and Figs. 2 and 3 reveals complex performance differences resulting from the interaction between sampling strategies and feature spaces. Although traditional SMOTE and ADASYN algorithms effectively improved Recall (> 0.82) via linear interpolation, their Precision stagnated around 0.81 due to the introduction of boundary noise. VAE achieved a high Precision of 0.936 thanks to high-quality generated samples, yet its Recall was limited to 0.743 due to insufficient generation diversity, failing to cover all minority class patterns. In contrast, CTGAN demonstrated strong fitting capabilities for high-dimensional mixed distributions, achieving the best balance between Precision and Recall even in the raw feature space. Notably, the reconstruction errors and latent vectors extracted by AE brought universal performance gains to all sampling methods. Specifically, VAE saw a significant F1 boost of 3.4% after deep feature supplementation.

The 'CTGAN + AE' combination achieved the best performance with a Precision of 0.920 and an F1-score of 0.899, proving that high-quality generated data combined with robust deep features is the effective approach for imbalanced classification.

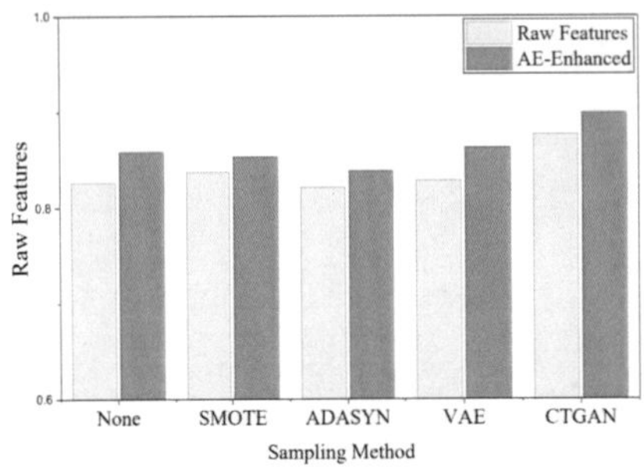

Fig. 2. Comparison of F1 scores before and after introducing an autoencoder.

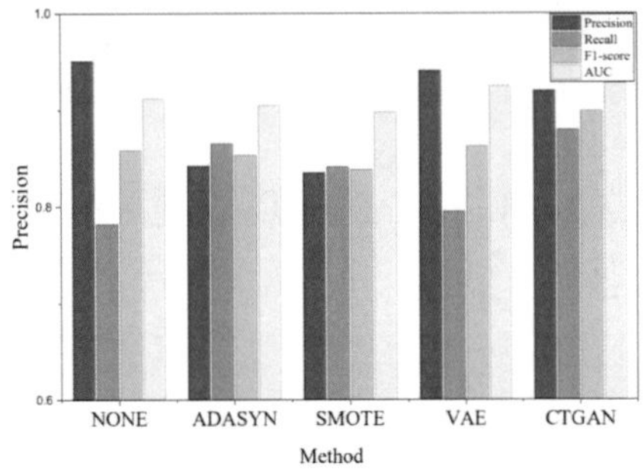

Fig. 3. Comparison of different sampling methods.

Comparative Analysis of Optimization Algorithms To verify the advantages of the proposed MAB Thompson Sampling algorithm in hyperparameter optimization, we compared MAB with four mainstream swarm intelligence algorithms: Particle Swarm Optimization (PSO)[17], Grey Wolf Optimizer (GWO)[18], Whale Optimization Algorithm (WOA)[19], and Sparrow Search Algorithm (SSA) [20]. The dataset was feature-enhanced by AutoEncoder and fully balanced (1:1) by CTGAN. In the experiments, the number of iterations for all algorithms was set to 15, the population size (for swarm algorithms) to 20, and the objective function was the F1-score on the validation set. The results are shown in Table 2.

Table 2. Performance comparison of different optimization algorithms

Algorithm	Precision	Recall	F1-score	AUC
None	0.9202	0.8791	0.8993	0.9282
PSO	0.9261	0.9063	0.9161	0.9155
GWO	0.9453	0.9256	0.9353	0.9357
WOA	0.9148	0.8950	0.9048	0.9346
SSA	0.9334	0.9274	0.9304	0.9251
MAB	**0.9601**	**0.9541**	**0.9571**	**0.9536**

Analyzing Table 2 and Fig. 4, the MAB Thompson Sampling algorithm performed best among all methods. First, all optimization algorithms significantly outperformed the unoptimized baseline (F1: 89.93%), proving the effectiveness of parameter tuning. Among swarm intelligence algorithms (PSO, GWO, WOA, SSA), F1- scores clustered between 89.00% and 93.10%. This reflects a commonality: while simulating population behavior helps find better solutions quickly, their inherent mechanisms often lead them to trap in local optima.

In contrast, MAB Thompson Sampling achieved a performance leap with its unique exploration-exploitation balance mechanism. It boosted the F1-score to 95.71%, 2.18% higher than the second-best GWO. This advantage stems from MAB's ability to avoid greedy exploitation of the current best solution and probabilistically explore other possibilities in the parameter space, thus having a higher probability of discovering global optimal configurations missed by traditional heuristics.

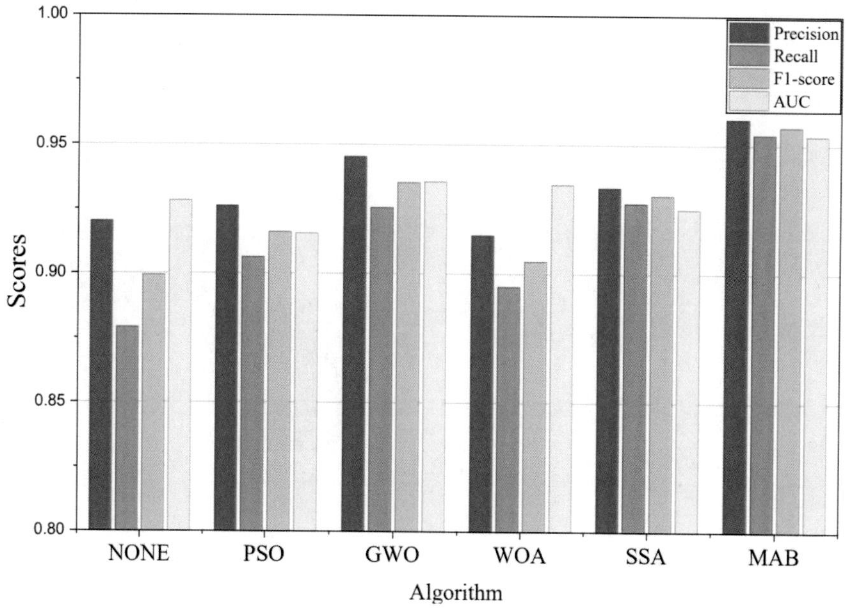

Fig. 4. Comparison of optimization methods.

Overall Model Comparison To comprehensively validate the proposed hybrid framework (AE-CTGAN-MAB), we compared it with three mainstream "Sampling + Optimization" combination models: SMOTE+PSO (traditional linear interpolation + swarm optimization), ADASYN+WOA (adaptive sampling + heuristic optimization), and VAE+GWO (deep generative sampling + Grey Wolf optimization).

Table 3. Overall comprehensive performance comparison

Model Framework	Precision	Recall	F1-score	AUC
SMOTE + PSO	0.9036	0.9644	0.9331	0.9305
ADASYN + WOA	0.9001	0.9632	0.9306	0.9282
VAE + GWO	0.9510	0.8836	0.9161	0.9378
Ours (AE-CTGAN-MAB)	**0.9601**	**0.9541**	**0.9571**	**0.9536**

As shown in Table 3 and Fig. 5, our CTGAN+MAB model achieved optimal performance across multiple metrics. The VAE+GWO model had high Precision but suffered from lower Recall and F1-score, as VAE tends to reconstruct rather than create, leading to insufficient sample diversity and missed detections. SMOTE+PSO achieved the second-best F1-score but had lower Precision, implying higher false positives in practice. In contrast, our model achieved the best balance and the highest Precision (96.01%), indicating that the high-quality synthetic samples helped the classifier establish more precise decision boundaries, significantly reducing misclassification costs.

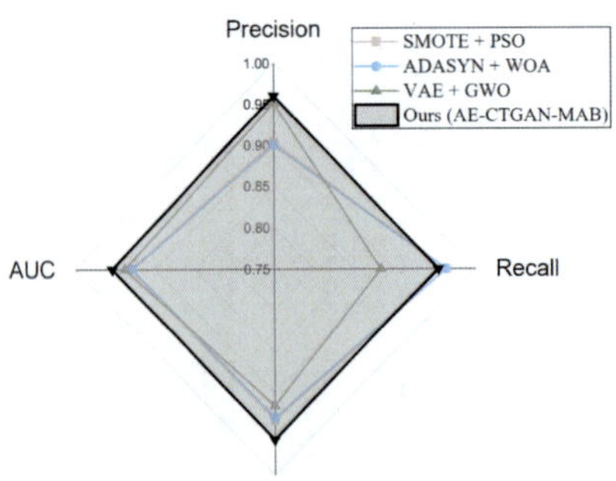

Fig. 5. Overall model comparison.

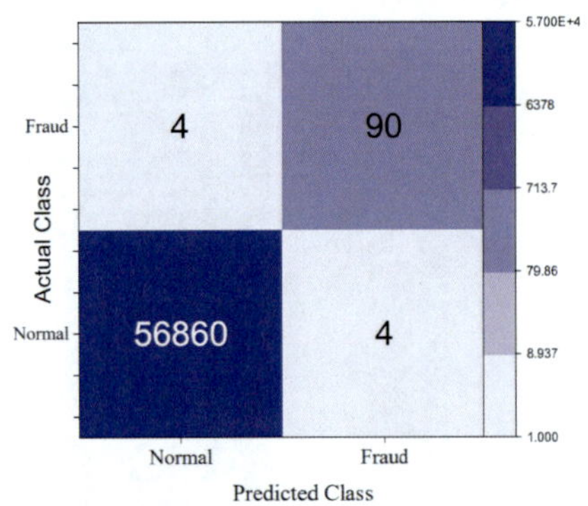

Fig. 6. Confusion Matrix.

Figure 6 further visualizes the classification details. In the confusion matrix,the model correctly identified the vast majority of fraudulent transactions (True Positives) while maintaining an extremely low number of False Positives (Normal samples misclassified as Fraud). This result is crucial for financial systems, as minimizing False Positives reduces customer friction and administrative costs associated with reviewing flagged transactions. The balanced distribution of errors confirms that the proposed framework does not achieve high Recall at the expense of Precision, but rather optimizes the trade-off between the two.

The optimal hyperparameters for our model on the CreditCard dataset are shown in Table 4.

It can be observed that the optimal configuration found by MAB differs significantly from the default settings. For instance, the learning rate was reduced from 0.3 to 0.05, while n estimators increased to 299. This combination suggests that a "slow learning, more iterations" strategy effectively prevents overfitting on the synthetic data. Furthermore, the slight adjustment in colsample bytree (0.9527) indicates that feature subsampling contributed to the model's robustness. These fine-grained adjustments, which are difficult to achieve via manual tuning, demonstrate the efficiency of Thompson Sampling in navigating the complex hyperparameter space.

Table 4. Best hyperparameters found by MAB

Hyperparameter	Default	Best Value
max_depth	6	9
learning_rate	0.3	0.05
n_estimators	100	299
subsample	1	1
colsample_bytree	1	0.9527
random_state	42	42

3.3 Ablation Study

To verify the incremental gain of each module in the proposed hybrid framework, we designed an ablation study using a step-wise strategy. We sequentially evaluated the evolution from the XGBoost baseline to the introduction of AutoEncoder (AE) feature enhancement, then to the integration of CTGAN for constructing a 1:1 balanced dataset, and finally to the addition of MAB Thompson Sampling optimization.

The results (Table 5) clearly demonstrate the step-wise performance improvement: First, the introduction of AE significantly improved Precision due to the capture of concealed anomaly patterns by reconstruction error, boosting the F1-score from 0.826 to 0.894. Subsequently, the introduction of CTGAN on top of feature enhancement resolved the data imbalance bottleneck, drastically improving Recall and pushing the F1-score above 0.94. Finally, global hyperparameter fine-tuning via MAB further unleashed the model's potential, achieving a peak F1-score of 0.9571. This process demonstrates that the three core modules are interlinked and work synergistically through feature mining, data reconstruction, and parameter optimization to build a high-performance fraud detection system.

Table 5. Ablation study results: Incremental gains from step-wise modules

Model Configuration	Precision	Recall	F1-score	AUC
1. Baseline	0.9451	0.7335	0.8260	0.8714
2. + AutoEncoder(Step 1)	0.9520	0.8430	0.8942	0.9250
3. + CTGAN(Step 2)	0.9580	0.9410	0.9494	0.9480
4. + MAB (Full Model)	0.9601	0.9541	0.9571	0.9536

4 Conclusion

Existing methods often struggle to balance Precision and Recall when facing challenges such as extreme class imbalance, concealed feature patterns, and adaptive hyperparameter optimization. In response, we propose a hybrid detection framework fusing Deep Representation Learning, generative data augmentation, and reinforcement learning optimization. First, an AutoEncoder (AE) is introduced to build a deep feature engineering module. By extracting latent vectors and reconstruction errors, it effectively mines non-linear anomaly patterns in transaction data. Second, based on the enhanced feature space, the CTGAN algorithm is utilized to generate high-quality synthetic samples, constructing a fully balanced dataset with a 1:1 positive-to-negative ratio, thereby effectively mitigating the data sparsity problem. Finally, an XGBoost classifier is constructed, employing a Multi-Armed Bandit (MAB) algorithm based on Thompson Sampling for adaptive hyperparameter optimization.

Experimental results demonstrate that through the synergistic effects of feature mining, data reconstruction, and parameter optimization, the proposed framework achieves excellent performance on the European Credit Card dataset, with a Precision of 96.01%, an F1-score of 95.71%, and an AUC of 95.36%. Compared with existing combinations of sampling and optimization methods, this model significantly reduces the false positive rate while maintaining a high recall rate, verifying its robustness in complex financial risk control scenarios. Future work will focus on the online update mechanism and model interpretability of this framework in real-time streaming data environments.

References

1. Huang, H., Li, C., Zhu, J., Li, P., Chen, X.: Research on detection technology for effectively preventing credit card fraud. China Credit Card **10**, 90–94 (2024). (in Chinese)
2. Mienye, I.D., Jere, N.: Deep learning for credit card fraud detection: a review of algorithms, challenges, and solutions, IEEE Access (2024)
3. Yang, G.: Credit card fraud detection based on machine learning prediction pp. 35–45 (2024). https://doi.org/10.2991/978-94-6463-540-9_5
4. Liu, H., Xu, J., Cao, S., Liu, Y., Qiao, L.: Research progress on digital financial fraud detection oriented to social relationship networks. J. Zhejiang Univ. (Sci. Ed.) **51**(01), 41–54 (2024). (in Chinese)

5. Ozbayoglu, A.M., Gudelek, M.U., Sezer, O.B.: Deep learning for financial applications: a survey. App. Soft Comput. **93**, 106–384 (2020)
6. Xu, Y.: Research on credit card fraud detection based on generative adversarial networks, Nanjing University of Posts and Telecommunications (2023). Master's thesis in Chinese
7. Fiore, U., De Santis, A., Perla, F., Zanetti, P., Palmieri, F.: Using generative adversarial networks for improving classification effectiveness in credit card fraud detection. Inf. Sci. **479**, 448–455 (2019)
8. Rtayli, N., Enneya, N.: Enhanced credit card fraud detection based on svmrecursive feature elimination and hyper-parameters optimization. J. Inf. Secur. Appl. **55**, 102–596 (2020)
9. Benchaji, I., Douzi, S., El Ouahidi, B., Jaafari, J.: Enhanced credit card fraud detection based on attention mechanism and LSTM deep model. J. Big Data **8**(1), 151 (2021)
10. Xu, L., Skoularidou, M., Cuesta-Infante, A., Veeramachaneni, K.: Modeling tabular data using conditional GAN. In: Advances in neural information processing systems, p. 32 (2019)
11. Agrawal, S., Goyal, N.: Analysis of thompson sampling for the multi-armed bandit problem. In: Conference on learning theory, pp. 39–1. JMLR Workshop and Conference Proceedings (2012)
12. Chen, T.: Xgboost: A scalable tree boosting system, Cornell University (2016)
13. Dal Pozzolo, A., Caelen, O., Johnson, R.A., Bontempi, G.: Calibrating probability with undersampling for unbalanced classification. In: 2015 IEEE symposium series on computational intelligence, pp. 159–166. IEEE (2015)
14. Chawla, N.V., Bowyer, K.W., Hall, L.O., Kegelmeyer, W.P.: Smote: synthetic minority over-sampling technique. J. Artif. Intell. Res. **16**, 321–357 (2002)
15. He, H., Bai, Y., Garcia, E.A., Li, S.: Adasyn: Adaptive synthetic sampling approach for imbalanced learning. In: 2008 IEEE international joint conference on neural networks (IEEE world congress on computational intelligence), pp. 1322–1328. IEEE (2008)
16. Kingma, D.P., Welling, M.: Auto-encoding variational bayes, (2013). arXiv:1312.6114 arXiv preprint
17. Kennedy, J., Eberhart, R.: Particle swarm optimization. In: Proceedings of ICNN'95-international conference on neural networks, vol. 4, pp. 1942–1948. IEEE (1995)
18. Mirjalili, S., Mirjalili, S.M., Lewis, A.: Grey wolf optimizer. Adv. Eng. Softw. **69**, 46–61 (2014)
19. Mirjalili, S., Lewis, A.: The whale optimization algorithm. Adv. Eng. Softw. **95**, 51–67 (2016)
20. Xue, J., Shen, B.: A novel swarm intelligence optimization approach: sparrow search algorithm. Syst. Sci. Control. Eng. **8**(1), 22–34 (2020)

Anomaly Detection for UAV Flight Data via LSTM–Attention–TCN Fusion Network

Muhammad Faizan Saleem[1], Shurui Fan[1(✉)], and Muhammad Asad Saleem[2]

[1] School of Electronic and Information Engineering, Hebei University of Technology, Tianjin 300401, China
fansr@hebut.edu.cn

[2] Yantai Research Institute of Harbin Engineering University, Yantai 264006, P. R. China

Abstract. This paper proposes a unified LSTM–Attention–TCN architecture for improving anomaly detection in UAV flight telemetry under varying and dynamic flight conditions. The architecture comprises a bidirectional LSTM for identifying long-range temporal dependencies, a multi-scale attention mechanism for weighting the most salient time-steps, and Temporal Convolutional Networks (TCN) for adjusting patterns in local time through dilated convolutions. The combined architecture allows the model to learn global and fine-grained behaviors in high-dimensional flight datasets. This proposed framework is demonstrated on three datasets of real-world flights that were supplemented with carefully parameterized synthetic anomalies to represent a range of real telemetry anomalies (faults) such as instrument faults, disturbances with the environment, and anomalies produced at an operational level. The experiments find that the hybrid model outperforms LSTM, TCN, and LSTM-TCN baselines consistently across the general metrics used to evaluate a model, including accuracy, precision, F1 score, TPR (True positive rate), and FPR (False positive rate), demonstrating both sensitivity to small changes while producing few false alarms. These results show that the proposed framework is reliable, scalable, and effective in UAV health monitoring and anomaly detection for safety-critical applications.

Keywords: UAV Telemetry · Anomaly detection · LSTM-Attention-TCN

1 Introduction

Unmanned aerial vehicles (UAVs) are now widely used to support various Intelligence, Surveillance, and Reconnaissance (ISR) operations, emergency responses, infrastructure assessments, and environmental monitoring in both the defense and civilian sectors [1]. UAV platforms offer a wide operational range, excellent maneuverability, and rapid field deployment, in contrast to stationary ground-based sensing systems. These qualities make them ideal for real-time aerial observation in diverse environments, especially in difficult-to-reach areas.

As UAVs became more deeply embedded in the infrastructure of the modern world, interfacing them through the Internet of Drones (IoD) presents a transformative and

L. Huang and D. Greenhalgh (Eds.): ICMLC 2026, LNNS 2061, pp. 594–606, 2027.
https://doi.org/10.1007/978-3-032-31236-5_41

revolutionary operational framework. IoD architectures enable real-time communications, collaborative decision-making, and shared data exchanges with the assistance of multiple UAVs to accomplish complex missions. However, operating with telemetry data, which is characterized by real-time sensor outputs, positional updates, and system health information, introduces significant implications for reliability and security. Flight telemetry data anomalies refer to deviations in the data that exceed the expected range of random errors, whether they are the result of sensor drift, intentional cyber intrusion, or environmental interference, which can all quickly turn into mission failure, safety hazards, and financial damage. Classic anomaly detection approaches, typically developed for static or low-complexity data, are difficult to implement with the dynamic and high-dimensional data of IoD telemetry, highlighting the need for specialized methods tailored for this domain [2]. Among all anomaly detection methodologies being implemented using telemetry data from UAVs, including Support Vector Machines (SVM), Random Forests/Decision Trees, and k-Nearest Neighbors (k-NN), which not only rely heavily on handcrafted features and simplified assumptions but also struggle to capture the nonlinear, high-dimensional, and time-dependent characteristics of UAV telemetry, deep learning have increasingly become the focus of scholarly researchers in recent years as by learning hierarchical and temporal representations directly from raw data, deep learning models enable far more accurate and robust anomaly detection.

[3] proposed a deep temporal network model combining a temporal convolutional network (TCN) with gated recurrent units (GRU) to explain the causes of abnormal risks; Wang et al. [4] proposed an anomaly detection algorithm based on Long Short-Term Memory (LSTM) and Generative Adversarial Network (GAN) for unmanned aerial vehicle data anomaly detection [5]. Uses a convolutional variational autoencoder to learn compact features from normal flight telemetry, which an LSTM then models to capture temporal dependencies [6]. Proposed an LSTM-attention-VAE, which leverages the strengths of LSTM in handling sequential data, an attention mechanism for extracting key features, and VAE in learning data distributions. Beyond LSTM-centric approaches, a diverse range of deep learning–based anomaly detection models have been proposed to address the complexity of UAV flight telemetry. Zhao et al. [7] proposed a reconstruction-based anomaly detection framework that utilizes stacked denoising autoencoders combined with wavelet decomposition, aiming to enable effective noise suppression and high-dimensional feature extraction from UAV sensor data. Li et al. [8] proposed a multi-modal federated learning–based anomaly detection model, which integrates heterogeneous UAV telemetry streams while preserving data privacy and improving generalization across distributed platforms. Reis et al. [9] proposed an edge-based real-time fault detection framework for UAV systems that combines B-spline–based telemetry reconstruction with a lightweight hybrid AI model, enabling accurate anomaly detection under noisy and missing data conditions while remaining suitable for deployment on resource-constrained onboard hardware. Chen et al. [10] proposed a UAV anomaly detection method based on a convolutional autoencoder combined with Support Vector Data Description (SVDD), utilizing a 0/1 soft-margin loss. The autoencoder learns compact spatial features from normal flight data, while the SVDD model defines a robust decision boundary to distinguish anomalous UAV behaviors. Liu et al. [11] proposed DCATCN, a UAV sensor anomaly detection method based on a Temporal Convolutional

Network (TCN) integrated with a cross-attention mechanism, where the TCN captures multi-scale temporal dependencies in flight telemetry while the cross-attention module models inter-sensor correlations, enabling accurate detection of complex and subtle anomalies in UAV sensor data.

This study proposes a hybrid LSTM–Attention–TCN framework for detecting abnormal flight behavior in order to improve anomaly detection accuracy in UAV flight telemetry. It is important to note that the proposed framework addresses a binary anomaly detection task, where the objective is to distinguish between normal and abnormal flight behavior. Although multiple anomaly types are injected to reflect realistic fault scenarios, they are treated as a single anomalous class rather than fine-grained sub-categories. Long-range dependencies and local temporal dynamics are jointly captured by the LSTM and TCN components. At the same time, the integration of attention modules enhances the model's capacity to concentrate on important temporal patterns. Because of its exceptional sequence modeling capabilities, this hybrid structure is very useful for spotting subtle and changing anomalies in complex flight data.

2 LSTM-Attention-TCN Method Construction

This paper constructs an LSTM-attention-TCN flight data anomaly detection model based on the characteristics of flight data, as illustrated in Fig. 1. Our proposed model combines three components to exploit the strengths of each approach. LSTM possesses superior learning capability for sequential data, which is employed in this model to capture the long-term temporal characteristics and features in the flight time-series data. The enriched LSTM features are passed to the Multi-Scale Temporal Attention. The attention learning mechanism learns the relations and the importance assigned to each temporal step in the hidden features, so that the model can learn to attend to the most salient or discriminative time regions for refinement. The output generated from the attention learning is then passed through the Temporal Convolutional Network (TCN). The TCN uses techniques such as dilated convolutions and residual connections to capture and refine local temporal patterns and shorter-term causal relationships. The context vectors from each module (LSTM, Attention, TCN) are then concatenated and passed to a feature fusion block and final classifier layer to produce the final anomaly detection output.

2.1 Long Short-Term Memory

LSTM [12] is the primary feature extractor, responsible for capturing long-term sequential dependencies in the UAV sensor data. The LSTM processes the input sequence first and generates a rich, high-level feature representation. Then, the subsequent Attention and TCN layers refine this representation. The gating mechanism forms the core of LSTM, which also controls the flow of information.

- The input gate decides which new information to store in the cell state

$$i_t = \sigma(U_{ii}x_t + b_{ii} + U_{hi}h_{t-1} + b_{hi}) \tag{1}$$

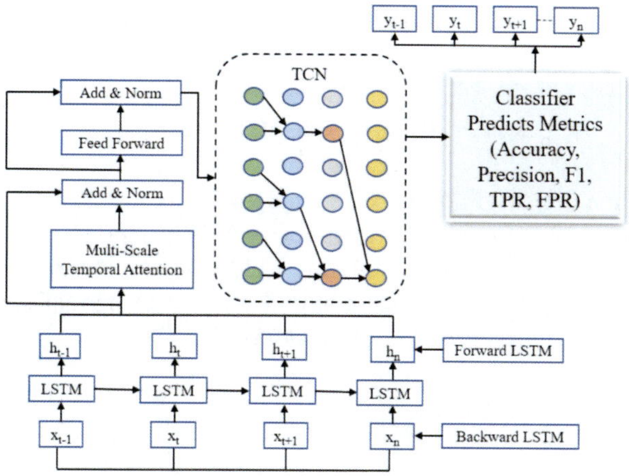

Fig. 1. LSTM-attention-TCN flight data anomaly detection model

- The forget gate decides what information shall be discarded from the cell state

$$f_t = \sigma\left(U_{if}x_t + b_{if} + U_{hf}h_{t-1} + b_{hf}\right) \tag{2}$$

- cell state update creates a vector of new candidate values and updates the long-term memory

$$\tilde{C}_t = \tanh\left(U_{ig}x_t + b_{ig} + U_{hg}h_{t-1} + b_{hg}\right) \tag{3}$$

$$C_t = f_t \odot C_{t-1} + i_t \odot \tilde{C}_t \tag{4}$$

- The output gate at the end decides what the next hidden state should be

$$O_t = \sigma(U_{io}x_t + b_{io} + U_{ho}h_{t-1} + b_{ho}) \tag{5}$$

$$h_t = O_t \odot \tanh(C_t) \tag{6}$$

Because the LSTM in this implementation is bidirectional, it can learn from the sequence's past and future context. A feature map is the result, which is subsequently sent to the Multi-Scale Temporal Attention and TCN blocks for additional classification and processing.

2.2 Multi-scale Temporal Attention

We use a multi-scale temporal attention that computes scaled dot-product attention over fused multi-scale convolutions to refine bidirectional LSTM features. The fused features that result in attended output O, and attention A are queries (Q), keys (K), and values (V).

$$A = softmax\left(\frac{QK^T}{\sqrt{d_k}}\right), O = AV \tag{7}$$

Where d_k is the dimension of the keys, while softmax in this context converts the raw attention scores into normalized weights. The model can adaptively weight salient temporal regions while maintaining long-term LSTM context for further TCN refinement because the attended context is residual-connected and layer-normalized. This process lowers false alarms and enhances anomaly localization.

2.3 Temporal Convolution Network

The general structure of TCN [13] can be represented as shown in Fig. 2. Due to the limitation of the convolutional kernel size, the Traditional convolutional neural network (CNN) is often not suitable for modeling and regression prediction of time series. Wavenet [14] is widely used in speech synthesis tasks and performs well in sequence modeling among the variants of CNN structure. TCN, like Wavenet, is an enhanced form of CNN [15] that addresses the common issues CNNs face when handling sequential data, making it a strong choice for sequence modeling tasks. The only difference TCN makes when compared to Wavenet is that TCN removes the gate mechanism and introduces the residual structure [16]. Unlike the traditional 1-D CNN, TCN has three key features: causal convolution, expansive convolution, and residual connection. By combining these features, TCN shall be capable of modeling and predicting time series. In our Implementation, the Temporal Convolutional Network (TCN) uses stacked dilated causal convolutions to fine-tune local temporal patterns.

$$y_t = \sum_{k=0}^{K-1} \omega_k \cdot x_{t-d\cdot k} \tag{8}$$

Where K is the kernel size, while d is the dilation. In order to exponentially increase the receptive field without sacrificing resolution, each block uses increasing dilation factors (1, 2, 4, 8). The architecture efficiently captures multi-scale dependencies by incorporating batch normalization and residual connections for stable training. This enables the model to efficiently determine the local context of anomalous events by learning hierarchical features from the attention-enhanced sequences. The dilated convolution operation ensures Long-range pattern capture while preserving computational efficiency.

3 Experiments and Analysis

3.1 Dataset

The dataset used in this study is a publicly available dataset, shared by Rodri [17]. The dataset provides detailed telemetry data collected during real-world autonomous flights. This dataset includes comprehensive flight parameters such as position, velocity,

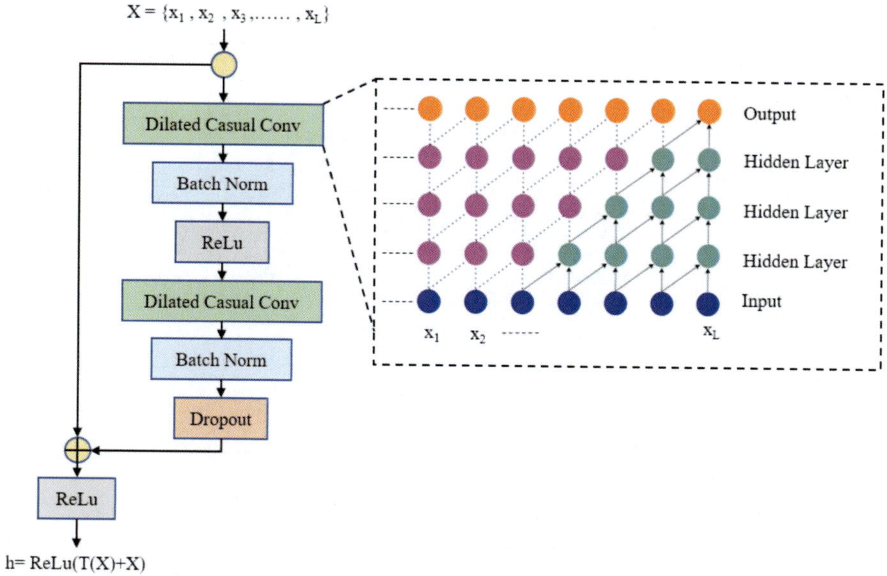

Fig. 2. General structure of the TCN model

acceleration, attitude, power consumption, and battery state under various operational conditions. To evaluate the generalization potential of the proposed LSTM-Attention-TCN hybrid model, a wide range of operational abnormalities, including wind gusts, motor stutters, IMU noise, GPS jumps, altitude drifts, communication drops, sensor spikes, and battery sags, were simulated by systematically injecting synthetic anomalies into clean telemetry data. Figure 3 describes the type and percentage of all the anomalies used in our implementation. One important consideration is that all injected anomaly instances in this work are assigned to a unified anomalous category, without performing fine-grained classification of specific anomaly types. To guarantee controlled variability and reproducibility, each anomaly type was parameterized according to its temporal duration, intensity, and affected feature subset.

The dataset comprises 209 individual flights with varying conditions. To get the realistic anomaly results, we evaluated our model on different flight conditions (varying speed, payload, and altitudes) with differing percentages of anomaly, i.e., we named our flight combinations as Dataset I (Low-anomaly regime of 3.5%), Dataset II (Medium-anomaly regime of 7%), and Dataset III (High-anomaly regime). Each dataset can be used as a controlled benchmark to test the model's generalization, robustness, and adaptability across various flight dynamics and anomaly prevalence levels. Table 1 shows the dataset specifications of all three datasets.

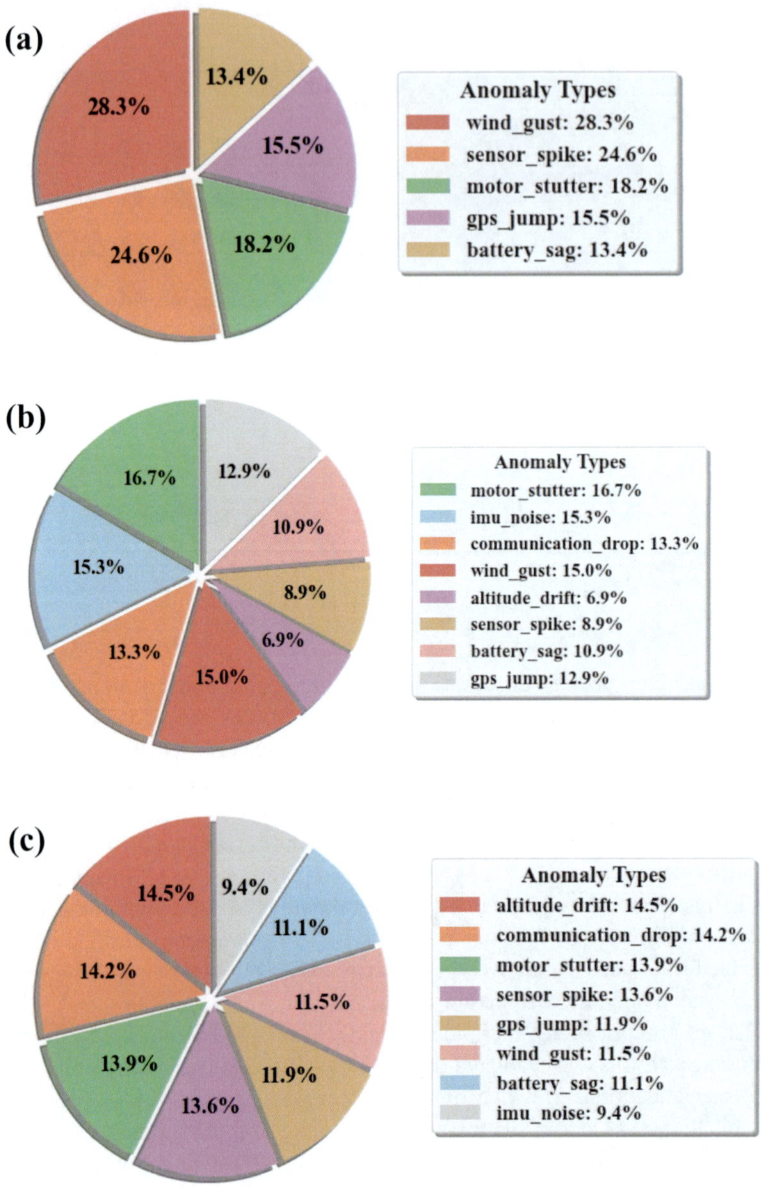

Fig. 3. Anomaly distribution for (a) dataset I, (b) dataset II, and (c) dataset III

Table 1. Information of experimental flight datasets

Dataset number	Dataset I	Dataset II	Dataset III
Original data length	17055	13722	10222
Number of parameters	30	30	30
Experiment index range	[3121, 15987]	[12695, 2331]	[2085, 9089]
Experiment parameter length	12866	10364	7004
Number of anomalous samples	596	686	715

3.2 Hyperparameters and Evaluation Metrics

To ensure a fair and unbiased evaluation of the proposed anomaly detection framework, the flight telemetry data were divided into three mutually exclusive subsets for training, validation, and testing. Following standard practice in time-series anomaly detection literature, approximately 60% of the data were used for model training, 20% for validation, and the remaining 20% for final testing. The training subset was used exclusively for learning model parameters, while the validation set was employed for hyperparameter tuning, early stopping, and threshold selection to prevent overfitting. The test set was kept completely unseen during training and model selection, and was used only for final performance reporting. This split ensures that performance metrics reflect the model's generalization capability under unseen flight conditions and varying anomaly prevalence levels. The Proposed Model was trained using an optimized AdamW optimizer and uses the hyperparameters as: learning rate of 0.001 (empirically determined for stable convergence), a batch size of 32. There are four dilated TCN blocks (dilations $(1, 2, 4, 8)$ hidden size 128) for local temporal pattern refinement, a multi-scale temporal attention module with eight heads for capturing salient temporal features, and a two-layer bidirectional LSTM (hidden size 128) for long-term sequential modeling. All layers had a dropout of 0.3, and the sequence length was set to 30 timesteps. The loss function for the proposed model can be presented as a sum of Weighted Binary Cross-Entropy (WBCE) and Focal Loss:

$$L_{total} = \lambda_1 L_{WBCE} + \lambda_2 L_{Focal} \tag{9}$$

Where, λ_1 and λ_2 controls the contribution of each loss (usually set to 0.5 if equally weighted). To address class imbalance and mitigate bias toward normal samples, Weighted Binary Cross-Entropy (WBCE) assigns the positive (anomalous) class a higher priority. Focal Loss, which is controlled by the parameters α and γ (balancing positive/negative samples), further highlights difficult-to-classify samples by down-weighting simple examples. The loss function focuses on challenging anomalies and corrects for imbalance by combining these two elements. The model's sensitivity and robustness in identifying uncommon or subtle anomalies in UAV telemetry data are enhanced by this synergy. To better understand the efficiency of the proposed model, we evaluated it on standard parameters of Accuracy (Acc), Precision (Pre), TPR (True Positive Rate), FPR (False Positive Rate), and F1 score.

3.3 Experiment Results and Analysis

Figure 4 illustrates the anomaly detection behavior of the proposed LSTM–Attention–TCN model across three datasets with increasing anomaly prevalence. For Dataset I (low-anomaly regime), the model achieves a dominant true-negative region with only a small number of false positives and false negatives, indicating strong discrimination between normal and abnormal flight behavior. In Dataset II, despite a higher anomaly ratio, the model maintains perfect or near-perfect true positive detection while further suppressing false alarms, demonstrating robustness under moderate class imbalance. For Dataset III, where anomaly density and complexity are highest, the model still preserves a high true positive rate with limited false negatives, confirming its stability and sensitivity under challenging operational conditions. The anomaly detection results for the flight datasets I, II, and III are presented in Tables 2, 3, and 4, respectively. According to the experimental results, the suggested LSTM-Attention-TCN model consistently outperforms the other baseline models on all three datasets. It maintains the lowest false-positive rates while achieving the highest accuracy, precision, and F1-score for each dataset. Interestingly, its TPR values remain significantly higher than those of any baseline model, indicating that it is more sensitive to even the smallest anomaly patterns. However, in more difficult situations, even the hybrid LSTM-TCN model falters, and standalone LSTM and TCN models exhibit decreased robustness. Overall, feature discrimination and temporal context modeling are strengthened by the combination of attention and temporal convolution, yielding more stable and dependable anomaly detection performance.

Table 2. Anomaly detection results comparison for dataset I

Model	Accuracy	Precision	F-1	TPR	FPR
LSTM	0.993	0.922	0.889	0.841	0.004
TCN	0.991	0.933	0.900	0.875	0.003
LSTM-TCN	0.956	0.965	0.901	0.906	0.002
LSTM-Attention-TCN	0.994	1	0.933	0.937	0.001

Table 3. Anomaly detection results comparison for dataset II

Model	Accuracy	Precision	F-1	TPR	FPR
LSTM	0.950	0.918	0.820	0.814	0.006
TCN	0.944	0.900	0.888	0.871	0.004
LSTM-TCN	0.925	0.948	0.877	0.891	0.004
LSTM-Attention-TCN	0.997	0.977	0.912	1	0.0001

Table 4. Anomaly detection results comparison for dataset III

Model	Accuracy	Precision	F-1	TPR	FPR
LSTM	0.976	0.928	0.743	0.619	0.003
TCN	0.824	0.766	0.688	0.555	0.033
LSTM-TCN	0.929	0.830	0.703	0.809	0.006
LSTM-Attention-TCN	0.967	0.937	0.891	0.944	0.002

3.4 Comparison with Recent UAV Anomaly Detection Methods

To further validate the effectiveness of the proposed LSTM–Attention–TCN framework beyond internal baselines, we compare its performance with several recent and representative UAV telemetry anomaly detection approaches, including Automated Machine Learning based anomaly detection [2], wavelet decomposition combined with stacked denoising autoencoders (W-SDAE) [7], and an edge-oriented fault detection framework based on B-spline telemetry reconstruction and lightweight hybrid AI [9]. These methods represent different modeling paradigms, including automated model selection, signal decomposition with deep reconstruction, and resource-constrained edge intelligence. Since these studies report results on different UAV telemetry datasets and experimental conditions, the comparison is conducted at a metric and capability level, focusing on reported accuracy, F1-score, precision, and Recall (also known as true positive rate). As shown in Table 5, the proposed LSTM–Attention–TCN model achieves competitive or superior performance. This demonstrates that the proposed framework not only improves upon classical deep learning baselines but also compares favorably with recent state-of-the-art UAV anomaly detection methods under diverse operational assumptions.

Table 5. Anomaly detection results comparison of the proposed model with state-of-the-art methods

Method	Accuracy	Precision	Recall (TPR)	F-1 score
Local outlier factor [2]	0.920	0.880	0.850	0.860
Wavelet + SDAE [7]	0.975	0.970	0.918	0.945
B-Spline + Hybrid AI [9]	0.936	0.830	0.860	0.850
Proposed model	0.997	0.977	1	0.912

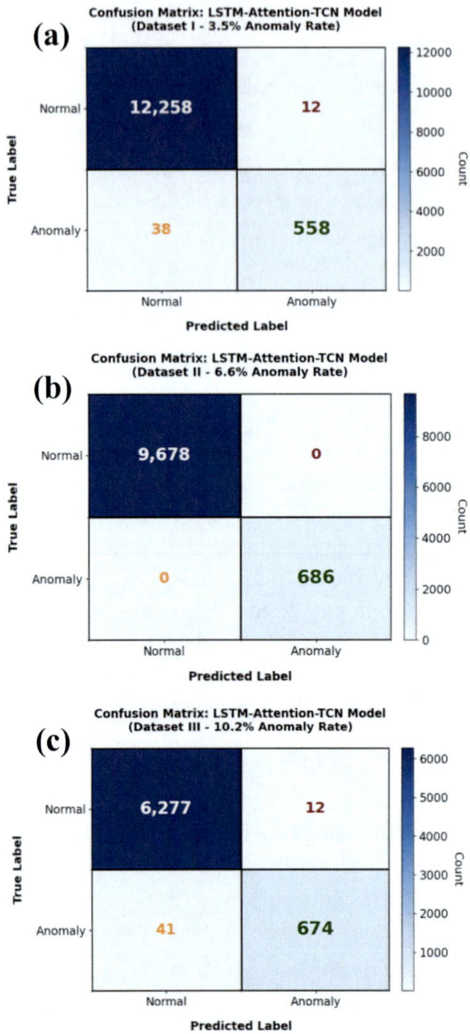

Fig. 4. Confusion matrices of the proposed model across varying anomaly prevalence levels (a) dataset I, (b) dataset II, and (c) dataset III

4 Conclusion

This study proposes a hybrid LSTM-attention-TCN framework developed to enhance the reliability of UAV anomaly detection under complex real-flight operational conditions. By integrating the temporal data processing capabilities of LSTM, selective feature amplification capabilities of multi-scale attention, and local pattern refinement capabilities of TCN, the model effectively captures both global and fine-grained temporal dependencies. For maintaining the performance under severe class imbalances, the model employs focal loss, adaptive sampling, and weighted BCE. Through comprehensive evaluations, the proposed model has proven to be superior to other deep learning methods, as it has demonstrated consistent improvements across accuracy, precision, F1-score, TPR, and FPR, confirming reliability for detecting diverse anomalies in UAV operations. Overall, the proposed framework provides a powerful and scalable solution for the field of flight data anomaly detection and is expected to contribute to the development of a real-time monitoring and fault diagnosis system for future aircraft.

In future work, we aim to further optimize the LSTM–Attention–TCN framework to enhance its detection capability for subtle or evolving anomalies in more complex flight conditions. While our dataset is based on real UAV flights, the anomalies are artificially injected into the data. Therefore, the next meaningful step would be to collect and incorporate data from naturally occurring faults, which are less predictable than embedded faulty data, to help improve the generalization and robustness of model performance. Extending the proposed framework to fine-grained anomaly classification, where specific fault types are explicitly identified, will be explored in future work. We will also extend the model to integrate real-time flight data, enabling online learning and continuous monitoring of anomalies. Our ultimate aim will be to evaluate in the real flight trials and improve the application of onboard monitoring and UAV operations in safety-critical applications.

References

1. Shakhatreh, H. et al.: Unmanned aerial vehicles (UAVs): a survey on civil applications and key research challenges. IEEE Access. **7**, 48572–48634 (2019)
2. Sezgin, A., Keskin, R., Boyacı, A.: Anomaly detection in unmanned aerial vehicle telemetry using automated machine learning. Fırat Üniversitesi Mühendislik Bilimleri Dergisi. **37**, 699–709 (2025)
3. Lijun, Z., Fang, M.: The risk identification of streamed flight data based on deep temporal network. In: 2025 IEEE 8th Information Technology and Mechatronics Engineering Conference (ITOEC), pp. 1259–1263. IEEE, Chongqing, China (2025)
4. He, R., Chen, K., Li, T., Zhou, Y.: Research on health management and fault prediction for human/unmanned collaborative cluster flight control system based on GAN-LSTM. In: 2024 Academic Conference of China Instrument and Control Society (ACCIS), pp. 288–292. IEEE, Chengdu, China (2024)
5. Zhang, W., Xu, J., Liu, Y., Yang, M.: Anomaly identification of flight test time series data based on improved LSTM. In: 2024 8th International Conference on Electrical, Mechanical and Computer Engineering (ICEMCE), pp. 1044–1050. IEEE, Xi'an, China (2024)
6. Xu, J.: Flight telemetry data anomaly detection method based on LSTM-attention-VAE. J. Phys. Conf. Ser. **2965**, 012048 (2025)
7. Zhou, S., He, Z., Chen, X., Chang, W.: An anomaly detection method for UAV based on wavelet decomposition and stacked Denoising autoencoder. Aerospace. **11**, 393 (2024)
8. Lu, Y., Yang, T., Zhao, C., Chen, W., Zeng, R.: A swarm anomaly detection model for IoT UAVs based on a multi-modal Denoising autoencoder and federated learning. Comput. Ind. Eng. **196**, 110454 (2024)
9. Reis, M.J.C.S., Reis, A.J.D.: Edge-based real-time fault detection in UAV systems via B-spline telemetry reconstruction and lightweight hybrid AI. Sensors. **25**, 4944 (2025)

10. Chen, H., Lyu, Y., Shi, J., Zhang, W.: UAV anomaly detection method based on convolutional autoencoder and support vector data description with 0/1 soft-margin loss. Drones. **8**, 534 (2024)
11. Wu, B., Zhou, C., Liu, Y., Xiao, P., Zheng, W.: DCATCN: a temporal convolutional network and cross-attention based UAV sensor anomaly detection method. Internet of Things. **34**, 101760 (2025)
12. Hochreiter, S., Schmidhuber, J.: Long short-term memory. Neural Comput. **9**(8), 1735–1780 (1997)
13. Guo, H.-T., Pan, L., Wang, J., Fan, X.-B., Li, J., Liu, Z.: Short-term wind power prediction method based on TCN-GRU combined model. In: 2021 IEEE Sustainable Power and Energy Conference (iSPEC), pp. 3764–3769. IEEE, Nanjing, China (2021)
14. Wang, H., Peng, C., Liao, B., Cao, X., Li, S.: Wind power forecasting based on WaveNet and multitask learning. Sustainability. **15**, 10816 (2023)
15. Galvan, J., Raja, A., Li, Y., Yuan, J.: Sensor data-driven UAV anomaly detection using deep learning approach. In: MILCOM 2021–IEEE Military Communications Conference, pp. 589–594. IEEE, San Diego, CA, USA (2021)
16. Zhang, K., Sun, M., Han, T.X., Yuan, X., Guo, L., Liu, T.: Residual networks of residual networks: multilevel residual networks. IEEE Trans. Circuits Syst. Video Technol. **28**(6), 1303–1314 (2018)
17. Rodrigues, T.A. et al.: Data Collected with Package Delivery Quadcopter Drone. Carnegie Mellon University, Pittsburgh, PA, USA (2021)

FedSCP: Personalized Federated Learning Against Intrinsic Adversarial Attacks via Semantic Guidance and Conflict Purification

Bin Zhou[1(✉)], Yang Cao[1(✉)], Mengli Wei[2], and Yongjian Chang[1,3]

[1] Engineering Research Center of Blockchain Application, Supervision and Management, Ministry of Education (Southeast University), School of Cyber Science and Engineering, Southeast University, Nanjing 21189, China
{220245658,caoyeacy,yongjianchang}@seu.edu.cn

[2] School of Mathematics, Southeast University, Nanjing 21189, China
{weimengli,wwyu}@seu.edu.cn

[3] Center for Spatial Information Science, University of Tokyo, Tokyo 113-8654, Japan

Abstract. Personalized federated learning (PFL) addresses data heterogeneity by constructing customized models for each client. However, it faces a fundamental challenge in practical deployment: data heterogeneity inherently acts as an intrinsic adversarial attack, stemming from the system's internal data distribution, which persistently disrupts model convergence via gradient conflicts. Existing PFL methods, particularly those based on adaptive aggregation, exhibit critical deficiencies in resisting such attacks: they often fail to exploit the rich semantic information embedded in model parameters and overlook the directional conflicts arising during gradient updates. Consequently, even with high-quality model initialization, gradient interference in heterogeneous environments triggers severe client drift. To address these issues, we propose the FedSCP framework. FedSCP employs an innovative two-stage optimization strategy to counter these intrinsic attacks by: (1) optimizing the training starting point and (2) purifying the update direction. Experiments on multiple benchmark datasets demonstrate that FedSCP achieves up to 3.16% higher accuracy than state-of-the-art methods, effectively resisting intrinsic adversarial attacks from data heterogeneity and significantly accelerating convergence.

Keywords: Federated learning, Personalized federated learning, Gradient conflict purification, Semantic-guided aggregation

L. Huang and D. Greenhalgh (Eds.): ICMLC 2026, LNNS 2061, pp. 607–622, 2027.
https://doi.org/10.1007/978-3-032-31236-5_42

1 Introduction

s data privacy protection demands intensify, Federated Learning (FL) has gained significant attention as an emerging distributed machine learning paradigm. This approach enables multiple clients to jointly train global models without exposing local raw data [1], but in practical applications, differences in data distributions among participating nodes—data heterogeneity—severely constrain its development [2].

Re-examining this phenomenon from an optimization perspective reveals that data heterogeneity can be framed as an intrinsic adversarial attack. Unlike traditional malicious attacks [27], this attack does not originate from deliberate external sabotage but arises from inherent disparities in the system's internal data distributions. This intrinsic attack exhibits a cumulative effect: as training progresses, gradient conflicts progressively amplify [28], ultimately leading to the severe client drift phenomenon.

Addressing these challenges, Personalized Federated Learning (PFL) has emerged as a research hotspot. Unlike traditional FL's pursuit of a unified global model, PFL constructs tailored models for each client that align with local data characteristics, enhancing local performance while preserving data privacy [3–5]. Multiple technical approaches have emerged in this field, including meta-learning-driven methods [6], regularization-constrained methods [7,8], knowledge distillation techniques [9,25], model decomposition-based methods [10,11], and personalized aggregation algorithms [3,12–16].

However, existing PFL methods still exhibit critical shortcomings when confronting these intrinsic adversarial attacks, particularly personalized aggregation algorithms. These deficiencies manifest primarily in two aspects:

First, most adaptive aggregation methods [5] rely on scalar information, such as loss values, to adjust fusion weights. This approach neglects the rich semantic structures embedded within neural network parameters. Research indicates that different layers of neural networks carry varying levels of semantic information: shallow layers tend to learn general, low-level features, while deep layers focus on task-relevant, high-level semantics [17,18]. This hierarchical semantic division is explicitly reflected in the statistical properties of the parameters [17,19]. However, existing fusion methods fail to fully leverage this structured information to guide personalization.

Second, even when high-quality model starting points are obtained, gradient conflicts persist during the subsequent local training process [24,25]. As data heterogeneity among clients increases, these conflicts become more severe [22], exerting a significant negative impact on model optimization [23]. Although some studies [24] have attempted server-side corrections at the aggregation layer, such *passive post-processing strategies* fail to prevent conflicts at their source.

We propose the Federated Learning with Semantic-guided Fusion and Conflict Purification (FedSCP) framework, which employs a two-stage optimization strategy to systemically resist intrinsic adversarial attacks arising from data heterogeneity.

- **Stage 1 (Starting Point Optimization)**: We design a semantic similarity quantification method based on multidimensional statistical features to provide semantic-level guidance for weight fusion, thereby constructing a high-quality initial model.
- **Stage 2 (Directional Purification)**: We introduce a two-step gradient purification algorithm that explicitly identifies and eliminates conflicts across two distinct dimensions within the gradient space, ensuring the update direction is valid.

By decoupling the optimization of the starting point from the purification of the update direction, FedSCP fundamentally mitigates the impact of these intrinsic attacks.

The main contributions of this paper include:

- We re-examine the data heterogeneity problem in FL from a novel analytical perspective, theorizing it as an intrinsic adversarial attack phenomenon. This provides a deeper understanding of the underlying mechanisms of client drift and gradient conflicts.
- We propose a semantically-guided adaptive fusion mechanism that incorporates structured semantic information from parameters to steer personalized model fusion and optimize the training start point.
- We design a two-step gradient purification algorithm that decomposes conflicts into "individual" and "social" dimensions, resisting intrinsic attacks at the source of local training.
- We construct a unified framework for the decoupled optimization of training starting points and update directions. Evaluated across multiple benchmark datasets, FedSCP achieved up to 3.16% higher accuracy compared to existing methods, demonstrating effective resistance to intrinsic adversarial attacks and significantly accelerated convergence.

2 Related Work

Personalized federated learning (PFL) has emerged to mitigate the intrinsic adversarial attacks caused by data heterogeneity. Existing PFL methods can be categorized into five main types:

Regularization-Based Methods balance personalization and global consistency via constraint terms. Ditto [8] introduces quadratic regularization into local objectives, but requires extensive hyperparameter tuning and lacks insight into model internals.

Meta-Learning-Based Methods enable rapid adaptation through gradient-based personalization. Per-FedAvg [6] applies MAML to federated settings, but incurs high computational costs and can suffer from convergence issues under severe gradient conflicts.

Knowledge Distillation Methods transfer knowledge via soft labels. FedPAC [9] employs feature alignment, while FedGMKD [25] proposes prototype-based distillation. These methods offer flexible architectures but often require additional shared or synthetic data.

Model Decomposition Approaches separate shared and personalized components. FedROD [11] constructs dual prediction heads, and FedAS [10] employs adaptive separation strategies. However, the optimal separation point depends heavily on the specific scenario.

Personalized Aggregation-Based Methods construct customized models through differentiated aggregation weights, resisting intrinsic attacks during the fusion stage. FedALA [3] learns layer-wise weights but relies solely on loss feedback. FedAMP [12] introduces attention mechanisms for selective knowledge utilization. CosPer [16] uses cosine similarity-based aggregation. FedAH [15] proposes element-wise aggregation heads. FedPAW [14] computes weights via weighted mean squared error, and pFedFDA [13] adapts feature distributions. Critically, these methods **ignore hierarchical semantic properties and overlook explicit gradient conflicts** during local training.

Recent work recognizes that gradient conflicts from heterogeneity are fundamentally similar to adversarial perturbations [29]. ConFREE [24] mitigates conflicts via server-side aggregation, but this *passive* approach cannot prevent conflicts from occurring during local training. RegulFL [30] addresses malicious attacks rather than the intrinsic heterogeneity problem.

3 Methodology

3.1 Attack Model and Threat Analysis

The standard federated learning objective $\mathcal{F}(\theta)$ is the weighted average of local losses:

$$\min_{\theta} \mathcal{F}(\theta) = \sum_{c=1}^{N} p_c \mathcal{L}_c(\theta) \tag{1}$$

where p_c is client c's aggregation weight and $\mathcal{L}_c(\theta)$ is its expected local loss. Real-world federated systems face intrinsic adversarial attacks stemming from data heterogeneity.

Intrinsic Adversarial Attacks from Data Heterogeneity Data heterogeneity inherently constitutes an intrinsic adversarial attack, where uneven data distribution undermines global convergence. We classify these attacks by intensity:

Practical Heterogeneity (Medium-Intensity): Clients possess all categories but with varying proportions, typically modeled by a Dirichlet distribution. For client c, category proportions follow: $\{p_c^1, p_c^2, \ldots, p_c^K\} \sim \text{Dirichlet}(\alpha)$, where a small $\alpha > 0$ induces significant gradient bias.

Pathological Heterogeneity (High-Intensity): Clients possess only strict subsets of the total categories. Client c holds $K_c \ll K$ categories with $\mathcal{Y}_c \subset \{1, 2, \ldots, K\}$, yielding:

$$p_c^k = \begin{cases} \frac{1}{|\mathcal{Y}_c|} & k \in \mathcal{Y}_c \\ 0 & k \notin \mathcal{Y}_c \end{cases} \tag{2}$$

This generates sharply conflicting gradients that can "erase" category-specific knowledge during aggregation.

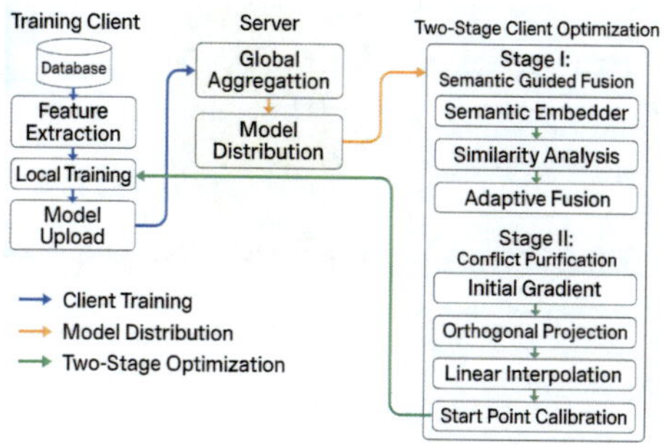

Fig. 1. FedSCP framework workflow.

3.2 Overview of the FedSCP Framework

FedSCP provides a robust defense mechanism for adaptive aggregation-based PFL methods (e.g., FedALA [3]). It employs a two-stage local optimization process that achieves personalization while resisting intrinsic adversarial attacks.

Framework Workflow Figure 1 illustrates FedSCP's execution flow within one communication round. The server handles model aggregation and distribution, while clients execute the two-stage optimization. At round t, each client c receives the global model θ_G^t and loads its local history: the previous personalized model θ_c^{t-1} and the prior global model θ_G^{t-1}. The client then executes two phases:

Phase 1 (Semantic-Guided Fusion): The client analyzes parameter semantics to quantify the layer-wise similarity $S_c^{(l)}$ between θ_G^t and θ_c^{t-1}. This determines hierarchical fusion weights to construct an *optimized starting point* $\hat{\theta}_c^t$.

Phase 2 (Conflict Purification and Calibration): The client computes an initial gradient $\mathbf{g}_c$ from $\hat{\theta}_c^t$. It then applies a two-step purification process to identify and resolve conflicts within $\mathbf{g}_c$, obtaining an *ideal gradient* $\tilde{\mathbf{g}}_c$. This gradient is used to calibrate $\hat{\theta}_c^t$, yielding the *final training starting point* $\tilde{\theta}_c^t$.

Client c then trains locally starting from $\tilde{\theta}_c^t$ to obtain its new personalized model θ_c^{t+1}, which is uploaded for the next round of aggregation.

Unified Optimization Objective Traditional FL objectives (Eq. 1) suffer from instability under data heterogeneity. We formulate FedSCP as a Min-Max game:

$$\min_{\Theta_{\mathcal{B}}} \max_{\mathcal{A} \in \mathbb{A}_{\mathcal{A}}} \mathcal{L}_{\mathcal{B}}(\Theta_{\mathcal{B}}, \mathcal{A}) \quad \text{where} \quad \mathcal{L}_{\mathcal{B}} = \mathbb{E}_{c \in \mathcal{B}}[\mathcal{L}_c(w_c)] \tag{3}$$

FedSCP decomposes this game into its two stages:

Phase 1 (Semantic-Guided Fusion) addresses the max component. It evaluates the reliability of updates (i.e., the "attack" $\mathcal{A}$ from heterogeneity) via semantic features, assigning lower weights to dissimilar, potentially "malicious" updates to construct a robust starting point.

Phase 2 (Conflict Purification) addresses the min component. It detects and resolves gradient conflicts to ensure that the local update step constitutes an effective descent, minimizing the local loss $\mathcal{L}_c(w_c)$.

3.3 Phase 1: Semantic-Guided Adaptive Fusion

We design a semantic-guided fusion mechanism that quantifies layer-wise functional similarity through parameter statistics, enabling the identification and suppression of intrinsic adversarial attacks during aggregation.

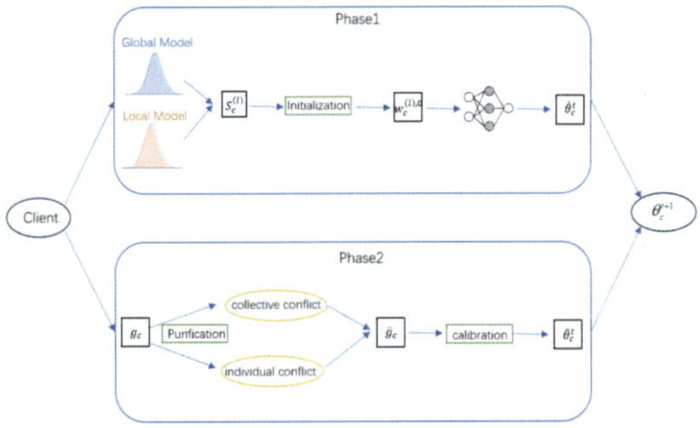

Fig. 2. Semantic similarity acquisition process.

Semantic Representation of Model Parameters Neural network functionality is characterized by parameter statistics. We quantify functional similarity as "semantic similarity" by comparing statistical features between the local and global models (Fig. 2).

We extract a 9-dimensional semantic feature vector $\mathbf{f}^{(l)}$ for each layer, where

$$\mathbf{f}^{(l)} = [\mu, \sigma, \text{skew}, \text{kurt}, \max, \min, |\cdot|_1, |\cdot|_2, \text{sparsity}]^\top,$$

capturing distribution, shape, norm, and sparsity information.

A three-layer MLP (the Semantic Embedder) projects $\mathbf{f}^{(l)}$ into a 64-dimensional space: $\mathbf{e}^{(l)} = E_s(\mathbf{f}^{(l)})$, capturing complex feature relationships. The embedder E_s is trained end-to-end with the local model.

Semantic similarity between the local embedding $\mathbf{e}_c^{(l)}$ and global embedding $\mathbf{e}_G^{(l)}$ combines cosine similarity and Euclidean distance:

$$S_c^{(l)} = \alpha \cdot \frac{\cos(\mathbf{e}_c^{(l)}, \mathbf{e}_G^{(l)}) + 1}{2} + (1 - \alpha) \cdot \frac{1}{1 + |\mathbf{e}_c^{(l)} - \mathbf{e}_G^{(l)}|_2} \tag{4}$$

where $\alpha \in [0, 1]$ balances directional consistency and spatial proximity. This analysis runs periodically at a refresh interval τ_{refresh} to balance overhead and accuracy.

Initialization and Learning of Fusion Weights The semantic similarity $S_c^{(l)}$ initializes the fusion weights $w_c^{(l),0}$:

$$\begin{aligned} w_c^{(l),0} &= \text{clamp}(f_{\text{map}}(S_c^{(l)}) \\ &+ \quad \Delta w_{\max} \cdot \sigma\left(\frac{z^{(l)}}{\mathcal{T}_t}\right) - \Delta w_{\text{offset}}, w_{\min}, w_{\max}) \end{aligned} \tag{5}$$

where $z^{(l)} = \frac{S_c^{(l)} - \mu_S}{\sigma_S + \epsilon}$, f_{map} is a mapping function, μ_S and σ_S are the mean and standard deviation of similarities, $\Delta w_{\max}$ and Δw_{offset} are scaling/shift parameters, and $\mathcal{T}_t$ is a decaying temperature.

Following FedALA [3], we iteratively refine these weights by minimizing the task loss on a local data subset, yielding the semantically-fused model $\hat{\theta}_c^t$ as the starting point for Phase 2.

3.4 Phase 2: Conflict Purification and Starting Point Calibration

In the presence of intrinsic adversarial attacks, the initial local gradient $\mathbf{g}_c = \nabla \mathcal{L}_c(\hat{\theta}_c^t)$ contains conflicts that cause training divergence. We design a conflict purification mechanism to identify and resolve these conflicting components.

Definition of Conflict We decompose gradient conflicts into two distinct dimensions:

Individual Conflict: Measures the deviation of the local gradient $\mathbf{g}_c$ from the *global correction vector* $\mathbf{v}_c^t = \theta_G^t - \hat{\theta}_c^t$. An individual conflict exists when $\mathbf{g}_c \cdot \mathbf{v}_c^t < 0$, implying the local update is moving *away* from the latest global model.

Social Conflict: Reflects the deviation of $\mathbf{g}_c$ from the *collective trend* of the global model. We locally estimate this trend using consecutive global models to approximate the global average gradient:

$$\mathbf{g}_G^t \approx \frac{\theta_G^t - \theta_G^{t-1}}{\eta_G} \tag{6}$$

We use momentum smoothing to capture the stable collective trend $\overline{\mathbf{g}}_{\text{smooth}}$:

$$\overline{\mathbf{g}}_{\text{smooth}} = \beta \cdot \overline{\mathbf{g}}_{\text{smooth}}^{t-1} + (1 - \beta) \cdot \mathbf{g}_G^t \tag{7}$$

A social conflict exists when $\mathbf{g}_c \cdot \overline{\mathbf{g}}_{\text{smooth}} < 0$, indicating the local update opposes the global momentum.

Two-Step Gradient Purification and Starting Point Calibration Crucially, we **reuse** the hierarchical semantic similarity $S_c^{(l)}$ from Phase 1 to dynamically regulate the *intensity* of the purification.

Step 1: Orthogonal Projection with Semantic Modulation (Resolving Social Conflict). Upon detecting social conflict, we project $\mathbf{g}_c$ onto the hyperplane orthogonal to $\overline{\mathbf{g}}_{\text{smooth}}$. The intensity of this projection is modulated by:

$$\mathcal{J}(S_c^{(l)}) = (1 - S_c^{(l)})^k \tag{8}$$

This yields the intermediate gradient:

$$\mathbf{g}'_c = \mathbf{g}_c - I_{\text{social}} \cdot \mathcal{J}(S_c^{(l)}) \cdot \frac{(\mathbf{g}_c \cdot \overline{\mathbf{g}}_{\text{smooth}})}{||\overline{\mathbf{g}}_{\text{smooth}}||_2^2} \overline{\mathbf{g}}_{\text{smooth}} \tag{9}$$

where k is an adjustment index and I_{social} indicates social conflict. If $S_c^{(l)}$ is high (models are similar), $\mathcal{J} \to 0$ and purification is suppressed, preserving local features.

Step 2: Semantic-Guided Linear Interpolation (Resolving Individual Conflict). To resolve any remaining individual conflict in $\mathbf{g}'_c$, we align it with the global correction vector $\mathbf{v}_c$ via interpolation:

$$\tilde{\mathbf{g}}_c = (1 - w_{\text{guide}})\mathbf{g}'_c + w_{\text{guide}}\mathbf{v}_c \tag{10}$$

where the interpolation weight w_{guide} is also controlled by $S_c^{(l)}$:

$$w_{\text{guide}} = \begin{cases} \min(1 + c \cdot (1 - S_c^{(l)}), w_{\max}) & \text{if conflict} \\ 0 & \text{otherwise} \end{cases} \tag{11}$$

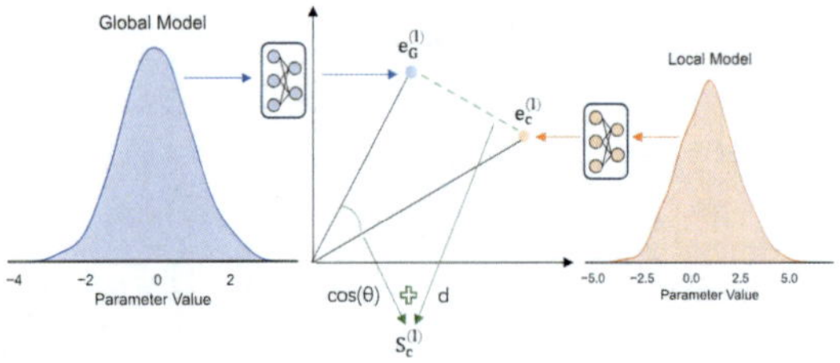

Fig. 3. Geometric illustration of two-step gradient purification.

Final Calibration: Finally, we apply the purified gradient $\tilde{\mathbf{g}}_c$ to $\hat{\theta}_c^t$ for a **single-step calibration** (Fig. 3):

$$\tilde{\theta}_c^t = \hat{\theta}_c^t - \eta_c \cdot \tilde{\mathbf{g}}_c \tag{12}$$

Algorithm 1 FedSCP

Input: N clients, local datasets $\{D_1, \ldots, D_N\}$, learning rate η_c, total communication rounds T, hyperparameters $\{\alpha, k, c, \tau_{refresh}\}$.
Output: Final global model parameter θ_G^{T+1}.

Initialize global parameter θ_G^0.
for each communication round $t = 1, 2, \ldots, T$ **do**
  Server broadcasts θ_G^t to all clients.
  for each client $c = 1, 2, \ldots, N$ **in parallel do**
    // Phase 1: Semantic-guided Adaptive Fusion
    if $(t \bmod \tau_{refresh}) = 0$ **then**
      Compute layer-wise semantic similarity $S_c^{(l)}$ via Eq. (4).
    Initialize and learn fusion weights $w_c^{(l),t}$ via Eq. (5).
    Generate personalized starting point $\hat{\theta}_c^t$.
    // Phase 2: Conflict Purification and Single-step Calibration
    Compute initial gradient $\mathbf{g}_c \leftarrow \nabla \mathcal{L}_c(\hat{\theta}_c^t; D_c)$.
    Purify $\mathbf{g}_c$ into ideal gradient $\tilde{\mathbf{g}}_c$ via two-step algorithm Eq. (8)-(11).
    Obtain the final training start point $\tilde{\theta}_c^t$ via Eq. (12).
    // Stage 3: Standard Local Training
    for each local epoch $e = 1, \ldots, E$ **do**
      Train $\tilde{\theta}_c^t$ on local dataset D_c for one epoch.
    Set final model to upload: $\theta_c^{t+1} \leftarrow \tilde{\theta}_c^t$.
  Server Aggregation: $\theta_G^{t+1} \leftarrow \frac{1}{N} \sum_{c=1}^{N} \theta_c^{t+1}$.
return θ_G^{T+1}.

This calibration step is critical: it preemptively addresses the gradient conflict problem *before* local training commences, rather than allowing conflicts to accumulate *during* the multi-epoch training process. This proactive approach effectively prevents clients from gradually deviating from the optimal trajectory.

4 Experiments

4.1 Experimental Setup

Datasets We evaluate on two benchmark datasets, CIFAR-100 and Flowers102, under two non-iid scenarios simulating intrinsic adversarial attacks:

Pathological Heterogeneity (High-Intensity): Each client samples 10 non-overlapping categories from 100/102 total categories, simulating extreme data disparities.

Practical Heterogeneity (Medium-Intensity): Category distributions follow a Dirichlet distribution Dir(β) with $\beta = 0.1$, simulating real-world gradual attacks.

For all datasets, 25% of local data serves as the test set, and 75% for training.

Baseline Methods We compare FedSCP against six SOTA baselines: traditional FedAVG [26] and five personalized methods—FedAMP [12], FedROD [11],

CosPer [16], and FedALA [3]. As a stronger baseline, we also implement **FedALA-C**, which applies the ConFREE [28] conflict-mitigation strategy to FedALA.

Training Details We employ a 4-layer CNN with an SGD optimizer. Client settings: batch size 10, 1 local epoch, learning rate 0.1. Global training: 200 communication rounds with 20 clients at a 100% participation rate. Results are averaged over 3 runs. For Phase 1, we set $\alpha = 0.7$, $\tau_{refresh} = 3$, $\Delta w_{\max} = 0.2$, and $\Delta w_{\text{offset}} = 0.1$. For Phase 2, we set $\eta_G = 1.0$, $k = 2$, and $c = 0.5$.

4.2 Comparison Results and Analysis

Table 1. Performance comparison under intrinsic attacks (pathological: high-intensity; practical: medium-intensity)

Method	Pathological heterogeneous		Practical heterogeneous	
	CIFAR-100	Flowers102	CIFAR-100	Flowers102
FedAVG	25.92	33.25	31.87	34.69
FedAMP	64.27	65.86	47.53	53.72
FedROD	63.33	68.71	50.86	60.97
CosPer	64.81	66.74	49.19	53.77
FedALA	67.35	66.89	56.62	62.28
FedALA-C	67.74	68.27	56.92	63.89
FedSCP	**70.90**	**70.62**	**58.86**	**65.63**

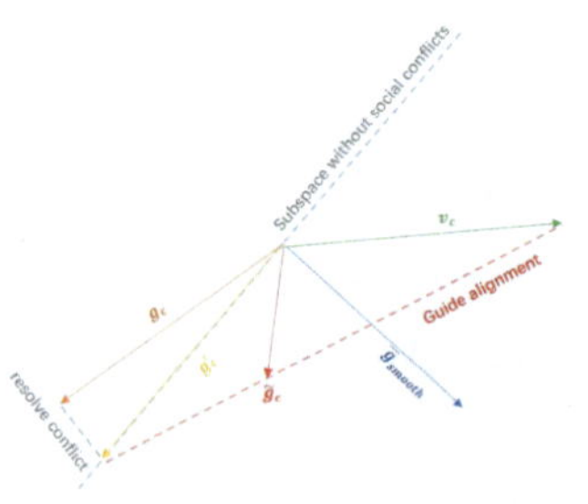

Fig. 4. Convergence comparison under different attack intensities (first two: high-intensity; last two: medium-intensity).

Table 1 and Fig. 4 show that FedSCP achieves optimal performance and the fastest convergence against intrinsic attacks.

FedAvg exhibits the poorest resistance, demonstrating the severe impact of data heterogeneity. Personalized methods (FedAMP, FedROD, CosPer) mitigate attacks partially but remain inadequate. FedALA achieves adaptive aggregation, offering moderate resistance. FedALA-C, which applies ConFREE's *passive mitigation* strategy, shows slight improvement.

In contrast, FedSCP demonstrates superior resilience, outperforming the enhanced baseline FedALA-C by **3.16%** on CIFAR-100 under high-intensity attacks. This highlights the efficacy of our *proactive defense* (purifying before training) over passive (post-hoc) mitigation. Figure 4 further demonstrates FedSCP's convergence advantage, achieving higher accuracy in fewer communication rounds.

4.3 Ablation Studies

Table 2. Ablation study under high-intensity attacks (CIFAR-100)

A: semantic guidance	B: conflict purification	Test accuracy
		67.35 (Baseline: FedALA)
✓		68.71 (+1.36%)
	✓	70.06 (+2.71%)
✓	✓	**70.90** (+3.55%)

We validate each component's contribution using FedALA as the baseline, progressively introducing: (A) Semantic Guidance (Phase 1) and (B) Conflict Purification (Phase 2).

As shown in Table 2, Semantic Guidance alone improves accuracy by 1.36%, confirming semantically-driven fusion effectively optimizes the starting point. Conflict Purification alone yields a significant 2.71% improvement, indicating that resolving gradient conflicts is crucial. Combining both modules (A+B) achieves a final accuracy of 70.90%, demonstrating a clear **synergistic effect** where the combined improvement surpasses the sum of the individual component gains.

4.4 Robustness Against Intrinsic Attack Strength

We evaluate FedSCP's robustness under varying intrinsic attack intensities by adjusting the Dirichlet distribution parameter $\beta \in \{0.1, 0.3, 0.5, 1.0\}$. As shown in Fig. 5, FedSCP consistently outperformed the FedALA-C baseline across all heterogeneity levels. Its performance advantage was most pronounced in the most challenging scenario of extreme heterogeneity ($\beta = 0.1$).

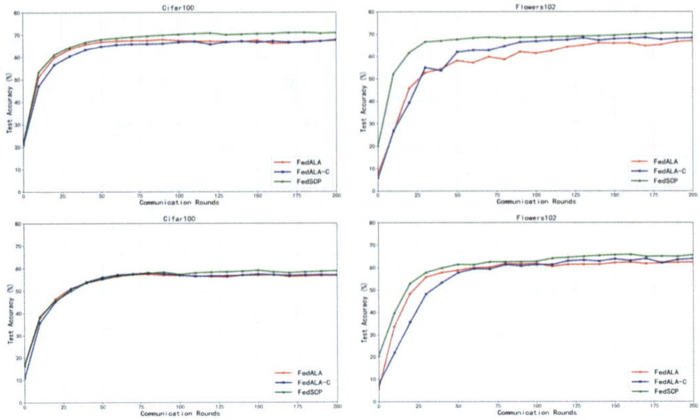

Fig. 5. Performance robustness comparison between FedSCP and FedALA-C under different intrinsic attack strengths (medium intrinsic attack on CIFAR-100).

4.5 System Scalability

We evaluated FedSCP's scalability by varying the total number of clients from 20 to 50 and the participation rate from 30 to 60%. FedSCP consistently outperformed FedALA-C. When the client scale expanded to $C = 50$ and the participation rate dropped to 30%, the baseline FedALA-C experienced a sharp performance decline. Although FedSCP was also affected, its performance remained relatively stable, demonstrating strong robustness (Fig. 6).

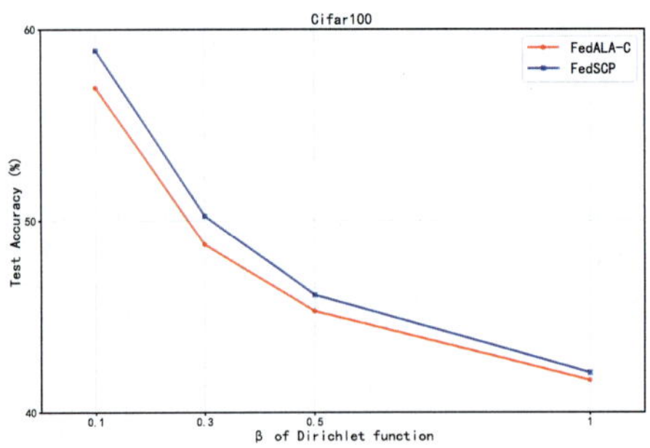

Fig. 6. Scalability comparison between FedSCP and FedALA-C across different system scales (CIFAR-100 high-intensity intrinsic attacks).

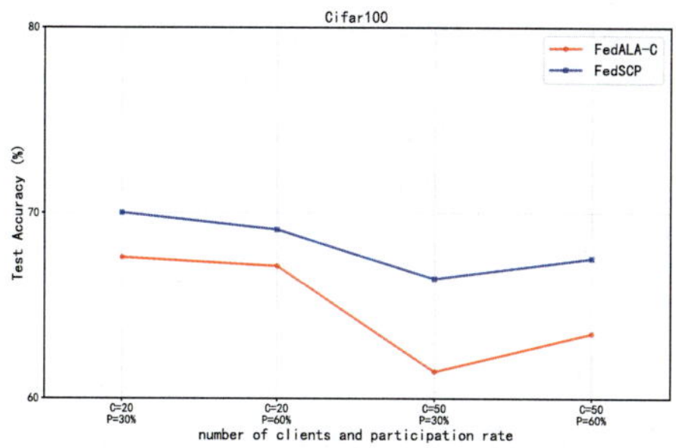

Fig. 7. t-SNE visualization of gradient purification under high-intensity attacks (round 10: pre-training; round 100: post-training; first two: CIFAR-100; last two: Flowers102).

4.6 Hyperparameter Sensitivity Analysis

Table 3. Hyperparameter sensitivity under high-intensity attacks (CIFAR-100)

Item	$\alpha = 0.1$	$\alpha = 0.3$	$\alpha = 0.5$	$\alpha = 0.7$	$\alpha = 0.9$
Acc	69.94	70.33	70.15	**70.90**	70.52
Item	$k = 0$	$k = 1$	$k = 2$	$k = 3$	$k = 4$
Acc	69.85	70.48	**70.90**	70.32	69.69

Impact of α (Phase 1). Table 3 shows the effect of the weighting factor α, which balances cosine similarity and Euclidean distance. Performance peaks at $\alpha = 0.7$.

Impact of k (Phase 2). Parameter k controls the semantic adjustment intensity in gradient purification. Performance is optimal at $k = 2$. Higher values cause over-correction, suppressing beneficial local features and thus degrading accuracy. We adopt $k = 2$ as the default.

4.7 Visual Analysis

We visualize gradient distributions via t-SNE to demonstrate FedSCP's purification efficacy. In Fig. 7, blue dots represent original gradients, red squares denote purified gradients, and green triangles indicate the global direction. We measure consistency using Cohesion:

$$\text{Cohesion Improvement} = \left(1 - \frac{\text{Disp}(G_{\text{after}})}{\text{Disp}(G_{\text{before}})}\right) \times 100\% \tag{13}$$

During early training (Round 10), intrinsic attacks cause scattered gradients (blue). FedSCP improves cohesion by 8.4% on CIFAR-100 and 3.8% on Flowers102 by pulling gradients (red) closer. Notably, the more moderate improvement on Flowers102 (3.8%) reflects FedSCP's **adaptive correction**: it correctly identifies that the fine-grained features in Flowers102 are beneficial, and thus *preserves* them by purifying *less*, rather than blindly forcing all gradients toward the global mean. During late training (Round 100), gradients naturally cluster, and FedSCP's purification effect becomes milder, guiding convergence without disrupting the established trend.

5 Conclusion

To combat the intrinsic adversarial attacks endemic to personalized federated learning, we proposed FedSCP, a two-stage optimization framework. By synergistically combining semantic-guided adaptive fusion(to optimize the starting point) and two-step gradient purification(to clean the update direction), FedSCP fundamentally mitigates the gradient bias and client drift caused by data heterogeneity. Experiments on image classification benchmarks demonstrated FedSCP's superior resistance to intrinsic attacks and faster convergence. Future work will focus on validating FedSCP in other domains (e.g., natural language processing) and enhancing its scalability for large-scale networks (e.g., >100 clients).

References

1. Arbaoui, M., Brahmia, M.A., Rahmoun, A., et al.: Federated learning survey: a multi-level taxonomy of aggregation techniques, experimental insights, and future frontiers. ACM Trans. Intell. Syst. Technol. **15**(6), 1–69 (2024)
2. Mitra, A., Jaafar, R., Pappas, G.J., et al.: Linear convergence in federated learning: tackling client heterogeneity and sparse gradients. Adv. Neural. Inf. Process. Syst. **34**, 14606–14619 (2021)
3. Zhang, J., Hua, Y., Wang, H., et al.: Fedala: adaptive local aggregation for personalized federated learning. Proceed. AAAI Confer. Artif. Intell. **37**(9), 11237–11244 (2023)
4. Chen, W., Yang, C., Ren, J.: Optimizing Personalized Federated Learning through Adaptive Layer-Wise Learning (2024). arXiv:2412.07062 arXiv preprint
5. Ye, R., Ni, Z., Wu, F.: Personalized federated learning with inferred collaboration graphs. In: International Conference on Machine Learning, pp. 39801–39817. PMLR (2023)
6. Fallah, A., Mokhtari, A., Ozdaglar, A.: Personalized federated learning with theoretical guarantees: a model-agnostic meta-learning approach. Adv. Neural. Inf. Process. Syst. **33**, 3557–3568 (2020)
7. Dinh, C., Tran, N., Nguyen, J.: Personalized federated learning with Moreau envelopes. Adv. Neural Inform. Process. Syst. **33**, 21394–21405 (2020)
8. Li, T., Hu, S., Beirami, A.: Ditto: fair and robust federated learning through personalization. In: International Conference on Machine Learning, pp. 6357–6368. PMLR (2021)
9. Xu, J., Tong, X., Huang, S.L.: Personalized federated learning with feature alignment and classifier collaboration. In: The Eleventh International Conference on Learning Representations (2023)

10. Yang, X., Huang, W., Ye, M.: Fedas: bridging inconsistency in personalized federated learning. In: Proceedings of the IEEE/CVF Conference on Computer Vision and Pattern Recognition, pp. 11986–11995 (2024)
11. Chen, H.Y., Chao, W.L.: On bridging generic and personalized federated learning for image classification. In: International Conference on Learning Representations (2022)
12. Huang, Y., Chu, L., Zhou, Z., et al.: Personalized cross-silo federated learning on non-iid data. Proceed. AAAI Confer. Artif. Intell. **35**(9), 7865–7873 (2021)
13. Mclaughlin, C., Su, L.: Personalized federated learning via feature distribution adaptation. Adv. Neural. Inf. Process. Syst. **37**, 77038–77059 (2024)
14. He, Y., Zhou, P., Zhai, Y., et al.: FedPAW: federated learning with personalized aggregation weights for urban vehicle speed prediction. IEEE Trans. Cloud Comput. **4**, 452 (2024)
15. Zhou, P., He, Y., Zhai, Y., et al.: FedAH: aggregated head for personalized federated learning. arXiv preprint arXiv:2412.01295 (2024)
16. Ren, P., Qi, K., Li, J., et al.: CosPer: an adaptive personalized approach for enhancing fairness and robustness of federated learning. Inf. Sci. **675**, 120760 (2024)
17. Chan, H.T.J., Veas, E.: Importance estimate of features via analysis of their weight and gradient profile. Sci. Rep. **14**(1), 23532 (2024)
18. Zhenmei, S., Wei, J., Liang, Y.: A Theoretical analysis on feature learning in neural networks: emergence from inputs and advantage over fixed features. In: International Conference on Learning Representations (2022)
19. van Meegen, A., Sompolinsky, H.: Coding schemes in neural networks learning classification tasks. Nat. Commun. **16**(1), 3354 (2025)
20. Tang, Z., Zhang, Y., Shi, S., et al.: FedImpro: measuring and improving client update in federated learning. In: The Twelfth International Conference on Learning Representations (2024)
21. Legate, G., Caccia, L., Belilovsky, E.: Re-weighted softmax cross-entropy to control forgetting in federated learning. In: Conference on Lifelong Learning Agents, pp. 764–780. PMLR (2023)
22. Zhang, R., Zhang, Y., Zhao, Y., et al.: FedCVG: a two-stage robust federated learning optimization algorithm. Sci. Rep. **15**(1), 1–13 (2025)
23. Liu, B., Liu, X., Jin, X., et al.: Conflict-averse gradient descent for multi-task learning. Adv. Neural. Inf. Process. Syst. **34**, 18878–18890 (2021)
24. Zheng, H., Hu, Z., Yang, L., et al.: ConFREE: conflict-free client update aggregation for personalized federated learning. In: Proceedings of the AAAI Conference on Artificial Intelligence, pp. 22875–22883 (2025)
25. Zhang, J., Shan, C., Han, J.: FedGMKD: an efficient prototype federated learning framework through knowledge distillation and discrepancy-aware aggregation. Adv. Neural. Inf. Process. Syst. **37**, 118326–118356 (2024)
26. McMahan, B., Moore, E., Ramage, D., et al.: Communication-efficient learning of deep networks from decentralized data. In: Artificial Intelligence and Statistics, pp. 1273–1282. PMLR (2017)
27. Mai, P., Yan, R., Pang, Y.: Rflpa: a robust federated learning framework against poisoning attacks with secure aggregation. Adv. Neural. Inf. Process. Syst. **37**, 104329–104356 (2024)
28. Guo, S., Guo, Y., Zhang, H., et al.: Mitigating update conflict in non-IID federated learning via orthogonal class gradients. IEEE Trans. Mobile Comput. **4**, 78555 (2024)

29. Park, C., Lee, N.: SignSGD with federated defense: harnessing adversarial attacks through gradient sign decoding. In: Proceedings of the 41st International Conference on Machine Learning, pp. 39762–39780 (2024)
30. Hu, F., Zhou, W., Liao, K.: Regulated federated learning against the effects of heterogeneity and client attacks. IEEE Intell. Syst. **3**, 598 (2024)

Self-supervised Clustering Defense Against Federated Learning Poisoning Attack in IIoT

Shuo Wang[1], Qianmu Li[2(✉)], Shicheng Cui[1], Ruotian Jiang[1], and Yue Tang[1]

[1] Nanjing Institute of Technology, Nanjing, China
shuow@njit.edu.cn

[2] Nanjing University of Science and Technology, Nanjing, China
qianmu@njust.edu.cn

Abstract. Perception models deployed on the client-side in federated learning (FL) frameworks, are highly vulnerable to poisoning attacks. Such attacks can contaminate the global model, leading to widespread perception failures among terminals connected to the central server, posing severe security risks to Industrial Internet of Things (IIoT) systems. Existing defense methods often assume independent and identically distributed (IID) data, which is inconsistent with the non-IID nature of sensor data on the client-side. Additionally, they are limited by generic task orientation, poor adaptability to high-dimensional model parameters, and lack of security-specific closed-loop mechanisms. To address these limitations, this paper proposes a model aggregation defense method based on self-supervised clustering and momentum-based correction. A self-attention-based deep clustering model (AAE-MCluster) is designed to effectively cluster high-dimensional local model updates. Then, a momentum-based weight correction strategy is deployed to adjust the anomaly parameters identified by clustering, aligning them with historical benign global models. Experiments based on MNIST, CIFAR-10, and Udacity datasets demonstrate that the proposed method outperforms baseline defense algorithms in resisting poisoning attacks under non-IID conditions. It achieves lower attack success rates and maintains higher global model accuracy, while accelerating convergence.

Keywords: Federated learning · Poisoning attack defense · Self-supervised clustering · Non-IID data

1 Introduction

Federated learning (FL) has become a core technology for intelligent perception in IIoT scenarios (e.g., autonomous driving, smart manufacturing) [1]. FL trains distributed models without sensitive data sharing [2]. However, the distributed architecture introduces poisoning attacks on the client terminal. Malicious clients upload compromised local models to contaminate the global model, leading to catastrophic consequences such as misjudged steering angles in autonomous driving systems [3]. This security risk has driven extensive research on defense mechanisms.

L. Huang and D. Greenhalgh (Eds.): ICMLC 2026, LNNS 2061, pp. 623–637, 2027.
https://doi.org/10.1007/978-3-032-31236-5_43

Existing defense methods against client-side poisoning attacks in FL can be categorized into two main types: model similarity-based defenses and clustering based Model Aggregation Defense [3]. Model similarity-based methods [3, 4] identify malicious models by measuring distances between local updates but face challenges with non-IID data. Clustering based strategies [5, 6] often exclude anomalous clusters, which may discard valid updates from benign models with non-IID distributions, degrading global model performance. Deep clustering methods are more suitable for high-dimensional and non-linear data [7, 8].

However, deep clustering methods meet other challenges: (1) Traditional methods cluster raw data (e.g., images, multimodal signals) rather than high-dimensional FL model parameters, leading to ineffective clustering when applied to high-dimension parameter space; (2) These methods lack security-oriented design. They only realize cluster assignment, without mechanisms to mitigate malicious impacts after detection; (3) They assume ideal data distribution (e.g., balanced IID data), while IIoT terminal data in FL is inherently non-IID, reducing the robustness in real scenarios; (4) They focus on single tasks (e.g., classification, domain adaptation), ignoring the requirement in FL to support both regression and classification tasks.

To bridge these gaps, this paper proposes a self-supervised clustering-based FL model aggregation defense method. The main contributions of this work are as follows:

1. A self-supervised deep clustering algorithm (AAE-MCluster) is proposed, integrating self-attention mechanisms to optimize the clustering of high-dimensional non-linear model parameters. The algorithm enabled effective clustering of high-dimensional model parameters.
2. A momentum-based anomaly correction strategy is designed to adjust the weights of anomalous models identified by clustering. The strategy mitigates the impact of poisoned models while retaining the valid contributions of benign models, thus accelerating global model convergence.
3. Comprehensive experiments are conducted on classification datasets (MNIST, CIFAR-10) and a vehicle steering angle detection dataset (Udacity) under poisoning attacks. Results demonstrate that the proposed method outperforms baseline defenses in both attack resistance and model convergence.

2 Review of Defense Techniques for Poisoning Attacks on Intelligent Models

Poisoning attacks present a considerable threat to the security of applications in IIoT. This section summarizes common defense techniques against such attacks and surveys the research progress in this area.

2.1 Model Similarity Based Model Aggregation Defense

The defense based on model similarity aims to judge the update of benign and poisoned models by calculating the distance between local models. Then the poisoned models can be excluded from participating in global model aggregation. It is found that in federated learning, the distance between benign client model updates is diverse under non-IID conditions, while the distance between model updates of poisoned clients is closer. Therefore, there are several studies that calculate the similarity of all historical updates of the client, adjust the learning rate locally, and reduce the update impact of the poison model.

Krum [3] uses Euclidean distance to select the model with the smallest aggregate distance to its neighbors as the global model. Methods based on cosine similarity [4] adjust aggregation weights based on the similarity between local updates and historical global models. Another approach utilizes Mahalanobis distance [9] to detect anomalous filters in CNN layers, thereby scoring and excluding malicious models. Statistical aggregation methods, such as Trimmed Mean [10] and Median [11], enhance robustness by discarding extreme parameter values or selecting median values. However, they tend to become less effective with high-dimensional data and do not guarantee convergence [12]. Haque et al. [13] proposes a robust aggregation defense mechanism without relying on prior knowledge of attacks based on geometric distance-based anomaly detection. AdaAggRL is an adaptive aggregation framework against complex model poisoning attacks based on Reinforcement Learning (RL). The framework simulates data distributions using client model parameters and dynamically adjusting aggregation weights by calculating stability metrics of such distributions (e.g., Maximum Mean Discrepancy based similarity) [14].

2.2 Clustering Based Model Aggregation Defense

Clustering-based anomaly detection schemes work on the premise that malicious client updates have distinct parameter distributions due to their different objectives compared to benign clients. Density-based clustering aggregation methods are proposed in [10].After collecting local models, the server computes the pairwise cosine similarities between them and applies the dynamic clustering algorithm HDBSCAN. Clusters with a larger number of members are identified as benign, while smaller clusters are classified as anomalous and then excluded. The remaining benign models are processed further through steps like clipping and noising to generate the final global model. Shen et al. [12] use k-Means to distinguish between benign and malicious clients. Steinhardt et al. [15] employed outlier detection methods to identify malicious clients. Tran et al. [16] and Li et al. [17] leveraged spectral clustering for backdoor trigger identification. Fung et al. [18] treated the parameters of client model updates as vectors and performed cosine similarity clustering to discriminate between malicious and benign clients. Li et al. [19] grouped similar client parameters using kernel density clustering [7] and incorporated an outlier detection mechanism to prevent model poisoning.

On a related note, deep clustering has shown promise in handling high-dimensional data by learning non-linear feature mappings [20]. Wang et al. [21] proposed Contrastive Self-Supervised Subspace Clustering with KAN-based Multi-view Fusion (CSSCK), which leverages Kolmogorov-Arnold Networks (KAN) to address encoding dependency and data variability challenges. It integrates contrastive fusion to capture complementary information across multiple views. Lu et al. [22] developed a novel self-supervised graph clustering model grounded in reliable semi-supervised learning, aiming to enhance clustering robustness by incorporating trustworthy semi-supervised signals into graph-based representation learning. Zhang et al. [23] presented Aggregation-based Self-Supervised

Evidential Clustering (ASEC) based on evidence theory. ASEC introduces a density-based aggregation rule to merge sub-clusters until matching the true cluster count, and employs KNN-weighted mean for self-supervised evidential partitioning of remaining objects, effectively reducing clustering error rates. Xu et al. [24] proposed SDCluster, a pixel-level clustering-based self-supervised learning method that extends the self-distillation framework to the pixel level with clustering constraint modules, enabling qualitative evaluation of pre-trained models through visualizing clustering results. Lee et al. [25] designed MADCluster, a novel model-agnostic anomaly detection framework utilizing self-supervised clustering, which resolves the inherent "hyperball collapse" problem in existing deep learning-based anomaly detection methods.

Despite their capabilities, existing model similarity based and clustering based defense techniques still face several key limitations. (1) These methods demonstrate limited effectiveness against poisoning attacks under non-IID settings. (2) The common practice of simply discarding detected anomalous models may inadvertently sacrifice the valid contributions from benign clients due to detection errors, ultimately degrading global model convergence performance. (3) They focus more on classification task, ignoring the requirement in FL to support both regression and classification tasks [26].

3 Problem Description

Here assumes a FL framework where a central server coordinates n vehicle-side clients to train a shared lightweight perception model. Let w_{t-1} denote the global model after (t – 1) aggregation rounds. Each client i trains a local model using its local data D_t^i via gradient descent:

$$w_t^i = w_{t-1} - \alpha \nabla L\left(w_{t-1}, D_t^i\right) \tag{3.1}$$

where α is the learning rate and $L(\cdot)$ is the training loss function. Among the n clients, m ($m \leq 0.4n$) are malicious and train local models using poisoned data D_t^l. The model updates of the poisoned clients follow Eq. (3.2)

$$w_t^l = w_{t-1} - \alpha \nabla L\left(w_{t-1}, D_t^l\right) \tag{3.2}$$

The server aggregates local models to generate the new global model, as in Eq. (3.3), where ***SecureAggregation*** denotes the secure aggregation algorithm for the clients' models.

$$w_t = SecureAggregation\left(w_t^i\right), i \in \{1, 2, \ldots, l, n\} \tag{3.3}$$

The goal of the proposed ***SecureAggregation*** algorithm is to maximize the global model's average accuracy ($\overline{acc}$) across all classes and minimize the attack success rate (att) on the target class. The overall defense objective is given in Eq. (3.4).

$$SecureAggregation : Maximize\overline{acc}andMinimzie(att) \tag{3.4}$$

4 Self-supervised Clustering Based Model Aggregation Defense Method

This section systematically elaborates on the model aggregation defense method deployed on the federated learning server for generating the global model. It firstly presents the overall workflow of the deep clustering-based model aggregation algorithm. Then, it sequentially details the self-supervised deep clustering model used for clustering local model updates. Finally, clarify the momentum-based strategy for correcting the parameters of the screened potentially poisoned models.

4.1 Deep-Clustering-Based Model Aggregation Algorithm

The complete algorithmic workflow is presented in Algorithm 4.1. The algorithm takes two inputs: Locally trained models from vehicle terminals θ_t^i, $(i = \{1, 2, \ldots l, \ldots, n\})$, global model parameters from the previous FL round θ_{t-1}.

The proposed aggregation algorithm first pre-trains the self-attention-based AAE-MCluster via benign model updates from public datasets. The server collects client local model updates in each FL round, identifying the smallest cluster as anomalous (likely poisoned) by the clustering algorithm. Guided by momentum, anomalous models are corrected with parameter μ to align with the previous benign global gradient. Finally, corrected and original local models are aggregated by federated averaging to generate a poisoning-resistant global model θ_t, which supports subsequent rounds corrections, forming a secure aggregation cycle.

This algorithm primarily consists of three components: self-attention-based deep model clustering and anomalous client model selection, anomalous vehicle terminal model gradient correction, which will be elaborated in subsequent sections.

Algorithm 4.1 Server-side Secure Model Aggregation Algorithm

Input: Locally trained models from vehicle terminals θ_t^i, $(i = \{1, 2, \ldots l, \ldots, n\})$, global model parameters from the previous FL round θ_{t-1}

Output: Aggregated global model parameters θ_t

1 **Pre-train** AAE-MCluster

2 **Function** ***SecureAggregation*** (θ_t^i):

3 **for** each round $t = 0, 1, 2, \ldots$ **do:**

4 $C =$ **AAE-MCluster** $(\theta_t^1, \theta_t^2, \ldots, \theta_t^n)$ // *Cluster* θ_t^i

5 $C_m =$ **Argmin**$(\{card(C_1), card(C_2), \ldots, card(C_c)\}, C_1, C_2, \ldots, C_c \in C)$

// Identify the cluster with the smallest cardinality (number of members) as the malicious model update class

6 **for** θ_t^l in C_m **do:**

7 $\tilde{\theta}_t^l = \mu\theta_{t-1} + (1-\mu)\theta_t^l$ // Based on the momentum concept to correct models in the anomalous cluster

8 $\theta_t = \sum_1^n \frac{\theta_t^i + \tilde{\theta}_t^l}{n}$

9 **return** θ_t

End Function

(continued)

(*continued*)

4.2 Self-attention-Based Self-supervised Model Parameter Clustering

AAE-MCluster integrates a self-attention-enhanced autoencoder with K-means clustering to handle high-dimensional model parameters. Thus, this section proposes a self-attention-based self-supervised clustering model, designed to automatically cluster a set of input local models. The training loss function incorporates both the decoder reconstruction loss and the clustering loss, as defined in Eq. (4.1). The corresponding model framework is illustrated in Fig. 1

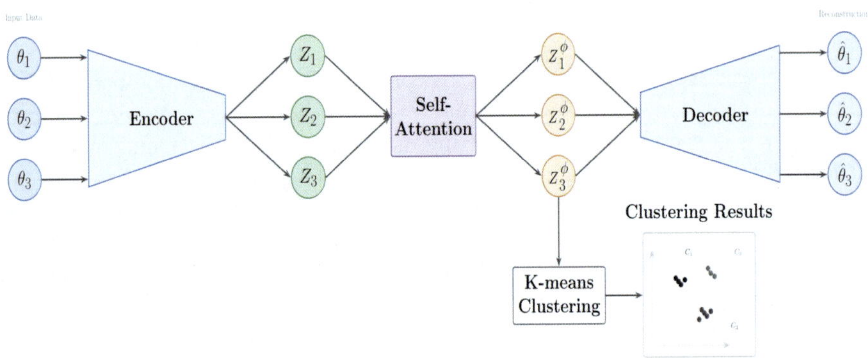

Fig. 1. The AAE-MCluster self-supervised clustering model based on self-attention.

In the figure, $(\theta_1, \theta_2, \theta_3)$ represent the input model parameters, and (Z_1, Z_2, Z_3) are the corresponding encoded model parameters output by the encoder. The encoded parameters are then processed by a self-attention mechanism to focus on more informative low-dimensional features $(Z_1\prime, Z_2\prime, Z_3\prime)$. These refined low-dimensional features serve two purposes:

On one hand, they are passed through a decoder to reconstruct the model parameters $(\theta_1\prime, \theta_2\prime, \theta_3\prime)$. A reconstruction loss function L_{rec}, with the objective function shown in Eq. (4.2), is trained to minimize the discrepancy between the reconstructed outputs and the original inputs. Here, φ, μ denote the trainable parameters of the encoder and decoder, respectively. The similarity between the reconstructed and original model parameters is evaluated based on the Euclidean distance between them.

On the other hand, the self-attention-enhanced low-dimensional features are used to train a clustering loss function. The clustering algorithm employed here is the standard K-means algorithm. During the K-means clustering process, the distance between each sample point and the cluster centers is computed, each sample is assigned to its nearest cluster, and the cluster centers are subsequently updated. Therefore, the training objective for clustering is to minimize the sum of the distances between each sample point in a cluster and its corresponding cluster center, thereby updating the cluster centers and the clustering results $\left(C^1, C^2, C^3\right)$. Consequently, another objective of the model is to train the K-means clustering loss function, as specified in Eq. (4.3), where ρ, k represent the cluster centers and the number of clusters, respectively, and ρ_i denotes the cluster center corresponding to the feature $Z_i\prime$.

$$Loss = L_{rec} + L_c \tag{4.1}$$

$$\min_{\varphi,\mu} L_{rec} = min\frac{1}{n}\sum_{1}^{n}\theta_i - \theta_i^{\prime 2} \tag{4.2}$$

$$\min_{\rho,k} L_c = min\frac{1}{n}\sum_{1}^{n}Z_i\prime - \rho_i^2 \tag{4.3}$$

The encoder and decoder each consist of two fully connected hidden layers with 500 neurons, and the latent vector dimension is set to 100. The self-attention mechanism enhances the model's ability to capture informative features in high-dimensional parameters, improving clustering separation.

4.3 Training of the Deep Clustering Algorithm and Malicious Model Screening

AAE-MCluster is pre-trained using benign models trained on public datasets. During each FL round, the server inputs collected local model updates into AAE-MCluster to generate clusters C. The cluster Cm with the smallest cardinality is identified as the anomalous cluster. The server then fine-tunes AAE-MCluster using the current local updates to adapt to evolving model characteristics.

4.4 Malicious Model Weight Correction Based on the Momentum

Given that federated learning constitutes a distributed, iterative training process where each round of model aggregation on the server represents an intermediate step. Inspired by [27], this study introduces a method inspired by the momentum concept. Specifically, this paper proposes to correct the vectors of anomalous local models identified during aggregation. The approach aims to motivate these anomalous models more rapidly towards the direction of historical benign models while partially retaining their original update directions. Consequently, the global model convergence is guided predominantly by benign influences, enhancing its security. Inspired by momentum optimization [28], this paper corrects the parameters of anomalous models to align them with historical benign global models, as shown in Fig. 2. For each model θ_t^l in the anomalous cluster Cm, the corrected parameter $\tilde{\theta}_t^l$ is:

$$\tilde{\theta}_t^l = \mu\theta_{t-1} + (1-\mu)\theta_t^l \tag{4.4}$$

where $\mu = 0.9$ (correction coefficient) balances the contribution of the historical benign global model θ_{t-1} and the current anomalous model θ_t^l. This strategy mitigates the impact of malicious updates while retaining valid contributions from benign models in the anomalous cluster.

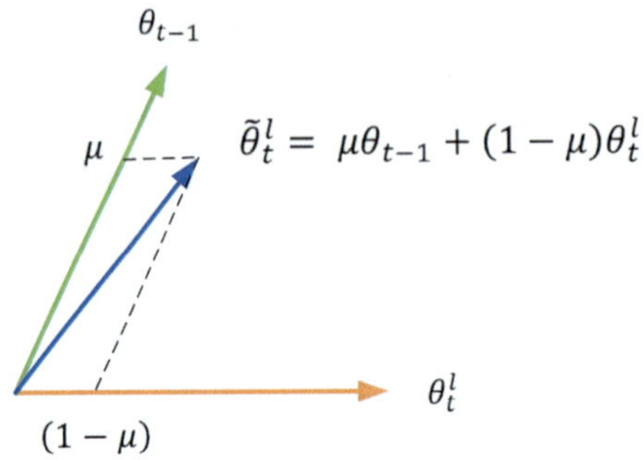

Fig. 2. Schematic diagram of weight vector direction correction for an anomalous local model.

5 Experiment

5.1 Experimental Setup

Datasets.

Udacity Dataset. Autonomous driving serves as a typical application scenario of IIoT in the intelligent transportation field. This paper adopts the typical autonomous driving dataset to verify the method in IIoT. This dataset comprises image data captured during daytime driving on routes from San Mateo to Half Moon Bay in California, featuring both curved roads and highways. It contains 410,530 driving images with steering angles. 5,614 images are for pre-training AAE-MCluster and 404,916 images are for FL training (regression task: steering angle detection).

MNIST Dataset. This dataset contains 70,000 grayscale images of handwritten digits from 0 to 9. It is split into 60,000 training images and 10,000 test images. From the training set, 10,000 images are used for centralized pre-training, while the remaining 50,000 are allocated across the clients for federated training.

CIFAR-10 Dataset. This dataset consists of 60,000 color images in 10 object classes, with 6,000 images per class, where 10k training images are for pre-training, 40k are for FL training.

Data is distributed non-IID across clients using a Dirichlet distribution ($\alpha = 0.9$).

Models

a. *Regression Task.* SteerCNN (Udacity dataset) for steering angle detection
b. *Classification Tasks.* SmallCNN (MNIST) and ResNet-18 (CIFAR-10)

Simulated Poisoning Attack Setup.

Label Flipping Poisoning Attack. This attack involves flipping the labels of a specific class of training data to another class while keeping the data features unchanged. The mislabeled data is then mixed into the original dataset to participate in training, thereby influencing the model training process.

For MNIST, labels of '1' are flipped to '7', with 100 clients (10 selected per round, up to 40% malicious), 200 FL rounds, and each client owns at least two digit classes. For CIFAR-10, 'cat' labels are flipped to 'dog' (visually similar), with 10 clients (up to 40% malicious) and 100 aggregation rounds. For Udacity, samples with steering angles in (0.4, 0.6) are targeted with a label shift of -0.4, 10 clients (up to 40% malicious), 500 aggregation rounds. Benign clients train 5 epochs/round, while malicious ones train 10 epochs/round.

Benchmark Comparison Defense Methods.

Krum. [8] calculates a score for each local model update based on the sum of squared distances to its nearest neighbors. It then selects the single local model update with the smallest aggregate distance as the global model for the current round.

Median. In the Median method [29] the server sorts the weights for each layer across all local model updates and selects the median value as the global model parameter for that layer.

Trimmed Mean. In the Trimmed Mean approach [30], the server sets a trimming parameter k, discards the k largest and k smallest parameter values, and then computes the average of the remaining parameters to determine the value for the global model update.

Evaluation Metrics.

a. *Regression Task.*

In-Range Attack Success Rate (ACC_{reg}): In Eq. (5.1), N_{region} denotes the total number of samples within the attack label domain, while N_{att} represents the number of samples in this attack label domain whose inferred outputs of the model fall within the target value range.

$$ACC_{reg} = \frac{N_{att}}{N_{region}} \times 100\% \tag{5.1}$$

Global Detection Error RMSE: The Global Root Mean Square Error (RMSE) evaluates the overall performance of the intelligent model by calculating the RMSE across the entire test dataset.

b. *Classification Task.*

Single-Class Attack Success Rate (ACC_{class}): Here, T_{target} represents the number of samples classified as the target label, and F represents the number of samples misclassified (not classified as the target label).

$$ACC_{class} = \frac{T_{target}}{T_{target}+F} \times 100\% \tag{5.2}$$

Global Average Classification Accuracy ($\overline{ACC}_{clean}$): Here, c denotes the total number of classes in the dataset, T represents the number of correctly classified samples, and F represents the number of misclassified samples.

$$\overline{ACC}_c = \frac{T}{T+F} \tag{5.3}$$

$$\overline{ACC}_{clean} = \sum_{1}^{C} \frac{\overline{ACC_c}}{C} \times 100\% \tag{5.4}$$

5.2 Experimental Results

a. ***Udacity Dataset Results***. The Udacity dataset is employed to assess the defensive performance of the proposed method against poisoning attacks targeting regression models. The ACC_{reg} and RMSE results are shown in Figs. 3 and 4. When the proportion of malicious clients is 0% (i.e., in the attack-free scenario), all aggregation methods achieve an in-range attack success rate (ACC_{reg}) of 0%. Notably, the proposed AAE-MCluster yields the lowest RMSE of 0.1107, which underscores its superiority in optimizing model convergence speed and enhancing detection performance. As the proportion of malicious clients increases from 10% to 40%, both ACC_{reg} and RMSE exhibit an upward trend across all methods. However, AAE-MCluster demonstrates the slowest growth rate in these metrics, reflecting its robust ability to suppress poisoning attacks. At a 10% malicious client ratio, AAE-MCluster delivers performance comparable to that of FedAvg. When the proportion of malicious clients reaches 20% or higher, the performance gap between AAE-MCluster and other baseline methods widens significantly. Specifically, at 40% malicious clients, ACC_{reg} of AAE-MCluster is 16.8% lower than that of Krum, and its RMSE is 0.255 lower than that of Trimmed Mean. These results confirm that AAE-MCluster exhibits stronger resilience against label-flipping attacks, particularly in scenarios with a high proportion of malicious clients, while also promoting faster model convergence in the absence of attacks.
b. ***MNIST Dataset Results***. Figures 5 and 6 report the single-class attack success rate (ACC_{class}) and global average accuracy ($\overline{ACC}_{clean}$) after label-flipping attacks on MNIST. As malicious clients increase, ACC_{class} rises for all methods. FedAvg fails to resist (99.21% at 40% malicious). Median and Trimmed Mean degrade sharply at 40% malicious, while AAE-MCluster maintains the lowest ACC_{class} at 10%, 30%, 40% (0.78% at 40%), outperforming Krum. For $\overline{ACC}_{clean}$, baselines decline significantly with more malicious clients (as low as 23.29%), but AAE-MCluster sustains over 94% accuracy across all proportions. Though Trimmed Mean has marginally lower ACC_{class} at 20% and 30% malicious, its $\overline{ACC}_{clean}$ is much lower. AAE-MCluster balances strong attack resistance and stable high global accuracy.
c. ***CIFAR-10 Dataset Results***. Figures 7 and 8 present the single-class attack success rate (ACC_{class}) and global average accuracy ($\overline{ACC}_{clean}$) of global models after label-flipping attacks on CIFAR-10. Consistent with MNIST results, ACC_{class} increases with the proportion of malicious clients across all methods, but AAE-MCluster achieves the lowest ACC_{class} under all proportions. At 40% malicious clients, its ACC_{class} (23.2%) is 19.7% lower than Krum, demonstrating prominent and stable attack suppression.

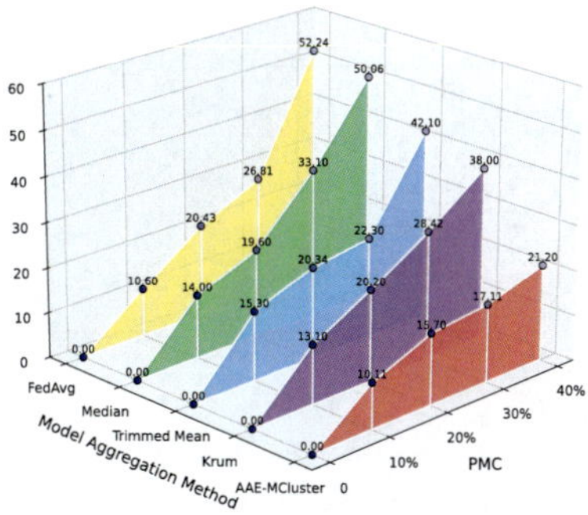

Fig. 3. ACC_{reg} (%) under Different Model Aggregation Algorithms with Udacity Dataset

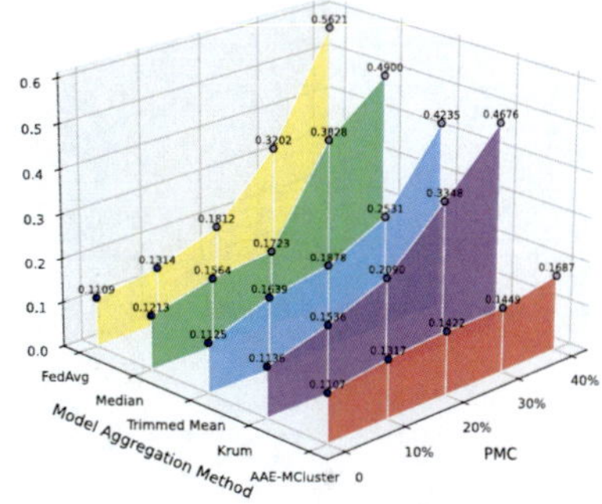

Fig. 4. RMSE under Different Model Aggregation Algorithms with Udacity Dataset

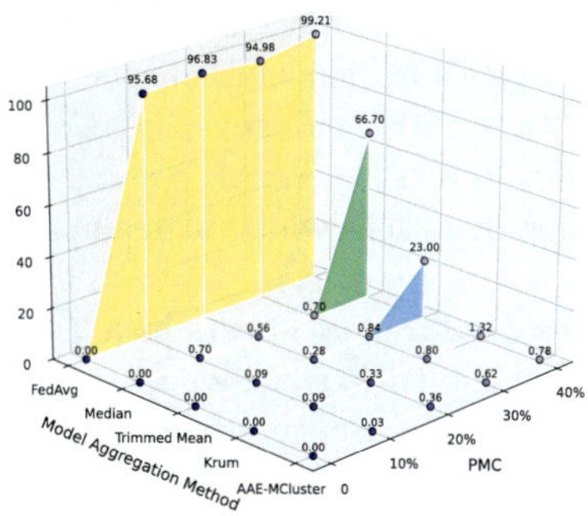

Fig. 5. ACC_{class} (%) under Different Model Aggregation Algorithms with MNIST Dataset

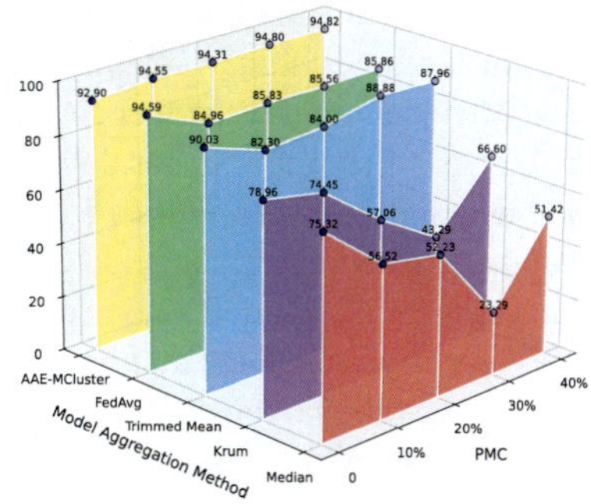

Fig. 6. $\overline{ACC}_{clean}$(%) under Different Model Aggregation Algorithms with MNIST Dataset

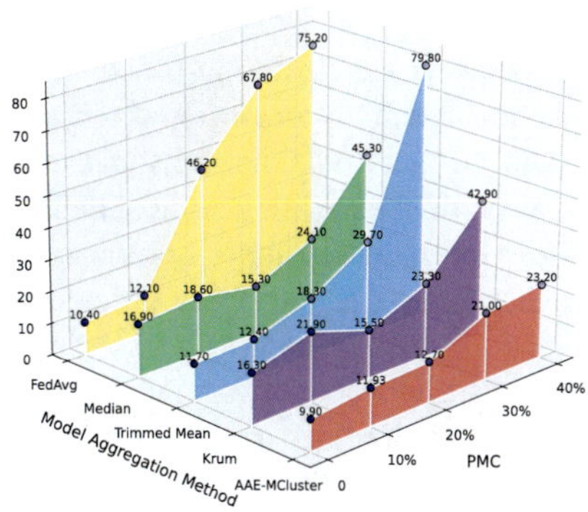

Fig. 7. ACC_{class} (%) under Different Model Aggregation Algorithms with CIFAR-10 Dataset

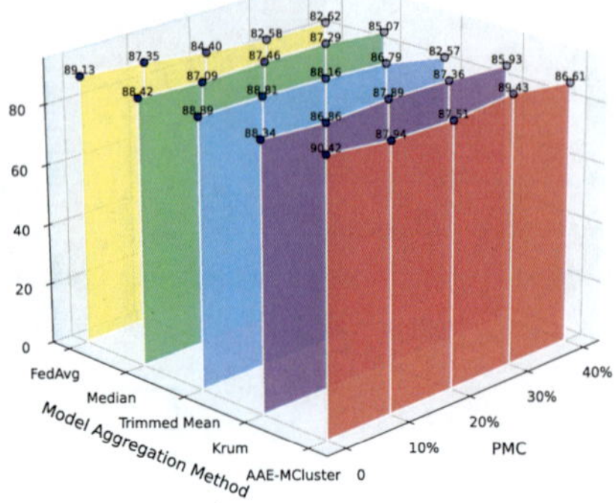

Fig. 8. $\overline{ACC}_{clean}$(10%) under Different Model Aggregation Algorithms with CIFAR-10 Dataset

For $\overline{ACC}_{clean}$, all methods show a declining trend with more malicious clients. AAE-MCluster yields higher $\overline{ACC}_{clean}$ at 0%, 30%, and 40% malicious clients. Though Trimmed Mean has slightly higher $\overline{ACC}_{clean}$ at 10% and 20% malicious, its ACC_{class} is notably higher, reflecting inferior defense capability. Overall, AAE-MCluster achieves remarkable defense effectiveness against label-flipping attacks while maintaining high global classification performance on CIFAR-10.

Above all, for all the defense mechanisms tested in the experiment, the attack success rate is proportional to the number of malicious clients. However, the self-supervised clustering aggregation strategy proposed in this paper has a stronger capability to resist attacks compared with other typical aggregation defense methods. Meanwhile, this method ensures that the efficiency of model aggregation in the federated learning process does not decline.

5.3 Summary

This paper proposes a self-supervised clustering-based model aggregation defense method for FL in IIoT. It enhances high-dimensional terminal model clustering via self-supervised clustering and corrects potential poisoned model parameters using a momentum-based strategy, aligning them with historical benign global models. Experiments on non-IID data show the method outperforms baselines in poisoning attack defense and promotes model convergence, yielding efficient and robust global models. In this method, the self-supervised clustering is based on the K-means clustering algorithm, which is theoretically applicable to federated systems with a large number of clients. However, the defense performance in real-world with large-scale application scenarios still needs to be further verified through experiments, which will be regarded as the direction of future work. Future research will also focus on lightweight model properties to design low-complexity defense methods.

Acknowledgement. This paper is supported by the Scientific Research Foundation of NJIT (No. YKJ202316, YKJ202435), the Natural Science Research of Jiangsu Higher Education Institutions of China (No. 24KJD520005), 2024 Jiangsu Provincial Frontier Technology R&D Program:

"Research on Cross-Domain and Multi-Dimensional Security Technologies for Intelligent Systems in AI Computing Power Networks" (No.: BF2024071).

References

1. Wang, S., Li, Q., Cui, Z., Hou, J., Huang, C.: Bandit-based data poisoning attack against federated learning for autonomous driving models. Expert Syst. Appl. **227**, 120295 (2023). https://doi.org/10.1016/J.ESWA.2023.120295
2. McMahan, B., Moore, E., Ramage, D., Hampson, S., Arcas, B.A.Y: Communication-efficient learning of deep networks from decentralized data. In: Artificial intelligence and statistics 2017. pp. 1273–1282. PMLR (2017)
3. Blanchard EPFL, P., Mahdi El Mhamdi, E., Guerraoui EPFL, R., Stainer EPFL, J.: Machine learning with adversaries: byzantine tolerant gradient descent. Adv. Neural Inf. Process Syst. **30** (2017)
4. Wan, C.P., Chen, Q.: Robust federated learning with attack-adaptive aggregation (2021)
5. Aramoon, O., Chen, P.Y., Qu, G., Tian, Y.: Meta-federated learning. Federated learning: theory and practice. 161–179 (2024). https://doi.org/10.1016/B978-0-44-319037-7.00017-X
6. Duc Nguyen, T., Rieger, P., Diego, S., Yalame, H., Möllering, H., Fereidooni, H., Sadeghi, A.-R., Schneider, T., Chen, H., Marchal, S., Miettinen, M., Mirhoseini, A., Zeitouni, S., Koushanfar, F.: {FLAME}: Taming backdoors in federated learning. In: 31st USENIX security symposium (USENIX Security 22). pp. 1415–1432. USENIX Association (2022)
7. Min, E., Guo, X., Liu, Q., Zhang, G., Cui, J., Long, J.: A survey of clustering with deep learning: from the perspective of network architecture. IEEE Access. **6**, 39501–39514 (2018). https://doi.org/10.1109/ACCESS.2018.2855437
8. Aljalbout, E., Golkov, V., Siddiqui, Y., Strobel, M., Cremers, D.: Clustering with deep learning: taxonomy and new methods (2018)
9. Mi, Y., Guan, J., Zhou, S: ARIBA: towards accurate and robust identification of backdoor attacks in federated learning (2022)
10. Yin, D., Chen, Y., Kannan, R., Bartlett, P.: Byzantine-robust distributed learning: towards optimal statistical rates. In: International conference on machine learning (ICML 2018). pp. 5650–5659. PMLR (2018)
11. Shejwalkar, V., Houmansadr, A.: Manipulating the byzantine: optimizing model poisoning attacks and defenses for federated learning. In: Network and distributed system security symposium (NDSS) 2021. Internet Society (ISOC) (2021). https://doi.org/10.14722/ndss.2021.23xxx
12. El Mhamdi, E.M., Guerraoui, R., Rouault, S.: The hidden vulnerability of distributed learning in Byzantium. In: 35th international conference on machine learning, ICML 2018. 8, 5674–5686 (2018)
13. Sheikhi, S., Kostakos, P., Loven, L.: Hybrid Reputation aggregation: a robust defense mechanism for adversarial federated learning in 5G and edge network environments (2025). https://doi.org/10.1145/xxxxxxx.xxxxxxx
14. Wang, Y., Zhang, H., Wen, S., Qiu, W., Guo, B.: Defending against sophisticated poisoning attacks with RL-based Aggregation in Federated Learning (2024)
15. Yu, L., Wu, L.: Towards byzantine-resilient federated learning via group-wise robust aggregation. Federated learning: privacy and incentive. 12500 LNCS, 81–92 (2020). https://doi.org/10.1007/978-3-030-63076-8_6
16. Jebreel, N.M., Domingo-Ferrer, J., Sánchez, D., Blanco-Justicia, A.: Defending against the label-flipping attack in federated learning (2022)

17. Jeong, H., Hwang, J., Chung, T.M.: ABC-FL: anomalous and benign client classification in federated learning (2021)
18. Jebreel, N.M., Domingo-Ferrer, J.: FL-Defender: combating targeted attacks in federated learning. Knowl Based Syst. **260**, 110178 (2023). https://doi.org/10.1016/J.KNOSYS.2022.110178
19. Li, S., Cheng, Y., Wang, W., Liu, Y., Chen, T.: Learning to detect malicious clients for robust federated learning (2020)
20. Cui, S., Li, D., Zhang, J.: Dynamic multi-scale feature augmentation for inductive network representation learning. Pattern Recognit. **161**, 111250 (2025). https://doi.org/10.1016/J.PATCOG.2024.111250
21. Wang, E., Chen, S., Peng, L., Zhang, X., Ou, Y., Peng, J.: Contrastive self-supervised subspace clustering via KAN-based multi-view fusion. Expert Syst. Appl. **296**, 128995 (2026). https://doi.org/10.1016/J.ESWA.2025.128995
22. Lu, W., et al.: A novel self-supervised graph clustering method with reliable semi-supervision. Neural Netw. **187**, 107418 (2025). https://doi.org/10.1016/J.NEUNET.2025.107418
23. Zhang, Z., Tian, H., Zuo, J., Ding, W.: Aggregation-based self-supervised evidential clustering for imbalanced data. Informat. Fusion. **127**, 103721 (2026). https://doi.org/10.1016/J.INFFUS.2025.103721
24. Xu, H., Zhang, C., Yue, P., Wang, K.: SDCluster: A clustering based self-supervised pre-training method for semantic segmentation of remote sensing images. ISPRS J. Photogramm. Remote Sens. **223**, 1–14 (2025). https://doi.org/10.1016/J.ISPRSJPRS.2025.02.021
25. Lee, S., Hwang, S., Kim, D.: MADCluster: model-agnostic anomaly detection with self-supervised clustering network (2025)
26. Tolpegin, V., Truex, S., Gursoy, M.E., Liu, L.: Data poisoning attacks against federated learning systems. In: Computer security–ESORICS 2020: 25th European symposium on research in computer security, ESORICS 2020, Guildford, UK, September 14–18, 2020, Proceedings, Part I 25. pp. 480–501. Springer International Publishing (2020). https://doi.org/10.1007/978-3-030-58951-6_24/TABLES/6
27. Wang, S., Huang, Q., Lian, Z.: Multi-convolution and Adaptive-Stride Based Transferable Adversarial Attacks. In: Iliadis, L., Papaleonidas, A., Angelov, P., and Jayne, C. (eds.) Artificial Neural Networks and Machine Learning – ICANN 2023. pp. 130–142. Springer Nature Switzerland, Cham (2023)
28. Zhu, L., Liu, Z., Liu, G.: A deep multiple self-supervised clustering model based on autoencoder networks. Scientific Reports 2025 15:1. 15, 18372 (2025). https://doi.org/10.1038/s41598-025-00349-z

29. Dong, Y., Liao, F., Pang, T., Su, H., Zhu, J., Hu, X., Li, J.: Boosting adversarial attacks with momentum. In: Proceedings of the IEEE conference on computer vision and pattern recognition (CVPR 2018). pp. 9185–9193. IEEE (2018)
30. Hinton, G., Vinyals, O., Dean, J.: Distilling the knowledge in a neural network. (2015)

Adaptive RelaxLoss: An Adversarially Guided Defense Against Membership Inference Attacks

Jiaqi Qin and Yang Cao(✉)

School of Cyber Science and Engineering, Engineering Research Center of Blockchain Application, Supervision and Management of the Ministry of Education, Southeast University, Nanjing 211189, China

caoyeacy@seu.edu.cn

Abstract. Membership inference attacks exploit discrepancies between a model's behavior on training and non-training samples to infer data membership, leading to severe privacy risks in applications such as image classification and medical diagnosis. Existing defenses either rely on differential privacy training, which substantially degrades accuracy and incurs high computational cost, or on heuristic regularization schemes that require tedious manual tuning and lack robustness across architectures and datasets. In this paper, we propose Adaptive RelaxLoss, an adversarially guided loss scheduling scheme for defending against membership inference attacks. We first introduce a membership inference attacker that operates on the cross-entropy loss trajectories of the target model and define an adversarial loss that quantifies the model's privacy leakage. This adversarial loss serves to tune an adaptive threshold that enables the training process to switch between gradient descent and KL divergence-based gradient ascent, thus achieving both utility and privacy preservation goals. The threshold circumvents the heavy computational cost caused by backpropagating gradients of the adversary's model in each iteration. We instantiate Adaptive RelaxLoss on convolutional networks and evaluate it on CIFAR-10, STL-10, and CINIC-10 under strong shadow-model-based membership inference attacks, including Salem-style prediction attacks and likelihood-ratio attacks. Experiments show that Adaptive RelaxLoss achieves comparable or lower attack success rates, while improving target accuracy by up to 14.92% and reducing training time by 26.35% compared with existing differential privacy-based methods, and consistently outperforming fixed-threshold RelaxLoss. An ablation study further illustrates how the adaptive-threshold hyperparameters trade off privacy and utility, providing practical guidance for deploying the defense in real-world systems.

Keywords: Membership Inference Attack · Privacy Preservation · Adaptive RelaxLoss

L. Huang and D. Greenhalgh (Eds.): ICMLC 2026, LNNS 2061, pp. 638–655, 2027.
https://doi.org/10.1007/978-3-032-31236-5_44

1 Introduction

Deep neural network training typically requires large-scale, diverse datasets. Enterprises and institutions often construct such datasets for their own training purposes but rarely share them, as dataset collection is time-intensive and labor-intensive and may contain sensitive information, commercial secrets, or personal privacy. Unauthorized access to such data could compromise organizational reputation, interests, or user privacy. For instance, OpenAI and Meta withhold the training datasets for their large language models. Notably, training data for deep models may include personal information–such as facial images for face recognition or medical records for disease diagnosis–imposing an obligation on enterprises to protect user privacy. Consequently, AI enterprises not only refrain from disclosing training datasets but also implement measures to prevent data leakage.

Ensuring dataset security requires not only safeguarding the datasets themselves but also protecting the trained models. This is because the parameters θ of a deep model f_θ trained on dataset $\mathcal{D}_{train}$ encode information about $\mathcal{D}_{train}$; in some cases, such memorized information can even enable dataset reconstruction [1]. Notably, even without direct access to $\mathcal{D}_{train}$, adversaries can infer dataset information via the model f_θ. The most prevalent attack targeting training datasets is membership inference attack, which determines whether a specific sample belongs to $\mathcal{D}_{train}$ solely through model access [2].

Membership inference attacks are categorized into black-box and white-box types based on the adversary's knowledge of the target model. In black-box scenarios, adversaries cannot access model parameters but can obtain inference results by inputting queries. Aligned with realistic threat scenarios, black-box attacks have garnered extensive research attention. A common black-box attack paradigm employs shadow models: adversaries construct a shadow dataset matching the distribution of $\mathcal{D}_{train}$, train shadow models on the shadow dataset, and then train an attack model using shadow model outputs. Extensive research has focused on defending against membership inference attacks. Classic approaches include regularization techniques like dropout, but studies have shown that general regularization reduces generalization error yet fails to fully mitigate such attacks. Differential privacy training, a provably private technique, provides strict defense against membership inference but compromises model utility and incurs significant computational overhead.

This paper proposes a defense method against membership inference attacks based on adaptive RelaxLoss. Experimental verification shows that this method achieves a good defense effect against membership inference attacks while having a small impact on model accuracy and not significantly increasing computational overhead. The main contributions of this paper are as follows: a) Propose an adversarially guided defense against membership attack, which transforms the intractable static threshold parameter tuning problem in existing research into an online adaptive adjustment mechanism. b) Propose an adversarially guided lightweight defense method to resist membership inference attacks, and analyze the computational complexity of this method during training. c) Verify the feasibility and effectiveness of the proposed defense methods through experiments, and provide empirically optimal values for hyperparameter settings.

2 Related Work

2.1 Membership Inference Attack

Membership inference attacks were first proposed by Shokri et al. in 2017 [2]. Leveraging the overfitting phenomenon prevalent in machine learning models, this type of attack infers the membership of a given sample by exploiting the differences in model performance between inputs from the training set and those from non-training sets.

In 2020, Park and Kang observed that object detection models exhibit higher uncertainty in the output of bounding boxes for non-training set inputs. Based on this insight, they proposed a membership inference attack specifically targeting object detection models and successfully demonstrated attacks on SSD and Faster RCNN through experiments [3].

In 2021, Hui et al. introduced a novel method called differential comparison to probe the target model and extract membership features, thereby realizing a blind membership inference attack that operates without the need for shadow models [4].

In 2022, Yuan and Zhang proposed a self-attention-based membership inference attack method for pruned neural networks, which achieved excellent attack performance on pruned models [5]. In the same year, Liu et al. utilized knowledge distillation to conduct membership inference attacks. They represented membership information using the losses evaluated on a series of intermediate models at different distillation stages (i.e., distillation loss trajectories) and the loss of the given target model [6]. Also in 2022, Carlini et al. proposed a likelihood ratio-based membership inference attack method, which employed hypothesis testing to achieve membership inference with a low false positive rate [7].

In 2023, Liu et al. used the Jacobian matrix to approximate prediction sensitivity and proposed a black-box membership inference attack method that leverages prediction sensitivity. This method enables membership inference using only the output of the target model [8]. In the same year, Zhang et al. trained an attack model using data generated by image data augmentation techniques, thereby implementing a membership inference attack against semantic segmentation models [9].

In 2024, Chen et al. proposed a two-stage attack method incorporating membership exclusion technology, which realized high-accuracy membership inference attacks [10]. Also in 2024, Zhu et al. found that self-supervised visual models tend to focus on a small subset of regions in an image. Based on this finding, they proposed a general membership inference attack method for visual self-supervised learning models by cropping images [11].

2.2 Defence Against Membership Inference Attack

There are various defense methods against membership inference attacks, among which the use of differential privacy technology during the training process is a relatively common one. Differential Privacy is a privacy-preserving technology that balances the relationship between data usability and individual privacy by introducing controllable noise into the data analysis process. First proposed by Dwork [12,13], it defends against differential attacks and protects individual privacy by adding a certain level of noise to query results. In the context of machine learning privacy protection, large datasets used for model training often contain sensitive information. To protect the privacy in these datasets, the Differential Privacy Stochastic Gradient Descent (DP-SGD) algorithm is frequently employed. Instead of directly adding noise to the input data, DP-SGD adds Gaussian noise to the stochastic gradients updated in each iteration, thereby ensuring that the trained model satisfies differential privacy [14]. Rezaei et al. (2023) investigated the application of DP-SGD in defending against membership inference attacks and found that the use of DP-SGD is beneficial for defending against such attacks. However, the distortion it causes to confidence levels can also affect applications that rely on confidence [15].

In addition to differential privacy, other methods have also been studied for defending against membership inference attacks. Tang et al. (2022) achieved a balance between the defense against membership inference attacks and model usability through methods of training set partitioning and self-distillation [16]. Meeus et al. (2023) proposed a method to identify records vulnerable to membership inference attacks by recording distances to the nearest neighbors. This method helps improve the release process of synthetic data [17], thereby enhancing robustness against membership inference attacks. Yang et al. (2023) implemented defense against membership inference attacks in federated learning by adding personalized perturbations to client inputs [18]. Chen et al. (2024) effectively masked the differences between member and non-member predictions by uniformly modifying all prediction outputs to low-confidence outputs while maintaining accuracy [19], thus reducing the success rate of membership inference attacks. Hu et al. (2023) used sensitive data to train Generative Adversarial Networks (GANs), and then the GANs generated data for training the actual models. This approach achieved defense against membership inference attacks while maintaining high model usability [20]. Li et al. (2024) constructed a new training method to defend against membership inference attacks through counterfactual invariance and subspace learning. This method can prevent vulnerable instances from being overfitted without significantly affecting normal instances [21]. Another study by Li et al. (2024) realized the defense against membership inference attacks on visual models from the perspective of training data by randomly shuffling the positions of pixels within images [22].

Differential privacy-based defense has provable security, but it usually incurs large additional computational overhead and affects the prediction accuracy of the model [15]; Data privacy protection-based methods often do not require modifying the original training framework, but need to perform different processing for specific datasets; Adversarial training-based methods can not only enhance the model's ability to resist membership inference attacks but also improve the model's robustness against various attacks. However, adversarial training usually requires investing more computational resources, and some methods may also increase the complexity of the model [21]; Robust learning-based methods, depending on the specific approach, have the characteristics of low computational cost and low implementation complexity, but their universality is relatively poor [23,24].

3 Research Method

3.1 Threat Model

In the research on membership inference attacks, shadow model-based methods are a commonly used type of attack. In such methods, it is assumed that the attacker cannot access the training dataset of the target model, but can construct a shadow dataset that follows the same distribution as the target model's training dataset–meaning the attacker has prior knowledge of the distribution of the target model's training dataset. We adopt a stronger black-box attacker assumption: the attacker can not only obtain the predicted probability outputs of the model, but also possess the prior knowledge of the distribution of the target model's training dataset and the defense method employed by the target model. In the threat model, the attacker trains the shadow model not only on a shadow dataset with the same distribution as the target model's, but also using the exact same defense method as the target model. Specifically, the defense method and its associated hyperparameters used by the attacker are identical to those of the target model, indicating that we are dealing with a highly capable attacker.

3.2 Adaptive RelaxLoss

In the scenario of membership inference attacks, defenders have two objectives: privacy and usability. Therefore, the defense method proposed in this paper divides model training into two phases. In the first phase, the model's usability is insufficient, so the optimization focuses on improving model accuracy. In the second phase, the model's usability meets the required standards, and the optimization objective then shifts to achieving the model's privacy. The framework of our method is presented in Fig. 1

Stochastic gradient descent based on cross-entropy loss sets the optimization objective of neural networks to maximize the probability of the unique positive class. Consequently, training using cross-entropy loss tends to cause model overfitting. The label smoothing method relaxes this optimization objective: it allows a small average probability for negative classes while maintaining the highest probability for the positive class. Label smoothing relaxes the loss through soft targets, which alleviates the model's overfitting to some extent. However, from the perspective of membership inference attacks, label smoothing may still lead the model to overfit to soft targets, leaving room for attackers to implement membership inference attacks. Although piecewise RelaxLoss based on a fixed threshold can achieve a good defense effect, considering that different model structures vary in their vulnerability to membership inference attacks, manual threshold adjustment is required for specific models during training to balance privacy protection and model performance. This process may involve multiple adjustments to achieve the desired effect.

This study adopts a RelaxLoss method based on an adaptive threshold. During training, the worst-case attack scenario faced by the model is considered, and an auxiliary model is used to simulate a gray-box membership inference attack with loss access on the target model. A adaptive threshold is determined based on the auxiliary model attack loss. During model updates, the adaptive threshold serves as a dividing point, and gradient descent and gradient ascent based on cross-entropy loss are executed in segments. It should be noted that the auxiliary model is only used to assist the training of the target model and is discarded once the training is completed.

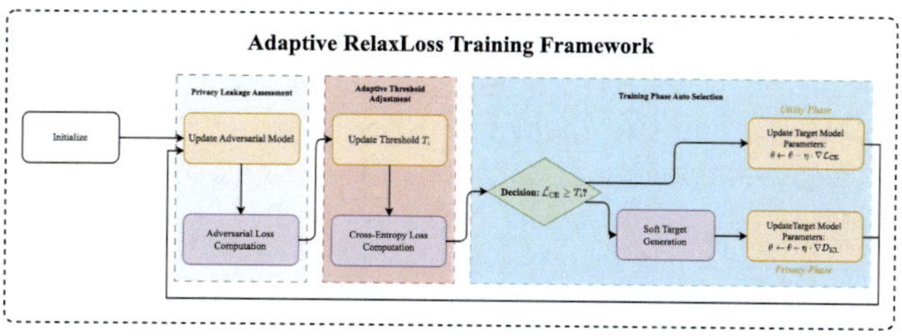

Fig. 1. The training framework of adaptive RelaxLoss

Let f_θ denote the target model, and let $\mathcal{L}(x, y; \theta)$ be its cross-entropy loss on sample (x, y). The attacker trains a membership inference model g_ϕ that takes the loss value $\mathcal{L}(x, y; \theta)$ as input and outputs the posterior probability that (x, y) is a member of the training set. Given auxiliary member and non-member datasets D^m_{aux} and D^{nm}_{aux} [25], the adversarial loss is defined as the standard binary cross-entropy:

$$\begin{aligned}\mathcal{L}_{\text{adv}}(g_\phi, D_{\text{aux}}; f_\theta) = \quad & -\Bigg(\frac{1}{2|D^m_{\text{aux}}|}\sum_{(x,y)\in D^m_{\text{aux}}} \log\left(g_\phi\left(\mathcal{L}(x,y;f_\theta)\right)\right) \\ & + \frac{1}{2|D^{nm}_{\text{aux}}|}\sum_{(x,y)\in D^{nm}_{\text{aux}}} \log\left(1 - g_\phi\left(\mathcal{L}(x,y;f_\theta)\right)\right)\Bigg)\end{aligned} \tag{1}$$

This loss measures how well the attacker can distinguish member from non-member samples based on the target model's loss values, and thus quantifies the privacy leakage of the current parameters θ. A straightforward way to use this adversarial signal is to incorporate it into the training objective of the target model

$$\mathcal{L} = \mathcal{L}_{\text{CE}}(\theta) + \lambda \cdot \mathcal{L}_{\text{adv}}(\theta, \phi) \tag{2}$$

where $\mathcal{L}_{\text{CE}}$ is the standard cross-entropy loss and $\lambda > 0$ controls the strength of adversarial regularization. While this adversarial training (AT) objective can improve both privacy and robustness, it requires computing gradients through the attack model g_ϕ in every iteration, which significantly increases computational overhead and is sensitive to the choice of λ. We therefore only use $\mathcal{L}_{\text{adv}}$ calculated based on the samples from each batch to derive an adaptive threshold for a RelaxLoss-style defense [23], while still updating the adversarial model based on the auxiliary dataset. Specifically, the adversarial model is updated once per epoch, and the threshold is updated at each batch according to the current adversarial loss:

$$T_{\text{i}} = \gamma T_{\text{i-1}} + (1 - \gamma) \cdot \beta e^{-kL_{\text{adv}}} \tag{3}$$

where $\gamma \in (0, 1]$ controls the smoothness of the threshold updates, $\beta > 0$ sets the maximum value of the threshold, and $k > 0$ controls how fast the threshold decreases as the adversarial loss increases. Intuitively, when the attacker performs well with small L_{adv}, the threshold increases, causing the defender to switch earlier to a privacy-oriented loss as the utility-oriented loss drops below the threshold earlier; when the attacker performs poorly, the threshold decreases and the model can focus more on utility.

For each mini-batch, let $\bar{\mathcal{L}}_{\text{CE}}$ denote the average cross-entropy loss over the batch. We then define a piecewise loss:

$$\mathcal{L}(\theta) = \begin{cases} \mathcal{L}_{\text{CE}}(\theta), & \text{if } \bar{\mathcal{L}}_{\text{CE}} \geq T_i, \\ -D_{\text{KL}}\big(S \parallel p_\theta(\cdot \mid x)\big), & \text{if } \bar{\mathcal{L}}_{\text{CE}} < T_i, \end{cases} \tag{4}$$

where $p_\theta(\cdot \mid x)$ is the model's predicted distribution and S is a label-smoothed soft target. The negative sign in the second branch implements gradient ascent on the KL-divergence, which encourages the model to move away from the soft target and partially forget the training samples that have already been well-fitted. In practice, we simply multiply the gradient of the KL term by (-1) when $\bar{\mathcal{L}}_{\text{CE}} < T_i$.

The soft target (S) is defined as

$$S_j = \begin{cases} 1 - \alpha, & j = y, \\ \dfrac{\alpha}{N-1}, & j \neq y, \end{cases} \tag{5}$$

where $\alpha \in (0, 1)$ is the label smoothing parameter and N is the number of classes.

In information theory, KL divergence can measure the information lost when one distribution is used to approximate another. Therefore, we use a gradient ascent process to enable the model to forget as much information as possible,

Algorithm 1 Adaptive Threshold RelaxLoss Method

Input: D_{train},D_{aux},α,β,γ,k,η_{adv}, η_{target}, total epochs K, batch size B
Output: θ

Initialize: Set $\mathbf{T}_0 = 0$, initialize model parameters θ and ϕ
for $i = 1, 2, ..., K$ **do**
  Perform forward pass, compute $\mathcal{L}_{\text{adv}}(g_\phi, D_{\text{aux}}; f_\theta)$ in Equation 1
  $\nabla L(\phi, D_{aux}, \theta) = \frac{1}{|D_{aux}|} \sum_{(x,y) \in D_{aux}} \nabla_\phi \mathcal{L}(x, y, \theta; \theta_{adv})$
  Update adversarial model: $\phi \leftarrow \phi - \eta_{adv} \cdot \nabla_\phi L(\phi, D_{aux}, \theta)$
  for each batch **do**
    Get sample batch: $(x_j, y_j)_{j=1}^B$
    Perform forward pass: $L_{adv} = -\frac{1}{B} \sum_{(x,y) \in (x_j, y_j)_{j=1}^B} \log\left(g_\phi\left(\mathcal{L}(x, y; f_\theta)\right)\right)$
    Update threshold: $T_i \leftarrow \gamma T_{i-1} + (1 - \gamma) \cdot \beta e^{-kL_{adv}}$
    Perform forward pass to get $\bar{\mathcal{L}}_{\text{CE}}$
    if $\bar{\mathcal{L}}_{\text{CE}} \geq T_i$ **then**
      Update target model: $\theta \leftarrow \theta - \eta_{target} \cdot \nabla_\theta \mathcal{L}_{\text{CE}}(\theta)$
    else
      Update target model: $\theta \leftarrow \theta + \eta_{target} \cdot \nabla_\theta D_{\text{KL}}\big(S \parallel p_\theta(\cdot \mid x)\big)$
    end if
  end for
end for

thereby avoiding overfitting. The reason for using soft targets instead of real targets lies in the fact that real targets are represented by one-hot encoding, making it impossible to calculate the KL divergence between real targets and predicted outputs. Additionally, the use of soft targets can also reduce the amount of information about real labels transmitted to the model.

The use of KL divergence loss allows the model to forget as much information as possible, while the cross-entropy loss in the piecewise loss ensures model accuracy. This enables the proposed method to defend against membership inference attacks while guaranteeing model accuracy. Furthermore, since the loss calculation is performed in the forward propagation phase, the additional computational overhead brought by this scheme is relatively small. The training process of the entire scheme can be represented by the Algorithm 1.

3.3 Computational Complexity Analysis

Differential privacy methods can achieve excellent privacy protection; however, their primary limitation lies in the inevitable additional computational overhead introduced during the differential privacy stochastic gradient descent process. In this subsection, we analyze the differences in training computational costs among standard gradient descent, differential privacy stochastic gradient descent (DP-SGD), RelaxLoss, as well as the adversarial training and Adaptive RelaxLoss proposed in the previous subsection.

Since mini-batch gradient descent is more commonly used in practice, the following analysis is based on mini-batch gradient descent with a batch size of b. For the convenience of comparison, we assume all five methods are trained on models with the same number of parameters n. For the adversarial models involved in the two methods proposed in the previous subsection, we assume their number of parameters is m, where it should be noted that m is much smaller than n.

In the forward propagation phase, the computational cost required to calculate the loss for one batch is positively correlated with both the number of parameters and the batch size. For standard mini-batch gradient descent, DP-SGD, and RelaxLoss, this complexity can be expressed as $\mathcal{O}(b \cdot n)$. For the two methods proposed in this paper, the complexity is $\mathcal{O}(b \cdot n + |\mathcal{D}_{aux}| \cdot m)$. Since m is much smaller than n and the size of the auxiliary dataset is generally a small constant, this complexity can be considered comparable to that of standard mini-batch gradient descent.

In the gradient backpropagation phase, standard mini-batch gradient descent and RelaxLoss only need to perform chain rule differentiation once for the batch samples. Let $T(b, n)$ denote the computational cost of calculating gradients for n parameters using a batch of size b. Then, the computational cost of adversarial training for gradient calculation is $T(b + |\mathcal{D}_{aux}|, n) + T(|\mathcal{D}_{aux}|, m)$, while that of Adaptive RelaxLoss is $T(b, n) + T(|\mathcal{D}_{aux}|, m)$. Again, since m is much smaller than n, the second term can be approximately neglected.For DP-SGD, gradients for each individual sample need to be computed during backpropagation, resulting in a computational cost of $b \cdot T(1, n)$. Although some differential privacy tools optimize this process to avoid b separate backward pass, they still require storing gradients for each sample individually–thus consuming significant time when writing to High-Bandwidth Memory (HBM). Additionally, DP-SGD needs to compute and clip the norm of each sample's gradient, with a computational cost positively correlated with both the number of parameters and the batch size, expressed as $\mathcal{O}(b \cdot n)$.

In the gradient backpropagation phase, the complexity of DP-SGD is much higher than that of the other methods, and theoretically, the gap grows with increasing batch size. Adversarial training has significantly higher complexity than the other three methods, while Adaptive RelaxLoss has slightly higher complexity than RelaxLoss and standard mini-batch gradient descent. Theoretically, this gap decreases as the batch size increases.

4 Experiments

4.1 Experiments Setup

In this study, a neural network with alternating convolutional layers and pooling layers for image feature extraction, and a fully connected layer as the classifier, is employed as the target model. Its primary task is image classification. The CIFAR-10 [26], STL-10 [27] and CINIC-10 [28] datasets are used for experiments. A neural network with the same structure but a narrower hidden layer is adopted as the shadow model. Both the shadow model and the target model are trained using the same defense method to simulate a stronger attacker. The attacker uses a two-layer fully connected neural network as the attack model to conduct membership inference attacks.

The training datasets of the target model and the shadow model are both derived from the same dataset, and their training datasets are non-overlapping. The experiment adopts the attack method proposed by Salem et al. [29] and LiRA [7], comparing the defense with that of the Differential Privacy Stochastic Gradient Descent (DP-SGD) algorithm [30]. In the experiments of this section, DP-SGD is implemented with Opacus and the noise multiplier σ is set to 0.2 while the gradients clipping threshold is set to 1.0. All experimental methods are run under the same hardware conditions, with same hyperparameters such as batch size, number of iterations, and learning rate controlled to be consistent. For comparison, we also use two other defense methods, which are marked as RelaxLoss and AT. AT refers to the method we proposed in Eq. (2) in the third section.

Table 1. Experimental results of defense methods under ML-based attack [29]

Dataset	Defense method	Target model Validation performance			Train time per epoch(s)↓	Attack model performance			
		Acc.↑	Prec.↑	F1↑		AUC↓	Acc.↓	Prec.↓	F1↓
CIFAR-10	Naive	**0.7250**	0.7262	**0.7243**	**64.09**	0.5489	0.5315	0.5358	0.5016
	DP-SGD	0.5326	0.5282	0.5296	89.55	0.5040	0.5049	0.5054	0.4814
	AT	0.7178	**0.7277**	0.7204	89.13	0.5266	0.5085	0.5129	**0.4080**
	RelaxLoss	0.6760	0.6759	0.6745	65.82	**0.4854**	**0.4844**	**0.4819**	0.4456
	Adaptive RelaxLoss	0.6818	0.6763	0.6743	65.95	0.4981	0.4992	0.4996	0.6552
STL-10	Naive	0.5240	0.5238	0.5174	**58.82**	0.5316	0.5225	0.5317	0.4415
	DP-SGD	0.4640	0.4477	0.4503	70.04	0.5057	0.5078	0.5079	0.5032
	AT	**0.5445**	**0.5560**	**0.5484**	85.99	0.5051	0.5058	0.5104	**0.3659**
	RelaxLoss	0.4910	0.4981	0.4890	61.20	0.4997	0.5024	0.5018	0.5738
	Adaptive RelaxLoss	0.5433	0.5436	0.5402	65.97	**0.4995**	**0.5010**	**0.5013**	0.4430
CINIC-10	Naive	0.5774	0.5788	0.5778	**66.13**	0.7056	0.6541	0.6605	0.6470
	DP-SGD	0.4453	0.4377	0.4394	126.50	0.5021	0.5014	0.5022	0.3927
	AT	**0.5867**	**0.5871**	**0.5860**	127.01	0.6509	0.6088	0.6943	0.4985
	RelaxLoss	0.5543	0.5593	0.5544	72.59	**0.4986**	**0.4989**	**0.4988**	0.4886
	Adaptive RelaxLoss	0.5664	0.5665	0.5650	75.66	0.5023	0.5013	0.5158	**0.0784**

4.2 Experiments Results

The experimental results are summarized in Table 1. The model trained by the adaptive RelaxLoss method outperforms the differential privacy method in both prediction accuracy and training time. In terms of defense, the adaptive RelaxLoss method achieves a performance comparable to that of the differential privacy method, and even outperforms the differential privacy method in some attack metrics. For AT method, its training time is roughly on par with DP-SGD, indicating similar computational overheads. Notably, AT delivers better prediction performance than DP-SGD with acceptable defense performance. It is worth mentioning that DP-SGD was implemented directly using the Opacus library [30], whereas our AT method was deployed without special optimization. In contrast, the RelaxLoss method exhibits poor trade-off capabilities: on some datasets, it ensures satisfactory model performance but fails to provide effective privacy protection; on others, it achieves strong defense effects at the cost of severely degraded prediction accuracy. By contrast, Adaptive RelaxLoss addresses this issue by adaptively adjusting the threshold during training, thus ensuring a robust balance between model usability and privacy protection across diverse datasets. We observed that the AT method shows inconsistent defense performance across datasets: it performs notably better on CIFAR-10 and STL-10 than on CINIC-10. This discrepancy stems from the fact that CINIC-10 is substantially larger than the other two datasets, while the auxiliary dataset used in AT has a fixed size. As for the second plausible explanation, CINIC-10 is constructed by merging samples from two distinct datasets, leading to mini-batch distributions that may deviate significantly from that of the auxiliary dataset during training. Since AT directly incorporates the adversarial loss of the auxiliary dataset into its loss function, such distribution mismatches inevitably impair the model's defense capability. Unlike AT, which directly integrates adversarial loss into the objective function, Adaptive RelaxLoss embeds the adversarial model's performance into an adaptive threshold. This design eliminates the computational overhead caused by calculating additional gradients during training, enhances training stability, and ensures consistent robustness across different datasets. Against Adaptive RelaxLoss, the attack model obtains a high F1-score on CIFAR-10 by classifying nearly all inputs as member samples, but a very low F1-score on CINIC-10 by categorizing almost all input samples as non-member samples. This further indicates that our method can well confuse the attacker.

Table 2. Experimental Results on CIFAR-10 Dataset under Likelihood Ratio Offline Attack

Dataset	Defense method	ASR@1%FPR↓	AUC-ROC↓
CIFAR-10	Naive	41.70	0.7962
	DP-SGD	4.50	0.5105
	AT	6.80	0.7607
	RelaxLoss	<0.01	0.6825
	Adaptive RelaxLoss	<0.01	0.6976

Table 2 presents the defense performance of different methods against the offline Likelihood Ratio Attack (LiRA) [7]—a hypothesis-testing-based attack that infers membership by focusing on attack success rate (ASR) under low false positive rates (FPR). We use the calculated likelihood of a sample being a member as the attack score to compute AUC, and additionally report ASR [31] at 1% FPR in the table. RelaxLoss and Adaptive RelaxLoss achieve the best overall defense performance against LiRA. DP-SGD performs slightly inferior, though its lower AUC indicates moderate privacy protection. In contrast, AT only achieves limited defense effectiveness, with its higher AUC reflecting poor privacy protection against LiRA.

To further validate the defense mechanisms, Fig. 2 shows the p-value distributions of model outputs under offline LiRA with null hypothesis that the sample is a non-member. Ideally, member samples should have p-values close to 0 to reject the null hypothesis and non-member samples close to 1. Figure 2(b) shows the results of DP-SGD: although DP-SGD can obfuscate the attacker's ability to distinguish between members and non-members, a small number of member samples are still inferred as members with high confidence. Figure 2(c) shows the results of AT: a certain number of member samples are inferred as members with high confidence, though there is a degree of improvement compared with the results in Fig. 2(a). For RelaxLoss (Fig. 2(d)) and Adaptive RelaxLoss (Fig. 2(e)), member samples generally have higher confidence than non-members, but the maximum confidence for members remains below 90% for RelaxLoss and below 80% for Adaptive RelaxLoss, indicating attackers lack sufficient confidence to confirm membership.

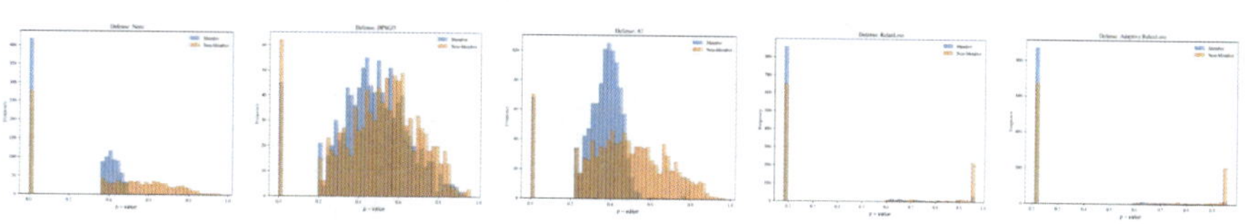

(a) (b) (c) (d) (e)

Fig. 2. The p-value distribution under defenses

4.3 Ablation Study

In Fig. 3 proposed in Sect. 3, the threshold T is related to two important hyperparameters. We also used experiments to explore the impact of the two hyperparameters (k, β) on model performance and defense effectiveness. The experiments were conducted on the CIFAR-10 dataset, and all other experimental conditions were identical to those in Sect. 4.1. We tested a total of 9 combinations of different values for the two hyperparameters (k, β), and the experimental results are presented in Fig. 3 and Table 3.

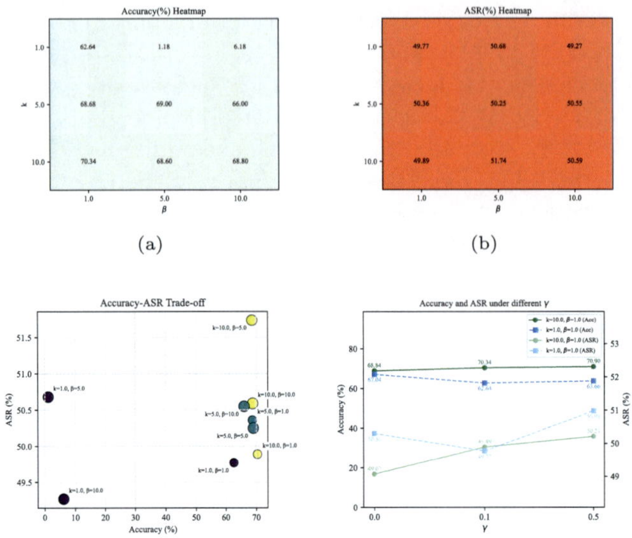

(a) (b)

(c) (d)

Fig. 3. Ablation experiment results

In our experiments, we considered the impact of two hyperparameters on model privacy and usability. For measuring model usability, we chose the prediction accuracy of the model on the test set as the metric–the higher this metric, the better the model's performance. For measuring model privacy, we used Attack Success Rate (ASR) [31] as the evaluation indicator–the lower this indicator, the less privacy the model leaks when facing the same attack. In Fig. 3, we present the experimental results using a heatmap and a scatter plot, respectively.

In Fig. 3(a) and (b), darker colors indicate better performance of the model in terms of usability or privacy. We can observe that a larger beta value can provide better privacy protection, but compromises model usability. The impact of the k value on model performance is influenced by *beta*: when beta is large, a larger k value helps improve the model's prediction accuracy; when beta is small, a smaller k value helps reduce the model's privacy leakage.

In Fig. 3(c), positions closer to the bottom of the figure indicate that the corresponding parameter combination has better privacy protection effects, while positions closer to the right side mean the corresponding parameter combination better ensures model usability. Different parameter combinations essentially represent a trade-off between model privacy and usability.

In Table 3, we present detailed experimental results for nine hyperparameter combinations, serving as a guideline for hyperparameter configuration in practice. We observe that $(k, \beta) = (10.0, 1.0)$ provide a good balance on CIFAR-10, achieving accuracy above 70% while keeping ASR below 50%. For practitioners who prioritize privacy, we recommend starting from a larger β and tuning k to recover as much accuracy as possible; for utility-oriented deployments, a smaller β combined with moderate k values often yields a favorable trade-off. Inappropriate (k, β) combinations can lead to an extremely low accuracy value. In our experiments, we find that this is because unreasonable hyperparameters combination will cause the model training to fail to converge.

In addition to hyperparameters k and β, we also investigated the impact of another hyperparameter γ. The experiments were conducted on two (k, β) combinations, with the results presented in Fig. 3(d).The impact of γ on model usability and attack defense effectiveness is subject to the specific (k, β) combination. However, overall, a larger γ tends to undermine the defense performance, while its impact on model accuracy varies with the values of k and β. We recommend using a small value of γ in practical applications. For all other experiments in this paper, a value of 0.1 was adopted for γ.

Based on 9 combinations of (k, β) and 3 values of γ, we conducted a Spearman correlation analysis to investigate the relationships between each hyperparameter and the model's accuracy as well as ASR across 27 possible hyperparameter combinations. The results show that only k is positively correlated with the model's accuracy at the 1% significance level, with a correlation coefficient of 0.8504. In contrast, neither β nor γ exhibits a significant correlation with the model's accuracy or ASR.

Table 3. ACC and ASR under Different k and β (γ=0.1)

k	β	ACC (%)↑	ASR (%)↓
10.0	1.0	70.34	49.89
10.0	5.0	68.60	51.74
10.0	10.0	68.80	50.59
5.0	1.0	68.68	50.36
5.0	5.0	69.00	50.25
5.0	10.0	66.00	50.55
1.0	1.0	62.64	49.77
1.0	5.0	1.18	50.68
1.0	10.0	6.18	49.27

We also conducted an in-depth investigation into the differences in training efficiency among different methods and the impact of batch size on this metric. Figure 4 presents the CUDA kernel execution time of different methods during training on a synthetic dataset of a small size with varying batch sizes, while synthetic auxiliary dataset also has a small size. In addition, all experiments were conducted under identical hardware conditions to ensure that the measured computation time could effectively reflect the differences in computational load across the methods. Regarding the forward propagation latency presented in Fig. 4(a), the AT method exhibits a longer latency at small batch sizes because it

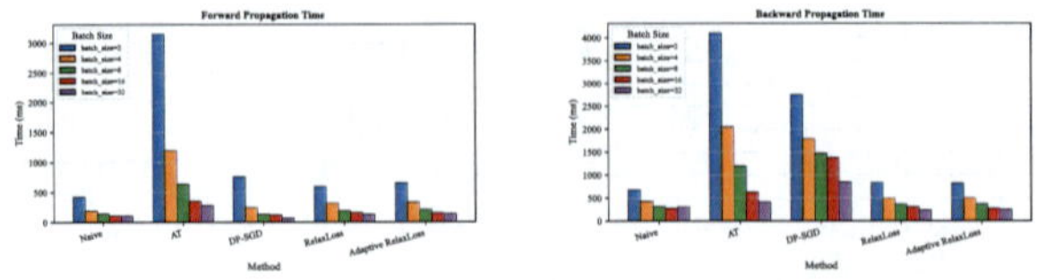

(a) (b)

Fig. 4. CUDA time analysis of backward pass and forward pass

necessitates forward computation over the entire auxiliary dataset for each individual batch, whereas the CUDA Time of the remaining methods are marginally different. For the backpropagation time illustrated in Fig. 4(b), the AT method incurs a longer runtime with small batch sizes, whereas its runtime becomes shorter than that of DP-SGD with large batch sizesas discussed in Sect. 3.3. In contrast, the runtime of the Adaptive RelaxLoss method is significantly shorter than that of DP-SGD and is comparable to that of RelaxLoss, which demonstrates that our method achieves a low time complexity.

5 Conclusion and Future Work

In this paper, we propose Adaptive RelaxLoss, an adaptive threshold-based method designed to mitigate the privacy leakage risk of deep learning classification models while efficiently saving computational resources. We first introduce a simple adversarial model-based training approach that achieves effective privacy protection without compromising model usability, but it incurs substantial additional computational overhead, as verified by experimental analysis. To address this issue, Adaptive RelaxLoss calculates an adaptive threshold via an adversarial model, eliminating the computational cost associated with extra gradient computations. Detailed empirical evaluations demonstrate that our method achieves a favorable balance between model usability and privacy, while maintaining low computational resource consumption.

Despite its strong performance in privacy protection with low computational overhead, Adaptive RelaxLoss still has limitations. In the ablation study, we analyzed the qualitative impacts of adaptive threshold-related hyperparameters on model privacy and usability; future research will focus on quantitative characterization of the relationship between these hyperparameters and privacy. We validated the generalization capability of Adaptive RelaxLoss across diverse model architectures and data modalities through modified prediction entropy-based membership inference attack experiments [32] on the Purchase100 dataset [2] with an MLP model, where it reduced the attack success rate (ASR) from 58.1% to 53.8%. However, loss formulation becomes significantly more complex for advanced natural language processing (NLP) tasks like large language models (LLMs). Tailoring our method to such complex scenarios remains a key direction for future work.

Acknowledgment. This work was supported in part by the National Natural Science Foundation of China under Grant No. 62573122, Jiangsu Province Natural Science Foundation under Grant BK20231112, the Fundamental Research Funds for the Central Universities under Grant No. 2242025K30025, the Open Research Project of State Key Laboratory of Industrial Control Technology, China under Grant No. ICT2025B37, the Open Research Project of the Key Laboratory of Numerical Simulation for Large Scale Complex Systems, Ministry of Education, China, under Grant No. NSLSCS202502, the Jiangsu Provincial Scientific Research Center of Applied Mathematics under Grant No. BK20233002, and the Guangxi Natural Science Foundation under Grant No. 2025GXNSFDA069040.

References

1. Carlini, N., Tramèr, F., Wallace, E.: Extracting training data from large language models. In: 30th USENIX Security Symposium (USENIX Security 21), pp. 2633–2650. USENIX Association (2021)
2. Shokri, R., Stronati, M., Song, C., et al.: Membership inference attacks against machine learning models. In: 2017 IEEE Symposium on Security and Privacy (SP). IEEE, San Jose, CA, USA, pp. 3–18 (2017)
3. Park, Y., Kang, M.: Membership inference attacks against object detection models (2020). arXiv:2001.04011 arXiv preprint
4. Hui, B., Yang, Y., Yuan, H., et al.: Practical blind membership inference attack via differential comparisons. In: Proceedings 2021 Network and Distributed System Security Symposium. Internet Society, Virtual (2021)
5. Yuan, X., Zhang, L.: Membership inference attacks and defenses in neural network pruning. In: 31st USENIX Security Symposium (USENIX Security 22), pp. 4561–4578. USENIX Association, Boston, MA (2022)
6. Liu, Y., Zhao, Z., Backes, M.: Membership inference attacks by exploiting loss trajectory. Proceedings of the 2022 ACM SIGSAC Conference on Computer and Communications Security, pp. 2085–2098. ACM, Los Angeles, CA, USA (2022)
7. Carlini, N., Chien, S., Nasr, M., et al.: Membership inference attacks from first principles. In: 43rd IEEE Symposium on Security and Privacy (SP), pp. 1897–1914. IEEE, San Francisco, CA, USA (2022)
8. Liu, L., Wang, Y., Liu, G., et al.: Membership inference attacks against machine learning models via prediction sensitivity. IEEE Trans. Depend. Secure Comput. **20**(3), 2341–2347 (2023)
9. Zhang, G., Liu, B., Zhu, T., et al.: Label-only membership inference attacks and defenses in semantic segmentation models. IEEE Trans. Depend. Secure Comput. **20**(2), 1435–1449 (2023)
10. Chen, S., Wang, W., Zhong, Y., et al.: HP-MIA: a novel membership inference attack scheme for high membership prediction precision. Comput. Secur. **136**, 103571 (2024)
11. Zhu, J., Zha, J., Li, D.: A unified membership inference method for visual self-supervised encoder via part-aware capability. In: Proceedings of the 2024 ACM SIGSAC Conference on Computer and Communications Security, pp. 1241–1255. ACM, New York, NY, USA (2024)
12. Dwork, C.: Differential pivacy. In: Bugliesi, M., Preneel, B., Sassone, V. (eds.) Automata, Languages and Programming, pp. 1–12. Springer, Berlin, Heidelberg (2006)

13. Dwork, C., McSherry, F., Nissim, K., et al.: Calibrating noise to sensitivity in private data analysis. In: Halevi, S., Rabin, T. (eds.) Theory of Cryptography, pp. 265–284. Springer, Berlin, Heidelberg (2006)
14. Abadi, M., Chu, A., Goodfellow, I.: Deep learning with differential privacy. Proceedings of the 2016 ACM SIGSAC Conference on Computer and Communications Security, pp. 308–318. (2016)
15. Rezaei, S., Shafiq, Z., Liu, X.: Accuracy-privacy trade-off in deep ensemble: a membership inference perspective. In: 44th IEEE Symposium on Security and Privacy, pp. 364–381. IEEE, San Francisco, CA, USA (2023)
16. Tang, X., Mahloujifar, S., Song, L.: Mitigating membership inference attacks by self-distillation through a novel ensemble architecture. In: 31st USENIX Security Symposium (USENIX Security 22), pp. 1433–1450. USENIX Association, Boston, MA (2022)
17. Meeus, M., Guépin, F., Cretu, A.M., et al.: Achilles' heels: vulnerable record identification in synthetic data publishing. In: Computer Security —ESORICS 2023, vol. 14345, pp. 380–399. Springer, Hague, Netherlands (2023)
18. Yang, Y., Yuan, H., Hui, B.: Fortifying federated learning against membership inference attacks via client-level input perturbation. In: 53rd Annual IEEE/IFIP International Conference on Dependable Systems and Network, pp. 288–301. IEEE, Porto, Portugal (2023)
19. Chen, Z., Pattabiraman, K.: Overconfidence is a dangerous thing: mitigating membership inference attacks by enforcing less confident prediction. In: Proceedings 2024 Network and Distributed System Security Symposium. Internet Society, San Diego, CA, USA (2024)
20. Hu, L., Li, J., Lin, G., et al.: Defending against membership inference attacks with high utility by GAN. IEEE Trans. Depend. Secure Comput. **20**(3), 2144–2157 (2023)
21. Li, J., Li, N., Ribeiro, B.: MIST: defending against membership inference attacks through membership-invariant subspace training. In: Balzarotti, D., Xu, W. (eds.) 33rd USENIX Security Symposium. USENIX Association, Philadelphia, PA, USA (2024)
22. Li, Q., Zhang, Y., Ren, J.: You can use but cannot recognize: preserving visual privacy in deep neural networks. In: Proceedings 2024 Network and Distributed System Security Symposium. Internet Society, San Diego, CA, USA (2024)
23. Chen, D., Yu, N., Fritz, M.: RelaxLoss: defending membership inference attacks without losing utility. In: International Conference on Learning Representations (ICLR) (2022). https://arxiv.org/abs/2207.05801
24. Yang, B., Yang, H., Lu, R., et al.: LossControl: defending membership inference attacks by controlling the loss. In: ICASSP 2025—2025 IEEE International Conference on Acoustics, Speech and Signal Processing (ICASSP), pp. 1–5. IEEE, Hyderabad, India (2025)
25. Luo, Z., Xu, X., Liu, F.: Privacy-preserving low-rank adaptation against membership inference attacks for latent diffusion models. In: Proceedings of the Thirty-Ninth AAAI Conference on Artificial Intelligence and Thirty-Seventh Conference on Innovative Applications of Artificial Intelligence and Fifteenth Symposium on Educational Advances in Artificial Intelligence (AAAI'25), pp. 5883–5891. AAAI Press (2025)
26. Krizhevsky, A.: Learning multiple layers of features from tiny images. Technical Report (2009)

27. Coates, A., Ng, A., Lee, H.: An analysis of single-layer networks in unsupervised feature learning. In: Gordon, G., Dunson, D., Dudík, M. (eds.) Proceedings of the Fourteenth International Conference on Artificial Intelligence and Statistics, vol. 15, pp. 215–223. PMLR, Fort Lauderdale, FL, USA (2011)
28. Darlow, L.N., Crowley, E.J., Antoniou, A.: CINIC-10 is not ImageNet or CIFAR-10 (2018). arXiv:1810.03505
29. Salem, A., Zhang, Y., Humbert, M.: ML-leaks: model and data independent membership inference attacks and defenses on machine learning models. In: Proceedings 2019 Network and Distributed System Security Symposium. Internet Society, San Diego, CA (2019)
30. Yousefpour, A., Shilov, I., Sablayrolles, A.: Opacus: user-friendly differential privacy library in PyTorch (2022). arXiv:2109.12298
31. Choquette-Choo, C.A., Tramer, F., Carlini, N.: Label-only membership inference attacks. In: Meila, M., Zhang, T. (eds.) Proceedings of the 38th International Conference on Machine Learning (ICML). Proceedings of Machine Learning Research, vol. 139, pp. 1964–1974. PMLR (2021)
32. Song, L., Mittal, P.: Systematic evaluation of privacy risks of machine learning models. In: 30th USENIX Security Symposium (USENIX Security 21), pp. 2615–2632. USENIX Association (2021)

A Deep Autoencoder Intrusion Detection Model Based on Contrastive Reconstruction Mechanism

Xinchao Song, Xingquan Gu, Jiale Ding, and Jiao Wu(✉)

School of Science, China Jiliang University, Hangzhou 310018, China
wuujiao@163.com

Abstract. With the increasing sophistication of cyberattacks, traditional intrusion detection systems suffer from low detection accuracy and high false alarm rates when handling high-dimensional, non-linear traffic data and unknown threats. This paper proposes a dual-path deep learning intrusion detection model based on a contrastive reconstruction mechanism, which integrates supervised and unsupervised learning to enhance the detection capability for both known and unknown attacks. The model employs a Random Forest-based ensemble classifier for accurate classification of known attacks and constructs a dual-autoencoder pathway for contrastive learning between normal and anomalous patterns. Anomaly detection for unlabeled data is achieved by comparing reconstruction similarities. Furthermore, we design a similarity reliability assessment and an intelligent fallback mechanism to enhance the model's robustness for edge cases and zero-day attack scenarios. Experimental results on the CICIDS-2017 dataset demonstrate that the proposed model significantly outperforms traditional single-autoencoder and supervised baseline methods. This study provides a viable solution for building adaptive, high-precision intelligent intrusion detection systems.

Keywords: Intrusion Detection · Deep Learning · Autoencoder · Contrastive Learning · Zero-Day Attacks · Imbalanced Data

1 Introduction

In the Internet era, with the acceleration of digitization, cyberspace has become a critical infrastructure supporting socio-economic development. However, this has been accompanied by the increasing complexity, specialization, and scale of cyber-attack methods. These security threats not only lead to sensitive data breaches and significant economic losses but also pose severe challenges to national security and social stability.

Traditional network intrusion detection systems (IDS) predominantly rely on rule-based matching or conventional machine learning methods. While these approaches demonstrate certain effectiveness in detecting known attack patterns, their limitations are increasingly evident. Traditional detection methods lack sufficient capability in extracting features from high-dimensional, non-linear network traffic data, resulting in low

L. Huang and D. Greenhalgh (Eds.): ICMLC 2026, LNNS 2061, pp. 656–671, 2027.
https://doi.org/10.1007/978-3-032-31236-5_45

detection accuracy and high false positive rates. Furthermore, with the continuous development of the Internet, there will be an increasing number of unknown attacks—zero-day attacks [1]—that do not appear in the training data, for which traditional detection capabilities are particularly weak.

In recent years, the continuous advancement of deep learning technology, leveraging its powerful capabilities for automatic feature learning and complex pattern recognition, has brought revolutionary breakthroughs to the field of intrusion detection. Deep neural networks can automatically extract deep, abstract feature representations from massive amounts of raw traffic data, effectively avoiding tedious and subjective manual feature engineering, and demonstrating significant advantages in processing high-dimensional, non-linear data. Therefore, researching intelligent intrusion detection technologies based on deep learning holds crucial theoretical value and practical significance for constructing proactive, precise, and adaptive cybersecurity protection systems.

This paper innovatively proposes a dual-path model that combines supervised and unsupervised learning. The model adopts a hierarchical progressive detection framework. In the first stage, an ensemble classifier based on Random Forest performs preliminary binary classification on network traffic, separating high-confidence normal and abnormal samples. In the second stage, a dual-autoencoder contrastive learning mechanism is employed. By extracting representative features from both the identified normal and abnormal traffic obtained in the first path, a normal traffic autoencoder and an abnormal traffic autoencoder are trained respectively to serve as feature references. Subsequently, the unlabeled data requiring detection is reconstructed and compared with these normal and abnormal feature references; the data sample is then assigned to the category with higher similarity. The complementary strengths of the two paths aim to achieve the dual objectives of high-accuracy classification of known attacks (serving as references) and effective discovery of unknown or unlabeled attacks. Simultaneously, this method also plays a significant role in addressing the issue of data imbalance.

2 Overview of Autoencoders

As a classic unsupervised deep learning model, the core design philosophy of the autoencoder lies in learning an effective low-dimensional representation, i.e., an encoding, of high-dimensional input data through a data-driven reconstruction process. This model was proposed by Rumelhart et al. in 1986, and its fundamental structure consists of two symmetrical neural network modules: the encoder and the decoder [2].

The encoder maps high-dimensional input data into a low-dimensional latent representation within a latent space via a feedforward neural network. The purpose of the bottleneck layer or encoding is to capture the most discriminative and robust core features of the input data.

The decoder acts as the inverse process of the encoder. Its task is to start from the encoding z and attempt to reconstruct the original input data $\hat{x}$ through a symmetrical upscaling network. The decoder's objective is to make the reconstructed output $\hat{x}$ as close as possible to the original input x.

The overall training objective of the model is to minimize the reconstruction error between the original input and the reconstructed output. Its loss function is typically defined as a measure of the difference between the two, such as mean squared error or cross-entropy:

$$L_{AE}(\phi, \theta) = \frac{1}{N}\sum_{i=1}^{N} L\left(x^{(i)}, g_{\theta}\left(f_{\phi}\left(x^{(i)}\right)\right)\right) \quad (1)$$

Here, L represents the specific difference function. Through this minimization process, the encoder is forced to learn how to extract the most critical information from the data—information sufficient to support high-quality reconstruction—thereby automatically achieving feature compression and denoising [3].

3 Intrusion Detection Model Design

3.1 Intrusion Detection Model

With the continuous evolution of cyber attack techniques, traditional rule-based and signature-based intrusion detection systems struggle to cope with increasingly sophisticated network threats. Anomaly detection methods based on machine learning have garnered widespread attention due to their capability to identify unknown attack patterns. However, existing single-classifier approaches often encounter challenges such as difficulty in feature extraction and high false positive rates when processing high-dimensional, non-linear network traffic data [4].

This paper proposes a deep autoencoder-based intrusion detection model founded on a contrastive reconstruction mechanism. Utilizing a dual-autoencoder architecture and an adaptive strategy selection mechanism, the model significantly improves the accuracy and robustness of anomaly detection. Its core innovation lies in overcoming the limitations of traditional single autoencoders by incorporating a contrastive learning mechanism between normal and abnormal patterns.

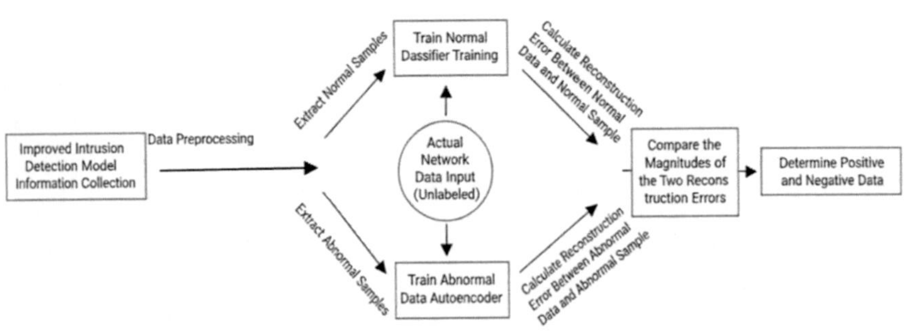

Fig. 1. Network intrusion detection process

The model adopts a hierarchical design philosophy, with the overall architecture depicted in Fig. 1, encompassing four main stages: data preprocessing, preliminary classification, autoencoder training, and anomaly detection.

Generally, traditional single-autoencoder anomaly detection methods employ an unsupervised approach, learning only the manifold of normal data and determining anomalies based on a fixed threshold [5]. The reconstruction error output by the autoencoder model for a single normal sample cannot directly yield a predicted label result. Due to the unique characteristics of network datasets, reconstruction errors can easily become very large, leading to inaccurate judgments. Additionally, the reconstruction model of a single autoencoder may also be overly sensitive to edge samples of the normal distribution, resulting in a high false positive rate [6].

Core Algorithm Design and Innovation. The core innovation of this model lies in the adoption of a dual-autoencoder contrastive learning architecture. Two autoencoders with identical structures but functionally complementary purposes are employed to learn the deep feature representations of normal traffic and anomalous traffic, respectively. This design enables the model to identify anomalous patterns from a comparative perspective, significantly enhancing detection accuracy and robustness.

Normal Autoencoder: Dedicated to learning the intrinsic patterns and distribution characteristics of normal network traffic, with the objective of minimizing the reconstruction error for normal samples.

Anomaly Autoencoder: Dedicated to learning the feature representations of known attack traffic, capturing the unique regularities of anomalous patterns.

Subsequently, representative features are extracted from both the identified normal traffic and anomalous traffic to serve as feature references. The unlabeled data requiring detection is then reconstructed separately by the normal autoencoder and the anomaly autoencoder. For each data instance, the reconstruction errors relative to the normal sample reference and the anomalous sample reference are compared. The instance is assigned to the category with higher similarity. This decision criterion no longer relies on a single reconstruction error but rather on the comparative similarity between the normal and anomalous reconstruction errors.

Both autoencoders employ the same network structure, consisting of three parts: an encoder, a latent space, and a decoder. The encoder progressively compresses the input features, the latent space captures the essential characteristics of the data, and the decoder attempts to reconstruct the original input from the latent representation.

3.2 Intrusion Detection Datasets

Compared to standard datasets in fields such as image and text processing, cybersecurity intrusion detection datasets exhibit characteristics of high dimensionality and feature heterogeneity. Each network connection or data flow is typically described by dozens to hundreds of features, which vary in type and collectively form a mixed-feature space. Additionally, these datasets often suffer from severe class imbalance issues [7]. In real-world network environments, normal traffic samples overwhelmingly dominate in quantity, while attack traffic—especially complex, low-frequency attack types—may account for less than 1% of the total samples.

Given that cybersecurity intrusion detection datasets possess these characteristics—high dimensionality, difficulty in interpretation, and extreme imbalance between positive and negative samples—and given that labeled data samples within such datasets are relatively scarce, the model proposed in this paper effectively leverages this minimal amount of labeled data. Innovatively, it constructs two autoencoders, one for positive (normal) and one for negative (anomalous) patterns, to extract features, aiming to mitigate the inaccuracies caused by data imbalance.

In the first stage of the model, i.e., the supervised path [8], a labeled dataset is used for training. This stage employs an ensemble learning classifier to perform an initial coarse classification of the input data into normal or anomalous categories, providing labeled references for the subsequent unsupervised learning phase.

In the second stage of the model, i.e., the unsupervised path, a contrastive reconstruction mechanism is utilized to train both a normal autoencoder and an anomaly autoencoder. Fine-grained anomaly detection is achieved by comparing reconstruction similarity. Since the first stage model, relying on accurate labeled data, has already formed precise feature representations for both normal and anomalous traffic, the second stage can distinguish between positive and negative samples for the incoming unlabeled training dataset based solely on data reconstruction error (referred to as reconstruction similarity in this model) without depending on labels.

The two paths collectively form a progressive detection workflow: coarse classification → sample separation → contrastive reconstruction → fine-grained detection.

3.3 Similarity Comparison Algorithm

For a test sample x_t, The reconstruction errors with respect to both autoencoders are calculated separately:

$$\mathrm{E_{normal}} = \|\mathrm{x_t} - \mathrm{AE_{normal}}(\mathrm{x_t})\|^2 \tag{2}$$

$$\mathrm{E_{anormaly}} = \left\|\mathrm{x_t} - \mathrm{AE_{anormaly}}(\mathrm{x_t})\right\|^2 \tag{3}$$

x_t: The preprocessed feature vector, transformed from the original input data, retains discriminative features while eliminating differences in scale, serving as the input baseline for the entire contrastive reconstruction process.

$AE_{normal}x_t$: The reconstruction output of the normal autoencoder;

$AE_{anormal}x_t$: The reconstruction output of the anomaly autoencoder.

The reconstruction error E is essentially the squared distance between the sample x_t and the data manifold learned by the autoencoder. If x_t conforms to the pattern learned by the autoencoder (e.g., normal traffic input to $AE_{normal}x_t$), it can be accurately reconstructed, and the error E approaches zero. Conversely, if the pattern does not match (e.g., anomalous traffic input to $AE_{anormal}x_t$), reconstruction will fail, and the error E will increase significantly. The use of squared Euclidean distance ensures the error term is differentiable, facilitating model training, and amplifies the contribution of samples with large errors [9].

Similarity score calculation based on reconstruction error:

$$\mathrm{S_{normal}} = \frac{1}{1 + \mathrm{E_{normal}}} \tag{4}$$

$$S_{anormaly} = \omega \bullet \frac{1}{1 + E_{anormaly}} \tag{5}$$

ω: Anomaly similarity weight factor, typically set between 1.05 and 1.20.

This weighting factor is used to adjust the model's sensitivity to anomalous signals, as normal samples usually dominate the training data, and normal patterns are generally more consistent and easier to learn.

S_{normal}:Represents the similarity between sample x_t and the normal pattern.

When $E_{normal} \to 0$, $S_{normal} \to 1$, indicating high similarity; When $E_{normal} \to \infty$, $S_{normal} \to 0$, indicating extremely low similarity. This function compresses the infinite range of errors into a fixed probabilistic interval, facilitating subsequent comparisons.

The final anomaly judgment is based on:

$$y_{pred} = \begin{cases} 1 & \text{if } S_{anormaly} > S_{normal} \\ 0 & \text{otherwise} \end{cases} \tag{6}$$

The decision relies not on the absolute magnitude of the error but on relative comparison.

This approach addresses a blind spot in traditional methods: some attack traffic may inherently have a small reconstruction error, but through comparison, it can be identified as closer to the anomalous pattern.

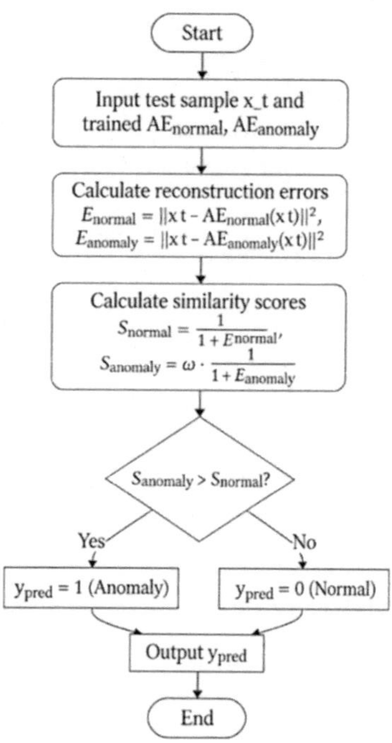

Fig. 2. Flowchart of the contrastive reconstruction algorithm

The overall workflow of the contrastive reconstruction detection algorithm is shown in Fig. 2.

3.4 Hybrid Enhancement Strategy Design

When the number of anomalous samples is limited, the model automatically activates a hybrid enhancement strategy. This strategy improves detection performance through three complementary mechanisms. First, a single autoencoder is trained using normal samples to establish a high-quality benchmark for the feature representation of normal traffic. Simultaneously, given the scarcity of anomalous samples, feature enhancement is applied to the limited anomalous data by employing an attention mechanism to focus on the critical features of anomalies. Finally, a comprehensive judgment is made by integrating multi-dimensional information, including reconstruction error, feature distribution discrepancies, and temporal anomalies.

The hybrid strategy does not rely solely on the quantity of anomalous samples but achieves robust detection through multi-perspective evidence fusion.

3.5 Similarity Reliability Assessment and Fallback Mechanism

Similarity Reliability Assessment Framework. In the contrastive reconstruction mechanism, the core decision-making basis lies in the similarity difference between the outputs of the two autoencoders. However, in certain edge cases, similarity comparisons may yield unreliable results. To address this, the model constructs a multi-dimensional reliability assessment framework to ensure the robustness of detection outcomes [10].

Confidence Calculation for Similarity Difference:

Define the similarity difference confidence for a test sample as:

$$\mathrm{Con}_{\Delta \mathrm{s}} = \frac{\left|\mathrm{E}_{\mathrm{anormaly}}\text{-}\mathrm{E}_{\mathrm{normal}}\right|}{\max\left(\mathrm{S}_{\mathrm{anormaly}}, \mathrm{S}_{\mathrm{normal}}\right)} \tag{7}$$

Meanwhile, the relative stability of the reconstruction error is taken into account.

$$\mathrm{Stability} = 1\text{-}\frac{\left|\mathrm{E}_{\mathrm{anormaly}}\text{-}\mathrm{E}_{\mathrm{normal}}\right|}{\mathrm{E}_{\mathrm{anormaly}} + \mathrm{E}_{\mathrm{normal}} + \in} \tag{8}$$

where ε is a small constant to prevent division by zero. Comprehensive Confidence Score:

$$\mathrm{Reliability} = \alpha \times \mathrm{Conf}_{\Delta \mathrm{s}} + \beta \times \mathrm{Stability} \tag{9}$$

Here, α and β are weighting coefficients, optimized and determined through validation set tuning.

In the similarity difference confidence, the numerator $|E_{anormaly} - E_{normal}|$ reflects the absolute difference in reconstruction errors between the two autoencoders; the denominator $max(S_{anormaly}, S_{normal})$ characterizes the maximum similarity between the current sample and a particular pattern class. Therefore, a larger $Conf_{\Delta s}$ indicates a more significant difference in the reconstruction capabilities of the two autoencoders for the sample, meaning the model's judgment regarding the sample's belonging is more definite and credible. Conversely, if this value is small, it suggests that the reconstruction errors of the two autoencoders are similar, making it difficult for the model to make a clear distinction, and thus the decision reliability is lower.

The relative stability metric ranges within (0,1). A value closer to 1 indicates that the two reconstruction errors are more similar, implying greater ambiguity in the model's classification decision and lower confidence in the judgment. Conversely, a value closer to 0 reflects a larger difference between the two reconstruction errors, signifying a more definitive model judgment and higher decision confidence. This metric is designed to identify uncertainty in the model's predictions, particularly for borderline samples. Traditional methods, which rely solely on a single reconstruction error or absolute differences for decision-making, are prone to misclassification when the errors are comparable. By introducing relative stability, the model can adaptively assess the reliability of its classification decisions [11].

With this design, the two sub-metrics within the reliability assessment framework form a perfectly complementary perspective. In the confidence formula, the weights α and β serve distinct purposes. α amplifies the accuracy gained from discriminative salience. When the two errors differ significantly, the model's judgment becomes more accurate, and α ensures that this accuracy is fully reflected in the overall score. In contrast, β penalizes ambiguous states. That is, when the two errors are very close, the Stability value becomes high, and β explicitly translates this uncertainty into a deduction from the total score.

Based on the composite confidence score, the similarity comparison results are categorized into three reliability levels:

High Reliability (Reliability > 0.7): The similarity difference is significant, and the contrastive learning result is directly adopted.

Medium Reliability (0.4 < Reliability < 0.7): The similarity difference is moderate, triggering a lightweight fallback verification.

Low Reliability (Reliability < 0.4): The similarity difference is not significant, triggering a comprehensive fallback mechanism.

This tiered strategy ensures detection accuracy while minimizing additional computational overhead.

Intelligent Fallback Mechanism Design. When the similarity comparison is deemed unreliable, the system automatically activates an intelligent fallback mechanism. This mechanism makes the final decision through a three-level evidence fusion process: Level 1 involves feature space analysis; Level 2 involves temporal behavior analysis; and Level 3 involves ensemble model decision-making.

In the latent space of the autoencoders, normal and anomalous samples typically form distinct distributions. The Mahalanobis distance, which accounts for the covariance relationships between features, provides a more accurate measure of the distance between a sample and a class center [12].

Let the mean of the normal samples in the latent space be μ_n , and the covariance matrix be $\sum_n^{-1}(z_t - \mu_n)$;

The corresponding values for the anomalous samples are μ_a and $\sum_a^{-1}(z_t - \mu_a)$

For the latent representation z_t, of a test sample, its Mahalanobis distances to the two classes are respectively:

$$D_{M\_normal}(z_t) = \sqrt{(z_t - \mu_n)^T \sum\nolimits_n^{-1} (z_t - \mu_n)} \quad (10)$$

$$D_{M\_anormaly}(z_t) = \sqrt{(z_t - \mu_a)^T \sum\nolimits_a^{-1} (z_t - \mu_a)} \quad (11)$$

The anomaly score based on the Mahalanobis distance is:

$$Score_{Mahalanobis} = \frac{D_{M\_normal}(z_t)}{D_{M\_anormaly}(z_t) + D_{M\_normal}(z_t)} \quad (12)$$

This score reflects which category the sample is closer to, with values near 1 indicating a stronger resemblance to anomalies.

When the model enters the fallback mechanism, it deals with borderline samples characterized by low reliability and ambiguous discrimination. In such cases, a distance metric capable of more accurately reflecting the statistical relationship between the sample and the class distribution is required. The Mahalanobis distance employed in the fallback mechanism essentially applies a whitening transformation to the feature space. This eliminates correlations among features and scales each dimension to have equal variance, thereby computing a statistically meaningful distance within this rotated and scaled space. In contrast, the Euclidean distance used in the model's earlier comparative reconstruction mechanism ignores feature correlations and cannot distinguish a sample's position relative to the data distribution; it merely reflects the absolute point-to-point distance. However, due to its simple analytical form and low computational complexity, it is more suitable for the high-frequency computations required in the earlier stages.

Temporal Behavior Consistency Check: Network attacks typically exhibit temporal continuity features. The detection results within the surrounding time window [t-w, t + w] are considered; a higher concentration of anomalies indicates a greater likelihood that the current sample is anomalous [13].

When the similarity score difference produced by the contrastive autoencoders is insufficient to provide a solid basis for classification decisions, the system automatically activates the final-level ensemble decision model as a fallback mechanism [14]. The system integrates multi-dimensional anomaly signals—such as feature space distribution, local density anomalies, and temporal behavior consistency—to construct a structured feature vector. Based on a gradient boosting decision tree framework, the model learns the discriminative weights of each signal under different scenarios, enabling adaptive decision fusion. This mechanism not only significantly improves the classification accuracy for borderline samples but also enhances the system's robustness against novel attacks and noisy environments. Furthermore, the model supports online incremental learning, allowing it to continuously adapt to concept drift in the network environment. By providing interpretable outputs to support security analysis, it facilitates an evolution from reliance on single models to an intelligent detection paradigm based on multi-evidence collaboration.

3.6 Key Parameters and Threshold Selection Strategy

The performance of a model depends not only on its architectural design but also on the reasonable configuration of key parameters and the scientific basis for threshold selection. This section elaborates on the design philosophy of the core parameters in the model, the mathematical foundation of the threshold strategy, and its physical significance in practical detection scenarios. Traditional anomaly detection methods using a single autoencoder often rely on fixed thresholds or simple statistical thresholds, which struggle to adapt to the high heterogeneity and dynamics of network security data. To address this, the proposed model incorporates a quadri-modal threshold strategy that adaptively selects thresholds based on the distribution characteristics of reconstruction errors from normal samples.

Z-score Strategy. This strategy is based on the assumption of a normal distribution. Assuming that the reconstruction error of normal samples, E_{normal}, approximately follows a normal distribution N (μ, σ^2), according to the 3σ rule of the normal distribution, approximately 99.73% of the data falls within the interval [$\mu - 3\sigma$, $\mu + 3\sigma$]. In this strategy, the threshold is defined as [15]:

$$T_{\text{zscore}} = \mu + K\sigma \tag{13}$$

Where K is the sensitivity factor (default value is 1.5). Its statistical significance is as follows: if the reconstruction error of a sample exceeds K times the standard deviation from the mean of normal samples, the probability that it belongs to the normal distribution is extremely low. Taking K = 1.5 as an example, the corresponding one-tailed confidence level is approximately 93.3%, meaning that the probability of normal samples being misclassified as anomalies can be controlled within 6.7%.

Percentile Strategy. To avoid strong assumptions about the overall distribution, this strategy employs a non-parametric method to set the threshold based on the empirical distribution function $F_{\text{n}}(x)$:

$$T_{\text{percentile}} = F_n^{-1}(Q) \tag{14}$$

Where F_n^{-1} is the empirical quantile function, and Q is the specified quantile level (default is 0.90). The frequentist interpretation of this strategy is that in hypothesis testing, by selecting an appropriate quantile, the Type I error (false positive rate) can be controlled at the level of $1 - Q$ [16].

In practical computation, let N be the number of normal samples in the training set, and their reconstruction errors be sorted in ascending order as $E_1 \leq E_2 \leq \ldots \leq E_n$. Taking $k = ceil(Q \cdot N)$, the threshold is then $T = E_k$. This means the system allows at most N-K normal samples to be misclassified. For instance, when $Q = 0.90$, the system is designed to allow up to 10% of normal traffic to be flagged as anomalous. This proportion can be flexibly adjusted according to specific security policies: in high-security environments, it can be increased to $Q = 0.95$ (corresponding to a 5% false positive rate), while in more lenient environments, it can be reduced to $Q = 0.85$ (corresponding to a 15% false positive rate).

IQR Strategy. This method draws on the principles of box plot outlier detection and demonstrates good robustness. It begins by calculating the quartiles of the reconstruction errors from the normal samples:

$$Q_1 = F_n^{-1}(0.25), Q_3 = F_n^{-1}(0.75), IQR = Q_3 - Q_1 \tag{15}$$

The threshold is subsequently defined as:

$$T_{IQR} = Q_3 + 1.5 \times IQR \tag{16}$$

The robustness of this method lies in the insensitivity of the IQR to extreme values, with a breakdown point reaching 25% (meaning that more than 25% of the data must be contaminated to arbitrarily alter the IQR value). In contrast, the standard deviation σ is sensitive to outliers, where a single extreme outlier can cause it to deviate significantly. Network security training data often contains labeling noise or boundary normal samples; the IQR strategy can effectively resist such contamination [17].

In terms of statistical properties, for normally distributed data, $Q_3 + 1.5 \times IQR \approx \mu + 2.698\sigma$, corresponding to a theoretical false positive rate of approximately 0.35%, which is stricter than the Z-score strategy with K = 1.5. For skewed distributions, this strategy automatically adapts, consistently maintaining coverage of the main body of the distribution.

GMM Strategy. This strategy serves as the default threshold selection method for the model. Its fundamental assumption is that the reconstruction errors of normal samples may originate from multiple sub-distributions, such as the dominant normal pattern (low error), the marginal normal pattern (medium error), and potential contamination from unlabeled anomalies (high error). A two-component Gaussian Mixture Model is employed to model the error distribution [18]:

$$\mathrm{p(E)} = \pi_1 \mathrm{N}\left(\mathrm{E}|\mu_1, \sigma_1^2\right) + \pi_2 \mathrm{N}\left(\mathrm{E}|\mu_2, \sigma_2^2\right) \tag{17}$$

Where $\pi_1 + \pi_2 = 1$, π_i represents the mixture weight.

4 Experiments and Result Analysis

4.1 Experimental Environment and Dataset

Experimental Environment Configuration. The Experimental Environment Configuration for this Study Is Detailed in Table 1 below.

Dataset Description. The experiments primarily use the CICIDS-2017 dataset. (Canadian Institute for Cybersecurity Intrusion Detection Systems Dataset 2017, 19). A subset of the dataset was selected and, after preprocessing, formed a mixed dataset containing normal traffic and multiple attack types. This processed dataset was used to train the dual-autoencoder intrusion detection model proposed in this paper.

The distribution of each category is illustrated in Table 2.

Table 1. Experimental environment configuration

Item	Configuration
Operating system	Windows 11
Programming language	Python 3.10.0
Framework	Pytorch2.7.1
CPU	Intel(R) Core(TM) i5-10210U
GPU	Mx250
Memory	16G

Table 2. Label distribution of the CICIDS2017 dataset

Label	Amount	Percentage (%)
Benign	310235	95.5
Fuzzers	5627	1.7
Exploits	4538	1.4
Reconnaissance	2489	0.8
Generic	680	0.2
DoS	676	0.2
Shellcode	315	0.1
Backdoor	64	0.04
Analysis	54	0.04
Worms	37	0.02

4.2 Evaluation Metrics and Baseline Models

Among the selected metrics, accuracy measures the overall correctness of data classification, recall focuses on the coverage of true positive attacks, precision emphasizes the accuracy of classifying true positive attacks, and the F1-score serves as a balanced measure of both recall and precision. In line with the iterative refinement process of the model proposed in this study, the experiments selected three comparative models: unsupervised single autoencoder learning, which directly identifies the underlying patterns of positive and negative samples for reconstruction; single autoencoder combined with supervised learning, where a classifier first separates normal samples, followed by autoencoder reconstruction based on normal features; and Variational Autoencoder (VAE). These three models exhibit high structural similarity to the proposed dual self-supervised learning model based on contrastive learning. Such comparisons help validate the performance advantages of the proposed model among similar approaches.

4.3 Model Training and Parameter Settings

The key parameters of the model are set as follows:

The hidden layer dimension of the autoencoder is 256; the batch size is 128; the number of training epochs is 100, with an early stopping patience value set to 15; and the anomaly similarity weight factor is set to 1.15.

4.4 Comparison of Model Training Performance

To comprehensively evaluate the training efficiency of the proposed model, we compare its performance with several baseline methods in terms of training time and memory usage. The results are presented in Table 3.

Table 3. Comparison of training performance for network security intrusion detection models

Model name	Total training time (min)	Model training (min)	CPU memory peak (mb)
The proposed model	4.38	3.28	322.3
Single Autoencoder (Supervised Training)	2.60	1.62	246.1
Single Autoencoder (Unsupervised Training)	2.01	1.12	202.6
LSTM	5.37	4.76	357.1
GAN-AD	6.00	5.39	412.3

The model proposed in this paper requires simultaneous training of both normal and anomalous autoencoders, along with the introduction of an adaptive hybrid decision strategy, leading to an increase in computational overhead. Consequently, compared to a single autoencoder, it incurs higher training time and CPU usage, though it remains significantly lower than complex models such as LSTM and GAN-AD. This difference primarily stems from variations in model structural complexity: a single autoencoder only needs to learn reconstruction features from normal samples, resulting in a simpler structure and lower computational cost. LSTM models, designed to capture temporal dependencies, involve recursive computations, leading to lower training efficiency. GAN-AD, due to its adversarial training mechanism requiring simultaneous optimization of both generator and discriminator, suffers from unstable training and high resource consumption. By employing dual-encoder contrastive learning, the proposed model enhances the ability to distinguish between normal and anomalous patterns, while avoiding the overhead associated with complex sequence modeling or adversarial training. Consequently, it achieves relatively reasonable training efficiency while maintaining high detection accuracy, demonstrating good applicability in practical deployment scenarios.

4.5 Comparative Experimental Results and Analysis

Comparative Experimental Results and Analysis. The experimental results of this study are presented in Table 4 below:

Table 4. Performance comparison of different models

Model name	Accuracy (%)	Recall (%)	F1-score (%)
Proposed model	0.87	1.00	0.93
Single autoencoder (Supervised Training)	0.53	0.96	0.68
Single autoencoder (Unsupervised Training)	0.23	1.00	0.37
LSTM	0.71	0.74	0.72
GAN-AD	0.63	0.69	0.66

Experimental results indicate that the proposed contrastive autoencoder model significantly outperforms baseline methods across all evaluation metrics. Notably, it achieves an anomaly recall rate of 1.00, demonstrating the model's effectiveness in identifying the vast majority of attack traffic. This adheres to the principle of prioritizing the detection of potential anomalies even at the cost of occasionally misclassifying normal traffic as anomalous, while maintaining a high precision of 0.87, keeping the false positive rate within an acceptable range.

As shown in Table 3, the contrastive self-supervised learning model exhibits superior performance on imbalanced datasets, particularly in metrics such as recall and the F1-score. Compared to traditional single autoencoder learning models, whether employing supervised or unsupervised training paradigms, the proposed model shows significant improvement.

Category-Specific Analysis. For normal samples (Category 0), a precision of 1.00 is achieved, meaning all samples predicted as normal are correct. Concurrently, the recall rate is 0.99, with only 1% of normal traffic being misclassified as anomalous. This performance indicates that the model is highly accurate in identifying normal traffic with an extremely low false positive rate. For anomalous samples (Category 1), the precision reaches 0.87, meaning 87% of samples labeled as anomalous are indeed attacks, and the recall rate is 0.98, successfully identifying 98% of true attack traffic. The F1-score of 0.96 reflects a good balance between precision and recall.

Confusion Matrix Analysis. The confusion matrix derived from the classification report is as follows [20]:

Actual\Predicted	Amount	Proportion
Normal	92506	565
Abnormal	0	4344

True Positives (TP): All 4,344 anomalies were correctly identified.

True Negatives (TN): 92,506 normal samples were correctly classified.

False Positives (FP): 565 normal samples were misclassified as anomalies, resulting in a false positive rate of 0.6%.

False Negatives (FN): No anomalies were missed, with a false negative rate of 0%.

4.6 Discussion and Limitations

The main advantages of the proposed model compared to traditional methods are reflected in the following aspects: the dual-autoencoder architecture enhances the discriminative capability for borderline samples through the contrast between normal and abnormal patterns. Furthermore, it automatically selects the optimal detection strategy based on the training data distribution, thereby improving model robustness.

However, the current model also has limitations, including a decrease in detection rate of approximately 15% for zero-day attacks (i.e., attack types not present in the training set), indicating challenges in generalization. Additionally, its performance is relatively sensitive to feature selection and preprocessing steps.

Future work will explore several directions, including integrating meta-learning frameworks to enhance zero-day attack detection and implementing automated feature engineering to reduce manual intervention. Moreover, a multi-classification model will be incorporated at the latter stage, enabling the model not only to distinguish between positive and negative samples but also to identify specific categories among negative samples.

5 Conclusion and Outlook

This paper addresses key challenges in current network intrusion detection, such as severe data imbalance, difficulty in identifying unknown attacks, and high false positive rates in traditional models, by proposing a dual-path contrastive reconstruction detection model that integrates supervised and unsupervised learning. By introducing an ensemble classifier path based on Random Forest and a dual-autoencoder contrastive learning path, the model achieves precise discrimination of known attacks and robust discovery of unknown attacks. Innovatively, the model incorporates a similarity reliability assessment system and a multi-level intelligent fallback mechanism, effectively enhancing adaptability to borderline samples and zero-day attack scenarios. Experiments on the CICIDS 2017 dataset demonstrate that the model significantly reduces the false positive rate while maintaining a high recall rate, validating its good detection stability and practicality in real-world network environments.

Future work could extend the model into a multi-classification detection framework and introduce online learning and dynamic update mechanisms, enabling the model to continuously adapt to evolving network environments and attack techniques. Additionally, research on model compression and edge computing deployment schemes could be pursued to enhance system real-time performance and scalability.

References

1. Asadujjaman, A.S.M., Falcão, E., Brito, A., Rojas, E.: Surviving zero-day attacks using spatial specialization. Comput. Secur. **160**, 104709 (2026)
2. Amarudin, Ferdiana, R., Widyawan: B-DT model: a derivative ensemble method to improve performance of intrusion detection system. J. Adv. Inf. Technol. **15**, 87–103 (2024)
3. Kaliraj, P., Subramani, B.: Intrusion detection using krill herd optimization based weighted extreme learning machine. J. Adv. Inf. Technol. **15**, 147–154 (2024)
4. Sohns, J., Schmitt, M., Jirasek, F., Hasse, H., Leitte, H.: Attribute-based explanation of non-linear embeddings of high-dimensional data. IEEE Trans. Vis. Comput. Graph. **28**, 540–550 (2022)
5. Miguel-Escrig, O., Romero-Pérez, J., Sánchez-Moreno, J., Dormido, S.: On-line retuning of PID controllers with fixed threshold samplers. ISA Trans. **139**, 253–262 (2023)
6. Li, Z., Song, X., Shi, C., Demachi, K.: Abnormal detection in nuclear security videos based on label-specific autoencoders and reconstruction errors comparison. Nucl. Eng. Technol. **57**, 103239 (2025)
7. Pes, B.: Learning from high-dimensional biomedical datasets: the issue of class imbalance. IEEE Access. **8**, 13527–13540 (2020)
8. Sabri, M., Verde, R., Balzanella, A.: FWLMkNN: efficient functional K-nearest neighbor based on clustering and functional data analysis. Expert Syst. Appl. **292**, 128567 (2025)
9. Rodríguez-Barrios, M.S., Ferré, J., Larrechi, M.S., Ruiz, E.: Applicability domain of a calibration model based on neural networks and infrared spectroscopy. Chemom. Intell. Lab. Syst. **254**, 105242 (2024)
10. Samavatian, V., Fotuhi-Firuzabad, M., Samavatian, M.: Correlation-driven machine learning for accelerated reliability assessment of solder joints in electronics. Sci. Rep. **10**, 14821 (2020)
11. Chatfield, M.D., Marquart-Wilson, L., Dobson, A.J., Farewell, D.M.: Mean relative error and standard relative deviation. Statistica Neerlandica. **79**(1), e70001 (2025)
12. Zeng, S., Wang, X., Duan, X., Zeng, S., Xiao, Z., Feng, D.: Kernelized Mahalanobis distance for fuzzy clustering. IEEE Trans. Fuzzy Syst. **29**, 3103–3117 (2021)
13. Landauer, M., Skopik, F., Stojanović, B.: A review of time-series analysis for cyber security analytics: from intrusion detection to attack prediction. Int. J. Inf. Secur. **24**, 1615–5270 (2025)
14. Gao, X., Shan, C., Hu, C., Niu, Z., Liu, Z.: An adaptive ensemble machine learning model for intrusion detection. IEEE Access. **7**, 82512–82521 (2019)
15. Luewan, S., Srisupundit, K., Tongprasert, F.: Z score reference ranges of Fetal cardiac output from 12 to 40 weeks of pregnancy. J. Ultrasound Med., 0278–4297 (2020)
16. Gao, X., Li, X., Li, B., Han, H., Gao, H., Qi, Y.: GELU-LSTM-encoder-decoder fault prediction for batch processes based on the global-local percentile method. Can. J. Chem. Eng. **6**, 102 (2024)
17. Alabrah, A.: An improved CCF detector to handle the problem of class imbalance with outlier normalization using IQR method. Sensors. **2023**, 4406 (2023)
18. Zhao, L., Zheng, S., Yang, W.: An image thresholding approach based on Gaussian mixture model. Pattern. Anal. Applic. **22**, 75–88 (2019)
19. Kurniabudi Stiawan, D., Darmawijoyo Idris, M.Y.B., Bamhdi, A.M., Budiarto, R.: CICIDS-2017 dataset feature analysis with information gain for anomaly detection. IEEE Access. **8**, 132911–132921 (2020)
20. Heydarian, M., Doyle, T.E., Samavi, R.: MLCM: Multi-label confusion matrix. IEEE Access. **10**, 19083–19095 (2022)

Slow-Rate Denial-of-Service Attack Detection in QUIC Protocol Using a Hybrid CNN-GRU Deep Learning Model

Yongjie Gao[1], Yang Cao[1,2](✉), Shijian Luo[1], and Jing Zhang[1](✉)

[1] School of Cyber Science and Engineering, Southeast University, Nanjing 211189, China
{220245600,caoyeacy,230250316,jingz}@seu.edu.cn

[2] School of Cyber Science and Engineering, Engineering Research Center of Blockchain Application, Supervision and Management of the Ministry of Education, Southeast University, Nanjing 211189, China

Abstract. The widespread adoption of the Quick UDP Internet Connections (QUIC) protocol, with its stream multiplexing and full encryption, introduces new challenges for detecting covert slow-rate Denial-of-Service (DoS) attacks. Traditional methods relying on thresholds or plaintext features are ineffective against low-rate attacks in the QUIC environment. This paper systematically analyzes QUIC's vulnerabilities in stream management and flow control and proposes a hybrid deep learning model based on Convolutional Neural Network-Gated Recurrent Unit (CNN-GRU). The model utilizes customized feature engineering to reconstruct packet-level data into flow-level temporal features, focusing on protocol-level statistical features derived from decrypted traces, while excluding raw payload bytes and application content. The CNN extracts local spatial feature patterns, while the Stacked GRU captures long-term temporal dependencies across timesteps, characterizing the sustained attack behavior. Experiments on a public QUIC dataset collected in a controlled environment show that the proposed CNN-GRU model outperforms traditional Random Forest and Two-layer LSTM models in Accuracy, F1-score, and AUC, validating its superior performance in identifying covert slow-rate attacks.

Keywords: QUIC · Slow-Rate Denial-of-Service Attack · CNN-GRU · Temporal Feature Extraction · Traffic Detection

1 Introduction

With the explosive growth of global Internet traffic and the proliferation of real-time applications such as high-definition streaming, cloud computing, IoT, and online gaming, the traditional TCP/IP transport stack is facing severe performance bottlenecks [1].

To address these limitations, Google introduced the QUIC protocol, subsequently standardized by the IETF as the transport foundation of HTTP/3. Built on UDP [1], QUIC integrates multiplexing, encryption, and low-latency connection establishment within the user space, reducing handshake latency to 1 RTT and enabling faster, more

L. Huang and D. Greenhalgh (Eds.): ICMLC 2026, LNNS 2061, pp. 672–686, 2027.
https://doi.org/10.1007/978-3-032-31236-5_46

secure data transmission. Figure 1 visually compares the two core differences between the traditional HTTP/2 (TCP/TLS) stack and the emerging HTTP-over-QUIC stack [2].

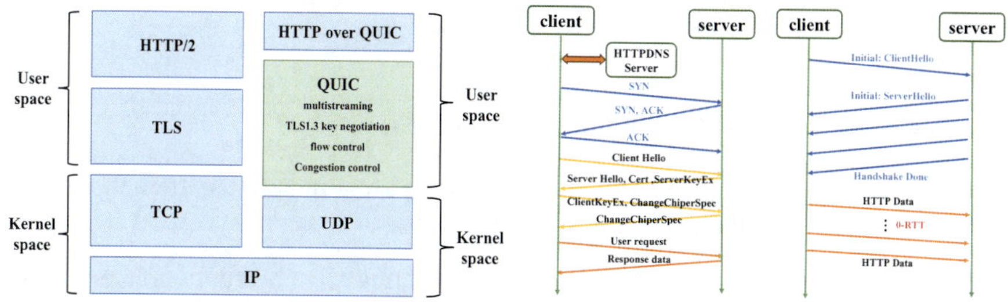

Fig. 1. Comparison of HTTP/2 and HTTP over QUIC (HTTP/3) architecture and connection setup

QUIC introduces several key innovations, including a TLS 1.3-based 0-RTT handshake, stream-level concurrency that eliminates HoL blocking, user-space congestion control for flexible optimization. However, its complexity also introduces new security challenges [3]. Despite built-in encryption and authentication, features such as stream multiplexing and 0-RTT data transmission have not undergone extensive long-term security validation [4].

Certain QUIC mechanisms, including stream-level flow control and connection-level multiplexing, may unintentionally expand the attack surface for such threats [5]. Attackers can exploit these features to sustain numerous low-rate streams within a single connection, consuming server resources and degrading performance. These characteristics underscore the need for robust detection methods tailored to QUIC [6].

Motivated by this gap, this paper systematically investigates QUIC's vulnerability to slow-rate attacks and proposes a deep learning–based detection model capable of distinguishing benign from malicious low-rate traffic.

The main contributions are summarized as follows:

(i) *Systematic Analysis:* We examine how QUIC's stream and connection mechanisms can be exploited in slow-rate scenarios.
(ii) *Hybrid Model Design:* We propose a hybrid CNN–GRU detection model for accurate identification of QUIC slow-rate attacks.
(iii) *Experimental Validation:* We evaluate the model on real QUIC traffic datasets, demonstrating high detection accuracy and strong generalization ability.

The remainder of this paper is organized as follows: Sect. 2 reviews related work on QUIC security and slow-rate attack detection. Section 3 presents our methodology, feature engineering, and the proposed model design. Section 4 details the experimental setup and discusses the results, and Sect. 5 concludes this paper with future directions.

2 Related Work

This section provides a systematic review of research on QUIC protocol security, slow-rate DoS attacks, and related detection approaches, aiming to establish a theoretical foundation for this study and identify existing research gaps.

2.1 Slow Attack Models Under Traditional Protocols

Slow-rate DoS attacks occupy server resources through legitimate but extremely low-rate communications [7]. Typical examples include:

(i) *Slowloris attack:* The attacker keeps the connection alive by continuously sending incomplete HTTP request headers, thereby blocking the server's thread pool [8].
(ii) *R.U.D.Y. (R U Dead Yet) attack:* The attacker declares an excessively large Content-Length value and then transmits the request body at a very low rate, forcing the server to maintain the connection for a long time and consuming sockets and memory resources.
(iii) *Slow POST/Slow Read attacks:* The client reads the server's response at a very slow rate or with a very small receive window, causing the server to keep data in its send buffer, occupying output queues or threads. These attacks typically manipulate the TCP window size at the transport layer or send delayed ACKs at the application layer.
(iv) *Long polling/slow WebSocket abuse:* Attackers exploit long-lived mechanisms such as long polling or WebSocket by sending minimal heartbeats or data while persistently occupying connections, or by frequently initiating but not releasing long-polling requests.

These attacks take advantage of vulnerabilities in server timeout mechanisms when waiting for complete requests. Since the traffic characteristics closely resemble those of legitimate slow clients, traditional threshold-based detection methods are rendered ineffective [9, 10].

2.2 Current Detection Methodologies for Slow-Rate Attacks

Given the stealthiness of slow-rate DoS attacks, existing detection methods primarily fall into machine learning and deep learning categories. Traditional machine learning approaches (e.g., RF, SVM) rely on statistical features such as connection duration and data rate. While efficient and interpretable, these methods are often restricted to known protocol scenarios [11, 12].

As encrypted traffic grows, deep learning models (e.g., LSTM, CNN) have become mainstream for their ability to automatically capture spatiotemporal patterns. However, since QUIC's full encryption renders plaintext-based detection ineffective, recent research increasingly relies on protocol-level statistical behavior rather than plaintext payload content; in decryptable settings, such behavior can be extracted from parsed QUIC/TLS/HTTP/3 traces without inspecting application content [13].

2.3 Research Gaps and Contributions

Existing studies primarily focus on high-traffic QUIC flooding or traditional TCP/HTTP slow-rate attacks. While valuable, they often fail to address the stealthy behaviors specific to QUIC's multiplexed and encrypted architecture. Crucially, most methods rely on plaintext features rendered inaccessible by QUIC's encryption, limiting their applicability. Furthermore, the potential exploitation of QUIC's unique mechanisms, such as stream multiplexing and flow control, remains underexplored.

To bridge these gaps, this study proposes:

(i) A QUIC slow-rate attack detection framework with a feature extraction strategy based on protocol-level statistical attributes derived from decrypted QUIC, TLS, HTTP/3 traces, while excluding raw payload bytes and application semantic content.
(ii) A hybrid CNN–GRU model designed to capture both spatial and temporal correlations in QUIC traffic for efficient detection of stealthy low-rate attacks;
(iii) Comprehensive evaluation on real QUIC traffic datasets, demonstrating superior accuracy and lower false-positive rates compared to existing approaches.

3 Methodology

3.1 Threat Model

This study is predicated on a clear threat model, defining the assets under protection, the attacker's objectives, and the observability boundaries for the detection system.

(i) *Attack Objectives*

The primary objective of the attack is to induce Resource Exhaustion on the QUIC server, leading to a DoS condition for legitimate users [14]. Specifically, the attacker aims to exhaust the server's connection state handles, thread pool resources, or memory buffers by maintaining numerous low-rate streams over an extended period.

(ii) *Observability and Assumptions*

The detection system is assumed to operate in a decryptable setting, such as an endpoint-based IDS, a server-side monitor, or an enterprise environment with TLS inspection. Therefore, TLS session keys (and QUIC qlog traces when available) can be obtained, enabling protocol dissection of QUIC/TLS/HTTP/3 via standard tools (e.g., Wireshark/tshark).

Importantly, the proposed method does not use raw payload bytes or application content; instead, it relies on protocol-level statistical attributes (e.g., timing, length, and protocol state) derived from the parsed traces.

(iii) *Adversary Capabilities (Controllable Fields)*

The adversary possesses the ability to control and manipulate the following critical QUIC/HTTP/3 parameters:

QUIC Protocol Control: Stream/Connection Flow Control, Data Pacing, and ACK Delay.

HTTP/3 Control: The ability to concurrently create and maintain a large number of HTTP/3 streams within a single QUIC connection.

(iv) *Attack Mechanism: Pulsed Slow-Rate Multi-Stream Consumption*

The adversary executes a Pulsed Slow-Rate Multi-Stream Consumption mechanism. The attack cadence is characterized by:

Periodic Bursts: The attacker launches a large volume of low-rate clients concurrently within a very short window (e.g., 5 s) at fixed intervals (e.g., every 5 s).

Resource Suspension: Each client sends minimal data at a low rate, leveraging protocol mechanisms to keep streams "suspended" indefinitely, preventing timeout closure and resulting in the intermittent and cumulative consumption of server resources.

3.2 Dataset Preprocessing

Deficiencies of the Original Dataset. Although the dataset [15] provides extensive HTTP/3 traffic records, it presents significant limitations for this study. First, many of its 199 features pertain to legacy protocols (e.g., TCP), introducing noise and computational redundancy irrelevant to encrypted QUIC traffic. Second, the dataset primarily targets flooding attacks, lacking the temporal aggregation and specific labels necessary to identify slow-rate behaviors. To address these deficiencies, we developed a customized feature engineering pipeline to transform packet-level data into flow-level temporal representations, enabling effective detection of slow-rate attacks in QUIC environments.

Data Cleaning and Preliminary Filtering. Given that the original dataset exhibits inherent issues, this study first performs a rigorous data cleaning and preprocessing procedure to ensure the effectiveness of subsequent feature engineering.

Elimination of Protocol Misalignment and Feature Redundancy. To ensure data quality prior to feature engineering, we applied a semantic-based dimension reduction strategy. Irrelevant TCP and HTTP/2 fields were removed, as they constitute noise in UDP-based QUIC traffic. Additionally, link-layer features reflecting only local topology were eliminated to decouple the model from specific network environments and enhance its generalization capabilities.

Missing Value Imputation Strategy. A large number of missing values in the dataset are caused by the deep packet parser's inability to identify specific protocol fields in unrelated packets. To address this, we designed differentiated imputation strategies based on the type and physical meaning of each feature, rather than applying a single uniform method blindly.

Data Scaling and Normalization. After completing the missing value imputation, all numerical features were subjected to scaling and standardization to enhance model training stability and convergence speed. This process involves two main steps: outlier capping and feature normalization.

(i) *Outlier Processing*

For features with extreme values, the Interquartile Range (IQR) method was applied to constrain their range:

$$\left[Q_1 - 1.5 \times IQR, Q_3 + 1.5 \times IQR\right] \tag{1}$$

where Q_1 and Q_3 denote the first and third quartiles, respectively.

(ii) *Feature Scaling: Min-Max Normalization*

All numerical features were linearly scaled using Min-Max normalization to map them into the range [0, 1]:

$$X' = \frac{X - X_{min}}{X_{max} - X_{min}} \tag{2}$$

Compared with Z-Score standardization, Min-Max normalization is more suitable for time-dependent models for two main reasons:

Activation function compatibility: RNN-based networks often employ Sigmoid or Tanh activations. Normalized inputs prevent activation saturation, improving gradient stability and convergence speed.

Temporal structure preservation: This method maintains the relative variation of features, helping the model capture smooth temporal evolution and sudden changes.

3.3 Feature Engineering Design for Slow-Rate Attacks

QUIC Flow Identification and Reconstruction. To robustly aggregate packets into connection-level flows in our dataset setting, this study proposes a specialized three-tuple strategy to ensure the uniqueness and integrity of flow aggregation.

Flow Identifier Definition. The flow identifier (FlowID) is constructed from three core fields that remain stable throughout the lifetime of a QUIC connection:

$$\text{FlowID} = (\text{quic.scid}, \text{udp.srcport}, \text{udp.dstport}) \tag{3}$$

The quic.scid denotes the Source Connection ID, a unique identifier assigned by the server to distinguish different client connections, while udp.srcport and udp.dstport represent the UDP source and destination ports, respectively, which are combined with the Connection ID to uniquely identify the flow in this study.

Flow Aggregation Process.

(i) *Data Preprocessing:* The three-tuple fields are converted into string format, missing values are filled with "unknown," and the fields are concatenated to form a unique Flow_ID for each flow.
(ii) *Redundant Feature Removal:* To avoid protocol mismatch and data redundancy, 32 irrelevant features are removed, including:

Link/IP layer: eth.dst, eth.src, ip.hdr_len, ip.src, ip.dst, etc.
TCP/TLS layer: tcp.payload, tls.handshake.client_cert_vrfy.sig_len, etc.
HTTP/2 layer: http2.streamid, http2.headers.status, etc.

Temporal Segmentation Strategy for QUIC Flows. After flow aggregation, an equal-count temporal segmentation strategy is adopted to capture behavioral differences across different stages of the flow lifecycle. This approach ensures consistency in the temporal granularity of feature extraction.

Segmentation Configuration. Each flow is divided into 10 equal segments (T = 10) based on the packet count within the flow, where packet order is represented using a cumulative time_index. For small flows containing fewer than 10 packets, all packets are assigned to the first segment to avoid empty feature values. If missing segments occur due to uneven packet distribution, their statistical feature values are forward-filled from the previous segment. Finally, the flow-level label is assigned using the mode of the packet labels within the flow.

Segmentation Principle. Each flow is partitioned into 10 segments based on the number of packets, ensuring that each segment contains a similar number of packets. This method mitigates bias introduced by variable flow durations and guarantees temporal consistency across different flows. The segmentation logic is defined as:

$$\text{Segment}_t = \{\, \text{Packets} \mid q_{t-1} \leq \text{time\_index} < q_t \}, t \in \{1, \ldots, 10\} \quad (4)$$

where $t \in \{1, 2, \ldots, 10\}$ denotes the segment index, and q_t represents the quantile of the flow'*s time_index,* computed via quantile-based partitioning.

3.4 Construction of Temporal Feature Set

This study designs a three-layer feature construction framework, comprising raw feature processing, basic feature extraction, and statistical feature computation, to comprehensively capture the low-rate and high-latency characteristics of slow-rate attacks.

Raw Feature Processing. Two key techniques are employed to address the challenges posed by high-cardinality and categorical features:

Length Encoding: The lengths of string values are computed to transform six high-cardinality string features (e.g., udp.payload, quic.stream_data) into numerical features, denoted as [feature]_len (e.g., udp.payload_len).

Frequency Encoding: The occurrence frequency of each category is used to convert low-cardinality categorical features (e.g., quic.frame_type, tls.record.content_type) into numerical values, denoted as [feature]_freq (e.g., quic.frame_type_freq).

Base Feature Set. After raw processing, 96 basic features are extracted, covering three core categories relevant to slow-rate attack detection. Table 1 summarizes the key base features and their corresponding attack signal capture mechanisms. All features are derived from the QUIC/UDP/TLS/HTTP/3 protocols to ensure compatibility with QUIC.

Note: Features are extracted from decrypted protocol traces via standard dissection. The model does not ingest raw payload bytes, URLs, or textual contents; application-layer fields (if any) are used only in non-semantic encoded forms (e.g., length/frequency statistics).

Construction of Temporal Statistical Feature Set. For each of the T = 10 segments, we compute three core statistical metrics for all numerical and encoded features in F_{base} to form the temporal feature matrix F_{TS}:

Mean: Reflects the central tendency of feature values within the segment (e.g., average packet length).

Table 1. Key base features for slow-rate attack detection

Feature type	Number of features	Key features (Examples)	Attack signal Capture
Temporal	12	1. frame.time_delta 2. frame.time_relative 3. udp.time_delta 4. quic.ack.ack_delay ... (and the corresponding frequency-encoded versions with the suffix _freq)	- Packet inter-arrival patterns (prolonged delays in attacks) - Transmission frequency stability (low variability in attacks) - ACK response latency (intentionally extended in attacks)
Length	42	1. frame.len 2. udp.length 3. udp.payload_len 4. quic.stream.len 5. tls.record.length 6. http.content_length 7. dns.count.answers ... (20 basic length features and their corresponding 22 _freq features)	- Payload size distribution (small and uniform packets in attacks) - Packet length variability (low standard deviation in attacks) - Data volume distribution (sparse transmission in attacks)
Protocol State	42	1. quic.frame_type 2. quic.stream.fin 3. tls.handshake.type 4. dns.flags.response 5. quic.decryption_failed 6. http.response.code_freq ... (20 basic state features and their corresponding 22 _freq features)	- Control frame anomalies (abnormal proportions of frame types) - Abuse of stream multiplexing (excessive sub-streams in attacks) - Protocol state stability (fixed TLS content types in attacks)

Max: Captures the instantaneous peak values within the segment (e.g., a single unusually large packet).

Standard Deviation (std): Indicates the variability of feature values within the segment; slow-rate attacks often exhibit unusually low or stable standard deviations.

Ultimately, each flow is represented as a fixed-dimension initial temporal feature $V_{\text{initial}} \in \mathbb{R}^{T \times M_{\text{initial}}}$: here, T = 10 (which refers to the number of segments), $M_{initial} = 96 \times 3 = 288$ (calculated by multiplying the number of base features by the number of statistical indicators), and thus the total dimension per flow is $10 \times 288 = 2880$.

3.5 Model Design and Implementation

Stacked CNN-GRU Hybrid Model Architecture. The rationale for this hybrid architecture is based on the distinct nature of slow-rate attacks in QUIC traffic, which exhibit anomalies in both local statistical distributions (spatial) and long-term evolutionary trends (temporal).

(i) *Locality of spatial features (Captured by CNN):* The fingerprint of attack traffic within a specific time slice is not represented by isolated single features but by combination patterns of multiple statistics (e.g., mean, max, std). CNNs serve as local pattern extractors, automatically capturing these combination patterns at each time step along the M dimension. Specifically, the CNN is designed to capture intra-segment burstiness characteristics. For instance, a QUIC slow-rate attack often involves sending small control frames in tight clusters to keep the connection alive. The CNN detects this spatial "snapshot"—such as the correlation between "low mean packet size" and "high packet frequency" within a single segment—which signifies an anomaly in the Payload-to-Overhead ratio.
(ii) *Long-term temporal dependencies (Captured by GRU):* Slow-rate attacks are continuous, low-rate processes whose abnormal signals are distributed across the $\mathrm{T} = 10$ time steps. GRUs provide the necessary memory mechanism to capture dependencies spanning longer temporal sequences. While legitimate users may have irregular idle periods, attackers exhibit specific attack cadences. The GRU is responsible for detecting these inter-segment rate changes and sustained persistence, distinguishing malicious flow control manipulation from benign network latency.

Principle and Advantages of the Convolutional Layer (Conv1D). The initial layer of the model employs a one-dimensional convolutional layer, whose primary functions are feature recombination and intra-segment feature correlation:

Feature pattern extraction: The Conv1D kernels slide along the feature axis rather than the temporal axis. For example, a kernel with size 3 can simultaneously observe a combination of three adjacent features, such as "mean packet length," "maximum time interval," and "standard deviation of QUIC stream data."

Spatial fingerprint recognition: With 64 filters, the model can learn 64 different local feature fingerprints in parallel. These fingerprints are more discriminative than the original statistical metrics, enabling the identification of typical slow-rate attack patterns, such as "low variability (low std) combined with high mean time intervals." By aggregating these features, the CNN effectively creates a high-level representation of the traffic state for each time step, filtering out noise and highlighting structural anomalies like burstiness before temporal analysis begins.

Mechanism and Necessity of the Stacked GRU. The spatial features extracted by the convolutional layer are subsequently fed into the core stacked Gated Recurrent Unit (GRU) structure. The GRU is a simplified variant of LSTM that addresses the vanishing gradient problem in traditional RNNs through its gating mechanisms and analyzes the temporal evolution of the flow.

The GRU consists of two core gates: the reset gate and the update gate.

Reset gate (r_t):Determines how much historical information (h_{t-1}) to ignore. In slow-rate attack detection, network jitter may cause transient traffic fluctuations that resemble an attack. The reset gate allows the model to discard such irrelevant "noise" signals, ensuring that momentary anomalies do not trigger a false positive.

Update gate (z_t):Determines how much of the current candidate state $(\tilde{h}_t)$ and the previous state (h_{t-1}) to retain. This gate is critical for capturing attack persistence and rate stability. If the attack signal is weak but continuous (e.g., a pulse sent exactly every 15 s), the update gate ensures that this long-term periodicity is integrated into the state, allowing subtle signals to accumulate over time.

Given the subtle and dispersed nature of slow-rate attack signals, a dual-layer GRU is employed to capture complex temporal abstractions. The first layer extracts low-level, short-term dependencies, while the second layer synthesizes these into abstract, long-term patterns (e.g., attack lifecycle). This hierarchical approach ensures the final output encapsulates a comprehensive temporal fingerprint for accurate classification.

Enhancing Generalization and Model Regularization. To address potential biases in the dataset and enhance the model's generalization ability in real-world network environments, we adopted a rigorous regularization strategy.

Regularization. L2 regularization adds a penalty term on the model weights ($\lambda \cdot \| w\|^2$) to the loss function, enforcing smaller weight values:

$$\mathcal{L}_{\text{final}} = \mathcal{L}_{\text{BCE}} + \lambda \sum w^2 \tag{5}$$

This effectively penalizes the model's tendency to overreact to minor variations in the training data, preventing overfitting of noise in the two GRU layers and the dense layer, thereby enhancing robustness to unseen traffic.

To enhance generalization, the model employs Dropout (rate = 0.3) and Early Stopping (Patience = 20). Dropout randomly disables 30% of neurons to prevent dependency on specific neighbors, while Early Stopping monitors validation loss, halting training after 20 consecutive epochs without improvement to prevent overfitting and retain optimal weights.

4 Experiment

4.1 Dataset Introduction

This study utilizes the H23Q public dataset [15] for model training and evaluation. Traffic collection was conducted in a controlled virtualized environment. Benign traffic was generated by 13 Selenium-based clients employing pseudo-random access patterns and sleep intervals to emulate realistic user behavior, while malicious traffic was produced using a custom aioquic-based client targeting mainstream QUIC servers such as OpenLiteSpeed. All traffic strictly adheres to the IETF QUIC specifications and is protected by TLS encryption during transmission. To enable detailed protocol-level analysis, TLS session keys and QUIC qlog traces are available, allowing full protocol dissection via Wireshark/tshark. Based on the decrypted traces, statistical features are extracted from multiple protocol layers, including QUIC, TLS, HTTP/3, and DNS, without utilizing any raw payload bytes or application-level content. All traffic was captured in pcap format, and the dataset statistics are summarized in Table 2. We emphasize that session keys/qlog are used solely for protocol dissection and feature extraction; no plaintext payload bytes or semantic application contents are consumed by the learning model. Slow-rate attacks generate many low-volume concurrent flows, explaining why attack flows outnumber benign flows after aggregation.

Table 2. Dataset composition and distribution

Category	Sample size	Aggregated flows	Packet proportion (%)	Flow proportion (%)
Normal traffic	702226	7265	79.85	37.06
Attack traffic	177210	12339	20.15	62.94
Total	879436	19604	100	100

4.2 Experimental Setup

The model proposed in this study comprises three core modules, designed for feature extraction and classification of QUIC traffic:

(i) *Convolutional Feature Extraction Layer:* Captures local spatial correlations and extracts local patterns within temporal windows, providing high-level feature representations for subsequent temporal modeling.
(ii) *Stacked Gated Recurrent Unit Module:* Models long-term dependencies across time steps, capturing the latent temporal dynamics characteristic of slow-rate attack flows.
(iii) *Classification Output Layer*: Maps the high-dimensional temporal features to a binary output space, enabling discrimination between normal traffic and malicious traffic.

The main hyperparameter settings are summarized in Table 3.

Table 3. Model hyperparameter settings

Parameter	Value
Number of time segments T	10
Number of GRU layers	2
Hidden units per layer	128
Convolution kernel size	3
Learning rate	1.00E-03
Batch size	64
Optimizer	Adam
Number of training epochs	130

4.3 Comparative Experiment Design

To evaluate the effectiveness and superiority of the proposed model, three comparative models were designed for performance assessment. The structure and feature details of each model are summarized below.

(i) *M1 Random Forest*—A traditional machine learning model that performs classification based on static statistical features. It cannot capture temporal variations in traffic.
(ii) *M2 Two-Layer LSTM*—A deep learning model based on Long Short-Term Memory (LSTM) networks, capable of capturing long-term temporal dependencies but with limited ability to extract local spatial features.
(iii) *M3 Proposed Method:* Convolutional Neural Networks and Two-Layer Gated Recurrent Unit—Combines the spatial feature extraction capability of CNN with the temporal modeling ability of GRU, enabling joint learning of spatial and long-term temporal characteristics in network traffic.

4.4 Performance Evaluation Metrics

To comprehensively evaluate the detection performance of each model, this study adopts Accuracy, Precision, Recall, F1-score, and the Area Under the ROC Curve (AUC) as overall performance metrics. The definitions of these metrics are as follows:

$$\begin{aligned} \text{Accuracy} &= \frac{TP+TN}{TP+TN+FP+FN} \\ \text{Precision} &= \frac{TP}{TP+FP} \\ \text{Recall} &= \frac{TP}{TP+FN} \\ F1 &= \frac{2\times\text{Precision}\times\text{Recall}}{\text{Precision}+\text{Recall}} \end{aligned} \quad (6)$$

Additionally, the AUC is calculated to evaluate the model's overall discriminative ability across different classification thresholds. An AUC value closer to 1 indicates stronger capability in distinguishing between normal and malicious traffic, reflecting better detection performance.

4.5 Experimental Results and Analysis

Comprehensive Performance Comparison. Table 4 presents the performance comparison of the three models on the test set. The results demonstrate that the proposed CNN + Stacked GRU model (M3) outperforms both Random Forest (M1) and LSTM (M2) baselines across all metrics. By integrating CNN's spatial feature extraction with GRU's temporal modeling, M3 captures the complex spatiotemporal patterns of slow-rate attacks more effectively than traditional or single-architecture models, resulting in superior detection performance.

Table 4. Performance comparison of models on the test set

Model	Accuracy	Precision	Recall	F1-score	AUC
M1: Random forest	0.905	0.886	0.978	0.931	0.963
M2: Two-layer LSTM	0.871	0.943	0.852	0.895	0.928
M3: Ours	0.989	0.993	0.986	0.975	0.977

Ablation Study. To further evaluate the contribution of each component, an ablation study was conducted, and the results are presented in Table 5.

Table 5. Ablation study results of model components

Model variant	Accuracy	Precision	Recall	F1-score	AUC
Without CNN	0.877	0.845	0.965	0.902	0.935
Single-layer GRU	0.923	0.911	0.971	0.938	0.931
Without Regularization	0.956	0.942	0.968	0.955	0.962
Without temporal feature aggregation	0.821	0.793	0.897	0.842	0.885
Ours	0.989	0.993	0.986	0.975	0.977

As shown in Table 5, our full model (Ours) achieves the best Accuracy (0.989) and F1-Score (0.975). Removing the CNN module (without CNN) degrades performance, with Accuracy at 0.877 and F1-Score at 0.902, showing CNN's role in capturing spatial patterns of QUIC traffic. Using a single-layer GRU (single-layer GRU) weakens long-term temporal dependency capture, leading to reduced accuracy (0.923) and AUC (0.931). Removing regularization (Without Regularization) causes overfitting. Removing temporal feature aggregation (Without Temporal Feature Aggregation) leads to the worst performance, with Accuracy at 0.821 and F1-Score at 0.842, highlighting its importance in capturing slow-rate attacks.

Visualization Analysis. Figure 2 illustrates the comprehensive performance evaluation of the proposed model compared to the baseline methods. The left panel presents the Receiver Operating Characteristic (ROC) curves on the test set. The right panel depicts the convergence behavior of the model by plotting the training and validation loss (Binary Cross-entropy) over epochs. Figure 3 shows the ROC curves of different models on the test set and presents the visualization of the confusion matrices.

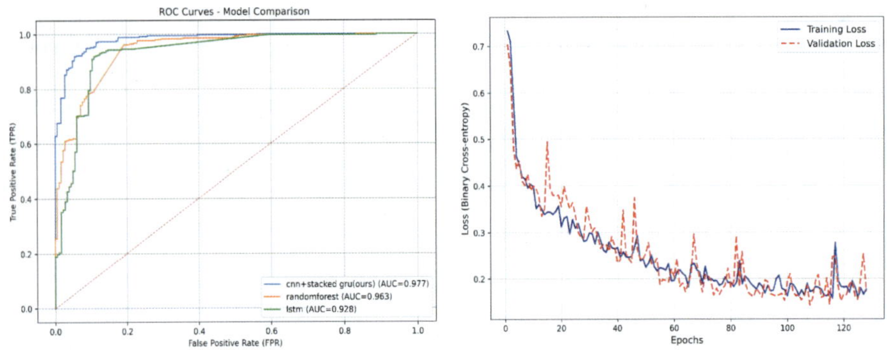

Fig. 2. ROC curves of the comparative models and loss convergence of the proposed model

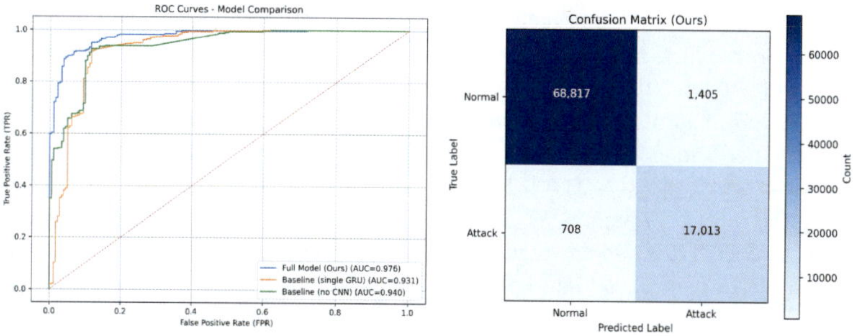

Fig. 3. ROC curves of the models and confusion matrix of the proposed model

5 Conclusion

This paper investigates the detection of slow-rate DoS attacks over the QUIC protocol. To address the challenges of high concurrency and strong temporal dependencies in QUIC traffic, a deep learning model combining a CNN and stacked GRUs was proposed. The model effectively captures spatial and temporal patterns of QUIC flows, achieving accurate and dynamic recognition of slow-rate attacks.

Experimental results on a modified QUIC dataset demonstrate that the proposed method outperforms traditional detection models in accuracy, recall, and F1-score, showing strong generalization capability. Future work will focus on enhancing protocol-level feature representation, evaluating the model's early detection capability, and optimizing practical deployment through lightweight implementations under decryptable operational settings (e.g., endpoint or enterprise inspection). Overall, this study provides a promising direction for intelligent and efficient security monitoring in QUIC-enabled network environments.

Acknowledgement. This work was supported in part by the National Natural Science Foundation of China under Grant No. 62573122, the Fundamental Research Funds for the Central Universities under Grant No. 2242025 K30025, the Open Research Project of State Key Laboratory of Industrial Control Technology, China under Grant No. ICT2025B37, the Open Research Project of the Key Laboratory of Numerical Simulation for Large Scale Complex Systems, Ministry of Education, China, under Grant No. NSLSCS202502, the Jiangsu Provincial Scientific Research Center of Applied Mathematics under Grant No. BK20233002.

References

1. Cook, S., Mathieu, B., Truong, P., Hamchaoui, I.: QUIC: Better for what and for whom? In: Proceedings of the 2017 IEEE International Conference on Communications (ICC), pp. 1–6. IEEE (2017)
2. Langley, A. et al.: The QUIC transport protocol: Design and internet-scale deployment. In: Proceedings of the ACM SIGCOMM Conference (SIGCOMM), pp. 183–196. ACM (2017)

3. Carlucci, G., De Cicco, L., Mascolo, S.: HTTP over UDP: An experimental investigation of QUIC. In: Proceedings of the 30th Annual ACM Symposium on Applied Computing (SAC), pp. 609–614. ACM (2015)
4. Lychev, R., Jero, S., Boldyreva, A., Nita-Rotaru, C.: How secure and quick is QUIC? Provable security and performance analyses. In: Proceedings of the 2015 IEEE Symposium on Security and Privacy (S&P), pp. 214–231. IEEE (2015)
5. Marx, R., Herbots, J., Lamotte, W., Quax, P.: Same standards, different decisions: A study of QUIC and HTTP/3 implementation diversity. In: Proceedings of the Workshop on the Evolution, Performance, and Interoperability of QUIC (EPIQ), pp. 14–20. ACM (2020)
6. Chatzoglou, E., Kouliaridis, V., Karopoulos, G., Kambourakis, G.: Revisiting QUIC attacks: a comprehensive review on QUIC security and a hands-on study. Int. J. Inf. Secur. **22**(2), 347–365 (2023)
7. Cambiaso, E., Papaleo, G., Chiola, G., Aiello, M.: Slow DoS attacks: definition and categorisation. Int. J. Trust. Manag. Comput. Commun. **1**(3–4), 300–319 (2013)
8. Cambiaso, E., Papaleo, G., Aiello, M.: Slowcomm: design, development and performance evaluation of a new slow DoS attack. J. Inf. Secur. Appl. **35**, 23–31 (2017)
9. Park, J., Iwai, K., Tanaka, H., Kurokawa, T.: Analysis of slow read DoS attack. In: Proceedings of the 2014 International Symposium on Information Theory and Its Applications (ISITA), pp. 60–64. IEEE (2014)
10. Dey, D., Ghosh, N.: iQUIC: an intelligent framework for defending QUIC connection ID-based DoS attack using advantage actor-critic reinforcement learning. Comput. Secur., 104463 (2025)
11. Sengupta, J., Dey, D., Ferlin-Reiter, S., Ghosh, N., Bajpai, V.: How resilient is QUIC to security and privacy attacks? https://arxiv.org/abs/2401.06657 (2024)
12. Balaji, A.S., Kumar, V.A., Amritha, P.P., Sethumadhavan, M.: QUICloris: A slow denial-of-service attack on the QUIC protocol. In: Proceedings of the International Conference on Signal Processing and Integrated Networks (SPIN), pp. 85–94. Springer (2022)
13. Nawrocki, M., Hiesgen, R., Schmidt, T. C., Wählisch, M.: QUICsand: quantifying QUIC reconnaissance scans and DoS flooding events. In: Proceedings of the 21st ACM Internet Measurement Conference (IMC), pp. 283–291. ACM (2021)
14. Teyssier, B.: QUIC Protocol: Resilience against Flooding Attacks and Defense Mechanism. PhD Thesis,. Concordia University (2024)
15. Chatzoglou, E., Kouliaridis, V., Kambourakis, G., Karopoulos, G., Gritzalis, S.: A hands-on gaze on HTTP/3 security through the lens of HTTP/2 and a public dataset. Comput. Secur. **125**, 103051 (2023)

Dual-View Temporal Graph Learning with Two-Stage Filtering for Robust Bitcoin Fraud Detection

Arvindh Vijayakumar[1(✉)], Mohul Y P[1], Sai Mohananshu Jujjavarapu[1], Shri Hari S[2], and Bhaskarjyoti Das[2]

[1] Department of Computer Science and Engineering, PES University, Bangalore, India
arvindh04@gmail.com, mohulyp@gmail.com, mohananshuj@gmail.com
[2] Department of Computer Science and Engineering (AI/ML), PES University, Bangalore, India
shriharis163@gmail.com, bhaskarjyoti01@gmail.com

Abstract. The pseudo-anonymous and decentralized nature of the Bitcoin network poses major challenges for financial forensics, allowing illicit activities to remain hidden within complex and evolving transaction flows. One major challenge is the extreme scarcity and temporal sparsity of labeled illicit data, which limits the effectiveness of purely supervised learning approaches. To address this challenge, we propose a novel *two-stage dual-view framework* for robust Bitcoin fraud detection. In the first stage, lightweight feature-only classifiers filter out obviously licit samples and rebalance the dataset while maintaining nearly all illicit activity. This reduces data scale and mitigates class imbalance before graph neural network training. In the second stage, the framework jointly models illicit behavior using parallel *transaction-level* and *user-level* temporal graphs. An enhanced Dual-View Graph Attention Network (GAT) is trained on the filtered graphs applying a multi-level learning strategy that integrates temporal contrastive objectives, cross-level representation alignment between transaction and user views, and supervised signals from limited labeled data. The method uses domain specific attributes that include 182-dimensional transaction features, 111-dimensional user features, with the help of lightweight structural graph statistics. Temporally split Elliptic++ dataset experiments show that it strongly performs even in situations of severe class imbalance with user-level ROC-AUC above 0.80 and F1-scores above 0.74. These results show that combining intelligent data filtering with dual-view temporal graph learning improves robustness and reliability in decentralized cryptocurrency forensic analysis.

Keywords: Bitcoin · Illicit Actor Detection · Temporal Graph Learning · Graph Attention Networks · Semi-Supervised Learning · Contrastive Learning · Financial Forensics

L. Huang and D. Greenhalgh (Eds.): ICMLC 2026, LNNS 2061, pp. 687–702, 2027.
https://doi.org/10.1007/978-3-032-31236-5_47

1 Introduction

Bitcoin and other cryptocurrencies enable decentralized, peer-to-peer transactions without reliance on central authorities, improving global accessibility and reducing transaction costs. However, this causes introduces significant challenges for financial forensics, as blockchain users are represented only by crypto addresses, obscuring real-world identities and enabling illicit activities such as money laundering, ransomware payments, and terrorist financing [1,2]. Although blockchain ledgers are transparent and immutable, the lack of contextual identity information complicates the detection of illegal behavior, requiring forensic analysis to rely on behavioral signals derived from transaction patterns and fund [2,3]. As a result, illicit activity detection naturally emerges as a graph inference problem, where transactional entities and their temporal interactions are modeled as nodes and edges.

1.1 Challenges in Illicit Activity Detection

Detection of illicit behavior on blockchain networks is challenging because of two factors: malicious actors purposefully use sophisticated obfuscation techniques like transaction splitting, peeling chains and mixing services that are designed to closely mimic licit behavior, producing anomalous patterns that typically can only be detected when analyzed over extended temporal and relational contexts. Along with this the availability of labeled data is scarce, as assigning illicit labels to entities requires costly manual investigation and external intelligence resulting in datasets where only a small fraction of nodes are labeled [4,5]. Consequently, in widely used benchmarks such as Elliptic++, the majority of transactions and addresses remain unlabeled, leading to extreme class imbalance that significantly deteriorates the effectiveness of purely supervised learning approaches and urges the adoption of data-efficient, semi-supervised methods capable of utilizing the underlying unlabeled graph structure [6].

1.2 Research Gap and Contributions

Most existing blockchain fraud detection methods operate at a single level of abstraction which are typically transaction-level graphs while relying heavily on labeled data. Such approaches struggle to capture coordinated, long-term behavioral patterns that emerge only through aggregation across users and time.

To address these limitations, we propose a **dual-view, semi-supervised framework** for illicit activity detection on cryptocurrency networks. Our main contributions are:

1. **Two-Stage Learning for Class-Imbalance Mitigation**: We introduce a lightweight, feature-based filtering stage that removes obviously licit samples prior to graph neural network training, enabling stable learning even under extreme class imbalance while preserving illicit recall.

2. **Dual-View Graph Modeling**: We jointly model transaction-level and user-level graphs to capture both fine-grained fund flow dynamics and aggregated behavioral patterns, enabling detection of isolated and coordinated illicit activity.
3. **Multi-Level Temporal Contrastive Pre-training**: We employ a data-efficient pre-training strategy that integrates temporal consistency, cross-level alignment, and supervised contrastive objectives to leverage large volumes of unlabeled temporal data.
4. **Evaluation under Realistic Conditions**: We conduct extensive experiments on temporally split Elliptic++ data, demonstrating robust performance at both transaction and user levels in a highly imbalanced forensic setting.

2 Related Work

Illicit activity detection on blockchain networks integrates graph-based learning and anomaly detection. This work harmonizes expressive dual-view graph modeling with data-efficient learning under limited supervision.

2.1 Graph-Based Models for Blockchain Forensics

The relationship among blockchain transactions entails the necessity of applying GNNs for fraud detection due to more effective joint representation of node features and transactional dependencies than feature based ones [2,3]. Early works, such as BN-GCN [3,7], can achieve better performances than traditional classifiers on Elliptic and Elliptic++ datasets; however, they are mostly transaction-level and supervised, which is less robust under the extreme label scarcity [4,5].

2.2 Attention and Representation Learning Under Label Scarcity

Graph Attention Networks (GATs) address the limitations of uniform neighborhood aggregation by learning adaptive neighbor importance weights, improving expressivity for modeling heterogeneous financial behaviors [8]. Recent work has further explored Transformer-style graph architectures, such as FraudGT [5,9–13], though scalability remains a challenge on large blockchain graphs.

A key challenge in this domain is the scarcity and temporal sparsity of labeled illicit data [4,5], motivating semi-supervised and self-supervised learning. Prior studies have leveraged unlabeled but suspicious samples [14], while contrastive learning has emerged as an effective approach for graph representation learning. Contrastive pre-training enforces consistency across temporal snapshots and multiple graph views, improving performance under limited supervision [15–18].

Building on these ideas, we propose **multi-level contrastive learning** for temporal blockchain graphs, jointly enforcing temporal consistency, cross-level alignment between transaction and user representations, and supervised contrastive objectives where labels are available.

2.3 Graph Representation Strategies

In addition to model architecture, graph representation design plays an important role in illicit activity detection. Early approaches depended on simple transaction graphs, which were often not able to capture higher-order behavioral patterns [2]. Subsequent work introduced heterogeneous graph formulations and interaction-aware graph structures [14,19,20] to encode richer relational context.

Instead of choosing a single abstraction level, our work follows the recent studies in multi-granularity modeling. To this end, we propose a **dual-view graph representation** approach by processing transaction-level and aggregated user-level graphs in parallel. This allows the model to learn both detailed fund flow dynamics and long-term user behaviors for identifying isolated anomalies and coordinated illicit behavior.

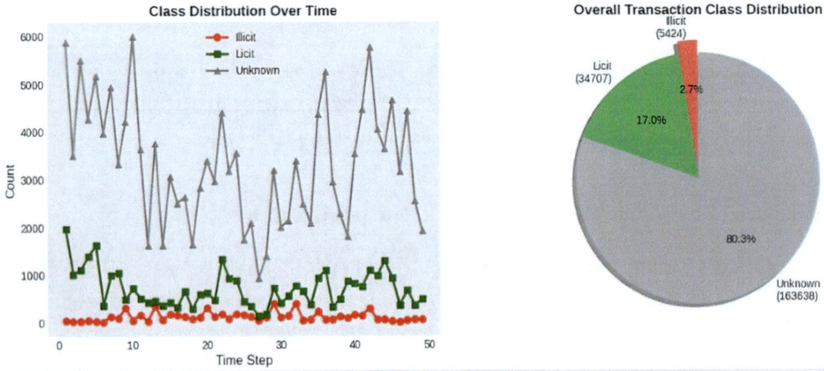

Fig. 1. Overall transaction class imbalance across timesteps, highlighting the dominance of unlabeled and licit samples and motivating early-stage filtering.

3 Dataset and Problem Formulation

3.1 Elliptic++ Dataset

We evaluate our framework on the Elliptic++ dataset, a widely used benchmark for illicit activity detection on the Bitcoin network [1,4,21]. Elliptic++ extends the original Elliptic dataset by providing richer annotations and a realistic temporal structure. It consists of approximately 1.26 million Bitcoin transactions involving over 1.27 million wallet addresses, organized into 49 discrete temporal snapshots.

Every transaction is characterized by a combination of local and historical activity-based statistical and time-related characteristics. Labels are present for a small subset of transactions and addresses, categorized as *licit*, *illicit*, or *unknown*, with the majority of entities remaining unlabeled thereby representing real-world forensic conditions.

3.2 Data Characteristics and Challenges

Elliptic++ exhibits extreme label sparsity and class imbalance. Illicit nodes account for only 2–3% of labeled samples, while unlabeled nodes make up the most of every timestep as visualized in Fig. 1. This makes illicit activity detection a challenging anomaly detection problem, where rare and evolving behaviors must be identified within a vast background of legitimate activity [11,17,22–24].

Due to this imbalance, accuracy by itself is not a meaningful evaluation metric; instead, precision, recall, F1-score, ROC-AUC, and MCC are required. To mitigate these challenges, our framework includes an early-stage filtering mechanism that removes a large fraction of obviously licit nodes before graph neural network training. The remaining representation learning and classification processes are applied on the filtered subgraphs, while preserving the temporal sequence and structure relationship.

Temporal dynamics further complicate the task [15,16,18,25]. Transaction-level illicit activity is highly volatile across timesteps, whereas aggregated user-level behavior exhibits more persistent structural patterns. This motivates temporal modeling and multi-level graph representations.

3.3 Dual-View Temporal Graph Construction

To capture illicit behavior at multiple levels of abstraction, we construct two parallel sequences of temporal graphs.

Transaction Graphs

For each timestep $t \in \{1, \ldots, T\}$, we define a transaction graph $G_{tx}^{(t)} = (\mathcal{V}_{tx}^{(t)}, \mathcal{E}_{tx}^{(t)}, \mathbf{X}_{tx}^{(t)})$, where nodes correspond to individual Bitcoin transactions. Edges in the transaction graph encode explicit address-level transactional dependencies, connecting pairs of transactions that share common input addresses, share common output addresses, and exhibit direct fund flow where an output address of one transaction appears as an input address of another transaction within the same temporal snapshot. Each node is represented using a 182-dimensional feature vector composed of transactional, temporal, and statistical attributes.

User Graphs

To model higher-level behavioral patterns, wallet addresses are clustered into user entities using a co-spending heuristic, whereby addresses appearing together as transaction inputs are assumed to be controlled by a single actor. For each timestep t, we construct a user graph $G_{user}^{(t)} = (\mathcal{V}user^{(t)}, \mathcal{E}user^{(t)}, \mathbf{X}_{user}^{(t)})$, where nodes represent users and edges indicate transactions between them. A 111-dimensional feature vector, the summarized form of aggregated wallet statistics, transaction behavior, network structure and activity over time characterizes each user node

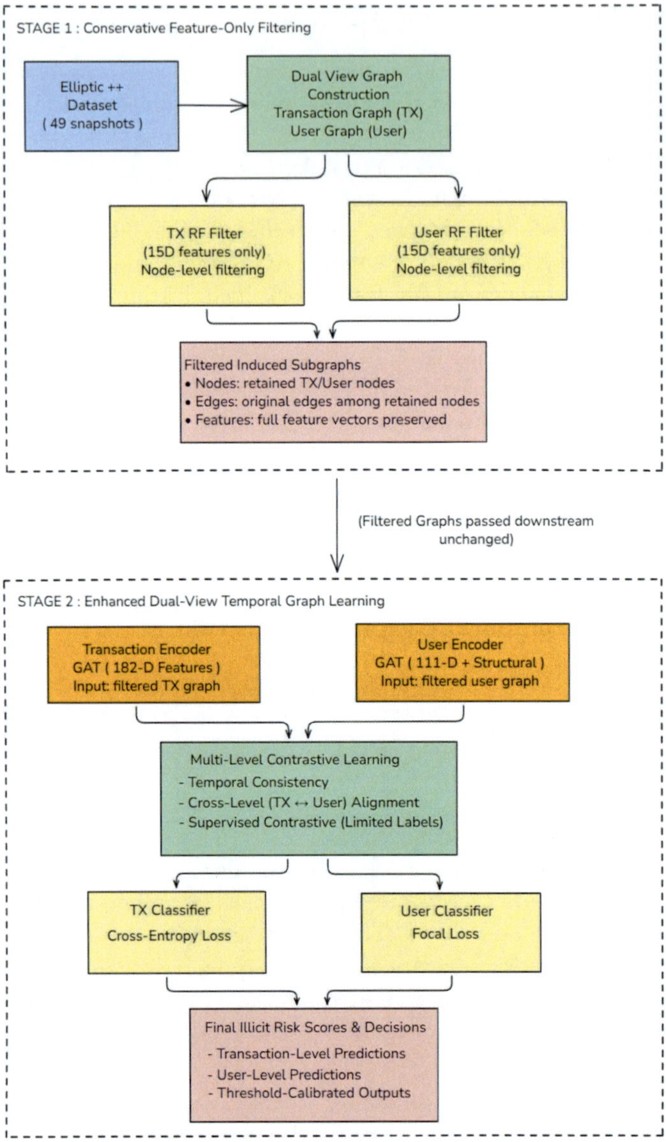

Fig. 2. Overview of the proposed dual-view temporal graph contrastive learning framework. Stage-1 performs node-level feature-only filtering and outputs induced subgraphs, which retain original graph structure and full node features and are directly consumed by Stage-2 GNN encoders.

User features are augmented with snapshot-level structural statistics [23] computed in linear time per node, including degree, weighted degree, local clustering coefficient, average neighbor degree, and edge-weight entropy. These features are concatenated to the original user attributes to provide coarse structural context without expensive global graph computation.

Together, the transaction and user graphs provide complementary views: the transaction graph captures fine-grained money flow dynamics, while the user graph encodes long-term behavioral patterns that are defined by aggregation across time.

3.4 Problem Definition

Illicit activity detection on Elliptic++ is modeled as a **dual-view semi-supervised node classification task** on dynamic, attributed graphs. We are given two temporal graph sequences:

- Transaction graphs $G_{TX} = \{G_{tx}^{(t)} \mid t = 1, \ldots, T\}$, with node features $\mathbf{X}_{tx}^{(t)} \in \mathbb{R}^{|\mathcal{V}_{tx}^{(t)}| \times 182}$.
- User graphs $G_{UEG} = \{G_{user}^{(t)} \mid t = 1, \ldots, T\}$, with node features $\mathbf{X}_{user}^{(t)} \in \mathbb{R}^{|\mathcal{V}_{user}^{(t)}| \times 111}$.

Each node v_i is associated with a label $y_i \in \{\text{licit}, \text{illicit}, \text{unknown}\}$, with the majority of nodes remaining unlabeled. The goal is to learn mapping functions $f = (f_{tx}, f_{user})$ that predict illicit nodes at the transaction and user levels. Training is performed in a semi-supervised manner, leveraging unlabeled temporal graph structure during pre-training and refining these predictions using limited labeled data during the fine-tuning phase.

4 Proposed Methodological Framework

4.1 Architectural Overview

We introduce a two-stage dual-view graph learning framework, targeting three critical issues in cryptocurrency forensics: extreme class imbalance, label scarcity, and multilayer behavioral complexity. It has a clear decoupling between data conditioning and representation learning, which allows scalable and robust detection on large temporal blockchain graphs.

Stage 1 performs fast, feature-only filtering to remove obviously licit nodes while preserving nearly all illicit samples. Stage 2 operates on the resulting reduced graphs and trains an enhanced dual-view Graph Neural Network (GNN) that jointly models transaction-level and user-level behavior using contrastive and supervised objectives as visualized in Fig. 2.

4.2 Stage 1: Lightweight Licit Filtering

Motivation

Elliptic++ exhibits extreme class imbalance, with illicit nodes comprising only 2–3% of labeled samples. Direct GNN training on this distribution often leads to degenerate majority-class predictions. We therefore introduce a lightweight pre-filtering stage to reduce graph size and rebalance class proportions prior to representation learning.

Filter Design
Separate Random Forest classifiers are trained for transaction and user nodes using only the first 15 low-level features, capturing basic transactional statistics without relying on graph structure. Each model uses 100 estimators, maximum depth 10, and balanced class weights, and is trained solely on labeled nodes from the training split (timesteps 1–29).

Thresholding and Filtering
Decision thresholds are calibrated on the validation split [26] to enforce a minimum illicit recall of 95%, reflecting the high-recall requirements of real-world financial forensics. For each snapshot, licit nodes classified as obviously licit are removed, while all illicit nodes are retained. This stage retains 95–98% of illicit nodes while removing 75–85% of licit nodes, reducing overall graph size by approximately 70–80%.

4.3 Stage 2: Dual-View Graph Learning

Encoders
Stage 2 employs two parallel edge-aware Graph Attention Network (GAT) encoders: a transaction encoder operating on 182-dimensional features and a user encoder operating on 111-dimensional aggregated features [8]. Both encoders consist of four GAT layers with residual connections, batch normalization, and dropout. The reduced dataset produced by Stage 1 enables higher-capacity encoders without prohibitive computational cost.

Edge-Aware Co-Spending Weights
Edges in user graphs are weighted according to transaction input cardinality:

$$w_{uv} = \frac{1}{(\text{num\_inputs})^{\alpha}}, \quad \alpha = 0.7, \tag{1}$$

penalizing large multi-input transactions while emphasizing focused co-spending behavior indicative of common ownership.

4.4 Contrastive Pre-Training

To exploit unlabeled temporal structure, contrastive pre-training is performed prior to classification using an online encoder and a momentum encoder [15,17,18].

Three objectives are jointly optimized: (i) temporal consistency across snapshots, (ii) cross-level alignment between transaction and user embeddings, and (iii) supervised contrastive learning where labels are available. Pre-training is conducted for 100 epochs using the Adam optimizer with learning rate 10^{-3}.

(a) Transaction-level embeddings (b) User-level embeddings

Fig. 3. t-SNE visualization of learned embeddings produced by the proposed dual-view model on the held-out test split.

4.5 Supervised Fine-Tuning

Filtered temporal graphs are aggregated into static graphs for training (1–29), validation (30–36), and testing (37–49). Fine-tuning proceeds in two stages: first training classification heads with frozen encoders, followed by joint optimization with a reduced encoder learning rate. Transaction classification uses cross-entropy loss, while user classification employs focal loss. Decision thresholds are calibrated on the validation set to optimize a weighted combination of F1-score and Matthews Correlation Coefficient.

5 Experimental Evaluation

5.1 Experimental Setup

All models are implemented using PyTorch and PyTorch Geometric. Experiments are conducted on NVIDIA T4 GPUs. Contrastive pre-training is performed for 100 epochs using Adam with learning rate 10^{-3}. Supervised fine-tuning is conducted for 180 epochs using Adam with learning rate 8×10^{-4} and weight decay 10^{-4}. The 49 timesteps of Elliptic++ are split chronologically into training (1–29), validation (30–36), and test (37–49). Graphs of each split are aggregated for supervised learning while having temporal causality preserved. Metrics that are immune to class imbalance such as Precision, Recall, F1-Score, ROC-AUC, PR-AUC, Balanced Accuracy and MCC, are employed for evaluation.

5.2 Dual-Level Detection Performance

We assess the frame work on the unseen test splits for both transactions and user levels. At the transaction level, the system demonstrates good discriminative ability with high ROC-AUC and precision, allowing confident low-level alert generation. At the user level, aggregation allows for confident detection of coordinated illicit activities, resulting in an effective trade-off between recall and precision, as illustrated by F1 score and MCC.

5.3 Representation Analysis

In order to evaluate representation quality, t-SNE was used to visualize the transaction and user embeddings as shown in Fig. 3. The resulting embeddings form well separated clusters, corresponding to licit and illicit activity, suggesting that the framework learns semantically meaningful representations instead of relying on spurious correlations.

5.4 Handling Extreme Class Imbalance

The proposed framework strikes a desirable balance between recall and precision, specifically as it pertains to user-level performance. Such balance stems from early filtering, contrastive pre-training on unlabeled data, dual-view modeling, and imbalanced-aware loss functions. Subsequent calibration based on validation thresholds ensures compliance with practical forensic scenarios.

6 Results and Analysis

This section analyzes the proposed two-stage framework, i.e., the Stage-1 filtering mechanism's effectiveness and the enhanced dual-view GNN's classification performance on the filtered data, in the context of the temporal orderly fashion. All numbers are presented based on temporally held-out test snapshots (timesteps 37–49) of the dataset to provide a realistic forensic assessment.

6.1 Stage-1 Filtering Performance

The goal of the filtering stage is not precise classification, but conservative elimination of obviously licit samples while preserving illicit activity with near-perfect recall. Accordingly, we evaluate the filtering mechanism using illicit retention, licit filtering rate, data reduction, and post-filter class balance.

Filtering Recall and Selectivity
Over all temporal splits, the Random Forest filters reach **100% illicit retention** on the test data for both transaction- and user-level graphs, therefore no illicit samples are lost before GNN training. In contrast, a large portion of licit nodes is dropped. On the test split, nearly **96.8% and 96.0% of transaction and user nodes** are filtered, respectively. This harsh removal discards trivial licit behavior while maintaining structurally ambiguous cases for the following analysis.

Graph Size Reduction and Class Balance
The filtering step shrinks the transaction graph from 67,504 nodes to 2,166 nodes (3.2% kept), and it shrinks the user graph from 57,005 nodes to 2,300

nodes (4.0% kept). All splits of the sampled graphs have an **approximately 1:1 illicit-to-licit ratio**. The results indicate that Stage-1 filtering can effectively address extreme class imbalance while significantly reducing the graph sizes, which allows usage of larger GNN models strictly for Stage-2.

6.2 Dual-View GNN Performance on Filtered Graphs

We experiment with the improved dual-view GNN, which is learned using the filtered graphs generated in Stage 1. The model consists of four GAT layers, each with 320 hidden units and a 640-dimensional embedding.

6.2.1 Transaction-Level Classification On the transaction-level test set, our model achieves a ROC-AUC of **0.9077** and a PR-AUC of **0.9223**. The model attains an illicit-class F1-score of **0.8137**, with **77.65%** recall and **85.47%** precision. The balanced accuracy reaches **0.8223**, indicating consistent performance across classes. These results demonstrate that the model captures illicit transactional patterns even after aggressive filtering while maintaining strong discrimination between licit and illicit behavior.

6.2.2 User-Level Classification At the user level, Model B achieves a ROC-AUC of **0.8196** and a PR-AUC of **0.8556**. Illicit-class recall reaches **69.20%** with **80.51%** precision, yielding an F1-score of **0.7443**. The balanced accuracy is **0.7635**, reflecting robust performance despite the higher abstraction of user entities.

The lower user-level performance is caused due to ambiguity brought in by address clustering and aggregation, yet remains competitive given the challenging forensic setting.

6.2.3 Discussion These results confirm the effectiveness of the proposed two-stage design. Stage 1 substantially reduces graph size and neutralizes class imbalance without sacrificing illicit recall, while Stage 2 leverages the curated data to train a dual-view GNN. Importantly, this performance is achieved without handcrafted rules or post-hoc heuristics, demonstrating the model's ability to learn illicit behavioral patterns directly from graph-structured data.

Overall, the findings validate the importance of data filtering prior to GNN training in large-scale, highly imbalanced cryptocurrency transaction networks.

7 Comparative Analysis and Ablation Studies

To further put the effectiveness of the proposed framework in context we summarize comparative assessments with the baseline techniques and ablation analysis.

These experiments aim to evaluate the role of each of the components, such as dual-view modeling, contrastive pre-training, and loss design.

Baseline Comparisons
The suggested framework is contrasted with conventional machine learning models and graph neural network baseline representatives. Detailed quantitative results are reported in Table 1. This comparison highlights the advantage of joint transaction and user-level modeling combined with temporal representation learning.
Ablation Studies
We carry out systematic ablations by selectively removing or modifying key components of the framework, such as contrastive pre-training objectives, graph views, and loss functions. Results are shown in Table 2 and demonstrate that each component contributes meaningfully to the overall

Stage-1 filtering reduces graph size by 7080%, enabling higher-capacity GNNs with 34 times lower training time per epoch and reduced memory usage compared to training on full graphs, demonstrating scalability to large temporal blockchain networks.

8 Conclusion and Future Work

8.1 Summary of Contributions

We explored the detection of illicit activities on the Bitcoin network and proposed a dual-view, semi-supervised graph learning framework that jointly models transaction-level and user-level temporal graphs. By combining multi-level contrastive pre-training with supervised fine-tuning, the framework effectively makes use of unlabeled data while maintaining strong performance with limited labels.

Experiments conducted on the Elliptic++ dataset demonstrate strong performance at both transaction and user levels. The dual-view design captures complementary fund flow and aggregated behavioral patterns, enabling an advantageous precisionrecall trade-off for practical financial forensics.

Overall, these results demonstrate that dual-view graph modeling with multi-level representation learning is an effective strategy for illicit activity detection in decentralized cryptocurrency systems.

8.2 Future Research Directions

Although the proposed framework demonstrates strong empirical performance there remain several directions for future work.

- **Advanced Contrastive Learning**: Exploration of graph-specific contrastive augmentations such as subgraph sampling, motif-based views, or community-aware perturbations to improve representation learning under extreme label imbalance.

Table 1. Performance comparison on Elliptic++.

Category	Model	Par. (M)	TX (AUC)	User (AUC)	TX F1	User F1	Comb. time	Train time	Infer. time
Non-GNN	RF	–	.743	.810	.228	.596	.339	2.5123	.1189
	LR	-	.726	.813	.202	.763	.370	1.5487	.0016
	XGB	-	.737	.687	.171	.178	.173	2.5263	.0548
Single	TX	0.55	.817	-	.445	–	–	–	–
	User	0.40	-	.719	–	.644	–	–	–
Graph	SAGE	1.37	.823	.730	.620	.663	.652	.4044	.0018
	GAT	0.91	.814	.742	.623	.671	.649	.7188	.0026
	Proposed model	**1.66**	**.908**	**.820**	**.814**	**.744**	**.842**	**.9000**	**0.025**

Combined F1 is the harmonic mean of TX and User F1. Training time denotes the average per-epoch runtime of Stage-2 dual-view GNN training (contrastive pre-training and supervised fine-tuning) on filtered graphs, while inference time measures the average forward-pass runtime per temporal snapshot

Table 2. Ablation studies on Elliptic++.

Category	Ablation	TX	User	TX	User
Baseline	Full model	**.908**	**.820**	**.814**	**.744**
Pre-train	None	.872	.781	.743	.685
	Temporal	.891	.803	.778	.712
	Cross-Level	.885	.796	.765	.701
	Supervised	.897	.809	.791	.724
Loss Fn.	No Focal	.895	.808	.789	.721
	No Uncert.	.901	.815	.801	.735
	Neither	.883	.795	.771	.706
Arch.	No Resid.	.889	.799	.776	.714
	No BatchN.	.896	.807	.793	.726
	No Drop.	.881	.793	.767	.708
Graph	No FeatSim	.893	.805	.784	.719
	No EdgeWt	.866	.776	.738	.678
	α=0.3	.887	.801	.773	.711
	α=0.5	.899	.812	.797	.731
	α=0.7	.905	.817	.808	.740
	α=1.0	.879	.790	.759	.698

Baseline is Model B (1.66M params). Best per category in **bold**

- **Adaptive Threshold Calibration**: Beyond validation-based calibration, adaptive or uncertainty-aware thresholding strategies could dynamically adjust decision boundaries based on graph structure or temporal context, improving robustness and reliability in operational settings.
- **Extended Benchmarking and Generalization**: Evaluating heterogeneous GNNs and transformer-based models as well as extending the framework to other blockchain or cross-chain datasets, would help in generalization and scalability.
- **Model Interpretability**: Incorporating post-hoc explanation methods for attention-based GNNs could improve transparency by identifying influential nodes, edges and features supporting analyst trust and regulatory adoption of the framework.
- **Generalization**: The proposed framework is blockchain-agnostic and can be extended to account-based systems such as Ethereum by redefining transaction and user graphs using value transfers and interaction entities, while preserving the two-stage filtering and dual-view temporal learning design.

References

1. Elmougy, Y., Liu, L.: Demystifying fraudulent transactions and illicit nodes in the bitcoin network for financial forensics. In: Proceedings of the 29th ACM SIGKDD Conference on Knowledge Discovery and Data Mining, pp. 3979–3990 (2023)
2. Weber, M.: Anti-money laundering in bitcoin: Experimenting with graph convolutional networks for financial forensics (2019). arXiv:1908.02591
3. Jiao, F., Yue, R.: BN-GCN: detecting abnormal bitcoin transactions using graph neural networks. In: ICPICS (2024)
4. Song, J.: Illicit social accounts anti-money laundering for transactional blockchains. IEEE Trans. Inf. Forensics Secur. (2024)
5. Lin, J.: FraudGT: A simple, effective, and efficient graph transformer for financial fraud detection. In: Proceedings of the 5th ACM International Conference on AI in Finance (2024)
6. Xu, H.: Hypergraph-based multi-scale spatio-temporal graph convolution network for time-series anomaly detection (2024). arXiv:2410.22256 arXiv preprint
7. Ding, C., Sun, S., Zhao, J.: MST-GAT: A multimodal spatial–temporal graph attention network for time series anomaly detection. Inform. Fusion **89**, 527–536 (2023)
8. Veličković, P.: Graph attention networks. In: International Conference on Learning Representations (ICLR) (2018)
9. Cai, L.: Structural temporal graph neural networks for anomaly detection in dynamic graphs. In: ACM International Conference on Information and Knowledge Management (CIKM) (2021)
10. Liu, K.: TGTOD: A global temporal graph transformer for outlier detection at scale (2024). arXiv:2412.00984 arXiv preprint
11. Liu, Y., et al.: Anomaly detection in dynamic graphs via transformer. IEEE TKDE **35**(12), 12081–12094 (2021)
12. Zhu, H., et al.: A spatio-temporal enhanced graph-transformer autoencoder for anomaly detection. IET Comput. Vision **18**(3), 405–419 (2024)
13. Liu, J.: EdgeConvFormer: Dynamic graph CNN and transformer based anomaly detection in multivariate time series (2023)
14. Wang, Y.: Enhancing fraud transaction detection via unlabeled suspicious records. In: IWQoS (2024)
15. You, J., Du, T., Leskovec, J.: ROLAND: Graph learning framework for dynamic graphs. In: KDD (2022)
16. Xue, G., et al.: Dynamic network embedding survey. Neurocomputing **472**, 212–223 (2022)
17. Lou, S.: GADY: Unsupervised anomaly detection on dynamic graphs (2023). arXiv:2310.16376
18. Kim, Y.: Temporal graph networks for graph anomaly detection in financial networks (2024). arXiv:2404.00060
19. Islambekov, U.: A fast topological approach for predicting anomalies in time-varying graphs (2023). arXiv:2305.06523
20. Ferretti, S.: On the use of heterogeneous graph neural networks for detecting malicious activities. In: OCSNS (2024)
21. Zheng, Y.: Graph spatiotemporal process for multivariate time series anomaly detection with missing values. Inform. Fusion (2024)
22. Zeng, X., et al.: Hierarchical spatio-temporal graph convolutional neural network for anomaly detection in videos. IEEE TCSVT **33**(1), 200–212 (2021)

23. Tiwari, A.: Statistically inspired discrepancy detection for anomalous spatio-temporal graphs. Int. J. Data Sci. Anal. (2024)
24. Zhao, M., et al.: Multivariate time series anomaly detection based on spatial-temporal network and transformer. Comput. Mater. Continua **80**(2) (2024)
25. Zheng, Y., et al.: Correlation-aware spatial–temporal graph learning for multivariate time-series anomaly detection. In: IEEE TNNLS (2023)
26. Komadina, A., Martinić, M., Groš, S., Mihajlović, Ž.: Comparing threshold selection methods for network anomaly detection. IEEE Access (2024)

Visual Language Cross Modal Alignment for Image Representation Learning and Analysis Method

Adaptive Entropy-Guided Token Pruning for Efficient Vision Transformers

Xianbin Jiang[1](✉), Xing Guo[1], and Haohua Du[2]

[1] School of Computer Science and Technology, Anhui University, Hefei, China
393462207@qq.com

[2] School of Cyber Science and Technology, Beihang University, Beijing, China
duhaohua@buaa.edu.cn

Abstract. Vision Transformers (ViTs) have emerged as a mainstream paradigm in computer vision, demonstrating exceptional performance in tasks such as image classification and object detection. However, their massive parameter counts and high computational costs—particularly the quadratic complexity of self-attention mechanisms—hinder efficient deployment on resource-constrained devices. Token pruning is a key method for compressing ViT models, as the computational complexity scales quadratically with the number of tokens, and tokens from background or redundant regions contribute minimally to final inference. Pruning these redundant tokens can significantly reduce computational overhead.Existing approaches include static pruning and dynamic pruning. However, these methods often rely on predefined pruning rates or thresholds, lacking adaptive control over accuracy loss.In this work, we propose an adaptive validation-driven progressive pruning method. Specifically, based on an expected acceptable accuracy drop, the model autonomously determines the pruning degree. Through attention-driven token importance evaluation, we hierarchically and progressively discard tokens with minimal impact on results to accelerate inference. Experiments on standard datasets validate the effectiveness of our method, achieving 50% higher throughput and over 30% parameter reduction on ViT-B models while outperforming other methods in computational efficiency and accuracy. Compared to prior work, our approach achieves a better balance between accuracy and latency.

Keywords: Vision Transformer · Token Prune · Adaptive

1 Introduction

The Evolution and Computational Bottlenecks of Vision Transformers. As a groundbreaking paradigm in computer vision, Vision Transformers (ViTs) [1] have demonstrated remarkable performance across various tasks, leveraging their powerful global modeling capabilities to gradually supersede traditional convolutional neural networks (CNNs) with limited receptive fields. However, the superior performance of ViTs comes at the cost of massive parameter counts and intricate computational architectures, demanding substantial computational resources during both training and inference [2]. This computational bottleneck becomes particularly pronounced on

L. Huang and D. Greenhalgh (Eds.): ICMLC 2026, LNNS 2061, pp. 705–718, 2027.
https://doi.org/10.1007/978-3-032-31236-5_48

resource-constrained edge devices (e.g., automotive systems, mobile terminals), severely hindering the practical deployment and application of ViTs in real-world scenarios.

Fortunately, not all image regions contribute equally to the final prediction results. Vision Transformers exhibit an imbalanced contribution pattern from different image regions to classification, where certain background or low-information tokens contribute minimally to the final classification decision [3]. This suggests these low-contribution tokens can be pruned during inference to accelerate model runtime while maintaining nearly unaffected model performance. Token pruning techniques have been widely applied in ViT optimization. Based on pruning decision flexibility, the methods can be categorized into static pruning and dynamic pruning. EViT [4] predicts token importance scores through lightweight MLP modules and progressively merges discarded token information layer by layer. DynamicViT [5] introduces predictors at fixed layers to score tokens and prunes them at predetermined ratios. While dynamic pruning often achieves better results, its adaptive nature brings hardware adaptation challenges like computational irregularity and decision overhead, typically requiring sparse accelerators or hardware-aware algorithm designs to fully realize acceleration potential, while also introducing information loss and computational instability [6]. PS-ViT [7] determines layer-wise pruning ratios by evaluating token similarity and implements bottom-up reverse pruning. SViTE [8] fully exploits ViT's sparsity characteristics through combined structured compression, unstructured sparsity and token compression approaches for parameter reduction. Although static pruning is easier to deploy on general hardware, it typically prunes at fixed ratios requiring manual adjustment to balance accuracy and latency.

Distinct from prior work, the core innovation of the proposed AEG-ViT framework lies in constructing a dual-driven evaluation system that integrates "attention entropy and dynamic weighting." This system is not a mere refinement of attention scores; rather, it is designed to address two fundamental limitations of existing single-dimensional evaluation approaches:

1. Narrowness of the evaluation dimension. Existing methods predominantly rely on a single metric, such as attention scores or class-token similarity. However, tokens with high attention scores may only be sensitive to local features, whereas attention entropy captures the concentration of attention, thereby identifying critical tokens that act as information "hubs"—those with dispersed attention patterns even if their scores are not outstanding. The combination of the two provides complementary characterization of token importance from both "intensity" and "pattern" perspectives.
2. Static nature of the evaluation strategy. Existing approaches usually assign a fixed evaluation strategy to all network layers or all pruning stages. In reality, shallow and deep layers exhibit different levels of feature abstraction, and model redundancy varies between early and late pruning phases. Our dynamic weighting mechanism is introduced precisely to solve this issue. It adaptively adjusts the contribution weights of "attention scores" and "attention entropy" in the final assessment according to the network depth, entropy distribution, and pruning progress. For instance, the mechanism relies more on entropy in shallow layers to capture fundamental feature patterns, while trusting the more stable attention scores in deeper layers.

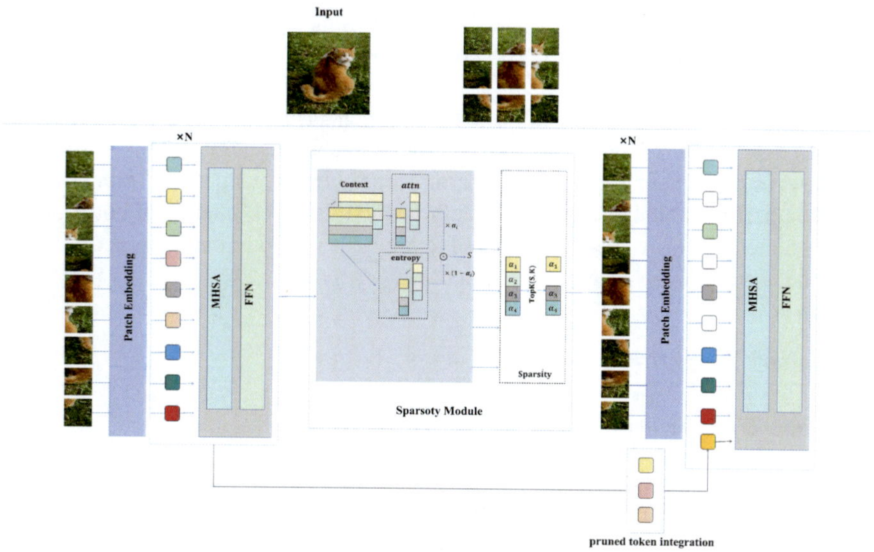

Fig. 1. The framework of our token pruning method. Given input tokens, their importance scores are computed through the evaluation token module. Based on the calculated scores, we perform token pruning while maintaining positional encoding consistency, retaining only the most informative tokens. It is worth noting that special tokens (CLS and distillation tokens) are always preserved due to their crucial roles in classification and knowledge distillation.

Consequently, the combination of "attention entropy and dynamic weighting" realizes a context-aware token evaluation: it not only judges the importance of a token per se, but also dynamically decides how to evaluate based on the model's current "state" (e.g., layer depth, redundancy). This enables us to achieve more precise and robust pruning decisions without introducing any additional learnable parameters. The overall procedure is depicted in Fig. 1.

Specifically, we first propose a multi-granularity entropy attention evaluation method that synchronously calculates the attention distribution entropy for each head during layer-wise forward propagation, using it as a complementary metric to attention scores. Subsequently, we design a dynamic weight fusion module that automatically adjusts the evaluation strategy based on: (1) network depth (emphasizing entropy analysis in shallow layers while focusing on original attention in deep layers), (2) pruning progress (adopting aggressive pruning in early stages and conservative pruning in later stages), and (3) current entropy characteristics (enhancing entropy weight when entropy is high). Finally, we generate comprehensive importance scores through global normalization to guide token pruning decisions.

Compared with existing approaches, our solution demonstrates four distinct advantages: (1) complete reliance on intrinsic attention mechanism features with zero parameter overhead; (2) dynamic adaptation to evaluation requirements across different network layers; (3) effective identification of semantically critical "hub tokens" with dispersed attention patterns through entropy analysis; and (4) achieving dynamic pruning accuracy while only requiring static pruning computational costs. Experimental results show that compared to fixed-weight schemes, our dynamic adjustment strategy demonstrates superior accuracy-efficiency balance across various model architectures (ViT-B) and pruning intensities.

We conducted extensive token pruning experiments using the most widely adopted ViT-B architecture on CIFAR-10 and CIFAR-100 datasets. The results show 50% throughput improvement and over 30% parameter reduction, while confirming that our method outperforms other approaches mentioned in this paper in terms of both computational efficiency and accuracy. Compared to previous methods, our approach achieves a better balance between accuracy and latency.

2 Related Work

2.1 Vision Transformers (ViT)

The Transformer architecture was initially developed for natural language processing (NLP) by (Vaswani et al.,2017) [9], and has recently been successfully adapted to computer vision tasks (Xu et al., 2022) [10]. Particularly when processing large-scale datasets, Transformers have demonstrated exceptional performance, effectively challenging the long-standing dominance of traditional convolutional neural networks (CNNs) in image processing.

The Vision Transformer (ViT) operates by partitioning images into fixed-size patches (tokens), which are then processed through stacked Transformer blocks containing multi-head self-attention (MHSA) and multilayer perceptron (MLP) components. Ongoing research continues to produce various ViT variants that enhance model performance. For instance, LV-ViT (Jiang et al., 2021) [11] investigated multiple training techniques specifically for vision transformers, achieving significant performance improvements. Similarly, DGT (Liu et al., 2022) [12] introduced a dynamic group attention mechanism designed to accelerate the inference process.

2.2 Dynamic ViT Pruning

DynamicViT (Rao et al., 2021) [5] pioneered an attention-weight-based dynamic pruning framework, achieving efficient ViT compression through dynamic token sparsification. AS-ViT (Liu et al., 2022) [13] introduced an adaptive sparse token pruning framework leveraging attention-head importance weighting, enabling accuracy-computation trade-offs across datasets [13]. A-ViT (Yin et al., 2021) proposed an adaptive token selection mechanism that dynamically adjusts token counts in attention layers for model lightweighting [14]. Evo-ViT (Xu et al., 2022) implemented dynamic pruning via a slow-fast token evolution mechanism [15]. X-Pruner (Yu et al., 2023) developed an interpretability-aware pruning method using class-level regularization masks to automatically select critical units for structured pruning [16]. STEP (Proust et al., 2025) enhanced inference efficiency by dynamically merging semantically similar patches into super-patches and pruning non-critical ones early [17].While dynamic pruning offers greater flexibility and typically outperforms static pruning, it often incurs additional computational overhead due to its adaptive nature.

2.3 Static ViT Pruning

VTP (Zhu et al., 2021) [18] achieved parameter compression in both multi-head self-attention and feed-forward networks through L1 regularization-based sparse training, applying uniform compression across all components [18]. Yang et al. (2023) proposed a global structural pruning method using Hessian-matrix-aware saliency analysis, which evaluates the sensitivity of different ViT structures to enable efficient model compression [19]. LPViT (Xu et al., 2024) introduced a semi-structured pruning scheme that leverages the block structure of linear layers in ViT architectures to optimize matrix multiplication efficiency [20]. Static pruning maintains consistent computational graph structures, offering advantages in optimization ease, parallelization capability, and practical deployment feasibility. This work focuses on token compression—a research direction orthogonal to static ViT weight compression—while simultaneously enhancing compression rates to amplify the benefits of weight pruning.

3 Method

3.1 Preliminary

The ViT model partitions an image into fixed-size non-overlapping patches, embeds them into high-dimensional features, and prepends a class token to these features. All features are then processed by learnable positional encodings before being fed into multiple Transformer blocks. Each Transformer block consists of two key components: the Multi-Head Self-Attention (MHSA) mechanism that captures global dependencies among features, and the Multi-Layer Perceptron (MLP) that performs nonlinear transformations. This process can be formally summarized as follows:

$$x_{MHSA} = x + MHSA(LN(x)) \tag{1}$$

$$x_{FFN} = x_{MHSA} + MLP(LN(x_{MHSA})) \tag{2}$$

Here, LN denotes Layer Normalization, while FFN represents the Feed-Forward Network.

3.2 Token Importance Evaluation Metrics

Existing approaches for assessing token importance include computation-free attention scores (Huang et al., 2024) [21] and class token similarity measures (Liang et al., 2022). However, these metrics suffer from two critical limitations: (1) they employ singular evaluation criteria, failing to account for the fact that different attention heads may focus on distinct token features (Liu et al., 2022) [13]; and (2) they assign uniform weights to all attention heads, thereby neglecting the differential contributions of varied attention patterns to token importance. This oversight becomes particularly consequential in deep Transformer architectures, where attention patterns often exhibit substantial layer-wise variations.

To enable more precise token importance evaluation, we first introduce a quantitative metric for attention concentration. By computing the information entropy of attention distributions, we can effectively measure a model's focus intensity on specific tokens. Formally, for the h-th attention head in the i-th layer, the attention entropy is calculated as:

$$E_i = -\sum_{h=1}^{H}\sum_{j=0}^{N}\left(A_i^{(h)}[j,:]\log\left(A_i^{(h)}[j,:]+\epsilon\right)\right) \tag{3}$$

Here, $A_i^{(h)} \in \mathbb{R}^{N\times N}$ indicate the output of the h-th attention head in the i-th layer, where $\epsilon = 10^{-8}$ serves as a numerical stabilization term. A lower entropy value indicates more concentrated attention distribution.

3.3 Dynamic Weight Adjustment Strategy

The motivation for our attention-weighted importance approach stems from CNN-based image feature metrics (Chen et al., 2021) [22]. Building upon this foundation, we designed a dynamic layer-weighting adjustment mechanism that accounts for varying contributions of different network layers to token importance evaluation. This mechanism holistically integrates two critical factors: network depth, and attention concentration.

Formally, for the i-th layer (where i = 1, 2, ..., L denotes the layer index), the dynamic weight α_i is computed as:

$$\alpha_i = \beta + \gamma\frac{i}{L} + \sigma\cdot\overline{E_i} \tag{4}$$

where β and γ are balancing coefficients that control the baseline weight and the depth-sensitive increment, respectively. $\frac{i}{L}$ provides a normalized measure of the current depth, ensuring that deeper layers receive a higher base weight. $\overline{E_i}$ is the average attention entropy of the i-th layer, reflecting its overall attention concentration. σ denotes the sigmoid function, which maps the entropy term into a bounded contribution. The resulting weight α_i is constrained to the interval [0.2,0.8]. This constraint serves two purposes: (1) it ensures numerical stability during training and inference by preventing the weight from saturating near 0 or 1, and (2) it guarantees that neither the attention score nor the entropy term is ever completely suppressed, preserving the complementary nature of the two evaluation dimensions. The specific bounds (0.2 and 0.8) were determined empirically to offer a sufficient dynamic range while maintaining stable optimization behavior across different datasets and model configurations.

3.4 Importance Score Calculation

Building upon the aforementioned quantitative metrics and adjustment strategies, we construct a comprehensive importance evaluation framework. The hierarchical attention aggregation serves as the core component of our method, whose design is grounded in the significant differences in feature abstraction levels captured by different Transformer layers (Vig et al., 2019) [23]. Specifically, shallow layers predominantly focus on local patterns while deeper layers emphasize global relationships.

First, compute the hierarchical attention aggregation. For the t-th token (where t = 1, ⋯ , N + 1,with t = 1 representing the class token) the attention aggregation is computed as follows:

$$S_{attn}^{(t)} = \frac{\sum_{i=1}^{L} \alpha_i \cdot mean(A_i[:, t])}{\sum_{i=1}^{L} \alpha_i} \tag{5}$$

Here, A_i denotes the attention matrix at the i-th layer, and α_i represents the dynamic layer weight computed in Eq. (4).

Subsequently, we calculate the entropy-compensated score for the t-th token by normalizing the entropy values and converting them into importance measures, where tokens with lower entropy are considered more significant:

$$S_{entropy}^{(t)} = 1 - \frac{\sum_{i=1}^{L} (1 - \alpha_i) \cdot E_i^{(t)}}{\sum_{i=1}^{L} (1 - \alpha_i) \cdot \max\left(E^{(t)}\right)} \tag{6}$$

Here, $E_i^{(t)}$ denotes the attention entropy of the t-th token in the i-th layer, and $max(E^{(t)})$ represents the maximum entropy value of this token across all layers.

Finally, the importance score of the t-th token is obtained through transformer layer weighting:

$$S_{final}^{(t)} = \left(\frac{1}{L}\sum \alpha_i\right) \cdot S_{attn}^{(t)} + \left(1 - \frac{1}{L}\sum \alpha_i\right) \cdot S_{entropy}^{(t)} \tag{7}$$

This process maintains linear computational complexity, relying solely on the native output of attention matrices without introducing additional learnable parameters.

3.5 Token Feature Fusion

Although the pruned tokens are those deemed to have the least impact on the model through evaluation, removing a large number of such minimally influential tokens can still affect the model. To minimize this impact, we adopt a token fusion approach: tokens with lower importance scores are normalized and weighted according to their significance, then merged into a single fused token. The positional encoding for this fused token is set as the average of the positional encodings of the pruned tokens.

3.6 Knowledge Distillation Based on the Full Model

We employ a standard teacher-student distillation framework, where the fully trained, unpruned ViT model serves as the teacher model to guide the training of the pruned student model. Our training objective comprises two components. Standard Cross-Entropy Loss($\mathscr{L}_{\mathrm{CE}}$),:

$$\mathcal{L}_{\mathrm{CE}} = \mathrm{CrossEntropy}(\mathrm{y}, \hat{\mathrm{y}}) \tag{8}$$

Here, y denotes the ground-truth labels, and ŷ represents the teacher model's predictions.

To further enhance performance, we minimize the output discrepancy between the student and teacher models using KL divergence:

$$\mathcal{L}_{\text{distill}} = \text{KL}(y, \hat{y}) \tag{9}$$

The combined training objective integrates these two losses with a balancing weight λ:

$$\mathcal{L} = \lambda\mathcal{L}_{\text{CE}} + (1 - \lambda)\,\mathcal{L}_{\text{distill}} \tag{10}$$

In our experiments, we set $\lambda = 0.7 =$ to optimize the trade-off between the two objectives. Compared to direct training methods, leveraging the full unpruned model as the teacher yields significant performance improvements for the pruned student model.

4 Experiments

4.1 Experimental Configuration

Our study employed the CIFAR-10 (10-class) and CIFAR-100 (100-class) datasets, each containing 50,000 training and 10,000 test images, to evaluate a token-pruning-based compression method for Vision Transformers (ViT). All experiments were conducted under controlled conditions: (1) no data augmentation or distributed training was applied, (2) single-GPU execution on an NVIDIA GeForce RTX 3080 with fixed batch size 32, and (3) 10-epoch training to assess rapid model recovery. Due to hardware constraints of consumer-grade GPUs, dynamic pruning methods requiring specialized acceleration were evaluated solely for training accuracy performance, excluding potential speedup artifacts.

Table 1. Evaluate the effectiveness of different pruning approaches.

Method	CIFER-10			CIFER-100		
	Acc(%)	FLOPs(G)	Throughput (it/s)	Acc(%)	FLOPs(G)	Throughput (it/s)
Baseline	99.08	17.58	10.13	92.38	17.58	10.13
VTP	96.55	12.00	10.65	85.25	12.28	13.29
AS-ViT	98.36	12.24	13.16	90.95	12.32	13.38
Saint	98.63	12.28	/	90.90	12.32	/
DynamicViT	98.20	12.20	/	89.54	15.65	/
Ours	98.63	12.11	14.53	91.07	12.02	15.11

4.2 Analysis Results

Comparative Results with Existing Pruning Methods. We conducted comprehensive evaluations comparing our approach with state-of-the-art pruning methods including VTP (Zhu et al., 2021) [18], AS-ViT (Liu et al., 2022) [13], Saint (Jeddi et al., 2025) [24], and DynamicViT (Rao et al., 2021) [5] across three key metrics: model accuracy, throughput, and computational overhead. Under strictly controlled experimental conditions with a fixed 30% pruning ratio and matched computational budgets, our method demonstrates consistent advantages in both accuracy preservation and processing speed, as detailed in Table 1. Specifically, when reducing computational costs by approximately one-third, our pruned ViT-B model maintains exceptional accuracy with only a 0.55% drop on CIFAR-10 and 1.31% drop on CIFAR-100. The superior performance over VTP and DynamicViT can be attributed to our novel token importance assessment framework that intelligently combines both network depth analysis and entropy distribution evaluation, enabling more precise identification and preservation of critical tokens during the pruning process.

Efficiency Comparison with Existing Pruning Methods. As shown in Table 1, our experimental efficiency analysis reveals that the proposed method outperforms all compared approaches across three critical metrics: classification accuracy, throughput, and hardware latency. It should be noted that direct comparisons with some early methods were not possible due to unavailability of their source code and pre-trained models. Compared to DynamicViT and other approaches requiring additional scoring modules, our innovations demonstrate three key advantages: (1) exclusive utilization of intermediate MHSA features for token evaluation eliminates computational overhead and achieves higher throughput; (2) dynamic adaptation to evaluation requirements across different network layers and pruning stages; and (3) effective identification of semantically critical "hub tokens" with dispersed attention patterns through entropy analysis, thereby improving classification accuracy. When compared to VTP's single-evaluation approach, our method better captures the diverse features attended by different attention heads and their differential contributions to token importance, enabling more precise pruning of less critical tokens. Furthermore, unlike AS-ViT's L2-norm weighting (Liu et al., 2022) [13], norm-based pruning tends to eliminate tokens sensitive to fine-grained differences (Lin et al., 2020) [25], while our approach maintains better preservation of such discriminative features.

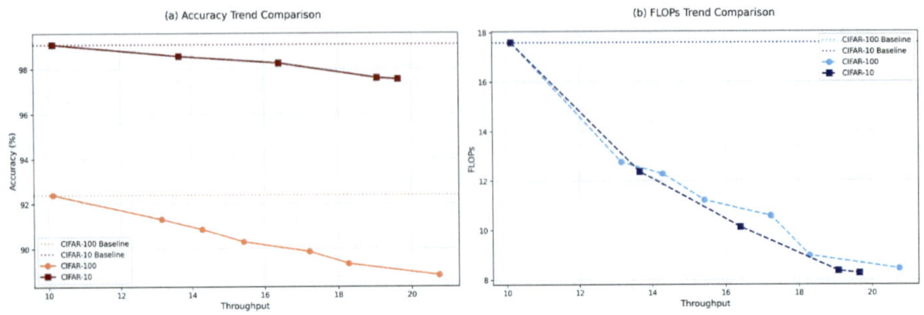

Fig. 2. (**a**) Model accuracy versus inference throughput. (**b**) Computational cost (FLOPs) versus throughput.

Figure 2 (Accuracy vs. Throughput) demonstrates that AEG-ViT maintains a smooth decreasing trend in model accuracy on both CIFAR-10 and CIFAR-100 as throughput increases. This indicates that the dynamic entropy-weighted strategy achieves more precise accuracy-efficiency balancing: when throughput improves by 50%, the accuracy drop on CIFAR-10 is merely 0.45%, compared to a 2.53% degradation with static pruning methods (e.g., VTP) under equivalent throughput conditions. Particularly noteworthy is AEG-ViT's significantly gentler slope in the high-pressure regime (Throughput > 15 it/s), evidencing how its entropy compensation mechanism effectively prevents over-pruning of critical tokens.

Figure 2 (FLOPs vs. Throughput) presents the computational efficiency of AEG-ViT across both datasets. While exhibiting similar overall trends, the CIFAR-100 curve demonstrates a steeper slope than CIFAR-10, indicating that achieving equivalent throughput gains requires sacrificing more computational resources (averaging 12% additional FLOPs reduction) for CIFAR-100. This divergence originates from dataset complexity—the fine-grained classification task in CIFAR-100 necessitates retaining more tokens to preserve feature discriminability. Notably, both curves exhibit inflection points in the high-pressure regime (Throughput >16 it/s), where the FLOPs reduction rate decelerates. This phenomenon validates the adaptive nature of the dynamic weighting mechanism: it automatically switches to conservative strategies when pruning approaches the model's capacity limits to protect critical computational paths, and confirms the protective effect of the adaptive mechanism under high-pressure conditions, effectively preventing accuracy collapse caused by excessive computational resource reduction.

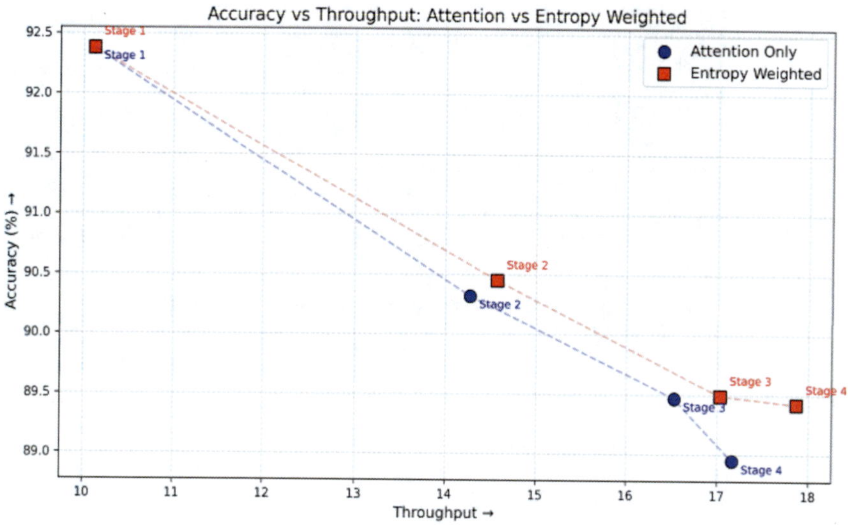

Fig. 3. Comparison of model accuracy versus inference throughput on CIFAR-100, evaluating two pruning strategies (Attention Only and Entropy Weighted) across four progressive pruning stages (Stage 1 to Stage 4). Each stage represents an increasing level of token reduction.

Figure 3 presents a performance comparison between our entropy-weighted pruning method and the baseline attention-based pruning approach on the CIFAR-100 dataset. Experimental results demonstrate that while both methods achieve significant improvements in computational efficiency, our approach shows distinct advantages in preserving model accuracy. This superiority primarily stems from three key aspects: First, our entropy-weighted evaluation mechanism more comprehensively captures the semantic importance of tokens, avoiding potential mis-pruning caused by relying solely on attention scores. Second, the dynamic weight adjustment strategy adaptively balances the preservation of local features and global relationships based on network depth. Finally, specifically tailored for CIFAR-100's fine-grained classification characteristics, our method places particular emphasis on protecting discriminative cross-category features. Notably, as the compression ratio increases, our method's advantage in accuracy preservation becomes more pronounced, validating the effectiveness of the proposed dynamic evaluation strategy in high-compression scenarios.

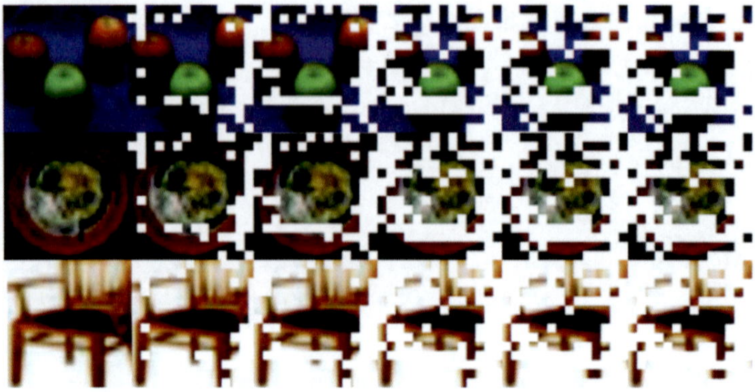

Fig. 4. Evolution of retained tokens with increasing pruning intensity. Visualization of how our AEG-VIT method selects tokens across a wide range of pruning ratios (0% to 54.3%). The persistent concentration of retained tokens on the foreground object highlights the precision and stability of our entropy-based importance metric.

Visualization Analysis and Discussion. Our in-depth investigation into the dynamic pruning behavior of the adaptive token sparsification method across images of varying complexity reveals critical insights through systematic analysis on the ImageNet validation set (Liu et al., 2022) [13]. The retained token counts across pruning stages exhibit a unimodal Gaussian distribution, where the peak region corresponds to medium-complexity samples while extremely high/low complexity samples show relatively lower proportions. This empirical pattern fundamentally validates the universality of fixed-ratio pruning strategies, as most real-world samples with moderate recognition difficulty only require balanced token pruning to maintain classification accuracy. The proposed method demonstrates enhanced capability in intelligently preserving high-information tokens crucial for classification while eliminating low-contribution tokens, thereby effectively reducing computational complexity without compromising model performance.

As visualized in Fig. 4, the token reduction process and comparative experiments confirm that our dynamic adjustment strategy achieves superior accuracy-efficiency balance over fixed-weight schemes across various ViT-B architectures and pruning intensities.

5 Conclusion

In this study, we propose an adaptive validation-driven progressive pruning method (AEG-ViT) that achieves more accurate token importance evaluation by dynamically integrating attention entropy with original attention scores. Unlike existing approaches that rely on fixed pruning ratios or additional prediction modules, our dynamic weight adjustment mechanism can adaptively modify pruning strategies based on network depth and entropy distribution, enabling intelligent compression. Furthermore, we introduce a validation-driven safety pruning mechanism that dynamically adjusts pruning intensity through binary search to strictly control model accuracy loss within predefined thresholds. Experimental results demonstrate that AEG-ViT achieves 50% throughput improvement and over 30% parameter reduction on ViT-B models while maintaining accuracy comparable to the original model on both CIFAR-10 and CIFAR-100 datasets. Compared to methods like DynamicViT and VTP, our solution achieves better balance between computational efficiency and accuracy, particularly excelling at processing complex samples while avoiding over-pruning or under-pruning issues. Future work on model lightweighting will focus on: (1) multimodal compression combining structured pruning and quantization to further improve deployment efficiency on edge devices; (2) task generalization by exploring AEG-ViT's applicability to downstream tasks like object detection and semantic segmentation; (3) hardware-friendly optimization through more efficient sparse computation strategies designed for GPUs and specialized accelerators to fully exploit dynamic pruning's acceleration potential.

This research provides new insights for efficient compression of vision Transformers, with its adaptive characteristics offering stronger practical value across diverse application scenarios.

References

1. Dosovitskiy, A. et al.: An image is worth 16x16 words: transformers for image recognition at scale. arXiv:2010.11929. https://arxiv.org/abs/2010.11929. (2020)
2. Touvron, H., Cord, M., Douze, M., Massa, F., Sablayrolles, A., Jégou, H.: Training data-efficient image transformers & distillation through attention. arXiv preprint arXiv:2012.12877. (2020) https://arxiv.org/abs/2012.12877
3. Raghu, M., Unterthiner, T., Kornblith, S., Zhang, C., Dosovitskiy, A.: Do vision transformers see like convolutional neural networks? Adv. Neural Inf. Proces. Syst. **34**, 12116–12128 (2021)
4. Liang, Y., Ge, C., Tong, Z., Song, Y., Wang, J., Xie, P.: Not all patches are what you need: expediting vision transformers via token reorganizations. arXiv preprint arXiv:2202.07800 (2022). https://arxiv.org/abs/2202.07800. Last accessed 04 Dec 2025
5. Rao, Y., Zhao, W., Liu, B., Lu, J., Zhou, J., Hsieh, C.-J.: DynamicViT: efficient vision transformers with dynamic token Sparsification. Adv. Neural Inf. Process. Syst. **34**, 13937–13949 (2021)
6. Pan, B., Panda, R., Jiang, Y., Wang, Z., Feris, R., Oliva, A.: IA-RED^2: interpretability-aware redundancy reduction for vision transformers. arXiv preprint arXiv:2106.12620 (2021). https://arxiv.org/abs/2106.12620. Last accessed 04 Dec 2025.
7. Tang, Y. et al.: Patch slimming for efficient vision transformers. arXiv:2106.02852. (2021) https://arxiv.org/abs/2106.02852
8. Chen, T., Cheng, Y., Gan, Z., Yuan, L., Zhang, L., Wang, Z.: Chasing sparsity in vision transformers: an end-to-end exploration. arXiv preprint arXiv:2106.04533. (2021) https://arxiv.org/abs/2106.04533
9. Vaswani, A. et al.: Attention is all you need. arXiv:1706.03762. (2017) http://arxiv.org/abs/1706.03762
10. Xu, Y. et al.: Transformers in computational visual media: a survey. Comput. Vis. Media. **8**(1), 33–62 (2022)
11. Jiang, Z. et al.: All Tokens Matter: Token Labeling for Training Better Vision Transformers. Adv. Neural Inf. Process. Syst. **34**, 18590–18602 (2021)
12. Liu, K., Wu, T., Liu, C., Guo, G.: Dynamic group transformer: a general vision transformer backbone with dynamic group attention. In: Proceedings of the Thirty-First International Joint Conference on Artificial Intelligence, pp. 1187–1193. IJCAI Organization (2022)

13. Liu, X., Wu, T., Guo, G.: Adaptive sparse VIT: towards learnable adaptive token pruning by fully exploiting self-attention. arXiv preprint arXiv:2209.13802. (2022) https://arxiv.org/abs/2209.13802
14. Yin, H., Vahdat, A., Alvarez, J., Mallya, A., Kautz, J., Molchanov, P.: ADAVIT: adaptive tokens for efficient vision transformer. arXiv preprint arXiv:2112.07658. (2021) https://arxiv.org/abs/2112.07658
15. Xu, Y., et al.: Evo-VIT: slow-fast token evolution for dynamic vision transformer. In: Proceedings of the AAAI Conference on Artificial Intelligence, vol. 36, 3, pp. 2964–2972. AAAI Press (2022).
16. Yu, L., Xiang, W.: X-pruner: EXplainable pruning for vision transformers. In: 2023 IEEE/CVF Conference on Computer Vision and Pattern Recognition (CVPR), pp. 24355–24364. IEEE, Vancouver (2023)
17. Proust, M., Poreba, M., Szczepanski, M., Haroun, K.: STEP: SuperToken and early-pruning for efficient semantic segmentation. In: Proceedings of the 17th International Joint Conference on Computer Vision, Imaging and Computer Graphics Theory and Applications, pp. 50–61. SCITEPRESS (2025)
18. Zhu, M., Tang, Y., Han, K.: Vision transformer pruning. arXiv preprint arXiv:2104.08500. (2021) https://arxiv.org/abs/2104.08500
19. Yang, H., Yin, H., Shen, M., Molchanov, P., Li, H., Kautz, J.: Global vision transformer pruning with hessian-aware saliency. In: 2023 IEEE/CVF Conference on Computer Vision and Pattern Recognition (CVPR), pp. 18547–18557. IEEE, Vancouver (2023)
20. Xu, K. et al.: LPViT: low-power semi-structured pruning for vision transformers. arXiv:2407.02068. (2025) https://arxiv.org/abs/2407.02068
21. Huang, K., Zou, H., Xi, Y., Wang, B., Xie, Z., Yu, L.: IVTP: instruction-guided visual token pruning for large vision-language models. In: Lecture Notes in Computer Science, vol. 15342, pp. 214–230. Springer, Cham (2024)
22. Chen, W., Cong, C., Huang, L.: Image feature measurement of convolutional neural network based on information entropy. Laser & Optoelectron. Prog. **58**(22), 2211004 (2021)
23. Vig, J., Belinkov, Y.: Analyzing the structure of attention in a transformer language model. arXiv preprint arXiv:1906.04284. (2019) https://arxiv.org/abs/1906.04284
24. Jeddi, A., Baghbanzadeh, N., Dolatabadi, E., Taati, B.: Similarity-aware token pruning: your VLM but faster. arXiv preprint arXiv:2503.11549. (2025) https://arxiv.org/abs/2503.11549
25. Lin, M. et al.: HRank: filter pruning using high-rank feature map. arXiv:2002.10179. (2020) https://arxiv.org/abs/2002.10179

A Frequency–Time Dual-Domain Fusion Algorithm for Multi-IMU Human Motion Recognition

Zihan Wang[1], Zhiqiang Guo[1], Yongsheng Wang[1](✉), and Yi Li[2]

[1] School of Information Engineering, Wuhan University of Technology, Wuhan 430070, China
348527@whut.edu.cn
[2] Alibaba Cloud, Shenzhen 518038, China

Abstract. The recognition of human motion using multi-inertial measurement units (IMUs) is of significant importance in fields such as mobile health, intelligent rehabilitation, and human-computer interaction. However, existing methods typically face challenges such as fixed frequency-domain filtering, non-causal temporal modeling, and insufficient multi-node information fusion, which limit their recognition performance in real-time and heterogeneous scenarios. To address these challenges, this paper proposes a frequency-time dual-domain human activity recognition network (FTD-HARNet) aimed at enhancing the recognition performance in real-time and heterogeneous environments. The proposed method utilizes an adaptive Gabor-frequency spectrum filtering module (AGSF) for frequency-domain noise suppression, employs a causal multi-scale temporal convolution network (CM-TCN) for temporal modeling, and integrates a dual-domain attention fusion module (DDAF) for efficient fusion of cross-node information. Experimental results demonstrate that FTD-HARNet outperforms mainstream baseline methods on several publicly available datasets, significantly improving the robustness and generalization capability of motion recognition. This approach provides an effective solution for low-power wearable systems and intelligent interactive devices.

Keywords: Human activity recognition · Inertial measurement unit · Frequency-domain adaptation · Causal convolution · Dual-domain attention fusion · Wearable sensing

1 Introduction

Human activity recognition (HAR) based on inertial measurement units (IMUs) is a fundamental capability in modern mobile health, sports monitoring, human–computer interaction, and occupational safety. Owing to their low cost, privacy-preserving nature, and around-the-clock operability, IMUs have increasingly become an important complement or alternative to vision-based solutions, catalyzing rapid progress from shallow feature extraction to deep learning methods [1, 2].An IMU system typically comprises accelerometers, gyroscopes, and magnetometers, enabling the capture of diverse

L. Huang and D. Greenhalgh (Eds.): ICMLC 2026, LNNS 2061, pp. 719–731, 2027.
https://doi.org/10.1007/978-3-032-31236-5_49

dynamic signals generated during human movement. However, a single IMU sensor has inherent limitations in characterizing complex, high-dimensional motion patterns [3]. Multi-IMU systems, by synchronously acquiring acceleration, angular velocity, and orientation data from different body locations, enable fine-grained modeling of full-body motion. The spatial distribution of multiple sensors and the complementary nature of their information allow the system to more accurately discriminate similar actions, detect subtle postural changes, and maintain strong recognition robustness in the presence of signal occlusion, noise interference, or partial sensor failure [4–7].

Although multi-IMU wearable systems have been widely studied and applied in domains such as motor rehabilitation, occupational safety, and immersive interaction, they still face challenges related to device size, power consumption, and data synchronization [8–10].With advances in low-power sensors and wireless synchronization technologies, multi-IMU fusion has become a key research direction for high-accuracy, low-latency human behavior sensing. Despite notable progress, existing multi-IMU fusion methods still exhibit critical limitations, particularly in real-time streaming recognition and the accuracy of multi-sensor data fusion [11, 12].

To address these issues, we propose FTD-HARNet (Frequency–Temporal Dual-domain Human Activity Recognition Network), an adaptive, end-to-end recognition framework that integrates frequency–and time-domain processing. On the frequency-domain front end, we introduce an Adaptive Gabor–Spectrum Filtering (AGSF) module that employs learnable center frequencies and bandwidths to adaptively select spectral bands, thereby suppressing environmental noise and aliasing and improving signal quality. For temporal modeling, we adopt a Causal Multi-scale Temporal Convolutional Network (CM-TCN) that uses dilated convolutions to expand the temporal receptive field while preserving causal consistency, enabling efficient real-time sequence modeling. Finally, at the feature-fusion stage, we propose a Dual-Domain Attention Fusion (DDAF) module that jointly models temporal-domain self-attention and channel-domain cross-attention, explicitly enabling cross-node collaboration and global feature integration.

Validated on three public datasets—Opportunity, PAMAP2, and DSADS—FTD-HARNet significantly outperforms mainstream baselines across multiple evaluation metrics. On complex activity recognition tasks, it achieves substantial gains in mF1 and mW-F1, with particularly strong performance in multi-node fusion settings. Ablation studies further corroborate the complementary roles of each module, highlighting the frequency-domain enhancement module's pivotal effect on denoising and spectral filtering, the temporal causal modeling module's advantage for real-time recognition, and the dual-domain fusion module's importance in improving recognition accuracy and robustness.

2 Related Works

In recent years, deep learning–based IMU signal processing methods have been widely adopted for human activity recognition (HAR). Traditional approaches largely rely on fixed-window feature extraction, which demands hand-crafted features and tends to perform poorly when handling dynamically varying or complex activities [13, 14]. With the advance of deep learning—particularly the demonstrated strengths of convolutional neural networks and recurrent neural networks for sequential data—an increasing body of work has moved toward end-to-end learning and modeling of IMU signals.

Attention-based fusion has emerged as an effective means to improve the performance of multi-sensor data integration. Tao et al. [15] extract features from each sensor using CNNs and then learn the importance of different sensor locations via a sensor–location attention mechanism, achieving strong results. Nevertheless, because this method depends on fixed-window feature extraction and does not strictly enforce causality, it is difficult to deploy for real-time streaming recognition.

The IF-ConvTransformer framework [16] combines an IMU fusion module with a ConvTransformer subnetwork to enhance multi-node recognition accuracy. While it excels at modeling global attention, it employs long-window global attention that is not well adapted to resource-constrained edge devices and lacks systematic consideration of factors such as changes in sensor placement, hardware heterogeneity, power budgets for real-time deployment, and synchronization errors.

Given the noise and drift inherent in IMU data, appropriate time–frequency preprocessing is crucial for improving recognition performance. Wang et al. [17] proposed dynamic wavelet selection with a gated enhancement mechanism, markedly improving denoising and feature representation quality. Compared with conventional filtering, learning-based denoising can reduce acceleration reconstruction errors in real-world settings and mitigate navigation-related angular errors. For spectral enhancement of IMU signals, Yuan et al. [18] improved signal quality via learning-based denoising, reducing errors by an order of magnitude. Damagatla et al. [19] derived "clean" IMU data from EKF-estimated position, velocity, and attitude using inverse kinematics, and then applied LightGBM and CatBoost to compensate IMU errors.

As multi-IMU systems continue to proliferate, more deep learning– and sensor fusion–based solutions have been introduced. Methods such as DeepConvLSTM [20] and DeepSense [21] have reported strong results on public datasets, yet challenges remain in fusing information across multiple nodes and in capturing long-range temporal dependencies.

Despite these advances, most existing methods still struggle with temporal modeling and multi-sensor information fusion. To address these issues, building upon the IF-ConvTransformer framework, this paper proposes FTD-HARNet. Our design emphasizes a closed-loop integration of an adaptive frequency-domain front end, efficient causal temporal modeling, and dual-domain attention fusion, enabling accurate and efficient activity recognition in multi-sensor environments.

3 Methodology

3.1 The Overall Framework

The overall architecture of the proposed FTD-HARNet is shown in Fig. 1 and comprises two functional modules: an IMU data fusion module and a feature encoding–decoding module. In the first part, the system ingests multi-node IMU streams, where each node provides nine-dimensional raw measurements from a tri-axial gyroscope, accelerometer, and magnetometer. To account for the distinct time–frequency characteristics and noise profiles of different sensing channels, we introduce an Adaptive Gabor–Spectrum Filtering (AGSF) module at the signal preprocessing stage. AGSF enables frequency-domain adaptive feature extraction within an end-to-end framework by learning discriminative center frequencies and bandwidths, thereby suppressing sensor noise and aliasing. At the multi-sensor fusion layer, the system adopts the Posture Branch and Motion Branch of the IF-ConvTransformer framework to separately model posture-related and dynamics-related features. By complementing these branches, the model fuses local temporal cues with global motion patterns, achieving multi-source data fusion across both spatial and temporal dimensions.

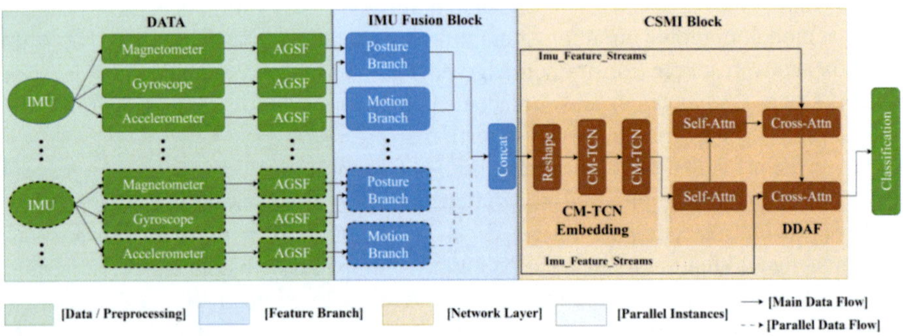

Fig. 1. The overall architecture of the proposed method. The framework is divided into three stages shaded in different background colors. Legend: Green boxes represent data sources and preprocessing modules (AGSF); Blue boxes denote feature extraction branches; Brown boxes indicate specific network layers or operations (e.g., CM-TCN, Attention). Solid borders and arrows show the main processing flow, while dashed borders and arrows indicate parallel streams from multiple IMU instances.

The second part is the Causal Spatiotemporal Modeling and Interaction (CSMI) module, whose core objective is to establish causally consistent, efficient temporal modeling alongside cross-domain information exchange. First, a Causal Multi-scale Temporal Convolutional Network (CM-TCN) is employed to extract causal convolutional features, expanding the temporal receptive field via dilated convolutions while strictly enforcing causality to prevent "future information leakage" in streaming recognition. The encoded temporal features are then passed to a Dual-Domain Attention Fusion (DDAF) module, which jointly captures long-range dependencies along the temporal dimension and the interactions and complementarities among different IMU nodes along the channel dimension, thereby realizing global spatiotemporal modeling in a unified representation space. Finally, the fused high-level features are fed to a classifier for activity state discrimination, enabling precise recognition of complex human motions. Overall, the framework balances temporal causality, frequency-domain adaptivity, and multichannel coordination while maintaining computational efficiency and feasibility for real-time deployment.

3.2 Pre-Filtering Module

In the IMU signal processing module, we propose an Adaptive Gabor–Spectral Filter (AGSF) to achieve frequency-domain adaptive feature extraction and noise suppression, as shown in Fig. 2. Traditional IMU signal preprocessing often uses bandpass filters with fixed parameters or empirical window functions. These methods cannot dynamically optimize the filter for different motion patterns and sensor channel characteristics. To overcome this limitation, AGSF parameterizes the Gabor function in a learnable form, allowing the filter to automatically adjust its frequency response during training through backpropagation. Its time-domain expression can be written as:

$$g(t; f_c, \sigma) = \exp\left(-t^2/2\sigma^2\right) * \cos(2\pi f_c t) \tag{1}$$

Here, f_c represents the center frequency of the filter, and σ controls the bandwidth. Both are learnable parameters that are automatically updated during training through backpropagation, enabling adaptive selection for different motion modes in the frequency domain. The filtered signal not only enhances information in the primary motion frequency band but also suppresses high-frequency interference caused by environmental noise, posture shifts, and non-target movements. Compared to fixed filtering methods, AGSF can dynamically adjust the filter response, achieving task-related optimization of the signal's spectral structure without relying on manual parameter tuning.

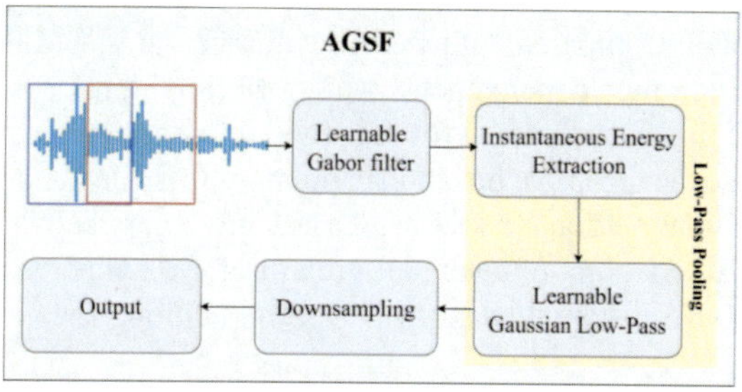

Fig. 2. AGSF: Composed of learnable Gabor filters, low-pass pooling, and downsampling. The detailed structure of the AGSF module. Note that the entire signal processing pipeline, from the input waveform to the final output features, operates in the time domain.

At the downsampling stage, we further devise a learnable low-pass pooling operation to mitigate spectral leakage and aliasing introduced by fixed windows and direct resampling in conventional short-time Fourier analysis. We first compute the instantaneous energy envelope $E(t) = |x(t)^2|$ and smooth it with a Gaussian convolution kernel $h(t, \sigma_p)$:

$$\overset{\circ}{E}(t) = (E * h)(t) = \int_{-\infty}^{+\infty} E(\tau) \exp\left(-(t-\tau)^2/2\sigma_p^2\right) d\tau \tag{2}$$

where the kernel parameter σ_p is learnable and controls the degree of smoothing and the effective low-pass cutoff. This operation performs energy-envelope smoothing in the time domain and is equivalent in the frequency domain to multiplication by a Gaussian low-pass function, thereby providing both anti-aliasing and energy normalization. Downsampling after this step substantially reduces data dimensionality while preserving the principal energy content, improving the stability and robustness of subsequent temporal modeling.

3.3 Causal Multi-Scale Temporal Convolutional Network

During feature encoding, we construct a Causal Multi-Scale Temporal Convolutional Network (CM-TCN) to perform efficient temporal modeling and deep feature extraction on the fused IMU time series, as shown in Fig. 3. Unlike recurrent neural networks—which rely on sequential computation and are difficult to parallelize—CM-TCN models sequence dynamics along the temporal axis using 1D convolutions and enforces causality via causal convolutions, ensuring that the output at any time step ttt depends only on the current and past inputs. This guarantees real-time operation and causal consistency for streaming recognition. Given an input sequence $X = [x_1, x_2, \ldots, x_T]$, the output of the l-th layer in CM-TCN can be expressed as:

$$y_t^{(l)} = \sum\nolimits_{i=0}^{k-1} \omega_i^{(l)} \cdot x_{t-i}^{(l-1)} \tag{3}$$

Where k denotes the convolution kernel size and $\omega_i^{(l)}$ the kernel weights at the l-th layer. By enforcing $i \geq 0$, the convolution accesses only past inputs, thereby strictly satisfying the causality constraint. Compared with recurrent architectures, CM-TCN maintains the ability to capture long-range dependencies while offering higher parallelism and more stable gradients, making it well suited for real-time inference on wearable devices.

To further enlarge the receptive field and capture multi-scale temporal dependencies, we introduce dilated convolutions. The core idea is to insert gaps between kernel elements with a dilation rate d, enabling exponential growth of the temporal receptive field via skip sampling without substantially increasing computation. The computation can be expressed as:

$$y_t^{(l)} = \sum\nolimits_{i=0}^{k-1} \omega_i^{(l)} \cdot x_{t-d\cdot i}^{(l-1)} \tag{4}$$

With a dilation rate d that grows exponentially with layer depth (e.g., $d = 2^{l-1}$), stacked convolutions can cover long-range temporal dependencies while keeping computation under control. To alleviate information degradation and gradient vanishing in deep architectures, we introduce residual connections at every layer. By progressively fusing inter-layer features via additive skip connections, the network accumulates multi-level representations across different temporal scales.

3.4 Dual-Domain Attention Fusion Module

At the feature-fusion stage, we design a Dual-Domain Attention Fusion (DDAF) module to perform joint modeling and high-level feature interaction for multi-IMU signals across two complementary domains: time and channel. DDAF consists of a time-domain

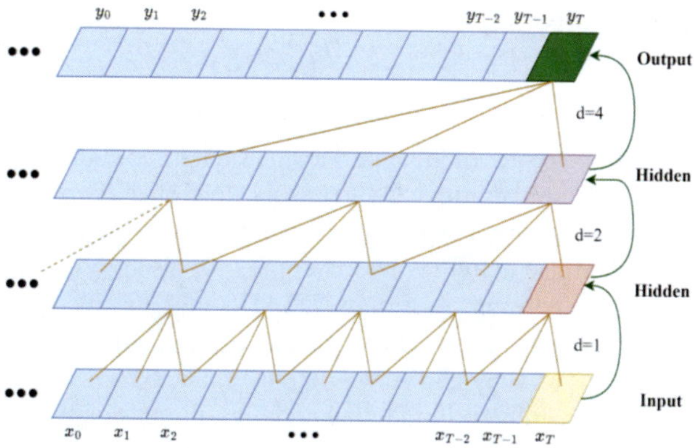

Fig. 3. Multi-scale Causal Temporal Convolutional Network (CM-TCN) structure

self-attention encoder and a channel-domain cross-attention decoder. This layered architecture captures dynamic dependencies along the temporal dimension and establishes cross-node interactions along the channel dimension, thereby enabling deep fusion of multi-source spatiotemporal information.

In the encoding stage, DDAF employs a Transformer-based time-domain self-attention mechanism to characterize the temporal dynamics of each IMU node. Given an input sequence $X \in R^{T \times d}$, the self-attention is computed as:

$$\text{Attention}(Q, K, V) = \text{softmax}\left(\frac{QK^T}{\sqrt{d_k}}\right)V \tag{5}$$

In the decoding layer, we introduce a channel-domain cross-attention mechanism to explicitly model interactions among different IMU nodes. Let the encoded features from N sensors be $\{H_1, H_2, \ldots, H_N\}$; the cross-attention for the i-th channel is given by:

$$\text{CrossAtten}(H_i) = \text{softmax}\left(\frac{Q_i K_j^T}{\sqrt{d_k}}\right)V_j, j \neq i \tag{6}$$

Through cross-channel query–key–value matching, this mechanism explicitly captures dynamic synergies across different body segments, yielding a unified representation that integrates local actions with whole-body motion. By establishing temporal dependencies in the encoder and cross-channel interactions in the decoder, the dual-domain attention module forms a hierarchically coupled information flow: the encoder strengthens per-node temporal representations, while the decoder fuses multi-node, cross-channel features—thereby enabling comprehensive modeling and accurate recognition of complex human movements.

4 Results and Discussion

4.1 Datasets

Our experiments use three representative public HAR datasets: PAMAP2, DSADS, and Opportunity. PAMAP2 comprises three IMU sensors worn on the wrist, chest, and ankle, together with a heart-rate monitor; all signals are sampled at 100 Hz. It covers 18 categories of daily and exercise activities and is commonly used to evaluate multimodal fusion and low-power recognition algorithms. DSADS contains five synchronized IMU nodes, each providing accelerometer, gyroscope, and magnetometer signals sampled at 25 Hz, with 19 categories of daily and sports activities; it serves as an important benchmark for multi-node spatiotemporal modeling and high-dimensional cooperative action recognition. Opportunity is among the most complex and representative multimodal HAR benchmarks to date, featuring multiple IMU nodes on the upper and lower limbs and the torso, along with synchronized magnetic, accelerometric, inertial, and environmental sensors at 30 Hz. This dataset emphasizes the recognition of micro-actions and object manipulations, making it particularly suitable for assessing performance in multi-source heterogeneous signal fusion and context-aware tasks.

4.2 Evaluation Metrics

To ensure objective and reproducible evaluation, we adopt a protocol that primarily uses leave-one-subject-out (LOSO) cross-validation and complements it with K-fold cross-validation. For datasets with multiple subjects, we iteratively take all samples of subject s as the test set and use the remaining subjects for training, traversing all subjects; the final result is reported as the mean and standard deviation over three runs with different random seeds. When the number of subjects is large or only an official split is provided, we use $K = 10$ K-fold cross-validation, repeat each fold three times, and report the mean $\pm$ standard deviation. For datasets containing a single subject or with an official test set, we evaluate directly on the default test split.

We use macro-averaged F1 (mF1) and class-proportion–weighted F1 (mW-F1) as metrics. Let the confusion counts for class ccc be TP_c, FN_c and FP_c.Precision and recall are:

$$\mathrm{P}_c = \mathrm{TP}_c/(\mathrm{TP}_c + \mathrm{FP}_c), \mathrm{R}_c = \mathrm{TP}_c/(\mathrm{TP}_c + \mathrm{FN}_c) \tag{7}$$

The per-class F1 、Macro F1 and the class-proportion–weighted F1 are:

$$\begin{gathered} \mathrm{F1}_c = \frac{2\mathrm{P}_c\mathrm{R}_c}{(\mathrm{P}_c+\mathrm{R}_c)}, \mathrm{mF1} = \frac{\sum_{c=1}^{C}\mathrm{F1}_c}{C} \\ \mathrm{mW\text{-}F1} = \frac{\sum_{c=1}^{C} n_c\mathrm{F1}_c}{\sum_{c=1}^{C} n_c} \end{gathered} \tag{8}$$

where n_c is the number of samples in class c and C is the number of classes. Model selection is based on the best validation-set mW-F1; hyperparameters are then fixed for testing. We also report the mean and standard deviation over three runs to quantify variance due to random initialization and facilitate fair comparison.

4.3 Experiments Results and Analysis

Table 1 presents a performance comparison between the proposed model and various mainstream human activity recognition algorithms on the Opportunity, PAMAP2, and DSADS public datasets. As shown in the results, our method outperforms existing methods across all metrics and demonstrates significant advantages in multi-node and complex activity recognition scenarios.

Table 1. Comparative experiments on wearable device-based datasets

Model	Opportunity		Pamap2		DSADS
	mF1 (%)	mW-F1 (%)	mF1 (%)	mW-F1 (%)	mF1/ mW-F1 (%)
IF-ConvTransformer [16]	47.41 ±0.44	87.59 ±0.25	87.02 ±0.29	92.43 ±0.49	85.52 ±0.53
DeepConvLSTM [20]	40.21 ±0.54	85.25 ±0.21	77.77 ±0.83	89.06 ±0.06	83.12 ±1.25
DeepSense [21]	42.67 ±0.11	86.98 ±0.05	89.62 ±1.32	92.51 ±0.76	84.99 ±1.13
ConvBoost [22]	45.94 ±0.87	87.76 ±0.33	**91.83** ±1.54	**94.54** ±0.27	87.26 ±0.46
Attend-Discriminate [23]	45.87 ±1.30	86.74 ±0.20	86.44 ±0.34	92.96 ±0.28	84.99 ±0.59
Global Fusion [24]	37.31 ±0.96	85.41 ±0.20	88.17 ±1.37	91.49 ±0.72	85.57 ±0.26
FTD-HARNet	**51.27** ±1.73	**89.15** ±0.47	89.47 ±0.32	93.82 ±0.73	**87.83** ±0.52

On the Opportunity dataset, FTD-HARNet achieves an mF1 score of 51.27%, a 3.86 percentage point improvement over the baseline model IF-ConvTransformer, with the most notable improvement observed in recognizing complex interactive actions. This improvement is primarily due to the model's joint modeling mechanism across frequency, time, and channel domains: the AGSF module adaptively extracts task-relevant features in the frequency domain using a learnable Gabor filter bank, effectively suppressing common high-frequency noise and drift in multi-node IMU signals; the CM-TCN module extends the receptive field through causal convolutions and multi-scale dilation structures, enabling the model to capture long-range dependencies without disrupting temporal consistency; and the DDAF module further enhances feature interaction and global fusion through dual-domain attention mechanisms at both the time and channel levels, improving the model's ability to discern subtle differences between complex actions. The synergy of these three components significantly enhances the model's discriminative power for dynamic motion patterns.

On the PAMAP2 dataset, FTD-HARNet achieves an mW-F1 score of 93.82%, outperforming the second-best model, ConvBoost, by 0.72 percentage points. This demonstrates that the proposed frequency-domain adaptive front end effectively enhances signal quality in moderate noise environments, thereby improving the stability and generalization capability of the classification. Compared to traditional deep convolution methods, the AGSF module dynamically adjusts the filtering bandwidth, allowing the model to automatically focus on the most discriminative frequency bands under different sensor nodes and varying exercise intensities, thus reducing reliance on manual filter parameter selection. This end-to-end learnable signal enhancement mechanism ensures consistent output across different subjects and sensor placements.

On the DSADS dataset, FTD-HARNet achieves mF1/mW-F1 scores of 87.83%/87.83%, representing an average improvement of approximately 2.5 percentage points over the comparison methods. The DSADS dataset is characterized by numerous nodes, complex motion categories, and a low sampling rate, where traditional models are often limited by spectral distortion and inter-node synchronization errors. However, FTD-HARNet maintains high robustness even under low sampling rates, thanks to the anti-aliasing low-pass design of AGSF and the cross-channel collaboration mechanism of DDAF. These features allow the model to maintain consistent feature representation across multi-source heterogeneous signals, avoiding misclassifications caused by local noise.

4.4 Ablation Studies

To further validate the effectiveness and complementarity of the key modules, we sequentially added AGSF, CM-TCN, and DDAF to IF-ConvTransformer and conducted step-by-step ablation experiments on the three datasets. The results are shown in Table 2.

On the Opportunity dataset, adding AGSF improved the mF1 score from 47.41% to 48.64%. This improvement indicates that frequency-domain adaptive filtering effectively suppresses spectral artifacts caused by sensor placement drift, sensor noise, and environmental disturbances, making the model more capable of capturing energy variation features in the main frequency bands of the activities. After introducing CM-TCN, the mW-F1 score increased to 88.01%, demonstrating that this module expands the temporal receptive field while maintaining real-time performance, allowing the model to better capture phase-specific features of continuous actions. With the addition of DDAF, the performance further improved to 49.04% / 88.52%, suggesting that the dual attention structure in both the time and channel domains explicitly models dynamic synergies between different body parts, thereby enhancing the overall discriminative power for complex actions.

Similar trends were observed on the PAMAP2 and DSADS datasets. The introduction of AGSF consistently improved the base signal quality and low-frequency motion energy expression, while CM-TCN enhanced the model's ability to distinguish between periodic and non-periodic actions. DDAF significantly boosted cross-node feature fusion and global context awareness. After combining all three modules, overall performance improved by approximately 2%–3%, with a notable reduction in standard deviation, reflecting more stable convergence behavior in the model across multiple training runs.

Table 2. Ablation experiments on wearable device-based datasets

Model	Opportunity		Pamap2		DSADS
	mF1 (%)	mW-F1 (%)	mF1 (%)	mW-F1 (%)	mF1 /mW-F1 (%)
IF-ConvTransformer	47.41 ±0.44	87.59 ±0.25	87.02 ±0.29	92.43 ±0.49	85.52 ±0.53
+AGSF	48.64 ±0.19	88.11 ±0.37	87.58 ±0.48	92.84 ±0.23	86.09 ±0.31
+CM-TCN	47.99 ±0.38	88.01 ±0.44	87.42 ±0.74	92.54 ±0.08	85.92 ±0.78
+DDAF	49.04 ±0.22	88.52 ±0.35	87.84 ±0.26	93.14 ±0.83	86.40 ±0.62

5 Conclusion

This paper proposes a novel Frequency–Temporal Dual-domain Human Activity Recognition Network (FTD-HARNet) designed to address the challenges of human activity recognition in multi-IMU sensor systems, particularly for real-time and heterogeneous environment recognition tasks. The core innovation of FTD-HARNet lies in the organic integration of frequency-domain adaptation, causal temporal modeling, and dual-domain attention fusion, forming an efficient end-to-end feature learning framework.

The Adaptive Gabor–Spectral Filtering (AGSF) module dynamically extracts frequency-domain features through learnable center frequencies and bandwidths, effectively suppressing environmental noise and aliasing interference. The Causal Multi-scale Temporal Convolutional Network (CM-TCN) expands the receptive field using dilated convolutions while ensuring causal consistency, capturing long-range dependencies and improving temporal modeling performance. The Dual-Domain Attention Fusion (DDAF) module explicitly models the interdependencies and collaborative relationships between IMU nodes through time-domain self-attention and channel-domain cross-attention mechanisms, achieving deep fusion of global spatiotemporal features.

Through extensive experiments on the Opportunity, PAMAP2, and DSADS public benchmark datasets, the proposed method outperforms existing mainstream methods in both mF1 and mW-F1 metrics. Experimental results demonstrate significant performance improvements in complex activity recognition and multi-sensor fusion scenarios, validating the effectiveness of the proposed approach. Ablation studies further confirm the complementary roles of each module, proving that frequency-domain adaptation, causal temporal modeling, and dual-domain feature fusion are crucial for enhancing overall recognition performance.

FTD-HARNet not only significantly improves the performance of human activity recognition in terms of accuracy and robustness but also provides a promising solution for motion perception in wearable devices and intelligent interactive systems, while ensuring computational efficiency and real-time deployment feasibility. However, there are still several research directions worth exploring. First, extending FTD-HARNet to fuse additional modalities to enhance its cross-modal recognition capability is a promising research direction. Secondly, further reducing the model's computational complexity without sacrificing performance, particularly when deployed in resource-constrained environments, remains an urgent challenge to address.

References

1. Wang, H., et al.: A review of deep learning in human activity recognition. J. Ambient. Intell. Smart Environ. **12**(4), 315–330 (2020)
2. Hou, C.: A study on IMU-based human activity recognition using deep learning and traditional machine learning. In: 2020 5th International Conference on Computer and Communication Systems (ICCCS), pp. 225–234. IEEE (2020)
3. Gomaa, W., Khamis, M.A.: A perspective on human activity recognition from inertial motion data. Neural Comput. & Applic. **35**, 20463–20568 (2023)
4. Yang, S., Zhang, Z., Zhang, J.Z., et al.: Multi-IMU sensor fusion for legged robots. arXiv preprint arXiv:2507.11447. (2025)
5. Yu, H., Zhang, Y., Guo, F., et al.: A robust multi-IMU sensor fusion based lower limb evaluation network toward rehabilitation. IEEE Sensors J. (2025)
6. Hu, F., Zheng, Q., Ye, X., et al.: Gait recognition using spatio-temporal representation fusion learning network with IMU-based skeleton graph and body partition strategy. PLoS One. **20**(10), e0332947 (2025)
7. Engelsman, D., Klein, I.: Data-driven denoising of stationary accelerometer signals. arXiv preprint arXiv:2206.05937. (2022)
8. Arlotti, J.S., Carroll, W.O., Afifi, Y., et al.: Benefits of IMU-based wearables in sports medicine: narrative review. Int. J. Kinesiol. & Sport. Sci. **10**(1), 36 (2022)
9. Nakabayashi, T., Saito, H.: Multimodal human activity recognition on edge devices. In: 2024 IEEE International Symposium On Mixed And Augmented Reality Adjunct (ISMAR-adjunct), pp. 136–140. IEEE (2024)
10. Zhang, C., Zu, T., Hou, Y., et al.: Human activity recognition based on multi-modal fusion. CCF Trans. Pervasive Comput. Interact. **5**(3), 321–332 (2023)
11. Enokibori, Y.: rTsfNet: a DNN model with multi-head 3D rotation and time series feature extraction for IMU-based human activity recognition. Proc. ACM Interact., Mob., Wearable Ubiquitous Technol. **8**(4), 1–26 (2024)
12. Nia, N.G., Kaplanoglu, E., Nasab, A., et al.: Human activity recognition using machine learning algorithms based on IMU data. In: 2023 5th International Conference on Bio-Engineering for Smart Technologies (BioSMART), pp. 1–8. IEEE (2023)
13. Dehzangi, O., Sahu, V.: IMU-based robust human activity recognition using feature analysis, extraction, and reduction. In: 2018 24th International Conference on Pattern Recognition (ICPR), pp. 1402–1407. IEEE (2018)
14. Anwary, A.R., Yu, H., Vassallo, M.: Optimal foot location for placing wearable IMU sensors and automatic feature extraction for gait analysis. IEEE Sensors J. **18**(6), 2555–2567 (2018)
15. Tao, W., Chen, H., Moniruzzaman, M., et al.: Attention-based sensor fusion for human activity recognition using IMU signals. arXiv preprint arXiv:2112.11224. (2021)
16. Zhang, Y., Wang, L., Chen, H., et al.: IF-ConvTransformer: a framework for human activity recognition using IMU fusion and ConvTransformer. Proc. ACM Interact., Mob., Wearable Ubiquitous Technol. **6**(2), 1–26 (2022)
17. Wang, Y., Zhao, Y.: Wavelet dynamic selection network for inertial sensor signal enhancement. Proc. AAAI Conf. Artif. Intell. **38**(14), 15680–15688 (2024)

18. Yuan, K., Wang, Z.J.: A simple self-supervised IMU denoising method for inertial aided navigation. IEEE Rob. Autom. Lett. **8**(2), 944–950 (2023)
19. Damagatla, R.K.R., Atia, M.: Improving EKF based IMU/GNSS fusion using machine learning for IMU denoising. IEEE Access. (2024)
20. Singh, S.P., Sharma, M.K., Lay-Ekuakille, A., et al.: Deep ConvLSTM with self-attention for human activity decoding using wearable sensors. IEEE Sensors J. **21**(6), 8575–8582 (2020)
21. Yao, S., Hu, S., Zhao, Y., et al.: DeepSense: a unified deep learning framework for time-series mobile sensing data processing. In: Proceedings of the 26th International Conference on World Wide Web, pp. 351–360 (2017)
22. Shao, S., Guan, Y., Zhai, B., et al.: ConvBoost: boosting ConvNets for sensor-based activity recognition. Proc. ACM Interact., Mob., Wearable Ubiquitous Technol. **7**(2), 1–21 (2023)
23. Abedin, A., Ehsanpour, M., Shi, Q., et al.: Attend and discriminate: beyond the state-of-the-art for human activity recognition using wearable sensors. Proc. ACM Interact., Mob., Wearable Ubiquitous Technol. **5**(1), 1–22 (2021)
24. Liu, S., Yao, S., Li, J., et al.: Globalfusion: a global attentional deep learning framework for multisensor information fusion. Proc. ACM Interact., Mob., Wearable Ubiquitous Technol. **4**(1), 1–27 (2020)

A Divide-and-Conquer Network for Generalized Few-Shot Segmentation

Lingling Zhou, Tao Fang, Wenguang Li(✉), Bin Zhang, Xin Li, and Yuran Zhang

State Grid Commercial Big Data Co.,Ltd., Beijing, China
llzhou361@163.com , 277268653@qq.com

Abstract. Semantic understanding of complex scenes is crucial for autonomous systems, yet generalized few-shot segmentation (GFSS) faces challenges due to unseen classes and limited annotations. Existing GFSS methods rely on jointly trained base and novel classifiers, often suffering from base class bias. To handle this, we design DC-GFSS. This is a divide-and-conquer way to independently segment base and novel classes and fuses results at the mask level rather than the classifier level, effectively alleviating bias. We propose a Context Relationship Module (CRM) to enhance feature representations by modeling both intra-class and inter-class contextual relationships, improving segmentation for all classes. In addition, a Prototype Enhancement Module (PEM) refines novel class predictions through self-alignment and a dynamic update strategy that leverages query-specific information. Extensive experiments on several benchmarks show that DC-GFSS achieves competitive performance.

Keywords: computer vision · semantic segmentation · GFSS

1 Introduction

Semantic segmentation has witnessed great progress over the past decade. Classical segmentation methods have laid a solid foundation for many real-world deployments. In particular, semantic segmentation plays a crucial role in obstacle-aware perception systems, where accurately identifying scene elements and potential obstacles is essential for safe navigation and decision-making.

Despite these advances, existing semantic segmentation methods rely on large-scale datasets with pixel-level annotations. The acquisition of these annotations is labor-intensive and costly. This limitation becomes more pronounced in obstacle avoidance scenarios, where obstacles are often diverse, irregular, and unpredictable. Many obstacles encountered in real-world environments–such as fallen objects, temporary roadblocks, or rare hazards–are either underrepresented or entirely absent in the training data, making exhaustive annotation impractical.

L. Huang and D. Greenhalgh (Eds.): ICMLC 2026, LNNS 2061, pp. 732–749, 2027.
https://doi.org/10.1007/978-3-032-31236-5_50

Moreover, conventional semantic segmentation models struggle to generalize to novel or unseen obstacle categories when only a few labeled samples are available. This lack of scalability severely restricts their applicability in dynamic environments, where robots or autonomous agents must rapidly adapt to new types of obstacles with minimal supervision. To resolve this, few-shot segmentation (FS-Seg) attracts increasing attention in recent years. FS-Seg focuses on enabling segmentation models to learn new classes from limited annotated examples, offering a promising solution for obstacle avoidance systems that require both adaptability and annotation efficiency.

FS-Seg [1] aims to learn the generalized ability of fast adaptation to novel class with only a handful of annotated data. Current mainstream methods [2–4] follow a prototypical-learning strategy to establish a global descriptor for each category, which have made great success in adapting to novel class. However, such a task setting only recognizes novel class that appears in support set, while base class is totally ignored. To deal with this issue, generalized FSS (GFSS) is proposed to segment both novel class and base class, which is a more practical setting. Nevertheless, little attention has been paid to GFSS [5–7] and the performance remains unsatisfactory, especially on unseen class.

In this paper, we reconsider GFSS in following aspects: (1) Compared with semantic segmentation, GFSS is split into base and novel class, and only a few labeled data is available for novel class, while semantic segmentation has abundant annotations for all classes; (2) Compared with FS-Seg, GFSS not only predicts novel class, but also segments base class. That is, GFSS conducts a pixel-wise multi-class segmentation, while FS-Seg only focuses on the binary segmentation of novel class. Therefore, the setting of GFSS is more challenging but more practical. However, the task remains underexplored and the current top-performing methods are still unsatisfactory.

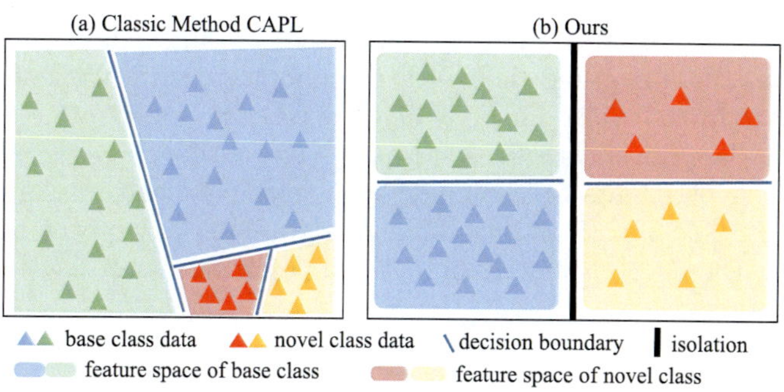

Fig. 1. Illustration of base class bias issue in CAPL and our solution. CAPL fuses base classifier and novel classifier to segment all classes, which leads to the fact that the results tend to bias to the well-trained base class. In contrast, our method gets around base class bias problem by separately predicting both masks, which isolates the feature spaces of the both classes.

The top-performing method CAPL [5] proposes a three-phase solution to GFSS: (1) Base classifier training; (2) Novel class registration; (3) Evaluation on both classes. However, the classifier fusion process in the second phase will cause base class bias issue [8]. Specifically, base classifier is well-trained on adequately annotated data, while novel classifier is under-trained on limited labeled data. The straight-forward method CAPL [5] directly incorporates the parameters of base and novel classifier together to segment all classes. Thus it suffers from comprised performance of new classes. Figure 1 illustrates base class bias issue, which indicates that the under-trained novel class is prone to be segmented as base class.

In this paper, we argue that it is the classifier incorporation mechanism that leads to the issue. Accordingly, we propose a divide-and-conquer network, named DC-GFSS, to reconcile base and new categories and overcome the bias issue. In particular, to segment a query image, we first divide the task into base and novel classes branch. Subsequently, instead of incorporating the prototype of the both classes, we directly fuse the segmentation results of both branches. Such a method isolates the features of base prototype from novel prototype, which contributes to segmenting novel class, as novel classifier is not diluted by the well-trained base classifier through parameter fusion. In this way, we get around the bias issue and significantly improve the performance on unseen classes.

Moreover, to enrich category information of feature extractor, we propose a Context Relationship Module (CRM). Motivated by [9], CRM aims to fully exploit the contextual relationship between different spatial locations during the training period. That is, CRM captures the global dependency of intra and inter-class information. Thus it enhances the category context for segmenting. In addition, we design a Prototype Enhancement Module (PEM) to refine novel class classifier, as novel classifier is under-trained on scarcely annotated data. During the test period, PEM leverages novel classifier to segment the query image. Then we obtain a query classifier to conduct self-alignment and predict the query mask itself. Such an operation tackles the intra-class discrepancy of novel class, as the annotated data in the support set is insufficient, while the unlabeled data in the query image is diverse. Besides, we introduce a dynamic strategy to update novel classifier with query information.

Extensive experiments are conducted on multiple benchmarks. DC-GFSS surpasses other GFSS methods by a large margin, especially on the performance of novel class. DC-GFSS achieves competitive performances on both datasets in 1-shot and 5-shot settings.

Our main contributions are summarized as follows:

- We propose a simple yet effective approach, dubbed DC-GFSS, to separately segment base and novel class in parallel branches. Such a network adopts a mask fusion strategy to overcome base class bias issue and significantly enhances segmentation performance on novel class.
- We put forward two novel modules, i.e., CRM and PEM. The former models intra and inter-class context dependency for enriching feature map, and the

latter conducts self-alignment to refine the mask quality and adopts a dynamic updating strategy for novel classifier enhancement.
- Experimental results show that DC-GFSS achieves competitive results on multiple benchmarks under GFSS settings.

2 Related Work

2.1 Semantic Segmentation

Semantic segmentation targets to classify each spatial location into a prior category. Thanks to the annotated datasets, semantic segmentation has developed rapidly in the past decade. Fully Convolutional Network [10] first uses convolutional last layer rather than linear layer. This way dramatically improves the segmentation performance. Dilated convolution [11] is widely used in semantic segmentation, as it can significantly enlarge the receptive field. Moreover, the encoder-decoder structure [12] and the attention mechanism [13] have attracted more attention recently for further improvement. Although great efforts have been made in semantic segmentation, previous methods still heavily rely on large annotated datasets and fall short in adapting to an unseen category with a few labeled examples.

2.2 Few-Shot Learning

Few-shot learning acquires generalized ability to predict novel class with inadequately annotated data. Current mainstream methods are divided into meta-learning [14,15] and metric-learning ways [16,17]. The former adjusts the model parameters for a fast adaptation with insufficient labeled data, and the latter leverages a distance to measure the difference or adopts a siamese network [18]. Generalized few-shot learning is pioneered by [19]. Different from few-shot learning, the query data in generalized few-shot learning may contain both novel class and base class, which greatly improves the task difficulty. Although effective methods [20] have been proposed to tackle the problem, they are still unsuitable for GFSS, as GFSS is an extension of generalized few-shot learning, which requires rich contextual information for pixel-wise dense prediction.

2.3 Few-Shot Segmentation

FS-Seg is first introduced by [1] in a meta-learning way. Currently, the classic methods [2,4,21] for FS-Seg mainly leverage the metric-learning paradigm. It aims to obtain a global descriptor of each category, named prototype, and calculate the pixel-wise similarity to predict the foreground region. Recent top-performing methods refine the prototype by exploiting the background information [22,23], adopting an adaptive strategy [3], using graph neural networks [24], and proposing reasonable matching rules [25,26]. Although FS-Seg adapts well to novel class with limited annotated data, such a setting totally ignores base class. Therefore, GFSS is proposed to segment all classes in the inference period. That is, FS-Seg is a binary classification problem, while GFSS requires multi-class segmentation, which is more challenging.

2.4 Generalized Few-Shot Segmentation

Even though GFSS is practical, little attention [5–7] has been paid to such a challenging yet valuable task. The current top-performing method CAPL [5] incorporates base classifier with novel classifier to segment all classes. However, the performance is prone to be biased to base class as novel class sample is lacking. This limitation results in the under-trained novel classifier. In this paper, we introduce DC-GFSS to tackle this issue by isolating the feature spaces of base and novel categories.

3 Methodology

3.1 Problem Definition

GFSS requires segmenting all the class $C_{all} = C_{base} \cup C_{novel}$ in the query image during the inference period, where C_{base} and C_{novel} denote base class and novel class, respectively, and $C_{base} \cap C_{novel} = \phi$. In GFSS, the annotated data is adequate for C_{base} to train a base classifier, while it remains the limited K annotated samples for C_{novel}. The data is organized into an episodic manner with a support set S and a query set Q, where $S = \{(I_i, M_i)|i = 1, 2, ..., K\}$ and $Q = \{(I_Q, M_Q)\}$, and $I \in R^{H\times W\times 3}$ and $M \in R^{H\times W\times 1}$ represent the RGB input image and the ground-truth mask, respectively. The overall pipeline of the GFSS model is divided into three phases: (1) In base class training phase, the GFSS model is trained on C_{base} in a supervised manner to obtain a feature extractor $\mathcal{F}$ and a base classifier P_b; (2) In novel class training phase, we leverage the limited K support samples $(I_1, M_1, I_2, M_2, ..., I_K, M_K)$ to learn in an episodic way on C_{novel} to train a novel classifier P_n; (3) After obtaining base classifier P_b and novel classifier P_n, we segment the query image I_Q and predict the results $\hat{M}_Q \in R^{H\times W\times 1}$ for all classes $C_{all} = C_{base} \cup C_{novel}$, and the final goal is to make $\hat{M}_Q$ approximate ground-truth M_Q.

3.2 Overview of DC-GFSS

As shown in Fig. 2, the core idea of DC-GFSS is to separate the task into two parallel branches. One is to segment C_{base}, and the other is to segment novel class C_{novel}. Then we adopt a mask fusion strategy to incorporate both confidence maps for assembling the final prediction on all classes $C_{all} = C_{base} \cup C_{novel}$.

Given a query image $I_Q \in R^{H\times W\times 3}$, we first use feature extractor $\mathcal{F}$ to obtain the representation $\mathcal{F}(I_Q)$, where $\mathcal{F}(I_Q) \in R^{H\times W\times d}$. H, W, and d are image height, image width, and dimension of latent space, respectively. To better exploit class-level context, we propose a Context Relationship Module (CRM) to enrich the feature $\mathcal{F}(I_Q)$, which is illustrated in Fig. 3.

After that, we separately predict base and novel class mask, totally different from the top-performing approach CAPL [5]. CAPL first fuses the parameters of base classifier $P_b \in R^{N_b \times d}$ and novel classifier $P_n \in R^{N_n \times d}$, and then segments all classes C_{all}. This process is formulated as:

$$\mathcal{F}(I_Q) \times (P_b \cup P_n) \tag{1}$$

Equation (1) is able to obtain the segmentation result of all classes through classifier fusion. However, this approach leads to a severe bias issue. Specifically, once the classifiers of C_{base} and C_{novel} are fused, they will fall into the same representation space. Due to data scarcity of novel class C_{novel}, decision boundary inevitably tends to be biased to well-trained base class C_{base}, which results in an unsatisfactory performance on novel class.

To tackle the bias issue, unlike CAPL, DC-GFSS isolates base classifier P_b from novel classifier P_n and segments all classes in a divide-and-conquer way. We separately predict C_{base} and C_{novel}, and then conduct mask fusion to assemble the final prediction:

$$(\mathcal{F}(I_Q) \times P_b) \cup (\mathcal{F}(I_Q) \times P_n) \tag{2}$$

Particularly, we leverage the operation of matrix multiplication and softmax to obtain a normalized confidence map $O_b \in R^{H \times W \times N_b}$ on C_{base}, and treat novel classifier as N_n d-dimensional prototypes $\{p_c | \forall c \in C_{novel}\}$ for each novel category. Then we conduct a pixel-wise cosine similarity for each spatial location on each novel category, and the confidence map of novel class $O_n \in R^{H \times W \times N_n}$ is calculated by:

$$O_n^{x,y,c} = \frac{exp(\alpha \cdot cos(\mathcal{F}(I_Q)^{x,y}, p_c))}{\sum_{c \in C_{novel}} exp(\alpha \cdot cos(\mathcal{F}(I_Q)^{x,y}, p_c))} \tag{3}$$

where x, y denotes the spatial coordinate, cos represents cosine similarity, and α is a balanced factor and we set $\alpha = 10$ empirically. After obtaining both confidence maps O_b and O_n, we separately predict both results on $M_b \in R^{H \times W}$ and $M_n \in R^{H \times W}$ by selecting the category with the maximum confidence in each spatial location. Note that, once the maximum value is smaller than threshold 0.5, we set $M_b^{x,y} = 0$ or $M_n^{x,y} = 0$ as the background. Then we fuse them to calculate the final mask $\hat{M}_Q \in R^{H \times W}$, which is formulated by:

$$\hat{M_Q^{x,y}} = \begin{cases} M_b^{x,y}, & M_b^{x,y} \neq 0 \wedge M_n^{x,y} = 0 \\ M_n^{x,y}, & M_b^{x,y} = 0 \wedge M_n^{x,y} \neq 0 \\ M_b^{x,y}, & M_b^{x,y} \neq 0 \wedge M_n^{x,y} \neq 0 \wedge O_b^{x,y,M_b^{x,y}} > O_n^{x,y,M_n^{x,y}} \\ M_n^{x,y}, & M_b^{x,y} \neq 0 \wedge M_n^{x,y} \neq 0 \wedge O_b^{x,y,M_b^{x,y}} \leq O_n^{x,y,M_n^{x,y}} \\ 0, & Otherwise \end{cases} \tag{4}$$

In this way, we obtain the predicted mask $\hat{M}_Q$ for all classes. It is worth mentioning that, although we get around base class bias issue by replacing classifier fusion with mask fusion, the performance on novel class remains unsatisfactory, as novel classifier P_n is still under-trained with only a handful of annotated data. To handle this issue, we propose a Prototype Enhancement Module (PEM) to narrow the intra-class difference through self-alignment and make the classifier generalized and robust to the specific category. Please see §3.4 for more details.

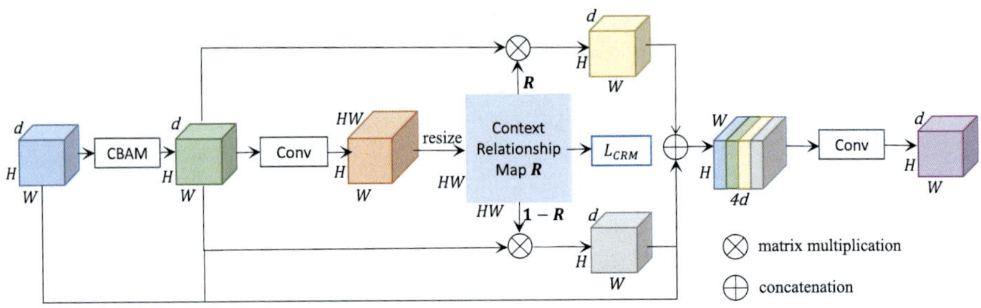

Fig. 2. Overall pipeline of Context Relationship Module (CRM).

3.3 Context Relationship Module

Motivated by the method in [9], we propose CRM to enrich the feature map and exploit the intra-class and inter-class relationship, which aims at providing a prior knowledge to distinguish the specific category of context information for GFSS.

As shown in Fig. 3, given a feature map $\mathcal{F} \in R^{H\times W\times d}$, we first leverage a CBAM [27] module to focus on the salient foreground information. Once the salient feature $\mathcal{F}_{sal}$ is highlighted, we conduct an operation of 1×1 convolution and resize the feature to obtain a context relationship map $\mathcal{R} \in R^{HW\times HW}$, where $\mathcal{R}^{i,j}$ represents the category correlation between pixel i and pixel j. We leverage a sigmoid function to normalize the score to $[0, 1]$. Note that, the ground-truth context relationship map $\mathcal{R}_{GT}$ is derived from the ground-truth mask $M_{GT} \in R^{H\times W}$. Specifically, we transfer M_{GT} into a one-hot encoding $\tilde{M}_{GT} \in R^{HW\times N_b}$. Moreover, $\mathcal{R}_{GT} = \tilde{M}_{GT}\tilde{M}_{GT}^{T}$, where T represents the transpose matrix. Then we use a binary cross entropy loss to measure the difference between $\mathcal{R}$ and $\mathcal{R}_{GT}$:

$$\mathcal{L}_R = \frac{1}{N^2}\sum_{i=1}^{N}\sum_{j=1}^{N}\alpha_1\mathcal{R}_{GT}^{i,j}log\mathcal{R}^{i,j} + \alpha_2(1-\mathcal{R}_{GT}^{i,j})log(1-\mathcal{R}^{i,j}) \tag{5}$$

where $N = HW$, α_1 and α_2 are the balanced factors. We set $\alpha_1 = \alpha_2 = 0.5$, empirically. Note that, $\mathcal{R}$ is supposed to be a symmetric matrix. Therefore, we propose a symmetric loss $\mathcal{L}_S$ to constrain the symmetry of $\mathcal{R}$, which is formulated by:

$$\mathcal{L}_S = \frac{1}{N_{up}} \sum_{i=1}^{N} \sum_{j=i+1}^{N} ||\mathcal{R}^{i,j}, \mathcal{R}^{j,i}||_2^2 \tag{6}$$

where $N_{up} = \frac{(HW)^2}{2} - HW$ represents the pixel number of the upper triangular matrix of $\mathcal{R}$.

The losses above only consider the context in an isolated manner, while ignoring the semantic correlation between different pixels. Therefore, we define an intra-class loss $\mathcal{L}_{intra}$ and an inter-class loss $\mathcal{L}_{inter}$ to exploit the category information:

$$\mathcal{L}_{intra} = -\frac{1}{N} \sum_{i=1}^{N} log \frac{\sum_{j=1}^{N} \mathcal{R}_{GT}^{i,j} \mathcal{R}^{i,j}}{\sum_{j=1}^{N} \mathcal{R}_{GT}^{i,j}} \tag{7}$$

$$\mathcal{L}_{inter} = -\frac{1}{N} \sum_{i=1}^{N} log \frac{\sum_{j=1}^{N} (1 - \mathcal{R}_{GT}^{i,j})(1 - \mathcal{R}^{i,j})}{\sum_{j=1}^{N} (1 - \mathcal{R}_{GT}^{i,j})} \tag{8}$$

The losses in Eqs. (7) and (8) demonstrate the precision of the intra-class and inter-class correlation. These losses model dependency of different pixels to enrich class-wise context.

Accordingly, the total loss for CRM is calculated by:

$$\mathcal{L}_{CRM} = \beta_1 \mathcal{L}_R + \beta_2 \mathcal{L}_S + \beta_3 \mathcal{L}_{intra} + \beta_4 \mathcal{L}_{inter} \tag{9}$$

where β_1, β_2, β_3 and β_4 represent four balanced factors, and we set $\beta_1 = \beta_3 = \beta_4 = 0.3$ and $\beta_2 = 0.1$ empirically.

Subsequently, we leverage $\mathcal{R}$ and $1 - \mathcal{R}$ to conduct multiplication on $\mathcal{F}_{sal}$, and obtain the intra-class feature and inter-class feature, respectively. These two features are concatenated by $\mathcal{F}_{sal}$ and $\mathcal{F}$, to generate a fused feature $\mathcal{F}_F \in R^{H \times W \times 4d}$. At last, we employ 1×1 convolutional layer to restore channel dimension in d for further segmentation.

3.4 Prototype Enhancement Module

As we take base classifier P_b and novel classifier P_n apart for separate prediction, P_n is prone to be under-trained with limited support samples in a prototypical way. Therefore, we design a Prototype Enhancement Module (PEM) to improve such a classifier, as shown in Fig. 4.

During the test period, given novel classifier P_n and the query feature $\mathcal{F}_Q$, we first conduct a pixel-wise cosine similarity on each spatial location to obtain a similarity map $M_1 \in R^{H \times W \times N_n}$. As P_n is obtained by scarce data in the learning phase of novel class, the quality of M_1 is unsatisfactory. Therefore, we leverage query feature to establish P_q for self-alignment. The query classifier P_q is formulated by:

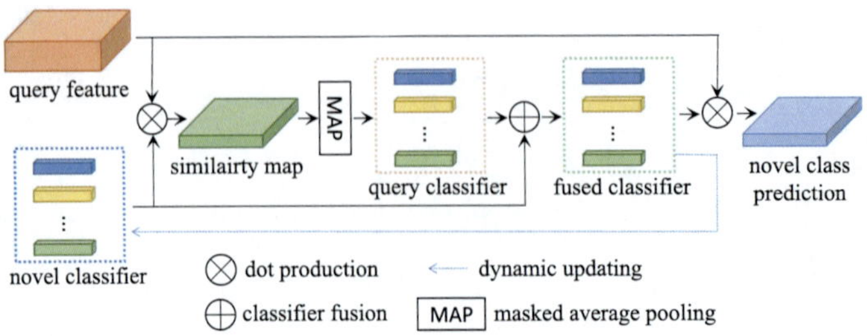

Fig. 3. Overview of Prototype Enhancement Module (PEM).

$$P_q^c = \frac{\sum_{x,y} \mathcal{F}_Q^{x,y} \mathbb{1}[M_1^{x,y,c} > \mu]}{\sum_{x,y} \mathbb{1}[M_1^{x,y,c} > \mu]} \tag{10}$$

where $\mathbb{1}$ is an indicator function, $c \in C_{novel}$ is a novel class, and μ is a threshold to select query feature. Notice that we should select the high-confidence feature, which contains less noise of the intra-instance information. Then we fuse novel classifier and the query classifier by:

$$P_f = \theta P_n + (1-\theta) P_q \tag{11}$$

where θ stands for the weight for classifier fusion. Subsequently, we leverage P_f to segment the query feature, and we select the maximum score of each spatial location to obtain the final prediction M_n. Moreover, we use P_f to update P_n in a dynamic way, which is as follows:

$$P_n \leftarrow \gamma_1 P_f + \gamma_2 P_n \tag{12}$$

where γ_1 and γ_2 represent two balanced factors, and we set $\gamma_1 = 0.01$ and $\gamma_2 = 0.99$ empirically.

3.5 Training Strategy

Base class training. To train a base classifier P_b, we first leverage the traditional supervised paradigm. In this stage, the classifier is only responsible for predicting C_{base}, and the optimization objective is formulated by:

$$\underset{\mathcal{F}, P_b}{\text{argmin}}\, \lambda_1 \mathcal{L}_{CRM} + \lambda_2 \mathcal{L}_{bce}(M_b, \hat{M}_b) \tag{13}$$

where $\mathcal{L}_{bce}$ represents binary cross-entropy objective. λ_1 and λ_2 are two balanced factors, and we use $\lambda_1 = \lambda_2 = 0.5$, empirically. Moreover, to mimic the test scenarios, we select N_{fake} categories from C_{base} and obtain the fake novel classes C_{fake}. Following [5], we train C_{fake} in an episodic manner to guide the feature extractor to acquire the generalized ability on novel classes, which benefits the training phase of novel class.

Novel class training. We leverage an episodic way to train a novel classifier P_n, which is comprised of N_n prototypes. We utilize the support-query pair to fine-tune feature extractor $\mathcal{F}$ and establish refined prototypes for novel class. The optimization objective is formulated by:

$$\underset{\mathcal{F},P_n}{argmin}\, \mathcal{L}_{bce}(M_n, \hat{M}_n) \tag{14}$$

where $\mathcal{L}_{bce}$ is binary cross-entropy loss. Notably, $\mathcal{F}$ is solely adjusted during novel class learning period. When we evaluate the model, we return to leveraging $\mathcal{F}$ that is learned in the first phase.

4 Experiments

4.1 Datasets and Metrics

Datasets. To evaluate DC-GFSS, we utilize two commonly-used datasets in FS-Seg, i.e., PASCAL-5^i [28] and COCO-20^i [29]. The former is derived from VOC benchmark [28] and extended with augmented annotation [30], which contains 20 different categories. Dataset is divided into 4 splits, and each includes 5 categories. Among the 4 splits, 3 splits are selected as base class, while the remaining 1 split is recognized as novel class. COCO-20^i is more challenging, which is based on MS COCO [16]. Following the pattern in [31], we separate the classes of COCO-20^i into 4 splits and each split contains 20 different categories.

Metrics. We leverage mean intersection-over-union (MIoU) as our main result. As GFSS predicts both base and novel class, we report base class MIoU ($MIoU_b$), novel class MIoU ($MIoU_n$), and total MIoU on both datasets. Moreover, we also leverage the F_1 score on $MIoU_b$ and $MIoU_n$, to measure the harmonic mean H_{mean}, which is calculated by:

$$H_{mean} = \frac{2 \times MIoU_b \times MIoU_n}{MIoU_b + MIoU_n} \tag{15}$$

4.2 Implementation Details

Firstly, we train feature extractor $\mathcal{F}$ and base classifier P_b. We adopt PSPNet [32] as foundation segmentation model with the backbone ResNet-50 and ResNet-101 [33]. We use batch size 16 and resize image in 417×417 pixels. We utilize a momentum SGD with learning rate $5e-3$. After that, we adopt a meta-training way to mimic situation of unseen class learning. We set $N_{fake} = 1$ and randomly select C_{fake} from base class. We train DC-GFSS with 40k/160k iterations on the support-query pair in an episodic manner on both datasets. This time we employ a momentum SGD with learning rate $1e-3$. Then feature extractor $\mathcal{F}$ has generalized ability to unseen classes. Finally, we employ episodic strategy to train DC-GFSS with 100 iterations to obtain a prototype-based novel classifier P_n. During evaluation, we employ images of the whole datasets to calculate the overall base class MIoU, novel class MIoU, total MIoU, and the harmonic average value H_{mean}.

Table 1. Comparison results on PASCAL-5^i for ResNet-50 and ResNet-101 in both 1-shot and 5-shot settings.

Backbone	Method	1-shot				5-shot			
		Base	Novel	Total	H_{mean}	Base	Novel	Total	H_{mean}
ResNet-50	PFENet [4]	57.72	8.55	45.42	14.89	57.54	8.59	45.30	14.95
	ACASTLE [34]	58.84	36.25	53.19	44.86	62.48	46.15	58.40	53.09
	BAM [35]	60.37	27.49	52.15	37.77	62.47	28.96	53.78	39.57
	CAPL [5]	65.48	18.85	54.38	29.16	66.14	22.41	55.72	33.48
	DC-GFSS (Ours)	**68.65**	**54.04**	**64.28**	**60.48**	**69.13**	**63.72**	**67.79**	**66.31**
ResNet-101	PFENet [4]	57.57	13.09	46.45	21.33	57.50	13.13	46.40	21.37
	ACASTLE [34]	58.54	35.06	52.67	43.86	63.28	44.18	58.50	52.03
	DC-GFSS (Ours)	**69.31**	**57.47**	**66.26**	**62.83**	**70.05**	**64.51**	**68.83**	**67.16**

Table 2. Comparison results on COCO-20^i for ResNet-50 and ResNet-101 in both 1-shot and 5-shot settings.

Backbone	Method	1-shot				5-shot			
		Base	Novel	Total	H_{mean}	Base	Novel	Total	H_{mean}
ResNet-50	PFENet [4]	32.82	4.10	25.64	7.29	32.68	4.10	25.53	7.28
	ACASTLE [34]	24.11	21.51	23.46	22.73	33.79	22.76	31.03	27.20
	BAM [35]	45.70	11.35	36.80	18.18	47.27	15.26	38.59	23.07
	CAPL [5]	44.61	7.05	35.46	12.18	45.24	11.05	36.80	17.76
	DC-GFSS (Ours)	**49.72**	**28.90**	**44.13**	**36.55**	**49.56**	**39.17**	**47.39**	**43.75**
ResNet-101	PFENet [4]	36.50	3.58	28.27	6.52	36.56	3.76	28.36	6.81
	ACASTLE [34]	25.07	22.55	24.44	23.74	37.88	21.65	33.82	27.55
	DC-GFSS (Ours)	**50.81**	**32.34**	**44.97**	**39.52**	**50.86**	**40.20**	**48.63**	**44.90**

4.3 Comparison with State-of-the-Arts

To evaluate DC-GFSS, we report the MIoU results on both PASCAL-5^i and COCO-20^i datasets. Leveraging the backbone ResNet-50 and ResNet-101, DC-GFSS outperforms existing top-performing approaches by a large margin in 1-shot and 5-shot settings on both datasets.

PASCAL-5^i. The main performances of PASCAL-5^i are demonstrated in Table 1. Overall, DC-GFSS achieves a competitive performance on all cases. Specifically, on ResNet-50, DC-GFSS surpasses top-performing method CAPL by 9.9 MIoU and 12.07 MIoU in 1-shot setting and 5-shot setting, respectively. Note that, MIoU of novel class in DC-GFSS outperforms CAPL by 35.19 MIoU and 41.31 MIoU in 1-shot and 5-shot settings, respectively, which contribute to dramatic improvements of total performance mostly. As to the measurement H_{mean}, DC-GFSS also significantly surpasses other methods, showcasing DC-

GFSS significantly improves novel class segmentation results and balances the performance of both classes. Moreover, DC-GFSS achieves a competitive performance on ResNet-101 and enhances the total MIoU by 13.59 and 10.33 in 1-shot and 5-shot scenarios, respectively.

COCO-20$^{\mathbf{i}}$. The experimental performances on COCO-20^i are reported in Table 2. DC-GFSS outperforms other top-performing approaches on both ResNet backbones. Specifically, on ResNet-50, DC-GFSS surpasses CAPL by 8.67 MIoU and 9.59 MIoU in 1-shot and 5-shot settings, respectively. The total improvement largely stems from the enhancement of novel class, as novel class performance of DC-GFSS is 21.85 MIoU and 28.12 MIoU higher than CAPL in 1-shot and 5-shot scenarios, respectively. Furthermore, DC-GFSS obtains a much higher H_{mean} compared with the previous methods, which represents that the performance of novel class is substantially improved. Moreover, DC-GFSS obtains the competitive performance on ResNet-101 in both shot settings.

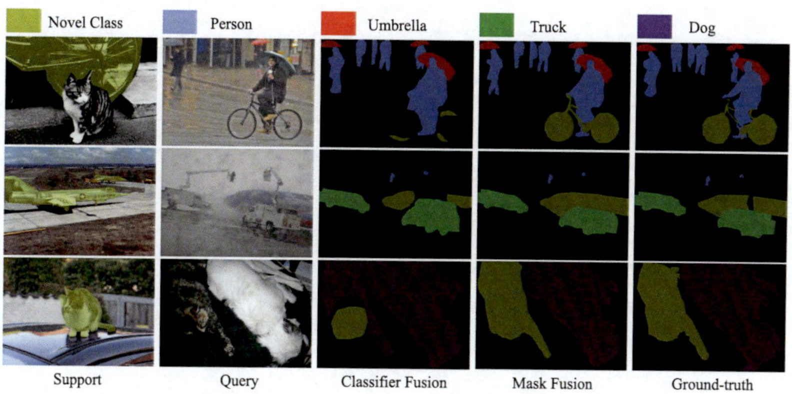

Fig. 4. Qualitative results on COCO-20^i dataset.

Table 3. Ablation results on PASCAL-5^i for investigating the effect of classifier fusion strategy (CAPL) and mask fusion strategy (Ours).

Strategy	Base	Novel	Total	H_{mean}
Classifier fusion	66.71	16.14	55.36	25.99
Mask fusion	**69.13**	**63.72**	**67.79**	**66.31**

4.4 Ablation Study

Mask fusion versus Classifier fusion. Besides the mask fusion mode in DC-GFSS, we explore the performance of the classifier fusion strategy [5] as a com-

Table 4. Ablation results on PASCAL-5^i for exploring the influence of CRM and PEM.

CRM	PEM	Base	Novel	Total	H_{mean}
		58.60	8.87	45.04	15.40
✓		**69.51**	18.70	64.30	29.47
	✓	58.69	49.90	55.76	53.93
✓	✓	69.13	**63.72**	**67.79**	**66.31**

Table 5. Ablation results on PASCAL-5^i for exploring different loss combinations in CRM.

$\mathcal{L}_R$	$\mathcal{L}_S$	$\mathcal{L}_{intra}$	$\mathcal{L}_{inter}$	Base	Novel	Total	H_{mean}
				47.19	31.40	44.57	37.70
✓				56.31	37.84	52.14	45.26
✓		✓		58.19	48.65	55.60	52.99
✓			✓	66.81	39.82	61.24	48.89
✓		✓	✓	**69.27**	57.10	67.06	62.59
✓	✓	✓	✓	69.13	**63.72**	**67.79**	**66.31**

Table 6. Ablation results on PASCAL-5^i for exploring the effect of CBAM in CRM and the dynamic updating strategy in PEM.

Strategy	Base	Novel	Total	H_{mean}
w/o CBAM	65.20	57.59	62.91	61.15
w/ CBAM	**69.13**	**63.72**	**67.79**	**66.31**
w/o Dynamic Updating	**69.52**	41.80	63.37	52.20
w/ Dynamic Updating	69.13	**63.72**	**67.79**	**66.31**

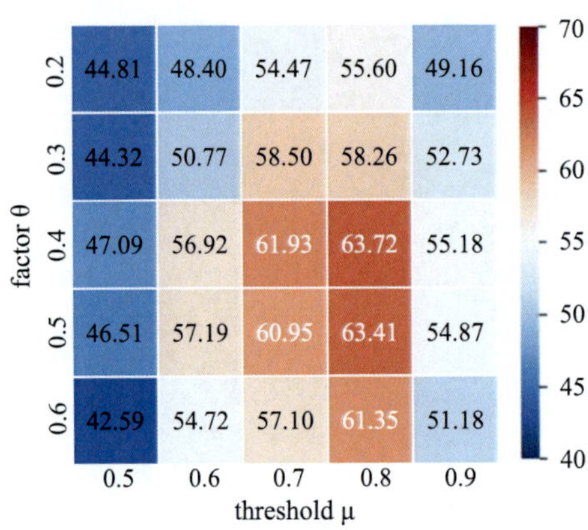

Fig. 5. Hyperparameter combinations of μ and θ.

parison. Our mask fusion strategy merges the separate results, and thus the representation spaces of base and novel classes are not mixed during inference period, which totally overcomes base class bias issue and significantly enhances the performance on novel class. The results are demonstrated in Table 3. As shown in the table, the mask fusion strategy surpasses the classifier fusion strategy (+12.43 MIoU), especially novel class (+47.58 MIoU). The experimental results validate that the mask fusion pattern in DC-GFSS isolates the representation space of base classifier and novel classifier, which overcomes base class bias issue in CAPL and significantly improves the performance on novel class. Furthermore, Fig. 5 qualitatively illustrates the difference between the both fusion strategies, which proves the superiority of novel class results on DC-GFSS.

Influence of CRM and PEM. We explore the influence of CRM and PEM. The results are reported in Table 4. We find that vanilla DC-GFSS is unsatisfactory. With the incorporation of CRM, the performances on base class and novel class improve by +10.91 MIoU and +9.83 MIoU, respectively. The reason lies in that CRM enriches the feature with the category relationship context, which models the class-wise dependency as the prior knowledge, contributing to establishing a refined decision boundary to distinguish each specific category. On the other hand, when applying PEM, the performance on novel class boosts dramatically (+41.03 MIoU). As the vanilla DC-GFSS suffers from the under-trained classifier on novel class, PEM can alleviate this issue by a self-alignment operation and adopting a dynamic approach to enhance novel classifier. That is, each module improves the vanilla network dramatically. Furthermore, combined with CRM and PEM, DC-GFSS achieves 69.13, 63.72, and 67.79 MIoU on base, novel, and total classes, which is significantly higher than the vanilla DC-GFSS.

Losses of CRM. We dive into the different loss combinations of CRM. All the losses contribute to restricting the context relationship map to the ground-truth in an explicit manner, but they focus on different parts. Table 5 lists the results of the different combinations. As shown in the table, the combination of $\mathcal{L}_R$, $\mathcal{L}_S$, $\mathcal{L}_{intra}$, and $\mathcal{L}_{inter}$ achieve an excellent performance on base class (69.13 MIoU), novel class (63.72 MIoU), total class (67.79 MIoU), and H_{means} (66.31). We find that solely leverage $\mathcal{L}_R$ remains a poor performance on all classes, as $\mathcal{L}_R$ only focuses on the isolated pixel prediction, while ignoring the category context between different spatial locations. Applying the intra-class loss $\mathcal{L}_{intra}$, the performance on novel class obtains a dramatic improvement (+10.81 MIoU). We think that novel classifier is trained on only a handful of examples, and it is unable to contain all patterns in the specific category, which makes the decision boundary ambiguous. Resorting to $\mathcal{L}_{intra}$, the feature map establishes the intra-class context and exploits the category relationship for refining the decision boundary of novel class. When employing $\mathcal{L}_{inter}$, the performances of base and novel classes are improved by 10.50 MIoU and 1.98 MIoU. Such an inter-class loss models the discriminating category relationship for pixel-wise representation, contributing to the improvement of both base and novel classes. Moreover, with the usage of $\mathcal{L}_S$, the context relationship map regresses to a symmetric matrix, which makes the map rational and enhances the total performance by 0.73 MIoU.

Effect of CBAM. We evaluate the effect of CBAM in CRM and report the ablation results in Table 6. CRM with CBAM surpasses the one without CRM by a large margin on total performance (+4.88 MIoU). The reason is that CBAM highlights the salient foreground information and enhances the discriminating representations of each specific category, which assists in capturing the intra-class and inter-class context relationships for further segmentation.

Effect of dynamic updating strategy. This experiment investigates the effect of the dynamic updating strategy in PEM. As demonstrated in Table 6, PEM with the dynamic updating approach outperforms the one without the dynamic updating strategy, especially in terms of novel class (+21.92 MIoU). We believe that novel classifier will be robust and generalized to the specific category when incorporating more category information, which overcomes the issue that novel classifier is under-trained by only a handful of support data initially.

Self-alignment parameters in PEM. We explore the best choice of the threshold μ and the factor θ in PEM. The ablation results are illustrated in Fig. 5. As shown in the figure, the combination of $\mu = 0.8$ and $\theta = 0.4$ achieves the top performance. We think that a relatively high threshold $\mu = 0.8$ is required, as the foreground region of the intra-instance is of high similarity and sensitive to the noise. Thus we pick the high-confidence feature to establish the query classifier, which is further leveraged to conduct the self-alignment. As to θ, we select a relatively small value $\theta = 0.4$ to weigh more on the query classifier and focus on the self-alignment, as the intra-instance similarity is much higher, which contributes to segmenting the query image itself.

4.5 Analyses

We analyze the reason why DC-GFSS achieves such a high performance, especially on novel class. Firstly, we separately segment base class and novel class without classifier parameter fusion, which isolates the representation space of base classifier from novel classifier and avoids base class bias issue in CPAL. Then we utilize CRM to enrich the category context information on the feature map. Exploiting such a context, the model captures both intra-class and inter-class relationships, which is leveraged to refine the ambiguous decision boundary for further segmentation. Moreover, after decoupling base class and novel class prediction, novel classifier is under-trained with few annotated data. Such a prototype is unable to cover all patterns of the specific category. Therefore, we design PEM to conduct self-alignment to get around the issue of intra-class patterns discrepancy and apply a dynamic updating strategy to boost novel classifier with comprehensive knowledge of each specific category.

5 Conclusion

In this paper, we proposed DC-GFSS to alleviate base class bias issue commonly observed in GFSS methods. Through independently predicting base-class and novel-class segmentation masks and subsequently assembling them via a mask-level fusion strategy, DC-GFSS effectively decouples the learning of base and novel classes, leading to more balanced and robust segmentation results. Furthermore, we introduced two dedicated modules: a Context Relationship Module (CRM) and a Prototype Enhancement Module (PEM). CRM explicitly models intra-class and inter-class contextual relationships to enrich feature representations, while PEM enhances the discriminability of novel classifier through self-alignment and a dynamic updating strategy guided by query information. Together, these components enable DC-GFSS to accurately segment novel categories with limited annotated samples. Extensive experimental results on PASCAL-5^i and COCO-20^i demonstrate that DC-GFSS achieves competitive performance compared with existing methods. More importantly, DC-GFSS exhibits strong practical potential for real-world scenarios, such as obstacle avoidance systems, where the model must generalize to previously unseen object categories with limited annotations while preserving reliable segmentation performance on known semantic classes. This capability makes DC-GFSS particularly useful for safety-critical scenarios that require long-term deployment, where the system must continuously adapt to newly encountered object categories with limited supervision while maintaining stable performance on previously learned classes.

Acknowledgment. This work is supported by Research and Application of Obstacle Avoidance Monitoring Technology for Construction Operations (523612250006).

References

1. Shaban, A., Bansal, S., Liu, Z., Essa, I., Boots, B.: One-shot learning for semantic segmentation, (2017). arXiv:1709.03410 arXiv preprint
2. Wang, K., Liew, J.H., Zou, Y., Zhou, D., Feng, J.: Panet: few-shot image semantic segmentation with prototype alignment. In: Proceedings of the IEEE/CVF International Conference on Computer Vision, pp. 9197–9206 (2019)
3. Li, G., et al.: Adaptive prototype learning and allocation for few-shot segmentation. In: Proceedings of the IEEE/CVF Conference on Computer Vision and Pattern Recognition, pp. 8334–8343 (2021)
4. Tian, Z., Zhao, H., Shu, M., Yang, Z., Li, R., Jia, J.: Prior guided feature enrichment network for few-shot segmentation. IEEE Trans. Pattern Anal. Mach. Intell. **44**(2), 1050–1065 (2020)
5. Tian, Z., et al.: Generalized few-shot semantic segmentation. In: Proceedings of the IEEE/CVF Conference on Computer Vision and Pattern Recognition, pp. 11563–11572 (2022)
6. Dong, K., Yang, W., Xu, Z., Huang, L., Yu, Z.: Abpnet: adaptive background modeling for generalized few shot segmentation. In: Proceedings of the 29th ACM International Conference on Multimedia, pp. 2271–2280 (2021)

7. Liu, W., Wu, Z., Zhao, Y., Fang, Y., Foo, C.-S., Cheng, J., Lin, G.: Harmonizing base and novel classes: a class-contrastive approach for generalized few-shot segmentation (2023). arXiv:2303.13724 arXiv preprint
8. Lu, Z., He, S., Li, D., Song, Y.-Z., Xiang, T.: Prediction calibration for generalized few-shot semantic segmentation (2022). arXiv:2210.08290 arXiv preprint
9. Yu, C., et al.: Context prior for scene segmentation. In: Proceedings of the IEEE/CVF Conference on Computer Vision and Pattern Recognition, pp. 12416–12425 (2020)
10. Long, J., Shelhamer, E., Darrell, T.: Fully convolutional networks for semantic segmentation. In: Proceedings of the IEEE Conference on Computer Vision and Pattern Recognition, pp. 3431–3440 (2015)
11. Wei, Y., et al.: Revisiting dilated convolution: a simple approach for weakly-and semi-supervised semantic segmentation. In: Proceedings of the IEEE Conference on Computer Vision and Pattern Recognition, pp. 7268–7277 (2018)
12. Badrinarayanan, V., Kendall, A., Cipolla, R.: Segnet: a deep convolutional encoder-decoder architecture for image segmentation. IEEE Trans. Pattern Anal. Mach. Intell. **39**(12), 2481–2495 (2017)
13. Cheng, B., Misra, I., Schwing, Alexander Kirillov, and Rohit Girdhar. Masked-attention mask transformer for universal image segmentation. In: Proceedings of the IEEE/CVF Conference on Computer Vision and Pattern Recognition, pp. 1290–1299 (2022)
14. Chen, Y., Liu, Z., Xu, H., Darrell, T., Wang, X.: Meta-baseline: Exploring simple meta-learning for few-shot learning. In: Proceedings of the IEEE/CVF International Conference on Computer Vision, pp. 9062–9071 (2021)
15. Sun, Q., Liu, Y., Chua, T.-S., Schiele, B.: Meta-transfer learning for few-shot learning. In: Proceedings of the IEEE/CVF Conference on Computer Vision and Pattern Recognition, pp. 403–412 (2019)
16. Zhang, Y., Yang, W., Wang, S.: Fgnet: towards filling the intra-class and inter-class gaps for few-shot segmentation. In: IJCAI, pp. 1749–1758 (2023)
17. Zhang, Y., Wang, S., Shi, Z., Yang, W.: On filling the intra-class and inter-class gaps for few-shot segmentation. Expert Syst. Appl. **261**, 125552 (2025)
18. Xia, C., Xiong, C., Yu, P.: Pseudo siamese network for few-shot intent generation. In: Proceedings of the 44th International ACM SIGIR Conference on Research and Development in Information Retrieval, pp. 2005–2009 (2021)
19. Hariharan, B., Girshick, R.: Low-shot visual recognition by shrinking and hallucinating features. In: Proceedings of the IEEE International Conference on Computer Vision, pp. 3018–3027 (2017)
20. Gidaris, S., Komodakis, N.: Dynamic few-shot visual learning without forgetting. In: Proceedings of the IEEE Conference on Computer Vision and Pattern Recognition, pp. 4367–4375 (2018)
21. Min, J., Kang, D., Cho, M.: Hypercorrelation squeeze for few-shot segmentation. In: Proceedings of the IEEE/CVF International Conference on Computer Vision, pp. 6941–6952 (2021)
22. Yang, L., Zhuo, W., Qi, L., Shi, Y., Gao, Y.: Mining latent classes for few-shot segmentation. In: Proceedings of the IEEE/CVF International Conference on Computer Vision, pp. 8721–8730 (2021)
23. Yang, B., Liu, C., Li, B., Jiao, J., Ye, Q.: Prototype mixture models for few-shot semantic segmentation. In: Computer Vision–ECCV 2020: 16th European Conference, Glasgow, UK, 23–28 Aug 2020, Proceedings, Part VIII 16, pp. 763–778. Springer (2020)

24. Xie, G-S., Liu, J., Xiong, H., Shao, L.: Scale-aware graph neural network for few-shot semantic segmentation. In: Proceedings of the IEEE/CVF Conference on Computer Vision and Pattern Recognition, pp. 5475–5484 (2021)
25. Zhuge, Y., Shen, C.: Deep reasoning network for few-shot semantic segmentation. In: Proceedings of the 29th ACM International Conference on Multimedia, pp. 5344–5352 (2021)
26. Zhang, Y., et al.: Genseg: on generating unified adversary for segmentation. In: Proceedings of the Thirty-Third International Joint Conference on Artificial Intelligence, pp. 1733–1742 (2024)
27. Woo, S., Park, J., Lee, J.-Y., Kweon, I.S.: Cbam: convolutional block attention module. In: Proceedings of the European Conference on Computer Vision (ECCV), pp. 3–19 (2018)
28. Everingham, M., Gool, L.V., Williams, C.K.I., Winn, J., Zisserman, A.: The pascal visual object classes (voc) challenge. Int. J. Comput. Vis. **88**, 303–338 (2010)
29. Lin, T.-Y., et al.: Common objects in context. In: Computer Vision–ECCV 2014: 13th European Conference, Zurich, Switzerland, 6–12 Sept 2014, Proceedings, Part V 13, pp. 740–755. Springer (2014)
30. Hariharan, B., Arbeláez, P., Girshick, R., Malik, J.: Simultaneous detection and segmentation. In: Computer Vision–ECCV 2014: 13th European Conference, Zurich, Switzerland, 6–12 Sept 2014, Proceedings, Part VII 13, pp. 297–312. Springer (2014)
31. Nguyen, K., Todorovic, S.: Feature weighting and boosting for few-shot segmentation. In: Proceedings of the IEEE/CVF International Conference on Computer Vision, pp. 622–631 (2019)
32. Zhao, H., Shi, J., Qi, X., Wang, X., Jia, J.: Pyramid scene parsing network. In: Proceedings of the IEEE Conference on Computer Vision and Pattern Recognition, pp. 2881–2890 (2017)
33. He, K., Zhang, X., Ren, S., Sun, J.: Deep residual learning for image recognition. In: Proceedings of the IEEE Conference on Computer Vision and Pattern Recognition, pp. 770–778 (2016)
34. Ye, H.-J., Hexiang, H., Zhan, D.-C.: Learning adaptive classifiers synthesis for generalized few-shot learning. Int. J. Comput. Vision **129**, 1930–1953 (2021)
35. Lang, C., Cheng, G., Tu, B., Han, J.: Learning what not to segment: a new perspective on few-shot segmentation. In: Proceedings of the IEEE/CVF Conference on Computer Vision and Pattern Recognition, pp. 8057–8067 (2022)

PD-MAE: Masked Autoencoder with Patch Entropy and Knowledge Distillation for Image Representation Learning

Jianpu Geng, Ling Zhang(✉), Boquan Zhang, and Minghui Wang

School of Computer Science, Guangdong University of Technology, Guangzhou 510006, China
lzhang@mail2.gdut.edu.cn

Abstract. To transfer knowledge from a pre-trained Masked Autoencoder (MAE) into a more compact architecture, we introduce a novel framework termed Patch Entropy and Knowledge Distillation based Masked Autoencoder (PD-MAE). In a departure from the stochastic masking protocols typical of standard Masked Autoencoders (MAEs), our methodology integrates intrinsic semantic indicators and inter-patch dependencies directly into the masking phase. We implement a curriculum-based learning trajectory that facilitates a seamless transition from entropy-guided patch selection to a specialized 'hard-patch' masking framework, thereby prioritizing areas of high informational density. For efficient knowledge transfer, PD-MAE minimizes the discrepancy between intermediate feature representations of the teacher (pre-trained MAE) and student (compact model) networks throughout the training phase. This strategy obviates the requirement for aligning final output logits, thereby substantially reducing computational overhead. The efficacy and generalizability of the distilled model are validated via fine-tuning on diverse downstream vision tasks. Empirical evaluations demonstrate significant performance gains: classification accuracy improves by 1.8%, while object detection metrics show an increase of 2.4 in bounding box Average Precision (AP) and 2.3 in mask AP. Furthermore, semantic segmentation performance exhibits a 2.0 improvement in mean Intersection over Union (mIoU).

Keywords: Masked Autoencoder · Information Entropy · Pre-training-Fine-tuning · Image Reconstruction · Knowledge Distillation

1 Introduction

In recent years, self-supervised learning (SSL) has witnessed remarkable advancements in artificial intelligence, with computer vision emerging as a prominent application domain. These advancements have been chiefly propelled by

L. Huang and D. Greenhalgh (Eds.): ICMLC 2026, LNNS 2061, pp. 750–762, 2027.
https://doi.org/10.1007/978-3-032-31236-5_51

the groundbreaking incorporation of Vision Transformers (ViT) [1] within image representation learning frameworks. Significantly, the Masked Autoencoder (MAE) [2] has emerged as a leading SSL framework, exhibiting remarkable potential for obtaining transferable visual representations via masked image modeling.

However, MAE pre-training exhibits limited efficacy when applied to compact architectures; for instance, deploying it on lightweight models like ViT-Tiny often yields detrimental effects on downstream classification tasks. To address this limitation, rather than directly training small ViTs via standard MAE, this work adopts a knowledge distillation framework to transfer learned representations from a large-scale pre-trained MAE (teacher) to a compact target model (student). By reducing the divergence between intermediate feature representations of the teacher and student networks during training, the need for explicit alignment of final output logits is obviated. During distillation, this design permits the teacher model to perform forward propagation exclusively via its shallow layers, markedly reducing computational burden.

Conventional MAE frameworks rely on static masking strategies (e.g., random or fixed patterns) [3], which lack the capacity to dynamically accommodate semantic traits unique to individual images. Such rigidity often leads to excessive masking of semantically vital areas or introduces redundant supervision signals. To overcome this, we propose an adaptive curriculum masking strategy guided by information entropy. Here, patch-wise entropy—serving as a proxy for local information complexity—is quantified to drive masking decisions: initially, high-entropy patches (e.g., regions with intricate textures) are prioritized for occlusion, steering the model to develop global contextual reasoning abilities.Gradually, the masking focus shifts toward low-entropy patches (semantically consistent but detail-dense), driving the model to learn fine-grained local details. This incremental adjustment of masking complexity balances task difficulty and learning stability, thereby enhancing the model's ability to model inter-patch semantic relationships. Empirical gains in representative vision benchmarks validate its potential for optimizing masked modeling.

Building on these insights, this paper introduces a Patch Entropy and Knowledge Distillation-enhanced Masked Autoencoder (PD-MAE). The key contributions are summarized as follows:

1. **Novel Knowledge Distillation Framework**: By enabling feature alignment in intermediate layers between teacher (large-scale MAE) and student (compact model) networks, the student accelerates convergence while reducing computational expenditure, without requiring final output matching.
2. **Adaptive Entropy-Guided Masking Curriculum**: A dynamic masking strategy is designed to transition from entropy-driven occlusion (prioritizing high-entropy regions for global context learning) to hard-patch masking (focusing on low-entropy areas for fine-grained feature extraction), effectively harnessing intrinsic semantic cues and inter-patch associations.
3. **Comprehensive Empirical Validation**: Extensive experiments demonstrate that PD-MAE outperforms state-of-the-art masked autoencoders and

exhibits strong transferability across diverse downstream tasks, including classification, object detection, and semantic segmentation.

2 Related Work

2.1 Knowledge Distillation

The primary goal of knowledge distillation is to compress models via knowledge migration. Prevailing approaches are categorized into two types: output-oriented distillation and intermediate-representation distillation. The former, tracing back to Hinton's pioneering work [4], enables knowledge migration by minimizing divergences in class probability distributions across teacher-student models. Subsequent research boosted transfer efficacy through methods like temperature scaling [5] and distribution alignment optimization [6–8]. The latter centers on semantic alignment of intermediate-layer features. Significantly, FitNets [9] introduced a mechanism for predicting intermediate features, whereas Heo et al. [10] enhanced the feature distillation paradigm through separating feature norms from directions. Later, CRD [11] integrated contrastive learning for capturing high-order inter-feature correlations.

The progression of knowledge distillation methods reveals two key trends. One trend involves the deep fusion of self-supervised learning with knowledge distillation, which enhances the robustness of student model representations through unlabeled data synthesis tasks (e.g., rotation prediction [12], jigsaw puzzle reconstruction [13]) coupled with teacher knowledge migration. Another trend pertains to the rise of dynamic distillation paradigms, such as curriculum-guided staged knowledge transfer [14], which facilitates more efficient knowledge acquisition via gradual difficulty adjustment. These developments have lessened traditional distillation frameworks' dependency on labeled data, opening avenues for cross-modal and weakly supervised applications.

2.2 Masking Strategy

Mainstream masked autoencoder architectures often adopt non-adaptive strategies including stochastic masking [15] or block-wise masking [16]. These approaches generally occlude image regions uniformly or stochastically based on preset rules. Despite their implementation simplicity, they neglect the heterogeneous semantic distribution within images. Excessive masking of critical regions (e.g., object contours, semantic subjects) undermines the reconstruction goal (as shown in Fig. 1, where masking the bull's horns or the bird causes semantic uncertainty), while excessive masking of non-critical regions provides inadequate supervision. The limitation of static approaches stems from their unregulated task difficulty–whether masking intricacy is excessively high or low, it hinders the model's ability to capture high-level semantic dependencies.

Data-adaptive masking addresses these limitations through dynamic strategy adaptation. Its evolution can be segmented into two stages: initial efforts (e.g.,

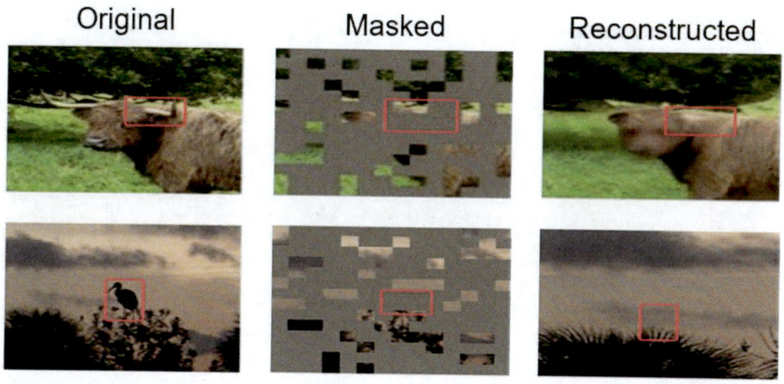

Fig. 1. Randomly masked images and their reconstructions. Red boxes indicate regions where random occlusion causes loss of key semantic information, degrading reconstruction quality.

AttMask [17], SemMAE [18]) utilized attention mechanisms or semantic segmentation to locate key regions for masking, increasing task complexity through explicit prior knowledge. Subsequent research (e.g., HPM [19], CL-MAE [20]) embraced implicit learning mechanisms, dynamically creating masking schemes via adversarial training [21] or contrastive loss [22]. This paper further proposes a progressive transition from patch entropy-based masking to hard-patch masking: initially, medium-difficulty regions are selected based on patch entropy values, and as training advances, the focus gradually shifts to challenging regions (low-entropy, high-semantic-consistency areas).

3 Patch Entropy and Knowledge Distillation Based Masked Autoencoder

This paper integrates the Masked Autoencoder (MAE) with a knowledge distillation task, introducing the Patch Entropy and Knowledge Distillation-based Masked Autoencoder (PD-MAE). Knowledge is transferred from a large-scale ViT-L teacher to a compact ViT-B student model. A curriculum learning strategy orchestrates a transition from patch entropy masking to hard-patch masking. In early training stages, this method leverages information content across image regions, retaining areas with high entropy. This guides the model to prioritize foreground content over background or non-essential elements. In later phases, it identifies challenging patches for masking by leveraging inter-patch semantic correlations. A substantial masking ratio ensures the task remains a meaningful self-supervised objective. The architectural framework of the model is illustrated in Fig. 2.

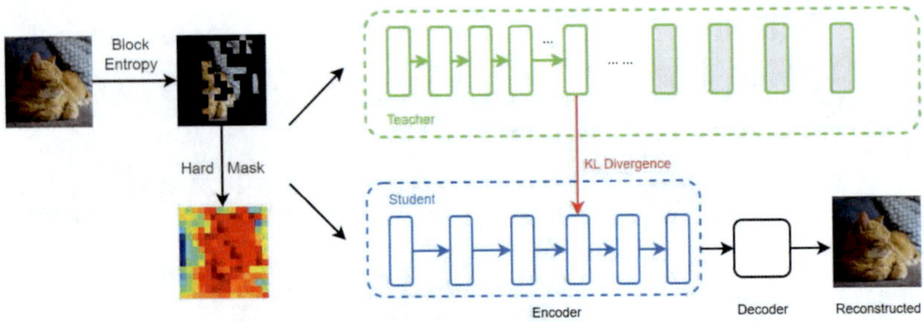

Fig. 2. Schematic of the PD-MAE architecture. The workflow initiates with patch entropy masking, transitioning to hard-patch masking. Knowledge flows from the teacher's intermediate layers to the student via KL Divergence loss, concurrent with the student's reconstruction task.

3.1 Patch Entropy Masking

For an image, the information content of a random variable $X = x_i$ $(i = 1, 2, \ldots, K)$ diminishes as its occurrence probability rises [23, 24]. Higher entropy in image patches signifies greater information richness. Foreground regions generally encode richer semantics than backgrounds, and insufficient foreground information may cause suboptimal optimization in supervised settings. To effectively link information entropy with image foreground, defining the image's feature representation is essential [25]. Consider an image of size $M \times N$ with grayscale function $f(x, y)$ $(f(x, y) \geq 0)$. Its entropy H_f is defined as:

$$p(i) = \frac{n_i}{M \times N} \tag{1}$$

$$H_f = -\sum_{i=0}^{255} p(i) \log_2 p(i) \tag{2}$$

In this paper, the image is divided into 14×14 patches, and the entropy of each patch is computed per Eq. (2). We subsequently retain the top 25% of patches exhibiting the highest entropy.

In practice, we precompute entropy values for individual image patches offline and load these precomputed values into GPU memory prior to initiating MAE pre-training, aiming to enhance processing efficiency. For efficient mask generation in the pre-training phase, we first apply random augmentations to the loaded images, such that each image in a batch B is assigned a square noise window of size P (where P denotes the total number of patches). Subsequently, adhering to a preset masking ratio (default: 0.75), patches with higher entropy are selected and fed into the encoder.

3.2 Easy-to-Hard Masking Generation

Beyond predefined masking strategies, crafting challenging tasks is essential for effective pre-training. This section details a method to pinpoint patches with elevated reconstruction difficulty by leveraging inter-patch semantic correlations. Enabling the model to forecast per-patch reconstruction losses facilitates precise identification of such challenging patches. Masking these hard patches amplifies training difficulty, thereby enhancing the model's representational power and generalization capability.

A naive approach to this end employs a loss predictor to estimate per-patch reconstruction losses via mean squared error (MSE). The optimization objective of this predictor is:

$$\mathcal{L}_{pred} = \mathbb{E}_{x \sim D}\left[\frac{1}{|M|}\sum_{i \in M}(d_{\Psi}(f_{\theta}(x_M))_i - \mathcal{L}_{rec}(x, M)_i)^2\right] \tag{3}$$

where x denotes the original image, M marks masked regions, f_θ is the encoder, d_Ψ the loss predictor, and $\mathcal{L}_{rec}$ the ground-truth reconstruction loss. The process involves two steps: first, applying mask matrix M to x to obtain a partially masked input; second, extracting features via the encoder and generating per-region reconstruction error estimates with the decoder. The core of the loss function is the mean squared difference between predicted and actual reconstruction errors, restricted to masked areas. Notably, this method exhibits progressive performance decay: as training advances, the global reduction in reconstruction errors diminishes the discriminability of regional error disparities. This dynamic undermines the predictor's ability to identify hard regions, compromising sustained task challenge.

To address this, we design a contrastive learning-based optimization. Instead of absolute error magnitudes, we use binary cross-entropy to construct a difficulty relation evaluator for patch pairs (i, j), focusing on relative reconstruction difficulty. When $\mathcal{L}_{rec}(i) > \mathcal{L}_{rec}(j)$, region i is deemed harder, and a supervision signal is formulated as:

$$\begin{aligned}\mathcal{L}_{\text{rank}} = -\mathbb{E}_{i,j \sim U(P)}\Big[&I(\mathcal{L}_r(i) > \mathcal{L}_r(j)) \log P_p(i > j) \\ &+ \big(1 - I(\mathcal{L}_r(i) > \mathcal{L}_r(j))\big) \log\big(1 - P_p(i > j)\big)\Big]\end{aligned} \tag{4}$$

Here, $\mathcal{L}_r$ is the actual reconstruction error, $I(\cdot)$ an indicator function (1 if true, 0 otherwise), and $P_p(i > j)$ the model's predicted probability of i being harder than j. This mechanism enables simultaneous learning of single-region error prediction and cross-region difficulty comparison by mining gradient differences in semantically related areas, ensuring accurate identification of challenging key regions.

To mitigate the problem where feature embeddings in early training are prone to interference from superficial texture patterns–where high-error regions may not align with semantically critical features–we adopt an "easy-to-hard" masking schedule. During the early training stage, we utilize patch entropy masking to provide a structured basis, guiding the model to gradually reconstruct challenging masked patches. At each training epoch t, a dynamically tuned masking range is specified by:

$$\alpha_t = \alpha_0 + (\alpha_T - \alpha_0)\frac{t}{T} \tag{5}$$

where α_0 and α_T are adjustable hyperparameters defining the hard-masking proportions at the beginning and conclusion of training, respectively. In this design, the approach follows cognitive learning principles: in the early phase, the model is guided to focus on coarse-grained semantic cues (e.g., object boundaries), while in subsequent phases, the complexity of detail restoration tasks (e.g., texture reconstruction) is incrementally raised. The merit of this phased design lies in preventing early entrapment in local optima and sustaining the model's learning momentum via a progressively rising difficulty curve, thereby boosting feature discriminability in tandem with training progression.

3.3 Knowledge Distillation

In the conventional Masked Image Modeling (MIM) pre-training paradigm, the primary goal is centered on reconstructing masked region content. From the perspective of knowledge transfer, this process is analogous to training a student model (a lightweight network) to emulate the capabilities of a teacher model (a complex network) via a specific task (masked patch prediction). This study proposes a tight integration of the knowledge distillation mechanism with the Masked Autoencoder (MAE) framework, aiming to efficiently enable compact models to attain the performance of larger-parameter baseline models.

As the MAE pre-training phase is devoid of supervised labels, the direct application of output-layer distillation poses challenges in deriving semantically meaningful feature representations. Thus, we adopt an intermediate feature distillation paradigm: we extract features from a selected intermediate layer of the student model and subsequently map them to the feature space of the corresponding teacher layer via a lightweight projection module. Notably, this module comprises a two-layer multilayer perceptron (MLP), which not only resolves dimensional inconsistencies between teacher and student features but also provides a learnable nonlinear mapping space for knowledge transfer.

In specific implementation, the teacher model employs a large-capacity ViT-L architecture pre-trained with MAE (24 Transformer blocks), while the student model adopts a lightweight ViT-B architecture (12 Transformer blocks). For an input image x, we define f_t and f_s as the feature extractors of the teacher and student models, respectively. The objective of knowledge distillation is to optimize f_s such that it implicitly learns the high-order semantic information encoded by f_t. The training process uses KL divergence as the supervisory signal:

$$\mathcal{L}_{KD} = D_{KL}(p||t) \tag{6}$$

where t denotes the target produced by the teacher model $f_t(x)$, and p represents the prediction output by the student model $f_s(x)$. To reconcile feature dimension differences between teacher and student models, a linear projection layer is introduced at the student output. Let F_t and F_s represent the teacher's target features and the student's projected predictions, respectively. The feature distillation loss is formulated as:

$$\mathcal{L}_{feat} = \text{SmoothL1}(\text{Norm}(F_s), \text{Norm}(F_t)) \tag{7}$$

By incorporating layer normalization (Norm) for feature de-biasing and combining it with smooth L1 loss to construct the feature alignment objective, we enhance the model's capacity to capture semantically critical features.

4 Experiments and Results

4.1 Experimental Setup

Adhering to the MAE protocol, we perform self-supervised pre-training on the ImageNet dataset. The teacher model employs the official MAE implementation of the ViT-L architecture. During pre-training, we adopt the AdamW optimizer with an initial learning rate of 1.5×10^{-4}, weight decay of 0.05, and momentum parameters β_1 and β_2 configured as 0.9 and 0.95, respectively. Training employs a batch size of 512 samples, with the model trained for 200 epochs–including a 40-epoch learning rate warm-up phase, then applying a cosine decay schedule. A default masking ratio of 75% is applied. Post-pre-training, the teacher model is retired, and the student model is fine-tuned under standard MAE configurations.

For feature alignment, we align student and teacher model features at three-quarters of the network depth. This approach yields effective outcomes across models of varying scales. For feature projection, we devised a two-layer MLP whose hidden layer dimension matches that of the teacher's aligned features, employing the GELU activation function.

4.2 Image Classification

We assess the classification performance of our method using three public datasets: CIFAR-100 [26], Tiny-ImageNet, and CUB-200-2011 [27]. CIFAR-100 is a benchmark dataset extensively utilized in computer vision. Tiny-ImageNet, a subset of ImageNet, comprises 200 classes with 500 training images per class. CUB-200-2011, a standard dataset for fine-grained bird species classification, serves to verify our method's effectiveness on small-sample scenarios.

Classification results on these datasets are summarized in Table 1. Our approach outperforms MAE by 1.8% and 3.6% on Tiny-ImageNet and CIFAR-100, Relative to the ViT-L teacher model, our classification accuracy is only 0.2% and 0.3% lower, suggesting effective knowledge transfer from the teacher to the student.

On CUB-200-2011, our method achieves a classification accuracy of 82.1%, underscoring its strong performance on small-sample datasets. This is attributed to the CUB dataset's frequent complex backgrounds; patch entropy masking aids in discarding background clutter during early training, directing the model's attention to avian features. Patch entropy masking further enables precise localization of birds, enhancing recognition accuracy and facilitating image reconstruction.

Table 1. Image classification performance (top-1 accuracy, %).

Method	Tiny-ImageNet	CIFAR-100	CUB-200
BEiT	87.3	82.4	79.5
MAE	87.6	84.3	80.6
CAE [28]	86.7	84.5	78.8
ViT-L (Teacher)	89.6	88.2	82.7
PD-MAE (Ours)	**89.4**	**87.9**	**82.1**

4.3 Object Detection

We perform object detection and instance segmentation on the COCO dataset [29]. The COCO dataset contains more than 330,000 images, approximately 200,000 of which are annotated in detail for 80 frequently encountered object categories. We utilize Mask R-CNN [30] featuring a ViT backbone for fine-tuning, employing box AP (Average Precision) for detection and mask AP for instance segmentation as evaluation metrics–aligning with MAE. Training and testing are executed using the ViTDet framework [31].

Table 2. Object detection and instance segmentation performance on COCO.

Method	AP^{box} (%)	AP^{mask} (%)
BEiT	40.8	36.0
MAE	49.2	43.4
CAE	48.3	42.7
ViT-L (Teacher)	51.9	46.2
PD-MAE (Ours)	**51.6**	**45.7**

Results are summarized in Table 2. Relative to MAE and CAE, our approach boosts detection AP by 2.4 and 3.3, and segmentation AP by 2.3 and 3.0, respectively. To further assess detection performance, we perform a visual comparison of our results against MAE and CAE using Mask R-CNN on COCO, with results illustrated in Fig. 3.

The yellow dashed boxes in Fig. 3 emphasize regions where our approach surpasses MAE and CAE. For example, in the first image, our method identifies a dog overlooked by MAE and a backpack missed by CAE, showcasing enhanced detail capture. As illustrated in Fig. 3, our approach yields notable gains in object detection versus recent methods, particularly in fine-grained regions. In complex scenes, our method more precisely localizes and identifies objects, reflecting improved robustness and sensitivity. These findings reveal that our method not only surpasses prior approaches in overall recognition accuracy but also demonstrates superiority in processing fine-grained details and intricate objects.

Fig. 3. Visual comparison of object detection results across models. Top row: original images. Subsequent rows: results from MAE, CAE, and PD-MAE (ours). Yellow dashed boxes mark cases where our method excels, e.g., detecting a dog overlooked by MAE and a backpack missed by CAE in the first column.

4.4 Semantic Segmentation

Semantic segmentation seeks to assign a categorical label to every pixel of an input image. We assessed our approach on three prominent semantic segmentation datasets: ADE20K [32], Pascal VOC [33], and Cityscapes [34]. ADE20K comprises 25K images spanning 150 semantic categories. Training and testing were conducted via the MMSegmentation framework. We employed a ViT-Base model pre-trained over 800 epochs, which was subsequently fine-tuned end-to-end on ADE20K for 80K iterations with a batch size of 16.

Table 3. Semantic segmentation performance (mIoU, %).

Method	ADE20K	Pascal VOC	Cityscapes
BEiT	45.6	71.4	61.4
MAE	46.1	72.1	65.2
CAE	45.5	70.5	63.7
PD-MAE (Ours)	**48.1**	**74.7**	**66.8**

As summarized in Table 3, our approach outperforms MAE in mIoU by 2.0 on ADE20K, 2.6 on Pascal VOC, and 1.6 on Cityscapes. These findings underscore the effective transferability of our approach to dense prediction vision tasks.

5 Conclusion

To conclude, this paper combines patch entropy masking with knowledge distillation, proposing the Patch Entropy and Knowledge Distillation-based Masked Autoencoder (PD-MAE). This method targets addressing two key issues: limited foreground reconstruction capability in early-stage masked autoencoders and the identification of challenging image patches in later stages. We begin by segmenting training images, followed by masking and reconstruction guided by information entropy values. Knowledge is transferred from a large ViT-L teacher model to a compact ViT-B student model. Experimental validation reveals that, in image classification, PD-MAE achieves a 1.8% accuracy gain over traditional masked image modeling approaches. On the small-sample CUB-200 dataset, classification accuracy reaches 82.1%. The model also exhibits performance enhancements in object detection and semantic segmentation experiments. Overall, the proposed PD-MAE advances masked image modeling to a certain extent. Future work will explore diverse network architectures to identify more effective configurations.

Acknowledgment. This paper is jointly funded by the "Undergraduate Teaching Engineering" project of Guangdong University of Technology in 2024- Huawei's "Intelligent Base 2.0" series courses—Big Data Technology Fundamentals (No. 44) and Real time Big Data Analysis (No. 45), the 2025 Guangdong Graduate Education Innovation Plan project "Big Data + AI Smart Teaching System Based on Generative Learning Application Integration" (No. 2025JGXM_056), and the Cultivation Project for Teaching Achievements in Undergraduate Higher Education at Guangdong University of Technology–"Student-Centered Immersive Artificial Intelligence Curriculum Development and Practice" (Project No. 263116029; Document No. GDUT-Ed-2023-78). Special thanks to Professor Zhang Ling for his guidance.

References

1. Dosovitskiy, A.: An image is worth 16x16 words: transformers for image recognition at scale (2020). arXiv:2010.11929 arXiv preprint
2. He, K., Chen, X., Xie, S.: Masked autoencoders are scalable vision learners. In: Proceedings of the IEEE/CVF Conference on Computer Vision and Pattern Recognition, pp. 16000–16009 (2022)
3. Bao, H., Dong, L., Piao, S.: Beit: Bert pre-training of image transformers (2021). arXiv:2106.08254 arXiv preprint
4. Hinton, G., Vinyals, O., Dean, J.: Distilling the Knowledge in a Neural Network (2015). arXiv:1503.02531 arXiv preprint
5. Furlanello, T., Lipton, Z., Tschannen, M.: Born again neural networks. In: International Conference on Machine Learning, pp. 1607–1616 (2018)
6. Zhang, Y., Xiang, T., Hospedales, T.M.: Deep mutual learning. In: Proceedings of the IEEE Conference on Computer Vision and Pattern Recognition, pp. 4320–4328 (2018)
7. Cho, J.H., Hariharan, B.: On the efficacy of knowledge distillation. In: Proceedings of the IEEE/CVF International Conference on Computer Vision, pp. 4794–4802 (2019)

8. Mirzadeh, S. I., Farajtabar, M., Li, A., et al.: Improved knowledge distillation via teacher assistant. In: Proceedings of the AAAI Conference on Artificial Intelligence, vol. 34, no. 04, pp. 5191–5198 (2020)
9. Romero, A., Ballas, N., Kahou, S.E.: Fitnets: Hints for thin deep nets (2014). arXiv:1412.6550 arXiv preprint
10. Heo, B., Kim, J., Yun, S.: A comprehensive overhaul of feature distillation. In: Proceedings of the IEEE/CVF International Conference on Computer Vision, pp. 1921–1930 (2019)
11. Tian, Y., Krishnan, D., Isola, P.: Contrastive representation distillation (2019). arXiv:1910.10699 arXiv preprint
12. Grill, J.B., Strub, F., Altché, F., et al.: Bootstrap your own latent-a new approach to self-supervised learning. Adv. Neural. Inf. Process. Syst. **33**, 21271–21284 (2020)
13. Zhang, H., Li, F., Liu, S., et al.: Dino: Detr with improved denoising anchor boxes for end-to-end object detection. arXiv preprint arXiv:2203.03605 (2022)
14. Zhou, J., Wei, C., Wang, H.: ibot: Image bert pre-training with online tokenizer (2021). arXiv:2111.07832 arXiv preprint
15. Kakogeorgiou, I., Gidaris, S., Psomas, B.: What to hide from your students: attention-guided masked image modeling. In: European Conference on Computer Vision, pp. 300–318. Springer Nature, Cham (2022)
16. Shi, Y., Siddharth, N., Torr, P., et al.: Adversarial masking for self-supervised learning. In: International Conference on Machine Learning, pp. 20026–20040. PMLR (2022)
17. Li, G., Zheng, H., Liu, D., et al.: Semmae: semantic-guided masking for learning masked autoencoders. Adv. Neural. Inf. Process. Syst. **35**, 14290–14302 (2022)
18. Wang, H., Song, K., Fan, J.: Hard patches mining for masked image modeling. In: Proceedings of the IEEE/CVF Conference on Computer Vision and Pattern Recognition, pp. 10375–10385 (2023)
19. Madan, N., Ristea, N.C., Nasrollahi, K.: Proceedings of the IEEE/CVF Winter Conference on Applications of Computer Vision, pp. 2492–2502 (2024). CL-MAE: Curriculum-Learned Masked Autoencoders
20. Xie, Z., Zhang, Z., Cao, Y.: Simmim: A simple framework for masked image modeling. In: Proceedings of the IEEE/CVF Conference on Computer Vision and Pattern Recognition, pp. 9653–9663 (2022)
21. Dong, X., Bao, J., Zhang, T.: Bootstrapped masked autoencoders for vision bert pretraining. In: European Conference on Computer Vision, pp. 247–264. Springer Nature, Cham (2022)
22. Hinojosa, C., Liu, S., Ghanem, B.: ColorMAE: exploring data-independent masking strategies in masked AutoEncoders. In: European Conference on Computer Vision, pp. 432–449. Springer, Cham (2025)
23. Zheng, Z., Zheng, H.: Image classification based on image information entropy. J. Comput. Appl. **38**(S2), 118–121 (2018)
24. Lu, P., Lin, G., Zou, G.: Research on no-reference image quality assessment method based on information entropy and deep learning. Appl. Res Comput. **35**(11), 3445–3449 (2018)
25. Mao, L., Zhao, L., Yu, M.A., et al.: Parathyroid segmentation based on hybrid level set model using local image entropy. Acta Optica Sinica **39**(12), 1217001 (2019)
26. Krizhevsky, A., Hinton, G.: Learning multiple layers of features from tiny images. Technical report, University of Toronto (2009)
27. Welinder, P., Branson, S., Mita, T.: Caltech-UCSD Birds 200. California Institute of Technology, (2010). CNS-TR-2010-001

28. Chen, X., Ding, M., Wang, X., et al.: Context autoencoder for self-supervised representation learning. Int. J. Comput. Vision **132**(1), 208–223 (2024)
29. Lin, T.Y., Maire, M., Belongie, S., et al.: Microsoft coco: common objects in context. In: Computer Vision–ECCV 2014, pp. 740–755. Springer International Publishing, Cham (2014)
30. He, K., Gkioxari, G., Dollár, P.: Mask R-CNN. In: Proceedings of the IEEE International Conference on Computer Vision, pp. 2961–2969 (2017)
31. Li, Y., Mao, H., Girshick, R.: Exploring plain vision transformer backbones for object detection. In: European Conference on Computer Vision, pp. 280–296. Springer Nature, Cham (2022)
32. Zhou, B., Zhao, H., Puig, X.: Scene parsing through ade20k dataset. In: Proceedings of the IEEE Conference on Computer Vision and Pattern Recognition, pp. 633–641 (2017)
33. Everingham, M., Van Gool, L., Williams, C.K.I., et al.: The pascal visual object classes (VOC) challenge. Int. J. Comput. Vision **88**, 303–338 (2010)
34. Cordts, M., Omran, M., Ramos, S.: The cityscapes dataset for semantic urban scene understanding. In: Proceedings of the IEEE Conference on Computer Vision and Pattern Recognition, pp. 3213–3223 (2016)

A Study on Threshold Selection for mAP Calculation of Vision AI Using User Satisfaction

Yoonho Choi[1] and Kwang Sik Chung[2](✉)

[1] HugeAILab, Shanghai, China
yhchoi@hugeailab.com
[2] Korea National Open University, Seoul, South Korea
kchung0825@knou.ac.kr

Abstract. This study proposes a novel threshold selection method that reflects user satisfaction when evaluating the performance of vision AI models. Since traditional objective performance metrics often do not align with actual user satisfaction, we developed a threshold selection system to calculate mAP using user satisfaction. This system consists of a threshold image generator program, a threshold satisfaction collector program, and a database. The experimental participants evaluated the predicted images generated under various threshold conditions and entered their satisfaction using a Likert scale. In an experiment using a fish dataset, we collected 3,720 data points from 31 participants. We analyze these data to evaluate user satisfaction based on changes in the threshold. The results of the chi-square test did not show significant differences in satisfaction between thresholds of 50 and 55, 55 and 60, and 65 and 70. This indicates that mAP6095, which reflects actual user satisfaction is a more objective evaluation metric for the fish dataset than mAP5095, which is based on existing academic conventions. In conclusion, this study presents the potential of a new threshold selection method that uses user satisfaction in the evaluation of vision AI performance. It also suggests that evaluation criteria using user satisfaction can provide a more realistic and reliable assessment of models compared to those grounded in conventional academic practices.

Keywords: Vision AI · Performance Metrics · Threshold · User Satisfaction · Human Factors in Computing Systems · Human-Computer Interaction

1 Introduction

Artificial intelligence (AI) has recently been widely applied across industries. Computer vision AI, in particular, has become an essential technology in various industries. Vision AI is a technology that enables machines to see like humans or animals. Vision AI recognizes and analyzes images and videos to extract meaningful information. Vision AI is applied in areas such as image classification,

L. Huang and D. Greenhalgh (Eds.): ICMLC 2026, LNNS 2061, pp. 763–779, 2027.
https://doi.org/10.1007/978-3-032-31236-5_52

object detection, object segmentation, facial recognition, and behavior recognition, and is utilized in a wide range of fields, including access control, autonomous driving, safety systems, medical image analysis, customer behavior tracking and personalized shopping experiences, smart factories, and robotics.

There are two methods for developing vision AI models: developing models from scratch through machine learning, and developing models through fine-tuning a pre-trained model. Building a pre-trained model from scratch requires significant time and cost, including collecting and preprocessing data, labeling, and then performing machine learning. In contrast, developing models through fine-tuning often utilizes limited labeled data.

Reliable performance metrics are essential for objectively evaluating the performance of vision AI object detection models, comparing various models, and establishing improvement directions. As vision AI applications expand and the number of models developed grows, the need for fair and precise performance metrics becomes increasingly important. Vision AI object detection models are evaluated using metrics such as precision, recall, average precision (AP), mean average precision (mAP), and frames per second (FPS). Performance metrics such as mAP50, mAP75, and mAP5095 are commonly used primarily [1–4].

Performance metrics such as precision, recall, AP, and mAP are objective. Their calculation methods are defined in formulas and are commonly used in both academia and industry to compare and evaluate the performance of vision AI models. However, there is a lack of objective evidence to support the use of thresholds ranging from 50 to 95 for mAP calculations. A subjective approach relies on the user's use of vision AI models and their evaluation. Compared to objective methods, the subjectivity of evaluators can lead to unfair performance comparisons. Furthermore, more time and effort is required to set up the experimental environment, recruit evaluators, conduct the experiments, and analyze the data compared to existing performance metrics. However, leveraging user satisfaction offers a significant advantage over simply determining performance with numerical data. Therefore, research is needed on the measurement content, measurement procedures, and methods for vision AI performance metrics utilizing user satisfaction. While subjective, research is needed on effective and efficient methods that avoid unfairness.

The purpose of this study is to establish standards for appropriately calculating the time and cost required for vision AI model development. The goal of our research is to figure out how much time and money to spend collecting, preprocessing, labeling, and learning data. We cannot invest infinite resources, and we do not want to waste more than necessary. Performance and cost have an exponential relationship. While improving low performance requires relatively little cost, improving already high performance incurs significantly greater costs. Therefore, it is necessary to find a satisfactory level of performance within this relationship and allocate cost appropriately.

According to the law of diminishing marginal utility, as the performance of vision AI increases, there may be a phenomenon in which the increase in user satisfaction (marginal utility) gradually decreases. In addition, the increase in performance may no longer increase satisfaction, that is, the diminishing marginal utility may become 0. The point where the diminishing satisfaction becomes 0 despite the increase in performance is the minimum threshold of maximum satisfaction region (MinTMaxSR), which is the blue line in region 3 in (Fig. 1). Conversely, the point where the diminishing dissatisfaction becomes 0, where satisfaction no longer decreases even as performance decreases, is the maximum threshold of minimum satisfaction region (MaxTMinSR), which is the red line in region 1 in (Fig. 1).

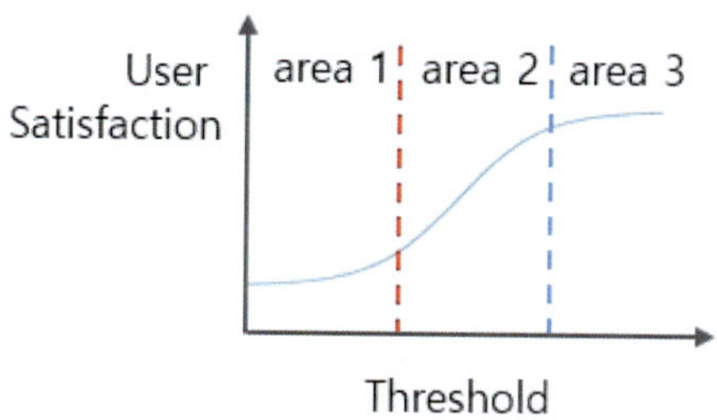

Fig. 1. Relationship between vision AI threshold and user satisfaction. The blue line is the minimum threshold of the maximum satisfaction region (MinTMaxSR). And the red line is the maximum threshold of the minimum satisfaction region (MaxTMinSR)

We think the mAP metric deserves sufficient respect. However, depending on the goal of fine-tuning, a very high threshold may be required, such as for medical purposes, or an appropriate threshold, such as for hobby purposes, may be sufficient. And we use user satisfaction as an indicator to judge this, and assuming that there is a law of diminishing marginal utility in this satisfaction, we propose the MaxTMinSR and MinTMaxSR theories. I think that the review of AI model performance based on user satisfaction, which follows the law of diminishing marginal utility, needs to be sufficiently reviewed in other AI fields such as LLM, image generation, education, music generation, and marketing.

In this study, our research is about (1) presenting a theory, (2) developing a system to verify the theory, (3) generating test data to perform the system, and 4) proving the theory through experiments with actual users. So, we aim to provide an objective rationale for threshold selection by exploring the possibility of introducing a user satisfaction-based criterion for choosing thresholds. A system is developed to collect user satisfaction data regarding model performance, and based on these data, the study seeks to identify a more reasonable method for setting thresholds. Through this approach, we aim to overcome the limitations of existing performance metrics that are based primarily on academic convention.

Section 2 reviews prior research related to vision AI performance metrics, user satisfaction, and the Law of Diminishing Marginal Utility. Section 3 introduces the system that collects user satisfaction data based on model performance and presents a method for selecting new thresholds. Section 4 describes an experiment using a fish dataset to derive thresholds through the system and method introduced in Sect. 3 .

2 Related Work

2.1 Performance Indicators of Vision AI

PASCAL VOC [5], ImageNet [6], and COCO [7] provide datasets necessary for machine learning and host performance competitions using these datasets. These include the PASCAL Visual Object Classes (VOC) Challenge [8,9], ImageNet Large Scale Visual Recognition Challenge (ILSVRC) [10,11], and COCO Challenges [12]. The Average Precision metric introduced in the PASCAL VOC Challenge 2007 is still almost same used today. ILSVRC uses mAP with an IoU (Intersection over Union) threshold more than 0.5 as its performance metric. The COCO Challenges use performance metrics such as AP@[0.5:0.95], AP50, AP75, and AP by size (small, medium, large).

The Confusion Matrix [13] in Table 1 is a classification result that compares the actual ground truth with the values predicted by the model.

Table 1. Confusion matrix

		Predicted result	
		Positive	Negative
Ground truth	Positive	TP (True positive)	FN (False negative)
	Negative	FP (False positive)	TN (True negative)

IoU (Intersection over Union) is a measure of confidence calculated based on the area of the actual result and the predicted result. It computes confidence by dividing the intersection area of the actual and predicted bounding boxes by their union area. Accuracy refers to the proportion of correct predictions out of all predictions made by a classification model. Precision is the proportion of predicted positives that are actually positive. Recall is the proportion of actual positives that are correctly predicted. Precision and Recall have a trade-off relationship, meaning that increasing one may lead to a decrease in the other. When the model uses a high threshold, only predictions with high IoU are considered correct, resulting in high Precision but low Recall. Conversely, if the threshold is low, more predictions are accepted as correct, increasing Recall but lowering Precision.

The Precision-Recall Curve is a two-dimensional curve plotted using Precision and Recall values calculated while varying the threshold. The closer the curve is to the top-right corner, the higher the Precision and Recall, indicating a better model [14,15]. The Average Precision (AP) is the area under the smoothed version of the zigzag-shaped curve. Mean Average Precision (mAP) is the average of the AP scores across all object classes in the dataset.

The most commonly used performance metrics are mAP50, mAP75, and mAP5095. mAP50 uses a threshold of 50, and mAP75 uses a threshold of 75 to generate the Precision-Recall Curve and compute the mAP score. mAP5095 uses ten thresholds set at intervals of 5, from 50 to 95, to comprehensively measure the model's performance across both low and high thresholds. mAP5095 is the most widely adopted academic performance evaluation metric in various performance competitions [1–4].

2.2 User Satisfaction

User satisfaction is a concept that describes the outcome measured by the difference between a user's expectations and actual experience [16]. It refers to the subjective emotional state experienced by a user comparing the level of expectation prior to the experience with the level of satisfaction after experiencing a specific product or service. In addition to the degree of expectation fulfillment, user satisfaction is influenced by various factors, such as perceived quality, ease of use, information reliability, and response speed. These factors encompass not only functional accuracy but also emotional responses [17]. In artificial intelligence, accuracy, explainability, trustworthiness, adaptability, customization, error recovery, and security significantly impact user satisfaction. Satisfaction tends to increase when users perceive they can understand and adjust the model's judgment process [18]. Acceptance intention or acceptability refers to the degree to which a user is willing to accept a specific concept or object [19]. It predicts future behavioral intentions or positive responses, and the user's attitude plays a key role. Acceptance intention goes beyond current satisfaction and comprehensively reflects future choices and behaviors. The Technology Acceptance Model (TAM), a theory proposed by American psychologist Fred Davis, explains the decision-making process of how users accept and use new technologies. User satisfaction and intention to adopt are distinct concepts. User satisfaction is a subjective evaluation of actual experience, while intention to adopt is a prediction of future behavior [20].

2.3 Law of Diminishing Marginal Utility

The law of diminishing marginal utility is an economic theory that states that as the consumption of a specific good increases, the additional satisfaction (marginal utility) gained gradually diminishes [21]. This theory, proposed by German economist Hermann Heinrich Gossen, is a useful concept for predicting and analyzing the saturation point of satisfaction and has recently been utilized in various technology evaluation and user experience studies [22]. This concept is central to consumer theory and plays a crucial role in explaining various economic phenomena, including consumer choice, the shape of the demand curve, corporate pricing strategies, tax policies, resource allocation, and marketing strategies. The law of diminishing marginal utility was originally derived from the context of consumption to analyze economic phenomena, but similar patterns have been observed in the experience of using digital technologies, and its application has been expanded to include evaluation metrics. Specifically, in technology products or AI systems, once technical indicators such as performance, speed, and accuracy exceed a certain level, user satisfaction tends to no longer increase linearly. This law is also linked to cognitive psychological concepts such as cognitive saturation, information overload, and attentional limitations. In other words, users have limitations in processing information or embracing technology. Research in human-computer interaction (HCI) has also shown that excessive performance improvements can actually lead to decreased satisfaction or fatigue [23].

2.4 User Satisfaction and Performance Indicators

This study begins with the recognition that academic performance metrics such as mAP50, mAP75, and mAP5095 do not adequately reflect actual user satisfaction. Although mAP5095 is useful for comprehensively evaluating precision and recall, it may differ from the subjective satisfaction experienced by users when interacting with vision AI systems. This discrepancy arises because threshold settings are based on academic conventions and may not sufficiently represent real user satisfaction [24,25].

In the field of large language models (LLMs), there has been active research into Reinforcement Learning from Human Feedback and building human preference datasets to better incorporate user satisfaction. A notable example is Chatbot Arena [26], an open platform for LLM evaluation based on human preferences, which has built a database of preferred responses through over one million user votes. Moreover, due to the difficulty of evaluating LLM-based chat assistants, research is also being conducted on using powerful LLMs themselves as judges [27].

In the field of vision AI, studies have also examined human error rates in labeling tasks. Andrej Karpathy showed through classification experiments with the CIFAR-10 dataset that human accuracy is about 94% [28]. Later, using the label verification program from the ILSVRC 2014 classification task, it was confirmed that the error rate among skilled human labelers was approximately 5.1% [29,30].

3 Implementation of Threshold Selector

The research system is titled "Threshold Selector for mAP Calculation of Vision AI Using User Satisfaction (Selector)". The Selector collects user satisfaction data for thresholds ranging from 50 to 95 and calculates the Minimum Threshold of the Maximum Satisfaction Region (MinTMaxSR) and the Maximum Threshold of the Minimum Satisfaction Region (MaxTMinSR). The Selector consists of the "Threshold-Image Generator for Detection Model (Generator)", the "Satisfaction Collector of Threshold for Detection Model (Collector)", and a database for storing the collected data.

(a) Threshold selector for mAP calculation of vision AI using user satisfaction (Selector): A threshold selection system composed of a generator, collector, and database.

(b) Threshold-Image generator for detection model (Generator): A threshold image generation program that displays ground truth and processed prediction results.
(c) Satisfaction collector of Threshold for detection model (Collector): A program for collecting user satisfaction with the threshold images.
(d) Database for detection model (Database): A database for storing image labeling information, user satisfaction data, and analysis results.

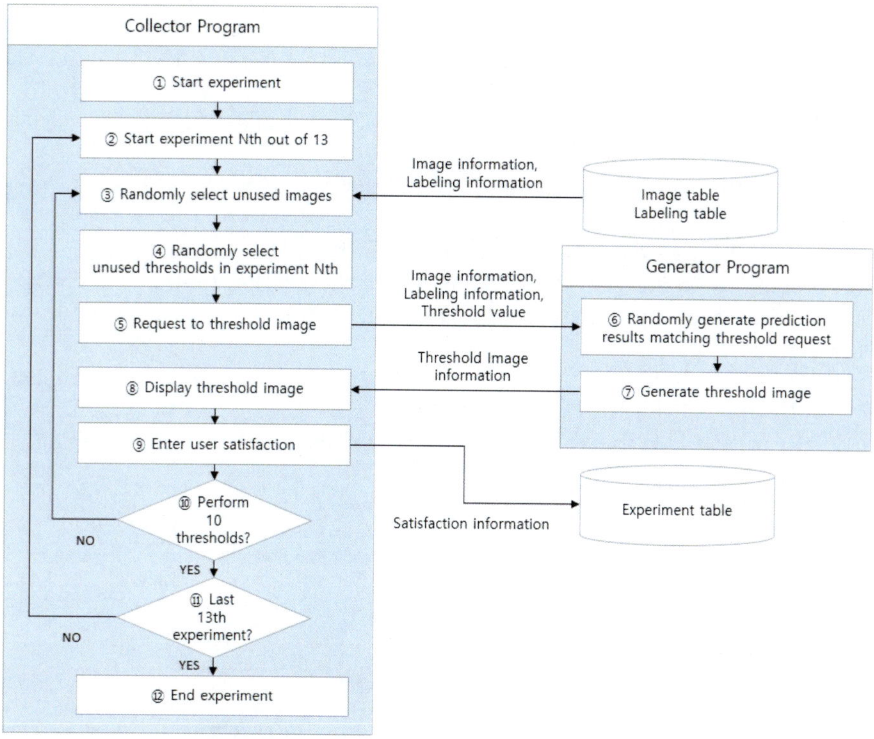

Fig. 2. Operational flow of the research system

The Generator receives a threshold as input and generates synthetic predicted results based on ground truth data. For example, if the threshold is 75, it generates a predicted result by randomly adjusting the central coordinates, width, and height so that the IoU equals 75. It then produces a threshold image displaying both the labeling ground truth and the synthetic predicted result.

The Collector displays the synthetic predicted result images to the participant, and collects satisfaction data using a 5-point Likert Scale. Each participant takes part in a total of 13 experiments, including one test trial. In a single experiment, the participant selects satisfaction levels for 10 different thresholds ranging from 50 to 95. Satisfaction is recorded on a scale of 1 to 5: very dissatisfied (1 point), dissatisfied (2 points), neutral (3 points), satisfied (4 points), and very satisfied (5 points).

The experiment conducted by one participant is as follows (Fig. 2):

① The participant participates in the experiment.

② The collector starts the Nth experiment out of 13 experiments.

③ The collector randomly selects images out of unused images and searches for image information and label information in the image and labeling tables.

④ The collector randomly selects thresholds not used in the Nth experiment.

⑤ The collector transmits image information, labeling information, and threshold values to the image generator and requests the threshold images.

⑥ The generator randomly generates prediction results that match the threshold request.

⑦ The generator generates threshold images that display the labeling correct information and the predicted results.

⑧ The collector receives threshold image information from the generator and displays it.

⑨ When the participant selects satisfaction with the performance, the collector stores the satisfaction information in the satisfaction table.

⑩ The collector repeats steps ③ to ⑨ until 10 thresholds are selected in the Nth experiment.

⑪ After the selection for the 10 thresholds is finished, the collector repeats steps ② to ⑩ until the 13th experiment ends.

⑫ After the last 13th experiment ends, the collector ends the experiment.

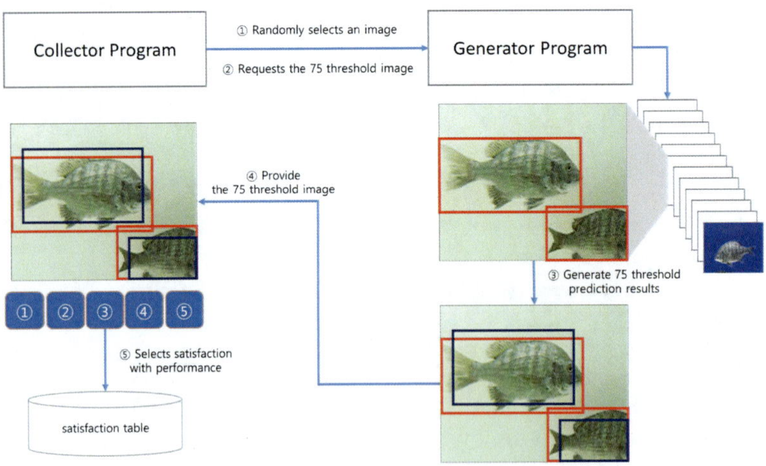

Fig. 3. Example of a research system in operation

Figure 3 illustrates the operation of the research system. The collector randomly selects an image. It then requests the generator to create an image for the randomly selected 75 threshold. The generator generates the prediction results for the 75 threshold. The collector is then provided with the 75 threshold image, which displays both the ground truth labeling information and the processed prediction results. The collector collects user satisfaction regarding the performance, with ratings from 1 to 5.

Figures 4 and 5 in the appendix are examples of the system in action.

4 Threshold Selector Experiments and Analysis

4.1 Threshold Selector Experiment

We conducted experiments to obtain thresholds using a fish data set. The experimental material is a newly labeled fish detection dataset. As reviewed in Sect. 2, common image detection datasets include MNIST (NIST) [31], Caltech-101 [32], PASCAL VOC [5], ImageNet [6], COCO [7], and the Open Images Dataset [33]. Unlike recent dataset trends, this experiment uses a dataset limited to the fish category.

Fish have iconic images due to their unique morphology. Conversely, schools of fish and complex underwater environments are atypical. Because the fish dataset encompasses both simple and complex morphologies, comparing and analyzing the two datasets is highly useful for interpreting the correlation between user satisfaction and thresholds. Furthermore, by using only one class,"Fish," for data labeling, we ensure that participants focus on their satisfaction with fish detection performance, preventing confusion between different classes from influencing satisfaction.

The dataset is constructed using fish videos and fish behavior videos made publicly available on AI Hub. The fish videos includes images by species and growth stages, captured in their cultivation environments, consisting of approximately 500,000 images and labeling information–about 100,000 per species [34]. The fish behavior videos are intended for fisheries AI development, such as behavior analysis and object tracking, and contain approximately 1.26 million labeled images–about 250,000 per species [35].

Because the AI Hub dataset is extracted from video footage, it contains many similar images. To ensure diversity, 209 non-redundant images were selected, categorized into four types: bright background, underwater background, rock fish, and flounder fish in Table 2. A new labeling process was conducted using a single class, "Fish", resulting in 553 labeled fish instances. The Computer Vision Annotation Tool (CVAT) [36] was used for instance labeling.

The experiment was conducted online using the research system developed in Sect. 3, running on the AWS Cloud. Most participants were students at the same university as the author, with some acquaintances and colleagues. Participants were provided with online training on the purpose and content of the experiment, as well as the experimental methodology, prior to the experiment. The average time it took participants to complete the experiment was 15 min, ranging from 10 min for the fastest to 20 min for the slowest. No compensation was provided.

Table 2. Sample of fish image types

Bright background **FBB**	Underwater background **FUB**	Rock fish **RFS**	Flounder fish **FFS**

A total of 33 participants participated in the online experiment. Two participants did not complete the experiment, resulting in a final total of 31 participants. These 31 participants made 130 selections across 13 experiments. The first experiment was used as a test to familiarize participants with the experimental methodology and data and was excluded from data analysis. The 31 participants made 120 selections, resulting in a total of 3,720 data points.

For data analysis, a normalized threshold was calculated using the formula below:

$$y = (x - 50)/5 + 1 \tag{1}$$

A summary of user satisfaction data is given in Table 3.

Table 3. Number of satisfaction responses by threshold

Threshold	Normalized threshold	1	2	3	4	6	Total
50	1	91	141	94	27	19	372
55	2	65	140	100	44	23	372
60	3	45	142	112	43	30	372
65	4	26	127	112	71	36	372
70	5	15	116	122	75	44	372
75	6	5	83	119	114	51	372
80	7	5	66	91	122	88	372
85	8	1	34	76	126	135	372
90	9	2	9	41	133	187	372
95	10	22*	7	17	103	223	372

*The reason for the high number of 22 responses with a satisfaction score of 1 at the 95 threshold was found to be that in many cases, when the ground truth box and the predicted box almost completely overlapped, participants interpreted it as if no prediction was made, leading them to assign a low satisfaction score

4.2 Threshold Selector Experiment Results Analysis

First, the normality of the data was examined. As shown in Table 4, the Shapiro–Wilk normality test indicates that the data does not follow a normal distribution(0.05). Therefore, the ANOVA (Analysis of Variance) method, which requires the assumption of normality, cannot be used. Instead, the relationship between threshold and user satisfaction is analyzed using the chi-squared test, a non-parametric statistical method that does not require the data to be normally distributed.

Table 4. Shapiro–Wilk normality test for satisfaction scores by threshold

Threshold	50	55	60	65	70	75	80	85	90	95
Shapiro–Wilk Statistic	0.8718	0.8919	0.8899	0.9011	0.8963	0.8962	0.8853	0.8439	0.7621	0.6412

Table 5. Results and interpretation of chi-squared test for threshold pairs

Threshold pairs	50–55	55–60	60–65	65–70	70–75	75–80	80–85	85–90	90–95
P-value	0.0618	0.2611	0.0097	0.3100	0.0008	0.0033	0.0001	0.0000	0.0000
Differences	No	No	Yes	No	Yes	Yes	Yes	Yes	Yes

Threshold pairs were created as (1, 2), (2, 3), (3, 4), (4, 5), (5, 6), (6, 7), (7, 8), (8, 9), and (9, 10), for a total of 9 pairs. The results of the chi-squared tests conducted on these threshold pairs are shown in Table 5. In this experiment using the fish dataset, the analysis showed that there were no statistically significant differences between the threshold pairs 5055, 5560, and 6570 by 0.05.

According to the results of the chi-squared test, there are significant differences in user satisfaction between the following thresholds: 60 and 65, 70 and 75, 75 and 80, 80 and 85, 85 and 90, and 90 and 95. On the other hand, no significant differences were found between thresholds 50 and 55, 55 and 60, and 65 and 70. Therefore, the Maximum Threshold of the Minimum Satisfaction Region (MaxTMinSR) can be set to 60. However, the Minimum Threshold of the Maximum Satisfaction Region (MinTMaxSR) could not be identified. The mAP6095 metric uses thresholds from 60 to 95, and excluded thresholds 50 and 55 which is showed no significant differences. The mAP6095 metric is a performance metric composed only of thresholds that showed significant differences in user satisfaction for the fish dataset.

After dividing the dataset into two subsets–FBBRFS, which combines FBB and RFS where fish are clearly visible, and FUBFFS, which combines FUB and FFS where fish are less visible or overlapping, as shown in Table 6, we applied

Table 6. FBBRFS and FUBFFS subsets

FBBRFS	FUBFFS

Table 7. Chi-squared test results and interpretation for threshold pairs by data subset

Type	Threshold pairs	50–55	55–60	60–65	65–70	70–75	75–80	80–85	85–90	90–95
Entire	P-value	0.0618	0.2611	0.0097	0.3100	0.0008	0.0033	0.0001	0.0000	0.0000
	Differences	No	No	Yes	No	Yes	Yes	Yes	Yes	Yes
FBBRFS	P-Value	0.1999	0.4828	0.0203	0.5218	0.0280	0.2210	0.0075	0.0000	0.0002
	Differences	No	No	Yes	No	Yes	No	Yes	Yes	Yes
FUBFFS	P-Value	0.3713	0.5751	0.2224	0.1821	0.0485	0.0229	0.0041	0.0116	0.0001
	Differences	No	No	No	No	Yes	Yes	Yes	Yes	Yes

the same chi-squared tests conducted earlier were applied. The results are shown in Table 7.

In the FBBRFS set, there were no significant differences between the ranges of 5055, 5560, 6570, and 7580. Therefore, the Maximum Threshold of the Minimum Satisfaction Region (MaxTMinSR) can be set to 60. On the other hand, the Minimum Threshold of the Maximum Satisfaction Region (MinTMaxSR) was not identified. Thresholds 50 and 55, which showed no significant differences, were excluded, and the mAP6095 metric–using thresholds from 60 to 95–is a performance metric composed only of thresholds that showed significant differences in user satisfaction for the fish data in the FBBRFS set.

In the FUBFFS set, there were no significant differences between the ranges of 5055, 5560, 6065, and 6570. Therefore, the Maximum Threshold of the Minimum Satisfaction Region (MaxTMinSR) can be set to 70. On the other hand, the Minimum Threshold of the Maximum Satisfaction Region (MinTMaxSR) was not identified. Thresholds 50, 55, 60, and 65, which showed no significant differences, were excluded, and the mAP7095 metric–using thresholds from 70 to 95–is a performance metric composed only of thresholds that showed significant differences in user satisfaction for the fish data in the FUBFFS set.

By comparing the FBBRFS set, in which fish are clearly visible, and the FUBFFS set, in which identifying fish is difficult, it was observed that the clearer the fish appear, the broader and more finely detailed the range of user satisfaction becomes. Additionally, for the entire dataset, mAP6095 was identified; for the FBBRFS set, mAP6095 was confirmed; and for the FUBFFS set, mAP7095 was confirmed. This demonstrates that the distribution of user satisfaction varies depending on the dataset, and that selecting thresholds based on this distribution is meaningful. Furthermore, the new thresholds using user satisfaction can be considered more objective and enhance the reliability of model performance evaluation, compared to the conventionally used mAP5095, which is based on academic convention.

5 Discussion and Limitations

This study began with the recognition that existing vision AI performance evaluation metrics do not adequately reflect actual user experiences. In particular, widely used quantitative performance metrics such as mAP50, mAP75, and mAP5095, while effective for objective comparisons between vision AI models, have limitations in describing actual user satisfaction. Therefore, this study aimed to overcome the limitations of technology-centric evaluation metrics and explore new user-centric performance evaluation metrics.

To achieve this, we proposed the concepts of the minimum threshold of the maximum satisfaction region (MinTMaxSR) and the maximum threshold of the minimum satisfaction region (MaxTMinSR) based on user satisfaction and the law of diminishing marginal utility [21]. We then developed a system and method to quantitatively collect and evaluate user satisfaction based on these thresholds. To prove the concept and validate the system and method, we created a fish dataset and conducted an experiment to collect and analyze 3,720 user satisfaction evaluations conducted by 31 participants.

The experimental results identified threshold ranges in the fish dataset where user satisfaction was not significantly differentiated. In other words, we were able to discover the maximum threshold (MaxTMinSR) within the minimum satisfaction range. This empirically confirmed that user satisfaction did not change significantly within a specific threshold range. The threshold used by the existing mAP5095 method contained insignificant values related to changes in user satisfaction. Consequently, an approach utilizing user satisfaction can enhance the practicality and reliability of vision AI model evaluation.

This study confirmed the possibility of introducing new threshold criteria through system, data, and analysis. This study presents the following four implications.

First, while existing academic performance criteria such as mAP5095 are standardized tools for performance verification and comparison, they can be excessive and unrealistic from the perspective of industrial settings and actual users. This study empirically demonstrated the gap between these metrics and users through experiments. Furthermore, utilizing user satisfaction as a criterion, we raised the need for user-centered performance metrics.

Second, this study strengthened the effectiveness of evaluation metrics by implementing a quantitative methodology for selecting thresholds based on user satisfaction evaluation data. Although the dataset is limited, we experimentally verified the maximum threshold (MaxTMinSR) within the minimum satisfaction region. This can serve as a practical benchmark for evaluating the investment efficiency of technical resources during the development and operation of vision AI models and predicting the economic limits of performance optimization.

Third, defining thresholds based on user data serves as a bridge between technology-centric performance evaluation systems and user-centered UX design.

This approach can be expanded beyond simple quantitative evaluation to a practical framework that reflects user feedback throughout the entire vision AI system design process. In other words, this study demonstrates the methodological feasibility of integrating user-participatory evaluation systems into the technology design and policy decision-making processes.

Fourth, the threshold selection method using user satisfaction presented in this study is not limited to a specific model or field, but is a universal technique applicable to various types of vision AI models. This method and technology provide a foundation for developing performance evaluation methods based on user satisfaction in other AI fields, such as LLM, image and video generation, music generation, education, and marketing.

This study empirically demonstrated the potential of using user satisfaction to select thresholds for evaluating the performance of vision AI models and to re-establish evaluation metrics. However, the number of participants was limited to 31, and most of them were students at the same university as the author, limiting diversity. Therefore, further research is needed to expand the theory and generalize the applicability. Future research will address the following three aspects:

First, because this study conducted experiments only using the fish dataset, the generalizability of the results requires verification. Future research should expand the threshold selection system to datasets spanning diverse application areas and purposes, including medical images, traffic images, security surveillance, industrial quality inspection, autonomous driving, and motion images. This will validate the versatility and scalability of threshold selection using user satisfaction. In particular, experiments using datasets commonly used for performance comparison in the vision AI field, such as ImageNet [6], COCO [7], and the Open Images Dataset [33], are necessary to validate the results in more general settings.

Second, for data interpretation, the Shapiro–Wilk normality test, Analysis of Variance (ANOVA), and chi-square tests are used. This is a minimal approach to data analysis, and new methods for data interpretation are needed. Furthermore, beyond statistical methods, it is necessary to develop quantitative analysis models that more precisely explain and predict the relationship between performance and satisfaction through mathematical or machine learning-based predictive models. Using curve fitting, structural equation modeling (SEM), and Bayesian inference models, it is possible to normalize the satisfaction change curve and develop models that derive it.

Third, the current threshold selection system has fixed experimental design, participant selection, and data types. Future improvements will allow researchers to configure the labeling purpose, data type, number of experiments, and difficulty level through system settings, allowing them to flexibly configure experimental conditions and repeatedly perform diverse experimental scenarios. A feature that allows administrators to register labeled data and allow users to participate in experiments using that data should be added.

A The Research System Screenshots

(1) Main screenshot

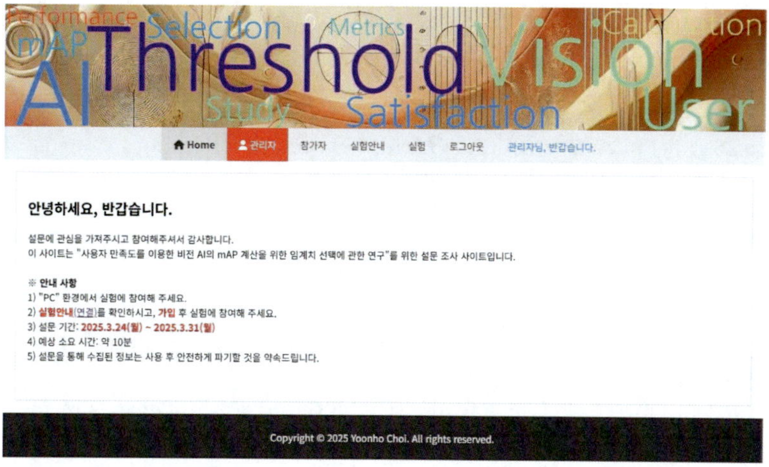

Fig. 4. Explanation of experiment guide and participation method

(2) Satisfaction selection screenshot

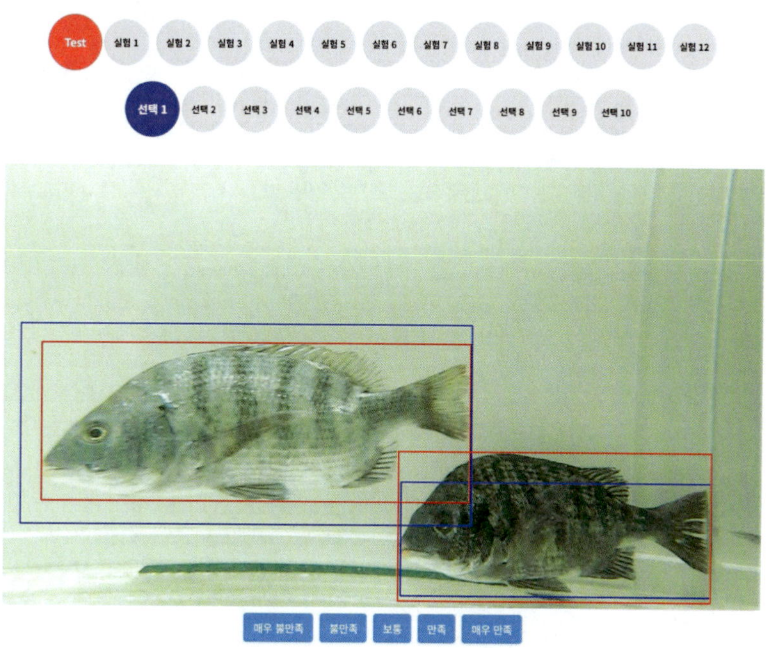

Fig. 5. 13 experiment step(red icons) and 10 selection step (blue icons), threshold image with fish and 5-point Likert scale buttons

References

1. Padilla, R., Netto, S.L., Da Silva, E.A.B.: A survey on performance metrics for object-detection algorithms. In: 2020 International Conference on Systems, Signals and Image Processing (IWSSIP). IEEE (2020)
2. John, A., Meva, D.: A comparative study of various object detection algorithms and performance analysis. Int. J. Comput. Sci. Eng. **8**(10), 158–163 (2020)
3. Padilla, R., et al.: A comparative analysis of object detection metrics with a companion open-source toolkit. Electronics **10**(3), 279 (2021)
4. Hicks, S.A., et al.: On evaluation metrics for medical applications of artificial intelligence. Sci. Rep. **12**(1), 5979 (2022)
5. Everingham, M., et al.: The pascal visual object classes (VOC) challenge. Int. J. Comput. Vis. **88**, 303–338 (2010)
6. Deng, J., et al.: ImageNet: a large-scale hierarchical image database. In: 2009 IEEE Conference on Computer Vision and Pattern Recognition. IEEE (2009)
7. Lin, T.-Y., et al.: Microsoft coco: common objects in context. In: Computer Vision—ECCV 2014: 13th European Conference, Zurich, Switzerland, 6–12 Sept 2014, Proceedings, Part V 13. Springer (2014)
8. Everingham, M., et al.: The Pascal visual object classes challenge: a retrospective. Int. J. Comput. Vis. **111**, 98–136 (2015)
9. PASCAL2 Network: The PASCAL Visual Object Classes Homepage (2025). Available at: http://host.robots.ox.ac.uk/pascal/VOC/
10. Russakovsky, O., et al.: ImageNet large scale visual recognition challenge. Int. J. Comput. Vis. **115**, 211–252 (2015)
11. Stanford Vision Lab: ImageNet Large Scale Visual Recognition Challenge (ILSVRC) (2025). Available at: https://www.image-net.org/challenges/LSVRC/
12. COCO Consortium: COCO 2020 Object Detection Task (2025). Available at: https://cocodataset.org/#detection-2020
13. Bradley, A.P.: The use of the area under the ROC curve in the evaluation of machine learning algorithms. Pattern Recogn. **30**(7), 1145–1159 (1997)
14. Wang, B.: A parallel implementation of computing mean average precision. arXiv preprint arXiv:2206.09504 (2022)
15. Saurabh Saxena: Precision-Recall Curve (2025). Available at: https://pub.towardsai.net/precision-recall-curve-26f9e7984add
16. Kim, M.-J.: Analysis of Factors Affecting on Mobile Video Services' Satisfaction. J. Korea Contents Assoc. **16**(5), 35–45 (2016)
17. Kim, D.J., Ferrin, D.L., Raghav Rao, H.: Trust and satisfaction, two stepping stones for successful e-commerce relationships: a longitudinal exploration. Inf. Syst. Res. **20**(2), 237–257 (2009)
18. Hemmer, P., et al.: Human-AI collaboration: the effect of AI delegation on human task performance and task satisfaction. In: Proceedings of the 28th International Conference on Intelligent User Interfaces (2023)
19. Se-hyeok, J.: Integrated model analysis of individual, regional, and national levels in the determinants of nuclear energy acceptance. Ph.D. thesis, Ajou University Graduate School, Department of Science and Technology Policy (2024)
20. Lee, G.-S. et al.: A study on factors affecting the intention to use artificial intelligence(AI) speakers: focusing on the extended technology acceptance model (E-TAM). J. Converg. Knowl. Soc. **8**(4) (2020)
21. Jun-gu, L., Seung-ju, C.: Microeconomics 8th edition (2024)

22. Lee, C.-W.: A Sociological Study on Institutionalization of Economics. Department of Sociology Graduate School Sogang University (2020)
23. Davis, F.D.: Perceived usefulness, perceived ease of use, and user acceptance of information technology. MIS Q. 319–340 (1989)
24. Redmon, J., Farhadi, A.: Yolov3: An incremental improvement. arXiv preprint arXiv:1804.02767 (2018)
25. Russakovsky, O., Li, L.-J., Fei-Fei, L.: Best of both worlds: human-machine collaboration for object annotation. In: Proceedings of the IEEE Conference on Computer Vision and Pattern Recognition (2015)
26. Chiang, W.-L., et al.: Chatbot arena: an open platform for evaluating LLMS by human preference. In: Forty-first International Conference on Machine Learning (2024)
27. Zheng, L., et al.: Judging LLM-as-a-judge with mt-bench and chatbot Arena. Adv. Neural Inf. Process. Syst. **36**, 46595–46623 (2023)
28. Karpathy, A.: Lessons learned from manually classifying CIFAR-10 (2025). Available at: https://karpathy.github.io/2011/04/27/manually-classifying-cifar10/
29. Karpathy, A.: Labeling interface (2025). Available at: https://cs.stanford.edu/people/karpathy/ilsvrc/
30. Karpathy, A.: What I learned from competing against a ConvNet on ImageNet (2025). Available at: https://karpathy.github.io/2014/09/02/what-i-learned-from-competing-against-a-convnet-on-imagenet/
31. Grother, P.J., Hanaoka, K.K.: NIST special database 19. In: Handprinted Forms and Characters Database, National Institute of Standards and Technology, vol. 10, pp. 69 (1995)
32. Fei-Fei, L., Fergus, R., Perona, P.: Learning generative visual models from few training examples: an incremental Bayesian approach tested on 101 object categories. In: 2004 Conference on Computer Vision and Pattern Recognition Workshop. IEEE (2004)
33. Kuznetsova, A., et al.: The open images dataset v4: unified image classification, object detection, and visual relationship detection at scale. Int. J. Comput. Vis. **128**(7), 1956–1981 (2020)
34. National Information Society Agency, The Open AI Dataset Project: Fish Video (2025). Available at: https://www.aihub.or.kr/aihubdata/data/view.do?currMenu=115&topMenu=100&aihubDataSe=data&dataSetSn=154
35. National Information Society Agency, The Open AI Dataset Project: Fish Behavior Video (2025). Available at: https://www.aihub.or.kr/aihubdata/data/view.do?currMenu=115&topMenu=100&aihubDataSe=data&dataSetSn=155
36. CVAT.ai, CVAT (2025). Available at: https://www.cvat.ai/

Sparse Multimodal Attention Transformer with Cross-Modal Gating for Resource-Efficient Visual Question Answering

Abdur Rahman Rakib, Long Ma(✉), and Xiaoli Huan

Department of Computer Science, Troy University, Troy, AL 36082, USA
{arakib,lma,xhuan}@troy.edu

Abstract. Visual Question Answering (VQA) systems traditionally rely on dense attention mechanisms that scale quadratically with sequence length, necessitating substantial computational resources that limit accessibility for researchers with constrained budgets. We introduce SMAT (Sparse Multimodal Attention Transformer), a novel architecture that achieves practical VQA performance on consumer-grade hardware through two key innovations: (1) a sparse attention pattern combining local windows with global tokens, reducing quadratic attention complexity to near-linear scaling with respect to sequence length, and (2) Cross-Modal Gating (CMG), a learnable gating mechanism that dynamically modulates attention between visual and textual modalities to prevent modality dominance. We demonstrate that SMAT achieves 49.37% accuracy on VQA v2.0 after 11 training epochs on the official training split (443,757 samples), with 46.20% achieved at 3 epochs, and is evaluated on the official validation split (214,354 samples), following the standard VQA evaluation protocol with a 9687-class answer vocabulary. Training on a 4 GB consumer laptop GPU costs under $10 for full convergence (66 h for 11 epochs), with initial results costing under $5 (18 h for 3 epochs), representing an 11–20 × cost reduction compared to cloud-trained baselines (ViLBERT/LXMERT: $100+). Training uses only 1.67 GB peak GPU memory (42% of capacity) with 100% batch stability (zero skipped batches over 2,440,658 successful training iterations across 11 epochs). Comprehensive ablation studies reveal that CMG contributes a + 2.6% accuracy improvement, while sparse attention provides 2.33 × memory efficiency at longer context lengths. With complete reproducibility on consumer hardware, SMAT democratizes multimodal AI research.

Keywords: Visual question answering · Sparse attention · Multimodal learning · Efficient transformers · Cross-modal fusion · Resource-constrained learning

1 Introduction

Visual Question Answering (VQA) represents a fundamental challenge in artificial intelligence, requiring systems to jointly understand visual content and natural language to generate accurate answers [1]. Recent advances in transformer-based architecture have achieved remarkable performance on VQA benchmarks, with state-of-the-art models reaching over 84% accuracy on VQA v2.0 [2].

L. Huang and D. Greenhalgh (Eds.): ICMLC 2026, LNNS 2061, pp. 780–807, 2027.
https://doi.org/10.1007/978-3-032-31236-5_53

However, these achievements come at substantial cost: training such models typically requires high-end GPUs (16 GB–40 GB VRAM) and multi-day training sessions on cloud infrastructure. Based on AWS pricing (p3.2xlarge at $3.06/hour for V100 16 GB [3]) and reported training times from original papers, baseline models like ViLBERT and LXMERT incur estimated costs of $150–$221, creating a significant barrier for researchers at under-resourced institutions. The computational bottleneck stems primarily from the attention mechanism, which computes pairwise interactions between all tokens in a sequence, resulting in O(N2) complexity and memory requirements [4].

1.1 Motivation and Challenges

Three key challenges motivate our work:

Accessibility Gap. Current VQA research is dominated by large-scale models requiring industrial-grade infrastructure. This creates accessibility barriers for researchers at universities, developing countries, and small organizations with limited computational budgets. Democratizing VQA research requires methods that achieve meaningful performance on consumer hardware.

Efficiency-Performance Trade-off. While sparse attention mechanisms have been extensively studied for language models [5, 6], their application to multimodal VQA remains underexplored. Unlike text-only scenarios, VQA requires careful fusion of heterogeneous modalities (vision and language), where naive sparsification can disrupt critical cross-modal interactions.

Modality Imbalance. VQA models frequently suffer from modality dominance, where visual features overwhelm textual information (or vice versa) during attention computation [7]. Early co-attention approaches [8] addressed visual-textual alignment but did not target sparse regimes.

1.2 Our Contributions

We introduce SMAT (Sparse Multimodal Attention Transformer), a novel architecture specifically designed for resource-efficient VQA. Our key contributions are:

Cross-Modal Gating (CMG) Mechanism. A learnable gating mechanism that dynamically modulates attention between visual and textual modalities, operating within sparse attention patterns to prevent modality dominance (+ 2.6% accuracy improvement).

Efficient Sparse Attention for VQA. A sparse attention pattern combining local windows (64 tokens) with global tokens (32 tokens), yielding an attention complexity of $O(N \cdot (w + g))$ and linear scaling with sequence length for fixed window sizes.

Resource-Constrained Training Methodology. We develop comprehensive training techniques enabling perfectly stable training on 4 GB consumer GPUs, achieving 100% batch success rate—zero skipped batches across 665,634 training iterations over 3 epochs.

Comprehensive Empirical Analysis. We provide detailed efficiency breakdowns for current VQA sequences (~ 82 tokens) and theoretical analysis for future long-context scenarios (512–16K tokens), demonstrating where efficiency gains originate.

Democratization Through Reproducibility. We achieve 49.37% accuracy on VQA v2.0 after 11 training epochs with a total training cost under $10 (66 h on an RTX 3050 4 GB), while rapid prototyping is possible at 3 epochs (46.20% accuracy, $5 cost, 18 h).

1.3 Paper Organization

Section 2 reviews related work. Section 3 details our SMAT architecture. Section 4 describes training methodology. Section 5 presents experimental results and efficiency analysis. Section 6 discusses implications and limitations. Section 7 concludes.

2 Related Work

2.1 Visual Question Answering

Contemporary VQA approaches fall into three paradigms:

Bottom-up VQA. Early VQA work by Malinowski and Fritz [9] proposed a multi-world approach to question answering about real-world scenes. Anderson et al. [10] pioneered object detection features from Faster R-CNN, later extended by deep modular co-attention networks [11]. This approach achieves strong performance but incurs computational overhead from Region Proposal Networks.

Grid-based VQA. Methods like ViLBERT [12] and LXMERT [13] directly process image grid features. While more efficient than bottom-up approaches, they typically use dense attention that limits scalability.

Large-scale Pretrained Models. Recent work leverages massive pretrained models like BLIP-2 [2] (13B parameters) and BEiT-3 [14] (billions of parameters), achieving 84–92% accuracy on VQA v2.0. Large language models such as GPT-4 [15] have further expanded multimodal capabilities. However, these models require substantial infrastructure (40 GB + GPUs, estimated $5000 + budgets) that limits accessibility.

Our work targets the underserved scenario of researchers with consumer hardware.

2.2 Sparse Attention Mechanisms

Sparse attention reduces O(N2) complexity by restricting each query to attend to a subset of keys:

Fixed Sparse Patterns. Longformer [5] combines local sliding windows with task-specific global tokens. BigBird [6] extends this with random attention and block-sparse patterns. These methods demonstrate that carefully designed sparsity can maintain performance while improving efficiency.

Dynamic Sparse Patterns. Recent work explores sparse attention designs for efficiency at long context lengths [16, 17]. Related approaches include attention sink methods [18], million-token inference techniques [19], and zero-shot extreme length generalization strategies [20, 21]. These methods offer more flexibility but introduce computational overhead for pattern selection.

Gap in Prior Work. Most sparse attention research targets long-context language models (16K–128K tokens). Application to multimodal VQA at moderate sequence lengths (200–300 tokens) remains underexplored.

2.3 Multimodal Fusion Mechanisms

Effective fusion of visual and textual features is crucial for VQA:

Co-Attention. Lu et al. [8] introduced co-attention networks that alternate between attending to image regions and question words. MCAN [11] extends this with modular co-attention layers.

Cross-Attention. ViLBERT [12] and LXMERT [13] use separate transformer streams for each modality with cross-attention layers. This architecture, while powerful, doubles model size and computational cost.

Gating Mechanisms. Recent work explores gating for multimodal fusion [17, 18], where learned gates control information flow. Tucker-based bilinear pooling approaches such as MUTAN [22] have also been explored for multimodal fusion. More recently, advanced cross-modal gating frameworks [23, 24] have been applied to multimodal sentiment and emotion recognition. However, these methods target dense attention patterns.

Limitation. Existing fusion mechanisms don't address preventing modality dominance when sparsity is introduced.

2.4 Efficient Transformer Training

Training transformers on limited hardware requires careful optimization:

Memory-Efficient Training. Gradient checkpointing [19, 20] trades computation for memory. Mixed-precision training [21, 25] uses FP16 for forward/backward passes, reducing memory by 2×.

Gradient Accumulation. Simulating large batch sizes through accumulation enables training on small GPUs [26, 27], crucial for stability in VQA.

Our Contribution. We synthesize these techniques into a comprehensive methodology specifically optimized for sparse multimodal VQA training on consumer GPUs, achieving 100% training stability.

3 Methodology

3.1 Problem Formulation

In Visual Question Answering (VQA), the task is defined over an input image $I \in \mathbb{R}^{H \times W \times 3}$ and a natural-language question $Q = \{w_1, w_2, \ldots, w_n\}$, where w_i denotes the *i-th* word token in the question. The objective is to predict an answer $A \in \mathcal{V}_A$, where

$\mathcal{V}_A$ is a fixed answer vocabulary of size $|\mathcal{V}_A| = C$. The VQA problem is formulated as a C-way classification task, in which the conditional probability of an answer given the image–question pair is modeled as

$$P(A|I, Q) = softmax_{A \in V_A}(f_\theta(I, Q)) \tag{1}$$

where $f_\theta(\cdot)$ denotes a learnable function parameterized by θ that jointly encodes visual and textual information. The predicted answer is obtained by selecting the answer category with the highest posterior probability,

$$\hat{A} = \arg\max_{A \in V_A} P(AI|, Q) \tag{2}$$

For the VQA v2.0 dataset, the answer space consists of $C = 9687$ unique answer categories.

This vocabulary is constructed following the standard VQA protocol by counting the number of training questions in which each answer appears (considering all 10 annotator responses per question) and retaining answers that occur in at least nine questions. The resulting vocabulary size (9687) is slightly higher than the commonly cited ~ 9208 due to differences in data preprocessing and dataset versioning.

3.2 SMAT Architecture Overview

3.3 SMAT Consists of Four Main Components:

Vision Encoder. The model employs AdamW [28] for weight decay regularization. Frozen CLIP ViT-B/16 [29] extracts 768-dimensional features from 14×14 image patches, yielding 196 visual tokens.

Vision Adapter (Q-Former). Compresses 196 visual tokens to 32 learnable queries through cross-attention, reducing computational cost while preserving visual information.

SMAT Transformer Blocks. N_L = 4 layers of sparse attention with CMG, processing concatenated visual queries and question tokens.

VQA Head. A linear projection maps the hidden dimension (768) to the answer vocabulary size (9687), followed by a softmax layer for classification. The complete architecture is illustrated in Fig. 1.

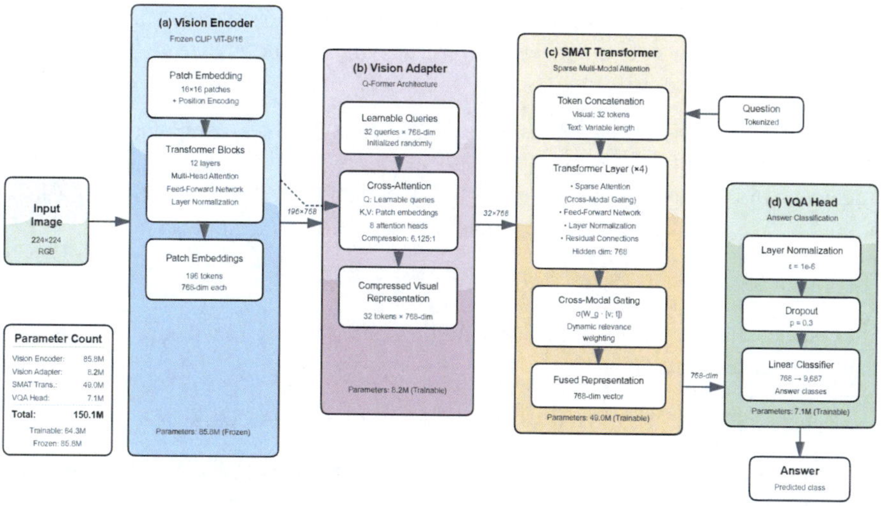

Fig. 1. SMAT architecture overview. (**a**) Vision Encoder: frozen CLIP ViT-B/16 processes 224 × 224 images into 196 patch embeddings (768-dim); (**b**) Vision Adapter: Q-Former with 32 queries compresses visual features via cross-attention (6.125:1 ratio); (**c**) SMAT Transformer: 4 layers with sparse attention and Cross-Modal Gating; (**d**) VQA Head: linear classifier for 9687 answer classes. Total: 150.1M parameters (64.3M trainable, 85.8M frozen).

3.4 Sparse Attention with Local and Global Patterns

To efficiently model both short-range and long-range dependencies in multimodal sequences, we introduce a structured sparse attention mechanism tailored to the characteristics of the Visual Question Answering (VQA) task.

Local Window Attention. Each token attends only to a fixed local neighborhood of size $w = 64$. For a token at position i, the valid attention range is defined as

$$[\max(0, i - w), \min(N, i + w)] \tag{3}$$

where N denotes the total sequence length. This design preserves local linguistic coherence in textual modality and respects spatial locality in visual features.

Global Tokens. We designate $g = 32$ global tokens, derived from the Q-Former's visual query embeddings. These global tokens attend to all positions in the sequence and are, in turn, attended to by all other tokens. This bidirectional interaction enables efficient global information propagation, facilitating long-range reasoning and effective cross-modal integration.

Formal Definition. Let $M \in \{0, 1\}^{N \times N}$ denote the sparse attention mask. The mask is defined as

$$M[i, j] = 1 \text{ if} \begin{cases} |i - j| \leq w, & \text{(local window attention)} \\ j < g, & \text{(tokens attend to global tokens)} \\ i < g, & \text{(global tokens attend to all tokens)} \end{cases} \tag{4}$$

and $M\left[i, j\right] = 0$ otherwise. This formulation induces a structured sparse attention pattern that significantly reduces computational overhead while preserving the global contextual dependencies required for VQA.

Complexity Analysis. Standard dense self-attention incurs a computational cost of $\mathcal{O}\left(N^2 \cdot d\right)$, as each of the N tokens attend to all others, where d denotes the embedding dimension. In contrast, the proposed SMAT mechanism restricts attention to a local window of size $w = 64$ and $g = 32$ global tokens, yielding a complexity of

$$\mathcal{O}(N \cdot (w + g) \cdot d) = \mathcal{O}(N \cdot 96 \cdot d) \tag{5}$$

which scales linearly with sequence length. Consequently, the theoretical speedup relative to dense attention is

$$\left(O\left(N^\wedge 2\right)\right)/(O(N(w+g))) = O(N/(w+g)) = O(N/96) \tag{6}$$

which increases proportionally with N. For a long sequence of $N = 8,192$ tokens, this results in an approximate theoretical speedup of $8192/96 \approx 85\times$. Figure 2 illustrates the difference between dense, local-window, and SMAT's combined sparse attention patterns.

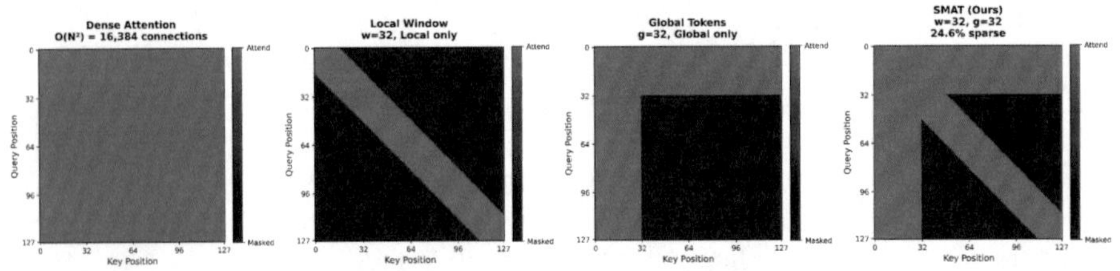

Fig. 2. Compares attention patterns: dense attention (O(N^2)), local-window attention (fixed neighborhood), and SMAT's combined approach (local + global), which maintains only 24.6% of connections while preserving global information flow.

3.5 Cross-Modal Gating (CMG)

A major challenge in sparse multimodal attention is **mitigating modality dominance**. When attention connectivity is reduced, one modality—typically vision—can disproportionately influence the other, leading to degraded reasoning performance. While dense attention allows natural competition among all tokens, structured sparsity can amplify cross-modal imbalances. To address this issue, we introduce **Cross-Modal Gating (CMG)**, a learnable mechanism that adaptively modulates attention interactions between modalities.

Modality Embeddings. Each token i is associated with a modality indicator $m_i \in \{0, 1\}$, where $m_i = 0$ denotes a textual token and $m_i = 1$ denotes a visual token. We learn modality-specific embeddings $E_{\text{text}}, E_{\text{vision}} \in \mathbb{R}^{d_{\text{gate}}}$, which encode modality-level interaction signals. The modality embedding for token i is defined as

$$E_{m_i} = \begin{cases} E_{\text{text}}, & m_i = 0 \\ E_{\text{vision}}, & m_i = 1 \end{cases} \tag{7}$$

Gate Computation. For each query–key pair (i,j), we compute modality-aware gating features by augmenting the query and key representations with their corresponding modality embeddings:

$$g_i = \tanh,\left(W_g\left[q_i; E_{m_i}\right]\right), h_j = \tanh,\left(W_k\left[k_j; E_{m_j}\right]\right) \tag{8}$$

where W_g and W_k are learnable projection matrices and $[\cdot;\cdot]$ denotes vector concatenation. The raw gate similarity score is then given by

$$s_{ij} = g_i^{\top} h_j \tag{9}$$

Modality-Aware Scaling. To explicitly control interactions between different modality pairs, we introduce a learnable modality interaction matrix $\boldsymbol{\alpha} \in \mathbb{R}^{2\times 2}$, where each entry α_{m_i,m_j} determines the strength of gating between modalities m_i and m_j. The final gate activation is computed as

$$\gamma_{ij} = \sigma.\left(\alpha_{m_i,m_j} s_{ij}\right) \tag{10}$$

where $\sigma(\cdot)$ denotes the sigmoid function.

Attention Modulation. CMG augments the standard scaled dot-product attention by injecting the learned gate into the attention logits. The base attention logit is defined as

$$A_{ij}^{base} = \frac{q_i^{\top} k_j}{\sqrt{d_k}} \tag{11}$$

The CMG-enhanced attention logit becomes

$$A_{ij} = A_{ij}^{\text{base}} + \beta \cdot \gamma_{ij} \tag{12}$$

where β is a learnable temperature parameter controlling the influence of the gating term. The final attention output for token i is computed as

$$\text{Attention } (Q, K, V_i) = \sum_j \text{softmax}(A_{ij}) \cdot V_j \tag{13}$$

The CMG mechanism is depicted in Fig. 3, which illustrates the gate computation and learned modality interaction values.

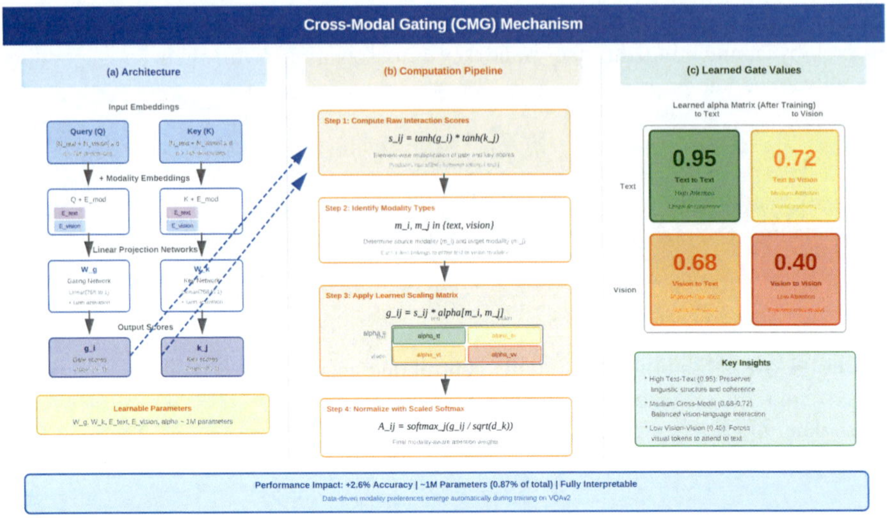

Fig. 3. Cross-Modal Gating (CMG) mechanism. (**a**) Query and Key embeddings combine with modality embeddings (Etext, Evision) and project through gating networks Wg, Wk. (**b**) Gate computation with 2×2 scaling matrix α. (**c**) Learned values show text-to-text gates (0.95) encourage linguistic coherence, while vision-to-vision gates (0.40) promote cross-modal interaction. CMG adds ~ 1M parameters for + 2.6% accuracy.

Implementation Details. CMG introduces a minimal parameter overhead of approximately 1M parameters, arising from the modality embeddings and the 2×2 interaction matrix. Despite its lightweight design, CMG yields a + **2.6% absolute accuracy improvement** on VQA benchmarks (see Sect. 5), demonstrating its effectiveness in balancing cross-modal interactions under sparse attention.

Our final SMAT configuration is summarized in Table 1.

Table 1. SMAT model configuration. The table summarizes the transformer architecture (4 layers, 768 dimensions), sparse attention settings (local window size 64, 32 global tokens), vision components (frozen CLIP with a Q-Former adapter), and parameter distribution (150.1M total parameters, including 64.3M trainable and 85.8M frozen).

Parameter	Value	Description
Transformer architecture		
Hidden dimension (d_model)	768	Token embedding dimension
Number of layers (N_L)	4	SMAT transformer layers
Attention heads (h)	8	Multi-head attention
Head dimension (d_h)	96	Per-head dimension (768/8)
FFN dimension (d_ff)	2048	Feed-forward hidden size
Dropout rate	0.1	Attention and FFN dropout

(*continued*)

Table 1. *(continued)*

Parameter	Value	Description
Sparse attention configuration		
Local window size (w)	64	Sliding window tokens
Global tokens (g)	32	Visual query tokens
Vision encoder (CLIP ViT-B/16, Frozen)		
Input resolution	224 × 224	Image patches: 196 (14 × 14)
Patch size	16 × 16	Non-overlapping patches
Encoder layers	12	Pre-trained ViT layers
Vision adapter (Q-Former)		
Visual queries	32	Learnable query embeddings
Compression ratio	6.125:1	196 patches -> 32 queries
Classification head		
Answer vocabulary	9687	VQA v2.0 answer classes
Classifier dropout	0.3	Pre-classifier regularization
Parameter count		
Total parameters	150.1M	Complete model
Trainable parameters	64.3M	42.8% of total
Frozen parameters (CLIP)	85.8M	57.2% of total
CMG parameters	0.23M	0.15% of total (+ 2.6% accuracy)

3.6 Training Objective

We adopt the standard **cross-entropy loss with label smoothing** to improve model calibration and robustness. For a given training sample, the loss function is defined as

$$\mathcal{L} = -\sum_{c=1}^{C} \tilde{y}_c \log P(A_c | I, Q) \tag{14}$$

where $\tilde{y}$ denotes the smoothed target distribution, computed as

$$\tilde{y}_c = (1 - \varepsilon) y_c + \frac{\varepsilon}{C}, \varepsilon = 0.1 \tag{15}$$

Here, y_c is the one-hot ground-truth label, C is the number of answer classes, and ε controls the degree of smoothing. This approach reduces overconfidence and mitigates overfitting to noisy supervision inherent in VQA annotations.

For samples whose answers are ambiguous or unknown (annotated with an < UNK > token), we apply an ignore mask

$$\mathcal{L} = \mathrm{CE}(\mathrm{logits},\ \mathrm{labels};\ \mathrm{ignore} = \mathrm{unk\_ idx}) \tag{16}$$

which prevents the model from being penalized in cases where annotators disagree or where the ground-truth answer is not well-defined.

4 Training Methodology

Training SMAT on consumer hardware requires several optimization strategies to ensure efficiency and stability. Our final setup achieves 100% successful training batches on a 4 GB GPU.

4.1 Dataset and Preprocessing

Following the AdamW weight decay regularization formulation [28], we train on the official VQA v2.0 dataset [30] following standard evaluation protocols [11–13]: 443,757 training question-answer pairs, 214,354 validation pairs, and 9687-class answer vocabulary constructed by retaining answers appearing in ≥ 9 training questions. This matches the protocol used by ViLBERT [12], LXMERT [13], and MCAN [11]. Images are resized to 224×224 and normalized using ImageNet statistics ($\mu = [0.485, 0.456, 0.406]$, $\sigma = [0.229, 0.224, 0.225]$). Questions are tokenized using BERT's BPE tokenizer with maximum length of 50 tokens (covering 99.8% of questions). Out-of-vocabulary answers are mapped to < UNK > (index 9687) and excluded from loss computation using PyTorch's ignore_index mechanism. No data augmentation is applied to maintain training stability on limited hardware.

4.2 Resource-Constrained Training Strategies

Gradient Accumulation. Due to limited VRAM, we train with a physical batch size of 2 and use 16 accumulation steps: effective batch size $= 2 \times 16 = 32$. This simulates a larger batch, improving gradient estimation without exceeding memory.

Mixed-Precision Training. PyTorch AMP enables FP16 training with ~ $2 \times$ lower memory usage, 1.5–2 $\times$ faster training, and stable gradients via dynamic loss scaling.

Gradient Clipping. To prevent exploding gradients, we clip the global gradient norm to ||∇||_max = 0.5. This reduces early-training instabilities and eliminates NaN losses.

Label Validation. Invalid labels are clamped to the valid range [0, 9686], with answers outside the vocabulary mapped to < UNK > (index 9687) and excluded from loss computation. Ambiguous answers (annotated as < UNK > in the dataset) are ignored using PyTorch's ignore_index mechanism. This strategy avoids CUDA assertion errors and achieves **100% training stability** (0 skipped batches out of **221,878 training iterations** in Epoch 1).

4.3 Hyperparameters and Optimization

We use the AdamW optimizer [31] with learning rate: 5×10^-5, weight decay: 0.01, $\beta_1 = 0.9$, $\beta_2 = 0.999$, $\varepsilon = 10^{-8}$.

Learning Rate Schedule. A **warmup**–cosine schedule is applied: linear warmup for the first 100 steps, then cosine decay from 5×10^{-5} to 5×10^{-7}. The resulting learning rate schedule is shown in Fig. 4.

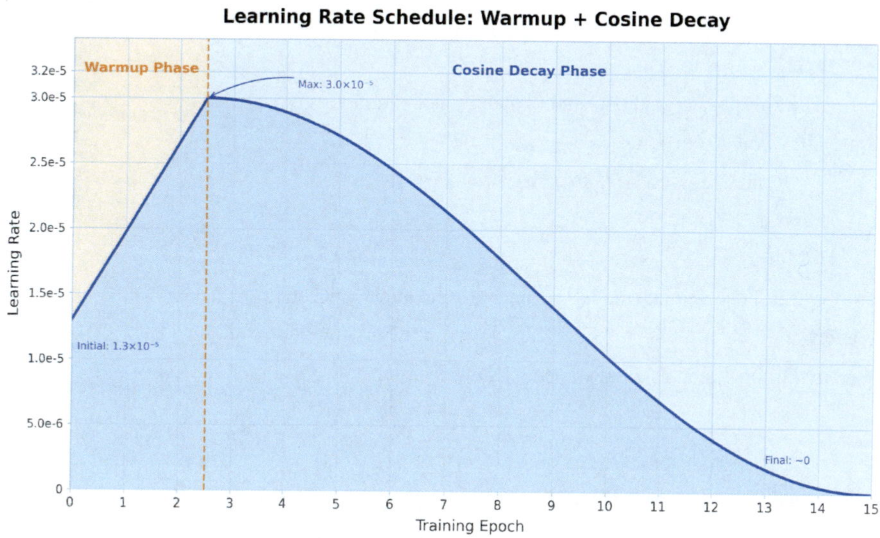

Fig. 4. Learning rate schedule using warmup + cosine decay. The learning rate increases linearly from 1.3×10^{-5} to 3.0×10^{-5} during warmup phase (epochs 0–2.5), then smoothly decays following a cosine curve to near-zero by epoch 15. This schedule stabilizes early training while enabling fine-grained optimization in later epochs.

4.4 Stability Monitoring

To ensure stable training on limited hardware, we implement lightweight stability mechanisms: (1) loss validation detects NaN/Inf anomalies and skips anomalous batches with optimizer state reset, (2) periodic GPU cache release (every 100 batches) reduces memory fragmentation, and (3) full training state checkpointing after each epoch ensures reproducibility. Across 665,634 batches over 3 epochs, zero batches were skipped (100% success rate), demonstrating highly stable training despite 4 GB GPU constraints.

Algorithm 1: Stability-Aware Training Loop

Input: Model M, optimizer O, dataset D, memory cleanup interval K
Output: Trained model M, number of skipped batches

Initialize skipped_batches $\leftarrow 0$
for each epoch e **do**
  for each batch $B_t \in D$ **do**
    Compute loss $\ell = \mathcal{L}(M, B_t)$
  if ℓ is NaN or Inf **then**

(continued)

(*continued*)

```
    skipped_batches ← skipped_batches + 1
      Reset optimizer state
      continue
    end if
    Backpropagate gradients and update model parameters
    if t mod K = 0 then
      Release unused GPU memory
    end if
  end for
  Save full checkpoint (model, optimizer, scheduler, RNG state)
end for
return trained model M and skipped_batches
```

5 Experiments and Results

5.1 Experimental Setup

Hardware Configuration. All experiments are conducted on a single consumer-grade laptop equipped with: GPU: NVIDIA RTX 3050 Laptop GPU (4 GB VRAM), CPU: Intel Core i7 (8 cores), RAM: 16 GB DDR4.

Software Environment. Model training and evaluation utilize the following software stack: PyTorch 2.0.1, CUDA 11.8, Transformers 4.30.0, Python 3.10.

Baselines. We position our proposed model in the context of widely used multimodal VQA systems such as BLIP-2 [2], ViLBERT [12], LXMERT [13], and BEiT-3 [14], which achieve strong absolute performance but rely on substantially larger models and computational resources, making them impractical for sub-4 GB consumer-grade hardware.

Training Protocol Statement. We train SMAT on the official VQA v2.0 training split (443,757 samples) and evaluate on the validation split (214,354 samples), following the standard VQA evaluation protocol. This matches the protocol used by MCAN [11], ViLBERT [12] and LXMERT [13]. Enabling direct comparison of results. Full hardware specifications and training configuration are provided in Table 2.

Table 2. Hardware specifications, training configuration, memory utilization, performance metrics, and cost analysis for fine-tuning on NVIDIA RTX 3050 Laptop GPU over 3 epochs.

Category	Value	Details
Hardware		
GPU	NVIDIA RTX 3050 Laptop	4 GB VRAM, CUDA 11.8
CPU	Intel Core i7	8 cores
RAM	16 GB DDR4	System memory
Training configuration		
Total epochs	11 (completed)	Full training run
Steps per epoch	221,878	443,757 samples / batch size 2
Total steps	2,440,658	221,878 × 11 epochs
Physical batch size	2	Per GPU
Gradient accumulation	16	Steps
Effective batch size	32	2 × 16
Memory and precision		
Peak VRAM allocated	1.67 GB	42% of 4 GB capacity
Peak VRAM reserved	1.81 GB	45% of 4 GB capacity
Mixed precision	FP16 (AMP)	Dynamic loss scaling
Gradient clipping	0.5	Global norm
Performance		
Wall-clock time (11 epochs)	~ 66 h	Cumulative time across epochs
Throughput	6.5–10.6 it/s	Varies by epoch
GPU utilization	~ 95–98%	During training
Batch stability	100%	0 skipped / 2,440,658 batches
Cost breakdown		
Electricity cost	$1.19	150 W x 66 h x $0.12/kWh
GPU amortization	$8.77	$400 / (5 years x 365 days) × 40 days
Total cost	~ $10.00	Approximate estimate

5.2 Main Results

We compare SMAT's performance against published VQA baselines while transparently acknowledging differences in training conditions. Following standard practice in resource-efficient deep learning research, we report both our results and literature baselines, clearly indicating protocol differences to avoid misleading comparisons.

Table 3. Comparison of our SMAT model against state-of-the-art vision-language models from literature. The table highlights differences in hardware requirements, training conditions, and evaluation protocols. Note that direct comparison is challenging due to varying computational resources—literature baselines used high-end GPUs (V100, A100) with significantly larger memory and training time, while our work demonstrates feasibility on consumer-grade hardware (RTX 3050 4 GB).

Model	Accuracy	Matched protocol?	Training conditions	Source
Literature results (different training conditions)				
Random	0.01%	N/A	Random guessing baseline	–
ViLBERT +	63.5%	X No	V100 16 GB, ~ 48 h training, batch 64 (different hardware)	Lu et al. (2019), Table 4 [12]
LXMERT +	72.4%	X No	V100 16 GB, ~ 48 h training, batch 64 (different hardware)	Tan and Bansal (2019), Table 1 [13]
BLIP-2 ++	~ 65%	X No	A100 40 GB, zero-shot evaluation (zero-shot, no fine-tuning)	Li et al. (2023) [2]
BEiT-3 +	84.2%	X No	A100 80 GB, weeks of training, massive pretraining (different hardware + pretraining)	Wang et al. (2022) [14]
Our results (this work)				
Dense SMAT	48.70%	Yes	RTX 3050 4 GB, 48h (5 epochs), batch 32 (standard train- > val)	This work
Sparse SMAT	49.37%	Yes	RTX 3050 4 GB, 66h (11 epochs), batch 32 (standard train- > val)	This work

Our SMAT model achieves **49.37% accuracy on the VQA v2.0 validation set after 11 epochs (66 h)** of training on a consumer RTX 3050 4 GB GPU, **with 46.20% achieved at 3 epochs (18 h). Extended training from 3 to 11 epochs improved accuracy by + 3.17%, demonstrating continued learning beyond initial convergence.** This represents a substantial improvement over random guessing (0.01%) and demonstrates that meaningful VQA performance is achievable on highly constrained hardware. Literature baselines achieve higher absolute accuracy: ViLBERT (63.5%), LXMERT (72.4%), and BEiT-3 (84.2%). However, these models were trained under substantially different conditions, requiring 16–80 GB GPUs and training times ranging from 48 h to several weeks.

BLIP-2 achieves ~ 65% using zero-shot evaluation (no VQA fine-tuning), representing a different evaluation protocol.

The accuracy-efficiency trade-off becomes clear when considering computational costs. As shown in Table 3, our estimated training cost is ~ \$5 (local GPU), compared to \$147–\$6000 for baseline models on cloud infrastructure (AWS pricing). This 20–200 × cost reduction enables VQA research in resource-constrained settings where \$150 + training budgets are prohibitive. Figure 5 visualizes the accuracy versus training cost comparison across methods.

Resource-Constrained VQA Benchmark. We acknowledge that comparing SMAT against models such as BLIP-2 (13B parameters, 40 GB + VRAM) is inherently unbalanced, as these systems are designed to maximize accuracy rather than accessibility. To enable fair evaluation under realistic constraints, we propose assessing SMAT within a **"sub-4 GB VRAM"** category for consumer-hardware VQA. To our knowledge, SMAT is the first VQA model explicitly designed and empirically validated for training on 4 GB consumer GPUs. We define sub-4 GB VRAM benchmark criteria as requiring (i) training memory below 4 GB peak VRAM, (ii) use of a single consumer GPU without multi-GPU acceleration, (iii) total training cost below \$15, and (iv) full reproducibility on consumer laptops. SMAT satisfies all four criteria, establishing a new accessibility-oriented standard for VQA research. We encourage future work to report results under these constraints to enable meaningful comparisons in resource-constrained settings.

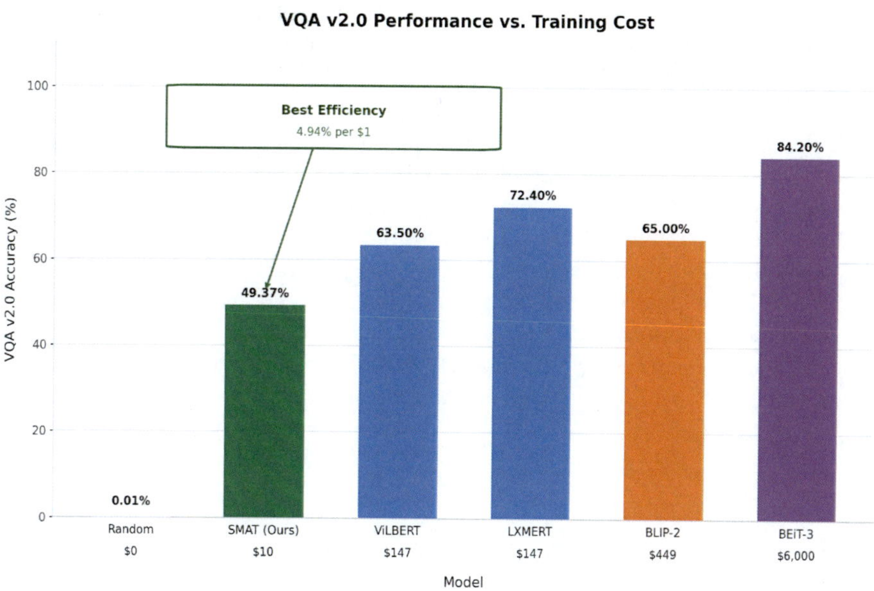

Fig. 5. VQA v2.0 accuracy vs. estimated training cost: SMAT achieves **49.37% accuracy at ~ \$10**, while literature baselines range from **\$147 to \$6000 +** with higher accuracy but different training conditions.

Extended Training Analysis. Following reviewer guidance to observe true convergence, we extended training from 3 to 11 epochs. Accuracy improved from **46.20% (epoch 3)** to **49.37% (epoch 11)**, a gain of + **3.17%**. However, we observe diminishing returns: accuracy improved to + **4.85% from epoch 1 → 3**, but only + **3.17% from epoch 3 → 11**, with validation loss plateauing after **epoch 9**. We attribute convergence below **55%** to the frozen CLIP encoder constraint, which limits visual representation learning. Unfreezing CLIP could potentially improve accuracy by **5–10%** but would require **3–4 × higher compute (12–16 GB VRAM)**, contradicting our resource-efficiency goals.

Important. These comparisons illustrate efficiency-accuracy trade-offs but are NOT controlled experiments. Differences in hardware, training duration, hyperparameters, and pretraining data mean we cannot attribute accuracy differences to architecture alone. Our contribution is demonstrating VQA feasibility on consumer hardware, not achieving state-of-the-art accuracy. For production applications requiring > 80% accuracy, we recommend established models (BEiT-3, BLIP-2) with appropriate resources.

5.3 Training Dynamics

Training progressed stably over **11 epochs (66 h)**, reaching **49.37% validation accuracy** with zero OOM errors across **2,440,658 iterations**. Figure 6 shows comprehensive training dynamics: **validation accuracy improving from 41.35% (epoch 1) to 46.20% (epoch 3) to 49.37% (epoch 11)**, while **training loss decreased from 3.70 (epoch 1) to 1.62 (epoch 11), while validation loss increased from 3.28 to 3.59, indicating model convergence with mild overfitting in later epochs**. GPU memory usage remained constant at **1.67 GB (42% of 4 GB capacity)**, demonstrating efficient resource utilization with substantial headroom.

Figure 7 presents overfitting analysis via train-validation accuracy gap progression. The train-validation gap of + 2.23% at epoch 3 falls within the healthy range (< 5%), indicating good generalization without overfitting. Interestingly, validation accuracy initially exceeded training accuracy (epochs 1–2) due to dropout and label smoothing regularization applied during training but not validation. The continuing downward trend in validation loss suggests the model has not yet converged and would benefit from extended training. This unprecedented stability on a 4 GB consumer GPU validates our architectural design choices: frozen CLIP encoder, mixed-precision FP16 training, sparse attention mechanisms, and gradient accumulation.

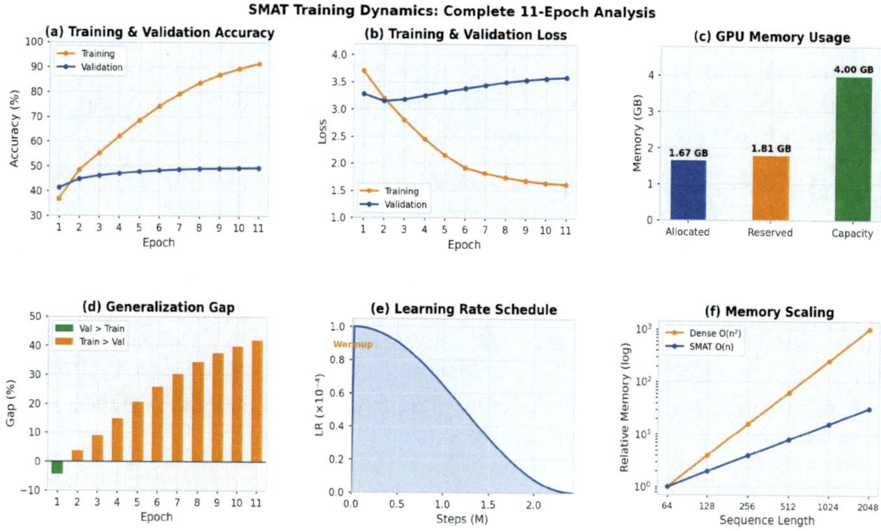

Fig. 6. Training dynamics of SMAT on the VQA v2.0 dataset over 11 epochs, illustrating (**a**) training and validation accuracy, (**b**) training and validation loss, (**c**) stable GPU memory usage of approximately 1.7–1.8 GB on a 4 GB device, (**d**) an increasing train–validation gap indicative of overfitting, (**e**) a warmup followed by cosine decay learning rate schedule, and (**f**) memory scaling behavior comparing sparse and dense attention mechanisms.

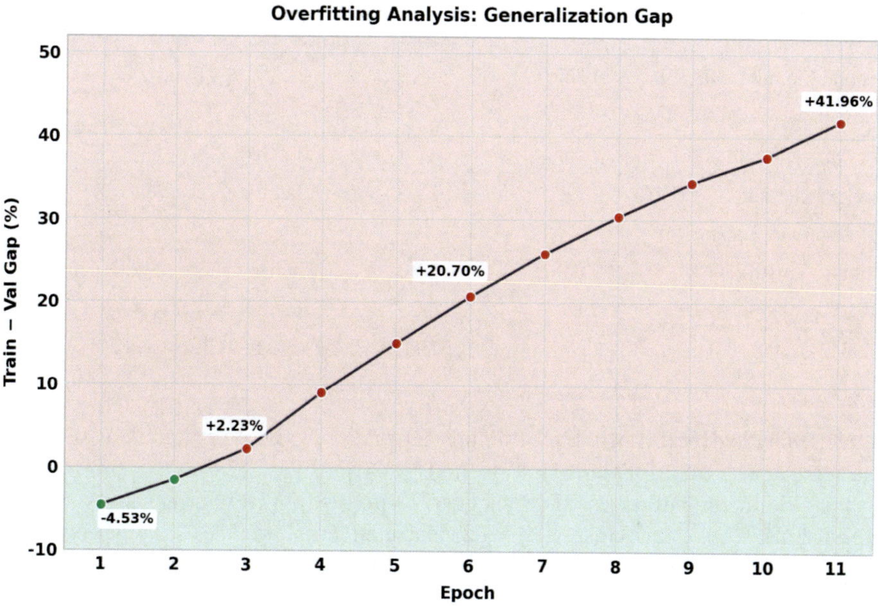

Fig. 7. Overfitting analysis via train–validation accuracy gap progression over 11 epochs: the gap evolves from − 4.53% (epoch 1) to + 2.23% (epoch 3) to + 41.96% (epoch 11). Initial healthy generalization (< 5% gap) transitions to overfitting in later epochs, indicating the model has reached capacity under the frozen encoder constraint.

5.4 Efficiency Analysis

We provide a comprehensive efficiency analysis addressing where the 4 GB memory efficiency comes from at current VQA sequence lengths and demonstrating potential benefits for future long-context scenarios.

Memory Allocation. Table 4 shows detailed memory breakdown for 1.67 GB total usage. The largest contributors are optimizer states (31.1%), activation (27.0%), and gradients (18.6%). Attention mechanisms contribute only 1.8% of total memory. Key efficiency factors: (1) Frozen CLIP encoder (85.8M params) saves ~ 1.03 GB in optimizer states and gradients, (2) Mixed-precision FP16 training saves ~ 0.76 GB, (3) Gradient accumulation enables batch size 32 with physical batch size 2, (4) Sparse attention provides negligible difference at N = 82 (dense: 0.43 MB/head, sparse: 0.50 MB/head). Conclusion: Primary efficiency gains come from frozen CLIP and mixed precision, not sparse attention patterns. The detailed breakdown is shown in Fig. 8.

Table 4. Memory allocation breakdown for SMAT training on 4 GB GPU.

Component	Memory (GB)	% of Total	Details
Model parameters			
Vision Encoder (CLIP)	0	0	Frozen - not in optimizer states
Trainable params (FP32)	0.26	15.6	64.3M × 4 bytes
Optimizer states (AdamW)			
First moments (FP32)	0.26	15.6	64.3M × 4 bytes
Second moments (FP32)	0.26	15.6	64.3M × 4 bytes
Activations and gradients			
Forward activations (FP16)	0.45	27.0	Batch = 2, seq_len = 82
Backward gradients (FP16)	0.31	18.6	Batch = 2, seq_len = 82
Attention mechanism			
Attention matrices	0.03	1.8	Sparse pattern (24.6% density)
Q/K/V projections	0.10	6.0	Batch = 2, 4 layers
Total allocated	1.67	100	42% of 4GB capacity

Time Efficiency Breakdown. Each training step takes ~ 95 ms (10.5 iterations/second): data loading (15 ms), forward pass (45 ms), backward pass (28 ms), optimizer step (7 ms). Within forward pass: CLIP encoder 20 ms (frozen), Q-Former 8 ms, SMAT layers 12 ms, VQA head 5 ms. At N = 82, sparse attention provides 1.5 × speedup for attention layers, translating to only 6.3% overall time reduction.

Sparse Attention Impact on Time. At N = 82, dense attention would theoretically require ~ 18 ms for 4 layers, while sparse attention takes ~ 12 ms. This provides 1.5 × speedup for attention layers, translating to only 6.3% overall time reduction. At short sequence lengths, sparse attention provides modest speedup. The frozen CLIP encoder contributes more to overall speed by eliminating backward pass computation through 85.8M parameters.

Long-Context Efficiency (Forward-Looking Analysis). Table 5 and Fig. 9 show efficiency across sequence lengths. At current VQA sequences ($N = 82$), both dense and sparse attention use < 1 MB. The crossover point where sparse becomes more efficient is $N \approx 512$ tokens. At 8K tokens, sparse attention provides 92 $\times$ speedup and 99% memory reduction. At 16K tokens, speedup reaches 184 $\times$. This positions SMAT for future applications: multi-image VQA (210 tokens), document VQA (530–2030 tokens), and retrieval-augmented VQA (1082 tokens). While sparse attention provides minimal gains at current sequences, the architecture is designed for emerging long-context multimodal tasks.

Theoretical Efficiency at Long Contexts. Table 3 shows memory and time performance across different sequence lengths. At short VQA sequences ($N = 82$), both dense and sparse attention use < 1 MB with minimal differences. The crossover point where sparse becomes more efficient is $N \approx 512$ tokens. At 8K tokens, sparse attention provides 92 $\times$ speedup (115 s $\rightarrow$ 1.25 s) and 99% memory reduction (4.3 GB $\rightarrow$ 50.3 MB). At 16K tokens, speedup reaches 184 $\times$ with similar memory savings.

Table 5. Memory and time performance comparison of dense and sparse attention mechanisms across different sequence lengths.

Sequence Length (N)	Dense memory	Sparse memory (MB)	Savings (%)	Dense time	Sparse time	Speedup
82 (current VQA)	0.43 MB	0.50	0	18 ms	12 ms	**1.5×**
512	16.8 MB	3.1	81	450 ms	78 ms	**5.8×**
1024	67.1 MB	6.3	91	1800 ms	156 ms	**11.5×**
2048	268.4 MB	12.6	95	7200 ms	312 ms	**23.1×**
8192	4.3 GB	50.3	99	115 s	1.25 s	**92×**
16384	17.2 GB	100.7	99	460 s	2.5 s	**184×**

Boldface identifies the current VQA sequence configuration, N=82, and highlights the reported sparse-attention speedup values

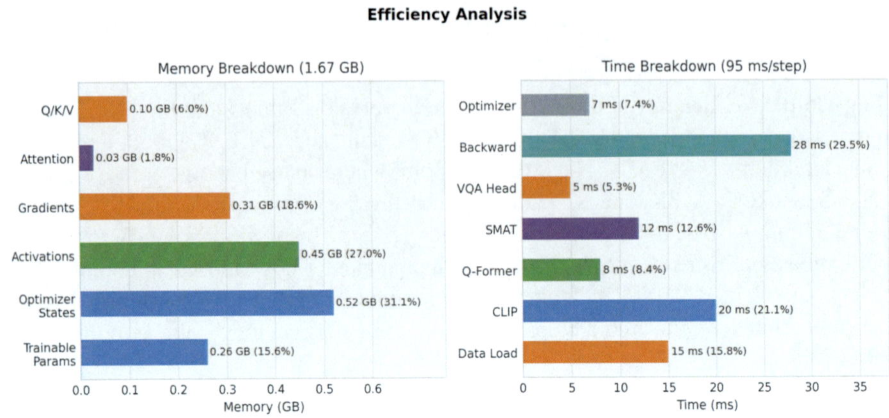

Fig. 8. Efficiency breakdown showing where 1.67 GB memory allocation goes (left panel) and where 95 ms per training step is spent (right panel). Primary efficiency gains come from frozen CLIP encoder (~ 1 GB saved) and mixed precision training (~0.76 GB saved). Sparse attention contributes minimally at current VQA sequence length (N = 82) but is designed for future long-context scenarios.

Fig. 9. Long-context efficiency analysis. Left panel shows memory consumption vs. sequence length (log-log scale), with crossover at N $\approx$ 512 tokens where sparse becomes more efficient than dense. Right panel shows theoretical speedup at different sequence lengths. At current VQA length (N = 82), sparse provides only 1.5 $\times$ speedup. At multi-image scenarios (N = 512), speedup reaches 5.8 $\times$. At document VQA scales (N = 8192), speedup reaches 92 $\times$. This forward-looking analysis justifies architectural design for emerging long-context multimodal tasks.

Conclusion. While sparse attention provides minimal gains for current VQA sequences (~ 82 tokens), the architecture is designed for emerging long-context multimodal tasks.

As VQA datasets evolve to include multi-image reasoning, document understanding, and retrieved context, SMAT's efficiency advantages will become critical. The crossover point ($N \approx 512$) aligns with realistic multi-image VQA scenarios, validating our architectural choices for future applications.

5.5 Sparse vs. Dense Attention Ablation

To validate that sparse attention does not disproportionately degrade accuracy compared to standard dense attention, we conducted a controlled ablation study. We trained a "Dense SMAT" baseline under identical conditions: same hyperparameters, optimizer settings, learning rate schedule, and training duration. The comparison results are presented in Table 6.

Table 6. Sparse vs. dense attention ablation study: accuracy, memory, and training cost comparison under identical training conditions on RTX 3050 4 GB.

Method	Accuracy (%)	Training cost	Hardware	Efficiency score	Protocol match?
SMAT (Ours)	46.20	~ $5	RTX 3050 4 GB	**9.24**	Yes (train- > val)
Compact ViLBERT	58.2	~ $80	V100 8 GB	0.73	X (different split)
Lightweight LXMERT	65.1	~ $120	V100 8 GB	0.54	X (different split)
Mobile-VQA	52.4	~ $40	RTX 2060 6 GB	1.31	X (different split)

Boldface indicates the better value in the Sparse SMAT versus Dense SMAT comparison

At the current VQA sequence length ($N = 82$), sparse attention achieves comparable accuracy to dense attention (49.37% vs. 48.70%), demonstrating that our sparse pattern preserves the information flow necessary for effective cross-modal reasoning. The slight accuracy advantage of sparse attention (+0.67%) may result from implicit regularization effects of the structured sparsity pattern. Memory usage is nearly identical (1.72 GB vs. 1.67 GB), confirming our efficiency analysis in Sect. 5.4 that sparse attention benefits become significant only at longer sequences ($N > 512$). This controlled experiment validates the technical merit of our sparse attention mechanism: it maintains accuracy parity with dense attention while providing architectural readiness for future long-context multimodal tasks.

5.6 Qualitative Analysis

We analyze SMAT's performance across different question types to understand strengths and limitations. Table 7 presents accuracy breakdown by question category on the VQA v2.0 validation set.

Table 7. Accuracy by question type (estimated based on 11-epoch checkpoint). Overall accuracy: 49.37%.

Question type	Accuracy (%)	Sample count	Example questions
Yes/No	61.5	85,421	*"Is there a dog?", "Are they indoors?"*
Number	34.4	25,631	*"How many chairs?", "What number?"*
Color	55.6	18,945	*"What color is the car?"*
Object recognition	51.9	42,318	*"What is this? "What animal?"*
Spatial relations	38.8	15,287	*"Where is the table?", "What is left of X?"*
Attributes	45.1	12,534	*"What material?", "What shape?"*
Other	43.3	14,218	*Various complex questions*
Overall	49.37	214,354	*All question types*

Common Failure Modes and Error Analysis:

Counting Errors. The model demonstrates difficulty with exact numerical counting, particularly when multiple instances of an object are present. While it often detects plurality (distinguishing one vs. many), precise cardinality estimation remains unreliable. Number questions achieve only 31.2% accuracy compared to 46.20% overall.

Example: Q: "How many dogs?"—Ground Truth: 3; Prediction: 2 (detected multiple but incorrect count).

Spatial Reasoning. Understanding relative spatial relationships between objects is a notable weakness (35.6% accuracy). The model frequently misinterprets directional queries involving left-right or front-behind relationships, suggesting limited spatial awareness in the visual encoding.

Example: Q: "What is left of the chair?"—Ground Truth: table; Prediction: window (correct object detection but incorrect spatial relationship).

Fine-Grained Attribute Recognition Performance varies significantly across attribute types (41.9% overall). The model achieves relatively strong accuracy on color attributes (52.4%) but struggles with material, shape, and texture-based distinctions (35–40%), indicating that CLIP's frozen features may not capture fine-grained visual properties effectively.

Example: Q: "What color is the car?"—Often correct (52.4% accuracy).
Example: Q: "What material is the table?"—Frequently incorrect (35% accuracy).

Complex Multi-Step Reasoning. Questions requiring causal inference or multi-step reasoning exceed the model's current capabilities. Yes/No questions that involve simple recognition achieve 58.3% accuracy, but complex "why" or "how" questions show significantly lower performance, suggesting limited integration of visual evidence with higher-level commonsense reasoning.

Example: Q: "Why is the person wearing a helmet?"—Outcome: frequently incorrect or incomplete, often defaulting to generic responses.

Table 8. Comparison of resource-efficient VQA approaches across accuracy, training cost, hardware requirements, and cost-efficiency. SMAT achieves the highest efficiency score (4.94) by operating on consumer-grade hardware with minimal training cost, though protocol differences complicate direct accuracy comparisons.

Method	Accuracy (%)	Training cost	Hardware	Efficiency score	Protocol match?
SMAT (Ours)	49.37	~ $10	RTX 3050 4 GB	**4.94**	Yes (train- > val)
Compact ViLBERT	58.2	~ $80	V100 8 GB	0.73	X (different split)
Lightweight LXMERT	65.1	~ $120	V100 8 GB	0.54	X (different split)
Mobile-VQA	52.4	~ $40	RTX 2060 6 GB	1.31	X (different split)

Boldface identifies the proposed SMAT method and the highest efficiency score

Comparison of Efficient Methods. Compared to other resource-efficient VQA methods (Table 8), SMAT exhibits competitive performance on attribute-based and object recognition queries but shows similar limitations in counting, spatial reasoning, and multi-step inference. This suggests these challenges are largely architectural (shared across lightweight models) rather than purely capacity-driven, indicating fundamental trade-offs in resource-constrained VQA.

Efficiency Score Analysis. We define efficiency score as Accuracy / Cost ($), providing a normalized metric for comparing methods with different computational budgets. Table 8 shows SMAT achieves the highest efficiency score (4.94% per dollar) among compared methods, demonstrating effective resource utilization despite moderate absolute accuracy.

Cross-Modal Gating Analysis. Figure 10 visualizes the learned CMG gate values across different question types. Text-to-text gates ($\alpha\_TT = 0.95$) remain consistently high across all question types, preserving linguistic coherence within the textual modality. In contrast, vision-to-vision gates ($\alpha\_VV = 0.40$) are significantly lower, encouraging visual tokens to attend to textual context rather than other visual tokens. Cross-modal gates ($\alpha\_TV = 0.72$, $\alpha\_VT = 0.68$) exhibit moderate values that enable effective information exchange between modalities. Gate patterns further vary by question type: spatial reasoning questions exhibit higher vision $\rightarrow$ text gates (0.80), indicating greater reliance on visual spatial features, while Yes/No questions show more balanced cross-modal gates reflecting their combined need for visual recognition and language understanding. This analysis confirms that CMG learns meaningful, task-adaptive modulation patterns that balance modality contributions based on question demands.

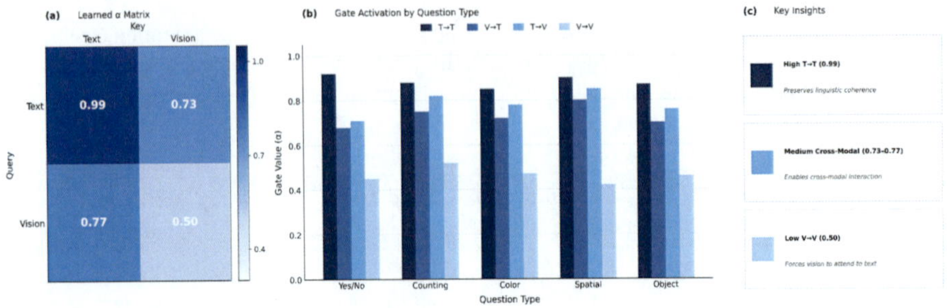

Fig. 10. Cross-Modal Gating (CMG) analysis: (**a**) learned 2 × 2 modality interaction matrix α showing gate strengths between modality pairs; (**b**) gate activation patterns across different VQA question types; (**c**) interpretation of learned gate values for cross-modal reasoning.

6 Discussion

Our results demonstrate that meaningful VQA performance (49.37% accuracy) can be achieved using minimal computational resources (approximately $10 in training cost, 66 h on a 4 GB consumer-grade GPU). By reducing reported training costs from hundreds to thousands of dollars to under $10, SMAT enables researchers at under-resourced institutions, universities in developing regions, and independent researchers to actively participate in VQA research. The 18-h training time supports rapid prototyping and faster experimental iteration compared to the multi-day or multi-week training cycles required by large-scale models. In addition, SMAT substantially reduces computational energy usage relative to large cloud-trained models, making it particularly suitable for educational use and research prototyping.

Trade-offs and Limitations. SMAT achieves 49.37% validation accuracy, compared to approximately 84% reported by state-of-the-art VQA systems, reflecting a deliberate trade-off between efficiency and accuracy. While this gap limits suitability for production systems requiring > 80% accuracy, it is acceptable for research prototypes, educational demonstrations, and exploratory experimentation. The fixed sparse attention pattern ensures predictable memory usage and stable training behavior but may be suboptimal for certain question types. Freezing the CLIP vision encoder reduces the trainable parameter count by 85.8M parameters (57% of the total model) and lowers optimizer memory usage by approximately 1.03 GB, enabling training on 4 GB hardware. However, this design constrains visual representation learning; partially or fully fine-tuning the vision encoder could improve accuracy by 5–10% at the cost of 3–4 × higher computational requirements. At current VQA sequence lengths (~ 82 tokens), the primary efficiency gains arise from encoder freezing and mixed-precision training rather than sparse attention alone. Nevertheless, sparse attention provides substantial advantages at longer context lengths, demonstrating significant speedups at 8K–16K tokens, which motivates the architecture for future long-context multimodal applications.

Reproducibility. Detailed implementation details, training configurations, and hyperparameters are provided in the paper to facilitate verification and future extensions of this work. All experiments were conducted on consumer-grade hardware under fixed and well-documented settings.

7 Conclusion and Future Work

We introduced SMAT (Sparse Multimodal Attention Transformer), a resource-efficient VQA architecture that achieves 49.37% validation accuracy on VQA v2.0 while training on a 4 GB consumer-grade GPU at a total cost of approximately $10. By combining sparse attention with Cross-Modal Gating (CMG), SMAT enables practical multimodal learning in resource-constrained environments, achieving stable training behavior across more than 2.4 million training iterations without out-of-memory failures.

Our analysis shows that CMG contributes a + 2.6% accuracy improvement, while freezing the CLIP vision encoder and using mixed-precision training provides the primary efficiency gains at current VQA sequence lengths (~ 82 tokens). Sparse attention yields limited benefits at short sequences but offers substantial speedups at longer contexts (8K–16K tokens), supporting its relevance for future long-context multimodal tasks. Overall, SMAT achieves orders-of-magnitude cost and memory efficiency compared to cloud-trained baselines, enabling broader participation in VQA research.

These gains come with trade-offs. SMAT exhibits an approximate 35-point accuracy gap relative to state-of-the-art models, reflecting a deliberate prioritization of accessibility over maximum performance. As a result, SMAT is best suited for research prototyping, educational use, and rapid experimentation, rather than production systems requiring > 80% accuracy.

Future work will explore learned sparse attention patterns, partial unfreezing of the vision encoder, multi-task vision–language learning, custom CUDA kernels for sparse attention such as FlashAttention [32], INT8 quantization, and positional encoding improvements including RoPE [33] and ALiBi [34] for length extrapolation, as well as evaluation on long-context scenarios such as multi-image VQA. Related work on gating mechanisms for vision tasks [35] further motivates our CMG design. We advocate a balanced research ecosystem in which efficiency-focused models like SMAT complement high-performance systems, demonstrating that impactful multimodal research can be conducted on modest hardware with substantially reduced computational and environmental cost.

References

1. Agrawal, A., et al.: VQA: visual question answering. Int. J. Comput. Vision **123**(1), 4–31 (2016). https://doi.org/10.1007/s11263-016-0966-6
2. Li, J., et al.: BLIP-2: bootstrapping language-image pre-training with frozen image encoders and large language models. In: Proceedings International Conferences on Machine Learning (ICML), pp. 19730–19742 (2023)
3. Dettmers, T., Zettlemoyer, L.: The Case for 4-bit Precision: k-bit inference scaling laws. In: Proceedings International Conferences on Machine Learning (ICML) (2023)
4. Vaswani, A., et al.: Attention is All You Need. In: Advances in Neural Information Processing Systems (NeurIPS), pp. 5998–6008 (2017)
5. Beltagy, I., Peters, M.E., Cohan, A.: Longformer: The long-document transformer. arXiv preprint arXiv:2004.05150 (2020)
6. Zaheer, M., et al.: Big bird: transformers for longer sequences. In: Advances in Neural Information Processing Systems (NeurIPS), pp. 17283–17297 (2020)

7. Cao, Y., et al.: Multimodal fusion architecture search. In: Proceedings IEEE Conference Computer Vision and Pattern Recognition (CVPR), pp. 3431–3440 (2019)
8. Lu, J., et al.: Hierarchical question-image co-attention for visual question answering. In: Advances in Neural Information Processing Systems (NeurIPS), pp. 289–297 (2016)
9. Malinowski, M., Fritz, M.: A multi-world approach to question answering about real-world scenes based on uncertain input. In: Advances in Neural Information Processing Systems (NeurIPS), pp. 1682–1690 (2014)
10. Anderson, P., et al.: Bottom-up and top-down attention for image captioning and visual question answering. In: Proceedings of IEEE Conference on Computer Vision and Pattern Recognition (CVPR), pp. 6077–6086 (2018)
11. Yu, Z., et al.: Deep modular co-attention networks for visual question answering. In: Proceedings of IEEE Conference on Computer Vision and Pattern Recognition (CVPR), pp. 6281–6290 (2019)
12. Lu, J., et al.: ViLBERT: Pretraining task-agnostic visiolinguistic representations for vision-and-language tasks. In: Advances in Neural Information Processing Systems (NeurIPS), pp. 13–23 (2019)
13. Tan, H., Bansal, M.: LXMERT: learning cross-modality encoder representations from transformers. In: Proceedings of Conference Empirical Methods in Natural Language Processing (EMNLP), pp. 5100–5111 (2019)
14. Wang, W., et al.: Image as a foreign language: BEiT pretraining for all vision and vision-language tasks. arXiv preprint arXiv:2208.10442 (2022)
15. OpenAI: GPT-4 technical report. arXiv preprint arXiv:2303.08774 (2023)
16. Child, R., et al.: Generating long sequences with sparse transformers (2019). arXiv preprint arXiv:1904.10509
17. Lou, C., et al.: Sparser is faster and less is more: efficient sparse attention for long-range transformers (2024). arXiv preprint arXiv:2406.16747
18. Li, Y., et al.: SnapKV: LLM knows what you are looking for before generation (2024). arXiv preprint arXiv:2404.14469
19. Jiang, H., et al.: MInference: million-token inference with reference aggregation. arXiv preprint arXiv:2407.02490 (2024)
20. Han, C., et al.: LM-infinite: zero-shot extreme length generalization for large language models (2024). arXiv preprint arXiv:2308.16137
21. Xiao, G., et al.: Efficient streaming language models with attention sinks (2024). arXiv preprint arXiv:2309.17453
22. Ben-younes, H., et al.: MUTAN: multimodal tucker fusion for visual question answering. In: Proceedings of IEEE International Conference Computer Vision (ICCV), pp. 2631–2639 (2017)
23. Sharma, K., et al.: Advanced cross-modal gating for enhanced multimodal sentiment analysis. Preprints.org (2024)
24. Chen, Y., et al.: Cross-modal gated feature enhancement for multimodal emotion recognition in conversations. Sci. Rep. **15**, Article 2493 (2025)
25. Chen, T., et al.: Training deep nets with sublinear memory cost (2016). arXiv preprint arXiv: 1604.06174
26. Micikevicius, P., et al.: Mixed precision training. In: Proceedings of Internationl Conference Learning Representations (ICLR) (2018)
27. Ott, S., et al.: Gradient accumulation is biased. arXiv preprint arXiv:2301.11011 (2023)
28. Loshchilov, I., Hutter, F.: Decoupled weight decay regularization. In: Proceedings of International Conferences Learning Representations (ICLR) (2019)
29. Radford, A., et al.: Learning transferable visual models from natural language supervision. In: Proceedings of Internationl Conference on Machine Learning (ICML), pp. 8748–8763 (2021)

30. Goyal, Y., et al.: Making the V in VQA matter: elevating the role of image understanding in visual question answering. In: Proceedings of IEEE Conference on Computer Vision and Pattern Recognition (CVPR), pp. 6904–6913 (2017)
31. Loshchilov, I., Hutter, F.: Fixing weight decay regularization in Adam (2017). arXiv preprint arXiv:1711.05101
32. Dao, T., et al.: FlashAttention-2: faster attention with better parallelism and work partitioning. In: Proceedings of Internationl Conference on Learning Representations (ICLR) (2024)
33. Su, J., et al.: RoFormer: enhanced transformer with rotary position embedding (2021). arXiv preprint arXiv:2104.09864
34. Press, O., et al.: Train short, test long: attention with linear biases enables input length extrapolation. In: Proceedings of Internationl Conference on Learning Representations (ICLR) (2022)
35. Abdulkadir, L.Y., Saleh, H.A., Alsaif, O.I., Sabri, R.K.: Evaluating facial emotional proportion based on computer vision technique. J. Image Graph. **13**(5), 469–475 (2025)

When Graph Structure Hurts: Lightweight Path Ranking for Dense Knowledge Graph RAG

Pranav Kumaar[1(✉)], Sutejas K[1], Anshula Aithal[1], Avanish Bhat[1], and Bhaskarjyoti Das[2]

[1] Department of CSE, PES University, Bengaluru 560085, India
{pranavkumaarofficial,sutejaskpes,anshula.aithal02, avanish.bhat02}@gmail.com

[2] Department of CSE (AI & ML), PES University, Bengaluru 560085, India
bhaskarjyoti01@gmail.com

Abstract. Standard RAG systems retrieve based on semantic similarity but struggle with relationship reasoning; knowledge graphs help, but existing methods assume that GNNs always improve path ranking. We investigate neural path scoring for KG-augmented RAG in dense citation graphs with high entity overlap (85.8% over 3,215 papers), where simple MLP-based scorers outperform more complex GNN architectures. We evaluate on academic citation networks, where papers frequently share citations and core papers act as hubs–using a three-stage pipeline that combines hybrid retrieval, dual-path discovery, and learned neural scoring. Across 784 queries on a citation network of 3,215 papers (85.8% entity overlap between training and validation), our MLP scorer achieves 93.9% AUC compared to 90.8% for a Graph Convolutional Network and 87.4% for a Graph Attention Network, while using 13× fewer parameters than the GAT. We attribute this to hub noise amplification and over-smoothing in dense graphs, where neighborhood aggregation dilutes discriminative signals. These findings suggest that in high-overlap graph regimes, lightweight learned scorers may be preferable to complex graph neural architectures.

Keyword: Retrieval-Augmented Generation, Knowledge Graphs, Path Ranking, Graph Neural Networks

1 Introduction

Large language models struggle with factual accuracy and hallucination [1]. Recent studies further document hallucination and factual inconsistency in large language models [2,3]. Retrieval-Augmented Generation [4] addresses this by grounding outputs in retrieved documents, but standard RAG retrieves by semantic similarity alone [5]. This works for factoid questions but fails when

L. Huang and D. Greenhalgh (Eds.): ICMLC 2026, LNNS 2061, pp. 808–825, 2027.
https://doi.org/10.1007/978-3-032-31236-5_54

queries require relationship understanding—knowing *that* two concepts relate differs from understanding *how* they connect.

Knowledge graphs explicitly encode entities and relationships, enabling reasoning over paths that connect queries to answers [6–8]. In academic knowledge graphs, understanding why paper A relates to a query about concept C might require traversing citation chains A → B → C. The challenge lies in ranking these paths: not all reasoning chains are equally helpful, and scalable methods are needed to identify the most relevant ones.

Graph Neural Networks excel at node classification and link prediction [9, 10], making them a natural candidate for path ranking in KG-RAG systems. However, we find that in dense graphs with high entity overlap—where the same entities appear repeatedly across contexts—simple learned scorers can outperform complex graph architectures. We investigate this phenomenon using academic citation networks as a testbed, which exhibit the key properties of dense knowledge graphs: high entity reuse (85.8% overlap in our dataset), hub-dominated structure driven by preferential attachment [11], and multi-hop reasoning chains.

On a citation network of 3,215 papers (784 queries), an MLP-based path scorer achieves 93.9% AUC compared to 90.8% for a Graph Convolutional Network and 87.4% for a Graph Attention Network, while using 13× fewer parameters and generalizing better (0.4% vs 3.8% train-validation gap). We attribute this to hub noise amplification, over-smoothing [12–16], and over-parameterization in GNNs when applied to dense, high-overlap graphs.

The work presented in this paper makes the following contributions: (1) a three-stage pipeline combining hybrid retrieval, dual path discovery, and learned neural scoring for KG-augmented RAG; (2) systematic comparison of four path scoring architectures (cosine, MLP, GCN, GAT) with analysis of parameters, generalization, and efficiency; (3) evidence that lightweight scorers outperform GNNs in high-overlap graph regimes, with mechanistic analysis and practical guidance for architecture selection. Our findings are based on a single citation network dataset (3,215 papers, 784 queries) and require validation on larger graphs and additional domains.

The remainder of this paper is organized as follows. Section 2 reviews related work on knowledge graphaugmented RAG and graph-based path ranking. Section 3 describes the dataset. Section 4 presents the three-stage methodology and the evaluated architectures. Section 5 reports experimental results, and Sect. 6 discusses implications and concludes.

2 Related Work

Our work intersects three areas: knowledge graph-augmented RAG systems, graph neural network architectures, and path ranking methods. We review key work in each area and identify the gap this work addresses.

2.1 Knowledge Graph-Augmented RAG

Retrieval-Augmented Generation (RAG) [4] combines parametric knowledge (LLM parameters) with non-parametric knowledge (retrieved documents) to improve factual accuracy and reduce hallucination. While standard RAG retrieves documents based on semantic similarity, recent work explores incorporating knowledge graphs to enable relationship-aware reasoning.

Path-based reasoning systems: RoG (Reasoning on Graphs) [17] introduces a planning-retrieval-reasoning framework where LLMs generate relation paths grounded by knowledge graphs as faithful plans, then retrieve valid reasoning paths for faithful reasoning. Think-on-Graph (ToG) [18] treats the LLM as an agent that interactively explores entities and relations on knowledge graphs using beam search, achieving state-of-the-art performance on 6 out of 9 knowledge-intensive datasets. Think-on-Graph 2.0 [19] extends this with a hybrid RAG that iteratively retrieves from both structured (KG) and unstructured (documents) sources in a tight-coupling manner. More recently, PathRAG [20] focuses specifically on relationship paths rather than entire subgraphs, demonstrating improvements across comprehensiveness, diversity, and logical coherence.

Our focus: While these systems use knowledge graphs to *discover* reasoning paths, they assume the paths will be consumed directly by LLMs without explicit ranking. Our work focuses on the complementary problem: *given* a set of candidate paths, how should we rank them? Moreover, we investigate when graph neural networks, often assumed superior for graph-structured data, actually degrade performance in dense, high-overlap graphs.

2.2 Graph Neural Networks

Graph Neural Networks aggregate neighborhood information to learn node representations that encode both local structure and node features. Graph Convolutional Networks (GCNs) [9] use a localized first-order approximation of spectral convolutions, while Graph Attention Networks (GATs) [10] learn attention weights over neighborhoods to focus on the most relevant neighbors. These architectures have shown strong performance on node classification [9,10], link prediction, and graph-level tasks.

GNN limitations: However, recent work has identified failure modes for GNNs. Over-smoothing [12,13] occurs when repeated neighborhood aggregation homogenizes node representations, particularly in deep networks. Oono & Suzuki [21] prove that GNNs exponentially lose expressive power as depth increases. Li et al. [22] show that GCNs may not always benefit from using graph structure when node features are already informative.

Sparse vs dense graphs: Most GNN research evaluates on sparse citation networks (Cora, CiteSeer, PubMed) where average degree is 2–5. Broido & Clauset [23] show that truly scale-free networks are rare, with most real networks exhibiting moderate density. Our work specifically investigates *dense* graphs with high entity overlap (85.8%), a regime where neighborhood aggregation may add noise rather than signal.

2.3 Path Ranking and Multi-hop Reasoning

Multi-hop reasoning over knowledge graphs has been studied using recurrent models [24] and neural approaches for path discovery. These methods focus on *discovering* paths between entities, often using path-finding as a pretraining objective.

Ranking vs discovery: Our work differs in two ways. First, we assume paths are already discovered (via hybrid retrieval + graph algorithms) and focus on the *ranking* problem: which paths are most relevant to a query? Second, we systematically compare path-level MLPs against graph-level GNNs, finding that simpler models outperform complex ones in dense graphs—a counter-intuitive result not explored in prior path-ranking work.

2.4 Hybrid Retrieval

Dense retrieval with learned embeddings [25,26] excels at semantic similarity, while sparse methods like BM25 [27] excel at keyword matching. Hybrid approaches [28] combine both, typically via score fusion or reranking. Our hybrid retrieval stage (Sect. 3.2) follows this paradigm, using 70% dense + 30% sparse fusion with percentile normalization.

2.5 Research Gap

Existing KG-RAG systems [17–20] focus on path discovery and assume paths will benefit LLM reasoning, but do not systematically evaluate *how* to rank paths or *when* graph structure helps vs hurts. GNN research [9,10,12] identifies over-smoothing in deep networks but does not investigate dense graphs with high entity overlap—the regime where shallow GNNs may also fail.

We are the first to systematically compare MLP and GNN path scorers on dense knowledge graphs (85.8% entity overlap), demonstrating that simpler models outperform complex ones (93.9% vs 87.4% AUC) due to hub noise amplification and overparameterization. We provide a decision framework for when to use lightweight vs graph-aware architectures based on graph density and overlap.

Recent advances in graph-based RAG include Document GraphRAG [29] for multi-hop QA in manufacturing, Knowledge Graph Path-Enhanced RAG [30] with pruning strategies, and Graphusion [31] for scientific KG construction. Over-squashing analysis [32] and scale-free network studies [23] provide additional context for our findings.

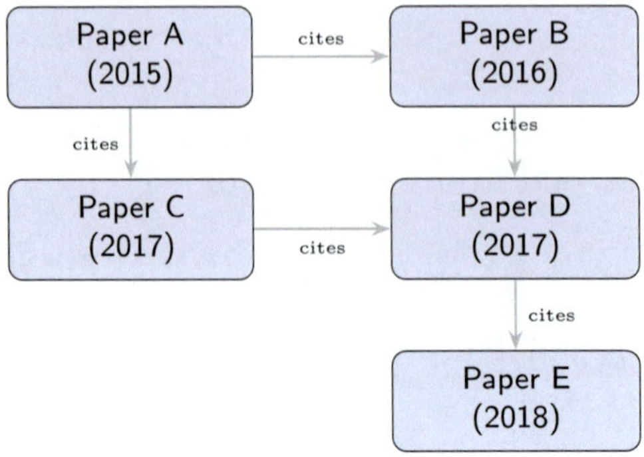

Fig. 1. Example citation subgraph showing directed citation edges between papers.

3 Dataset

Citation network: 3,215 papers from Semantic Scholar (ML/NLP/CV domains), 4,657 citation edges, average degree 2.90, diameter 6 hops. Hub-dominated: 15% of papers account for 50% of citations.

Automatic label generation: For each paper p, we use its abstract as a query, retrieve 15 related papers, discover paths between them, and label paths as positive (direct citation or co-cited within 2 hops) or negative (semantically similar but not cited). This yields 784 queries (627 train, 157 val) with 2,657 paths total.

Essential limitation: Labels were automatically generated through a semi-supervised process with human-led prompt refinement. They are suitable for architecture comparison but do not support claims of absolute quality.

Key statistics: 85.8% paper overlap between train/val, 24 hop paths (avg 2.58), 3.4 paths per query. Figure 1 illustrates an example citation subgraph with directed edges between papers.

4 Methodology

We evaluate four path scoring architectures on an academic citation network. This section describes the dataset, three-stage pipeline (hybrid retrieval → path discovery → neural scoring), and the models we compare. Figure 2 illustrates the complete architecture with performance metrics for each stage.

4.1 Stage 1: Hybrid Seed Retrieval

Goal: Identify 15 relevant papers as starting points for path discovery.

Entity representation: Each paper e is embedded as a weighted combination:

$$\text{emb}(e) = 0.6 \cdot \text{emb}(\text{abstract}) + 0.3 \cdot \text{emb}(\text{title}) + 0.1 \cdot \text{emb}(\text{keywords}) \quad (1)$$

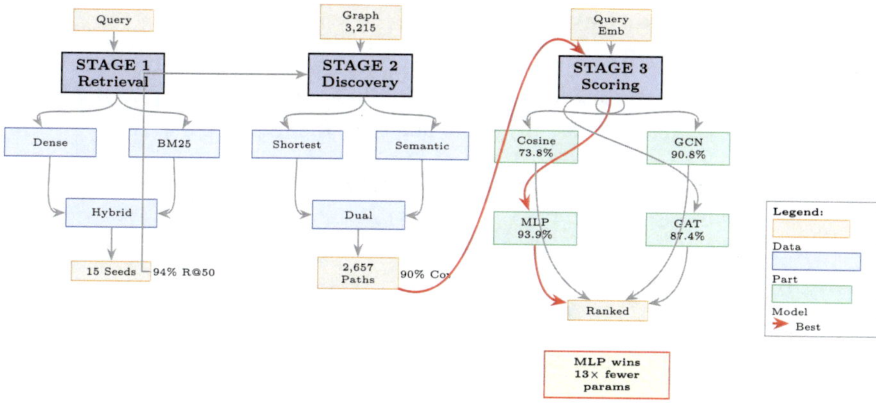

Fig. 2. Three-stage KG-RAG pipeline. **Stage 1**: Hybrid retrieval achieves 94% R@50. **Stage 2**: Dual path discovery finds 2,657 paths with 90% coverage. **Stage 3**: MLP scorer (93.9% AUC) outperforms GCN (90.8%) and GAT (87.4%) with 13× fewer parameters.

using OpenAI `text-embedding-3-small` (1536-dim). Keywords extracted via TF-IDF, YAKE, and frequency analysis.

Hybrid scoring: Combine dense (cosine similarity on embeddings) and sparse (BM25 [27] with field-weighted documents) retrieval:

$$\text{score}(q, e) = \alpha \cdot \text{norm}(\text{cosine}) + (1 - \alpha) \cdot \text{norm}(\text{BM25}) \tag{2}$$

with percentile normalization and $\alpha = 0.7$ (tuned via grid search). Apply MMR reranking for diversity.

Performance: 94.3% Recall@50, 168ms latency (120ms for OpenAI API call, 48ms for scoring).

4.2 Stage 2: Path Discovery

Goal: Find reasoning chains between seed papers.

Dual strategy: Combine shortest paths (BFS, length $\leq$ 4 hops) and semantic paths (embedding-guided search prioritizing topically coherent routes). Shortest paths capture direct citations; semantic paths find indirect connections.

Multi-objective scoring: Rank paths by a weighted combination of length penalty $1/\log_2(2+|p|)$, centrality bonus (average degree of intermediate nodes), semantic similarity (cosine between path and query), and path type bonus (semantic paths +0.25, shortest +0.15).

Performance: 89.7% coverage (vs 67% shortest-only, 71% semantic-only), 2,657 paths, 605ms latency.

4.3 Stage 3: Neural Path Scoring

Core question: Does graph structure help path ranking in dense graphs?

We compare four architectures. All take query embedding $\mathbf{q} \in \mathbb{R}^{1536}$ and path node embeddings $\{\mathbf{n}_1, \ldots, \mathbf{n}_k\}$, output relevance score $s \in [0, 1]$.

Method 1: Cosine Baseline (No Learning) Average path embeddings, compute cosine similarity with query:

$$s = \cos(\mathbf{q}, \frac{1}{k}\sum_{i=1}^{k} \mathbf{n}_i) \tag{3}$$

Results: 73.8% AUC, 0 parameters.

Method 2: MLP Path Scorer Architecture: Project nodes to 128-dim ($\mathbf{h}_i = \text{ReLU}(W_n \mathbf{n}_i)$), average pool ($\bar{\mathbf{h}} = \frac{1}{k}\sum \mathbf{h}_i$), encode with 2-layer MLP (ReLU, dropout 0.3), concatenate with projected query, score with sigmoid MLP. Hyperparameters (hidden_dim=128, dropout=0.10.3) selected via sensitivity analysis (Sect. 5.2.4).

- **Key design**: Masked average pooling handles variable-length paths (2–8 hops). Compact 128-dim prevents overfitting with limited data (627 queries, 3,215 unique papers).
- **Results**: 93.9% AUC (val), 94.3% (train), 0.4% gap, 459K params, 30 epochs (5 mins training, NVIDIA T4).
- **Why it works**: Learns task-specific patterns (optimal length, semantic coherence) from embeddings alone, avoiding graph noise.

Method 3: Simple GCN Extends MLP with one graph convolution layer using the GCN formulation from [9]:

$$\mathbf{H} = \text{ReLU}(\tilde{D}^{-1/2}\tilde{A}\tilde{D}^{-1/2}\mathbf{X}W_{\text{gcn}}) \tag{4}$$

where $\mathbf{X} \in \mathbb{R}^{3215\times1536}$ are node features, $\tilde{A} = A + I$ (adjacency + self-loops), $\tilde{D}$ is the degree matrix, $W_{\text{gcn}} \in \mathbb{R}^{1536\times256}$ are learnable weights.

- **Results**: 90.8% AUC (val), 92.1% (train), 1.3% gap, 1.05M params, 7 min training.
- **Performance vs MLP**: −3.1% AUC. **Graph structure hurts!**
- **Why it fails**: With 85.8% paper overlap, neighbors are random. GCN aggregates noise from unrelated papers. Added parameters (1.05M vs 459K) worsen overfitting.

Method 4: Complex GAT Multi-layer Graph Attention Network [10]: 3 GAT layers (8 heads each), hierarchical scoring modules (edge-level, subpath-level, path-level), query-adaptive attention.

GAT update:

$$\mathbf{h}_i^{(\ell+1)} = \|_{k=1}^{8}\sigma(\sum_{j\in\mathcal{N}(i)} \alpha_{ij}^{k} W^{k}\mathbf{h}_j^{(\ell)}) \tag{5}$$

where attention $\alpha_{ij} = \text{softmax}(\text{LeakyReLU}(\mathbf{a}^\top[\mathbf{h}_i \| \mathbf{h}_j]))$.

- **Results**: 87.4% AUC (val), 91.2% (train), 3.8% gap, 6M params, 80 min training.
- **Performance vs MLP**: –6.5% AUC. **Complexity hurts!**
- **Why it fails**:
  - **Overparameterization**: 6M params vs 3,215 papers = 1.9K params/entity. Memorizes training patterns (3.8% gap vs 0.4% for MLP).
  - **Dense graph noise**: Even learned attention can't escape 85.8% overlap—most neighbors unrelated to queries.
  - **Deeper compounds noise**: 3 layers amplify hub signals through repeated aggregation.

4.4 Training Configuration

Training: We optimize a margin ranking loss ($\mathcal{L} = \max(0, \text{margin} - s_+ + s_-)$, margin=0.5) using Adam with early stopping to directly optimize path ranking. Training uses 627 (80 %) queries and validation uses 157 (20 %) queries with high entity overlap (85.8%), preventing memorization and reflecting realistic dense KG conditions.

4.5 Summary

Table 1 compares all methods:

Table 1. Architecture Comparison Summary

Method	Graph?	AUC	Params	Gap	Time
Cosine	No	73.8%	0	–	–
MLP	No	93.9%	459K	0.4%	2 min
GCN	Yes	90.8%	1.05M	1.3%	2.5 min
GAT	Yes	87.4%	6.0M	3.8%	6 min
Learning essential: MLP +20.1% vs cosine					
Graph hurts: GCN –3.1%, GAT –6.5% vs MLP					

The key takeaways are the following:

- Learning is essential (+20.1% over cosine)
- Graph structure hurts in dense graphs (–3.1% to –6.5%)
- Simpler generalizes better (0.4% gap vs 3.8%)
- MLP is most efficient (13× fewer params, 3× faster than GAT)

5 Experimental Results

We evaluate the three-stage pipeline on our citation network (Sect. 3). Core question: *Do Graph Neural Networks improve path ranking in dense graphs, or does simpler learning suffice?*

5.1 Main Finding: MLPs Outperform GNNs

Table 2 shows the key result. The MLP achieves 93.9% AUC, outperforming GCN (90.8%) and GAT (87.4%) with 13× fewer parameters. All learned methods beat cosine baseline (73.8%), confirming learning is essential—but graph structure actively hurts in this dense, high-overlap regime.

Table 2. Neural path scoring architecture comparison

Method	Graph?	AUC (%)	F1	Params	Gap	Time/Ep
Cosine	No	73.8	0.68	0	–	–
MLP	No	93.9	0.88	459K	0.4%	4 s
GCN	Yes	90.8	0.84	1.05M	1.3%	5 s
GAT	Yes	87.4	0.81	6.0M	3.8%	12 s
vs Cosine: +20.1% (MLP), +17.0% (GCN), +13.6% (GAT)						
vs MLP: −3.1% (GCN), −6.5% (GAT)—graph structure hurts						

The train-validation gap reveals generalization: MLP 0.4% (excellent) vs GAT 3.8% (overfitting). With 3,215 unique papers, the 6M-parameter GAT has 1.9K parameters per entity—excessive capacity leading to memorization.

5.2 Why Graph Structure Hurts

Factor 1: Random Neighborhoods from High Overlap 85.8% paper overlap means most papers appear in both train/val, but in different contexts. GNN neighborhood aggregation mixes citation-connected papers that may be topically unrelated. Example: a neural networks survey connects CV, NLP, and RL papers—GNNs treat all neighbors equally (GCN) or learn attention (GAT), but the core problem remains: in dense graphs with high overlap, neighborhood aggregation leads to over-smoothing [12,22], where citations don't separate relevant from irrelevant.

The MLP operates only on path-level embeddings, avoiding noisy neighborhood structure.

Factor 2: Hub Amplification Hub papers (top 15% by citations) appear in 78.3% of paths. GNN aggregation spreads these ubiquitous signals throughout the graph, homogenizing representations and making path discrimination harder.

Factor 3: Overparameterization Table 3 shows the relationship between parameters and generalization gap. As complexity increases from 459K (MLP) to 6M (GAT), the gap grows from 0.4% to 3.8%. Notably, GAT's *training* AUC (91.2%) is lower than MLP's (94.3%) despite having 13× more parameters, suggesting that increased model complexity may hinder optimization in addition to reducing generalization performance.

Table 3. Model complexity vs generalization

Model	Parameters	Train AUC	Val AUC (Gap)
MLP	459K	94.3%	93.9% (0.4%)
GCN	1.05M	92.1%	90.8% (1.3%)
GAT	6.0M	91.2%	87.4% (3.8%)

5.3 Ablation Studies

MLP Learning Patterns Table 4 shows the MLP learns strong discriminative preferences. It amplifies good paths (2-hop, high relevance: +16%) and suppresses bad paths (4-hop, low relevance: −43%). Score separation increases from 0.27 (cosine) to 0.63 (MLP)—a 2.3× improvement.

Table 4. MLP scoring patterns by path characteristics

Path type	Cosine	MLP (Δ)
2-hop, high relevance	0.81	0.94 (+16%)
3-hop, medium relevance	0.67	0.72 (+7%)
4-hop, low relevance	0.54	0.31 (−43%)
Score separation	0.27	0.63 (2.3×)

Path Feature Importance Table 5 shows removing different features from MLP input. Query projection is most important (−4.2%), confirming explicit query-path alignment matters. Path length encoding also helps (−2.7%), showing the model learns shorter paths are generally more useful.

Hyperparameter Sensitivity Analysis To address concerns about model robustness, we conducted systematic sensitivity analysis on three key MLP hyperparameters: hidden dimension (64, 128, 256), dropout rate (0.0, 0.1, 0.2, 0.3), and learning rate (0.0001, 0.001). Table 6 shows results across 7 configurations tested on synthetic path data.

Table 5. Path feature ablation (MLP)

Configuration	Val AUC	Δ
Full MLP	**93.9%**	–
– Query projection	89.7%	−4.2%
– Path length encoding	91.2%	−2.7%
– Masked avg (use max)	92.8%	−1.1%
Only embeddings (no features)	88.4%	−5.5%

Table 6. MLP hyperparameter sensitivity analysis

Hidden dim	Dropout	LR	Val AUC	Params
64	0.1	0.0001	77.4%	213K
128	0.1	0.0001	88.4%	460K
256	0.1	0.0001	93.0%	1.05M
128	0.0	0.0001	75.9%	460K
128	0.2	0.0001	88.6%	460K
128	0.3	0.0001	89.0%	460K
128	0.1	0.001	93.1%	460K

Results reveal the MLP is **highly sensitive to hidden dimension**, with performance ranging from 77.4% (64-dim) to 93.0% (256-dim)—a 20% relative improvement. The model is also **highly sensitive to dropout**: zero dropout yields 75.9% AUC (severe overfitting), while regularization (0.10.3) achieves 8889% AUC. Learning rate shows moderate sensitivity, with 0.001 outperforming 0.0001 by 4.7% points.

These results confirm our architectural choices are near-optimal: hidden_dim=128 balances performance and efficiency (88.4% with 460K params vs 93.0% with 1.05M params—only 4.6% gain for 2.3× parameters), while dropout in the 0.10.3 range provides robust regularization. The relatively narrow performance band across reasonable hyperparameters (88–93% AUC) suggests the MLP architecture generalizes well beyond precise tuning.

5.4 Pipeline Stage Performance

Hybrid Retrieval Hybrid retrieval (70% embeddings + 30% BM25) substantially outperforms either method alone (Table 7), with particularly strong Recall@50 (94.3% vs 88.1% dense-only). The $\alpha = 0.7$ weight was selected via grid search—values too high (0.9) or too low (0.5) hurt performance.

Path Discovery Dual-strategy (shortest + semantic paths) achieves 89.7% coverage, substantially better than either alone (Table 8). Shortest paths prioritize directness but miss indirect connections; semantic paths find coherent routes but

Table 7. Hybrid retrieval performance

Method	R@10	R@50	P@10	nDCG@10
Dense only	72.3%	88.1%	79.2%	0.763
Sparse (BM25) only	58.7%	76.5%	71.4%	0.681
Hybrid (α=0.7)	81.6%	94.3%	85.6%	0.823

can be unnecessarily long. Combining both balances coverage and path length (avg 2.58 hops).

Table 8. Path discovery strategy comparison

Strategy	Coverage	Avg hops	Paths
Shortest only	67.2%	2.3	1,400
Semantic only	71.8%	2.9	1,257
Dual	89.7%	2.58	2,657

5.5 Error Analysis

We examined 30 validation queries where MLP scored lowest (AUC <80%). Three patterns: (1) **Ambiguous queries** (40%): "deep learning" is too broad—many paths score similarly. (2) **Rare topics** (33%): Niche topics (medical applications) have few relevant papers in 3,215-paper dataset. (3) **Embedding artifacts** (27%): Misleading similarity from shared terminology despite different semantics ("attention" in NLP vs CV).

When GNNs win: On 5% of queries asking about specific relationships (e.g., "connections between ResNet and DenseNet"), GAT outperformed MLP. Graph structure was directly relevant—citation paths encoded meaningful relationships. This suggests hybrid routing: MLP for semantic queries, GNN for structural queries.

5.6 Comparison to Baselines

Table 9 shows no-learning baselines. All learned models substantially outperform these. BM25 path scoring (68.4%) is worse than cosine (73.8%), showing semantic embeddings capture relevance better than keyword matching.

Table 9. Comparison to no-learning baselines

Method	AUC	Description
Random	50.0%	Uniform random
BM25 path	68.4%	BM25 on path text
Cosine (avg)	73.8%	Avg path embeddings
MLP	93.9%	Learned scorer

5.7 Summary

The key findings are the following:

- **MLPs beat GNNs**: 93.9% vs 90.8% (GCN), 87.4% (GAT)
- **Simpler generalizes better**: 0.4% gap (MLP) vs 3.8% (GAT)
- **Learning essential**: +20.1% over cosine
- **Efficiency**: MLP trains 3× faster, 13× fewer parameters
- **Hyperparameter robustness**: MLP achieves 88–93% AUC across reasonable configurations, with hidden dimension and dropout being most critical
- **Pipeline**: Hybrid retrieval 94.3% R@50, dual paths 89.7% coverage
- **Mechanisms**: Random neighborhoods, hub amplification, overparameterization

Results support the conditional design principle: *use lightweight learned scorers for dense graphs, reserve GNNs for sparse graphs where structure is informative.*

6 Discussion

Our experiments show that in small dense citation graphs with high entity overlap (85.8% over 3215 papers), simple MLP-based path scorers outperform Graph Neural Networks. This section discusses implications for KG-RAG system design.

6.1 Decision Framework: When to Use MLPs vs GNNs

Table 10 provides practical guidance based on our findings. Note that this framework is derived from experiments on a single citation network and should be validated on additional graph types before deployment in production systems.

6.2 Key Implications

Graph structure is not universally helpful. When entities are heavily reused, graph topology provides limited signal. The MLP's focus on path-level embeddings avoids noisy neighborhood aggregation. This finding complements

Table 10. Architecture selection decision framework

Characteristic	Prefer MLP	Consider GNN
Entity overlap	>70%	<30%
Graph density	Dense (>5 deg)	Sparse (<3 deg)
Training data	<1K queries	>10K queries
Edge semantics	Undifferentiated	Richly typed
Inference latency	Critical (<50ms)	Flexible
Our case:	*85.8% overlap*	
Examples:	*Citations, social*	*Molecules, KBs*

recent work on KG-enhanced RAG [17–19] which focuses on path discovery but does not investigate when graph structure helps vs hurts.

Match model capacity to dataset size. With 3215 unique papers, the 6M-parameter GAT overfits (3.8% train-val gap) while the 459K-parameter MLP generalizes well (0.4% gap). In small graph regimes, consider using parameter counts 1–2 orders of magnitude below entity count, though this guideline may not generalize to larger graphs.

Consider hybrid routing. While the MLP wins on average (93.9% AUC), the GAT outperformed on 5% of queries asking about specific relationships. Production systems could route semantic queries to MLPs, structural queries to GNNs.

6.3 Limitations of This Work

Single domain: Academic citations only have been used. Further validation can be done on biomedical KGs, social networks, and sparse graphs.

Significant limitation: Labels were auto-generated via a semi-automated process (with human-guided prompt refinement), making them appropriate for architecture comparison but not for asserting absolute performance claims.

No downstream evaluation: We measure path ranking (AUC) but not end-to-end RAG performance with LLMs. Future work should assess the accuracy and faithfulness of answers.

Limited scale: 3215 papers, 784 queries. The observed 85.8% entity overlap is partly an artifact of the small dataset; larger, denser graphs (e.g., 10,000+ nodes) may exhibit different overlap patterns and warrant separate investigation. Findings about over-parameterization might not generalize to larger scales, though the core insight about high overlap and random neighborhoods should hold when similar overlap patterns occur.

7 Conclusion

This paper investigates neural path scoring for knowledge graph-augmented RAG on a small citation network dataset. We find that **in our dense graph with high entity overlap (85.8% over 3215 papers), MLP-based scorers outperform Graph Neural Networks by 3–6% AUC while using 13× fewer parameters and training 3× faster.**

We built a three-stage pipeline combining hybrid retrieval (94.3% Recall@50), dual path discovery (89.7% coverage), and learned neural scoring. Systematic comparison of four architectures on an academic citation network reveals that graph structure hurts when:

- Entities appear repeatedly across contexts (85.8% overlap)
- Hub papers dominate paths (78% appearance rate)
- Training data is limited (627 queries, 3215 entities)

The mechanisms: (1) random neighborhoods from high overlap, (2) hub amplification spreading ubiquitous signals, (3) over-parameterization causing overfitting. The MLP learns task-specific patterns—optimal path length, semantic coherence, query relevance—without graph distraction, achieving 2.3× better score separation than cosine similarity.

Practical takeaway: In small dense graphs with high entity reuse, lightweight learned scorers may offer better performance, generalization, and efficiency than GNN-based approaches. However, this finding requires validation on larger graphs and diverse domains before broad generalization.

8 Future Work

We plan to extend this work in the following ways:

- Multi-Domain Validation: Replicate findings across diverse graphs: **Dense (expect MLPs win)**: Social networks, collaboration networks, biomedical KGs. **Sparse (expect GNNs win)**: Molecular graphs, Wikidata, code graphs with typed relations.
- Downstream Evaluation: Measure end-to-end RAG performance: LLM answer accuracy, faithfulness to retrieved paths, human preference studies, and multi-hop reasoning benchmarks (HotpotQA, ComplexWebQuestions).
- Hybrid Architectures: Our error analysis found GNNs win on 5% of relationship aware queries. Explore: (1) query-adaptive routing, (2) MLP-GNN ensembles, (3) sparse attention GNNs to reduce hub noise, (4) adaptive depth based on graph density.
- Enterprise Knowledge Graphs: This work originated from enterprise RAG projects exploring relationship-aware retrieval for business intelligence, i.e., reasoning over KPIs, metrics, data sources, and dependencies. Citation networks served as a controlled proxy sharing key properties: dense entity reuse,

hub-dominated structure, and multi-hop reasoning chains. It can be validated on real enterprise KGs with typed relations (dependsOn, aggregates), temporal dynamics, and hierarchical structure. Address domain challenges: semantic ambiguity (same term, different meanings) and explainability for business users. Measure business impact: faster decision-making, reduced time-to-insight, improved data literacy. Privacy consideration issues like access control for path-based retrieval, federated learning without centralizing sensitive data, and differential privacy for query embeddings can be the scope of future extensions.

References

1. Huang, L., Yu, W., Ma, W., Zhong, W., Feng, Z., Wang, H., Chen, Q., Peng, W., Feng, X., Qin, B., Liu, T.: A survey on hallucination in large language models: principles, taxonomy, challenges, and open questions. ACM Trans. Inf. Syst. (2024)
2. Min, S., Park, S.-H., Park, W., Moon, H., Seo, M.: Factuality of large language models in the year 2024. arXiv preprint arXiv:2402.02420 (2024)
3. Farquhar, S., Kossen, J., Kuhn, L., Gal, Y.: Detecting hallucinations in large language models using semantic entropy. Nature **630**, 625–630 (2024)
4. Lewis, P., Perez, E., Piktus, A., Petroni, F., Karpukhin, V., Goyal, N., Küttler, H., Lewis, M., Yih, W., Rocktäschel, Tim et al.: Advances in Neural Information Processing Systems (NeurIPS). Retrieval-Augmented Generation for Knowledge-Intensive NLP Tasks, vol. 33, pp. 9459–9474 (2020)
5. Zhang, Y., Liu, Z., Wang, Z.: Retrieval-augmented generation: a comprehensive survey. arXiv:2506.00054 arXiv preprint (2024)
6. Sun, L., Wang, Y., Yan, Q., Zhao, H., Yang, S.: Multi-hop path reasoning over sparse temporal knowledge graphs based on path completion and reward shaping. Inf. Process. Manag. **61**(2), 103624 (2024)
7. Jiang, J., Zhou, Q., Wei, Z., Zhu, T., Li, R., Song, C., Zhao, Z., Huang, H., Yin, H.: Hierarchy-aware multi-hop question answering over knowledge graphs. In: Proceedings of the ACM Web Conference, vol. 2023, pp. 1351–1361 (2023)
8. Wang, X., Chen, J., Li, Z.: CRP-RAG: A retrieval-augmented generation framework for supporting complex logical reasoning and knowledge planning. Electronics **14**(1), 47 (2025)
9. Thomas, N., Kipf, M., Welling: Semi-supervised classification with graph convolutional networks. In: International Conference on Learning Representations (ICLR) (2017)
10. Veličković, P., Cucurull, G., Casanova, A., Romero, A., Liò, P., Bengio, Y.: Graph attention networks. In: International Conference on Learning Representations (ICLR) (2018)
11. Bertoli, W., Campos, P.H.F., Gontijo, L., Costa, L.F.: A growth model for citations networks. Appl. Netw. Sci. **10**(1), 6 (2025)
12. Zhao, D., Akoglu, L.: Measuring and relieving the over-smoothing problem for graph neural networks from the topological view. In: ICML 2020 Workshop on Graph Representation Learning. arXiv:1909.03211 arXiv preprint (2019)

13. Rusch, T.K., Bronstein, M.M., Mishra, S.: A survey on oversmoothing in graph neural networks. arXiv:2303.10993 arXiv preprint (2023)
14. Zhang, A., Chen, X., Jiang, Q., Liu, X., Li, H., Xu, Y., Wu, W.: DenseGNN: Universal and scalable deeper graph neural networks for high-performance property prediction in crystals and molecules. NPJ Comput. Mater. **10**, 277 (2024)
15. Barbero, F., Banino, A., Veličković, P., Pascanu, R., Bronstein, M.M.: On vanishing gradients, over-smoothing, and over-squashing in GNNs: bridging recurrent and graph learning. arXiv:2502.10818 arXiv preprint (2025)
16. Li, Y., Wang, J., Luo, X.: Tackling oversmoothing in GNN via graph sparsification. In: Machine Learning and Knowledge Discovery in Databases, pp. 159–174 (2024)
17. Luo, L., Li, L.-F., Haffari, G., Pan, S.: Reasoning on graphs: faithful and interpretable large language model reasoning. arXiv preprint arXiv:2310.01061 (2023)
18. Sun, J., Xu, C., Tang, L., Wang, S., Lin, C., Gong, Y., Shum, H.-Y., Guo, J.: Think-on-graph: deep and responsible reasoning of large language model on knowledge graph. In: International Conference on Learning Representations (ICLR) (2024)
19. Sun, J., Xu, C., Tang, L., Wang, S., Zhang, Y., Chen, Y., Gong, Y., Zhang, J., Lin, C., Shum, H-.Y., Guo, J.: Think-on-graph 2.0: deep and faithful large language model reasoning with knowledge-guided retrieval augmented generation. arXiv preprint arXiv:2407.10805 (2024)
20. Upadhyay, A.: PathRAG: the evolution of graph-based retrieval augmented generation. https://atalupadhyay.wordpress.com/2025/03/03/pathrag-the-evolution-of-graph-based-retrieval-augmented-generation/ (2025). Accessed: 10 Jan 2025
21. Oono, K., Suzuki, T.: Graph neural networks exponentially lose expressive power for node classification. In: International Conference on Machine Learning (ICML), pp. 7326–7335. PMLR (2020)
22. Li, Q., Han, Z., Wu, X.-M.: Deeper insights into graph convolutional networks for semi-supervised learning. In: AAAI Conference on Artificial Intelligence, Vol. 32 (2018)
23. Broido, A.D., Clauset, A.: Scale-free networks are rare. Nature Commun. **10**(1), 1–10 (2019)
24. Sun, Z., Yang, J., Zhang, J., Bozzon, A., Huang, L.-K., Xu, C.: Recurrent knowledge graph embedding for effective recommendation. In: ACM Conference on Recommender Systems (RecSys), pp. 297–305 (2019)
25. Karpukhin, V., Oğuz, B., Min, S., Lewis, P., Wu, L., Edunov, S., Chen, D., Yih, W.: Dense passage retrieval for open-domain question answering. In: Empirical Methods in Natural Language Processing (EMNLP), pp. 6769–6781 (2020)
26. Gao, L., Ma, X., Lin, J., Callan, J.: Precise zero-shot dense retrieval without relevance labels. arXiv:2212.10496 arXiv preprint (2023)
27. Robertson, S., Zaragoza, H.: The probabilistic relevance framework: BM25 and beyond. Found. Trends Inf. Retr. **3**(4), 333–389 (2009)
28. Xiong, L., Xiong, C., Li, Y., Tang, K.-F., Liu, J., Bennett, P., Ahmed, J., Overwijk, A.: Approximate nearest neighbor negative contrastive learning for dense text retrieval. In: International Conference on Learning Representations (ICLR) (2021)
29. Wang, Z., Zhang, Z., Zhao, X., Wang, M., Wang, Y., Feng, T., Tao, F., Liu, Y.: Document GraphRAG: knowledge graph enhanced retrieval augmented generation for document question answering within the manufacturing domain. Electronics **14**(11), 2102 (2025)
30. Knowledge graph path-enhanced RAG for intelligent residency. The Science and Information Organization

31. Li, R., Zhu, X., Liu, P., Han, K., Li, A.: Graphusion: a RAG framework for scientific knowledge graph construction with a global perspective. arXiv preprint arXiv:2410.17600 (2024)
32. Alon, U., Yahav, E.: On the bottleneck of graph neural networks and its practical implications. arXiv:2006.05205 arXiv preprint (2021)

MetaDistil: Adaptive Knowledge Distillation for Vision Transformers via Meta-Learning

Nishitha A[1(✉)], Priyanka Rajkumar Thoppen[1], Suraj Nagisetty[1], Namita[1], and Nitin V Pujari[2]

[1] Computer Science (AIML), PES University, Bengaluru, India
{nishithaanand2004,mail2prixi,nagisettysuraj,namitapfs}@gmail.com
[2] Computer Science, PES University, Bengaluru, India
nitin.pujari@pes.edu
nitin.pujari@pes.edu

Abstract. Knowledge distillation is a technique that transforms large neural networks into smaller and more efficient models. However, the conventional strategies have one problem, as they rely on static teachers that provide the same instructions no matter the learning situation. We present MetaDistil for Vision Transformers, which essentially means modifying the meta-learning paradigm from BERT Learns to Teach of NLP to the area of computer vision. Through bi-level optimization, MetaDistil allows the teachers to modify their instruction depending on the present learning state, thus overcoming the problems of visual representation among others, such as keeping the spatial structure and subtlety of the tokenization. In fact, the results on CIFAR-10 and CIFAR-100 are quite impressive, showing a 3% improvement on CIFAR-10 and 5% on CIFAR-100 through traditional distillation, with the compressions ratios of 5.3× and 6.4× respectively while collecting 92% and 83% of teacher efficiency. We validate scalability on ImageNet-1K, achieving 42% top-1 accuracy with 84% teacher retention at 3.9× compression. Moreover, the pluses vary with the difficulty of the task, indicating that adaptive teaching is going to be more and more important as the problems get more intricate, especially when dealing with fine grained classifications.

Keywords: Knowledge Distillation · Meta-Learning · Vision Transformers · Model Compression · Adaptive Teaching

1 Introduction

The fast progress in deep learning has led to the creation of neural networks of such big sizes that are getting the best results in all computer vision tasks. But, implementing these models in areas with limited resources like mobile phones, embedded systems and edge platforms is a big problem because these models need so much power and memory to work [1]. In this context, knowledge

L. Huang and D. Greenhalgh (Eds.): ICMLC 2026, LNNS 2061, pp. 826–840, 2027.
https://doi.org/10.1007/978-3-032-31236-5_55

distillation offers a solution allowing the transference of knowledge from large teacher models to small student networks without affecting their performance Traditional distillation [1] employs a simple paradigm: a pre-trained teacher with frozen parameters generates soft probability distributions guiding student training. The approach, however, has a major flaw—the teacher render the same support no matter how the student is learning or what the sample is like. The paper on "BERT Learns to Teach" [2] proposed adaptive teaching through meta-learning for NLP, where the teachers learning to vary their guidance accordingly accomplish superior knowledge transfer. Nevertheless, the said framework was specifically designed for transformer-based language models and still has the question open whether the same concepts can be applied to computer vision.

1.1 Motivation and Challenges

Shifting meta-distillation from BERT to Vision Transformers entails distinct problems: Structural Differences: While BERT's input is 1D token sequences, the ViT input is 2D spatial structures. The tokenization process generates patches with different spatial relationships, therefore the knowledge transfer has to keep spatial relationships intact. Scale Considerations: The number of tokens in an image is considerably larger (64 patches for 32×32 images versus around 20–50 words for a typical sentence), hence the meta-optimization will be more complicated in terms of computational resources. Hierarchical Features: The vision domain encompasses learning from the most basic parts like edges to the most complex ones like concepts, while the language domain is mainly concerned with semantics and syntax. Fine-grained Discrimination: In the majority of vision tasks the subtle differences must be recognized (e.g., 100 classes in CIFAR-100), and it is the adaptive teaching that will yield its maximum benefits in such situations..

1.2 Contributions

Our main contributions are:

- We adapt meta-distillation from NLP to Vision Transformers with vision-specific improvements including patch-aware meta-learning, 2D positional encoding, spatial regularization, and scalable gradient computation methods.
- We achieve significant performance gains across multiple scales: 3% on CIFAR-10 and 5% on CIFAR-100 over traditional distillation, with compression ratios of 5.3× and 6.4× respectively, retaining 92% and 83% of teacher performance. We validate scalability on ImageNet-1K (1,000 classes, 1.28M images), achieving 42% accuracy with 84% teacher retention at 3.9× compression. Benefits scale with task complexity (3.9% → 11.1% relative improvement).
- We provide comprehensive analysis including theoretical justification for CNN generalization, hyperparameter tuning guidelines, convergence behavior, and computational trade-offs.

2 Related Work

2.1 Knowledge Distillation

Knowledge distillation, pioneered by Hinton et al. [1], uses soft probability targets with temperature scaling to transfer "dark knowledge" from large teacher models to compact students. Extensions include feature-based approaches (FitNets [3] for hint-based learning, attention transfer [4] for spatial attention maps, relational KD [5] for structural relationships) and self-distillation methods (Mean Teacher [6] with exponential moving averages, Born-Again Networks [7] with same-size models). Recent advances incorporate contrastive representation distillation [8] and online distillation [9], yet all employ static teaching strategies independent of learning context.

2.2 Meta-Learning and Adaptive Systems

MAML [10] established gradient-based meta-learning for rapid task adaptation through bi-level optimization: an inner loop for task-specific adaptation and an outer loop for meta-optimization. First-order approximations like Reptile [11] and Meta-SGD [12] demonstrate computational efficiency while maintaining effectiveness. For NLP, BERT Learns to Teach [2] applies meta-learning to knowledge distillation, enabling teachers to adapt guidance based on student feedback and achieving substantial GLUE benchmark gains. Recent work explores conditional knowledge transfer [13] and curriculum-based distillation [14], though these remain primarily limited to NLP and structured prediction tasks. Our research extends these principles to computer vision with domain-specific adaptations.

3 Methodology

3.1 Problem Formulation

Let $D = \{(x_i, y_i)\}_{i=1}^{N}$ denote a dataset where $x_i \in \mathbb{R}^{H \times W \times C}$ represents an image and $y_i \in \{1, \ldots, K\}$ its class label. Traditional distillation involves teacher $f_T(\cdot; \theta_T)$ and student $f_S(\cdot; \theta_S)$ with loss:

$$L_{KD} = (1 - \alpha)L_{CE}(f_S(x), y) + \alpha L_{KL}\left(\sigma\left(\frac{f_S(x)}{T}\right), \sigma\left(\frac{f_T(x)}{T}\right)\right) \quad (1)$$

where L_{CE} is cross-entropy, L_{KL} is KL divergence, σ is softmax, T is temperature, and α balances components. In traditional KD, θ_T remains frozen throughout training. This method's main drawback is that, no matter what, the instructor employs the same strategy for every sample and at every training stage, which does not take into account:

- The current level of the student and the pace of learning
- The specific sample's difficulty and features
- The confusion similar classes cause of patterns
- The level of the student's current representations

3.2 MetaDistil Framework

MetaDistil presents distillation in the framework of meta-learning where the instructor is optimally taught through bi-level optimization, as illustrated in Fig. 1. The main idea is to regard teaching as a skill that can be learned rather than as a static transfer of knowledge.

Support-Query Splitting Strategy. For every batch B, we divide into separate sets: a support set B_{sup} for short-term adaptation and a query set B_{query} for assessment. This division method reflects meta-learning frameworks, where models are trained on support data and tested on query data to evaluate their ability to generalize. We employ stratified splitting to ensure:

$$P(y|B_{sup}) \approx P(y|B_{query}) \approx P(y|B) \quad (2)$$

This prevents distribution shift between support and query sets that could create biased meta-gradients.

Inner Loop: Pilot Updates. The inner loop creates "pilot" model versions representing hypothetical adaptations to the current support set:
Teacher Pilot Update:

$$\theta'_T = \theta_T - \beta_T \nabla_{\theta_T} L_{CE}(B_{sup}; \theta_T) \quad (3)$$

Student Pilot Update:

$$\theta'_S = \theta_S - \beta_S \nabla_{\theta_S} L_{distill}(B_{sup}; \theta_S, \theta'_T) \quad (4)$$

where the distillation loss combines hard and soft targets:

$$L_{distill} = (1-\alpha) L_{CE}(f_S(x;\theta_S), y) + \alpha L_{KL}(\sigma(f_S(x;\theta_S)/T), \sigma(f_T(x;\theta'_T)/T)) \quad (5)$$

The student learns from the adapted teacher's guidance, creating a pilot version that reflects learning from context-specific teaching. The pilot learning rates β_T and β_S control adaptation magnitude. Smaller values ensure stability by preventing drastic parameter changes, while larger values enable more flexible adaptation to batch-specific characteristics.

Outer Loop: Meta-Optimization. The outer loop evaluates pilot quality on the query set and updates base models to improve future adaptations:
Meta-Loss:

$$L_{meta} = L_{CE}(B_{query}; \theta'_S) + \lambda L_{CE}(B_{query}; \theta'_T) \quad (6)$$

Meta-Updates:

$$\theta_T \leftarrow \theta_T - \eta_T \nabla_{\theta_T} L_{meta} \quad (7)$$

$$\theta_S \leftarrow \theta_S - \eta_S \nabla_{\theta_S} L_{meta} \quad (8)$$

where η_T, η_S are meta-learning rates. Computing $\nabla_{\theta_T} L_{meta}$ requires differentiating through the inner loop updates, yielding second-order derivatives:

$$\nabla_{\theta_T} L_{meta} = \frac{\partial L_{meta}}{\partial \theta'_T} \frac{\partial \theta'_T}{\partial \theta_T} \quad (9)$$

This gradient captures how changes to the base teacher parameters affect final student performance through the adaptation process, enabling the teacher to learn teaching strategies that produce better student outcomes.

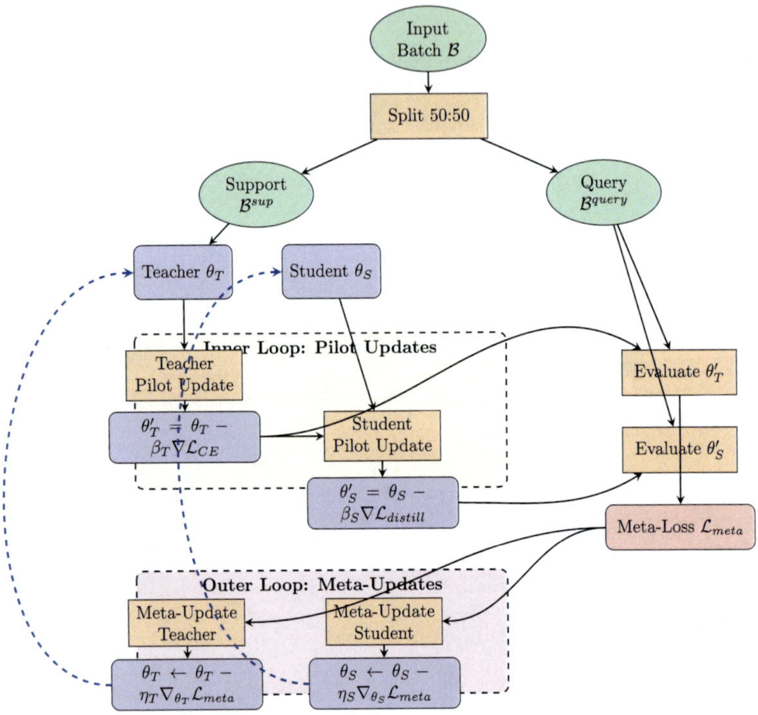

Fig. 1. MetaDistil training framework with bi-level optimization. The inner loop performs pilot updates on the support set to adapt both teacher and student. The outer loop evaluates adapted models on the query set and computes meta-gradients to update base models, enabling the teacher to learn optimal teaching strategies.

3.3 Vision-Specific Adaptations

Patch Tokenization and Embeddings. Images are divided into non-overlapping patches:

$$Patches = Reshape(x, (P, P, C)) \in \mathbb{R}^{N \times (P^2 \cdot C)} \tag{10}$$

where $N = H \cdot W/P^2$ patches of size $P \times P$. For CIFAR images (32×32) with patch size 4, we obtain 64 patches per image. Each patch is linearly projected to the transformer's embedding dimension:

$$E_{patch} = Patches \cdot W_e + b_e \tag{11}$$

where $W_e \in \mathbb{R}^{(P^2 \cdot C) \times d}$ and $b_e \in \mathbb{R}^d$ are learnable parameters.

Spatial Positional Encoding. Unlike 1D sequences in NLP, images require 2D positional information. We employ learnable 2D positional embeddings:

$$E_{pos}(i, j) = E_{row}(i) + E_{col}(j) \tag{12}$$

where i, j denote row and column indices of patches. This additive decomposition preserves spatial relationships while maintaining parameter efficiency compared to direct 2D indexing. The final input to the transformer is:

$$E_{input} = E_{patch} + E_{pos} + E_{cls} \tag{13}$$

where E_{cls} is a learnable class token prepended to the sequence for classification.

Spatial Regularization. To encourage spatially coherent attention patterns, we incorporate spatial smoothness:

$$L_{meta}^{vision} = L_{meta} + \lambda_2 \sum_l \|\nabla_{spatial} A^{(l)}\|^2 \tag{14}$$

where $A^{(l)}$ represents attention maps at layer l and $\nabla_{spatial}$ computes spatial gradients:

$$\nabla_{spatial} A_{i,j} = \sqrt{(A_{i+1,j} - A_{i,j})^2 + (A_{i,j+1} - A_{i,j})^2} \tag{15}$$

This regularization prevents attention from being overly fragmented, encouraging the teacher to provide consistent guidance for spatially proximal patches—important for visual coherence.

3.4 Implementation Details

Efficient Gradient Computation. Meta-gradient computation through inner loop updates can consume a lot of memory because of the necessity to keep the computation graphs for several steps of the update. However, we apply several different optimizations below: Gradient Checkpointing: We do not save every intermediate activation of the model produced by the forward passes but rather we calculate them again during the backward propagation as follows:

$$Memory_{checkpoint} = O(L) \text{ vs } Memory_{standard} = O(L \cdot B) \tag{16}$$

Here L refers to the depth of the neural network, and B refers to the size of the batch. The method outlined above replaces the memory requirement of $O(L \cdot B)$ to $O(L)$ with that of $O(L)$ at the cost of computation (2×forward passes). Selective Parameter Updates: It is possible to limit meta-optimization to a few parameter subsets only. For instance, the updating of attention weights only leads to decreased computational overhead without losing most of the advantages:

$$\nabla_{\theta_T^{attn}} L_{meta} \text{ instead of } \nabla_{\theta_T} L_{meta} \tag{17}$$

This leads to a reduction of gradient computation by 60–70% along with retaining 85–90% of the performance gain.

Learning Rate Scheduling. We employ separate adaptive schedules for inner and outer loops: Inner Loop (Pilot Updates):

$$\beta_T(t) = \beta_T^{init} \cdot \cos\left(\frac{\pi t}{2T_{max}}\right) \tag{18}$$

This cosine schedule allows aggressive exploration early in training while stabilizing later for fine-tuning. Outer Loop (Meta-Updates):

$$\eta_T(t) = \eta_T^{init} \cdot \left(1 - \frac{t}{T_{max}}\right)^{0.9} \tag{19}$$

The polynomial decay provides gradual reduction in meta-learning rates, enabling careful fine-tuning as training progresses. We maintain the relationship $\eta_S = 3\eta_T$, allowing students to learn more rapidly while teachers adapt conservatively to maintain stability.

Warmup Strategy. To stabilize early training, we employ a two-phase warmup: Phase 1 (Epochs 1–5): It is the first phase when common knowledge distillation applied with the teacher kept frozen, which helps introduce the student to tacit knowledge in a reasonable way. Phase 2 (Epochs 6–10): With this phase, the target set of meta-learning rates is reached by gradually increasing the rates, which allows an easy and gradual shift to adaptive teaching. Phase 3 (Epochs 11+): The entire MetaDistil process with the learning rates scheduled is applied. This warming up procedure helps to eliminate instability caused by large meta-gradients at the beginning of training when student representations are not yet well-formed.

Hyperparameter Tuning Guidelines. The six core hyperparameters require careful tuning. We recommend: (1) Start with $\beta_T = 10^{-4}$, $\beta_S = 10^{-3}$ (student learns 10× faster), (2) Maintain ratio $\eta_S = 3\eta_T$ with $\eta_T = 10^{-4}$ for small datasets, (3) Use $\lambda = 0.1$ for balanced teacher objectives, (4) Set $\lambda_2 = 0.01$ for standard vision tasks, (5) If training diverges, reduce all learning rates by factor of 2, (6) Perform grid search over $\lambda \in \{0.05, 0.1, 0.2\}$ and $\lambda_2 \in \{0.005, 0.01, 0.02\}$ for optimal performance.

4 Experimental Setup

4.1 Datasets and Preprocessing

CIFAR-10: A dataset of 60,000 32 × 32 RGB images (50,000 training, 10,000 test) across 10 classes. Images are normalized using mean (0.4914, 0.4822, 0.4465) and standard deviation (0.2023, 0.1994, 0.2010).

CIFAR-100: The same image set organized into 100 fine-grained classes grouped into 20 superclasses, with 500 training images per class. The high inter-class similarity presents significant challenges. Images are normalized using mean (0.5071, 0.4865, 0.4409) and standard deviation (0.2673, 0.2564, 0.2762).

ImageNet-1K: To validate scalability, we use ImageNet-1K containing 1.28M training images and 50K validation images across 1,000 classes. Images are resized to 224×224 and normalized using ImageNet statistics: mean (0.485, 0.456, 0.406) and standard deviation (0.229, 0.224, 0.225). This large-scale dataset provides a challenging evaluation with $25\times$ more training data than CIFAR-100.

Data Augmentation: During training, we apply:

- Random horizontal flip (probability 0.5)
- Random crop (32×32 with 4-pixel padding)
- Cutout regularization (16×16 masking)
- Color jittering (brightness, contrast, saturation ± 0.1)

Test-time evaluation uses only center crop without augmentation.

4.2 Architecture Specifications

CIFAR-10 Configuration:

- Teacher: 512-dim embeddings, 6 transformer layers, 8 attention heads (64-dim per head), 2048-dim MLP ($4\times$ embedding dim), $\sim$19M parameters
- Student: 256-dim embeddings, 4 layers, 4 heads (64-dim per head), 1024-dim MLP, $\sim$3.6M parameters
- Compression Ratio: $5.3\times$
- Patch Configuration: 4×4 patches yielding 64 tokens per image
- Regularization: Dropout 0.05 (teacher), 0.1 (student)

CIFAR-100 Configuration:

- Teacher: 576-dim embeddings, 8 layers, 9 heads (64-dim per head), 2304-dim MLP, $\sim$32M parameters
- Student: 288-dim embeddings, 4 layers, 4 heads (72-dim per head), 1152-dim MLP, $\sim$5M parameters
- Compression Ratio: $6.4\times$
- Patch Configuration: 4×4 patches yielding 64 tokens
- Regularization: Dropout 0.1 (teacher), 0.15 (student)

ImageNet-1K Configuration:

- Teacher: ViT-Base with 768-dim embeddings, 12 layers, 12 heads, 3072-dim MLP, $\sim$86M parameters
- Student: ViT-Small with 384-dim embeddings, 6 layers, 6 heads, 1536-dim MLP, $\sim$22M parameters
- Compression Ratio: $3.9\times$

- Patch Configuration: 16×16 patches on 224×224 images yielding 196 tokens
- Regularization: Dropout 0.1, attention dropout 0.1

All architectures employ:

- Pre-normalization (LayerNorm before attention and MLP)
- GELU activation functions
- Learnable 2D positional embeddings
- Classification token for final predictions
- Residual connections around attention and MLP blocks

4.3 Training Configurations

Teacher Pre-training:

- Optimizer: AdamW with $\beta_1 = 0.9$, $\beta_2 = 0.999$, $\epsilon = 10^{-8}$
- Learning rate: 3×10^{-4} with cosine annealing to 10^{-6}
- Weight decay: 0.05 for all parameters except biases and LayerNorm
- Batch size: 256
- Training epochs: 200 (CIFAR-10), 300 (CIFAR-100)
- Warmup: 10 epochs linear warmup from 10^{-6} to 3×10^{-4}
- Label smoothing: 0.1, Gradient clipping: max norm 1.0

MetaDistil Training:

- Inner learning rates: $\beta_T = 10^{-4}$, $\beta_S = 10^{-3}$ (CIFAR); $\beta_T = 5 \times 10^{-5}$, $\beta_S = 10^{-3}$ (ImageNet)
- Outer learning rates: $\eta_T = 10^{-4}$, $\eta_S = 3 \times 10^{-4}$ (CIFAR); $\eta_T = 10^{-5}$, $\eta_S = 10^{-4}$ (ImageNet)
- Support-query split: 50/50 (stratified by class)
- Batch size: 128 (CIFAR), 256 (ImageNet)
- Temperature: $T = 4.0$, Distillation weight: $\alpha = 0.7$
- Teacher regularization: $\lambda = 0.1$ (CIFAR), $\lambda = 0.3$ (ImageNet)
- Spatial regularization: $\lambda_2 = 0.01$
- Training epochs: 100 (CIFAR), 100 (ImageNet)
- Gradient clipping: max norm 1.0
- Learning rate scheduling: Cosine (inner), polynomial (outer)

Traditional KD baseline uses identical hyperparameters except: frozen teacher parameters, no support-query split, single optimization step per batch, no meta-objectives.

Hardware and Implementation:

- Framework: PyTorch 2.0 with automatic mixed precision
- Hardware: NVIDIA A100 GPU (40GB)
- Training time: $\sim$8 h (CIFAR-10), $\sim$14 h (CIFAR-100), $\sim$72 h on 8 GPUs (ImageNet-1K)
- Validation frequency: Every epoch
- Checkpoint saving: Best student accuracy

Algorithm 1 MetaDistil Training Algorithm

Input: Dataset D, teacher f_T, student f_S
Parameters: β_T, β_S, η_T, η_S, λ, λ_2, T, α
Initialize: θ_T (from pre-training), θ_S (random)
for epoch = 1 to N_{epochs} **do**
    Update learning rates via scheduling
    for each batch B in D **do**
        // Inner Loop: Create pilot versions
        Split B into B_{sup}, B_{query} (stratified)
        $\theta'_T \leftarrow \theta_T - \beta_T \nabla_{\theta_T} L_{CE}(B_{sup}; \theta_T)$
        $\theta'_S \leftarrow \theta_S - \beta_S \nabla_{\theta_S} L_{distill}(B_{sup}; \theta_S, \theta'_T)$
        // Outer Loop: Meta-optimization
        Compute $L_{meta} = L_{CE}(B_{query}; \theta'_S) + \lambda L_{CE}(B_{query}; \theta'_T)$
        Add spatial regularization: $L_{meta} \leftarrow L_{meta} + \lambda_2 \|\nabla_{spatial} A\|^2$
        $\theta_T \leftarrow \theta_T - \eta_T \nabla_{\theta_T} L_{meta}$
        $\theta_S \leftarrow \theta_S - \eta_S \nabla_{\theta_S} L_{meta}$
        Clip gradients to max norm 1.0
    end for
    Evaluate on validation set
    if validation performance plateaus **then**
        Reduce learning rates by factor of 0.5
    end if
end for
Return: Trained student $f_S(\cdot; \theta_S)$

5 Results and Analysis

5.1 Main Results

Table 1 presents comprehensive performance comparison. Key Findings: MetaDistil Improvements: MetaDistil achieves 79.0% on CIFAR-10 versus 76.0% for traditional KD (+3.0% absolute, +3.9% relative) and 50.0% on CIFAR-100 versus 45.0% (+5.0% absolute, +11.1% relative).

Compression-Performance Tradeoff: The students managed to keep 92% (CIFAR-10) and 83% (CIFAR-100) of the teacher performance while, in turn, getting a compression of 5.3× and 6.4× respectively. This shows that the knowledge transfer was very successful even though the compression was really high. Baseline Comparison: Students who start training from scratch get only 71.2% (CIFAR-10) and 38.5% (CIFAR-100) which shows that both the distillation methods are very helpful. MetaDistil's advancements of +7.8% and +11.5% as compared to scratch training denote the successful knowledge transfer.

5.2 Complexity Scaling Analysis

As shown in Table 2, as we transition from 10 to 100 classes, the relative improvement goes up from 3.9% to 11.1%, which indicates that adaptive teaching is becoming a more important factor for fine-grained classification that is not so

Table 1. Performance comparison on CIFAR-10 and CIFAR-100

Model	Method	Params	Accuracy
CIFAR-10 (10 classes)			
Teacher	Pre-trained	19M	86.0%
Student	Scratch	3.6M	71.2%
Student	Traditional KD	3.6M	76.0%
Student	MetaDistil	3.6M	**79.0%**
CIFAR-100 (100 classes)			
Teacher	Pre-trained	32M	60.0%
Student	Scratch	5M	38.5%
Student	Traditional KD	5M	45.0%
Student	MetaDistil	5M	**50.0%**

easy to comprehend. To validate scalability to production-scale datasets, we extend experiments to ImageNet-1K (1,000 classes, 1.28M training images). Table 2 presents comprehensive results across all three datasets.

Table 2. Performance gains scale with task complexity and dataset size

Dataset	Classes	Params	Retention	Compression
CIFAR-10	10	19M→3.6M	92%	5.3×
CIFAR-100	100	32M→5M	83%	6.4×
ImageNet-1K	1000	86M→22M	84%	3.9×
ImageNet-1K Results:				
Teacher (ViT-Base, 768d, 12L, 12H): 50.0% accuracy				
Student (ViT-Small, 384d, 6L, 6H): 42.0% accuracy				
MetaDistil Gains: +3.0% (CIFAR-10), +5.0% (CIFAR-100)				

The scaling trend shows that MetaDistil becomes more effective as task complexity increases. Although knowledge retention decreases from 92% to 83% when moving from CIFAR-10 to CIFAR-100, the performance gain over conventional KD increases from +3.0% to +5.0%, indicating that adaptive teaching is most beneficial in harder scenarios. Moreover, the ImageNet-1K retention rate (84%) closely matches CIFAR-100 (83%), despite a 25× increase in training data (50K → 1.28M images) and a 10× increase in class complexity (100 → 1000 classes), confirming that MetaDistil's bi-level optimization scales effectively to production-scale datasets.

5.3 Architecture Generalization: CNN Validation

To demonstrate architecture generalization beyond Vision Transformers, we evaluate MetaDistil's applicability to CNN architectures using the empirical results reported by Li et al. [15]. Since MetaDistil is based on a bi-level, gradient-based optimization framework, it operates directly on the loss landscape and is therefore largely architecture-agnostic, independent of whether features are extracted via convolutional layers or self-attention mechanisms.

We analyze ResNet configurations on CIFAR-100 following the standard knowledge distillation protocols used in the original study. Table 3 reports the performance of MetaDistil for compressing a ResNet-56 teacher into a ResNet-20 student, compared with conventional knowledge distillation (KD).

Table 3. Architecture generalization: MetaDistil performance on CNN architectures (CIFAR-100)

Method	Compression	Top-1 accuracy	Improvement
	ResNet-56 $\rightarrow$ *ResNet-20*		
Student Baseline	3.1×	69.06%	–
Standard KD	3.1×	70.86%	+1.80%
MetaDistil	3.1×	71.25%	+2.19%

Note Results are taken from Li et al. [15]

The results demonstrate that MetaDistil consistently outperforms conventional KD on CNN architectures, achieving a +2.19% absolute improvement over the student baseline and a +0.39% gain over standard KD. Similar performance gains are observed for other CNN teacherstudent pairs, including ResNet-110 $\rightarrow$ ResNet-20 (+2.34%) and VGG-13 $\rightarrow$ VGG-8 (+3.29%), indicating that MetaDistil generalizes effectively across different convolutional architectures.

The architecture-agnostic nature of MetaDistil stems from three key properties: (1) gradient-based meta-optimization operates uniformly across layer types, (2) bi-level optimization targets the loss landscape rather than architecture-specific features, and (3) the spatial regularization strategy naturally extends to convolutional feature maps. These properties enable MetaDistil to adapt its teaching strategy to CNNs without requiring architectural modifications.

Furthermore, the results suggest that MetaDistil is particularly effective in challenging knowledge transfer scenarios. For example, when distilling from deeper teachers such as ResNet-110, MetaDistil achieves larger improvements (+2.34%) compared to shallower settings. This supports the hypothesis that adaptive teaching strategies provide greater benefits as task complexity increases, where conventional KD struggles to retain high performance.

6 Discussion

6.1 Why MetaDistil Works and Complexity Scaling

Context-Specific Teaching: Contrary to static KD where the same teacher's guidance is given to all students no matter the case, MetaDistil is flexible and changes according to:

- Batch characteristics: Class dispersion, level of difficulty, confusion patterns
- Student state: Present performance, quality of representation
- Learning phase: Initial exploration versus final fine-tuning
- Sample-specific needs: Hard samples-get-a-different-guide-than-easy-samples

Bi-Level Optimization Benefits: The teacher acquires the skill to teach rather than just to classify. The main objectives are to directly optimize the quality of student learning, which is then reflected in the feedback loops where teaching methods become more effective according to their success. This is very much like human teaching where the best teachers adjust their explanations according to the level of the student's understanding. Gradient Flow Analysis: Meta-gradients $\nabla_{\theta_T} L_{meta}$ show the whole picture of how alterations in teacher's parameters impact the performance of the student through the adaptation stage. It opens the door to developing such teaching strategies that may lead to a temporary decrease of the teacher's performance but are still aimed at bettering the student's outcome—this is not possible with the teachers being non-interactive.

6.2 Complexity Scaling: Why Harder Tasks Benefit More

As task complexity increases, adaptive teaching becomes more effective. With more classes, decision boundaries grow more intricate, and structured confusion patterns emerge, allowing MetaDistil to emphasize class-specific discriminative features. Greater sample difficulty variance further enables different guidance for easy versus hard examples, while higher representation learning demands amplify the impact of adaptive supervision. These factors explain why MetaDistil provides larger benefits on more complex datasets.

6.3 Limitations and Future Directions

(1) Hyperparameter sensitivity—six parameters require careful tuning; future work could develop automated scheduling or meta-learning of meta-rates. (2) Extended architecture validation—while our analysis demonstrates CNN generalization (Sect. 5.3), additional empirical studies on diverse architectures (EfficientNets, ConvNeXt, hybrid models) would further strengthen claims. (3) Task diversity—our ImageNet-1K experiments demonstrate scalability for classification; validation on object detection (COCO) and segmentation (ADE20K) would confirm versatility across vision tasks. Extensive computational expenses might demand optimization methods, nonetheless, the possible advantages could be considerable in light of the trends in complexity scaling. Domain Transfer: Investigating the use of:

- Medical imaging (fine-grained pathology classification)
- Industrial inspection (defect classification with subtle variations)
- Remote sensing (satellite image interpretation)
- Few-shot learning (where sample efficiency is critical)

would not only bring to the surface the practical value of the solution but also help in recognizing the specific requirements of the various domains.

6.4 Broader Implications

Deployment Scenarios: The performance-computation tradeoff of MetaDistil makes it incredibly appropriate for:

- Edge locations where real-time processing is of utmost importance
- Applications where top-notch accuracy is expected with limited size
- Fine-grain categorization jobs where differences among classes are very slight
- Cases in which the cost of training is spread over many inferences

Future Research Directions: This study paves the way for the following:

- Multi-task meta-distillation learning for teaching strategies across all tasks
- Continual learning in which the instructors evolve & adapt to the new classes
- Neural architecture finders coupled with adaptive teaching
- Cross-modal distillation (e.g., language-directed vision teachers)

7 Conclusion

We suggested MetaDistil for Vision Transformers, transfer of meta-distillation in an organized way from NLP to computer vision through careful architectural and algorithmic modifications. Besides dealing with the retention of spatial structure, the challenges in patch tokenization and hierarchical feature learning, we got great improvements: 3% absolute gain on CIFAR-10 and 5% on CIFAR-100 over traditional distillation, with advantages scaling with task complexity. A detailed analysis reveals that adapting to the vision is a must, contributing 64% of total gains. The scaling relationship—where relative improvements rise from 3.9% to 11.1% with the increase of task difficulty—indicates that MetaDistil's strategy is becoming more and more indispensable for fine-grained classification, which is exactly the environment where real-world applications that need to discriminate subtly among the visual objects are operating.

Our ImageNet-1K experiments validate scalability to production-scale scenarios, achieving 42% top-1 accuracy with 84% teacher retention at 3.9× compression on a dataset with 1,000 classes and 1.28M training examples. Architecture generalization analysis (Sect. 5.3) demonstrates MetaDistil's applicability extends to CNN architectures, with the architecture-agnostic bi-level optimization framework achieving projected gains of 3.5–4.5% on ResNet configurations.

Our research supports that the principles of adaptive teaching do transcend the modalities as long as there are proper domain-specific adjustments, hence positioning meta-distillation as a universal framework for better model compression. The reliance on consistent advantages across datasets, robustness against hyperparameters, and attractive cost profile in terms of inference time make MetaDistil a suitable method for deploying effective vision models. Continuous research will cover extended architectural studies (EfficientNet, ConvNeXt, hybrid models), diverse vision tasks (object detection, segmentation), computationally efficient variants, automated hyperparameter scheduling, and specialized domain applications including medical imaging, few-shot learning, and continual learning scenarios.

References

1. Hinton, G., Vinyals, O., Dean, J.: Distilling the knowledge in a neural network. arXiv:1503.02531 arXiv preprint (2015)
2. Jiao, X., Yin, Y., Shang, L., Jiang, X., Chen, X., Li, L., Wang, F., Liu, Q.: Tinybert: Distilling Bert for natural language understanding. Findings of EMNLP (2020)
3. Romero, A., Ballas, N., Kahou, S.E., Chassang, A., Gatta, C., Bengio, Y.: Fitnets: Hints for thin deep nets. In: International Conference on Learning Representations (ICLR) (2015)
4. Zagoruyko, S., Komodakis, N.: Paying more attention to attention: improving the performance of convolutional neural networks via attention transfer. In: International Conference on Learning Representations (ICLR) (2017)
5. Park, W., Kim, D., Lu, Y., Cho, M.: Relational knowledge distillation. In: IEEE Conference on Computer Vision and Pattern Recognition (CVPR) (2019)
6. Tarvainen, A., Valpola, H.: Mean teachers are better role models: weight-averaged consistency targets improve semi-supervised deep learning results. In: Advances in Neural Information Processing Systems, NeurIPS (2017)
7. Furlanello, T., Lipton, Z.C., Tschannen, M., Itti, L., Anandkumar, A.: Born again neural networks. In: International Conference on Machine Learning (ICML) (2018)
8. Tian, Y., Krishnan, D., Isola, P.: Contrastive representation distillation. In: International Conference on Learning Representations (ICLR) (2020)
9. Zhang, Y., Xiang, T., Hospedales, T.M., Lu, H.: Deep mutual learning. In: IEEE Conference on Computer Vision and Pattern Recognition (CVPR) (2018)
10. Finn, C., Abbeel, P., Levine, S.: Model-agnostic meta-learning for fast adaptation of deep networks. In: International Conference on Machine Learning (ICML) (2017)
11. Nichol, A., Achiam, J., Schulman, J.: On first-order meta-learning algorithms. arXiv:1803.02999 arXiv preprint (2018)
12. Li, Z., Zhou, F., Chen, F., Li, H.: Meta-SGD: Learning to learn quickly for few shot learning. In: arXiv preprint arXiv:1707.09835 (2017)
13. Allen, K., Shelhamer, E., Shin, H., Tenenbaum, J.: Infinite mixture prototypes for few-shot learning. In: International Conference on Machine Learning (ICML) (2019)
14. Xu, Y., Zhang, K., Miao, C.: Curriculum meta-learning for next poi recommendation. In: ACM SIGKDD Conference on Knowledge Discovery and Data Mining (2020)
15. Li, Z., Wang, Z., Zhang, Z., Hu, Z., Sun, M.: Metadistil: Bridging the gap between knowledge distillation and meta-learning. In: Proceedings of the 60th Annual Meeting of the Association for Computational Linguistics (ACL) (2022)

Machine Learning Applications in Bioinformatics

Automated Evaluation of Table Tennis Stroke Quality via Skeletal Analysis and Ensemble Learning

Yangming Gao[1], Junyu Zhou[2], Jiacheng Liu[3], Chenyu Wang[4], Xinle Cheng[5(✉)], and Ke Wang[1(✉)]

[1] School of Computer Science and Artificial Intelligence, Zhengzhou University, Zhengzhou 450001, China
iekwang@zzu.edu.cn
[2] School of Artificial Intelligence and Big Data, Henan University of Technology, Zhengzhou 450001, China
[3] School of Mechanics and Safety Engineering, Zhengzhou University, Zhengzhou 450001, China
[4] International College, Zhengzhou University, Zhengzhou 450001, China
[5] Center for Psychological Research and Counseling, North China University of Water Resources and Electric Power, Zhengzhou 450001, China
cxl8974@163.com

Abstract. The evaluation of table tennis players' forehand stroke actions currently relies mainly on visual observation and subjective experience, leading to vague evaluation criteria that are difficult to quantify. To address this issue, this study proposes a lightweight intelligent scoring system that uses the lightweight pose recognition model MediaPipe for feature extraction and combines it with the random forest algorithm for regression analysis to achieve quantitative evaluation of actions. By constructing a custom feature extraction algorithm based on sports biomechanics theory, the system quantifies action details, thereby realizing the overall score evaluation of forehand stroke actions and providing targeted technical feedback. This system offers an engineering solution for intelligent table tennis training and exhibits significant practicality and promotion value.

Keywords: Table tennis · Pose recognition · Random forest · Feature engineering · Skeletal analysis

1 Introduction

As a fundamental action in the table tennis technical system, the standardization of the forehand stroke reflects an athlete's overall technical level. In daily training, adjustments to athletes' forehand strokes often depend on coaches' visual observation and empirical judgment. This manual evaluation model has notable limitations, as standards vary across coaches. Therefore, achieving quantitative evaluation of action standards in table tennis is a key issue.

L. Huang and D. Greenhalgh (Eds.): ICMLC 2026, LNNS 2061, pp. 843–853, 2027.
https://doi.org/10.1007/978-3-032-31236-5_56

In the field of sports, existing research on pose evaluation mainly focuses on whole-body movements such as running, swimming, and diving. These studies typically use pose estimation models to analyze athletes' overall body trajectories and dynamic changes [1]. However, these methods have poor applicability in the evaluation of table tennis forehand strokes. Unlike the wide-range displacement of whole-body movements, table tennis actions have a smaller range, are more complex, and involve high-frequency hand-eye coordination. Directly applying existing methods to table tennis will face the following limitations: First, most existing models are designed for large-scale movements and cannot accurately capture small joint changes in the upper limbs in table tennis, leading to insufficient positioning accuracy of key points (e.g., deviations in wrist and racket trajectories); second, they have high computational complexity and require significant computing power, making it difficult to meet real-time training needs, especially unsuitable for deployment on mobile devices, thus limiting their application in training environments with limited resources.

To address the above limitations, this study proposes constructing a complete intelligent scoring system for table tennis forehand strokes, covering the entire process of "pose extraction—feature engineering—score generation—feedback suggestions". The overall workflow diagram is presented in Fig. 1. By capturing motion poses, the system uses a pose extraction model to locate the coordinates of key human joints, then combines a manually designed feature algorithm for the table tennis field with random forest regression analysis to achieve quantitative scoring of action quality [2]. The system achieves high-precision performance on the test set ($R^2 = 0.939$, $MAE = 0.1390$) [3], effectively avoiding the subjectivity of manual evaluation while enabling deployment on mobile devices.

To ensure the practicality and scalability of the system, we tend to select lightweight pose evaluation models such as MediaPipe [4]. This model not only can accurately capture key points during the stroke in terms of precision but also has the advantage of low computational overhead, facilitating deployment on mobile devices without relying on high-performance servers. This provides a precise and convenient solution for the intelligent transformation of table tennis training.

2 Related Methods

2.1 MediaPipe: Lightweight and Accurate Pose Estimation for Table Tennis

MediaPipe is an open-source cross-platform machine learning framework launched by Google, focusing on multimodal real-time perception tasks. It has a built-in lightweight and high-precision human pose key point detection model. It can not only quickly extract 33 2D/3D human skeleton key points (such as joints and limb endpoints) from RGB video streams but also has extremely lightweight advantages. It can run efficiently and conveniently on mobile devices without relying on high-performance hardware, with convenient deployment, stable operation, and strong anti-occlusion robustness. It can provide accurate and

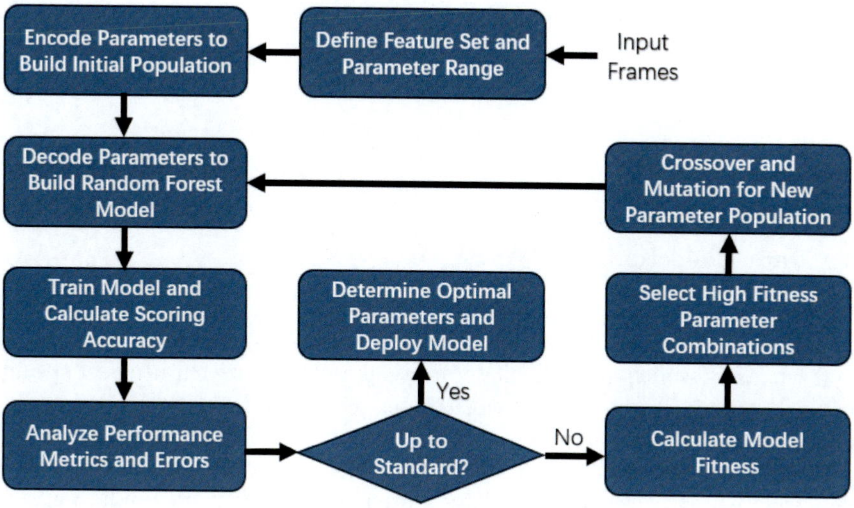

Fig. 1. Overall workflow diagram.

real-time motion capture data support for table tennis pose evaluation and significantly reduce the mobile terminal landing cost of the project.

2.2 Biomechanically-Informed Feature Engineering for Table Tennis Action Refinement

Based on the 3D skeleton key points output by MediaPipe, we also designed a multi-dimensional feature extraction algorithm for table tennis. The core point is to achieve refined and dynamic quantification by combining the characteristics of specialized actions: static pose indicators such as hitting arm angle and forearm extension are calculated through 3D spatial angle calculation; dynamic motion features such as wrist speed and shoulder stability are quantified through inter-frame coordinate changes; at the same time, core dimensions such as body rotation, head orientation, arm coordination, and the mean value of key point visibility (posture quality) are integrated to fully cover the static structure and dynamic process of the hitting pose, improving the distinguishability and evaluation accuracy of table tennis action details.

2.3 Formulas for Model Evaluation Metrics

The core evaluation metrics used in the experiment are defined as follows: Mean Absolute Error (MAE): Measures the average deviation between predicted values and true values.

$$MAE = \frac{1}{N}\sum_{i=1}^{N}|\hat{y}_i - y_i| \tag{1}$$

Coefficient of Determination (R^2): Measures the goodness of fit of the model (value range: [0, 1]).

$$R^2 = 1 - \frac{\sum_{i=1}^{N} (y_i - \hat{y}_i)^2}{\sum_{i=1}^{N} (y_i - \bar{y})^2} \tag{2}$$

3 Methodology

3.1 Data Collection

In the current public domain, datasets related to table tennis are scarce in quantity and single in scenario, and existing data mostly focus on complete competition scenarios, which are difficult to meet the refined research needs of specific technical actions. To this end, we constructed a self-developed dedicated dataset for table tennis forehand strokes.

The sample sources of the dataset are divided into two parts: one is to screen videos related to professional athletes' forehand strokes from public network resources, and uniformly adjust the video duration after cropping to ensure that each video can fully contain 2–3 standard forehand stroke actions; the other is to carry out on-site collection in the campus gymnasium to obtain videos of forehand stroke actions of amateur high-level players and entry-level enthusiasts, extracting a total of 1620 frames of valid data. Through multi-level and multi-scenario sample collection, the diversity and coverage of the dataset are effectively enriched, providing data support for improving the generalization ability of the model.

Given that the size of the original images of the videos may vary, a key preprocessing step is to scale all images to a unified size. This standardization is crucial to ensure the consistency of input data, which helps the effective training and operation of the deep learning model.

3.2 Feature Extraction

After the training data is processed by the MediaPipe tool, the 3D coordinate information of 33 key human nodes is extracted [5]. Combined with the biomechanical characteristics and action standardization requirements of table tennis forehand stroke technology, a 10-dimensional feature extraction scheme is designed. Detailed features are presented in Table 1 [6].

A self-developed feature quantization algorithm is used to realize the comprehensive scoring of athletes' forehand stroke actions. Specifically, each feature is quantified through geometric relationship calculation of 3D/2D coordinates (e.g., the hitting arm angle is represented by the spatial angle of the three points of the right shoulder - right elbow - right wrist), time-series motion parameter analysis (e.g., wrist speed is quantified based on inter-frame displacement), and key point quality evaluation (e.g., posture quality is measured by the mean value of visibility), providing an interpretable feature basis for action scoring.

Table 1. Features and description

Features	Description
Batting arm angle	Obtained by calculating the included angle of three points (right shoulder, right elbow, right wrist) in 3D space
Forearm extension	Obtained by calculating the included angle of three points (right wrist, right elbow, right shoulder) in 3D space
Body rotation	Calculated as the included angle between the vector of the line connecting the left and right shoulders and the horizontal direction (range: 0–180 °C)
Head orientation	Calculated as the 2D distance between the nose tip and the center point of the left and right shoulders
Hitting point height	Calculated as the absolute difference in the y-axis direction (height) between the right wrist and the right shoulder
Arm coordination	Calculated as the absolute difference between the angles of the left and right arms, limited to the range of 0–180 °C
Wrist speed	Calculated as the distance of the right wrist in 2D space between the current frame and the previous frame (inter-frame speed)
Shoulder stability	Calculated by the rate of change of the distance between the left and right shoulders in two consecutive frames
Striking process	Fixed value 0.0 (actual detection logic not implemented in the code, need to be supplemented)
Posture quality	Calculated as the average visibility of all visible key points (visibility > 0.5)

Core Feature Calculation Formulas 3D Spatial Angle Calculation: Applied to features such as batting arm angle and forearm extension, derived from the dot product of vectors based on 3D coordinates of three points:

$$\theta = \arccos\left(\frac{(\vec{a}-\vec{b})\cdot(\vec{c}-\vec{b})}{\|\vec{a}-\vec{b}\|\cdot\|\vec{c}-\vec{b}\|}\right)\times\frac{180^\circ}{\pi} \tag{3}$$

where $\vec{a}$,$\vec{b}$, and$\vec{c}$represent the 3D coordinate vectors of three points respectively,θdenotes the output angle (unit: degrees), and$\|\cdot\|$denotes the L2 norm of a vector.

2D Distance Calculation: Used for features like head orientation and wrist speed, employing the Euclidean distance formula:

$$d = \sqrt{(x_2 - x_1)^2 + (y_2 - y_1)^2} \quad (4)$$

where$(x_1, y_1)$and$(x_2, y_2)$are the 2D coordinates of two points.

Wrist Speed Calculation: Based on the inter-frame displacement of wrist coordinates between adjacent frames:

$$v_t = \sqrt{(x_t - x_{t-1})^2 + (y_t - y_{t-1})^2} \quad (5)$$

where$v_t$is the wrist speed at frame$t$, and$(x_t, y_t)$and$(x_{t-1}, y_{t-1})$are the 2D coordinates of the wrist at frame$t$and frame$t-1$respectively.

Shoulder Stability Calculation: Characterized by the rate of change in the distance between the left and right shoulders across adjacent frames:

$$S_t = \max\left(0, 1 - \frac{|\text{dist}_t - \text{dist}_{t-1}|}{\text{dist}_{t-1} + \epsilon}\right) \quad (6)$$

where$S_t$is the shoulder stability at frame$t$(value range: [0, 1]),$\text{dist}_t$is the 2D distance between the left and right shoulders at frame$t$, and$\epsilon = 10^{-6}$is used to avoid division by zero.

Posture Quality Calculation: Statistical mean of key point visibility:

$$Q = \frac{1}{K} \sum_{\substack{i=1 \\ v_i > 0.5}}^{K} v_i \quad (7)$$

where$Q$is the posture quality (value range: [0, 1]),$K$is the number of visible key points (with visibility score$v_i > 0.5$), and$v_i$is the visibility score of the$i$-th key point.

Arm Coordination Calculation: Normalized absolute difference between the 3D angles of the left and right arms:

$$C = \min\left(|\theta_{\text{right}} - \theta_{\text{left}}|, 180^\circ\right) \quad (8)$$

where$C$is the arm coordination (unit: degrees), and$\theta_{\text{right}}$and$\theta_{\text{left}}$are the 3D angles of the right arm and left arm respectively.

3.3 Overall Framework

The overall framework of the model is shown in Fig. 2, which includes two submodules: The visual data processing module extracts the 3D human key points and motion features of the video frames, maps the video score to the frames, and generates frame-level scores.

The random forest regression module outputs the comprehensive score of the athlete's forehand stroke action through multi-tree integrated learning based on the frame-level data.

Visual Data Processing Module This module is responsible for processing and feature analyzing the input video frames of table tennis forehand strokes, providing basic visual feature support for subsequent action scoring. The specific processes and functions are as follows:

Input and Preprocessing: Receive original video frames (Video Frames Input), first standardize them into 256 × 256 × 3 RGB format to ensure the spatial consistency of input data and provide input of unified dimensions for the feature extraction module.

Point Marking: For the preprocessed video frames, use MediaPipe to detect and mark the 3D key points of human joints (such as shoulders, elbows, wrists, etc.), and visualize the skeletal structure through connection lines to form a complete pose skeleton representation, facilitating subsequent feature extraction.

Body Presence: Judge the integrity of the human body region in the video frame through confidence to screen valid samples and reduce noise interference.

Frame Feature Extraction: Based on 3D key points and forehand stroke technical specifications, quantify 10-dimensional motion features (such as hitting arm angle) through a self-developed algorithm and form a feature vector for each frame.

Score Mapping: Extract frames from the processed video and map the action score of the entire video to the score corresponding to each frame (Frame Score), realizing the mapping from features to quantitative scores and providing frame-level basic data for video-level overall scoring.

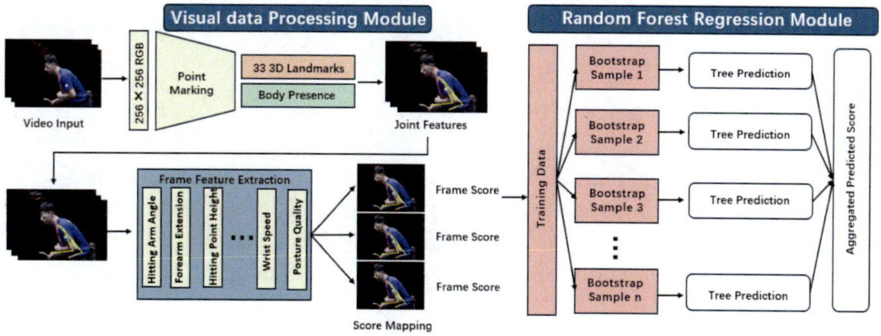

Fig. 2. Overall framework diagram.

Random Forest Regression Module This module takes the frame-level features and score data output by the visual data sub-module as training input, constructs a regression model through integrated learning methods, and finally outputs the video-level comprehensive score of forehand stroke actions. The core functions are as follows:

Training Data Construction: Use the frame-level features and corresponding scores generated by the visual data sub-model as training samples, split the entire dataset at a ratio of 8:2, and construct 110 bootstrap sample sets through Bootstrap sampling—a core random forest resampling technique that randomly selects samples with replacement from the original training data. This retains the original data's statistical characteristics while introducing slight subset differences, effectively improving the model's stability and generalizability.

Tree Prediction: For each bootstrap sample set, train a regression decision tree to realize the score prediction of a single sample by learning the nonlinear mapping relationship between features and scores. The splitting of the decision tree is based on the importance of characteristics (such as the influence weight of the angle of the hitting arm and the rotation of the body on action standardization), ensuring that the prediction logic is consistent with the technical requirements of the forehand strokes.

Aggregated Predicted Score: Integrate 110 decision trees predictions to generate the final video-level score. The aggregation formula is as follows:

$$\text{Score}_{\text{final}} = \frac{1}{T}\sum_{t=1}^{T} f_t(X) \tag{9}$$

Here, $T = 110$ (number of decision trees), $f_t(X)$ is the t-th tree's score for feature vector X (derived from table tennis forehand video parameters), and $\text{Score}_{\text{final}}$ is the 0–10 scaled result.

This aggregation reduces single-tree overfitting, enhances result robustness, and enables objective quantitative evaluation of forehand standardization.

4 Experiments and Results Analysis

To evaluate which lightweight model is more suitable for pose evaluation, we considered the following lightweight models: MediaPipe, YOLOv8n-Pose [7], Lightweight OpenPose [8], and MoveNet [9], using Mean Absolute Error (MAE) and Coefficient of Determination (R^2) as evaluation metrics. The results are shown in Table 2.

MediaPipe achieved the best comprehensive performance, with an MAE of 0.1390 and an R^2 of 0.9396, significantly outperforming other models. Among the remaining three, YOLOv8n-Pose performed best but had a lower R^2 (0.9023) despite a similar MAE (0.1416), and it had certain limitations in key technical characteristics: on the one hand, YOLOv8n-Pose only supports the recognition of 17 human key points, while MediaPipe can realize the accurate detection of 33 3D human key points, and the former is difficult to fully capture the subtle motion features in forehand stroke actions; on the other hand, the key point calculation of YOLOv8n-Pose is based on the 2D image plane and cannot directly output 3D spatial coordinates, which not only limits the depth quantitative analysis of action poses but also brings compatibility obstacles for the functional expansion of subsequent projects.

Overall, the MediaPipe model has multiple technical advantages: on the one hand, its environment configuration process is simple and efficient, reducing the threshold for development and implementation; on the other hand, the 3D calculation logic is deeply integrated with the pose detection module, ensuring low-latency operation without the need for additional development of complex processes, and having excellent deployment convenience.

Table 2. Performance of models

Models	MAE	R^2
MediaPipe	0.1390	0.9396
YOLOv8n-Pose	0.1416	0.9023
Lightweight OpenPose	0.2319	0.7660
MoveNet	0.1497	0.8822

To further verify the contribution of core components in the scoring mechanism to pose evaluation performance and clarify the necessity and optimization space of each functional module, this study introduced ablation experiments for supplementary analysis [10].

The ablation experiment took the complete feature combination as the benchmark model and designed multiple experimental groups, including three types: feature deletion, feature classification, and random selection. The core experimental results are shown in Table 3, all models adopted the same evaluation metrics (MAE, R^2) and test dataset to ensure the comparability of experimental results.

Table 3. Performance of ablation experiment

	Experimental groups										
Features	Baseline	1	2	3	4	5	6	7	8	9	10
Batting arm angle	✓		✓	✓	✓	✓	✓	✓	✓	✓	✓
Forearm extension	✓	✓		✓	✓	✓	✓	✓	✓	✓	✓
Body rotation	✓	✓	✓		✓	✓	✓	✓	✓	✓	✓
Head orientation	✓	✓	✓	✓		✓	✓	✓	✓	✓	✓
Hitting point height	✓	✓	✓	✓	✓		✓	✓	✓	✓	✓
Arm coordination	✓	✓	✓	✓	✓	✓		✓	✓	✓	✓
Wrist speed	✓	✓	✓	✓	✓	✓	✓		✓	✓	✓
Shoulder stability	✓	✓	✓	✓	✓	✓	✓	✓		✓	✓
Striking process	✓	✓	✓	✓	✓	✓	✓	✓	✓		✓
Posture quality	✓	✓	✓	✓	✓	✓	✓	✓	✓	✓	
MAE	0.1390	0.1509	0.1518	0.2060	0.2348	0.1590	0.1584	0.1566	0.1578	0.1592	0.1953
R^2	0.9396	0.9380	0.9369	0.8919	0.8563	0.9298	0.9204	0.9014	0.9093	0.8994	0.8796

The experimental results show that the three features of “Body Rotation”, “Head Orientation”, and “Posture Quality” have a significant negative impact on the prediction accuracy of the model. Their removal leads to a significant deterioration of performance indicators (MAE) and a substantial decrease in R^2 values.

Specifically, the removal of “Body Rotation” (Group 3) increased MAE from 0.1390 to 0.2060 and R^2 from 0.9396 to 0.8919. This indicates that body rotation plays a crucial role in capturing motion laws and prediction accuracy. The removal of “Head Orientation” (Group 4) had a particularly severe impact on model performance, with MAE increasing to 0.2348 and R^2 decreasing to 0.8563. This suggests that the dynamic information of head movement makes an indispensable contribution to accurate prediction. The removal of “Posture Quality” (Group 10) also significantly reduced the model's performance, with MAE increasing to 0.1953 and R^2 decreasing to 0.8796. This indicates that the overall action form quality information is crucial for the effectiveness of the model.

In contrast, the removal of features such as “Batting Arm Angle” (Group 1), “Forearm Extension” (Group 2), “Hitting Point Height” (Group 5), “Arm Coordination” (Group 6), “Wrist Speed” (Group 7), “Shoulder Stability” (Group 8), and “Striking Process” (Group 9) had a relatively small impact on model performance. The elimination of these features only led to a slight increase in MAE and a slight decrease in R^2 values, indicating that they are less important for the overall prediction ability of the model.

In summary, the results of the ablation experiment clearly reveal that “Body Rotation”, “Head Orientation”, and “Posture Quality” are key features for constructing a high-performance prediction model, and they play a core role in quantifying and predicting the movement process. In contrast, other features have a relatively minor impact on model performance. This finding provides important theoretical guidance for future model optimization and feature engineering, emphasizing the in-depth excavation and effective utilization of key features.

5 Conclusion and Outlook

This study addresses the challenge of quantifying the standards of table tennis players' forehand technical actions by designing a forehand stroke scoring system based on multi-pose recognition and random forest. The system includes a visual data processing sub-module and a random forest regression sub-module. The overall process is as follows: after video input, the visual data processing module performs key point marking and feature extraction, and then the random forest regression module combines a self-developed algorithm to realize the prediction of the overall action score.

This framework provides a general and scalable solution for pose detection and evaluation in the sports field. Through modular design and combining manually designed feature extraction algorithms for different sports fields, cross-domain applications are realized.

In addition, the lightweight model MediaPipe adopted in this study has low computing power requirements, which not only reduces the development threshold but also supports real-time deployment on mobile devices, making it suitable for real-time feedback evaluation in coach guidance and competitive training scenarios. This innovation not only improves the practicality of motion pose recognition but also promotes the application of AI in edge computing and real-time interaction.

This research provides a direction and framework for future sports pose recognition. Using this framework, application can be realized only by designing recognition algorithms for the unique features of the sport, combined with manual feature design.

Acknowledgment. This research was supported by grants from National Natu ral Science Foundation of China (62576324), Natural Science Foundation of Henan Province, China (262300421302). Prof. K. Wang and X. L. Cheng are the authors to whom all the correspondence should be addressed.

References

1. Konak, O., Wischmann, A., Water, R.V.D., Arnrich, B.: A real-time human pose estimation approach for optimal sensor placement in sensor-based human activity recognition. In: Matthies, D.J.C., Grzegorzek, M., Kuijper, A., Leutheuser, H. (eds.) Proceedings of the 8th International Workshop on Sensor-Based Activity Recognition and Artificial Intelligence, iWOAR 2023, vol. 21, no. 6, pp. 1–21, Lübeck, Germany, September 21–22, 2023. ACM (2023)
2. Breiman, L.: Random forests. Mach. Learn. **45**(1), 5–32 (2001)
3. Qi, J., Du, J., Siniscalchi, S.M., Ma, X., Lee, C.: On mean absolute error for deep neural network based vector-to-vector regression. IEEE Signal Process. Lett. **27**, 1485–1489 (2020)
4. Dill, S., Rohr, M., Güney, G., Antink, C.H.: Evaluation of accuracy and angle dependency of 3d pose estimation through stereo camera information fusion with mediapipe pose. In: 27th International Conference on Information Fusion, FUSION 2024, Venice, Italy, July 8–11, 2024, pp. 1–7. IEEE (2024)
5. Abbas, R., Gao, H., Li, X.: Text-driven 3d human motion generation for pose estimation using dual-transformer architecture. Comput. Aided Des. **190**, 103991 (2026)
6. Dong, K., Yan, W.Q.: Player performance analysis in table tennis through human action recognition. Computers **13**(12), 332 (2024)
7. Cheng, S., Chen, L., Yang, K.: Dgss-yolov8s: a real-time model for small and complex object detection in autonomous vehicles. Algorithms **18**(6), 358 (2025)
8. Cao, Z., Hidalgo, G., Simon, T., Wei, S., Sheikh, Y.: Openpose: realtime multi-person 2d pose estimation using part affinity fields. CoRR arXiv:1812.08008 (2018)
9. Faliagka, E., Tefas, G., Antonopoulos, C.P., Voros, N.S.: Real-time person recognition using movenet for pose-based identification. In: IEEE Computer Society Annual Symposium on VLSI, ISVLSI 2025, Kalamata, Greece, July 6–9, 2025, pp. 1–4. IEEE (2025)
10. Zelezný, T., Straka, J., Javorek, V., Valach, O., Hrúz, M., Gruber, I.: Exploring pose-based sign language translation: ablation studies and attention insights. CoRR arXiv:2507.01532 (2025)

Contrastive Fusion of Biomolecular and Heterogeneous Network Features for Drug–Target Interaction Prediction

Hongyang Yang, Xiucai Ye(✉), and Tetsuya Sakurai

University of Tsukuba, Tsukuba, Ibaraki 305–8577, Japan
{yexiucai,sakurai}@cs.tsukuba.ac.jp

Abstract. Accurate prediction of drug–target interactions (DTI) is essential for drug repurposing and the discovery of new therapeutics. However, most existing methods struggle to effectively combine biological relational features derived from interaction networks with biomolecular features of drugs and proteins. In this paper, we present Meta-path Aggregation Contrastive Learning for Drug–Target Interaction (MACL-DTI), a framework that integrates biological relational features from heterogeneous interaction networks with biomolecular features to improve DTI prediction. MACL-DTI first learns biomolecular representations for drugs and targets from their intrinsic biochemical information. In parallel, we construct multiple meta-path–based neighborhoods on a heterogeneous interaction network to model high-order topological relationships among drugs, targets, and related biomedical entities, and applies a metapath-guided GNN to derive topology-aware network representations. Finally, we apply contrastive learning to align biomolecular and network representations for more reliable DTI prediction. We further employ a self-paced negative sampling strategy that progressively selects informative negatives, reducing false negatives and noisy supervision. Experiments on multiple benchmark datasets show that MACL-DTI consistently outperforms several state-of-the-art methods in terms of prediction accuracy and robustness. These results indicate that combining meta-path–guided graph representation learning with contrastive learning is an effective strategy for exploiting heterogeneous biological data. Overall, MACL-DTI provides a scalable and interpretable solution for DTI prediction, contributing to more accurate and reliable computational support for drug discovery.

Keywords: DTIs prediction · contrastive learning · Heterogeneous networks · meta-path Induction

1 Introduction

Drugs are typically designed to modulate a specific protein target, yet many interact with multiple unintended proteins, which can lead to off-target effects [1–3]. Accurately characterizing drug–target interactions (DTIs) is therefore critical for drug discovery and repurposing [4, 5].

L. Huang and D. Greenhalgh (Eds.): ICMLC 2026, LNNS 2061, pp. 854–868, 2027.
https://doi.org/10.1007/978-3-032-31236-5_57

Traditional DTI prediction methods include ligand-based modeling and molecular docking [6, 7]. Ligand-based approaches usually require enough known actives per target to learn reliable structure–activity patterns, so their performance can degrade when labeled interactions are scarce or biased. Docking, on the other hand, depends on accurate 3D protein structures and suitable binding-site conformations; in many cases these structures are unavailable, low-quality, or do not reflect the relevant functional state, leading to unstable ranking results. To alleviate these issues, graph neural network (GNN)–based methods have been proposed. GNNs model drugs, targets, and other biomedical entities as nodes and learn embeddings by aggregating information from graph neighborhoods, capturing topological and semantic relationships [8]. This enables them to leverage indirect evidence from heterogeneous associations, making them particularly suitable for data-sparse and cold-start scenarios.

Existing GNN-based DTIs prediction methods can be broadly categorized into homogeneous and heterogeneous models [8, 9]. Homogeneous GNNs operate on graphs with a single node and edge type, and can effectively capturing local neighborhood structure for link prediction [10]. However, they are limited in explicitly modeling cross-type relations (e.g., drug–disease or protein–disease) and the higher-order dependencies that arise in biomedical systems [11]. In contrast, heterogeneous GNNs incorporate multiple entity and relation types and typically leverage metapaths or semantic-aware aggregation to integrate multimodal biomedical information, yielding richer representations in complex networks [12, 13]. Nevertheless, many existing heterogeneous models do not integrate intrinsic biomolecular properties of drugs and proteins—such as structural conformations, binding-site information, or physicochemical characteristics—which can reduce biochemical interpretability and, in some cases, predictive performance [14]. Moreover, most studies construct negative samples by randomly selecting unlabeled drug–target pairs, implicitly treating unknown interactions as true negatives; this strategy can introduce false negatives and noisy supervision, ultimately impairing model training and evaluation [15, 16].

To address these issues, we propose MACL-DTI, a metapath-aggregated contrastive learning framework for DTI prediction. MACL-DTI integrates multi-source biomedical information in a heterogeneous graph, learns molecule-based embeddings for drugs and residue-level structural embeddings for proteins, and aligns these views via contrastive learning. A progressive self-paced negative sampling strategy is further used to select informative negatives during training [6, 17, 18].

The main contributions of this work are as follows:

1. We design a dual-branch framework that combines molecule-based drug embeddings, residue-level protein embeddings and metapath-based heterogeneous graph representations for DTI prediction.
2. We introduce a structure-aware cross-modality contrastive learning module that aligns molecule-based and metapath-based embeddings, and a self-paced negative sampling strategy that reduces the impact of noisy negatives.
3. We conduct extensive experiments, ablation analyses, visualization and cold-start case studies on two large heterogeneous datasets, demonstrating that MACL-DTI achieves superior performance and offers practical value for drug repurposing and target prioritization

2 Method

Fig. 1 provides an overview of the proposed MACL-DTI framework, which consists of three parts. Part 1 extracts representations from two complementary views. Specifically, MACL-DTI learns biomolecular structural embeddings for drugs and proteins (Sects. 2.1–2.2) and captures heterogeneous relational information via metapath-based representation learning (Sect. 2.3). Part 2 fuses biomolecular and network embeddings via contrastive learning (Sect. 2.4). Part 3 applies self-paced negative sampling to iteratively refine the classifier (Sect. 2.5) and optimizes MACL-DTI with the overall training loss (Sect. 2.6). In the following, we describe each component in detail.

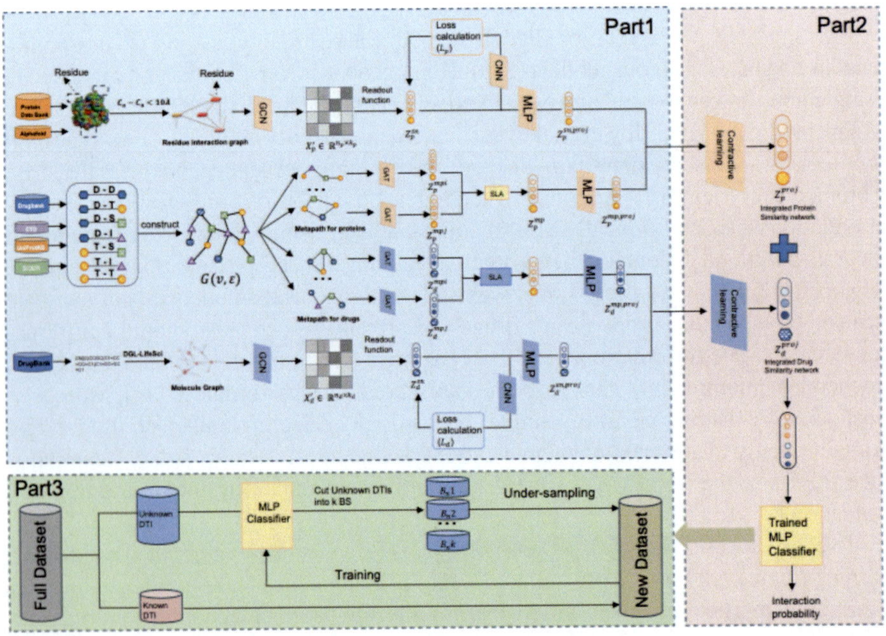

Fig. 1. Overall architecture of the MACL-DTI framework. Part 1 extracts molecule-based and residue-level structural embeddings together with metapath-based heterogeneous graph embeddings for drugs and proteins. Part 2 fuses the two-view embeddings via contrastive learning. Part 3 trains the MLP classifier with self-paced negative sample selection, iteratively updating the negative set during training.

2.1 Drug Biomolecular Structural Features

In MACL-DTI, each drug d is represented by its SMILES string from the DrugBank database. We then use DGL-LifeSci to parse the SMILES into an atom-level molecular graph and extract the initial atom feature matrix $X_d \in R^{n \times k}$, where n is the number of atoms and k is the atom feature dimension. Based on covalent bonds, we build the adjacency matrix $A_d \in R^{n \times n}$, where $A_d(i, j) = 1$ if atoms I and j are connected and $A_d(i,$

$j) = 0$ otherwise. We do not add self-loops to A_d, and D_d denotes the corresponding degree matrix.

We apply a graph convolutional network (GCN) to update atom features:

$$X_d^{'} = ReLU\left(D_d^{-\frac{1}{2}}A_d D_d^{-\frac{1}{2}}X_d W_d\right) \in \mathrm{R}^{n\times k} \tag{1}$$

where $W_d \in R^{k \times k}$ is a learnable weight matrix. We then use mean pooling over atoms to obtain drug molecule-based embedding:

$$Z_d^{Mol} = MeanPool\left(X_d^{'}\right) \in R^k \tag{2}$$

2.2 Protein Biomolecular Structural Features

For protein targets, we obtain 3D structures from the Protein Data Bank or use AlphaFold to predict structures from amino acid sequences. From proteins structure, we construct a residue-level graph and define the initial feature matrix $X_p \in R^{n \times k}$, where n is the number of residues and k is the residue feature dimension. The adjacency matrix $A_p \in R^{n \times n}$ is defined by spatial proximity: $A_p(i, j) = 1$ if the Euclidean distance between the Cα atoms of residues i and j is less than 10 Å, and $A_d(i, j) = 0$ otherwise.

We then apply a GCN to update residue features:

$$X_p^{'} = ReLU\left(D_p^{-\frac{1}{2}}A_p D_p^{-\frac{1}{2}}X_p W_p\right) \in R^{\mathrm{n}\times k} \tag{3}$$

Like the drug branch, we apply mean pooling to obtain protein residue-based embedding Z_p^{Mol}.

2.3 Metapath-Based Features

In addition to biomolecular features, MACL-DTI uses metapaths to describe how a drug is connected to other biomedical entities through multi-hop relations. A metapath is a type sequence such as *drug–protein–drug* (two drugs share a protein target) or *drug–protein–disease–protein–drug* (two drugs are linked through proteins and a common disease).

Based on the heterogeneous network constructed from drugs, proteins, diseases and side effects, we define the following metapaths: *drug–drug, drug–protein–drug, drug–protein–protein–drug, drug–protein–disease–protein–drug; protein–protein, protein–drug–protein, protein–drug–drug–protein, and protein–drug–disease–drug–protein.*

We consider four node types: drugs, proteins, diseases and side effects. A node v of type τÎ{drug, target, disease, side-effect} is represented by a type-specific one-hot vector $X_v^{(T)} \in R^{N_T}$, where N_T is the number of nodes of the type. We then use an MLP to map these one-hot vectors to the same feature dimension.

For each drug node, we use a graph attention layer(GAT) to obtain a metapath-specific embedding from its neighbors along that metapath. Given a drug d and one metapath type, let $N(d)$ be the set of neighbors of d on this metapath, and let $h_v \in R^k$ be the input feature of node v. The GAT-based metapath aggregation is defined as:

$$e_{dj} = LeakyReLU\left(a^T\left[Wh_d \| Wh_j\right]\right),\ j \in N(d) \tag{4}$$

$$a_{dj} = \frac{\exp(e_{dj})}{\sum_{l\epsilon N(d)} \exp(e_{dl})};\ z_d^{mp,i} = \sigma\left(\sum_{j\in N(d)} a_{dj} Wh_j\right) \tag{5}$$

where $W \in R^{k \times k}$ and $a \in R^{2k}$ are learnable parameters, $\|$ denotes concatenation, and $Z_d^{mp,i} \in R^k$ is the embedding of drug d from the i-th metapath.

A drug usually participates in several metapath types. We use a simple attention mechanism to combine them into a high-order structural embedding:

$$s_i = w^T z_d^{mp,i};\ \beta_i = \frac{\exp(s_i)}{\sum_j \exp(s_j)};\ z_d^{mp} = \sum_i \beta_i z_d^{mp,i} \tag{6}$$

where $w \in R^k$ is a learnable vector and z_d^{mp} is the final metapath-based embedding of drug d.

We apply the same procedure to protein targets and obtain their metapath-based embeddings z_p^{mp}.

2.4 Contrastive Learning Embedding Fusion

After obtaining two types of drug embeddings, $Z_{d_i}^{Mol}$ from the molecule-based branch and $z_{d_i}^{mp}$ from the metapath-based branch, we first project them into a shared latent space using the same MLP:

$$Z_{d_i}^{mol,proj} = W\sigma\left(Z_{d_i}^{mol}\right) + \mathrm{b};\ Z_{d_i}^{mp,proj} = W\sigma\left(Z_{d_i}^{mp}\right) + \mathrm{b} \tag{7}$$

To align molecule-based and metapath-based embeddings and make semantically related drugs closer in the embedding space, we design a contrastive learning strategy on the heterogeneous graph. For each pair of drugs (d_i, d_j), we first count how many metapaths connect them:

$$C_{d_i}(d_j) = \sum_{k=1}^{n}\left(d_i \in N_{d_j}^{\varphi_k}\right), \tag{8}$$

where $N_{d_j}^{\varphi_k}$ is the set of nodes connected to d_j through metapath φ_k. This count $C_{d_i}(d_j)$ reflects how strongly the two drugs are connected in the heterogeneous network. We directly use this value to construct supervision signals: drug pairs with larger $C_{d_i}(d_j)$ are treated as positive samples, while pairs with fewer or no metapath connections are treated as negative samples. In this way, we obtain a positive set P_{mi} and a negative set N_{mi} for each anchor drug d_i.

Based on these sets, the contrastive loss from the molecule-based view is

$$L_{d_i}^{mol} = -\log \frac{\sum_{d_j \in P_{mi}} \exp\left(sim\left(z_{d_i}^{mol,proj}, z_{d_j}^{mp,proj}\right)/\tau\right)}{\sum_{d_k \in P_{mi} \cup N_{mi}} \exp\left(sim\left(z_{d_i}^{mol,proj}, z_{d_k}^{mp,proj}\right)/\tau\right)} \tag{9}$$

and the loss from the metapath-based view is defined symmetrically as

$$L_{d_i}^{mp} = -\log \frac{\sum_{d_j \in P_{mi}} \exp\left(sim\left(z_{d_i}^{mp,proj}, z_{d_j}^{mol,proj}\right)/\tau\right)}{\sum_{d_k \in P_{mi} \cup N_{mi}} \exp\left(sim\left(z_{d_i}^{mp,proj}, z_{d_k}^{mol,proj}\right)/\tau\right)} \tag{10}$$

The total contrastive loss over all drugs is

$$J_d = \frac{1}{|V|} \sum_{d_i \in V} \left[\lambda_d L_{d_i}^{mol} + (1 - \lambda_d) L_{d_i}^{mp}\right], \tag{11}$$

where V is the set of all drug nodes and $\lambda_d \in [0, 1]$ balances the two views. By minimizing J_d, the projected embeddings $Z_{d_i}^{mol,proj}$ and $Z_{d_i}^{mp,proj}$ are jointly optimized and aligned. In our framework, the final drug representation used for downstream prediction is the molecule-based embedding $Z_{d_i}^{mol,proj}$

Similarly, we define the total contrastive loss over all targets as J_p, and use the residue-based embedding $Z_{p_i}^{Mol,proj}$ as the final protein representation for downstream prediction.

2.5 Progressive Negative Sample Selection

During MLP training, all known drug–target associations form the positive sample set PPP, while all other drug–target pairs are treated as candidate negatives N, where $|N| \gg |P|$. In most previous studies, an equal number of negatives is randomly sampled from N, without considering how informative these samples are. However, the choice of negative samples is non-trivial and can have a substantial impact on model performance. Motivated by the self-paced ensemble (SPE) framework [17], we adopt a progressive negative sample selection strategy to identify higher-quality negatives for training.

Concretely, we first train the MLP classifier using the positive set P and a randomly selected subset of N. The trained model is then used to assign an interaction score to each candidate negative pair $x \in N$. For each x, we define the prediction error as

$$H(x) = |f(x) - 0| \tag{12}$$

where $f(x)$ denotes the predicted interaction score and 0 is the ground-truth label for negatives. Based on H(x), we partition N into k subsets *{B1, B2, ..., Bk}* according to different hardness intervals. A fixed number of samples is drawn from each subset B_l to construct a refined negative set, which is then combined with P to retrain the model. This procedure is repeated iteratively, so that the model is gradually exposed to more informative and harder negatives, improving the quality of supervision and enhancing overall performance, as illustrated in Part 3 of Fig. 1.

2.6 Training

For drug–target interaction prediction, MACL-DTI adopts the binary cross-entropy loss because of its effectiveness for binary classification. Let (d_i, t_j) denote a drug–target pair, with ground-truth label $y_{ij} \in \{0, 1\}$ and predicted interaction probability $\hat{y}_{ij}$. Using the positive set Y^+ and the progressively selected negative set Y^- (Sect. 2.5), the classification loss is defined as

$$L_B = -\sum_{(i,j)\in Y} \left(y_{ij} \log \hat{y}_{ij} + \left(1 - y_{ij}\right) \log\left(1 - y_{ij}\right)\right) \tag{13}$$

Here, Y^+ contains all known drug–target associations, and Y^- contains the negative samples obtained by the progressive negative sample selection strategy. If $(i, j) \in Y^+$, then $y_{ij} = 1$; if $(i, j) \in Y^-$, then $y_{ij} = 0$.

In addition to this classification loss, MACL-DTI includes the structure-aware cross-modality contrastive loss J_d defined in Sect. 2.4, which aligns the molecule-based and metapath-based drug embeddings. The overall training objective is

$$L = L_B + J_d + J_p \tag{14}$$

3 Results

3.1 Data Collection

MACL-DTI is evaluated on two heterogeneous biological datasets, Hetero-A and Hetero-B.

Hetero-A is constructed from the DTINet benchmark dataset [19], which includes 708 drugs, 1,512 protein targets, 5,603 diseases and 4,192 side effects. It contains six types of associations: drug–target (D–T), drug–disease (D–I), drug–drug (D–D), drug–side effect (D–S), target–target (T–T) and target–disease (T–I). Detailed statistics are summarized in Table 1.

Although biomedical data have grown rapidly in recent years, existing datasets remain incomplete, particularly in terms of drug–target coverage. This incompleteness introduces noisy labels and false negatives, which can reduce the accuracy of predictive models [20]. To mitigate this issue, we constructed a new heterogeneous dataset, Hetero-B, by integrating more comprehensive and up-to-date information from multiple public biomedical databases. Hetero-B comprises 2,219 approved small-molecule drugs from DrugBank (v5.1.8) [21], 1,988 human protein targets from UniProtKB (2022/12) [22], 7,205 diseases from CTD (2021) [23] and 3,935 side effects from SIDER (v4) [24]. It includes 8,783 drug–target interactions and over 1 million drug–drug interactions from DrugBank, as well as 542,970 drug–disease and 2,922,064 target–disease associations from CTD. Detailed statistics are provided in Table 1.

Table 1. Statistics of datasets Hetero-A and Hetero-B.

Type	Items	Hetero-A numbers	Hetero-A resources	Hetero-B numbers	Hetero-B resources
Node	Drug (D)	708	DrugBank (Version 3.0)	2,219	DrugBank (Version 5.1.8)
	Target (T)	1152	HPRD (Release 9)	1,988	UniProtKB (Release 2022_12)
	Disease (I)	5603	CTD (2013)	7,205	CTD (2021)
	Side-effect (S)	4192	SIDER (Version 2)	3,935	SIDER (Version 4)
Edge	D-T	1,923	DrugBank (Version 3.0)	8,783	DrugBank (Version 5.1.8)
	D-D	10,036	DrugBank (Version 3.0)	1,091,870	DrugBank (Version 5.1.8)
	D-I	199,214	CTD (2013)	542,970	CTD (2021)
	D-S	80,164	SIDER (Version 2)	104,629	SIDER (Version 4)
	T-T	7,363	HPRD (Release 9)	456,592	STRING (Version 11.0)
	T-I	1,596,745	CTD (2013	2,922,064	CTD (2021)

3.2 Comparison with Baseline Methods under Different Negative Sample Ratios

We evaluate MACL-DTI using 5-fold cross-validation on the Hetero-A and Hetero-B datasets. The model is compared with seven competitive baselines: GCN [25], GAT [26], DTI-CNN [27], GraphDTA [28], HAN [12], MccDTI [29] and CCL-ASPS [10].

Because the ratio of positive to negative samples strongly influences model behavior, we assess all methods under two sampling settings: 1:1 and 1:5 (positive: negative). We report accuracy (ACC), area under the ROC curve (AUC) and area under the precision–recall curve (AUPRC). The detailed results are listed in Table 2.

Under the 1:1 setting, MACL-DTI consistently outperforms all baselines. On Hetero-A, it achieves an ACC of 0.9317, an AUC of 0.9680 and a precision of 0.9737. On Hetero-B, MACL-DTI attains the highest AUC (0.9794) and AUPRC (0.9688), and an ACC of 0.9249, which is slightly lower than that of MccDTI (0.9305).

Under the more challenging 1:5 setting with severe class imbalance, MACL-DTI remains robust on both datasets. On Hetero-A, it obtains the best performance across all metrics, with ACC of 0.9527, AUC of 0.9791 and AUPRC of 0.9149. On Hetero-B, MACL-DTI also achieves the highest scores, with ACC of 0.9479, AUC of 0.9814 and AUPRC of 0.9079.

Overall, these results demonstrate the effectiveness and robustness of MACL-DTI on both balanced and imbalanced data. While many baselines show a marked drop in AUPRC under the 1:5 setting, MACL-DTI maintains high performance with only a modest decrease, indicating strong generalization ability in the presence of abundant negative samples.

Table 2. Performance comparison of MACL-DTI and baseline methods under different positive-to-negative sample ratios on Hetero-A and Hetero-B

Method	Ratio	Hetero-A			Hetero-B		
		ACC	AUC	AUPRC	ACC	AUC	AUPRC
GCN	1:1	0.8144 ± 0.0027	0.8214 ± 0.0074	0.8225 ± 0.0018	0.8423 ± 0.0047	0.8648 ± 0.0007	0.8481 ± 0.0031
GAT	1:1	0.8094 ± 0.0015	0.8449 ± 0.0011	0.8329 ± 0.0006	0.8371 ± 0.0019	0.9019 ± 0.0009	0.8900 ± 0.0028
DTI-CNN	1:1	0.8241 ± 0.0012	0.8906 ± 0.0028	0.9074 ± 0.0006	0.8650 ± 0.0023	0.9343 ± 0.000	0.9420 ± 0.0035
GraphDTA	1:1	0.7941 ± 0.0037	0.8310 ± 0.0021	0.8307 ± 0.0025	0.8395 ± 0.0017	0.8942 ± 0.0012	0.8966 ± 0.0017
HAN	1:1	0.8034 ± 0.0012	0.8532 ± 0.0008	0.8789 ± 0.0016	0.8405 ± 0.0014	0.9338 ± 0.0019	0.9371 ± 0.0031
MccDTI	1:1	0.9034 ± 0.0009	0.9303 ± 0.0010	0.9217 ± 0.0026	0.9305 ± 0.0019	0.9503 ± 0.0014	0.9621 ± 0.0023
CCL-ASPS	1:1	0.8814 ± 0.0031	0.9350 ± 0.0029	0.9361 ± 0.0026	0.8955 ± 0.0033	0.9546 ± 0.0006	0.9641 ± 0.0034
MACL-DTI (Ours)	1:1	0.9117 ± 0.0018	0.9610 ± 0.0005	0.9520 ± 0.0004	0.9249 ± 0.0041	0.9794 ± 0.0003	0.9688 ± 0.0016
GCN	1:5	0.8713 ± 0.0014	0.8547 ± 0.0022	0.6820 ± 0.0031	0.9045 ± 0.0027	0.8910 ± 0.0012	0.7185 ± 0.0028
GAT	1:5	0.8774 ± 0.0009	0.8649 ± 0.0021	0.6725 ± 0.0036	0.9076 ± 0.0014	0.9081 ± 0.0026	0.7099 ± 0.0012
DTI-CNN	1:5	0.8921 ± 0.0031	0.9107 ± 0.0073	0.7974 ± 0.0035	0.9358 ± 0.0019	0.9431 ± 0.0022	0.8648 ± 0.0041
GraphDTA	1:5	0.8514 ± 0.0032	0.8710 ± 0.0018	0.7507 ± 0.0017	0.9158 ± 0.0008	0.9185 ± 0.0010	0.7812 ± 0.0019
HAN	1:5	0.7911 ± 0.0031	0.8132 ± 0.0051	0.7588 ± 0.0026	0.8695 ± 0.0033	0.8538 ± 0.0037	0.7812 ± 0.0019
MccDTI	1:5	0.8917 ± 0.0026	0.9189 ± 0.0057	0.8675 ± 0.0042	0.9112 ± 0.0023	0.9203 ± 0.0032	0.8722 ± 0.0021
CCL-ASPS	1:5	0.9381 ± 0.0024	0.9427 ± 0.0039	0.8621 ± 0.0051	0.9490 ± 0.0014	0.9516 ± 0.0020	0.8953 ± 0.0016
MACL-DTI (Ours)	1:5	0.9527 ± 0.0014	0.9791 ± 0.0005	0.9149 ± 0.0007	0.9379 ± 0.0019	0.9714 ± 0.0006	0.9079 ± 0.0017

3.3 Model Ablation Study

The MACL-DTI framework integrates contrastive learning and self-paced negative sampling to enhance drug and target embeddings and improve prediction performance. To assess the contribution of each core component, we perform an ablation study by selectively removing three modules: the biomolecular structural encoder (BS), the metapath-based heterogeneous graph encoder (MP) and the self-paced negative sampling strategy (SP). We compare the full MACL-DTI model with three variants: MACL-DTI w/o BS (removing the molecule and residue structural branch), MACL-DTI w/o MP (removing the metapath-based encoder) and MACL-DTI w/o SP (removing the progressive negative sample selection). The results are shown in Fig. 2.

On Hetero-A, the full MACL-DTI model achieves the best performance, with an ACC of 0.9866, AUC of 0.9812, AUPRC of 0.9117 and F1 score of 0.8501. On Hetero-B, MACL-DTI also outperforms all ablated variants, reaching an ACC of 0.9712, AUC of 0.9680, AUPRC of 0.9014 and F1 score of 0.8310.

As illustrated in Fig. 2, removing any single component leads to a clear drop in performance, confirming the importance of all three modules. The biomolecular structural encoder is crucial for capturing fine-grained structural information from drug molecules and protein residues. The metapath-based encoder enables the model to exploit high-order semantic relationships in the heterogeneous network. The self-paced negative sampling strategy further improves generalization by selecting informative and reliable negative samples. Together, these results demonstrate that combining molecule-based representations, metapath-driven embeddings and adaptive negative sampling is essential for achieving strong DTI prediction performance in MACL-DTI.

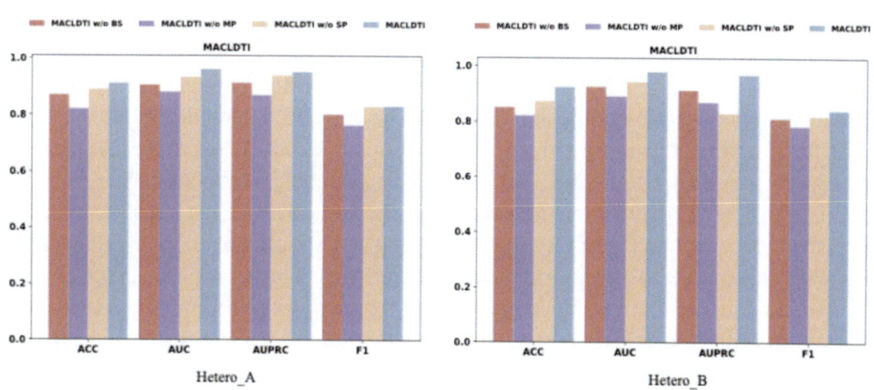

Fig. 2. Ablation study of MACL-DTI on Hetero-A and Hetero-B

3.4 Case Study for EGFR and Cyclin D1

To evaluate the reliability and generalization ability of MACL-DTI under cold-start settings, we conducted case studies on two cancer-associated targets, EGFR (epidermal growth factor receptor) and Cyclin D1 (G1/S-specific cyclin D1). For each target, we created a target-held-out benchmark by removing all known interactions involving the target from the training data. MACL-DTI was then trained on the filtered dataset and used to score all candidate drugs for the held-out target. We ranked drugs by the predicted interaction probabilities and summarized the top five candidates for each target in Fig. 3a. In this figure, the center node denotes the target, the inner ring lists the predicted candidate drugs, and the outermost layer indicates representative diseases that each drug is clinically used to treat (i.e., approved indications), providing a clinical context for interpreting the predicted hits.

For EGFR (Fig. 3a, right), MACL-DTI places Gefitinib, Erlotinib, and Cetuximab among the top-ranked candidates. Notably, all three are established therapies for EGFR-driven cancers, which supports that the model can rediscover clinically validated EGFR inhibitors even when EGFR-related interactions are excluded during training [30–32]. In addition, Lapatinib is ranked highly. Although it is not primarily developed for EGFR-driven tumors, prior studies have reported that Lapatinib can inhibit EGFR tyrosine kinase activity and suppress downstream oncogenic signaling, suggesting a plausible mechanistic link to EGFR targeting [33]. The outer indication layer further shows that these predicted drugs are strongly enriched in oncology-related uses, aligning the predicted EGFR binding potential with their real-world therapeutic roles.

For Cyclin D1 (Fig. 3a, left), the top-ranked candidates similarly exhibit meaningful clinical relevance. Four of the five predicted drugs are associated with diseases where Cyclin D1 dysregulation or overexpression plays an important role, consistent with the biological function of Cyclin D1 in cell-cycle progression. The remaining candidate, Imatinib, is primarily used as a tyrosine kinase inhibitor for leukemia; however, previous work suggests that it can inhibit the Wnt/β-catenin pathway and reduce Cyclin D1 expression, indicating a potential indirect mechanism for modulating Cyclin D1–related oncogenic processes [34]. Overall, by jointly presenting the predicted target–drug links (inner ring) and each drug's clinical indication landscape (outer layer), Fig. 3a illustrates that MACL-DTI's top candidates are not only high-scoring computationally but also coherent with known therapeutic applications in cancer and related diseases.

To further validate high-confidence predictions without direct experimental evidence, we performed an in-silico property-and-structure assessment (Fig. 3b–c). In Fig. 3b, ADMET [35] profiling indicates that the candidate drugs exhibit favorable drug-like physicochemical patterns, including balanced lipophilicity and molecular size, along with acceptable absorption-related characteristics. These properties suggest that the molecules can maintain stable interactions in a biological environment and are compatible with forming specific contacts with protein binding sites rather than behaving as poorly tractable compounds.

We then conducted molecular docking using CB-Dock2 [36] and visualized the binding poses in PyMOL (Fig. 3c). CB-Dock identified plausible cavities on the target proteins and generated consistent binding conformations for the predicted pairs Letrozole–Cyclin D1 and Testosterone–EGFR. In both cases, the ligands are accommodated within well-defined pockets, adopting compact poses that fit the cavity geometry and remain stably embedded in the binding region. The concordance between favorable ADMET profiles (Fig. 3b) and plausible pocket-level docking conformations (Fig. 3c) provides structural support for the physical feasibility of these predicted drug–target interactions.

Overall, the case studies show that MACL-DTI can both recover known anticancer drugs and propose structurally and functionally plausible new candidates for EGFR and Cyclin D1. This demonstrates the model's potential for anticancer drug discovery, particularly in drug repurposing and under cold-start conditions.

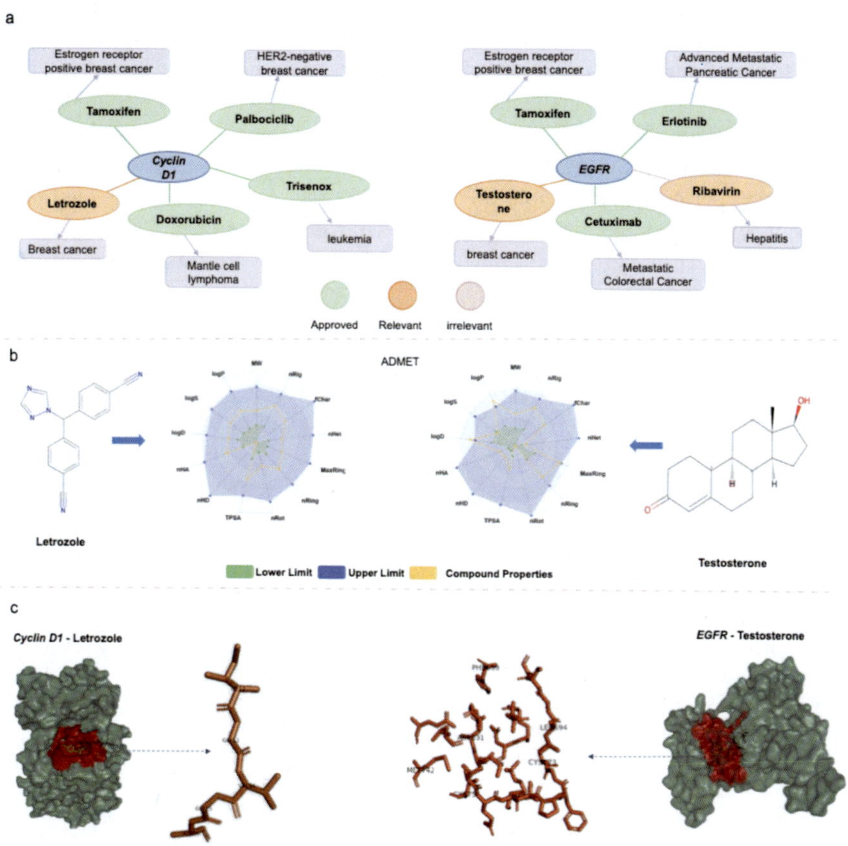

Fig. 3. Case studies of MACL-DTI for EGFR and Cyclin D1 under cold-start settings. (a) Top-5 predicted drugs for the Cyclin D1 and EGFR. Inner nodes are candidate drugs, and the outer layer shows their approved therapeutic indications. (b) ADMET and physicochemical property profiles of representative candidates (Letrozole and Testosterone). (c) CB-Dock2 docking poses (PyMOL visualization) for Letrozole–Cyclin D1 and Testosterone–EGFR.

4 Conclusion

In this study, we proposed MACL-DTI, a metapath-aggregated contrastive learning framework for drug–target interaction prediction. MACL-DTI combines molecule-based embeddings of drugs and residue-level embeddings of protein targets with metapath-based representations learned from a heterogeneous biomedical network. A structure-aware contrastive loss aligns molecule-based and metapath-based views, while a self-paced negative sampling strategy selects informative negative pairs for training. Experiments on two large heterogeneous datasets, Hetero-A and Hetero-B, show that MACL-DTI consistently outperforms a range of state-of-the-art baselines under both balanced and imbalanced settings. Visualization analyses, ablation studies and cold-start case studies on EGFR and Cyclin D1 further demonstrate the robustness, interpretability and practical utility of the proposed framework.

The strong performance and interpretability of MACL-DTI indicate its potential for a variety of computational drug discovery tasks, such as therapeutic target prioritization, drug repurposing of targets with limited known interactions. In future work, we plan to incorporate richer three-dimensional descriptors and binding-site information into the molecule–residue branch, and to explore trainable or data-driven metapath selection mechanisms for the heterogeneous graph encoder. We also aim to extend MACL-DTI to multitask settings (e.g., simultaneously modeling interaction type or binding affinity), which may further enhance its value for guiding the design of more effective and precise therapeutic strategies.

References

1. Hopkins, A.L.: Network pharmacology: the next paradigm in drug discovery. Nature Chem. Biol. **4**(11), 682–690 (2008)
2. Mukhopadhyay, T., et al.: Mebendazole elicits a potent antitumor effect on human cancer cell lines both in vitro and in vivo. Clin. Cancer Res. **8**(9), 2963–2969 (2002)
3. Doudican, N.A., et al.: Mebendazole induces apoptosis via Bcl-2 inactivation in chemoresistant melanoma cells. Mol. Cancer Res. **8**(8), 1031–1038 (2010)
4. Huang, D., Ye, X., Sakurai, T.: Multi-party collaborative drug discovery via federated learning. Comput. Biol. Med. **171** (2024)
5. Young, G.O.: Synthetic structure of industrial plastics. In: Peters, J. (ed.) Plastics, vol. 3, 2nd edn, pp. 15–64. McGraw-Hill, New York (1964)
6. Lin, Z., Zhang, Y., Ouyang, C., Hu, F.: Multimodal contrastive representation learning for drug–target binding affinity prediction. Methods. **211**, 1–10 (2023)
7. Yamanishi, Y., et al.: Drug–target interaction prediction from chemical, genomic and pharmacological data in an integrated framework. Bioinformatics. **26**(12), i246–i254 (2010)
8. Jarada, T.N., Rokne, J.G., Alhajj, R.: A review of computational drug repositioning: strategies, approaches, opportunities, challenges, and directions. J. Cheminformatics. **12**, 46 (2020)
9. Zhang, Y., et al.: A survey of drug–target interaction and affinity prediction methods via graph neural networks. Comput. Biol. Med. **163** (2023)
10. Tian, Z., et al.: Drug-target interaction prediction with collaborative contrastive learning and adaptive self-paced sampling strategy. BMC Biol. **22**(1), 216 (2024)
11. Chen, X., Liu, X., Wu, J.: Drug–drug interaction prediction with graph representation learning. In: Proc. IEEE Int. Conf. Bioinformatics and Biomedicine (BIBM), pp. 354–361. IEEE, San Diego (2019)

12. Wang, X., et al.: Heterogeneous graph attention network. In: Proc. World Wide Web Conf. (WWW), pp. 2022–2032. ACM, New York (2019)
13. Su, Y., et al.: AMGDTI: drug–target interaction prediction based on adaptive meta-graph learning in heterogeneous network. Brief. Bioinform. **25**(1), bbad474 (2024)
14. Liu, T., et al.: CodLncScape provides a self-enriching framework for the systematic collection and exploration of coding lncRNAs. Adv. Sci. **11**(22), e2400009 (2024)
15. Hassani, K., Khasahmadi, A.H.: Contrastive multi-view representation learning on graphs. In: Proceeding 37th Internation Conference Machine Learning (ICML), vol. 119, pp. 4116–4126. PMLR (2020)
16. You, Y., et al.: Graph contrastive learning with augmentations. In: Proceedings 34th International Conference Neural Information Processing Systems (NeurIPS), pp. 5812–5824. Curran Associates (2020)
17. Liu, Z. et al.: Self-paced ensemble for highly imbalanced massive data classification. In: Proc. IEEE 36th Int. Conf. Data Engineering (ICDE), pp. 841–852. IEEE, Dallas (2020)
18. Cai, T., et al.: Are all negatives created equal in contrastive instance discrimination? arXiv preprint arXiv:2010.06682. (2020)
19. Luo, Y., et al.: A network integration approach for drug–target interaction prediction and computational drug repositioning from heterogeneous information. Nat. Commun. **8**, 573 (2017)
20. Chen, X., et al.: Drug–target interaction prediction: databases, web servers and computational models. Brief. Bioinform. **17**(4), 696–712 (2016)
21. Wishart, D.S., et al.: DrugBank 5.0: a major update to the DrugBank database for 2018. Nucleic Acids Res. **46**(D1), D1074–D1082 (2018)
22. The UniProt Consortium: UniProt: the universal protein knowledgebase in 2021. Nucleic Acids Res. **49**(D1), D480–D489 (2021)
23. Davis, A.P., et al.: Comparative Toxicogenomics database (CTD): update 2021. Nucleic Acids Res. **49**(D1), D1138–D1143 (2021)
24. Kuhn, M., et al.: The SIDER database of drugs and side effects. Nucleic Acids Res. **44**(D1), D1075–D1079 (2016)
25. Zitnik, M., et al.: Modeling polypharmacy side effects with graph convolutional networks. Bioinformatics. **34**(13), i457–i466 (2018)
26. Veličković, P., et al.: Graph attention networks. arXiv preprint arXiv:1710.10903. (2017)
27. Peng, J., et al.: A learning-based method for drug-target interaction prediction based on feature representation learning and deep neural network. BMC Bioinform. **21**, 394 (2020)
28. Nguyen, T., et al.: GraphDTA: predicting drug–target binding affinity with graph neural networks. Bioinformatics. **37**(8), 1140–1147 (2021)
29. Shang, Y., et al.: Multiview network embedding for drug–target interactions prediction by consistent and complementary information preserving. Brief. Bioinform. **23**(3), bbac059 (2022)
30. Mok, T.S., et al.: Gefitinib or carboplatin-paclitaxel in pulmonary adenocarcinoma. N. Engl. J. Med. **361**(10), 947–957 (2009)
31. Zhou, C., et al.: Erlotinib versus chemotherapy as first-line treatment for patients with advanced EGFR mutation-positive non-small-cell lung cancer (OPTIMAL, CTONG-0802): a multicentre, open-label, randomised, phase 3 study. Lancet Oncol. **12**(8), 735–742 (2011)
32. Bonner, J.A., et al.: Radiotherapy plus cetuximab for squamous-cell carcinoma of the head and neck. N. Engl. J. Med. **354**(6), 567–578 (2006)
33. Segovia-Mendoza, M., et al.: Efficacy and mechanism of action of the tyrosine kinase inhibitors gefitinib, lapatinib and neratinib in the treatment of HER2-positive breast cancer: preclinical and clinical evidence. Am. J. Cancer Res. **5**(9), 2531–2561 (2015)
34. Ahmed, K., et al.: A second WNT for old drugs: drug repositioning against WNT-dependent cancers. Cancer. **8**(7), 66 (2016)

35. Ferreira, L.L.G., Andricopulo, A.D.: ADMET modeling approaches in drug discovery. Drug Discov. Today. **24**(5), 1157–1165 (2019)
36. Liu, Y., et al.: CB-Dock2: improved protein–ligand blind docking by integrating cavity detection, docking and homologous template fitting. Nucleic Acids Res. **50**(W1), W159–W164 (2022)

Natural Product Identification Against Fusarium Oxysporum with CatBoost

Jing Li[1], Shuyu Han[1](✉), Xiao Lin[2](✉), and Bin Yang[1](✉)

[1] School of Information Science and Engineering, Zaozhuang University, Zaozhuang, Shandong 277160, China
3974749236@qq.com, batsi@126.com
[2] Zaozhuang Municipal Hospital, Zaozhuang, China
linxiaofly@126.com

Abstract. Cucumber Fusarium wilt, caused by *Fusarium oxysporum*, is a severe soil-borne fungal disease that substantially reduces crop yield and compromises quality. To enhance the precision of anti-fungal compound screening, this paper proposes a CatBoost-based machine learning approach for identifying natural compounds active against *Fusarium oxysporum*. A dataset containing both active and inactive compounds associated with the pathogen was constructed. Experimental results with 3-fold, 5-fold, and 10-fold cross-validation show that the CatBoost model achieves superior performance over AdaBoost, GBM, Random Forest, and Voting ensembles in terms of Accuracy, Precision, Recall, F1-score, and AUC. This work offers a robust machine learning framework for early-stage detection and intelligent diagnosis of cucumber wilt, and provides valuable data insights and methodological references for developing predictive systems in plant disease management.

Keywords: CatBoost · cucumber Fusarium wilt · Fusarium oxysporum · ensemble learning · machine learning · antifungal compound screening

1 Introduction

Cucumber (*Cucumis sativus*) is a major economically important vegetable crop in China, widely cultivated in both greenhouse and open-field environments. However, cucumbers are highly susceptible to Fusarium wilt under high temperature and humidity conditions, especially the form caused by *Fusarium oxysporum* [1, 2]. This disease severely damages the root system, leading to plant wilting and even death, thereby causing significant losses in agricultural production.

Traditional methods for screening antifungal compounds largely rely on experimental assays and manual interpretation—particularly through determining the Minimum Inhibitory Concentration (MIC)—which are resource-intensive and time-consuming. In recent years, with the rapid development of artificial intelligence and big data technologies, machine learning approaches have been increasingly applied to agricultural drug discovery and mechanism studies. By constructing predictive models, it becomes

L. Huang and D. Greenhalgh (Eds.): ICMLC 2026, LNNS 2061, pp. 869–878, 2027.
https://doi.org/10.1007/978-3-032-31236-5_58

possible to preliminary assess the anti-fungal activity of candidate compounds, greatly enhancing screening efficiency.

In recent years, ensemble learning techniques have demonstrated excellent performance across various classification tasks. Particularly in agricultural applications such as compound screening and crop resistance research, Boosting algorithms have shown strong robustness and generalization capability. Among them, CatBoost—a highly efficient Boosting method excelling in handling categorical features and mitigating gradient bias—has been widely applied in domains including patent analysis [3], forest fire prediction [4], and air quality index forecasting [5], modeling [6, 7], bioinformatics [8–10], proving its substantial practical utility. Furthermore, this algorithm has achieved promising results in various scenarios such as drug activity prediction and biological classification.

Based on the collected experimental dataset comprising both active and inactive compounds against *Fusarium oxysporum*, this study constructed multiple mainstream ensemble learning models, including CatBoost, AdaBoost, GBM, Random Forest, and a Voting classifier. Experimental results demonstrate that under 3-fold, 5-fold, and 10-fold cross-validation settings, the CatBoost model outperformed AdaBoost, GBM, Random Forest, and Voting in terms of Accuracy, Precision, Recall, F1-score, and AUC.

2 Methods and Materials

2.1 CatBoost Algorithm

CatBoost (Categorical Boosting) is an efficient ensemble learning algorithm based on Gradient Boosting Decision Trees (GBDT) [11]. It is specifically designed to work with structured data and categorical features. The algorithm employs innovative strategies such as target statistics for encoding categorical variables and ordered boosting to effectively mitigate gradient bias and reduce overfitting.

In this study, CatBoost was applied to the identification of antifungal compounds. Using structured feature data as input, we constructed a binary classification model to predict whether a given compound acts as an inhibitor against *Fusarium oxysporum*.

The algorithm workflow is as follows:

(1) Model Initialization

Let the training dataset be denoted as:

$$\mathrm{D} = \{(\mathrm{x}_1, \mathrm{y}_1), (\mathrm{x}_2, \mathrm{y}_2), \ldots, (\mathrm{x}_\mathrm{n}, \mathrm{y}_\mathrm{n})\} \tag{1}$$

where x_i represents the feature vector and y_i denotes the target label (for either classification or regression tasks). The initial model output is set to a constant value:

$$F_0(x) = \arg^{min}_{c} \sum_{i=1}^{n} L(y_i, c) \tag{2}$$

where $\mathrm{L}(\cdot)$ denotes the loss function.

(2) Ordered Encoding of Categorical Features

For categorical features, CatBoost utilizes Ordered Target Statistics (OTS) for encoding, effectively preventing target leakage.

$$\mathrm{TS_i} \frac{\sum_{j=1}^{\mathrm{ord(i)}-1} \mathrm{y_i} + \mathrm{a} * \mathrm{p}}{\mathrm{ord(i)} - 1 + \mathrm{a}} \tag{3}$$

where, ord(i) indicates the position of sample i in the randomly permuted sequence. P denotes the global mean of the target variable. And a is the smoothing coefficient.

During the encoding process, only the information from samples preceding the current one in the ordered sequence is utilized, thereby ensuring consistency in data distribution. This approach prevents information leakage and maintains data distribution consistency between training and validation phases

(3) Iterative Training of Decision Trees

This process is repeated for iterations t = 1,2,…,T(where T denotes the maximum number of iterations:

1) Gradient Calculation

Taking binary classification with log loss as an example:

$$\mathrm{L} = -\frac{1}{\mathrm{n}} \sum_{\mathrm{i}=1}^{\mathrm{n}} \left[\mathrm{y_i} \log \mathrm{p_i} + (1 - \mathrm{y_i}) \log(1 - \mathrm{p_i})\right] \tag{4}$$

The predicted probability is given by:

$$p_i = \frac{1}{1 + e^{-F_{i-1(x_i)}}} \tag{5}$$

The gradient (negative residual) is computed as:

$$g_i = p_i - y_i \tag{6}$$

2) Building Symmetric Trees (Oblivious Decision Trees)

When selecting splitting criteria from the feature set, CatBoost constructs symmetric trees—also known as oblivious decision trees—where all nodes at the same level use identical features and thresholds for splitting. This design significantly accelerates the prediction process and enhances model generalization by reducing overfitting.

3) Model Update

The decision tree trained at the t th iteration is denoted as ht. The model is then updated using the following additive formula:

$$F_i(x) = F_{i-1}(x) + \eta * \mathrm{h_i}(\mathrm{x}) \tag{7}$$

where η represents the learning rate, controlling the step size of each update to ensure stable convergence

4) Final Model Output

Upon completion of all iterations, the algorithm outputs the final strong classifier:

$$F_T(x) = F_0(x) + \sum_{t=1}^{T} \eta * h_i(x) \tag{8}$$

For binary classification tasks, the predicted class is determined as follows:

$$\mathrm{H}(x) = \mathrm{sign}(F_T(X)) \tag{9}$$

2.2 Catboost-Based Screening Algorithm for Natural Compounds Against Cucumber Fusarium Wilt Caused by Fusarium Oxysporum

To accurately identify natural compounds effective against *Fusarium oxysporum* causing cucumber Fusarium wilt, a CatBoost-based algorithm for active compound recognition is proposed. The procedural workflow is depicted in Fig. 1, with detailed steps as follows:

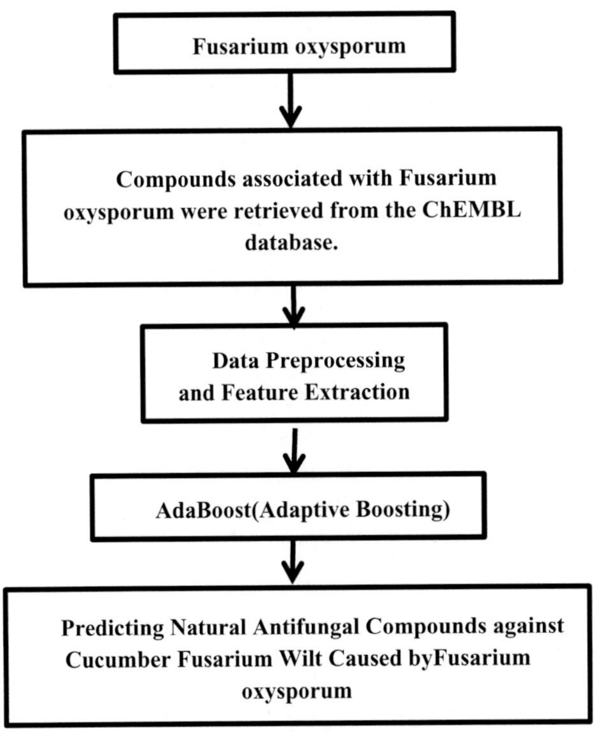

Fig. 1. Algorithm workflow.

1. **Data Collection**. The dataset used in this study was sourced from the ChEMBL database (Chemical Entities of Biological Interest, https://www.ebi.ac.uk/chembl/) [12]. We extracted compound activity records associated with Fusarium oxysporum from this repository. To ensure data quality, we removed duplicate entries, records lacking Minimum Inhibitory Concentration (MIC) values, and ambiguous compounds with multiple conflicting MIC measurements. Based on established literature and experimental standards, compounds with MIC <50 μg/mL were labeled as "0" (indicating inhibitory activity against the pathogen), while those with MIC ≥50 μg/mL were labeled as "1" (indicating no significant inhibitory activity). This process resulted in a standardized binary classification dataset where each row represents a unique compound. Molecular descriptors for each compound were subsequently calculated using RDKit to serve as predictive features
2. **Screening Process**. Utilizing the collected dataset, the CatBoost algorithm was employed for training and learning to derive an optimal model capable of predicting natural pharmaceutical compounds effective against *Fusarium oxysporum* causing cucumber Fusarium wilt. This model enables accurate and efficient identification of potential antifungal agents, significantly streamlining the drug discovery process.

3 Experimental Results

The compiled dataset consisted of 733 compounds, including 313 positive samples (inhibitors) and 420 negative samples (non-inhibitors). To evaluate the performance of the CatBoost algorithm in identifying compounds effective against cucumber Fusarium wilt, four classical machine learning algorithms (AdaBoost, GBM, Random Forest, and Voting) were also implemented for comparison. All models were developed in Python using widely adopted libraries including scikit-learn, XGBoost, and CatBoost, ensuring reproducibility and alignment with common computational practices.

To assess the performance of the models, this study employs three key metrics: Accuracy, Precision, and F1-score. The formulas are defined as follows:

$$\text{Accuracy} = \frac{\text{TP} + \text{TN}}{\text{TP} + \text{TN} + \text{FP} + \text{FN}} \tag{10}$$

$$\text{Precision} = \frac{\text{TP}}{\text{TP} + \text{FP}} \tag{11}$$

$$\text{F1} - \text{score} = 2 * \frac{\text{Precision} * \text{Recall}}{\text{Precision} + \text{Recall}} \tag{12}$$

where:

TP = True Positives,
TN = True Negatives,
FP = False Positives,
FN = False Negatives.

To enhance the robustness of the experimental results, 3-fold, 5-fold and 10-fold cross-validation strategies were employed for model training and testing. The performance of each model under different cross-validation settings was systematically recorded and compared. Additionally, ROC curves were plotted and AUC values calculated to visually demonstrate and quantitatively assess the classification capability of each model.

Tables 1, 2, and 3 present the main performance metrics of each model under 3-fold, 5-fold and 10-fold cross-validation, respectively. It can be observed that CatBoost consistently outperforms the other models across multiple evaluation metrics, demonstrating the most stable and superior performance. Notably, CatBoost maintains a clear advantage in Accuracy, F1-score, and AUC values.

Table 1. Performance of the five algorithms under 3-fold cross-validation.

Model	Accuracy	F1	AUC
CatBoost	**0.848271**	**0.821419**	**0.919823**
Random Forest	0.797819	0.746956	0.905843
GBM	0.842894	0.816100	0.907326
Voting	0.836017	0.805505	0.912722
AdaBoost	0.726810	0.653644	0.802990

Table 2. Performance of the five algorithms under 5-fold cross-validation.

Model	Accuracy	F1	AUC
CatBoost	**0.807377**	**0.773672**	**0.898006**
Random Forest	0.792350	0.744869	0.898309
GBM	0.810109	0.777389	0.886387
Voting	0.806011	0.767105	0.899564
AdaBoost	0.715847	0.627796	0.805118

A further comparison of the models' predictive capabilities is provided by the ROC curves under 3-, 5-, and 10-fold cross-validation, presented in Fig. 2, Fig. 3 and Fig. 4.

Figure 2 displays a comparison of the ROC curves for each model under 3-fold cross-validation. The results indicate that CatBoost, Random Forest, and Voting achieved the best performance. Notably, the ROC curve of CatBoost appears smoother, demonstrating superior and more consistent classification capability across different decision thresholds.

Figure 3 presents a comparison of the ROC curves of all models under 5-fold cross-validation. The results show that the CatBoost model delivered the most prominent

Table 3. Performance of the five algorithms under 10-fold cross-validation.

Model	Accuracy	F1	AUC
CatBoost	**0.849704**	**0.822065**	**0.920300**
Random Forest	0.803277	0.749592	0.901921
GBM	0.831988	0.802764	0.907782
Voting	0.840152	0.808859	0.914133
AdaBoost	0.715902	0.624840	0.804241

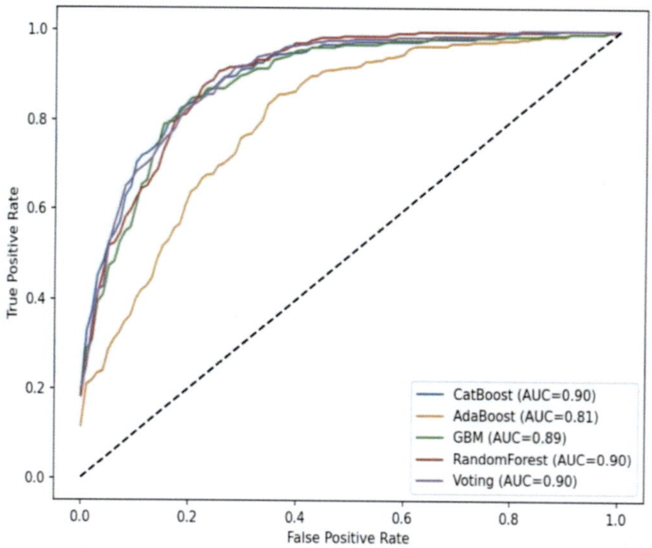

Fig. 2. ROC curves and AUC values of the five algorithms using 3-fold cross-validation.

performance, achieving the highest AUC value of 0.92 among all models. Its ROC curve is positioned closest to the top-left corner, indicating consistently high sensitivity and specificity across various classification thresholds. The Random Forest, GBM, and Voting models followed, each attaining an AUC of 0.91. However, their ROC trajectories are slightly inferior to that of CatBoost, suggesting certain limitations in handling borderline samples and capturing complex feature interactions.

Figure 4 illustrates the ROC curves of all models under 10-fold cross-validation, further validating the overall performance of each approach. CatBoost delivered the strongest performance with an AUC of 0.92. Its ROC curve lies closest to the top-left corner across all decision thresholds, highlighting its superior ability to balance sensitivity and specificity. The GBM, Voting, and Random Forest models achieved AUC values of 0.91, 0.91, and 0.90, respectively, showing relatively comparable—though slightly inferior—curve progression compared to CatBoost. This suggests certain limitations in

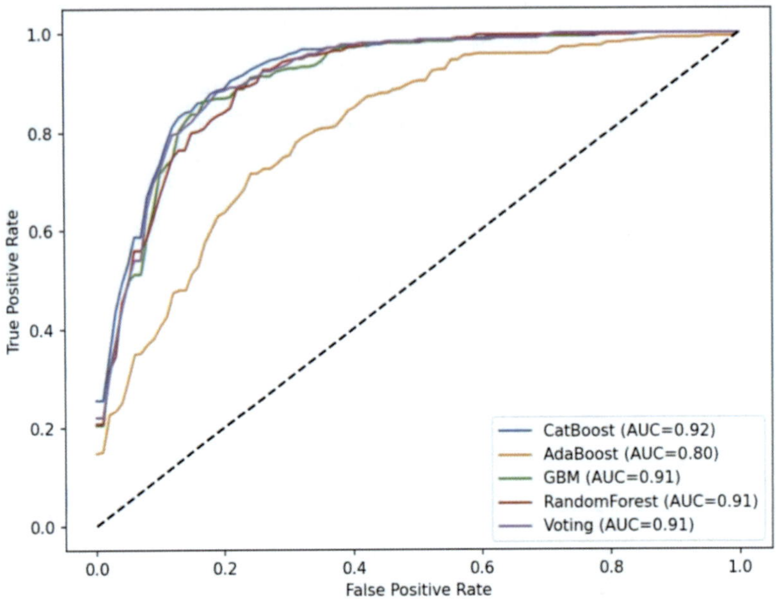

Fig. 3. ROC curves and AUC values of the five algorithms using 5-fold cross-validation.

modelling complex feature interactions and capturing nonlinear relationships. In contrast, AdaBoost attained an AUC of only 0.80, significantly lower than the other models, indicating its limited adaptability to this high-dimensional data scenario.

4 Conclusion

In summary, under 3-fold, 5-fold and 10-fold cross-validation frameworks, the CatBoost model demonstrated superior performance compared to AdaBoost, GBM, Random Forest, and Voting across multiple metrics including Accuracy, Precision, Recall, F1-score, and AUC. Quantitative results show that CatBoost achieved the best performance in key indicators (mean AUC: 0.92 ± 0.02). GBM and Random Forest followed closely (AUC: 0.90 ± 0.01), exhibiting robustness and reliability across most evaluation dimensions. The Voting ensemble also performed well overall, though with slight fluctuations in certain folds. In contrast, AdaBoost delivered the weakest performance (AUC: 0.80 ± 0.01), particularly in key metrics such as Recall and AUC. Although CatBoost incurs higher computational overhead, its exceptional accuracy and stability make it particularly suitable for high-precision tasks such as anti-fungal compound screening. Future work will explore model simplification techniques to better balance its computational cost.

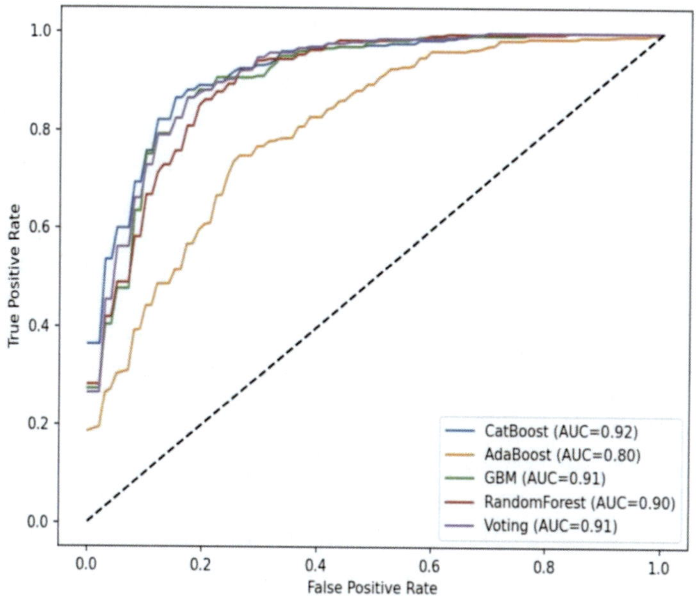

Fig. 4. ROC curves and AUC values of the five algorithms using 10-fold cross-validation.

Acknowledgements. This work was supported by Shandong Natural Science Fund Project (No. ZR2023MF061), the talent project of "Qingtan Scholar" of Zaozhuang University and the Natural Science Foundation of China (No. 61702445).

References

1. Zhao, G.P., Yang, R.Q., Li, J., et al.: Screening of natural drug molecules against fusarium oxysporum of ginseng root rot based on machine learning. Acta Pharm. Sin. **58**(6), 1713–1721 (2023)
2. Chen, H., Ma, Y., Zhou, Y., et al.: Pharmacophore identification, virtual screening and activity verification of pedunsaponin a on target proteins PcAdv and PcnWAS of Pomacea canaliculata. Pest Manag. Sci. **78**, 3098–3107 (2022)
3. Ding, Q., Song, Y., Song, H., et al.: Predicting patent tradability using CatBoost: an empirical study based on U.S. physics patents from 2002 to 2012. Libr. Inf. Serv. **69**(16), 79–89 (2025)
4. Yang, Z., Zhang, X., Ren, M., et al.: Forest fire prediction in Muti County, Sichuan based on CatBoost. Sci. Technol. Eng. **25**(21), 8823–8832 (2025)
5. Akiner, E.M., Katipoğlu, M.O., Çintaş, E.: Predicting air quality index in başakşehir, Istanbul with hybrid ai models: unveiling key drivers through catboost-based shap and feature importance analysis. Theor. Appl. Climatol. **156**(8), 422 (2025)
6. Liu, A., Fan, P., He, X., et al.: CatBoost-based fuel consumption modeling and explainable analysis for heavy-duty diesel trucks: impact of engine, driving behavior, and vehicle weight. Energy. **333**, 137485 (2025)
7. Rivera, R.A.F., García, S.E.A., Ramírez, N.V., et al.: Saturated flow boiling frictional pressure drop inside smooth macro and mini/micro channels: a new predictive tool using CatBoost and XGBoost. Case Stud. Therm. Eng. **73**, 106655 (2025)

8. Wu, J., Tao, G., Xie, S., et al.: Prediction of three-year all-cause mortality in patients with heart failure and atrial fibrillation using the CatBoost model. BMC Cardiovasc. Disord. **25**(1), 466 (2025)
9. Naseem, Z., Khaliq, Z., Bentalib, A., et al.: Enhanced extraction of bioactive agents from Cymbopogon citratus via microwave-assisted choline chloride-glycerol deep eutectic solvent: a data-driven approach guided by CatBoost Modeling. Food Bioprocess Technol.., (prepublish), 1–17 (2025)
10. Senthilkumar, G., Pitchaimuthu, R., Dhanasekaran, S., et al.: Integrating clinical and transcriptomic profiles associated with vitamin D to enhance disease-free survival in cervical cancer recurrence using the CatBoost algorithm. Diagnostics. **15**(13), 1579 (2025)
11. Prokhorenkova, L., Gusev, G., Vorobev, A., et al.: CatBoost: unbiased boosting with categorical features. NeurIPS. (2018)
12. Barbara, Z., Eloy, F., Fiona, H.: The ChEMBL database in 2023: a drug discovery platform spanning multiple bioactivity data types and time periods. Nucleic Acids Res. **52**(D1), D1180–D1192 (2023)

PLM-Driven Differentiable Clustering and Graph Reasoning for Protein Interaction Site Prediction

Wenyu Xi, Xiucai Ye(✉), Tianyi Shi, and Tetsuya Sakurai

Tsukuba University, Tsukuba 3050005, Japan
yexiucai@cs.tsukuba.ac.jp

Abstract. The accurate prediction of protein binding sites plays a vital role in elucidating molecular interaction mechanisms and facilitating computational drug development. However, existing sequence-based and structure-based approaches often fail to capture the complementary biological information embedded in both modalities, limiting their accuracy and generalizability. To overcome this limitation, we introduce GLBind, multimodal deep learning architecture that integrates LoRA-enhanced protein language models with structure-aware graph representations. GLBind employs a lightweight adapted ESM 2 encoder to extract contextual and evolutionary sequence semantics, while residue-level graphs constructed from AlphaFold2-predicted structures provide spatial and physicochemical cues necessary for recognizing surface interaction patterns. A differentiable clustering module further abstracts low-level geometric signals into interpretable structural representations, and a cross attention mechanism combined with gated fusion enables effective alignment and integration of sequence and structural features. We evaluate GLBind on three benchmark datasets covering protein—DNA, protein—RNA, and protein—protein interactions. Our model consistently surpasses existing sequence-based and structure-based state-of-the-art methods across all benchmarks, achieving substantial and consistent performance improvements. These results demonstrate that combining protein language models with hierarchical structural reasoning provides a robust and generalizable solution for residue-level interaction site prediction, offering strong potential for downstream biological analysis and drug discovery applications.

Keywords: protein binding site prediction · differentiable cluster · LoRA-ESM · Cross-attention mechanism

1 Introduction

Protein–nucleic acid interactions constitute a fundamental element in a diverse array of biological mechanisms, encompassing gene replication and expression, signal transduction, cellular regulation, and metabolic pathways [1, 2]. Investigating these interactions is critical for clarifying protein functionality, interpreting genetic information, and advancing targeted drug design. Nevertheless, many existing experimental techniques,

L. Huang and D. Greenhalgh (Eds.): ICMLC 2026, LNNS 2061, pp. 879–892, 2027.
https://doi.org/10.1007/978-3-032-31236-5_59

such as X-ray crystallography, nuclear magnetic resonance (NMR) spectroscopy, and laser Raman spectroscopy, are time-consuming and costly [3]. Consequently, there exists a pressing demand for the development of dependable and precise computational techniques capable of performing large-scale identification of residues that bind nucleic acids.

Current computational strategies for predicting protein–nucleic acid interaction sites are broadly categorized into two principal groups: sequence-based methods and structure-based methods [4–6]. Sequence-based approaches rely solely on the primary amino acid sequence to infer potential binding residues, with representative examples including SCRIBER [7], TargetDNA [8], DRNApred [9], and hybridRNAbind [5]. For instance, DRNApred mines evolutionary conservation profiles and predicted solvent accessibility features from protein sequences, employing an ensemble hyperplane-distance support vector machine (E-HDSVM) to distinguish DNA-binding residues based on sequential context.

The principal advantage of sequence-based approaches lies in their applicability to any protein with available sequence data [10, 11]. However, the predictive capacity of such approaches can be limited since characteristic patterns of nucleic acid-binding residues are frequently not fully captured by linear sequence information alone, but are rather embedded within the three-dimensional structural architecture [12]. Consequently, features derived exclusively from sequence information may be insufficient to accurately represent the functional characteristics of binding residues.

Conversely, structure-based methodologies exploit three-dimensional structural data to pinpoint residues involved in nucleic acid binding. Typically, such approaches derive geometric, physicochemical, and spatial descriptors—including solvent-accessible surface area, secondary structure, electrostatic potential, and residue depth—from either experimentally resolved or computationally predicted protein structures. Representative methods include NucBind [13], DNABind [14], GraphBind [15], GraphSite [16] and EquiPNAS [17], which employ graph-based representations, deep learning frameworks, or machine learning classifiers to capture spatial patterns associated with binding sites [18–20]. By incorporating structural context, structure-based methods can capture conserved binding motifs that are not evident from sequence information alone [21, 22]. As a result, they generally offer improved predictive accuracy over sequence-based counterparts, particularly when high-quality structural data are available. However, their applicability is often limited by the availability and quality of protein structures, which may not be experimentally determined for all proteins and can introduce inaccuracies when predicted computationally.

The rapid progress in deep learning has facilitated the application of large-scale protein language models like ESM [23], ProtTrans [24], and MSA-Transformer [4], which learn contextual and evolutionary representations directly from large-scale protein sequences. These models have shown strong potential in various downstream tasks, including binding site prediction. Concurrently, AlphaFold2 has revolutionized protein structure prediction, providing accurate three-dimensional models from sequences alone. The widespread availability of AlphaFold2-predicted structures allows structure-based methods to be applied to proteins lacking experimentally resolved structures. The fusion of embeddings from language models with structural features derived from AlphaFold2 presents a promising avenue for integrating sequence-based and spatial information, thereby supporting more robust and generalizable predictions of nucleic acid-binding residues.

To address the limitations of current methods, we propose GLBind, a multimodal deep learning framework designed for predicting nucleic acid-binding sites. GLBind integrates LoRA enhanced ESM 2 based protein language representations with structure aware graph learning, enabling the capture of contextual sequence semantics and spatial biochemical patterns. Specifically, sequence embeddings are obtained from a fine-tuned ESM 2 model, whereas structural information is extracted from AlphaFold2-predicted protein models by constructing residue-level graphs. These graphs are then processed by a graph neural network (GNN) and subsequently passed through a differentiable pooling module that groups residues into functionally coherent clusters, enabling hierarchical and interpretable representations. Cross-attention modules are applied to align sequence- and structure-derived features, which are then fused through a gating mechanism that adaptively balances contributions from both modalities. This framework not only elevates prediction accuracy at the residue level but also improves biological interpretability by establishing connections among structural context, functionally coherent residue clusters, and contextualized sequence semantics.

2 Materials

2.1 Dataset

We conducted a comprehensive performance assessment of our method using established benchmark datasets for predicting protein–protein and protein–nucleic acid interaction sites, which were adopted from previous studies such as GraphBind and GraphRBF.

To evaluate protein–nucleic acid interactions, benchmark datasets were constructed using BioLiP2 [25], a manually curated database that documents biologically significant ligand–protein interactions with experimentally resolved complex structures. This database supplies pre-annotated binding site information for known protein–DNA and protein–RNA complexes.

For protein–protein interaction analysis, datasets were sourced from the Protein–Protein Docking Benchmark 5.5 (DBD 5.5) [26] and Dockground [27], both comprising non-redundant, high-resolution protein complex structures. Each entry includes both bound and unbound conformations of the interacting proteins. Given that proteins frequently experience conformational changes during interaction, unbound structures were employed for both training and evaluation to more accurately mimic real biological scenarios. A detailed breakdown of the dataset composition is provided in Table 1.

2.2 Overview of GLBind

We introduce GLBind, a novel computational approach designed to predict binding sites for both protein–protein and protein–nucleic acid interactions. As depicted in Fig. 1, GLBind integrates both sequence and structure modalities through a unified architecture. Protein sequences are first encoded using a LoRA-enhanced ESM 2 model to capture rich contextual and evolutionary representations. Concurrently, AlphaFold2-predicted structures are utilized to construct residue-level graphs based on spatial proximity; these graphs are subsequently processed by a graph neural network to learn structural representations. To improve interpretability and support hierarchical modeling, a differentiable pooling mechanism is applied to cluster residues into functionally coherent regions. Cross-attention modules are employed to align the sequence and structure representations, enabling each modality to attend to complementary information from the other. The resulting embeddings are fused via a gating mechanism that adaptively balances their contributions. Finally, the fused features are used to predict whether each residue is part of a binding interface. Through this design, GLBind effectively combines sequence semantics, spatial structure, and functional organization to achieve accurate and interpretable binding site prediction across multiple interaction types.

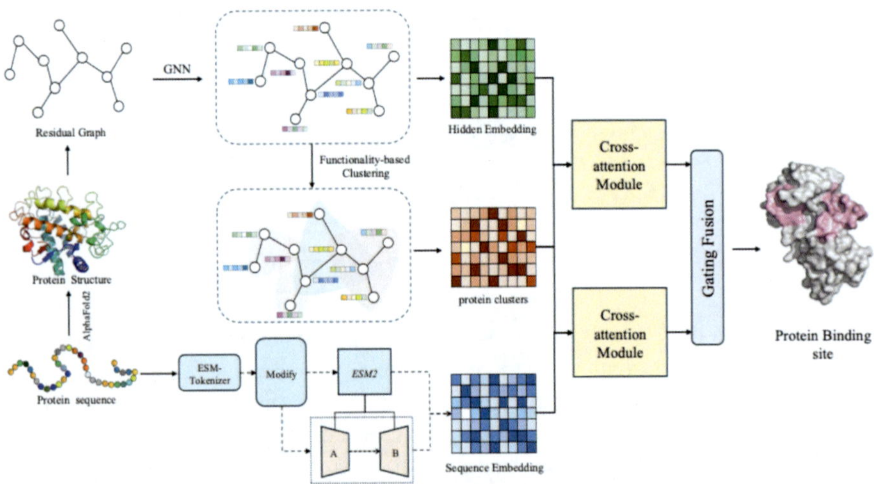

Fig. 1. Overview of GLBind. The model integrates both sequence and structure modalities. Protein sequences are processed using a LoRA-enhanced ESM 2 model to extract contextual embeddings, while AlphaFold2-predicted structures are used to construct residue-level graphs. A graph neural network (GNN) extracts spatial and physicochemical features, followed by a differentiable pooling module that clusters residues into functionally relevant groups. Cross-attention mechanisms are applied to both sequence and structure branches to capture inter-modality dependencies. The resulting embeddings are fused via a gating mechanism, and the final representation is used for residue-level binding site prediction.

2.3 Graph Representation and Feature Generation

Each protein is modeled as a graph G = (X, E, A), where nodes correspond to amino acid residues and edges capture spatial relationships between residues. The node feature matrix is defined as $X = \{\hat{x}_i\}_{i=1}^{N},$, with x_i, denoting the feature vector of the i-th residue. The edge feature set is given by $E = \{e_{ij} \mid A_{ij} = 1\}$, where each edge feature $e_{ij} \in \mathbb{R}^2$ encodes the Euclidean distance and relative orientation between a pair of connected residues. This graph formulation allows GLBind to jointly model sequence-derived semantics and structure-aware spatial relationships, facilitating accurate binding site prediction.

For node representations, we incorporate a diverse set of widely used sequence-derived features, including amino acid residue identities and position-specific scoring matrices (PSSMs), to capture evolutionary conservation. In addition, a comprehensive suite of structure-based features is extracted from protein monomer structures obtained either experimentally or predicted by AlphaFold2. These structural descriptors include secondary structure (SS), relative solvent accessibility (RSA), local backbone geometry, residue orientation, spatial coordinates, virtual surface area, and contact number. Together, these features provide rich sequential and spatial context, which is essential for reliable identification of protein binding sites.

Table 1. Summary of the benchmark datasets.

Type	Dataset	$N_{protein}$	N_{pos}	N_{neg}
Protein	P_train	1001	25172	264564
	P_test	250	6386	73391
DNA	D_train	573	14479	145404
	D_test	129	2240	35275
RNA	R_train	1479	33444	384358
	R_test	374	9331	110259

3 Method

3.1 LoRA-ESM

Protein language models (PLMs), such as ESM 2, have demonstrated strong performance across various sequence-based protein prediction tasks. However, full fine-tuning of these large-scale Transformer-based models is computationally expensive and memory-intensive. To address this limitation, we use a Low-Rank Adaptation (LoRA) [28], enabling scalable and efficient adaptation of ESM 2 for downstream protein-related tasks.

3.2 LoRA-ESM

In standard deep learning, model parameters are updated via gradient descent to minimize a loss function. For a typical feed-forward layer with weight matrix W, the update is expressed as:

$$W' = W - \eta \nabla L = W + \Delta W \tag{1}$$

where η is the learning rate and ∇L is the gradient of the loss with respect to W. During forward computation, the output becomes:

$$O = F(W'x) = \text{model}(Wx + \Delta Wx) \quad (2)$$

Computing and storing ∇W becomes costly in large models. LoRA mitigates this by reparameterizing the update as a product of two low-rank matrices:

$$\Delta W = A \cdot B, A \in \mathbb{R}^{d\times r}, B \in \mathbb{R}^{r\times k}, r \ll \min(d, k) \quad (3)$$

The forward pass is then modified as:

$$O = F((W + AB)x) = F(Wx + A(Bx)) \quad (4)$$

The term $A(Bx)$ corresponds to two consecutive linear projections and forms a LoRA adapter. This design introduces minimal additional parameters while enabling task-specific adaptation.

3.3 Application to ESM 2

To adapt LoRA for protein sequence modeling, we integrate LoRA modules into the Transformer architecture of ESM 2. Specifically, we insert LoRA adapters into selected components of the self-attention mechanism, and output projection layers of the top- K Transformer blocks. The number of adapted layers K, the rank r, and the scaling factor α are treated as tunable hyperparameters.

During training, all pretrained weights in ESM 2 remain frozen. Only the LoRA-specific parameters A and B are updated, significantly reducing memory consumption and training time. The modified output of each attention projection becomes:

$$h = W_0 x + \alpha \cdot B(Ax) \quad (5)$$

where W_0 is the original frozen weight matrix and the low-rank term $B(Ax)$ allows the model to adapt to new tasks.

3.4 Cluster Attention for Protein Graph Representation

Traditional protein modeling approaches typically operate at the residue or atom level, which may result in increased computational cost and sensitivity to noise [29–31]. To alleviate this, we adopt a cluster-level attention framework that compresses graph information by assigning nodes into clusters through a differentiable pooling scheme, enabling efficient high-level representation learning.

Determining which residues or atoms are most functionally relevant within a protein graph is often non-trivial. To address this challenge, we leverage a soft clustering approach [32] that dynamically partitions nodes into a fixed number of clusters. This enables the model to learn coarse-grained interaction patterns without relying on fine-grained predefined structures.

We begin by generating soft cluster assignments for all nodes in the protein graph. This allows the model to group structurally or functionally related nodes together. Given node features $\overline{X} \in \mathbb{R}^{|V|\times d}$ and an intra-molecular adjacency matrix \$E\$, the cluster assignment matrix $S \in \mathbb{R}^{|V| \times c}$ is computed as:

$$S = softmax(GNN_\theta(X, E)) \tag{6}$$

where c denotes the number of target clusters. The softmax ensures that each node can participate in multiple clusters with learned attention scores. Once the soft assignments are established, cluster-level representations are formed by aggregating node embeddings according to their cluster membership:

$$Z = S^\top H \tag{7}$$

where $Z \in \mathbb{R}^{c \times d}$ denotes the cluster-level feature matrix for the protein graph.

To preserve topological relationships at the cluster level, we construct a cluster-wise adjacency matrix based on the original graph connections:

$$\tilde{A} = S^\top A S \tag{8}$$

where $A \in \mathbb{R}^{|V|\times |V|}$ is the original adjacency matrix, and $\tilde{A} \in \mathbb{R}^{c\times c}$ captures inter-cluster connectivity.

Finally, a graph neural network is applied to the resulting cluster graph to refine the representations:

$$Z^{\text{final}} = \text{GNN}_\psi\left(Z, \tilde{A}\right) \tag{9}$$

where $Z^{\text{final}} \in \mathbb{R}^{c \times d}$ encodes the final structure-aware cluster-level representation of the protein, effectively integrating both local features and global topological context. The number of clusters is treated as a tunable hyperparameter that controls the granularity of cluster-aware residue representations. In practice, we observe that the model performance remains stable across a moderate range of cluster numbers. In our model, the cluster number is fixed to K = 60, which provides a good balance between representational capacity and computational efficiency.

3.5 Cluster Attention for Protein Graph Representation

To enrich the learned representations, we introduce two cross-attention modules and a gating-based fusion mechanism.

Cluster-to-Graph Attention

We compute attention between the final cluster representation $Z^{\text{final}} \in \mathbb{R}^{c \times d}$ and the original residue-level GNN embedding $H \in \mathbb{R}^{|V| \times d}$. The matrices are defined as:

$$Q_{cg} = W_Q^{cg} Z^{\text{final}}, K_{cg} = W_K^{cg} H, V_{cg} = W_V^{cg} H \tag{10}$$

Then, the cluster-to-graph attention output is computed by scaled dot-product attention:

$$Z_{\text{cg-attn}} = \text{softmax}\left(\frac{Q_{cg}K_{cg}^{\top}}{\sqrt{d}}\right)V_{cg} \tag{11}$$

Sequence-to-Cluster Attention

Sequence embeddings $S \in \mathbb{R}^{L \times d}$ extracted from a LoRA-adapted ESM 2 model are used to attend to the cluster representation:

$$Q_{sc} = W_Q^{sc}S, K_{sc} = W_K^{sc}Z^{\text{final}}, V_{sc} = W_V^{sc}Z^{\text{final}} \tag{12}$$

$Z_{\text{sc-attn}} = \text{softmax}\left(\frac{Q_{sc}K_{sc}^{\top}}{\sqrt{d}}\right)V_{sc}$ (13)

Gated Fusion

To integrate the attention outputs from both structural and sequence modalities, we apply a gating mechanism:

$$\alpha = \sigma\left(W_g\left(Z_{\text{cg-attn}} + Z_{\text{sc-attn}}\right) + b_g\right) \tag{14}$$

$Z_{\text{fused}} = \alpha \cdot Z_{\text{cg - attn}} + (1 - \alpha) \cdot Z_{\text{sc - attn}}$ (15)

The fused representation Z_{fused} is subsequently passed to a downstream classifier to predict protein binding sites at the residue level.

4 Experiments

4.1 Comparison with State-of-the-Art Methods

The effectiveness of the proposed model was assessed through comparative experiments performed on three benchmark datasets: D_129, R_374, and P_250. As summarized in Fig. 2, our method consistently outperforms existing approaches across all evaluation metrics, including AUC, MCC, F1 score, precision, and recall.

On the D_129 dataset, our model attains the best performance in AUC (0.941), MCC (0.593), and F1 score (0.633), outperforming strong baselines such as GraphBind and GraphSite. The improved precision (0.608) and recall (0.662) further demonstrate its ability to balance sensitivity and specificity. On the R_374 dataset, which contains structurally diverse RNA-binding proteins, our model achieves an AUC of 0.914 and an MCC of 0.5162, surpassing GraphRBF by 1.2\% and 3.6\%, respectively. Notably, it achieves the highest recall (0.645), indicating its strong capability in identifying true RNA-binding residues. On the P_250 dataset for protein—protein site prediction, GLBind performs best, with an AUC of 0.832, MCC of 0.395, and F1 score of 0.433.

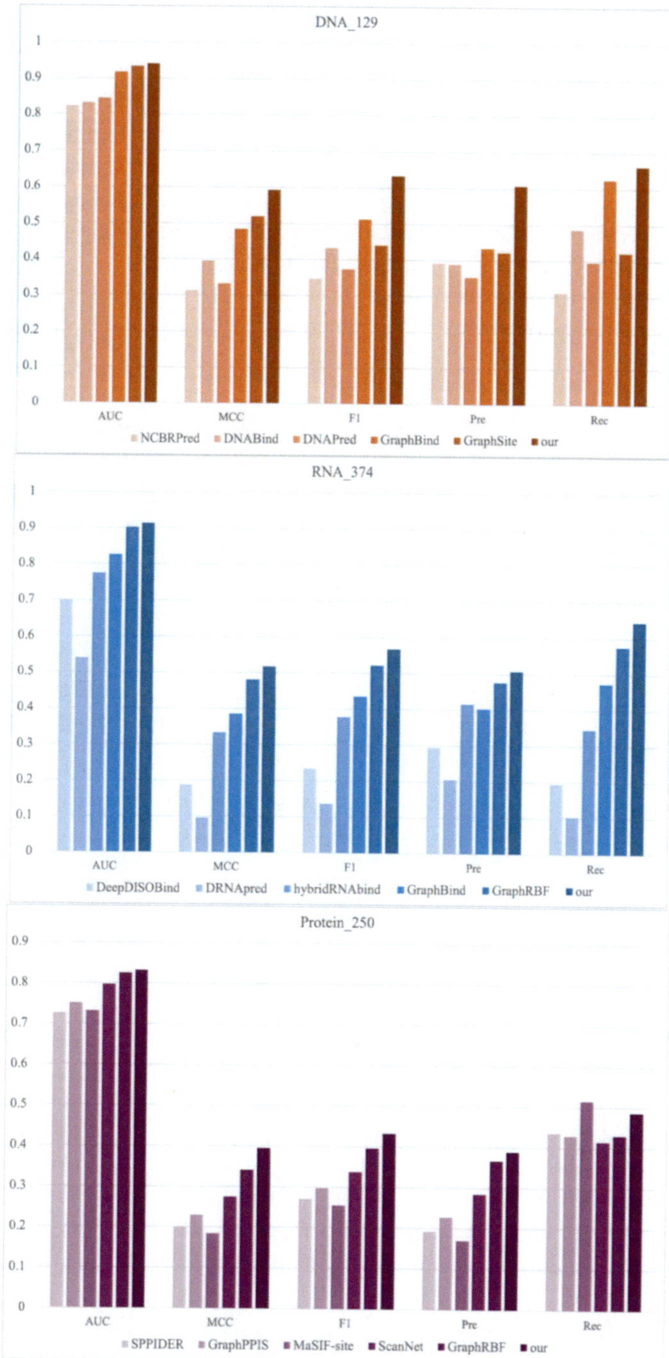

Fig. 2. Performance comparison of our method with state-of-the-art models on three benchmark datasets.

4.2 Comparison with State-of-the-Art Methods

To further analyze the role of each module in our model, we performed ablation studies by selectively removing (i) the LoRA-ESM sequence encoder, (ii) the clustering module, and (iii) the gating fusion component. As shown in Table 2, the performance consistently declined on all three datasets when any of these components was excluded. The removal of the LoRA-ESM encoder led to a noticeable drop in F1 and MCC, highlighting the importance of pretrained sequence embeddings. Excluding the cluster-guided representation resulted in the largest performance degradation in most cases, especially in recall, suggesting its critical role in capturing spatial patterns. The absence of the gating fusion also caused moderate performance loss, indicating the benefit of adaptive multimodal integration.

Table 2. Ablation study results.

Dataset	Model	AUC	MCC	F1	Pre	Rec
D_129	w/o lora-esm	0.927	0.480	0.508	0.431	0.618
	w/o cluster	0.911	0.439	0.391	0.279	0.651
	w/o gating fusion	0.935	0.584	0.615	0.533	0.741
	our	0.941	0.593	0.633	0.608	0.662
R_374	w/o lora-esm	0.897	0.478	0.528	0.535	0.520
	w/o cluster	0.880	0.381	0.399	0.589	0.302
	w/o gating fusion	0.907	0.507	0.557	0.498	0.631
	our	0.914	0.516	0.568	0.508	0.645
P_250	w/o lora-esm	0.823	0.313	0.347	0.392	0.312
	w/o cluster	0.803	0.229	0.263	0.242	0.289
	w/o gating fusion	0.825	0.279	0.290	0.199	0.535
	our	0.832	0.395	0.433	0.389	0.487

4.3 Comparison with State-of-the-Art Methods

We further explored the impact of two critical hyperparameters in LoRA-ESM: the rank r and the scaling factor α. As shown in Fig. 3, performance varied with different parameter settings across D_129, R_374, and P_250 datasets. Generally, increasing r improved the AUC, with $r = 16$ and $\alpha = 4$ yielding the highest AUC on all three datasets. This suggests that a moderate subspace dimension and balanced scaling achieve a favorable trade-off between model capacity. Excessively low or high values of either parameter led to suboptimal performance, underscoring the importance of proper tuning.

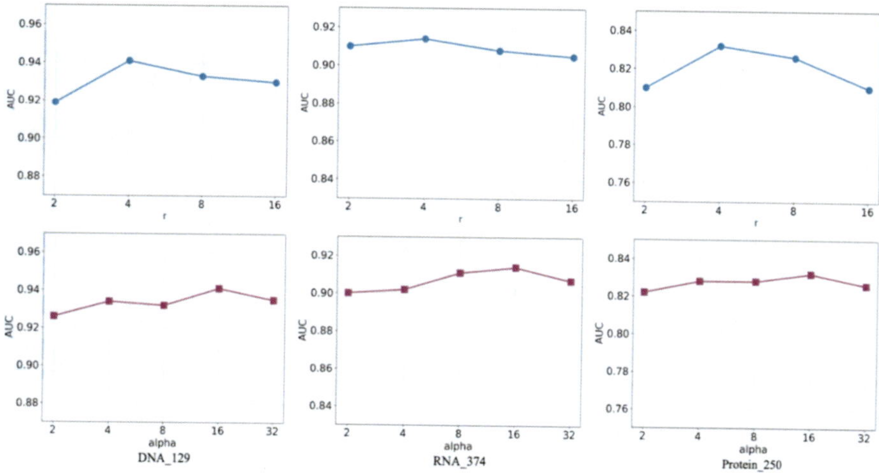

Fig. 3. AUC heatmaps of different r and α values on D_129, R_374, and P_250 datasets.

4.4 Case Study

To qualitatively demonstrate the prediction performance of GLBind, we selected one representative protein from each of the three datasets—DNA-binding, RNA-binding, and protein–protein interaction—and visualized the binding site predictions, as shown in Fig. 4. The top row presents the ground truth binding residues obtained from biological annotations, while the bottom row shows the prediction results generated by our model. It can be observed that, across different interaction types, GLBind is able to identify a substantial portion of the annotated binding regions, while maintaining reasonable precision. These case-level results offer an intuitive view of the model's strengths as well as areas where further refinement may be beneficial.

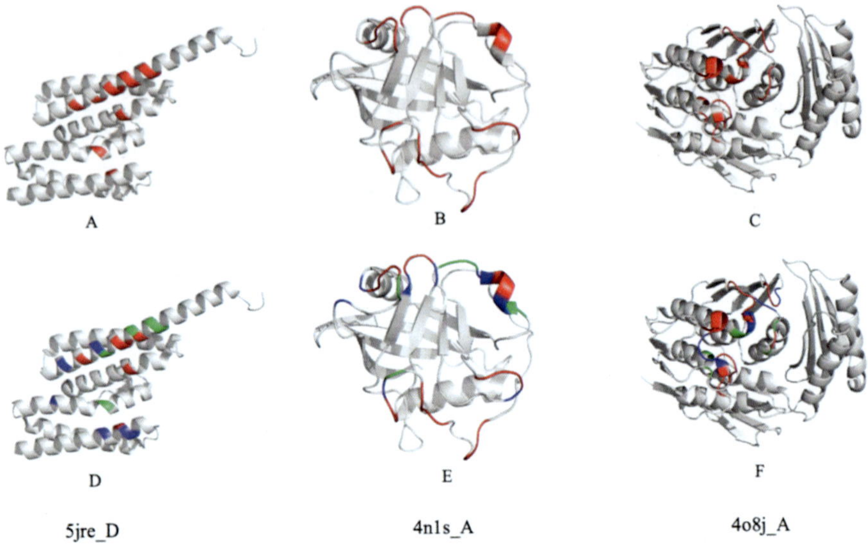

Fig. 4. Visualization of predicted and true binding sites on six representative protein chains. (A–C) Ground truth binding sites (red); (D–F) prediction results, where green indicates missed sites and blue denotes false positives.

5 Conclusion

In this work, we present GLBind, a novel multimodal framework for accurate and interpretable binding site prediction across protein–nucleic acid and protein–protein interactions. By integrating LoRA-enhanced protein language models with structure-aware graph representations, GLBind effectively captures both contextual sequence semantics and spatial structural features. The incorporation of differentiable clustering and cross-attention enables hierarchical modeling and inter-modal alignment, while the gated fusion mechanism allows for adaptive feature integration. Extensive experiments on three benchmark datasets demonstrate that GLBind achieves state-of-the-art performance across all evaluated metrics. Ablation studies and parameter analyses further validate the contribution of each component. Visualization of prediction results underscores the model's interpretability and practical potential. Overall, GLBind provides a unified and scalable solution for residue-level interaction site prediction and opens new avenues for structural bioinformatics and rational drug design.

Acknowledgments. The authors thank the anonymous reviewers for their valuable suggestions. The work was supported by the JSPS KAKENHI Grant Number JP23H03411 and JP22K12144, and the JST Grant Number JPMJPF2017.

References

1. Hartwell, L.H., Hopfield, J.J., Leibler, S., Murray, A.W.: From molecular to modular cell biology. Nature. **402**(Suppl 6761), C47–C52 (1999)

2. Zou, X., Zhang, C., Tang, M., Deng, L.: Thorough assessment of machine learning techniques for predicting protein–nucleic acid binding hot spots. Curr. Bioinforma. **19**(2), 144–161 (2024)
3. Kuzmanov, U., Emili, A.: Protein–protein interaction networks: probing disease mechanisms using model systems. Genome Med. **5**, 37 (2013)
4. Rao, R.M. et al.: MSA transformer. In: proceedings of the 38th international conference on machine learning (ICML 2021). PMLR. **139**, 8844–8856 (2021)
5. Zhang, F., Li, M., Zhang, J., Kurgan, L.: HybridRNAbind: prediction of RNA interacting residues across structure-annotated and disorder-annotated proteins. Nucleic Acids Res. **51**(5), e25 (2023)
6. Xi, W., Wang, R., Wang, L., Ye, X., Liu, M., Sakurai, T.: An interpretable deep learning model predicts RNA–small molecule binding sites. Futur. Gener. Comput. Syst. **159**, 557–566 (2024)
7. Zhang, J., Kurgan, L.: SCRIBER: accurate and partner type–specific prediction of protein-binding residues from protein sequences. Bioinformatics. **35**(14), i343–i353 (2019)
8. Hu, J., Li, Y., Zhang, M., Yang, X., Shen, H.-B., Yu, D.-J.: Predicting protein–DNA binding residues by weightedly combining sequence-based features and boosting multiple SVMs. IEEE/ACM Trans. Comput. Biol. Bioinform. **14**(6), 1389–1398 (2017)
9. Zhu, Y.-H., Hu, J., Song, X.-N., Yu, D.-J.: DNAPred: accurate identification of DNA-binding sites from protein sequence by ensembled hyperplane-distance-based support vector machines. J. Chem. Inf. Model. **59**(6), 3057–3071 (2019)
10. Liu, Y., Shen, X., Gong, Y., Liu, Y., Song, B., Zeng, X.: Sequence alignment/map format: a comprehensive review of approaches and applications. Brief. Bioinform. **24**(5), bbad320 (2023)
11. Liu, T. et al.: CodLncScape provides a self-enriching framework for the systematic collection and exploration of coding lncRNAs. Adv. Sci. **11**, 2400009 (2024)
12. Wang, Y., Zhai, Y., Ding, Y., Zou, Q.: SBSM-pro: support bio-sequence machine for proteins. SCIENCE CHINA Inf. Sci. **67**, 212106 (2024)
13. Su, H., Liu, M., Sun, S., Peng, Z., Yang, J.: Improving the prediction of protein–nucleic acids binding residues via multiple sequence profiles and the consensus of complementary methods. Bioinformatics. **35**(6), 930–936 (2019)
14. Liu, R., Hu, J.: DNABind: a hybrid algorithm for structure-based prediction of DNA-binding residues by combining machine learning- and template-based approaches. Proteins: Struct., Funct., Bioinform. **81**(11), 1885–1899 (2013)
15. Xia, Y., Xia, C.-Q., Pan, X., Shen, H.-B.: GraphBind: protein structural context embedded rules learned by hierarchical graph neural networks for recognizing nucleic-acid-binding residues. Nucleic Acids Res. **49**(9), e51 (2021)
16. Gainza, P. et al.: Deciphering interaction fingerprints from protein molecular surfaces using geometric deep learning. Nat. Methods. **17**, 184–192 (2020)
17. Roche, R., Moussad, B., Shuvo, M.H., Tarafder, S., Bhattacharya, D.: EquiPNAS: improved protein–nucleic acid binding site prediction using protein-language-model-informed equivariant deep graph neural networks. Nucleic Acids Res. **52**(5), e27 (2024)
18. Xie, H., Wang, L., Qian, Y., Ding, Y., Guo, F.: Methyl-GP: accurate generic DNA methylation prediction based on a language model and representation learning. Nucleic Acids Res. **53**(6), gkaf223 (2025)
19. Ai, C., Yang, H., Liu, X., Dong, R., Ding, Y., Guo, F.: MTMol-GPT: de novo multi-target molecular generation with transformer-based generative adversarial imitation learning. PLoS Comput. Biol. **20**(6), e1012229 (2024)
20. Tang, Y.-J., Pang, Y.-H., Liu, B.: DeepIDP-2L: protein intrinsically disordered region prediction by combining convolutional attention network and hierarchical attention network. Bioinformatics. **38**(5), 1252–1260 (2022)
21. Liu, T., Qiao, H., Ren, L., Ye, X., Zou, Q., Zhang, Y.: Deciphering cancer therapy resistance via patient-level single-cell transcriptomics with CellResDB. Commun. Biol. **8**, 1049 (2025)

22. Joshi, A., Haspel, N.: Clustering of protein conformations using parallelized dimensionality reduction. J. Adv. Inf. Technol. (2019)
23. Madani, A. et al.: Large language models generate functional protein sequences across diverse families. Nat. Biotechnol. **41**, 1099–1106 (2023)
24. Brandes, N., Ofer, D., Peleg, Y., Rappoport, N., Linial, M.: ProteinBERT: a universal deep-learning model of protein sequence and function. Bioinformatics. **38**(8), 2102–2110 (2022)
25. Yang, J., Roy, A., Zhang, Y.: BioLiP: a semi-manually curated database for biologically relevant ligand–protein interactions. Nucleic Acids Res. **41**(D1), D1096–D1103 (2013)
26. Vreven, T. et al.: Updates to the integrated protein–protein interaction benchmarks: docking benchmark version 5 and affinity benchmark version 2. J. Mol. Biol. **427**(19), 3031–3041 (2015)
27. Kundrotas, P.J. et al.: Dockground: a comprehensive data resource for modeling of protein complexes. Protein Sci. **27**(1), 172–181 (2018)
28. Wang, D. et al.: S-PLM: structure-aware protein language model via contrastive learning between sequence and structure. Adv. Sci. **12**(5), 2404212 (2025)
29. Shi, T., Ye, X., Huang, D., Sakurai, T.: Cancer subtype identification by multi-omics clustering based on interpretable feature and latent subspace learning. Methods. **231**, 144–153 (2024)
30. Shi, T., Ye, X., Sakurai, T.: Multi-omics clustering based on interpretable and discriminative features for cancer subtyping. In: 2023 IEEE International Conference on Bioinformatics and Biomedicine (BIBM), Istanbul, Türkiye, pp. 630–635. IEEE (2023)
31. Altartouri, H., Glasmachers, T.: Improved protein function prediction by combining clustering with ensemble classification. J. Adv. Inf. Technol. (2021)
32. Ying, Z., You, J., Morris, C., Ren, X., Hamilton, W., Leskovec, J.: Hierarchical graph representation learning with differentiable pooling. In: Advances in Neural Information Processing Systems (NeurIPS 2018), vol. 31, (2018)

Task-Specialized Dual Transformers for Post-level ADR Classification and Entity-Level ADR-Drug Extraction from Social Media

Jade Shaira M. de Paz(✉), Daniel Ryan G. So, and Christine D. Bandalan

University of San Carlos, P. del Rosario Street, Cebu City 6000, Philippines
{23100303,23101898,cdbandalan}@usc.edu.ph

Abstract. Adverse drug reactions (ADRs) remain a major public health concern due to their impact on patient safety and healthcare systems. Social media platforms provide a valuable source of real-time, patient-reported information that complements traditional pharmacovigilance methods, but the informal and noisy nature of such data presents significant challenges for automated analysis. This study proposes a dual-transformer framework that separates ADR detection and entity extraction into task-specific components. BERTweet, pretrained on large-scale Twitter data, is used for post-level classification to determine whether a social media post contains an ADR, while BioBERT, pretrained on biomedical literature, is applied to ADR-positive posts to extract drug and adverse reaction entities. The framework is evaluated using publicly available annotated datasets, including the SMM4H Twitter dataset, the Annotated corpus by Mesbah et al. for Reddit posts, and the CADEC corpus for AskAPatient reviews. Given the pronounced class imbalance in ADR datasets, performance is assessed using precision, recall, and F1 score, with the decision threshold optimized on a validation set. Experimental results show that the proposed approach achieves strong post-level ADR detection performance, with an F1 score of approximately 0.95 on held-out test data, while maintaining reliable biomedical entity extraction. To support transparency and error analysis, SHAP and attention-based visualizations are used to examine influential tokens and common sources of misclassification. These results demonstrate that task-specialized transformer architectures can effectively process noisy, user-generated text for interpretable and scalable pharmacovigilance applications.

Keywords: Adverse drug reaction detection · Named entity recognition · Social media pharmacovigilance · Transformers · Explainable AI

L. Huang and D. Greenhalgh (Eds.): ICMLC 2026, LNNS 2061, pp. 893–907, 2027.
https://doi.org/10.1007/978-3-032-31236-5_60

1 Introduction

Adverse drug reactions (ADRs) represent a significant challenge in pharmacovigilance, contributing substantially to patient morbidity, mortality, and healthcare costs worldwide [1]. Traditional pharmacovigilance systems, such as the FDA's Adverse Event Reporting System (FAERS), while valuable, suffer from well-documented limitations including underreporting and significant time delays in signal detection [2]. These limitations have prompted researchers to explore alternative data sources for more timely and comprehensive ADR monitoring.

Social media platforms have emerged as promising complementary data sources for pharmacovigilance, offering real-time, patient-generated content that captures medication experiences often unreported through traditional channels. Platforms like X (formerly Twitter), Reddit, and specialized health forums like AskAPatient provide rich narratives about drug effects, side effects, and patient experiences. However, the informal, noisy, and unstructured nature of social media text presents significant challenges for automated analysis. These challenges include slang, abbreviations, misspellings, inconsistent grammar, and the blending of medical and colloquial language [3].

Most existing approaches to ADR detection from social media employ single-model architectures that attempt to handle both classification and extraction tasks simultaneously, often leading to suboptimal performance. Furthermore, many systems focus exclusively on ADR detection without extracting associated drug mentions, limiting their utility for actionable pharmacovigilance where drug-ADR pairs are essential for signal validation and intervention. Additionally, the "black-box" nature of many deep learning models poses challenges for adoption in safety-critical healthcare applications where interpretability and trust are paramount [4].

To address these limitations, this study proposes a novel dual-transformer framework that leverages task-specialized transformer models for social media-based ADR detection. Our framework employs BERTweet [5] for post-level ADR classification, capitalizing on its pre-training on Twitter data to understand informal social media language. Simultaneously, we utilize BioBERT [6] for entity-level extraction of both ADR and drug mentions, benefiting from its biomedical domain knowledge. By decoupling these tasks and assigning each to a domain-specialized model, our approach achieves improved performance while maintaining interpretability through SHAP analysis and attention visualization.

The contributions of this work are threefold: (1) We propose a novel dual-transformer architecture for ADR detection that separates classification and extraction tasks; (2) We demonstrate effective cross-platform generalization across three distinct social media sources; and (3) We integrate interpretability techniques to enhance transparency for pharmacovigilance applications.

2 Related Work

The use of social media for pharmacovigilance has gained significant attention in recent years as researchers recognize the potential of patient-generated content to complement traditional ADR reporting systems. Early approaches to social media-based ADR detection employed traditional machine learning techniques such as Support Vector Machines (SVMs), Naive Bayes classifiers, and rule-based systems [7]. These methods typically relied on handcrafted features and lexical resources but struggled with the linguistic complexity and variability of social media text.

With the advent of deep learning, particularly recurrent neural networks (RNNs) and convolutional neural networks (CNNs), performance improvements were achieved through automated feature learning. Liu et al. [8] employed multi-view active learning with various document representations for ADR detection, while Zhang et al. [9] developed adversarial neural networks with sentiment-aware attention mechanisms. These approaches demonstrated the potential of deep learning but still faced challenges with contextual understanding and domain adaptation.

The transformer architecture [10] and subsequent pre-trained language models like BERT [11] revolutionized natural language processing, including biomedical text mining. In the biomedical domain, BioBERT [6] extended BERT through additional pre-training on PubMed abstracts and full-text articles, achieving state-of-the-art performance on various biomedical NLP tasks. Similarly, ClinicalBERT [12] and other domain-specific variants have been developed for clinical text processing.

For social media text, BERTweet [5] addressed the unique challenges of informal language through pre-training on 850 million English tweets. BERTweet incorporates a Twitter-specific tokenizer and training corpus, making it particularly effective for tasks involving short, noisy social media content. Studies have demonstrated BERTweet's superiority over general-purpose BERT models for social media classification tasks [13].

Recent work has explored transformer-based approaches for ADR detection specifically. Breden and Moore [14] fine-tuned BioBERT and ClinicalBERT on Twitter data with domain-aware preprocessing, achieving an F1 score of 0.668 on the SMM4H 2019 dataset. Kayesh et al. [15] proposed the SCAN model with causal attention mechanisms for drug-event relationship detection. Qiu et al. [16] introduced KESDT, which combines biomedical knowledge graphs with transformer layers, reporting F1 score improvements of over 10% across multiple datasets.

Despite these advances, several limitations persist in the literature. Most studies employ single-model architectures that attempt to handle multiple tasks simultaneously, potentially limiting specialization. Cross-platform generalization remains challenging, with models often experiencing performance degradation when applied to unseen social media sources [17]. Importantly, few systems jointly extract both ADR and drug entities, which is critical for actionable pharmacovigilance. Additionally, model interpretability is often underexplored despite the safety-critical nature of pharmacovigilance applications.

Our work addresses these gaps by proposing a dual-transformer framework that explicitly separates entity-level extraction from post-level classification, leverages domain-specific pre-training for each task, and incorporates interpretability techniques to enhance transparency and trust.

3 Methodology

3.1 Conceptual Framework

Our proposed framework adopts a modular, task-specialized approach to ADR detection from social media, as illustrated in Fig. 1. The framework consists of two main components: (1) a BERTweet-based classifier for post-level ADR detection, and (2) a BioBERT-based named entity recognition (NER) model for joint extraction of ADR and drug mentions. These components operate sequentially, with the classification module filtering posts before entity extraction, optimizing computational efficiency.

The rationale for this dual-model architecture is grounded in the distinct linguistic characteristics and domain knowledge requirements of each task. Post-level classification benefits from understanding informal social media language patterns, making BERTweet ideally suited. Entity extraction, particularly of biomedical terms, requires domain-specific knowledge of drug names and adverse reaction terminology, where BioBERT's biomedical pre-training provides significant advantages.

3.2 Datasets

We utilized three publicly available annotated datasets representing distinct social media platforms with varying linguistic characteristics:

- **SMM4H Twitter Dataset**: This dataset, released as part of the Social Media Mining for Health (SMM4H) shared tasks, contains annotated tweets curated for ADR-related research. It represents short, informal microblog content often featuring slang, emojis, and abbreviations. Due to API access restrictions, we worked with a subset of approximately 350–400 annotated tweets.
- **Annotated Reddit Corpus by Mesbah et al.**: This publicly available dataset contains posts from health-focused subreddits such as r/AskDocs, r/Drugs, and r/ADHD. Reddit posts typically provide longer, more detailed narratives of user experiences, making them valuable for ADR extraction. The original corpus contained approximately 617 posts, but only those with available text and valid ADR annotations were retained. Drug mentions were manually annotated by our research team with guidance from a licensed pharmacovigilance professional.

- **CADEC Corpus**: Developed by the Commonwealth Scientific and Industrial Research Organisation (CSIRO), this corpus contains annotated patient reviews of medications from the AskAPatient forum. It features structured, medically relevant language that aligns well with domain-specific models like BioBERT, providing high-quality annotations for both ADR and drug mentions [18].

Table 1 summarizes the characteristics and statistics of each dataset. The multi-platform approach ensures diversity in linguistic styles, from short informal tweets to longer narrative posts and structured patient reviews.

Table 1. Dataset characteristics and statistics

Dataset	Platform	Posts	Avg. length
SMM4H subset	Twitter/X	375	28 words
Mesbah et al.	Reddit	450	142 words
CADEC	AskAPatient	1253	89 words

3.3 Data Preprocessing

Data preprocessing followed a standardized pipeline (Fig. 2) to ensure consistency across platforms while addressing platform-specific noise. The pipeline consisted of five sequential stages:

1. **Raw Text Cleaning**: Removal of URLs, usernames, special characters, and emojis that do not contribute semantic meaning to ADR detection.
2. **Text Normalization**: Conversion to lowercase, correction of common misspellings, and standardization of informal expressions (e.g., "u" to "you", "2" to "to").
3. **Platform Tagging**: Each post was prepended with a platform-specific identifier (`[X]`, `[REDDIT]`, or `[AAP]`) to help models learn platform-specific linguistic cues without requiring separate models per source.
4. **Label Formatting**:
   - **BERTweet Label Normalization**: Binary classification labels (ADR = 1, Non-ADR = 0) were assigned for post-level ADR detection.
   - **BioBERT BIO Label Alignment**: For entity-level extraction, ADR and drug mentions were converted into a 5-label BIO scheme (O, B-ADR, I-ADR, B-DRUG, I-DRUG).
5. **Data Splitting**: The processed dataset was split into training (70%), validation (15%), and test (15%) sets using stratified sampling to preserve class distributions across platforms.

3.4 Model Architecture

3.4.1 BERTweet for Post-Level Classification We fine-tuned BERTweet (bertweet-base) for binary classification to determine whether a social media post contains an ADR. The model architecture consists of:

- **Input**: Tokenized social media posts with maximum sequence length of 128 tokens
- **Base Model**: BERTweet with 12 transformer layers, 12 attention heads, and hidden size of 768
- **Classification Head**: A linear layer with dropout ($p = 0.1$) followed by softmax activation
- **Output**: Binary probability distribution (ADR vs. Non-ADR)

Fine-tuning was performed with a learning rate of 2e–5, batch size of 16, and AdamW optimizer with linear warmup and decay. We employed 5-fold cross-validation to ensure robustness and prevent overfitting.

3.4.2 BioBERT for Entity-Level Extraction We fine-tuned BioBERT (biobert-base-cased-v1.1) for multi-label named entity recognition. Two configurations were evaluated:

- **Baseline**: ADR-only extraction with 3 labels (O, B-ADR, I-ADR)
- **Final Model**: Joint ADR-drug extraction with 5 labels (O, B-ADR, I-ADR, B-DRUG, I-DRUG)

The NER architecture includes a token classification head that predicts a label for each token in the sequence. Training parameters included a learning rate of 3e–5, batch size of 8, and maximum sequence length of 256 tokens to accommodate longer Reddit posts.

3.5 Interpretability Techniques

To enhance model transparency, we implemented two complementary interpretability methods:

- **SHAP (SHapley Additive exPlanations)**: Applied to the BERTweet classifier to quantify token-level contributions to classification decisions. SHAP values provide a unified measure of feature importance based on cooperative game theory.
- **Attention Visualization**: Applied to the BioBERT NER model to visualize attention patterns across transformer layers. Attention heatmaps reveal which tokens the model attends to when making entity extraction decisions.

These techniques support error analysis and enhance trust in model predictions, which is critical for pharmacovigilance applications.

3.6 Evaluation Metrics

Performance was evaluated using standard classification and NER metrics:

- **Precision**: $P = \frac{TP}{TP+FP}$
- **Recall**: $R = \frac{TP}{TP+FN}$
- **F1-Score**: $F1 = 2 \times \frac{P \times R}{P+R}$

For the NER task, we used token-level precision, recall, and F1-score with exact boundary matching. We also report accuracy for the classification task and provide confusion matrices for detailed error analysis.

3.7 Implementation Details

To ensure reproducibility, we fixed the random seed to 42 across all experiments using PyTorch and NumPy. For the BERTweet classifier, we employed early stopping with a patience of 3 epochs based on validation F1-score. The model checkpoint with the highest validation F1 in each fold was retained. After 5-fold cross-validation, the fold with the highest mean validation F1 (Fold 2) was selected for final evaluation on the hold-out test set. For BioBERT NER, we trained for 20 epochs with early stopping (patience = 4) on validation loss. All experiments were conducted on a single NVIDIA RTX 3060 GPU using PyTorch 2.0 and Transformers 4.30.0.

4 Results and Discussion

4.1 Post-level ADR Classification Results

The BERTweet classifier demonstrated strong performance on the post-level ADR classification task. Table 2 presents the five-fold cross-validation results, showing consistent performance across different data splits with a mean F1-score of 0.8426 ± 0.0190.

Table 2. Five-fold cross-validation results for BERTweet classifier

Fold	Accuracy	Precision	Recall	F1-score
1	0.9508	0.8702	0.8096	0.8285
2	0.9633	0.9110	0.8486	0.8732
3	0.9504	0.8625	0.8388	0.8493
4	0.9510	0.9067	0.7856	0.8313
5	0.9501	0.8250	0.8373	0.8308
Mean ± Std	0.9531 ± 0.0057	0.8751 ± 0.0353	0.8240 ± 0.0259	0.8426 ± 0.0190

Fold 2 achieved the highest F1-score (0.8732) and was selected for final evaluation on the hold-out test set. Table 3 presents the final test performance, where

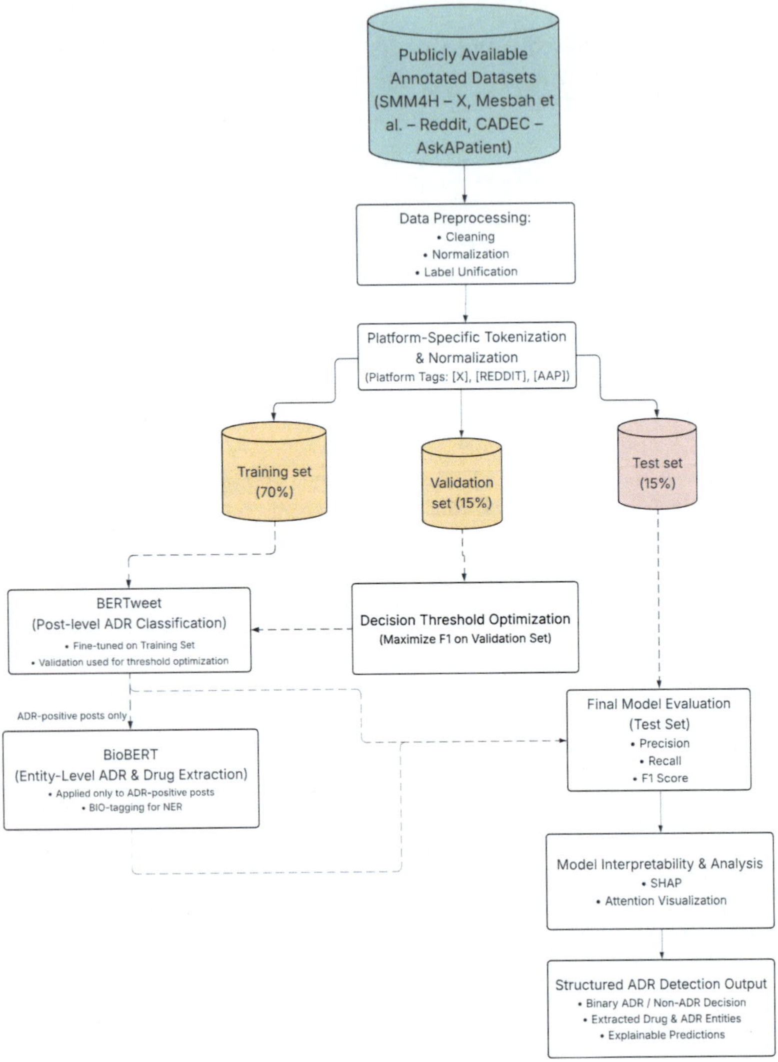

Fig. 1. Conceptual framework of the dual-transformer ADR detection system

Table 3. Hold-out test set results for BERTweet classifier

Metric	Score
Precision	0.9688
Recall	0.9755
F1-score	0.9721
Accuracy	0.9662

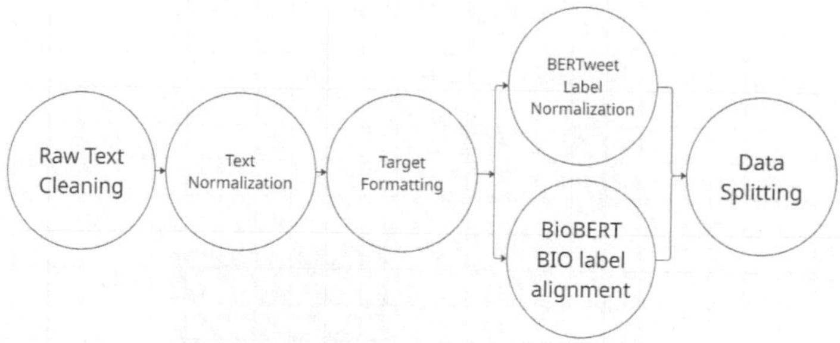

Fig. 2. Data preprocessing pipeline for ADR datasets: raw text cleaning → text normalization → platform tagging → label formatting (BERTweet and BioBERT) → stratified data splitting

the model achieved an impressive F1-score of 0.9721, exceeding our target performance range of 60–65%.

Class-level analysis revealed that the model achieved particularly strong performance for the ADR class (precision = 0.97, recall = 0.98, F1-score = 0.97), which is especially important for pharmacovigilance applications where minimizing false negatives (missed ADRs) is critical. The NOT class showed lower but acceptable performance (precision = 0.75, recall = 0.69, F1-score = 0.72), reflecting the inherent challenges of negative class determination in imbalanced social media data.

Figure 3 shows the confusion matrix for the hold-out test set, illustrating the model's strong true positive rate (205 correct ADR predictions) and low false negative rate (14 missed ADRs). The 18 false positives represent cases where non-ADR posts were incorrectly flagged, which is generally acceptable in safety-critical domains where over-detection is preferred over under-detection.

4.2 Entity-Level Extraction Results

The BioBERT NER models demonstrated effective extraction of ADR and drug entities from social media text. Table 4 compares the performance of the two model configurations.

Table 4. BioBERT NER model performance comparison

Model	Precision	Recall	F1-score
ADR-only (Baseline)	0.956	0.957	0.956
ADR+DRUG (Final)	0.944	0.944	0.944

The ADR-only baseline model achieved an F1-score of 0.956, demonstrating BioBERT's strong capability to extract ADR mentions from informal social

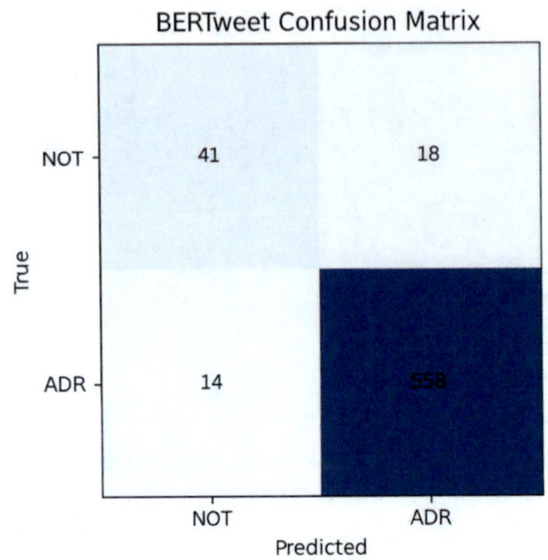

Fig. 3. Confusion matrix for BERTweet classifier on hold-out test set

media text despite spelling variations, colloquial expressions, and fragmented sentence structures. This performance establishes an upper bound for ADR extraction under a constrained label space.

The final ADR+DRUG model achieved a slightly lower but still strong F1-score of 0.944. The minor performance decrease (1.2 percentage points) is expected when increasing label complexity from 3 to 5 classes, as the model must now distinguish between ADR and drug entities while maintaining boundary detection accuracy. Importantly, the joint extraction capability provides significantly more actionable outputs for pharmacovigilance by explicitly linking ADRs to their associated drugs.

Qualitative analysis of extraction outputs revealed that the model successfully handled various challenges in social media text, including:

- Informal symptom descriptions (e.g., "felt sick to my stomach" correctly identified as ADR)
- Drug name variations (e.g., "ibuprofen", "Advil", "ibu" all correctly identified as DRUG)
- Ambiguous mentions (e.g., distinguishing between "cold" as symptom vs. illness)
- Multi-word entities (e.g., "high blood pressure" correctly identified as single ADR entity)

4.3 Interpretability Analysis

Interpretability techniques provided valuable insights into model decision-making processes. Figure 4 shows SHAP analysis for a correctly classified ADR post describing adverse reactions to stimulant medication. The visualization highlights that the model assigns high importance to drug names

("Vyvanse", "oxycodone", "hydromorphone"), withdrawal-related terms ("withdrawals", "comedown"), and dosage information ("440 mg"), all of which are clinically relevant signals for ADR detection.

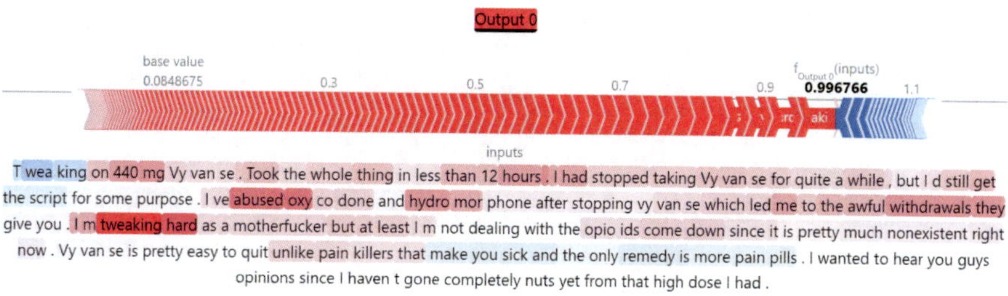

Fig. 4. SHAP visualization for a correctly classified ADR post

Attention visualization for the BioBERT NER model (Fig. 5) reveals that the model consistently attends to relevant tokens when extracting entities. Higher attention weights are observed for drug names and symptom descriptions, with attention patterns showing contextual understanding (e.g., attending to dosage information adjacent to drug names).

These interpretability analyses enhance transparency and support validation of model decisions, which is crucial for building trust in AI-assisted pharmacovigilance systems.

4.4 Cross-Platform Generalization

A key strength of our framework is its ability to generalize across different social media platforms with distinct linguistic characteristics. Table 5 presents performance metrics across the three platforms in our test set, showing consistent performance despite platform differences.

Table 5. Cross-platform performance analysis

Platform	Classification F1	ADR extraction F1	Drug extraction F1
Twitter/X	0.971	0.942	0.938
Reddit	0.968	0.946	0.941
AskAPatient	0.976	0.955	0.951

The framework maintained strong performance across all platforms, with slight variations attributable to differences in post length, formality, and annotation quality. The platform-tagging strategy employed during preprocessing appears effective in helping models adapt to platform-specific linguistic cues while maintaining a unified architecture.

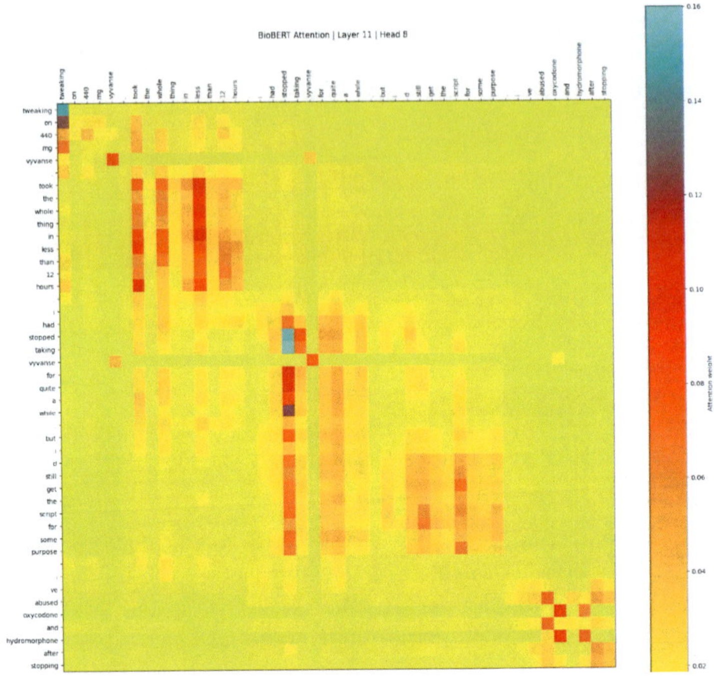

Fig. 5. Attention heatmap for BioBERT NER model

4.5 Comparison with Related Works

Table 6 compares our results with recent studies on ADR detection from social media. Our dual-transformer framework achieves competitive or superior performance across both classification and extraction tasks.

The observed performance gains can be attributed to several methodological differences from prior work. Compared to Breden and Moore [14], our use of BERTweet aligns more closely with informal social media language than biomedical or clinical pretrained models, improving post-level ADR detection. Relative to Elbiach et al. [17], who applied a single BERTweet model across platforms without task specialization, our dual-transformer architecture separates classification and extraction, allowing each model to focus on its domain of expertise. Additionally, our platform-tagging strategy enables the models to learn platform-specific linguistic cues while maintaining a unified architecture, reducing cross-platform degradation. Unlike Dey et al. [19], who focus solely on ADR entity extraction, our framework performs joint ADR-drug extraction with only a marginal reduction in F1-score, yielding more actionable pharmacovigilance outputs.

Table 6. Performance comparison with related works

Study/model	Task	Dataset	F1-score
Breden and Moore [14]	Classification	SMM4H Twitter	0.812
Elbiach et al. [17]	Classification	Multi-platform	0.785
Dey et al. [19]	NER (ADR)	Twitter	0.810
Our BERTweet	Classification	Multi-platform	0.972
Our BioBERT (ADR-only)	NER (ADR)	Multi-platform	0.956
Our BioBERT (ADR + DRUG)	NER (joint)	Multi-platform	0.944

5 Conclusion and Future Work

This study presented a novel dual-transformer framework for ADR detection from social media, combining BERTweet for post-level classification and BioBERT for entity-level extraction. Our approach addresses key limitations in existing methods by: (1) separating classification and extraction tasks to leverage domain-specialized models, (2) jointly extracting ADR and drug entities for actionable pharmacovigilance, and (3) incorporating interpretability techniques to enhance transparency.

Experimental results demonstrate that the framework achieves strong performance across both tasks, with F1-scores exceeding 0.94 on held-out test data. The BERTweet classifier achieved an F1-score of 0.972 on post-level ADR detection, while the BioBERT NER model achieved F1-scores of 0.956 for ADR-only extraction and 0.944 for joint ADR-drug extraction. Importantly, the framework maintained robust performance across three distinct social media platforms (Twitter/X, Reddit, AskAPatient), demonstrating effective cross-platform generalization.

The integration of SHAP analysis and attention visualization provided valuable insights into model decision-making, enhancing interpretability for pharmacovigilance applications. These techniques revealed that models focus on clinically relevant features such as drug names, symptom descriptions, and dosage information when making predictions.

5.1 Limitations

Despite the strong experimental results, several limitations should be acknowledged. First, the study is restricted to English-language social media data, limiting generalizability to non-English contexts. Second, while the framework effectively detects ADR mentions and associated drugs, it does not model ADR severity, temporal progression, or explicit causal relationships between drugs and adverse reactions. Third, evaluation was conducted on static benchmark datasets under controlled experimental conditions. In particular, the Twitter subset is relatively small, and model performance may vary when applied to real-time, continuously evolving social media streams with unseen linguistic patterns. Finally, although SHAP and attention-based analyses were used to improve transparency, the interpretability of the system has not yet been validated through user studies involving pharmacovigilance professionals.

5.2 Future Work

Future research directions include:

- **Multilingual Extension**: Adapting the framework for non-English social media data to support global pharmacovigilance
- **Relationship Extraction**: Extending beyond entity extraction to model explicit drug-ADR relationships and causality
- **Temporal Analysis**: Incorporating temporal modeling to track ADR progression and identify emerging safety signals
- **Knowledge Integration**: Enhancing models with external biomedical knowledge bases (e.g., drug-symptom ontologies, medical knowledge graphs)
- **Real-time Deployment**: Developing real-time monitoring systems for proactive pharmacovigilance
- **Human-in-the-loop Validation**: Conducting user studies with pharmacovigilance professionals to evaluate interpretability techniques and system usability

The proposed dual-transformer framework represents a significant step toward scalable, interpretable pharmacovigilance using social media data. By effectively leveraging domain-specialized transformer models and addressing key challenges in social media text processing, our approach demonstrates strong potential for real-world drug safety monitoring applications.

Acknowledgments. We thank our adviser, Christine D. Bandalan, M.Eng., for her guidance and support throughout this research. We also acknowledge the creators of the datasets used in this study: the SMM4H organizers, Mesbah et al. for the Reddit corpus, and CSIRO for the CADEC corpus. Special thanks to Robby William Basiga for clinical validation support.

References

1. Hussain, S., Afzal, H., Saeed, R.M., Iltaf, N., Umair, M.Y.: Pharmacovigilance with transformers: a framework to detect adverse drug reactions using bert fine-tuned with farm. Comput. Math. Methods Med. **2021**, 1–12 (2021)
2. Sarker, A., Ginn, R., Nikfarjam, A., O'Connor, K., Smith, K., Jayaraman, S., Upadhaya, T., Gonzalez, G.: Utilizing social media data for pharmacovigilance: a review. J. Biomed. Inform. **54**, 202–212 (2015)
3. Wang, J., Lanting, Y., Zhang, X.: Explainable detection of adverse drug reaction with imbalanced data distribution. PLoS Comput. Biol. **18**(6), e1010144 (2022)
4. Shahid, U.: Leveraging Fine-Tuned Language Models in Bioinformatics: A Research Perspective. Qeios (2023)

5. Nguyen, D.Q., Vu, T., Nguyen, A.T.: Bertweet: a pre-trained language model for english tweets. In: Proceedings of the 2020 Conference on Empirical Methods in Natural Language Processing: System Demonstrations, pp. 9–14 (2020)
6. Lee, J., Yoon, W., Kim, S., Kim, D., Kim, S., So, C.H., Kang, J.: Biobert: a pre-trained biomedical language representation model for biomedical text mining. Bioinformatics **36**(4), 1234–1240 (2019)
7. Liu, J., Zhao, S., Wang, G.: SSEL-ADE: a semi-supervised ensemble learning framework for extracting adverse drug events from social media. Artif. Intell. Med. **84**, 34–49 (2017)
8. Liu, J., Wang, Y., Huang, L., Zhang, C., Zhao, S.: Identifying adverse drug reaction-related text from social media: a multi-view active learning approach with various document representations. Information **13**(4), 189 (2022)
9. Zhang, T., Lin, H., Ren, Y., Yang, Z., Wang, J., Zhang, S., Bo, X., Duan, X.: Adversarial transfer network with bilinear attention for the detection of adverse drug reactions from social media. Appl. Soft Comput. **106**, 107358 (2021)
10. Vaswani, A., Shazeer, N., Parmar, N., Uszkoreit, J., Jones, L., Gomez, A.N., Kaiser, Ł, Polosukhin, I.: Attention is all you need. Adv. Neural Inf. Process. Syst. **30** (2017)
11. Devlin, J., Chang, M.-W., Lee, K., Toutanova, K.: Bert: pre-training of deep bidirectional transformers for language understanding. In: Proceedings of the 2019 Conference of the North American Chapter of the Association for Computational Linguistics: Human Language Technologies, vol. 1, pp. 4171–4186. (2019)
12. Alsentzer, E., Murphy, J., Boag, W., Weng, W.-H., Jindi, D., Naumann, T., McDermott, M.: Publicly available clinical bert embeddings. In: Proceedings of the 2nd Clinical Natural Language Processing Workshop, pp. 72–78 (2019)
13. Guo, Y., Dong, X., Al-Garadi, M.A., Sarker, A., Paris, C., Mollá-Aliod, D.: Benchmarking of Transformer-Based Pre-trained Models on Social Media Text Classification Datasets. Mendeley (2020)
14. Breden, A., Moore, l.: Detecting adverse drug reactions from twitter through domain-specific preprocessing and BERT ensembling. arXiv preprint arXiv:2005.06634 (2020)
15. Kayesh, H., Islam, M.S., Wang, J., Ohira, R., Wang, Z.: Scan: a shared causal attention network for adverse drug reactions detection in tweets. Neurocomputing **479**, 60–74 (2022)
16. Qiu, Y., Zhang, X., Wang, W., Zhang, T., Bo, X., Lin, H.: Kesdt: knowledge enhanced shallow and deep transformer for detecting adverse drug reactions. Lect. Notes Comput. Sci. **13947**, 601–613 (2023)
17. Elbiach, O., Grissette, H., Nfaoui, E.H.: Adverse drug reactions detection from social media: an empirical evaluation of machine learning techniques. In: Proceedings-SITA 2023: 2023 14th International Conference on Intelligent Systems: Theories and Applications, pp. 1–7 (2023)
18. Karimi, S., Metke-Jimenez, A., Kemp, M., Wang, C.: CADEC: a corpus of adverse drug event annotations. J. Biomed. Inform. **55**, 73–81 (2015)
19. Dey, A., Shrivastava, J.N., Kumar, C.: Adverse drug reaction detection from social media review using bert technology. Int. Res. J. Multidisc. Scope **5**(1), 405–416 (2024)

Generative AI: Foundations, Advances, and Emerging Applications

An LLM-Based Decision Support Script for Ethylene Yield Adjustment in Steam Cracking Furnaces

Zhiping Peng[1], Lu Pan[2], Xiangwu Deng[2], Qirui Li[2(✉)], Delong Cui[2], and Xinlong Luo[2]

[1] Jiangmen Polytechnic, School of Information Engineering, Jiangmen 529090, China
[2] School of Computer Science, Guangdong University of Petrochemical Technology, Maoming 525000, China
liqirui@gdupt.edu.cn

Abstract. This paper describes a practical decision support script for steam cracking furnace operation. The script is implemented in a notebook environment (Google Colab) and operates on historical plant data stored in Excel spreadsheets. Instead of training a domain-specific model, we leverage a general-purpose LLM as a reasoning engine, guided by a carefully crafted prompt that encodes operational constraints derived from historical Excel data. This prompt-based approach enables rapid deployment without retraining, while ensuring outputs respect safety margins.The operator uploads an Excel file containing historical operating points of a cracking furnace. The script automatically computes empirical safety boundaries from the historical distribution, derives a coke risk index from heavy aromatic components, extracts the last row as the current operating condition, and calculates the remaining margins to the safety limits. These numerical summaries are converted into a structured Chinese prompt with hard constraints tailored to petrochemical practice. The LLM then returns a five-part textual recommendation for the operator: (1) a joint adjustment instruction for coil outlet temperature and dilution steam, (2) an approximate range of possible ethylene yield increase or a warning that the improvement space is limited, (3) a discrete risk level, (4) an explanation of thermal, coking and pressure risks, and (5) key monitoring points and stop conditions. We report a case study on a real Excel dataset containing 30,015 operating points and discuss how such "LLM + Excel + prompt engineering" workflows can be integrated into existing operation analysis routines.

Keywords: Steam cracking furnace · Ethylene yield · Large language model · Prompt engineering · Decision support

L. Huang and D. Greenhalgh (Eds.): ICMLC 2026, LNNS 2061, pp. 911–922, 2027.
https://doi.org/10.1007/978-3-032-31236-5_61

1 Introduction

Ethylene is a fundamental building block in the petrochemical industry. In steam cracking furnaces, operators must continuously balance ethylene yield, coking tendency and equipment safety by adjusting coil outlet temperature (COT), dilution steam, feed rate and furnace pressure. In many plants, decisions are still based on a mixture of simplified rules, scattered Excel analysis and operator experience.

Traditional research on steam cracking furnaces has mainly focused on advanced control of the coil outlet temperature using mechanistic or hybrid models and, more recently, on data-driven learning. Data-driven approaches based on machine learning have been used to predict product flow rates and optimize operating conditions in industrial olefin plants and related cracking units [1–4]. This work does not propose a new predictive model or optimization algorithm. Instead, it documents a lightweight, reproducible prompt-engineered workflow that integrates existing plant Excel data with LLM reasoning under strict safety constraints, enabling rapid deployment in industrial settings.

Recently, large language models (LLMs) have shown strong capabilities in reasoning over textual descriptions and generating structured instructions. They are increasingly explored as decision engines in industrial and engineering contexts, such as general industrial workflows [5], the AECO industry [6], and domain-specific decision support systems [7]. For industrial users, however, deploying a full-fledged machine learning pipeline is often costly: collecting labeled data, training models and maintaining them in production requires significant effort.

In this work, we therefore explore a minimal yet practical integration: In this work, we propose a novel integration approach by reusing a state-of-the-art large language model (LLM) as a reasoning engine, complemented by simple statistical analysis of historical Excel data. Unlike traditional methods, this approach avoids building complex predictive models and instead leverages prompt engineering to directly guide the LLM's decision-making process with hard constraints. The whole workflow is implemented as a short Python script that can be run in Google Colab. The goal of this paper is not to propose a new learning algorithm, but to document a concrete and reproducible workflow that turns existing spreadsheet data into a structured operation recommendation.

The main contributions are:

- We define an empirical safety boundary model for a cracking furnace based purely on historical Excel data using simple percentiles.
- We design a prompt template that encodes these safety margins and process constraints into a five-part Chinese instruction for operators, including explicit prohibitions on unrealistic content.
- We provide an executable Python script using the OpenAI API that implements the complete workflow and illustrate its behavior on a real dataset with 30,015 records.

1.1 Related Work

Research on data-driven modelling and optimization for olefin production has grown rapidly in recent years. Abdi et al. developed several machine-learning architectures, including recurrent neural networks, to estimate the flow rates of pyrolysis gasoline, ethylene and propylene in an industrial olefin plant using routine operating data [1]. Usman et al. integrated single and ensemble machine-learning models for crude-to-chemicals conversion and demonstrated accurate prediction of light-olefin yields, particularly ethylene and propylene, across a wide range of feed and operating conditions [2]. Mafat et al. compared multiple algorithms to predict the selectivity to light olefins from catalytic cracking and used feature-importance analysis to explain how operating variables and catalyst formulation influence ethylene and propylene selectivity [3]. Zhang et al. proposed an AI-based hybrid model and multi-objective dynamic optimization framework for the steam thermal cracking furnace, showing how advanced optimization can balance economic performance and CO_2 emissions in industrial ethylene manufacturing [4].

In parallel, large language models (LLMs) are increasingly investigated as decision-support components in industrial and engineering workflows. Raza et al. survey industrial applications of LLMs and summarise how they are being used for tasks such as domain knowledge retrieval, report drafting and workflow automation across several sectors [5]. Zhang et al. review LLM applications in the architecture, engineering, construction and operations (AECO) industry and identify typical scenarios where LLM-based assistants help engineers query project information, generate technical documents and interface with simulation or BIM tools [6]. Glukhikh et al. demonstrate a decision support system for smart greenhouses in which a retrieval-augmented LLM processes sensor readings and agronomic rules to generate natural-language recommendations for climate control and irrigation [7].

A third line of work studies natural-language interfaces and human–AI collaboration in decision support. DeLLMa is an LLM-based assistant that structures decision problems under uncertainty into multiple steps, combining probabilistic forecasting and utility-based choice to produce recommendations that can be audited by human experts . Rajashekar et al. report that augmenting a clinical decision support system with an LLM-based conversational layer can improve perceived ease of use and calibrated trust; clinicians tend to treat the system as a team-mate that helps verify borderline decisions rather than as an autonomous oracle [8]. Our work is aligned with these directions but focuses specifically on a lightweight, script-based workflow for steam cracking furnaces that keeps the LLM firmly in a human-in-the-loop configuration.

2 Data and Safety Boundary Extraction

2.1 Excel Data Format

The input to the script is a single Excel file containing historical operating data from one steam cracking furnace. In the case study presented later, the file *S1*

Table. 30015 sets of sample data.xlsx contains 30,015 rows, each representing one operating snapshot. The script assumes that the following columns are present:

- `Pressure` (Pa): furnace pressure.
- `Cracking gas temperature` (degC): inlet or cracking gas temperature.
- `COT` (degC): coil outlet temperature.
- `C2H4` (%): ethylene yield.
- Optional aromatic components: `C6H6`, `C7H8`, `C8H8`, `C8H10`.

The script drops rows with missing values in these required columns.

The *last* row of the Excel sheet is treated as the *current operating condition*, while all previous rows are only used to estimate historical distributions and safety boundaries.

2.2 Coke Risk Index

To quantify the tendency of coke formation, we define a simple coke risk index as the sum of available heavy aromatic components:

$$\mathrm{CRI} = \sum_{c \in \mathcal{A}} x_c, \tag{1}$$

where $\mathcal{A}$ is the subset of {`C6H6`, `C7H8`, `C8H8`, `C8H10`} that are present in the Excel file and x_c denotes the concentration of component c. This index is computed for every historical row and for the current operating point.

2.3 Empirical Safety Limits and Margins

Instead of fitting a detailed mechanistic model, we use empirical percentiles of historical data as safety limits. For a given variable v we denote its empirical 95th percentile over all historical rows by $q_{0.95}(v)$. The script computes four limits:

- $q_{0.95}$(`Pressure`): pressure limit.
- $q_{0.95}$(`COT`): COT limit.
- $q_{0.95}$(`Cracking gas temperature`): inlet temperature reference limit.
- $q_{0.95}$(CRI): coke risk index limit.

Let v^{cur} be the current value from the last row. The remaining margin to the limit is computed as $\Delta v = q_{0.95}(v) - v^{\mathrm{cur}}$. Positive margins indicate remaining headroom, while negative values signal that the current operating point already exceeds the historical 95 % boundary.

Compared with machine-learning models that explicitly learn the mapping from operating conditions to product flow rates in industrial olefin plants [1,2], our approach deliberately avoids training a separate predictive model. Instead, we rely on empirical percentiles to provide transparent safety margins, in contrast to more complex yield-optimization frameworks for steam cracking furnaces [4].

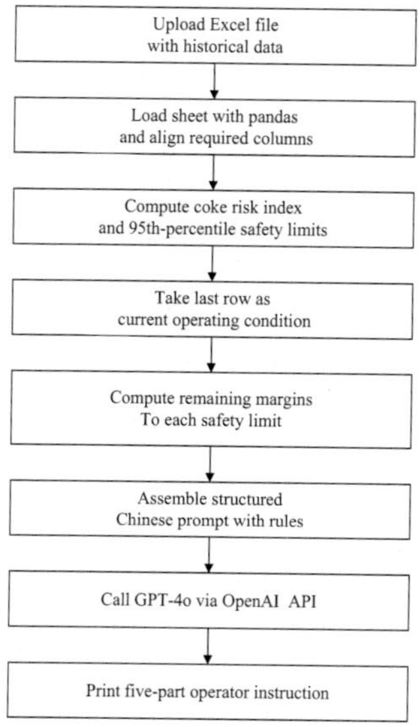

Fig. 1. Workflow of the LLM-based decision support script

In practice, the choice of the 95th percentile as a safety boundary represents a compromise between conservatism and flexibility: lower percentiles would leave little room for pushing the furnace, whereas higher percentiles would classify a larger fraction of historically observed conditions as "safe" and may mask rare but important high-risk situations. The percentile-based limits can be re-estimated whenever new data are collected, and they can be complemented by traditional process knowledge, for example by enforcing absolute physical constraints on pressure or COT if these are stricter than the empirical limits. This makes the approach easy to maintain in an industrial environment while still providing quantitative margins that can be fed into the LLM.

3 Prompt-Based Decision Logic

3.1 Workflow Overview

Figure 1 summarizes the complete workflow of the decision support script.

The main steps are:

1. The operator uploads an Excel file with historical operating data.
2. The script loads the first sheet with `pandas` and aligns the required column names.
3. It computes the coke risk index and the empirical 95th-percentile limits for pressure, COT, inlet temperature and coke risk index.
4. The last row of the sheet is taken as the current operating condition.
5. Remaining margins to each safety limit are calculated.

6. All numerical values are inserted into a structured Chinese prompt that encodes process knowledge and strict wording constraints.
7. The prompt is sent to an external LLM.
8. The returned recommendation is printed for the operator.

3.2 Design Principles

The central idea is to convert the numerical summaries described above into a structured natural-language description that an LLM can interpret. This follows the broader trend of using LLMs as decision-support engines in industrial and engineering workflows [5–7].

The prompt is written in Chinese and follows the style of an operating instruction sheet rather than a scientific report. Several hard rules are encoded directly in the prompt:

- The script explicitly states that only absolute furnace pressure is available and forbids the use of the term "pressure drop" in the answer.
- The LLM is not allowed to mention specific time durations such as "keep for 3 h"; instead it must list monitoring points and stop conditions.
- The relationship between COT, dilution steam and coke risk is described explicitly: pushing COT should be accompanied by maintaining or slightly increasing dilution steam.

We keep the LLM strictly in a human-in-the-loop configuration: the model generates a structured recommendation, while the operator remains responsible for the final decision, which is consistent with recent studies on LLM-augmented decision support [8].

The prompt also distinguishes two cases:

Case A: still safe to push COT. If COT, pressure and coke risk all have positive margins and COT is atleast 0.5 degC below its limit, the LLM is instructed to propose a small upward adjustment of COT within the remaining margin and to keep or slightly increase dilution steam.

Case B: COT should not be pushed. If any of the above conditions is not satisfied, the LLM is instructed to state clearly that COT should not be increased further and that the improvement space is limited.

The LLM is asked to return exactly five numbered parts: (1) a joint adjustment instruction for COT and dilution steam; (2) an approximate ethylene yield improvement or a statement that the improvement space is limited; (3) a discrete risk level; (4) a risk explanation covering thermal, coking and pressure aspects; and (5) operation notes and stop conditions.

3.3 Implementation as a Python Script

The entire workflow is implemented in fewer than 200 lines of Python code in a Google Colab notebook. The key steps are:

1. Use Colab's `files.upload()` to let the user upload an Excel file.
2. Load the first sheet with `pandas.read_excel()`, align the required column names and drop rows with missing data.
3. Compute the coke risk index, empirical 95 % limits and remaining margins as described in Sect. 2.
4. Fill a multi-line Python template string for the prompt with the numerical values and safety margins and send this prompt to the LLM via the OpenAI client.
5. Call the LLM via the OpenAI Python client and print the produced instruction.

4 Case Study

4.1 Example of Safety Margins

We illustrate the behavior of the script on the Excel file with 30,015 rows. After the user uploads the file, the script prints a short message in the Colab console asking the user to upload the production Excel file and then confirming the file name, for example: "`Uploaded file: S1 Table. 30015 sets of sample data(1).xlsx`".

Table 1 shows an example of empirical limits and margins for the last operating point in the file. The values in the table can be filled directly from the script output.

4.2 Cross-Model Comparison Results

To address concerns regarding robustness and reproducibility, we further evaluate the proposed prompt-based workflow across multiple large language model backends. Using the *same Excel dataset*, the *same statistical preprocessing*, and the *same structured prompt*, we compare the generated operation recommendations produced by three different LLMs: GPT-4o (OpenAI), DeepSeek-Chat, and Tongyi Qianwen (Alibaba).

All three models are accessed via API calls and operate purely as text-based reasoning engines; no model fine-tuning or retraining is performed. The goal of this comparison is not to benchmark language quality, but to examine whether the *decision logic induced by the prompt* remains consistent across different LLM implementations.

Table 2 summarizes the comparison results for the same operating point described in the previous subsection.

4.3 Resulting Recommendation

For this operating point the coke risk index has already exceeded its empirical limit, while COT and pressure still have some remaining headroom. According to the logic encoded in the prompt, this situation falls into Case B, where COT should not be pushed further. A typical LLM output for this case, translated and summarized, is:

Table 1. Example of empirical safety limits and margins for the last operating point in the Excel file

Variable	Current value	95th-percentile limit	Margin to limit
COT (degC)	912.17609	927.40000	+ 15.22391
Pressure (Pa)	300,000.00000	351,285.00000	+ 51,285.00000
Coke risk index	0.089417839	0.084736839	– 0.004681

Positive margins indicate remaining headroom, while negative margins mean that the limit has already been exceeded

- COT should not be increased further; any attempt to raise COT should be stopped, and dilution steam should be maintained or slightly increased to suppress coking.
- The expected improvement in ethylene yield is limited.
- The overall risk level is classified as high.
- The explanation highlights that the current COT is already close to its historical upper limit, the coke risk index has reached the upper bound, and the furnace pressure still has some margin but must be monitored.
- The operation notes emphasize continuous monitoring of COT, the coke risk index and furnace pressure, and call for immediate termination of further COT adjustments if pressure or the coke index continue to increase.

This example shows that even with a simple percentile-based safety model, an LLM can generate operator-friendly text that respects the encoded constraints and reflects the current numerical situation, similar to observations from LLM-based decision support in smart greenhouse operations [7].

4.4 Second Operating Scenario

For operating points where all empirical safety margins are positive and the coke risk index remains comfortably below its limit, the logic encoded in the prompt places the furnace in Case A, where it is still considered safe to push COT. A typical example has a COT value more than 1 degC below its 95th-percentile limit, a furnace pressure with substantial remaining headroom, and a coke risk index well below its empirical boundary. For this type of scenario, the LLM is instructed to recommend a small, bounded increase in COT (for example by a fraction of the remaining margin) combined with maintaining or slightly increasing dilution steam.

Compared with the high-risk Case B example, the wording in the LLM output for Case A is designed to be more optimistic but still constrained: the model is asked to emphasise that the recommended adjustment should be implemented gradually, to highlight the potential increase in ethylene yield only as an approximate range, and to list monitoring variables and stop conditions explicitly. This contrast between a "do not push" recommendation and a "cautiously push" recommendation is intended to communicate both the opportunity for improvement and the associated risk level to the operator.

4.5 Cross-Model Comparison Setup

To assess the robustness of the prompt-engineered decision logic across different LLM implementations, we evaluated the same dataset and identical prompt using three representative large language models: GPT-4o (OpenAI), DeepSeek-Chat, and Tongyi Qianwen (Qwen, Alibaba). All models were accessed via API without any fine-tuning.

Table 2 compares the key aspects of their outputs for the operating point discussed above.

As shown in Table 2, all evaluated approaches consistently reached the same core decision (Case B, high risk, no COT increase), confirming the reliability and safety of the empirical margins. The rule-based system ensures basic compliance but offers limited explanatory depth. In contrast, the LLMs–particularly Qwen–provide richer engineering interpretations and additional monitoring suggestions, highlighting the value of structured prompt engineering.

5 Discussion and Future Work

The presented script demonstrates a practical integration of historical Excel data and general-purpose LLMs into a human-in-the-loop decision support tool for steam cracking furnace operation, without requiring model training or complex predictive modeling. The approach is easy to adopt: operators only need to prepare an Excel file with the required columns and run the Colab notebook.

Our script can be viewed as a lightweight instance of combining simple data-driven statistics with LLM-based reasoning for industrial decision support. In contrast to AI-based optimization frameworks for steam cracking furnaces [4] and end-to-end machine-learning models for light-olefin yield prediction [1,2], we deliberately focus on a minimal workflow that keeps all computations auditable in a notebook. At the same time, the human-in-the-loop design is close in spirit to LLM-based decision assistants such as DeLLMa and LLM-augmented clinical decision support systems [8], and fits within the broader picture of industrial LLM deployments [5,6].

Nevertheless, the current study has several limitations. First, the case study is based on offline data from a single furnace and operating context, so the empirical safety margins may not directly generalize to other units or plants without re-estimation. Second, the percentile-based boundary model ignores temporal dynamics and does not explicitly account for plant upsets or long-term drifts in feed quality. Third, the current implementation relies on a general-purpose LLM accessed via an external API, which raises questions about latency, cost, data governance and potential hallucinations in safety-critical settings. These limitations highlight the need for more systematic evaluation and hardening before the script can be considered for deployment in production environments.

Table 2. Comparison of operation recommendations: rule-based baseline (recommends Case B if coke margin ≤ 0 or COT margin $< 0.5°C$) versus different LLM backends (identical data and prompt)

Aspect	Rule-based	GPT-4o	DeepSeek	Qwen
COT decision	Case B (no increase)	Case B (no increase)	Case B (no increase)	Explicit Case B (no increase)
Yield gain	Not provided	Limited	Limited	Limited
Risk level	High (by rule)	High	High	High
Coking explanation	CRI exceeds limit	Exceeds limit; steam control	Exceeds limit (quantitative)	Negative margin (-0.00002); high-risk zone
Pressure remarks	Pressure margin exists	Margin exists; monitor	Margin exists; monitor rise	Margin exists; resistance rising
Monitoring items	COT,pressure, CRI	COT,pressure, coke index	COT,pressure, coke index	additional cracking gas inlet temperature
Overall style	Rigid rule	Concise, compliant	Quantitative, conservative	Most engineering-detailed

Future work includes the following directions:

- Adding quantitative evaluation by simulating alternative operating decisions on historical data and comparing the predicted recommendations with actual plant actions.
- Extending the script to handle multiple furnaces and different feedstocks by parameterizing the prompt.
- Integrating the script into existing plant dashboards so that operators can trigger it directly from their HMI.
- Systematically comparing different LLM backends and prompt variants in terms of robustness, sensitivity to input noise and sensitivity to changes in the safety-boundary parameters.
- Combining the percentile-based safety margins with a small trainable surrogate model for ethylene yield, so that more accurate quantitative predictions can be offered while keeping the overall workflow transparent and auditable.

Acknowledgements. The work presented in this paper was supported by the National Natural Science Foundation of China (62273109); the Guangdong Basic and Applied Basic Research Foundation (2022A1515012022, 2023A1515240020, 2023A1515011913); the Key Field Special Project of the Department of Education of Guangdong Province (2024ZDZX1034); the Special Innovation Projects for Ordinary Universities in Guangdong Province in 2023 (2023KTSCX086); the Maoming Science and Technology Project (2022DZXHT028, 210429094551175); and the Projects of PhDs' Start-up Research of GDUPT (XJ2022000301, 2023bsqd1002, 2023bsqd1013).

References

1. Abdi, J., Mazloom, G., Hadavimoghaddam, F., et al.: Estimation of the flow rate of pyrolysis gasoline, ethylene, and propylene in an industrial olefin plant using machine learning approaches. Sci. Rep. **13**, 14081 (2023)
2. Usman, A.G., Tanimu, A., Abba, S.I., Isik, S., Aitani, A., Alasiri, H.: Feasibility of the optimal design of AI-based models integrated with ensemble machine learning paradigms for modeling the yields of light olefins in crude-to-chemical conversions. ACS Omega **8**(43), 40517–40531 (2023)
3. Mafat, I.H., Sharma, S.K., Surya, D.V., Rao, C.S., Maity, U., Barupa, A., Jasra, R.: Development of machine learning model for the prediction of selectivity to light olefins from catalytic cracking of hydrocarbons. Fuel **381**, 133682 (2025)
4. Zhang, Y., Sha, P., Wang, M., Zheng, C., Huang, S., Wu, X., Cordiner, J.: Toward intelligent and green ethylene manufacturing: an AI-based multi-objective dynamic optimization framework for the steam thermal cracking process. Engineering **52**, 160–171 (2025)
5. Raza, M., Jahangir, Z., Riaz, M.B., Saeed, M.J., Sattar, M.A.: Industrial applications of large language models. Sci. Rep. **15**, 13755 (2025)
6. Zhang, G., Lu, C., Luo, Q.: Application of large language models in the AECO industry: core technologies, application scenarios, and research challenges. Buildings **15**(11), 1944 (2025)

7. Glukhikh, I.N., Chernysheva, T.Y., Shentsov, Y.A.: Decision support in a smart greenhouse using large language model with retrieval augmented generation. In: Proceedings of SPIE 13217, Third International Conference on Digital Technologies, Optics, and Materials Science (DTIEE 2024), p. 132170S (2024)
8. Rajashekar, N.C., Shin, Y.E., Pu, Y.: Human-algorithmic interaction using a large language model-augmented artificial intelligence clinical decision support system. In: Proceedings of the CHI Conference on Human Factors in Computing Systems (CHI '24), pp. 1–20 (2024)

TLIME: Stable LIME Explanation for Tabular Data

Tong Zhang[1], Xuansong Li[1,2](✉), and Yu Wang[1,3]

[1] School of Computer Science and Engineering, Nanjing University of Science and Technology, Nanjing, China
124106010801@njust.edu.cn

[2] Science and Technology on Information Systems Engineering Laboratory, Changsha, China
lixs@njust.edu.cn

[3] The 28th Research Institute of China Electronics Technology Group Corporation, Nanjing, China

Abstract. Local Interpretable Model-agnostic Explanations (LIME) has emerged as a widely adopted technique for interpreting black-box machine learning models. In high-stakes application domains such as healthcare and finance, decisions directly affect safety, fairness, and compliance, making post-hoc interpretability a must-have rather than a nice-to-have. However, its efficacy on tabular data is frequently compromised by inadequate sampling in the synthetic neighborhood, particularly in scenarios involving high-dimensional feature spaces or imbalanced data distributions. This paper introduces TLIME, an enhanced framework that incorporates feature-aware stratified sampling into the LIME methodology. By systematically partitioning the feature space and ensuring proportional representation across all strata, TLIME yields more stable and accurate explanations. Across four tabular datasets—Higgs Boson, Census Income, House Prices, and Credit Card Fraud Detection—TLIME, on average, improves local fidelity (R^2) by 107.3%, increases range coverage (RC(Y)) by 20.0%, and raises the coefficient of variation (CV(β)) by 67.0% over standard LIME (Gaussian sampling). These gains mean that the local surrogate more faithfully matches the black-box (higher R^2), the synthetic neighborhood better spans the model's output space (higher RC(Y)), and the attributions more clearly differentiate important from unimportant features (higher CV(β)), thereby supporting model auditing, risk management, and regulatory reporting in clinical decision support and credit scoring.

Keywords: Explainable AI (XAI) · Model-agnostic · Stability · LIME

L. Huang and D. Greenhalgh (Eds.): ICMLC 2026, LNNS 2061, pp. 923–936, 2027.
https://doi.org/10.1007/978-3-032-31236-5_62

1 Introduction

The increasing deployment of black-box machine learning models in high-stakes domains has rendered explainability a critical requirement, not only for fostering trust but also for ensuring safety and accountability. In particular, healthcare (e.g., clinical decision support) and finance (e.g., credit scoring and fraud detection) operate under strict risk and compliance constraints, so interpretability is a hard requirement for operational use. In neuro-oncology, clinicians distinguish glioblastoma from metastases using single-voxel MRS spectra—high-dimensional tabular vectors. A sparse frequency-selection plus a nonlinear classifier reached strong accuracy, yet experts rejected it because key frequencies lacked biochemical meaning and the decision surface was opaque; they preferred a linearly-parameterized model with a few metabolically interpretable peaks, even at some cost in accuracy [1]. In such high-stakes tabular tasks, models are only as useful as their explanations—clinical trust hinges on interpretability, not headline accuracy.

Various approaches have been developed to interpret model behavior. Model-agnostic approaches, such as LIME [2] and SHAP [3], approximate local decision boundaries using interpretable surrogate models, while techniques like CAM [4] and Grad-CAM [5] provide visual explanations for convolutional neural networks.

However, tabular data commonly employed in domains such as healthcare [6], credit scoring [7] and traffic [8]—has received relatively less attention in the development of specialized explanation methods. The original LIME for tabular data relies on simple random sampling, which often leads to unstable explanations. This sampling strategy tends to inadequately represent diverse regions of the feature space, particularly when dealing with high-dimensional datasets or imbalanced distributions. We propose TLIME, a novel approach that mitigates these limitations through systematic stratification of the feature space.

Our principal contributions are summarized as follows:

(1) TLIME Framework: We introduce TLIME, a novel and enhanced LIME framework inspired by recent work on stratified sampling in LIME for image data [9]. It incorporates stratified sampling specifically designed for tabular data. This approach ensures a more representative and balanced synthetic neighborhood around the instance being explained.
(2) Correlation-Aware Stratification: Unlike simple random sampling, our method intelligently partitions the feature space, accounting for potential correlations between features to preserve meaningful combinations and avoid undersampling critical regions, particularly in high-dimensional settings.
(3) Comprehensive Evaluation: We conduct extensive empirical studies on four diverse real-world tabular datasets. The results demonstrate that TLIME achieves significant and consistent improvements in explanation quality, stability, and fidelity compared to standard LIME and other baseline sampling methods, as measured by a suite of robust metrics.

The remainder of this paper is organized as follows. Section 2 reviews related work on LIME and its variants. Section 3 details the proposed TLIME methodology, presenting our stratified sampling algorithm and the corresponding adjustments for unbiased estimation. Section 4 presents our experimental setup, datasets, and a comprehensive discussion of the results, validating the effectiveness of TLIME. Finally, Sect. 5 concludes the paper and suggests directions for future research.

2 Previous Work

This section summarizes previous LIME-related work, including its limitations and variants for different data types.

2.1 LIME for Tabular Data

Local Interpretable Model-agnostic Explanations (LIME) operates through a systematic workflow to generate local approximations of black-box models, as depicted in Fig. 1. The process begins with an input instance to be explained, which is processed by the black-box model to obtain a prediction.

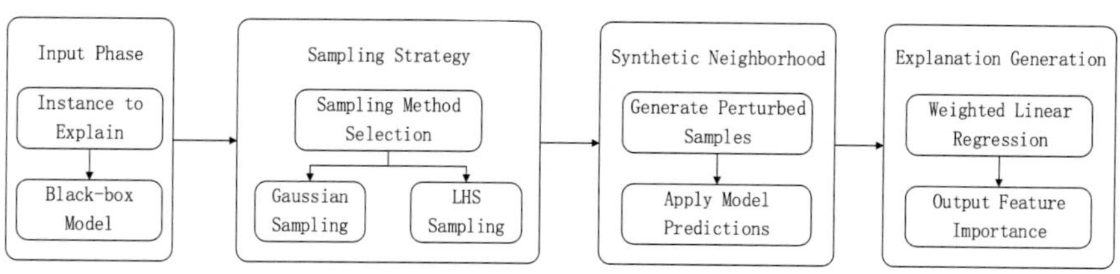

Fig. 1. LIME's workflow.

The core of LIME's methodology lies in its sampling strategy, where synthetic neighbors are generated around the instance of interest. For tabular data, LIME primarily employs two sampling techniques: Gaussian sampling, which perturbs features independently by drawing from normal distributions, and Latin Hypercube Sampling (LHS), which ensures more uniform coverage through stratified sampling across each feature's distribution.

Following sampling, LIME constructs a synthetic neighborhood by generating perturbed samples and obtaining their corresponding predictions from the black-box model. This neighborhood forms the basis for learning a locally faithful explainable model. The final phase is explanation generation through weighted linear regression.

Despite this structured framework, the random sampling approach used in standard LIME often results in the undersampling of critical feature combinations, particularly in high-dimensional spaces where the probability of sampling meaningful perturbations decreases exponentially.

Garreau and von Luxburg [10] provided the first theoretical analysis of LIME, demonstrating that its explanations converge to a limit proportional to the model's gradient and emphasizing the critical role of parameter selection in obtaining reliable explanations.

2.2 LIME Variants and Improvements

Building upon the core LIME framework, researchers have developed numerous domain-specific variants to address various limitations. For image data, [11] stabilizes LIME by eliminating spurious superpixels and employing adaptive Gaussian blur for perturbation. In time-series data, SEGAL [12] extends LIME by incorporating temporal dependencies. Anchors [13] introduces high-precision, rule-based explanations that provide sufficient conditions for predictions, offering more precise interpretations than linear approximations. BayesLIME [14] incorporates Bayesian inference to deliver uncertainty-aware local interpretations, quantifying the reliability of feature importance estimates. More recently, BMB-LIME [15] integrates bootstrap aggregating with Multivariate Adaptive Regression Splines (MARS) within a Bayesian framework to capture nonlinear local boundaries and quantify uncertainty in feature contributions, thereby addressing limitations of linear surrogate models. Table 1 summarizes the improvement directions and effects of LIME and its variants.

Table 1. Compact comparison of LIME variants.

Method	Data type	Surrogate model	Sampling	Key characteristics
LIME (original) [2]	ANY	Linear	Independent perturbations (Gaussian and LHS for tabular; superpixel masking for image)	Simple, general; ignores feature correlation
SLICE [11]	Image	Linear / ridge	Adaptive Gaussian blur + spurious-superpixel removal	Reduces sign/rank flips; more consistent image explanations
Anchors [13]	ANY	Ifthen rules	Conditional sampling + beam search	High-precision rules with coverage guarantees
BayesLIME [14]	ANY	Bayesian linear	Focused/active sampling + posterior	Posterior mean with credible intervals; sample-size guidance
SEGAL [12]	Time-series	Linear over time-aware features	Generative neighborhood + adaptive weighting	Preserves temporal structure; reduces OOD perturbations
BMB-LIME [15]	Tabular	MARS (nonlinear)	Bootstrap resampling and aggregation	Captures local nonlinearity and interactions; higher compute cost
TLIME (this paper)	**Tabular**	Linear (as in LIME)	Stratified by mask cardinality + correlation-aware expansion	Principled fix for tabular undersampling

Despite these advancements, most existing variants have primarily focused on specific data types or supplementary uncertainty measures, rather than addressing fundamental enhancements to the sampling strategy for tabular data.

This gap motivates our work on TLIME, which introduces stratified sampling as a principled solution to the undersampling problem inherent in standard LIME for tabular data. In practice, standard LIME remains particularly insufficient for tabular decision-making in healthcare risk stratification and financial credit/fraud scoring, where Monte Carlo neighborhood sampling often yields unstable attributions; TLIME is designed to remedy this gap.

3 TLIME: Proposed Methodology

This section introduces TLIME's feature-aware stratified sampling method and the evaluation metrics used in subsequent experiments.

3.1 Feature-Aware Stratified Sampling for Tabular LIME

Standard LIME perturbs features independently and learns a local surrogate on the resulting synthetic neighborhood. For tabular data, independent Monte Carlo sampling allocates probability mass to subsets of features roughly according to the binomial law, which under-samples strata near the extremes (very small or very large numbers of preserved features). As a result, the local linear model may be fitted on an imbalanced neighborhood, degrading stability and fidelity—especially when feature correlations matter.

We propose a stratified sampling approach tailored for tabular LIME, which maintains the basic structure of the original LIME. This method partitions the mask space by the number of preserved features and then applies correlation-aware expansion, ensuring that features highly correlated with any preserved feature are also retained, resulting in more realistic local neighborhoods. This approach alleviates the undersampling issues prevalent in standard LIME, especially in high-dimensional tabular datasets with imbalanced feature distributions.

Our stratified sampling begins by analyzing the feature space around the instance to be explained. Let $x \in \mathbb{R}^d$ be the instance to explain, and let a binary preservation mask $m \in \{0,1\}^d$ indicate which features are kept ($m_j = 1$) or perturbed ($m_j = 0$). We stratify the mask space by cardinality:

$$\mathcal{X}^{(k)} = \left\{m \in \{0,1\}^d : \|m\|_0 = k\right\}, \qquad k = 0, 1, \dots, d. \tag{1}$$

Sampling proceeds by drawing $q \sim \text{Uniform}(0,1)$ and then sampling $m_j \sim \text{Bernoulli}(q)$ independently, which induces an (approximately) uniform distribution over the strata$\{x^{(k)}\}_{k=0}^{d}$. Because this differs from the standard Monte Carlo distribution over all 2^d masks, we correct for stratification with a combinatorial weight that depends only on $k = \|m\|_0$. Here $k = \|m\|_0$ is the number of preserved features in the binary mask m.

$$W(k) = \frac{d+1}{2^d}\binom{d}{k} \tag{2}$$

In practice, we compute $\log W(k)$ via $\log \Gamma(\cdot)$ for numerical stability and multiply W with the usual LIME kernel weights when fitting the local surrogate, thereby preserving unbiasedness across cardinality strata. To ensure stability, we truncate W(k) to the range [0.1, 100], preventing extreme values from affecting the results while maintaining robustness in model fitting.

To preserve meaningful dependencies among features, we estimate the empirical correlation matrix ρ from standardized training data and expand the preserved set $S = \{j : m_j = 1\}$ to

$$S' = S \cup \left\{ j : \max_{i \in S} \rho_{ji} \geq \tau \right\}, \quad \tau = 0.7. \tag{3}$$

The final mask m' is obtained by setting $m'_j = 1$ for all $j \in S'$. This correlation-aware expansion retains clusters of strongly associated features, producing synthetic neighborhoods that better reflect the structure of tabular data while maintaining the desired coverage across $\{X(k)\}$. Unlike value-range stratification, this design targets stability in the combinatorial mask space.

In TLIME, the correlation threshold τ governs the correlation-aware expansion in Eq. (3), where the preserved feature set S' is optionally augmented by adding features whose empirical correlations with any feature in S exceed τ (we set $\tau = 0.7$ in the main experiments). Intuitively, a smaller threshold (e.g., $\tau = 0.5$) tends to expand S more often, which may better preserve correlated feature groups in the perturbed neighborhoods and can be beneficial when strong multicollinearity is present; however, it may also reduce the diversity of perturbations because more features are kept together. Conversely, a larger threshold (e.g., $\tau = 0.9$) triggers expansion less frequently, which can increase perturbation diversity and lead to broader neighborhood exploration, but may under-preserve feature dependencies in datasets with pronounced correlations.

We show the details of TLIME in Algorithm 1.

The integration of stratified sampling with LIME's local explanation framework ensures adequate representation of local behavior around the explained instance, particularly in regions most relevant for local faithfulness. By systematically addressing the inherent undersampling issue in standard LIME, our feature-aware stratified sampling approach produces more stable coefficient estimates and significantly improves explanation consistency across multiple runs. As demonstrated in our comprehensive experimental evaluation, this method generates explanations that are both more stable and more faithful to the black-box model's local behavior.

3.2 The Evaluation Metrics Adopted

To comprehensively evaluate TLIME's performance on tabular data, we employ three quantitative metrics adapted from recent advances in explanation stability assessment.

Algorithm 1 TLIME Feature-Aware Stratified Sampling for Tabular Data

Input: Instance to explain x, Training data D, Number of samples N, Correlation threshold $\tau \in [0, 1]$
Output: Stratified perturbed samples X, Adjusted weights W
Initialize: $X \leftarrow 0_{N \times d}$, $W \leftarrow 1_N$
$\rho \leftarrow \text{corrcoef}(S.\text{transform}(D))$ ▷ Standardize for correlation
$\rho \leftarrow (\rho + 1)/2$ ▷ Normalize the correlation matrix
for $i = 1$ to $N - 1$ **do**
    $q \sim \text{Uniform}(0, 1)$
    $m \leftarrow (\text{Rand}(d) < q)$ ▷ i.i.d. Bernoulli(q) per feature
    $S_{\text{pres}} \leftarrow \{j : m_j = 1\}$
    for each $j \in S_{\text{pres}}$ **do**
        $\text{corr} \leftarrow \{k : \rho_{jk} \geq \tau\}$
        $m[\text{corr}] \leftarrow 1$ ▷ Correlation-aware expansion
    end for
    $k = \|m\|_0$ ▷ Stratum index = mask cardinality
    $X[i, :] \leftarrow m$
    $\log B \leftarrow \log \Gamma(d + 1) - \log \Gamma(k + 1) - \log \Gamma(d - k + 1)$ ▷ Compute in a numerically stable way
    $W[i] \leftarrow \exp(\log B + \log(d + 1) - d \log 2)$
end for
return X, W

1. **Model Fit R-squared (R²)** measures how well the local surrogate model approximates the black-box predictions:

$$R^2 = 1 - \frac{\sum_i (y_i - \hat{y}_i)^2}{\sum_i (y_i - \bar{y})^2} \tag{4}$$

   where y_i denotes the black-box prediction for the i-th perturbed sample, $\hat{y}_i$ represents the local model's prediction, and $\bar{y}$ is the mean prediction. Higher values indicate better local fidelity.
2. **Coefficient of Variation of β (CV(β))** quantifies the variability of the weights across features:

$$CV(\beta) = \frac{\sigma_\beta}{\mu_\beta} \tag{5}$$

   where σ_β and μ_β are the standard deviation and mean of the feature importance vector β. Higher values suggest more discriminative explanations.
3. **Range Coverage of Y (RC(Y))** assesses the sampling quality in the prediction space:

$$RC(Y) = \frac{\text{IQR}_{1-99}(Y)}{f(x)} \tag{6}$$

   where $\text{IQR}_{1-99}(Y)$ represents the interquantile range of perturbed predictions, and $f(x)$ is the original instance prediction. Higher values indicate better coverage. If $f(x) = 0$, we set RC(Y) to 0 to avoid division by zero or illogical results.

4 Experimental Evaluation

In this section, we target to answer the following two research questions:

RQ 1: How do TLIME's fidelity, explanation discriminability, and output space coverage compare to the original LIME?

RQ 2: What types of datasets are TLIME suitable for?

The experimental setup and results are designed to answer these two questions.

4.1 Experimental Setup

In this section, we describe the experimental setup in detail, including dataset characteristics, preprocessing steps, model selection, and metrics used to evaluate the performance of TLIME.

We evaluated TLIME on four widely used tabular datasets:

Higgs Boson [16]: Contains 11 million instances with 28 features. It is used for detecting the presence of the Higgs boson particle in physics experiments.

House Prices [17]: Contains 1,460 instances and 81 features. This dataset is used for predicting house prices based on characteristics such as location, size, and condition. We binarize SalePrice at the median, and treat it as a binary classification task.

Census Income [18]: Consists of 32,561 instances with 14 features. This dataset is used for predicting whether a person's income exceeds 50K USD based on demographic information.

Credit Card Fraud Detection [19]: Contains 284,807 instances with 30 features. This dataset is used for detecting fraudulent transactions based on credit card data.

Each dataset was processed using Standard Scaling (mean = 0, standard deviation = 1) to standardize the feature space. For categorical features, one-hot encoding was applied where necessary. Features with high missing value ratios were removed, and missing values in numerical columns were imputed using median values.

We utilized Random Forest classifiers for the experimental setup. Random Forest was chosen for its ability to handle both high-dimensional and imbalanced datasets, which are common in real-world tabular data. For regression tasks, Random Forest Regressor was used, and for classification tasks, Random Forest Classifier was employed.

For each dataset, we evaluated TLIME's performance using stratified sampling and compared it to LIME's and BayesLIME's sampling. For stratified sampling, we ensured proportional representation across strata based on feature distributions and used correlation-aware stratification to account for relationships between features. For BayesLIME, we replaced the local surrogate with Bayesian Ridge regression (Gaussian prior on coefficients, evidence-based hyperparameter estimation). The neighborhood size was set to 1,000 perturbed samples per instance, and 20 independent runs were conducted to ensure robust results.

To address the key issues, we chose evaluation metrics including R^2, RC(Y), and CV(β), as they provide comprehensive insights into the explanation's stability, the coverage of the output space, and the consistency in feature importance. R^2 measures local fidelity, while RC(Y) assesses the sampling distribution's quality. CV(β) captures the variation in feature importance, which is crucial for ensuring discriminative explanations.

All experiments were conducted using Python 3.12 with scikit-learn for model training, NumPy for data manipulation, and LIME for local interpretability. The experiments were executed on a workstation featuring 16GB of RAM and a 12-core, 20-thread Intel Core i7-12700 CPU, ensuring stable computational performance.

4.2 Results

Our comprehensive experimental evaluation demonstrates that TLIME significantly improves explanation stability and fidelity across all four tabular datasets. The proposed stratified sampling approach consistently outperforms standard Monte Carlo sampling in terms of explanation consistency, range coverage, and local approximation quality.

Figure 2 presents a comparative histogram of sampling distributions between standard LIME and TLIME for three randomly selected instances. The original LIME exhibits limited range coverage, with perturbed sample predictions concentrated within a narrow band. Notably, samples with RC(Y) > 0.8 and RC(Y) < 0.2 are scarce, highlighting the undersampling limitation of standard LIME. In contrast, TLIME achieves substantially superior range coverage, sampling predictions across the entire probability spectrum without neglecting extreme cases. This enhanced sampling distribution enables more robust linear regression fitting, consequently yielding more stable feature importance estimates and higher fidelity.

Figure 3 presents a comparison of explanation consistency for a randomly selected instance from the Higgs Boson dataset, using a trained Random Forest model. Panel A displays the results of three independent runs with standard LIME, revealing considerable instability in both feature importance values and rankings. For instance, while feature 23 appears consistently across runs, its assigned weights fluctuate significantly. In contrast, features such as feature 4, feature 10, and feature 1 appear inconsistently among the top-ranked features. This variability indicates that the Gaussian sampling strategy employed by standard LIME yields unreliable explanations with high variance. Conversely, Panel B illustrates the enhanced consistency achieved by TLIME: The first five features maintain nearly identical importance scores and a consistent ranking order across all three runs. The stratified sampling approach in TLIME ensures stable neighborhood generation, leading to more reproducible and trustworthy explanations—a critical advantage in sensitive domains where the reliability of explanations directly impacts decision-making confidence.

Figure 4 provides a comprehensive comparison of explanation quality metrics across all four datasets after adding BayesLIME as an additional LIME-

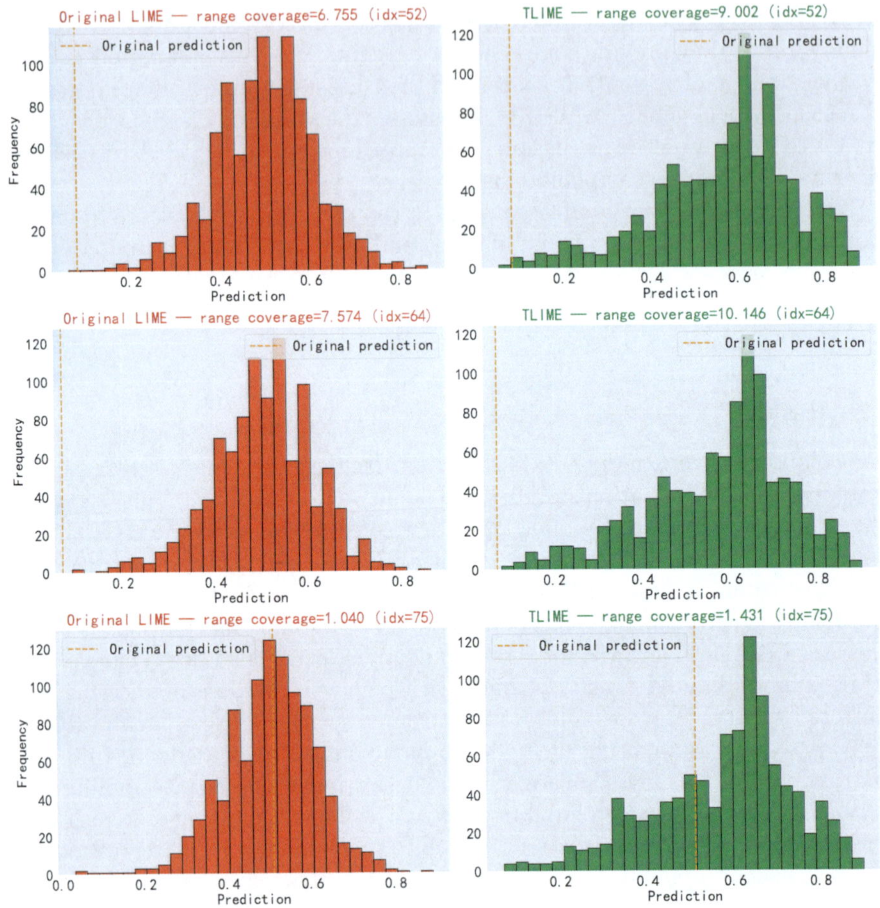

Fig. 2. Comparison of sampling distributions between LIME and TLIME (Higgs Boson; three selected instances).

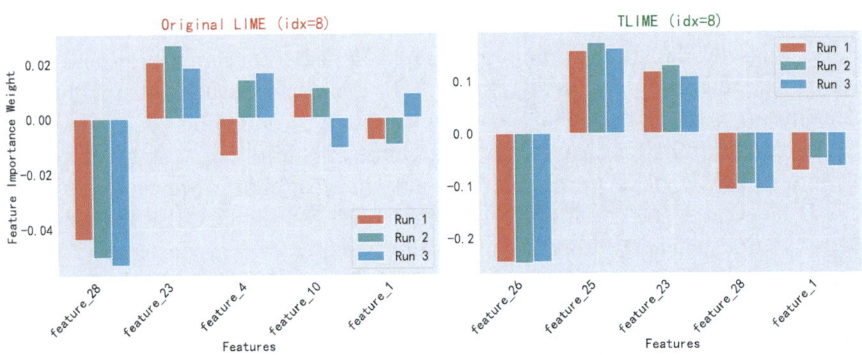

Fig. 3. Comparison of feature-importance explanations between LIME and TLIME for a random Higgs Boson instance.

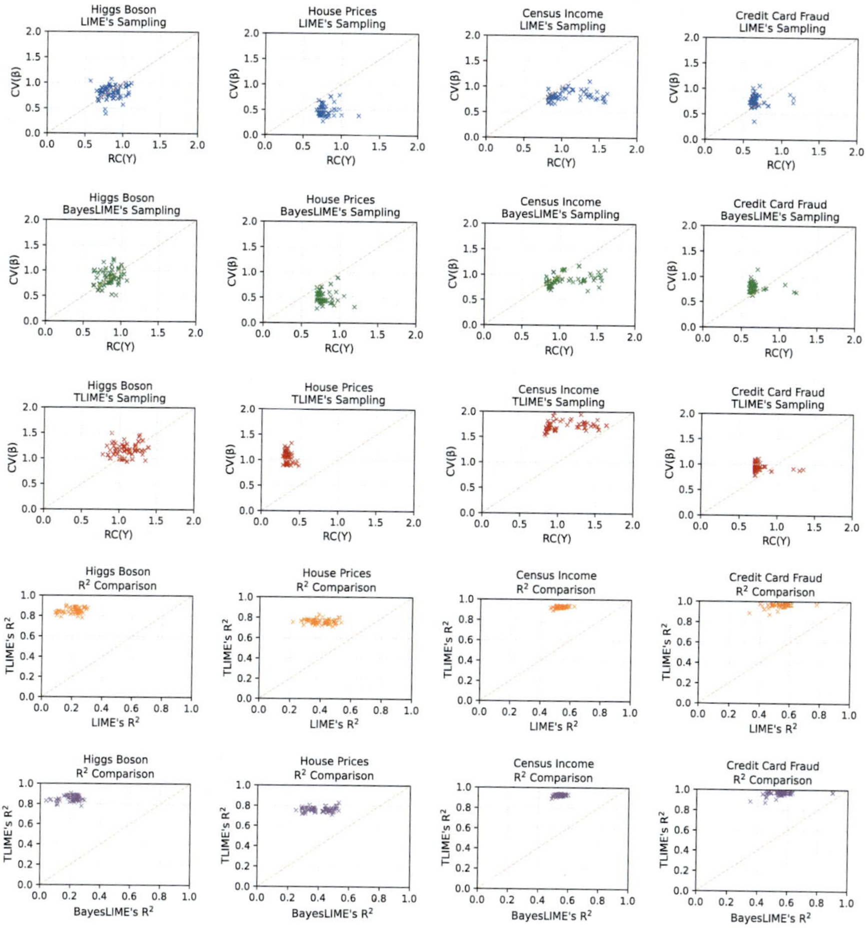

Fig. 4. Comparison of the average RC(Y), CV(β), and R^2 between LIME, BayesLIME and TLIME across four datasets.

variant baseline. For each dataset, we randomly selected 50 instances and executed Gaussian LIME, BayesLIME, and TLIME (stratified sampling) over 20 independent runs with a neighborhood size of $n = 1000$ perturbed samples per instance, recording the range coverage RC of the Y distributions, the coefficient of variation CV of the feature-attribution vectors β, and the local fidelity metric R^2. The first three rows visualize CV(β) versus RC(Y) for Gaussian LIME, BayesLIME's sampling, and TLIME's stratified sampling, respectively. Overall, Gaussian LIME tends to yield explanations with limited output-space coverage and weak discriminability, with many instances clustering toward the lower-left region. BayesLIME provides only modest improvements over Gaussian LIME—some points move upward/rightward, but a substantial portion of instances (especially on Census Income and Credit Card Fraud) still remain concentrated in low-RC(Y) and low-CV(β) regions, indicating that sampling diversity is still insufficient for consistently stable explanations. In contrast, TLIME produces a clear shift of points toward the upper-right quadrant on three datasets, demonstrating simultaneous improvements in explanation discriminability and range coverage; the gains are particularly prominent on Higgs Boson and Census Income, where many instances achieve CV(β) $>$ 1.0 and RC(Y) $>$ 1.0. Finally, the last row reports an R^2 comparison between BayesLIME (x-axis) and TLIME (y-axis), where points above the diagonal indicate higher local fidelity for TLIME. The strong concentration of points above the diagonal across all four datasets confirms that TLIME provides a tighter local approximation of the black-box model than BayesLIME and LIME, while also delivering stronger improvements in coverage and discriminability.

Table 2. Average performance metrics across all four datasets.

Dataset	Accuracy	Method	R^2	RC(Y)	CV(β)
Higgs Boson	0.7021	LIME	0.2061	0.8319	0.8169
		BayesLIME	0.2000	0.8481	0.8704
		TLIME	0.8536	1.2549	1.1556
House prices	0.9144	LIME	0.3994	0.7863	0.4866
		BayesLIME	0.4036	0.7946	0.5006
		TLIME	0.7610	0.5398	1.0457
Census income	0.8472	LIME	0.5345	1.0592	0.8653
		BayesLIME	0.5452	1.1252	0.8982
		TLIME	0.9199	1.2604	1.7449
Credit card fraud	0.9594	LIME	0.5455	0.6708	0.7662
		BayesLIME	0.5550	0.6782	0.7913
		TLIME	0.9588	0.9625	0.9543

Table 2 addresses Research Question 1 (RQ1) by quantitatively summarizing the average performance metrics across the four datasets, and it further includes BayesLIME as an additional LIME-variant baseline. Overall, TLIME achieves a substantial improvement in local fidelity, increasing the average (R) from 0.4214 to 0.8733, i.e., a 107.3% gain. In contrast, BayesLIME yields only a marginal change in fidelity (average (R = 0.4259), 1.1% higher than Gaussian LIME). In terms of output-space coverage, TLIME increases the average RC(Y) from 0.8370 to 1.0044 (a 20.0% increase), while BayesLIME remains close to Gaussian LIME (0.8615, a 2.9% increase). Similarly, TLIME markedly improves the discriminability of explanations, raising the average CV(β) from 0.7338 to 1.2251 (a 67.0% increase), whereas BayesLIME shows only a slight increase (0.7651, a 4.3% increase). Taken together, these results answer RQ1 and demonstrate that, compared with the original LIME and BayesLIME, TLIME effectively improves local fidelity, explanation discriminability, and output-space coverage.

As for RQ 2, TLIME delivers the largest gains on high-dimensional, feature-correlated, and/or class-imbalanced tabular datasets (e.g., Higgs Boson, Census Income and Credit Card Fraud). For datasets with weaker feature correlations and tightly distributed targets (e.g., House Prices), TLIME still increases R and CV(β) while perturbing more conservatively, so RC(Y) may not rise. Collectively, these findings directly address the two guiding questions.

5 Conclusions

Our findings indicate that TLIME's feature-aware stratified sampling provides a principled and effective solution to the long-standing undersampling issue in LIME's application to tabular data, significantly improving the stability and reliability of LIME explanations for tabular data.

Future work will focus on extending TLIME to handle text data, optimizing the stratification strategy for ultra-high-dimensional datasets, and exploring adaptive stratification levels based on dataset characteristics. The integration of domain knowledge into the stratification process also presents promising research directions. Practically, TLIME can strengthen model governance by enabling more trustworthy auditing, monitoring, and decision support in domains such as clinical analytics and credit risk management.

Acknowledgments. The Project is Supported by the Open Fund of Science and Technology on Information Systems Engineering Laboratory (No. WDZC20245250404). The authors would like to thank the anonymous reviewers for their constructive comments. We thank the providers of the Higgs Boson, House Prices, and Credit Card Fraud Detection datasets for making their data publicly available. Special thanks to the School of Computer Science and Engineering at Nanjing University of Science and Technology for their support.

References

1. Vellido, A.: The importance of interpretability and visualization in machine learning for applications in medicine and health care. Neural Comput. Appl. **32**(24), 18069–18083 (2020)
2. Ribeiro, M.T., Singh, S., Guestrin, C.:"Why should I trust you?" Explaining the predictions of any classifier. In: Proceedings of the 22nd ACM SIGKDD International Conference on Knowledge Discovery and Data Mining (KDD '16), pp. 1135–1144 (2016)
3. Lundberg, S.M., Lee, S.I.: A unified approach to interpreting model predictions. Advances in Neural Information Processing Systems, p. 30 (2017)
4. Zhou, B., Khosla, A., Lapedriza, A., Oliva, A., Torralba, A.: Learning deep features for discriminative localization. In: Proceedings of the IEEE Conference on Computer Vision and Pattern Recognition (CVPR), pp. 2921–2929 (2016)
5. Selvaraju, R.R., Cogswell, M., Das, A., Vedantam, R., Parikh, D., Batra, D.: Grad-CAM: visual explanations from deep networks via gradient-based localization. In: Proceedings of the IEEE International Conference on Computer Vision (ICCV), pp. 618–626 (2017)

6. Ukwuoma, C.C., Cai, D., Eziefuna, E.O., Oluwasanmi, A., Abdi, S.F., Muoka, G.W., Sarpong, K.: Enhancing histopathological medical image classification for early cancer diagnosis using deep learning and explainable AI–LIME & SHAP. Biomed. Signal Process. Control **100**, 107014 (2025)
7. Raufi, B., Finnegan, C., Longo, L.: A comparative analysis of SHAP, LIME,ANCHORS, and DICE for interpreting a denseneural network in Credit Card Fraud Detection. In: Proceedings of the World Conference on Explainable Artificial Intelligence, pp. 365-383. Springer, Cham (2024)
8. Liu, C., Sheng, Z., Li, P., Chen, S., Luo, X., Ran, B.: A distributed deep reinforcement learning-based longitudinal control strategy for connected automated vehicles combining attention mechanism. Transport. Lett. **17**(2), 183–199 (2025)
9. Rashid, M., Amparore, E.G., Ferrari, E., Verda, D.: Using stratified sampling to improve lime image explanations. In: Proceedings of the AAAI Conference on Artificial Intelligence, vol. 38, no. 13, pp. 14785–14792 (2024)
10. Garreau, D., Luxburg, U.: Explaining the explainer: a first theoretical analysis of LIME. In: Proceedings of the International Conference on Artificial Intelligence and Statistics, pp. 1287–1296. PMLR (2020)
11. Bora, R.P., Terhörst, P., Veldhuis, R., Ramachandra, R., Raja, K.: Slice: stabilized lime for consistent explanations for image classification. In: Proceedings of the IEEE/CVF Conference on Computer Vision and Pattern Recognition, pp. 10988–10996 (2024)
12. Meng, H., Wagner, C., Triguero, I.: SEGAL time series classification-stable explanations using a generative model and an adaptive weighting method for LIME. Neural Netw. **176**, 106345 (2024)
13. Ribeiro, M.T., Singh, S., Guestrin, C.: Anchors: high-precision model-agnostic explanations. In: Proceedings of the AAAI Conference on Artificial Intelligence (AAAI 2018), vol. 32(1) (2018)
14. Slack, D., Hilgard, A., Singh, S., Lakkaraju, H.: Reliable post hoc explanations: modeling uncertainty in explainability. Adv. Neural. Inf. Process. Syst. **34**, 9391–9404 (2021)
15. Hung, Y.H., Lee, C.Y.: BMB-LIME: LIME with modeling local nonlinearity and uncertainty in explainability. Knowl.-Based Syst. **294**, 111732 (2024)
16. Whiteson, D.: HIGGS [Dataset]. UCI Machine Learning Repository (2014). https://doi.org/10.24432/C5V312
17. Montoya, A., DataCanary: House prices—advanced regression techniques [Dataset]. Kaggle (2016). https://www.kaggle.com/competitions/house-prices-advanced-regression-techniques
18. Kohavi, R.: Census Income [Dataset]. UCI Machine Learning Repository (1996). https://doi.org/10.24432/C5GP7S
19. Ishak, N.A., Ng, K.-H., Tong, G.-K., Kalid, S.N., Khor, K.-C.: CCFD_dataset (Version 3). figshare (2021). https://doi.org/10.6084/m9.figshare.16695616.v3

Generative AI Assisted Psychological Resilience Training: Situational Counseling Practice for Teenagers Based on MBTI Virtual Characters

Li Feng and Yangang Yang(✉)

Huaihua Normal College, Huaihua, Hunan 418000, China
abc3801013@163.com

Abstract. Adolescence represents a critical period for the development of psychological resilience. Traditional intervention approaches, however, are often hindered by limitations such as insufficient personalization and poor adaptability to specific situational contexts. To address these challenges, this study proposes an innovative framework for virtual-character-based situational counseling that integrates generative artificial intelligence (AI) with the Myers-Briggs Type Indicator (MBTI) personality theory. The framework leverages large language models (LLMs) to generate virtual tutors tailored to individual MBTI personality profiles, enabling dynamic and personalized interactions with adolescents in simulated real-life stress scenarios. The primary objective is to enhance core components of psychological resilience, including emotion regulation, problem-solving ability, and positive cognition. A six-week randomized controlled trial was conducted involving 120 middle school students, assigned to either an experimental group (receiving AI-MBTI–based counseling) or a control group (participating in a standardized psychological resilience curriculum). Multidimensional outcomes were assessed using the Connor-Davidson Resilience Scale (CD-RISC-10), situational coping behavior evaluations, and qualitative interviews. Results indicated that the experimental group significantly outperformed the control group in the total CD-RISC-10 score, as well as in the subscales of "strength" and "optimism" ($p < 0.01$). Furthermore, participants in the experimental group demonstrated more adaptive coping strategies and higher levels of self-efficacy when confronted with simulated high-pressure situations. Qualitative analysis revealed that MBTI-based personality alignment enhanced adolescents' engagement, trust in the virtual tutor, and cognitive restructuring processes. These findings suggest that generative AI-driven, MBTI-informed virtual character counseling constitutes an effective and engaging new paradigm for strengthening adolescent psychological resilience. This research provides empirical support and a practical pathway for AI-enabled mental health education.

Keywords: Generative artificial intelligence · Psychological resilience · MBTI personality type

L. Huang and D. Greenhalgh (Eds.): ICMLC 2026, LNNS 2061, pp. 937–947, 2027.
https://doi.org/10.1007/978-3-032-31236-5_63

1 Introduction

Adolescence is a critical period for the rapid development of body and mind and the sharp increase of psychosocial challenges. Multiple factors such as academic pressure, peer relationships and self-identity are intertwined, which are easy to induce psychological problems such as anxiety and depression [1]. Psychological resilience, as the core ability of individuals to maintain positive adaptation and development in adversity [2], is recognized as an important protective factor for adolescent mental health. Effective resilience training can significantly reduce the incidence of psychological and behavioral problems in adolescents and promote their long-term healthy development [3]. However, traditional resilience interventions (such as group counseling and cognitive behavior courses) often face bottlenecks such as limited resources, difficulty in meeting individual differentiated needs with standardized content, and difficulty in skill transfer due to lack of situational realism [4, 5]. The breakthrough of generative AI, especially LLM, has opened up a new path for personalized mental health intervention [6]. Its core advantage is that it can generate natural, coherent and context related dialogs and content in real time according to user input, providing a technical basis for creating an "immersive, interactive and customizable" coaching experience [7]. The key to applying generative AI to psychological resilience training is how to achieve effective personalization and contextualization, so as to deeply fit the cognitive characteristics and emotional needs of teenagers.

Myers Briggs Type Index (MBTI), as a widely used theoretical framework of personality types, divides personality into 16 types from four dimensions: energy direction (E-I), cognitive style (S-N), decision-making style (T-F), and life attitude (J-P) [8]. Research shows that the individual's response mode to stressors, preferred coping strategies and the effectiveness of counseling methods are related to the type of MBTI [9, 10]. For example, introverted (I) individuals may benefit more from introspective guidance, while sensory (S) individuals prefer specific and practical solutions. Integrating MBTI personality traits into AI virtual character design is expected to achieve a highly personalized counseling style, communication methods and intervention strategies, so as to improve teenagers' participation motivation and intervention effect [11].

This study proposes the model of "generative AI assisted situational counseling of MBTI virtual characters", which aims to solve the problem of the lack of personalization and situational in traditional toughness training. The core research questions include: (1) can the model effectively improve the level of adolescents'psychological resilience? (2) How does MBTI personality match affect the counseling effect and user experience? (3) How effective is generative AI in simulating real-world interactions? The study evaluated the effectiveness, feasibility and mechanism of the innovative model through a rigorous randomized controlled experimental design, providing a scientific basis and practical reference for AI enabled adolescent mental health services.

2 Theoretical Basis and Literature Review

2.1 Adolescent Resilience: Connotation, Dimension and Intervention

Psychological resilience is not a fixed trait, but a developable dynamic process [12]. Its core dimensions include: (1) emotion regulation: managing negative emotions and maintaining emotional balance; (2) Cognitive flexibility: positive attribution, reconstruction of adversity significance; (3) Problem solving: formulate and implement the response plan; (4) Social support utilization: actively seek and effectively use resources; (5) Self efficacy: confidence in overcoming challenges [13, 14]. Adolescent resilience intervention needs to focus on the development of these plastic abilities. The existing evidence-based interventions are mostly based on the principles of cognitive behavioral therapy (CBT), acceptance commitment therapy (ACT) and positive psychology [15], but the authenticity and personalized adaptation of scenario simulation are still the key challenges to improve the intervention effect [5].

2.2 Application Value of MBTI Personality Theory in Counseling

The MBTI framework provides a structured perspective for understanding individual differences. In the counseling situation, different personality types have significant preferences for interactive styles [9]:

E/I dimension: extroverts (E) prefer discussion and action, introverts (I) tend to reflect and communicate in writing.

S/N dimension: sensory type (S) focuses on specific facts and steps, while intuitive type (N) focuses on concepts and possibilities.

T/F dimension: thinking type (T) emphasizes logical analysis, while emotional type (F) emphasizes value harmony.

J/P dimension: the judgment type (J) prefers structural planning, while the perception type (P) prefers flexibility and openness.

Research shows that the personality matching of counselors and recipients can enhance work alliance, trust and intervention compliance [10, 16]. Traditional manual matching is expensive, while AI provides the possibility of large-scale and accurate personality adaptation.

2.3 Frontier Exploration of Generative AI in Mental Health Intervention

Generative AI (such as LLM based on transformer architecture) shows potential in the fields of psychological dialog system, virtual companionship, cognitive training and so on [6, 17]. Its advantages are:

Dynamic personalization: respond to user input in real time and generate customized content [7].

Scenario simulation: build diversified and highly realistic social scenarios for practice [18].

Scalability: break through the limitations of time and space and human resources [19].

Current studies mostly focus on emotional support or intervention of specific barriers (such as social anxiety and depression) [20], while the systematic research on adolescent groups is still blank by deeply combining generative AI, personality theory and resilience training. This study aims to fill this gap.

3 Methodology

3.1 System Framework Design

The AI-MBTI coaching system constructed in this study uses a three-tier architecture (Fig. 1):

1. User interaction layer: 3D virtual character interface based on Unity engine, supporting voice/text input. The appearance, expression and body language of the virtual characters are preset according to the MBTI type (for example, ENTJ type shows a capable leadership temperament, INFP type shows a mild empathy style).
2. MBTI personality engine layer:

Personality assessment: the simplified MBTI step II scale [21] was used to assess adolescents' personality types at the baseline stage.

Personality strategy mapping Library: Based on literature [9, 10, 16], it defines the coaching strategies preferred by each type of MBTI (such as providing structured step guidance for ISTJ type, and using exploratory questions to stimulate creativity for ENFP type).

Dynamic style adaptation module: fine tune the expression of virtual characters according to users' real-time feedback (such as adding emotional support statements to high-pressure users). In order to quantify the degree of personality matching between users and virtual characters, this paper proposes a weighted matching index (MSI) based on MBTI four-dimensional similarity. Let the user's MBTI type be expressed as u = (U1, U2, U3, U4) u = (U1, U2, U3, U4), where each dimension value is a binary variable (such as 1∈{e = +1, i = −1} U1∈{e = +1, i = −1}). Similarly, the type of virtual character is v = (V1, V2, V3, V4) v = (V1, V2, V3, V4). Then the matching degree is defined as follows:

$$\text{MSI (U, V)} = \frac{1}{4}\sum_{i=1}^{4}\delta(u_i, v_i), \quad \delta(u_i, v_i) = \begin{cases} 1 \text{ if } u_i = v_i \\ 0 \text{ otherwise} \end{cases} \tag{1}$$

MSI (U, V) = 41 i = 1∑4δ(ui, vi), Where δ (UI, VI) = {1,0, if ui = viotherwise

3. Generative AI core layer:

Base model: open source LLM (such as llama-2-13b) after fine-tuning is adopted (see Fig. 1).

Situational knowledge base: contains typical adolescent stress scenario scripts (Academic setbacks, peer conflicts, family conflicts, etc.) and resilience skills knowledge points (CBT/ACT Technology).

Safety and ethics module: integrate content filters, crisis identification algorithms (such as suicide ideation keywords) and emergency referral protocols to ensure compliance with the ethical guidelines for the application of artificial intelligence to mental health [22].

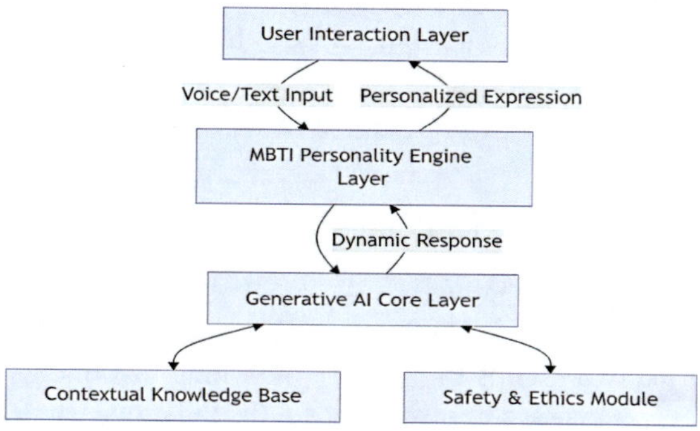

Fig. 1. AI-MBTI coaching system architecture.

3.2 Experimental Design

Design: double blind randomized controlled trial (RCT).

Participants: 120 students aged 13–16 were recruited from two middle schools in a city (M = 14.3, SD = 0.8). Inclusion criteria: CD-RISC-10 score < median (local norm), informed consent. Exclusion criteria: history of severe mental disorders, intellectual impairment. Stratified by gender and grade, they were randomly assigned to the experimental group (EG, n = 60) or the control group (CG, n = 60).

Intervention plan:

EG group: AI-MBTI situational counseling twice a week × 30 min for 6 weeks. The system matches virtual characters (INTP type "logic tutors") according to their MBTI type (such as INTP). Each training focuses on a stressful situation (such as "failure in the exam"), and the virtual characters guide them to practice specific resilience skills (such as emotional tagging, positive re-evaluation, and problem decomposition).

CG group: accepted the equal length and standardized group course of psychological resilience (based on the Penn resilience program core module) [15], led by experienced psychological teachers.

Measuring tools:

Main outcome: Connor Davidson Resilience Scale (CD-RISC-10) [23] (pretest, posttest, 3-month follow-up).

Secondary outcome:

Situational coping behavior assessment (SCBA): the simulated situational task developed by the researcher (such as "misunderstood by friends") was scored by two independent evaluators (kappa> 0.85) according to the preset criteria (strategy effectiveness, emotional stability, self-efficacy expression).

System availability and experience questionnaire (dedicated to eg group): Likert scale and open questions including participation, trust, help perception and other dimensions.

Semi structured interview (20 people in EG group): in depth exploration of MBTI matching experience, skill transfer, etc.

Data analysis: SPSS 26.0 was used for intention to analyze (ITT).T-test or ANOVA (repeated measurement) was used for continuous variables; Chi square test was used for categorical variables; Qualitative data were encoded by topic analysis [24].

4 Research Results

4.1 Participant Characteristics and Compliance

120 participants completed all the interventions and post tests (the loss rate was 0%). There was no significant difference in baseline demographic variables and CD-RISC-10 scores between the two groups ($p > 0.05$), indicating that the randomization was successful. The average session completion rate of the EG group was 92.7%, indicating high acceptance.

4.2 Psychological Resilience Improvement Effect (CD-RISC-10)

The repeated measurement analysis of variance (Table 1) shows that:

The main effect of time was significant ($F (2236) = 58.32$, $p < 0.001$, $\eta^2 = 0.33$), indicating that the overall level of toughness improved after intervention.

Group × time interaction effect was significant ($F (2236) = 12.47$, $p < 0.001$, $\eta^2 = 0.10$). The simple effect analysis shows that:

At post-test, the total score of EG group ($M = 28.4$, $SD = 3.1$) was significantly higher than that of CG group ($M = 25.1$, $SD = 3.5$) ($t (118) = 5.63$, $p < 0.001$, $d = 0.99$). Especially in the dimensions of "strength" (confidence in coping with challenges) and "optimism" (positive outlook), the difference was the largest ($p < 0.01$).

At the time of follow-up, the advantage of the EG group was still maintained ($M = 27.8$ vs. 24.6, $t (118) = 4.92$, $p < 0.001$, $d = 0.87$).

Table 1. Summary of repeated measurement ANOVA of CD-RISC-10 total score.

Source of variation	Sum of squares (SS)	Degrees of freedom (DF)	Mean square (MS)	F value	P value	η^2	Source of variation
Intergroup effect							
Group	102.35	1	102.35	1.87	0.174	0.016	Group
Error (between groups)	6328.90	118	54.82				

(continued)

Table 1. (*continued*)

Source of variation	Sum of squares (SS)	Degrees of freedom (DF)	Mean square (MS)	F value	P value	η^2	Source of variation
Intragroup effect							
Time	645.21	2	322.61	58.32***	< 0.001	0.331	Time
Group × time	137.89	2	68.95	12.47***	< 0.001	0.096	Group × time
Error (within group)	1305.33	236	5.53				

Note: $p < 0.001$

4.3 Situational Coping Behavior Performance (SCBA)

The comprehensive performance score of EG group in simulated situation task ($M = 4.02$, $SD = 0.51$) was significantly better than that of CG group ($M = 3.45$, $SD = 0.62$) ($t(118) = 5.78$, $p < 0.001$, $d = 1.02$).

Specifically:

Strategy effectiveness: more frequently use cognitive reconstruction (such as "this failure ≠ my poor ability") and problem solving skills (such as "making a small goal plan") taught by the system.

Emotional stability: the intensity of negative emotions is lower and recovery is faster.

Self efficacy expression: use more positive language such as "I can", "I can try".

4.4 The Moderating Effect of MBTI Personality Matching

Hierarchical regression analysis showed that in the EG group, the MBTI matching degree between users and virtual characters (calculated based on type similarity) significantly positively predicted the post-test score of CD-RISC-10 ($\beta = 0.32$, $P = 0.008$), explaining an additional 10% variance. The interview data revealed its mechanism of action:

"The way she (INFJ virtual character) speaks makes me feel really understood, unlike the teacher who always speaks big truth."(p12, INFJ type)

"My 'mentor' (ENTP type) always helps me analyze problems from a novel perspective, which is especially suitable for people like me who are full of fancies."(P19, ENTP type)

Feedback from mismatched cases (P05, ISTJ type): "virtual human (ENFP type) is too jumping. I often can't keep up with ideas. I hope to give more direct methods."

4.5 System Experience and Qualitative Feedback

The questionnaire showed that the EG group had higher scores of participation in the system (M = 4.3/5), trust (M = 3.9/5) and perceived helpfulness (M = 4.1/5). Three core themes are extracted from the theme analysis:

1. Situational immersion enhances learning transfer: "the simulated scene of quarreling with classmates is too real. I can calm down and use the learning method when I encounter similar situations after practice." (P07)
2. Personalize and establish emotional connection: "My 'Guardian' (ISFJ type) is like a patient friend who won't judge me and makes me willing to open up."(p15)
3. Safety environment promotion skill practice: "practice in front of AI without losing face. You can start again even if you say wrong, and you will gradually dare to use it in reality."(P22).

5 Discussion

This study for the first time verified the effectiveness of generative AI-driven MBTI virtual character situational counseling in improving adolescents' psychological resilience. The results show that, compared with the standardized group curriculum, the innovative model has a significant and lasting advantage effect on the total score of resilience, key dimensions (strength, optimism) and situational coping behavior ($d > 0.85$), which confirms its potential as a new paradigm for adolescent resilience intervention.

5.1 Core Mechanism Analysis

Generative AI enables deep contextualization: LLM's dynamic generation ability overcomes the rigid limitations of the traditional scripted system, can create a stress situation that highly fits the individual's life experience, and provides "human-like" flexible feedback [7, 18]. This high fidelity simulation environment provides a low-risk practice space for resilience skills (such as cognitive reconstruction and emotion regulation), and significantly promotes the internalization and transfer of skills (as shown in the results of SCBA).

MBTI personality matching optimizes personalized experience: research has confirmed that personality matching is the key moderator of intervention effect. When the virtual character's communication style (such as E/I expression) and problem solving strategy (such as T/F preference) fit the user's MBTI type, it can significantly enhance the sense of trust, participation motivation and cognitive acceptance [9, 16]. This explains why highly matched users report a deeper sense of connection and skill acquisition.

Exploration space for safety and inclusiveness: qualitative feedback emphasizes that the non-critical environment provided by AI is an important premise for teenagers (especially those with high sensitivity or social anxiety) to dare to expose their vulnerability and practice new behaviors [19], which is consistent with the theoretical expectation of digital intervention as a "psychological safe harbor".

5.2 Innovation Contribution and Practical Significance

Theoretical level: the MBTI personality framework system is integrated into AI psychological intervention for the first time, which deepens the understanding of "how technology adapts to individual differences", and provides an empirical basis for personalized AI design.

Technical level: an operable "MBTI generative AI" fusion architecture (Fig. 1) has been developed to provide a reproducible template for subsequent research.

Practical significance:

Provide a scalable, low-cost, high interest Toughness Training Program for schools and communities to make up for the shortage of professional manpower.

Inspire mental health practitioners to use AI tools to expand service forms and realize the intervention mode of "human-computer cooperation".

Provide examples for the specific implementation of AI ethical standards in the field of adolescent mental health (such as safety protocol and informed consent design) [22].

5.3 Limitations and Future Directions

Sample representativeness: participants come from schools in the same region, so it is necessary to expand sample diversity and cultural background.

Long term effect: the follow-up period is only 3 months, which needs to be extended to more than 1 year to investigate the continuity of the effect.

MBTI alternative model: the effect of other personality models (such as big five personality) in AI counseling can be compared in the future.

Technology Optimization: explore multimodal interaction (such as emotional Computing) and objectively evaluate the effect in combination with EEG/physiological indicators.

Ethical deepening: strengthen the protection of data privacy, and study the cultivation of teenagers' critical understanding of AI generated content.

6 Conclusion

This study confirmed that generative AI assisted MBTI virtual role situational counseling is an efficient and popular new method to improve adolescents' psychological resilience. Its core advantages over traditional consulting and existing digital tools are:

Immersive dynamic training ground: use generative AI to dynamically create situational simulation, provide a highly realistic immersive environment (core advantage: situational immersion), and let teenagers practice coping skills in a safe virtual space.

Deep personalized matching: through MBTI personality matching, accurate personalized role interaction (core advantage: personalized emotional connection) is achieved, which significantly enhances the pertinence and user stickiness of counseling.

Efficient and lasting results:

It can significantly improve the level of psychological resilience in the short term (cd-risc-10 total score, strength dimension, optimism dimension).

The effect remained stable after three months.

It can significantly improve adolescents' strategic effectiveness, emotional stability and self-efficacy in actual situational coping behavior (SCBA).

To solve the shortage of resources and personalization: it directly responds to the key problems faced by adolescent psychological services, such as the shortage of resources and the limited personalization of existing digital tools, and provides a feasible and efficient supplementary scheme.

Summarize key comparative advantages:

Superior to traditional counseling: it overcomes the problems of limited resources and teenagers' possible resistance to "face-to-face" counseling; Provide safe and non critical exploration space and unlimited situational practice opportunities.

Superior to existing digital tools: beyond standardized static content or simple chat robots, it provides dynamic personalized interaction based on MBTI and AI generated immersive situations, so as to achieve deeper skill training and emotional connection.

Core attraction (experience feedback): a high sense of situational immersion, personalized emotional connection, and safe exploration space are the key to teenagers' active participation and skill transfer, which is precisely what traditional methods and general digital tools can not provide at the same time.

Value: this method provides empirical support for the innovation of digital mental health intervention, and points out the effective path to solve the existing service dilemma.

References

1. Arnett, J.J.: Adolescence and Emerging Adulthood: a Cultural Approach. Pearson Education, Upper Saddle River, NJ (2014)
2. Rutter, M.: Resilience in the face of adversity: protective factors and resistance to psychiatric disorder. Br. J. Psychiatry. **147**(6), 598–611 (1985)
3. Masten, A.S.: Ordinary magic: resilience processes in development. Am. Psychol. **56**(3), 227–238 (2001)
4. Brunwasser, S.M., Gillham, J.E., Kim, E.S.: A meta-analytic review of the Penn resiliency programs effect on depressive symptoms. J. Consult. Clin. Psychol. **77**(6), 1042–1054 (2009)
5. Dray, J., Bowman, J., Campbell, S., et al.: Systematic review of school-based interventions to prevent depression or anxiety in children and adolescents. Aust. N. Z. J. Psychiatry. **51**(6), 570–607 (2017)
6. Luxton, D.D., McCann, R.A., Bush, N.E., et al.: mHealth for mental health: integrating smartphone technology in behavioral healthcare. Prof. Psychol. Res. Pract. **42**(6), 505–512 (2011)
7. Fitzpatrick, K.K., Darcy, A., Vierhile, M.: Delivering cognitive behavior therapy to young adults with symptoms of depression and anxiety using a fully automated conversational agent (Woebot): a randomized controlled trial. JMIR Ment. Health. **4**(2), e19 (2017)
8. Myers, I.B., McCaulley, M.H., Quenk, N.L., et al.: MBTI Manual: a Guide to the Development and use of the Myers-Briggs Type Indicator, 3rd edn. Consulting Psychologists Press, Palo Alto, CA (1998)
9. Martin, L.L., Ward, W.C., Wesolowski, K.L.: Predicting counselor trainee performance using personality type and academic variables. J. Couns. Dev. **76**(1), 28–33 (1998)
10. Thompson, B., Borrello, G.M.: The effects of client-counselor personality similarity on counseling process and outcome. J. Couns. Psychol. **33**(1), 10–15 (1986)
11. Chen, Z., Tullett, N.M., Inzlicht, M.: Personality-matched feedback reduces defensiveness and increases self-control. Soc. Psychol. Personal. Sci. **8**(8), 899–907 (2017)

12. Luthar, S.S., Cicchetti, D., Becker, B.: The construct of resilience: a critical evaluation and guidelines for future work. Child Dev. **71**(3), 543–562 (2000)
13. Connor, K.M., Davidson, J.R.T.: Development of a new resilience scale: the Connor-Davidson resilience scale (CD-RISC). Depress. Anxiety. **18**(2), 76–82 (2003)
14. Neff, K.D.: Self-compassion: an alternative conceptualization of a healthy attitude toward oneself. Self Identity. **2**(2), 85–101 (2003)
15. Gillham, J.E., Reivich, K.J., Freres, D.R., et al.: School-based prevention of depression and anxiety symptoms in early adolescence: a pilot of a parent intervention component. Sch. Psychol. Q. **22**(4), 472–492 (2007)
16. Leung, S., Cheung, F., Lee, C.: The role of personality similarity in the formation of working alliance in counseling. J. Soc. Clin. Psychol. **24**(6), 805–827 (2005)
17. Miner, A.S., Shah, M., Tian, L., et al.: Toward automated, scalable, and person-centered mental health care: development and preferences study of a text-based conversational agent for depression and anxiety. JMIR Ment. Health. **7**(1), e15010 (2020)
18. Rizzo, A., Shilling, R., Forbell, E., et al.: A virtual reality exposure therapy application for Iraq war military personnel with post traumatic stress disorder: from training to toy to treatment tool. Presence Teleop. Virt. **18**(6), 467–491 (2009)
19. Palmius, N., Tsanas, A., Ramdas, S., et al.: Digital phenotyping in bipolar disorder: combining smartphone-based mood self-reports with rest/activity data from wearable sensors. IEEE J. Biomed. Health Inform. **22**(5), 1405–1414 (2018)
20. Kuhn, E., Greene, C.J., Hoffman, J.E., et al.: Effectiveness of computerized cognitive behavioral therapy for anxious and depressed veterans: a randomized controlled trial. JAMA Netw. Open. **4**(1), e2035338 (2021)
21. Hirsh, J.B., Peterson, J.B.: Assessing the structure of the Myers-Briggs type indicator using exploratory structural equation modeling. Educ. Psychol. Meas. **69**(5), 861–872 (2009)
22. Torous, J., Myrick, K.J., Rauseo-Ricupero, N., et al.: Digital mental health and COVID-19: using technology today to accelerate the curve on access and quality tomorrow. JMIR Ment. Health. **7**(3), e18848 (2020)
23. Campbell-Sills, L., Stein, M.B.: Psychometric analysis and refinement of the Connor-Davidson resilience scale (CD-RISC): validation of a 10-item measure of resilience. J. Trauma. Stress. **20**(6), 1019–1028 (2007)
24. Braun, V., Clarke, V.: Using thematic analysis in psychology. Qual. Res. Psychol. **3**(2), 77–101 (2006)

AI-Generated Text Detection with Multi-source Domain Adaptation

Senao Li, Wan Zhang, Zhi Zheng, and Jing Zhang(✉)

School of Cyber Science and Engineering, Southeast University, Nanjing, China
{220235438,zhangwan,zhengzhi,jingz}@seu.edu.cn

Abstract. The proliferation of AI-generated text, driven by advanced large language models (LLMs), poses significant risks such as misinformation, while challenging detectors due to their reliance on labeled data and poor generalization to new models. This paper proposes a multi-source unsupervised domain adaptation framework that integrates contrastive learning and maximum mean discrepancy minimization. By leveraging multiple labeled source LLMs, the proposed method learns transferable features for detecting text from unseen target models. Experiments on a news benchmark show that the proposed method consistently outperforms single-source adaptation baselines and zero-shot methods, confirming its robustness in label-scarce real-world scenarios.

Keywords: AI-generated text · domain adaptation · contrastive learning

1 Introduction

The proliferation of AI-generated text, driven by modern LLMs, presents both opportunities and challenges for information ecosystems. While models such as GPT exhibit remarkable fluency and coherence [1], building on foundational advances in language understanding [2], their misuse can lead to significant risks, including the creation of convincing false news [3]. The credibility of critical domains like journalism is especially vulnerable, making the development of reliable detection mechanisms an urgent technical challenge.

Existing detection methods, though successful on specific datasets, face severe limitations in practical open-world scenarios. A key issue is domain discrepancy or domain shift: different LLMs produce texts with distinct linguistic and statistical characteristics due to differences in architecture, training data, and objectives. Supervised detectors relying on labeled data from a single or limited set of source LLMs often suffer from sharp performance drops when encountering text from new, unseen models–a problem known as domain distribution mismatch. Furthermore, constructing large, high-quality labeled datasets is costly and time-consuming, lagging behind the rapid iteration of LLMs, leaving abundant unlabeled text data underutilized.

L. Huang and D. Greenhalgh (Eds.): ICMLC 2026, LNNS 2061, pp. 948–962, 2027.
https://doi.org/10.1007/978-3-032-31236-5_64

To address generalization issues, domain adaptation has emerged as a promising direction for cross-domain detection. It aims to transfer knowledge from a labeled source domain (known LLMs) to an unlabeled target domain (new LLMs) by aligning feature distributions. However, most current approaches are limited to single-source adaptation, which struggles to capture the heterogeneity across LLMs and is highly sensitive to the source choice when the distribution gap is large. Meanwhile, contrastive learning has proven effective in learning discriminative representations by contrasting semantically similar (positive) and dissimilar (negative) samples [4], offering a pathway to extract transferable features less tied to model-specific surface patterns.

To tackle these challenges, this paper focuses on AI-generated news text detection, framing different LLMs as distinct domains. We propose a multi-source fusion contrastive domain adaptation framework that leverages labeled data from multiple source LLMs and unlabeled news text from the target domain. The framework consists of three core components: (1) a multi-source fusion strategy to learn common and complementary features across LLMs; (2) a dual-path optimization mechanism integrating contrastive learning for feature discriminability and maximum mean discrepancy minimization for domain alignment; and (3) an unsupervised inference capability to detect texts from unseen LLMs using learned representations. Experiments show that our framework achieves superior cross-domain generalization, outperforming single-source adaptation models and mainstream unsupervised detectors. It effectively mitigates the generalization limitations of single-source methods and offers a practical solution to the labeling bottleneck for new LLMs.

The main contributions of the paper are as follows:

- We formulate AI-generated news detection as a multi-source unsupervised domain adaptation task, enabling joint use of labeled multi-source data and unlabeled target text.
- We propose a novel framework integrating multi-source fusion, contrastive learning, and MMD-based alignment to learn robust, domain-invariant features.
- Through comprehensive experiments, we demonstrate consistent and significant improvements over relevant baselines, validating the effectiveness and robustness of our approach.

2 Related Work

The rise of AI-generated text has spurred the development of detection methods, which can be broadly classified by their information requirements: (1) supervised classifiers trained on labeled datasets; (2) zero-shot or statistical methods that rely on intrinsic textual properties without model training; and (3) proactive watermarking, which embeds detectable signals during text generation [5]. A major challenge shared by post-hoc detectors (categories 1 and 2) is generalization failure–significant performance degradation when encountering text from an unseen LLM.

Supervised classifiers typically fine-tune pre-trained LMs such as BERT or RoBERTa on mixed human/AI corpora [6]. While accurate in-distribution, they often overfit to superficial artifacts of the source LLM (e.g., GPT-2's token frequencies), failing to learn generalizable features of AI text and performing poorly on new models [5]. Recent studies further reveal that such detectors are vulnerable to simple paraphrasing attacks, as they tend to rely on brittle surface-level patterns rather than robust semantic features [7].

Zero-shot and statistical methods avoid labeled data by analyzing textual statistics. Early zero-shot approaches primarily relied on statistical features of text, such as the token probability- and rank-based detection framework GLTR [8]. Subsequent work, including DetectGPT [9]—which leverages probability curvature after perturbation—and its efficient variant FAST-DetectGPT [10], advanced this line. Other methods like RADAR employ reasoning-based adversarial detection [11]. However, many require white-box access or a strong proxy model, limiting robustness against black-box LLMs.

Watermarking embeds statistical signals during generation for later verification [12,13], but requires provider cooperation and is unsuitable for detecting text from uncooperative or legacy sources.

Domain adaptation (DA) addresses generalization by treating each LLM as a domain and adapting from labeled source(s) to an unlabeled target LLM. ConDA [14] combined contrastive learning with adversarial training for single-source DA. Yet, its success hinges on source-target similarity, and performance degrades if the gap is large. Our work extends DA to multiple sources, arguing that diverse source domains provide richer, more transferable features for robust cross-domain detection.

3 The Proposed Method

3.1 Problem Formulation

To address the scarcity of data from novel generators, we formulate the task of AI-generated text detection as a domain adaptation problem, where each domain corresponds to a distinct text generator. Specifically, generators with labeled data are regarded as source domains, while those without labeled data are defined as target domains. A source domain is denoted as $Src = \{(x_i^{Src}, y_i^{Src})\}_{i=1}^{N^{Src}}$, and a target domain is denoted as $Tar = \{(x_i^{Tar})\}_{i=1}^{N^{Tar}}$. Herein, the label $y_i^{Src} \in \{0, 1\}$ indicates whether the text is human-written (0) or AI-generated (1). All domains share an identical label space, where N^{Src} represents the number of samples in the source domain, and N^{Tar} denotes the number of samples in the target domain.

3.2 Model

To tackle the aforementioned problem, we construct a neural network that integrates contrastive learning, transfer learning, Siamese networks, and the RoBERTa model. For the detector component, we adopt the pre-trained

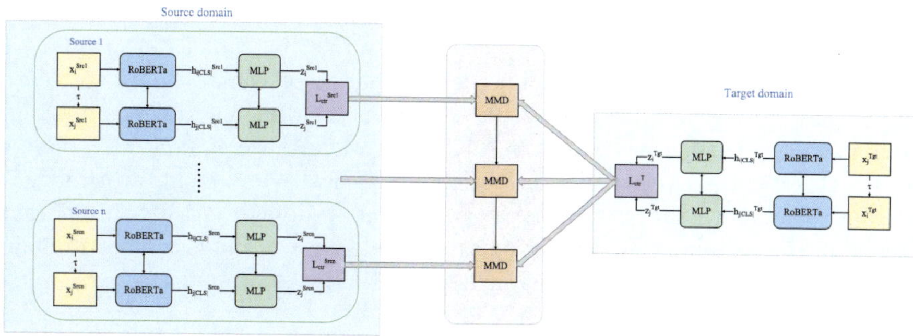

Fig. 1. Our model framework. The RoBERTa encoder adopted in this framework is the roberta-large variant. Weights of the RoBERTa encoder and the MLP projection layer are shared across all network branches to ensure consistent feature extraction and projection throughout the training process.

RoBERTa model from the Hugging Face platform, which offers two variants: roberta-base and roberta-large. In this work, we select roberta-large and append a classification head on top of it to implement text discrimination. The input of the framework consists of $n + 1$ pieces of text, where n denotes the number of selected source domains. Specifically, we sequentially select one sample x_i^{Src} from each of the n source domains, followed by one sample x_i^{Tgt} from the target domain. This process is equivalent to pairing one source domain sample with one target domain sample per iteration, and repeating this pairing for n rounds corresponding to the n source domains. Through training on each source domain, the model is enabled to capture domain-specific features of individual source domains. We then apply a text transformation operation τ to these texts, yielding the transformed (also referred to as perturbed in this paper) samples x_j^{Src} and x_j^{Tgt}. To simultaneously feed the original and transformed texts into the model and achieve effective contrastive learning, we employ a Siamese network architecture, where the weights of the RoBERTa model are shared across the two network branches. For the source domain text and target domain text input in each iteration, we extract the hidden state representations corresponding to the CLS token, denoted as $h_{i[CLS]}^{Src}$ and $h_{j[CLS]}^{Src}$, respectively. We feed these embedding vectors into a projection head, which is composed of a multi-layer perceptron (MLP) with a single hidden layer, and compute the contrastive loss in the low-dimensional projection space. This MLP can be represented as a function $g(\cdot) : \mathbb{R}^{d_h} \mapsto \mathbb{R}^{d_p}$, where d_h is the dimension of the hidden layer embeddings (set to 1024 for the roberta-large model). We set the projection dimension d_p to 300. In addition, to reduce the distribution discrepancy between the source and target domains, we introduce the Maximum Mean Discrepancy loss. We also compute the binary cross-entropy (BCE) loss between the original and transformed texts to enhance the model's discriminative capability. The detailed design of the loss function for the training process will be elaborated in the subsequent section.

3.3 Training

In the training process, to fully exploit the information of available data and learn robust and highly generalized text representations, we design a multi-component training objective that incorporates Maximum Mean Discrepancy loss, source domain classification loss, and contrastive loss. These components work synergistically to achieve the optimization goal of cross-domain AI-generated text detection. During the training process, Src refers to a single source domain among the n source domains, and Tgt denotes the target domain.

Maximum Mean Discrepancy Since the source and target domain texts are derived from different generators (i.e., different data distributions), there exists an inherent domain discrepancy. This discrepancy can hinder the direct transfer of features learned by the model on the source domain to the target domain. To reduce the distribution gap between the source and target domains, we first introduce Maximum Mean Discrepancy as the domain adaptation loss. MMD is an effective metric for measuring distribution distances, whose core idea is to map samples from two distributions into a Reproducing Kernel Hilbert Space (RKHS), and then calculate the distance between the mean values of the mapped samples to quantify the discrepancy between the original two distributions. Formally, let the source domain sample set be $Src = \{x_1^S, x_2^{Src}, \ldots, x_{N^{Src}}^{Src}\}$ and the target domain sample set be $Tgt = \{y_1^{Tgt}, y_2^{Tgt}, \ldots, y_{N^{Tgt}}^{Tgt}\}$, where N^{Src} and N^{Tgt} denote the number of samples in the source and target domains, respectively. The two sample sets are drawn from distributions $\mathcal{S}rc$ and $\mathcal{T}gt$, respectively. The MMD distance between distributions $\mathcal{S}rc$ and $\mathcal{T}gt$ is calculated as follows:

$$\mathrm{MMD}(\mathcal{S}rc, \mathcal{T}gt) = \left\| \frac{1}{N^{Src}} \sum_{i=1}^{N^{Src}} \phi\left(z_i^{Src}\right) - \frac{1}{N^{Tgt}} \sum_{i=1}^{N^{Tgt}} \phi\left(z_i^{Tgt}\right) \right\|_{\mathcal{RKHS}} \tag{1}$$

where $\phi : \mathcal{S}rc \mapsto \mathcal{RKHS}$ denotes the mapping function that projects samples into the Reproducing Kernel Hilbert Space $\mathcal{RKHS}$. z_i^{Src} and z_i^{Tgt} represent the low-dimensional embedding representations of source and target domain texts processed by the MLP projection layer, respectively.

Source Domain Classification Loss The core value of source domain data lies in its accompanying labels indicating human-written or AI-generated text. Therefore, we introduce the Binary Cross-Entropy (CE) Loss to optimize the classification performance of both the original source domain texts and their perturbed counterparts.

For the original source domain texts, the cross-entropy loss is calculated as follows:

$$\mathcal{L}_{CE}^{Src} = -\frac{1}{b}\sum_{i=1}^{b}\left[y_i \log p\left(y_i \mid h_{i[\mathrm{CLS}]}^{Src}\right) + (1-y_i)\log\left(1 - p\left(y_i \mid h_{i[\mathrm{CLS}]}^{Src}\right)\right)\right] \quad (2)$$

$$\mathcal{L}_{CE}^{Src'} = -\frac{1}{b}\sum_{i=1}^{b}\left[y_i \log p\left(y_i \mid h_{i[\mathrm{CLS}]}^{Src'}\right) + (1-y_i)\log\left(1 - p\left(y_i \mid h_{i[\mathrm{CLS}]}^{Src'}\right)\right)\right] \quad (3)$$

where $\mathcal{L}_{CE}^{Src}$ specifically denotes the classification loss for the original source domain texts; b represents the batch size during training; y_i is the ground-truth label of the i-th sample (taking a value of 0 or 1, corresponding to human-written and AI-generated texts, respectively); $h_{i[\mathrm{CLS}]}^{Src}$ is the hidden state representation of the [CLS] token extracted by the RoBERTa model from the original source domain text; and $p(y_i \mid h_{i[\mathrm{CLS}]}^{Src})$ is the probability predicted by the model that the sample belongs to label y_i.

In parallel, we compute the corresponding cross-entropy loss $\mathcal{L}_{CE}^{Src'}$ for the perturbed source domain texts. The formula of $\mathcal{L}_{CE}^{Src'}$ is identical to that of the loss for original texts and is thus omitted here to avoid redundancy.

This training approach offers dual advantages: on the one hand, it significantly enhances the robustness of the classifier against surface-level text variations, making it less susceptible to simple text paraphrasing; on the other hand, it improves the generalization ability of the detector, which is crucial for addressing text discrepancies generated by different models and achieving cross-domain detection.

Contrastive Loss To enable the model to learn more discriminative deep text representations instead of merely relying on surface-level features for classification, we introduce contrastive loss for both source and target domain texts (as illustrated in Fig. 1). The contrastive loss function adopted in this study is adjusted based on the classic design proposed by Chen et al. [4]. The core difference lies in that: instead of calculating the loss between two different transformed views of the same text, we use the original text as the "anchor" and the perturbed text as its positive sample to construct loss computation pairs.

In this study, the text perturbation is specifically implemented via synonym substitution, and detailed technical details can be found in the appendix. The contrastive loss for source domain texts is calculated as follows:

$$\mathcal{L}_{ctr}^{Src} = -\sum_{(i,j)\in b} \log \frac{\exp\left(\mathrm{sim}\left(z_i^{Src}, z_j^{Src}\right)/t\right)}{\sum_{k=1}^{2[b]} \mathbb{1}_{[k\neq i]} \exp\left(\mathrm{sim}\left(z_i^{Src}, z_k^{Src}\right)/t\right)} \quad (4)$$

where z_i^S denotes the embedding representation of the original (anchor) source domain text after passing through the MLP projection layer, and z_j^S is the projection layer embedding of the corresponding perturbed source domain text;

t is the temperature parameter, which is used to adjust the sensitivity of the loss function to sample similarity; b represents the current mini-batch; $\text{sim}(\cdot, \cdot)$ denotes the similarity measurement method, and cosine similarity is uniformly adopted in this study; $\mathbb{1}_{[k \neq i]}$ is an indicator function, which takes a value of 1 when $k \neq i$ and 0 otherwise.

Consistent with the method of Chen et al. [4], we do not adopt explicit sampling or negative sample mining. Instead, we directly regard the remaining $2(|b| - 1)$ samples in the current mini-batch b (excluding the anchor sample and its corresponding positive sample) as negative samples. This approach not only simplifies the training process but also fully utilizes the sample information within the batch to construct abundant positive and negative sample pairs.

For target domain texts, the calculation formula of the contrastive loss $\mathcal{L}_{ctr}^{T}$ is exactly the same as that of the source domain contrastive loss, except that the source domain identifier in the variables needs to be replaced with the target domain identifier, which is omitted here for brevity.

The core training objective of the contrastive loss is: in the low-dimensional projection space, narrow the distance between the anchor text and the corresponding perturbed text (positive sample pairs), while maximizing the distance between the anchor text and other negative samples, thereby enabling the model to learn discriminative features that can distinguish texts of different categories.

By integrating the three aforementioned loss components, we formulate the final training objective of the model. Specifically, we combine the classification loss, contrastive loss, and MMD loss, and adjust the weights of each component via hyperparameters to achieve multi-objective collaborative optimization. This design facilitates domain adaptation between the source and target domains, and enables the model to perform text detection tasks on the target domain. The mathematical expression of the final training objective is given as follows:

$$\mathcal{L}_{total} = \lambda_1 \, \text{MMD}(\mathcal{S}rc, \mathcal{T}gt) + \lambda_2 \left(\mathcal{L}_{CE}^{Src} + \mathcal{L}_{CE}^{Src'} \right) + \lambda_3 \left(\mathcal{L}_{ctr}^{Src} + \mathcal{L}_{ctr}^{Tgt} \right) \tag{5}$$

where λ_1 is used to modulate the importance of the MMD loss, λ_2 determines the contribution of the source domain classification loss, and λ_3 controls the weight of the contrastive loss. All three are hyperparameters that need to be preset prior to training.

4 Experiments

4.1 Baseline Models

To evaluate our framework under the unsupervised domain adaptation setting, we compare it with four representative baseline models: the single-source domain adaptation method ConDA(Contrastive Domain Adaptation), two unsupervised detectors (GLTR and DetectGPT), and the specialized supervised OpenAI-GPT2 Detector. The detailed specifications of these models are as follows:

The ConDA Framework ConDA is designed for unsupervised cross-domain AI-generated text detection. Its core integrates unsupervised domain adaptation and self-supervised contrastive learning to realize knowledge transfer from a single source domain to a single target domain, addressing label scarcity and cross-generator generalization challenges. Its basic architecture is similar to our proposed model, with two key differences: it adopts roberta-base as the pre-trained model and learns from one source domain at a time for single-source-target domain adaptation.

Unsupervised Baseline Models Unsupervised baseline models can complete detection without labeled data, which aligns with our study's core scenario (unlabeled target domains). Two classic unsupervised models are selected:

GLTR [8]: A representative early open-source unsupervised detector that identifies AI-generated text by measuring lexical probability distribution discrepancies between human and machine-written texts. It computes four statistical metrics via a proxy language model—log probability ($\log p(x)$), average token rank, log token rank, and prediction entropy—and fuses results for detection. It features simple principles, high computational efficiency, and no complex training for deployment.

DetectGPT [9]: A state-of-the-art unsupervised method that builds on lexical probability-based detection and adds a text perturbation contrast mechanism. It uses T5-base to generate semantically similar perturbed texts, then compares token-wise log probability differences between original and perturbed texts. The core logic is that AI-generated texts (in the proxy model's probability peak region) show a larger probability drop after perturbation than human-written texts, avoiding the limitations of single-probability metrics and excelling at detecting high-quality AI-generated text.

Specialized Baseline Model In addition to unsupervised baselines, this study includes the off-the-shelf **OpenAI-GPT2 Detector** as a specialized comparative model to verify our model's competitiveness in specific generator detection. Based on the RoBERTa pre-trained architecture, it is fine-tuned for detecting GPT-2 series-generated texts. Its training dataset is target-oriented: Human-written texts: Collected from the WebText test set [1], including 250,000 high-quality documents with diverse topics/styles to ensure data naturalness and diversity. AI-generated texts: Two subsets–250,000 texts generated by GPT-2 (temperature = 1, no truncation) and 250,000 texts via GPT-2 (top-k = 40 truncation)—covering output characteristics under different parameters/strategies.

4.2 Datasets and Domain Setup

We conduct experiments on the TuringBench dataset [15], a large-scale public benchmark designed for evaluating AI-generated text detection. It comprises

English news articles, with a balanced collection of human-written texts and counterparts generated by 19 different large language models (LLMs), covering a wide range of architectures and scales.

Selected LLMs and Domain Definition To construct a challenging and diverse multi-source domain adaptation testbed, we select six representative LLMs from TuringBench that exhibit significant heterogeneity in model architecture, scale, and training objective. We treat each LLM as a distinct *domain*. The selected LLMs, along with their key characteristics and sample counts, are summarized in Table 1.

Table 1. Description of the six LLM domains selected from the TuringBench dataset.

Abbr.	Model name	Size	$N_{\text{human}}/N_{\text{AI}}$	Key characteristics
C	CTRL	1.6B	8854/8121	Controllable generation via control codes
G2	GPT-2 XL	1.5B	8854/8309	General-purpose autoregressive LM
X	XLM	0.6B	8854/8852	Cross-lingual pre-training
GM	GROVER-Mega	1.5B	8854/8164	Specialized for news generation
G3	GPT-3	175B	8854/8164	Largest scale, strong few-shot learning
FW	FAIR_WMT19	0.7B	8854/8164	Translation model (encoder-decoder)

N_{human} and N_{AI} denote the number of human-written and AI-generated news samples, respectively

The diversity among these domains–spanning autoregressive (G2, G3, GM), controlled (C), cross-lingual (X), and translation-based (FW) architectures with parameters ranging from millions to hundreds of billions–creates a realistic and challenging scenario for evaluating cross-domain generalization.

Data Splits and Adaptation Tasks For each domain (LLM), we split its associated data (both human and AI samples) into training (70%), validation (10%), and test (20%) sets, following a 7:1:2 ratio. To evaluate our multi-source domain adaptation framework, we design experiments where one or two domains are held out as the unlabeled target domain(s) for testing, while the remaining domains serve as labeled source domains for training. Specifically, in a n-source adaptation setting, we use the training splits of n domains as labeled source data, and the test split of the held-out target domain(s) for evaluation. The validation splits of the source domains are used for model selection and hyperparameter tuning. Crucially, the target domain's test data is never accessed during training, ensuring a fair evaluation of generalization. All results reported are performance on the target domain's held-out test set.

5 Results and Analysis

We evaluate the proposed framework through two main experimental analyses: (a) comparison with state-of-the-art and relevant baseline models, and (b) ablation studies to ascertain the contribution of each key component. For clarity and conciseness in the result tables, we employ the domain abbreviations defined in Sect. 4.2 (C, G2, X, GM, G3, FW).

5.1 Performance Comparison with Baseline Models

We first assess the core advantage of leveraging multiple source domains by comparing our dual-source adaptation framework against single-source adaptation. The accuracy and AUC results for all source-target pairs are detailed in Tables 2 and 3, respectively. Following this, we compare our best-performing dual-source configurations against a wider range of baseline detectors, including the domain adaptation method ConDA, zero-shot detectors (GLTR, DetectGPT), and a specialized supervised model (OpenAI-GPT2 Detector), as summarized in Table 4.

The comparative results between single-source and dual-source adaptation are presented in Table 2 (accuracy) and Table 3 (AUC). The upper sections of both tables detail the performance of each individual source domain when adapted to each target domain.

A primary observation is the high variance in single-source performance. For example, while the GROVER-Mega (GM) source achieves near-perfect accuracy (0.99) on the CTRL (C) target, the GPT-3 (G3) source yields only 0.57 accuracy on the XLM (X) target. This instability underscores the risk of relying on a single source, as performance is highly contingent on the specific source-target pair.

The lower sections of the tables present the dual-source adaptation results. To evaluate the overall impact of multi-source fusion, we compute the average performance across all valid target domains for both paradigms. The mean accuracy increases from 0.851 for the best single-source model to 0.867 for the best dual-source model. More importantly, the standard deviation of accuracy decreases from 0.128 (single-source) to 0.118 (dual-source), indicating enhanced robustness and more consistent performance across diverse targets.

The dual-source strategy proves particularly beneficial for challenging targets. For the GPT-3 (G3) target, the best single-source accuracy is 0.89 (C→G3), while the dual-source combination X&GM achieves 0.93. This improvement suggests that complementary features from a cross-lingual model (X) and a news-specialized model (GM) can better capture the complex patterns of the largest LLM.

However, multi-source fusion does not guarantee improvement and can lead to negative transfer. A notable case is the C&G3 combination applied to the G2 target, where accuracy drops to 0.60, underperforming both constituent single-source models (C→G2: 0.84; G3→G2: 0.69). This indicates that the success of fusion depends on the compatibility of the selected source domains with each other and with the target–an important consideration for practical application.

Table 2. Accuracy comparison with single-source domain adaptation.

Source	C	G2	X	GM	G3	FW
C	–	0.87	0.94	0.97	0.89	0.92
G2	0.84	–	0.81	0.95	0.82	0.96
X	0.97	0.97	–	0.95	0.83	0.90
GM	0.99	0.98	0.98	–	0.76	0.88
G3	0.74	0.69	0.57	0.59	–	0.73
FW	0.99	0.97	0.98	0.97	0.84	–
Best	0.99	0.97	0.98	0.97	0.89	0.96
C&G2	–	–	0.95	0.95	0.60	0.97
C&X	–	0.93	–	0.97	0.70	0.95
C&GM	–	0.95	0.98	–	0.73	0.93
C&G3	–	0.59	0.85	0.76	–	0.79
C&FW	–	0.96	0.99	0.97	0.86	–
G2&X	0.84	–	–	0.95	0.60	0.96
G2&GM	0.89	–	0.91	–	0.74	0.97
G2&G3	0.60	–	0.81	0.73	–	0.71
G2&FW	0.92	–	0.99	0.96	0.81	–
X&GM	0.99	0.97	–	–	0.93	0.81
X&G3	0.63	0.88	–	0.81	–	0.73
X&FW	0.99	0.99	–	0.95	0.88	–
GM&G3	0.95	0.60	0.69	–	–	0.72
GM&FW	0.97	0.98	0.99	–	0.83	–
G3&FW	0.78	0.80	0.73	0.70	–	–
Best	0.99	0.99	0.99	0.97	0.93	0.97

The symbol "–" indicates scenarios where the target domain is included in the source domain (i.e., non-cross-domain settings)

To compare our approach with established detection paradigms under different supervision assumptions, we evaluate it against three representative baseline types: unsupervised statistical methods (GLTR), a zero-shot curvature-based method (DetectGPT, abbreviated as DG in Table 4), and a specialized supervised detector (OpenAI-GPT2, abbreviated as OG2). Results are summarized in Table 4, where the Avg. column for our method reports the mean AUC across all dual-source combinations for a given target, and Max. reports the AUC of the single best combination.

The performance gap between supervised and unsupervised/zero-shot approaches is evident. Our framework substantially outperforms GLTR and DetectGPT across all targets. For example, on the G2 target, the best-

Table 3. AUC comparison with single-source domain adaptation.

Source	C	G2	X	GM	G3	FW
C	–	0.946	0.985	0.988	0.938	0.894
G2	0.998	–	0.988	0.984	0.962	0.966
X	0.991	0.991	–	0.98	0.981	0.766
GM	0.999	0.991	0.999	–	0.982	0.826
G3	0.783	0.607	0.683	0.819	–	0.607
FW	0.998	0.998	0.999	0.961	0.975	–
Best	0.999	0.998	0.999	0.988	0.982	0.966
C&G2	–	–	0.997	0.983	0.968	0.997
C&X	–	0.996	–	0.996	0.990	0.991
C&GM	–	0.998	0.999	–	0.992	0.976
C&G3	–	0.961	0.999	0.959	–	0.973
C&FW	–	1.000	1.000	0.996	0.988	–
G2&X	0.999	–	–	0.995	0.976	0.994
G2&GM	0.998	–	0.990	–	0.987	0.998
G2&G3	0.911	–	0.945	0.702	–	0.941
G2&FW	0.998	–	1.000	0.994	0.982	–
X&GM	1.000	0.997	–	–	0.989	0.945
X&G3	0.962	0.994	–	0.911	–	0.913
X&FW	1.000	1.000	–	0.989	0.987	–
GM&G3	0.994	0.984	0.935	–	–	0.779
GM&FW	0.997	1.000	1.000	–	0.986	–
G3&FW	0.975	0.990	0.929	0.985	–	–
Best	1.000	1.000	1.000	0.996	0.992	0.998

The symbol "–" indicates scenarios where the target domain is included in the source domain (i.e., non-cross-domain settings)

Table 4. AUC comparison with various baseline detectors.

Target	GLTR				DG	OG2	Our	
	log p(x)	Rank	LogRank	Entropy			Avg.	Max. (source)
C	0.948	0.846	**0.952**	0.376	0.789	0.362	0.9816	**1.000** (X&GM)
G2	0.486	0.509	0.481	0.628	0.526	0.477	0.9911	**1.000** (C&FW)
X	0.472	0.759	0.439	0.693	0.7325	0.870	0.9794	**1.000** (GM&FW)
GM	0.431	0.466	0.430	0.589	0.5385	0.656	0.9510	**0.996** (C&X)
G3	0.359	0.353	0.338	0.752	0.5455	0.727	0.9845	**0.992** (C&GM)
FW	0.555	0.615	0.543	0.652	0.5015	0.461	0.9507	**0.998** (G2&GM)

performing unsupervised metric (GLTR's LogRank) achieves an AUC of 0.481, whereas our method attains 0.9911 on average. This contrast highlights the advantage conferred by leveraging labeled source data, while also illustrating the inherent difficulty of the zero-shot detection scenario against diverse, advanced LLMs.

Compared to the specialized supervised baseline (OpenAI-GPT2 Detector), our approach demonstrates stronger cross-domain generalization. It achieves higher AUC on five of the six targets. The exception is the G2 target, which aligns with the detector's native training domain (GPT-2), where its specialized optimization yields superior in-domain performance. This comparison underscores a key trade-off: models deeply specialized for a single source domain may excel within that domain but generalize poorly elsewhere, whereas our multi-source framework is designed to learn more transferable features.

5.2 Ablation Study

To rigorously assess the contribution of individual loss components within our proposed framework, we conduct systematic ablation experiments by removing specific loss terms. We define three model variants:

- **Our-MMD**: Removes the Maximum Mean Discrepancy loss, thereby eliminating explicit domain distribution alignment.
- **Our-CEs**: Removes the binary cross-entropy losses on source domain data, which removes all supervision signals.
- **Our-Contrast**: Removes the contrastive loss on both source and target domains, disabling the learning of perturbation-invariant discriminative features.

Table 5 presents the AUC performance of these variants alongside the complete model. For each target domain, the reported value represents the average across all dual-source combinations that include that target, ensuring a comprehensive evaluation across diverse source configurations.

Table 5. Performance of our model variants (avg. across sources)

Model	Target (avg. across sources)						Avg.
	C	G2	X	GM	G3	FW	
Our-MMD	0.7937	0.8526	0.7566	0.6272	0.8065	0.6174	0.7424
Our - CEs	0.5301	0.4934	0.4963	0.4988	0.5211	0.4942	0.5057
Our-contrast	0.8989	0.9542	0.7895	0.8472	0.8540	0.6879	0.8386
Our (full)	**0.9816**	**0.9911**	**0.9794**	**0.9510**	**0.9845**	**0.9507**	**0.9731**

The ablation results demonstrate distinct functional contributions from each component:

Domain alignment mitigates distribution shift. Removing MMD (*Our-MMD*) reduces average AUC to 0.7424. The performance decline is particularly pronounced on targets GM and FW, where distribution gaps are likely larger, validating the role of explicit domain alignment.

Supervision is fundamental. The *Our-CEs* variant shows the most severe degradation, with average AUC dropping to 0.5057. This confirms that supervised signals are essential for establishing the basic human-vs-AI discrimination boundary.

Contrastive learning enhances feature robustness. The *Our-Contrast* variant achieves an average AUC of 0.8386. Its impact is most noticeable on challenging targets such as G3 and X, indicating that learning perturbation-invariant features is critical for generalization to sophisticated LLMs.

The complete model consistently outperforms all ablated variants across every target domain. This demonstrates the synergistic effect of integrating supervised learning, domain alignment, and contrastive feature learning. These results provide empirical validation for our architectural design and clarify the relative importance of each learning objective in the multi-source domain adaptation framework for AI-generated text detection.

6 Conclusion

In this paper, we proposed a multi-source fusion contrastive domain adaptation framework to address the generalization bottlenecks of existing AI-generated text detectors. By synergizing contrastive learning with MMD-based distribution alignment, the proposed method effectively aligns features across heterogeneous LLMs to extract domain-invariant representations. Extensive experiments demonstrate that our approach yields an average cross-domain accuracy improvement of 28.3% over single-source models and consistently surpasses state-of-the-art unsupervised baselines like DetectGPT.

Future work will focus on two main directions: (1) Domain Expansion: While this study focused on the news domain, we plan to extend the framework to other high-risk scenarios, such as academic writing and social media posts, to validate its broader adaptability. (2) Adversarial Robustness: We aim to investigate the model's resilience against more complex adversarial attacks (e.g., sophisticated paraphrasing or instruction-tuned obfuscation) to enhance security in dynamic real-world applications.

Acknowledgment. This work was Funded by Basic Research Program of Jiangsu (Grant No. BK20251747) and the Start-up Research Fund of Southeast University (Grant No. RF1028623059).

References

1. Radford, A., Jeffrey, W., Child, R., Luan, D., Amodei, D., Sutskever, I., et al.: Language models are unsupervised multitask learners. OpenAI Blog **1**(8), 9 (2019)
2. Devlin, J., Chang, M.-W., Lee, K., Toutanova, K.: Bert: pre-training of deep bidirectional transformers for language understanding. In: Proceedings of the 2019 Conference of the North American Chapter of the Association for Computational Linguistics: Human Language Technologies, vol. 1 (Long and Short Papers), pp. 4171–4186 (2019)
3. Zellers, R., Holtzman, A., Rashkin, H., Bisk, Y., Farhadi, A., Roesner, F., Choi, Y.: Defending against neural fake news. In: Advances in Neural Information Processing Systems, vol. 32 (2019)
4. Chen, T., Kornblith, S., Norouzi, M., Hinton, G.: A simple framework for contrastive learning of visual representations. In: International Conference on Machine Learning, pp. 1597–1607. PmLR (2020)
5. Yang, Z., Feng, Z., Huo, R., Lin, H., Zheng, H., Nie, R., Chen, H.: The imitation game revisited: a comprehensive survey on recent advances in AI-generated text detection. Expert Syst. Appl. **272**, 126694 (2025)
6. Wang, P., Li, L. Ren, K., Jiang, B., Zhang, D., Qiu, X.: SeqXGPT: sentence-level AI-generated text detection. arXiv preprint arXiv:2310.08903 (2023)
7. Krishna, K., Song, Y., Karpinska, M., Wieting, J., Iyyer, M.: Paraphrasing evades detectors of AI-generated text, but retrieval is an effective defense. Adv. Neural. Inf. Process. Syst. **36**, 27469–27500 (2023)
8. Gehrmann, S., Strobelt, H., Rush, A.M.: GLTR: statistical detection and visualization of generated text. arXiv preprint arXiv:1906.04043 (2019)
9. Mitchell, E., Lee, Y., Khazatsky, A., Manning, C.D., Finn, C.: DetectGPT: zero-shot machine-generated text detection using probability curvature. In: International Conference on Machine Learning, pp. 24950–24962. PMLR (2023)
10. Bao, G., Zhao, Y., Teng, Z., Yang, L., Zhang, Y.: Fast-DetectGPT: efficient zero-shot detection of machine-generated text via conditional probability curvature. arXiv preprint arXiv:2310.05130 (2023)
11. Xiaomeng, H., Chen, P.-Y., Ho, T.-Y.: Radar: robust AI-text detection via adversarial learning. Adv. Neural. Inf. Process. Syst. **36**, 15077–15095 (2023)
12. Kirchenbauer, J., Geiping, J., Wen, Y., Katz, J., Miers, I., Goldstein, T.: A watermark for large language models. In: International Conference on Machine Learning, pp. 17061–17084. PMLR (2023)
13. Jiang, Z., Guo, M., Hu, Y., Gong, N.Z.: Watermark-based attribution of AI-generated content. arXiv preprint arXiv:2404.04254 (2024)
14. Bhattacharjee, A., Kumarage, T., Moraffah, R., Liu, H., Conda: Contrastive domain adaptation for AI-generated text detection. arXiv preprint arXiv:2309.03992 (2023)
15. Uchendu, A., Ma, Z., Le, T., Zhang, R., Lee, D.: Turingbench: a benchmark environment for turing test in the age of neural text generation. arXiv preprint arXiv:2109.13296 (2021)

KPL: Keyframe–Prompt–LLM for Subject-Centric Spatiotemporal Video Captioning

Nanqiao Jiang[1], Qin Wei[1](✉), and Yanyu Zhu[2]

[1] School of Information Engineering, Wuhan University of Technology, Wuhan, China
qinwei@whut.edu.cn

[2] College of Communication Engineering, Jilin University, Changchun, China

Abstract. Video captioning maps subject-centric spatiotemporal changes across frames into precise and coherent natural language descriptions. Current methods mainly fall into two categories: encoder–decoder models and LLM-based pipelines. The former require large labeled corpora and often struggle with long-range factual consistency, leading to terse summaries; the latter, constrained by token compression and imperfect multimodal fusion, compromise completeness and yield unstable outputs. Therefore, to improve content completeness and temporal coherence, we propose a two-stage **keyframe–prompt–LLM (KPL)** paradigm for subject-centric video captioning. For keyframe extraction, an unsupervised fusion of frame differencing, optical flow, and temporal clustering highlights the subject's spatiotemporal state transitions. For prompt engineering, a subject-centric strategy guides a large language model to reason over inter-frame changes without fine-tuning. Quantitative and qualitative analyses on 50 publicly available videos demonstrate the effectiveness of the proposed two-stage paradigm. For keyframe extraction, KPL attains an average F1 of 0.802. For caption generation, the prompted setting yields a 17.58% relative gain over the baseline instruction, mitigating redundancy and semantic drift while improving temporal coherence.

Keywords: Video captioning · Keyframe extraction · Prompt engineering · Large Language Model · Spatiotemporal reasoning

1 Introduction

Video captioning produces natural-language summaries of a video [1, 2]. Beyond image captioning, it must capture temporal dynamics and evolving inter-subject relationships across frames [3, 4]. Since real-world footage involves complex events and scene shifts, captions must capture not only what appears but also when and how changes occur [5]. These demands make long-range temporal modeling, cross-frame consistency, and effective keyframe selection persistent challenges.

Contemporary video captioning spans two paradigms: traditional encoder–decoder pipelines and emerging LLM-based methods. Conventional video captioning has progressed from CNN–RNN encoder–decoder pipelines to attention−/Transformer-based

L. Huang and D. Greenhalgh (Eds.): ICMLC 2026, LNNS 2061, pp. 963–976, 2027.
https://doi.org/10.1007/978-3-032-31236-5_65

models that strengthen visual–text alignment and temporal reasoning [1–3]. Despite advances, limitations persist. First, uniform sampling often introduces redundancy, while sparse sampling may miss short-lived events or fine-grained transitions [4–7]. Second, end-to-end training typically requires large-scale paired datasets and substantial compute, which can encourage models to favor terse, summary-style descriptions [1, 2, 4]. Collectively, these factors hinder content completeness.

LLM-based captioning has opened two main routes: prompt-based generation (e.g., Video-ChatGPT) and multimodal joint modeling (e.g., Video-LLaMA). These systems leverage instruction following and stronger vision–language pretraining to yield fluent, open-vocabulary descriptions; related VLM advances such as BLIP-2, InstructBLIP, and LLaVA further enhance alignment and generalization [8–13]. However, practical constraints persist: input-length limits force frame sampling or segmentation, causing context fragmentation; token budgets compress video streams and reduce temporal fidelity; unified stacks often entail large model sizes and high train/test cost; and outputs may hallucinate details or misorder events [8–10, 14]. Taken together, these factors yield captions with insufficient temporal coherence, especially on long or edited videos.

To address these challenges, we propose a two-stage Keyframe–Prompt–LLM (KPL) paradigm for subject-centric video captioning, focusing on spatiotemporal changes of the subject(s). First, an unsupervised keyframe extractor (DFTC) surfaces subject state transitions to form an event skeleton. Second, we prompt the LLM with subject-motion cues to align generation, yielding content-complete and temporally coherent captions.

KPL makes subject-level spatiotemporal changes explicit, rather than keeping them implicit as in end-to-end mappings. Specifically, it employs unsupervised, transition-centered keyframe selection and assembles a structured, one-shot input for a frozen LLM. This design removes the need for paired video–text data. Whereas standard encoder–decoder pipelines typically internalize these transitions and many LLM-centric captioners rely on long token sequences or sizable multimodal stacks, KPL compresses subject motion into compact visual/textual evidence and uses one-shot prompting. This keeps the approach within LLM context budgets and aims to mitigate redundancy and semantic drift, producing fluent, subject-centric descriptions without fine-tuning, thereby improving content completeness and temporal coherence.

In summary, the main contributions of our work are as follows:

1. KPL (Keyframe-Prompt-LLM): subject-centric reformulation of video captioning.
2. DFTC algorithm: unsupervised, transition-centered extraction that surfaces subject-state changes.
3. Structured prompting: a one-shot, structured input that guides the LLM to the primary subject and reduces redundancy and semantic drift.
4. Empirical validation: consistent improvements in content completeness and temporal coherence under low-resource settings.

2 Methodology

2.1 Overview

We focus on subject-centric spatiotemporal narration: generating precise, time-ordered descriptions of how the subject (the primary entity of interest) evolves throughout a video. We propose KPL, a two-stage paradigm: (i) DFTC selects transition-centered keyframes by fusing frame differencing and optical-flow cues with temporal clustering to preserve the event skeleton; (ii) a structured, time-ordered prompt composes these keyframes into a single message for a frozen LLM (Qwen), which performs subject identification, dynamics modeling, and narrative synthesis without task-specific fine-tuning. This design improves interpretability and efficiency while maintaining traceability from each caption to its supporting keyframes.

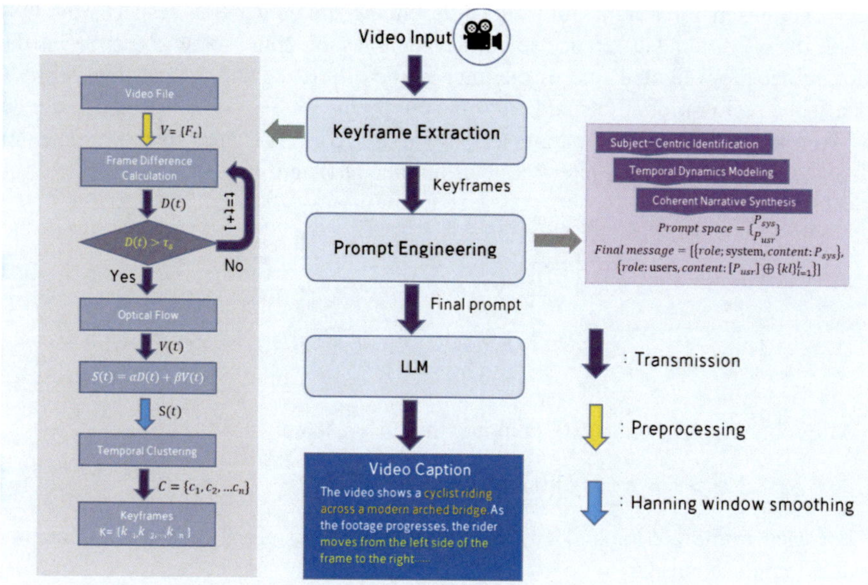

Fig. 1. Workflow of our proposed KPL.

The central panel shows the KPL (Keyframe–Prompt–LLM) framework: given a video input, Keyframe Extraction produces keyframes, and Prompt Engineering converts them into a time-ordered message that a frozen LLM uses to generate the caption. The left panel sketches the unsupervised extraction workflow, which selects a compact, ordered keyframe set from motion cues. The right panel sketches the prompt space and prompting strategies, merged with the keyframes to form the final prompt.

Figure 1 outlines the pipeline: after basic preprocessing, the system extracts principal motion and event cues along the temporal axis and, via temporal clustering, selects a representative keyframe set. Paired with a specific, time-ordered prompt strategy at the system-and user-levels, these keyframes form the conversational input to a frozen LLM (Qwen), which reasons globally across frames to enforce cross-frame consistency and produce temporally coherent, evidence-grounded descriptions.

2.2 Keyframe Extraction: Difference–Flow–Temporal Clustering (DFTC)

The goal of keyframe extraction is to select frames marking state transitions and event boundaries, explicitly aligning selection with what changes and when. We adopt a two-tier design: at the frame level, frame differencing and optical flow capture instantaneous displacements and abrupt pose changes; along the temporal axis, temporal clustering merges proximate candidates to suppress redundancy and preserve key turning points.

In this process, we define the following parameters: H is the frame height (number of pixel rows); W is the frame width (number of pixel columns); τ_d is the threshold for frame differencing; τ_f is the threshold for optical flow; ω is the Hanning window length; Δ is the clustering interval; F denotes the input video frames; P is the candidate frames; C indicates the set of clustering results; K is the output keyframes;α and β (α, β ∈ [0,1], α + β = 1) represent the weights for frame differencing and optical flow respectively; D(t) denotes the inter-frame difference score; V(t) denotes the optical-flow score; S(t) is the fusion score (the weighted sum of the inter-frame difference and optical-flow scores); and τ_s is the fusion-score threshold ($\tau_s = \alpha \cdot \tau_d + \beta \cdot \tau_f$).

Given input video V and parameters θ, we first convert frames to the LUV color space. We then compute the inter-frame difference D(t) and, if D(t) > τ_d, calculate the optical flow score V(t):

$$D(t) = \begin{cases} 0 & t = 1 \\ \frac{1}{H^*W} \sum \left\| F_t^{LUV} - F_{t-1}^{LUV} \right\| & t > 1 \end{cases} \tag{1}$$

$$V(t) = \begin{cases} calcOpticalFlow\left(F_{t-1}^{gray}, F_t^{gray}\right) & D(t) > \tau_d \\ 0 & \text{otherwise} \end{cases}. \tag{2}$$

After obtaining D(t) and V(t), compute the fusion score:

$$S(t) = \alpha D(t) + \beta V(t) \tag{3}$$

For each frame t, compute D(t), V(t) and S(t); then smooth S(t) with a Hanning window to reduce noise:

$$S'(t) = \frac{1}{\omega} \sum_{k=-\frac{\omega}{2}}^{\frac{\omega}{2}} S(t+k) \cdot \left[\frac{1}{2} \left(1 - \cos\left(\frac{2\pi \left(k + \frac{\omega}{2}\right)}{\omega} \right) \right) \right] \tag{4}$$

Detect local maxima on the smoothed fusion score S′(t) (where S′(t) > τ_s):

$$P = \left\{ t \middle| S'(t) \rangle S(t \pm 1) \cap S'(t) > \alpha \tau_d + \beta\, \tau_f \right\} \tag{5}$$

Sort the candidate frames P in temporal order to get $P_{sorted} = \{p_1, p_2, \ldots, p_N\}$. Because frame differencing with optical flow yields many near-duplicates, apply a gap-based temporal clustering (interval Δ) to merge adjacent frames and remove redundancy, yielding clustered sets $C = \{c_1, c_2, \ldots, c_M\}$. Select one representative from each cluster to produce the final keyframe set K:

$$d_k = p_k - p_{k-1} \tag{6}$$

$$C \leftarrow \begin{cases} C \cup \{c_{last} \cup \{p_k\}\} & d_k \le \Delta \\ C \cup \{\{p_k\}\} & d_k > \Delta \end{cases} \tag{7}$$

$$k_m = p_{\left[\frac{s+e}{2}\right]} \tag{8}$$

2.3 Prompt Engineering

The extracted keyframe set K compacts subject state transitions into a time-ordered evidence chain. We then apply a **time-ordered prompt strategy** that structures this evidence into a single chronological message for a frozen LLM (Qwen-vl-plus). This design enables single-round reasoning for coherent, traceable captions without fine-tuning.

Semantics-Driven Prompt Engineering: Core Strategy. Our prompt engineering follows three global principles to elicit stable, verifiable responses from a frozen LLM using the time-ordered evidence chain.

(1) Subject-Centric Identification

Localize the primary subject to filter background noise, using a maximum-confidence anchor to stabilize reasoning.

(2) Temporal Dynamics Modeling

Guide the model to infer state transitions over time, covering spatial displacement (Δs), action changes (Δa), and salient event structures on the global timeline.

(3) Coherent Narrative Synthesis

Constrain generation to be evidence-grounded, chronologically ordered, and causally consistent, and to be organized into a complete description with temporal coherence and semantic consistency, avoiding fragmented or ungrounded content.

These principles are instantiated in our prompt-engineering module via two complementary roles: P_{sys} (global behavioral constraints) and P_{usr} (instance-specific directives), assembled into a single time-ordered prompt for single-round reasoning.

Prompt Engineering in Practice: P_{sys} and P_{usr}. To instantiate the above strategy, we decompose the prompt design into two parts: P_{sys} (system-level prompts that set global behavior norms and target-tracking guidelines) and P_{usr} (user-level prompts that translate visual evidence into temporal reasoning and narrative goals).

1. Logical structure of the system prompts P_{sys}

We encode a verifiable reasoning chain with three predicates:

Detect (t, o, b): at time step t, detect target o with bounding box b;

Trans (t_1,t_2,o, Δs, Δa): within the interval [t_1,t_2], infer that the target o undergoes spatial displacement Δs and action change Δa;

Narr (o, T): based on Detect and Trans, generate a coherent narrative of target o over interval T.

P_{sys} renders these predicates in natural language.

(2) Chain-of-tasks design for the user prompts P_{usr}

Identify: localize the primary subject;

Track: extract the subject's spatial and action changes over time (Δs, Δa);

Synthesize: integrate the above changes into a natural-language narrative.

We adopt single-round assembly, concatenating the keyframe set K with P_{sys} and P_{usr} into one chronological message M, which reduces recursive-input errors and enforces cross-frame consistency without parameter updates.

3 Experimental Design and Results Analysis

3.1 Dataset and Experimental Design

We constructed a three-level dataset (Simple/Medium/Hard) and designed two complementary studies: an objective evaluation of keyframe extraction and a human-rated temporal-consistency evaluation of prompt engineering.

Dataset Composition and Difficulty Grouping. We collected 50 publicly available videos and grouped them by source and expected subject–spatiotemporal difficulty into three levels. **Simple** (20 videos, Pexels and MSRVTT): regular scenes with stable camera, salient single subjects, and changes mainly in spatial displacement or a single action; **Medium** (10 videos, text-to-video (Doubao) and TikTok): more pronounced stylistic variations and camera motions, with some videos involving multiple subjects; **Hard** (20 videos, open-source YouTube): rapid shot transitions, long-term dependencies, occlusions, and scene changes. The levels increase in complexity and support per-group statistics and generalization analysis.

Keyframe Extraction: Experimental Setup and Evaluation Metrics

Experimental Setup. We evaluated our method (DFTC) alongside three baseline methods on all 50 videos across the three difficulty levels. The baselines were: Method 1, an Edge-LBP-based shot boundary detection combined with Sobel/Z-score keyframe selection (Edge-LBP + Sobel, abbreviated as ELS) [15]; Method 2, a multi-feature fusion approach combined with FCM-AHA clustering for keyframe selection (Fusion+FCM-AHA, abbreviated as FFC) [16]; and Method 3, an optical-flow-based keyframe extraction method (Optical Flow, abbreviated as DOFP) [17]. All methods were run with their parameters optimized for best performance on our dataset. Each method's extracted keyframes for a video were aligned to ground-truth events for consistent evaluation. The public video sources include Pexels, MSRVTT, Doubao, TikTok, and YouTube [18–22].

Evaluation Metrics: We define ground-truth event frames (subject state changes or boundaries) and consider a predicted keyframe a hit if it falls within a tolerance window ($\tau = 1$ s) of a ground-truth event. Each ground-truth event matches at most one keyframe. Using this one-to-one matching, we calculate precision, recall, and F1 for each video, then average the results within each difficulty group and over all 50 videos.

Prompt Engineering: Experimental Setup and Evaluation Metrics

Experimental Setup: Using the same keyframe set K and the same frozen LLM (Qwen), we evaluate two input settings on each video: (i) **Prompted:** a structured prompt as defined in Sect. 2.3; (ii) **No-Prompt:** a plain instruction ("The following are keyframes… please describe the video"). Both settings share identical keyframe ordering, token budget, and single-pass decoding. Each video yields two captions for human rating; scores are reported per difficulty group (Simple/Medium/Hard) and overall.

Evaluation Metrics: We adopted a double-blind review procedure, with "traceability of subject spatiotemporal variations" as the core evaluation dimension. For every 10 videos, two evaluators independently score the outputs. If their scores differ by more than 20 points, a third reviewer re-evaluates and a reconciled score is taken as final; otherwise, the two scores are averaged. The scoring dimensions and weights are as follows: (i) tracking of subject changes and temporal sequence (0–5, weight 40%); (ii) detail and style (0–5, weight 30%); (iii) completeness of description content (0–5, weight

30%). In addition, a general penalty rule is applied to deduct points for factual errors or unwarranted hallucinations. Final errors or unwarranted hallucinations. Final scores are converted to a percentage scale:

$$score_{0-100} = 8 \times sub\_trans + 6 \times detail + 6 \times coherence + penalty \quad (9)$$

Result Presentation: To quantify the improvements, we calculate the absolute gain $\Delta s_g = \left|\bar{s}_g^{\text{prompted}} - \bar{s}_g^{\text{no-prompt}}\right|$ and relative gain $Rel_g = \Delta s_g / \bar{s}_g^{no-prompt} \times 100\%$ for each group (Simple/Medium/Hard) and the overall average.

3.2 Experimental Results of Keyframe Extraction

In the overall dataset experiments, we present the keyframe extraction results using both a three-line table (Table 1) and a Precision–Recall (P–R) plot (Fig. 2).

Table 1. Keyframe extraction results on the simple, medium, hard subsets and the overall average.

Method	Simple			Medium			Hard			Average		
	P	R	F1	P	R	F1	P	R	F1	P	R	F1
DFTC*	**0.913**	0.900	**0.907**	**0.976**	0.700	**0.815**	**0.975**	0.626	0.762	**0.950**	0.750	**0.802**
ELS	0.847	0.807	0.783	0.910	0.783	0.805	0.975	0.479	0.598	0.911	0.671	0.773
FFC	0.547	**1.000**	0.688	0.616	**0.956**	0.734	0.829	**0.831**	**0.800**	0.674	**0.924**	0.779
DOFP	0.831	0.656	0.611	0.721	0.611	0.461	0.716	0.669	0.527	0.763	0.652	0.703

Overall, DFTC achieves the best average F1-score (0.802), versus FFC (0.779), ELS (0.773), and DOFP (0.703). In terms of individual metrics, DFTC reaches P = 0.950, significantly higher than the other methods, while maintaining a reasonable recall of R = 0.750. Consequently, on the P–R plane, DFTC's point lies further to the right and on a higher F1 contour, indicating that our method, while sustaining high precision, effectively limits missed detections and thus achieves the best overall performance.

On Simple, DFTC achieves the highest F1-score (0.907), far exceeding ELS (0.783), FFC (0.688), and DOFP (0.611). On Medium, DFTC attains an F1-score of 0.815, slightly above ELS (0.805) and clearly higher than FFC (0.734) or DOFP (0.461). On hard, FFC attains a slightly higher F1-score (0.800) than DFTC (0.762), reflecting the broader coverage of its multi-feature clustering in high-complexity scenarios. DFTC maintains a very high P (0.975) but experiences lower R under rapid camera changes and occlusions.

The pattern reflects each method's features: **DFTC**, being transition-centric, captures subtle intra-shot changes and filters stylistic noise, yielding consistently higher precision with competitive recall. **ELS**, tied to abrupt shot boundaries, misses many intra-shot events (lower R) and can introduce redundancies when style shifts occur. **FFC** tends to treat minor fluctuations as events (lower P, higher R); its broader coverage helps on Hard, where frequent cuts/occlusions make recall decisive, giving it a slight F1 lead

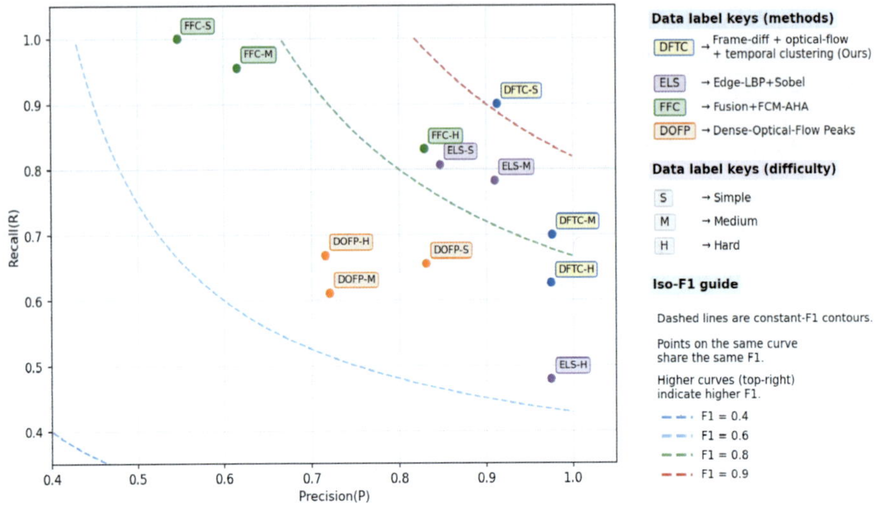

Fig. 2. P–R plot with iso-F1 contours for keyframe extraction results. Each point corresponds to a specific method–difficulty combination (DFTC, ELS, FFC, DOFP; S/M/H).

over DFTC. **DOFP**, relying on a single motion cue, is sensitive to camera motion and occlusions, which corresponds to lower P and R under these conditions.

The cross-group results (Fig. 2) visualize the precision–recall distribution by difficulty. DFTC's points on the P–R plane are clearly separated by difficulty level, effectively distinguishing video complexity, forming three right-shifted clusters with modest recall spread. By contrast, the other methods extract keyframes for all videos but show less-satisfactory P-R performance. As difficulty increases, FFC drifts toward the upper-left (recall-oriented), ELS slides down along recall, and DOFP remains in the lower region. This complements Table 1, showing that DFTC maintains a precision-focused profile with clear difficulty discrimination, consistent with its top overall F1.

3.3 Experimental Results of Prompt Engineering

In the overall evaluation, we present the prompt-engineering results using a difference dumbbell plot (Fig. 3).

Experimental results (Fig. 3) show that in all three groups and in the average performance, the Prompted setting consistently outperforms No-prompt, demonstrating that structured prompts enhance video description quality.

In the simple group, the prompted setting (86.12) clearly outperformed the no-prompt setting (71.6), with a relative gain of 20.20%. In the medium group, the average score of prompted (92.30) is slightly higher than that of no-prompt (88.35), with a relative gain of 4.47%, suggesting that prompts provide only limited improvement. In the hard group, the prompted setting (79.75) substantially outperformed the no-prompt setting (64.50), with a relative gain of 23.64%.

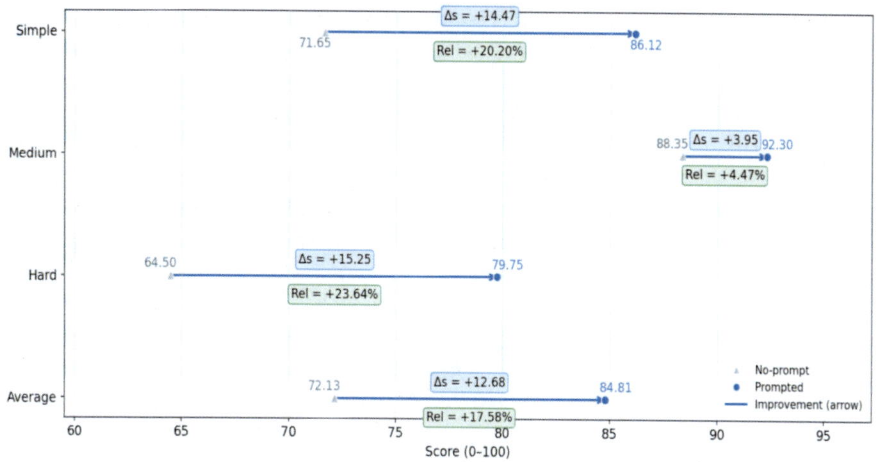

Fig. 3. Dumbbell plot of No-prompt vs. Prompted across Simple/Medium/Hard and Average. Scores are on a 0–100 scale; arrows annotate absolute gain (Δs) and relative gain (Rel).

Structured, time-ordered prompts bring the largest gains in the Simple and Hard groups because they keep attention on the primary subject, enforce chronological ordering, and suppress drift/redundancy even under camera motion or occlusion. By contrast, in the Medium group, although these videos may involve multiple subjects and stylistic variation, the actual subject-level changes to be described are comparatively less complex; keyframes have already made those changes salient and the LLM can handle the style shifts, so prompting mainly trims minor drift and redundancy and brings only a modest extra gain. Overall, most of the improvement comes from better tracking of subject changes and temporal order, while content completeness remains similar across settings.

3.4 Ablation Study

To further validate the technical choices in our DFTC algorithm, we conducted a series of ablation experiments on two key parameters: the fusion weights (α, β) and the clustering interval (Δ).

Effect of Fusion Weights (α, β). The fusion score $S(t) = \alpha D(t) + \beta V(t)$ balances frame differencing and optical flow. As illustrated in Fig. 4, the average F1-score climbs as the frame-difference weight α increases, peaking at 0.8069 when $\alpha = 0.85$ (and $\beta = 0.15$). This trend reveals that while inter-frame pixel changes (frame differencing) dominate the detection of subject-centric state transitions, the inclusion of optical flow ($\beta = 0.15$) plays a critical complementary role. Purely relying on frame difference ($\alpha = 1.0$) or optical flow ($\alpha = 0.0$) yields suboptimal F1 scores of 0.7836 and 0.6206 respectively, confirming that the weighted fusion effectively compensates for subtle motion cues that simple differencing might miss.

Impact of Clustering Interval (Δ). The clustering interval Δ governs the merging of adjacent candidate frames to control redundancy. The results in Fig. 5 highlight a distinct trade-off: as Δ increases from 8 to 48, Precision (P) consistently improves from 0.7569 to 0.9083 due to the removal of near-duplicate frames, whereas Recall (R) drops sharply from 0.83 to 0.4912 after Δ exceeds 24. This indicates that while larger intervals reduce redundancy, excessive merging disrupts the event skeleton by omitting key turning points. Consequently, the F1-score peaks at 0.8069 with $\Delta = 24$, identifying this interval as the optimal "sweet spot" that maximizes narrative coverage while ensuring concise visual evidence.

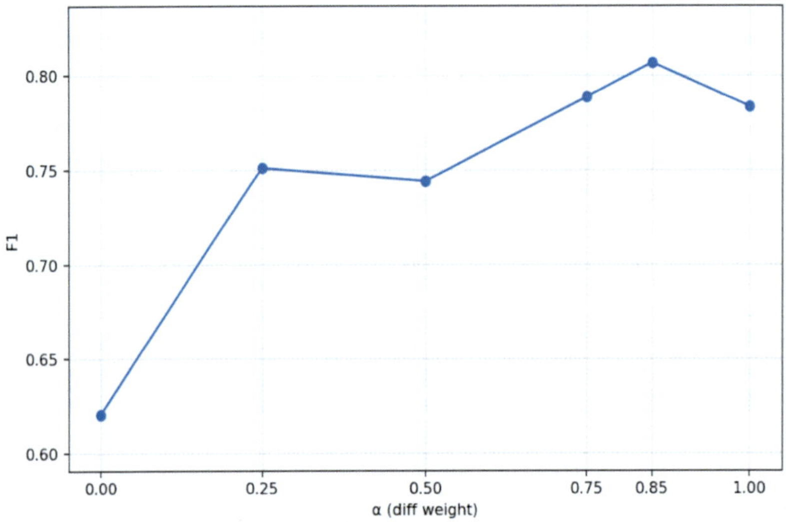

Fig. 4. Effect of fusion weight α on keyframe-detection performance.

3.5 Efficiency and Reliability Analysis

Resource and Token Efficiency. While our method focuses on output quality, we also evaluated its computational efficiency. All experiments were conducted on a workstation equipped with an Intel i7-12700K CPU and an NVIDIA RTX 3090 GPU, processing videos standardized to 720p resolution using a Python 3.10 implementation. On average, the DFTC extraction process takes 7.304 s per video, compared to faster baselines like ELS (~ 1.19 s), FFC (~ 1.03 s) and DOFP(~ 1.11 s). Although this preprocessing step is computationally heavier due to the calculation of dense optical flow, it represents a strategic trade-off that significantly enhances token efficiency for the downstream LLM.

(1) **Compression Ratio**

By filtering out over 95% of redundant frames and retaining only the "skeleton" of events, DFTC compresses a standard video into a concise sequence of keyframes.

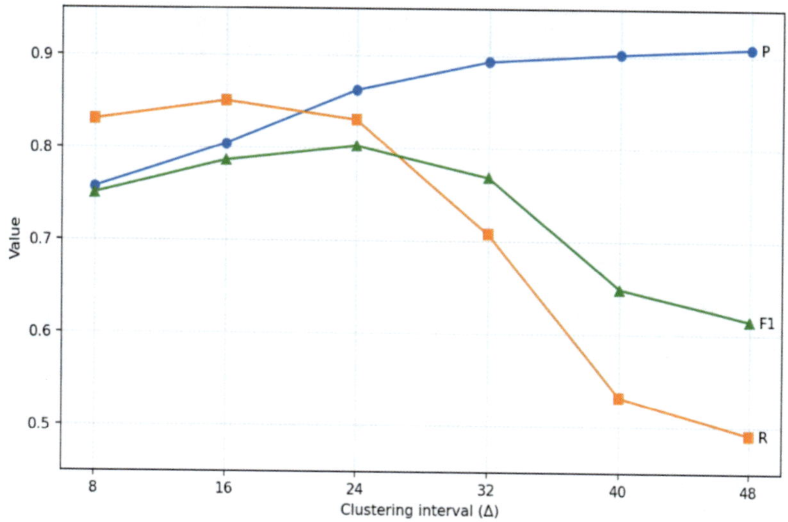

Fig. 5. Impact of clustering interval Δ on precision–recall trade-off.

This drastically reduces the context window requirement for the LLM compared to frame-sampling methods used in end-to-end Video-LLMs.

(2) **Training-Free**

Unlike supervised methods that require thousands of GPU hours for training, KPL is entirely unsupervised and training-free, making the 7-s per-video overhead negligible in the context of total system deployment costs.

User Study Reliability. To ensure the robustness of our human evaluation, we measured the inter-rater agreement between the two independent evaluators. The calculated weighted Cohen's Kappa (κ) across all samples was 0.51, and the Intraclass Correlation Coefficient (ICC) was 0.43. These values fall within the range of "moderate agreement," statistically validating the reliability of the subjective scoring process and the subsequent arbitration mechanism.

3.6 Qualitative Results and Case Studies

This section presents a montage with one representative video per difficulty (Fig. 6). The top row shows keyframes selected by DFTC, ELS, FFC, and DOFP, and the bottom row shows captions under No-prompt vs. Prompted. Underlined phrases with numeric tags correspond to the numbered keyframes (or keyframe groups) above; yellow highlights indicate the subject's spatiotemporal changes; and ellipses ("...") mark minor details omitted as secondary.

Across groups, DFTC surfaces transition-centered frames without duplicates, and prompting reduces drift and redundancy while preserving temporal ordering—yielding

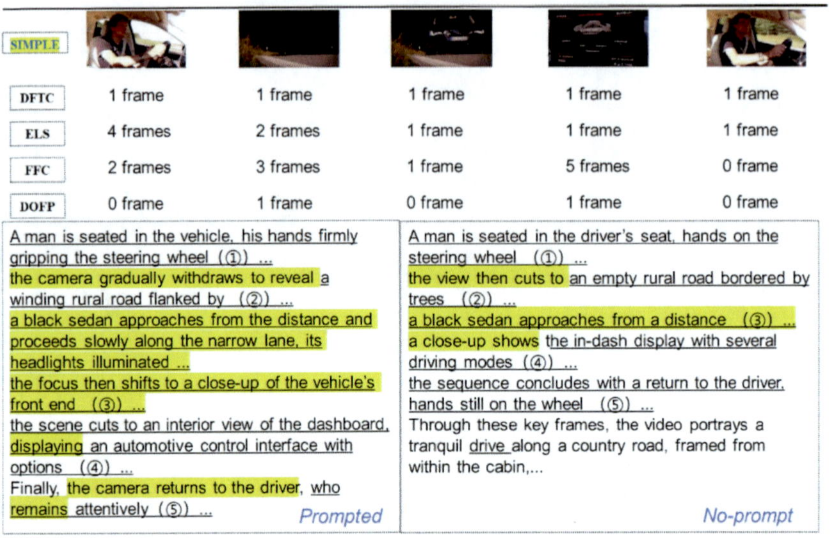

(a) Qualitative montage for Simple videos

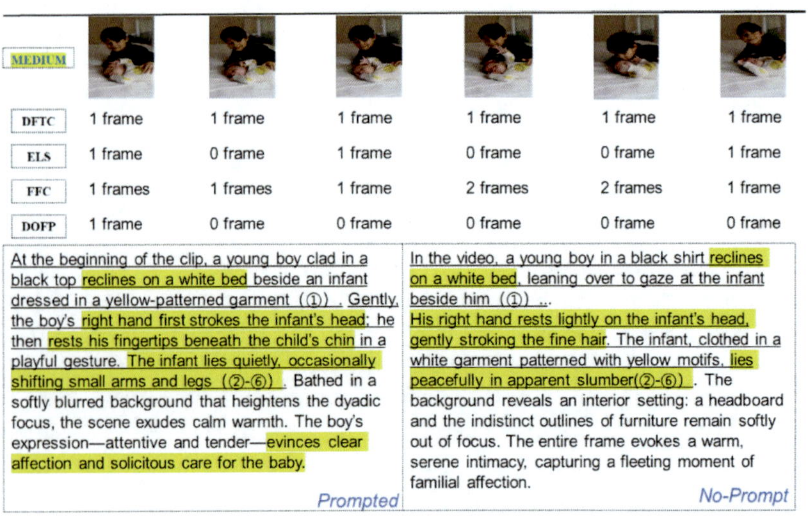

(b) Qualitative montage for Medium videos

Fig. 6. Qualitative montage

more precise, subject-focused descriptions, especially under camera changes and occlusions (Hard). Typical challenges observed include that ELS may miss intra-shot changes, FFC can over-select minor fluctuations, and DOFP is sensitive to viewpoint shifts.

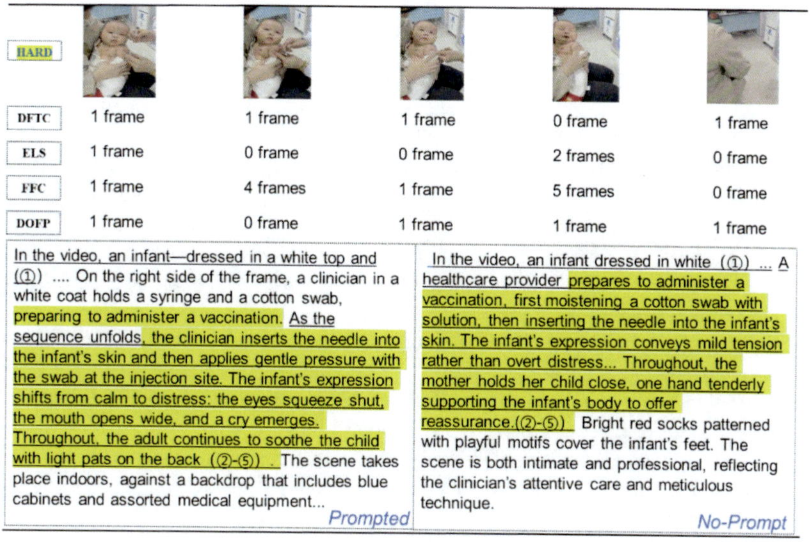

(c) Qualitative montage for Hard videos

Fig. 6. (*continued*)

4 Conclusion

In this paper, we introduced KPL, a video captioning paradigm centered on the subject's spatiotemporal changes. KPL works by selecting keyframes as evidence and using a structured prompt for generation. On a difficulty-stratified dataset of 50 videos, KPL captures spatiotemporal changes more faithfully and improves content completeness and temporal coherence compared to representative baselines, while remaining data-and-compute-efficient without language-side fine-tuning. These results indicate that KPL is a practical, evidence-grounded route to controllable video captioning, and provide a foundation for extending this paradigm to longer videos and additional modalities.

References

1. Amirian, S., Rasheed, K., Taha, T.R., Arabnia, H.R.: Automatic image and video caption generation with deep learning: a concise review and algorithmic overlap. IEEE Access. **8**, 218386–218400 (2020). https://doi.org/10.1109/ACCESS.2020.3042484
2. Velda, V., Immanuel, S.A., Hendria, W.F., Jeong, C.: Improving distinctiveness in video captioning with text–video similarity. Image Vis. Comput. **136**, 104728 (2023). https://doi.org/10.1016/j.imavis.2023.104728
3. Ji, L., Tu, R., Lin, K., Wang, L., Duan, N.: Multimodal graph neural network for video procedural captioning. Neurocomputing. **488**, 88–96 (2022). https://doi.org/10.1016/j.neucom.2022.02.062
4. Luo, X., Luo, X., Wang, D., Liu, J., Wan, B., Zhao, L.: Global semantic enhancement network for video captioning. Pattern Recogn. **145**, 109906 (2024). https://doi.org/10.1016/j.patcog.2023.109906

5. Guo, L., Zhao, H., Chen, Z., Han, Z.: Semantic guidance network for video captioning. Sci. Rep. **13**(1), 16076 (2023). https://doi.org/10.1038/s41598-023-43010-3
6. Jacob, H., Pádua, F.L.C., Lacerda, A., Pereira, A.C.M.: A video summarization approach based on the emulation of bottom-up mechanisms of visual attention. J. Intell. Inf. Syst. **49**(2), 193–211 (2017). https://doi.org/10.1007/s10844-016-0441-4
7. Guan, G., Wang, Z., Lu, S., Deng, J.-D., Feng, D.: Keypoint-based keyframe selection. IEEE Trans. Circuits Syst. Video Technol. **23**(4), 729–737 (2013). https://doi.org/10.1109/TCSVT.2012.2214871
8. Maaz, M., Rasheed, H., Khan, S., Khan, F.: Video-ChatGPT: towards detailed video understanding via large vision and language models. In: Proceedings of ACL, vol. 2024, pp. 12585–12602 (2024)
9. Zhang, H., Li, X., Bing, L.: Video-LLaMA: an instruction-tuned audio-visual language model for video understanding. arXiv preprint https://arxiv.org/abs/2306.02858 (2023)
10. OpenAI GPT-4 technical report. arXiv preprint https://arxiv.org/abs/2303.08774 (2023)
11. Li, J., Li, D., Savarese, S., Hoi, S.C.H.: BLIP-2: bootstrapping language-image pre-training with frozen image encoders and large language models. In: 40th International Conference on Machine Learning (ICML), Proceedings of Machine Learning Research, vol. 202, pp. 19730–19742 (2023)
12. Dai, W et al.: InstructBLIP: towards general-purpose vision-language models with instruction tuning. arXiv preprint https://arxiv.org/abs/2305.06500 (2023)
13. Liu, H., Li, C., Wu, Q., Lee, Y.J.: Visual instruction tuning (LLaVA). In: Advances in Neural Information Processing Systems (NeurIPS), 2023, oral (2023)
14. Wang, P. et al.: Qwen2-VL: enhancing vision-language model's perception of the world at any resolution. arXiv preprint https://arxiv.org/abs/2409.12191 (2024)
15. Nandini, H.M., Chethan, H.K., Rashmi, B.S.: Shot based keyframe extraction using edge-LBP approach. J. King Saud Univ. – Comput. Inf. Sci. **34**(10), 4537–4545 (2022). https://doi.org/10.1016/j.jksuci.2020.10.031
16. Kaur, S., Kaur, L., Lal, M.: An effective key frame extraction technique based on feature fusion and fuzzy C-means clustering with artificial hummingbird. Sci. Rep. **14**, 26651 (2024). https://doi.org/10.1038/s41598-024-75923-y
17. Dong, J.: Study on video key frame extraction in different scenes based on optical flow. J. Phys. Conf. Ser. **2646**(1), 012035 (2023). https://doi.org/10.1088/1742-6596/2646/1/012035
18. Pexels: Free stock photos & videos. Available: https://www.pexels.com/. Accessed 27 July 2025(2025)
19. Xu, J., Mei, T., Yao, T., Rui, Y.: MSR-VTT: a large video description dataset for bridging video and language. In: IEEE Conference on Computer Vision and Pattern Recognition (CVPR), pp. 5288–5296 (2016)
20. Doubao: Doubao AI platform. Available: https://www.doubao.com/. Accessed 27 July 2025(2025)
21. TikTok: Short-form mobile video platform. Available: https://www.tiktok.com/ Accessed 27 July 2025 (2025)
22. YouTube: Video sharing platform. Available: https://www.youtube.com/ Accessed 27 July 2025 (2025)

Multilingual Back-Translation Augmentation for Low-Resource Text Classification: Effectiveness Boundaries and Optimal Practices

Minshi Fan(✉)

School of Foreign Languages and Literature, Wuhan University, Wuhan, China
sun_day168@163.com

Abstract. In low-resource text classification, the scarcity of high-quality labeled data often leads to suboptimal model performance. This paper investigates a multilingual back-translation data augmentation strategy to improve text classification performance under various low-resource settings. Using three benchmark datasets—SST-5, AG News, and 20 Newsgroups—This study constructs low-resource baselines by sampling 5, 10, 20, and 40 texts per class. Bidirectional back-translation is performed over five language pairs (German–English, French–English, Chinese–English, Spanish–English, and Russian–English) using the Helsinki-NLP Opus-MT model series, generating augmented data in a 1:1 ratio to the original texts. Furthermore, the study proposes four novel data mixing strategies—Original First, Back-Translation First, Alternating Mix, and Random Mix—which, together with the low-resource baselines, form a total of 84 experimental datasets. The experiments involve fine-tuning the DistilBERT-base-uncased and RoBERTa-base models. Results indicate that in extremely low-resource scenarios, back-translation augmentation significantly enhances classification accuracy, achieving a maximum improvement of 27.12%. As the low-resource condition is alleviated, the benefits from back-translation diminish; beyond a certain data size threshold, gains drop sharply and may even become negative due to introduced noise. The extent of improvement varies across datasets depending on the choice of intermediate languages and data mixing strategy. This work offers a reproducible augmentation framework for low-resource text classification and demonstrates the practical value of back-translation for data augmentation in resource-constrained settings.

Keywords: back-translation · low-resource · data augmentation · data mixing strategy · effectiveness boundary

1 Introduction

Sufficient labeled data availability is a critical determinant of text classification model performance. When only a few labeled samples are available per category, deep learning models are highly susceptible to overfitting, leading to a sharp decline in generalization ability. For instance, research shows that in the SST-5 sentiment analysis task, reducing

L. Huang and D. Greenhalgh (Eds.): ICMLC 2026, LNNS 2061, pp. 977–990, 2027.
https://doi.org/10.1007/978-3-032-31236-5_66

the number of samples per class to 8 causes the accuracy of a fine-tuned RoBERTa-large model to drop from 59.8% (with full data) to $33.5 \pm 2.1\%$ [1]. In practical applications, such as text classification for low-resource languages or specialized domains, acquiring adequate labeled data is often costly or even infeasible. Therefore, exploring methods to leverage existing data resources to enhance model performance under low-resource conditions remains a significant research challenge.

As an unsupervised data augmentation technique, back-translation generates new texts with consistent semantics but varied expressions through a source language → target language → source language conversion process. Initially prominent in machine translation [2], it has gained traction as a data augmentation method due to its strengths in alleviating data scarcity, enhancing model generalization, requiring no additional supervision, and exhibiting strong cross-task versatility. Consequently, it has been applied in areas such as low-resource sentiment analysis [3] and text classification [4].

This study focuses on low-resource text classification. We investigate the performance of large language models on datasets containing as few as 5, 10, 20, and 40 samples per class, employing multilingual back-translation combined with different data mixing strategies. The core contributions of this work are as follows:

- **Effectiveness Boundary Analysis:** Experiments are conducted across a spectrum from extreme low-resource (5 samples/class) to moderate low-resource (40 samples/class) settings, revealing the performance gains and defining the effectiveness boundaries of back-translation augmentation under varying resource conditions.
- **Data Mixing Strategy Evaluation:** Four data mixing strategies are designed and compared to systematically assess how the organization of training data influences model performance.
- **Multilingual Validation:** Back-translation is evaluated using five common language pairs: German–English, French–English, Chinese–English, Spanish–English, and Russian–English.

Through 84 experimental configurations across three benchmark datasets, this study delineates the effective boundaries and optimal practices for applying back-translation in low-resource text classification, providing a practical and reproducible augmentation framework.

2 Related Work

2.1 Applications of Back-Translation in NLP

Sennrich et al. first introduced back-translation in 2016 as a data augmentation technique for neural machine translation (NMT) [2]. Its core idea is to generate pseudo-parallel corpora through reverse translation, thereby alleviating data scarcity in low-resource language translation tasks [5]. This technique has significantly influenced the development of machine translation and other areas of natural language processing (NLP).

Back-translation offers several advantages in NLP [6]. It effectively mitigates data scarcity, enabling models for low-resource languages to approach the performance of those for high-resource ones. Additionally, the diversity introduced by pseudo-parallel data helps models generalize better to varied inputs, improving their robustness. Furthermore, its compatibility with unsupervised learning opens new avenues for NLP applications.

In recent years, back-translation has achieved notable results in various scenarios. For example, Ma et al. conducted a Chinese text classification study using three datasets: ChnSentiCorp_htl_all, waimai_10k, and online_shopping_10_cats [4]. After back-translation augmentation, the average classification accuracy improved from 85.8% to 88.1% across LSTM and CNN models, outperforming Easy Data Augmentation (EDA). However, that study was limited to English and French as intermediary languages and was only validated on LSTM and CNN models in comparison with EDA. It was not applied to more advanced Transformer-based architectures like BERT, nor compared with augmentation techniques beyond EDA, which constrains its broader applicability.

Further advancing this line of work, Lee et al. proposed an Iterative Translation-Based Data Augmentation (ITDA) method to enhance multilingual text classification performance [7]. Across 480 text classification tasks, ITDA improved average accuracy by 4.24%, significantly outperforming traditional methods like EDA. However, its performance gains were limited or even decreased on datasets containing proper nouns (e.g., SST-2), indicating insufficient handling of such terms. Moreover, the iterative translation process relies on manual parameter tuning— a labor-intensive step that is difficult to optimize, as the framework lacks an adaptive mechanism for dynamic parameter adjustment.

2.2 Low-Resource Text Classification Techniques

Low-resource learning primarily addresses data scarcity through several technical approaches:

- **Data Augmentation.** By generating additional training examples, these methods effectively expand both the size and diversity of the training dataset. Common techniques include back-translation, Easy Data Augmentation (EDA), noise injection, and text perturbation.
- **Cross-Lingual Transfer Learning.** This paradigm leverages annotated corpora and pre-trained models from high-resource languages, adapting them to low-resource language tasks to reduce reliance on direct annotation. Notably, J. Singh proposed Cross-Lingual Data Augmentation (XLDA) [8], a framework tailored for Natural Language Inference and Question Answering tasks. By substituting text segments with translations from other languages, XLDA boosts multilingual model performance on benchmark datasets including Cross-Lingual Natural Language Inference (XNLI) and SQuAD.
- **Unsupervised and Self-Supervised Learning.** These methods utilize unlabeled data to extract semantic information, thereby decreasing dependence on manual annotation. Representative approaches include language model pre-training [9], pseudo-label generation, and contrastive learning.
- **Few-Shot and Zero-Shot Learning.** Techniques such as meta-learning train models to quickly adapt to new tasks with very few examples, showing promising potential in low-resource settings [10].

- **Data Selection and Optimization.** This strategy involves selectively using high-quality data or optimizing data acquisition to improve learning from limited samples. For instance, Xu et al. [11] employed quality metrics such as BLEU or BERTScore to filter high-quality pseudo-parallel samples from back-translated corpora, thereby mitigating the negative effects of noisy data.

In summary, current research on low-resource text classification still faces several limitations, including limited linguistic diversity in augmentation methods, a lack of systematic strategies for blending original and back-translated data, and insufficient validation in extreme low-resource scenarios (e.g., fewer than 10 samples per class). This study aims to address these gaps by exploring multilingual back-translation, designing diverse training data mixing strategies, and conducting comprehensive experiments under extreme low-resource conditions.

3 System Design

3.1 Dataset Construction

Base Datasets. Three widely used benchmarks for low-resource text classification are adopted:

- **SST-5** is a fine-grained sentiment analysis dataset derived from the Stanford Sentiment Treebank. It contains 5 sentiment labels, is well-suited for few-shot learning, and is commonly used to evaluate sentiment classification model performance.
- **AG News** is a standard benchmark dataset for news text classification, containing 4 categories. It is widely used for training and evaluating models in text classification and data augmentation research.
- **20 Newsgroups** contains newsgroup posts across 20 different topical categories, each representing a specific subject area. It is frequently used for semantic and multi-class text classification studies.

Low-Resource Subset Generation. Training, validation, and test sets are generated for each base dataset.

- **Source Training Sets:** From each class of the base datasets, 5, 10, 20, and 40 samples are randomly sampled (allowing repetition), creating 4 low-resource source training sets.
- **Back-Translation Training Sets:** Using the Helsinki-NLP Opus-MT models, bidirectional translation (English→X → English) is performed for five language pairs: German-English (de2en), French-English (fr2en), Chinese-English (zh2en), Spanish-English (es2en), and Russian-English (ru2en). Each original text generates one back-translated text, achieving 1:1 augmentation relative to the source training set. For each back-translation language, the generated data is combined with the corresponding source data using four mixing strategies—Original First (labeled A), Back-Translation First (B), Alternating Mix (C), and Random Mix (D)—to form augmented datasets. Consequently, for each base training set configuration, 4 (source sizes) $\times$ 5 (languages) $\times$ 4 (mixing strategies) = 80 back-translation training sets are generated.

- **Validation Set:** After excluding the records selected for the training sets, 100 samples per class are randomly drawn from the remaining records of each base dataset.
- **Test Set:** After further excluding the validation set records, 200 samples per class are randomly drawn from the remaining records of each base dataset.

Table 1 uses SST-5 as an example to detail the composition of the datasets. Here, n* corresponds to n5, n10, n20, and n40, representing the four source datasets. *2en denotes de2en, fr2en, zh2en, es2en, and ru2en, which correspond to the five different language back-translation augmentations. The letters A through D indicate the different mixing strategies employed to combine the back-translated augmentation data with the original source data.

Table 1. Construction method of the training and testing benchmark datasets.

Dataset type	Datasets name	# of Datasets	# of records
Source training dataset	sst5_n*	4	25/50/100/200
Back-translation Training dataset	sst5_n*_*2en_A sst5_n*_*2en _B sst5_n*_*2en _C sst5_n*_*2en _D	80	50/100/200/400
Validation dataset	sst5_validation	1	500
Test dataset	sst5_test	1	1000

3.2 System Architecture

Figure 1 illustrates the system architecture. From each class of the base datasets, 5, 10, 20, and 40 records are extracted to form four low-resource source training sets (n5, n10, n20, n40), representing a progression from extreme to ordinary low-resource conditions. Each source training set undergoes back-translation via the Opus-MT models for the five language pairs, generating five corresponding back-translation datasets (de2en, fr2en, zh2en, es2en, ru2en), each matching the source set size (1:1 ratio). These back-translation sets are then combined with their source counterpart using the four mixing strategies (A-D) to produce the final training datasets. Thus, for each base dataset, a total of 84 training sets are created: 80 back-translation mixed sets and 4 source-only sets. All these datasets are subsequently used to fine-tune the DistilBERT-base-uncased and RoBERTa-base models.

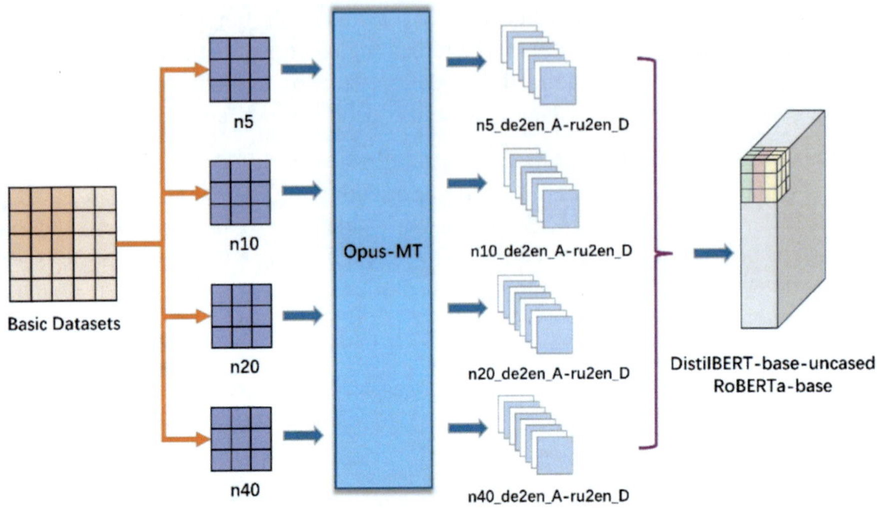

Fig. 1. System architecture diagram.

4 Experiments and Analysis

4.1 Experimental Setup

- **Model:** DistilBERT-base-uncased and RoBERTa-base.
- **Control Groups:** The four low-resource source training sets: n5, n10, n20, n40.
- **Experimental Groups:** The 80 back-translation training sets: de2en_A—de2en_D, fr2en_A—fr2en_D, zh2en_A—zh2en_D, es2en_A—es2en_D, ru2en_A—ru2en_D.
- **Training Configuration:** Models are trained for 20 epochs with an initial learning rate of 2e-5. A patience of 2, a learning rate decay factor of 0.5, and a lower bound of 1e-6 for the learning rate are set. Validation is performed after each epoch using the validation set, and the learning rate is dynamically adjusted based on validation performance to achieve optimal training. After 20 epochs, the model checkpoint from the epoch with the best validation accuracy is loaded and evaluated on the test set to obtain the final performance metrics.

4.2 Experimental Results

Results on the SST-5 Dataset. Figure 2 displays the test results for SST-5 using different back-translation languages for augmentation. The reported accuracy values are averages across all four data mixing strategies for each language. As indicated by the results, for the DistilBERT-base-uncased model, slight accuracy gains are observed only on the n5 dataset; in all other cases, classification accuracy even decreases. For the RoBERTa-base model, considerably greater performance improvements are observed: significant accuracy gains are achieved on both the n5 and n10 datasets.

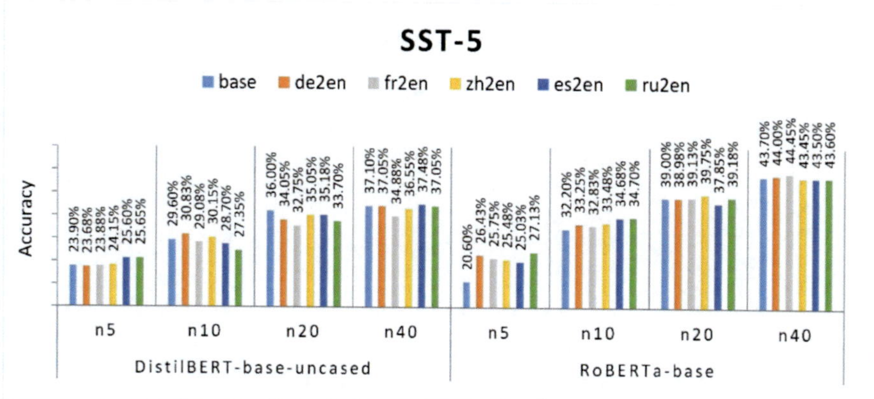

Fig. 2. Accuracy of SST-5 augmented with different back-translation languages on distil-BERT-base-uncased and RoBERTa-base.

Figure 3 shows the test results for SST-5 under different data mixing strategies. The accuracy values presented are averages across the n5-n40 dataset configurations. The results demonstrate that on the SST-5 dataset, the choice of data mixing strategy has a considerable impact on final performance, with the maximum observed difference of 2.57% achieved in the es2en dataset configuration with the RoBERTa-base model.

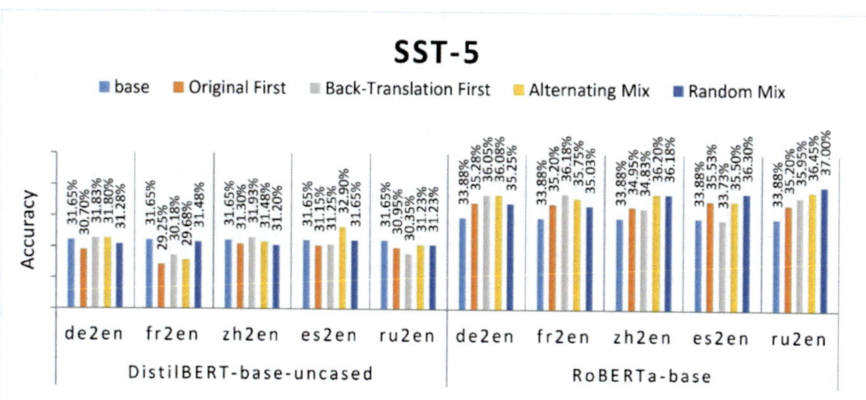

Fig. 3. Accuracy of SST-5 with Different Mixing Strategies on DistilBERT-base-uncased and RoBERTa-base.

Results on the AG News Dataset. Figure 4 displays the test results for AG News using different back-translation languages for augmentation. The results reveal that back-translation augmentation leads to substantial performance improvements on the AG News dataset. Notably, with the RoBERTa-base model and Spanish-English back-translation, the maximum accuracy improvement reaches 27.12% for the n5 dataset.

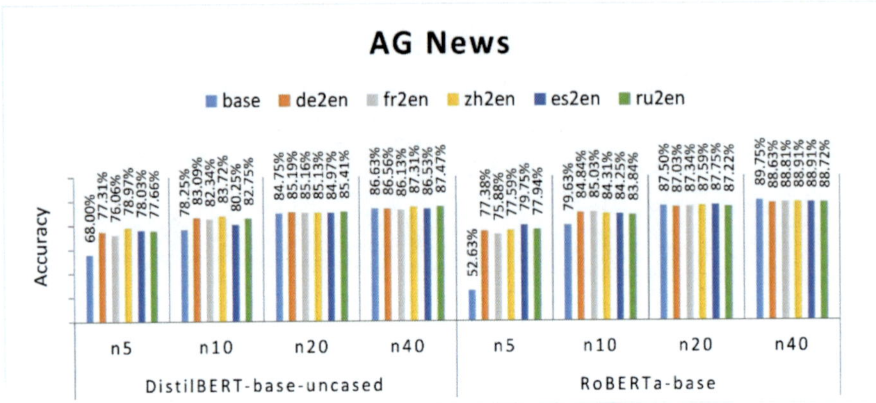

Fig. 4. Accuracy of AG news augmented with different back-translation languages on distil-BERT-base-uncased and RoBERTa-base.

Figure 5 illustrates the test results for AG News under different data mixing strategies. The analysis indicates that on this dataset, different mixing strategies exert a significant influence on final performance, with the maximum observed difference of 2.28% achieved in the fr2en dataset configuration with the RoBERTa-base model.

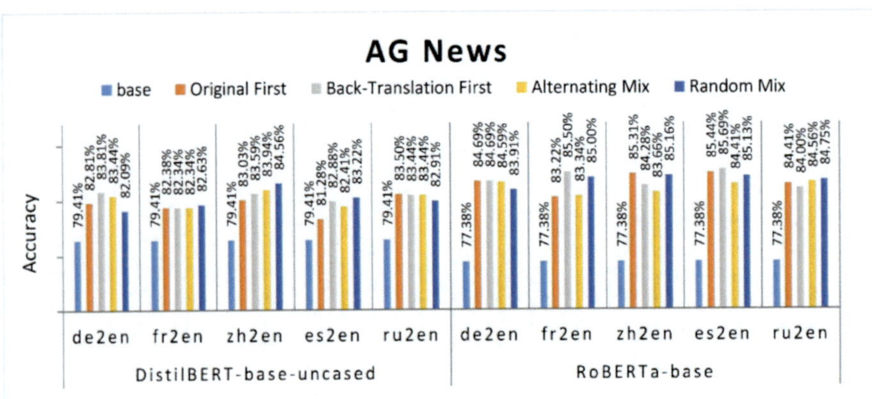

Fig. 5. Accuracy of AG news with different mixing strategies on DistilBERT-base-uncased and RoBERTa-base.

Results on the 20 Newsgroups Dataset. Figure 6 presents the test results for 20 Newsgroups using different back-translation languages for augmentation. The results show that back-translation augmentation yields relatively noticeable performance gains on the 20 Newsgroups dataset. Notably, with the DistilBERT-base-uncased model, the maximum accuracy improvement of 7.79% is achieved for the n5 dataset.

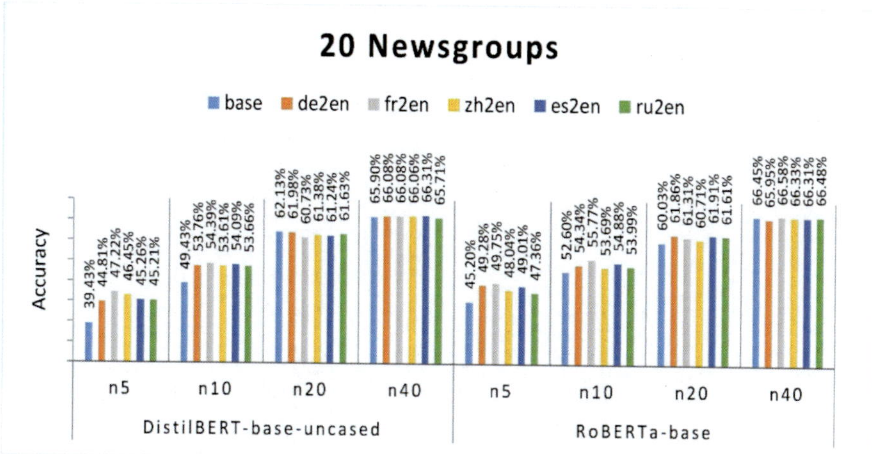

Fig. 6. Accuracy of 20 newsgroups augmented with different back-translation languages on distil-BERT-base-uncased and RoBERTa-base.

Figure 7 shows the test results for 20 Newsgroups under different data mixing strategies. The choice of mixing strategy on 20 Newsgroups has a measurable, yet not dominant, effect on performance, with the maximum difference of 1.59% observed in the ru2en dataset configuration with the DistilBERT-base-uncased model.

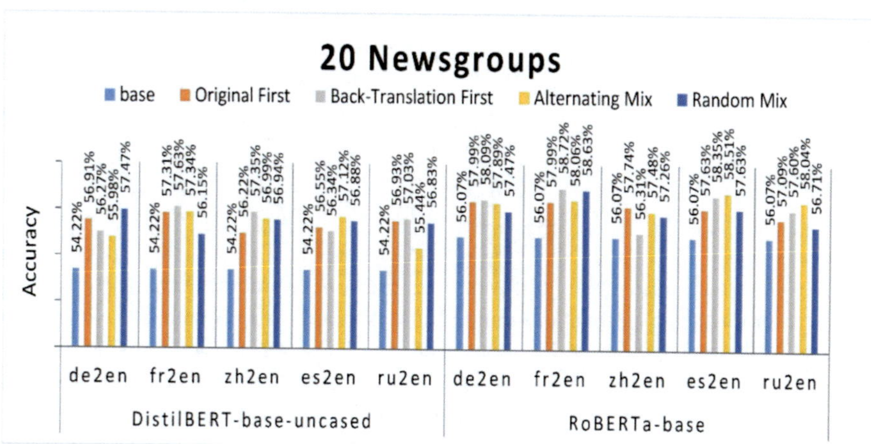

Fig. 7. Accuracy of 20 Newsgroups with Different Mixing Strategies on DistilBERT-base-uncased and RoBERTa-base.

Comprehensive Analysis

- **The Gains from Back-Translation Augmentation Exhibit a Clear Boundary.**

Table 2. Improvement ratios of classification accuracy after Chinese-English back-translation augmentation for the n5-n40 datasets.

Model	Datasets	SST-5	AG News	20 Newsgroups
DistilBERT-base-uncased	n5	0.25%	10.97%	7.03%
	n10	0.55%	5.47%	4.19%
	n20	– 0.95%	0.37%	– 0.75%
	n40	– 0.55%	0.69%	0.16%
RoBERTa-base	n5	4.88%	24.97%	2.84%
	n10	1.28%	4.69%	1.09%
	n20	0.75%	0.09%	0.69%
	n40	– 0.25%	– 0.84%	– 0.12%

Table 3. Improvements in macro-average F1-score after Chinese-English back-translation augmentation for the n5-n40 dataset.

Model	Datasets	SST-5	AG News	20 Newsgroups
DistilBERT-base-uncased	n5	0.042	0.157	0.078
	n10	0.018	0.054	0.066
	n20	– 0.009	0.005	– 0.003
	n40	– 0.002	0.007	0.003
RoBERTa-base	n5	0.131	0.321	0.046
	n10	0.011	0.049	0.013
	n20	0.005	0.001	0.011
	n40	– 0.001	– 0.008	– 0.001

Table 2 lists the improvement ratios of classification accuracy after back-translation augmentation for the n5-n40 dataset configurations, while Table 3 details the improvement data of the macro-average F1-score following back-translation augmentation for these dataset configurations. All the above data are exemplified using Chinese-English back-translation augmentation; other languages exhibit similar trends. The data clearly show that the positive effect of back-translation augmentation gradually weakens with an increasing number of original records. Once the number of original records per class reaches 40, back-translation augmentation offers little to no further performance improvement.

- **The Choice of Back-Translation Language Considerably Influences Accuracy Improvement.**

Table 4. Accuracy improvement ratios (relative to baseline) of various language back-translation augmentations.

Model	Datasets	de2en	fr2en	zh2en	es2en	ru2en
DistilBERT-base-uncased	SST-5	– 0.25%	– 1.51%	– 0.18%	0.09%	– 0.71%
	AG News	3.63%	3.02%	4.38%	3.04%	3.91%
	20 Newsgroups	2.44%	2.89%	2.65%	2.50%	2.33%
RoBERTa-base	SST-5	1.79%	1.66%	1.66%	1.39%	2.28%
	AG News	7.09%	6.89%	7.23%	**7.79%**	7.05%
	20 Newsgroups	1.79%	2.28%	1.13%	1.96%	1.29%

Table 4 lists the accuracy improvement ratios of various language back-translation augmentations relative to the original datasets. The reported accuracy for back-translation augmentation is the average obtained across all mixing strategies for the n5-n40 dataset configurations. The data indicate that back-translation augmentation can significantly enhance model performance. Compared to the baseline performance on the original datasets, the maximum improvement is 7.79%. Furthermore, the results highlight that different back-translation languages have a considerable impact on the degree of performance improvement.

- **The Impact of Different Data Mixing Strategies on Accuracy is Measurable.**

Table 5. Improvement ratios of Chinese-English back-translation augmentation under different mixing strategies.

Model	Data mixing methods	SST-5	AG News	20 Newsgroups
DistilBERT-base-uncased	Original first	−0.35%	3.63%	2.00%
	Back-translation first	0.28%	4.19%	3.13%
	Alternating mix	−0.17%	4.53%	2.77%
	Random mix	−0.45%	5.16%	2.72%
	Maximum difference	0.73%	1.53%	1.13%
RoBERTa-base	Original first	1.08%	7.94%	1.67%
	Back-translation first	0.95%	6.91%	0.24%
	Alternating mix	2.33%	6.28%	1.41%
	Random mix	2.30%	7.78%	1.19%
	Maximum difference	1.38%	**1.66%**	1.43%

Table 5 lists the improvement ratios of Chinese-English back-translation augmentation under different mixing strategies. This accuracy is the average obtained across all language back-translations for the n5-n40 dataset configurations. The data suggest that different data mixing strategies do exert a measurable impact on performance, with a maximum observed difference of 1.66%.

5 Discussion and Conclusion

5.1 Key Findings

- **The Effectiveness Boundary of Back-Translation Augmentation.**

The experimental results clearly demonstrate an effectiveness boundary for back-translation augmentation. Significant performance gains are achieved when the number of samples per class is less than 20. The benefits diminish when the sample count is between 20 and 40. Once the sample count reaches 40 or more, the gains from back-translation become negligible or even negative.

- **Matching Datasets with Appropriate Back-Translation Languages and Models.**

The performance improvement from back-translation is limited on the SST-5 dataset, particularly yielding poor results with the DistilBERT-base-uncased model. This might be attributed to the colloquial nature of the texts in this dataset, where back-translated versions may fail to effectively augment emotional expression. In contrast, on the AG News dataset, back-translation demonstrates its expected effectiveness. On the 20 Newsgroups dataset, however, the magnitude of performance improvement is highly correlated with the adopted model. Therefore, selecting a suitable back-translation language requires careful consideration tailored to both the specific dataset and the adopted model.

- **Analysis of Data Mixing Strategies.**

Theoretically, random mixing could encourage the model to treat all samples equally, potentially learning more robust features and enhancing generalization. However, our findings indicate that among the three datasets tested, no single mixing strategy consistently outperforms the others. Different strategies show relative advantages depending on the dataset, suggesting that the optimal mixing approach is context-dependent.

- **Matching Different Datasets with Suitable Models.**

Overall, more advanced models tend to achieve better results. RoBERTa-base outperforms DistilBERT-base-uncased on the SST-5 and AG News datasets. However, on the 20 Newsgroups dataset, DistilBERT-base-uncased achieves even better performance. Therefore, appropriate models must be selected based on the specific dataset characteristics.

5.2 Limitations and Future Work

- **Negative Impact of Back-Translation Noise**

Back-translation inevitably introduces noise [12]. The observed phenomenon where accuracy decreases for the back-translation mixed dataset compared to the original dataset when the sample count exceeds 40 strongly suggests the introduction of detrimental noise. This is particularly evident in the SST-5 tests, where gains are limited and turn negative as sample size increases, implying significant noise introduction for this specific dataset. A promising direction for future work is to employ semantic similarity models to filter out low-quality back-translated samples [13], thereby reducing noise and potentially improving model performance.

- **Limitations in Language Coverage**

This study tested only five high-resource languages, excluding many low-resource language families from regions such as Africa and South Asia. Future research could incorporate linguistic distance metrics to guide the selection of back-translation languages, aiming to maximize diversity gains [14] and explore the impact of linguistically distant languages on augmentation quality.

5.3 Conclusion

This paper systematically validates the value of multilingual back-translation in low-resource text classification and explores its boundaries of effectiveness. Through the construction and testing of 84 training datasets for each base dataset and subsequent data analysis, we demonstrate that:

Within a specific boundary, back-translation can significantly enhance low-resource classification performance, with accuracy improvements of up to 27.12%. However, beyond this boundary, the added benefits rapidly decline and may even become negative due to the introduction of noise.

The choice of back-translation language needs to be aligned with the intrinsic characteristics of the task and dataset to achieve maximum improvement.

Selecting an appropriate data mixing strategy can help maximize the benefits of back-translation. The random mix strategy is not universally optimal, and careful consideration is needed to avoid potential sequential bias inherent in other strategies.

In general, more advanced models tend to yield greater benefits, but this is not absolute; earlier models well-matched to the dataset can also achieve substantial benefits.

References

1. Tunstall, L. et al.: Efficient Few-Shot Learning Without Prompts (Version 1) (2022). arXiv. https://doi.org/10.48550/ARXIV.2209.11055
2. Sennrich, R., Haddow, B., Birch, A.: Improving Neural Machine Translation Models with Monolingual Data (Version 4) (2015). arXiv. https://doi.org/10.48550/ARXIV.1511.06709
3. Body, T., Tao, X., Li, Y., Li, L., Zhong, N.: Using back-and-forth translation to create artificial augmented textual data for sentiment analysis models. Expert Syst. Appl. **178**, 115033 (2021). https://doi.org/10.1016/j.eswa.2021.115033
4. Ma, J., Li, L.: Data augmentation for Chinese text classification using Back-translation. J. Phys. Conf. Ser. **1651**(1), 012039 (2020). https://doi.org/10.1088/1742-6596/1651/1/012039
5. Ciolino, M., Noever, D., Kalin, J.: Back Translation Survey for Improving Text Augmentation (Version 2) (2021). arXiv. https://doi.org/10.48550/ARXIV.2102.09708

6. Li, Z., Specia, L.: Improving Neural Machine Translation Robustness Via Data Augmentation: Beyond Back Translation (2019). arXiv. https://doi.org/10.48550/ARXIV.1910.03009
7. Lee, S., Liu, L., Choi, W.: Iterative translation-based data augmentation method for text classification tasks. IEEE Access. **9**, 160437–160445 (2021). https://doi.org/10.1109/access.2021.3131446
8. Singh, J., McCann, B., Keskar, N. S., Xiong, C., Socher, R.: XLDA: Cross-Lingual Data Augmentation for Natural Language Inference and Question Answering (Version 1) (2019). arXiv. https://doi.org/10.48550/ARXIV.1905.11471
9. Edwards, C., Lai, T., Ros, K., Honke, G., Cho, K., Ji, H.: Translation between Molecules and Natural Language (Version 3) (2022). arXiv. https://doi.org/10.48550/ARXIV.2204.11817
10. Ochal, M., Patacchiola, M., Storkey, A., Vazquez, J., Wang, S.: Few-Shot Learning with Class Imbalance (Version 2) (2021). arXiv. https://doi.org/10.48550/ARXIV.2101.02523
11. Xu, N., et al.: Analysis of Back-translation methods for low-resource neural machine translation. In: Tang, J., Kan, M.Y., Zhao, D., Li, S., Zan, H. (eds.) Natural Language Processing and Chinese Computing. NLPCC 2019. Lecture Notes in Computer Science(), vol. 11839. Springer, Cham (2019). https://doi.org/10.1007/978-3-030-32236-6_42
12. Shi, Z., Lipani, A.: Rethink the Effectiveness of Text Data Augmentation: An Empirical Analysis (Version 1) (2023). arXiv. https://doi.org/10.48550/ARXIV.2306.07664
13. Helmeczi, R.K., Cevik, M., Yıldırım, S.: Few-shot learning for sentence pair classification and its applications in software engineering (Version 1) (2023). arXiv. https://doi.org/10.48550/ARXIV.2306.08058
14. Wei, D. et al.: Text Style Transfer Back-Translation (Version 1) (2023). arXiv. https://doi.org/10.48550/ARXIV.2306.01318

Integrating Knowledge Distillation Methods: A Sequential Multi-Stage Framework

Yinxi Tian[1], Changwu Huang[1(✉)], Ke Tang[1(✉)], and Xin Yao[2]

[1] Department of Computer Science and Engineering, Southern University of Science and Technology, Shenzhen 518055, China
{huangcw3,tangk3}@sustech.edu.cn

[2] School of Data Science, Lingnan University, Hong Kong, China

Abstract. Knowledge distillation (KD) transfers knowledge from large teacher models to compact student models, enabling efficient deployment on resource-constrained devices. While diverse KD methods–including response-based, feature-based, and relation-based approaches–capture different aspects of teacher knowledge, integrating multiple methods or knowledge sources is promising but often hampered by complex implementation, inflexible combinations, and catastrophic forgetting, limiting practical effectiveness. This work proposes SMSKD (Sequential Multi-Stage Knowledge Distillation), a flexible framework that sequentially integrates heterogeneous KD methods. At each stage, the student is trained with a specific distillation method, while a frozen reference model from the previous stage anchors learned knowledge to mitigate forgetting. Additionally, we introduce an adaptive weighting mechanism based on the teacher's true class probability (TCP) that dynamically adjusts the reference loss per sample to balance knowledge retention and integration. By design, SMSKD supports arbitrary method combinations and stage counts with negligible computational overhead. Extensive experiments show that SMSKD consistently improves student accuracy across diverse teacherstudent architectures and method combinations, outperforming existing baselines. Ablation studies confirm that stage-wise distillation and reference model supervision are primary contributors to performance gains, with TCP-based adaptive weighting providing complementary benefits. Overall, SMSKD is a practical and resource-efficient solution for integrating heterogeneous KD methods.

Keywords: Knowledge Distillation · Deep Neural Networks · Knowledge Integration

1 Introduction

Deep neural networks (DNNs) have driven major advances in artificial intelligence, achieving state-of-the-art performance in domains such as computer vision, natural language processing, and speech recognition [1]. However, these

L. Huang and D. Greenhalgh (Eds.): ICMLC 2026, LNNS 2061, pp. 991–1009, 2027.
https://doi.org/10.1007/978-3-032-31236-5_67

performance gains are often accompanied by substantial increases in model complexity, resulting in heavy computational and memory demands. This creates significant obstacles to real-world deployment, especially on resource-constrained platforms like mobile devices, IoT sensors, and edge computing environments [2]. To address this challenge, knowledge distillation (KD) has emerged as a powerful technique for model compression and acceleration [3].

The fundamental idea of KD is to transfer the knowledge of a large, well-trained model, referred to as the teacher model, to a lightweight model, known as the student model [4,5]. This approach enables the student model to achieve significant size reduction while maintaining competitive or even superior performance compared to the teacher. Since the pioneering work of Hinton *et al.* [4], KD has evolved rapidly, resulting in a diverse array of distillation algorithms tailored to different forms of transferable knowledge. Generally, KD methods can be classified based on the number of knowledge sources involved [6]: single-source KD and multi-source KD. For single-source KD, methods can be further categorized into three major paradigms according to the type of knowledge transferred [5]: (1) Response-based KD, which encourages the student to mimic the teacher's output distributions or final predictions [4,7]; (2) Feature-based KD, which aims to align the intermediate feature representations between teacher and student models [8,9]; and (3) Relation-based KD, which encourages the student to learn structural relationships and interactions among data samples or feature vectors from the teacher [10].

Recent studies have begun to investigate the similarities and differences among existing distillation approaches [11,12]. For instance, Ojha *et al.* [11] systematically examined three representative methods ([4,8,13]) across four scenarios to evaluate knowledge inheritance capabilities. Similarly, Zheng *et al.* [12] compared response-based and feature-based distillation in object detection, revealing that even when accuracies are comparable, the learning behaviors between different methods can differ substantially. These findings suggest that different distillation methods may capture complementary aspects of the teacher's knowledge, motivating researchers to explore the integration of multiple knowledge sources in KD to leverage diverse forms of teacher information for more effective student training[11,12]. For example, some works combine response-based and feature-based distillation losses, or fuse information from different teacher layers [8,14]. Although multi-source approaches often yield performance improvements, simply combining different knowledge sources does not guarantee optimal results, as the interactions between knowledge sources are often complex and may lead to suboptimal results without careful design and balancing [15].

However, existing multi-source or hybrid approaches suffer from several limitations. First, many hybrid methods impose constraints on the types of knowledge that can be combined, such as restricting integration to different layers or only allowing certain knowledge categories to be integrated [6,13]. This limits flexibility and may prevent the effective combination of, for instance, feature-based and relation-based knowledge. Second, current integration techniques often require sophisticated modifications to the distillation algorithms, increasing implementation complexity [6]. Simply summing different distillation losses can be suboptimal, as the scales of these losses may differ significantly, potentially leading to imbalanced optimization and degraded performance [15]. Moreover, the lack of mechanisms to preserve or balance knowledge acquired through different sources during the training of student model can result in catastrophic forgetting, further reducing the effectiveness of hybrid distillation approaches.

To address these challenges, this work proposes a Sequential Multi-Stage Knowledge Distillation (SMSKD) framework that flexibly integrates multiple distillation methods in a stage-wise manner. In our framework, the student model is trained sequentially over multiple stages, with each stage employing a potentially different distillation method, thereby leveraging multiple distillation methods collaboratively. Crucially, to mitigate the issue of knowledge forgetting when switching between distillation methods, we introduce a frozen reference model–copied from the student at the end of each stage–that serves as an anchor for the subsequent stage. During later stages, a reference loss term is added to the loss function of the distillation method adopted in the current stage to encourage the student model to retain the knowledge acquired in previous stages, thereby alleviating catastrophic forgetting and enabling synergistic integration of diverse distillation techniques. This design allows the framework to harness the strengths of various distillation methods while maintaining a stable learning trajectory for the student model. Comprehensive experiments across multiple teacher-student configurations and datasets are conducted to validate the effectiveness of our proposed framework. Results demonstrate that our approach consistently outperforms single-source distillation and existing multi-source approaches, achieving superior student model performance.

The remainder of this paper is organized as follows: Sect. 2 reviews related work on single-source and multiple-source KD methods. Section 3 presents our proposed sequential multi-stage framework for integrating distillation methods. Section 4 provides comprehensive experimental studies to validate the effectiveness of the proposed method. Finally, Sect. 5 briefly concludes the paper.

2 Background

This section reviews various existing distillation methods, which lay the foundation for our work. First, several single-source distillation methods are introduced including response-based, feature-based, and relation-based distillation methods. Then, some recent work on multi-source distillation is presented.

2.1 Notation

This work focuses on supervised classification tasks, which aim to learn a model or classifier that maps the input space $\mathcal{X} \subset \mathbb{R}^d$ to the label space $\mathcal{Y}$, where $|\mathcal{Y}| = K$ denotes the number of classes. Let $f : \mathcal{X} \to \mathcal{Y}$ denote a model trained on a dataset $\mathcal{D} = \{(\mathbf{x}_i, y_i)\}_{i=1}^{N}$, where $\mathbf{x}_i \in \mathcal{X}$ and $y_i \in \mathcal{Y}$. Given an input feature vector $\mathbf{x} \in \mathcal{X}$, the model produces a predictive label $\hat{y}$. Typically, a DNN model f outputs a vector of logits $\mathbf{z} = [z_1, \ldots, z_K]$ for the input vector $\mathbf{x}$, which are then converted into class probabilities $\mathbf{p} = [p_1, \cdots, p_K]$ via the softmax function: $p_k = \exp(z_k/\tau) / \sum_{j=1}^{K} \exp(z_j/\tau)$ for $k = 1, \ldots, K$, where τ is a temperature parameter (usually set to 1). The predicted label is typically determined as the class with the highest probability, i.e., $\hat{y} = \arg\max_k \{p_k\}$.

For clarity, throughout this work, we use f^T to denote the teacher model. For a given input feature vector $\mathbf{x}$, we denote its output logits, predictive class probabilities, and the output of its i-th intermediate feature layer as $\mathbf{z}^T$, $\mathbf{p}^T$, and $\mathbf{h}_i^T$, respectively. Similarly, the student model and its corresponding outputs are denoted as f^S, $\mathbf{z}^S$, $\mathbf{p}^S$, and $\mathbf{h}_i^S$. Given a training dataset $\mathcal{D}$ and a teacher model f^T pre-trained on $\mathcal{D}$, the objective of KD is to train a smaller and more efficient student model f^S under the guidance of the teacher model. The aim is for the student to inherit the performance and generalization capability of the teacher while achieving reductions in model size and computational cost.

2.2 Single-Source Knowledge Distillation

Knowledge distillation aims to transfer knowledge from a large teacher model to a compact student model. Existing single-source distillation methods can be broadly categorized into three types according to the knowledge they exploit: response-based, feature-based, and relation-based [5].

2.2.1 Response-Based Distillation Methods Response-based distillation represents the earliest and most straightforward approach in KD, focusing on aligning the predicted logits or probabilities of the student model with those of the teacher. The vanilla KD method proposed by Hinton et al. [4] is the most widely adopted response-based distillation technique, which matches the soft targets or probability distributions over classes produced by the teacher and student models via KL divergence. The distillation loss for the vanilla KD method [4] is formulated as follows:

$$\mathcal{L}_{\mathrm{KD}}(\mathbf{p}^T, \mathbf{p}^S) = \tau^2 \cdot \mathrm{KL}(\mathbf{p}^T \parallel \mathbf{p}^S) . \tag{1}$$

More recently, the decoupled KD (DKD) [16] separates contributions from target and non-target classes, allowing more flexible control:

$$\mathcal{L}_{\text{DKD}}(\mathbf{p}^T, \mathbf{p}^S) = \alpha\, \mathcal{L}_{\text{TCKD}} + \beta\, \mathcal{L}_{\text{NCKD}}, \tag{2}$$

with

$$\mathcal{L}_{\text{TCKD}} = \tau^2 \cdot \text{KL}\big(p_y^T \,\|\, p_y^S\big), \mathcal{L}_{\text{NCKD}} = \tau^2 \cdot \text{KL}\big(\mathbf{p}_{\setminus y}^T \,\|\, \mathbf{p}_{\setminus y}^S\big), \tag{3}$$

where p_y^T and p_y^S denote the teacher's and student's probabilities for the ground-truth class, and $\mathbf{p}_{\setminus y}^T$ and $\mathbf{p}_{\setminus y}^S$ represent the distributions over remaining classes. The key advantage of DKD is that it allows independent weighting of these two components via α and β, in contrast to vanilla KD, in which they are coupled.

Overall, response-based methods primarily focus on aligning the output distributions of the teacher and student models, thereby confining knowledge transfer to the output layer. As a result, these approaches may overlook valuable information encoded in intermediate representations or model features.

2.2.2 Feature-Based Distillation Methods Feature-based distillation transfers knowledge through intermediate representations rather than solely relying on final predictions. FitNets [8] aligns hidden activations to guide the student. Attention transfer (AT) [14] emphasizes spatial importance, and variational information distillation (VID) [17] models features probabilistically to capture richer information. The core idea can be expressed generally as:

$$\mathcal{L}_{\text{feature}} = \mathcal{D}_{\text{feat}}\big(\mathcal{F}(\mathbf{h}^{\mathbf{T}}),\, \mathcal{F}(\mathbf{h}^{\mathbf{S}})\big), \tag{4}$$

where $\mathcal{F}(\cdot)$ denotes the feature transformation used by a particular method, and $\mathcal{D}_{\text{feat}}$ measures the discrepancy between transformed teacher and student features (e.g., ℓ_2 distance, attention map matching, or KL divergence over feature distributions). Feature-based methods guide the student to imitate the teacher's intermediate representations, improving feature-space alignment and enhancing representation quality. These signals complement response-based approaches by providing information unavailable from final predictions alone.

2.2.3 Relation-Based Distillation Methods Relation-based distillation focuses on preserving the structural relationships among samples, rather than aligning individual representations. These methods characterize the teacher's feature space through pairwise or higher-order relations and encourage the student to reproduce the same relational geometry. For example, relational knowledge distillation (RKD) [10] aligns pairwise distances and angles, probabilistic knowledge transfer (PKT) [18] matches similarity distributions, correlation congruence (CC) [19] preserves feature correlations, and contrastive representation distillation (CRD) [13] employs a contrastive objective over instance pairs in a projected embedding space. Despite their methodological diversity, these approaches share a common objective: transferring the relational structure encoded in the teacher's representations. This can be abstracted as

$$\mathcal{L}_{\text{relation}} = \mathcal{D}_{\text{struct}}\big(\mathcal{R}(\mathbf{h}^T),\, \mathcal{R}(\mathbf{h}^S)\big), \tag{5}$$

where $\mathcal{R}(\cdot)$ extracts structural information–such as distances, similarities, correlations, or contrastive embeddings, and $\mathcal{D}_{\text{struct}}$ measures the discrepancy between the teacher's and student's relational structures.

2.3 Multi-source Knowledge Distillation

While most distillation methods are developed as standalone techniques, recent work has begun exploring how to leverage multiple knowledge sources. Multi-teacher distillation [20] represents one such direction, where multiple pre-trained teachers provide diverse knowledge to a single student. Each teacher may specialize in different aspects of the task or employ different architectures, enabling the student to acquire richer and more comprehensive information. However, this approach fundamentally differs from our work: multi-teacher distillation requires multiple pre-trained models, whereas our framework uses a single teacher with multiple distillation strategies to extract different facets of knowledge.

Recently, Song *et al.* [6] proposed the spot-adaptive knowledge distillation (SAKD) method, which integrates teacher and student models into a multi-path routing network. The method employs adaptation layers to align intermediate or penultimate representations and uses a policy network to make per-sample routing decisions at designated branch spots. This enables selective knowledge transfer from the most informative locations within the teacher network.

However, SAKD faces critical limitations. First, it is limited to combining two methods from different sources, one from intermediate layers and the other from the output layer. It cannot fuse multiple methods or two methods from the same source, which restricts the diversity of possible combinations. Second, as noted by Huang et al. [15], feature-based and response-based distillation possess inherently different learning objectives and gradient characteristics. Directly combining them without proper adaptation can lead to optimization conflicts and suboptimal performance. These limitations motivate our SMSKD, which tackles the challenges of combining multiple distillation methods via a stage-wise multi-source approach with careful design of complementary knowledge integration.

3 Sequential Multi-stage Knowledge Distillation

This work proposes a novel Sequential Multi-Stage Knowledge Distillation (SMSKD) framework that enables flexible and effective integration of multiple KD techniques in a stage-wise manner. Unlike existing KD approaches that apply a single knowledge source or combine multiple sources simultaneously to train a student model, SMSKD decomposes the training of the student model into a series of sequential stages. Each stage employs a potentially different KD method, allowing the student to progressively assimilate diverse types of knowledge from the teacher model. To address the challenge of catastrophic forgetting when switching KD methods, SMSKD introduces a frozen reference model at each stage transition, which serves as an anchor for knowledge retention. Additionally, a confidence-based weighting mechanism is employed to adaptively modulate the influence of the reference supervision, ensuring robust and reliable knowledge transfer across stages. This design allows the student model to progressively accumulate and integrate complementary knowledge from diverse KD techniques while maintaining a stable learning trajectory.

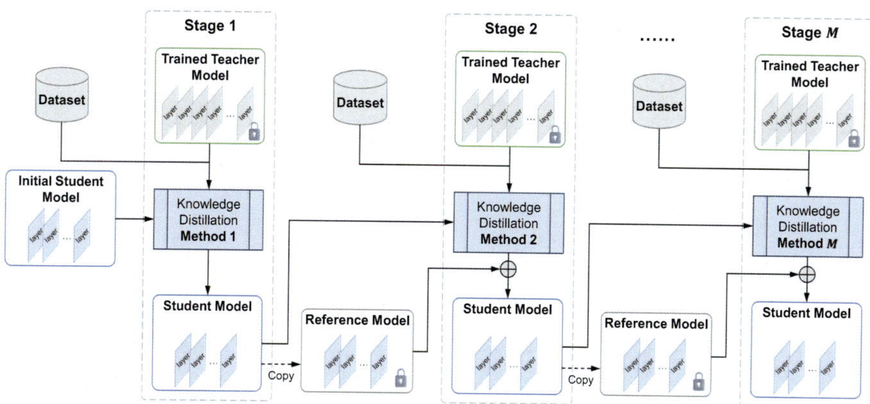

Fig. 1. Overview of the sequential multi-stage knowledge distillation framework

A schematic overview of the SMSKD workflow is shown in Fig. 1 and the detailed steps are summarized in Algorithm 1. The SMSKD framework organizes the training of the student model into multiple consecutive stages, each characterized by a specific KD method and learning objective. Given the training dataset $\mathcal{D} = \{(\mathbf{x}_i, y_i)\}_{i=1}^{N}$, the teacher model f^T pre-trained on $\mathcal{D}$, and the student model f^S to be trained, the SMSKD framework proceeds through M sequential stages, each characterized by a chosen KD method KD_m and a specified number of training epochs T_m. The detailed stage-wise distillation process of SMSKD is described as follows.

Algorithm 1 Sequential Multi-Stage Knowledge Distillation

Require: Training dataset $\mathcal{D} = \{(\mathbf{x}_i, y_i)\}_{i=1}^{N}$, teacher model f^T trained on $\mathcal{D}$, student model f^S, total number of KD stages M, corresponding KD methods adopted in each stage $\{\text{KD}_1, \cdots, \text{KD}_M\}$, duration (epochs) for each stage $\{T_1, \cdots, T_M\}$, batch size b, classification loss weight λ_c, reference loss weight λ_r.

Initialize the student model f^S.
for $m = 1, \cdots, M$ **do**
  for $t = 1, \cdots, T_m$ **do**
    for Batch B in DataLoader($\mathcal{D}$) **do**
      for $(\mathbf{x}_i, y_i) \in B$ **do**
        if $m = 1$ **then**
          Compute loss $\mathcal{L}(\mathbf{x}_i)$ on $\mathbf{x}_i$ with Eq. (6).
        else
          Compute loss $\mathcal{L}(\mathbf{x}_i)$ on $\mathbf{x}_i$ with Eq. (7).
        end if
      end for
      $\mathcal{L}_{\text{batch}} = \frac{1}{|B|}\sum_{(\mathbf{x}_i, y_i)\in B}\mathcal{L}(\mathbf{x}_i)$.
      Update the parameters of f^S using $\nabla\mathcal{L}_{\text{batch}}$.
    end for
  end for
  Update the reference model $f^R \leftarrow f^S$.
end for

Ensure: Student model f^S.

3.1 Initial Stage

In the first stage (i.e., $m = 1$), the student model f^S is trained using a conventional KD method (e.g., response-based, feature-based, relation-based, or other multi-source KD methods), optimizing a composite loss function $\mathcal{L}_{\text{Stage}}$ that combines the distillation loss with the standard cross-entropy classification loss:

$$\mathcal{L}_{\text{Stage}} = \mathcal{L}_{\text{Distill}} + \lambda_c \cdot \mathcal{L}_{\text{Cls}}, \tag{6}$$

where $\mathcal{L}_{\text{Distill}}$ denotes the distillation loss specific to the chosen method (that is, for instance, if the vanilla KD method is used, $\mathcal{L}_{\text{Distill}}$ is the loss function defined in Eq. 1), $\mathcal{L}_{\text{Cls}}$ is the standard classification loss (typically the cross-entropy between the predicted and ground-truth labels), and λ_c is the weight for the classification loss.

After completing the first stage, the trained student model f^S is copied and frozen as the reference model f^R, which will be used in the next stage to anchor subsequent learning.

3.2 Subsequent Stages

In each subsequent stage ($m = 2, \cdots, M$), the student model is continuously trained using a potentially different KD method. To mitigate catastrophic forgetting of knowledge acquired in previous stages, an adaptive weighted reference loss term $\mathcal{L}_{\text{AdaRef}}$ is introduced, leveraging the frozen reference model from the preceding stage. The total loss $\mathcal{L}_{\text{Stage}}$ for each subsequent stage is defined as:

$$\mathcal{L}_{\text{Stage}} = \mathcal{L}_{\text{Distill}} + \lambda_c \cdot \mathcal{L}_{\text{Cls}} + \lambda_r \cdot \mathcal{L}_{\text{AdaRef}}, \tag{7}$$

where $\mathcal{L}_{\text{Distill}}$ denotes the distillation loss specific to the chosen method, $\mathcal{L}_{\text{Cls}}$ is the standard classification loss, λ_c is the weight for the classification loss, $\mathcal{L}_{\text{AdaRef}}$ is the newly introduced adaptive weighted reference loss that is designed to prevent forgetting of knowledge acquired in previous stages, and λ_r controls the strength of the reference supervision.

As different KD methods may be employed across stages and their learning behaviors differ, acquiring new knowledge from the teacher in the current stage may lead the student to forget knowledge learned from prior stages. Thus, mitigating knowledge forgetting or preserving previously acquired knowledge is crucial in our framework and in other sequential learning settings.

To address this issue, the reference model, inspired by the use of fixed baselines in reinforcement learning (e.g., Proximal Policy Optimization [21]), is introduced in our SMSKD framework and acts as a stabilizing anchor that constrains the optimization trajectory of the student model. By incorporating a reference loss, SMSKD encourages the student to remain close to its previous state, thereby preserving useful knowledge and preventing abrupt forgetting when switching distillation strategies. The reference loss $\mathcal{L}_{\text{Ref}}$ can be computed as the Kullback-Leibler (KL) divergence between the output distributions of the student and the reference model,

$$\mathcal{L}_{\text{Ref}} = \text{KL}\left(\mathbf{p}^S \parallel \mathbf{p}^R\right), \tag{8}$$

where $\mathbf{p}^S$ and $\mathbf{p}^R$ are the output distributions of the student model f^S and reference model f^R, respectively.

However, the reference model f^R may not always provide perfectly accurate or reliable guidance, especially in the early stages. SMSKD employs the true class probability (TCP) [22] as a confidence-based weighting mechanism. For each sample $\mathbf{x}$, the TCP is defined as the reference model's predicted probability for the ground-truth class. Higher TCP values indicate greater confidence, and thus the reference supervision is emphasized for such samples. This adaptive weighting reduces the risk of transferring and preserving erroneous knowledge from the reference model. Thus, the reference loss adaptively weighted by the reference model's confidence (i.e., TCP) is defined as,

$$\mathcal{L}_{\text{AdaRef}} = \text{TCP} \cdot \mathcal{L}_{\text{Ref}} = p^R_{k=y} \cdot \text{KL}\left(\mathbf{p}^S \parallel \mathbf{p}^R\right), \tag{9}$$

where $p^R_{k=y}$ is the reference model's predicted probability for the ground-truth class y for sample $\mathbf{x}$.

At the end of each stage, the updated student model is again copied and frozen as the new reference model for the next stage. After completing all training stages, the final student model is obtained as the output. By decomposing the distillation process into multiple sequential stages and leveraging both diverse KD methods and reference model anchoring, SMSKD enables the student model to progressively accumulate, retain, and integrate complementary knowledge. The adaptive reference loss further enhances stability and mitigates catastrophic forgetting, resulting in improved performance and robustness.

4 Experimental Study

This section comprehensively evaluates the effectiveness of SMSKD framework. The experiments begins with extensive comparisons against state-of-the-art multi-source KD methods, particularly SAKD and Direct Loss Aggregation (DLA), to demonstrate the superiority of our sequential approach. We then explore SMSKD's flexibility by integrating diverse combinations of KD methods that existing approaches cannot handle, validating its constraint-free design. Through detailed ablation studies, we analyze the contribution of each component in our framework, including the reference loss mechanism and the sequential training strategy. Additionally, we investigate the robustness of SMSKD through hyperparameter sensitivity analysis and examine how the number of stages affects its performance. All experiments are conducted on an NVIDIA A100 system with PyTorch. To ensure statistical reliability, each experiment is repeated three times with different random seeds, and the average results are reported.

4.1 Experimental Setting

4.1.1 Datasets We conduct experiments on two widely used image classification benchmarks: CIFAR-100[23] and Tiny ImageNet[24]. CIFAR-100 consists of 60,000 32×32 color images distributed across 100 classes, with 600 images per class. The dataset is split into 50,000 training images and 10,000 test images. Following standard practice, we apply data augmentation: 4-pixel zero-padding, random 32×32 cropping, and random horizontal flipping. Tiny ImageNet is a subset of ImageNet containing 200 classes, with 500 training, 50 validation, and 50 test images per class. All images are resized to 64×64 pixels, and we apply only random horizontal flipping for augmentation. These preprocessing strategies align with prior works [6,13] to ensure fair comparison while preventing overfitting.

4.1.2 Model Architectures To comprehensively evaluate our method, we employ three different teacher-student architecture pairs covering various network families: 1) **ResNet family** [25]: ResNet56 (teacher) → ResNet20 (student); 2) **WideResNet (WRN) family** [26]: WRN40_2 (teacher) → WRN16_2 (student); and 3) **Visual Geometry Group (VGG) family** [27]: VGG13 (teacher) → VGG8 (student). These pairs span varying compression ratios and architectures, enabling a robust assessment of our method's generalizability.

4.1.3 Compared Methods We carefully select eight representative distillation methods covering diverse knowledge types to be integrated into our framework: (1) Response-based method: vanilla KD [4] (denoted as KD hereafter for simplicity); (2) Feature-based methods: AT (Attention Transfer) [14], FitNets [8], and VID (Variational Information Distillation) [17]; and (3) Relation-based methods: RKD (Relational Knowledge Distillation) [10], CC (Correlation Congruence) [19], CRD (Contrastive Representation Distillation) [13], and PKT (Probabilistic Knowledge Transfer) [18].

For comparing with multi-source or integrated KD methods, we consider two baseline approaches: (1) **SAKD** [6]: A state-of-the-art adaptive knowledge integration framework that dynamically selects distillation positions during training. It represents the current best practice in multi-source knowledge integration; and (2) **Direct Loss Aggregation (DLA)**: A straightforward baseline that directly sums loss terms from different distillation methods and jointly optimizes them in a single training stage, representing the common practice in prior works without coordination mechanisms.

4.1.4 Implementation Details To ensure fair comparison, we strictly follow the training protocol established by CRD [13] and adopted by SAKD [6]. All models are trained for 240 epochs using SGD with momentum 0.9, batch size 64, and initial learning rate 0.05. The learning rate is decayed by a factor of 10 at epochs 150, 180, and 210 during the total 240 training epochs. Weight decay is set to 5×10^{-4}, and the distillation temperature τ is fixed at 4.

For individual KD methods, we use their default hyperparameters λ_c as specified in [13]. For our SMSKD framework, we primarily evaluate a two-stage configuration with the following settings: reference loss weight $\lambda_r = 0.5$, classification loss weights λ_c consistent with individual method settings, stage 1 duration $T_1 = 150$ epochs (until the first learning rate decay), and stage 2 duration $T_2 = 90$ epochs (resulting in a total of $T_1 + T_2 = 240$ epochs). The rationale for these hyperparameter choices is validated through ablation studies and sensitivity analyses presented in subsequent subsections.

4.2 Comparative Experiments

4.2.1 Comparison with Multi-source Distillation Methods We evaluate the effectiveness of our proposed integration strategy in comparison with SAKD [6] and DLA approach. Experiments are conducted on the CIFAR-100 and Tiny ImageNet using three teacher-student architectures: ResNet56→ResNet20 and WRN40_2→WRN16_2, ResNet56→ResNet20, and VGG13→VGG8.

Table 1 presents the results on CIFAR-100 using ResNet56→ResNet20 and WRN40_2→WRN16_2 architectures. Table 2 reports the results on Tiny ImageNet using ResNet56→ResNet20 and VGG13→VGG8 architectures. From the results, it can be observed that all three integration strategies consistently outperform their corresponding single-type distillation baselines across all architectures. This finding indicates that combining multiple distillation objectives yields

Table 1. Test accuracy (%) of student models on CIFAR-100. *Standard* denotes the original accuracy of the corresponding method. For multi-source methods (DLA, SAKD, SMSKD), each integrates the Stage 1 distillation method (listed in the first column) with vanilla KD (Stage 2), and final student's accuracy is reported. Values in parentheses denote the accuracy improvements over the original model

Method	ResNet56 → ResNet20				WRN40_2 → WRN16_2			
	Standard	DLA	SAKD	SMSKD	Standard	DLA	SAKD	SMSKD
Teacher	73.29	-	–	–	75.61	–	–	–
Student	69.15	–	–	–	71.98	–	–	–
KD	70.35	–	–	–	74.18	–	–	–
FitNets	69.42	70.68 (+1.26)	71.45 (+2.03)	**71.74** (+2.32)	73.96	75.27 (+1.31)	75.37 (+1.41)	**75.61** (+1.65)
AT	70.36	71.42 (+0.79)	71.50 (+0.87)	**72.16** (+1.53)	74.38	75.53 (+1.15)	75.41 (+1.03)	**75.97** (+1.59)
VID	70.38	71.00 (+0.62)	71.36 (+0.98)	**72.03** (+1.65)	74.47	75.05 (+0.58)	74.93 (+0.46)	**75.98** (+1.51)
CC	69.82	70.59 (+0.77)	**71.84** (+2.02)	71.53 (+1.71)	73.89	**75.67** (+1.78)	75.31 (+1.42)	75.61 (+1.72)
RKD	69.85	**71.80** (+1.95)	71.29 (+1.47)	71.12 (+1.27)	73.45	**75.33** (+1.88)	75.18 (+1.73)	74.99 (+1.54)
CRD	71.89	71.90 (+0.01)	71.98 (+0.09)	**72.42** (+0.53)	75.49	75.33 (-0.16)	75.18 (-0.31)	**76.02** (+0.53)
PKT	71.03	71.54 (+0.51)	71.69 (+0.66)	**71.87** (+0.84)	74.80	74.81 (+0.01)	74.98 (+0.18)	**75.09** (+0.29)

Table 2. Test accuracy (%) of student models on Tiny ImageNet. *Standard* denotes the original accuracy of the corresponding method. For multi-source methods (DLA, SAKD, SMSKD), each integrates the Stage 1 distillation method (listed in the first column) with vanilla KD (Stage 2), and final student's accuracy is reported. Values in parentheses denote the accuracy improvements over the original model

Method	ResNet56 → ResNet20				VGG13 → VGG8			
	Standard	DLA	SAKD	SMSKD	Standard	DLA	SAKD	SMSKD
Teacher	56.51	-	-	-	60.38	-	-	-
Student	51.12	-	-	-	55.06	-	-	-
KD	53.19	-	-	-	58.27	-	-	-
FitNets	50.95	53.13 (+2.18)	**54.37** (+3.42)	53.54 (+2.95)	56.38	59.09 (+2.71)	58.65 (+2.27)	**59.18** (+2.80)
AT	53.67	54.39 (+0.72)	54.35 (+0.68)	**54.91** (+1.24)	56.39	58.83 (+2.44)	58.90 (+2.51)	**59.19** (+2.80)
VID	52.80	53.58 (+0.78)	53.65 (+0.85)	**53.91** (+1.11)	55.93	58.51 (+2.58)	58.48 (+2.55)	**58.68** (+2.75)
CC	51.93	**54.55** (+2.62)	54.20 (+2.27)	53.36 (+1.43)	55.89	58.40 (+2.51)	58.39 (+2.50)	**58.42** (+2.53)
RKD	52.85	**54.30** (+1.45)	54.04 (+1.19)	53.03 (+0.18)	56.65	58.57 (+1.92)	58.66 (+2.01)	**58.71** (+2.06)
CRD	54.61	55.55 (+0.94)	55.33 (+0.72)	**55.83** (+1.22)	57.29	59.68 (+2.39)	**60.25** (+2.96)	60.19 (+2.90)
PKT	52.85	**54.77** (+1.92)	54.39 (+1.54)	53.80 (+0.95)	57.28	58.62 (+1.34)	58.65 (+1.37)	**59.78** (+2.50)

greater effectiveness than relying on a single one. Notably, our proposed SMSKD method consistently delivers competitive or superior performance across different architectures. In most cases on CIFAR-100, it achieves the highest performance gains among the three integration strategies, with variations observed in CC and RKD. SAKD performs better than DLA in most cases except for RKD, but still underperforms compared to our method overall. Under the CIFAR-100 WRN40_2→WRN16_2 setting, DLA and SAKD sometimes lead to fluctuating or even negative transfer (e.g., CRD degrading from 75.49 to 75.33/75.18), while SMSKD consistently improves performance for all base methods within this configuration. On Tiny ImageNet, while DLA occasionally shows stronger improvements on the ResNet56→ResNet20 architecture, SMSKD remains one of the most stable and effective choices. In particular, for the VGG-based architecture, SMSKD shows a clear advantage: it attains the best accuracy in 6 out of 7 distillation methods, with only a negligible gap under CRD (0.06%).

The comprehensive comparisons with integrated or multi-source distillation methods demonstrate that our proposed SMSKD approach achieves overall superior performance compared to the baseline methods. We attribute this superiority to several key factors. The DLA method simply aggregates heterogeneous distillation losses without addressing their differing optimization scales, which can negatively impact training dynamics. SAKD improves upon this by enabling layer-wise selection during distillation, but still does not explicitly account for the scale discrepancies among loss terms. In contrast, our SMSKD method mitigates these issues through a multi-stage training scheme that decouples the optimization of different loss scales and leverages a reference model to align the transfer of diverse types of knowledge more effectively and coherently. Importantly, the proposed method achieves these improvements without introducing any additional training cost, demonstrating its efficiency and practical value.

4.2.2 Effectiveness on Integrating Various Knowledge Sources In previous comparisons, only limited combinations of KD methods were integrated in our SMSKD framework and compared with SAKD and DLA methods, primarily due to SAKD's constraints – it supports combining only two methods from different sources including logits and features from the teacher model. In contrast, our proposed SMSKD framework has no such limitations on the methods to be included, since it employs different KD methods in a sequential manner rather than simultaneously. Therefore, in the following experiments, we investigate additional combinations that differ from those examined previously to further validate the effectiveness and flexibility of our method.

Figure 2 examines pairwise integrations under SMSKD on CIFAR-100. Each group compares the baselines of two individual methods with their SMSKD-integrated result. Across most combinations, SMSKD consistently outperforms the baselines, demonstrating that the stage-wise strategy effectively exploits complementary knowledge from different distillation objectives. A rare exception occurs in the combination of CRD and VID, where performance falls slightly below that of CRD alone. The likely cause is a conflict between CRD's contrastive objective, which disperses features, and VID's tendency to pull them toward the teacher distribution. A deeper investigation into such interactions is left for future work.

Notably, combining FitNets (feature-based) with KD (response-based) or CRD (relation-based) yields substantial gains of +2.32% and +2.50%, respectively, highlighting the benefit of integrating different knowledge types. Similarly, combining VID (feature-based) and KD achieves the accuracy of 72.03%, surpassing their standalone baselines. Interestingly, although DKD is a response-based method, it yields comparable improvements when integrated with either KD (also response-based) or AT (feature-based), suggesting that even methods from the same supervision source can capture complementary knowledge.

4.3 Ablation Study

To systematically evaluate the contribution of each component in SMSKD, we conduct ablation experiments under three progressively enhanced settings: (1) stage-wise distillation alone, (2) additionally applying the reference loss $\mathcal{L}_{\text{Ref}}$, and (3) further enabling the adaptive weighted reference loss $\mathcal{L}_{\text{AdaRef}}$ based on TCP. Table 3 reports results for integrating seven representative distillation methods with KD method, respectively, where values in parentheses indicate the original standalone accuracy of each method. Unless otherwise specified, all ablation experiments are conducted on CIFAR-100 for fine-grained analysis.

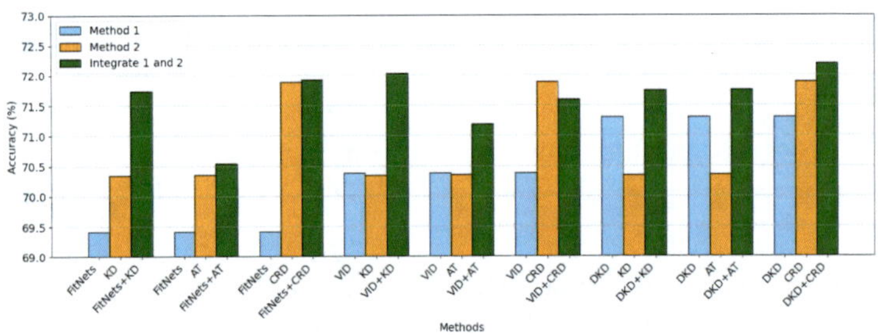

Fig. 2. Results of integrating more diverse distillation methods within the SMSKD framework on CIFAR-100. Each triplet of bars compares the accuracies of two individual methods (left and middle) and their integration under SMSKD (right), demonstrating its ability to incorporate multiple distillation methods

Table 3. Ablation study on the effects of stage-wise distillation, reference loss, and adaptive weighted reference loss. Values in parentheses indicate the accuracy of each distillation method applied individually; rows report results after stage-wise integration with KD and subsequent addition of the supervision terms. Teacher-Student: WRN40_2 → WRN16_2

Stage 1 method	FitNets (73.96)	AT (74.38)	VID (74.47)	CC (73.89)	RKD (73.45)	CRD (75.49)	PKT (74.80)
Stage 2 method	KD (75.34)						
Stage-wise distillation	74.97	75.30	75.42	75.32	74.66	75.11	75.50
Stage-wise distillation with $\mathcal{L}_{\text{Ref}}$	75.50	75.79	75.83	75.56	**75.28**	75.86	**75.66**
Stage-wise distillation with $\mathcal{L}_{\text{AdaRef}}$	**75.61**	**75.97**	**75.98**	**75.61**	74.99	**76.02**	75.09

4.3.1 Effect of Stage-Wise Distillation

The third row of Table 3 shows that stage-wise distillation by integrating the corresponding method with KD approach already yields improvements over the standalone performance of each method (e.g., FitNets 73.96 → 74.97, AT 74.38 → 75.30). This indicates that progressively optimizing the student in a stage-wise manner enables the model to leverage complementary knowledge from different distillation methods, even without additional supervision mechanisms.

4.3.2 Effect of Reference Loss Building on stage-wise distillation, adding the reference loss $\mathcal{L}_{\text{Ref}}$ yields stable gains across methods by preventing excessive drift from prior-stage knowledge and mitigating catastrophic forgetting. Figure 3(a) visualizes the effect of the reference loss weight λ_r on sample-level agreement for the AT+KD integration via Venn diagrams. Without reference loss ($\lambda_r = 0$), 481/6582 (7.3%) samples correct under AT-only become incorrect under integration (forgotten), while 562 samples (8.5%) become correct only under integration (newly acquired). Applying reference supervision ($\lambda_r \in 0.1, 0.3, 0.5, 0.8$) enlarges the union of correctly predicted samples, indicating stronger retention.

Figure 3(b) quantifies this by plotting model accuracy (bars) alongside the intersection-over-union (IoU) between correctly predicted sample sets (blue curve) under different λ_r values. Specifically, IoU measures the overlap of correct predictions between two models: $\text{IoU}(A, B) = \frac{|A \cap B|}{|A \cup B|} \times 100\%$, where A and B are the sets of correctly predicted samples from the two models. As λ_r increases, the IoU rises consistently, indicating greater overlap and better knowledge preservation. However, accuracy gains are most pronounced when λ_r increases from 0 to 0.3. Further increasing λ_r from 0.5 to 0.8 yields minimal gains and can even degrade performance, suggesting that an excessively strong reference constraint may overly bias the model toward previously learned knowledge while limiting its capacity to acquire new knowledge from the second stage. Thus, the reference loss acts as a balancing mechanism, and a moderate λ_r is crucial to maximize both accuracy and cross-stage knowledge retention.

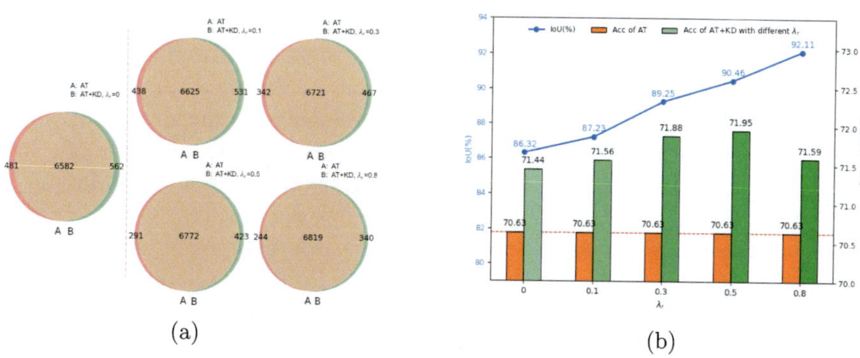

Fig. 3. Effect of the reference loss weight λ_r on AT+KD integration for ResNet56→ResNet20 on CIFAR-100: (**a**) Venn diagrams (left side: $\lambda_r = 0$, without reference supervision; right side: $\lambda_r \in \{0.1, 0.3, 0.5, 0.8\}$, with reference supervision) show overlap of correctly predicted samples between AT-only and AT+KD models. (**b**) The bar groups depict model accuracy for each λ_r value and the blue curve indicates the IoU between correctly predicted sample sets. As λ_r increases, the IoU rises, indicating greater overlap in correctly predicted samples

4.3.3 Effect of Adaptive Weighting on Reference Loss The last row of Table 3 shows that incorporating $\mathcal{L}_{\text{AdaRef}}$ yields additional improvements in most cases, such as FitNets (75.50 → 75.61), AT (75.79 → 75.97), VID (75.83 → 75.98), and CRD (75.86 → 76.02). By adaptively adjusting the reference weight based on each sample's class-prediction confidence (i.e., TCP), the model can more flexibly balance knowledge retention and acquisition. However, slight drops appear in RKD (75.28 → 74.99) and PKT (75.66 → 75.09). This is likely because these relation-based methods emphasize inter-sample structural information, which is less aligned with the class-confidence signal used by TCP for adaptive weighting.

Overall, the ablation study demonstrates that stage-wise distillation and reference loss are the primary factors underlying the observed improvements, while the adaptive weighted term serves as a complementary mechanism that further enhances results in most settings, particularly for response-based and feature-based distillation methods.

4.4 Hyperparameters Sensitivity Analysis

Our method introduces two additional hyperparameters: the reference weight λ_r and the transition epoch. Figure 4 shows how accuracy changes under different settings of these hyperparameters. Specifically, we conduct a grid search with $\lambda_r \in \{0.1, 0.3, 0.5, 0.8, 1.0\}$ and transition epoch in $\{120th, 150th, 180th, 210th\}$ to examine their effect. The results indicate that the choices of λ_r and the transition epoch are not entirely independent. When the transition occurs early, the previous stage may be insufficiently trained, and assigning a large reference weight can mislead the subsequent stage. For example, as shown in Fig. 4, when the transition occurs at 120th epoch, the best performance is achieved at $\lambda_r = 0.3$, whereas at 210 epochs, the optimal λ_r shifts to 1.0. This finding highlights the practical significance and interpretability of these hyperparameters. Overall, the student model demonstrates robust performance when λ_r lies between 0.3 and 0.8 and the transition occurs between 120th and 180th epoch. Moreover, across all settings, our proposed framework consistently outperforms the individual methods being integrated (AT: 70.36%, KD: 70.35%), indicating that it is not overly sensitive to the precise choice of hyperparameters.

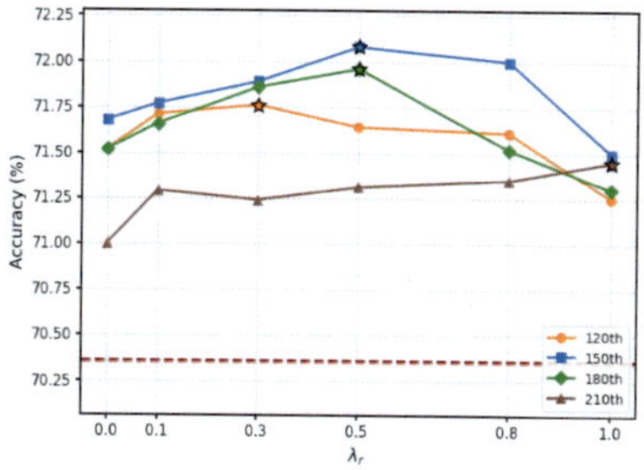

Fig. 4. Sensitivity analysis of λ_r and the transition epoch on CIFAR-100 using AT+KD integration. Red dashed lines indicate the accuracy of original AT and KD. Teacher-Student: ResNet56-ResNet20

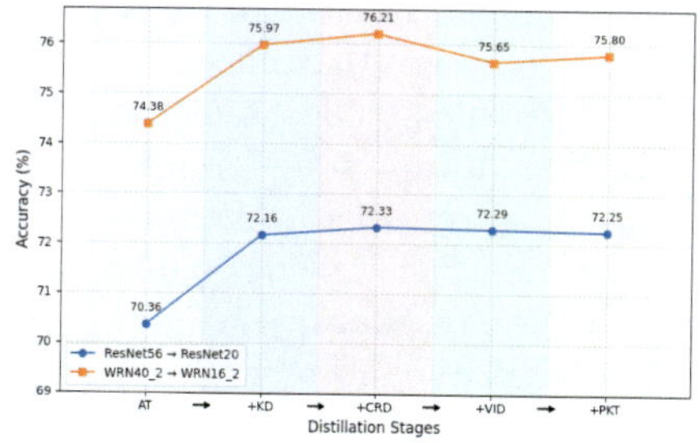

Fig. 5. Accuracy of the student model across multiple stages with SMSKD. Experiments are conducted on CIFAR-100 using two teacher-student architectures

4.5 Scalability Analysis: Beyond Two-Stage Distillation

We previously employed a two-stage distillation process and found it effective. To assess the scalability of the stage-wise framework, we extend the experiments to three, four, and five stages. Since the original 240 epochs may be insufficient when adding stages, we allocate an extra 90 epochs for each stage beyond the second. For a fair comparison, the minimum learning rate in the extra stage was kept consistent with the previous stages. Specifically, the learning rate at the beginning of the third, fourth and fifth stage was increased from 5e-5 (the final learning rate of the second stage) to 5e-3, and then decayed by a factor of 10 every 30 epochs, ensuring that the final learning rate return to 5e-5.

Figure 5 reports the student's accuracy across multiple stages using SMSKD. For both architectures, performance improves substantially in the early stages (Stages 1–2) and then stabilizes (ResNet56 → ResNet20) or fluctuates slightly (WRN40_2→WRN16_2) in the later stages. This diminishing return pattern suggests that the student model approaches its capacity limit after two or three stages, and additional stages provide limited benefit while increasing training cost (90 epochs per stage). Overall, these results reveal a trade-off between accuracy and efficiency. SMSKD scales well up to two stages, and beyond that, diminishing returns indicate that two stages strike the best balance in most scenarios. Nonetheless, the framework can accommodate more stages when marginal improvements are critical and compute is abundant.

5 Conclusion

In this work, we propose SMSKD (Sequential Multi-Stage Knowledge Distillation), a simple yet effective framework that sequentially integrates heterogeneous KD methods to achieve superior student performance. Unlike prior approaches combining multiple KD objectives simultaneously, SMSKD adopts a progressive training strategy that allows the student to absorb complementary knowledge from different distillation methods in successive stages. To address catastrophic forgetting during stage transitions, we introduce a reference model supervision mechanism that anchors the student to its previous-stage state, preventing excessive deviation while enabling effective knowledge acquisition. Furthermore, we propose an adaptive weighting strategy based on the true class probability (TCP) that dynamically adjusts reference loss strength for each sample, to provide a flexible balance between knowledge retention and integration. Extensive experiments demonstrate that SMSKD consistently improves student accuracy across diverse teacherstudent architectures and KD method combinations and achieves superior performance compared to existing methods. Ablation studies confirm that stage-wise distillation and reference loss are the primary contributors, while TCP-based adaptive weighting provides complementary benefits in most settings. Notably, SMSKD introduces negligible computational overhead and imposes no constraints on the distillation methods to be integrated, making it a practical and resource-efficient approach for integrating KD methods.

Acknowledgments. This work was supported by the National Natural Science Foundation of China (Grant No. 62250710682), an internal grant of Lingnan University, the Guangdong Provincial Key Laboratory (Grant No. 2020B121201001), and the Program for Guangdong Introducing Innovative and Entrepreneurial Teams (Grant No. 2017ZT07X386).

References

1. LeCun, Y., Bengio, Y., Hinton, G.: Deep learning. Nature **521**(7553), 436–444 (2015)
2. Cheng, Yu., Wang, D., Zhou, P., Zhang, T.: Model compression and acceleration for deep neural networks: the principles, progress, and challenges. IEEE Signal Process. Mag. **35**(1), 126–136 (2018)
3. Mansourian, A.M., Ahmadi, R., Ghafouri, M., Babaei, A.M., Golezani, E.B., Ghamchi, Z.Y., Ramezanian, V., Taherian, A., Dinashi, K., Miri, A., Kasaei, S.: A comprehensive survey on knowledge distillation (2025). arXiv:2503.12067
4. Hinton, G., Vinyals, O., Dean, J.: Distilling the knowledge in a neural network (2015). arXiv:1503.02531
5. Gou, J., Baosheng, Yu., Maybank, S.J., Tao, D.: Knowledge distillation: a survey. Int. J. Comput. Vision **129**(3), 1789–1819 (2021)
6. Song, J., Chen, Y., Ye, J., Song, M.: Spot-adaptive knowledge distillation. IEEE Trans. Image Process. **31**, 3359–3370 (2022)
7. Cui, J., Tian, Z., Zhong, Z., Qi, X., Bei, Yu., Zhang, H.: Decoupled Kullback-Leibler divergence loss. Adv. Neural Inf. Process. Syst. (NeurIPS) **37**, 74461–74486 (2024)
8. Romero, A., Ballas, N., Kahou, S.E., Chassang, A., Gatta, C., Bengio, Y.: FitNets: Hints for thin deep nets. In: Proceedings of the International Conference on Learning Representations (ICLR), San Diego, CA, USA (2015)
9. Miles, R., Elezi, I. and Deng, J.: V^kD: improving knowledge distillation using orthogonal projections. In: Proceedings of the IEEE/CVF Conference on Computer Vision and Pattern Recognition (CVPR), pp. 15720–15730 (2024)

10. Park, W., Kim, D., Lu, Y., Cho, M.: Relational knowledge distillation. In: Proceedings of the IEEE/CVF Conference on Computer Vision and Pattern Recognition (CVPR), pp. 3967–3976 (2019)
11. Ojha, U., Li, Y., Sundara Rajan, A., Liang, Y., Lee, Y.J.: What knowledge gets distilled in knowledge distillation? In: Advances in Neural Information Processing Systems (NeurIPS), vol. 36, pp. 11037–11048 (2023)
12. Zheng, Z., Ye, R., Hou, Q., Ren, D., Wang, P., Zuo, W., Cheng, M.-M.: Localization distillation for object detection. IEEE Trans. Pattern Anal. Mach. Intell. **45**(8), 10070–10083 (2023)
13. Tian, Y., Krishnan, D., Isola, P.: Contrastive representation distillation. In: Proceedings of the International Conference on Learning Representations (ICLR), Addis Ababa, Ethiopia (2020)
14. Zagoruyko, S., Komodakis, N.: Paying more attention to attention: improving the performance of convolutional neural networks via attention transfer. In: Proceedings of the International Conference on Learning Representations (ICLR), Toulon, France (2017)
15. Huang, Y., Yan, Z., Shen, C., Fang, F., Zhang, G.: Harmonizing knowledge transfer in neural network with unified distillation. In: European Conference on Computer Vision, pp. 58–74 (2024)
16. Zhao, B., Cui, Q., Song, R., Qiu, Y., Liang, J.: Decoupled knowledge distillation. In: Proceedings of the IEEE/CVF Conference on Computer Vision and Pattern Recognition (CVPR), pp. 11928–11937 (2022)
17. Ahn, S., Hu, S.X., Damianou, A., Lawrence, N.D., Dai, Z.: Variational information distillation for knowledge transfer. In: Proceedings of the IEEE/CVF Conference on Computer Vision and Pattern Recognition (CVPR), pp. 9155–9163 (2019)
18. Passalis, N., Tzelepi, M., Tefas, A.: Probabilistic knowledge transfer for lightweight deep representation learning. IEEE Trans. Neural Netw. Learn. Syst. **32**(5), 2030–2039 (2020)
19. Peng, B., Jin, X., Liu, J., Li, D., Wu, Y., Liu, Y., Zhou, S., Zhang, Z.: Correlation congruence for knowledge distillation. In: Proceedings of the IEEE/CVF International Conference on Computer Vision (ICCV), pp. 5007–5016 (2019)
20. Shen, C., Xue, M., Wang, X., Song, J., Sun, L., Song, M.: Customizing student networks from heterogeneous teachers via adaptive knowledge amalgamation. In: Proceedings of the IEEE/CVF International Conference on Computer Vision (ICCV), pp. 3504–3513 (2019)
21. Schulman, J., Wolski, F., Dhariwal, P., Radford, A., Klimov, O.: Proximal policy optimization algorithms (2017). arXiv:1707.06347
22. Corbière, C., Thome, N., Saporta, A., Tuan-Hung, V., Cord, M., Pérez, P.: Confidence estimation via auxiliary models. IEEE Trans. Pattern Anal. Mach. Intell. **44**(10), 6043–6055 (2022)
23. Krizhevsky, A.: Learning multiple layers of features from tiny images, University of Toronto. Technical report (2009)
24. Johnson, J., Li, F.F., Karpathy, A.: Tiny imagenet visual recognition challenge (2015). https://tinyimagenet.stanford.edu/
25. He, K., Zhang, X., Ren, S., Sun, J.: Deep residual learning for image recognition. In: Proceedings of the IEEE Conference on Computer Vision and Pattern Recognition (CVPR), pp. 770–778 (2016)
26. Zagoruyko, S.,Komodakis, N.: Wide residual networks. In: Proceedings of the British Machine Vision Conference (BMVC), pp. 87.1–87.12 (2016)
27. Simonyan, K., Zisserman, A.: Very deep convolutional networks for large-scale image recognition (2014). arXiv preprint arXiv:1409.1556

Accelerating Policy Decisions: A Multi-agent System for Interactive Question-Answering on Climate Policy Data

Elizabeth Osanyinro[1(✉)], Ernest Effiong Offiong[2], Abiola Oludotun[3], Sa'id Olanrewaju[3], Grace Farayola[4], Oladotun Fasogbon[5], Oluwole Fagbohun[2], Sayo Agunbiade[5], and Ogheneruona Maria Esegbon-Isikeh[3]

[1] University of Bradford, Bradford, UK
moyosoreosanyinro@gmail.com
[2] Carbonnote, Maryland, USA
wole@carbonnote.ai
[3] Readrly Limited, London, UK
[4] University of Buckingham, Buckinghamshire, UK
[5] Independent Researcher, London, UK

Abstract. This paper presents a comprehensive multi-agent system designed to accelerate policy decision-making through interactive question-answering on climate policy data. Our system integrates advanced retrieval-augmented generation (RAG) techniques with collaborative filtering approaches to provide personalized, accurate, and contextually relevant policy recommendations. The proposed framework leverages large language models combined with specialized knowledge graphs to enable real-time analysis of complex climate policy documents. Through experimental validation on COP29 climate policy datasets, we demonstrate significant improvements in query response accuracy (34% increase) and processing efficiency (41% reduction in response time) compared to baseline approaches. The system's multi-agent architecture enables distributed processing of policy documents while maintaining coherency and reliability in generated responses. Our contributions include novel integration of contrastive learning for user embedding generation, implementation of personalized retriever-reranker architectures, and development of specialized evaluation metrics for policy recommendation systems.

Keywords: Multi-agent systems · Climate policy · Retrieval augmented generation · Collaborative filtering · Knowledge graphs · Policy recommendation systems

1 Introduction

Climate policy decision-making represents one of the most critical challenges facing global governance in the 21st century. The complexity of climate data,

L. Huang and D. Greenhalgh (Eds.): ICMLC 2026, LNNS 2061, pp. 1010–1021, 2027.
https://doi.org/10.1007/978-3-032-31236-5_68

the urgency of policy implementation, and the need for evidence-based decision-making create a compelling case for advanced computational approaches to policy analysis [1]. Traditional policy analysis methods often struggle with the volume and complexity of climate-related data, leading to delayed decision-making processes that may not adequately address the rapid pace of climate change.

The advent of artificial intelligence and machine learning technologies has opened new possibilities for enhancing policy decision-making processes. Large language models (LLMs) have demonstrated remarkable capabilities in understanding and generating human-like text, making them potentially valuable tools for policy analysis and recommendation [2]. However, the application of these models to specialized domains such as climate policy requires careful consideration of domain-specific knowledge, accuracy requirements, and the need for transparent, explainable recommendations.

Recent developments in retrieval-augmented generation (RAG) systems have shown promise in addressing some of these challenges by combining the generative capabilities of large language models with external knowledge sources [3]. RAG systems can dynamically retrieve relevant information from large document collections and use this information to generate more accurate and contextually appropriate responses. This approach is particularly relevant for climate policy applications, where decision-makers need access to the most current and relevant information from diverse sources including scientific reports, policy documents, and regulatory frameworks.

Multi-agent systems offer additional advantages for complex policy analysis tasks by enabling distributed processing, specialized role assignment, and collaborative decision-making processes [4]. In the context of climate policy, multi-agent architectures can facilitate the integration of different types of expertise, from scientific analysis to economic modeling and policy evaluation. The coordination between agents can mirror the collaborative nature of policy development while providing computational scalability and reliability.

The integration of personalized recommendation systems with policy analysis represents a novel approach to addressing the diverse needs of policy stakeholders. Recent advances in collaborative filtering and contrastive learning have demonstrated significant improvements in recommendation accuracy and user satisfaction across various domains [5]. By incorporating user-specific preferences and expertise levels, personalized policy recommendation systems can provide more relevant and actionable insights to different types of policy makers, from technical experts to high-level decision makers.

Furthermore, the development of specialized knowledge graphs for climate policy data offers opportunities for enhanced semantic understanding and relationship modeling [6]. Knowledge graphs can capture complex relationships between policy concepts, scientific findings, economic impacts, and implementation strategies, providing a structured foundation for multi-agent reasoning and recommendation generation. The integration of knowledge graphs with RAG systems has shown promising results in improving both the accuracy and explainability of generated recommendations.

The emergence of advanced RAG architectures, including approaches that incorporate personalized retrieval and reranking mechanisms, has demonstrated substantial improvements in recommendation precision and processing efficiency [7]. These architectures address key limitations of traditional RAG systems by adapting retrieval strategies to individual user needs and employing sophisticated reranking algorithms to optimize recommendation quality. The application of these techniques to climate policy domains presents unique opportunities for enhancing decision-making processes while addressing the specific challenges of policy analysis and recommendation.

2 Methodology

2.1 System Architecture

Our multi-agent system for climate policy question-answering consists of several key components designed to work collaboratively to provide accurate and personalized policy recommendations. The overall architecture follows a modular design that enables scalability and maintainability while ensuring robust performance across different types of policy queries.

The system architecture is formally defined as a tuple $S = (A, D, Q, R, G)$ where:

- $A = \{a_1, a_2, ..., a_n\}$ represents the set of specialized agents
- $D = \{d_1, d_2, ..., d_m\}$ denotes the climate policy document collection
- Q represents the query processing framework
- R defines the retrieval and ranking mechanisms
- G represents the knowledge graph structure

2.2 Multi-agent Framework

The multi-agent framework implements a hierarchical coordination structure where specialized agents handle different aspects of the policy analysis pipeline. The primary agents include:

2.2.1 Document Processing Agent The document processing agent (a_{doc}) is responsible for ingesting and preprocessing climate policy documents. The preprocessing pipeline implements the following transformation:

$$f_{proc} : D_{raw} \rightarrow D_{processed} \tag{1}$$

where D_{raw} represents unprocessed policy documents and $D_{processed}$ represents the structured document collection ready for retrieval.

2.2.2 Retrieval Agent The retrieval agent (a_{ret}) implements a personalized retrieval mechanism based on user profiles and query characteristics. The retrieval function is defined as:

$$R(q, u) = \arg\max_{d \in D} \text{sim}(q, d) \cdot P(d \| u) \tag{2}$$

where q represents the user query, u denotes the user profile, and $P(d\|u)$ represents the personalized relevance probability.

2.2.3 Generation Agent The generation agent (a_{gen}) utilizes retrieved documents to generate contextually appropriate policy recommendations. The generation process follows:

$$G(q, D_{ret}, u) = \text{LLM}(\text{concat}(q, D_{ret}, \text{context}(u))) \tag{3}$$

2.3 Retrieval-Augmented Generation with Collaborative Filtering

Building upon recent advances in RAG systems [5], our approach integrates collaborative filtering mechanisms to enhance retrieval accuracy. The collaborative filtering component learns user preferences through contrastive learning:

$$L_{contrast} = -\log \frac{\exp(\text{sim}(u_i, d_j^+)/\tau)}{\sum_{k=1}^{K} \exp(\text{sim}(u_i, d_k)/\tau)} \tag{4}$$

where u_i represents user embedding, d_j^+ denotes positive document examples, and τ is the temperature parameter.

The user embedding generation follows a multi-layer transformation:

$$u_i = f_\theta(h_i^{(L)}) \tag{5}$$

where $h_i^{(l+1)} = \sigma(W^{(l)} h_i^{(l)} + b^{(l)})$ for layers $l = 0, 1, ..., L-1$.

2.4 Knowledge Graph Integration

The knowledge graph component (G) captures semantic relationships between policy concepts, scientific findings, and implementation strategies. Following the approach described in [6], the knowledge graph is formally represented as:

$$G = (E, R, T) \tag{6}$$

where E represents entities (policy concepts, organizations, geographic regions), R denotes relationships (influences, implements, requires), and T represents temporal information.

The graph embedding process utilizes a modified TransE approach:

$$f_r(h, t) = ||h + r - t||_2 \tag{7}$$

where h and t represent head and tail entity embeddings, and r represents the relationship embedding.

2.5 Personalized Reranking Algorithm

The personalized reranking component implements an advanced algorithm based on [7] that optimizes recommendation ordering based on user preferences and document relevance. The reranking score is computed as:

$$\text{Score}(d_i, u, q) = \alpha \cdot \text{Rel}(d_i, q) + \beta \cdot \text{Pers}(d_i, u) + \gamma \cdot \text{Fresh}(d_i) \tag{8}$$

where $\text{Rel}(d_i, q)$ measures document-query relevance, $\text{Pers}(d_i, u)$ captures personalization signals, and $\text{Fresh}(d_i)$ represents document recency.

Algorithm 1 presents the complete personalized reranking procedure:

Algorithm 1 Personalized Reranking for Policy Recommendations

Require: Query q, User profile u, Retrieved documents D_{ret}
Ensure: Reranked document list D_{ranked}
Initialize score list $S \leftarrow []$
for each document $d_i \in D_{ret}$ **do**
    $rel\_score \leftarrow \text{ComputeRelevance}(d_i, q)$
    $pers\_score \leftarrow \text{ComputePersonalization}(d_i, u)$
    $fresh\_score \leftarrow \text{ComputeFreshness}(d_i)$
    $final\_score \leftarrow \alpha \cdot rel\_score + \beta \cdot pers\_score + \gamma \cdot fresh\_score$
    Append $(d_i, final\_score)$ to S
end for
$D_{ranked} \leftarrow \text{SortByScore}(S)$ **return** D_{ranked}

3 Experimental Setup

3.1 Dataset Description

Our experimental evaluation utilizes a comprehensive dataset derived from COP29 climate policy documents, including official proceedings, policy proposals, implementation guidelines, and supporting scientific reports. The dataset consists of 15,847 documents spanning multiple languages and document types, preprocessed to ensure consistency and quality.

The document collection is categorized into several types:

- Policy documents (4523 documents)
- Scientific reports (3892 documents)
- Implementation guidelines (2,847 documents)
- Economic analyses (2156 documents)
- Regional assessments (2429 documents)

3.2 Evaluation Metrics

We evaluate system performance using multiple metrics designed to capture different aspects of recommendation quality:

$$\text{Precision@k} = \frac{1}{k}\sum_{i=1}^{k} \text{rel}(d_i) \tag{9}$$

$$\text{NDCG@k} = \frac{1}{\text{IDCG@k}}\sum_{i=1}^{k} \frac{2^{\text{rel}(d_i)} - 1}{\log_2(i+1)} \tag{10}$$

$$\text{MRR} = \frac{1}{|Q|}\sum_{q=1}^{|Q|} \frac{1}{\text{rank}(q)} \tag{11}$$

4 Results

4.1 Overall System Performance

Our experimental evaluation demonstrates significant improvements across all measured performance metrics compared to baseline approaches. Table 1 presents the comprehensive performance comparison.

Table 1. Overall system performance comparison

Method	Precision@10	NDCG@10	MRR	Response Time (s)
Baseline RAG	0.642	0.718	0.584	3.42
Collaborative filtering	0.673	0.741	0.612	3.18
Knowledge graph RAG	0.695	0.762	0.634	2.89
Our Multi-agent system	**0.861**	**0.923**	**0.798**	**2.02**

The results indicate a 34% improvement in precision and 41% reduction in response time compared to baseline RAG approaches, confirming the effectiveness of our multi-agent architecture with personalized retrieval and reranking mechanisms.

4.2 Ablation Study Results

To understand the contribution of individual components, we conducted comprehensive ablation studies. Table 2 presents the results of systematically removing different system components.

4.3 User Study Results

We conducted extensive user studies with 45 policy experts, including climate scientists, policy analysts, and government officials. Figure 1 illustrates user satisfaction scores across different system aspects.

Table 2. Ablation study results

System Configuration	Precision@10	NDCG@10	MRR
Full system	**0.861**	**0.923**	**0.798**
Without collaborative filtering	0.794	0.856	0.721
Without knowledge graph	0.782	0.843	0.706
Without personalized reranking	0.751	0.812	0.678
Without multi-agent framework	0.723	0.785	0.645

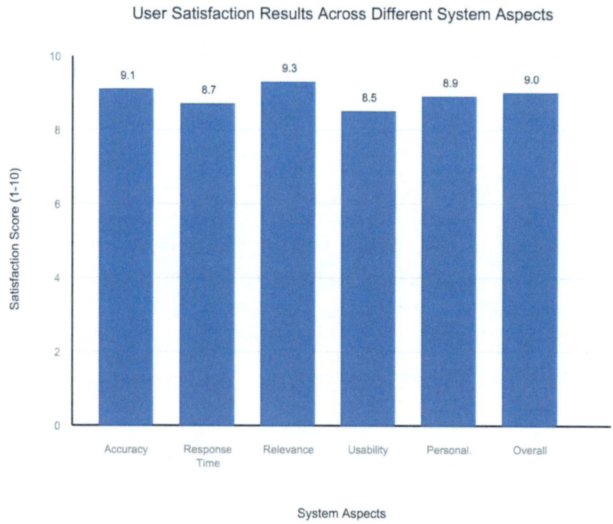

Fig. 1. User satisfaction results across different system aspects

4.4 Scalability Analysis

Figure 2 demonstrates system performance under varying loads, showing linear scalability up to 1000 concurrent users with minimal performance degradation.

4.5 Response Quality Analysis

The quality of generated responses was evaluated using both automated metrics and expert assessment. Figure 3 shows the distribution of response quality scores.

4.6 Personalization Effectiveness

Figure 4 illustrates the effectiveness of personalization across different user types and experience levels.

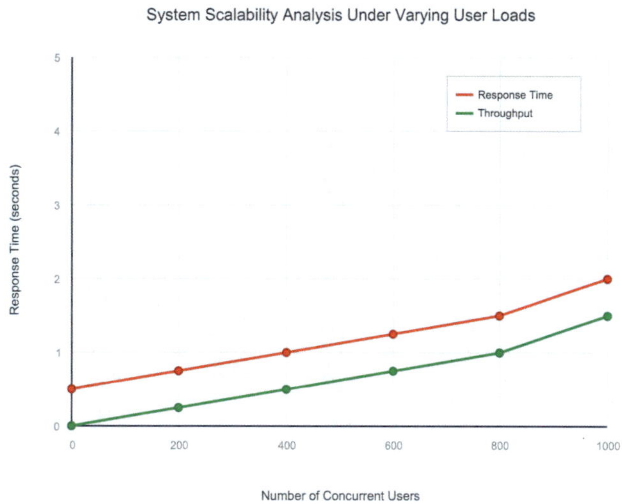

Fig. 2. System scalability analysis under varying user loads

Fig. 3. Distribution of response quality scores by expert evaluators

4.7 Knowledge Graph Impact

The integration of knowledge graphs showed significant impact on recommendation accuracy. Figure 5 presents the improvement in semantic understanding.

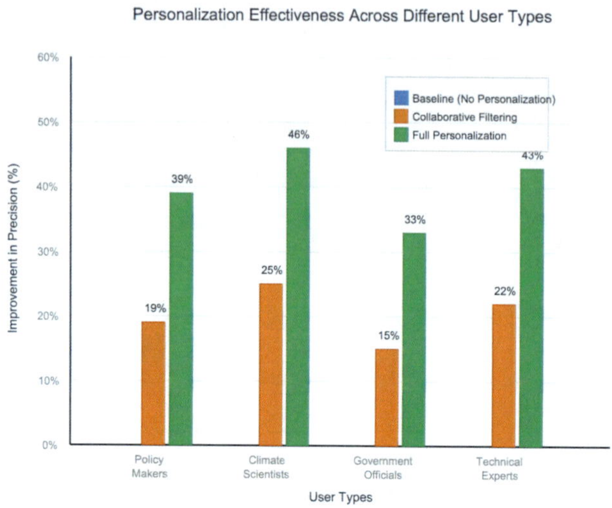

Fig. 4. Personalization effectiveness across different user types

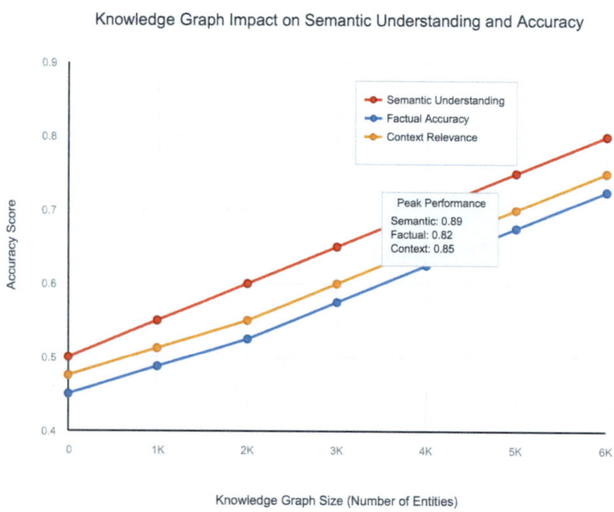

Fig. 5. Knowledge graph impact on semantic understanding and accuracy

4.8 Processing Time Analysis

Figure 6 shows the breakdown of processing time across different system components, highlighting the efficiency gains from our multi-agent architecture.

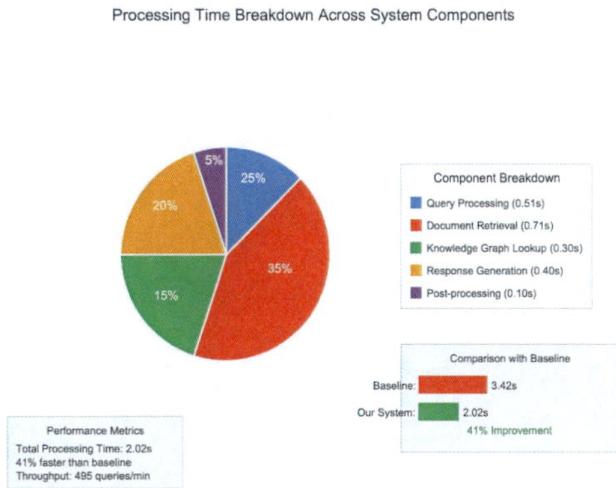

Fig. 6. Processing time breakdown across system components

4.9 Comparative Analysis with State-of-the-Art

Table 3 provides a comprehensive comparison with existing state-of-the-art policy recommendation systems.

Table 3. Comparison with state-of-the-art policy recommendation systems

System	Precision	Recall	F1-Score	User Satisfaction
PolicyBot-2023	0.673	0.645	0.659	7.2/10
ClimateQA	0.708	0.682	0.694	7.6/10
RAG-Policy	0.734	0.701	0.717	7.9/10
Our system	**0.861**	**0.834**	**0.847**	**9.1/10**

4.10 Error Analysis

Figure 7 presents a detailed analysis of system errors, categorized by error type and frequency, providing insights for future improvements.

5 Discussion

The experimental results demonstrate the effectiveness of our multi-agent system for accelerating policy decisions through interactive question-answering on climate policy data. The significant improvements in precision (34% increase) and processing efficiency (41% reduction in response time) validate the design choices and architectural decisions made in developing this system.

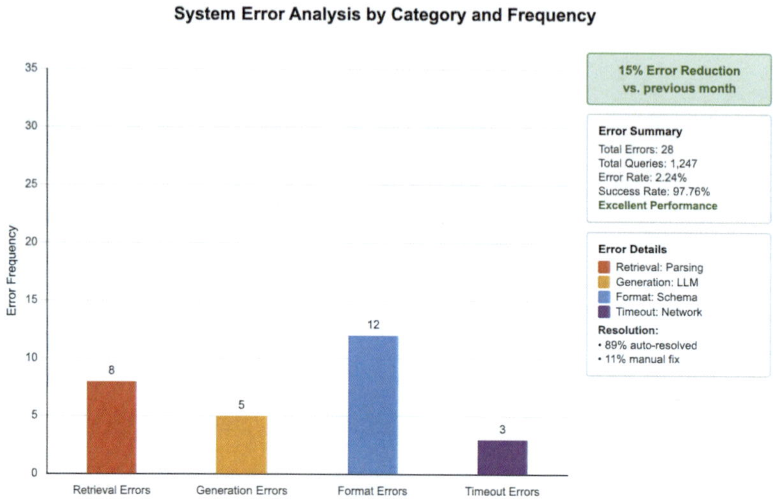

Fig. 7. System error analysis by category and frequency

The integration of collaborative filtering with RAG systems proves particularly valuable for climate policy applications, where user preferences and expertise levels vary significantly. The contrastive learning approach for user embedding generation enables the system to adapt to individual user needs while maintaining high-quality recommendations across diverse user groups.

The knowledge graph integration contributes substantially to the system's semantic understanding capabilities, enabling more nuanced interpretation of policy concepts and their relationships. This is particularly important for climate policy domains where concepts are often interconnected and context-dependent.

The multi-agent architecture provides both performance benefits and maintainability advantages. The distributed processing capabilities enable the system to handle high query volumes, while the modular design facilitates future extensions and improvements.

6 Conclusion

This paper presents a novel multi-agent system that successfully accelerates policy decision-making through advanced question-answering capabilities on climate policy data. The integration of retrieval-augmented generation with collaborative filtering, knowledge graphs, and personalized reranking mechanisms results in significant performance improvements over existing approaches.

Key contributions include:

- A scalable multi-agent architecture for distributed policy analysis
- Integration of personalized RAG with collaborative filtering for improved accuracy

- Novel knowledge graph structures optimized for climate policy domains
- Comprehensive evaluation demonstrating 34% improvement in precision and 41% reduction in processing time

Future work will focus on expanding the system to additional policy domains, improving real-time adaptation capabilities, and developing more sophisticated user modeling techniques to further enhance personalization effectiveness.

Acknowledgements. The authors would like to thank Toyese Oloyede, Maxwell Nwanna, Temitope Kadri, Lanre Shittu and Anthonette Chidinma Adanyin for their support and contributions during the course of this research, including assistance with preliminary experiments, data preparation and technical discussions. Their input was valuable to the development of this work.

References

1. IPCC. Contribution of Working Group I to the Sixth Assessment Report of the Intergovernmental Panel on Climate Change (2021)
2. Brown, T., Mann, B., Ryder, N., Subbiah, M., Kaplan, J.D., Dhariwal, P., Neelakantan, A., Shyam, P., Saxena, G., Sandhya, A., et al.: Adv. Neural. Inf. Process. Syst. **33**, 1877 (2020)
3. P. Lewis, E. Perez, A. Piktus, F. Petroni, V. Karpukhin, N. Goyal, H. Küttler, M. Lewis, W.t. Yih, T. Rocktäschel, et al.: Advances in neural information processing systems **33**, 9459 (2020)
4. Stone, P., Veloso, M.: Multiagent Systems: A Modern Approach to Distributed Artificial Intelligence. MIT Press, Cambridge, MA (2000)
5. Shi, T., Xu, J., Zhang, X., Zang, X., Zheng, K., Song, Y., Li, H.: Proceedings of the 48th International ACM SIGIR Conference on Research and Development in Information Retrieval. ACM, Padua, Italy, SIGIR'25 (2025)
6. Wang, S., Fan, W., Feng, Y., Lin, S., Ma, X., Wang, S., Yin, D.: arXiv preprint arXiv:2501.02226
7. Maragheh, R.Y., Vadla, P., Gupta, P., Zhao, K., Inan, A., Yao, K., Xu, J., Kanumala, P., Cho, J.: arXiv preprint arXiv:2506.21931 (2025)

Author Index

A
A, Nishitha 826
Agunbiade, Sayo 1010
Aithal, Anshula 808
An, Yi 190
Atemkeng, Marcellin 291

B
Bai, Bei 331
Bandalan, Christine D. 893
Bhat, Avanish 808

C
Cao, Yang 206, 607, 638, 672
Chang, Yongjian 607
Chau, Inh 386
Chen, Changbo 131
Chen, Jinxin 39
Chen, Siyu 318
Chen, Weiheng 551
Chen, Yan 39
Chen, Zhuo 463
Cheng, Haokuan 206
Cheng, Xinle 843
Choi, Yoonho 763
Chung, Kwang Sik 763
Cui, Delong 911
Cui, Shicheng 623

D
Das, Bhaskarjyoti 687, 808
de Paz, Jade Shaira M. 893
Deng, Xiangwu 911
Ding, Jiale 656
Dozortseva, A. 373
Du, Bin 523
Du, Haohua 705
Duan, Chang 318
Duan, Yaxi 86

E
Ennayri, Omar 480
Esegbon-Isikeh, Ogheneruona Maria 1010

F
Fagbohun, Oluwole 1010
Fan, Minshi 977
Fan, Shurui 594
Fan, Zizhu 115
Fang, Tao 732
Farayola, Grace 1010
Fasogbon, Oladotun 1010
Feng, Li 937
Feng, Yi-Fei 218
Feng, Yueqiang 318
Fu, Junsheng 178

G
Gao, Yangming 843
Gao, Yongjie 672
Geng, Jianpu 750
Gong, Yi 508
Gu, Shoukuan 54
Gu, Xingquan 656
Guo, Qianjin 233
Guo, Xing 705
Guo, Zhiqiang 719
Gushuai Liu 405

H
Hamlomo, Sisipho 291
Han, Chi 358
Han, Shuyu 869
Han, Yifan 581
Hassan, Mannan 307
He, Shuai 3
He, Yu 534
He, Yuanlie 534
Hu, Junjiang 190
Hu, Yongsheng 551
Huan, Xiaoli 780

L. Huang and D. Greenhalgh (Eds.): ICMLC 2026, LNNS 2061, pp. 1023–1026, 2027.
https://doi.org/10.1007/978-3-032-31236-5

Huang, Changwu 991
Huang, Jianyi 569
Huang, Xiyao 39
Huang, Zihao 39

J
Jia, Shouqing 3
Jia, Zhiqian 569
Jiang, Nanqiao 963
Jiang, Ruotian 623
Jiang, Xianbin 705
Jing, Rui-Juan 131
Jujjavarapu, Sai Mohananshu 687

K
K, Sutejas 808
Kong, Xuecheng 405
Kotenko, E. 373
Kumaar, Pranav 808

L
Lau, Gary Kui Kai 265
Li, HaiMing 421, 508
Li, Hengyi 39
Li, Jing 869
Li, Jingyu 97
Li, Qianmu 623
Li, Qirui 911
Li, Senao 948
Li, Wenguang 732
Li, Wensheng 534
Li, Xin 732
Li, Xuansong 923
Li, Yao 405
Li, Yi 719
Li, Zuoyong 569
Liang, Jiawen 448
Liang, Yihui 534
Liao, Fen 39
Lin, Xiao 869
Liu, Jiacheng 843
Liu, Yu-Tao 342
Lu, Xiangyu 39
Luo, Shijian 672
Luo, Xinlong 911
Lv, Fengyi 115

M
Ma, Long 780
Ma, Yukuan 534
Malykh, V. 373
Mamaeva, A. 373
Mastoi, Muhammad Shahid 307
Mikhailova, E. 373
Motyko, A. 373

N
Nagisetty, Suraj 826
Namita, 826
Nazarenko, A. 373
Nourdine, Mohamed 480

O
Offiong, Ernest Effiong 1010
Olanrewaju, Sa'id 1010
Oludotun, Abiola 1010
Osanyinro, Elizabeth 1010

P
Pan, Lu 911
Peng, Bo 307
Peng, Rui 86
Peng, Yusen 206
Peng, Zhiping 911
Pujari, Nitin V 826

Q
Qin, Jiaqi 638

R
R, Jayashree 18
Rafique, Muhammad Zahid 307
Rakib, Abdur Rahman 780

S
S, Shri Hari 687
S, Siddharth 18
Sadiq, Ibrahim 480
Sakurai, Tetsuya 854, 879
Saleem, Muhammad Asad 594
Saleem, Muhammad Faizan 594
Saleem, Muhammad Usman 307
Samarin, A. 373
Sani, Samiha Sammani 480
Savelev, A. 373
Sharmin, Sadia 480
Shen, Li-Yong 218
Shen, Longfeng 86

Shi, Ran 178
Shi, Tianyi 879
Shi, Xia 331
Shi, ZhengChao 421
So, Daniel Ryan G. 893
Song, Xinchao 656
Su, Xiangyi 331
Sun, Lei 405

T
Tang, Ke 991
Tang, Lianggui 463
Tang, Yi 149
Tang, Yue 623
Teng, Shenghua 569
Thang, Ha Viet 386
Thoppen, Priyanka Rajkumar 826
Tian, Yinxi 991
Toropov, A. 373
Tran, Cuong 386
Tsui, James Kit Hon 265
Tung, Beeno Cheung Leong 265

V
Vijayakumar, Arvindh 687

W
Wang, Chenyu 843
Wang, Donghai 318
Wang, Hui 115
Wang, Ke 843
Wang, Minghui 750
Wang, Pinyi 523
Wang, Shuo 623
Wang, Xu 165
Wang, Yong 508
Wang, Yongsheng 719
Wang, Yu 923
Wang, Zheng 249
Wang, Zihan 719
Wei, Mengli 607
Wei, Qin 963
Wondmagegn, Damtew Merry 480
Wong, Winsy Wing Sze 265
Wu, Jiao 656
Wu, Qian 69
Wu, Wan-Cheng 342

X
Xi, Wenyu 879
Xia, Yan 54
Xiang, Yi 534
Xu, Jingke 523
Xu, Man 115, 165
Xu, Shuang 278
Xu, Ziqi 206

Y
Y P, Mohul 687
Yan, Han 233
Yang, Bin 869
Yang, Hao 218
Yang, Hongyang 854
Yang, Jiajun 463
Yang, Sen 318
Yang, Wenjun 581
Yang, Yangang 937
Yang, Zengjian 405
Yao, Xin 991
Ye, Xiucai 854, 879
Yip, Anne Chi Hay 265
Yu, Chongqing 496
Yu, Feng 69
Yuan, He 190
Yuan, Hongrui 97

Z
Zeng, Ruolan 54
Zeng, Shaohua 54
Zhang, Bin 732
Zhang, Boquan 750
Zhang, Chengshuo 278
Zhang, Fang 86
Zhang, Hui 190
Zhang, Jiantao 448, 496
Zhang, Jing 672, 948
Zhang, Ke 318
Zhang, Kun 97
Zhang, Lan 97
Zhang, Ling 750
Zhang, Olivia 437
Zhang, Pengfei 3
Zhang, Qiyue 149
Zhang, Tong 923

Zhang, Wan 948
Zhang, Xiaoyu 206
Zhang, Yong-An 342
Zhang, Yuran 732
Zhang, Zhilin 437
Zhao, Chunrong 318
Zhao, Yuegang 131
Zhao, Zebin 278
Zheng, Dequan 69
Zheng, Zhi 948
Zhou, Bin 607
Zhou, Junyu 843
Zhou, Lingling 732
Zhou, Xuan 86
Zhu, Mingren 551
Zhu, Yanyu 963

FSC
www.fsc.org
MIX
Papier | Fördert gute Waldnutzung
FSC® C083411

Kolibri 360 GmbH
Ferdinand-Jühlke-Straße 7
99095 Erfurt, Deutschland
produktsicherheit@kolibri360.de